CANADA
NORTH DAKOTA
MINNESOTA
SOUTH DAKOTA
WISCONSIN
MICHIGAN
IOWA
MIDWEST and GREAT PLAINS STATES
NEBRASKA
ILLINOIS
KANSAS
MISSOURI
OKLAHOMA
ARKANSAS
SOU
TEXAS
LOUISIANA
GULF
VT.
NEW YORK
NEW ENGLAND
N.H.
MASS.
R.I.
CONN.
MIDDLE ATLANTIC STATES
PENN.
N.J.
DEL.
MD.
ATLANTIC OCEAN

THE OXFORD ILLUSTRATED LITERARY GUIDE TO THE UNITED STATES

Henry Wadsworth Longfellow and his daughter Edith, in a photograph taken about two years before Longfellow died. From 1837 until his death in 1882, the poet lived in this house, 105 Brattle Street, in Cambridge, Massachusetts, and here wrote much of his poetry.

THE OXFORD ILLUSTRATED LITERARY GUIDE TO THE UNITED STATES

EUGENE EHRLICH
and
GORTON CARRUTH

A Hudson Group Book

New York Oxford
OXFORD UNIVERSITY PRESS
1982

Library of Congress Cataloging in Publication Data
Ehrlich, Eugene
The Oxford illustrated literary guide to the United States
"A Hudson Group book"
Includes indexes.
1. Literary landmarks—United States.
2. United States—Description and travel—1981- —Guidebooks.
3. Authors American—Homes and haunts.
I. Carruth, Gorton.
II. Title.
PS141.E74 1982 810'.9 82-8034
ISBN 0-19-503186-5

Printing (last digit): 9 8 7 6 5 4 3 2 1
Printed in the United States of America

For our wives, Norma and Gisèle

PREFACE

We compiled *The Oxford Illustrated Literary Guide to the United States* to help travelers find places associated with the lives and works of writers. Many other books have been written for literary travelers. Some focus on the lives of individual writers or on writers identified with particular cities, states, or regions. Others cover only the major literary shrines, for example, Poe Cottage in New York City, William Faulkner's stately home in Oxford, Mississippi, and Jack London's favorite haunt, the First and Last Chance Saloon in Oakland, California. Such books are helpful when planning a trip to a particular region or a major literary shrine. Unfortunately, travelers with an uncertain itinerary have had to carry an assortment of guidebooks, or risk overlooking places they would have found amply rewarding.

We have, therefore, assembled information about the homes and work places of more than fifteen hundred literary figures from Colonial times to the present and in all parts of the country. In order not to intrude on privacy, we have not included the present addresses of authors, but we have otherwise been as specific and helpful as possible. Literary travelers now can find their way to the Bethlehem, Pennsylvania, gravesite of the Indian who was the model for James Fenimore Cooper's heroic Mohican; the Columbus, Ohio, penitentiary in which an athletic field is named for O. Henry; and the Salem, Massachusetts, home in which a young writer named Nathaniel Hawthorne worked on *The Scarlet Letter,* which proved to be his masterpiece.

Thousands of literary sites are identified and located in 1,586 hamlets, villages, towns, and cities. Some of these places are associated with only one author, but many have multiple literary associations. Needless to say, Boston, Chicago, New Orleans, San Francisco, and our other major cities are bonanzas for the literary traveler. New York City alone, by virtue of its size and dominance in the publishing life of our country, can hold one's attention for days.

Less populous places have their own attractions. Some towns have established themselves on the literary map because of a single writer: Hannibal, Missouri, for Mark Twain; Hillsboro, West Virginia, for Pearl Buck; Red Cloud, Nebraska, for Willa Cather; West Salem, Wisconsin, for Hamlin Garland; and Sauk Centre, Minnesota, for Sinclair Lewis. Other places, such as New Haven, Cambridge, and Princeton, have rich literary associations because of the presence of major universities. Finally, there is the special interest of places that are identified with literary works themselves. In Scituate, Massachusetts, the literary traveler may see the Old Oaken Bucket Homestead; in Charleston, South Carolina, the area where DuBose Heyward's Porgy begged; in Monterey, California, once-thriving Cannery Row; in Santa Fe, New Mexico, the cathedral Willa Cather's Bishop Jean Latour worked so hard to build.

In some towns and cities of our country, residents have marked authors' former homes, and sites where authors' homes once stood; in others, newspaper archives, files of historical societies, and public libraries must be searched to turn up the necessary clues. Such literary detective work can be an exciting part of any holiday—we have found it so.

While no illustration can give the satisfaction of an actual visit, we have included many photographs and drawings in the *Oxford Guide.* They can help literary travelers find houses they are looking for, and serve also as a reminder of places visited. Some of the photographs were supplied by authors or their families. Many are snapshots taken by literary travelers, who willingly share their souvenirs with our readers. Still others are postcard views of literary sites. In this sense, then, the *Oxford Guide* has somewhat the immediacy of a family picture album. These illustrations serve another function, that of showing us how varied have been the fortunes of our poets and writers. Some houses that appear in the following pages are almost palatial, but many suggest straitened circumstances or even temporary shelter without amenities.

During the four years in which we worked on the *Oxford Guide,* we were struck by the restlessness of American writers. To follow Mark Twain, for example, as he moved from place to place is a staggering task, but his continual moving was characteristic of many of our poets and authors. They appear to have moved mostly in search of work, work as a writer when possible, but too often just any kind of work. A reading of *The Oxford Literary Guide to the British Isles* (1977), the excellent guidebook written by Dorothy Eagle and Hilary Carnell, reveals far less of this moving about by British writers. Can it be that American wandering manifests an intense restlessness of spirit, a conviction that one's literary destiny lies always in the place beyond, rather than at home?

A word should be said about how the *Oxford Guide* was compiled. Much of our research was initiated through questionnaires sent to authors, librarians, local historians, public officials, and such organizations as local Chambers of Commerce. The material we obtained in this way yielded precise information we would otherwise have been unable to find. We name each of these many helpful people and organizations in the Acknowledgments Section of this book.

We also acquired considerable information from our own local public libraries. In addition, these libraries provided access to an extensive interlibrary system, enabling us to read books we would otherwise have found difficult to obtain. The names of our local librarians appear in the Acknowledgments Section, but we wish to express here as well our appreciation for the help given us by the librarians of the Mount Pleasant Public Library, Pleasantville, New York, who never failed to assist us to the limits of their resources, and always with ample patience.

We wish also to express our appreciation to Pam Forde Graphics for developing a handsome design into an attractive finished book, and to Christopher Barbieri, the artist whose fine maps of our country appear in this book. Finally, we convey our deep appreciation to Raymond V. Hand, Jr., of The Hudson Group. With never-failing devotion, Ray assisted in our research and editorial work from early in its inception until completion. Without his diligence, loyalty, persistence, and good judgment, *The Oxford Illustrated Literary Guide to the United States* would lack a good measure of its authority and completeness. We thank him for his professionalism and for his unfailing good humor.

EUGENE EHRLICH
GORTON CARRUTH

Pleasantville, New York
June 1982

We invite correspondence providing information about literary sites in your town.
Please write us at The Hudson Group, 74 Memorial Plaza, Pleasantville, New York 10570.

TABLE OF CONTENTS

STATES

CITIES & TOWNS

THE OXFORD ILLUSTRATED LITERARY GUIDE TO THE UNITED STATES

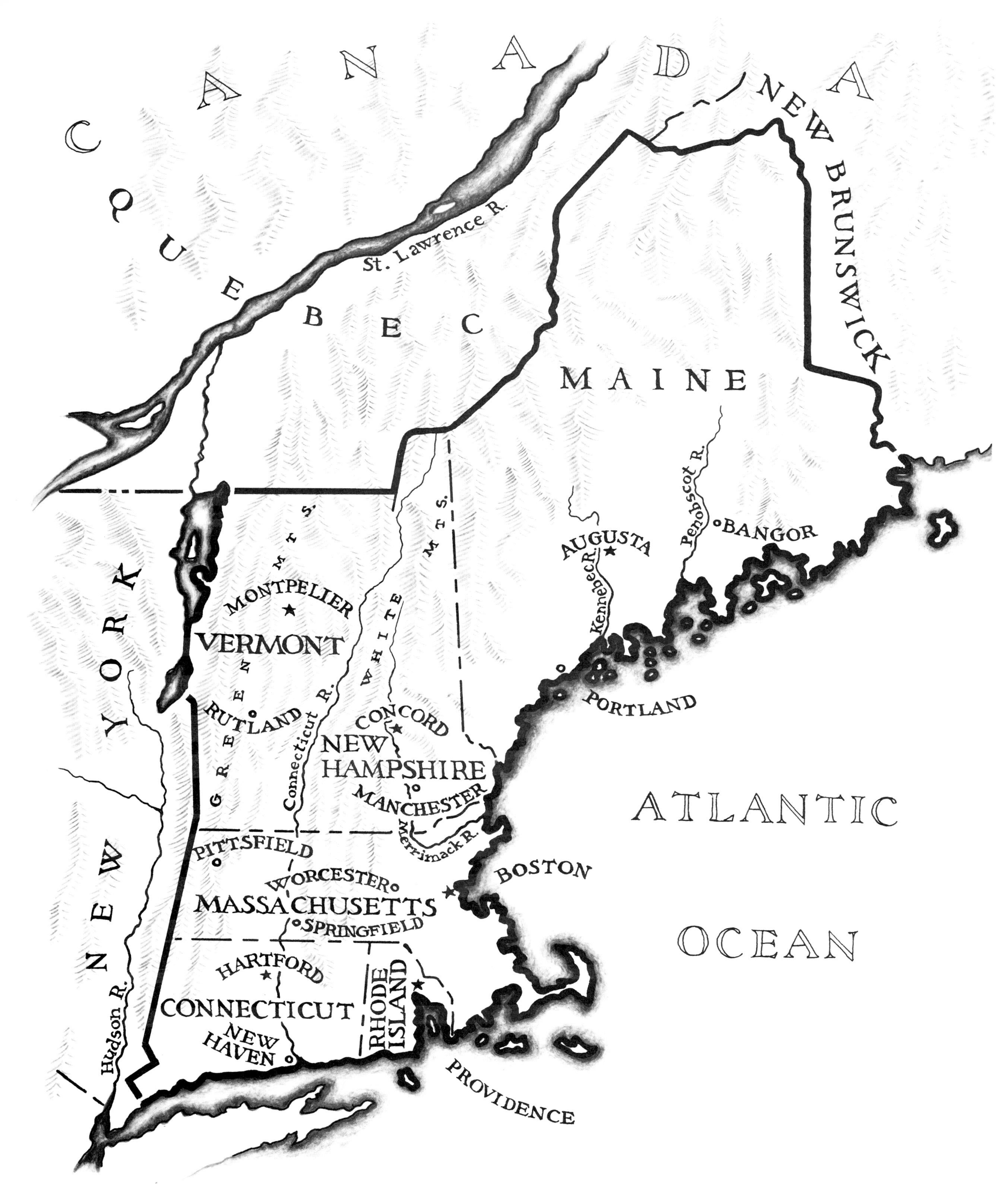

CANADA
QUEBEC
St. Lawrence R.
NEW BRUNSWICK
MAINE
Penobscot R.
BANGOR
AUGUSTA
Kennebec R.
PORTLAND
NEW YORK
VERMONT
MONTPELIER
GREEN MTS.
RUTLAND
Connecticut R.
WHITE MTS.
CONCORD
NEW HAMPSHIRE
MANCHESTER
Merrimack R.
PITTSFIELD
WORCESTER
MASSACHUSETTS
BOSTON
SPRINGFIELD
HARTFORD
CONNECTICUT
NEW HAVEN
RHODE ISLAND
PROVIDENCE
Hudson R.
ATLANTIC OCEAN

NEW ENGLAND

MAINE

ALNA

The village of Head Tide in Alna, a town on the Sheepscot River about fifteen miles south of Augusta, was the birthplace of **Edwin Arlington Robinson,** born on December 22, 1869. The poet's father, Edward Robinson, ran a store nearby and sold timber to shipbuilders; with the wealth he attained he moved his family to Gardiner less than a year after Edwin was born. The poet did come back to Head Tide on visits, however, and he used material from these visits in such poems as "Stafford's Cabin" (1916). After Robinson's death in 1935, a group of his friends bought the house he was born in and undertook its repair and maintenance. In 1973, the house was given as a gift to Colby College, which now offers it each summer as a vacation home to a member of the college faculty. The house is occupied during the winter months as well, and literary visitors are welcome if prior arrangements are made, or if someone staying in the house is at home.

AUBURN

This city across the Androscoggin River from Lewiston was for many years the home of **Holman F. Day,** who moved here after his marriage in the mid-1890s. Day was born in Maine and lived in the state during most of his life. Many of his

Birthplace of Edwin Arlington Robinson in Alna. Robinson, who did not live here very long, once wrote to a subsequent owner of the house: "I cannot tell you much except that I was born in it in December, 1869. Six months later we all went to Gardiner."

Maine is still the closest to Heaven that I've yet come, and the place I'd be willing to settle on for a celestial abode.

—Hodding Carter, Jr.

books, whether volumes of poetry such as *Up in Maine* (1900) or novels such as *King Spruce* (1908), dealt with Maine subjects. His home here, the Holman Day House, 2 Goff Street, has changed little since Day lived here. Now privately owned, the house has been nominated for inclusion in the National Register of Historic Places.

BANGOR

Owen Davis was born in Bangor, Maine's third largest city, on January 29, 1874. His childhood home, a brick house at 166 Union Street now known as the Isaac Farrar House, is open to the public daily. Davis won the Pulitzer Prize for his play *Icebound* (1923).

BATH

MacDonald Clarke, called the mad poet of Broadway, was born on June 18, 1798, in Bath, a town about thirty-five miles up the coast from Portland. Clarke eventually took his own life in 1842 while he was being held in the insane asylum on Blackwell's Island, New York. His poems were collected in such volumes as *Elixir of Moonshine by the Mad Poet* (1822) and *The Belles of Broadway* (1833). Today Clarke is remembered almost exclusively for a single delicate fragment:

Now twilight lets her curtain down
And pins it with a star.

BLUE HILL

Mary Ellen Chase, the novelist, essayist, and biographer who often wrote about Maine, was born in this town on the coast, south of Bangor, on February 24, 1887. The Chase family home on Union Street, though still standing, is privately owned and not open to visitors. Chase attended services at the Congregational Church, on Main Street, and was graduated from the George Stevens Academy, on Union Street. Among her many books is a biography of another Blue Hill resident, Jonathan Fisher (1768–1847), a minister whose house on Main Street is now a museum. Of Chase's novels, *Mary Peters* (1934), the story of a New England family, and *Windswept* (1941) were the best known. Chase died on July 28, 1973, and is buried in Seaside Cemetery, a short walk east from the Blue Hill Post Office. The inscription on her gravestone is from Isaiah: "They shall mount up with wings, as eagles."

Postcard view of Goff Street in Auburn at the turn of the century. The Holman Day house, at the left, is easily identifiable by its three-story tower and covered porch.

BOOTHBAY HARBOR

Toward the end of July 1929, **Thomas Wolfe** rented a cottage at Ocean Point in this popular coastal resort and here read the proofs of *Look Homeward, Angel* (1929), his first novel. The cottage he rented is still standing but not open to the public.

BRIDGTON

Seba Smith, best known for his series of satirical pieces written under the name of **Major Jack Downing,** lived in Bridgton, west of Lewiston, from 1799 until about 1815. Smith studied and taught in the local public schools, for a brief time attended Bridgton Academy, which is still functioning.

BRUNSWICK

This town northeast of Portland is the home of Bowdoin College, founded in 1794 and alma mater of a number of literary men. **Jacob Abbott,** who was graduated in 1820, was best known for his series of Rollo books, written for children. **Hodding Carter, Jr.,** who was graduated in 1927, was a distinguished newspaperman and author of the novels *Winds of Fear* (1944) and *Flood Crest* (1947). He often returned to Maine later in life for vacations. **Elijah Kellogg,** who was graduated in 1840, was an author of children's stories. He is remembered still for his poem "Spartacus to the Gladiators" (published originally in *School Reader,* 1846). **Seba Smith,** the humorist and satirist, was graduated in 1818. **Robert P[eter] Tristram Coffin,** who was graduated from Bowdoin in 1915, was born in Brunswick on March 18, 1892. He returned to Bowdoin in 1934, after graduate work at Oxford and a period as a member of the English faculty at Wells College, to serve as Pierce Professor of English. His residence at 44 Harpswell Street, not far from his birthplace, is still standing but not open to visitors. Coffin won a Pulitzer Prize in 1936 for his collection of verse called *Strange Holiness,* and many of his other books, such as *Saltwater Farm* (1937), had Maine as their subject.

Of all Bowdoin's literary alumni, the two most famous were graduated together in 1825, in a class that also included Franklin Pierce, a future U.S. President. One of them, **Nathaniel Hawthorne,** had come here unwillingly and made conventional but poor use of his undergraduate years. He gambled a little, enjoyed drinking with his friends, and generally devoted himself to the pursuit of light-hearted good fun. He used Bowdoin later on as the setting for *Fanshawe* (1828), his first novel. The dormitory he lived in and Massachusetts Hall, where he attended classes, are both still standing. So is an off-campus building he lived in, at 25 Federal Street. The college library, not surprisingly, maintains a large collection of the author's works. While here, Hawthorne was barely acquainted with **Henry Wadsworth Longfellow,** the other noteworthy man of letters in the class of 1825.

Longfellow entered at age fourteen and spent his time productively. He published a number of poems and worked hard at his courses. By his senior year, he felt confident enough to write to his father: "I most eagerly aspire after future eminence in literature." After graduation, Longfellow was offered a chair in modern languages on condition that he study further in Europe. This he did, from 1826 to 1829, and then returned to Bowdoin, for appointment both as professor of modern languages and college librarian. In July of 1831, he married Mary Potter, and the couple lived at 76 Federal Street, still standing but not open to visitors. Dissatisfied with his position at the college, Longfellow left for Europe in 1835 with his wife. Like Hawthorne, Longfellow attended classes at Massachusetts Hall and lived at 25 Federal Street. He also attended Brunswick's First Parish Church. The Bowdoin Library has one of the most extensive collections of Longfellow materials to be found anywhere.

As though this list of literary residents were not long enough, Brunswick also claims **Harriet Beecher Stowe,** who moved from Ohio in 1850 to a house at 63 Federal Street after her husband was appointed to the Bowdoin faculty. It was in the First Parish Church that Stowe had the vision that inspired her to write *Uncle Tom's Cabin* (1852). The novel, published first in serial form, sold some 10,000 copies in the first week of its publication as a book, and more than 300,000 copies within a year. Stowe's other books include *A Key to Uncle Tom's Cabin* (1853) and *Dred* (1856). Her home in Brunswick is now used as an inn and has been designated a National Historic Landmark. A plaque marks her pew at the First Parish Church.

BUCKFIELD

Seba Smith was born on September 14, 1792, in this town northwest of Auburn. His family moved to Bridgton when Seba was seven. Smith's humorous works included *May-Day in New York* (1845), *'Way Down East* (1854), and *My 30 Years Out of the Senate* (1859).

BUXTON

This town in the southern tip of the state entered literary history through the work of **Kate Douglas Wiggin,** who lived for most of her life in Hollis, the next town west. Tory Hill Meeting House at Buxton Lower Corner, on Route 112, is the setting for her 1917 play based on her book *The Old Peabody Pew* (1907) which is presented in Buxton every year by the Dorcas Society. A marble cross erected in Wiggin's memory in the Tory Hill Cemetery carries the final words of her autobiography, *My Garden of Memory* (1923): *The Song Is Never Ended.*

CALAIS

Harriet Prescott Spofford was born on April 3, 1835, in a house on Main Street in this Down East town on the St. Croix River. Her birthplace was torn down several years ago. Only a flat stone, believed to have been at the base of the steps, remains on the vacant lot. Spofford's works included novels, poetry, essays, and autobiography, but she was best known for her romantic stories, collected in such volumes as *The Amber Gods* (1863) and *New-England Legends* (1871).

CAMDEN

Edna St. Vincent Millay came to this town on Penobscot Bay with her mother, recently divorced, who found work as a housekeeper. Often they lived in the houses where Mrs. Millay was employed, and one of them at least, at 82 Washington Street, is still standing but privately owned. Millay attended Sunday School at the First Congregational Church, now the United Church of Christ on the corner of Elm and Free streets. She took piano lessons at the Tufts-Cushing residence on Chestnut Street and was graduated from Camden High School. The high-school building, though still in use, has been condemned. The young poet, perhaps in 1910, climbed the hills around nearby Mount Battie, a favorite spot of hers. Out of the experience she wrote her famous poem "Renascence" (1912), which had an enormous impact on her future. Sitting in the audience when she read the poem at Camden's Whitehall Inn was Caroline B. Dow, head of the National Training School of the YWCA. Dow was so impressed that she raised money to send the young poet to college. In gratitude the poet dedicated to Dow the volume *Second April* (1921), Millay's second book of poems. The Whitehall Inn is still standing and has an Edna St. Vincent Millay Reading Room, which contains such items as Millay's high-school diploma, photographs, and a scrapbook of her press clippings. Mount Battie has a plaque commemorating the poet's climb in the nearby hills.

Another literary resident of Camden was **William Gilbert Patten,** known for the Frank Merriwell novels he wrote under the name **Burt L. Standish.** Patten spent many summers at his vacation house, Overocks, at 2 Limerick Street, and spent longer periods of time here in his later years when he turned away from his popular fiction and tried, unsuccessfully, to do more serious work. After his wife's death in 1938, Patten suffered a nervous breakdown. He moved to California in 1941 and lived there until his death in 1945.

CASTINE

Ellen Glasgow came to this village on Penobscot Bay, about thirty-five miles south of Bangor, every summer from 1935 to 1945. She rented two houses in her time here: Littleplace, on Battle Avenue, and Appledoor—this was the novelist's name for the house—on the corner of Perkins and Madockawando streets. Both houses are privately owned and closed to visitors. Another summer resident was **Robert Lowell,** who made reference to Maine in a number of poems.

CORINNA

William Gilbert Patten was born on October 25, 1866, in this town northwest of Bangor. Patten wrote Western stories under the pen name **William West Wilder** but adopted the pen name **Burt L. Standish** for his Frank Merriwell novels, the first of which appeared in 1896 and at the

All I could see from where I stood
Was three long mountains and a wood;
I turned and looked the other way,
And saw three islands in a bay.
So with my eyes I traced the line
Of the horizon, thin and fine,
Straight around till I was come
Back to where I'd started from;
And all I saw from where I stood
Was three long mountains and a wood.

—Edna St. Vincent Millay,
in "Renascence"

In the fall of 1829, I took it into my head I'd go to Portland. . . . I up and told father, and says: "I am going to Portland, whether or no, and I'll see what this world is made of yet."

Father stared a little at first, and said he was afraid I would get lost, but when he see I was bent upon it . . . he stepped up to his chest and opened the till, and took out a dollar and gave it to me; and says he:

"Jack, this is all I can do for you. Go and lead an honest life, and I believe I shall hear good of you yet."

—Seba Smith,
in "My First Visit to Portland"

Backward, turn backward, O Time,
in your flight,
Make me a child again just for
tonight.

—Elizabeth Chase Akers,
in "Rock Me to Sleep, Mother"

Our worthy Captain Lovell among
them there did die,
They killed Lieutenant Robbins,
and wounded good young Frye,
Who was our English Chaplain;
he many Indians slew,
And some of them he scalped
when bullets round him flew.

—From the anonymous ballad
"Lovell's Fight"

end numbered more than two hundred volumes. The Stewart Free Library in Corinna maintains a nearly complete collection of the Frank Merriwell books.

DARK HARBOR

James Weldon Johnson had a summer home in this town on Warren Island and was staying here at the time of his death in an automobile accident on June 26, 1938, in Wiscasset.

EAST ORLAND

Walter Van Tilburg Clark was born on August 3, 1909, in this tiny village south of Bangor. Clark grew up in Nevada, so it is not surprising that his fiction was laid in the West. Clark's best-known work is *The Ox-Bow Incident* (1940).

FARMINGTON

Jacob Abbott, who wrote many inspirational books for children, summered at his father's home in Farmington, about forty miles north of Lewiston, from 1836 to 1870. From 1870 until his death in 1879, Abbott lived year round at Fewacres, a house at 93 Main Street, now privately owned and listed in the National Register of Historic Places. The Mantor Library of the University of Maine, in Farmington, has a large collection of Abbott's work.

FARMINGTON FALLS

Elizabeth Chase Akers came to live in this town near Farmington in the 1830s, when her father, Thomas Chase, ran a sawmill here. The mill burned down in 1838, and the family moved away soon afterward, only to return a few years later, when Mr. Chase bought the Farmington Falls Hotel. Elizabeth, who had begun writing poetry at an early age, briefly supplemented the family's income by running a school across the street from the hotel. She moved away in 1851. Her only widely known poem, "Rock Me to Sleep, Mother," was published in the Philadelphia *Saturday Post* ten years later, but some long-time Farmington Falls residents have claimed that Chase wrote it while living here.

Fewacres, home of Jacob Abbott in Farmington, as it appeared in 1910. The author of the Rollo books lived here during the last decade of his life.

FRYEBURG

Not far from this town in southern Maine, near the New Hampshire border, is Lovell's Pond, scene of a day-long battle on May 8, 1725, between Indians and volunteer scouts led by Capt. John Lovell that found its way into nineteenth-century literature. Two survivors of the battle appear in the story "Roger Malvin's Burial" (1832), by Nathaniel Hawthorne, and the battle itself was the subject of "The Battle of Lovell's Pond" (1820), the first published poem by Henry Wadsworth Longfellow, thirteen years old at the time. The battle was also the subject of "Lovell's Fight," an anonymous ballad once widely known.

Clarence E. Mulford moved to Fryeburg in 1926 from Brooklyn, where he had spent twenty of his most productive years as a Western writer. His most famous creation was Hopalong Cassidy, introduced in 1910 in a novel of that name and later becoming a staple of motion pictures and television. Mulford lived in Fryeburg until his death on May 11, 1956, and is buried in Pine Grove Cemetery, Route 302. His house, now privately owned and not open to visitors, stands at 102 Main Street. The Fryeburg Public Library exhibits books from Mulford's personal collection as well as copies of most of his works and assorted memorabilia.

Fryeburg's only known literary native son is **Charles Gamage Eastman,** sometimes called the Burns of New England, who was born here on June 1, 1816. In addition to writing poetry, Eastman founded two periodicals, *Spirit of the Age* (1840) and *Vermont Patriot* (1846). His collected *Poems* appeared in 1848.

GARDINER

Laura Elizabeth Richards lived for much of her life at The Yellow House, 3 Dennis Street, in this town just south of Augusta. She married an Augusta native, Henry Richards, in 1871, and five years later they came to live here. The author remained until her death in 1943. It was while living here that Richards began to write poetry. She branched out later into other literary fields, writing more than eighty books in all and winning a Pulitzer Prize in 1917 for a biography—written in collaboration with others—of her mother, Julia Ward Howe. Richards's other works included *Snow White* (1900), *Abigail Adams and Her Times* (1917), and *Honor Bright* (1920). She died in Gardiner on January 14, 1943, about six weeks short of her ninety-third birthday, and is buried in the cemetery of Christ Church, on Dresden Avenue. Today The Yellow House is privately owned and not open to visitors.

One of Richards's other biographical subjects was **Edwin Arlington Robinson,** who lived here from 1870 until 1891, when he was twenty-one. Robinson lived in childhood in a Greek Revival house, still standing, at 67 Lincoln Avenue. In his small room at the back of the house he wrote early poems, and from his experiences in Gardiner absorbed much that would influence his mature work. One lasting impression came from the funeral processions that frequently passed on their way to a nearby cemetery. Gardiner is said to be a partial model for Tilbury Town, the setting for a number of Robinson's poems. Robinson returned to Gardiner from Harvard in 1893 to care for his father, whose health had begun to fail, and stayed until 1896, the year he published, with his own funds, his first volume of poetry. Originally entitled *The Torrent and the Night Before,* the book was reissued in 1897 as *The Children of the Night.* The house on Lincoln Avenue was sold in 1903 and remained out of family hands until about twenty years ago, when it was bought by the poet's niece and her husband. Today the house is still a private residence, but a plaque identifies it as a National Historic Landmark. The cemetery near the house contains the poet's ashes, and a monument to him in the common on Church Square carries an inscription written by Laura Elizabeth Richards.

> *"Right through the forest, where none can see,*
> *There's where I'm going, to Tilbury Town.*
> *The men are asleep,—or awake, may be,—*
> *But the women are calling John Evereldown.*
> *Ever and ever they call for me,*
> *And while they call can a man be free?*
> *So right through the forest, where none can see,*
> *There's where I'm going, to Tilbury Town."*
>
> —Edwin Arlington Robinson,
> in "John Evereldown"

HALLOWELL

Jacob Abbott, author of the twenty-eight-volume series of Rollo books, was born in this town just south of Augusta on November 14, 1803. He was educated at Hallowell Academy and lived here until 1828, when he moved with his bride to Boston. His residence here, at 53 Winthrop Street, is still standing but not open to visitors. Another Hallowell native is **Osborne Russell,** born here on June 12, 1814. Russell's years in the Far West provided material for *Journal of a Trapper, or Nine Years in the Rocky Mountains 1834–43,* an important account of life on the frontier that, for various reasons, went unpublished until 1914.

HARPSWELL

Elijah Kellogg, preacher and poet, often visited this coastal town south of Brunswick during his undergraduate years, 1837 to 1840, at Bowdoin College. When he entered Andover Theological Seminary in 1840, the people of Harpswell asked him if he would become their minister after graduating. Kellogg joked that he would, but only if they built him a church. Three years later, on his graduation, he learned that they had taken him seriously: a church was already under construction. A man of his word, Kellogg turned down other offers and moved here in 1844. The home he built on a hill overlooking Middle Bay was completed in 1849, and there he lived until his death in 1901. In 1846, Kellogg's poem "Spartacus to the Gladiators" was published in Epes Sargent's *School Reader* and soon became a favorite recitation piece for schoolchildren. Kellogg also wrote a number of children's books, usually in series and with Maine settings and characters. Among them were the Elm Island series and the Whispering Pine series. His Harpswell home, on Route 123, is privately owned and reported to be in disrepair, but it has recently been nominated for inclusion in the National Register of Historic Places.

HIRAM

In recognition of service during the Revolutionary War, Gen. Peleg Wadsworth in 1790 was granted a large tract of land near Hiram, about thirty-five miles northwest of Portland. Here Wadsworth built the large mansion he called Wadsworth Hall, to which, after serving in Congress, he came to retire. Among his visitors here was his young grandson **Henry Wadsworth Longfellow.** From an early age Henry spent his summer vacations at Wadsworth Hall. He would listen to his grandfather's stories about the Revolutionary War, learning from him the story of the Battle of Lovell's Pond, which the young poet used as the basis for his first published poem (1820). Peleg Wadsworth died in 1829 and is buried on the grounds of Wadsworth Hall. The house itself was subsequently used over the years as a meeting hall and indoor drill area and is now a private residence.

HOLLIS

This town in the southern tip of Maine was for many years the home of **Kate Douglas Wiggin.** Wiggin, born in 1856, lived here for most of her childhood. At seventeen she moved away, first to California and later to New York City, but returned here in the summer of 1889, following the success of her book *The Birds' Christmas Carol* (1888) and the death of her husband. She

Portrait of Edwin Arlington Robinson by Lilla Cabot Perry. The painting hangs in the Robinson Memorial Room of the Colby College library, in Waterville, Maine.

Home of Edwin Arlington Robinson in Gardiner, where the three-time Pulitzer Prize-winner first began writing poetry.

Rebecca's eyes were like faith,—"the substance of things hoped for, the evidence of things not seen." Under her delicately etched brows they glowed like two stars, their dancing lights half hidden in lustrous darkness. Their glance was eager and full of interest . . . and had the effect of looking directly through the obvious to something beyond. . . . They had never been accounted for, Rebecca's eyes.

—Kate Douglas Wiggin,
in *Rebecca of Sunnybrook Farm*

remarried in 1895 and, some ten years later, moved into a house that still stands on Salmon Falls Road near the Saco River, at Salmon Falls. She had admired the house since childhood. In choosing a name for the house, Wiggin decided that if a dove's home is a dovecote, a writer's home must be a quillcote. And Quillcote she named the house. She also bought another nearby building, near the Salmon Falls Bridge, in which she started a kindergarten. She and her sister later converted the building for use as a library and donated it to the village. Wiggin's books, most of them written for children, reflect her love of Maine and its people. *The Birds' Christmas Carol* is sometimes considered her finest work, but the one known best is still unquestionably *Rebecca of Sunnybrook Farm* (1903). Wiggin's autobiography, *My Garden of Memory,* was published in 1923, the year of her death. Although Wiggin died in England, her ashes were brought back to Hollis and scattered over the waters of the Saco River, near Quillcote. The house today is privately owned and included in the National Register of Historic Places.

KENNEBUNK

Kenneth Roberts was born on December 8, 1885, in the Storer Mansion on Storer Street in Kennebunk, a town near the coast about midway between Portland, Maine, and Portsmouth, New Hampshire. Kennebunk formerly was called Arundel, and Roberts used the town as the setting for one of his best historical novels, *Arundel* (1930), the story of Benedict Arnold's unsuccessful expedition against Quebec in the Revolutionary War. The novel was the first in a projected series initially called Chronicle of Arundel. Whether fiction or nonfiction, many of Roberts's books showed his fondness for his native state. His nonfiction works included *Trending into Maine* (1938) and *Good Maine Food* (1939), written with Marjorie Mosser. Among his other novels were *Rabble in Arms* (1933), *Oliver Wiswell* (1940), and his best-known work, *Northwest Passage* (1937), a story about Rogers' Rangers. The Storer Mansion, Roberts's birthplace, is still standing but is not open to visitors.

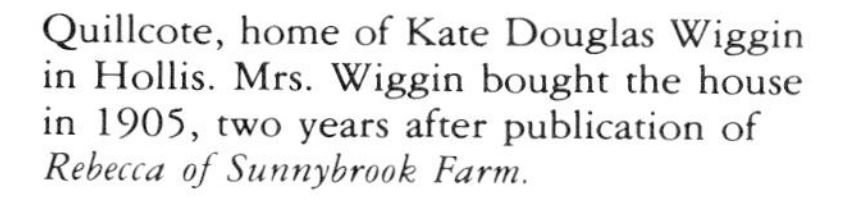

Quillcote, home of Kate Douglas Wiggin in Hollis. Mrs. Wiggin bought the house in 1905, two years after publication of *Rebecca of Sunnybrook Farm.*

KENNEBUNKPORT

This coastal town near Kennebunk has attracted several literary figures. **Margaret Deland,** the prolific novelist and short-story writer, who was known best for *The Awakening of Helena Richie* (1906) and *The Iron Woman* (1911), spent nearly every summer here from her marriage in 1880 until her death in 1945. Her home, Greywood, was not far from that of her good friend **Kenneth Roberts,** who lived in Kennebunkport year round. When he was not writing, Roberts spent his time pursuing a variety of interests, such as painting, raising ducks, reclaiming land, and dowsing, in which arcane practice he believed fervently. He also hunted and fished at places throughout Maine. Roberts built his Kennebunkport home, Rocky Pastures, in 1938 and died there on July 21, 1957, not long after publication of his third book on dowsing, *Water Unlimited.* His home, on Wildes District Road, was severely damaged by fire several years ago, but has since been rebuilt completely.

Roberts had another good friend and neighbor in **Booth Tarkington,** who first came here in 1915 while working with a theater group and writing a play, *Clarence* (1919), for Helen Hayes. Tarkington returned to Kennebunkport in 1917 and built a colonial frame house, now private, which he called Seawood, and which he referred to as "the house that Penrod built." Seawood, located on South Maine Street, on a hill overlooking the Kennebunk River, was to be the author's summer home for the rest of his life. He enjoyed sailing on his schooner here and going out by motorboat to look for whales. Tarkington worked on his books in an oak-paneled room at Seawood, as well as in a studio on Ocean Avenue, in a building called the Floats. Sometimes he worked in the cabin of his schooner, the *Regina,* which was for many years to be found tied up at the Floats, its bowsprit pointing out over Ocean Avenue. The schooner has since gone the way of many old vessels, its stripped hulk sunk at sea.

Another literary resident of Kennebunkport was **John Townsend Trowbridge,** the author and editor, who spent summers in a house he called Spouting Rock Cottage. In fact, it was Trowbridge who named two local scenic points—Spouting Rock and Blowing Cave.

KITTERY POINT

William Dean Howells bought a house in 1902 on Pepperell Road in Kittery Point, in the southern tip of the state, and summered here until his wife's death in 1910. Howells used the house occasionally after that until 1919, when he gave it to his son. Today the place is owned by Harvard University, which uses it as a conference center and as a vacation house for faculty members. It is not open to the general public.

MERCER

Frank A. Munsey, founder in 1889 of the magazine *Munsey's,* was born in Mercer, northwest of Waterville, on August 21, 1854. The family moved to Gardiner six months later, and their house in Mercer was destroyed by fire some time afterward. Among Munsey's novels, none much read today, were *Derringforth* (1894) and *A Tragedy of Errors* (1899).

MOUNT DESERT

Margaret Barnes spent summers on Mount Desert Island, whose best-known town is Bar Harbor. Her home was in a rural district without street names or numbers. Barnes won a Pulitzer Prize for her novel *Years of Grace* (1930).

MOUNT VERNON

Erskine Caldwell lived from 1929 to 1933 at his wife's parents' home in Mount Vernon, a town on Route 41, northwest of Augusta. He had come here to improve his skills as a short-story writer, and one of his efforts, "Country Full of Swedes," eventually won for him the 1933 Yale Review Award for Fiction. During the same period Caldwell wrote five novels, including the immensely popular *Tobacco Road* (1932) and *God's Little Acre* (1933). The novelist's wife, from whom he was later divorced, still lives in the house, which is not open to visitors.

NOBLEBORO

Henry Beston lived with his wife, **Elizabeth J[ane] Coatsworth,** for many years at Chimney Farm, in Nobleboro, off Route 1, northeast of Bath. Beston's works included children's books and nature writings, of which *The Outermost House* (1928) and *American Memory* (1937) were the best known. His *White Pine and Blue Water* (1950) was written about the state of Maine. Beston died here on April 15, 1968, and is buried in the Chimney Farm Burying Ground. Elizabeth Coatsworth, a poet, novelist, and author of books for children, is known especially for *The Cat Who Went to Heaven* (1930). Her novel *Here I Stay* (1938) and her book of sketches *Country Neighbor* (1944) were set in Maine. Chimney Farm is still owned by Elizabeth Coatsworth.

NORRIDGEWOCK

Rebecca Sophia Clarke, who wrote children's books under the pen name **Sophie May,** was born in Norridgewock, fifteen miles northwest of Waterville, in 1833. Educated at Norridgewock's Female Academy, she moved with her family into what is now called the Sophie May House. She returned here in 1857, after a seven-year stay in Illinois, and lived in the house until her death in 1906. Sophie May set many of her stories here, drawing material from the people and places she knew so well, and was known best for her *Dolly Dimple Stories* (1867–69). In recognition of her attachment to Norridgewock, she gave the town a building for its library, and the town reciprocated by naming for her the street on which her house stands. The house itself, now privately owned, was nominated in 1976 for inclusion in the National Register of Historic Places.

NORTH YARMOUTH

Elizabeth Oakes Smith was born on August 12, 1806, in North Yarmouth, Route 9, about twenty miles north of Portland. She wrote novels and poetry under several pen names, including **Ernest Helfenstein,** and achieved a popularity comparable to that of her husband, **Seba Smith,** the satirist. Her many novels included *The Newsboy* (1854) and *The Bald Eagle* (1867).

OGUNQUIT

Sinclair Lewis lived in a rented cottage in this village in the southern tip of Maine during the summer of 1939. He worked as stage manager on a production of *Our Town* (1938), by Thornton Wilder, and wrote parts of *Bethel Merriday* (1940), his novel about a young woman's struggle to achieve recognition as an actress.

PORTLAND

Portland is the premier literary city in Maine. **Elizabeth Chase Akers** lived here in three distinct periods of her career in the second half of the nineteenth century while editing newspapers, including the Portland *Daily Advertiser,* and writing poetry. It was while she was here that she wrote "Rock Me to Sleep, Mother," the poem with which she is identified today. **Elizabeth Oakes Smith** and her husband, **Seba Smith,** spent many years in Portland both before and after their marriage in 1823. Elizabeth Oakes had moved here with her family in 1814, when she was eight. Smith, in about 1821, came to edit the *Eastern Argus.* To boost the circulation of the Portland *Courier,* which he founded in 1829, he began writing his famous letters from a character called Major Jack Downing. The pieces were an immediate success. His wife's success as an author did not come until after 1839, when they moved to New York City. **Ann Sophia Stephens** edited and contributed poetry to her husband's *Portland Magazine* from 1831 to 1837.

A number of writers were born in Portland. One of them is **Harry Brown,** who was born on April 30, 1917. His novels include *A Walk in the Sun* (1944), *Wake of the Red Witch* (1949), and *Stars in Their Courses* (1960). **Owen Davis,** who was born here on January 29, 1874, won a Pulitzer Prize for his play *Icebound* (1923) and

The truth is I love the place; and if I seem to talk overmuch of it, it is because I would like those who read about it to see it as I saw it, and to know the sweet smell of it and to love it as I do.

—Steven Nason, narrator of Kenneth Roberts's *Arundel,* speaking of Maine

A man never knows what's going to happen next in the State of Maine; that's why I wish sometimes I'd never left the Back Kingdom to begin with. I was making sixty a month, with the best of bed and board, back there in the intervale; but like a God-damn fool I had to jerk loose and came down here near the Bay. I'm going back where I came from, God-helping.

—Erskine Caldwell, in "Country Full of Swedes"

Seawood, home of Booth Tarkington in Kennebunkport. Here Tarkington entertained such literary guests as Alexander Woollcott and Alice Duer Miller.

The first letter written by Henry Wadsworth Longfellow: "Dear Papa, Ann wants a little bible like little Betsy's. Will you please to buy her one if you can find any in Boston. I have been to school all the week and got only seven marks. I shall have a billet[?] on monday. I wish you to buy me a drum. Henry W. Longfellow, Portland, Jan. 1814."

dramatized a number of novels, including *Ethan Frome* (1936), by Edith Wharton. **Joseph Holt Ingraham,** who was born here on January 25, 1809, wrote serials for newspapers and historical novels, including *Lafitte; or The Pirate of the Gulf* (1836) and *The Prince of the House of David* (1855). **Elijah Kellogg,** who was born here on May 20, 1813, is remembered for his poem "Spartacus to the Gladiators" (1846).

William Henry Thomes, born here on May 5, 1824, was the author of *On Land and Sea* (1883), an account of his life aboard a hide-trader. Thomes is said to have gone to sea because of his fascination with *Two Years Before the Mast* (1840), by Richard Henry Dana. *Lewey and I* (1884) was based on Thomes's experiences in the California Gold Rush. **N[athaniel] P[arker] Willis,** born here on January 20, 1806, wrote for and edited the New York *Mirror.* **Sara Payson Willis,** a sister of N. P. Willis, was also born here, on July 9, 1811. Using the pen name **Fanny Fern,** she produced essays, sketches, novels, and other books for children. Her *Fern Leaves from Fanny's Portfolio* (1853) sold 70,000 copies in its first year. **Edward Sylvester Ellis,** the dime novelist, died on June 20, 1916, while vacationing on Cliff Island, in Casco Bay, off Portland.

By far Portland's most famous literary native son is **Henry Wadsworth Longfellow,** born on February 27, 1807, at his aunt's house, which stood on Fore Street until it was demolished in the 1950s. Longfellow grew up here and in the house of his grandfather, Peleg Wadsworth, at 487 Congress Street. He was educated at various schools, including Portland Academy, where one of his teachers was Jacob Abbott. The family's place of worship was the First Parish Church, 425 Congress Street, where the family pew is marked by a brass plate. Longfellow's first poem, "The Battle of Lovell's Pond," was published in the Portland *Gazette* on November 21, 1820. A year later he went off to Bowdoin College, and after that never lived in Portland for more than short periods. Nonetheless, for years the poet considered his grandfather's house in Portland to be his true home, and he returned for visits at least until 1881. On one visit, in 1847, he walked with his friend William Pitt Fessenden to "the gallows," a hillside at Congress and Vaughan streets so called because two men had once been hanged there. Ten years later the hill suggested the setting for Longfellow's poem "Changed" (1858). The Wadsworth-Longfellow House, at 487 Congress Street, is open as a memorial to the poet from June to September each year, and its fifteen rooms display Longfellow memorabilia.

Another distinguished Portland literary native was **John Neal,** born on August 25, 1793.

Wadsworth-Longfellow House, boyhood home of Henry Wadsworth Longfellow in Portland. On display for visitors are the poet's cradle and schoolbooks and the physician's bill for attending Longfellow's birth.

From the outskirts of the town,
Where of old the mile-stone stood,
Now a stranger, looking down,
I behold the shadowy crown
Of the dark and haunted wood.

Is it changed, or am I changed?
Ah! the oaks are fresh and green,
But the friends with whom I ranged
Through their thickets are estranged
By the years that intervene.

—Henry Wadsworth Longfellow, in "Changed"

Neal was much admired in his own day: Edgar Allan Poe received encouragement from Neal—"the very first . . . I remember to have heard"—and ranked him among the great American writers. Neal was one of few Americans to win the esteem of *Blackwood's Edinburgh Magazine,* the prominent literary journal, and the first writer to attempt a history of American literature. Neal's novels included *Errata; or, The Works of Will. Adams* (1823) and *Seventy-six* (1823). The latter is usually considered his finest. Neal had been living overseas as well as elsewhere in America when he decided, in 1827, to return to Portland. At 173–175 State Street he built two connected buildings that later served him as a residence and are still standing. Neal died in Portland on June 20, 1876, and is buried in Western Cemetery, at Danforth and Vaughan streets.

Westbrook College, in Portland, has a large collection of literary works by women associated with Maine, including Rebecca Sophia Clarke, Margaret Deland, Sarah Orne Jewett, Edna St. Vincent Millay, Kate Douglas Wiggin, and Harriet Beecher Stowe.

RANGELY

Elizabeth Chase, known later in life as the poet **Elizabeth Chase Akers,** lived as a young girl in this town on Rangely Lake in western Maine. Her father worked as an itinerant preacher and later ran a gristmill in Rangely, but the family moved to Farmington Falls because young Elizabeth's mother was frightened by life in the wilderness that was nineteenth-century western Maine.

ROCKLAND

Edna St. Vincent Millay was born in Rockland, a town on the western shore of Penobscot Bay, on February 22, 1892, and lived here for the next few years. Her birthplace at 198/200 Broadway is marked by a plaque. Her collection *The Harp-Weaver and Other Poems* (1923) won a Pulitzer Prize, and her other noteworthy books of poetry include *Conversation at Midnight* (1937) and *Huntsman, What Quarry?* (1939). Her *Collected Poems* was published posthumously in 1956.

ST. GEORGE

[Charles] Wilbert Snow was born on April 6, 1884 (or 1883, according to his older brother—there are no official records of his birth) on White Head, an eighty-acre island at the western entrance of Penobscot Bay, near this tiny coastal village on Route 131, below Rockland. Snow taught for many years at colleges and universities and later went on to become Maine's lieutenant governor. His love for the state was apparent in much of his poetry, collected in such volumes as *Maine Coast* (1923), *Down East* (1932), and *Maine Tides* (1940).

SEAL HARBOR

Walter Lippmann, the Pulitzer Prize-winning political journalist and commentator, had a home in this coastal community about forty-five miles southeast of Bangor for thirty years until his death in a New York City nursing home on December 14, 1974. Lippmann's later books included *The Public Philosophy* (1955) and *The Communist World and Ours* (1959).

SEARSMONT

Ben Ames Williams had a summer home in this village, about ten miles southwest of Belfast. The home was left to the novelist in 1931 by his friend A. L. McCorrison, who appears in some of Williams's stories as Bert McAusland. Williams wrote several books with Maine settings, including *Audacity* (1924) and *Immortal Longings* (1927). His home here, on Route 131, is still standing but privately owned and not open to visitors.

SHEEPSCOTT

Sheepscott, a tiny community on Route 27, south of Augusta, was the birthplace of **Ola Elizabeth Winslow,** born about 1885. Winslow kept a home in Sheepscott for much of her life and, after retiring in 1950 from full-time teaching, spent much of the year here. Best known as an authority on Puritan life, she won a Pulitzer Prize for her biography *Jonathan Edwards* (1940). Her other works included *Master Roger Williams* (1957) and *John Bunyan* (1961).

SHIRLEY MILLS

Edgar Wilson Nye, later to become widely known as **Bill Nye,** was born in this town on Route 6, south of Moosehead Lake, on August 25, 1850. Of the house where the humorist was born, situated on the north side of Main Street, only the cellar hole remains. As a tribute to the author, the hole is marked by a plaque—a touch of humor Nye himself might have appreciated. Nye achieved international recognition through his humorous writings, which were published in newspapers and in such bestselling books as *Bill Nye and Boomerang* (1881) and *Bill Nye's History of the U.S.* (1894).

SOUTH BERWICK

This coastal town in the southern tip of the state will forever be identified with the name of **Sarah Orne Jewett,** whose stories and novels about Maine life place her in the first rank among America's regional authors. She was born here on September 3, 1849, in what is now called the Sarah Orne Jewett House, an eighteenth-century building at 101 Portland Street. Except for the nearby Theodore Eastman Jewett House, built by her father in the 1850s, where she spent a few years of her childhood, the house at 101 Portland Street was the only one she ever called home. Jewett was educated at Miss Olive Rayne's School and the Berwick Academy, from which she was graduated in 1865. Much later, in 1901, she became the first woman to receive a Litt. D. from Bowdoin College, her father's alma mater. Jewett also did some traveling, spending time in Boston and Europe, but most of her life was passed in South Berwick—in and near the Portland Street house and in the rural area surrounding the town. Her father was a doctor, and as a child she would ride with him when he made his calls. From those early experiences came her love of the area and

Maine is noted for being the easternmost State in the Union, and has been utilized by a number of eminent men as a birthplace. White birch spools for thread, Christmas-trees, and tamarack and spruce-gum are found in great abundance. It is the home of an industrious and peace-loving people. Bar Harbor is a cool place to go to in summer-time and violate the liquor law of the State.

—Bill Nye,
in *Bill Nye's History of the U.S.*

I know some poison I could drink;
I've often thought I'd taste it;
But Mother bought it for the sink,
And drinking it would waste it.

—Edna St. Vincent Millay,
in "From a Very Little Sphinx"

Birthplace of Edna St. Vincent Millay in Rockland. Millay's family moved to Union, Maine, soon after she was born.

Drawing of boyhood home of Nathaniel Hawthorne in South Casco, by W. Fallon. Hawthorne later wrote: "I lived in Maine like a bird of the air, so perfect was the freedom I enjoyed. But it was there I first got my cursed habit of solitude."

her feeling for its inhabitants, and from that feeling came her books.

Jewett began to write stories when she was young, and had her first story accepted by *Atlantic Monthly* when she was nineteen. More stories followed and in 1877, at the urging of William Dean Howells, she collected them as her first book, *Deephaven.* The books that came later included *A Country Doctor* (1884), a novel that portrays a New England woman who rejects marriage to pursue a career as a physician; the novel *A Marsh Island* (1885); *A White Heron and Other Stories* (1886); and *The Country of the Pointed Firs* (1896). This last named, a portrait of a seaport town in Maine, is unquestionably Jewett's masterpiece. In an oft-quoted statement, the author's friend Willa Cather ranked it as one of the three American novels she expected to "have the possibility of a long, long life. . . ."

Sarah Orne Jewett died at home on June 24, 1909, and is buried in the Portland Street Cemetery. Her house is maintained as a memorial by the Society for the Preservation of New England Antiquities and is open to visitors. Among the memorabilia on display in the author's upstairs bedroom, which remains today as it was when she lived there, are her writing desk, riding crops, skating lantern, and two slates she used as a child.

When one really knows a village like this and its surroundings, it is like becoming acquainted with a single person. The process of falling in love at first sight is as final as it is swift in such a case, but the growth of true friendship may be a lifelong affair.

—Sarah Orne Jewett,
in *The Country of the Pointed Firs*

SOUTH CASCO

Nathaniel Hawthorne was nine years old in 1813 when he came to South Casco to live in a house on land owned by his uncle, Richard Manning. The preceding years had not been good ones for Hawthorne: his father had died in 1808, and an injury suffered while playing had restricted Nathaniel to sedentary pursuits. In this town on Sebago Lake, in southern Maine, Hawthorne nevertheless had a splendidly active time. He skated, fished, hunted, and learned to enjoy the pleasures of solitude and open countryside, and his experiences here had a marked influence on his later life. The house he lived in came to be used later as a tavern and as a meeting house, and today is owned by the Hawthorne Community Association. Although still used as a meeting place, the house is open to the public only during July and August.

STRONG

Elizabeth Chase Akers was born in this village on Route 149, north of Farmington, on October 9, 1832. Her family stayed here only a few months before moving to Rangely.

VASSALBORO

Holman F. Day was born on November 6, 1865, in this town on Route 201, north of Augusta. Day's works, many of which had Maine locales and subjects, included a collection of poems, *Pine Tree Ballads* (1902), and the novels *King Spruce* (1908) and *The Ramrodders* (1910).

VINALHAVEN

Harold Vinal was born in Vinalhaven, on Vinal Island in Penobscot Bay, on October 17, 1891. A descendant of the man for whom the island was named, Vinal went on to found and edit the quarterly magazine *Voices* (1921). Among his volumes of verse were *White April* (1922), *A Stranger in Heaven* (1927), and *Hurricane* (1936). His birthplace on Chestnut Street is still standing, but is privately owned and not open to visitors.

WATERFORD

Charles Farrar Browne, who became the widely read humorist **Artemus Ward,** was born on April 26, 1834, on a farm near the village of Waterford, in southern Maine. He later moved with his family into Waterford proper, where he lived until he was thirteen. Browne's celebrity came in 1857, when the Cleveland *Plain Dealer* published the first Artemus Ward letters. Although they were well received for their wild misspellings, grammatical errors, and puns, they are

I was born in the State of Maine of parents. As an infant I abstracted a great deal of attention. The nabers woud stand over my cradle for hours and say, "How bright that little face looks! How much it nose!"

—Artemus Ward

memorable more for their witty observations on American life and current events. In his lifetime Browne published only two books, *Artemus Ward, His Book* (1862) and *Artemus Ward, His Travels* (1865). He died in England on March 6, 1867, but is buried in the village cemetery in Waterford.

WISCASSET

James Weldon Johnson, the black poet and author, died in Wiscasset, on Boothbay Harbor, on June 26, 1938. The automobile in which he was a passenger was struck by a railroad train at a grade crossing on Main Street. Johnson died moments after being pulled from the wreckage, and his wife was seriously injured. The poet was dressed for burial—by his testamentary request—in his working outfit: a lounging robe and formal trousers. His hands held a copy of *God's Trombones* (1927), one of his books of poetry. Because Johnson's wife remained in serious condition after the accident, the poet's brother was the only relative to attend the funeral services, which were held in New York City, but among the many wreaths sent was one from Eleanor Roosevelt.

NEW HAMPSHIRE

ALLENSTOWN

Robert Frost honeymooned in 1896 in a cottage near the Suncook River, outside this village south of Concord. Two of Frost's poems, "Flower-Gathering," published in *A Boy's Will* (1913), and "The Self-Seeker," published in *North of Boston* (1914), were based on the poet's stay here.

AMHERST

Near this town, southwest of Manchester, is the one-story frame house in which **Horace Greeley** was born on February 3, 1811. Greeley learned to read while living here, soon becoming so proficient that he could devour anything from the Bible to the locally published *Farmer's Cabinet.* A memorial to Greeley stands in the town common. On the east side of the local church is the site of the Second Courthouse, where **Daniel Webster** once spoke. The building itself has been moved to Foundry Street and is now a private home.

BETHLEHEM

Anna Hempstead Branch, a poet known for such volumes as *Rose of the Wind* (1910) and *Sonnets from a Lock Box* (1929), often spent summers in her small house here, about thirty miles southwest of Berlin.

CANTERBURY

This tiny village north of Concord was the home of a thriving Shaker community during the nineteenth century. Among those who visited the community were **Nathaniel Hawthorne,** in 1831, and **Herman Melville,** in 1850 and 1851.

CENTER HARBOR

During the many summers **John Greenleaf Whittier** spent in New Hampshire in the 1870s and 1880s, he often came to stay in this town on Lake Winnipesaukee. On some visits he stayed in the Sentor House hotel, on others at the Sturtevant House, where friends of his lived. The Sentor House burned down in 1887; on its site now stand the Nichols Memorial Town Library, the fire station, the municipal building, and the post office. The Sturtevant House, though still standing, is privately owned. Near it is the site of the Whittier Pine, an old pine tree that once stood here and under which the poet often wrote. Among the poems he is said to have written under the tree are "The Wood Giant," whose title refers to the Whittier Pine, and "An Outdoor Reception." The land on which the tree stood is now privately owned.

Play lightly on his slender keys,
O wind of summer, waking
For hills like these the sound of seas
On far-off beaches breaking!

And let the eagle and the crow
Find shelter in his branches,
When winds shake down his winter snow
In silver avalanches.

—John Greenleaf Whittier,
in "The Wood Giant"

CHARLESTOWN

Joseph Dennie, the essayist and editor, served from 1791 to 1793 as a law clerk in the southern New Hampshire town of Charlestown,

Birthplace of Horace Greeley in Amherst, as it appeared about 1890, some eighteen years after the editor's death. A view of the house as it looks today.

Constance Fenimore Woolson, great-niece of James Fenimore Cooper.

near the Connecticut River. Benjamin West, the lawyer he clerked for, had recently built a large house on the east side of Main Street, near the center of town, and Dennie may have studied and lived there. The home, built in 1784, is still standing but is a private residence. The Hoyt House, also on Main Street, was for several years the summer home of **Charles Hale Hoyt,** the dramatist, whose play *A Temperance Town* (1893) was set in a town modeled after Charlestown.

CHOCORUA

William James, the philosopher and psychologist, had a summer home here, in the Ossipee Mountains of eastern New Hampshire. It was to Chocorua that he came with his brother **Henry James, Jr.,** the great novelist, after they returned from England in the summer of 1910. William had been in ill health for some time and, not long after arriving here, died in his sleep on August 26, at the age of sixty-eight. Henry accompanied William's widow to a memorial service at Harvard and returned afterward with his brother's ashes, which then were scattered over the waters of a nearby stream. Henry stayed on for a few weeks in deep and desolate mourning.

Unutterable, unforgettable hour—
with those that have followed . . .
all unspeakable.

—Henry James,
in a notebook entry written in Chocorua after his brother's death

CLAREMONT

Constance Fenimore Woolson, great-niece of James Fenimore Cooper, was born here on March 3, 1840, in a house since burned down. She went on to write novels, including *Anne* (1882) and *Horace Chase* (1894). The Woolson family moved on to Ohio shortly after Woolson was born, so it is not surprising that her work does not deal with Claremont.

CONCORD

Ralph Waldo Emerson was preaching in New Hampshire's capital city in December 1827 when he met seventeen-year-old Ellen Louisa Tucker, whom he described as "beautiful by universal consent." A year later they became engaged and, on September 30, 1829, were married in Concord in the home of Ellen's stepfather, Col. William A. Kent, on Pleasant Avenue. Their marriage ended tragically in 1831, when Ellen died. In 1835 the Kent home provided a welcome haven for **John Greenleaf Whittier** and George Thompson, the English abolitionist. Walking down the street together toward the courthouse, where Thompson was to lecture, they were set on by a mob throwing rocks and debris, but found refuge in the Kent home. The house still stands today, but not at its original address. In the 1860s it was moved to 24 Spring Street, where it remains to this day. The house is privately owned, but visitors who have made appointments have occasionally been admitted for a look at the interior.

Concord was also the home of **Charles Hale Hoyt,** who was born here on July 26, 1860. His birthplace may have been the American House, a hotel here, but this has not been confirmed. The greatest commercial success of Hoyt's twelve plays was *A Trip to Chinatown,* produced in 1891. After the death of his wife, in 1898, Hoyt suffered an irreversible mental breakdown and never wrote again.

"What does it think it's doing running west
When all the other country brooks flow east
To reach the ocean? It must be the brook
Can trust itself to go by contraries. . . ."

—Robert Frost,
in "West-Running Brook"

CONWAY

Bernard De Voto came to Conway with his wife in 1929 to spend the summer. De Voto made considerable progress on a manuscript he had started work on in March that would eventually be published as *Mark Twain's America* (1932). He returned alone the following summer to escape the heat of Cambridge, where he was professor of English at Harvard, and stayed until August.

CORNISH

Ever since 1885, when Augustus Saint-Gaudens, the renowned sculptor, came to live and work here, this town just north of Claremont and near the Vermont border has been home to numerous artists and writers. Among those who have lived or stayed here are **Witter Bynner,** author of such volumes as *Grenstone Poems* (1917) and *Take Away the Darkness* (1947); **Herbert D[avid] Croly,** founder and first editor of *The New Republic,* and author of such nonfiction works as *The Promise of American Life* (1909) and *Progressive Democracy* (1914); **Finley Peter Dunne,** creator of Mr. Dooley, an eminently quotable Irish saloon keeper, who spent the summer of 1903 here with Saint-Gaudens; **Maud Howe Elliott,** daughter of Julia Ward Howe and co-author of a Pulitzer Prize-winning biography of her mother (1915); **Norman Hapgood,** journalist and author of such realistic novels as *An Anarchist Woman* (1909) and *Types from City Streets* (1910); **Robert Herrick,** whose novels of social criticism included *The Master of the Inn* (1908) and *Chimes* (1926); **Percy Mackaye,** poet and playwright, whose *Yankee Fantasies* (1912) dealt with New England life; **Langdon Mitchell,** playwright, whose most important play was *The New York Idea* (1906); **William Vaughn Moody,** playwright and poet, who is thought to have worked on his play *A Sabine Woman* (1906) while living here; and **Maxwell Perkins,** the great editor, whose authors included Hemingway and Fitzgerald.

The Cornish literary resident of longest standing was **Winston Churchill.** Churchill retired to Cornish in 1917, after completing work on his novel *The Dwelling Place of Light* (1917), and lived here until his death in 1947. Churchill wrote little here. He dabbled in carpentry and painting, and kept to a small circle of friends.

CROYDON

This town boasts a general store called Coniston, after the 1906 novel by **Winston Churchill,** who lived in nearby Cornish. *Coniston* depicts ethical conflicts in the politics of nineteenth-century New England.

DERRY

In 1900 **Robert Frost** was told by his doctor to live in the country, and the poet moved to a one-man farm near Derry, on Route 93. Frost tried to run the place as a working farm for five years, keeping a cow, a horse, hens, and a garden and orchard. He failed in his farm venture, and for much of his time here made a living by teaching at Derry's Pinkerton Academy. In 1909 he gave up on the farm and moved into town. The farm, known as Frosty Acres, now is owned and operated by the state of New Hampshire and is open to the public. Frost's "Mending Wall," first published in *North of Boston* (1914), was based on this farm. Nearby is West-Running Brook, which gave its name to another of Frost's poems; the poem in turn gave its name to the volume *West-Running Brook* (1928), in which the poem was first published.

DUBLIN

Snippets of information about Dublin's literary association tantalize the curious and, in the end, prove unsatisfying. In 1905 **Mark Twain** summered in this town southeast of Keene, in a house on the south side of Dublin Lake. The house is now privately owned and carries no markers. Twain returned the following summer to Upton Farm, on Upper Jaffrey Road. That house too is unmarked and privately owned. Another summer resident was **Amy Lowell.** When **Randolph Bourne,** the essayist, was working in a nearby house, Lowell invited him over to dine—but only once. Just as Bourne arrived, he was attacked by her sheepdogs, fortunately without inflicting great harm.

EXETER

This town southwest of Portsmouth is the home of the Phillips Exeter Academy, founded in 1783. Among its graduates are **Robert Anderson, Robert Benchley, Booth Tarkington,** and **Gore Vidal.** But the town has also had its share of literary residents. **Bliss Perry,** editor of the *Atlantic Monthly* from 1899 to 1909 and author of such literary studies as *The American Mind* (1912), lived for some years at The Exeter Inn, 90 Front Street, and died at Exeter Hospital on February 13, 1954.

Exeter claims among its natives **Granville Hicks,** the critic and novelist, born here on September 9, 1901, at 43 Main Street. The house is still standing but is privately owned.

Henry Augustus Shute, author of *The Real Diary of a Real Boy* (1902), was born in Exeter on November 17, 1856, and lived here for most of his life. It is thought that he was born in a house still standing at 1 High Street. Shute attended the Court Street School and Phillips Exeter Academy. After returning from Harvard in 1879, he read law in the offices of Judge William W. Stickney. He went on to practice law here until 1926, when he was seventy years old. He was long an active member of the Exeter Brass Band. Shute's boyhood home has been demolished; the house he bought at 3 Pine Street, with income from his popular *Real Diary,* still stands at the same address but is privately owned and not open to visitors. Shute died on January 25, 1943, and was buried in Exeter Cemetery, on Linden Street.

Tabitha Tenney, author of the popular satirical novel *Female Quixotism* (1801), was born here on April 7, 1762, and lived at 65 High Street, in a home still standing but private and unmarked.

FRANCONIA

This lovely resort town on the northwest edge of the White Mountain National Forest has long attracted literary visitors, including **Washington Irving, Henry Wadsworth Longfellow,** and **Nathaniel Hawthorne.** Hawthorne memorialized the famous Old Man of the Mountains, about six miles south of here, in his tale "The Great Stone Face" (1851).

Franconia's most famous literary resident was **Robert Frost,** who brought his family here in the winter of 1914, not long after he returned from his three-year stay in England. His first two books—*A Boy's Will* (1913) and *North of Boston* (1914)—had established Frost's reputation as a poet, and he sought in Franconia to isolate himself in well-loved surroundings. Throughout the winter of 1914–15, Frost and his family stayed at the farm home of John Lynch, where they had visited before. In April the poet began looking for a farm to buy and quickly found one just outside the village, on what is now Ridge Road, off Route 116. He arranged to pay $1,000 for the property, only to find himself being charged more when the farmer selling it to him got wind of his literary fame. Nonetheless, he moved his family to Ridge Road in June 1915 and soon settled down to good-humored, unremunerative farming, at which he excelled. In the five years he lived here, Frost took long walks up nearby Mount Lafayette, served as president of the local PTA, and wrote one book of verse, *Mountain Interval* (1916), and most of another, *New Hampshire* (1923). The Frost Place, as the farm is now known, is owned by the Town of Franconia and open to visitors during summers and early fall. In addition to serving as a memorial to the poet, it provides a center for the arts: concerts and poetry readings are presented in summer in the barn, which is now a lecture hall. Among the Frost memorabilia on display are a worn leather Morris chair, and some

> *It was a happy lot for children to grow up to manhood or womanhood with the Great Stone Face before their eyes, for all the features were noble, and the expression was at once grand and sweet, as if it were the glow of a vast, warm heart, that embraced all mankind in its affections, and had room for more.*
>
> —Nathaniel Hawthorne,
> in "The Great Stone Face"

> *Well, if I have to choose one way or the other,*
> *I choose to be a plain New Hampshire farmer*
> *With an income in cash of, say, a thousand*
> *(from, say, a publisher in New York City).*
> *It's restful to arrive at a decision,*
> *And restful just to think about New Hampshire.*
> *At present I am living in Vermont.*
>
> —Robert Frost,
> in "New Hampshire"

The Frost Place, home of Robert Frost in Franconia. A community center since 1977, the house has two rooms containing Frost memorabilia, including handwritten copies of his poems and signed editions of his books.

That is happiness, to be dissolved into something complete and great.

—The gravestone inscription of Willa Cather, a quotation from *My Ántonia*

of his manuscript poems. In a plot of woods near the house is a nature trail marked by twelve of Frost's poems, including "Mending Wall" and "The Road Not Taken."

Among Frost's friends here was **Ernest Poole,** who owned a summer home, now said to be remodeled as a multiple dwelling, on Birch Road in nearby Sugar Hill. While staying here, Poole took extensive walks and made a daily trip to the local post office. Although best known for his portrayals of city life in novels such as *The Harbor* (1915) and the Pulitzer Prize-winning *His Family* (1917), Poole also wrote of the Franconia area, including his home and his friendship with Robert Frost, in *The Great White Hills of New Hampshire* (1946), a nonfiction bestseller.

FRANKLIN

Daniel Webster was born on January 18, 1782, in Franklin, north of Concord, in a one-story farmhouse his father had built. Franklin then was known as Salisbury.

HAMPTON FALLS

Franklin B[enjamin] Sanborn was born in Hampton Falls, southwest of Portsmouth, on December 15, 1831. Sanborn was a friend of some of the most distinguished American literary figures of his day, and he wrote about them in such books as *The Personality of Thoreau* (1901) and *Hawthorne and his Friends* (1908). It is not known whether his birthplace, on Exeter Road near Applecrest Farm, is still standing.

John Greenleaf Whittier (*standing left*) on the balcony of Gove House in Hampton Falls. This is the last photograph taken of Whittier.

A distinguished literary visitor here was **John Greenleaf Whittier,** who spent the summer of 1892 at Elmfield, a house owned by one of his cousins. It was here, on August 24, that Whittier wrote the poem that proved to be his last. It was a tribute to his friend Oliver Wendell Holmes, and not long after writing it, on September 7, 1892, the poet died of a stroke. Elmfield, or Gove House as it is now called, is still standing but privately owned.

The hour draws near, howe'er delayed and late,
When at the Eternal Gate
We leave the words and works we call our own,
And lift void hands alone
For love to fill. Our nakedness of soul
Brings to that Gate no toll;
Giftless we come to Him, who all things gives,
And live because He lives.

—John Greenleaf Whittier, in "To Oliver Wendell Holmes"

HANOVER

This town on the Connecticut River is best known as the home of Dartmouth College, founded in 1769. Among the literary men who studied here are **Richard Eberhart,** who was graduated in 1926 and returned as a professor of English and poet in residence in 1956; **Robert Frost,** who entered in 1892 but left before the term was out; **A. J. Liebling,** who attended from 1920 to 1923 and was expelled for refusing to attend chapel; **Budd Schulberg,** who was graduated in 1936 and is known best for his satirical novel *What Makes Sammy Run?* (1941); **George Ticknor,** who was graduated in 1801 and whose *Life, Letters, and Journals* was published in 1876; and **Daniel Webster,** who was also graduated in 1801, and later successfully argued before the Supreme Court that the State of New Hampshire could not forcibly make his alma mater a state institution. Webster is said to have torn up his diploma on learning that he was not to be named class valedictorian. The slight appears not to have dulled Webster's loyalty to Dartmouth.

Percy Marks taught at Dartmouth while working on material he would later use in *The Plastic Age* (1924), his novel of life at an American college. **Lewis Mumford,** who taught at Dartmouth from 1931 to 1935 and from 1945 to 1948, is thought to have lived at 46 North College Street or 46 South College Street. The latter site was razed in the years after 1948 to make room for expansion of the college. Among the many distinguished visitors here was **Walt Whitman,** who read "As a Strong Bird on Pinions Free" at commencement here in 1872. The college today owns one of the finest collections of material on Robert Frost, and also has a special collection devoted to Vachel Lindsay.

Perhaps the unhappiest aspect of Hanover's literary associations relates to the notorious stay of **F. Scott Fitzgerald** and **Budd Schulberg** at the Hanover Inn during the 1940 Dartmouth Winter Carnival. The writers had been at work in Hollywood on a motion-picture script based on the carnival and thought Fitzgerald could benefit from visiting the actual site. Fitzgerald, who had been on the wagon, fell to drinking and became abusive. His condition became so acute that he had to be taken to a hospital in New York City for treatment. Schulberg later wrote scenes recalling the incident for his novel *The Disenchanted* (1950), about the last year of an author's life. The Hanover Inn has been rebuilt since the unfortunate episode.

ISLES OF SHOALS

This group of islands, which includes Appledore, Star, and White islands, lies about ten miles southeast of Portsmouth. Isles of Shoals is the literary domain of **Celia [Laighton] Thaxter,** who spent the better part of her life here. Her father, a businessman and state legislator, moved his family to White Island in 1839, after being disappointed in his attempt to become governor of New Hampshire. He ran the island's lighthouse, still standing, for about six years, moving in 1848 to Appledore Island, where he opened and ran the successful resort hotel called the Appledore Hotel. On White Island Celia and her brothers had had few playmates, but on Appledore their enforced isolation was relieved by the guests who stayed at the hotel. Among them were **John Greenleaf Whittier, William Morris Hunt,** and **Childe Hassam.** Celia herself remained on Appledore only three years: she left for Star Island in 1851 after marrying her tutor, Levi Lincoln Thaxter. They stayed on Star Island until 1860, when they moved to the mainland. Celia began writing verse, and her first published work appeared in 1861 in *The Atlantic Monthly.* Encouraged by Whittier, among others, she continued to write and brought out a number of volumes over the next thirty years. These volumes included *Poems* (1872), *Among the Isles of Shoals* (1873), and *An Island Garden* (1894). Her illustrator for the third volume was Childe Hassam. In the meantime she and her husband had moved back, in 1866, to Appledore Island, where her brothers continued to run the family hotel after their father's death. Celia bought a cottage nearby, keeping a garden and holding literary salons, whose guests sometimes included Whittier. The Appledore Hotel was destroyed by fire in 1914, but Celia's cottage still stands.

JAFFREY

This town in southern New Hampshire, site of beautiful Mount Monadnock, has been a favorite spot for several writers. **Willa Cather** came here over a twenty-year period, staying at the Shattuck Inn, which burned down in 1909 and was rebuilt as a religious conference center. Cather loved the area so much that she wanted to be buried here. After her death in 1947, she was brought to the Old Burying Yard in nearby Jaffrey Center. **William Dean Howells** stayed at the Shattuck Inn for a summer when it was a privately owned farm called, simply, Shattuck Farm. **Henry David Thoreau,** drawn by the beauty of Mount Monadnock, came here several times. Thoreau spent a night at the summit in 1843 or 1844, crossed the mountain again in September 1852, and camped near the summit for several nights in June 1858. In August 1860, on his last visit, he camped here for about a week.

Photograph of Celia Thaxter. The poet and short-story writer was encouraged by John Greenleaf Whittier to write her volume of sketches *Among the Isles of Shoals.*

LITTLETON

This town in the northwest part of the state is the birthplace of children's author **Eleanor Hodgman Porter** (born 1868). Porter's books were all moderately successful until she published *Pollyanna* (1913), which not only won the hearts of millions of Americans but enriched the English language—at least for a time—by giving it a new word.

pollyanna n *one having a disposition or nature characterized by irrepressible optimism and a tendency to find good in everything*

—from *Webster's Third New International Dictionary,* 1961

MANCHESTER

Robert Choquette was born in New Hampshire's largest city on April 22, 1905. The poet moved with his family to Montreal when he was ten and lived and worked there from then on. Choquette's books of verse include *A Travers les Vents* (1925) and *Metropolitan Museum* (1931).

MONT VERNON

G[eorge] W[ilkins] Kendall was born in Mont Vernon, thirteen miles southwest of Manchester, on August 22, 1809. Kendall was cofounder of the New Orleans *Picayune* in 1837,

Old photograph of Appledore Hotel, once located on the Isles of Shoals. The hotel was owned by Celia Thaxter's family, who entertained many literary guests here.

still published as the New Orleans *Times-Picayune*. The newspaper's field reports from the Mexican War, often written by Kendall himself, were among the pioneering efforts in modern war reporting.

NELSON

May Sarton lived in Nelson, nine miles northeast of Keene, from 1957 to 1972. Here Sarton wrote many books, including the novel *The Small Room* (1961) and the books of verse *As Does New Hampshire* (1967) and *A Grain of Mustard Seed* (1971).

NEWPORT

Sara Josepha [Buell] Hale was born on October 24, 1788, on her great-grandfather's farm near here, midway between Lake Sunapee and the Vermont border. She lived here for a good part of her life, devoting herself to domestic pursuits before taking up the literary activities that made her famous through America in her lifetime. Today her memory, if not her name, is preserved by one small work, the poem "Mary's Lamb" (1830), better known by its first line: "Mary had a little lamb." Hale's birthplace has long since been torn down, and there are today no buildings on the site. Newport's Richards Free Library, 58 North Main Street, has a large collection of material on her, and each year the Friends of the Library give the Hale Award to a New England literary figure.

NORTH HAMPTON

Ogden Nash died in Baltimore in 1971 but was buried in the Little River Cemetery on Atlantic Avenue in North Hampton, eight miles southwest of Portsmouth.

ORFORD

Charles Jackson, known best for his novel *The Lost Weekend* (1944), had a home from 1946 to 1954 in this beautiful town just north of Hanover. Today the house is known as the Vanderbilt House and is privately owned. Built in 1825–28, it stands at the northern end of Orford's row of historical houses known as The Ridge.

Home of May Sarton on the village green in Nelson. In the fifteen years she lived here, the poet and novelist completed a number of books, including her autobiography, *Plant Dreaming Deep* (1968).

PETERBOROUGH

Few towns the size of Peterborough have played such a large part in the literary life of the nation. Peterborough, about fifteen miles east of Keene, is the home of the MacDowell Colony, founded in 1908 as a memorial to the composer Edward MacDowell, who made his summer home here. The colony provides artists with a haven in which to pursue their work undisturbed: each resident is given room and board, a working studio, and free access to the estate's 400 acres. And many literary figures have made good use of the facilities. **Stephen Vincent Benét** stayed at the colony for three weeks in 1926 while finishing *Spanish Bayonet* (1926) and making notes for *John Brown's Body* (1928). **Willa Cather** wrote *Death Comes for the Archbishop* (1927) while staying here. **Edwin Arlington Robinson** spent every summer here from 1911 to 1935. Among his many works published in this period was *The Man Against the Sky* (1916), a volume that contains many of his best-known poems. **Thornton Wilder** was staying at the colony when he wrote much of his novel *The Bridge of San Luis Rey* (1927) and his play *Our Town* (1938). A few of the many other fine writers who have stayed at MacDowell over the years are **James Baldwin, Louise Bogan, DuBose Heyward, Stanley Kunitz,** and **W. D. Snodgrass;** and **Amy Lowell** lectured here from time to time.

PLYMOUTH

Robert Creeley attended the Holderness School in Plymouth, central New Hampshire, about eighteen miles northwest of Laconia. The school is still functioning here today. Another short-term resident of Plymouth was **Robert Frost,** who lived in a cottage at the corner of Highland Avenue and School Street during 1911–12 while he was a teacher at Plymouth Normal School, now Plymouth State College. Frost taught in Room 24 of Rounds Hall, overlooking the town's Main Street and, beyond that, the Pemigewasset River Valley. While living here Frost worked on the poems that would be included in his first two books, *A Boy's Will* (1913) and *North of Boston* (1914).

The Pemigewasset Inn, no longer standing, was once the town's chief landmark. In May 1864 two distinguished personages stayed there: Franklin Pierce, who had served as president of the United States, and **Nathaniel Hawthorne.** Hawthorne, often a guest at the inn, died there on May 19, 1864. A plaque on the village green, erected by students at Plymouth High School, pays tribute to him.

PORTSMOUTH

Francis Parkman spent many summers with his daughter and son-in-law at the Wentworth Mansion in this seaport city. Parkman wrote part of *Montcalm and Wolfe* (1884) and finished *A Half*

Century of Conflict (1892) while staying here. The house, on Little Harbor Road at the end of the cemetery on Sagamore Avenue, is open to the public from May 3 to October 12. The building is marked, and there are signs directing visitors at the connecting roads.

Portsmouth was also the birthplace of four distinguished literary figures. **Thomas Bailey Aldrich** was born on November 11, 1836, in a house at 45 Court Street. He spent much of his childhood in New York and New Orleans, but returned to Portsmouth about 1849. He spent three happy years, writing verse that was published in the Portsmouth *Journal.* Aldrich described his Portsmouth house at length in his semi-autobiographical work *The Story of a Bad Boy* (1870), in which he called his home Nutter House. The house, sometimes referred to by the name Aldrich gave it, is open to the public between June 15 and September 15 and is part of Strawbery Banke, a historical restoration area in Portsmouth. The Aldrich house has been restored and furnished much as it was in Aldrich's day.

James Thomas Fields was born here on December 31, 1817. His place in literary history was ensured when he became a founder of the eminent publishing house of Ticknor, Reed and Fields, which is known today as Ticknor & Fields and is located in New Haven, Connecticut. He also is remembered as editor of *The Atlantic Monthly* from 1861 to 1871 and author of volumes of poetry and reminiscence. His birthplace at 83–85 Gates Street, formerly 12 Gates Street, is still standing but privately owned. A plaque outside pays tribute to Fields.

Benjamin Penhallow Shillaber, who wrote humor in the form of discourses by a character called Mrs. Partington, was born here on July 12, 1814. His home, at the corner of Langdon and McDonough streets, is no longer standing. In addition to writing his own humorous pieces, Shillaber founded *The Carpet-Bag* (1851–53), a humorous weekly published in Boston, in which appeared the first published story by **Mark Twain,** "The Dandy Frightening the Squatter" (1852).

Celia [Laighton] Thaxter was born here on June 29, 1835, in a Federal-style building at 48 Daniel Street. The building has been altered considerably and converted for use as a navy uniform store. Thaxter lived here until the age of four, when she moved with her family to the Isles of Shoals off New Hampshire's coast.

ROCHESTER

Gladys Hasty Carroll was born on June 26, 1904, at 58 Portland Street in Rochester, in southeastern New Hampshire. The novelist is remembered particularly for *As the Earth Turns* (1933).

SALEM DEPOT

Robert Frost completed grammar school at Salem District School Number Six, in this town in the southeastern tip of the state. The Frost family lived in rented rooms on the farm of Loren Bailey, for whom Robert worked briefly. His poem "The Grindstone," published in *New Hampshire* (1923), described chores he had to perform here.

SILVER LAKE

Joy Farm in Silver Lake, in the central part of the state, was owned by the parents of **E. E. Cummings.** The poet inherited the farm and continued to come here throughout his life. He was staying here in the summer of 1962 when he died, on September 3, at Memorial Hospital in North Conway.

WALPOLE

Louisa May Alcott, the creator of *Little Women* (1868–69), lived in 1855 and 1856 in this town on the Connecticut River, northwest of Keene. The house she lived in stood formerly on Main Street, on the present site of the Bridge Memorial Library. The house, since moved to 83 High Street, is used now as an apartment building. Louisa and her sisters formed a dramatic club and performed some of their plays at the Elmwood, an inn on the Common. Later named the Colony and now a private home, it was there that **James Michener** began work on his novel *Hawaii* (1959).

Joseph Dennie lived in Walpole from 1796 to 1798, staying part of the time in the home of the Reverend Thomas Fessenden. The house, still standing, is now owned by Hubbard Farms, Inc. Dennie lived here while editing the Federalist newspaper *Farmer's Weekly Museum,* for which he wrote his celebrated "Lay Preacher" essays. Today the building in which the *Museum* was published is a restaurant, the Thomas Ryder House. Dennie's close friend **Royall Tyler** published his picaresque novel *The Algerine Captive* (1797) in Walpole, and it became the first American novel to be republished in England.

Thomas Bailey Aldrich.

Nutter House, home of Thomas Bailey Aldrich in Portsmouth.

A lover's claim is mine on all
I see to have and hold,—
The rose-light of perpetual hills,
And sunsets never cold!

—John Greenleaf Whittier,
in "Sunset on the Bearcamp"

The whole place is even more beautiful than I had remembered, with spring here. When are Ellen and you coming? . . . You head for Woodstock, and ask at the post office how to get to Barnard; at Barnard, the general store, ask how to get here.

—Sinclair Lewis,
in a letter to Alfred Harcourt,
May 26, 1929

The property the Lewises bought, traditionally called "Twin Farms" and soon to be given that name again, was a mountain, a hill and a valley, with the great sweep of view toward Mount Ascutney from the highest point. . . . [The Big House] was a wreck. The framework was good enough but nobody had taken any care of it for many decades and the roofs and floors all needed a good deal of attention. Alongside it, to the left as one looked down and up toward Ascutney, there was a vast barn, also derelict and falling apart.

—Vincent Sheean,
in *Dorothy and Red*

WEST OSSIPEE

John Greenleaf Whittier spent many summers throughout the 1870s and 1880s at the Bearcamp House hotel here, named for a river nearby, in the Ossipee Mountains. The entire area can legitimately be called Whittier country: there is even a mountain named after him about six miles northwest of here. Bearcamp House is no longer standing, but Whittier memorialized the Bear Camp River in his poem "Sunset on the Bearcamp."

WILTON

This small town about ten miles from the Massachusetts border has attracted such literary visitors as **Henry Wadsworth Longfellow, Oliver Wendell Holmes,** and **William Dean Howells.** Among its full-time residents was **Eleanor Abbott,** who moved here in 1908. Abbott was best known for her children's novel *Molly Make-Believe* (1910). Today her home is the site of the Anne Jackson Memorial, owned by the Girl Scouts of New Hampshire and not open to the public.

WINDHAM

Robert Frost stayed a week in 1891 as a working boarder at the farm of Joseph Dinsmoor on Cobbetts Pond in this township ten miles northeast of Nashua. The farm is still there, but the present house may not be the same one that Frost stayed in. In any case, the farm is not open to the public. Frost's experience here is said to have provided the basis for "The Tuft of Flowers," which was included in *A Boy's Will* (1913).

VERMONT

ARLINGTON

An appropriate way to open a discussion of Vermont's literary history is with **Ethan Allen,** the leader of the Green Mountain Boys, who wrote his religious tract *Reason the Only Oracle of Man* (1784) in a house later called the Studio Tavern, in Arlington, a town north of Bennington. Authorship of the volume, sometimes referred to as Ethan Allen's Bible, is in dispute. Some scholars attribute most of the work to Thomas Young (1732–77).

Dorothy Canfield [Fisher] spent much of her youth on her parents' estate just north of town on Red Mountain and, after she married John Fisher in 1907, moved with him into a small house on the estate, which remains in the family and is not open to the public. Nearby is a pine forest that the Fishers planted. When the forest was still in its seedling stage, **Robert Frost** and Dorothy Canfield would walk together here. In Canfield's many short stories and in her novels *The Brimming Cup* (1921), *Bonfire* (1933), and *Seasoned Timber* (1939), she demonstrated the independence and close personal ties that characterize life in Vermont villages. Canfield died in Arlington on November 9, 1958, and was buried beside her husband in the northwest corner of St. James Cemetery, on Main Street. The Martha Canfield Library, also on Main Street, is housed in a building donated by Dorothy Canfield. The literary visitor will especially value a portrait of the author and a collection of memorabilia.

Sarah N[orthcliffe] Cleghorn, the poet and novelist, visited her close friend Dorothy Canfield here and later took a house on Manchester Street, next to the Dellwood Greenhouse and near the Dellwood Cemetery. **Arthur Guiterman,** the journalist and author of several collections of poems, including *The Laughing Muse* (1915), also had a home, called Hillhouse, in Arlington.

Wide and shallow in the cowslip marshes
Floods the freshet of the April snow.
Late drifts linger in the hemlock gorges,
Through the brakes and masses trickling slow
Where the Mayflower,
Where the painted trillium, leaf and blow.

—Sarah N. Cleghorn,
in "Vermont"

BARNARD

Sinclair Lewis and **Dorothy Thompson** in the fall of 1928 bought a 300-acre farm located halfway between Barnard and Pomfret, in eastern Vermont. They called their property Twin Farms because it had two farmhouses, one of which they reserved for house guests. Their visitors included Alfred Harcourt, the publisher, H. L. Mencken, and Louis Untermeyer. It was at this farm that Lewis worked on his novel *Dodsworth* (1929) and, on November 5, 1930, learned by telephone that he had won a Nobel Prize. In 1942 Lewis and Thompson divorced, and Thompson in the following year married Maxim Kopf, the painter, at the Universalist Church here. Dorothy Thompson later returned to live at Twin Farms. She died in 1961 and was buried in the Barnard Village Cemetery. **[James] Vincent Sheean,** the journalist and biographer who wrote a memoir of Lewis and his wife, *Dorothy and Red* (1963), also lived for a time at Twin Farms, beginning in 1938.

BARTON

Edward Hoagland lived in this northern Vermont town, about thirteen miles south of Newport, and worked here on three of his books of essays, *The Courage of Turtles* (1970), *Walking the Dead Diamond River* (1973), and *Red Wolves and Black Bears* (1976), as well as on a travel book, *African Calliope* (1979).

BENNINGTON

Bennington College, founded here in 1932 for women only, has had a number of distinguished literary figures among its faculty and student body. **W. H. Auden** taught here during the spring term of 1946. **Bernard De Voto,** who was editor of *The Saturday Review of Literature* from 1936 to 1938, lived in the Shingled Cottage here during the summer of 1937. **Ralph Ellison** lectured at the college on black American culture. **Francis Fergusson,** the drama critic, taught here and served as director of the College Theater from

1934 to 1947. **Barbara Howes,** author of several collections of poems, including *The Undersea Farmer* (1948), was a student here from 1933 to 1937. **Stanley J. Kunitz,** the poet, was professor of English here from 1946 to 1949.

Bernard Malamud worked, while teaching English here, on his novels *Idiots First* (1963), *The Fixer* (1967), *Pictures of Fidelman* (1969), *The Tenants* (1971), *Rembrandt's Hat* (1973), and *Dubin's Lives* (1979). **Howard Nemerov** has taught at the college for many years, beginning in 1948. During these years he has produced various works of fiction as well as a great many poems, collected in such volumes as *Guide to the Ruins* (1950), *The Salt Garden* (1955), and *Mirrors and Windows* (1958). **Theodore Roethke,** the poet, met Beatrice O'Connell during the time he was teaching here. They later married. **Allan Seager** and **Genevieve Taggard** also taught here.

BOMOSEEN

Alexander Woollcott, the journalist and drama critic, and a number of his friends organized a club in 1926 to purchase Neshobe Island on Lake Bomoseen, in western Vermont, as a place for members to stay during summers. Woollcott, the unofficial head of this club, later had a stone house built near the clubhouse for his own permanent residence. Other literary figures who were members or guests of members included **S. N. Behrman, Moss Hart, Ben Hecht, Charles MacArthur, Cornelia Otis Skinner,** and **Thornton Wilder.**

BRATTLEBORO

This city has a distinguished literary history. In 1892, for example, **Rudyard Kipling,** having just married a Vermont woman, Caroline Balestier, visited his new in-laws here. Kipling was so taken with the scenery that he and his wife decided to live on the Balestier estate, first in Bliss Cottage, now a private residence across the street from its original site on Kipling Road, and then in Naulakha, off Kipling Road, about three miles northeast of town. Kipling himself planned Naulakha and had it built. The house still looks much as it did when Kipling lived here and wrote two volumes of stories—*The Jungle Book* (1894) and *Second Jungle Book* (1895)—and the novel *Captains Courageous* (1897), the story of the schooner *We're Here* and the men who sailed her.

With **[Charles] Wolcott Balestier,** Kipling wrote *The Naulahka* (1892), a novel about an American speculator in India. Balestier, who also wrote novels and short stories on his own, wrote material for the American chapters of *The Naulahka.* Although Kipling did a considerable amount of writing here and found the area beautiful, he was not inspired to write many pieces about Vermont. One he did write, the short story "A Walking Delegate" (1898), had horses as its main characters. The Kipling family left Vermont in 1896 to return to England. Eleven years later, Kipling became England's first Nobel Prize-winner in literature. Naulakha is open to the public only on special occasions.

Henry David Thoreau was another writer who enjoyed the Vermont countryside. While visiting friends in Brattleboro in September 1856, he made botanical notes in his journal and met with Charles Christopher Frost, a leading authority on Vermont flora.

The first literary figure associated with Brattleboro was **Royall Tyler,** author of *The Contrast* (1787), the first comedy written by an American and performed by a professional company in this country. In 1801, Tyler moved his family here to a house, no longer standing, on Meeting House Hill, then the center of town. The site, on what now is Orchard Street, is occupied by another house and has a view of the Connecticut River Valley and Wantastiquet Mountain, in New Hampshire. During Tyler's years in Brattleboro, he served on the Supreme Court of Vermont and in 1807 was made its Chief Justice. After he published his *Yankee in London* (1809), a series of letters concerning English life and customs, many of his readers were astonished to discover that he had never been to England but had relied on information obtained from English visitors. The

My workroom in the Bliss Cottage was seven feet by eight, and from December to April the snow lay level with its window-sill. It chanced that I had written a tale about Indian Forestry work which included a boy who had been brought up by wolves. In the stillness, and suspense, of the winter of '92 some memory of the Masonic Lions of my childhood's magazine, and a phrase in Haggard's Nada the Lily, *combined with the echo of this tale. After blocking out the main idea in my head, the pen took charge, and I watched it begin to write stories about Mowgli and animals, which later grew into the* Jungle Books.

—Rudyard Kipling,
in *Something of Myself*

I must explain something about this island. It has several levels. On one of the high ones is built my own stout all-year-round house with stone walls two feet thick—a fortress successfully designed to defeat a Vermont winter. When it's 30 below outside—it was 18 below the day I went over the mountain to help bury Otis Skinner—this house is cozy and will be as long as it is possible to get oil.

—Alexander Woollcott,
in letter to Edward Sheldon,
April 24, 1942

Naulakha, home of Rudyard Kipling in Brattleboro. The house, named for a priceless Indian jewel in *The Naulahka,* a novel Kipling wrote with Wolcott Balestier—Kipling later married Balestier's sister—looks across the Connecticut River valley to Mount Monadnock, in New Hampshire.

Tyler family lived in three other houses in Brattleboro before buying, in 1820, the house now occupied by the Department of Social Welfare, at Putney Road and Park Place, near the Common. It was in this last house that Tyler wrote *The Chestnut Tree* (not published until 1931), a long poem foretelling the rise of industrialism, and *The Back Bay* (not published until 1971), an autobiographical account of life in eighteenth-century Boston. Tyler died on August 16, 1826, and was buried in Prospect Hill Cemetery.

BURLINGTON

Ethan Allen lived his last years in Burlington, on a farm now called Ethan Allen Park, off North Avenue. He died here on February 12, 1789, and like many other Revolutionary War soldiers was buried in Greenmount Cemetery, on Colchester Avenue.

John Dewey, the philospher, educator, and author of many influential books, was born here on October 20, 1859.

The presence in Burlington of the University of Vermont is responsible for other literary associations of the city. **Walter Van Tilburg Clark,** author of *The Ox-Bow Incident* (1940), received an M.A. degree here in 1934. **Lloyd C. Douglas,** known particularly for the novels *Magnificent Obsession* (1929) and *The Robe* (1942), received a D.D. degree here in 1931. The Royall Tyler Theater, named after the playwright, who also served as a teacher and trustee of the university, is the site of many campus productions. Two other authors connected with Burlington are **John Godfrey Saxe,** the poet, who edited the Burlington *Sentinel* from 1850 to 1856, and **Rex Stout,** the detective story writer, who wrote four short stories while living in a boarding house here in 1913.

The Royall Tyler Theater, on the campus of the University of Vermont in Burlington.

DANBY

Pearl Buck lived for extended periods in this town, fifteen miles south of Rutland, during the last three years of her life. The novelist was active in efforts to restore houses and shops on Main Street. She bought some of the buildings herself and formed a construction and decorating company for the remodeling effort. The home in which she lived until her death on March 6, 1973, is still standing but not open to the public.

DERBY LINE

Beginning in 1931, a cooperative Vermont farmer whose sugarbush in this small town was adjacent to the Canadian border—and remote from customs inspectors—bootlegged liquor year round for **Bernard De Voto,** who spent summers in Morgan Center, eleven miles to the southeast, and drank whiskey from time to time throughout the year. By such chance and chancy activities do some men enter literary history. A letter in carefully worked-out code would set in motion the transport vehicles of the irregular Quebec and Southeast Transportation Company. Its only stockholder was De Voto, and its only cargo was good cheap liquor. The southern terminus of the Q.S.T. Company was Harvard College, Cambridge, Massachusetts, where De Voto taught and was generous in distributing his illegally obtained bottled spirits among close friends. The enterprising farmer, transformed by De Voto's typewriter into a country philosopher, was immortalized by De Voto in several of his "Easy Chair" columns for *Harper's Magazine.*

DORSET

Stuart P[ratt] Sherman, the literary critic and historian and one of the editors of the *Cambridge History of American Literature,* was buried in the cemetery of this southwestern Vermont village on August 25, 1926, after a funeral service in the Congregational Church, where his grandfather once had been the pastor.

EAST POULTNEY

Horace Greeley, from 1826 to 1830, long before he founded and edited the New York *Tribune,* worked in this western Vermont village as an apprentice printer for *The Northern Spectator.* Greeley, a teenager, lived in a boarding house and spent most of his free time reading, but he did take the time to attend meetings of the local debating society in the schoolhouse, where he made his first speech. *The Northern Spectator* stopped publishing in 1830, but the building it occupied is still standing. It houses The Horace Greeley Museum, open to visitors in July and August for a small admission charge.

FERRISBURG

Rowland Evans Robinson, known best for his stories and illustrations of nineteenth-century Vermont villages, was born on May 14, 1833, at Rokeby, his family's farmhouse on Route 7 north of this western Vermont town. Robinson died on October 15, 1900, in the same room in which he was born. In his long career Robinson

OLD BUILDING WHERE "NORTHERN SPECTATOR" WAS PRINTED, EAST POULTNEY, VT., AND WHERE HORACE GREELEY LEARNED HIS TRADE

HORACE GREELEY

APPRENTICESHIP
NORTHERN SPECTATOR
1826 – 1830

The town house was an unpainted, weatherbeaten, clapboarded building of one story, with one rough, plastered room, furnished with rows of pine seats, originally severely plain, but now profusely ornamented with carved initials, dates, and strange devices. A desk and seat on a platform at the farther end, for the accommodation of the town officers, and a huge box stove, so old and rusty that it seemed more like the direct product of a mine than of a furnace, completed the furniture of the room, wherein were now gathered a majority of the male inhabitants of the town.

—Rowland Evans Robinson, in "An Old-Time March Meeting"

gave readers a picture of Vermont life, insight into Vermont character and myth, and an accurate account of Vermont speech and folkways. His many works are collected in a seven-volume Centennial Edition, published between 1933 and 1936. Rokeby, now a museum, is open to the public. The building was originally a small house but was enlarged by the Robinson family to serve as a Quaker Meeting place, way station on the Underground Railroad, town clerk's office, local library, and cultural center as well. Rokeby is furnished with the original eighteenth- and nineteenth-century furniture and artifacts of the Robinson family and exhibits art works produced by family members. Admission is charged.

GUILFORD

In 1791 **Royall Tyler** left the Boston area to practice law in this village just south of Brattleboro. By 1797 his house had been built, near Guilford Center, and he arranged for his wife and child to join him. It was while living in Guilford that Tyler wrote *The Algerine Captive* (1797). It was the first American-produced novel to be reprinted in England. The novel satirized higher education, medical quackery, and slavery. Another Tyler work written here was the lost play *The Georgia Spec; or, Land in the Moon* (1797), which satirized land speculation in Yazoo, Georgia. The land fever was later revealed as a swindle, and is known

Rowland Evans Robinson, the Vermont author, and Rokeby, his home for most of his life. Now a museum, the house contains a library of materials dealing with state and local history.

We are indeed placed in a cold and rough country, far in the interior, with no great marts of commerce, manufactures or trade. But we all have, among these green hills, the high health and contentment, which combine to make up almost the sum total of all animal happiness. We have the habits of industry and frugality, which best ensure us pecuniary independence as a separate community, and, at the same time, best assure, with our admitted intelligence and virtue, our *part of the duty of perpetuating our great national blessings of freedom and equal rights.*

—Daniel Pierce Thompson, in *History of the Town of Montpelier*

Daniel Pierce Thompson. This portrait of the Vermont author and lawyer by Thomas Waterman Wood hangs in the offices of the Vermont Historical Society in Montpelier.

to historians as the Yazoo land fraud. Tyler's home is no longer standing, but a marker near the Guilford Country Store memorializes the author.

HIGHGATE CENTER

John Godfrey Saxe was born on June 2, 1816, on a farm just north of the little town of Highgate Center, about eight miles northeast of St. Albans. Saxe, in addition to operating a grist mill on Rock River, was the author of popular verse. His work included *Progress: A Satirical Poem* (1846), *Humorous and Satirical Poems* (1850), and *Leisure-Day Rhymes* (1875). Saxe's home is still standing but not open to the public. A monument to Saxe can be seen nearby, on St. Armand Road.

JAMAICA

Genevieve Taggard, the poet and teacher at Mount Holyoke and Sarah Lawrence, had a summer home in this south central Vermont village, where she eventually came to live permanently. She died here soon after, on November 8, 1948.

MANCHESTER

In 1885, when **Sarah N[orthcliffe] Cleghorn** was nine, she and her younger brother were sent by their widowed father to this town, now known for its stately homes, to live with two maiden aunts. She and her brother stayed first at Sans Souci, a cottage on the west road behind North Main Street, and later, in April 1890, at the Elijah Littlefield house. The latter stood near the cemetery and was a place where the town children could pick strawberries in summer and skate in winter. After her college years Cleghorn returned to this community, where she wrote two novels, *The Turnpike Lady* (1907) and *The Spinster* (1916), as well as much of her poetry, which she later classified as either her "sunbonnet poems"—those on rural Vermont—or her "burning poems"—those on social reform. "The Golf Links," found in the collection *Portraits and Protests* (1917), is one of the latter:

The golf links lie so near the mill
That almost every day
The laboring children can look out
And see the men at play.

Walter Rice Hard, poet, folklorist, and columnist for the Rutland *Herald,* was born here on May 3, 1882. When grown, Hard managed his family's drugstore, which his childhood friend Sarah Cleghorn described as a place where townspeople met and talked politics. Hard, with his wife, also ran the Johnny Appleseed Bookshop, which is still in business in Manchester. Hard's volumes of verse, *Salt of Vermont* (1931), *A Mountain Township* (1933), and *Vermont Valley* (1939), faithfully reflected the character, speech, and customs of his fellow Vermonters. He died here on May 21, 1966.

MIDDLEBURY

Middlebury College Library has important literary collections, including some 300 volumes relating to Robert Frost, in the Robert Frost Room, and memorabilia of Henry David Thoreau, in the Abernathy Collection. Two of Middlebury's alumni achieved literary recognition: **John Godfrey Saxe,** the poet, B.A. 1839, M.A. 1842; and **Daniel Pierce Thompson,** the novelist, B.A. 1820. **William Hazlett Upson,** known best for his stories about the misadventures of Alexander Botts, salesman for the Earthworm Tractor Company in *Alexander Botts, Earthworm Tractors* (1929), had a residence at 24 Daniel Chipman Park, still standing but not open to visitors. One can visit the William Hazlett Upson Memorial Bandstand on the village green and his burial plot at the West Cemetery here. The humorist died at Porter Medical Center here on February 5, 1975.

MONTPELIER

Daniel Pierce Thompson grew up in the late 1700s on a farm on the Stevens branch of the Winooski River, three miles south of Vermont's capital city. A marker at 370 River Street, Montpelier, indicates the former site of Thompson's boyhood home. Thompson lived later in his life at 10 Barre Street, in a house that also is gone. He practiced law in this city, was active in county and state politics and from 1849 to 1865 served as editor of the *Green Mountain Freeman,* an antislavery publication. Thompson's considerable knowledge of the history of Vermont led him to write novels and short stories based firmly on historical events. *The Green Mountain Boys* (1839) celebrates the adventures of Ethan Allen, the larger-than-life Vermont hero. *Locke Amsden; or, The Schoolmaster* (1847) depicts in great detail life on the Vermont frontier. Thompson died here on June 6, 1868. The people of Vermont have honored him by placing his portrait and a bronze plaque in the State House.

MORGAN CENTER

Bernard De Voto and his family summered in 1931 near this town ten miles south of the Canadian border. The place gave De Voto the seclusion he needed for his writing as well as ample room for a steady stream of guests from Cambridge and New York City. They held many boozy picnics here and on the Canadian shore of Lake Memphremagog. The summer spent so idyllically prompted De Voto to write for *Harper's Magazine* an article entitled "New England: There She Stands."

NEWPORT

Stewart H[all] Holbrook, whose books on Vermont include *Ethan Allen* (1940) and *Down on the Farm* (1954), was born on August 22, 1893, on a farmstead on Lake Road in this northern Vermont town. The house, still standing, is not open to the public.

NORTH BENNINGTON

Shirley Jackson, the fiction writer, lived here with her husband, **Stanley Edgar Hyman,** the literary critic and member of the Bennington College faculty. Jackson, whose most widely read short story is "The Lottery," died here on August 8, 1965. Her classic story of psychopathic behavior in a realistic setting, which originally appeared in *The New Yorker,* is the title story of a collection of her stories that was published in 1949.

OLD BENNINGTON

The grave of **Robert Frost** is in the cemetery next to the Old First Church (Congregational) here on Route 9. The cemetery is open to the public.

PAWLET

Upton Sinclair worked as a clerk during the summer of 1895 at the Crescent Valley Hotel, then standing in this western Vermont community fifteen miles south of Poultney. An elementary school now occupies the site, which is across the street from the town hall.

PLYMOUTH

On September 12, 1928, **Sinclair Lewis** and his wife **Dorothy Thompson** visited this village, birthplace of President Calvin Coolidge, fifteen miles southwest of Woodstock. While here, Lewis sent a postcard message to H. L. Mencken: "Stopping in this idyllic spot today we bared our heads and in your behalf as well as our own, sent up a prayer for the welfare of our Pres. and the success of the party." In 1934 they visited here again, this time with **Carl Van Doren,** to see if they could catch a glimpse of the former President on his front porch. They failed to see Coolidge, but enjoyed talking with tourists making the same pilgrimage.

PUTNEY

Norman Mailer completed his novel *Barbary Shore* (1958) while staying in this village north of Brattleboro.

QUECHEE

Susan Bogert Warner, whose pen name was Elizabeth Wetherell, set her popular children's novel *Queechy* (1852) in this village just east of Woodstock.

RIPTON

The countryside surrounding this town, ten miles east of Middlebury, has many literary associations, mainly because of the Bread Loaf School of English, founded in 1920, and the Bread Loaf Writers' Conference, founded in 1926. Both operate each summer on Middlebury College's mountain campus, about three miles east of town. The School of English is a degree program, while the Writers' Conference is a two-week session of lectures and workshops in poetry, fiction, and nonfiction. While the number of literary figures who have participated in these programs is too great to cite here, writers who also lived in the area can be mentioned. **Robert Frost** joined the staff of Bread Loaf in 1921 and taught here almost every summer for the rest of his life. In 1940, when he was sixty-six, Frost bought a farm two miles east of Ripton, on Route 125. It was the former Homer Noble Farm, and he was to own it until his death in 1963. The farm included three houses and a log cabin, and it was in the cabin that Frost chose to live. Frost shared his simple accommodations with his dog, Gillie. Kathleen Morrison, who served as Frost's personal secretary, and **Theodore Morrison,** the poet and novelist, who was director of the Writers' Conference, lived in the main house of the farm. Theodore Morrison wrote several novels about university intellectual life, including *To Make a World* (1957) and *The Whole Creation* (1962). On many days Frost would write from about ten o'clock in the morning until three in the afternoon, leaving the rest of the day and evening for doing chores, taking walks, and socializing. It was while Frost lived here that he was awarded his fourth Pulitzer Prize, for *A Witness Tree* (1942), and it was here that he prepared several other collections of his verse: *Steeple Bush* (1947), *Complete Poems* (1949), and *In the Clearing* (1962). The Frost farm, now owned by Middlebury College, is not open to the public, but the literary visitor can stop at the nearby Robert Frost Wayside Recreation Area, as the poet often did, and go for a walk on the Memorial Nature Trail, just off Route 125. Also, while traveling from the recreation area east toward the Bread Loaf Inn, one can stop along the way to read seven of Frost's poems, which are on plaques off the highway.

> *In Vermont, where I was born and where my people have lived since Colonial times, the name Ethan Allen is well known, even if few Vermonters know much about the man. In foreign parts . . . Allen, if heard of at all, is at best a misty character. . . . This is unfortunate, for not in the American scene has there been a livelier, lustier character than the late and profane General Allen.*
>
> —Stewart H. Holbrook, in *Ethan Allen*

Robert Frost on his farm near Ripton.

I bought myself a farm for Christmas. One hundred and fifty-three acres in all, fifty in woods. The house a poor little cottage of five rooms, two ordinary fireplaces and one large kitchen fireplace all in one central chimney as it was in the beginning. The central chimney is the best part of it—that and the woods. . . . We have no trout brook, but there is a live spring that I am told should be made into a trout pond.

—Robert Frost,
in a letter to Louis Untermeyer,
January 6, 1929

William Hazlett Upson, who was a frequent participant in the Writers' Conference, lived for many years at Earthworm Manor, just west of the Bread Loaf Inn, on Route 125. This white farmhouse, now owned by the college, is not open to the public.

RUTLAND

Julia C. R. Dorr, poet and author of the novel *Farmingdale* (1854), lived on Dorr Road here. In her home, The Maples, she entertained such literary figures as Ralph Waldo Emerson, Oliver Wendell Holmes, Henry Wadsworth Longfellow, and James Russell Lowell. She died here on January 18, 1913.

Sinclair Lewis and **Dorothy Thompson** on September 23, 1929, were dinner guests of the Rutland Rotary Club at the Community House, now the Lawrence Recreation Center. Lewis, who was asked to speak about Vermont, delighted the large audience by praising the state at some length, but cautioned those listening to protect Vermont against "the terrific speed to make things grow," lest this proclivity destroy Vermont's pastoral beauty.

O woman-form, majestic, strong and fair,
Sitting enthroned where in upper air
Thy mountain-peaks in solemn grandeur rise,
Piercing the splendor of the summer skies,—
Vermont! Our mighty mother, crowned to-day
In all the glory of thy hundred years,
If thou dost bid me sing, how can I but obey?

—Julia C. R. Dorr,
in "Vermont: A Centennial Poem"

ST. ALBANS

Frances Frost, poet and author of the novels *Innocent Summer* (1936) and *Uncle Snowball* (1940), was born in this city north of Burlington on August 3, 1905, in a house at 10 Stebbins Street, still standing. Her poems reflect a concern for nature, the land and people of Vermont, the lives of women, and the complexity of modern life.

SOUTH SHAFTSBURY

Robert Frost in 1919 bought a farm here, called The Stone House. After renovating the house, Frost found it a good place in which to write and here completed part of his collection *New Hampshire* (1923), for which he won his first Pulitzer Prize. His children often stayed here with him, and the entire family took pleasure in the outdoor life. In July 1922, for example, they accomplished a hundred-mile hike through the Vermont mountains. Frost developed a sore foot during the outing and had to hobble all the way home. The Stone House eventually proved too crowded. In 1930, Frost gave the house to his son and daughter-in-law, and the poet and his wife, Elinor, moved nearby to a white farmhouse called The Gully. It was in The Gully that Frost wrote "A Drumlin Woodchuck" (1936), after watching the activities of woodchucks in his yard. Frost received two additional Pulitzer Prizes during his residence here, for *Collected Poems* (1930) and for *A Further Range* (1936), but the years in The Gully were sad ones: his daughter Marjorie died in 1934; his wife, in 1938; and his son Carol, in 1940.

STARKSBORO

Robert Lewis Taylor, known especially for the historical novel *The Travels of Jamie McPheeters* (1958), which won a Pulitzer Prize, lived on a farm from 1955 to 1958 in this village south of Burlington.

WILMINGTON

Robert E. Spiller wrote *The Americans in England During the First Half-Century of Independence* (1926), *Fenimore Cooper, Critic of His Times* (1931), and *The Cycle of American Literature* (1955) at the Spiller family house east of Wilmington, about seventeen miles west of Brattleboro.

Beginning in 1917, **William Carlos Williams,** along with his wife and two boys, spent summers near Mt. Olga, which is south of Lake Raponda. They stayed on the farm owned by Mrs. Williams's aunt and uncle.

H[alsey] W[illiam] Wilson, founder of the H. W. Wilson publishing company in 1898, was born on May 12, 1868, in a house on East Main Street, still standing but not open to the public. He established many important indexes and other reference works, including the *Cumulative Book Index,* begun in 1898, and the *Readers' Guide to Periodical Literature,* begun in 1901.

WINDSOR

Maxwell Perkins, the celebrated editor, spent his childhood summers in Windsor, where his family had a large brick house at 26 North Main Street. After Perkins was married, he and his family continued to come to this house and to others nearby for weekends and for longer vaca-

Summer home of Maxwell Perkins in Windsor, built by his grandfather in 1825. Perkins spent summers all through childhood in this house and later returned for summers here with his own children.

tions. The North Main Street house is not open to the public.

WOODSTOCK

Sinclair Lewis was divorced by **Dorothy Thompson** in the Windsor County Court House here on January 2, 1942.

George Seldes in 1934 bought a farm in Woodstock and here wrote *Sawdust Caesar* (1935), *Lords of the Press* (1938), and *You Can't Do That* (1938).

MASSACHUSETTS

AMESBURY

This town in northeastern Massachusetts is celebrated as the home for four decades of one of our country's best-known poets, **John Greenleaf Whittier.** Whittier's house, now known as the John Greenleaf Whittier Home, is at 86 Friend Street. A National Historic Landmark, it is open to the public. A small admission fee is charged. Whittier lived at this house from 1836 to 1876, when he moved to Danvers, but was a frequent visitor to his Amesbury home until his death in 1892. For forty years, the poet did most of his writing in the Garden Room of the house, which is maintained as it was in Whittier's day. The house contains many of its original furnishings, as well as Whittier's books, some of his manuscripts, and the desk on which he wrote his masterpiece, the long poem *Snow-Bound* (1866). During his years at Amesbury Whittier published *The Panorama and Other Poems* (1856), which includes "Maud Muller"; *Home Ballads, Poems and Lyrics* (1860), in which appears "Skipper Ireson's Ride"; and *In War Time and Other Poems* (1864), which contains the schoolchild's classic "Barbara Frietchie." Whittier is buried in Union Cemetery, on Route 110 (Haverhill Road), not far from his home. The Amesbury Public Library, at 149 Main Street, has a collection of items by and about Whittier, including editions of the poet's books, manuscript letters, and microfilms of newspapers edited by Whittier.

AMHERST

This town in western Massachusetts has been the home of several well-known writers, but the author most closely identified with Amherst is **Emily Dickinson,** who was born on December 10, 1830, in the red brick house at 280 Main Street. In 1840 she moved with her family to a house on North Pleasant Street, but the Dickinsons returned to the Main Street house in 1855. There Emily spent her remaining years, wrote most of her poems and, on May 15, 1886, died. Dickinson attended the local school, Amherst Academy, and

> *There we went summer after summer for a month at a time: once in April to a sugaring off in old man Aldrich's woods, helping to collect the sap buckets for him, watching the steaming pan and all the rest of it. But I never had any real desire to live the year through in Vermont.*
>
> —William Carlos Williams, in *The Autobiography of William Carlos Williams*

The Garden Room in the Whittier Home in Amesbury. Here John Greenleaf Whittier wrote many of his works.

Faith is a fine invention
For gentlemen who see;
But Microscopes are prudent
In an emergency.

—Emily Dickinson,
in *Poems*, Second Series

Photograph of Emily Dickinson at age seventeen. This is the only known picture of the poet.

made several close friends, including **Helen Maria Fiske,** who later wrote under her married name, **Helen Hunt Jackson.** As Emily Dickinson grew older, she became more reclusive, spending almost all of her time in her house, screened from the outer world, quite willingly, by her family. Of her multitude of poems, only a few were published in her lifetime. After Emily Dickinson's death, her sister found a sheaf of her poems and, disobeying the poet's request to have all her papers destroyed, made the poems available for publication. Today Emily Dickinson is considered one of the leading American poets, and her verses are frequently anthologized. The house at 280 Main Street, now a National Historic Landmark, is used as a faculty residence by Amherst College and is open to the public by appointment. The poet is buried in the family plot in West Cemetery in Amherst. The Jones Library, at 43 Amity Street, has a display of Emily Dickinson's manuscripts that will interest the literary visitor.

Amherst's other literary residents include **Eugene Field,** who lived from 1856 to 1865 with his cousin Mary Field French at her home, still standing, at 219 Amity Street. **Helen [Maria Fiske] Hunt Jackson,** known also as **H. H.,** born here on October 15, 1830, was best known for her novel *Ramona* (1884). Her childhood home at 249 South Pleasant Street, a two-and-a-half story frame house, is still standing.

Writers associated with Amherst College include **Jacob Abbott,** the clergyman, author, and teacher, who taught mathematics and natural philosophy here from 1824 to 1830; and **Clyde Fitch,** the playwright, who was graduated from Amherst in 1886. As a student, Fitch demonstrated a talent for literature and the stage. He was a steady contributor to, and for a time editor of, the Amherst *Student.* **Robert Frost** was a professor of English at Amherst in 1916–20, 1923–25, and 1926–38. **Noah Webster,** the lexicographer, came to the town of Amherst from New Haven in 1812 and lived here until 1822. During that time he helped to found Amherst College. **Richard Wilbur,** the poet, was graduated from Amherst in 1942; and **Stark Young,** the poet and novelist, taught here from 1915 to 1921.

". . . I'm Disko Troop o' the 'We're Here' o' Gloucester, which you don't seem rightly to know."

"I don't know and I don't care," said Harvey. "I'm grateful enough for being saved and all that of course; but I want you to understand that the sooner you take me back to New York the better it'll pay you."

—Rudyard Kipling,
in *Captains Courageous*

ANDOVER

A number of writers are associated with this town in northeastern Massachusetts through the several academies that have existed here. Among these institutions is Abbott Academy, run by **Jacob Abbott** and his brothers in the nineteenth century. Abbott had studied at Andover Theological Seminary in 1821–22 and in 1824. This school was moved in 1908 to Cambridge, and is now part of the Andover Newton Theological School, in Newton.

Its former buildings in Andover are now used by the town's third school of wide reputation, Phillips Academy. Phillips, founded in 1778 and later located on Main Street, where it now stands, was merged with Abbott Academy in 1973. Phillips Academy is the only school of the three mentioned still fulfilling its function in Andover. **Edgar Rice Burroughs** was a student at Phillips Academy in 1891–92. Other writers associated with Andover include **John Gould Fletcher,** the poet, who studied at Phillips Academy in the early 1900s, and **Alice French,** who was born on March 19, 1850, in the Double Brick House, on Main Street, now used by Phillips as a dormitory. The novelist and short-story writer was graduated from Abbott Academy in 1868 and later spent much of her life in Iowa and Arkansas. Under the pen name **Octave Thanet,** she wrote a number of books, set mostly in the Midwest and South. Among her volumes were the collection *Stories of a Western Town* (1893) and the novel *The Man of the Hour* (1905).

Munro Leaf, author of numerous books for children, lived for thirteen years in the 1950s and 1960s at 56 Salem Street, still standing. The family swimming pool, also still there, was the inspiration for *The Wishing Pool* (1960), a book for children. **Elizabeth Stuart Phelps [Ward],** author of the religious novel *The Gates Ajar* (1868), was born in Boston on August 31, 1844, but grew up in the house now used as the headmaster's residence at Phillips Academy.

Harriet Beecher Stowe moved to Andover in 1853, when her husband joined the faculty of Andover Theological Seminary. The Stowes remained in Andover until 1864 and lived in a stone house on Chapel Avenue, since moved to 80 Bartlett Street. During these years Mrs. Stowe wrote a number of books, including *Dred: A Tale of the Great Dismal Swamp* (1856) and *The Minister's Wooing* (1859). Mrs. Stowe is buried in the cemetery of the trustees of Phillips Academy in Andover. **Kate Douglas Wiggin** studied for a time at Abbott Academy in the 1870s but did not stay long enough to be graduated.

ANNISQUAM

Writers associated with this village, part of Gloucester, in northeastern Massachusetts, include **Russel Crouse,** the playwright, who owned a home here; and **Rudyard Kipling,** who visited Gloucester in the 1890s and gathered material for his classic novel *Captains Courageous* (1897), the story of a spoiled boy named Harvey, who is washed overboard from a luxury liner on which he is a passenger. The skipper and crew of the Gloucester fishing schooner that saves him teach Harvey the values of honesty, friendship, hard work, and courage. Kipling stayed for a time during 1896 in a house, still standing, at the corner of Leonard Street and Cambridge Avenue. A small Annisquam boat named *I Am Here* gave Kipling the idea for the name of the fishing schooner *We're Here* in *Captains Courageous.*

ARLINGTON

Robert Creeley, the poet, was born on May 21, 1926, at Sims Hospital in this town about six miles northwest of Boston. His many collections of verse include *A Form of Women* (1959) and *Selected Poems* (1976). **John Townsend Trowbridge,** one of the most widely read authors of books for boys in the second half of the nineteenth century, lived in Arlington from 1865 until his death on February 12, 1916. Among the books Trowbridge wrote while living here were the novel *Coupon Bonds* (1866), which was set in New England; *The Vagabonds and Other Poems* (1869), which included his poem "Darius Green and His Flying Machine"; and his autobiography, *My Own Story* (1903).

BELCHERTOWN

This town fifteen miles northeast of Springfield was named for Jonathan Belcher (1682–1757), colonial governor of Massachusetts and New Hampshire, and later of New Jersey. **Josiah Gilbert Holland,** the novelist, poet, and editor, was born in Belchertown on July 24, 1819. Holland's novel *Sevenoaks* (1875) was set in a New England industrial town run by an unscrupulous financier named Robert Belcher.

BELMONT

William Dean Howells, seeking a home in the quiet countryside where he could work and watch his children grow up, lived from 1878 to 1881 at Redtop, a three-story brick and redwood shingle house he built at 90 Somerset Street. Howells and his wife wanted the house to be spacious and well lighted. When Howells wrote to his architects expressing his feeling that redwood paneling would make the library too dark, the architect suggested that the paneling be painted. A similar incident occurs in *The Rise of Silas Lapham* (1885). While living here, Howells was editor of *The Atlantic Monthly* and worked on several novels, including *The Lady of the Aroostook* (1879) and *A Modern Instance* (1882). Among the literary people who visited Howells at Redtop were **Thomas Bailey Aldrich, Henry James, Henry Wadsworth Longfellow, Mark Twain,** and **Charles Dudley Warner.**

In late 1881 Howells suffered a physical collapse that left him permanently weakened, unable even to climb the hill to his house. His own medical expenses and those for his daughter, who was afflicted with an unknown ailment, coupled with his inability to sell his home in Cambridge, led Howells to give up his beloved home in the country. Redtop, now private, is a National Historic Landmark. The house, whose exterior has been modified over the years, contains a number of items from Howells's day, including several carved fireplace mantels, and the author's library, complete with original bookshelves and, over the library door, the motto "From Venice as far as Belmont." The red shingles that covered the upper exterior walls of the house have been replaced with stucco in recent years, but they may be restored in the future.

> *"Paint it?" gasped the Colonel.*
> *"Yes," said the architect quietly. "White, or a little off white."*
> *Lapham dropped the plum he had picked up from the table. His wife made a little move toward him of consolation or support.*
> *"Of course," resumed the architect, "I know there has been a great craze for black walnut. But it's an ugly wood; and for a drawing-room there is really nothing like white paint. . . ."*
>
> —William Dean Howells, in *The Rise of Silas Lapham*

BEVERLY

This city just northeast of Salem has been the home of a number of well-known authors. **Henry Adams** made his summer home in Beverly in the late nineteenth century, when it was known as Beverly Farms; and **Oliver Wendell Holmes,** the physician, teacher, and author, summered at 868 Hale Street in the 1800s. This two-and-a-half-story frame house was also the summer home of his son Oliver Wendell Holmes, Jr., chief justice of the Supreme Court. The elder Holmes wrote of his Beverly home and other places in the area in the poem "The Broomstick Train."

Beverly was the birthplace, on March 5, 1824, of **Lucy Larcom,** the poet, teacher, abolitionist, and author of the autobiography *A New England Girlhood* (1889). The Larcom family home at 415 Hale Street was torn down in 1832. At the age of seven, Larcom moved with her family to Lowell, where she worked in the mills and launched her literary career by contributing, under various pseudonyms, to a publication known as the *Lowell Offering,* written and produced by the girls who worked in the mills. In later years she lived in Beverly, which she wrote of in both her autobiography and poetry, and spent winters in Boston. One of her poems, "Mistress Hale of Beverly," concerns the wife of John Hale, the Beverly minister active in the witch trials of 1692. Hale's wife was accused of witchcraft; the minister, who knew of her innocence, rose to her defense and in so doing ended the mob hysteria whipped up against so-called witches—a hysteria he had helped create. Lucy Larcom died on April 17, 1893, and is buried in Beverly. The Beverly Historical Society, 117 Cabot Street, has a room devoted to Larcom's life and writings.

Redtop, the Belmont home of William Dean Howells, completed in July 1878. The upper stories of the house were shingled originally in California redwood.

My feet, they haul me Round the House,
They Hoist me up the Stairs;
I only have to steer them, and
They ride me Everywheres.

—Gelett Burgess,
in "My Feet"

Another Beverly native was **Philip [Gordon] Wylie,** born on May 12, 1902, at the Congregational Church, no longer standing, at 65 Dodge Street. Wylie coined the term "momism" in his best-known book, *Generation of Vipers* (1942).

BILLERICA

Elizabeth Palmer Peabody, born on May 16, 1804, in this town in northeastern Massachusetts, was the sister-in-law of Nathaniel Hawthorne and of the educator Horace Mann, and was a prominent figure in Boston literary life. She owned a celebrated bookstore in Boston and wrote in *Record of a School* (1835) of her association with the Temple School, conducted by Bronson Alcott.

BOSTON

Boston, since its founding in the 1600s, has been the home and workplace of numerous writers, and the city has also been the birthplace of a number of authors, including **Charles Follen Adams,** born on April 21, 1842, in Dorchester, now a section of Boston. Adams was known for his humorous dialect verse, published in such volumes as *Leedle Yawcob Strauss and Other Poems* (1878). **Henry Adams,** author of *The Education of Henry Adams,* was born on February 16, 1838, in a house on Hancock Avenue, on Beacon Hill near the Boston State House. In 1842 the Adams family moved to a house on Mount Vernon Street. Adams spent his boyhood there and in Quincy, Massachusetts.

The Robert Gould Shaw Memorial in Boston Common, sculpted by Augustus Saint-Gaudens. Shaw, a Civil War hero, is also recalled in poems by Emerson and William Vaughn Moody.

William Taylor Adams, born in Boston on July 30, 1822, wrote under the name **Oliver Optic** more than one hundred books for boys. **[Frank] Gelett Burgess,** the humorist known best for his rhyme beginning "I never saw a Purple Cow," was born here on January 30, 1866. **William Ellery Channing,** the poet and essayist, was born in Boston on November 29, 1817, and attended Boston Latin School. **John Ciardi,** the poet, was born on June 24, 1916 in a house on Sheafe Street. Ciardi has stated, "I was literally born there—midwived at home." **James B[rendan Bennet] Connolly,** author of many books set in or describing Gloucester, Massachusetts, was born here in 1868, and **Joseph Dennie,** the author and editor, was born here on August 30, 1768.

Ralph Waldo Emerson was born on May 25, 1803, in a house, no longer standing, on Summer Street. His father, the Reverend William Emerson, was pastor of the First Church of Boston, located at the corner of Summer and Chauncy streets. Emerson attended Boston Latin School, and served from 1829 to 1831 as pastor of the Second Church of Boston, located on Hanover Street but demolished in 1844. Emerson lived at several addresses, including 60 Essex Street, 24 Franklin Place, 276 Washington Street, and Franklin Place. Emerson composed his poem "Voluntaries" (1863) to honor a recently dead Civil War hero, Col. Robert Gould Shaw, who commanded the Fifty-fourth Massachusetts, the first enlisted Black regiment. Shaw's statue stands in Boston Common across from the State House. This statue also inspired **William Vaughn Moody** to write his poem "An Ode in Time of Hesitation" (1900).

Before the solemn bronze Saint Gaudens made
To thrill the heedless passer's heart with awe,
And set here in the city's talk and trade
To the good memory of Robert Shaw,
This bright March morn I stand,
And hear the distant spring come up the land;
Knowing that what I hear is not unheard
Of this boy soldier and his Negro band,
For all their gaze is fixed so stern ahead,
For all the fatal rhythm of their tread.
The land they died to save from death and shame
Trembles and waits, hearing the spring's
great name,
And by her pangs these resolute ghosts are stirred.

—William Vaughn Moody,
in "An Ode in Time of Hesitation"

Benjamin Franklin was born on January 17, 1706, in a house at 17 Milk Street, now the site of an office building, which bears a plaque honoring Franklin's birth. Franklin spent the first seventeen years of his life in Boston before leaving to establish himself as a printer in Philadelphia. The city honors Franklin with a statue in front of City Hall; the 520-acre Franklin Park in the West Roxbury section of Boston; and a Benjamin Franklin Collection in the Boston Public Library, 666 Boylston Street.

Other writers born in Boston include **Edward Everett Hale,** author of *The Man Without* *a Country* (1863), born here on April 3, 1822. Hale lived at various places in and around Boston during his life, including a house at 39 Highland Street in Roxbury, his home in the late 1850s. He also lived for forty years in a house at 12 Morley Street and died there on June 10, 1909. Hale, a Unitarian minister, served as pastor of Boston's South Congregational Church from 1856 to 1899. It was during this period that Hale wrote his best-known work, *The Man Without a Country,* a classic tale in patriotic American literature. Hale's home on Morley Street is used by the Research Institute of Africa and African Diaspora Arts, Inc. A memorial to Hale stands in the city's Public Gardens, adjacent to Boston Common. Also born in Boston was Hale's sister, **Lucretia Peabody Hale,** the novelist and author of books for children, on September 2, 1820. She was known best for her collection of humorous stories, *The Peterkin Papers* (1880). Miss Hale spent most of her life in Boston, and in later years lived in an apartment at 127 Charles Street. She died in Boston on June 12, 1900, and funeral services were held for her at the South Congregational Church.

Barbara Howes, the poet, who was born in Boston on May 1, 1914, has stated that in childhood she lived at 45 Woodland Road, where she wrote "just junky verse; hopeless." **Sarah Kemble Knight,** the diarist whose *Journal of Madame Knight* (1935) is considered a valuable source for information about colonial life, was born here on April 19, 1666. **Robert Lowell,** the poet, was born in Boston on March 1, 1917, spent his youth here, and returned to live in Boston during the 1950s. His addresses included 170 Marlborough Street, 33 Commonwealth Avenue, and 239 Marlborough Street.

In Copp's Hill Burying Ground, off Hull Street in Boston's North End, are buried three distinguished members of the Mather family: Cotton, Increase, and Samuel Mather. **Cotton Mather,** the clergyman and author and one of the leading figures of seventeenth-century America, was born in Boston on February 12, 1663. The son of the clergyman and historian Increase Mather, Cotton Mather wrote more than four hundred books in his lifetime, including such works as *The Wonders of the Invisible World* (1693), *Magnalia Christi Americana* (1702), and *The Christian Philosopher* (1721). For much of his life Mather served as minister at the Second Church in Boston, then located at the head of North Square. For some time between the mid-1660s and 1676, Mather lived with his father in a house on North Square. In 1688 Cotton bought a house on the west side of Hanover Street, believed to be on the site of the present 298 Hanover Street. He owned the house until 1718 but may have lived on there as a tenant until his death on February 13, 1727 (or 1728). Neither of the houses still stands.

Increase Mather was born in Dorchester, now part of Boston, on June 21, 1639. A theologian and author, Increase Mather was for sixty years minister of the Second Church of Boston, then at the head of North Square, the later site of the Old North Church. He lived in a house in North Square, where Paul Revere's house now stands, but a plaque there mistakenly identifies the place as the site of Cotton Mather's house, which was two blocks away, near the corner of Hanover and Prince streets. In 1677, after Increase Mather's house burned, he built a new house on the west side of Hanover Street near North Bennet Street. This remained his home until his death on August 23, 1723. **Samuel Mather,** Cotton Mather's son and biographer, who also was a clergyman, was born in Boston and lived from 1736 to his death on June 27, 1785, in a house, no longer standing, on the east side of Moon Street.

THE MATHERS

Cotton was the first son of Increase. At the age of 11, Cotton could write and speak Latin, could read the New Testament in Greek and had begun Hebrew grammar. He entered Harvard at 12 years and was a candidate for the ministry at 18. His youthful energy never abated. He published 444 books, collected New England's largest library of its day, and did scientific research that won him membership in the prestigious Royal Society of London. Many of his endeavors were ahead of their time. He was an early champion of innoculation against smallpox, but his advanced ideas won him only ridicule. Cotton was not dismayed. He was convinced his extraordinary ability was a personal sign from God. Perhaps, as a result, he was often overbearing, and always overzealous, traits that won him respect but rarely affection from his fellowman.

Despite his voluminous writing and constant activity, Cotton Mather's advice to visitors was inscribed over his library door, "Be Short."

Samuel Mather followed his father and grandfather to Harvard and into the ministry. Unlike them however, he did not venture outside the ministry into politics and with his death in 1785, the Mather Dynasty ended.

Plaque at the Mather tomb in Copp's Hill Burying Ground, burial site of Increase, Cotton, and Samuel Mather.

Writers born in Boston also include **Samuel Eliot Morison,** the Pulitzer Prize-winning historian and biographer, born on July 9, 1887; and **Francis Parkman,** the historian, born on September 16, 1823, who lived for many years at 50 Chestnut Street. **Sylvia Plath,** the poet, was born at Memorial Hospital in Boston on October 27, 1932, and lived in the late 1950s at 9 Willow Street in Louisburg Square. In 1982, many years after her death, she won a Pulitzer Prize for *The Collected Poems* (1981).

Edgar Allan Poe, not thought of as a New Englander, was born in Boston on January 19, 1809. Christened Edgar Poe—the middle name was added after his informal adoption by John Allan—he was the son of David and Elizabeth Arnold Poe, actors in the employ of the Boston and Charleston Players. A few months after Poe's birth the family moved to New York, Poe was orphaned before he was three, and Boston thereafter played only a slight role in Poe's life. Poe, who had broken with his foster father, came to Boston in the spring of 1827 and secured publication of his first volume, *Tamerlane and Other Poems,* at a job printing shop at 70 Washington Street. The slim volume did not bear Poe's name—the title page declared that the book was "By a Bostonian"—but it was Poe's first book and now is an

> *"But under no circumstances is he ever to hear of his country or to see any information regarding it; and you will specially caution all the officers under your command to take care, that in the various indulgences which may be granted, this rule, in which his punishment is involved, shall not be broken.*
>
> *"It is the intention of the Government that he shall never again see the country which he has disowned. Before the end of your cruise you will receive orders which will give effect to this intention.*
>
> *"Respectfully yours,*
> *"W. Southard, for the*
> *Secretary of the Navy."*
>
> —Edward Everett Hale,
> in *The Man Without a Country*

> *We grew in age—and love—together—*
> *Roaming the forest and the wild;*
> *My breast her shield in wintry weather—*
> *And, when the friendly sunshine smiled,*
> *And she would mark the opening skies,*
> *I saw no Heaven—but in her eyes.*
>
> —Edgar Allan Poe,
> in "Tamerlane"

Benjamin Franklin's birthplace, drawn by an eighteenth-century artist.

A. Bronson Alcott and Louisa May Alcott.

exceedingly rare collector's item. The book went unnoticed after publication and Poe, penniless, was forced to enlist in the U.S. Army. From May to October 1827 he was stationed at Fort Independence, in Boston Harbor. In late October his unit was reassigned to Fort Moultrie in Charleston, South Carolina, and Poe's days in the city of his birth came to an end.

Also born in Boston was **Laura Elizabeth Richards,** the poet and biographer, on February 27, 1850. Richards received a Pulitzer Prize for her biography of her mother, *Julia Ward Howe* (1916), written with Maud Howe Elliott and Florence Howe Hall. **Henry Morton Robinson,** author of the best-selling novel *The Cardinal* (1950), was born here on September 7, 1898. **Horace E[lisha] Scudder,** the author, and editor from 1890 to 1898 of *The Atlantic Monthly,* was born here on October 16, 1838. **Mary Stolz,** author of books for girls, was born in Boston on March 24, 1920; and **Daniel Pierce Thompson,** the novelist whose books dealt with early Vermont history, was born here on October 1, 1795.

Royall Tyler, the lawyer and writer, was born here on July 18, 1757, and practiced law in Boston. While living here Tyler wrote the play *The Contrast* (1787) and the picaresque novel *The Algerine Captive* (1797), said to be the first American novel republished in England. **Theodore H[arold] White,** the journalist-historian, was born on May 6, 1915, in a house on Erie Street. White also lived in a house on Greenwood Street and attended Boston Latin School. He won a Pulitzer Prize for *The Making of the President* (1961), a dramatic account of the 1960 election.

Writers who have lived or worked in Boston include **Jacob Abbott,** who came to Boston in 1928 and established the Mt. Vernon School for Girls. Abbott wrote of his progressive educational ideas in a book entitled *The Young Christian* (1832). In 1833 Abbott became minister of the Eliot Congregational Church, and in 1834 there appeared the first volume of the Rollo Books, the series of juveniles that made him famous.

Boston was a frequent home for **A. Bronson Alcott** and his family in the nineteenth century. Bronson Alcott first visited Boston in 1828 and opened a modest school on Salem Street. On May 23, 1830, he and Abigail May were married in King's Chapel, at the corner of Tremont and School streets. After living in Philadelphia for a time, Alcott moved his family back to Boston in 1834 and established his celebrated Temple School in the Masonic Temple that stood at the corner of Tremont Street and Temple Place. Here, with the assistance of Elizabeth Palmer Peabody, Alcott applied his progressive theories of education and, with Miss Peabody, wrote an account of the school in *The Record of a School* (1835). Alcott was visited at the school in 1835 by Ralph Waldo Emerson, who admired Alcott. In that same year Elizabeth Peabody left the school over a difference of opinion with Alcott, and Margaret Fuller soon after came to teach in her place.

Publication of Alcott's *Conversations on the Gospels* (1837) brought much criticism from those less progressive in their views, and Alcott's school began to lose pupils. The killing blow came when Alcott had the audacity to admit a Black student: the school was forced to move and in 1839 closed down. In 1840 Alcott moved his family to Concord, Massachusetts, the town with which he is most closely identified. But he continued to live in Boston at different times and different places. For example, the Alcott family lived in the 1850s at 20 Pinckney Street. There Alcott's daughter **Louisa May Alcott** wrote her first published story, "The Rival Prima Donnas," which appeared as the work of **Flora Fairfield** in the *Saturday Evening Gazette* of November 11, 1854. At 20 Pinckney Street, Louisa also saw publication of her first book, *Flower Fables* (1854), written six years earlier. This marked the beginning of her remarkable career as a writer. In the next thirty years she and the rest of her family were to divide their time largely between Concord and Boston, and they were to live in a bewildering number of Boston houses. For example, in 1868 and 1869 Louisa Alcott worked on the second volume of her popular novel *Little Women* (1868, 1869) in rented rooms on Brookline Street and at the Bellevue Hotel on Beacon Street. She frequently stayed at the Bellevue during visits to Boston in the 1870s and there worked on *A Modern Mephistopheles* (1877) and *Jack and Jill* (1880). In the fall of 1885 Louisa Alcott moved to 10 Louisburg Square and there prepared for publication a three-volume collection of stories, *Lulu's Library* (1886–89), and completed her novel *Jo's Boys* (1886). It was at this house as well that Bronson Alcott died on March 4, 1888. Toward the end of her life, Louisa Alcott suffered greatly from ill health and spent much time at 2 Dunreath Place, the Roxbury home of Dr. Rhoda Lawrence. Here she worked on her last book, *A Garland for Girls* (1888), and here she died on March 6, 1888.

Thomas Bailey Aldrich lived in Boston from 1865 until his death on March 19, 1907. During this period Aldrich served as editor of *Every Saturday* magazine from its inception in 1866 until its demise in 1874, and as editor of *Atlantic Monthly* from 1881 to 1890. Here, too, Aldrich produced almost all of his books, including *The Story of a Bad Boy* (1870), for which he is remembered today. Aldrich lived at 84 Pinckney Street, where he wrote *The Story of a Bad Boy;* at 131 Charles Street, from 1871 to 1881; and for many years at 59 Mt. Vernon Street. He described the area near his last home in the book *Two Bites at a Cherry* (1894).

> *The next day was Sunday, and a goodly troop of young and old set forth to church—some driving, some walking, all enjoying the lovely weather and the happy quietude which comes to refresh us when the work and worry of the week are over. Daisy had a headache; and Aunt Jo remained at home to keep her company, knowing very well that the worst ache was in the tender heart struggling dutifully against the love that grew stronger as the parting drew near.*
>
> —Louisa May Alcott, in *Jo's Boys*

Robert W. Anderson lived at 135 Marlborough Street and 14 Carver Street while a student at Harvard in the late 1930s, and there wrote some early plays. In 1940 Anderson moved to 66 Chestnut Street with his first wife. Anderson used Boston as a partial setting for his novel *After* (1973). **Charles Angoff** lived in the Dorchester and West End sections of Boston in his youth, worked on Boston newspapers in 1923, and used the city as a setting for the stories in his collection *When I Was a Boy in Boston* (1947). **Isaac Asimov** lived at 42 Worcester Square in 1949 while teaching at Boston University School of Medicine. **Nathaniel Benchley** died at New England Deaconess Hospital here on December 14, 1981. **Elizabeth Bishop,** the poet, died at her home here on October 6, 1979, and **Kay Boyle** wrote her autobiographical *Being Geniuses Together* (1968) in Boston.

Anne Bradstreet, the first woman poet to write in English in the New World, lived in Boston in the mid-1600s. **James Fenimore Cooper** came to Boston to gather material for his novel *Lionel Lincoln* (1825), set in Boston during the Revolutionary War. In the novel, Gage, the British general, planned strategy for the Battle of Bunker Hill in a house on Hull Street in Boston's North End. Lionel Lincoln watched the battle from Copp's Hill Burying Ground, off Hull Street. **Richard Henry Dana, Jr.** on August 14, 1834, sailed from Boston Harbor as an able seaman aboard the *Pilgrim,* bound for California. Two years later he returned to Boston aboard the ship *Alert,* resumed his studies at Harvard, and in 1840 was admitted to the bar. The same year saw publication of his classic volume *Two Years Before the Mast,* based on his experiences as a sailor, and the following year, publication of *The Seaman's Friend* (1841), for years the standard reference on maritime law. Dana lived and practiced law in Boston for many years, and his addresses included a house at 361 Park Street, his home from 1874 to 1880.

Margaret Deland lived much of her adult life in Boston and Cambridge. In Boston, toward the end of the nineteenth century, she lived at 76 and at 112 Mt. Vernon Street. In the early 1900s she lived at 35 Newbury Street. During her years in Boston, Mrs. Deland wrote most of her novels and collections, including *Old Chester Tales* (1898)—she was brought up in a Pennsylvania town called Manchester, now part of Pittsburgh—and it is for the fictional town of Old Chester that she is remembered today. In Deland's last years she lived at the Hotel Sheraton in Boston, and it was there that she died on January 13, 1945. She is buried in Forest Hills Cemetery, 95 Forest Hills Avenue.

James T. Fields, the author, editor, and partner in the Boston publishing firm of Ticknor and Fields, lived for thirty years in a house, no longer standing, at 148 Charles Street, and died there on April 24, 1881. Fields came to Boston at the age of fourteen and took a job at The Old Corner Bookstore, then located at the northwest corner of Washington and School streets, but later moved to Bromfield Street. He became one of the store's proprietors, and with John Reed and **W. D. Ticknor**—Ticknor was also a partner in the bookstore—in 1832 launched Ticknor, Reed, and Fields, known from 1854 on as Ticknor and Fields, which became one of the most prestigious publishing firms of the nineteenth century. The imprint still is used today. Fields's wife, Annie Adams Fields, was known as one of the leading literary hostesses of her day, and visitors to the Fields home included **Willa Cather, Charles Dickens,** and **Sarah Orne Jewett.** Jewett lived with the Fields family for a time after the death of her

WALDEN;

OR,

LIFE IN THE WOODS.

BY HENRY D. THOREAU,

AUTHOR OF "A WEEK ON THE CONCORD AND MERRIMACK RIVERS."

I do not propose to write an ode to dejection, but to brag as lustily as chanticleer in the morning, standing on his roost, if only to wake my neighbors up. — Page 92.

BOSTON:

TICKNOR AND FIELDS.

M DCCC LIV.

Title page of Henry David Thoreau's most famous work, published in 1854 by Ticknor and Fields.

Let Greeks be Greeks, and women
what they are.
Men have precedency, and still excel.
It is but vain unjustly to wage war:
Men can do best, and women know
it well.
Preëminence in all and each is yours—
Yet grant some small acknowledgment
of ours.

—Anne Bradstreet,
in "The Prologue"

Thomas Bailey Aldrich's study in his home on Mt. Vernon Street. The author lived in this house from 1883 until his death in 1907.

Etching of Oliver Wendell Holmes, the physician and man of letters.

father in 1878. James T. Fields served from 1861 to 1871 as editor of *The Atlantic Monthly* and is known best as an author for his collection of reminiscences *Yesterdays With Authors* (1872). Mrs. Fields also wrote several books, including *Authors and Friends* (1896).

Mary Hallock Foote spent the last ten years of her life in Boston, where she died on June 25, 1938. **Robert Frost** died in Peter Bent Brigham Hospital, Boston, on January 29, 1963. **Margaret Fuller** lived in Boston in the 1830s and 1840s. At her home on West Street, Miss Fuller held "conversations" with women on such subjects as art, literature, and music. During this period she worked as a teacher in the Temple School of **A. Bronson Alcott,** and served as editor of *The Dial,* the literary magazine founded by Emerson and others to publish the writings of the Transcendentalists. The Boston Public Library has a major collection of Margaret Fuller's manuscripts.

Hamlin Garland came to Boston in 1884 to complete his education. He spent much time reading in the Boston Public Library and studied and taught at the Boston School of Oratory. Garland lived here until the early 1890s and contributed articles and stories to such publications as the Boston *Transcript, Harper's Weekly,* and *The Arena.* He also wrote the stories collected in *Main-Travelled Roads* (1891) and made the acquaintance of William Dean Howells and other influential literary figures. Garland used Boston as the setting for his novel *Jason Edwards* (1892) and wrote of his experiences here in *A Son of the Middle Border* (1917).

William Lloyd Garrison lived in Boston for many years and here founded and edited the abolitionist magazine *The Liberator* (1831–65)—in 1835, for his attacks on slavery, he was dragged through the streets by an angry crowd. Garrison lived from 1864 to 1879 at 125 Highland Street in Roxbury. His two-and-a-half story frame house now is a National Historic Landmark. Garrison died on May 24, 1879, and was buried in Forest Hills Cemetery.

Home of William Lloyd Garrison on Highland Street. After *The Liberator* ceased publication in 1865, Garrison went into semiretirement here.

Nathaniel Hawthorne was appointed weigher and gauger at the Boston Custom House and moved to the city in 1839. He lived at 8 Somerset Place and 54 Pinckney Street, and worked at the Custom House until January 1841. A few months later he joined George Ripley's Brook Farm experiment in West Roxbury, now a part of Boston. In July 1842 Hawthorne married Sophia Peabody at the home of her parents on 13 West Street, and the couple went to live in Concord. Hawthorne used Boston as the setting for several of his short stories, including "My Kinsman, Major Molineux" and "Lady Eleanore's Mantle." He also wrote, in *The Scarlet Letter* (1850), of Boston's Old Prison, which stood on Prison Lane, now Court Street.

While living at 65 Pinckney Street in 1949, **Ruth Herschberger** wrote *A Dream Play* (1957), a play about Conrad Aiken. **Edward Hoagland** worked on his first novel, *Cat Man* (1956), at 14 Pinckney Street, and also in Cambridge.

Oliver Wendell Holmes, Sr., the physician and author, lived in Boston for most of his life. While living in a house, no longer standing, on Bosworth Street, off Montgomery Street, Holmes wrote the poem he considered his best: "The Chambered Nautilus" appeared in his volume *The Autocrat of the Breakfast-Table* (1858). Holmes also lived from 1859 to 1871 at 164 Charles Street and, for the last thirteen years of his life, at 296 Beacon Street. There he wrote *Over The Teacups* (1891) and was host to such literary friends as **Ralph Waldo Emerson, James Russell Lowell,** and **John Greenleaf Whittier.** Holmes used Boston as the inspiration or setting for a number of works. The U.S.S. *Constitution,* now at Boston Naval Shipyard and a National Historic Site, was the inspiration for Holmes's "Old Ironsides" (1830). Publication of this poem helped save the ship from destruction. Boston is also the setting for Holmes's poem "The Last Leaf" (1831), inspired by a Major Melville, fire warden of Boston for forty years; and for the novels *The Guardian Angel* (1867) and *A Moral Antipathy* (1885). Holmes was for decades one of Boston's leading literary figures. It was he who gave *The Atlantic Monthly* its name in 1857, and it was for the *Atlantic* that he wrote his several "Breakfast-Table" books, beginning with *The Autocrat of the Breakfast-Table.* Holmes died at his Beacon Street home on October 7, 1894.

> *Build thee more stately mansions, O my soul,*
> *As the swift seasons roll!*
> *Leave thy low-vaulted past!*
> *Let each new temple, nobler than the last,*
> *Shut thee from heaven with a dome more vast,*
> *Till thou at length art free,*
> *Leaving thine outgrown shell by life's unresting sea!*
>
> —Oliver Wendell Holmes, Sr. in "The Chambered Nautilus"

Julia Ward Howe, the author and reformer, lived in the mid-1800s at 13 Walnut Street. This four-story brick mansion, now private, was also the home of the artist John Singer Sargent and is a National Historic Landmark.

If Oliver Wendell Holmes Sr., gave *The Atlantic Monthly* its name, then certainly **William Dean Howells,** as editor in chief from 1871 to 1881, gave the magazine much of its deserved reputation as a publication of high literary merit. Howells came to work in Boston in 1866 to take a position as subeditor of *The Atlantic Monthly,* then located at 124 Tremont Street. He rented a house on Bullfinch Street, then moved to a house at 13 Boylston Street. During Howells's years with *The Atlantic Monthly,* he wrote a number of novels, including *Their Wedding Journey* (1872), *A Chance Acquaintance* (1873), and *The Lady of the Aroostook* (1879). In 1884, three years after resigning from *The Atlantic Monthly,* Howells moved to a house at 4 Louisburg Square, and there began to write his celebrated novel *The Rise of Silas Lapham* (1885), which he completed at 302 Beacon Street, in the Back Bay section of Boston. Back Bay is the setting for much of the novel. It was here that the fictional Silas Lapham built his mansion, and Howells used his own experience here and in Belmont, Massachusetts, to provide background for Lapham's life. Howells used Boston as the setting for a number of his novels, including *The Undiscovered Country* (1880), *A Modern Instance* (1882), *A Woman's Reason* (1883), *The Minister's Charge* (1887), and *April Hopes* (1888). It is thought that Howells modeled Nankeen Square in *The Rise of Silas Lapham* after Chester Square in Boston.

Henry James lived in Boston in the 1860s and 1870s, and used the city as a setting for several of his novels, including *The American* (1877), *The Europeans* (1878), and *The Bostonians* (1886). In 1865 and 1866 James lived at 13 Ashburton Place. **Ross Lockridge** taught at Simmons College in Boston from 1941 to 1946. During part of that time he worked on his novel *Raintree County* (1948) in apartment 19 at 46 Mountfort Street.

Henry Wadsworth Longfellow courted his second wife, Fanny Appleton, at her home at 39 Beacon Street and married her there on July 13, 1843. This house, along with the adjacent building at 40 Beacon Street, is owned by the Women's City Club, and is open to visitors. A longtime resident of Cambridge, Longfellow was one of the leading literary figures of nineteenth-century Boston and used the city as a setting for a number of poems. Longfellow wrote of Boston's Old North Church, which stood at Salem and Hull streets, in his ballad "Paul Revere's Ride" (1861). Legend has it that lanterns set in the church tower would warn the Minutemen of the route taken by British troops en route to Lexington and Concord—one lantern if by land, two if by sea. Whatever the truth of the story, it is certain that during the British occupation of the city, much of the old wooden church was dismantled for firewood.

Amy Lowell and her family spent winters in the late nineteenth century at a townhouse at 97 Beacon Street. Her relative, **James Russell Lowell,** was the first editor in chief of *The Atlantic Monthly,* from 1857 to 1862. **J. P. Marquand** in the 1920s lived in a house at 43 West Cedar Street. Marquand used Boston as a setting for *The Late George Apley* (1937), the novel for which he won a Pulitzer Prize, and for *H. M. Pulham, Esq.* (1941).

In a much later time, after World War II, **Edwin O'Connor** worked as a writer and producer at radio station WNAC in Boston. It is thought that Frank Skeffington, the fictional mayor and political boss making his last campaign for reelection in O'Connor's novel *The Last Hurrah* (1956), was modeled after James M. Curley, a long-time mayor of Boston. Boston is also the model for the setting of O'Connor's novel *The Edge of Sadness* (1961), for which he won a Pulitzer Prize. After establishing himself as a successful novelist, O'Connor moved to an apartment on Commonwealth Avenue and, on September 2, 1962, married Veniette Caswell at Holy Cross Cathedral, on Washington Street. On March 23, 1968, O'Connor died at New England Baptist Hospital, at 91 Parker Hill Avenue in Roxbury. Funeral services were held at Holy Cross Cathedral, and O'Connor was buried in Holyhood Cemetery, off Boylston Street in Brookline.

Eugene O'Neill spent his last, illness-wracked days at the Shelton Hotel, which was located on Bay State Road in Boston, and died there on November 27, 1953. Shortly before his death, he is reported to have said: "Born in a hotel room—and God damn it—died in a hotel room." O'Neill was buried in Forest Hills Cemetery, at 95 Forest Hills Avenue.

Francis Parkman lived in the Boston area much of the time from 1859 until his death. In 1859 he moved with his family to 8 Walnut Street and lived there until 1864, when he moved to 50 Chestnut Street. In the 1860s he spent summers in his own house on Jamaica Pond, which now is part of Olmstead Park, just north of the Jamaica Plain section of Boston. Parkman spent winters with his family in Boston and summered at Jamaica Pond every year until his death on November 8, 1893. He was buried in Mount Auburn Cemetery in Cambridge. During his Boston years, Parkman completed many of his historical studies, including *The Jesuits in North America* (1867), *The Discovery of the Great West* (1869), *Montcalm and Wolfe* (1884), and *A Half-Century of Conflict* (1892).

George William Apley was born in the house of his maternal grandfather, William Leeds Hancock, on the steeper part of Mount Vernon Street, on Beacon Hill, on January 25, 1866. He died in his own house, which overlooks the Charles River Basin and the Esplanade, on the water side of Beacon Street, on December 13, 1933. This was the frame in which his life moved, and the frame which will surround his portrait as a man. He once said of himself: "I am the sort of man I am, because environment prevented my being anything else."

—J. P. Marquand,
in *The Late George Apley*

Old photograph of 4 Louisburg Square, home of William Dean Howells.

"Yes, sir, it's about the sightliest view I know of. I always did like the water side of Beacon. Long before I owned property here, or ever expected to, m'wife and I used to ride down this way, and stop the buggy to get this view over the water. When people talk to me about the Hill, I can understand 'em. It's snug, and it's old-fashioned, and it's where they've always lived. But when they talk about Commonwealth Avenue, I don't know what they mean. It don't hold a candle to the water side of Beacon. You've got just as much wind over there, and you've got just as much dust, and all the view you've got is the view across the street. No, sir! When you come to the Back Bay at all, give me the water side of Beacon."

—William Dean Howells,
in *The Rise of Silas Lapham*

He always turned to the left, for never, except to funerals, did Mr. Nathaniel Alden walk down *Beacon Hill. He didn't go out for exercise. The need of exercise, he said, was a modern superstition, invented by people who ate too much, and had nothing to think about. Athletics didn't make anybody either long-lived or useful. An abstemious man could take plenty of sunshine and fresh air by his chamber window, while engaged in reading or writing. And if some impertinent relative wondered how he could take the sun in his house, with the blinds closed, he begged you to observe that all the slats in his blinds were movable, and that by inclining them inwards, or making them level, he could let in all the light and air desired, without exposing himself to public view.*

—George Santayana, in *The Last Puritan*

Elizabeth Palmer Peabody, one of the Transcendentalists, was a leading figure in nineteenth-century Boston literary life. She taught at Bronson Alcott's Temple School in the 1830s, and in 1839 opened a bookstore on West Street that became a gathering place for Boston writers. There George Ripley's Brook Farm commune was planned, and the literary magazine *The Dial* was published in 1842 and 1843. Peabody is thought to have been the model for Miss Birdseye in Henry James's novel *The Bostonians* (1886). Peabody died in Jamaica Plain on January 3, 1894, and was buried in Concord, Massachusetts.

William Hickling Prescott, the historian noted for his studies of Spanish colonization in the New World, lived for fourteen years before his death on January 28, 1859, at 55 Beacon Street. **William Makepeace Thackeray** visited Prescott here in the 1850s during Thackeray's American tours. It is said that the history relating to two crossed swords over the mantelpiece at Prescott's house gave Thackeray the idea for his novel *The Virginians* (1857–59). These swords now are kept by the Massachusetts Historical Society, at 1154 Boylston Street.

George Ripley, the editor and literary critic, and founder of Brook Farm in West Roxbury, was minister of Boston's Purchase Street Church from 1826 to 1841, when he resigned to become president of Brook Farm. The communal experiment lasted from 1841 to 1847 and included among its members **George William Curtis, Ralph Waldo Emerson, Margaret Fuller, Nathaniel Hawthorne,** and **Henry David Thoreau.** Hawthorne used his Brook Farm experiences as the basis for his novel *A Blithedale Romance* (1852). The site of Brook Farm, 670 Baker Street in West Roxbury, now a part of Boston, is a National Historic Landmark but is not accessible to visitors.

George Santayana, the philosopher, critic, and poet, came to Boston at the age of nine in 1872. Santayana lived with his family at 802 Beacon Street until 1881, during that time attending Boston Latin School. Santayana also lived in a house in Roxbury, address unknown. He taught philosophy at Harvard from 1889 to 1912 and then moved to Europe. Among the books Santayana wrote while living in Boston and Cambridge were *The Sense of Beauty* (1896) and the five-volume work *The Life of Reason* (1905–06). Santayana's only novel, *The Last Puritan* (1935), used Boston's Back Bay section as one of its settings.

Epes Sargent, the author and editor, moved in 1818 with his family to Roxbury. Sargent grew up there, attended Boston Latin School and, in the 1830s, worked for the Boston *Daily Advertiser* and Boston *Daily Atlas.* Sargent's plays *The Bride of Genoa* (1837) and *Velasco* (1839) were first produced at the Tremont Theatre in Boston. After living in New York City for some years, Sargent returned to Boston to work as editor of the Boston *Transcript* from 1847 to 1853. **Ellery Sedgwick,** editor from 1908 to 1938 of *The Atlantic Monthly,* lived for many years at 14 Walnut Street. **Benjamin Penhallow Shillaber** worked for the Boston *Post* in the 1830s and 1840s, and there wrote the first sketches featuring the character Mrs. Partington, whose humorous sayings and doings gave Shillaber a national reputation as a humorist. From 1851 to 1853 he edited the *Carpet-Bag,* a humor weekly that in 1852 published a story called "The Dandy Frightening the Squatter." It was the first published story signed with the name **Samuel L. Clemens. Charles Farrar Browne,** later to be known as **Artemus Ward,** set type on the *Carpet-Bag* and had some of his early bits of humor published in it. Upon the demise of this highly respected weekly, Shillaber returned to the *Post* and worked there until 1856, and then for the *Saturday Evening Gazette* from 1856 to 1866.

Upton Sinclair wrote of Boston and of the notorious Sacco-Vanzetti case in the novel *Boston* (1928), and **Jean Stafford** completed her novel *Boston Adventure* (1944) here. **Robert Louis Stevenson** is commemorated in Boston by a statue by Augustus Saint-Gaudens in the Nichols House Museum, at 55 Mt. Vernon Street. This four-story brick building, the former home of Saint-Gaudens's niece, Rose Standish Nichols, is open to visitors on a limited schedule.

Harriet Beecher Stowe lived with her family from 1826 to 1832 at 42 Green Street.

Henry David Thoreau lived in Boston with his family in the 1820s, first in a house in the South End of Boston, and then in a house at 4 Pinckney Street. **George Ticknor,** the educator and author of the scholarly *History of Spanish Literature* (1849) and *Life of William Hickling Prescott* (1863), lived in a house on the corner of Park and Beacon streets. **Benjamin Tompson** taught from 1666 to about 1671 at the free school in Boston, which became the Boston Latin School. In 1676 Tompson had published in Boston his best-known work, the narrative poem *New Englands Crisis,* an epic about King Philip's War. In the early 1700s, Tompson taught for several years in Roxbury.

Mark Twain spent much time in Boston in the late 1860s and early 1870s. Twain often conferred with James Redpath, his lecture agent, whose Boston Lyceum Bureau was located on School Street. Twain gave his first lecture in Boston in November 1869, and about that time met **William Dean Howells,** then a subeditor of *The Atlantic Monthly.* Twain and Howells soon became close friends, and Twain became a frequent contributor to *The Atlantic Monthly.* As Twain's reputation grew, so did his list of acquaintances. In Boston he met such prominent authors as **Oliver Wendell Holmes, Sarah Orne Jewett,** and **John Greenleaf Whittier.**

It was at the Hotel Brunswick on December 17, 1877, at a dinner celebrating the seventieth birthday of Whittier, that Twain gave a speech that would haunt him for the rest of his life: Twain poked fun at Emerson, Holmes, and Longfellow, honored guests at the evening's entertainment. Early in his talk, Twain realized he had blundered in mocking these eminent New Englanders, yet there was nothing to do but finish the talk. Howells was horrified by Twain's lack of sense, and Twain later sent letters of apology to the three men he had burlesqued. Some accounts of Twain's speech suggest that it was not as much of a disaster as Twain thought: there were some smiles and chuckles from the audience, even from Twain's targets. Emerson was an exception, but there are those who suggest that he may not have known what was happening—Emerson then was seventy-four years old.

Boston also enters literary tradition through an unlikely genre: Mother Goose of nursery tale and rhyme is said to have been modeled on a Boston woman named Elizabeth Vergoose. Evidence for this is far from conclusive, but Elizabeth Vergoose may have been the mother-in-law of one **Thomas Fleet,** who in 1719 published a volume entitled *Mother Goose's Melodies.* The Massachusetts volume of the Federal Writers' Project American Guide Series states that the real name of Fleet's mother-in-law was Mary Goose and that she is buried in the Old Granary Burying Ground, off Tremont Street just northeast of Boston Common. Also buried there are a number of notable historical figures of the Revolutionary War period, including Samuel Adams, John Hancock, and Paul Revere.

Boston literary figures of a more substantial nature than Mother Goose include **Daniel Webster,** the lawyer and orator, who lived at 57 Mt. Vernon Street and also at 138 Summer Street; and **John Greenleaf Whittier,** who spent many winters in Boston at the home of his publisher **James T. Fields,** where Whittier developed a close and lasting friendship not only with Fields's wife, Mrs. Annie Fields, but also with **Sarah Orne Jewett.**

Owen Wister worked for the Union Safe Deposit Bank in the 1880s and, during that time, became a member of one of this city's numerous clubs, the Tavern Club. Wister was made president of the club in 1929.

Boston is also the home of Boston University, whose Mugar Memorial Library, 771 Commonwealth Avenue, houses collections of the papers of Charles Angoff, Gladys Hasty Carroll, Thomas Duncan, John Clellon Holmes, Max Lerner, Frank G. Slaughter, Grace Zaring Stone, Elswyth Thane, Agnes Sligh Turnbull, Edward Wagenknecht, and Elizabeth Yates.

Other Boston literary sites include the Parker House, at 60 School Street, on the site of the first public Latin school in America. **Charles Dickens** stayed at the Parker House in November 1867, during his second tour of America. The hotel was also the meeting place for many years of the Saturday Club, a literary society founded in 1856. It included such literary figures as William Ellery Channing, Richard Henry Dana, Jr., Ralph Waldo Emerson, Nathaniel Hawthorne, Oliver Wendell Holmes, Henry James, Henry Wadsworth Longfellow, James Russell Lowell, and John Greenleaf Whittier.

It was the Magazine Club, a group of writers who also belonged to the Saturday Club, that in November 1857 founded *Atlantic Monthly,* which for over a century has published articles and fiction by many of this country's best writers. The magazine's editors have included **Thomas Bailey Aldrich, James T. Fields, William Dean Howells, James Russell Lowell, Horace E. Scudder,** and **Ellery Sedgwick.** Offices of the magazine, now known as *The Atlantic,* were for many years located at 4 Park Street, and now are located at 8 Arlington Street.

Other sites of literary interest include the Boston Public Library, on Copley Square, which has collections of the writings of Margaret Fuller and Thomas Wentworth Higginson, letters and manuscripts of Cotton and Samuel Mather, and George Ticknor's collection of Spanish and Portuguese books. The library facade bears a list of eminent men of science, arts, and letters. When the building was being constructed it was discovered that the first letters of each name, reading downward, formed the words "McKim Mead and White the architects." The names were later rearranged to thwart the scheme.

The Boston Athenaeum, 10½ Beacon Street, was founded in the early 1800s, and its present building dates to 1845. This private library, containing such items as George Washington's book collection, is now a National Historic Landmark.

There is an ancient tradition among those who deal with the law that a judge must not discuss a case in public, at least not while it is actually before him. But in handling Sacco and Vanzetti, "Web" Thayer could not keep within these traditional limits. He would talk about the case in a club dining-room, until all the men would leave the table; he would approach a Dartmouth professor on the football field, asking in a loud voice, "Did you see what I did to those anarchistic bastards the other day?" The horror of these words became such in Massachusetts that the Commonwealth had to send its policemen with hickory clubs to crack the skulls of demonstrants who carried the words on a banner. Because there was no way to keep them from being spoken aloud, Boston Common had to be closed to public speakers for the first time in its three hundred years of history.

—Upton Sinclair, in *Boston*

The Parker House, in the nineteenth century a meeting place of the Boston literary society known as the Saturday Club. Richard Henry Dana (*left*) and John Greenleaf Whittier (*below*) were members of the group.

All that tread
The globe are but a handful to the
tribes
That slumber in its bosom.

—William Cullen Bryant,
in "Thanatopsis"

My words are little jars
For you to take and put upon a shelf.
Their shapes are quaint and beautiful,
And they have many pleasant colours
and lustres
To recommend them.

—Amy Lowell,
in "A Gift"

The Old Corner Bookstore, northwest corner of Washington and School streets, was the home of the publishing firm of Ticknor and Fields, and was a popular gathering place for nineteenth-century writers. The building, moved to Bromfield Street and restored, contains many items of literary interest, including a first edition of Nathaniel Hawthorne's *The Scarlet Letter* (1850), and Oliver Wendell Holmes's desk. Another popular Boston bookstore, the Brattle Book Shop, was located at the corner of Franklin Avenue and Brattle Street for many years, but now is located at 5 West Street. The story goes that George Gloss, the proprietor, once opened his store to anyone who wanted a book, whether they could pay or not. By the end of the day, not a book was left on his shelves.

The Old Corner Bookstore as it appeared in 1865. Originally run by Ticknor and Fields, the bookstore was operated later by E. P. Dutton & Co.

BREWSTER

Conrad Aiken lived from 1940 to 1967 in a home called Forty-one Doors, at 457 Stony Brook Road, in this town on Cape Cod Bay. The poet and novelist also spent summers here from 1967 to 1971. Among Aiken's numerous works were the poem *The Kid* (1947), the autobiographical novel *Ushant* (1952), *Collected Poems* (1953), and the *Collected Short Stories of Conrad Aiken* (1960). Cape Cod was the setting of Aiken's novel *Conversation: Or, Pilgrim's Progress* (1940).

Another Brewster resident was **Horatio Alger, Jr.**, who served as pastor of the Unitarian Universalist Church, on Route 6-A, from 1864 to 1866. **Joseph C[rosby] Lincoln,** who wrote more than forty novels about life on Cape Cod, was born in Brewster on February 13, 1870, and lived for many years in Chatham. Among his novels set on the Cape was *Cap'n Eri* (1904), perhaps his best-known work.

BRIGHTON

John [Thomas] Gould, born in this community just northwest of Boston on October 22, 1908, is the author of a number of books, most of them dealing with life in Maine, as well as author of the weekly "Dispatch from the Farm," which appears in *The Christian Science Monitor. Among his books are a historical work, New England Town Meeting* (1945), and the humorous *The Farmer Takes a Wife* (1945).

BROCKTON

William Cullen Bryant lived in 1814 and 1815 in this city, then the north parish of the town of Bridgewater, about twenty miles south of Boston, at the home of his grandfather, 815 Belmont Street. He studied law in the nearby town of West Bridgewater. It is believed that Bryant composed his poems "Inscription for the Entrance to a Wood" and "The Yellow Violet" here and worked on the final draft of his best-known poem, "Thanatopsis."

Brockton was the birthplace, on February 6, 1879, of **Katharine Fullerton Gerould,** the novelist and short-story writer. After studying in Boston until 1893, she returned to Brockton to live in her uncle's home at 195 Newbury Street, still standing. Among Mrs. Gerould's works were the story collections *Vain Oblations* (1914) and *The Great Tradition* (1915); and the novel *Lost Valley* (1922).

BROOKFIELD

Mary Jane [Hawes] Holmes, born on April 5, 1825, in this town in central Massachusetts, was the author of thirty-nine sentimental novels, which together sold more than two million copies. Best known of these novels was *Lena Rivers* (1856), a romantic tale that was very popular in its day.

BROOKLINE

This town just west of Boston was the birthplace of **Louise Andrews Kent,** born on May 25, 1886. Kent wrote a number of novels for children as well as several humorous books, including *Mrs. Appleyard's Year* (1940) and *With Kitchen Privileges* (1953).

Amy [Lawrence] Lowell was born on February 9, 1874, at her family's home, Sevenels, now private, at 70 Heath Street. Lowell, one of the Imagist poets, spent most of her life in Brookline and here received such literary visitors as **Maxwell Bodenheim, Malcolm Cowley, Robert Frost,** and **William Carlos Williams.** Miss Lowell died on May 12, 1925, and is buried in Mount Auburn Cemetery in Brookline.

Other writers associated with Brookline include **Ring Lardner,** who lived in an apartment at 16 Park Drive in 1911, when he worked on the staff of the Boston *American;* and **Ben Ames Williams,** the novelist, who died of a heart attack at Brookline Country Club, 191 Clyde Street, on February 4, 1953. Williams, born in Mississippi, lived in Boston in his early years.

CAMBRIDGE

Writers who were born or lived in Cambridge include **Eleanor Hallowell Abbott,** author of books and stories for children, who was born on September 22, 1872. The granddaughter of Jacob Abbott, she wrote the popular children's book *Molly Make-Believe* (1910) and the autobiographical *Being Little in Cambridge When Everyone Else Was Big* (1936).

Conrad Aiken moved to Cambridge in 1915 in order to be the neighbor of John Gould Fletcher, the poet, whom Aiken considered his literary mentor. Aiken later had his own turn as mentor when he was visited in the summer of 1929 at his apartment, 8 Plympton Street, by Malcolm Lowry, the English-born novelist. Lowry, then on his first trip to America, wanted to pay Aiken for the privilege of living with and learning from him. Lowry saw Aiken not only as a friend but also as a literary model and father figure. Aiken's influence on the future author of *Under the Volcano* (1947) can only be guessed at, but it is interesting to note that Lowry's first novel, *Ultramarine* (1933), which he worked on at 8 Plympton Street, may owe its title in part at least to Aiken's own first novel, *Blue Voyage* (1927), which Lowry greatly admired. Aiken has yet another association with Boston. He wrote of William Blackstone, a legendary Boston figure, in his poem "The Kid" (1947). Blackstone is said to have welcomed Boston's first settlers on their arrival here, on what now is Boston Common.

Margaret Barnes, the novelist and playwright, died at her home in Cambridge on October 25, 1967. **E[dward] E[stlin] Cummings,** the poet and artist, was born on October 14, 1894, in his family's house at 104 Irving Street. He grew up in Cambridge, and attended Harvard University, from which he received his B.A. in 1915 and M.A. in 1916. Cummings returned to live in Cambridge, at 6 Wyman Road, in 1952, when he served for a year as Charles Eliot Norton Professor of Poetry at Harvard.

Richard Henry Dana, the critic and poet, was born here on November 15, 1787. Cambridge was also the birthplace of his son **Richard Henry Dana, Jr.,** born on August 1, 1815, in a house on Green Street. The younger Dana grew up in Cambridge and attended Harvard but left because of failing health. He joined the crew of a hide-trading ship bound for California. After two years Dana returned to Harvard, was graduated in 1837, and established himself in Boston as a lawyer. In 1840 appeared the book for which he is known, *Two Years Before the Mast,* and in 1852 Dana moved to a house at 4 Berkeley Street, Cambridge.

Margaret Deland lived in Boston and Cambridge in the late nineteenth and early twentieth centuries. **Robert Frost** lived in a house on Brewster Street for a time. **[Sarah] Margaret Fuller,** the Transcendental critic and poet, was born on May 23, 1810, in a house at 71 Cherry Street, which is used now as a settlement house. After living away from Cambridge for a time, the Fuller family returned in 1826 and soon after moved to a house, no longer standing, on the approximate site of 5 Dana Street. In 1832 they moved to the Brattle house, at 42 Brattle Street.

Cambridge was also the birthplace of **Robert Herrick,** the novelist and short-story writer, born on April 26, 1868; and **Thomas Wentworth Higginson,** the clergyman, writer, and reformer, born on December 22, 1823. Higginson grew up in Cambridge, enrolled at Harvard at the age of thirteen, and was graduated in 1841. He was graduated from Harvard Divinity School in 1847, whereupon he left Cambridge to serve as Unitarian minister in Newburyport, Massachusetts. In 1878 Higginson returned to Cambridge and made

Conrad Aiken, the poet and novelist.

Forty-one Doors, for many years the home of Conrad Aiken in Brewster. In a letter dated May 21, 1940, Aiken described the ramshackle house as "a fine wreck of a house . . . the delight of our hearts."

The alarm clocks tick in a thousand
furnished rooms,
tick and are wound for a thousand
separarate dooms;
all down both sides of North Infinity
Street
you hear that contrapuntal pawnshop
beat.

—Conrad Aiken,
in "North Infinity Street"

The mind-curers . . . have demonstrated that a form of regeneration by relaxing, by letting go, psychologically indistinguishable from the Lutheran justification by faith and the Wesleyan acceptance of free grace, is within the reach of persons who have no conviction of sin and care nothing for the Lutheran theology. It is but giving your little private convulsive self a rest, and finding that a greater Self is there.

—William James,
in *The Varieties of Religious Experience*

it his home until his death on May 9, 1911. During this time Higginson wrote many books, including the autobiographical *Cheerful Yesterdays* (1898); the collection of sketches *Old Cambridge* (1899); and the biographies *Henry Wadsworth Longfellow* (1902) and *John Greenleaf Whittier* (1902).

John Hollander, a member from 1954 to 1957 of Harvard's Society of Fellows, wrote some of the poems in his first collection, *A Crackling of Thorns* (1958), while living at 20 Prescott Street.

Oliver Wendell Holmes, Sr., the physician and author, was born in Cambridge on August 29, 1809. Holmes was graduated from Harvard in 1829 and received his medical degree there in 1836. He lived for many years in Boston, and was a leading figure in that city's literary circle. He died on October 7, 1894, and is buried in Mt. Auburn Cemetery, 580 Mt. Auburn Street in Cambridge, where also are buried **Louis Agassiz, William Ellery Channing, James Russell Lowell, Francis Parkman,** and **William Hickling Prescott. William Dean Howells,** another member of the Boston literary circle, became subeditor of *The Atlantic Monthly* in 1866 and moved to 41 Sacramento Street, where he lived for a time before moving to Boston. Howells returned to Cambridge, moving to 3 Berkeley Street, and about 1872 built a new house at 37 Concord Avenue, where he lived until 1878, when he moved to Belmont, Massachusetts.

Henry James [Jr.] enrolled at Harvard Law School in 1862. By 1864 his family had moved to Boston, and by 1866 to Cambridge. Thereafter James considered Cambridge his home when he was not living abroad. Except for an extended trip to Europe, he spent much of the years from 1866 to 1872 here and wrote articles and stories for *Atlantic Monthly,* where he became close friends with **William Dean Howells.** Howells encouraged James's literary efforts and was a frequent guest at the James house at 20 Quincy Street. After 1872 James spent most of his life abroad. His last trip to Cambridge was in 1910, shortly after the death of his brother William James. Henry stayed at his brother's house, 95 Irving Street, and during that visit had dinner there with **Somerset Maugham.**

William James, the philosopher and psychologist, received his medical degree from Harvard

Henry Wadsworth Longfellow with his daughter Edith in front of his Brattle Street home. The photograph was taken about two years before the poet's death in 1882.

Under a spreading chestnut tree
The village smithy stands:
The smith, a mighty man is he,
With large and sinewy hands;
And the muscles of his brawny arms
Are strong as iron bands.

—Henry Wadsworth Longfellow,
in "The Village Blacksmith"

in 1869 and taught there from 1870 to 1907. In 1889 he built the house at 95 Irving Street, his home until his death on August 26, 1910. A frequent traveler to Europe, James became thoroughly acquainted with European advances in psychology and was a leader in the development of experimental psychology in the United States. Perhaps his best remembered work is *The Varieties of Religious Experience* (1902).

Henry Wadsworth Longfellow, newly appointed professor of modern languages at Harvard, came to Cambridge in January 1837 and took rooms at the Foxcroft House, at the corner of Kirkland and Oxford streets. In August 1837 he rented a room at the home of Mrs. Elizabeth Shaw Craigie, at 105 Brattle Street. Mrs. Craigie died in 1841, and in 1843 Longfellow married Frances Appleton of Boston. Her father bought the house for Longfellow as a wedding present, and it remained Longfellow's home until his death here on March 24, 1882. The house, used in 1775 and 1776 by George Washington as headquarters during the siege of Boston, is administered now by the National Park Service and is open to the public daily. Here Longfellow wrote much of his poetry, which appeared in such volumes as *Hyperion* (1839), *Ballads and Other Poems* (1842), *Evangeline* (1847), *Hiawatha* (1855), *The Courtship of Miles Standish* (1858), *Tales of a Wayside Inn* (1863, 1872, 1874, collected edition 1886), *Ultima Thule* (1880), and *In the Harbor* (1882). One of Longfellow's best-known poems is "The Village Blacksmith" (1840), and one of the many items of interest at Longfellow's house is a chair made from the wood of the chestnut tree under which stood the prototype of Longfellow's smith. The chair was presented to Longfellow by the children of Cambridge on the poet's seventy-second birthday, February 27, 1879, and on that day Longfellow, to thank them, wrote the poem "From My Arm-Chair." Whenever children came to visit him, Longfellow would have them sit in the chair, and would give them a printed copy of the poem.

Other sites of interest to admirers of Longfellow are Longfellow Park, which stretches from the Longfellow House to the Charles River and contains a monument to the poet; and Longfellow Bridge, originally called Cambridge Bridge and the oldest bridge connecting Boston with Cambridge. The Dexter Pratt House, 56 Brattle Street, was the home of the village blacksmith who was the model for Longfellow's village smithy. A plaque near the corner of Brattle and Story streets reads:

Near this spot stood
the "spreading chestnut tree"
and the smithy referred to in
Longfellow's poem
"The Village Blacksmith."

Longfellow is buried in Mt. Auburn Cemetery, 580 Mt. Auburn Street.

Modern photograph of Longfellow's study in the house on Brattle Street. The room is little changed since Longfellow's death.

I du believe the people want
A tax on teas an' coffees,
Thet nothin' aint extravygunt,—
Purvidin' I'm in office;
Fer I hev loved my country sence
My eye-teeth filled their sockets,
An' Uncle Sam I reverence,
Partic'larly his pockets.

—James Russell Lowell,
in *The Biglow Papers*

James Russell Lowell, the poet, editor, and teacher, was born on February 22, 1819, at Elmwood, the three-story Georgian mansion at 33 Elmwood Avenue now used by Harvard University as a dean's residence. Except for short periods, Lowell lived in this house all his life and here wrote his best-known works. In the year 1848 alone, Lowell saw publication of three books that made his literary reputation secure: the poem *The Vision of Sir Launfal;* the collection of poems and prose pieces, *The Biglow Papers, First Series;* and *A Fable for Critics,* a critique in verse on the writers and leading figures of his day. Lowell, who had been graduated from Harvard in 1838 and Harvard Law School in 1840, taught at Harvard from 1855 to 1886. In 1857 he became the first editor of the new *Atlantic Monthly,* but resigned in 1861 to devote his energies to writing and teaching. In 1864 he became editor of the *North American Review,* one of the most highly respected journals of his day.

Lowell is remembered also for his "Ode Recited at the Commemoration to the Living and Dead Soldiers of Harvard University" (1865), commonly called the "Commemoration Ode," which he wrote to honor Harvard students who had died in the Civil War. Lowell also wrote about Cambridge in the sketch "Cambridge Thirty Years Ago" (1854). Lowell died at Elmwood on August 12, 1891, and was buried in Mt. Auburn Cemetery in Cambridge.

We sit here in the Promised Land
That flows with Freedom's honey and milk;
But 't was they won it, sword in hand,
Making the nettle danger soft for us as silk.
We welcome back our bravest and our best;—
Ah me! not all! some come not with the rest,
Who went forth brave and bright as any here!

—James Russell Lowell,
in "Commemoration Ode"

Birthplace of James Russell Lowell, built between 1763 and 1767 by Thomas Oliver, the last lieutenant governor of Massachusetts under English rule.

Other writers associated with Cambridge include **Philip Rahv,** the critic and editor, who died at his home here on December 22, 1973, and **George Santayana,** who came to live in Cambridge in the 1880s and taught philosophy at Harvard from 1889 to 1912. Harvard figures in Santayana's only novel, *The Last Puritan* (1935).

May Sarton came to Cambridge with her family in 1916, when she was four years old. In childhood she lived at 10 Avon Street, on Agassiz Street, and at 103 Raymond Street. At 5 Channing Place, her home from 1936 to 1944, she wrote her collection of poems *Encounter in April* (1937), the novel *The Single Hound* (1938), and the collection of poems *Inner Landscape* (1939). At 139 Oxford Street, her home from 1945 to 1950, she wrote the novel *The Bridge of Years* (1945), the collection *The Lion and the Rose* (1948), and the novel *Shadow of a Man* (1950). At 14 Wright Street, Sarton's home from 1950 to 1957, she wrote *The Leaves of the Tree* (1950), poems; *A Shower of Summer Days* (1952), a novel; *The Land of Silence* (1953), poems; *Faithful Are the Wounds* (1955), a novel; *In Time Like Air* (1957), poems; *The Birth of a Grandfather* (1957), a novel; *The Fur Person* (1956), a novel; and *I Knew a Phoenix* (1959), her autobiography. May Sarton also taught at Harvard from 1950 to 1953 and at Radcliffe in 1957 and 1958.

Horace E. Scudder, from 1890 to 1898 editor of *Atlantic Monthly,* lived at 17 Buckingham Street. **Harriet Prescott Spofford,** the prolific and versatile author, lived at 34 Cambridge Street in the 1860s; and **Mabel Loomis Todd,** the novelist, poet, and editor of *The Letters of Emily Dickinson* (1894), was born in Cambridge on November 10, 1856. **Richard Wilbur,** while living at 22 Plympton Street from 1945 to 1947, completed his collection of poems *The Beautiful Changes* (1947). At 37 Kirkland Street, Wilbur's home from 1947 to 1950, he wrote the poems collected in *Ceremony and Other Poems* (1950). Wilbur studied at Harvard University in 1947 and 1948, and was a Junior Fellow from 1947 to 1950.

Harvard University in Cambridge includes twelve constituent schools, of which two, Harvard College and Radcliffe College, are noted for the number of writers who have studied or taught there. Among the authors who have studied at Harvard and Radcliffe are **James Agee** (A.B. 1932), who was on the editorial board of the *Harvard Advocate* and wrote many poems and stories for it; **Conrad Aiken** (A.B. 1912); **Thomas Bailey Aldrich** (A.B. 1883, A.M. 1896); **Horatio Alger, Jr.** (graduated 1852); **Philip Barry,** who from 1919 to 1922 studied theater in George Pierce Baker's 47 Workshop; **Nathaniel Benchley** (B.S. 1938); **Robert Benchley** (A.B. 1912); **Van Wyck Brooks** (A.B. 1908); and **James Gould Cozzens,** who as a student from 1922 to 1924 wrote his first novel, *Confusion* (1924). Other Harvard students include **Countée Cullen** (A.M. 1926); **John Dos Passos** (A.B. 1916); and **W. E. B. Du Bois** (A.B. 1890, A.M. 1891, Ph.D. 1895).

The list goes on: **Ralph Waldo Emerson** studied in 1817, at age fourteen, under Harvard president Kirkland and lived in Wadsworth House, facing Massachusetts Avenue, and also studied at Harvard Divinity School in 1825. **Rachel Field** attended Radcliffe from 1914 to 1918 and studied drama in George Pierce Baker's 47 Workshop. While a student, she wrote her one-act play *Three Pills in a Bottle* (1918). **Robert Frost** studied at Harvard from 1897 to 1899. **Edward Hoagland** (A.B. 1954) worked on his first novel, *Cat Man* (1956), at Adams House. **Charles Fletcher Lummis** entered Harvard in 1877 but contracted brain fever three days before graduation in 1881. He was awarded his bachelor's degree in 1906, twenty-five years late.

Harvard's literary alumni also include **Norman Mailer** (A.B. 1943), who won *Story* magazine's annual college contest in 1941 with "The Greatest Thing in the World." **John P. Marquand** (A.B. 1915) lived for a time in the 1950s at One Reservoir Street. The three Mathers were Harvard men: **Cotton Mather** received his A.B. in 1678 and his A.M. in 1681; **Increase Mather** received his A.B. in 1656; and **Samuel Mather** received his A.B. in 1723. **Frank Norris** attended Harvard in 1894–95. While living at 47 Gray's Hall, Norris worked on his novels *McTeague* (1899) and *Vandover and the Brute* (1914). **Eugene O'Neill** studied in George Pierce Baker's 47 Workshop in 1914. **John Reed,** who was graduated in 1910, was editor of the very successful Harvard *Lampoon* and helped found Harvard's dramatic club.

George Ripley was graduated from Harvard in 1823 and from Harvard Divinity School in 1826. **Edwin Arlington Robinson** studied at Harvard from 1891 to 1893. **Gertrude Stein** attended Radcliffe from 1893 to 1897 but did not take a degree. **Wallace Stevens** attended Harvard from 1897 to 1900. **Henry David Thoreau** was graduated in 1837. **Barbara Tuchman** attended Radcliffe College from 1930 to 1933. **Royall Tyler** was graduated in 1776, and **John Hall Wheelock** received his A.B. in 1908. While here, Wheelock and his classmate **Van Wyck Brooks** published their anonymous collection, *Verses by Two Undergraduates* (1905). **Theodore H. White** was graduated from Harvard in 1938. **Owen Wister** was graduated from Harvard College in 1882 and Harvard Law School in 1888. While at Harvard, Wister established lifelong friendships with Oliver Wendell Holmes and William Dean Howells. **Thomas Wolfe** (A.M. 1922) studied in George Pierce Baker's 47 Workshop, for which Wolfe wrote the play *Welcome to Our City* (produced at Harvard in 1923). Wolfe's addresses in Cambridge included 48 Buckingham Street, 42 Kirkland Street, 67 Hammond Street, 21 Trowbridge Street, and 10 Trowbridge Street.

Harvard has had many distinguished writers among its faculty, many of them also Harvard graduates. Writers include **Henry Adams** (A.B. 1858, Ph.D. 1876), who taught here from 1870 to 1877; **George Pierce Baker** (A.B. 1887), whose 47 Workshop in playwriting, which he conducted here from 1905 to 1925, was a powerful influence on the numerous twentieth-century playwrights and authors who took the course; **John Berryman,** who taught here from 1940 to 1943; and **Bernard DeVoto** (A.B. 1920), who taught at Harvard from 1929 to 1936 and lived at several addresses in Cambridge, including 64 Oxford Street, 10 Mason Street, Coolidge Hill Road, and 8 Berkeley Street. **T. S. Eliot** (A.B. 1909, A.M. 1910) was Charles Eliot Norton Professor of Poetry here in 1932–33; and **William James** taught psychology and philosophy here from 1872 to 1907.

George Lyman Kittredge, a prolific writer and authority on Shakespeare and English literature, taught at Harvard from 1888 to 1936. Among his many books were *The Old Farmer and His Almanack* (1904), *Shakspere* (1916), and *Witchcraft in Old and New England* (1929). **Archibald MacLeish** (LL.B. from Harvard Law School, 1919) taught at Harvard from 1949 to 1962. **Lewis Mumford** taught at Harvard from 1965 to 1971, and at the Massachusetts Institute of Technology from 1957 to 1960 and 1973 to 1975. During this time Mumford lived at 50 Follen Street (1956), 14 Francis Avenue (1957–58), 19 Lowell Street (1958–59), 10 St. John's Place (1959–60), Leverett House at Harvard (1965 to 1971), and Eastgate at M.I.T. (1973 to 1975).

Other writers who have taught at Harvard include **Francis Parkman,** who was graduated from Harvard in 1844, received his LL.B. here in 1846, and in 1871 was appointed professor of horticulture; and **Bliss Perry,** who became professor of literature here in 1907—taking up the chair vacated by **James Russell Lowell** forty years earlier. Perry was Francis Lee Higginson Professor of English Literature from 1926 to 1935, and an overseer of Harvard from 1935 to 1941.

Harvard's former faculty also includes **Mark Schorer,** who taught here from 1937 to 1945; **Delmore Schwartz,** who studied at Harvard from 1935 to 1937 and taught here from 1940 to 1947; **Odell Shepard,** who received a Ph.D. from Harvard in 1916 and taught English here in 1916 and 1917; and **Wallace Stegner,** who wrote the novels *Fire and Ice* (1941) and *The Big Rock Candy Mountain* (1943), and the nonfiction book *Mormon Country* (1942) during the period in which he taught at Harvard, 1939 to 1945.

Inside, out of the rain, the lunch wagon was hot and sticky. Al Groot stopped in front of the doorway, wiped his hands and wrung his hat out, and scuffed his shoes against the dirt-brown mat. He stood there, a small, old, wrinkled boy of eighteen or nineteen, with round beady eyes that seemed incapable of looking at you unless you were in back of him.

—Norman Mailer,
in "The Greatest Thing in the World"

Home of William Dean Howells on Concord Avenue, where he entertained many literary figures. Mark Twain was a frequent visitor.

Edward Bellamy's writing table in his house in Chicopee.

Harvard University's library, which contains more than nine million volumes, is the largest university library in the world. The library maintains, among many others, collections of the books and manuscripts of Ralph Waldo Emerson, William Dean Howells, Herman Melville, and Henry David Thoreau.

CHARLEMONT

Charles Dudley Warner lived during his childhood in this community in northwestern Massachusetts, about fifteen miles east of Adams. Warner, his mother, and his brother lived from 1837 to 1841 in the home, still standing, of their guardian Jonas Patch.

CHATHAM

This town on the south shore of Cape Cod has attracted a number of writers as summer residents or visitors. **Bernard De Voto** spent the summer of 1927 here with his wife, and **Sinclair Lewis** stayed in Chatham with his wife and child for two weeks in May of 1918. **Sylvia Plath,** the poet, lived here from July to September of 1952. **Robert Lewis Taylor,** the newspaperman and author, rented houses in Chatham for the summers of 1949–51, and owned a house on Cotchpinicutt Road from 1952 to 1955. At the latter home, Taylor worked on his novel *The Bright Sands* (1954).

Chatham's best-known writer is **Joseph C[rosby] Lincoln,** author of more than forty popular novels set on Cape Cod. Lincoln experienced immediate success with his first novel, *Cap'n Eri* (1904). It was the story of three sea captains who lived under one roof and became so tired of keeping house that they decided to take a drastic action—one of them would marry. *Cap'n Eri* was followed by many other tales describing the life and people of Cape Cod in times gone by. Lincoln spent his summers in a cottage, still standing, in Chatham. He died at his apartment in Winter Park, Florida, on March 10, 1944, but is buried in Chatham. The Chatham Historical Society has a Joseph C. Lincoln collection.

CHICOPEE

Edward Bellamy, author of *Looking Backward: 2000–1887* (1888), was born on March 26, 1850, in Chicopee Falls, now part of Chicopee, in western Massachusetts. Bellamy was born at 41 East Street and later lived at 91–93 Church Street, now a National Historic Landmark known as the Edward Bellamy Homestead. From his writing table, located in front of the bay window in this house, Bellamy could look out upon the mills of the city, which symbolized to him the social injustice he sought to address in his best-known work and in his later writings. *Looking Backward* was the focus of his career: Bellamy had worked for the New York *Evening Post* and the Springfield *Union* before helping to found, in 1880, the Springfield *Daily News.* The working and living conditions of the people in European cities, which he had seen during a trip abroad, turned him from journalism and inspired him to write his novel, the account of a man who awakens in the year 2000 to find a world much improved over the one he left in 1887. *Looking Backward* attracted a great deal of attention when it was published, and numerous Bellamy Clubs were formed in America and abroad by people eager to discuss the implications of the author's work. In 1891 Bellamy founded *The New Nation* in Boston and spent large sums financing the journal, in which he elaborated his social ideas. Poor health eventually forced him to abandon the publication, and in his final years he concentrated his energies on writing *Equality* (1897), the sequel to *Looking Backward.* He worked against time and the tuberculosis that was killing him and, after traveling to the Southwest to seek a climate that he hoped would save him, returned to Chicopee, where he died on May 22, 1898. Bellamy is buried in Fairview Cemetery here. His home is open to visitors by appointment. Collections of Bellamy material are maintained at Our Lady of the Elms College, 291 Springfield Street, and at the Chicopee Public Library, on Market Square. A school on Pendleton Street in Chicopee is named in Bellamy's honor.

> *". . . I must know a little more about the sort of Boston I have come back to. You told me when we were upon the house-top that though a century only had elapsed since I fell asleep, it had been marked by greater changes in the conditions of humanity than many a previous millennium. With the city before me I could well believe that, but I am very curious to know what some of the changes have been. To make a beginning somewhere, for the subject is doubtless a large one, what solution, if any, have you found for the labor question? It was the Sphinx's riddle of the nineteenth century, and when I dropped out the Sphinx was threatening to devour society, because the answer was not forthcoming. It is well worth sleeping a hundred years to learn what the right answer was, if, indeed, you have found it yet."*
>
> *"As no such thing as the labor question is known nowadays," replied Dr. Leete, "and there is no way in which it could arise, I suppose we may claim to have solved it . . . ."*
>
> —Edward Bellamy, in *Looking Backward: 2000–1887*

CONCORD

This town northwest of Boston, scene of the opening battle of the American Revolution, was the home during the nineteenth century of the group of writers who came to be known as the Transcendentalists. Among the most celebrated were **A[mos] Bronson Alcott, William Ellery Channing, Ralph Waldo Emerson, Nathaniel Hawthorne,** and **Henry David Thoreau.** The Transcendentalists believed that man could attain knowledge without relying solely on information perceived through the senses. They believed in inspiration from God and nature and in the inner spiritual being of man, which they termed the oversoul. Transcendentalism was a reaction to the restrictive theology of the time that tended to deny man's creative and spiritual vitality. The Transcendentalists and other writers helped make Concord one of the leading literary centers of nineteenth-century America, and well worth a visit today. Most of the major literary sites in Concord are open to visitors, including the homes of the Alcotts, Emerson, and Hawthorne, and most charge a nominal admission fee. An annual "Points of Interest" brochure, listing all the sites and their hours, is available from the office of Minute Man National Historic Park, Post Office Box 160, Concord, Massachusetts, 01742.

A. Bronson Alcott, a teacher and philosopher whose progressive and unorthodox ideas kept his family close to the edge of poverty, first moved to Concord with his family in 1840. The Alcotts settled into the Hosmer Cottage, near the town's old South Bridge, and lived there for a time before leaving Concord. They returned in late 1844, and in April of 1845 moved into the Hillside, on Lexington Road, which remained their home until the fall of 1848, when the Alcotts moved to Boston. The Hillside was later the home of **Nathaniel Hawthorne,** who called it the Wayside, the name by which it is known today. In 1857 the Alcotts returned to Concord, to a house no longer standing, while their new home, Orchard House, on Lexington Road next door to the Hillside, was being prepared for them. In 1858 they moved to Orchard House, and lived there until 1884.

In Orchard House **Louisa May Alcott,** one of Bronson Alcott's daughters, wrote her best-known novel, *Little Women* (1868), based on her family and their life in Concord. Here too she worked on a number of other books, including the collection *Hospital Sketches* (1863), based on letters she had written to her family while working during the Civil War as a nurse in a Union hospital in Washington, D.C. This volume gave her literary recognition, but it was publication of *Little Women* that brought economic stability to the Alcott household: her father, an enlightened thinker and prolific writer, was an impractical man.

Bronson Alcott's greatest abilities were in the art of conversation, through which he developed and tested his ideas and greatly impressed such literary contemporaries as Emerson and Thoreau. Much of Bronson Alcott's writing is still unpublished, but the volume *Concord Days,* based on his journals, was issued in 1872, and a selection of his journals was published in 1938. Alcott presided over the Concord School of Philosophy, an annual four-week series of lectures and discussions held from 1879 to 1887 in a building still standing on the grounds of Orchard House.

In 1884 Louisa May Alcott sold Orchard House. Her final years were overshadowed by worsening illness, the result of the typhoid pneumonia she had contracted as a nurse, and the even more devastating medicine she had been given to cure it. Her condition was aggravated by the overwork she put herself through, long after financial matters—the reason she first turned to writing—had ceased to be a problem. She died in West Roxbury, now a suburb of Boston, on March 6, 1888, having outlived her father by only two days. Both are buried in Sleepy Hollow Cemetery, on Route 62 in Concord, where also are buried **William Ellery Channing, Ralph Waldo Emerson, Nathaniel Hawthorne, Henry David Thoreau,** and **Elizabeth Palmer Peabody,** the teacher and author, and sister-in-law of Hawthorne. The graves of the Alcotts, Emerson, Hawthorne, and Thoreau are in a section of the cemetery known as Authors' Ridge.

William Ellery Channing, the poet and essayist who was a friend of Emerson and Thoreau, made Concord his home from 1843 to his death in 1901. Channing, often called Ellery Channing to distinguish him from his namesake uncle, a noted clergyman, married Ellen Fuller, the younger sister of Margaret Fuller, and settled in Concord in order to be near Emerson, who had written favorably of Channing's poetry. Here Channing became the close friend of Thoreau and often accompanied Thoreau in his wanderings. Channing wrote *Thoreau, the Poet-Naturalist* (1873), the first biography of the man. After Thoreau's death in 1862, Channing began to retreat into solitude. He moved from his home on Main Street, near Thoreau's house, to a house at 25 Middle Street, just off Academy Lane, and spent his last years in the home of Franklin B. Sanborn on Elm Street, where he died on December 23, 1901.

Ralph Waldo Emerson first came to Concord during childhood to visit his grandmother and step-grandfather at the Old Manse on Monument Street. This was the parsonage built by his grandfather, the Reverend William Emerson, who died during the American Revolution while serving as a chaplain in the Continental army. The parson's successor, Rev. Ezra Ripley, who married

"I said we'd got to have a woman, and we have. One of us'll have to git married, that's all."

"Married!" *roared the two in chorus.*

"That's what I said, married, and take the others to board in this house. Look here now! When a shipwrecked crew's starvin' one of 'em has to be sacrificed for the good of the rest, and that's what we've got to do. . . ."

—Joseph C. Lincoln, in *Cap'n Eri*

(*below*) Marker in Sleepy Hollow Cemetery, Concord, directing visitors to Authors' Ridge, burial site of several of New England's best-known literary figures. (*left*) Orchard House, home of the Alcotts in Concord, built about 1650.

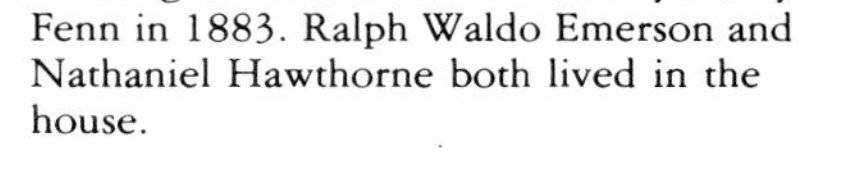

Painting of the Old Manse made by Harry Fenn in 1883. Ralph Waldo Emerson and Nathaniel Hawthorne both lived in the house.

Reverend Emerson's widow, welcomed the visits of Ralph Waldo Emerson to the Old Manse. Emerson later lived in the Old Manse from October 1834 to September 1835, and there worked on his first book, *Nature* (1836). In 1835 he moved with his bride, Lydia Jackson, to a spacious house on Lexington Road, which remained his home until his death there on April 27, 1882. In this house Emerson was visited by many literary figures of his day, including his neighbor Bronson Alcott, Margaret Fuller, William Dean Howells, and Mark Twain.

Henry David Thoreau lived in Emerson's house from time to time in the 1840s. Emerson's home had become the center of the Transcendental movement in American literature. It was there that Emerson wrote his volume *Essays* (1841) as well as a second series of essays in 1844, which together established his reputation in America and abroad. In subsequent volumes Emerson collected more essays, mostly based on the lectures he gave across the country. He also published two volumes of poetry: *Poems* (1846) and *May-Day and Other Pieces* (1867). Perhaps his best-known poem is "Concord Hymn," written for the dedication of the Concord Monument in 1836. Today, a monument featuring the well-known statue of a Minuteman, the work of Daniel Chester French, another illustrious resident of Concord, bears the lines of Emerson's poem. The statue stands on the site of the clash that occurred between American patriots and British soldiers on April 19, 1775.

In July 1872, a servant accidentally set fire to the Emerson house while rummaging in the attic. Neighbors helped extinguish the blaze, and carried Emerson's books, letters, and manuscripts to safety. The books and papers now are kept across the street at the Concord Antiquarian Society. Here visitors can see some of Emerson's letters that were singed by the fire. After the fire, Emerson left for Europe on a lecture tour; when he returned in 1873 he was greeted by the citizens of Concord, who had taken up a collection to repair his house, and he was escorted to his newly restored home. The Emerson house is open to visitors, as is the Concord Antiquarian Society.

In 1842 **Nathaniel Hawthorne** and his bride Sophia Peabody came to Concord and moved into Reverend Emerson's Old Manse. Hawthorne lived at the Old Manse until 1845, and there wrote many of the tales and sketches collected in his volume *Mosses from an Old Manse* (1846). In the fall of 1845, the Hawthornes left Concord. In the seven years following, Hawthorne wrote his most famous works, *The Scarlet Letter* (1850) and *The House of the Seven Gables* (1851). In 1852 he returned to Concord an established and respected author, and bought the Alcotts' old home, the Hillside, renaming it the Wayside. After only a short time, however, Hawthorne set off again for a seven-year stay abroad. When he returned in 1860, Hawthorne added to the house the three-story tower study in which he hoped to find the peace and comfort he needed for his writing, but it was not enough, and his sense of isolation and failure prevented him from writing anything comparable to his earlier works. Hawthorne died on May 19, 1864, while on a trip to the White Mountains near Plymouth, New Hampshire, and was buried in Sleepy Hollow Cemetery in Concord.

Portrait of Ralph Waldo Emerson. The poet and essayist lived most of his adult life in Concord.

By the rude bridge that arched the flood,
Their flag to April's breeze unfurled,
Here once the embattled farmers stood
And fired the shot heard round the world.

—Ralph Waldo Emerson,
in "Concord Hymn"

The Wayside, now open to visitors, claims among its literary residents not only Hawthorne and the Alcotts but also **Harriet Lothrop,** who, under the pen name **Margaret Sidney,** wrote the still-popular children's book *Five Little Peppers and How They Grew* (1881). Lothrop is also buried in Sleepy Hollow Cemetery.

A prominent figure in the Concord literary group of the nineteenth century was **Franklin B[enjamin] Sanborn,** who lived in a house, now private, at 49 Sudbury Road, and there ran a school for the children of Emerson, Hawthorne, Horace Mann, and John Brown the abolitionist. Brown visited Sanborn here shortly before his abortive raid on Harpers Ferry, Virginia. Sanborn was accused of complicity in planning the raid and in arranging the escape of some of Brown's followers, but when Federal authorities attempted to seize him the residents of Concord came to his defense and protected him. This dramatic incident in Sanborn's career contrasts with his otherwise quiet life and lifelong devotion to literature, history, education, and the lives of his Concord friends. Among Sanborn's books were *Henry David Thoreau* (1882); *A. Bronson Alcott: His Life and Philosophy* (1893); and, with William Torrey Harris, *Ralph Waldo Emerson* (1901) and *Hawthorne and His Friends* (1908). Sanborn also had the house on Elm Street in which William Ellery Channing spent his final years.

Of Concord's major Transcendentalist writers, only **Henry David Thoreau** was born here. The site of his birthplace, on Virginia Road near the village of Concord, is marked, and the house in which he was born, on July 12, 1817, has been moved east of its original site, but still on Virginia Road. Henry David Thoreau lived with his family for most of his life. They lived in the village of Concord in a bewildering succession of homes. Thoreau attended Concord Academy on Academy Lane and in the summer of 1838 opened his own school there, which he ran with his brother until 1841. In 1839 Thoreau traveled along the Concord and Merrimack rivers, north and east of Concord, and gathered the impressions he later used for his book *A Week on the Concord and Merrimack Rivers* (1849). In the early 1840s Thoreau lived for extended periods as an unofficial member of the family in the home of his friend Emerson, who was frequently away on lecture tours. In 1844 the Thoreau family built its own house, on what is now Belknap Street. The site of this house is marked by a bronze plaque.

Thoreau made the town of Concord and its environs his personal microcosm, a laboratory wherein he could discover and observe the nature of life everywhere. Perhaps his greatest experiment was his adventure at Walden Pond, where he built a small cabin and spent much of a two-year period in solitude. At the cabin he wrote *A Week on the Concord and Merrimack Rivers* and kept the journals that formed the basis of *Walden* (1854). In September 1847 he moved back to the Emerson house, where he remained until July 1848. Finally, after a period at the house on Belknap Street, Thoreau moved with his family to the house at 255 Main Street, his home from 1850 until his death on May 6, 1862. In an addition built at the back of this house Thoreau helped his father operate the family pencil business.

The house remained the Thoreau family home until 1877, when it was bought by **Louisa May Alcott** for her sister Anna. Here Louisa stayed in the 1880s, after Orchard House was sold, and here she wrote her book *Jo's Boys* (1886). The Thoreau-Alcott House, as it is now known, is a private home. Visitors to Concord can walk a footpath off Route 126 to Walden Pond and the site of Thoreau's cabin. A replica of the Walden cabin stands at the Thoreau Lyceum, 156 Belknap Street, which also maintains a natural-history room, a research library, exhibits relating to Concord history and literature, and a collection of Thoreau memorabilia. A modest admission is charged.

The Concord Antiquarian Society Museum, on Lexington Road, contains furniture from Thoreau's cabin. Admission is charged. The Concord Free Public Library, at 129 Main Street, has a collection of Thoreau's books and memorabilia. Here also are housed the Thoreau Society archives, the Fred Hosmer collection, and Thoreau's surveys and surveying equipment. Another Thoreau site is accessible in Concord. The writer was arrested in 1846 for not paying his poll tax and spent a night in jail before being bailed out by his aunt, who paid the tax for him. He later mentioned this experience in his essay "Civil Disobedience" (1849), which he first wrote as a lecture and delivered at the Concord Lyceum in February 1848. The approximate site of the jail, at the corner of Main Street and Lowell Road, opposite Monument Square, is now marked.

Other writers associated with Concord include **Conrad Aiken,** who prepared for college at the Middlesex School, about three miles north of Concord, in the early 1900s. There he was editor of the school newspaper, the *Anvil.* **George Horatio Derby,** the humorist and author, lived for a time in Concord and worked in a store here before joining the U.S. Army in 1842. **John Augustus Stone,** the actor, poet, and playwright, was born on December 15, 1800, in a frame house, now a gift shop, at 12 Walden Street. Stone was

A lake is the landscape's most beautiful and expressive feature. It is earth's eye; looking into which the beholder measures the depth of his own nature.

—Henry David Thoreau, in *Walden*

House in Concord where Henry David Thoreau was born. The present owners of the house have stated that Thoreau was born in a second-floor bedroom, to the right as one views the house from Virginia Road.

best known for his tragedy *Metamora, or, The Last of the Wampanoags* (1829), which he wrote for Edwin Forrest, the actor.

CUMMINGTON

William Cullen Bryant was born on November 2, 1794, in this town in northwestern Massachusetts about twenty miles east of Pittsfield. A monument now stands at the site of his birthplace, a log cabin that overlooked the north branch of the Westfield River about two miles from Cummington. When Bryant was five years old, the family moved to a house on Route 112 now known as the William Cullen Bryant Homestead. The building, a National Historic Landmark, was a one-and-a-half-story Dutch Colonial house when Bryant lived there, although a new lower floor was added to the structure many years later. This house was Bryant's home until he was twenty-two years old, and here, in 1811, he wrote the first version of his best-known poem, "Thanatopsis."

> *So live, that when thy summons comes to join*
> *The innumerable caravan, that moves*
> *To that mysterious realm, where each shall take*
> *His chamber in the silent halls of death,*
> *Thou go not, like the quarry-slave at night,*
> *Scourged to his dungeon, but, sustained and soothed*
> *By an unfaltering trust, approach thy grave,*
> *Like one who wraps the drapery of his couch*
> *About him, and lies down to pleasant dreams.*
>
> —William Cullen Bryant,
> in "Thanatopsis"

Bryant wrote of the area around Cummington in a number of his poems, including "The Rivulet," "Lines on Revisiting the Country," "Inscription for the Entrance to a Wood," and "A Lifetime." In 1865 Bryant, by then a successful newspaperman, bought his boyhood home and renovated it. In his later years he divided his summers between this house and his country estate in Roslyn, New York.

DANVERS

John Greenleaf Whittier spent winters in his last years, 1876 to 1892, at Oak Knoll, the home of his cousins, which stood on Sumner Street in this town twenty miles northeast of Boston.

DEDHAM

George Horatio Derby, the Western humorist who wrote under the pen names **John Phoenix** and **Squibob,** was born on April 3, 1823, in this city southwest of Boston. His humor was collected in *Phoenixiana* (1855) and *Squibob Papers* (1865).

DEERFIELD

Richard Hildreth, the historian and author, was born in this town in northwestern Massachusetts on June 28, 1807. His popular novel *The Slave: or, Memoirs of Archy Moore* (1836), which was republished as *The White Slave* and as *Archy Moore,* is said to have been the first antislavery novel. Hildreth also wrote a six-volume *History of the United States* (1849–52).

John McPhee attended Deerfield Academy, a preparatory school here that became a leading institution in its field largely through the efforts of Frank Boyden, Deerfield headmaster for many years. McPhee wrote *Headmaster* (1966), a biography of Boyden and a chronicle of Deerfield Academy from 1902 to the 1960s.

EASTHAM

In the late summer of 1927 **Henry Beston,** author of many nonfiction works, moved into a sixteen-by-twenty-foot shack built for him on the beach near this town on Cape Cod. He had intended to stay only a few weeks but became so enamored of the area and its natural beauty that he remained for a year. He wrote of the experience in what may be his most admired book, *The Outermost House* (1928). Beston donated the cabin in which he stayed to the National Audubon Society, and it was designated a National Historic Site in 1964. The cabin has since been washed away by the ocean, but Beston's description of it and of life here still attracts many readers. Other literary residents of Eastham include **Sylvia Plath** and her husband **Ted Hughes,** the poet, who lived in the Beston cabin here in the summer of 1957.

> *I lingered on, and as the year lengthened into autumn, the beauty and mystery of this earth and outer sea so possessed and held me that I could not go. . . . The flux and reflux of ocean, the incomings of waves, the gatherings of birds, the pilgrimages of the peoples of the sea, winter and storm, the splendour of autumn and the holiness of spring—all these were part of the great beach.*
>
> —Henry Beston,
> in *The Outermost House*

EDGARTOWN

Many writers vacation in this town on Martha's Vineyard, but those most closely associated with Edgartown include **Edward Hoagland,** the novelist and essayist, whose papers are stored in his mother's home here; and **Henry Beetle Hough,** editor for many years of the newspaper the *Vineyard Gazette,* and author of such books as *Country Editor* (1940) and *Once More the Thunderer* (1950), accounts of his experiences as editor of the *Gazette;* and of the biography *Thoreau of Walden* (1956). Hough has lived in Edgartown since 1920. The Dukes County Historical Society in Edgartown has a collection of items relating to Hough. **Thornton Wilder,** toward the end of his life, owned a house, still standing, in Edgartown. In that house Wilder worked on *Theophilus North* (1973), a series of connected stories based on his life and memories.

FALL RIVER

Victoria Lincoln, the novelist, wrote of this city in southeastern Massachusetts in her first novel, *February Hill* (1934). **Gladys Hasty Carroll** lived in Fall River in the late 1920s, first at 407 Hanover Street, and later at 387 High Street, where she wrote her first book for girls, *Cockatoo* (1929). Both buildings she lived in are now used for apartments.

GAY HEAD

Max Eastman, author of such works as *Enjoyment of Poetry* (1913) and *Love and Revolution* (1965), was for many years a resident of Gay Head, on the western end of Martha's Vineyard. Eastman first came to the island in 1929. For a time he lived in a rented cottage in Chilmark, just east of Gay Head. Then, about 1941, Eastman and his wife Eliena Krylenko bought a hilltop in Gay Head and built the house in which they summered until Krylenko's death in 1956. Eastman also spent summers here with his second wife, Yvette, until his death in Barbados on March 26, 1969. Yvette Eastman now lives in Gay Head during most of the year. Max Eastman is buried on Martha's Vineyard.

GLOUCESTER

This city on the coast of Cape Ann has a long and remarkable history as a fishing port, and the city and its surrounding communities have for many years attracted summer visitors, including writers and artists who have found inspiration in the character and history of Gloucester and its people. **Louisa May Alcott,** a frequent summer visitor to Gloucester, stayed several times in the late 1860s and early 1870s at the Fairview Inn, at 52 Eastern Point Road. **William Rose Benét** spent many summers in Gloucester, and the Gloucester Public Library has books that came originally from the poet's library. **John Berryman** lived in Gloucester in the 1920s. **James B. Connolly** wrote of Gloucester and its fishermen in a number of books, including *Out of Gloucester* (1902), *Book of the Gloucester Fishermen* (1927), and *The Port of Gloucester* (1940). **Richard Henry Dana, Sr.,** the poet, essayist, and editor, who was a founder of *North American Review,* divided his time in his later years between his home in Boston and his estate on the shore of Cape Ann near here. **Rudyard Kipling,** the British poet and novelist, stayed at the Fairview Inn in the 1890s while collecting material for his sea story *Captains Courageous* (1897).

Henry Wadsworth Longfellow wrote of Norman's Woe, a reef off the shore of West Gloucester, in his poem "The Wreck of the Hesperus,"

> *And fast through the midnight dark and drear,*
> *Through the whistling sleet and snow,*
> *Like a sheeted ghost, the vessel swept*
> *Tow'rds the reef of Norman's Woe.*
>
> —Henry Wadsworth Longfellow,
> in "The Wreck of the Hesperus"

and **William Vaughn Moody,** who lived here for a time after 1900, wrote of the Gloucester area in his poem *Gloucester Moors* (1901). **Elizabeth Stuart Phelps [Ward],** who owned a summer house in East Gloucester, wrote of Gloucester in *The Madonna of the Tubs* (1886), *Jack the Fisherman* (1887), and *A Singular Life* (1894).

Gloucester has been the birthplace of several writers, including **Jeremy [Mildred Dodge] Ingalls,** who was born on April 2, 1911, in a house on Wells Street. Ingalls lived in a house at 6 Harold Avenue until she was nine, and there began to write poetry. In 1920 her parents bought a house, now private, at 97 Washington Street, which remained the poet's nominal place of residence until 1941, although after 1928 she also lived in other places. Ingalls taught at Gloucester High School from 1935 to 1941, and the area of Gloucester and Cape Ann figures in her poetry.

Epes Sargent, the poet, playwright, and novelist, was born in Gloucester on September 27, 1813. Sargent's many works included the novels *The Bride of Genoa* (1837) and *The Stain of Birth* (1845). **William Winter,** the drama critic and poet, was born in Gloucester on July 15, 1836. Winter wrote for Henry Clapp's New York *Saturday Press* and was identified with the New York literary group known as the Bohemians. He later wrote for the New York *Tribune* and *Harper's Weekly,* and produced biographies of prominent theatrical figures, including Henry Irving and Joseph Jefferson, as well as volumes of verse and books dealing with the theater.

GOSHEN

Bert Leston Taylor, the poet and newspaper columnist who signed his column "A Line o' Type or Two" with his initials **B. L. T.,** was born on November 13, 1866, in this town in northwestern Massachusetts, just east of Cummington. His collection *Line o' Type Lyrics* (1902) was just one of several volumes based on his newspaper column, which appeared in the Chicago *Tribune* for many years and was continued by others after his death.

GREAT BARRINGTON

Writers associated with this town on the Housatonic River eighteen miles south of Pittsfield include **William Cullen Bryant,** who practiced law and served as town clerk here from 1815 to 1825. In Great Barrington he met Frances Fairchild, the inspiration for his poems "The Future Life," "The Life That Is," "Oh, Fairest of the Rural Maids," and "October." The two were married in 1821 in Bryant's two-and-a-half-story

> *The sweet calm sunshine of October, now*
> *Warms the low spot; upon its grassy mould*
> *The purple oak-leaf falls; the birchen bough*
> *Drops its bright spoil like arrow-heads of gold.*
>
> —William Cullen Bryant,
> in "October"

Henry Beetle Hough photographed at the Linotype of the *Vineyard Gazette* in Edgartown.

Her rest is quiet on the hill,
Beneath the locust's bloom;
Far off her lover sleeps as still
Within his scutcheoned tomb.

The Gascon lord, the village maid,
In death still clasp their hands;
The love that levels rank and grade
Unites their severed lands.

—John Greenleaf Whittier, in "The Countess"

house, which later was used as a tearoom addition to the Berkshire Inn. **Ellis Parker Butler,** best known for his humorous short story "Pigs Is Pigs" (1906), lived for two years in a village near Great Barrington and died there on September 13, 1937. **W[illiam] E[dward] B[urghardt] Du Bois,** the educator, editor, and author of such influential volumes as *The Suppression of the Slave Trade* (1896) and *The Souls of Black Folk* (1903), was born in a house on Church Street here on February 23, 1868. Du Bois taught at Atlanta University from 1897 to 1910 and edited the magazine *The Crisis* from 1910 to 1934. Another prominent Black author, **James Weldon Johnson,** bought and remodeled an old farmhouse near here in 1926, but information is not available on its location.

GREENFIELD

George Ripley, born in this town in northwestern Massachusetts on October 3, 1802, was a minister, writer, Transcendentalist, and one of the founders of *The Dial,* a literary journal. He also helped establish Brook Farm, the Utopian community near West Roxbury, now part of Boston, and served as its president. **Frederick Goddard Tuckerman** left his law practice to spend a life of seclusion in Greenfield. In his lifetime he published only one volume, *Poems* (1860), but Tuckerman has been rediscovered in the twentieth century. His *Complete Poems,* published in 1965, is now regarded by many as an important addition to American literature.

GROTON

Christopher La Farge, the poet and novelist, was graduated in 1916 from the Groton School, the preparatory school located in this town in northeastern Massachusetts.

HAMPDEN

Thornton W. Burgess, author of such children's books as *The Adventures of Reddy Fox* (1913) and *The Old Briar Patch* (1947), bought a home in this town a few miles east of Springfield in 1925. Burgess originally purchased the house, at 789 Main Street, as a summer home, but made it his permanent home after 1957. His house and 260-acre estate, Laughing Brook, were purchased by the Massachusetts Audubon Society after the author's death in 1965 and converted into a wildlife sanctuary and teaching center. The house is closed temporarily, but the grounds and other buildings are open to visitors for a modest admission charge.

Interior view of the John Greenleaf Whittier Homestead in Haverhill. Whittier described the fireplace in this room in *Snow-Bound.*

HARVARD

This town about fifteen miles northwest of Concord was the site, in the early 1840s, of an experiment in communal living led by **A[mos] Bronson Alcott.** In June of 1843 the Alcotts moved from Concord to a farmhouse on one hundred acres that they called Fruitlands—perhaps because there were no fruit trees on the property. Alcott was one of the two central figures in the experiment, the other being Charles Lane, an English reformer who financed the project and had, like Alcott, his own strong views on how the colony should operate. The community lasted only seven months. Lane moved to a Shaker community located a few miles away and in January 1844 the Alcotts moved to the nearby community of Still River. Today Fruitlands is a museum, listed in the National Register of Historic Places, and counts among its exhibits the Indian arrowhead collection of Henry David Thoreau.

HAVERHILL

Winfield Townley Scott was born in this city near the Merrimack River in northeastern Massachusetts on April 30, 1910. Scott is best known for his poems with New England themes and subjects. Among his books are *Elegy for Robinson* (1936), *Mr. Whittier* (1948), and *Collected Poems* (1962).

John Greenleaf Whittier was born on December 17, 1807, at his family's house about four miles east of Haverhill. The poet's birthplace and home for twenty-nine years is now known as the John Greenleaf Whittier Homestead and is open to visitors. Nominal admission is charged. Visitors may tour the setting of Whittier's classic poem *Snow-Bound* (1866) and see such items as the desk on which he wrote his early poems as well as his last poem. Whittier grew up in this area and set many of his poems in or near Haverhill. Among these were "Fernside Brook," "The Barefoot Boy," and "The Sycamores." The subject of Whittier's poem "The Countess" was a woman named Mary Ingalls, who was born in 1786 in the nearby community of Rocks Village, on the Merrimack River, and who married Count François de Vipart. Her grave is located off East Broadway near Rocks Village.

Whittier attended a district school as a boy and later wrote of it in the poem "In School-Days." He also studied at Haverhill Academy for two terms, in 1827 and 1828. He wrote an ode for the school's dedication and paid his tuition by teaching and making shoes. Whittier later worked, in 1830 and in 1836, as editor of the *Essex Gazette,* published at the corner of Merrimack and Main streets in Haverhill. After his second stint as editor of the *Gazette,* Whittier sold the family farm and moved to Amesbury, but he kept throughout his life an affection for Haverhill and the other places he knew in his youth. Whittier later wrote the poem "Haverhill" for the 250th anniversary of the town's founding, and also wrote the dedication hymn "The Library" for the Haverhill Public Library, which has a substantial Whittier collection.

HINGHAM

Henry Beston in the 1930s owned a summer home in this resort town, about ten miles southeast of Boston, and lived here with his wife **Elizabeth Coatsworth,** the poet, novelist, and author of children's books. **Mary Hallock Foote,** the novelist, died on June 25, 1938, at her daughter's home in Hingham. **Richard Henry Stoddard,** the poet best known for his poem *Abraham Lincoln: An Horatian Ode* (1865), was born in Hingham on July 12, 1825.

HOLYOKE

John Clellon Holmes, a poet and novelist of the Beat generation, was born in this city north of Springfield on March 12, 1926. Holmes is the author of the novels *Go* (1952), *The Horn* (1958), and *Get Home Free* (1964).

HUNTINGTON

Louise Dickinson Rich, author of *We Took to the Woods* (1942) and other books about life in the woodlands of New England, was born on June 14, 1903, in this community west of Holyoke.

IPSWICH

Anne Bradstreet, the first woman author in the American colonies to write in English, was both in England but lived in this town in northeastern Massachusetts for about nine years in the 1630s and 1640s. She lived at 33 High Street with her husband Simon, who was later to serve as governor of the Massachusetts Bay Colony, and with her growing family. It is believed that during her years here she wrote many of the poems that later appeared in her volume *The Tenth Muse Lately Sprung Up in America* (1650). The manuscript of this book was taken to England and there published without Mrs. Bradstreet's knowledge. The preface to *The Tenth Muse* was written by another Ipswich resident, Nathaniel Ward, the clergyman and writer, known best for his *The Simple Cobler of Aggawam* (1647). The site of Mrs. Bradstreet's home is identified by a marker.

> *If ever two were one, then surely we.*
> *If ever man were lov'd by wife, then thee.*
> *If ever wife was happy in a man,*
> *Compare with me, ye women, if you can.*
> *I prize thy love more than whole Mines of gold,*
> *Or all the riches that the East doth hold.*
> *My love is such that Rivers cannot quench,*
> *Nor ought but love from thee give recompence.*
> *Thy love is such I can no way repay;*
> *The heavens reward thee manifold I pray.*
> *Then while we live, in love let's so persever,*
> *That when we live no more, we may live ever.*
>
> —Anne Bradstreet,
> "To My Dear and Loving Husband"

Other former Ipswich residents include **Helen [Maria Fiske] Hunt Jackson,** who attended Ipswich Female Academy, no longer in existence, and **John Updike,** who lived in Ipswich from 1957 to 1975. Updike lived in three houses here, the two principal addresses being 26 East Street, where he lived longest, and 50 Labor-in-Vain Road, his home from 1970 to 1975. While in Ipswich, Updike published the novels *Rabbit, Run* (1960) and *The Centaur* (1963); a story collection, *The Music School* (1966); and the novels *Couples* (1968) and *Rabbit Redux* (1971). Updike won a Pulitzer Prize in 1982 for *Rabbit Is Rich* (1981).

Nathaniel Ward, originally a lawyer and then a minister, lived in Ipswich from 1634 to 1646. The town was originally known as Aggawam, and was the setting for Ward's satire *The Simple Cobler of Aggawam,* published in 1647, the year of Ward's return to England. Ward was also the author of "The Body of Liberties," a code of one hundred laws that anticipated, in its defense of the rights of individuals, the philosophical beliefs that later were reflected in the Declaration of Independence and the Constitution of the United States.

LANCASTER

Caroline Lee [Whiting] Hentz, born in this town in central Massachusetts on June 1, 1800, spent much of her adult life in the South. Mrs. Hentz wrote poetry, sketches, short stories, and plays, but was known best for her novels with Southern settings, such as *Linda, or The Young Pilot*

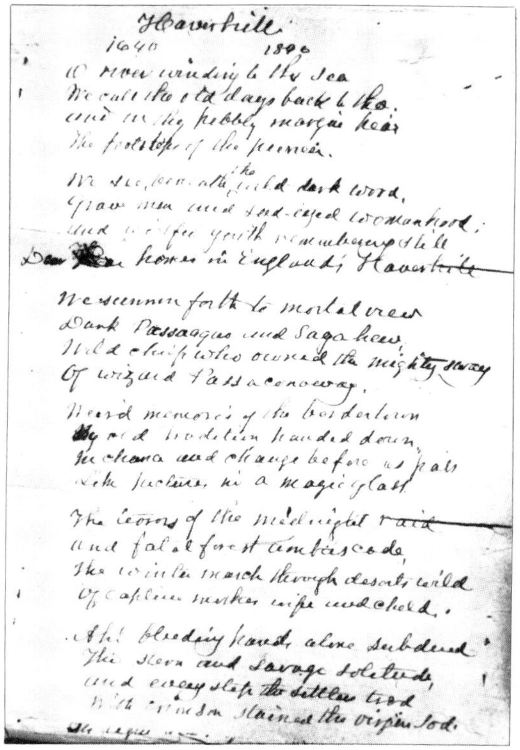

Haverhill
1640 1890
O river winding to the sea
We call the old days back to thee
And on thy pebbly margin hear
The footsteps of the pioneer.

We see, beneath the wild dark wood,
Grave men and sad-eyed womanhood;
And wistful youth remembering still
[illegible] homes in England; ~~Haverhill~~

We summon forth to mortal view
Dark Passaquo and Saggahew,
Wild chiefs who owned the mighty sway
Of wizard Passaconaway.

Weird memories of the border town
By old tradition handed down,
In chance and change before us pass
Like pictures in a magic glass

The terrors of the midnight ~~raid~~
And fatal forest ambuscade,
The winter march through deserts wild
Of captive mother wife and child.

Ah! bleeding hands alone subdued
The stern and savage solitude,
And every step the settler trod
With crimson stained the virgin sod.

Facsimile of the first page of the poem "Haverhill," written by John Greenleaf Whittier in 1890.

It was shaped something like a lengthened egg, but flattened more; and, at the ends, pointed more; and yet not pointed, but irregularly wedge-shaped. Somewhere near the middle of its under side, there was a lateral ridge; and an obscure point of this ridge rested on a second lengthwise-sharpened rock, slightly protruding from the ground. Beside that one obscure and minute point of contact, the whole enormous and most ponderous mass touched not another object in the wide terraqueous world.

—Herman Melville, in *Pierre*

of the Belle Creole (1850) and *The Planter's Northern Bride* (1854). Her birthplace, an old farmhouse on Main Street, opposite Creamery Road, is not open to visitors. **Edmund H. Sears,** who wrote the hymn "It Came Upon a Midnight Clear," was pastor of the Lancaster Unitarian Church from 1840 to 1847.

Unitarian Church in Lancaster where Edmund H. Sears served as pastor from 1840 to 1847. Two years after leaving Lancaster for Wayland, Sears wrote the hymn "It Came Upon a Midnight Clear."

LANESBORO

This town a few miles north of Pittsfield was the birthplace, on April 21, 1818, of **Henry Wheeler Shaw,** who became, under the name **Josh Billings,** one of America's leading nineteenth-century humorists. Shaw grew up in Lanesboro and, after a period out West, returned to work as a farmer in Lanesboro for several years in the 1840s. After another period in the West, Shaw settled in Poughkeepsie, New York, where he developed the style of writing that drew attention to his wit and humor. His writing, marked by outrageous misspelling and occasionally brilliant observation, was very popular in its day. It now seems labored and unsatisfying, but readers can still see the wit in the best of his sketches. Among his books were *Josh Billings, His Sayings* (1865) and *Josh Billings on Ice, and Other Things* (1868).

Near Lanesboro, not far from the north shore of Pontoosuc Lake, is Balance Rock, a giant boulder resting atop a smaller rock in such a way that it can be moved but not toppled. Herman Melville called it the Memnon Stone (after the Colossus of Memnon in ancient Egypt), or the Terror Stone, in his novel *Pierre* (1852).

LAWRENCE

Robert Frost, when he was ten years old, moved with his mother to this city in northeastern Massachusetts. The poet was graduated from Lawrence High School in 1892. Lawrence was the birthplace, on August 14, 1863, of **Ernest Lawrence Thayer,** the newspaperman and poet known in our time only as the author of the timeless "Casey at the Bat" (1888).

LENOX

This town seven miles south of Pittsfield has been the home of a number of prominent literary figures. **Henry Ward Beecher,** the clergyman and writer, lived on a farm, no longer in existence, on Lee Road near Lenox. There, it is reported, he wrote *Star Papers* (1855), a nature study of the Berkshire Mountains region, and the novel *Norwood; or, Village Life in New England* (1868), a series of related sketches. The novel was not a great success. "People used to accuse me of being the author of *Uncle Tom's Cabin,*" Beecher once said, "until I wrote *Norwood.*" **Maria Susanna Cummins,** the novelist, attended Mrs. Charles Sedgwick's School for Girls here in the 1800s. The school's buildings are now used by the Lenox Boys' School.

Nathaniel Hawthorne visited Lenox in the fall of 1849, seeking a place to live and write. By June 1850 he and his wife Sophia had moved from Salem to a small red farmhouse on an estate near Lenox. Here Hawthorne wrote *The House of the Seven Gables* (1851) and a collection of children's stories, *A Wonder-Book* (1852). He also began to plan his novel *The Blithedale Romance* (1852), which he wrote after his departure from Lenox in November 1851. It was during Hawthorne's stay in Lenox, however, that he made the acquaintance of **Herman Melville,** then rapidly approaching the height of his literary powers. The two writers first met at a friendly outing in the Berkshires, and Melville subsequently visited the Hawthornes at their cottage in September 1850. Melville found in Hawthorne someone who understood the meaning of his writings, and dedicated his novel *Moby-Dick* (1851) to him.

The Little Red House, as Hawthorne's Lenox cottage came to be known, was destroyed by fire in 1890 but has since been reconstructed. The reconstructed house, which stands on Hawthorne Avenue, is now owned by the Tanglewood-Berkshire Music Center. The Hawthorne cottage is open to visitors free of charge during the summer.

Other Lenox residents of an earlier time include **Fanny Kemble,** the actress and dramatist who, in the 1850s and 1860s, made her home at The Perch, a house no longer standing. The site of her home, on Kemble Street, is now identified by a plaque mounted on a boulder. Lenox was also the home for a number of years of **Catharine Maria Sedgwick,** who wrote of New England in such novels as *A New-England Tale* (1822) and *Hope Leslie* (1827). Mrs. Sedgwick was born and lived in nearby Stockbridge and in later years came to live in Lenox in the home, no longer standing, of her brother Charles and his wife, who ran Mrs. Charles Sedgwick's School for Girls. **Henry Wheeler Shaw,** the humorist, better known as **Josh Billings,** attended Lenox Academy on Main Street. The building has been restored for use by community organizations.

Edith Wharton lived from about 1902 to 1912 at The Mount, a twenty-three-room mansion she had built for her in Lenox. Here she wrote her novel *The House of Mirth* (1905) and the novelette *Ethan Frome* (1911), whose climactic scene is said to have been based on an actual sleigh ride down Courthouse Hill in Lenox. **Henry James, Jr.** visited Wharton at The Mount in 1904, and the building is now headquarters of Shakespeare and Company, an acting group, and is undergoing restoration as a theater center.

LEOMINSTER

This city in central Massachusetts claims as its native son **John Chapman,** the Swedenborgian ascetic who came to be known as **Johnny Appleseed.** Chapman was born, probably in 1774, the son of Nathaniel Chapman, one of the original Minutemen. The site of Chapman's birthplace, a log cabin overlooking the Nashua River, is now identified by a marker on the east side of the river, just across the bridge, on a rural road running south off Mechanic Street. Johnny Appleseed has been the subject of many stories and poems, notably the poem *In Praise of Johnny Appleseed* (1923) by Vachel Lindsay.

> *"Johnny Appleseed, Johnny Appleseed,"*
> *Chief of the fastnesses, dappled and vast,*
> *In a pack on his back,*
> *In a deer-hide sack,*
> *The beautiful orchards of the past,*
> *The ghosts of all the forests and the groves—*
>
> —Vachel Lindsay,
> in *In Praise of Johnny Appleseed*

LEXINGTON

This town northwest of Boston was the site of the initial confrontation, on April 19, 1775, between American Minutemen and the British soldiers sent from Boston to seize war supplies in Concord. James Fenimore Cooper wrote of the events at Lexington and Concord in his novel *Lionel Lincoln* (1825). **Joseph Dennie,** the essayist and editor, came to live in Lexington in 1775 when his parents moved here from Boston. The seven-year-old Dennie first attended school in Lexington. In more recent times, 1956–57, **Donald Hall,** the poet and editor, lived at 39 Paul Revere Road.

LINCOLN

Bernard De Voto moved in the summer of 1932 to a seventeen-room house on Weston Road in this community just southeast of Concord. While here, until 1936, De Voto wrote many stories and articles, and the novel *We Accept With Pleasure* (1934). He also taught at Harvard University and, in summer, at Bread Loaf Writers' Conference in Vermont.

LOWELL

This city in northeastern Massachusetts is associated with the lives of several major writers. **Jack Kerouac,** the Beat poet and novelist, was born on March 12, 1922, in a house, still standing, at 9 Lupine Street. Shortly after his birth, the Kerouac family moved to a house, also still standing, at 34 Beaulieu Street. Over the next few years the family moved several times more before settling at 16 Phebe Avenue when Jack was ten years old. From 1935 to 1938 the Kerouacs lived in a house at 33–35 Sarah Avenue. Kerouac attended Bartlett Junior High School, and later Lowell High School, on Kirk Street. His family again moved, this time to an apartment at 736 Moody Street. After high school Kerouac left Lowell but returned in the early 1940s briefly to work for the Lowell *Sun.* He found life in Lowell unsatisfying and, after a few months, left once again. By the time he returned to live in the city again, Kerouac had written all the books for which he was to become well known. In 1966 he moved with his wife and mother to a house on Sanders Street and there worked on *Vanity of Duluoz* (1968), a novel he had begun writing years earlier. One of Kerouac's favorite hangouts during these years was Nicky's Bar, on Gorham Street, owned by his brother-in-law. Kerouac tired of Lowell once again and in 1968 packed up and took his family to Florida, where he died on October 21, 1969. His funeral was held in Lowell, and he was buried in Edson Cemetery, on Gorham Street.

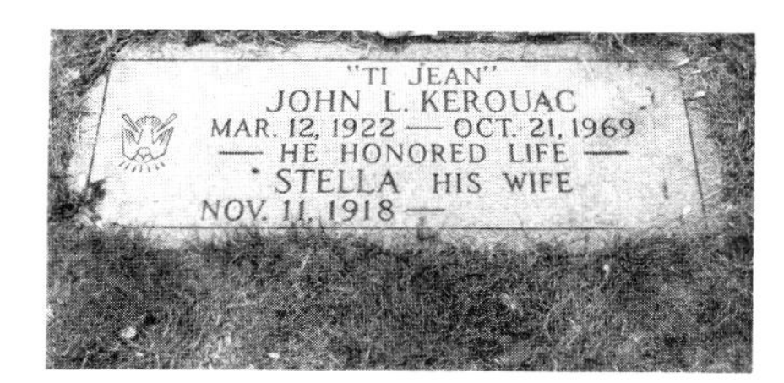

33–35 Sarah Avenue in Lowell, one of the many houses in which Jack Kerouac lived with his family in the 1930s. Headstone of Kerouac's grave in Edson Cemetery.

Other writers associated with Lowell include **Lucy Larcom,** who came to live here as a girl in the 1830s and worked in the textile mills. Her poetry first appeared in the *Operatives' Magazine* and later in the *Lowell Offering,* two publications that afterward merged. The contributors to these publications were the mill workers, girls who for one reason or another had left their rural homes to work in the mills. **Charles Dickens** visited Lowell and was impressed by the mills and by the *Offering,* both of which he described in a chapter of his *American Notes* (1842). Lucy Larcom wrote of her life here in her autobiography, *A New England Girlhood* (1889).

Edgar Allan Poe visited Lowell in July 1848 and lectured on "The Poetic Principle" here. In Lowell, he met and fell in love with Mrs. Annie Richmond, the inspiration for his poem "For Annie" (1849). Poe also wrote of her in his story "Landor's Cottage." **John Greenleaf Whittier** was editor of the Middlesex *Standard,* a Lowell newspaper, for several months in 1844 and 1845. During this time he met Lucy Larcom, who became a close friend.

> *There was ease in Casey's manner as*
> *he stept into his place,*
> *There was pride in Casey's bearing*
> *and a smile on Casey's face,*
> *And when responding to the cheers he*
> *lightly doft his hat,*
> *No stranger in the crowd could doubt,*
> *'t was Casey at the bat.*
>
> —Ernest Lawrence Thayer,
> in "Casey at the Bat"

LYNN

Charles Fletcher Lummis, the author and editor, was born on March 1, 1859, in this city

> *The big tree loomed bigger and closer,*
> *and as they bore down on it he thought:*
> *"It's waiting for us: it seems to know."*
> *But suddenly his wife's face, with*
> *twisted monstrous lineaments, thrust*
> *itself between him and his goal, and*
> *he made an instinctive movement to*
> *brush it aside. The sled swerved in*
> *response, but he righted it again, kept*
> *it straight, and drove down on the*
> *black projecting mass. There was a last*
> *instant when the air shot past him*
> *like millions of fiery wires; and then*
> *the elm. . . .*
>
> —Edith Wharton,
> in *Ethan Frome*

ten miles northeast of Boston. His birthplace, at the corner of King and Ocean streets, is private and unmarked. Lummis lived in Lynn only until the age of two but later, as a student at Harvard in the 1880s, he published his first poetry in the Lynn *Transcript.* Lynn was also the home of Moll Pitcher, the subject of the long poem *Moll Pitcher* (1832) by John Greenleaf Whittier.

MALDEN

Erle Stanley Gardner, creator of the lawyer-detective Perry Mason and one of the most successful writers of crime fiction in history, was born on July 17, 1889, in this city north of Boston. Malden was also the birthplace, on February 13, 1891, of **Elliot [Harold] Paul,** author of *The Life and Death of a Spanish Town* (1937) and *The Last Time I Saw Paris* (1942), both based on his experiences as an expatriate living in Europe. Paul was a founder, with Eugène Jolas, of the little magazine *transition,* published from 1927 to 1938.

MARBLEHEAD

H. P. Lovecraft used this town on the Atlantic Ocean northeast of Boston as a model for the fictional town of Kingsport, which appears in several of his tales of supernatural horror, including "The Festival" and "The Terrible Old Man." But Marblehead has other claims to literary fame. In 1948 **Eugene O'Neill** moved to a frame cottage on Point O' Rock Lane, where he lived until 1951. During this time, O'Neill was so ill that he was unable to complete any of his unfinished plays. He led a life of isolation and seclusion, punctuated only by occasional visits from old friends. **Anya Seton** set her novel *The Hearth and the Eagle* (1948) at the Hearth and the Eagle Inn, at 30 Franklin Street. It is said that Seton, considered to be a psychic, did not enter the inn, a composite of several houses, until after she had written her book. When she did go inside, she declared that the inn was exactly as she had imagined it.

Isaac Story, born on August 7, 1774, in a house, now private, at 1 Mechanic Street, was a lawyer, essayist, and poet. He wrote under various pseudonyms, and it is difficult to say how much he did produce in his short lifetime. Among his known works were *An Eulogy on the Glorious Virtues of the Illustrious Gen. George Washington* (1800) and the collection of verse *A Parnassian Shop, Opened in the Pindaric Stile; By Peter Quince, Esq.* (1801). His cousin, **Joseph Story,** born on September 10, 1779, in the house, now private, at 102 Washington Street, became associate justice of the Supreme Court in 1811 and a key figure in establishing the Court's authority and powers under the Constitution. He wrote a number of books on the law, as well as an early volume of poetry, *The Power of Solitude* (1804), which he later tried to suppress. His *Miscellaneous Writings* were collected by his son in 1852. John Greenleaf Whittier wrote of Marblehead in his poem "Skipper Ireson's Ride."

Of all the rides since the birth of time,
Told in story or sung in rhyme,—
On Apuleius's Golden Ass,
Or one-eyed Calender's horse of brass,
Witch astride of a human back,
Islam's prophet on Al-Borák,—
The strangest ride that ever was sped
Was Ireson's, out from Marblehead!
Old Floyd Ireson, for his hard heart,
Tarred and feathered and carried in a cart
By the women of Marblehead!

—John Greenleaf Whittier,
in "Skipper Ireson's Ride"

The anchor was up, the sails were set, and off we glided. It was a short, cold Christmas; and as the short northern day merged into night, we found ourselves almost broad upon the wintry ocean, whose freezing spray cased us in ice, as in polished armor. The long rows of teeth on the bulwarks glistened in the moonlight; and like the white ivory tusks of some huge elephant, vast curving icicles depended from the bows.

—Herman Melville,
in *Moby-Dick*

MEDFORD

Writers associated with this city about five miles north of Boston include **John Ciardi,** the poet and teacher, who was graduated from Tufts College here in 1938 and lived for a time at 84 South Street. Medford is the principal locale of his book *Lives of X* (1971). **Francis Parkman** lived on his grandfather's farm in Medford from age eight to thirteen. **Thomas Wolfe,** while a student at Harvard in the early 1920s, was a frequent visitor to the Medford home of his uncle, a Unitarian minister.

Tufts College in Medford has attracted other poets in addition to John Ciardi: **Robert P. Tristram Coffin** was Phi Beta Kappa Poet at Tufts, **Jeremy Ingalls** received her B.A. in 1932 and her M.A. here in 1933, **Denise Levertov** taught at Tufts, and **Archibald MacLeish** received his M.A. from Tufts in 1932. But it was a novelist, **Nathanael West,** a student at Tufts in the fall term of 1921, who set an all-time record for rapid advancement. Advised to leave the Tufts campus because of his terrible grades, he transferred to Brown University and, through a bureaucratic bungle, was awarded fifty-seven college credits that had been earned by another student whose name was identical with West's real name, **Nathan Weinstein.**

METHUEN

Robert Rogers, born on November 7, 1731, in this town in northeastern Massachusetts, was an explorer who fought in the French and Indian wars, and drew on his experiences for *A Concise Account of North America* (1765) and the verse drama *Ponteach* (1766), the first drama to use American Indians as its subject. The site of the cabin in which he was born is now the intersection of Cross and Hampshire streets and is marked.

NAHANT

Writers associated with this community northeast of Boston include **Cleveland Amory,** born on September 2, 1917, in a house, now private, at 2 Prospect Street. Amory is an editor, columnist, and author, whose books include *The Proper Bostonians* (1947), *Home Town* (1950), and *Who Killed Society?* (1960). **Henry Wadsworth Longfellow** owned a summer home, no longer standing, on Willow Road. The Nahant Public Library at 340 Nahant Road, has a handwritten copy of Longfellow's poem "The Bells of Lynn."

NANTUCKET

Thomas Beer, the novelist, short-story writer, and biographer, once owned a large summer house on this island off the southeast coast of Massachusetts. **Nathaniel Benchley** lived in Nantucket from 1971 until his death in Boston on December 14, 1981. **Robert Benchley** is buried in the Benchley family plot in Prospect Hill Cemetery in Nantucket. **Carson McCullers** worked on the play *The Member of the Wedding* (1951), an adaptation of her novel of the same name, while staying with **Tennessee Williams** here in the 1940s. At Williams's cottage, 31 Pine Street, McCullers wrote at one end of a table, while Williams worked on his play *Summer and Smoke* (1948) at the other end. **Herman Melville** wrote of the village of Nantucket and of its rival whaling port, New Bedford, in his novel *Moby-Dick* (1851). The whaler *Pequod* in the novel sailed from Nantucket.

NATICK

Horatio Alger, Jr. lived toward the end of his life in this town about fifteen miles west southwest of Boston. He lived at his sister's house, still standing, at 29, formerly 31, Florence Street. Alger died on July 18, 1899, and his funeral service was held at the Eliot Church, 16 Pleasant Street, South Natick. This was the church in which his father, a Unitarian minister, had preached. It is said that Alger also lived for a time at 6 Auburn Street, South Natick.

NEW BEDFORD

Conrad Aiken lived during childhood with his aunt in New Bedford, in southeastern Massachusetts. He came to New Bedford from Savannah, Georgia, after his father had killed Aiken's mother and then committed suicide. Aiken wrote of New Bedford and his aunt's house, now private, at 549 County Street, in his first novel, *Blue Voyage* (1927), and also in his autobiographical novel *Ushant* (1952).

Louisa May Alcott owned a summer house in the coastal community of Nonquit, just south of New Bedford. **William Ellery Channing,** the poet, served as an editor of the New Bedford *Mercury* in 1856. The newspaper's offices were in a building, no longer standing, at the corner of Union and South Second streets. **Frederick Douglass** lived in New Bedford after fleeing from slavery in Maryland.

Henry Beetle Hough, born on November 8, 1896, in a house, now private, at 85 Campbell Street, was for many years editor of the *Vineyard Gazette.* His novel *Long Anchorage* (1947) was based on his youth here, and Hough also wrote of New Bedford in his book *The New England Story* (1957) and in the novel *Lament for a City* (1960).

In January 1841, **Herman Melville** sailed aboard the whaler *Acushnet* from the community of Fairhaven, at one time part of New Bedford, bound for the South Seas. Melville's adventures on this voyage would later be used in his novels *Typee* (1846), *Omoo* (1847), and *Moby-Dick* (1851). In the 1850s, Melville often visited his sister at her home at 100 Madison Street, now used by the Swain School of Design. He also frequented the Seamen's Bethel, at 15 Johnny Cake Hill.

Albert Bigelow Paine, born on July 10, 1861, at the Bourne House, 716 County Street, was a poet, novelist, and biographer. He is best known today for his long association with Mark Twain. His three-volume *Mark Twain, A Biography* (1912) was for many years the definitive work on the writer. He also served as executor of the Mark Twain estate.

NEWBURYPORT

This city in northeastern Massachusetts has been the home of a number of writers, including **William Lloyd Garrison,** who was born on December 10, 1805, in a house at 5 School Street, now identified by a plaque. Garrison became a leader of the Abolitionist movement, expressing his views in the Newburyport *Free Press* and later in *The Liberator,* which he published in Boston from 1831 to 1865. As editor of the *Free Press,* he published the poems "The Exile's Departure" and "The Deity" by John Greenleaf Whittier, and encouraged the young poet in his literary and political interests. A statue of Garrison stands at Brown Square in Newburyport. **Thomas Wentworth Higginson,** the clergyman and author, was minister of the Unitarian Church here from 1847 to 1849, and lived at the church parsonage at 18 Essex Street. After resigning from the ministry, Higginson lived for a time at Curzon's Mill, just outside of the city.

Another resident of Curzon's Mill was **J[ohn] P[hillips] Marquand,** who divided his time as a boy between the home of his aunts here, and his family's home in Rye, New York. His aunts' Federal-style house at Curzon's Mill was the model for the main house in Marquand's novel *Wickford Point* (1939). He also used Newburyport as the model for the town of Clyde, Maine, in his novel *Point of No Return* (1949), the story of a man who returned to his hometown while waiting for an expected promotion at the bank where he worked. Marquand wrote of an actual Newburyport figure in two books, *Lord Timothy Dexter* (1925) and the posthumous *Timothy Dexter Revisited* (1960). Timothy Dexter, who dubbed himself Lord Timothy, was an eccentric whose actions convinced many of his townsmen that he was a lunatic. He filled his property with wooden statues of famous men—including a bust of himself—and in 1802 pub-

> *The main farmhouse was close to the Wickford River. It had been built in the late seventeen hundreds by one of the Macey family in Boston as a hunting lodge. The front of the house was square and handsome, painted an even white. Toward the rear a subsequent owner had added to the summer kitchen ell on different levels, so that in moving from one room to another you walked up and down steps through narrow passageways.*
>
> —J. P. Marquand, in *Wickford Point*

Curzon's Mill near Newburyport. John Greenleaf Whittier wrote of the mill in his poem "To __________ : Lines Written After a Summer Day's Excursion."

Half-way up the stairs it stands,
And points and beckons with its hands
From its case of massive oak,
Like a monk, who, under his cloak,
Crosses himself, and sighs, alas!
With sorrowful voice to all who pass,—
"Forever—never!
Never—forever!"

—Henry Wadsworth Longfellow,
in "The Old Clock on the Stairs"

lished a twenty-four-page pamphlet (reprinted in Marquand's 1925 biography) entitled *A Pickle for the Knowing Ones.* The pamphlet was entirely unpunctuated and erratically spelled. In an 1838 edition, Dexter supplied pages of commas and periods at the end of the book, along with the injunction to readers to "peper and solt it as they plese." Dexter was a shrewd businessman and speculator, and his house at 201 High Street was Newburyport's grand attraction until the wooden statues eventually were destroyed by weather. In later years Marquand lived on nearby Kent's Island, where he died on July 15, 1960, and is buried.

Harriet [Prescott] Spofford was a graduate, in 1852, of the Putnam Free School, which stood at the corner of Green and High streets. An essay she wrote at this school came to the attention of Thomas Wentworth Higginson, who encouraged her to become a writer and later introduced her to the Boston literary world. Her early literary contributions to magazines, in the 1850s, were intended to help family finances, and her first major success came in February 1859, when she sold her story "In a Cellar" to *The Atlantic Monthly.* In December 1865 she married Richard Smith Spofford, Jr. of Newburyport. In 1874 the couple bought Deer Island, in the Merrimack River not far from Newburyport, where Mrs. Spofford spent the rest of her life and wrote the stories, sketches, and poems collected in such volumes as *New-England Legends* (1871) and *In Titian's Garden* (1897). Mrs. Spofford died at her Deer Island home on August 14, 1921. **Henry David Thoreau,** in December 1850, visited **Thomas Wentworth Higginson**—presumably living then at Curzon's Mill—and spoke at Market Hall in Newburyport on December 6 of that year.

NEWTON

Nathaniel Benchley, son of Robert Benchley, was born in this Boston suburb on November 13, 1915. Nathaniel Benchley became a humorist, novelist, and author of many children's books. His novels included *The Off Islanders* (1961) and *Catch a Falling Spy* (1963). His books for young readers included *Gone and Back* (1971) and *Kilroy and the Gull* (1977).

S[amuel] Foster Damon, the poet, playwright, and biographer, was born here on February 22, 1893. His home at 98 Washington Street is now private. Damon's works included the collections of poetry *Astrolabe* (1927) and *Tilted Moons* (1929); and the biography *Thomas Holley Chivers, Friend of Poe* (1930).

Ralph Waldo Emerson lived in Newton briefly in 1833, at his mother's house, no longer standing, at 227 Woodeard Street.

NORTH ADAMS

Nathaniel Hawthorne visited this city in northwest Massachusetts in the summer of 1838 and several times climbed Mount Greylock, just to the west of the city. His experiences here, especially a midnight walk he took to see a burning limekiln, were the inspiration for his classic story "Ethan Brand," which he originally titled "The Unpardonable Sin."

Will[iam James] Durant, the historian known best for his eleven-volume work *The Story of Civilization* (1935–75), written with his wife Ariel Durant, was born in North Adams on November 5, 1885.

The kiln . . . was a rude, round, tower-like structure about twenty feet high, heavily built of rough stones, and with a hillock of earth heaped about the larger part of its circumference; so that the blocks and fragments of marble might be drawn by cart loads, and thrown in at the top . . . it resembled nothing so much as the private entrance to the infernal regions, which the shepherds of the Delectable Mountains were accustomed to show to pilgrims.

—Nathaniel Hawthorne,
in "Ethan Brand"

NORTHAMPTON

George Washington Cable lived from 1885 until his death in this city about fifteen miles north of Springfield. The short-story writer, novelist, and historian lived at 55 Elm Street in 1887 and 1888, and at 61 Paradise Road from 1889 to 1891. In 1892 Cable moved to 23 Dryads Green, now private, his home for the rest of his life. While in Northampton, Cable taught at Smith College and spent much of his time establishing the Home-Culture Clubs, now called the Northampton People's Institute, with the help of a grant from a friend, Andrew Carnegie. Cable died on January 31, 1925, in St. Petersburg, Florida, where he and his wife had gone for the winter. He is buried in the Bridge Street Cemetery, in Northampton. Among Cable's later books were *The Negro Question* (1888), one of several books reflecting his concern for social justice, *"Posson Jone" and Père Raphael* (1909), and *Gideon's Band* (1914).

William Ellery Channing attended the Round Hill School, whose buildings at 46 Round Hill Road are now used by the Clarke School for the Deaf. **Timothy Dwight** was born in Northampton on May 14, 1752. The poet and clergyman was a member of the literary group known as the Connecticut Wits. His poem *The Conquest of Canaan* (1785) is considered to be the first American epic poem. **Sylvester Judd III,** the clergyman and novelist, born in nearby Westhampton, lived in Northampton as a youth.

Many writers have studied or taught at Smith College, which first opened its doors in Northampton in 1875. Among them are **W. H. Auden,** who was a research professor here in 1953, and **Mary Ellen Chase,** the novelist and essayist, who taught at Smith from 1926 to 1955 and was professor emeritus from 1955 until her death on July 28, 1973, at Pine Rest Nursing Home, 25 Franklin Street. Chase lived for many years at 16 Paradise Road. During these years she wrote her autobiographical book *A Goodly Heritage* (1932), which described her girlhood in Maine; and the novels *Mary Peters* (1934), *Silas Crockett* (1935), *Windswept* (1941), and *The Edge of Darkness* (1957), all of which drew on her experiences in Maine and her understanding of New England life. She also wrote *A Goodly Fellowship* (1939), a continuation of her autobiography, in which she described her first years at Smith College.

Anne Morrow Lindbergh received her B.A. from Smith in 1928 and her M.A. in 1935. **Margaret Mitchell** studied briefly here in 1918. **Sylvia Plath** was graduated from Smith in 1955 and taught here in 1956 and 1957. When Plath was a student, she lived on campus at Haven House and later at Lawrence House. When she was a teacher, she lived with her husband, **Ted Hughes,** in an apartment on Elm Street, across from the Northampton High School.

NORTH ANDOVER

Anne Bradstreet lived in this town in northeastern Massachusetts in the 1660s. Her husband,

Simon Bradstreet—later to be governor of the Massachusetts Bay Colony—in 1667 built the house that stands across from the Phillips mansion, 148 Osgood Street, the home of the founder of Phillips Academy in nearby Andover.

NORTON

Lucy Larcom, the poet, taught at Wheaton Seminary, now Wheaton College, in this town in southeastern Massachusetts from 1849 to 1854. She continued as guest lecturer until 1863.

PIGEON COVE

Ross Lockridge spent the summer of 1943 in this community just north of Rockport. The novelist and his family stayed at a place called the Cleaves' barn, at 8 Pasture Road, now remodeled into apartments. Another summer guest at Pigeon Cove was **Richard Wright,** the novelist, who vacationed here in September 1941.

PITTSFIELD

This city in western Massachusetts has been the home of several writers. **Rose Terry Cooke,** the poet and short-story writer, lived in Pittsfield from 1887 until her death here on July 18, 1892. **Oliver Wendell Holmes, Sr.,** summered from 1849 to 1856 at Canoe Meadow, a small villa he had inherited from his great-grandfather. The house, at 497 Holmes Road, is now substantially altered and is private. Holmes wrote of Rattlesnake Mountain, known to Pittsfield residents as Snake Hill, in his novel *Elsie Venner* (1861), in which the mountain appears as Rattlesnake Ledge.

Henry Wadsworth Longfellow first visited Pittsfield in 1843, on his wedding trip. In 1846, he stayed with relatives in a house on East Street, at what is now the site of Pittsfield High School. In the house stood an old clock that inspired Longfellow to write his beloved poem "The Old Clock on the Stairs."

Pittsfield's best-known literary resident was **Herman Melville,** who lived from 1850 to 1863 at Arrowhead, a two-and-a-half-story frame house at 780 Holmes Road. The Herman Melville House, as it is also known, is a National Historic Landmark and headquarters of the Berkshire County Historical Society. It is open to visitors, and a modest admission fee is charged. Here Melville completed his novel *Moby-Dick* (1851). The effort required to write his greatest work depleted his physical and psychic reserves to such a degree that Melville never really recovered. *Moby-Dick* was misunderstood by critics and ignored by the public, and its appearance marked the beginning of Melville's slow but steady decline into obscurity. Only after Melville's death was the work recognized as a masterpiece of American literature.

At Arrowhead, Melville also wrote his autobiographical and allegorical novel *Pierre* (1852), and many short stories and sketches. Visitors can see the fireplace Melville described in "I and My Chimney," and the piazza where he wrote his *Piazza Tales* (1856). From Arrowhead Melville could look north and see the looming form of Mount Greylock, to which he dedicated *Pierre.* Before moving to Arrowhead, Melville had lived at different times at Broadhall, the estate of his

> *"This is a most remarkable structure, sir," said the master-mason, after long contemplating it in silence, "a most remarkable structure, sir."*
>
> *"Yes," said I, complacently, "every one says so."*
>
> *. . ."Twelve feet square; one hundred and forty-four square feet! Sir, this house would appear to have been built simply for the accommodation of your chimney."*
>
> —Herman Melville,
> in "I and My Chimney"

Arrowhead, home of Herman Melville in Pittsfield, where Melville completed *Moby-Dick.*

Whither, midst falling dew,
While glow the heavens with the last steps of day,
Far, through their rosy depths, dost thou pursue
Thy solitary way?

—William Cullen Bryant, in "To a Waterfowl"

Why don't you speak for yourself, John?

—Henry Wadsworth Longfellow, in *The Courtship of Miles Standish*

uncle Thomas Melville and now the Pittsfield Country Club. Broadhall appears as Saddle Meadows in *Pierre*. Other works completed by Melville during his years at Arrowhead include *Israel Potter* (1855) and the satirical novel *The Confidence-Man* (1857). Melville was visited at Arrowhead by a number of fellow writers, including **Nathaniel Hawthorne, Oliver Wendell Holmes, Sr.,** and **Catharine Maria Sedgwick.** The Berkshire Historical Society library here contains a collection of books by and about Melville, as does the Berkshire Athenaeum, at 1 Wendell Avenue.

Henry Dwight Sedgwick, the lawyer and writer, died at Pittsfield General Hospital on January 5, 1957.

PLAINFIELD

After being admitted to the bar in 1815, **William Cullen Bryant** decided, for reasons of economy, to begin practicing law in this town in northwestern Massachusetts, about seven miles north of his hometown of Cummington. While walking from Cummington to Plainfield one day in December of 1815, Bryant noticed a lone bird flying across the horizon and was so moved by the sight that he wrote the poem "To a Waterfowl" that evening. Bryant remained in Plainfield for eight months before moving to Great Barrington.

Charles Dudley Warner, born in Plainfield on September 12, 1829, was an editor, essayist, and novelist best known for his volume *My Summer in a Garden* (1871) and for his collaboration with Mark Twain on the novel *The Gilded Age* (1873). His birthplace, on Warner Hill Road, has been added to, torn down, and rebuilt. It is now private.

PLYMOUTH

This town on Massachusetts Bay, first settled by the Pilgrims on December 21, 1620, has been the setting for a number of literary works based on its early history, most notably Longfellow's poem *The Courtship of Miles Standish* (1858). **Henry David Thoreau** visited Plymouth several times in the 1850s. In July 1850, at the end of one of his famous walking tours, Thoreau attempted to walk the three miles from shore out to Clark's Island at low tide. About halfway across he was nearly drowned when the tide began to come in. A local fishing boat saved him and took him to the island, where Thoreau visited with Marston Watson and his family. In February 1852, Thoreau lectured at Leyden Hall as a favor to Watson and, in 1854, surveyed the Watson estate. Thoreau visited the Watsons again several times in 1857 and 1858.

Captain Jack's Wharf in Provincetown. Tennessee Williams lived here briefly in 1940, at the beginning of his career.

PROVINCETOWN

This town at the tip of Cape Cod has been associated with American drama since 1915, when **Susan Glaspell** and her husband **George Cram Cook** founded the Provincetown Players. The first production of the new theater group was the play *Suppressed Desires* (1915), written by Glaspell and Cook and first performed in a private home in Provincetown. After 1916 the group moved to Greenwich Village, in New York City, where it continued to produce the plays of promising newcomers until 1929, and where the Provincetown Playhouse still stands. **Eugene O'Neill** got his start with the company in Provincetown and is without question the theater's greatest alumnus. In Provincetown, at the Wharf Theater, O'Neill's play *Bound East for Cardiff* (1916) was produced, and his brilliant career as a dramatist was launched.

During the summer of 1916, **John Reed,** the poet and radical journalist, and Louise Bryant, later Reed's wife, stayed in Provincetown and met O'Neill. Both Reed and Bryant were part of the theater group's workshop, and Bryant played a part in the performance of O'Neill's play *Thirst* (1916). Reed, during that summer, was plagued by illness. Bryant and O'Neill were drawn together and had an affair, apparently with Reed's acquiescence if not his blessing, and the three worked together all summer. But things were not always quiet when O'Neill was around. In a drunken fit, O'Neill smashed all the crockery in the house Reed and Bryant were renting. O'Neill was trying to kill a green mouse he had seen.

John Dos Passos, a frequent visitor to Cape Cod, stayed in the fall of 1929 with his wife Katy at 571 Commercial Street, still standing. They bought a summer cottage in nearby South Truro but usually spent their summers at their Commercial Street home. Dos Passos summered here until 1947, when Katy died in a tragic automobile accident. Unable thereafter to bear the loneliness of the house, Dos Passos returned to Provincetown only for brief visits.

Alfred Kazin, the literary critic, lived for a time in Provincetown, but **Harry Kemp,** the tramp poet of the dunes, lived on Cape Cod for forty-one years, until his death here on August 8, 1960. Kemp lived at 577 Commercial Street, now a condominium; in a shack on the dunes; and finally in a small cottage built for him by friends and still standing at 15 Howland Street. A replica of Kemp's shack on the dunes now stands in the Provincetown Heritage Museum, at 356 Commercial Street.

Sinclair Lewis visited Provincetown in the summer of 1912 and here completed the first draft of his novel *Our Mr. Wrenn* (1914). Lewis lived on Avellar's Wharf in 1912. He returned to Provincetown in later years and became a friend of George Cram Cook and Susan Glaspell. In 1939 Lewis came to Provincetown to play a part in *Ah, Wilderness!* (1933) by Eugene O'Neill.

I broke my heart because of you, my dear;
I wept full many an unmanly tear—
But as in agony I lay awake
I thought, "What lovely poems this will make!"

—Harry Kemp,
"Literary Love"

Norman Mailer lived in the 1960s at 565 Commercial Street, still standing. **Frank Shay,** the founder of Frank Shay's Traveling Bookshop and author of such books as *My Pious Friends and Drunken Companions* (1927), edited several books dealing with the theater, and was a founder of the Provincetown Players. Shay owned a bookstore in Provincetown. **Wilbur Daniel Steele,** the short-story writer, completed his story "A White Horse Winter" (1912) in Provincetown, and lived for a time in the Avellar Shack, at 437–439 Commercial Street. **Henry David Thoreau** and his friend **William Ellery Channing,** the poet, stayed in July 1855 at Gifford's Union House, still standing, at 9 Carver Street. Thoreau visited Provincetown several times. He is known to have stayed at a hotel called the Pilgrim House. **Tennessee Williams** lived in a shack on Captain Jack's Wharf for a time during the summer of 1940. **Edmund Wilson** was a visitor to Provincetown in 1920. During that summer he often visited Edna St. Vincent Millay at her summer cottage near Truro.

QUINCY

Henry Adams spent summers during his childhood in the 1840s in this city south of Boston. He lived at the home of his grandfather, John Quincy Adams, at 135 Adams Street, now a National Historic Site and open to visitors. He also lived in another house nearby in his later youth. Adams wrote of his boyhood here in the opening chapter of his autobiography, *The Education of Henry Adams* (1918).

Quincy was the birthplace, on June 1, 1888, of **Henry Beston,** the writer best known for his account of his year on the beach of Cape Cod, *The Outermost House* (1928). Beston's home at 12 School Street is no longer standing. **John Cheever** was born in Quincy on May 27, 1912. His family then lived at 43 Elm Street, still standing, and later moved to a house on Winthrop Avenue in Wollaston, a section of Quincy. In later years Cheever's mother lived at 67 Spear Street, where the writer visited her frequently. **Benjamin Tompson,** our first native American poet, was born in Quincy on July 14, 1642, when the town was part of Braintree. He is known best for his poem "New Englands Crisis" (1676), based on King Philip's War, 1675–76.

RANDOLPH

Mary E[leanor] Wilkins [Freeman], born on October 31, 1852, in this town in eastern Massachusetts, lived in Randolph for much of her life until her marriage in 1902. The novelist and short-story writer often used eastern Massachusetts as a setting for her stories, collected in such volumes as *A Humble Romance* (1887) and *A New England Nun* (1891) and considered to be her best books.

REVERE

Horatio Alger, Jr. was born in this city northeast of Boston on January 13, 1834. The clergyman and novelist wrote nearly 130 novels for boys, all demonstrating that hard work and clean living—combined with a lot of luck—lead to success. This view of life was very popular in the nineteenth century, when countless boys and young men looking for wealth and happiness were leaving rural areas and moving to the cities.

ROCKPORT

Eleanor Estes, author of books for young people, summered in this town in northeastern Massachusetts from 1937 to 1947. Rockport was the setting for her book *The Sun and the Wind and Mr. Todd* (1943). **Elliot Paul,** the journalist and writer, is buried in Beech Grove Cemetery in Rockport. **Katherine Anne Porter** completed her novel *Ship of Fools* (1962) while staying at the Yankee Clipper Inn, 127 Granite Street.

SALEM

This city about fifteen miles northeast of Boston, proud of its association with the life of **Nathaniel Hawthorne,** has been the home or birth-

It was late in the afternoon, and the light was waning. There was a difference in the look of the tree shadows out in the yard. Somewhere in the distance cows were lowing and a little bell was tinkling; now and then a farm-wagon tilted by, and the dust flew; some blue-shirted laborers with shovels over their shoulders plodded past; little swarms of flies were dancing up and down before the people's faces in the soft air. There seemed to be a gentle stir arising over everything for the mere sake of subsidence—a very premonition of rest and hush and night.

—Mary E. Wilkins,
in "A New England Nun"

(*below left*) 571 Commercial Street, summer home of John Dos Passos in the 1930s and 1940s. (*below*) Norman Mailer lived at 565 Commercial Street in the 1960s.

Half-way down a by-street of one of our New England towns stands a rusty wooden house, with seven acutely peaked gables, facing towards various points of the compass, and a huge, clustered chimney in the midst.

—The opening sentence of *The House of the Seven Gables*

place of a number of well-known writers, and the setting of numerous works describing the life and history of colonial Salem. The seventeenth-century Salem witchcraft trials have been the subject of many literary works, including the poem "Giles Corey of the Salem Farms," by Henry Wadsworth Longfellow. H. P. Lovecraft, well versed in the history and folklore of Salem, used the city as a model for the fictional city of Arkham in several of his tales of supernatural horror, including "The Colour Out of Space" and "The Dreams in the Witch House." Salem also appears by its own name in his stories, which merge historical fact and the author's rich imagination so effectively that Lovecraft, despite his awkward style, manages in his best stories to shock and disturb the reader. Other works that deal with the witchcraft trials include the play *The Crucible* (1953) by Arthur Miller, and the novel *Rachel Dyer* (1828) by John Neal.

Anne Bradstreet and her husband Simon lived in Salem in the 1630s, only a few years after the town was settled. Their house stood on the site of the present Essex Institute, at 126 Essex Street. **Maria Susanna Cummins,** born in Salem on April 9, 1827, was a novelist whose most renowned work was *The Lamplighter* (1854), a best-selling, moralistic novel about a girl who was separated from her family in childhood, befriended by a lamplighter, adopted by an old woman, and finally reunited with her father. Miss Cummins's home, at 312 Essex Street, is no longer standing.

Jonathan Mitchell Sewall was born in Salem in 1748. A merchant, lawyer, and poet, Sewall wrote a popular song, "War and Washington," at the beginning of the American Revolution, and later wrote the poem *Eulogy on the Late General Washington* (1800). His poetry was collected in *Miscellaneous Verse* (1801). **Samuel Sewall,** a great-uncle of Jonathan Mitchell Sewall, served as a commissioner in the Salem witchcraft trials, but later publicly declared his sense of shame for his participation in the trials. Sewall wrote many pamphlets on a variety of subjects, but his three-volume *Diary,* which was published by the Massachusetts Historical Society in 1878–82, was the only work of enduring interest. **William Wetmore Story,** the lawyer, sculptor, and poet, a son of Joseph Story, was born in Salem on February 12, 1819. Another poet, **Jones Very,** was born here on August 28, 1813, and lived in Salem for much of his life. From 1833 until his death on May 8, 1880, Very lived at 154 Federal Street. His first published volume, *Essays and Poems* (1839), was edited by Ralph Waldo Emerson and was the only one of Very's works to be published in his lifetime.

Nathaniel Hawthorne was born on July 4, 1804, in a house that stood at 21 Union Street and later was moved to the grounds of the House of the Seven Gables, at 54 Turner Street. Hawthorne's series of homes in Salem is not easy to follow: he lived in the Union Street house until his father's death in 1808. The Hawthornes then took up residence at Nathaniel's grandfather's house at 10 Herbert Street, now private. From 1821 to 1825, Nathaniel attended Bowdoin College, in Maine. When he returned to Salem in 1825, he settled into the house on Herbert Street, resolved to become a professional writer. In 1828, Hawthorne published his first novel, *Fanshawe,* which showed little promise of the quality Hawthorne was to achieve when he reached his full powers as a writer—*Fanshawe* is thought to have been written in the Herbert Street house. About 1829, the Hawthorne family moved to a house on Dearborn Street, but moved back to Herbert Street in 1832. Hawthorne remained in Salem, with interruptions for travel and for a brief term in 1836 as editor of a Boston magazine, and wrote the stories that were collected in the first volume of *Twice-Told Tales* (1837).

In 1839, unable to support himself through writing, Hawthorne left town to work in the Boston Custom House, not to return to Salem until 1845. In that year, three years after his marriage to Sophia Peabody, he was back in Salem, this time with his wife and child, and living in the house on Herbert Street. In a short while, he and his young family moved to 16 Chestnut Street. By 1847, the Nathaniel Hawthornes were at 14 Mall Street, the house in which Hawthorne was to write his greatest novel, *The Scarlet Letter* (1850). Hawthorne was unhappy in Salem. He had to work at jobs he disliked: he was surveyor of the port of Salem and an agent in its custom house. Notwithstanding, Hawthorne went on writing and even though he wrote little, what he did write here was among his best work: in addition to *The Scarlet Letter,* he wrote his short story "Ethan Brand."

In June 1849 Hawthorne lost his job at the custom house and in July his mother died. But in winter of the same year, James T. Fields, of the Boston publishing house of Ticknor, Reed, and Fields, visited Hawthorne and pressed him for a publishable manuscript. Hawthorne maintained that he had nothing worth considering, but as Fields began to leave, Hawthorne had a change of heart and gave Fields an unfinished manuscript

House of the Seven Gables in Salem, where Nathaniel Hawthorne visited in his youth. In his novel *The House of the Seven Gables* Hawthorne wrote of the accursed Pyncheon family, who lived in their ancestral home in Salem.

of a story he had been working on. Fields reported later that Hawthorne had said: "It is either very good or very bad—I don't know which." Fields was impressed when he read the story and persuaded Hawthorne to complete it as a novel. After Hawthorne had completed the novel—it was, of course, *The Scarlet Letter*—and it had been published, Hawthorne and his family, eager to quit Salem, moved to Lenox. There, far from Salem, he wrote *The House of the Seven Gables* (1851), a novel set in a Salem house believed to be modeled after the house at 54 Turner Street, where his cousin Susannah Ingersoll lived. This house, which Hawthorne visited often as a boy, has become one of the country's leading literary shrines. Adjacent to it is Hawthorne's birthplace, moved here from its original site on Union Street. These houses are open to visitors.

Other Salem sites associated with Hawthorne include the Salem Maritime National Historic Site, at 178 Derby Street, which was the custom house in which Hawthorne worked from 1846 to 1849. Here visitors can see Hawthorne's office as it was in his day. The custom house is open free of charge daily. The Essex Institute, at 126 Essex Street, has a collection of Hawthorne books, manuscripts, and letters, and other exhibits of historical interest. It stands on the site of the home of Anne Bradstreet. The institute is open daily and admission is charged. A statue of Hawthorne stands on Hawthorne Boulevard in Salem.

SALISBURY BEACH

John Greenleaf Whittier visited this community on the coast of extreme northeastern Massachusetts in 1867. In the company of friends, Whittier pitched a tent on the beach and spent several days fishing, talking with people, and enjoying the scenic beauty of the coast. At Salisbury Beach he wrote the title poem of his collection *The Tent on the Beach* (1867).

SANDISFIELD

Edmund H[amilton] Sears, the clergyman and hymnist remembered as the author of the hymn "It Came Upon a Midnight Clear," was born on April 6, 1810, in this community in the southwestern corner of Massachusetts.

SANDWICH

Thornton W[aldo] Burgess, author of numerous books about nature and small animals that are still read by and to children, was born on January 4, 1874, in this town on Cape Cod. The writer's birthplace, at 6 School Street, is now private. Burgess lived in Sandwich until his graduation in 1891 from Sandwich Academy, no longer standing, and later returned to spend summers in Sandwich. In his childhood he lived in ten different houses here, but the one associated with him today is the former home of his aunt, Arabella Eldred Burgess, at 4 Water Street. It is now the Thornton

> *Once, when the sunset splendors died,*
> *And, trampling up the sloping sand,*
> *In lines outreaching far and wide,*
> *The white-maned billows swept to land,*
> *Dim seen across the gathering shade,*
> *A vast and ghostly cavalcade,*
> *They sat around their lighted kerosene,*
> *Hearing the deep bass roar their every*
> *pause between.*
>
> —John Greenleaf Whittier,
> in "The Tent on the Beach"

Birthplace of Nathaniel Hawthorne in Salem, built in 1750. Statue of Nathaniel Hawthorne in Salem.

Well here's your box. Nearly everything I have is in it, and it is not full. Pain and excitement are in it, and feeling good or bad and evil thoughts and good thoughts—the pleasure of design and some despair and the indescribable joy of creation. . . .

—John Steinbeck,
in a letter to Pascal Covici, used as the dedication of *East of Eden*

W. Burgess Museum, and here visitors can see, free of charge, the most complete collection of the author's writings known to exist, as well as a number of original illustrations used in his books.

Burgess's aunt, herself a naturalist, figures in Thornton Burgess's *Aunt Sally's Friends in Fur* (1955), and Burgess wrote of his life in Sandwich in his autobiography, *Now I Remember* (1960). The town's historic district contains a small park named in honor of Burgess, and the Sandwich elementary school library has a mural made by students that depicts a scene from a Burgess story, and a bronze plaque in tribute to Burgess. Many of Burgess's stories and books, especially his early volumes, were set in nearby East Sandwich, and the area he described is now known as The Old Briar Patch, after one of his best-known books, *The Old Briar Patch* (1947). This wildlife area includes fifty-seven acres of woodland and nature trails and is open to the public. It is located on Gully Lane, about one mile east of Sandwich.

How dear to this heart are the scenes of my childhood
When fond recollection presents them to view!
The orchard, the meadow, the deep-tangled wild-wood,
And every loved spot which my infancy knew!
The wide-spreading pond, and the mill that stood by it,
The bridge, and the rock where the cataract fell,
The cot of my father, the dairy-house nigh it,
And e'en the rude bucket that hung in the well—
The old oaken bucket, the iron-bound bucket,
The moss-covered bucket which hung in the well.

—Samuel Woodworth,
in "The Old Oaken Bucket"

SCITUATE

Samuel Woodworth, the poet and playwright best known for his poem "The Old Oaken Bucket" (1817)—sometimes found in anthologies under the title "The Bucket"—was born on January 13, 1784, in this town about fifteen miles southeast of Boston. Woodworth's poem was inspired by the well that served the white frame building here on Old Oaken Bucket Road and now known as the Old Oaken Bucket Homestead.

SHELBURNE FALLS

Horace Gregory, the poet and translator, whose last book of verse was *Another Look* (1976), died in a nursing home in Shelburne Falls on March 11, 1982. His wife, Marya Zaturenska, the Pulitzer Prize-winning poet, died on January 19, 1982, in the house here that she and her husband shared in the final years of their lives.

SIASCONSET

Robert Benchley summered in Siasconset, a resort on Nantucket, many times with his family. Benchley's son, **Nathaniel Benchley,** in his stays here as a boy came to know the New England fishing islands he later wrote about in *Sail a Crooked Ship* (1960) and other works.

In the summer of 1951, **John Steinbeck** and his wife Elaine rented a house, now private, next to Sankaty Light. Steinbeck continued work on the novel he had called *The Salinas Valley.* In Siasconset he decided on the title under which the novel was to be published, *East of Eden* (1952). Each day that Steinbeck worked on the novel, both here and in New York City, he began writing by composing a letter to his publisher, Pascal Covici. These letters, detailing the day-by-day progress of Steinbeck's novel, were published in 1969 as *Journal of a Novel: The East of Eden Letters.* While living at Siasconset, Steinbeck immersed himself

Sign marking the Thornton W. Burgess Museum in Sandwich. Home of Henry David Thoreau in South Chelmsford, where Thoreau's father made an unsuccessful attempt to run a grocery.

in his writing but took time to carve a mahogany box in which he intended to send the manuscript of *East of Eden* to Covici. The letter accompanying the manuscript was published as the dedication of the book.

SOMERVILLE

Isaac Asimov lived at 762 Broadway in this city outside of Boston from 1949 to 1951. **Jones Very,** the poet who said he drew his religious verse from visions and voices, spent some time at McLean Asylum here, after being talked into committing himself by his Harvard colleagues. Ralph Waldo Emerson thought Very "profoundly sane" and later helped the reclusive poet publish his collection of *Essays and Poems* (1839), the only volume of Very's work issued in his lifetime. While at the asylum, since moved to 115 Mill Street in nearby Belmont, Very continued to write, and Emerson helped him have his work published in the literary journal *The Western Messenger.* Somerville has an additional claim to literary immortality. One section of town is believed to be the model for Mudville in the poem "Casey at the Bat" (1888) by **Ernest Lawrence Thayer.**

SOUTHBOROUGH

W. H. Auden taught briefly during the spring of 1939 at Saint Mark's School in this town in central Massachusetts. **James Russell Lowell** came to live at Deerfoot Farm, his daughter's home in Southborough, after returning from his diplomatic post in England in 1885. In his final years Lowell was away from Southborough a good deal of the time: summer trips to England and Europe, lectures in various American cities, and periods of residence in his homes in Boston and Cambridge. At Deerfoot Farm he worked on his volume of poems *Heartsease and Rue* (1888) and on some of the poems that were to appear in his posthumous volume *Last Poems* (1895). He also collected the pieces that were published in *Political Essays* (1888).

SOUTH CHELMSFORD

Henry David Thoreau lived during childhood, from 1818 to 1821, in this town just southwest of Lowell. The house in which his family is believed to have lived is known as the Proctor house, now private, at 47 Proctor Road.

SOUTH HADLEY

This town north of Springfield is the home of Mount Holyoke College, which includes among its former faculty and students **W. H. Auden,** who taught here in 1950; **Emily Dickinson,** who studied here in the late 1840s, when the school was known as the South Hadley Female Seminary; and **Jeannette Marks,** the poet and playwright. In 1916 she organized Play and Poetry Shop Talk at Mount Holyoke for the purpose of sponsoring readings and lectures by visiting authors. In 1928 Marks founded the Laboratory Theater at Mount Holyoke, which she directed from 1929 to 1941. She also taught here from 1901 to 1910 and from 1913 to 1939. Her home was at 1 Dunlop Place. **Genevieve Taggard,** the poet, taught English literature at Mount Holyoke from 1929 to 1931. **John Clellon Holmes,** the novelist, lived in South Hadley as a child.

SOUTH LINCOLN

Richard Wilbur lived from 1950 to 1954 in a house on South Great Road in South Lincoln, about thirteen miles northwest of Boston. Here he worked on his collection of poems *Things of this World* (1956) and on *The Misanthrope* (1955), his translation of the play by Molière.

SOUTH YARMOUTH

Conrad Aiken owned a house on the Bass River in this Cape Cod community. Many of the poet's ancestors, who originally spelled their name Akin, are buried in the village cemetery. **Malcolm Lowry** stayed with Aiken in the summer of 1929.

SPRINGFIELD

This city in southwestern Massachusetts has a number of literary associations. **Edward Bellamy,** who lived in nearby Chicopee Falls, served in the 1870s as editor of the Springfield *Republican.* In 1880 he and his brother founded the Springfield *Daily News,* but by the time of Edward Bellamy's marriage in 1882, the novelist had decided to give up newspaper work and devote his time to literature.

Thornton W. Burgess came to Springfield in 1895 to work as an office boy for a publishing company. He later worked as a subeditor and as a researcher for *Good Housekeeping* and other magazines published in Springfield. Burgess from 1907 to 1957 owned a home, now private, at 61 Washington Road. The author died on June 5, 1965, in Hampden, Massachusetts, and was buried in Springfield Cemetery.

Rachel Field lived as a child in the early

Home of Conrad Aiken in South Yarmouth, at 10 Pleasant Street, about one-half block from Route 28.

1900s at 384 Union Street. The poet and novelist attended Springfield High School, where she earned her first money as a writer by winning a $20 prize in a school essay contest.

Springfield includes among its native sons **Theodore Seuss Geisel,** better known as **Dr. Seuss.** The author of many popular children's books with such titles as *How the Grinch Stole Christmas* (1957), was born on March 2, 1904, in a house, now private, at 162 Sumner Avenue. **Enos Hitchcock,** born on March 7, 1744, in Springfield, was the author of *Memoirs of the Bloomsgrove Family* (1790), said to be the second American novel published in book form. **Franklin B. Sanborn,** the teacher and biographer associated with the writers of Concord, lived in Springfield from 1868 to 1872 and edited the Springfield *Republican.* His several addresses during this period included 41 East Union Street and 190 Maple Street. Catharine Maria Sedgwick used Springfield of the 1630s as a locale for her novel *Hope Leslie* (1827).

STILL RIVER

In January of 1844 **A[mos] Bronson Alcott** moved his family to this town near Harvard, after the failure there of Fruitlands, his experiment in communal living. In Still River the Alcotts lived first in a house owned by their friend Edmund Hosmer, and in April moved to a house known as Brick Ends, on Route 110, near the town post office, where they lived until they returned to Concord in November 1844.

Brick Ends, home of the Alcott family in Still River.

STOCKBRIDGE

This rustic New England town is best known through the work of the artist Norman Rockwell, who lived here for many years until his death in 1978 and used village scenes and people as settings and models for his works. But Stockbridge also has a notable literary history. In the summer of 1886, **Matthew Arnold,** the English poet and critic, stayed at Laurel Cottage, which stood on Main Street, where town tennis courts now are located. The Historical Room of the Stockbridge Library Association, also on Main Street, has a plaque that once was affixed to a tree planted in Stockbridge by Arnold.

William Cullen Bryant wrote of Monument Mountain, a landmark near Stockbridge, in his poem "Monument Mountain." The poem tells of an Indian maiden who, because of an unlawful love, threw herself off the mountain to her death. Her people buried her on the slope of the mountain and placed stones in a cone-shaped monument over her grave. This monument of stones, according to Bryant's poem, was the origin of the name of the mountain.

And Indians from the distant West, who come
To visit where their fathers' bones are laid,
Yet tell the sorrowful tale, and to this day
The mountain where the hapless maiden died
Is called the Mountain of the Monument.

—William Cullen Bryant,
in "Monument Mountain"

Jonathan Edwards, the theologian and philosopher, wrote the treatise *A Careful and Strict Enquiry into the Modern Prevailing Notions of Freedom of Will* (1754), considered to be his greatest work, at a house on Main Street, where he lived from 1752 to 1758. The house was later named Edwards Hall in his honor. It became an inn and then a boy's school before being torn down in 1902. A sundial on the site and a plaque at the Congregational Church honor Edwards.

Rachel Field, near the turn of the twentieth century, spent part of her childhood in Stockbridge. The poet and novelist lived with her grandmother in several different houses, including the Old Corner House, on Main Street, which now is used for the Norman Rockwell Museum. Field is buried in Stockbridge Cemetery.

In August of 1850, **Nathaniel Hawthorne, Oliver Wendell Holmes, Sr.,** and **Herman Melville** met for the first time at a party at Laurel Cottage, no longer standing, on Main Street. The three writers climbed nearby Laurel Hill and Monument Mountain, and this outing marked the beginning of the close friendship that developed between Hawthorne, then living in Lenox, and Melville, who lived in Pittsfield.

Owen Johnson, the novelist, lived in the early years of this century in Stockbridge, in a house known as Ingleside. He died in Vineyard Haven but is buried in Stockbridge Cemetery. **Sinclair Lewis** stayed for a short time during 1937 at the Austen Riggs Center, recovering from an accident he suffered during a drinking spree. After Lewis had dried out a bit, he lived in a rented house on Clark Road for a time. **J. P. Marquand** and his wife lived in a house on Main Street for a few years in the early 1930s.

Catharine Maria Sedgwick, one of the most popular novelists of the first half of the nineteenth century, was born in Stockbridge on December 28, 1789, and lived much of her life in the family mansion on Main Street, still owned by the Sedgwick family. Here she was visited by many of the writers of her day. Her novels included *Hope Leslie* (1827), her most popular book, which used Laurel Hill as a locale, and *The Linwoods; or, 'Sixty Years Since' in America* (1835). In later years the novelist lived in nearby Lenox and wintered in New York City. She died on July 31, 1867, in West Roxbury (now part of Boston), and was buried in Stockbridge Cemetery. Miss Sedgwick was the great-aunt of **Henry Dwight Sedgwick,** the lawyer and writer, who was born at the Sedgwick mansion on September 24, 1861. His books included the collection *Essays on Great Writers* (1902). Sedgwick was the older brother of **Ellery Sedgwick,** the editor and publisher, who lived in Stockbridge during childhood. Ellery Sedgwick died in Washington, D.C. on April 21, 1960, and is buried in Stockbridge Cemetery. Henry Dwight Sedgwick died in Pittsfield on January 5, 1957, and also is buried in Stockbridge Cemetery.

TAUNTON

Robert Treat Paine, the poet, editor, and critic—son of Robert Treat Paine, a signer of the Declaration of Independence—was born in this city in southeastern Massachusetts on December 9, 1773. Paine was originally given the name Thomas but had his name legally changed in 1801. When he was seven years old his family moved to Boston. The poet's best-known works included the poem "The Invention of Letters" (1795) and the popular patriotic song "Adams and Liberty" (1798), which was sung to the tune later used for "The Star-Spangled Banner."

TRURO

Writers associated with this town on Cape Cod include **John Dos Passos,** who had a home in North Truro. **Waldo Frank,** the editor and writer, owned a home here for many years until his death in White Plains, New York, on January 9, 1967. Frank is buried in Truro. **Edna St. Vincent Millay** spent the summer of 1920 in a rented house, still standing, on a knoll overlooking an area known as the Long Nook Valley in South Truro. The poet wrote of Cape Cod in a number of her poems, including "Memory of Cape Cod." **Edmund Wilson** frequently bicycled here from nearby Provincetown to see Millay.

In May 1919, **Eugene O'Neill** and his wife Agnes moved into a converted Coast Guard station that had been bought for O'Neill by his father. This was the first house O'Neill owned, and it stood at the edge of the Atlantic Ocean—which greatly pleased the playwright. The house, three miles away from Truro, was cut off by ever-shifting sand dunes that made it all but inaccessible. Here O'Neill summered in the 1920s and worked on such plays as the Pulitzer Prize-winning *Beyond the Horizon* (1920), *The Emperor Jones* (1920), and *Anna Christie* (1921), which also won a Pulitzer Prize. The house, located on a fierce stretch of beach known as Peaked Hill Bar, was destroyed by the ocean in January 1930.

TYRINGHAM

Sidney [Coe] Howard, the dramatist, once owned a dairy farm on Jerusalem Road near this town about fifteen miles south of Pittsfield. Howard died here on August 23, 1939, when a tractor he was working on slipped into gear and crushed him against a garage wall. A short time earlier he had said to Dorothy Thompson that the tractor had "a life of its own." Howard, who won a Pulitzer Prize for his play *They Knew What They Wanted* (1924), is buried in a cemetery on Church Road, not far from his farm.

VINEYARD HAVEN

John Hersey lived in the 1960s in a house on Hatch Road in this community on Martha's Vineyard, where he worked on parts of his nonfiction collection *Here to Stay* (1963) and his novels *White Lotus* (1965) and *Too Far to Walk* (1966). In the 1970s Hersey lived at 167 Main Street, where he wrote his novels *My Petition for More Space* (1974) and *The Walnut Door* (1977).

Owen Johnson, author of the Lawrenceville stories, lived in Vineyard Haven late in his life. His house on Williams Street is still standing and is owned by Johnson's wife's nephew. Johnson died here on January 27, 1952, and was buried in Stockbridge.

WALTHAM

Isaac Asimov lived at 265 Lowell Street in this city west of Boston from 1951 to 1956. During this time Asimov published a flood of books, including his Foundation trilogy of science fiction novels: *Foundation* (1951), *Foundation and Empire* (1952), and *Second Foundation* (1953). **Edmund Gilligan,** the novelist whose works included *Voyage of the Golden Hind* (1945) and *My Earth, My Sea* (1959), was born in Waltham on June 7, 1899. **Saul Bellow, Max Lerner, Lewis Mumford, Howard Nemerov,** and **Philip Rahv** have taught at Waltham's Brandeis University.

WATERTOWN

Robert Benchley and his wife took an apartment in Watertown, at 117 Church Street, in September of 1914, and **Robert Creeley,** the poet, lived here for two years in the late 1920s. Watertown is west of Boston.

WAYLAND

Wayland's two literary figures, both active in the nineteenth century, are remembered today for single works. **Lydia Maria Child,** the writer and abolitionist who wrote the poem "Thanksgiving Day" (1857), lived in this town fifteen miles west of Boston from 1852 until her death on October 20, 1880. Her home at 91 Old Sudbury Road is now private. Child is buried in North Cemetery in Wayland.

Edmund H. Sears was made pastor of the Unitarian church, now the First Parish Church, at 225 State Road East, in 1839, before serving in Lancaster for seven years. In December 1849 he wrote the poem, later made into a hymn, by which he is recalled: "It Came Upon a Midnight Clear." Sears read his poem at church services in Wayland that Christmas. The poem was put to music and was published in the magazine *The Christian Register.*

WELLESLEY

This town twelve miles west of Boston, the home of Wellesley College, is associated with a

And ne'er shall the sons of Columbia
be slaves,
While the earth bears a plant, or the
sea rolls its waves.

—Robert Treat Paine,
in "Adams and Liberty"

Over the river and through the wood,
To grandfather's house we go;
The horse knows the way
To carry the sleigh
Through the white and drifted snow.

—Lydia Maria Child,
in "Thanksgiving Day"

And the essence of New Leeds was a kind of exaggeration. Everything here multiplied, like the jellyfish in the harbor. There were three *village idiots, grinning, in the post office; the average winter resident who settled here had had three wives; there were eight young bohemians, with beards, leaning from their pickup trucks; twenty-one town drunkards. . . . Nothing in New Leeds happened once only.*

—Mary McCarthy,
in *A Charmed Life*

Grave marker of Edmund Wilson in Wellfleet. The Hebrew maxim on the stone is חזק חזק ויתחזק
[be strong, be strong, and we shall strengthen one another]. To find Wilson's grave, travel east off Route 6A on Gross Hill Road, then take the first left to the Congregational Cemetery.

number of writers. **Marcia Davenport,** the novelist and biographer, attended Wellesley from 1921 to 1923. **Sylvia Plath** in 1942 moved with her mother to a house, still standing, at 23 Elmwood Road. The young poet attended local schools.

May Sarton taught at Wellesley College in the early 1960s. **Richard Wilbur** lived at 117 Crest Road from 1954 to 1957, working on his poetry collection *Things of This World* (1956) and on *The Misanthrope* (1955), his translation of the play by Molière. He also collaborated with Lillian Hellman and Leonard Bernstein on the musical *Candide* (1957).

WELLFLEET

Wellfleet has had many literary summer people including, in the 1960s, **Philip Roth, Gilbert Seldes,** and **Allen Tate. Mary McCarthy** lived in this town on Cape Cod Bay in the 1940s with her husband **Edmund Wilson.** Wellfleet appears as New Leeds in McCarthy's novel *A Charmed Life* (1955). Wilson, from the 1960s on, lived in a house on Route 6 in Wellfleet with his fourth wife, Elena Thornton. The house is still standing. Wilson died on June 12, 1972, and is buried in Wellfleet.

WEST ACTON

Robert Creeley, the poet, lived during childhood in this town about fifteen miles southwest of Lowell. His parents had a house on Elm Street and, later, one on Willow Street.

WESTBOROUGH

Esther Forbes was born on June 28, 1891, in this town east of Worcester, at 33 Church Street, now private. Forbes's historical novels included *A Mirror for Witches* (1928), set in New England in the 1600s; *The General's Lady* (1938), set during the American Revolution; and *The Running of the Tide* (1948), about clipper ships and early Salem. She won a Pulitzer Prize for her biography *Paul Revere and the World He Lived In* (1942).

WESTFIELD

Edward Taylor, who lived in this city west of Springfield in the 1600s, wrote poetry that went unpublished in his lifetime. His recognition as a poet came centuries later with the publication, in 1939, of the *Poetical Works of Edward Taylor.* **T[homas] B[angs] Thorpe** was born in Westfield on March 1, 1815. Thorpe was known best for his humorous tales of the Old Southwest, including "The Big Bear of Arkansas" (1841).

WESTFORD

Richard Henry Dana, Jr., author of *Two Years Before the Mast* (1840), attended the Westford School, in this town northwest of Cambridge, in the early 1820s. The school building, on Boston Road, is being converted into a museum.

WESTHAMPTON

Sylvester Judd III, born on July 23, 1813, in this western Massachusetts town, was a clergyman, poet, and novelist. While Judd was pastor of the Unitarian church in Augusta, Maine, he wrote several novels, including *Margaret* (1845) and *Richard Edney and the Governor's Family* (1850), both of which combined Judd's religious and social views with colorful descriptions of New England life. Judd grew up in Westhampton and nearby Northampton.

WEST NEWTON

Isaac Asimov lived from 1956 to 1970 at 45 Greenough Street in this Boston suburb. Among the many books Asimov published during these years were *The Intelligent Man's Guide to Science* (1960), the novel *Fantastic Voyage* (1966), and the two volumes of *Asimov's Guide to the Bible* (1968, 1969). In November 1851, **Nathaniel Hawthorne** moved his family from Lenox to Horace Mann's home in West Newton. Mann's wife was Hawthorne's sister-in-law, Mary Peabody. While in West Newton, Hawthorne wrote his novel *The Blithedale Romance* (1852), which was based on his experiences at Brook Farm, the utopian community in West Roxbury, now part of Boston. The Hawthornes remained at the Mann house until the spring of 1852, when they returned to Concord.

WEYMOUTH

Abigail Adams, born on November 11, 1744, in this town southeast of Boston, was the wife of John Adams and the mother of John Quincy Adams. Her *Letters of Mrs. Adams, The Wife of John Adams* (two vols., 1840) and *Familiar Letters of John Adams and His Wife During the Revolution* (1876), which were collected by her grandson, Charles Francis Adams, are a valuable source for historians. Another Weymouth native was **Bradford Torrey,** born on October 9, 1843. Torrey was an ornithologist and naturalist whose books included *Birds in the Bush* (1885) and *A Florida Sketch Book* (1894). He also edited the *Journals* (14 vols., 1906) of his friend Henry David Thoreau.

WILLIAMSTOWN

James Gould Cozzens, who won a Pulitzer Prize for his novel *Guard of Honor* (1948), lived in this town in northwestern Massachusetts from 1958 to 1971. Shadowbrook, his home at 170 Torrey Woods Road, is now private. **Max Lerner** lived in Williamstown while teaching at Williams College from 1938 to 1943. During this period, Lerner wrote his books *Ideas are Weapons* (1939) and *Ideas for the Ice Age* (1941).

Sinclair Lewis stayed at the Williams Inn, on the Williams College campus, first in June of 1943 and again in early 1946, when he began work on his novel *Kingsblood Royal* (1947). In 1946 Lewis rented, then bought, Thorvale Farm, a large estate on Oblong Road, now used for the Novitiate of the Carmelite Fathers' Chapel. At Thorvale Farm, Lewis revised the draft of *Kingsblood Royal* and wrote the novel *The God-Seeker* (1949), the story of Aaron Gadd, a young missionary from this area of Massachusetts. Lewis also worked on an unpublished novel, *Lucy Jade,* whose elements he later used in his last novel, *World*

So Wide (1951). Lewis put Thorvale Farm up for sale in 1949, but it was not sold until after his death, in Italy, in 1951. Among Lewis's literary visitors here were **Barnaby Conrad,** who wrote his novel *The Innocent Villa* (1948) while staying at Thorvale Farm, **Norman Mailer,** and **Carl Van Doren.**

Williamstown had a distinguished native son in **Bliss Perry,** the teacher, editor, and writer, who was born here on November 25, 1860. His birthplace was located at 57 Spring Street, but was moved later to 31 Bank Street. In 1889 Perry built a house, since destroyed by fire, behind a house now standing at 35 Bank Street. He taught at Williams College from 1888 to 1893, before leaving to teach at Princeton. While living here Perry wrote stories and essays as well as his first novel, *The Broughton House* (1890). He died in 1954 and was buried in Williams College Cemetery. **Chard Powers Smith,** the poet and novelist, died on October 31, 1977 at the Sweetbrook Nursing Home in Williamstown.

Other writers associated with Williams College include **William Cullen Bryant,** who studied here in 1810; **Max Eastman,** who was graduated in 1905; **Eugene Field,** who studied here in 1868 and 1869; and **Carl Jonas,** the novelist, who was graduated from Williams in 1936. **Stuart P. Sherman,** the teacher, editor, and critic, attended Williamstown High School before entering the sophomore class at Williams in 1900. He was graduated in 1903.

WINTHROP

Sylvia Plath lived in this town northeast of Boston during childhood in the late 1930s and early 1940s, and attended the Edward B. Newton School, still in use. She lived at 92 Johnson Avenue, in a stucco house facing Boston Harbor, with Massachusetts Bay as a backyard.

WORCESTER

This city in central Massachusetts has long been a major industrial, cultural, and political center. In the nineteenth century its public lecture halls echoed with the voices of such men as **Matthew Arnold, Charles Dickens, Frederick Douglass, Ralph Waldo Emerson, Abraham Lincoln,** and **Henry David Thoreau.** Worcester is the home of six major institutions of higher learning and numerous museums and cultural centers, including the American Antiquarian Society, at Salisbury Street and Park Avenue. The Society owns the largest known collection of material relating to **Cotton** and **Samuel Mather.**

Worcester was also the home, in the late 1800s and early 1900s, of **Eva March Tappan,** the teacher, biographer, and historian. In 1897 Tappan became head of the English department of English High School, at 20 Irving Street, in the building now used as a school administration building. She held this post until 1905, when she began to devote her full energies to writing. Among her historical books, written especially to interest young readers, were *In the Days of Alfred* (1900), *Our Country's Story* (1902), and *The Story of Our Constitution* (1922). The author lived at several addresses in Worcester, but her longest period of residence was at 15 Monadnock Road, her home from about 1911 until her death on January 29, 1930.

Worcester's greatest contributions to American literary history come from its native sons. **S[amuel] N[athaniel] Behrman,** the playwright, biographer, and short-story writer, was born on June 9, 1893, in a house, no longer standing, on Water Street. He later lived in a house, still standing, at 33 Providence Street. Behrman attended Classical High School, on Walnut Street, in a building since razed, and studied at Clark College, now Clark University, from 1912 to 1914. **Robert Benchley,** the humorist, was born on September 15, 1889, in a house, now private, at 14 Kingsbury Street. Benchley lived in several places in Worcester, including 6 King Street, still standing; 37 King Street; 3 Shepard Street; and 2 King Street. He attended South High School, at 14 Richards Street, for three years.

Worcester was also the birthplace, on July 23, 1838, of **Sam Hall,** an editor and author who

To be picked out as a missionary within half an hour after his conversion! To go west! To bring order and civilization to the aborigines, under the word of God! . . . Perhaps Aaron would be the first to understand them, love them, unite them with the whites in a titanic new race of men.

—Sinclair Lewis,
In *The God-Seeker*

(*left*) The Westford School in Westford, attended by Richard Henry Dana, Jr. (*below*) The American Antiquarian Society in Worcester.

wrote dime novels under the pen names **Major Sam S. Hall** and **Buckskin Sam. Stanley J[asspon] Kunitz,** winner of the Pulitzer Prize for his *Collected Poems* (1958), was born here on July 29, 1905. His birthplace is believed to have been 133 Green Street, no longer standing. Kunitz later lived in a house, also no longer standing, on Providence Street, and attended Classical High School when it was located at 20 Irving Street, where he was editor of the school newspaper, *The Argus.* Another Worcester native was **Charles [John] Olson,** the poet, born on December 27, 1910. Olson's essay *Projective Verse* (1959) is said to have influenced many poets in the 1960s, including Robert Creeley, Robert Duncan, and Denise Levertov.

RHODE ISLAND

BLOCK ISLAND

Long admired by visitors for its natural beauty and unhurried pace, this island nine miles off the southern coast of Rhode Island has several literary associations. **Richard Henry Dana, Jr.,** the novelist, came here often in the 1800s. **F. Hopkinson Smith,** before embarking on his career as novelist and short-story writer, worked during the latter half of the nineteenth century as an engineer in constructing the Block Island breakwater. The waters around Block Island served as a setting for scenes in a novel by John Hersey, *Under the Eye of the Storm* (1967), but the best-known use of the locale is in a poem by John Greenleaf Whittier. "The Palatine" tells the dramatic story of the wreck of the *Palatine* in the waters off the island, and the subsequent looting and burning of the ship by a band of Block Islanders. Island legend, known to Whittier, held that as long as the islanders involved in destroying the ship and murdering its passengers were alive, a phantom ship—sails set and awash in flame—would return each year to haunt the guilty. After the last looter died, the phantom *Palatine* was seen no more, but is recalled still in Whittier's poem.

Into the teeth of death she sped:
(May God forgive the hands that fed
The false lights over the rocky Head!)

O men and brothers! what sights were there!
White upturned faces, hands stretched in prayer!
Where waves had pity, could ye not spare?

Down swooped the wreckers, like birds of prey
Tearing the heart of the ship away,
And the dead had never a word to say.

And then, with ghastly shimmer and shine
Over the rocks and the seething brine,
They burned the wreck of the Palatine.

—John Greenleaf Whittier, in "The Palatine"

View of the Mohegan Bluffs and Southeast Light on Block Island. The area enters literary history principally through John Greenleaf Whittier's poem "The Palatine."

BRISTOL

This town on Narragansett Bay, southeast of Providence, was the birthplace on August 28, 1864, of **M[ark] A[ntony] DeWolfe Howe,** son of an Episcopal bishop of the same name. Howe grew up in Pennsylvania, and later moved to Boston to pursue an editorial career. In 1895, when his eyesight began to fail, he returned here to farm the family holdings and to write. A number of articles he wrote here for *The Bookman* magazine were collected as *The American Bookman* (1898). By 1899 he was able to return to Boston, where he edited such magazines as *The Youth's Companion* and the *Harvard Alumni Bulletin.* Although Howe made Boston his home, he often summered in Bristol in later years. His numerous books included biographies of many illustrious Americans, among them *Life and Letters of George Bancroft* (1908); *Barrett Wendell and His Letters* (1924), for which Howe won a Pulitzer Prize; and *Holmes of the Breakfast Table* (1939). He also wrote *Bristol, Rhode Island, A Town Biography* (1930) and the autobiographical *A Venture in Remembrance* (1941).

CRANSTON

Elliot Paul, author of *The Life and Death of a Spanish Town* (1937) and *The Last Time I Saw Paris* (1942), lived in this city south of Providence for a time in the 1950s. Paul's home was at 70 Victory Avenue, called Rue de la Victoire by his friends. Paul died at the Veterans Administration Hospital in Providence on April 7, 1958.

EAST GREENWICH

Literary visitors to this town in east-central Rhode Island include **Henry Wadsworth Longfellow,** who lived for two weeks in a house, still standing, at 144 Division Street. Longfellow bought the house for his friend George Washington Greene and later arranged to have a nearby windmill moved and attached to the house. A frequent visitor to the Greene household, Longfellow penned a copy of his poem "The Windmill" in a letter to Greene. Another visitor to East Greenwich was **Thomas Paine,** who stayed at the General James Mitchell Varnum House at 57 Pierce Street. Varnum, a brigadier general during the American Revolution, was also visited by such important figures as the Marquis de Lafayette and George Washington. The Varnum House is open to the public free of charge.

JAMESTOWN

Literary visitors to this island town in southern Narragansett Bay include **Maurice Maeterlinck,** the Belgian poet and playwright, who was here during the summer of 1945, and **Booth Tarkington,** the Indiana novelist, who spent the summer of 1893 here.

KINGSTON

This village in southern Rhode Island, whose buildings, some dating to the 1700s, remain largely untouched by the ravages of progress, is the home of the University of Rhode Island, founded in 1892 as the Rhode Island College of Agriculture and Mechanical Arts and later called Rhode Island State College. **Maurice Maeterlinck** stayed at the Kingston Inn here, 1320 Kingstown Road, during the summer of 1945, and **R. W. Stallman,** the poet and biographer, taught at Rhode Island State College in 1942–43. Thomas Robinson Hazard, better known to his contemporaries as Shepherd Tom Hazard, used the Kingston Inn as a setting for his tales about Rhode Island, collected in the volume *The Jonny-Cake Letters* (1882).

LITTLE COMPTON

In the Common Burial Ground in this town in southeastern Rhode Island lie the remains of Elizabeth Alden Pabodie, the first European girl to be born in New England. *Betty Alden* (1891), a novel for young people, by Jane G. Austin, tells the story of Elizabeth Pabodie's life. Her parents were John and Priscilla Alden, two of the central characters in the long poem by Henry Wadsworth Longfellow, "The Courtship of Miles Standish" (1858).

MATUNUCK

For twenty-five years before his death in 1909, **Edward Everett Hale** was a summer resident of this community on Rhode Island's southern coast. Hale's home, New Sybaris, was near Wash Pond and commanded a view of Matunuck Beach. Hale enjoyed the solitude of this house, a full seven miles from the nearest railroad station. Among Hale's books with Rhode Island settings were *Christmas in Narragansett* (1884) and the satirical *Sybaris and Other Homes* (1869).

MIDDLETOWN

From 1728 to 1731, **George Berkeley** lived at Whitehall, a home on Berkeley Avenue in this town north of Newport. Berkeley, a major figure in British intellectual life, had become the moving force in a plan to establish a college in Bermuda and, while in Middletown, began work on the project. While waiting for funds for his project, Berkeley helped to found the Philosophical Society of Newport. Here also he wrote the poem "On the Prospect of Planting Arts and Learning in America," in which appears the prophetic line, "Westward the course of empire takes its way!" A further achievement was his writing of *Alciphron, or the Minute Philosopher* (1733), a series of dialogues on the existence of God. Whitehall is open to visitors; an admission fee is charged.

Ogden Nash, the satirical poet, was a student from 1917 to 1920 at St. George's School, off Purgatory Road, and later taught at the school for a year.

NARRAGANSETT

Thomas Robinson Hazard, called **Shepherd Tom Hazard** because of his involvement in sheep-raising, was born in Narragansett on January 3, 1797. His birthplace, his grandfather's house on the east slope of what was known as Tower Hill, is no longer standing. Hazard achieved a lasting place in Rhode Island literary history primarily on the strength of two books:

Behold! a giant am I!
Aloft here in my tower,
With my granite jaws I devour
The maize, and the wheat, and the rye,
And grind them into flour.

I look down over the farms;
In the fields of grain I see
The harvest that is to be,
And I fling to the air my arms,
For I know it is all for me.

—Henry Wadsworth Longfellow, in "The Windmill"

Grave of Elizabeth Pabodie, daughter of John and Priscilla Alden, in the Common Burial Ground, Little Compton.

> *The Beaufort house was one that New Yorkers were proud to show to foreigners, especially on the night of the annual ball. The Beauforts had been among the first people in New York to own their own red velvet carpet and have it rolled down the steps by their own footmen, under their own awning, instead of hiring it with the supper and the ball-room chairs. They had also inaugurated the custom of letting the ladies take their cloaks off in the hall, instead of shuffling up to the hostess's bedroom and recurling their hair with the aid of the gas-burner; Beaufort was understood to have said that he supposed all his wife's friends had maids who saw to it that they were properly* coiffées *when they left home.*
>
> —Edith Wharton,
> in *The Age of Innocence*

Recollections of Olden Times (1879), which included detailed information about life in southern Rhode Island, and his better-known *The Jonny-Cake Letters* (1882), a collection of sketches. Hazard spent much of his life in Rhode Island, became wealthy through sheep-raising and textile manufacture, and was a leader in movements to improve education in the state and to reform treatment of the poor and the insane.

NEWPORT

This city once served as a playground for America's millionaires, who built spacious and elegant summer palaces here in the late eighteenth and early nineteenth centuries. These mansions, many of them open at times to visitors, and the beauty of Newport's older and humbler structures, have helped make Newport one of the main tourist attractions in the state, with ample interest for the literary visitor.

The Redwood Library, 50 Bellevue Avenue, was established in 1747 by the Philosophical Society of Newport, the organization **George Berkeley** helped to found while living in nearby Middletown. The library has a portrait of Berkeley as well as works by and about him. Trinity Church, 141 Spring Street, where Berkeley preached, has a silver pot and an organ case he owned. Berkeley presented an organ to Trinity Church in 1733, after his gift had been rejected by the town fathers of Berkeley, Massachusetts, as being "an instrument of the devil."

James Gordon Bennett, founder of the New York *Herald,* lived here in the late 1900s. The house, Stone Villa, stood across the street from the present Newport Casino and the Tennis Museum and Hall of Fame at 194 Bellevue Avenue. While living in Newport, **Natalie S. Carlson** wrote the children's books *The Talking Cat and Other Stories of French Canada* (1952), *Alphonse, That Bearded One* (1954), and *The Empty Schoolhouse* (1965). **Richard Henry Dana, Jr.** lived with his grandfather, William Ellery, in the latter's home, no longer standing, on the west side of Thames Street near Poplar Street. **Maud Howe Elliot** who, with her collaborators, Laura Elizabeth Richards and Florence Howe Hall, won a Pulitzer Prize for the biography *The Life and Letters of Julia Ward Howe* (1916), once lived at 150 Rhode Island Avenue, now private. **Bret Harte** summered in the 1870s at the Newbold Edgar villa, now the Village House Convalescent Home on Harrison Avenue. It was there he wrote the poems "A Newport Romance" and "A Grey-port Legend, 1797." In 1878, his popularity waning, Harte departed for Europe, and never returned to the United States.

> *And I hear no rustle of stiff brocade,*
> *And I see no face at my library door;*
> *For now that the ghosts of my heart are laid,*
> *She is viewless forevermore.*
>
> *But whether she came as a faint perfume,*
> *Or whether a spirit in stole of white,*
> *I feel, as I pass from the darkened room,*
> *She has been with my soul to-night!*
>
> —Bret Harte,
> in "A Newport Romance"

Thomas Wentworth Higginson, author and editor, lived from 1864 to 1877 at Hunter House, 54 Washington Street. The pediment over the front door features a wooden pineapple, a symbol of hospitality dating back to an early Newport tradition. When a sea captain returned from a voyage, he would place a pineapple on a pike near his door to let his friends know they were welcome to visit. Over the years, pineapples were incorporated into the design of several Newport homes. Hunter House is open daily; an admission fee is charged.

Other Newport residents included **Helen Hunt Jackson,** the poet and novelist, who lived here in 1866. **Henry James,** the novelist, in the years before the Civil War attended school at the Berkeley Institute, Classical and Commercial School at 10 Washington Square, now the site of the Savings Bank of Newport. His family lived at 463–465 Spring Street, now the site of a funeral home, and at 13–15 Kay Street. **Henry Wadsworth Longfellow** stayed at Hazard's Boarding House on Bath Road, now Memorial Boulevard, in 1838 and 1852, and visited the Joshua Perry house on Perry Street in 1855. **Clement C. Moore** lived in Newport from 1851 to 1863. Moore's home, The Cedars, where he died on July 10, 1863, is now a museum of children's toys and features copies of Moore's famous poem "A Visit from St. Nicholas" (1823). It has been announced that the Clement C. Moore House will be open on weekends, with admission charged.

Solomon Southwick, known for his long poem "The Pleasures of Poverty" (1823), was born in Newport on December 25, 1773. His father, also named **Solomon Southwick,** who took over the press of Benjamin Franklin's nephew James Franklin in 1768, lived at 23 Third Street.

Home of Henry James on Spring Street in Newport. Thirty-five years after leaving Newport, James wrote of the city that he knew it "too well and loved it too much for description or definition."

The elder Southwick edited the Newport *Mercury*, one of Rhode Island's earliest newspapers and still publishing. The editor was obliged to flee Newport during the American Revolution because of his newspaper's protests against British tyranny.

Edith Wharton, as a girl, spent many summers at Pencraig, her family's Harrison Avenue summer home, no longer standing. She later lived at Land's End, a house that challenged her talents in interior decorating and inspired her to write, with an architect named Ogden Codman, *The Decoration of Houses* (1897). Other Newport sites associated with Edith Wharton include the Chase and Chase Bookstore, then at 202 Thames Street, where her first book of verse was printed in 1878, when she was sixteen years old; and Oaklawn, the Charles H. Russel house, which may have been the model for Julius Beaufort's New York City home in *The Age of Innocence* (1920), for which Wharton won a Pulitzer Prize.

Literary visitors to Newport include the two sisters, **Alice Cary** and **Phoebe Cary,** both poets. Phoebe Cary died here on July 31, 1871. Alice Cary has the distinction of being the subject of "The Singer," a poem by John Greenleaf Whittier. **John Pendleton Kennedy,** the novelist, died here on August 18, 1870. **Jean Stafford** visited Newport in August 1947, looking for new settings for her stories. **Robert Louis Stevenson** stayed for two weeks in 1887 at the home of Mr. and Mrs. Clark Fairchild, which stood at 94 Washington Street. **John Greenleaf Whittier** was a guest at Cliff Lawn, now Cliff Walk Manor and a year-round restaurant, at 82 Memorial Boulevard. **Thornton Wilder,** who served in 1918 in the Coast Artillery Corps at Fort Adams, on Fort Adams Drive off Harrison Avenue, frequented Christie's Restaurant, at 351 James Street, where he could often be seen at work on his writing.

Newport sites used in literary works include the Samuel Hopkins house, 46 Division Street, home of the Reverend Dr. Samuel Hopkins, the main character in the novel by Harriet Beecher Stowe called *The Minister's Wooing* (1859). The Old Stone Mill, in Touro Park, on Mill Street off Bellevue Avenue, has played a part in several works. This structure is said to date back to the days of Viking exploration of North America but more likely was built in the seventeenth century by Gov. Benedict Arnold, grandfather of the unfortunate Revolutionary War general. James Fenimore Cooper wrote about the Old Stone Mill in his novel *The Red Rover* (1827), and Henry Wadsworth Longfellow wrote of it and the Viking legend in "The Skeleton in Armor." The mill appears again in a novel by Edison Marshall, *The Viking* (1951). Longfellow also wrote one of his most memorable poems about another Newport site, the Jewish cemetery located near Broadway.

PEACE DALE

Leonard Bacon, known as the Poet of Peace Dale, lived for a time on Broad Rock Lane here in southern Rhode Island, just north of the village of Wakefield. It was to Bacon's house that **Stephen Vincent Benét** and his wife came to live upon their return from Europe just after Benét had won a Pulitzer Prize for *John Brown's Body* (1928). Bacon himself received a Pulitzer Prize for his volume *Sunderland Capture* (1940).

Alice Cary, sister of Phoebe Cary.

How many lives we live in one,
And how much less than one, in all.

—Alice Cary,
in "Life's Mysteries"

There are eyes half defiant,
Half meek and compliant;
Black eyes, with a wondrous, witching charm
To bring us good or to work us harm.

—Phoebe Cary,
in "Dove's Eyes"

PORTSMOUTH

Julia Ward Howe, known best for her poem "The Battle Hymn of the Republic" (1862), spent summers in this town northeast of Newport. Oak Glen, her home at 745 Union Street, was a gathering place for literary figures of her day, and Mrs. Howe died at Oak Glen on October 17, 1910.

PROVIDENCE

Of the many writers associated with Rhode Island's capital city, two authors—both of whom failed while living to receive the recognition they deserved—now are acclaimed: **Edgar Allan Poe** and **H. P. Lovecraft,** whose stories of the supernatural are still growing in popularity. Poe, whose wife had died in 1847, made his second visit to this city late in the following year: he was infatuated with **Sarah Helen Whitman,** also a poet, who was born in Providence on January 19, 1803, and lived as an adult at 88 Benefit Street. While on an earlier visit to Providence, Poe is said to have seen Mrs. Whitman, long widowed, for the

The Old Stone Mill, in Touro Park, Newport. The mill figures in Longfellow's "The Skeleton in Armor" and other literary works.

Three weeks we westward bore,
And when the storm was o'er,
Cloud-like we saw the shore
Stretching to leeward;
There for my lady's bower
Built I the lofty tower,
Which, to this very hour,
Stands looking seaward.

—Henry Wadsworth Longfellow,
in "The Skeleton in Armor"

The Mansion House in Providence, where Edgar Allan Poe stayed during 1848. This engraving by W. D. Terry appeared in the Providence *Patriot* on September 1, 1832.

first time. She was picking flowers by moonlight in her garden, and he was on one of his midnight strolls. He met her formally when he attended a literary gathering at her home, now used by the Episcopal Church as housing for the elderly. Poe courted her at 88 Benefit Street and at the Providence Athenaeum, 251 Benefit Street. The Athenaeum, which now has a collection of portraits and letters of Poe and Mrs. Whitman, is open to visitors. During this visit to Providence, Poe stayed at the Mansion House, 81 Benefit Street. In its long service as an inn and tavern, the Mansion House was host to such luminaries as George Washington and Thomas Jefferson. The main structure was torn down in 1941 but a section is still used for apartments. Poe is said to have proposed marriage to Mrs. Whitman in the churchyard of St. John's Cathedral, 271 North Main Street. She consented only after receiving many letters, impassioned pleas, and visits from Poe. Shortly before the wedding, planned for Christmas Day, 1848, Mrs. Whitman came to the conclusion that the marriage would prove disastrous for both of them, and Poe returned alone to his cottage in New York City. Mrs. Whitman was the inspiration for Poe's second poem entitled "To Helen," and some believe that his poem "Annabel Lee" contains allusions to the events of this period of his life. After Poe's death in 1849, Mrs. Whitman defended him against his detractors in *Edgar Poe and His Critics* (1860). She died in Providence on June 27, 1878, and is buried in the North Burial Ground, North Main Street near Rochambeau Avenue.

> *It was many and many a year ago,*
> *In a Kingdom by the sea,*
> *That a maiden there lived whom you may know*
> *By the name of Annabel Lee;*
> *And this maiden she lived with no other thought*
> *Than to love and be loved by me.*
>
> —Edgar Allan Poe,
> in "Annabel Lee"

H[oward] P[hillips] Lovecraft was born on August 20, 1890, at the home of his maternal grandfather, Whipple Phillips. This gabled, three-story house stood for many years at 454 Angell Street. In 1893, after his father became insane, Lovecraft and his mother went back to live at 454 Angell Street and remained there until the grandfather's death in 1904. Lovecraft, a precocious reader, is said to have read through his grandfather's library of 2,000 books. After the grandfather's death, the house was sold, and Lovecraft and his mother moved to less spacious, rented quarters on the ground floor at 598 Angell Street. Lovecraft entered Hope Street High School in 1905, but illness and an inability to master algebra kept him from graduating. While living at 598 Angell Street, Lovecraft had letters published in the Providence *Journal* and in *Scientific American*. He also wrote science columns for the Pawtuxet Valley *Gleaner* and the Providence *Evening Tribune*, and such stories as "Dagon," "Beyond The Wall of Sleep," "The Terrible Old Man," "The Outsider," and "The Rats in the Walls." Lovecraft lived here until he moved to New York in 1924, and married Sonia Haft Greene. His marriage had turned out badly by April 1926, when he returned to Providence. Here he lived with an aunt, Lillian Clark, in half of the double house, still standing, at 10–12 Barnes Street. His years there, 1926 to 1933, proved highly productive. During this time he wrote many of his best stories, including "The Call of Cthulhu," "Cool Air," "Pickman's Model," "The Dream-Quest of Unknown Kadath," "The Colour Out of Space," "The Dunwich Horror," and "The Whisperer in Darkness," all tales of fantasy or supernatural horror.

The Case of Charles Dexter Ward, also written during this time and destined to become one of Lovecraft's best-known works, was not published until 1941, several years after his death. He made use of local settings and easily verifiable New England history to lend credibility to this horror tale.

> *I have often wondered if the majority of mankind ever pause to reflect upon the occasionally titanic significance of dreams, and of the obscure world to which they belong. Whilst the greater number of our nocturnal visions are perhaps no more than faint and fantastic reflections of our waking experiences—Freud to the contrary with his puerile symbolism—there are still a certain remainder whose immundane and ethereal character permits of no ordinary interpretation, and whose vaguely exciting and disquieting effect suggests possible minute glimpses into a sphere of mental existence no less important than physical life, yet separated from that life by an all but impassable barrier.*
>
> —H. P. Lovecraft,
> in "Beyond the Wall of Sleep"

The home of Charles Dexter Ward, the main character, is modeled after the Thomas Halsey house—according to local lore, a haunted house—at 140 Prospect Street, now the site of Halsey Estate Apartments. Many of Lovecraft's stories are rich with the early history of New England, especially of Providence and its environs, and many of his more than 100,000 letters describe the atmosphere and architecture of the city's older neighborhoods and structures. Perhaps seeking literary and psychic strength in long walks through old neighborhoods, Lovecraft frequented such places as the Market House, at the corner of North Main and College streets, which dates from the 1770s and now is owned by the Rhode Island School of Design; and the churchyard of the Cathedral of St. John, 271 North Main Street. In May 1933, Lovecraft moved with another of his aunts, Annie Gamwell, to apartments on the second floor of a house at 66 College Street, since moved to 65 Prospect Street. There he wrote his final stories, including "The Thing on the Doorstep," "The Haunter of the Dark," and "The Shadow Out of Time." Lovecraft died at Jane Brown Memorial Hospital on March 15, 1937. He is buried in the family plot at Swan Point Cemetery, on Blackstone Boulevard. Through the years his stories have attracted a wider and wider audience. Today he is undoubtedly Rhode Island's most important native writer.

Providence also includes among its native sons **George Pierce Baker.** Baker made his contribution to literary history through his 47 Workshop for playwrights, given at Harvard University from 1905 to 1925. He was born here on August 4, 1866, in a house at 241 Broad Street, no longer standing, and is buried in Swan Point Cemetery. Baker's students included Eugene O'Neill, John Dos Passos, and Thomas Wolfe. Professor Hatcher, in Wolfe's *Of Time and the River* (1935), is modeled after Professor Baker. **George M. Cohan,** the playwright, was born on July 3, 1878, in a house at 536 Wickenden Street, replaced later by a building that now serves the Fox Point Neighborhood Center. **George William Curtis,** the editor and author, was born on February 24, 1824, in a house on Williams Street. He later lived in a house, no longer standing, at 149 North Main Street, and attended Classical High School, still at 770 Westminster Street. **Stephen W. Meader** was born here on May 2, 1892. His numerous novels for boys included *Down the Big River* (1924), *The Long Trains Roll* (1944), and *Voyage of the Javelin* (1959). **Frederick William Thomas,** the novelist and poet, and a friend of Edgar Allan Poe's, was born here on October 25, 1806.

Writers who have lived in Providence include **James Fenimore Cooper,** who in 1823 wrote part of his romance *Lionel Lincoln* (1825) while living in the John Whipple house at 54 College Street, now used by Brown University's Music Department. **A. J. Liebling,** long-time contributor to *The New Yorker* magazine, worked as a writer and reporter for the Providence *Journal* in the late 1920s. **Edwin O'Connor,** author of the best-selling novel *The Last Hurrah* (1956), attended La Salle Academy, at 612 Academy Avenue, in the 1930s and worked in 1940 as an announcer for radio station WPRO, now on Route 114 in nearby East Providence. O'Connor's experience in radio here and elsewhere was the basis for his first novel, *The Oracle* (1951), which caricatured the life of a radio broadcaster.

Providence is part of me—I am Providence. . . . Providence would always be at the back of my head as a goal to be worked toward—an ultimate Paradise to be regain'd at last.

—From a letter by H. P. Lovecraft

Cathedral of St. John in Providence, built in 1810. Edgar Allan Poe and H. P. Lovecraft, masters of the supernatural horror story, are known to have wandered through the churchyard.

From a private hospital for the insane near Providence, Rhode Island, there recently disappeared an exceedingly singular person. He bore the name of Charles Dexter Ward, and was placed under restraint most reluctantly by the grieving father who had watched his aberration grow from a mere eccentricity to a dark mania involving both a possibility of murderous tendencies and a peculiar change in the apparent contents of his mind. Doctors confess themselves quite baffled by his case, since it presented oddities of a general physiological as well as psychological character.

—H. P. Lovecraft, in *The Case of Charles Dexter Ward*

> *A few nights ago I strolled into our Pompeian living room in my stocking feet, bedad, with a cigar in my mouth and a silk hat tilted back on my head, to find Maggie, with osprey plumes in her hair and a new evening cape, pulling on long white gloves. A little cluster of exclamation points and planets formed over her head as she saw me.*
>
> —S. J. Perelman,
> in "A Pox on You, Mine Goodly Host"

S. J. Perelman spent his early years here. His father operated a dry goods store in Providence in the early 1900s. He attended Classical High School, was an habitué of the Providence Public Library, and later attended Brown University. **Philip Rahv,** the critic, lived here as a boy in the 1920s. **Winfield Townley Scott,** the poet, worked for the Providence *Journal* from 1931 to 1951. He studied and later taught at Brown University. His home at 312 Morris Avenue is now private. **Richard Steere,** the poet remembered, if at all, for *The Daniel Catcher* (1713), lived here for a time in the late 1600s.

One of the main literary attractions in Rhode Island is Brown University, at 79 Waterman Street. The University's John Hay Library contains the prestigious Harris Collection of American Poetry and Plays; the Lincoln Collection, which contains more than 700 manuscripts of Abraham Lincoln; a collection of letters and manuscripts of H. P. Lovecraft; and the Henry S. Saunders Whitman Collection, a major source for material on Walt Whitman. Among the many writers associated with Brown University, two of the most colorful were **Nathanael West,** known best for his surrealistic novel *The Day of the Locust* (1939), and **S. J. Perelman.** West was advised in 1922 to leave Tufts University, in Massachusetts, because of his poor grades. He transferred to Brown and, through an administrative error, was awarded academic credits he had not earned, enough to make him a second-semester sophomore in good standing. Although he continued to treat college as a lark, he was graduated in 1924, a year ahead of Perelman, with whom he had become close friends. Perelman, who early in his life exhibited a talent for satirical word play, joined the staff of the campus humor magazine, *Brown Jug,* and eventually became its editor. This experience led to a job with the comic weekly *Judge* and, eventually, to his association with *The New Yorker,* which endured until his death in 1979. Perelman's friendship with West—Perelman was married to West's sister Laura—continued until West's untimely death in 1940.

Other writers were associated with Brown University. **John Berryman,** the poet, taught at Brown in 1962–63. **S. Foster Damon** taught here from 1927 to 1963, and lived at 24 Thayer Street. He was made curator of the Harris Collection of American Poetry and Plays in 1929. Damon's books included the collections of poems *Astrolabe* (1927) and *Tilted Moons* (1929), and the biographies *William Blake—His Philosophy and Symbols* (1924) and *Thomas Holley Chivers* (1930). **John Hawkes,** the novelist, has taught at Brown since 1958. **Percy Marks** taught here while writing *The Plastic Age* (1924), his controversial novel about life on a college campus. **Winfield Townley Scott,** the poet, edited the *Brown Daily Herald* while he was a student here. He was graduated in 1931 and went to work for the Providence *Journal,* remaining on the staff for twenty years. It was while he worked for the *Journal* that he published such volumes of poetry as *Elegy for Robinson* (1936) and *Wind the Clock* (1941).

SAUNDERSTOWN

Oliver La Farge and his brother, **Christopher La Farge,** spent childhood summers in the early 1900s at The Waterway, the La Farge family's summer home, which was destroyed by fire in 1944. By 1928 the family had begun to spend summers at River Farm, off Tower Hill Road (Route 1), near this village on the western side of Narragansett Bay and just south of the Jamestown Bridge. River Farm later became Christopher La Farge's year-round home, and his love for the southern part of Rhode Island was reflected in many of his books, notably in his first novel, *Hoxie Sells His Acres* (1934), written in verse; and in his best-selling prose novel, *The Sudden Guest* (1946). Other books he wrote here include the verse novel *Each to the Other* (1939), *Poems and Portraits* (1940), and the collection of stories *All Sorts and Kinds* (1949). He died in Providence on January 5, 1956, at Jane Brown Memorial Hospital.

Oliver La Farge, known best for *Laughing Boy* (1929), his Pulitzer Prize-winning novel about life among the Navajos, spent most of his time in the Southwest, pursuing his lifelong interest in anthropology, but also summered at River Farm as an adult. His novel *Long Pennant* (1933) was a tale of Rhode Island privateers in the War of 1812. He also wrote about his experiences in Saunderstown in an autobiographical work, *Raw Material* (1945).

Owen Wister had a summer home here, called Crowfield and located off Boston Neck Road (Route 1-A), about a mile and a half north of Saunderstown. The home is now known as Champs des Corbeaux—French for Crowfield. The novelist summered and wrote a great deal here from 1913 until he died at Crowfield on July 21, 1938. His principal work during this period was *Roosevelt, the Story of a Friendship* (1930).

SOUTH PORTSMOUTH

Thomas Robinson Hazard lived in South Portsmouth, northeast of Newport, from the 1840s until his death in 1886. Hazard retired about 1840 to Vaucluse, a spacious and comfortable estate that remained his home for the rest of his life. All that remains of Hazard's house is the doorway, which is owned by New York City's Metropolitan Museum of Art. In his later years, Hazard wrote *Recollections of Olden Times* (1879) and *The Jonny-Cake Letters* (1882). Hazard died in New York City on March 26, 1886, and was buried at Vaucluse.

WOONSOCKET

This city in northern Rhode Island was the birthplace, on July 29, 1918, of novelist **Edwin O'Connor.** His home at 247 Gaskill Street, near the Massachusetts border, is now private. O'Connor grew up in Woonsocket's North End, attending the Summer Street School, 219 Summer Street; the Boyden Street School, no longer standing; the Harris School, 60 High School Street; and Woonsocket Junior High School, 357 Park Place. As a boy, O'Connor played football and baseball in a vacant lot known as Hoyle's Field, still to be found off Gaskill Street near his home. O'Connor received a Pulitzer Prize in 1962 for his novel *The Edge of Sadness* (1961), which deals with the lives of Irish people in a city that is evocative of Boston.

CONNECTICUT

BRIDGEWATER

While living on Curtis Road in this rural town in western Connecticut, east of the Still River and north of Danbury, **Robert W. Anderson,** the playwright, worked on *I Never Sang for My Father* (1968), *After* (1973), and *Getting Up and Going Home* (1978).

On July 1, 1949, **Van Wyck Brooks** and his wife moved to a home on Main Street in which they were to entertain many literary visitors. A Sunday supper they gave in 1957, for example, was attended by Katherine Anne Porter, Norman Mailer, and William Styron. The Malcolm Cowleys lived only ten miles away and were frequent visitors. Several of Brooks's books were published while he was living here, including *Scenes and Portraits: Memories of Childhood and Youth* (1954), *Helen Keller: Sketch for a Portrait* (1956), and *From the Shadow of the Mountain: My Post-Meridian Years* (1961). Van Wyck Brooks died in Bridgewater on May 2, 1963, and is buried in the Center Cemetery here. In the Burnham Library there is a plaque that reads:

This North Reading Area of the Burnham Library
is dedicated to the memory of
Van Wyck Brooks 1886–1963—
Critic and Historian of American Letters

The house on Main Street remains private today.

BRISTOL

A[mos] Bronson Alcott, the Yankee peddler, taught school from November 1827 until March 1828 in this city, which is situated fifteen miles southwest of Hartford.

BROOKFIELD

Virgil Geddes, the poet and dramatist whose plays include *Native Ground* (1932) and *Pocahontas and the Elders* (1933), lived for many years in Brookfield, just east of Candlewood Lake. In 1941 he became town postmaster and later wrote the book *Country Postmaster* (1952). **Elliot Paul,** the journalist and writer, lived on Pumpkin Hill Farm here about 1940. *The Life and Death of a Spanish Town* (1937) was one of his better-known books.

BROOKFIELD CENTER

Joseph Hayes, playwright and novelist, wrote stories and plays while living in this small town north of Danbury. **Joshua Logan** and **Tom Heggen** worked together on the dramatization of *Mister Roberts* (1947) at Logan's home in Brookfield Center in 1947. **Clifford Odets** was traveling with the Group Theatre, which he had helped found, when the company moved here. He played a bit part in its first production, *The House of Connelly* (1931), by Paul Green.

BROOKLYN

On August 8, 1827, **A. Bronson Alcott** met Abigail May at the home of Reverend Samuel J. May in this town in the northeastern corner of the state. They were married in May 1830 in Boston.

CHESHIRE

At the age of thirteen, **A. Bronson Alcott** went to live with his mother's brother, Reverend Tillotson Bronson, who conducted the Cheshire Academy. The academy was located on Main Street, four houses behind St. Peter's Church, in this town midway between Waterbury and New Haven. It is unlikely that Cheshire Academy then

The Connecticut mind, as travellers often noted, was keen, strong and witty, but usually narrow, educated rather than cultivated. It abounded in prejudices that were often small, like most of the "Yankee notions." It lacked, as a rule, the power of generalizing, which had been marked in New York, in Virginia and in Boston. It was a village mind, in short, that had never breathed a larger atmosphere. It bore few of the fruits that spring from an intercourse and collision with other minds from other mental regions.

—Van Wyck Brooks,
in *The Flowering of New England*

Home of Van Wyck Brooks in Bridgewater, on Main Street across from the post office and to the right of the town hall.

They went out through the revolving doors that made a faintly derisive whistling sound when you pushed them. It was two blocks to the parking lot. At the drugstore on the corner she said, "Wait for me here. I forgot something. I won't be a minute." She was more than a minute. Walter Mitty lighted a cigarette. It began to rain, rain with sleet in it. He stood up against the wall of the drugstore, smoking. . . . He put his shoulders back and his heels together. "To hell with the handkerchief," said Walter Mitty scornfully. He took one last drag on his cigarette and snapped it away. Then, with that faint, fleeting smile playing about his lips, he faced the firing squad; erect and motionless, proud and disdainful, Walter Mitty the Undefeated, inscrutable to the last.

—James Thurber,
in "The Secret Life of Walter Mitty"

was run according to the educational principles Alcott would later espouse to such effect: the honor system, the conversational method of instruction, and athletics in the curriculum. Young Alcott, homesick, soon returned to his family's home in Wolcott. Founded in 1794, Cheshire Academy is still functioning at 10 Main Street.

CORNWALL

Van Wyck Brooks and his wife rented a large white eighteenth-century house opposite the stone library on the green in Cornwall, near Torrington, where they lived in 1948 and 1949. **Mark Van Doren** lived in this small town for most of his life and here wrote many works of poetry and criticism, including *Jonathan Gentry* (1931) and *The Happy Critic* (1961). He died in Torrington and was buried from the First Church of Christ Church in Cornwall Plains on December 3, 1972. **Robert Lewis Taylor,** while living at Dark Entry Farm here from 1946 to 1948, wrote his first book, *Adrift in a Boneyard* (1947).

James Thurber, celebrator, in "The Secret Life of Walter Mitty" (1939), of the undefeated and inscrutable hero present within all defeated husbands, lived for a time in Cornwall. At first, for the summer of 1942, he took a small house in the center of town. Thurber liked the house so much that he stayed until January 1943 and returned for the next summer. He worked on his story "The Catbird Seat" (1943) while here, as well as on *Men, Women and Dogs* (1943), *My World—and Welcome To It* (1942), *The Thurber Carnival* (1945), and many other books. In 1945 Thurber bought his dream house in Cornwall, a fourteen-room Colonial adjacent to the Mohawk State Forest. It was to be his home for the rest of his life. He died in New York City on November 3, 1961.

Home of James Thurber in Cornwall. In the preface to an edition of *The Thurber Carnival* he wrote on September 1, 1957, Thurber said he planned to spend his days in the house "reading *Huckleberry Finn,* raising poodles, laying down a wine cellar, playing boules, and talking to the little group of friends which he has managed to take with him into his crotchety middle age."

CORNWALL BRIDGE

Robert McCloskey wrote *Make Way for Ducklings* (1941) in this town in northwestern Connecticut, on Route 43. This delightful children's story tells of a family of mallards who make their home in the Public Gardens of Boston.

DANBURY

James Bailey, the newspaperman and humorist sometimes called the Danbury News Man, became well known in the 1860s for his humorous sketches. He lived in Danbury at 24 Granville Avenue, at the corner of Osborne Street. In 1865, Bailey and a partner bought the Danbury *Times* and, in 1870, the Danbury *Jeffersonian.* They merged the two newspapers into the Danbury *News* and, in 1871, established the *Evening News.* Bailey bought out his partner's interest in 1878 and was sole proprietor until he died in 1894.

The Danbury area appears in the plays *The Third Day* (1964) and *The Hours After Midnight* (1958), by Joseph Hayes.

Rose Wilder Lane, the novelist and short-story writer, died in 1968 at 23 King Street. Her body was transported to Missouri for burial near her parents' graves. **Robert Lowell** was sentenced in 1943 to a year and a day in the federal penitentiary here for disobeying the Selective Service Act: Lowell was a pacifist. His poem "In the Cage," from *Lord Weary's Castle* (1944), describes the experience of those who languish in jail.

In 1927, **Rex [Todhunter] Stout,** creator of Nero Wolfe, bought eighteen acres of hilly Danbury countryside astride the New York and Connecticut border to build a house modeled on a bey's palace he had seen outside Tunis. He called his place High Meadow, and lived in it for almost fifty years. Although the mailing address was Brewster, New York, the house itself was 1600 feet into Connecticut, in Danbury township. Stout supervised every aspect of the construction himself, and the final product was considered newsworthy. In January 1933, the New York *Times* ran an article about the house. Stout married his second wife, Pola Hoffman, at High Meadow on December 31, 1932. One of Stout's best friends in Danbury was Dr. "Than" Selleck, Jr., who attended the births of Stout's two daughters, Barbara and Rebecca. (Barbara Stout later married Dr. "Than" Selleck III.) Stout and Selleck often enjoyed a game of pool in the basement of Selleck's home at 25 Delta Avenue. Rex Stout wrote many books about his gourmet and orchidologist detective, Nero Wolfe, including *The Hand in the Glove* (1937), *Too Many Cooks* (1938), *Not Quite Dead Enough* (1944), and *Murder by the Book* (1951). Stout died in Danbury on October 27, 1975.

DARIEN

From 1939 to 1942, **Erskine Caldwell** and his second wife, Margaret Bourke-White, the photographer, lived on Point O'Woods Road. Caldwell wrote *Trouble in July* (1940) here. Their house still stands and is privately owned. When **Sinclair Lewis** visited Caldwell here, the two authors commiserated over the lonely life of a writer.

DERBY

Ann Sophia [Winterbotham] Stephens was born in 1813 in this town eight miles west of New Haven. She was popular in the nineteenth century as the author of flamboyant historical novels, such as *The Diamond Necklace, and Other Tales* (1846) and *The Rejected Wife* (1863). Her best-known book, *Malaeska; The Indian Wife of the White Hunter* (1860), has been called the first dime novel.

EASTON

Treasure Hill, the celebrated home of **Edna Ferber,** the novelist, stood on Maple Road in Easton, six miles east of Bridgeport. She lived there from 1938 to 1949. Two of her novels appeared during her years in Easton, *Saratoga Trunk* (1941) and *Great Son* (1945).

EAST WINDSOR

Jonathan Edwards, the mystic and clergyman, and probably the most talented writer anywhere in the Colonies, was born in East Windsor, north of Hartford, on October 5, 1703. He entered Yale at the age of thirteen and while there read Locke and Berkeley. He was graduated in 1720, but stayed on for postgraduate work until 1726, when he became pastor of a Congregational church in Northampton, Massachusetts. His many published works included *Divine and Supernatural Light* (1734), *Charity and Its Fruits* (1738), and *An Humble Attempt To Promote Visible Union of God's People* (1747).

FAIRFIELD

On July 20, 1783, **Timothy Dwight,** the poet, clergyman, and scholar, accepted the pastorate at the Congregational church in Fairfield, southwest of Bridgeport. Dwight stayed here for twelve years and established a coeducational school, which attracted students from northern and southern states because of Dwight's reputation as a Calvinist and a Federalist. He left in 1795 to become president of Yale. A leading member of the Connecticut Wits, Dwight wrote many sermons, political pamphlets, and poems: *The Conquest of Canaan* (1785), dedicated to George Washington, was of epic length and compared heroes of the Revolutionary War with biblical characters. *Greenfield Hill* (1794), dedicated to John Adams, described the New England community he served as pastor for twelve years.

John Hersey lived for a time in Southport, a subdivision of Fairfield, in a house on Hull's Farm Road. It was here that he wrote two of his novels, *A Single Pebble* (1956), which is laid in China, where Hersey was born and lived for ten years, and *The War Lover* (1959).

FARMINGTON

William Gillette, the playwright and actor, died in Hartford in 1937 and was buried in the Hooker family plot in a cemetery in Farmington, in northern Connecticut. Through his mother he was a direct descendant of Thomas Hooker, the founder of Hartford. **Archibald MacLeish,** the poet and playwright, often visited the Farmington home of his father-in-law, William Hitchcock, for extended periods in the 1920s. The house, at 27 Main Street, still stands and is privately owned. When MacLeish returned from France in 1928, he and his wife, Ada Hitchcock, settled on a farm here. His volume *New Found Land* (1930) appeared while MacLeish lived in Farmington.

GEORGETOWN

Anne Parrish, who wrote more than twenty novels during her long career, including *The Perennial Bachelor* (1925), *Loads of Love* (1932), and *A Clouded Star* (1948), was living in Georgetown at the time of her death on September 5, 1957, in Danbury Hospital. Parrish had moved in 1939 to Georgetown, where she established the home she called Quantness, on Peaceable Street.

A poem should be motionless in time
As the moon climbs

A poem should be equal to:
Not true

For all the history of grief
An empty doorway and a maple leaf

For love
The leaning grasses and two lights
above the sea—

A poem should not mean
But be

—Archibald MacLeish,
in "Ars Poetica"

Quantness, the Georgetown home of novelist Anne Parrish.

"I wish I could get a man to foot my bills. I'm sick and tired, cooking my own breakfast, sloshing through the rain at 8AM, working like a dog. For what? Independence? A lot of independence you have on a woman's wages. I'd chuck it like that for a decent, or an indecent, home."

—Clare Boothe, in *The Women*

GREENS FARM

Harold [Lincoln] Gray, the newspaper artist who created Little Orphan Annie, lived in this town south of Westport. **John Hersey,** the novelist, lived for a time on Turkey Road in Greens Farms.

GREENWICH

This New York City bedroom suburb *par excellence* has been home for many well-known writers. One of them was **Truman Capote,** the novelist, who lived in the Milbrook section of Greenwich for a time in his youth. He attended Greenwich High School but not long enough to be graduated. The school is currently being renovated for use as a town hall. The Bush-Holley House, once called the Holley House, on Strickland Road, was a gathering place for artists and writers early in this century. **Willa Cather** had a high-ceilinged room that overlooked Cos Cob harbor. She worked for *McClure's* magazine at the time, as did **Lincoln Steffens,** who also stayed in the Holley House. The house, now headquarters of the Greenwich Historical Society, is open to visitors.

Ring Lardner and his family rented a house on Otter Rock Drive, in the Belle Haven section of Greenwich, in time for the start of the 1919 school year. En route to their new home, Lardner's son, Ring Lardner, Jr., asked his father the travel question that would become classic through Lardner's story "The Young Immigrants," published in the *Saturday Evening Post* in January 1920: "'Are you lost daddy,' I asked tenderly. 'Shut up,' he explained." The Lardners lived here for only a year, moving in 1920 to a new home on Long Island.

Hendrick Van Loon, the journalist and author, died on March 11, 1944, at his home Nieuw Veere, on Cove Road, Old Greenwich, near Long Island Sound. The house is actually located on Van Loon Lane, which is a driveway and privately owned. Van Loon had ten writing projects in progress when he died. One was his autobiography, *Report to St. Peter,* published posthumously in 1947. Other works of the best-selling author were *The Story of Mankind* (1921), *The Story of the Bible* (1923), and *America* (1927). Funeral services for Van Loon were held at the First Congregational Church on March 14, 1944. Among the mourners were Mayor and Mrs. Fiorello La Guardia; President Franklin D. Roosevelt sent a message of condolence. Van Loon is buried in the First Congregational Church Cemetery, 108 Sound Beach Avenue, in Old Greenwich.

Clare Boothe [Luce], who grew up in the Old Sound Beach area of Greenwich and is best known for her play *The Women* (1936), in 1923 married her first husband, George Brokaw, at Christ Church, on East Street. She was divorced from Brokaw in 1929 and, in 1935, married **Henry Luce,** co-founder of *Time* and other magazines, in the First Congregational Church in Old Greenwich. They took a house at 1275 King Street in Greenwich. Their home was modestly called The House at the time. Here Boothe wrote the melodrama *Europe in the Spring* (1940). The House is now headquarters for the Avco Corporation. The face of the building remains the same, but additions have been made. Greenwich has yet another association with the life of Clare Boothe. When the playwright was traveling to New York City from Greenwich by train one day, she happened to sit next to a middle-aged man who looked vaguely familiar. He asked her questions about herself and her family, and then told her that he was her missing father. Understandably, she failed to ask all the questions she might have—and never saw him again.

Ernest Thompson Seton, the naturalist and writer of animal stories, first visited Greenwich as a guest at the Holley House, on Strickland Road. He lived later in Cos Cob and then on Round Hill, both in Greenwich, until he left for Santa Fe, New Mexico, in the late 1920s. The Bruce Museum, on Steamboat Road, has a collection of Seton's books as well as a display of his memorabilia. **Lincoln Steffens** bought a home, Little Point, in Riverside, across the harbor from Cos Cob, in the early 1900s. At the time, Steffens was employed as researcher by **Walter Lippman,** who also lived in the area. In 1915, Steffens rented the house to Owen D. Young, the industrialist, and in 1920 sold Little Point to Young.

The Bush-Holley House in Greenwich, where Willa Cather, Lincoln Steffens, and Ernest Thompson Seton stayed.

GUILFORD

Harriet Beecher [later **Stowe**] was sent to Guilford, on Route 95, after the death of her mother in 1815, to live with her aunt and her grandmother on her grandmother's farm, Nut Plains, on Long Island Sound. It was at the farm that Harriet Beecher, then four years old, first met blacks. Dinah and Harry were her grandmother's indentured servants, and because of them she faced one of the first dilemmas of her young life: if, as her father had told her, all children were equal, why were Dinah and Harry, who were also children, treated less well than herself, or even white servants? Harriet Beecher lived here for a year before returning to her native Litchfield in 1816.

Fitz-Greene Halleck, the poet and satirist, was born in a house at the corner of Whitfield and Water streets on July 8, 1790. The address is now 87 Whitfield Street, but the house is no longer standing. Halleck wrote the satiric *Croaker Papers* (1819) with Joseph Rodman Drake. Along with other members of the Knickerbocker Group, Halleck was an admirer of, and imitated the work of, Sir Walter Scott and Lord Byron. His *Marco Bozzaris* (1825), a stirring poem about the Greek fight for independence, owed much to Byron. Halleck died at 15 Water Street in Guilford. The house is still a private residence, and the number has been changed to 25 Water Street. Halleck is buried in the Alderbrook Cemetery, on Boston Street.

Strike—for your altars and your fires;
Strike—for the green graves of your sires;
God—and your native land!

—Fitz-Greene Halleck,
in *Marco Bozzaris*

HADLYME

William Gillette in 1913—at the height of his fame as creator and star actor in the play *Sherlock Holmes*—planned to build a home on Long Island. While cruising up the Connecticut River, however, he so admired the countryside around Essex that he soon decided that he would build his castle at Hadlyme. He bought 122 acres, including river frontage of three-quarters of a mile, and had transported from Long Island all the iron and steel he had gathered for his castle. His home, Seventh Sister, was built atop a hill commanding a view of the river and surrounding countryside. Construction began in 1914 and continued for five years. Southern white oak and native granite were used. Gillette himself was responsible for both exterior and interior design, including granite walls three feet thick, thick oaken doors with intricate locking devices—no two were the same—and some sliding furniture on metal tracks. A system of mirrors enabled him to keep an eye on the front door from his bedroom. Gillette entertained his guests by taking them for rides on his private railroad. The track, extending a little over three miles, began in New York City—all stations were Gillette's own designations—at Grand Central and passed through 125th Street Station. In 1943, in accordance with Gillette's wishes, the castle was acquired by the State Park and Forest Commission. Gillette Castle is open daily from Memorial Day to November 1. Admission is charged for visitors over twelve.

HAMDEN

Donald Hall, the poet and editor, lived at Coram and Winet streets in this town near New Haven. He also lived on Whitney Avenue, but it was at 160 Ardmore Street that he began writing poetry. The house on Ardmore Street still stands but is private. **Thornton Wilder,** the playwright and novelist, in 1929 built himself a brown-shingled house at 50 Deepwood Drive, where he lived for some years.

HARTFORD

Hartford, since its founding in 1636, has had its share of literary associations. Many talented writers were born and worked here and, it must be remembered, Hartford was an important center of intellectual life until its eclipse by Boston and New York City. In the eighteenth century, Hartford was the meeting place for a brilliant group known as the Connecticut Wits and as the Hartford Wits. In the latter half of the nineteenth century, a remarkable group of writers, artists, and intellectuals built their homes in Nook Farm, a section of the city. In the following pages the Connecticut Wits will appear first, then Nook Farm, and finally the Hartford writers not associated with either group.

The Connecticut Wits, all graduates of Yale University, first banded together with the intention of using satirical verse to further their political ideas. They met at the Black Horse Tavern in Hartford, address unknown, and among the Wits were a college president, a statesman, an editor, and a physician, all of them versifiers and writers. **Joel Barlow,** one of the leading members of the group, arrived here in 1782 and spent five years in Hartford. He published a newspaper called the *American Mercury* and worked on a poem that he later developed into *The Columbiad* (1807). This ponderous epic was popular with the citizens of the new nation. Barlow's other popular poem was "Hasty Pudding," a portrayal of New England home life. **Timothy Dwight,** who was to be president of Yale University for twenty-two years, was another member of the Wits. Dwight's best-known work was the song "Columbia," but his literary output included many volumes of sermons, travel sketches, and philosophy.

Colonel David Humphreys, an aide-de-camp to General Washington, was also a member. His many works included *Essay on the Life of the Honorable Major-General Israel Putnam* (1788), the earliest biography of the Revolutionary War hero. **Elihu Hubbard Smith,** a physician and poet as well as a Wit, compiled *American Poems, Original and Selected* (1793), perhaps the first anthology of poetry published in America. Another leader of the group was **John Trumbull,** whose best-known work was *The Progress of Dulness* (1772–73). Trumbull became known as the Father of American Burlesque for his mock-epic "M'Fingal," an attack on the Tories. *The Anarchiad: A Poem on the Restoration of Chaos and Substantial Night* (1786–87), *The Echo* (1791–1805), and *The Political Greenhouse for the Year 1798* (1799) were all written and published jointly by the Connecticut Wits.

Nook Farm became prominent in the second half of the nineteenth century. This area of literary fame consisted of about 140 acres on the western side of Hartford. Only a small portion of the tract is left as it was in its heyday, but the houses that still stand are rich in literary memories, and a trip to the intersection of Forest Street and Farmington Avenue is well worth the literary visitor's time.

William Gillette, the playwright and actor, was born on July 24, 1855, in a farmhouse on Forest Street in Nook Farm. When he was two years old, his parents built the elegant home on Forest Street where Gillette would spend his childhood. As a boy, he was interested in all aspects of theater. He attended Hartford Public High School and there entered an elocution contest. He lost because he gave his speech in the naturalistic way suggested by his father, rather than in the

I sing the sweets I know, the charms
I feel,
My morning incense, and my evening
meal,
The sweets of Hasty-Pudding.

—Joel Barlow,
in "Hasty Pudding"

What has posterity done for us,
That we, lest they their rights should
lose,
Should trust our necks to gripe of
noose?

—John Trumbull,
in "M'Fingal"

Harriet Beecher Stowe, photographed in 1882. Photograph taken in about 1865–70 of Oakholm, first Hartford home of Harriet Beecher Stowe. The man sitting on the veranda, reading a newspaper, is her husband, Calvin Stowe. The identity of the woman sitting with him is not known.

usual declamatory manner. Between 1888 and 1893, Gillette wrote plays at Nook Farm, but none of them received the recognition won by his best-known play, a dramatization of Conan Doyle's *Sherlock Holmes* (1899). He acted in that play almost until his death on April 29, 1937, in Hartford Hospital.

Harriet Beecher [Stowe] first came to Hartford in the fall of 1824 to attend the Hartford Female Seminary, which had just been founded by Catharine Beecher, Harriet's eldest sister and surrogate mother. Harriet remained at the seminary until 1832, both as a student and teacher. At the age of fourteen she wrote the long poem "A Description of the Twelve Months of the Year." Her unfinished blank verse tragedy *Cleon*, written soon after, is considered by her biographer to be a more fitting example of the juvenilia of the author who would, in 1852, burst on the literary scene with *Uncle Tom's Cabin*. With her friend, Georgiana May, Harriet explored the environs of Hartford, and they often picnicked in an oak grove along the Park River.

Harriet Beecher Stowe returned to Hartford in 1864 with her husband, Calvin Stowe, a clergy-

Mark Twain (*left*) and his friend Charles Dudley Warner (*right*), a frequent visitor at Twain's home in Hartford.

man and author of biblical interpretations. Incorporating many of Mrs. Stowe's suggestions, the architect Octavius Jordan designed Oakholm for the Stowes. In *The American Woman's Home* (1869), Mrs. Stowe discussed many of her ideas about home design and interior decoration, and her influence could be seen in the homes of other Nook Farm residents, particularly in their inclusion of conservatories, a favorite element of Mrs. Stowe's. The Stowes sold Oakholm in 1870, and for a while traveled and lived at their home in Florida. In 1873 they bought a home, now the Harriet Beecher Stowe House, on Farmington Avenue at Forest Street, where they lived from then on. After Mr. Stowe died in 1886, Mrs. Stowe was to live there alone until her death in 1896.

Her last ten years were quiet ones for Harriet Beecher Stowe. On her wanderings about Nook Farm, she was known to her friends and neighbors for her habit of picking flowers—roots and all—from other people's gardens. During her Hartford years, Mrs. Stowe wrote and published *Oldtown Folks* (1869), *Lady Byron Vindicated* (1870)—it created a scandal by charging that Lord Byron had had incestuous relations with his sister—*Sam Lawson's Oldtown Fireside Stories* (1872), *Poganuc People* (1878), and *Footsteps of the Master* (1877), a collection of her religious verse. Mrs. Stowe's home has been faithfully restored and is open for guided tours, as is the adjacent home, which belonged to Mark Twain. Current information about hours and cost of admission can be obtained from the Nook Farm Visitors Center, 77 Forest Street, Hartford, Connecticut, 06105.

In 1868 **Mark Twain** arranged with Elisha Bliss, Jr., of the American Publishing Company, to have Twain's *The Innocents Abroad* (1869) published. Twain was delighted to report to his family that he would get a five percent royalty on the subscription price, a very favorable agreement. In January 1868 he stayed for a few days with John and Isabella Beecher Hooker, in their Nook Farm home at the corner of Forest and Hawthorne streets. In July he stayed with Bliss, in his home at 821 Asylum Avenue, and read proofs in one of the upper rooms. Of his hosts, he wrote to his fiancée, Olivia Langdon: "Puritans are mighty straight-laced and they won't let me smoke in the parlor, but the Almighty don't make any better people." His opinion of the city of Hartford was equally favorable. So in 1871, the year after his marriage, Twain came to Hartford with his wife and their young son, Langdon, to look for a house. They first rented the Hooker home, at the corner of Forest and Hawthorne streets, where he had once been a guest and where young Langdon died, on July 2, 1872. At dinner one night in Nook Farm, **Twain** and a friend, **Charles Dudley Warner,** were criticizing novels their wives had been discussing, and the two women suggested that their husbands write a better one. The two men decided to do just that and, between February and April of 1873, came up with *The Gilded Age* (1873), the novel about corruption after the Civil War that gave its name to post-Civil War America. The novel was dramatized later, and William Gillette played in it when it was produced at the Hartford Opera House in May of 1875.

Soon after renting the Hooker house, Twain and his wife bought, with $31,000 of Langdon family money, a five-acre tract on Farmington Avenue and began planning their ideal home. Twain engaged an architect, who drew plans for a mansion that would soon amaze and delight the rest of Hartford. While the Twains were traveling in Europe in 1874, the house was built at a cost of $70,000. On their return, the family was forced to live upstairs because work was still going on downstairs. Twain's imagination showed itself throughout the house, particularly in the chimneys curving around a window that enabled Twain to watch snowflakes falling into the fireplace. The library boasted an old carved mantel salvaged from a ruined Scottish castle. And, of course, there was a conservatory of the design laid down by Harriet Beecher Stowe. Although Twain had intended to use one of the second-floor rooms as his study, he did his writing in the billiard room

In Paris we often saw in shop windows the sign "English Spoken Here," just as one sees in the windows at home the sign "Ici on parle français." *We always invaded these places at once—and invariably received the information, framed in faultless French, that the clerk who did the English for the establishment had just gone to dinner and would be back in an hour—would Monsieur buy something? We wondered why those parties happened to take their dinners at such erratic and extraordinary hours, for we never called when an exemplary Christian would be in the least likely to be abroad on such an errand. The truth was, it was a base fraud—a snare to trap the unwary—chaff to catch fledglings with. They had no English-murdering clerk.*

—Mark Twain,
in *The Innocents Abroad*

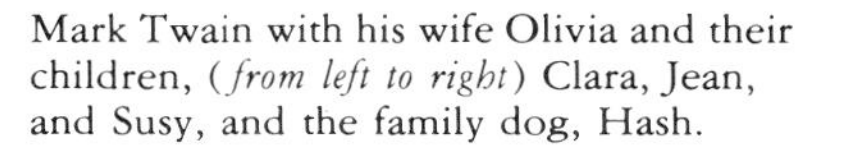

Mark Twain with his wife Olivia and their children, (*from left to right*) Clara, Jean, and Susy, and the family dog, Hash.

on the third floor. The exterior architecture of the house has been variously described, but the total price is indisputable. With the furnishings and the land, the new home cost $122,000, an impressive figure for that day.

Accounts of Twain and his family during their Hartford years picture them as happy and united, with the three daughters enjoying the stories their father made up for them. There was friendly intercourse among all the neighbors of Nook Farm and frequent literary visitors to the Twain home, among them **George W. Cable, William Dean Howells,** and **Bret Harte**—Twain collaborated with Harte and with Howells on various works. Amateur theatricals were performed in the conservatory, and the house had the further distinction

Mark Twain's home on Farmington Avenue, in Hartford, designed largely by Twain himself.

of being the first to have its own telephone: Twain strung a wire from his house to his office at the *Courant,* the distinguished Hartford newspaper.

One of Twain's best friends in Hartford was Joseph Twichell, pastor of the Asylum Hill Congregational Church. Twichell went along when the Twains traveled in Europe in 1878–79, and was the model for Harris in *A Tramp Abroad* (1880). Twain wrote *A Connecticut Yankee in King Arthur's Court* (1889) at Twichell's house. The book is about an arms factory superintendent in Hartford who goes back to King Arthur's time and amazes the people with his mechanical and meteorological knowledge. Other works written by Twain during his years in Hartford include *Old Times on the Mississippi* (1875), *The Adventures of Tom Sawyer* (1876), *The Prince and the Pauper* (1882), and his masterpiece, *The Adventures of Huckleberry Finn* (1884). In 1891, Twain's finances were in chaos because of his family's extravagance and his heavy investment in the Paige typesetter. The family went to live in Europe with the intention of economizing and never again lived in Hartford, although Twain returned for a short time in 1893. In 1900 Twain came back once more to serve as a pallbearer for his friend and neighbor Charles Dudley Warner. Only then did he decide to see the house on Farmington Avenue again; he sold it in May of 1903.

Charles Dudley Warner came to Hartford in 1860 to become associate editor of the Hartford *Evening Press.* He and his wife Susan first lived in a small cottage in Nook Farm. They moved in 1866 to a house on the south side of Hawthorne Street. When the *Evening Press* merged with the Hartford *Courant* in 1867, Warner became editor of the new *Courant.* During the 1860s, he wrote humorous articles about his gardening experiences in Nook Farm, and these were later collected and published as *My Summer in a Garden* (1870). Warner also edited the thirty-volume *The Library of the World's Literature* (1896–97) and wrote *The Golden House* (1895), *A Little Journey in the World* (1889), and *That Fortune* (1899). In 1884, the Warners moved to their last home on Forest Street, which had belonged to Warner's brother. Like their neighbors in Nook Farm, the Warners entertained celebrated visitors, including **Helen Keller, William Gillette, Thomas Bailey Aldrich,** and **William Dean Howells.** In 1884, the neighborhood children performed *The Prince and the Pauper* as a Christmas surprise for Mark Twain. The first production was held in the Warner house, and the cast included Warner and Twain children.

Oakholm, Harriet Beecher Stowe's first home in Nook Farm, was torn down in 1905. By the 1950s, most of the other homes on Forest Street had also been torn down, and in 1963 a new high school was erected there. Mark Twain's Victorian mansion and Mrs. Stowe's last house, both on the corner of Farmington Avenue and Forest Street, were saved and are open to the public year-round.

But the literary associations of Hartford do not end with Nook Farm. **Edward Albee,** the playwright, entered Trinity College here in 1946 but within two years was asked to leave after he refused to attend math classes and chapel. **Rose Terry Cooke,** the poet, was graduated from the Hartford Female Seminary in 1843, and **Louise Field Cooper,** the short-story writer, was born in Hartford on March 8, 1905.

Sinclair Lewis, when he was a Yale undergraduate, saw Hartford for the first time while hitchhiking. Lewis declared it "the prettiest town we saw," so it is not surprising that he returned in 1922 with his first wife, Grace Livingstone Hegger, and rented a house at 25 Belknap Road, where they entertained frequently. One of their houseguests was the English novelist **Hugh Walpole.** The house had a minstrel gallery, a dry fountain, and a guest room with a shrine, but these were not enough to keep the Lewises happy in Hartford for long. On September 13, 1922, Lewis wrote disparagingly of the city in a letter to his friend H. L. Mencken, and there were signs of domestic discord. On September 16, for example, the Hartford *Daily News* reported, inaccurately, that Mrs. Lewis had deleted portions of a typescript of Lewis's *Babbitt* (1922)! Late that year, the Lewises moved from Hartford.

May Sarton, the poet and novelist, worked for the Associated Actors, Inc. at the Wadsworth Atheneum about 1936. The Atheneum, a two-story granite structure featuring three-story towers, was built to house a fine-arts gallery. It is still located at 600 Main Street and is open to the public for an admission charge. **Odell Shepard** taught English at Trinity College, on Summit Street, from 1917 to 1946. He was the author of *The Harvest of a Quiet Eye* (1927), a description of a walking trip through northern Connecticut; and *Pedlar's Progress, The Life of Bronson Alcott* (1937), for which Shepard won a Pulitzer Prize for history. Shepard served as lieutenant-governor of Connecticut in 1940 and 1941.

Lydia Huntley Sigourney wrote anonymously for newspapers here and, in her time, was very popular for the sentimental and pious verses she wrote on the deaths of prominent citizens and small children, whom she always regarded as potential angels. She was called the Sweet Singer of Hartford. Her works, numbering about sixty books, included *Moral Pieces in Verse and Prose* (1815) and *Letters of Life* (1866), her last book and an account of her life. Sigourney, a pioneer in the cause of higher education for women, is remembered in an oil portrait in the Stowe House in Nook Farm. **Winchell Smith,** the playwright and actor, was born here on April 5, 1871. He was known best for his sentimental comedy *Lightnin'* (1918), written with Frank Bacon, which had a very long run. **Edmund Clarence Stedman** was born in Hartford on October 8, 1833, and lived here until he was two. Stedman, a Wall Street broker as well as a poet, wrote his most famous poem, "Pan in Wall Street," in 1867. He edited *Poets in America* (1885) and *An American Anthology* (1900).

Wallace Stevens lived and worked in Hartford for most of his life. In 1916 he began working for the Hartford Accident and Indemnity Company and was a vice president at the time he retired. Something of an outsider in Hartford, Stevens was known for walking in all weathers to and from work at his office at 690 Asylum Street. Walking to work was not considered unusual, but Stevens always refused an offer of a ride. It was while walking home with a student one evening that Stevens, according to Hugh Kenner in *A Homemade World* (1975), spoke of his recent poem "Notes Toward a Supreme Fiction" (1942): "I said that I thought that we had reached a point

> *Most of the readers of* The Times, *doubtless, have had at least an external view of the structure, which already has acquired something beyond a local fame; and such persons, we think, will agree with us in the opinion that it is one of the oddest looking buildings in the State ever designed for a dwelling, if not in the whole country.*
>
> —Hartford *Daily Times,* March 23, 1874

> *There was constant running in and out of friendly houses where the lively hosts and guests called one another by their Christian names or nicknames, and no such vain ceremony as knocking or ringing at doors. Clemens was then building that stately mansion in which he satisfied his love of magnificence as if it had been another sealskin coat, and he was at the crest of the prosperity which enabled him to humor every whim or extravagance.*
>
> —William Dean Howells, recalling Nook Farm

> *I am an American. I was born and reared in Hartford, in the State of Connecticut—anyway, just over the river, in the country. So I am a Yankee of the Yankees—and practical; yes, and nearly barren of sentiment, I suppose—or poetry, in other words. My father was a blacksmith, my uncle was a horse doctor, and I was both, along at first.*
>
> —Mark Twain, in *A Connecticut Yankee in King Arthur's Court*

Though Mr. Lewis himself was born (in 1885) in a Minnesota prairie hamlet, where his father was a quite typical country physician, that father and his ancestors for eight or nine generations were born in Connecticut. . . . It was natural, then, that he should have settled in Connecticut, being weary of travel and of what he himself once called . . . "the chronic wanderer's discovery that he is everywhere such an Outsider that no one will listen to him even when he kicks about the taxes and the beer."

—Sinclair Lewis,
in "The Death of Arrowsmith"

Georges Simenon in the yard of his Lakeville home, Shadow Rock Farm, in 1953.

at which we could no longer really believe in anything unless we recognized it was a fiction." And this was the doctrine of Wallace Stevens, poet and insurance man. His works include *Harmonium* (1923, reprinted with additions in 1931), the volume *Notes Toward a Supreme Fiction* (1942), and *The Necessary Angel* (1951). Stevens died on August 2, 1955, at 118 Westerly Terrace. He is buried in Cedar Hill Cemetery, 453 Fairfield Avenue.

. . . The steeple at Farmington
Stands glistening and Haddam shines and sways.

It is the third commonness with light and air,
A curriculum, a vigor, a local abstraction . . .
Call it, once more, a river, an unnamed flowing,

Space-filled, reflecting the seasons, the folk-lore
Of each of the senses; call it, again and again,
The river that flows nowhere, like a sea.

—Wallace Stevens,
in "The River of Rivers in Connecticut"

Noah Webster, the great American lexicographer, was admitted to the bar in Hartford in 1781. He did not practice until 1789, and then only until 1793, when he returned to New York City. **John Greenleaf Whittier,** in 1830, edited the *New England Weekly Review* here and, in 1831, published his first book of poetry, *The Legends of New England.*

JEWETT CITY

Samuel Adams Hammett, a southwestern-frontier humorist, was born on February 4, 1816, in this town north of Norwich. His collections of stories included *A Stray Yankee in Texas* (1853) and *Piney Woods Tavern, or Sam Slick in Texas* (1858). Sam Slick, eventually a stock character in American writing, was based on a Thomas Chandler Haliburton (1796–1865) character of the same name. Haliburton was a Canadian humorist.

KENSINGTON

James Gates Percival was born in Kensington, south of Hartford, on September 15, 1795. Because he knew languages, he was a valuable assistant to Noah Webster, the lexicographer, but Percival achieved his reputation as a poet. His "Prometheus," published in his *Poems* (1821), was said by some of his contemporaries to be the equal of Byron's *Childe Harold.*

KENT

James Gould Cozzens in 1922 was graduated from Kent School in this town in northwestern Connecticut. Cozzens's novel *The Last Adam* (1933), the story of a country doctor, was set in Connecticut. **Robert Lewis Taylor,** the newspaperman and novelist, lived from 1967 to 1972 in a house on Spectacle Mountain in Kent.

LAKEVILLE

This town in northwestern Connecticut has several literary associations. **Thomas B. Costain,** the novelist and historian, had a home here in 1953. **Sinclair Lewis,** while summering in a rented house here in 1941, ten years before his death, wrote "The Death of Arrowsmith," in the form of an obituary for himself. He represented the piece as having been written in 1971 and portrayed himself as having been settled for many years in a "small country place in northwestern Connecticut." "The Death of Arrowsmith" was published in *Coronet* magazine in July 1941.

Archibald Macleish, the poet and playwright, attended Hotchkiss School here before World War I and "hated" it. **John Hersey,** the novelist, also attended Hotchkiss, between 1928 and 1932. **Enid LaMonte Meadowcraft [Wright],** who wrote stories for young people, such as *The Adventures of Peter Whiffen* (1936) and *When Nantucket Men Went Whaling* (1966), lived in Lakeville. From September 1950 until January of 1955, Shadow Rock Farm was home to **Georges Sim,** who writes under the name **Georges Simenon.** Simenon, the Belgian mystery novelist, is internationally known as the creator of the imperturbable and implacable Inspector Maigret. Simenon's literary output has always been prodigious, and while living here he wrote, among other works, *Les mémoires de Maigret* (1951), *Maigret et l'homme du banc* (1953), and *Maigret et le corps sans tête* (1955). Simenon's books have been translated into English and many other languages.

LEDYARD

Samuel Seabury, the clergyman and pamphleteer who wrote in support of British rule, was born on November 30, 1729, in what is now the town of Ledyard, near Block Island Sound. In Seabury's time Ledyard was a section of Groton. His birthplace is not standing, but a marker in Ledyard Center commemorates his birth. Seabury wrote *Free Thoughts on the Proceedings of the Continental Congress* (1774), which was published in 1930, along with other pamphlets by Seabury, as *Letters of a Westchester Farmer 1774–1775.* In 1775, Seabury was seized by a mob and imprisoned. In 1783, he became the first Protestant Episcopal bishop in the U.S. His later works included *Discourses on Several Important Subjects* (three volumes, 1791–98).

LITCHFIELD

Harriet Beecher Stowe was born here on June 14, 1811, when Litchfield, in western Connecticut, was the state's fourth largest town. She was the seventh child of Roxanna Foote and Lyman Beecher, the pastor of the Congregational church here. He was a just and thoughtful man, known to his parishioners for his strong Calvinistic ideals. His church no longer stands, but the site is marked. Beecher brought his children up to believe in the betterment of mankind, equality, and social justice. In 1816, on returning from her grandmother Foote's house in Guilford, Harriet was enrolled in "Ma'am" Kilbourne's School on West Street. At the age of ten, she began attending John Brace's Litchfield Academy, where she excelled in her studies. Her essay "Can the Immortality of the Soul Be Proved by the Light of Nature?" written when she was twelve years old, won first prize and was read aloud at the graduation exercises, to the great delight and pride of her father. Stowe's *Poganuc People* (1878) was in-

Over the shoulders and slopes of the dune
I saw the white daisies go down to the sea,
A host in the sunshine, an army in June,
The people God sends us to set our heart free.

—Bliss Carman,
in "Daisies"

Sunshine House, where Bliss Carman stayed in New Canaan, sketched by Keith Ward.

spired by Litchfield Village and its Bantam River. Her birthplace is marked, and there is a bas-relief of Harriet Beecher Stowe and her brother, Henry Ward Beecher, the abolitionist preacher.

Elihu Hubbard Smith, the playwright and poet, was born here on North Street on September 4, 1771. The house still stands and is private.

MADISON

Leane Zugsmith lived for more than twenty years in Madison, east of New Haven, at 88 East Wharf Road, and died there on October 13, 1969. Her two-story house, with a small vegetable and flower garden, is now privately owned. No burial was held for Zugsmith because she had willed her body to the Yale Medical School. Two of her novels were *The Reckoning* (1934) and *A Time to Remember* (1936), about the labor movement and unionization in New York City.

MERIDEN

Philip Dunning was born on December 11, 1890, in Meriden, northeast of New Haven. His play *Broadway* (1926), written with George Abbott, was chosen as one of the ten best plays of the 1926 season. He also wrote *The Understudy* (1927). **George Sklar** was also born here, on June 1, 1908. In the 1930s, he studied with George Baker Pierce at the Yale Drama School. Among his works were the play *Stevedore* (1934), written with Paul Peters, and the novel *Two Worlds of Johnny Truro* (1947).

MIDDLETOWN

Richard Alsop, one of the Connecticut Wits, was born here on January 23, 1761, in a house at the northeast corner of Main and Parsonage (now College) streets. In 1839 a new owner moved it two doors east. In the 1950s the house was demolished.

Middletown is the home of Wesleyan University, which has attracted literary figures over the years. In the 1850s, **George William Curtis** spoke at Wesleyan on *The Duty of the American Scholar to Politics and the Times,* published in 1856. **Paul Horgan** served first as a fellow at Wesleyan, from 1960 to 1967, and later became permanent writer-in-residence. While living in Middletown, he wrote the novels *Mountain Standard Time* (1962), *Songs After Lincoln* (1965), *Whitewater* (1970), and *The Thin Mountain Air* (1977), among other works. The Conversation Club here has been a favorite haunt of the novelist. **Arthur Kopit,** the playwright, while a visiting fellow at the Center for Humanities, lived at 329 Washington Terrace. **Lewis Mumford,** the writer and architectural historian, lived at 35 Home Avenue while he worked at the Center for Advanced Studies at Wesleyan in the fall semester of 1963. **[Charles] Wilbert Snow** taught at Wesleyan in 1921. His books of poetry included *Maine Coast* (1923), *Down East* (1923), and *Maine Tides* (1940). **Richard Wilbur,** the poet, has taught at Wesleyan since 1955, winning a Pulitzer Prize for his *Things of This World* (1957).

NEW CANAAN

[William] Bliss Carman lived in New Canaan from 1908 until his death on June 8, 1929. During this period the poet produced *Echoes from Vagabondia* (1912), *April Airs: A Book of New England Lyrics* (1916), and *Wild Garden* (1929). Carman visited friends at 289 New Norwalk Road, known then as Sunshine House. He also took meals there and used their studio. The house now

It was eight-twenty-five in the morning when Maigret rose from the breakfast-table still drinking his last cup of coffee. Though it was only November, the lights were on. At the window, Madame Maigret was peering through the fog at the passers-by, who were hurrying to work, shoulders hunched and hands in pockets.

"You'd better put on your heavy overcoat," she said.

For it was by watching people in the street that she decided what the weather was like. This morning, everyone was walking fast and many wore scarves; they had a particular way of stamping along the pavement to warm their feet, and she had noticed several blowing their noses.

"I'll go and get it for you."

—Georges Simenon,
in *Maigret's Mistake*

Merrimack River was no longer the highway that it had been for Indians. The men using it now as a route north and south were trappers and hunters, and only occasionally settlers. Connecticut River carried the settlers. As for east and west traffic, there wasn't much anyway.

—LeGrand Cannon, Jr., in *Look to the Mountain*

serves as the clubhouse for the Canaan Close condominium complex. **Edward Hoagland,** the novelist, grew up at 340 Oenoke Ridge, in New Canaan, attending the New Canaan Country School and Ponus Ridge School through the ninth grade. The house on Oenoke Ridge and the Country School are still standing.

Maxwell Perkins had a home in the center of New Canaan, and it was there that the editor introduced **Thomas Wolfe** to **Van Wyck Brooks.** Wolfe was an unknown talent at the time. In 1936, while attending the wedding here of one of Perkins's daughters, Wolfe interrupted the ceremony with a characteristically loud comment about the perspiration in his hatband. **Sloan Wilson,** while living in his home at 216 Whiteoak Shade Road, wrote stories for *The New Yorker.* He later moved to a house at 8 (now 113) Harrison Avenue. The house on Whiteoak Shade Road, still standing, is near the center of town. **Elinor [Hoyt] Wylie** and her husband, **William Rose Benét,** lived at 151 Park Street from 1925 to 1927. The house was later torn down to make room for an apartment house. Wylie is thought to have written the novel *The Venetian Glass Nephew* (1925) while living here, and Benét's selected poems, *Man Possessed* (1927), was published during this time.

I shall lie hidden in a hut
In the middle of an alder wood,
With the back door blind and bolted shut,
And the front door locked for good.

I shall lie folded like a saint,
Lapped in a scented linen sheet,
On a bedstead striped with bright-blue paint,
Narrow and cold and neat.

The midnight will be glassy black
Behind the panes, with wind about
To set his mouth against a crack
And blow the candle out.

—Elinor Wylie, "Prophecy"

NEW HAVEN

For literary purposes New Haven is two cities. First there is the city that goes about its daily business of making a living, bearing children, and burying the dead. The second New Haven is Yale University, which educates the young and, quite often, sends its graduates on to literary careers. This second New Haven employs authors and poets who take time from their literary work to give instruction in their craft as well as teachers who, while fulfilling their academic responsibilities, also create works of literature. This second New Haven maintains libraries to meet its instructional needs and to provide material for literary scholarship. In the following pages we first encounter writers native to New Haven or residents of the city, then Yale students and teachers who have made names in literature, and finally the great Yale libraries.

Seven authors are known to have been born in New Haven. **LeGrand Cannon, Jr.,** the novelist who wrote *A Mighty Fortress* (1937), *The Kents* (1938), and *Look to the Mountain* (1942)—all about New Hampshire—was born in New Haven on December 1, 1899. **Edward Ellsberg,** author of works of fiction and nonfiction that derived from his career in the U.S. Navy, was born here on November 21, 1891. His books included *On the Bottom* (1929), *Under the Red Sea* (1946), and *The Far Shore* (1960). **Donald Hall** was born here on September 20, 1928. One of his books of poems is *Exiles and Marriages* (1955).

Mary Holley, the biographer and author, was born here on October 30, 1784, but most of her work concerned Texas. **Harriet Lothrop,** pen name **Margaret Sidney,** author of the beloved *Five Little Peppers and How They Grew* (1881), was born here on June 22, 1844. **Theodore Winthrop,** the poet and novelist whose works all were published posthumously, was born in New Haven on September 22, 1828, and lies buried in Grove Street Cemetery. Winthrop, who was killed in a Civil War battle in 1861, was the author of *Life in the Open Air* (1863), *The Life and Poems of Theodore Winthrop* (1884), and *Mr. Waddy's Return* (1904). **Eleanor Estes,** who lived in nearby West Haven, in her youth often visited Savin Rock, at the mouth of the harbor in New Haven. In her "Moffat" series, she refers to Savin Rock as Plum Beach. Between 1932 and 1940, Estes served as children's librarian at the New Haven Free Public Library, 133 Elm Street. **Percy Marks,** the novelist whose first published work, *The Plastic Age* (1924), achieved a success far beyond that of any of his many later novels, died at the Grace–New Haven Hospital on December 27, 1956.

Perhaps the most widely read-about Yale undergraduate of all time is Frank Merriwell, the wholesome college athlete. His exploits in all sports were followed avidly by American boys steeped in the novels of **Burt L. Standish,** pen name of **William Gilbert Patten,** who wrote over two hundred Frank Merriwell novels in all, beginning in 1896 and ending in 1941, when Frank was a grown man, yet dauntless as ever. But Yale, since its beginnings in the eighteenth century, has also been known for the great number of literary men—and now women—who studied there. **Louis Auchincloss,** who combines successful careers in law and literature, attended Yale from 1935 to 1939, when he left without taking an undergraduate degree. His decision to leave was based on the failure of two of his novels. Scribner's had rejected the first; he himself had deposited the second in the wastebasket. Auchincloss's novels include *Sybil* (1952), *Portrait in Brownstone* (1962), *The Embezzler* (1966), and *A World of Profit* (1968). **Joel Barlow,** the poet and satirist, was graduated from Yale in 1778. **Philip Barry** was a senior at Yale in 1919 when the Yale Dramatic Club produced his only one-act play, a work that has never been published. Two of his best-known plays are *The Animal Kingdom* (1932) and *The Philadelphia Story* (1939).

And now Frank Merriwell, though naturally a friendly, peace-loving person, felt again the strange thrill of eagerness, almost of exaltation, that invariably came upon him when danger threatened or battle impended. Never one to seek trouble, he likewise was one who never turned his back upon it when it came to him. No cowering pacifist who would not fight under any circumstances, was he. In his veins the red blood of self-respecting manhood ran full and strong.

—Gilbert Patten, in *Mr. Frank Merriwell*

Yale was alma mater for the two Benét brothers. **Stephen Vincent Benét** was graduated from Yale in 1919. As a student he lived at Pierson Hall and wrote two books of poetry, published as *Five Men and Pompey* (1915) and *Young Adventure* (1918). Benét later won a Pulitzer Prize for *John Brown's Body* (1928), his long narrative poem about the Civil War. Stephen's older brother,

I have fallen in love with American names,
The sharp names that never get fat,
The snakeskin-titles of mining-claims,
The plumed war-bonnet of Medicine Hat,
Tucson and Deadwood and Lost Mule Flat.

. . .

I shall not rest quiet in Montparnasse.
I shall not lie easy at Winchelsea,
You may bury my body in Sussex grass,
You may bury my tongue at Champmedy.
I shall not be there. I shall rise and pass.
Bury my heart at Wounded Knee.

—Stephen Vincent Benét,
in "American Names"

William Rose Benét, was graduated from Yale in 1907. He too won a Pulitzer Prize, for *The Dust Which Is God* (1941). Other books of his were *Merchants from Cathay* (1913), *The Falconer of God* (1914), *The Great White Wall* (1916), and *Day of Deliverance* (1944).

In 1802, **James Fenimore Cooper,** then thirteen years old, was sent to Yale, where he studied Latin, Hebrew, and Greek. His dismissal from Yale proved no handicap to success in literary life: Cooper went on to become famous as the creator of Natty Bumppo in *The Last of the Mohicans* (1826) and the other novels of the Leather-Stocking Tales, beginning with *The Pioneers* (1823) and ending with *The Deerslayer* (1841). **Clarence Day,** who attended Yale in the late nineteenth century, is known best for his autobiographical *Life With Father* (1935) and *Life With Mother* (1937). *Life With Father* was dramatized in 1939 by Howard Lindsay and Russel Crouse and ran on Broadway for years. Clarence Day was instrumental in the establishment of the Yale Series of Younger Poets.

Timothy Dwight early proved his brilliance by entering Yale in 1765, when he was thirteen, and graduating with honors in 1769. He went on to serve as a tutor from 1771 to 1777, and was president of Yale from 1795 to 1817. His five-volume work *Theology, Explained and Defended* (1818–19) is the series of 173 sermons he delivered while he was at Yale. Dwight died on January 11, 1817, at 380 Farnum Hall on College Street, and was buried at the Grove Street Cemetery. No less precocious than Dwight was **Jonathan Edwards,** the clergyman and mystic, who entered Yale in 1716, before he was thirteen, and was graduated in 1720. Perhaps his greatest work was *Freedom of Will* (1754), which won him prominence as a philosopher.

Waldo Frank, the novelist and critic, received a B.A. and M.A. from Yale. In 1911, during his undergraduate senior year, he was drama critic for the New Haven *Courier-Journal.* **John Hersey,** graduated from Yale in 1936, became secretary to Sinclair Lewis for a time and, later, correspondent for *Time* magazine. After becoming established as a novelist, he returned to Yale as master of Pierson College and lived at 231 Park Street, the Pierson master's house. While there he wrote *Under the Eye of the Storm* (1967) and *The Algiers Motel Incident* (1968). New Haven appears in his novels *Too Far to Walk* (1966), *My Petition for More Space* (1974), and *The Walnut Door* (1977).

While an undergraduate at Yale, **Owen Johnson** was chairman for a year of the *Yale Literary Magazine.* He was graduated in 1901 and a decade later scandalized Yale students with *Stover at Yale* (1911), his novel poking fun at senior societies and college undergraduates in general. Johnson's Lawrenceville stories about the prodigious Hickey, the Tennessee Shad, *et al.* made him a favorite of many young readers of an earlier generation. **Manuel Komroff** studied engineering and music at Yale for a time, leaving in 1912 without taking a degree. His novels included *Echo of Evil* (1948) and *Jade Star* (1951). **Max Lerner,** the essayist and political scientist, received his B.A. degree from Yale in 1932.

Sinclair Lewis first arrived at Yale as a freshman in autumn of 1903, taking rooms at 124½ Park Street, now the site of a bank. Later that fall he moved to 79 South Middle Street, considered a better location for meeting one's classmates. At the beginning of his sophomore year, in 1904, Lewis was back on Park Street again, this time at 105, but soon moved to a new college room at 232 Durfee. Lewis's first published stories appeared in the *Yale Literary Magazine,* and he sometimes wrote for the Hartford *Courant.* By 1906, bored with Yale, Lewis left college and became a janitor in Englewood, New Jersey. In December 1907, he returned to Yale and boarded in a ramshackle house at 14 Walley Avenue. There he enjoyed gossiping with his landlady about the other boarders. He also appreciated the low rent and the release his residence afforded him from such college routines as compulsory chapel. Lewis received his B.A. in 1908.

Augustus Longstreet, the writer and editor, was graduated from Yale in 1813. He served as president of several colleges in the South and became known as a writer for his *Georgia Scenes, Characters, and Incidents* (1835), a collection of humorous sketches of life in his native state. **Archibald MacLeish,** who won a Pulitzer Prize for his play *J.B.* (1958), received a B.A. from Yale in 1915. As an undergraduate, MacLeish edited the *Yale Literary Magazine.* **James Gates Percival,** who was graduated from Yale in 1815, wrote and acted in the tragedy *Zamor* while he was a senior. His best-known poem was "Prometheus" (1821). Several of Percival's poems were published in a New

"The trouble with me is just that. I'm impractical; have strange ideas. I'm not satisfied with Yale as a magnificent factory on democratic business lines; I dream of something else, something visionary, a great institution not of boys, clean, lovable and honest, but men of brains, of courage, of leadership, a great center of thought, to stir the country and bring it back to the understanding of what man creates with his imagination and dares with his will. It's visionary—it will come."

—Owen Johnson,
in *Stover at Yale*

The Pierson master's residence at Yale University, where John Hersey lived when he was master from 1965 to 1970.

"I'd better explain," I told her. "Mr. Wolfe is in the middle of a fit. It's complicated. There's a fireplace in the front room, but it's never lit because he hates open fires. He says they stultify mental processes. But it's lit now because he's using it. He's seated in front of it, on a chair too small for him, tearing sheets out of a book and burning them. The book is the new edition, the third edition, of Webster's New International Dictionary, Unabridged, published by the G. & C. Merriam Company of Springfield, Massachusetts. He considers it subversive because it threatens the integrity of the English language. In the past week he has given me a thousand examples of its crimes. . . ."

—Rex Stout,
in *Gambit*

Haven magazine, *The Microscope,* and for a time he edited the New Haven *Connecticut Herald.* **Frederick Prokosch,** the novelist and poet, received his Ph.D. degree from Yale in 1933 and taught English here for a brief time. His novels include *The Asiatics* (1935) and *The Conspirators* (1943). **Frederic [Sackrider] Remington** studied at the Yale School of Fine Arts. Remington wrote several books, including *Pony Tracks* (1895) and *Crooked Trails* (1898), but is best known for his action illustrations and sculpture of Western subjects. **Samuel Seabury,** the first Protestant Episcopal bishop in the U.S., was graduated from Yale in 1748. During the war he was a Loyalist pastor, and his *Letters of a Westchester Farmer 1774–1775* vigorously opposed proposals of the Continental Congress. His opposition to the Revolution inflamed his opponents, causing Seabury to seek protection by British troops.

Another editor of the *Yale Literary Magazine* was **Edward Rowland Sill,** the poet and essayist, who was graduated from Yale in 1861. *The Hermitage and Other Poems* (1868) and *The Venus of Milo and Other Poems* (1883) were two of his many books. **George Sklar,** while a member of George Pierce Baker's drama workshop at Yale, wrote *Merry-Go-Round* (1933) with Albert Maltz. Another student of Baker's was **Betty Smith,** the playwright and novelist, who attended Yale Drama School from 1930 to 1934. Smith is known best for her novel *A Tree Grows in Brooklyn* (1934). **Edmund Clarence Stedman,** the poet and critic, arrived at Yale in 1849 as the youngest member of his class. His poem "Westminster Abbey" won the sophomore prize in English composition, but he was neglecting his studies for "beer, skittles and other amusements." Stedman was expelled in 1851 when he was discovered touring New England with another student, calling themselves "the well-known tragedian Alfred Willoughby and his sister Miss Agnes Willoughby." He later received an honorary degree from Yale. **Donald Ogden Stewart,** the humorist and playwright, was graduated from Yale in 1916. His books included *A Parody Outline of History* (1921), *The Crazy Fool* (1925), and *Father William* (1929).

This I beheld, or dreamed it in a dream:—
There spread a cloud of dust along a plain;
And underneath the cloud, or in it, raged
A furious battle, and men yelled, and swords
Shocked upon swords and shields. A prince's banner
Wavered, then staggered backward, hemmed by foes.
A craven hung along the battle's edge,
And thought, "Had I a sword of keener steel—
That blue blade that the king's son bears,—but this
Blunt thing—!" he snapt, and flung it from his hand,
And lowering crept away and left the field.
Then came the king's son, wounded, sore bestead,
And weaponless, and saw the broken sword,
Hilt-buried in the dry and trodden sand,
And ran and snatched it, and with battle-shout
Lifted afresh he hewed his enemy down,
And saved a great cause that heroic day.

—Edward Rowland Sill,
"Opportunity"

John Trumbull passed the Yale entrance examinations when he was seven but came to Yale five years later. In 1770 he received an M.A. degree from Yale and, in 1772, became a tutor here. He attacked deficiencies in the Yale curriculum, particularly through his satirical poem *The Progress of Dulness* (1772–73). Trumbull practiced law in New Haven and Hartford after 1774, and joined several Yale classmates in organizing the Connecticut Wits. **Mark Twain** was granted an honorary degree by Yale in 1888. In his acceptance speech he defined his trade, that of a "funny man," as "the deriding of shams, the exposure of pretentious falsities, the laughing of stupid superstitions out of existence; and that whoso is by instinct engaged in this sort of warfare is the natural enemy of royalties, nobilities, privileges and all kindred swindles, and the natural friend of human rights and human liberties."

Royall Tyler received B.A. degrees from Yale and from Harvard in 1776. Tyler later was to write *The Contrast* (1787), which was the first comedy and only the second play of any kind written by an American. **Robert Penn Warren,** the novelist and poet, attended the Yale Graduate School in 1927 and 1928. With Cleanth Brooks and John Purser, Warren edited *An Approach to Literature* (1932), a textbook that became a classic in the teaching of literature to high-school and college students.

Noah Webster entered Yale in 1774 and was graduated in 1778 even though he had to take time from his studies to fight in the American Revolution. Webster returned to New Haven to live after a brief residence in New York City. While in New Haven he published *A Compendious Dictionary of the English Language* (1806), a forerunner of the dictionary for which he was to become famous. He had already embarked on his landmark career in American letters with *A Grammatical Institute of the English Language* (1783–85), the first part of which was subsequently published as the famous *Spelling Book* and did so much toward standardizing American spelling. The *Spelling Book,* with all its revisions, had a total sale of more than sixty million copies by 1890. Webster left New Haven in 1812 for Amherst, Massachusetts, but returned in 1822 to stay for the rest of his life. Webster's work on his great unabridged dictionary began in 1807, while he was in New Haven, and continued for twenty years. Called *An American Dictionary of the English Language* (1828), it contained about 38,000 entries and reflected American usage and spelling. In the 1840 revision, the number of entries had grown to some 70,000, and the quality and authority of Webster's work were firmly established. Webster died on May 28, 1843, and was buried in Grove Street Cemetery.

Thornton Wilder, the novelist and playwright, moved to New Haven in 1915 when his father took a position at Yale. Wilder entered Yale in 1917 and was graduated in 1920. Late in the 1920s Wilder, by then married, rented a frame house at 75 Mansfield Street in New Haven, where he and his family lived for a short time. He had already won a Pulitzer Prize for his novel *The Bridge of San Luis Rey* (1927). **N[athaniel] P[arker] Willis,** the editor and writer, was graduated from Yale in 1827. In that same year, he published *Sketches* (1827), a collection of his verse. In his time Willis was highly regarded as a poet.

In addition to a great many visiting lecturers from the literary world, Yale has had many distinguished writers, critics, and thinkers on its faculty whose presence has served to attract undergraduates interested in literature. **George Pierce Baker,** who conducted his 47 Workshop in playwriting so successfully at Harvard, moved his work to Yale in 1925. The Yale School of Drama was established as an outgrowth of Baker's work. **Joseph Heller** taught at Yale while writing *We Bombed in New Haven* (1968), his play about U.S. military involvement in Vietnam. The play was first presented by the Yale School of Drama Repertory Theatre on December 4, 1967. **John Hollander,** while teaching at Yale from 1959 to 1966, lived at 11 Mansfield Street and 240 Lawrence Street. It was here that Hollander wrote *Movie-Going and Other Poems* (1962), *Visions from the Ramble* (1965), and *Types of Shape* (1969). Other literary men and women who have taught at Yale include **Marc Connelly, Ralph Ellison, Paul Horgan, Stanley Kunitz, Ayn Rand, R. W. Stallman,** and **Allen Tate.**

The Beinecke Rare Book and Manuscript Library, on the northeast corner of Wall and High streets, houses Yale's magnificent collection of illuminated manuscripts, old manuscripts, first editions, and the papers of many American authors, including Joel Barlow, Philip Barry, Stephen Vincent Benét, William Rose Benét, James Fenimore Cooper, John Hersey, Sinclair Lewis, Archibald MacLeish, Eugene O'Neill, James Gates Percival, Mark Twain, Robert Penn Warren, and Thornton Wilder. The American collection in 1962 acquired a first edition of Melville's *The Whale* (3 vols. London, 1851), better known as *Moby-Dick.* Also in the American collection are the papers of Alice B. Toklas, writer and companion to Gertrude Stein, and author of *The Alice B. Toklas Cookbook* (1955). The Van Sinderen collection, given to Yale in 1940, consists of letters, manuscripts, and first editions of Walt Whitman.

NEW LONDON

Anna Hempstead Branch was born in 1875 in the Old Hempstead House, at 11 Hempstead Street, New London, where her family had lived since the city's earliest days. She grew up in New York City, but soon returned to New London, where she died on September 8, 1937. The poet was known best for *The Heart of the Road* (1901), *The Shoes That Danced* (1905), and *Nimrod and Other Poems* (1910). The Old Hempstead House is maintained by the Connecticut Antiquarian Society. **H[enry] C[uyler] Bunner,** the poet, short-story writer, and novelist, who died in 1896, is buried here at Cedar Grove Cemetery, 638 Broad Street. **John Hollander** lived at 61 Maeaug Avenue from 1957 to 1959 while serving as lecturer at Connecticut College here. The site of Hollander's home is now occupied by the U.S. Coast Guard Academy Library. While living in New London, Hollander wrote *The Untuning of the Sky: Ideas and Music in English Poetry 1500–1700* (1961).

By far the most brilliant of New London's literary lights is **Eugene O'Neill.** When O'Neill

Sometimes when all the world seems gray
and dun
And nothing beautiful, a voice will cry,
"Look out, look out! Angels are drawing nigh!"
Then my slow burdens leave me one by one,
And swiftly does my heart arise and run
Even like a child when loveliness goes by—
And common folk seem children of the sky,
And common things seem shaped of the sun,
Oh, pitiful! that I who love them, must
So soon perceive their shining garments fade!
And slowly, slowly, from my eyes of trust
Their flaming banners sink into a shade!
While this earth's sunshine seems the golden dust
Slow settling from the radiant cavalcade.

—Anna Hempstead Branch,
"While Loveliness Goes By"

was young, his family spent summers at Monte Cristo, a two-story, eight-room wooden cottage at 325 Pequod Avenue, overlooking the Thames River. The house, which the O'Neill family occupied from 1884 until 1920, was so named because James O'Neill, the playwright's father, spent much of his acting career playing the role of the Count of Monte Cristo. The ground floor of the cottage is well known to lovers of theater as the setting for two of O'Neill's best plays, *Ah, Wilderness!* (1933) and *Long Day's Journey into Night,* published posthumously in 1956. *Ah, Wilderness!* was O'Neill's only comedy, significantly set in the same new New England town that was the setting

Monte Cristo, boyhood home of Eugene O'Neill.

The wolf of winter
Devours roads and towns
In his white hunger.

The wolf of winter
Sticks his paw into the city's
rancid pot,
Wanly stirring its soup of whores
and suicides.

—Kenneth Patchen,
in "The Wolf of Winter"

There is rust upon locks and hinges,
And mould and blight on the walls,
And silence faints in the chambers,
And darkness waits in the halls.

—Louise Chandler Moulton,
in "House of Death"

If it possessed no other
Distinction, Norfolk would be the
world's creditor
Each June for its rhododendrons and
mountain laurel.
In fact, however, it enjoys other marks
of favor. Although the town
cherishes its genteel repose,
It contains, for instance, three notable
oddballs,
Brendan Gill, James Laughlin, and
the undersigned.

—Hayden Carruth,
in "A Short-Run View"

for his great tragedy of the Tyrone family, *Long Day's Journey into Night.* The Tyrones, in the cottage living room, bare their agonies and the agonies of O'Neill's own youth in a manner never to be forgotten by theater audiences. The Monte Cristo cottage has been repaired and converted into a museum and meeting-place by the Eugene O'Neill Theater Center of Waterford, Connecticut, and is open to the public. O'Neill spent one year studying at Princeton, six years traveling about the world—all the time drinking too much. He returned to New London when he was twenty-four and worked for a few months as a reporter and contributor to the poetry column of the New London *Telegraph,* which published his first creative writing. O'Neill soon came down with tuberculosis, and in December 1912 entered a sanitarium in Wallingford. Following his "rebirth" in 1913, O'Neill returned to New London, where he wrote most of his one-act sea play *Bound East for Cardiff* (1916).

Bishop Samuel Seabury, the clergyman and pamphleteer, was buried from St. James Episcopal Church at 76 Federal Street in 1796, and his remains were placed under the chancel in 1849. The church is still in use, and further information about it can be obtained from the Historical Society of New Haven.

NORFOLK

Hayden Carruth lived in the summer of 1961 in a small cottage near Tobey Pond, Norfolk. His volume *The Norfolk Poems of Hayden Carruth* (1962) reflects Carruth's stay here. The house in which Rose Terry Cooke, the poet and short-story writer, lived in Winsted was moved more than thirty years ago to Doolittle Pond in Norfolk. Norfolk is on Route 182, in northwestern Connecticut.

NORWALK

Kay Boyle wrote *Seagull on the Step* (1955) and *Generation Without Farewell* (1959) while living in Rowayton, a section of Norwalk. **George Seldes** wrote *Witch Hunt* (1940), *Facts and Fascism* (1943), and *The People Don't Know* (1949) while living in South Norwalk. **Philip Van Doren Stern** lived at 44 Parkhill Avenue in Norwalk when he wrote *When the Guns Roared: World Aspects of the American Civil War* (1965). **Sloan Wilson** was born on May 8, 1920, in a house on Whiteoak Shade Road, on the Norwalk–New Canaan line. Wilson is best known for the novel *The Man in the Gray Flannel Suit* (1955), a sardonic portrayal of the world of Madison Avenue advertising.

NORWICH

Donald Grant Mitchell, known as **Ik Marvel** or **Ike Marvel,** was born in Norwich, north of New London, on April 12, 1822. His first popular work was a book of essays, *Reveries of a Bachelor* (1850). His novel *Dr. Johns* (1866), which was set in a New England village, contrasted the strictness of Connecticut Calvinism with the outside influence of Catholicism. Another native of Norwich was **Lydia Howard Sigourney,** born on September 1, 1791. She attended school in Norwich and wrote anonymously for the newspapers.

OLD LYME

Kenneth Patchen and his wife, Miriam, moved to rural Old Lyme in the 1940s to escape the demands of literary life in New York City. Patchen was victim of a painful and crippling spinal disease. The Patchens rented a house on what is now Dunn's Lane, just off Library Lane. When *The Selected Poems of Kenneth Patchen* was published in 1947, Patchen donated a copy to the Old Lyme–Phoebe Griffin Noyes Library, but the book was placed on a reserved shelf because it was considered to be unsuitable for young people. The sexual imagery, antiwar sentiment, and profanity scattered throughout the poems were considered offensive. Patchen wrote a letter of complaint to the librarian. The secretary to the president of the library replied to Patchen that although he was a poet "who has been given some considerable critical attention . . . and students of modern poetry would probably be interested in his works . . . they may not be desirable for young people." There followed a barrage of letters, complaining of the censorship, from such well-known writers as Louis Untermeyer, Archibald MacLeish, and William Rose Benét. To Archibald MacLeish, who then was living in Paris, the secretary replied, "This is not censorship in the usual sense, but merely a practical adaptation to the needs of our particular community." Three years later, in 1951, after a tag sale on the front lawn of the house on Dunn's Lane, the Patchens left Old Lyme forever. *The Selected Poems* was placed in general circulation in the 1950s and is reported to have enjoyed that status ever since.

OLD SAYBROOK

Ann Petry, the novelist, was born here on October 12, 1912. Petry is known best for her novel *The Street* (1946), which was set in New York City's Harlem.

PLAINFIELD

Van Wyck Brooks, the critic and biographer, was born on February 16, 1886, in this town in northeastern Connecticut. Brooks won a Pulitzer Prize for his literary history *The Flowering of New England* (1936).

PLYMOUTH

A. Bronson Alcott came to Plymouth, north of Waterbury, in the spring of 1814 to work at the Seth Thomas Clock Factory. He stayed for a year before returning to his home in Wolcott, Connecticut.

POMFRET

[Ellen] Louise Chandler Moulton was born in Pomfret on April 10, 1835. Her works included *Bedtime Stories* (1873, 1874, 1880) and *Poems and Sonnets* (1909). Pomfret is near Putnam in northeastern Connecticut.

PORTLAND

Richard Wilbur lived at Hillcroft in Portland from 1957 to 1973 while teaching in nearby

Middletown. It was here that he wrote, among others, the poems collected in *Advice to a Prophet* (1961) and *Walking to Sleep* (1969), and translated Molière's *Tartuffe* (1963) and *The School for Wives* (1971).

REDDING

Joel Barlow was born in 1754 on his father's farm here on Route 107 and Dayton Road. The house is no longer standing, and the marker once in place has disappeared. Joel Barlow's brother had a farm on Station Road off Umpaug Road, where Joel worked for a while. The colonial saltbox that was the farmhouse is the second oldest house in Redding but is not open to the public. Barlow returned to the area during the Revolutionary War—he fought in the battles of Long Island and White Plains—when he visited soldiers encamped in Redding's Putnam Park and read poetry to them. Barlow's papers, letters, and first editions are on display in the Mark Twain Library here, on Route 53. The Joel Barlow High School, built in 1959 on Black Row Turnpike, has an oil portrait of Barlow. The original painting was by Robert Fulton. Several members of the Barlow family are buried in the family plot in the Old Burying Ground on Great Pasture Road, but Joel Barlow died in Poland while on a mission to negotiate a treaty with Napoleon, and is buried there.

Etching of Joel Barlow, one of the Connecticut Wits.

War after war his hungry soul requires,
State after State shall sink beneath his fires,
Yet other Spains in victim smoke shall rise
And other Moskows suffocate the skies,
Each land lie reeking with its people's slain
And not a stream run bloodless to the main.
Till men resume their souls, and dare to shed
Earth's total vengeance on the monster's head,
Hurl from his blood-built throne this king of woes,
Dash him to dust, and let the world repose.

—Joel Barlow,
in "Advice to a Raven in Russia, December, 1812"

Joseph Wood Krutch, the critic, in 1931 bought a house on Limekiln Road, in which he was to spend long weekends while teaching at Columbia University in New York City. Krutch wrote "The Colloid and the Crystal" here, an essay he particularly liked. The Limekiln house is not open to visitors. When **Sinclair Lewis** came to Redding in 1936 to visit his literary agent, Ann Watkins, he arrived waving a bottle of Scotch out the car window. Watkins's home on West Redding Road was just off the Danbury-Redding town line.

Albert Bigelow Paine, novelist and biographer of Mark Twain, lived from 1905 to 1920 on Diamond Hill Road, about a mile west of the Mark Twain Library. The house, described by Paine in *Dwellers in Arcady* (1919), burned down about 1975. After Mark Twain moved to Redding in 1908, he gave his friend Paine some property adjacent to Stormfield, Twain's own house. Paine, who called the property Markland, found that he was able to write there undisturbed. From 1935 until his death, Paine lived on Redding Road (Route 107), two miles north of the Mark Twain Library. That house is still standing. Paine died in Smyrna, Florida, on April 9, 1937, but is buried in the old cemetery on Umpawaug Hill, at the corner of Diamond Hill Road, two miles west of the Mark Twain Library.

The last home of **Mark Twain** was Stormfield, the house he built on Redding Road, where he lived with his daughter Jean until her death. He played billiards with Albert Paine, worked on his autobiography, and entertained such literary visitors as **William Dean Howells** and **Helen Keller.** On April 21, 1910, Twain died, and on July 25, 1923, Stormfield burned down. The house was replaced later with a house in the same Italianate style, but the property of two hundred acres carries no plaques or markers. The Mark Twain Library, on Route 53, was founded by Twain in memory of his daughter Jean, who acted as his secretary at Stormfield until her death in 1909 during an epileptic seizure. Twain donated books and land and encouraged his friends to follow suit and to raise funds. The library was dedicated in 1911, one year after Twain's death, and is open to the public. Its collections include autographed Twain first editions, letters, photographs, a self-portrait on copperplate, and a model of a Mississippi River paddleboat.

RIDGEFIELD

In 1942, **S[amuel] N[athaniel] Behrman,** a prolific playwright and short-story writer, bought a forty-six acre estate here in Western Connecticut on the New York border. It included a frame house and a large barn. Among Behrman's best-known plays are *Amphitryon 38* (1937) and *No Time for Comedy* (1939). **Konrad Bercovici,** the novelist, after 1955 lived off West Lane on Route 35A, in a house that is still standing. Bercovici's works included the novels *The Marriage Guest* (1925), *The Volga Boatman* (1926), and *The Exodus* (1947). Bercovici, who was born in Rumania, entitled his autobiography *It's the Gypsy in Me* (1941). **Howard Fast** lived on Harvey Road when he wrote the novel *The Crossing* (1962) and the play *The Hessian* (1972).

One glance was enough to show that it was all that the other old house was not. It did not sag, or lurch, or do any of those disreputable things. It stood up as straight and was as firm on its foundations as on the day when its last hand-wrought nail had been driven home, a century or so before. No mistaking its period or architecture—it was the long-roofed salt-box type, the first Connecticut habitation that followed the pioneer cabin; its vast central chimney had held it unshaken during the long generations of sun and storm.

—Albert Bigelow Paine,
in *Dwellers in Arcady*

> *She had never observed his face more composed and she grabbed his hand and held it to her heart. It was resistless and dry. The outline of a skull was plain under his skin and the deep burned eye sockets seemed to lead into the dark tunnel where he had disappeared. She leaned closer and closer to his face, looking deep into them, trying to see how she had been cheated or what had cheated her, but she couldn't see anything. She shut her eyes and saw the pin point of light but so far away that she could not hold it steady in her mind. She felt as if she were blocked at the entrance of something. She sat staring with her eyes shut, into his eyes, and felt as if she had finally got to the beginning of something she couldn't begin, and she saw him moving farther and farther away, farther and farther into the darkness until he was the pin point of light.*
>
> —Flannery O'Connor, in *Wise Blood*

S[amuel] G[riswold] Goodrich, known as **Peter Parley,** was born on West Lane on August 10, 1793. The house is no longer standing. When the author was four years old, his family moved to a house at 15 High Ridge Avenue, which may still be standing. Parley was known for his series of children's stories that began with *The Tales of Peter Parley About America* (1827) and went on to a total of more than one hundred volumes. He was also responsible for introducing the work of Nathaniel Hawthorne in *The Token,* his annual gift book (1827 to 1842). Sugar Hill, the Ridgefield home of **Clare Boothe [Luce],** and her husband **Henry R. Luce,** is still standing. Mrs. Luce had library equipment installed in the home so that she and her secretaries could do their research here.

From September 1949 to December 1950, **Flannery O'Connor** lived with her friends **Robert Fitzgerald** and his wife **Sally Fitzgerald** at 70 Acre Road here. Robert Fitzgerald was a poet and the respected translator of *The Odyssey* (1961) and other classical Greek works. Sally Fitzgerald is also a writer. She edited and collected Flannery O'Connor's letters, *The Habit of Being* (1979), and now is writing O'Connor's official biography. In August 1949, O'Connor wrote to a friend: "Me & novel are going to the rural parts of Connecticut. I have some friends named Fitzgerald who have bought a house on top of a ridge, miles from anything you could name. An exaggeration. . . ." She settled in and was able to work for about four hours a day on her first novel, *Wise Blood* (1949). Soon after Flannery O'Connor left Ridgefield, she was found to have the crippling disease that eventually caused her death. O'Connor died on August 3, 1964, in Milledgeville, Georgia, her principal home since childhood.

Eugene O'Neill lived at Brook Farm, 845 North Salem Road, in 1922 and 1923, while he was married to Agnes Boulton. The farm included thirty-one acres of wide lawns, pastures, and woodland. From 1948 to 1957, **Robert Lewis Taylor,** newspaperman and novelist, lived on Old Branchville Road. He wrote several books here, including *W. C. Fields, His Follies and Fortunes* (1949), *Center Ring: The People of the Circus* (1956), and *The Travels of Jamie McPheeters* (1958), for which Taylor won a Pulitzer Prize.

Brook Farm, the house in Ridgefield where Eugene O'Neill lived for a short time.

ROXBURY

Manfred B. Lee, half of the highly successful team that wrote mystery novels under the name of **Ellery Queen,** lived on South Street in Roxbury, east of Waterbury, from the mid-1950s until his death on April 3, 1971. Lee is buried in Roxbury. While living in Roxbury, **William Styron** wrote his novels *Set This House on Fire* (1960) and *Confessions of Nat Turner* (1967). While working on *Confessions of Nat Turner,* for which Styron won a Pulitzer Prize, he was given encouragement and help by **James Baldwin,** who worked on his own novel, *Another Country* (1962), at the Styron house.

SALISBURY

On his last visit to the United States, **Henry James** stayed at Matiff Farm in Salisbury. James was impressed by the northwestern Connecticut countryside, especially the mountains and lakes.

SAYBROOK

Mark Twain in 1872 came to Fenwick Hall in Saybrook, southwest of New London, with his wife, Olivia Langdon, and their daughter. Their only son, Langdon, had died that June. Twain was involved in putting the final touches on his invention, the Mark Twain Self-Pasting Scrapbook, which he patented the next year. Fenwick Hall, which stood on Long Island Sound on what is now Fenwick Avenue, has burned down. A golf course occupies the site today.

SEYMOUR

John William De Forest was born on March 31, 1826, in Seymour, northwest of New Haven and known in De Forest's time as Humphreysville. He attended the just-opened Miss Stoddard's Select School here. Among De Forest's many works were *Miss Ravenel's Conversion from Secession to Loyalty* (1867), a realistic novel about the Civil War, and *Honest John Vane* (1875), depicting the corruption of the Grant administration. De Forest Street in Seymour is named after the author's father, who was a successful businessman.

SHARON

Robert Lewis Taylor lived in Penoyer House on South Main Street here from 1957 to 1967. The house, still standing and privately owned, carries the sign "Penoyerhouse" over its front door. While living here, Taylor wrote several books, including *A Journey to Matecumbe* (1961) and *Vessel of Wrath: The Life and Times of Carrie Nation* (1966). Sharon is south of Salisbury.

SHELTON

Betty Smith, known best for her novel *A Tree Grows in Brooklyn* (1943), died on January

17, 1972, at the Hewitt Memorial Convalescent Home, at 230 Coram Avenue, Shelton, a town west of New Haven.

SIMSBURY

Sarah Pratt McLean was born in Simsbury, Route 10, north of Hartford, on July 3, 1856. Her stories and novels were set in New England, particularly Cape Cod. Her novels included *Cape Cod Folks* (1881), *Last Chance Junction* (1889), and *Everbreeze* (1913).

SOUTHBURY

Katherine Anne Porter lived in 1957 in a house near Southbury, off Route 84. **Gladys Taber,** who brought Southbury to life for her loyal readers, lived for many years at Stillmeadow Farm here, which she described in countless magazine articles and many books. Her novels included *Give Me the Stars* (1945) and *Spring Harvest* (1960). She died on March 11, 1980, in Hyannis, Massachusetts.

SOUTHINGTON

Joseph [Hopkins] Twichell, the biographer, was born in Southington, west of Waterbury, on May 27, 1838. The building he was born in stood in the western part of town, but was torn down much later to make way for Route 84. Twichell was a member of the literary coterie that gathered round Mark Twain in Nook Farm, in Hartford. Twichell traveled through Europe with the Twain family, and was called Harris in Twain's *A Tramp Abroad* (1880). In 1909, Reverend Twichell performed the marriage ceremony for Twain's daughter, Clara, and Ossip Gabrilowitsch at Twain's home in Redding.

STAMFORD

In July 1955 **Maxwell Anderson** bought a home at 141 Downs Avenue. There, on February 26, 1959, the dramatist suffered a stroke while making plans for the production of his unpublished play *Madonna and Child,* which he had completed in 1956. Two days later he died in the Stamford Hospital on Shelbourne Road. **Sholem Asch,** the novelist, maintained a home on Sky Meadow Drive, near High Ridge Road here in the 1940s. **Heywood Broun** lived on Hunting Ridge Road here from 1928 until his death on December 18, 1939. Broun bought part of an old farm here, and wrote and meditated in a studio he fixed for himself out of an old woodshed. He named the place Sabine Farm. Broun died of pneumonia at the Columbia-Presbyterian Medical Center in New York City. He was known best for his syndicated newspaper column, "It Seems to Me." **John Hawkes,** the novelist and playwright, was born on August 17, 1925, at the St. Elizabeth Hospital in Stamford. His novels include *The Beetle Leg* (1951), *The Lime Twig* (1961), and *Second Skin* (1964).

Eugene O'Neill attended Betts Academy, a boarding school for boys, from 1902 to 1906. The academy, which is now closed, was at 77 Strawberry Hill Avenue. **Maxwell Perkins,** the well-known Scribner's editor, died at the Stamford Hospital on June 17, 1947. **Elmer Rice** moved in the early 1960s to a house, still standing, at 815 Long Ridge Road. The playwright lived there until shortly before his death in England on May 8, 1967.

Edmund Wilson, the literary critic, rented Trees, a house outside Stamford, in 1936, the year before his *To the Finland Station* (1940) began to appear in *The New Yorker* and *Partisan Review.* Wilson wrote of his stay here: "Stamford: I hardly ever go to New York and am rural as any agrarian." Wilson married **Mary McCarthy,** the novelist, in April 1938, and they lived at 233 Stamford Avenue until 1941, when they moved to a house in Wellfleet, Massachusetts. While living in Stamford, McCarthy wrote "Cruel and Barbarous Punishment," which Robert Penn Warren published in *The Southern Review* (1939). The house on Stamford Avenue is still standing.

> *It is all too easy to idealize a social upheaval which takes place in some other country than one's own. . . . The remoteness of Russia from the West evidently made it even easier for American socialists and liberals to imagine that the Russian Revolution was to get rid of an oppressive past, to scrap a commercial civilization and to found, as Trotsky prophesied, the first really human society. We were very naïve about this. We did not foresee that the new Russia must contain a good deal of the old Russia: censorship, secret police, the entanglements of bureaucratic incompetence, an all-powerful and brutal autocracy. This book of mine assumes throughout that an important step in progress has been made, that a fundamental "breakthrough" has occurred, that nothing in our human history would ever be the same again. I had no premonition that the Soviet Union was to become one of the most hideous tyrannies that the world had ever known, and Stalin the most cruel and unscrupulous of the Russian tsars.*
>
> —Edmund Wilson, in the 1971 introduction to *To the Finland Station*

Penoyer House, once the home of Robert Lewis Taylor in Sharon. Penoyer House stands on the east side of South Main Street and south of the clock tower.

STONINGTON

In 1941 **Stephen Vincent Benét** bought the Amos Palmer House, a Georgian structure at the northwest corner of Main and Wall streets in Stonington. Built in 1787, the house was once the boyhood home of James McNeil Whistler, the artist. The Benét family spent summers here in southeastern Connecticut and winters at their home in New York City, where the poet died on March 13, 1943. Benét wrote in a small third-story room here, using a lapboard for his desk, and propping his feet against a chair.

STORRS

R. W. Stallman has taught, since 1949, at the University of Connecticut, in Storrs, on Route 195, north of Willimantic. He lived in the Faculty Apartments between 1949 and 1955. Most of his books have been written in Storrs, including *Stephen Crane: An Omnibus* (1952, 1954). Some of his papers are deposited in the University Library, Special Collections Section.

TORRINGTON

Carl Van Doren died on July 18, 1950, in the Charlotte Hungerford Hospital in Torrington. His brother, **Mark Van Doren,** died in that same hospital on December 10, 1972. Mark Van Doren, like his brother, had been a professor at Columbia, and among his students were Jack Kerouac and Thomas Merton. After receiving an A in a Shakespeare class, Kerouac quit football so that he could have more time to study with Van Doren. Torrington is in the northwestern part of the state, close to Cornwall, where Mark Van Doren spent much of his life.

The Amos Palmer House in Stonington, which Stephen Vincent Benét bought as a summer home in 1941.

WALLINGFORD

Choate School, located at 185 Christian Street in Wallingford, was founded in 1896. It has been known as Choate-Rosemary Hall since 1971. This first-rank preparatory school has had at least three students who have gone on to careers in literature. While a student here in 1944, **Edward Albee** wrote verse, short stories, a novel, and the play *Schism,* which was published in the *Choate Literary Magazine.* **John Dos Passos** attended Choate School about 1912, and **Walter Dumaux Edmonds,** the historical novelist, attended from 1918 to 1921.

Wallingford is also the home of Gaylord Hospital, where **Eugene O'Neill** spent six months in 1912–13, recuperating from tuberculosis. After his time there, which O'Neill referred to as his "rebirth," he turned from writing verse to writing plays, and *The Straw* (1921) was inspired by his experience at Gaylord. Some of the administration buildings are the same now as they were in O'Neill's time, but the cottages for patients have been torn down.

WASHINGTON

Phil Stong, who lived in Washington, on Route 47 north of Danbury, died on April 26, 1957. Stong's best-known works were *State Fair* (1932) and *Gold in Them Hills* (1957). His home on Sabbaday Lane, near the intersection of Wykeham Road, still stands.

WATERBURY

A. Bronson Alcott, who lived in nearby Wolcott, was confirmed on October 18, 1816, in the Episcopal Church in Waterbury. It was most likely at St. John's Episcopal Church, but the church records were destroyed in a fire. **Hayden Carruth,** the poet, was born at 58 Central Avenue on August 3, 1921. The building has since been converted into an apartment house. Carruth's poems have been collected in a number of volumes, the most recent being *Brothers, I Loved You All* (1978). On December 11, 1869, **Henry David Thoreau** gave a lecture entitled "Autumnal Tints" at Hotchkiss Hall in Waterbury. A newspaper account of the time referred to Thoreau as an "author with a deservedly high reputation, but [because of his monotonous delivery] one out of his element as a popular lecturer." It is thought that Thoreau stayed at the Scovill House, then a hotel on West Main Street on the south side of the Green. Hotchkiss Hall, where Thoreau lectured, was most likely in the Hotchkiss block, on the corner of North Main and East Main streets. Neither building still stands.

WATERFORD

The Eugene O'Neill Theater Center, on Route 213, has a collection of letters of Eugene

O'Neill, a desk he owned, and historical material. Each summer, the National Playwrights Conference is held at the Center. Actors, playwrights, and directors come to advise, participate in workshops, and produce their plays. The Center also owns the Monte Cristo cottage in New London, just east of Waterford. Monte Cristo was owned at one time by O'Neill's father.

WESTBURY

John Trumbull, the poet who led the Connecticut Wits, was born on April 12, 1750, in Westbury. He wrote *M'Fingal* (1782) and *The Progress of Dulness* (1772–73).

WEST CORNWALL

Wickwire, in West Cornwall, Route 128, was the summer home of **Carl Van Doren.** After he died on July 18, 1950, in Torrington, his ashes were scattered over his property here. West Cornwall has a less somber literary association: the epistolary story "File and Forget," by **James Thurber,** collected in *Thurber Country* (1953), took readers through the troubled relations between an author and his publisher. As letter after letter was fruitlessly exchanged, the misunderstandings inevitably multiplied. The writer in the story was James Thurber, and his return address was West Cornwall, Connecticut.

WEST HARTFORD

Rose Terry Cooke was born on a farm here on February 17, 1827. It stood on the south side of the Albany Turnpike, just west of Steele Lane, where St. Mary's Home for the Aged now is located. Beginning in 1861, Miss Cooke's stories of life in rural New England were published in *The Atlantic Monthly.* Two of her books were *Somebody's Neighbor* (1881) and *Huckleberries Gathered from New England Hills* (1891). **Wallace Stevens** lived with his family in West Hartford from 1924 to 1932. His daughter, Holly Stevens, has written of that period: "On summer days when I was small, my mother and I used to sit in the sun on the west side of the house where we lived at 735 Farmington Avenue, in West Hartford, near our front door (which was really the side door of the house). One of my earliest memories is of being there and watching her brush and comb her long blonde hair in the sunshine: it resembled strands of pure gold." Wallace Stevens was writing poetry, and working for an insurance company in Hartford at the time. **Noah Webster,** the lexicographer, was born at 227 S. Main Street on October 16, 1758. The family farmhouse, in a section of Hartford then known as West Division, is now the Noah Webster House, operated by the Noah Webster Foundation and Historical Society of West Hartford. It is open to the public, and admission is charged. The exhibits at the Noah Webster

Birthplace of Noah Webster in West Hartford, now a museum.

Eleanor Estes, creator of the fictional Moffats, and Peter De Vries, sardonic portrayer of suburban mores.

House are changed from time to time and do not always deal with Webster's life. Previous exhibits have included nineteenth-century costume and textiles, and nineteenth-century advertising.

WEST HAVEN

Eleanor Estes was born in this city, just west of New Haven, on May 9, 1906. From 1911 to 1915, she attended Wood Street School at 370 School Street. That building now houses Baker's Equipment Company, Inc. From 1915 to 1923, she studied at the Union School, 176 Center Street, which now is a meeting place for the elderly. Two of Estes's homes, at 57 Fourth Avenue and Twelve Ward Place, are both standing and private, as is a house at 264 Union Avenue, the scene of *The Echoing Green* (1947), an adult novel by Estes. Eleanor Estes was the creator of the Moffat family in popular children's books such as *Ginger Pye* (1951), *Pinky Pye* (1958), and *The Moffats* (1941). These books were all set in the town of Cranbury, modeled after West Haven.

WESTON

Franklin P[ierce] Adams, known to older generations as F. P. A., lived on Lyons Plains Road in Weston from 1932 to 1950. F.P.A. wrote "The Conning Tower," a daily newspaper column made up principally of satirical verse and, once a week, the lesser events in the life of the writer, and given under the title "The Diary of Our Own Samuel Pepys." These accounts, written in the style of the seventeenth-century diarist, were collected in two volumes in 1935 as *The Diary of Our Own Samuel Pepys.* Other works by F.P.A. included *Christopher Columbus* (1931) and *The Melancholy Lute* (1936). While living on Old Weston Road here, in a house previously owned by **Van Wyck Brooks, John Hersey** wrote *Here to Stay* (1963), *White Lotus* (1965), and *Too Far to Walk* (1966). Weston is on Route 57, off Route 15.

> *The house on the outskirts of Decency we lived in was built around a silo, which became my father's library. Swiss cheese, except for the silo, comprised the principal masonry, as the dank airs which continually stirred the draperies attested; the wiring was, to put it no lower, shocking; the fireplace drew briskly but in the wrong direction, sending out ashes which settled like a light snow on our family and on the strangers within our gates, for in those years my parents loved to entertain. They had lived originally in town but had begun to drift apart and needed more room. The capacious new house did in fact ease their relations, getting them out of one another's pockets I suppose, and I can still hear my mother wailing over some new kitchen crisis, "Oh, God," and my father answering cozily from the silo, "Were you calling me, dear?"*
>
> —Peter De Vries, in *Comfort Me with Apples*

WESTPORT

This city in southwestern Connecticut has long attracted writers, editors, and publishers who wish to be near but not part of New York City. **Van Wyck Brooks** from 1920 to 1946 lived on King's Highway here. Brooks was literary editor of *The Freeman* magazine for four years, beginning in 1920. His autobiographical *Days of the Phoenix* (1957), dealing with the 1920s, was set mainly in Westport. Several of Brooks's books were published during his years here, including *The Pilgrimage of Henry James* (1925); *Emerson and Others* (1927); *Sketches in Criticism* (1932); *The Flowering of New England* (1936), for which he won a Pulitzer Prize; and *The World of Washington Irving* (1944).

Peter De Vries, the quintessential chronicler of Westport life, while living on Old Road in the 1950s, wrote many pieces for *The New Yorker.* **Dylan Thomas,** the Welsh poet, stayed here with De Vries on one of his many visits to the United States. Between 1952 and 1956, De Vries lived on North Avenue, where he wrote *Tunnel of Love* (1954), featuring Westport as Avalon, and *Comfort Me with Apples* (1956).

In May 1920, just after their honeymoon in California, **F. Scott** and **Zelda Fitzgerald** rented the gray-shingled Wakeman cottage at 244 South Compo Road. The house is almost within view of Long Island Sound. While there, Fitzgerald wrote several short stories: "The Jelly Bean," "His Russet Witch," and "Two For a Cent." The Fitzgeralds joined the local beach club and frequently gave wild weekend parties for friends up from New York and Princeton. Even so, Westport was not exciting enough for the couple, and Zelda set off a fire alarm one evening in order to liven things up. By September 1920, Fitzgerald told his editor, Maxwell Perkins, "the duller Westport becomes, the more work I do." But an impending winter promised too dull a life, and the couple moved to New York City in November.

The morning of June 27th was clear and sunny, with the fresh warmth of a full-summer day; the flowers were blossoming profusely and the grass was richly green. The people of the village began to gather in the square, between the post office and the bank, around ten o'clock; in some towns there were so many people that the lottery took two days and had to be started on June 26th, but in this village, where there were only about three hundred people, the whole lottery took less than two hours, so it could begin at ten o'clock in the morning and still be through in time to allow the villagers to get home for noon dinner.

—Shirley Jackson,
in "The Lottery"

Shirley Jackson, author of the classic story "The Lottery," lived on Indian Hill Road here for a time in 1952. **Sinclair Lewis** and his second wife, **Dorothy Thompson,** rented the Franklin Pierce Adams house in Westport in 1930. Lewis was alone on November 5, 1930, when he received a telephone call from a Swedish newspaper correspondent in New York City telling him that he had won the Nobel Prize for Literature. The skeptical Lewis thought the call was a prank and began to imitate the man's accent; finally, an American friend had to get on the telephone to confirm the news. Following his trip to Sweden in 1931 to receive his award, Lewis returned to Westport. An interview in the Boston *Evening Transcript* at the time described him as "calm and wistful."

Ruth McKenney, the journalist, and her husband, Mike Lyman, lived for a time on Easton Road here in a frame house built in 1821. They had returned to Westport following an unhappy stay in Hollywood. **Paul Rosenfeld,** the music critic and novelist, had a summer home on Woodside Avenue in the early 1920s. Rosenfeld's only novel was *By Way of Art* (1928). It is believed that **J. D. Salinger,** the elusive novelist and short-story writer, once had a home here, but no one is sure. **Jerome Weidman,** the novelist, did live at 82 King's Highway from 1948 to 1961. His work during that time included *The Price Is Right* (1949) and *The Enemy Camp* (1958). The house is still standing. The names of the characters and particulars of their lives have changed, but it still is Westport life that **Sloan Wilson** described in *The Man in the Gray Flannel Suit* (1955).

WILTON

George Seldes wrote *Freedom of the Press* (1935) and parts of *Lords of the Press* (1938) and *The Catholic Crisis* (1940) while living in Wilton, on Route 33, off Route 15.

WINDSOR

The building that now is the Senior Citizen Center at 114 Palisado Avenue was the birthplace of **Edward Rowland Sill** on April 29, 1841. Sill's works included *The Hermitage and Other Poems* (1868) and *The Venus of Milo and Other Poems* (1883). Windsor is north of Hartford, off Route 91.

WINSTED

Rose Terry Cooke, the poet and short-story writer, married in 1873 and moved to Winsted. She and her husband lived here in what is now called the Rose Terry Cooke Mansion. In 1903, the house was moved to Norfolk. **Edmund Clarence Stedman,** the poet and critic, worked for a time in journalism; for a few years he owned and edited the *Mountain County Herald* in Winsted.

WOLCOTT

The birthplace of **Amos Bronson Alcott,** on November 29, 1799, in this small town northeast of Waterbury, is no longer standing, but there is a plaque on Beach Road, at the intersection of Spindle Hill Road, that reads, in part:

Near this spot stood the birthplace of
the educator and philosopher
AMOS BRONSON ALCOTT (1799–1888).
He was the father of
LOUISA MAY ALCOTT (1832–1888)
author of "Little Women."

When Alcott was one year old, the family moved to a new house in Petuker's Ring, a clearing about one mile from his birthplace. Alcott lived there with his family, in a house long since gone, until he was five years old. In 1805, while his father was working away from home, Amos lived with Col. Streat Richards, and attended school for the first time. The senior Alcott returned the next year, and the family moved to another house—made up of two houses put together—on eighty acres of land near the summit of Spindle Hill. Between the ages of six and ten, Amos attended school in a one-room schoolhouse, heated with a single wood stove. He made his own ink by steeping maple and oak bark in a mixture of indigo and alum, and practiced penmanship by copying "Avoid Alluring Company." Alcott soon left Wolcott to make his way as a peddler and as a teacher, returning to Wolcott only twice. By 1852, as an educator of some reputation, he went back to Wolcott to give what he termed "conversations" and later, with **Louisa May Alcott,** to visit following publication of his *Concord Days* (1872).

WOODBRIDGE

The area around this town just northwest of New Haven provided the inspiration for many of the pieces **Louise Field Cooper** did for *The New Yorker.* Among her works with Woodridge settings are *The Lighted Box* (1942) and *Summer Stranger* (1947).

WOODBURY

Hayden Carruth, the poet, lived in his youth in a house at 595 Main Street South in Lower Woodbury and attended the local public school until 1935, when his family moved to Pleasantville, New York. **James Thurber** moved to Woodbury in the fall of 1938 and spent the winter here, with interruptions for visits to Columbus, Ohio, and New York City. Here Thurber began to plan the play *The Male Animal* (1940), which he later wrote with Elliot Nugent.

Awoke this morning with the thought that Christmas is but two months away, and very happy to think I had so many friends that I might give things to, and what a vast sum of money I could save if I gave no gifts soever. So up, and read "Color," a book of poems by Countee Cullen, some of them very fine, and I proud of having printed many of them first in my journall. So up, and did on my Sunday suit, very brave and fine, and so to call upon Miss Edna Millay, and very glad of seeing her, too. So home, and read Sherwood Anderson's "Dark Laughter," which I did like enormously, and deem the best tayle ever he hath wrote.

—Franklin P. Adams,
in the Sunday, October 25, 1925, entry in *The Diary of Our Own Samuel Pepys*

Papa and Mama came to Cranbury to live so that Papa could study the birds of the marshes and the woods and the fields, and because Cranbury was in the middle between New York and Boston. Perhaps they, too, could not make up their minds which was more important, New York or Boston, and had to settle halfway.

After a while Gramma and Grampa moved to Cranbury too, so that Uncle Bennie would grow up knowing his niece and nephew, Rachel and Jerry, and none of them be strangers to any of them. Grampa was a piano tuner and he said he'd just as lief tune pianos in Cranbury as where he was and moreover he could have a boat in Cranbury, which he couldn't in New York. Sometimes Rachel and Jerry asked Grampa which he thought was more important, New York or Boston, and between plinking the piano keys, he'd say, "New York." But then, naturally, being from there he could not be a traitor and say, "Boston."

—Eleanor Estes,
in *Ginger Pye*

CANADA
QUEBEC
ONTARIO
LAKE ONTARIO
LAKE ERIE
St. Lawrence R.
VT.
MASS.
CONN.
NEW YORK
BUFFALO
ROCHESTER
SYRACUSE
UTICA
SCHENECTADY
ALBANY
BINGHAMTON
Hudson R.
ERIE
SCRANTON
Susquehanna R.
Allegheny R.
PENNSYLVANIA
Delaware R.
PITTSBURGH
HARRISBURG
TRENTON
NEW YORK
NEWARK
PHILADELPHIA
NEW JERSEY
OHIO
WHEELING
Ohio R.
Monongahela R.
BALTIMORE
Potomac R.
WILMINGTON
ANNAPOLIS
DOVER
WEST VIRGINIA
CHARLESTON
WASHINGTON D.C.
DELAWARE
MARYLAND
HUNTINGTON
VIRGINIA
James R.
ROANOKE
RICHMOND
NORFOLK
PORTSMOUTH
ATLANTIC OCEAN
NORTH CAROLINA

MIDDLE ATLANTIC STATES

NEW YORK

ADAMS

Marietta Holley, a popular humorist and novelist, was born on July 16, 1836, in the community of Ellisburgh, three miles south of Adams, a town in northern New York. Bonnie View, her home for most of her life, was built on the site of her birthplace on Route 11, about fifteen miles southwest of Watertown. Holley led a secluded life at Bonnie View and there wrote books under the name **Samantha Allen, Josiah Allen's Wife.** And it was at Bonnie View that she died on March 1, 1926. Among her books were *Samantha at the Centennial* (1877), *Samantha at Saratoga* (1887), and *Samantha on Women's Rights* (1913). Her novels, sketches, and poetry reflected her interest in the causes of temperance and women's rights. Holley is buried at Pierrepont Manor Cemetery, south of Adams, on Route 193, just west of Route 11.

ALBANY

This city, the capital of New York State, has been the home or birthplace of many writers. Among Albany's native sons are **James M[ontgomery] Bailey,** born on September 25, 1841. His popular humorous columns in the Danbury (Connecticut) *News* earned him the nickname the Danbury News Man, and he is sometimes called the first newspaper columnist.

By far the most famous writer to be born in Albany is **[Francis] Bret[t] Harte,** born on

Bonnie View, home of Marietta Holley, near Adams.

> *Jacques had been very temperate and industrious. He would have gained success if there had been half a chance. But mining in the Adirondack Mountains was hard. So long as a man was young, and new at the business, he could endure the disappointments, after a fashion; but when he was turned of forty years of age, and had learned that mining in the Adirondacks was contending against great commercial odds, if not natural laws, he was apt to think that he needed a change. One's best strength for "fighting the rocks" was likely to be impaired before middle life. Jacques, however, was still strong. It was the check upon earnest purposes and honest hopes that wrung his heart. He had tried so many times, he said, and so fair, and every time a failure.*
>
> —Philander Deming, in "Benjamin Jacques"

August 25, 1836. There is some controversy over the location of Harte's birthplace. Some believe he was born at 18 Beaver Street, and for many years a plaque marked the site, which came to be used as offices by the Albany *Knickerbocker Press.* The plaque is now in the possession of Capital Newspapers, at 645 Albany-Shaker Road, in the adjacent village of Colonie. Capital Newspapers is the descendant of the *Knickerbocker Press.* Another possible birthplace site is 15 Columbia Street, listed as the Harte family home in the Albany directory for 1835–36. At any rate, neither place still exists, and Harte had no further contact with the city. His father, who ran a school in Albany, moved the family from the city when Bret was very young.

Another writer born in Albany was **Charles King,** on October 12, 1844. King, a soldier and author, wrote many adventure novels dealing with military life in the West. His novels included *The Colonel's Daughter* (1883) and *Under Fire* (1894). As commandant of a military academy in Michigan, he was an early influence on a cadet named Edgar Rice Burroughs, who went on to a literary career.

Writers who have visited or lived in Albany include **James Fenimore Cooper,** who studied here from 1799 to 1803 under Reverend William Ellison, rector of St. Peter's Episcopal Church. **Philander Deming,** a lawyer, pioneer court stenographer, and writer, lived in Albany for many years and died at his home at 12 Jay Street on February 9, 1915. Deming's sketches of Adirondack life appeared in *Adirondack Stories* (1880) and other volumes. The pieces were somewhat autobiographical, as were the early pages of his book *The Story of a Pathfinder* (1907).

Charles Dickens visited and gave readings in Albany in early 1868, during his second tour of America. **Anne McVickar Grant,** the Scottish author of *Memoirs of an American Lady* (1810), an excellent portrait of colonial life in New York, lived for a time before the American Revolution in Albany, at the home of Philip Schuyler and his wife Margarite. Mrs. Grant's book is thought to have influenced a number of writers, including James Fenimore Cooper and James Kirke Paulding. **Isaac Mitchell,** the novelist and newspaper editor, was born near Albany about 1759 and edited the Albany *Republican Crisis* from about 1806 to 1812. **John Godfrey Saxe,** the poet and journalist, lived here from 1860 to 1872. After a series of tragedies that brought him great unhappiness, he spent his last years at his son's home in Albany, where he died on March 31, 1887.

Solomon Southwick came to Albany in the late 1700s and became active in newspaper publishing and in politics. He was associated with the Albany *Register* until 1817, and became editor in 1819 of the agricultural newspaper *The Ploughboy,* his own contributions appearing under the pen name **Henry Homespun, Jr.** His books included *The Pleasures of Poverty* (1823) and *Five Lessons for Young Men: by a Man of Sixty* (1837). Southwick died in Albany on November 18, 1839. Another well-known political newspaperman was **Thurlow Weed,** who was a major figure in state politics and important in national politics. Weed published the Albany *Evening Journal* from 1830 to 1860.

One of Albany's leading citizens in the early 1800s was William James, the millionaire merchant and banker. His son, **Henry James, Sr.,** wrote numerous books on religious and moral topics. Henry James, Sr. was born in Albany on June 3, 1811. He lived from 1845 to 1847 at 50 North Pearl Street with his growing family, which included his sons: **Henry James, Jr.,** the novelist, and William James, the psychologist. The house on North Pearl Street was located near the home of Henry James, Jr.'s grandmother. These houses no longer stand, but James described his grandmother's house, which he visited often, in the opening pages of *The Portrait of a Lady* (1880). In the back garden of his grandmother's house stood peach trees, and the young boy developed a great taste for peaches; in fact, peaches became closely identified in James's mind with his memories of Albany. In *The Portrait of a Lady,* James transferred his experiences to the heroine, Isabel Arden: ". . . the house offered to a certain extent the appearance of a bustling provincial inn kept by a gentle old landlady who sighed a great deal and never presented a bill. . . . Isabel had stayed with her grandmother at various seasons, but somehow all her visits had a flavour of peaches." James also wrote of his Albany experiences in his autobiographical *A Small Boy and Others* (1913).

Other sites associated with the remarkable James family include Albany Academy, still standing in Academy Park at the end of Washington Avenue. William James, the grandfather of Henry James, Jr., was president of the academy's board of trustees from 1826 to 1832. Henry James, Sr. attended the academy, and it was here, at age thirteen, that he was severely injured in a fire on the school green caused by kerosene used to inflate balloons. James's leg was so damaged that it had to be amputated. Another student at Albany Academy was **Herman Melville,** who came to the city with his family as a boy of twelve. The Melvilles took a house at 338 North Market Street in 1831

Marker in Clinton Square in Albany, a small, triangular park across from the home of Herman Melville, shown in the second photograph.

and later lived at Steuben Street, 282 North Market Street, and 3 Clinton Square. Clinton Square has been nearly obliterated by approaches to a superhighway, but Melville's house is still standing, and a state historical marker notes his residence there. While living in Albany, young Melville saw the publication of his first writing—letters to the editor of the Albany *Microscope.* Melville's mother was a Gansevoort, one of the early families to settle in the Albany area, and Melville described the Gansevoort family home in his novel *Pierre* (1852).

ALBION

Carl Carmer grew up in this village about thirty-two miles west of Rochester. In 1910 Carmer was graduated from Albion High School, now headquarters for Albion Educational Services. The author's home at the northeast corner of Main Street and East Avenue (Routes 31 and 98) has been replaced by a gasoline station, but more than thirty Albion buildings dating from Carmer's day have been designated for inclusion in the National Register of Historic Places.

ANNANDALE-ON-HUDSON

This village about twenty miles north of Poughkeepsie is the home of Bard College. Writers who have taught here include **Ralph Ellison,** who taught Russian and American literature from 1958 to 1961, and **Mary McCarthy,** who taught English in 1945 and 1946. McCarthy used her experiences as a college teacher in her satirical novel *The Groves of Academe* (1952).

AUBURN

Samuel Hopkins Adams, the journalist and author, owned a summer home for a time in this city at the northern end of Owasco Lake, about twenty-five miles west of Syracuse. **Edward Sandford Martin,** the humorist, essayist, and editor, was born on January 2, 1856, at Willowbrook, his family's home on Owasco Lake. Martin helped to found the *Harvard Lampoon* and the national humor magazine *Life* (1883–1936). He wrote for *Life* from 1887 to 1933, and was associated with *Harper's Weekly.*

AUSTERLITZ

Near this community about thirty miles southeast of Albany is Steepletop, the home of **Edna St. Vincent Millay** from 1925 until her death. The poet and her husband renovated the two farmhouses that stood on the property when they bought it and added several new structures, including a cabin studio, nestled in a pine grove a few hundred feet from the main house. It was there that Millay wrote her poetry. Millay's large literary output during her years at Steepletop included *The Buck in the Snow* (1928), *Huntsman, What Quarry?* (1939), and *Murder of Lidice* (1942). Millay died at Steepletop on October 19, 1950, and her ashes are buried here. Steepletop, a National Historic Landmark, is now the site of the Millay Colony for the Arts, a place for writers and artists to live and create amid the solitude and beauty of the area's rolling wooded hills. The colony is not accessible to visitors, but a state marker, incorrectly noting the years of Millay's residence, is located in Austerlitz.

BABYLON

Walt Whitman, during the winter of 1836–37, taught school near this community on Long Island's southern shore, about forty miles east of New York City. It is said that an incident at the school was the inspiration for his short story "Death in the Schoolroom" (1841).

BALLSTON SPA

This village about thirty miles north of Albany was the setting in December 1842 for a celebrated literary court case between **James Fenimore Cooper** and **Horace Greeley.** Cooper successfully sued Greeley and Thomas McElrath, owners of the New York *Tribune,* for libel in attacks on him and his writing. The court awarded Cooper two hundred dollars in damages. Greeley later published an account of the trial in the *Tribune.* The story was widely reprinted and reissued as a pamphlet, with the predictable result that Cooper sued Greeley once again. Here in Ballston Spa Cooper was also involved in litigation against Thurlow Weed, editor of the Albany *Evening Journal.* Weed, with Greeley and other newspapermen, had maintained a vicious barrage of words against Cooper for criticizing American life in his novels *Homeward Bound* (1838) and *Home as Found* (1838).

BARRYTOWN

Gore Vidal lived from 1950 to 1970 at Edgewater, his large, old house on the east bank of the Hudson River here, a few miles upriver from Kingston. At Edgewater Vidal wrote the novels *The Judgment of Paris* (1952) and *Messiah* (1954), the plays *Visit to a Small Planet* (1956) and *The Best Man* (1960), and many television plays.

BEACON

Writers associated with this city, about forty-five miles up the Hudson River from New York

Sweet sounds, oh, beautiful music, do not cease!
Reject me not into the world again.
With you alone is excellence and peace,
Mankind made plausible, his purpose plain.
Enchanted in your air benign and shrewd,
With limbs a-sprawl and empty faces pale,
The spiteful and the stingy and the rude
Sleep like scullions in the fairy-tale.
This moment is the best the world can give:
The tranquil blossom on the tortured stem.
Reject me not, sweet sounds! oh, let me live,
Till Doom espy my towers and scatter them,
A city spell-bound under the aging sun.
Music my rampart, and my only one.

—Edna St. Vincent Millay,
"On Hearing a Symphony of Beethoven"

. . . a large, square, double house, with a notice of sale in the windows of one of the lower apartments. There were two entrances, one of which had long been out of use but had never been removed. They were exactly alike—large white doors, with an arched frame and wide sidelights, perched upon little "stoops" of red stone, which descended sidewise to the brick pavement of the street.

—Henry James,
in *The Portrait of a Lady*

Marker on Route 203 in Austerlitz. The information it provides is incorrect: Edna St. Vincent Millay lived at Steepletop from 1925 until 1950.

And then Clyde, with the sound of Roberta's cries still in his ears, that last frantic, white, appealing look in her eyes, swimming heavily, gloomily and darkly to shore. And the thought that, after all, he had not really killed her. No, no. Thank God for that. He had not. And yet (stepping up on the near-by bank and shaking the water from his clothes) had he? Or, had he not?

—Theodore Dreiser, in *An American Tragedy*

City, include **Zelda Fitzgerald,** who in March 1934 was a patient at the Craig House Hospital, on Howland Avenue; and **Gertrude Knevels,** born on April 2, 1881, in a house still standing at 75 Knevels Avenue. Knevels became an author of plays and stories for children, as well as mystery stories for adults, including *The Octagon House Mystery* (1926).

BEDFORD

Sloan Wilson lived in the early 1960s in Bedford, which is about thirty-five miles northeast of New York City. The Victorian house he lived in is on Pound Ridge Road, between the Presbyterian Church and a pond at the edge of the village. While living in Bedford, Wilson wrote his novel *Georgie Winthrop* (1962).

BIG MOOSE

Just east of this town in the Adirondack Mountains, about fifty-five miles above Utica, is Big Moose Lake, the site of a murder that **Theodore Dreiser** was to use in writing his novel *An American Tragedy* (1925). In July 1906, a young man named Chester Gillette brought Grace Brown, made pregnant by Gillette, to Big Moose Lake, on the promise that he would marry her here. Instead, as was later brought out in Gillette's trial for murder, he drowned her in the lake and tried unsuccessfully to represent the tragedy as a boating accident. The motive for the murder was understood to be money: Grace Brown had stood in the way of Gillette's plans to marry into wealth and social position, and Gillette was executed for the crime. In the summer of 1923, while working on *An American Tragedy,* Dreiser visited the scene of the murder in the company of Helen Richardson, whom he was later to marry. They stayed at the Glenmore Hotel on the shore of Big Moose Lake, the same hotel at which Gillette and Brown had stayed. Dreiser was deeply involved with the story of the murder, and Richardson—who hoped the novelist would divorce his wife and marry her—is said to have wondered whether Dreiser might repeat the crime. Fortunately, both returned from this adventure, and Dreiser completed his best-known novel.

Hotel Glenmore in Big Moose. Theodore Dreiser stayed at the hotel in June of 1906 while writing *An American Tragedy.*

BOONVILLE

Walter D[umaux] Edmonds was born on July 15, 1903, in this town about thirty miles north of Utica. Edmonds spent much of his boyhood in Boonville, once a busy town on the Black River waterway and canal, and later wrote some of his books here. Many of his novels are set in northern New York State, and his first novel, *Rome Haul* (1929), is about the Boonville area. Among his books are *Drums Along the Mohawk* (1936), *Chad Hanna* (1940), *The Matchlock Gun* (1941), and *Bert Breen's Barn* (1975). Since he established himself as a writer, Edmonds has spent much of each year at Northlands, his 1,000-acre farm here.

Thomas S. Jones, Jr., a poet whose work was collected in the volumes *The Path of Dreams* (1905) and *The Voice in the Silence* (1911), was born in Boonville on November 6, 1882.

BREWSTER

Rex Stout, creator of the fictional detective Nero Wolfe, built a fourteen-room house off Milltown Road near Brewster, which is about fifty miles north of New York City. The house, called High Meadow, stands just across the state border, in Connecticut. Stout died at High Meadow, now private, on October 27, 1975, and his ashes were spread over the property. A bench dedicated to his memory by his publisher stands in a minipark on Main Street in Brewster. Plants in the park were contributed by the Stout family. (See DANBURY, CONNECTICUT.)

BRIARCLIFF MANOR

From 1925 to 1928, **John Hersey** lived with his parents at 44 Valentine Road, in a house still standing, in this village about thirty miles north of New York City. Hersey attended Briarcliff Junior and Senior High schools, at 1031 Pleasantville Road, now the site of the middle school.

BRIDGEHAMPTON

Bobby Van's, a restaurant and drinking place on Main Street in Bridgehampton, has been a favorite gathering place for many writers summering on eastern Long Island, including Truman Capote, James Jones, Willie Morris, Irwin Shaw, and Wilfrid Sheed.

BROADALBIN

Shortly after the publication of his popular novel *In the Quarter* (1894), **Robert W. Chambers** moved to his family's estate in Broadalbin, about thirty-five miles northwest of Albany. The novelist and short-story writer worked on his books here and, in winter, in a study in New York City. In 1912 Chambers commissioned his brother to redesign the family home, at 2 North Street, as a Colonial-style mansion, and Chambers made other improvements in the 800-acre estate while producing many more historical romances, from *The Red Republic* (1895) to *The Drums of Aulone* (1927). Chambers lived in Broadalbin until his death on December 16, 1933. He is buried in Broadalbin Mayfield Rural Cemetery. His house, somewhat altered, is now owned by St. Joseph's Roman Catholic Church and is not open to the public.

BROCKPORT

Mary Jane Holmes, one of the most popular American novelists in the second half of the nineteenth century, made her home for many years in this village eighteen miles west of Rochester. She was married in 1849 to David Holmes, a Brockton lawyer. After living for a short time in Versailles, Kentucky, where she gathered many impressions she used later in her books, Mrs. Holmes returned with her husband to live here. She spent much of her time in travel and in writing her thirty-nine sentimental novels, the one known best being *Lena Rivers* (1856). Two million copies of her books were sold in her lifetime, but they have not survived the test of time. In 1907, while returning from her summer home in Oak Bluffs, Massachusetts, Mrs. Holmes became ill. She died in Brockport on October 6 and is buried here.

BRONXVILLE

Writers associated with this village just north of New York City include **Carl Carmer,** who died at Lawrence Hospital here on September 11, 1976; **Lawrence Ferlinghetti,** the poet, who spent part of his boyhood growing up in a mansion owned by a member of the Lawrence family, founders of Sarah Lawrence College; and **Sinclair Lewis,** who bought a large Tudor-style house at 17 Wood End Lane and spent winters here with his wife **Dorothy Thompson** until April 1937, when Lewis decided that their disintegrating marriage had ended. Lewis left and Thompson retained the home, later renting it to **Vincent Sheean.** Sheean was living here when, on the night of February 9, 1941, the house caught fire. Sheean and his family escaped unharmed, and a thoughtful fireman saved the manuscript of Sheean's novel *Bird of the Wilderness* (1941), which Sheean later completed while living in a New York City hotel.

Sarah Lawrence College, in Bronxville, includes among its former faculty members **Horace Gregory, Edward Hoagland, Max Lerner, Mary McCarthy,** and **Harvey Swados.** McCarthy's fictional Jocelyn College in *The Groves of Academe* (1952) evokes thoughts of Sarah Lawrence College and of Bard College, in Annandale-on-Hudson, where the novelist taught before turning to writing full-time.

BUFFALO

This city at the eastern tip of Lake Erie is the birthplace of a number of distinguished writers. **Elizabeth J[ane] Coatsworth,** poet, novelist, and author of numerous books for children, was born here on May 31, 1893. Her poetry included *Fox Footprints* (1923) and *Compass Rose* (1929), and her best-known children's books were *The Cat Who Went to Heaven* (1930) and the five Sally books. **Paul Horgan,** born here on August 1, 1903, has set many of his books in the Southwest, but the cities of Buffalo and Rochester are models for the fictional city of Dorchester, which appears in his novel *The Fault of Angels* (1933). Horgan lived at 133 St. James Place as a young child, and at 45 Middlesex Road from age eight to twelve. In his early twenties, Horgan lived at 534 Delaware Avenue. **[James Morrison] Steele MacKaye,** born in Buffalo on June 6, 1842, was an actor and playwright. Of a score of plays that he wrote, perhaps the best known was *Hazel Kirke* (1880). His birthplace, an imposing white limestone structure built in 1837 and called The Castle, was razed in 1953 to make way for the approach road to the Peace Bridge linking Buffalo with Fort Erie, Ontario.

Harvey Swados, the novelist, short-story writer, and essayist, was born in Buffalo on October 28, 1920. His novels include *False Coin* (1959) and *The Will* (1963). **Elizabeth Yates** was born on December 6, 1905, in a house at 1243 Delaware Avenue, where the Unity Church of Practical Christianity now stands. Yates is known best for her books for children, including *Amos Fortune, Free Man* (1950).

Buffalo has been the home of authors born elsewhere, including **[Janet] Taylor Caldwell,** who came to America from her native England with her parents in 1907, grew up here and, in 1931, received a B.A. degree from the University of Buffalo. Caldwell lived for many years in Eggertsville, a suburb of Buffalo, and now lives in Buffalo. *Dynasty of Death* (1938), *The Eagles Gather* (1940), and her many other dramatic novels have been tremendously popular. **F. Scott Fitzgerald** lived with his parents here from 1898 to 1901 and from 1903 to 1908. During the period from 1901 to 1903, the family lived in Syracuse.

> *Jocelyn College . . . had a faculty of forty-one persons and a student-body of two hundred and eighty-three—a ratio of one teacher to every 6.9 students, which made possible the practice of "individual instruction" as carried on at Bennington (6:1), Sarah Lawrence (6.4:1), Bard (6.9:1), and St. John's (7.7:1). It had been founded in the late Thirties by an experimental educator and lecturer, backed by a group of society-women in Cleveland, Pittsburgh, and Cincinnati who wished to strike a middle course between the existing extremes, between Aquinas and Dewey, the modern dance and the labor movement.*
>
> —Mary McCarthy, in *The Groves of Academe*

Bronxville home of Sinclair Lewis and his wife, Dorothy Thompson. Lewis bought the house in 1933, but left in 1937 when the marriage broke up.

Marker at the site of Mark Twain's home in Buffalo. Twain and his wife moved into the house on February 3, 1870, a day after their wedding, but put the house up for sale in March of 1871.

While in Buffalo, young Fitzgerald attended Holy Angels Academy, and his family lived at several addresses, including 29 Irving Place and 71 Highland Avenue.

Charles Austin Fosdick, who wrote nearly sixty adventure books for boys, grew up here in the 1840s and 1850s. Fosdick wrote under the name **Harry Castleman. Edwin O'Connor** was a radio announcer for some time at station WBEN, which is still broadcasting. O'Connor used his experiences in radio here and elsewhere for his novel *The Oracle* (1951), which satirized radio broadcasting. **Edward Streeter** spent much of his youth in Buffalo, devoting a good deal of his time to reading books from the family library. At the age of nine he began—and abandoned—his first novel. From 1914 to 1916, after his graduation from Harvard, the humorist and novelist worked for the Buffalo *Express.*

The *Express,* now the Buffalo *Courier-Express* at 785 Main Street, is better known through its association with the career of **Mark Twain,** who was part owner and general editor of the newspaper from August 1869 to April 1871. The year 1869 saw publication of Twain's book *The Innocents Abroad,* and in the following year Twain married Olivia Langdon, the daughter of Jervis Langdon, a coal and iron magnate. Langdon had lent Twain a substantial sum of money with which to buy into the *Express,* then located at 14 Swan Street. Both men considered the newspaper a useful property for advancing their separate interests. Upon the marriage of his daughter to Twain, Langdon demonstrated his generosity by buying the newlyweds a three-story brick mansion, no longer standing, at 472 Delaware Avenue, which now is the address of the Cloister restaurant. A marker at the site commemorates Twain's residence in Buffalo.

Twain's life in Buffalo was burdened by tragedy. When Jervis Langdon died, in August 1870, Olivia suffered a nervous collapse. Twain's early burst of enthusiasm for the *Express* waned, and he had trouble completing *Roughing It* (1872), the account of his adventures in the West and in Hawaii. In November 1870, Olivia gave birth prematurely to a son, Langdon. Both mother and child became ill, and in February 1871 Olivia contracted typhoid fever. By that time Twain had decided to leave Buffalo for Elmira, Olivia's hometown, and to turn to writing full time. Twain sold his share of the Buffalo *Express* in April, and the family moved to Elmira.

Writers associated with the University of Buffalo include **Robert Creeley,** the poet, who taught here in 1966 and 1967; **Imamu Amiri Baraka,** original name **LeRoi Jones,** who was a visiting professor at the university in 1964; and **William Carlos Williams.** The collection of Williams papers here is one of the two largest in the country,

Home of F. Scott Fitzgerald at 29 Irving Place in Buffalo. Fitzgerald lived here from 1903 to 1905.

> *Tom put on his best suit, a freshly cleaned and pressed gray flannel. On his way to work he stopped in Grand Central Station to buy a clean white handkerchief and to have his shoes shined. During his luncheon hour he set out to visit the United Broadcasting Corporation. . . . There were eighteen elevators in the lobby of the United Broadcasting building. They were all brass colored and looked as though they were made of money.*
>
> —Sloan Wilson,
> in *The Man in the Gray Flannel Suit*

the other being at Yale. The University of Buffalo also has collections of the papers of the poets Jeremy Ingalls and Vachel Lindsay and, for students of best-selling novels, the original manuscript of the novel *The Man in the Gray Flannel Suit* (1955), by **Sloan Wilson,** who taught at the university in the early 1950s and lived in Snyder, a Buffalo suburb. Wilson wrote his biting portrayal of life in the executive suite while living at 204 Burbank Drive, in Snyder.

CANANDAIGUA

This city southeast of Rochester was the birthplace, on January 4, 1883, of **Max [Forrester] Eastman,** the poet, critic, and a leading force in the intellectual American leftist movement of the early twentieth century. Eastman helped found and edit the radical journals *Masses* (1913–17) and *The Liberator* (1918–22), but his disillusion with the system of government in the Soviet Union led him to criticize Marxism in several books, including *Marxism: Is It a Science?* (1940) and *Reflections on the Failure of Socialism* (1955). His most successful book was his first, *The Enjoyment of Poetry* (1913), an analysis of metaphor and simile. The book has gone through many editions.

CANTON

Irving Bacheller grew up in this village in northern New York. At an early age the novelist moved here from Pierrepont, his birthplace. He attended Canton Academy and enrolled in 1878 at St. Lawrence University in Canton. He was graduated in 1882. Bacheller used his childhood experiences in his first novel, *Eben Holden* (1900), set in the St. Lawrence valley. Canton is proud of its association with **Frederic [Sackrider] Remington,** born here on October 4, 1861, and buried here. Remington is known best for his sculptures and paintings of Western subjects, but he also wrote and illustrated a number of books, including *Pony Tracks* (1895), *Stories of Peace and War* (1899), and *The Way of an Indian* (1906). His birthplace at 55 Court Street is now private, but bears a commemorative plaque.

CARDIFF

In October 1869, workmen digging a well on a farm near this town, about ten miles south of Syracuse, unearthed what appeared to be the petrified body of a man more than ten feet tall. The farmer on whose land the giant was unearthed set up a tent and charged admission to view the creature, and soon a vigorous debate arose over whether the giant was a hoax. Eventually it was learned that the Cardiff Giant, as the curiosity came to be called, had been sculpted from a block of Iowa gypsum, and that the farmer had been party to the hoax. This revelation did not keep the attraction from drawing big crowds, and a number of people invested in the Cardiff Giant, seeing in it a chance to earn profits. One of the investors was David Hannum, who is said to have been the model for the main character in the commercially successful novel *David Harum: A Story of American Life* (1898), written by E[dward] N[oyes] Westcott, a native of Syracuse. **Carl Carmer** wrote of the Cardiff Giant hoax in his book *Listen for a Lonesome Drum* (1936), which recounted New York State folklore. The Cardiff Giant is now on display in Cooperstown.

CATSKILL

Catskill, a village about thirty miles south of Albany and just west of the Hudson River, stands in the Catskill Mountain region made famous by Washington Irving as the locale for his story "Rip Van Winkle" (1820). Spanning the Hudson River near Catskill is the Rip Van Winkle Bridge.

CAZENOVIA

One of the writers associated with this village southeast of Syracuse is **Hervey Allen,** who served as trustee of Cazenovia Junior College and who married Annette Hyde Andrews, daughter of Syracuse attorney Charles W. Andrews, in Cazenovia on June 30, 1927. The novelist, soon to write *Anthony Adverse* (1933), lived for a time in a house on Rippleton Road, near the Andrews family home. **Walter Van Tilburg Clark** spent ten years as a teacher in the public schools in Cazenovia, and rented rooms in various houses that are still standing. His best-known novel, *The Ox-Bow Incident* (1940), a psychological portrayal of a Nevada lynching, was published while Clark was living in Cazenovia. **Charles Dudley Warner,** the novelist, editor, and travel writer, lived in Cazenovia as a boy from 1842 to 1845, and was graduated from Cazenovia Seminary in 1845.

CHAPPAQUA

Lawrence Ferlinghetti, the poet, spent part of his childhood in an orphanage in this community north of New York City, but Chappaqua's principal literary association is with **Horace Greeley.** Greeley's books included *The American Conflict* (2 volumes, 1864–6) and *Recollections of a Busy*

Whoever has made the voyage up the Hudson must remember the Kaatskill Mountains. They are a dismembered branch of the great Appalachian family, and are seen away to the west of the river, swelling up to a noble height, and lording it over the surrounding country.

—Washington Irving, in "Rip Van Winkle"

Birthplace of Frederic Remington in Canton.

For the Thanksgiving and Christmas feasts four heavy turkeys were bought and fattened for weeks: Eugene fed them with cans of shelled corn several times a day, but he could not bear to be present at their executions, because by that time their cheerful excited gobbles made echoes in his heart. Eliza baked for weeks in advance: the whole energy of the family focussed upon the great ritual of the feast. A day or two before, the auxiliary dainties arrived in piled grocer's boxes—the magic of strange foods and fruits was added to familiar fare: there were glossed sticky dates, cold rich figs, cramped belly to belly in small boxes, dusty raisins, mixed nuts—the almond, pecan, the meaty nigger-toe, the walnut, sacks of assorted candies, piles of yellow Florida oranges, tangerines, sharp, acrid, nostalgic odors.

—Thomas Wolfe,
in *Look Homeward, Angel*

Life (1868), but he is remembered principally as a journalist. Greeley founded the New York *Tribune* in 1841 and served as editor until 1872, when he died. Greeley bought a farm in Chappaqua in the 1850s and ran it until his death. In 1864 Greeley moved his family from its Chappaqua home, known as the House in the Woods, which no longer exists, to a two-story farmhouse at 100 King Street, now the Greeley House Gift Shop and listed in the National Register of Historic Places. Greeley was an active and progressive farmer, and the three-story barn he designed and built in 1856 is believed by some to be the first concrete barn ever erected. Converted into a residence in 1892 and named Rehoboth, the structure stands at 33 Aldridge Road and is also listed in the National Register. Two other Chappaqua sites associated with the Greeley family are included in the National Register. St. Mary the Virgin Church on South Greeley Avenue, a copy of a Gothic-style church in England, was built in 1906 by Greeley's daughter. Beside the church is a grove of towering evergreen trees planted by Greeley in the 1850s as a windbreak. The second site is the Chappaqua Railroad Station and Depot Plaza, built on property donated by Greeley's daughter. Other points of interest include a statue of Greeley, which stands near the southbound exit of the Saw Mill River Parkway, and a room in the New Castle Town Hall, 200 South Greeley Avenue, which contains a desk used by Greeley, and other memorabilia. The room is open by appointment on Wednesdays. Chappaqua's high school is named in honor of Greeley.

Thomas Wolfe had Christmas dinner in 1937 at the home of friends living at 1177 Hardscrabble Road. The dinner was prepared to correspond to Wolfe's recollections of his childhood Christmas feasts. Wolfe later referred to the holiday as his last Christmas, the happiest since his boyhood. He died in 1938.

CHAUTAUQUA

The Chautauqua Institution was founded in 1874 in this resort village northwest of Jamestown and on the northwestern shore of Chautauqua Lake. The Chautauqua Institution originally offered training for Sunday-school teachers but soon began to conduct cultural and educational programs for the public. Tens of thousands of people have attended Chautauqua's summer programs each year ever since, and the great attraction of the programs has always been the appearances of guest lecturers, speaking on a broad range of topics. The Chautauqua idea spread widely in the nineteenth century, with the result that communities throughout the United States established their own meetings after the style of Chautauqua. In addition, traveling tent chautauquas toured the nation in summertime, providing welcome cultural stimulation for people in outlying communities before the advent of radio and motion pictures. A popular lecturer at the Chautauqua Institution in the late nineteenth century was **Isabella [MacDonald] Alden,** author of more than seventy-five books, most of them for young people and written under the name **Pansy.** Mrs. Alden spent her summers from 1873 to 1895 in Chautauqua, and all three of the houses she lived in—6 Ames Avenue, 37 Janes Avenue, and 20 Forest Avenue—are still standing. The house on Janes Avenue contains a large desk believed to have been built for Mrs. Alden. Among the books written by Mrs. Alden were a number of extremely popular novels set in Chautauqua, including *Four Girls at Chautauqua* (1876).

CHITTENANGO

L[yman] Frank Baum, best known for his Oz books, particularly *The Wonderful Wizard of Oz* (1900), was born on May 15, 1856, in this

Home of Horace Greeley at 100 King Street in Chappaqua. The editor, shown in the photograph below, lived here from 1864 until his death in 1872.

> *The room was carpeted, and furnished with convenient, substantial furniture, some of which was brought from the city, and the remainder having been manufactured by the mechanics of Templeton. There was a sideboard of mahogany, inlaid with ivory, and bearing enormous handles of glittering brass, and groaning under the piles of silver plate. Near it stood a set of prodigious tables, made of the wild cherry, to imitate the imported wood of the sideboard, but plain, and without ornament of any kind. Opposite to these stood a smaller table, formed from a lighter-colored wood, through the grains of which the wavy lines of the curled maple of the mountains were beautifully undulating.*
>
> —James Fenimore Cooper, in *The Pioneers*

Engraving of rear view of Otsego Hall, home of James Fenimore Cooper in Cooperstown. The statue of Cooper shown below stands where the house once stood.

village sixteen miles east of Syracuse. Baum wrote some sixty books, most of them for children, and used the pen name **Edith Van Dyne** for the books he wrote for girls.

CLAVERACK

Stephen Crane attended Claverack College and Hudson River Institute in this community about thirty miles south of Albany. The school was closed in 1902 and destroyed by fire in 1978, but its site is identified by a state marker.

CLINTON

Writers associated with Hamilton College, located in this village southwest of Utica, include **Samuel Hopkins Adams,** the novelist and journalist, who received a B.A. here in 1891; **Henry Wheeler Shaw,** better known as **Josh Billings,** who enrolled at Hamilton in 1832 but remained for only one year; **Carl Carmer,** who was graduated in 1914 and served as head of the college's public-speaking department in 1919; **Charles Dudley Warner,** who received a B.A. in 1851; and **Alexander Woollcott,** who was graduated in 1909. Hamilton's most distinguished literary alumnus is **Ezra Pound,** who received his Ph.B. in 1905. While here, from 1903 to 1905, Pound lived at 17 Hungerford Street.

Clinton Scollard, born in Clinton on September 28, 1860, grew up here, and was educated at Hamilton College: B.A., 1881; M.A., 1884. Scollard taught English at Hamilton from 1888 to 1896 and in 1911 and 1912. His works published during his lifetime include *Pictures in Song* (1884) and *Songs of a Sylvan Lover* (1912). In addition, some of Scollard's poems were collected in *The Singing Heart* (1934). The poet's birthplace on Fountain Street, just south of the Clinton Baptist Church, is now private, as are his boyhood home at 1 Marvin Street and his later home at 70 College Street. Scollard died on November 19, 1932, and was buried in Sunset Hill Cemetery in Clinton.

COLD SPRING

Paul Corey lived between 1931 and 1947 in two different homes in this village twenty miles south of Poughkeepsie. During this period he wrote the novels *Three Miles Square* (1939), *County Seat* (1941), and *Acres of Antaes* (1946), as well as other books, articles, and short stories.

COOPERSTOWN

This village, about forty miles west of Albany, was founded in the 1780s by William Cooper, father of **James Fenimore Cooper.** The novelist's father owned a large estate on the shore of Otsego Lake. His large frame house, known as Manor House, was later moved so as not to obstruct the view of the lake from Otsego Hall, the family's new, larger home completed about 1798. Manor House, no longer standing, is described in *The Pioneers* (1823), and Judge Marmaduke Temple in the book is modeled after William Cooper. Cooper grew up in this region, which was then at the edge of the western frontier, and after his

schooling and a stint in the U.S. Navy, returned to Cooperstown in 1814 with his wife Susan to build his own home here, which he called Fenimore Farm, or Fenimore Cottage.

In 1817, unable to afford the expense of living in the house, the Coopers left Cooperstown. They returned in 1834, and Cooper spent most of his later life here, dying in Cooperstown on September 14, 1851. He is buried in the graveyard near Christ Church. A statue of his character Natty Bumppo, or Leather-Stocking, stands in Lakewood Cemetery just outside Cooperstown, and there is a statue of Cooper in Cooper Park, on the site of Otsego Hall. Cooper drew the inspiration for the best of his fiction from his memories of the frontier he knew as a boy. His Leather-Stocking Tales, which comprise *The Pioneers* (1823), *The Last of the Mohicans* (1826), *The Prairie* (1827), *The Pathfinder* (1840), and *The Deerslayer* (1841), derived their popular literary power from Cooper's romantic view of frontier life of the early 1800s. Otsego Lake is just one of the local places Cooper worked into his novels. It appears as Glimmerglass in *The Deerslayer.*

Fenimore House, built in 1932 on the site of Fenimore Cottage, is now the headquarters of the New York State Historical Association, which maintains the Cooper Room, containing paintings, manuscripts, and other Cooper memorabilia. It is open daily, and an admission fee is charged, with combination rates available for those who also wish to visit Cooperstown's National Baseball Hall of Fame and Museum, or the Farmer's Museum and Village Crossroads, which exhibits the Cardiff Giant, moved from its site in Cardiff, New York.

CORNWALL-ON-HUDSON

Djuna Barnes, known best for her avant-garde novel *Nightwood* (1936), was born in this village on the Hudson River, south of Newburgh, on June 12, 1892. **E[dward] P[ayson] Roe,** the novelist, retired from the Presbyterian ministry when his second novel, *Opening a Chestnut Burr* (1874), repeated the commercial success of his first, *Barriers Burned Away* (1872). He moved to a two-story frame house, no longer standing, on a twenty-three-acre estate in Cornwall-on-Hudson that has since been subdivided. Payson Road now winds through Roe's former estate, where he wrote such novels as *Near to Nature's Heart* (1876) and *The Earth Trembled* (1887). Roe died at home here on July 19, 1888, and is buried in Willow Dell Cemetery on Church Street. In the village, across from Payson Road, is E. P. Roe Park, which contains a bronze tablet commemorating the author.

N[athaniel] P[arker] Willis, a prolific author and poet, but known principally as an editor, lived from 1853 until his death on January 20, 1867, at Idlewild, his much admired three-story brick home in Cornwall-on-Hudson. The first floor of the house, at 20 Idlewild Park Drive, is still standing and in use. It was in Idlewild that Willis wrote his only novel, *Paul Fane* (1857).

Frequently Grandfather had to admonish us against snobbishness. He was not wholly free from it, himself, however. A visiting New England lady to whom he was presented at a church festival, said, "Adams? Adams? Do you claim kinship with the Boston Adamses?"

"There is a Boston branch, I believe," he answered cautiously.

"I refer to the Presidential Adamses," the lady said haughtily.

"Ah! I was personally acquainted with the Honorable John Quincy Adams. A very respectable gentleman. He may well have been a connection of our line, though, being no braghard, he would naturally not press the claim. . . . May I fetch you a glass of water, madam?"

—Samuel Hopkins Adams, in *Grandfather Stories*

CORTLAND

Carl [Lamson] Carmer was born in this city south of Syracuse on October 16, 1893, while his parents were visiting Carmer's grandfather's farm at Dryden, a town southwest of Cortland. Carmer, a teacher and writer, wrote a number of books about New York State, including *Listen for a Lonesome Drum* (1936); *The Hudson* (1939), a volume in the Rivers of America series; and the novel *Genesee Fever* (1941). **David Ross Locke,** the humorist who became known as **Petroleum V. Nasby,** began his newspaper career here in the 1840s. At the age of ten, Locke apprenticed himself to the Cortland *Democrat* and worked for the newspaper for seven years.

CROTON FALLS

Robert McCloskey, the artist and author, wrote *Blueberries for Sal* (1948), *One Morning in Maine* (1952), and *Time of Wonder* (1957) while living in Croton Falls, about forty-five miles north of New York City.

CROTON-ON-HUDSON

Many twentieth-century writers have been attracted to this village, which lies about thirty-five miles north of New York City. **John Dickson Carr,** the prolific mystery writer, lived at 71 Mt. Airy Road in the 1930s. This house was just across the street from the home of Max Eastman. **Floyd Dell,** the novelist and playwright, lived in the 1920s and 1930s at 75 Mt. Airy Road, and **Max Eastman** lived nearby, at 70 Mt. Airy Road, at the junction of Mt. Airy Road and Riverview Trail. Eastman bought the house, a converted cider mill, about 1917. **Waldo Frank,** the editor and writer, lived in a house on Glengary Road. **Edna St. Vincent Millay,** while visiting friends here, met Eugen Boissevain, who had helped to pay for John Reed's trip to Russia. Boissevain was renting the house next door to Max Eastman's home. Millay and Boissevain were married there on July 18, 1923, and lived there until November, when they moved to Greenwich Village. **John Reed** lived here, first with **Mabel Dodge [Luhan],** and later with his wife, Louise Bryant. Reed and Bryant in late 1916 bought a cottage on Mt. Airy Road, about half a mile from Eastman's home. Reed worked on his poetry, short stories, and articles in a shed near the house. **Upton Sinclair** lived during 1914 in a rented house in Croton-on-Hudson. Many of the houses mentioned are still standing, but they are privately owned and unmarked.

DANSVILLE

Carl Carmer, as a child, lived in Dansville, about forty-five miles south of Rochester. Carmer's father was principal of Dansville High School.

DUNKIRK

Samuel Hopkins Adams, the novelist and journalist, was born in this city in southwestern New York on January 21, 1871. Adams worked for *McClure's, Collier's,* and the New York *Tribune* in the early twentieth century, and became known as a muckraking journalist before establishing himself as a novelist. Among his books are the novel *Canal Town* (1944) and the collection *Grandfather Stories* (1955), tales told to Adams as a boy by his grandfather, who lived in Rochester.

EAST AURORA

This village southeast of Buffalo was the home in the late nineteenth and early twentieth centuries of the Roycrofters, a group of artisans working in a community founded by **Elbert Hubbard.** The Roycrofters produced well-crafted articles of copper, leather, and wood, and espoused the values of family, community, and workmanship—qualities Hubbard saw lacking in the machine age. Hubbard expressed his views in magazines he edited in East Aurora, *The Philistine* (1895–1915) and *The Fra* (1908–1917). His best-known essay, "A Message to Garcia" (1899), first published in *The Philistine,* was reprinted widely. Hubbard's writings include *Little Journeys,* a series of 170 biographical sketches issued as pamphlets from 1895 to 1909 and collected later in fourteen volumes. The Roycroft community was a victim of the Great Depression, but many of its buildings are still standing, designated in the National Register of Historic Places as the Roycroft Campus.

EAST HAMPTON

This summer resort on eastern Long Island has attracted many writers. **Philip Barry,** the playwright, owned a summer home off Jeffreys Lane in the 1940s. The playwright died on December 3, 1949, and was buried in Saint Philomena's Cemetery. East Hampton was the home of Gerald and Sara Murphy, friends of **F. Scott Fitzgerald,** who dedicated to them his novel *Tender Is The Night* (1934). Fitzgerald claimed that the characters in the novel, Dick and Nicole Diver, were modeled partly on the Murphys, but they denied this. Their mansion is still standing but is private.

Ring Lardner moved to East Hampton in the late 1920s, living off West End Road in a thirteen-room house he called Still Pond. Next door was the home of Lardner's friend Grantland Rice, the sports writer. Lardner spent his last years at Still Pond, and died there on September 25, 1933. **A. J. Liebling,** the writer and expert on boxing, and his wife **Jean Stafford,** the novelist, first came to East Hampton in the early 1960s. Liebling died on December 28, 1963, and is buried in Green River Cemetery in East Hampton. After his death, Jean Stafford continued living in their house at 929 Fireplace Springs Road. She died in White Plains, New York, on March 26, 1979.

John Howard Payne, author of the song "Home Sweet Home" (1823), lived as a boy in the house at 14 James Lane, now the Home Sweet Home Museum, which contains a collection of china and furniture as well as some Payne memorabilia. It is open daily and a modest admission fee is charged. **John Hall Wheelock,** the poet, summered here for many years in a house, still standing, which his father built on Georgica Road. He wrote of East Hampton in many of his poems, including the poem "Bonac," the name old residents use for East Hampton. He wrote in one sketch that his poems celebrated "the spell cast by East Hampton, its sea, its dunes, its beaches—its glades and gardens."

This is enchanted country, lies under
a spell,
Bird-haunted, ocean-haunted—
land of youth.
Land of first love, land of death also,
perhaps, . . .

—John Hall Wheelock,
in "Bonac"

EGGERTSVILLE

[Janet] Taylor Caldwell wrote her first novel, *Dynasty of Death* (1938), at the kitchen table of her home in Eggertsville, a Buffalo suburb. She lived here for a number of years and has lived in the Buffalo area for most of her life, all the while writing her many popular novels.

ELIZABETHTOWN

Louis Untermeyer in 1928 bought Stony Water, a 160-acre farm in the Adirondack Mountains near Elizabethtown, a village about thirty-two miles south of Plattsburgh. He used the farm as a summer place for many years.

Home of Jean Stafford and A. J. Liebling in East Hampton.

Spring came early to the Genesee Country that new year of 1795. Hardly had Nathan and Whirl realized that they needed a new almanac when the first bluebird was a promise of spring sky above the frozen savannahs of withered grass. There was the late February day when the trees on the mountains across the river stood blackly silhouetted in warm sunlight above the snow. On the next morning bits of bare ground spotted the far-spread whiteness and little streams were foaming down the steep slopes into the swelling river. At once Whirl gave up his job at Springfield House and joined Nathan in the task of tapping the hard maples in their upland acres.

—Carl Carmer,
in *Genesee Fever*

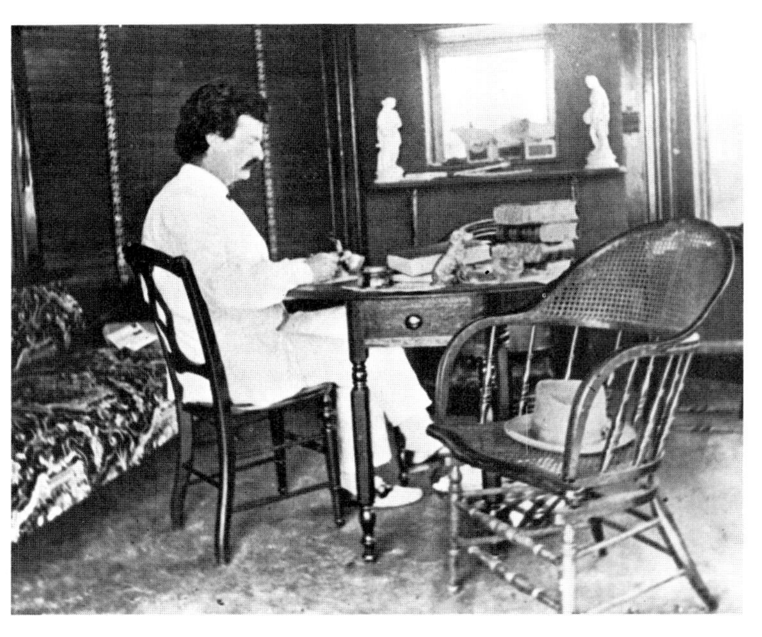

Mark Twain's study at its original site, Quarry Farm in Elmira, and Twain at work in the study in 1874. (*below right*) The main house at Quarry Farm. Twain family memorial in Elmira. The monument bears reliefs of Mark Twain and Ossip Gabrilowitsch, who married Twain's daughter Clara. At the base is the inscription:

Death is the starlit strip
Between the companionship
Of yesterday and the reunion
Of tomorrow
To the loving memory of
My father and my husband
C. C. G. 1937

ELMIRA

Max Eastman in the 1890s attended the Elmira Free Academy, then located at 600 Lake Street and now at 933 Hoffman Street, and **Clyde [William] Fitch,** whose many plays included *Beau Brummell* (1890) and *Captain Jinks of the Horse Marines* (1901), was born in Elmira on May 2, 1865.

The writer most closely associated with Elmira is **Mark Twain.** He first came here on the invitation of Charles Langdon, whom he had met on a steamboat trip in 1867. He fell in love with Langdon's sister Olivia, and married her on February 2, 1870, in the Langdon home. For many years Twain and his wife and family spent their summers in Elmira, at Quarry Farm, the home of Olivia's sister and brother-in-law, west of Crane Road and now private. At Quarry Farm, in an octagonal study, Twain worked on most of his best-known books. Twain described the study, now located on the grounds of Elmira College, as "octagonal, with a peaked roof, each face filled with a spacious window, and it sits perched in complete isolation on the very top of an elevation that commands leagues of valley and city and retreating ranges of distant blue hills." Twain worked from morning to late afternoon, tossing each page of manuscript to the floor as it was completed. At the peak of his literary power in these years, Twain found the solitude of his study here infinitely more conducive to writing than the study of his home in Hartford, Connecticut, where he could easily be dis-

tracted. Twain had four watering troughs, fed by springs, installed on the road from Elmira. Each was named for one of his children: Langdon, Susy, Clara, and Jean. The trough marked Clara is now on the Elmira College campus, next to Twain's study. At Quarry Farm, in the summer of 1889, a then-unknown writer named **Rudyard Kipling** came to see Twain, the visit marking the beginning of a lifelong friendship. Twain and his family are buried in the Langdon-Clemens plot in Woodlawn Cemetery, 1200 Walnut Street. Between Water Street in Elmira and the Chemung River is the Mark Twain Riverfront Park, named in honor of the writer, who once had suggested that Elmira raise a monument to Adam. The suggestion was not acted on, but Twain recalled the proposal in a short piece entitled "A Monument to Adam."

FAR ROCKAWAY

William Vaughn Moody, the poet and playwright, died in Colorado on October 17, 1910, and his ashes were cast over the waters of Jamaica Bay, off Far Rockaway, near Kennedy Airport. **John Hall Wheelock,** the poet and editor, was born here on September 9, 1886. Wheelock won a Bollingen Prize for his volume *The Gardener* (1961). A posthumous volume of Wheelock's poetry, *This Blessed Earth,* appeared in 1978.

FIRE ISLAND

This barrier beach off Long Island's southern shore attracts writers in summer. **W. H. Auden** spent summers here in the early 1940s in a cabin he called Bective Poplars. **Theodore H. White** lived on Fire Island while he wrote his novel *The Mountain Road* (1958), and **Herman Wouk,** at his summer home on Beachwold Road, worked on his play *The Caine Mutiny Court-Martial* (1953) and the novel *Marjorie Morningstar* (1955).

But not all the literary associations of Fire Island have been happy ones. **Margaret Fuller,** the nineteenth-century editor and critic, died in a shipwreck off Fire Island on July 19, 1850. She was returning with her husband and young child from a trip to Italy, where she had witnessed the revolution of 1849. Only the body of her child was recovered, and her manuscript account of the revolution was lost. Fuller is remembered in a monument on the beach at Fire Island, erected in 1901 through the efforts of Julia Ward Howe. Another Fire Island tragedy was the fatal accident suffered by **Frank O'Hara,** the poet and art critic. He was struck by a taxicab here on July 24, 1966, and died the next day at a hospital in Mastic Beach, Long Island.

FONDA

James Fenimore Cooper may hold the all-time record for literary litigiousness. The Old Court House in this village about forty miles northwest of Albany was the scene of a number of lawsuits brought by the novelist in the 1830s and 1840s. Cooper sued newspaper editors and publishers for their responses to opinions he had expressed in his novel *Home as Found* (1838). For his views on American life, Cooper had been called everything from a lunatic to a traitor by such editors as James Watson Webb of the New York *Enquirer* and Thurlow Weed of the Albany *Evening Journal.* Cooper brought Webb and Weed to trial here and won libel judgments against them. Editors may have learned not to slander Cooper, but the bad publicity Cooper received damaged his popularity. (See BALLSTON SPA, NEW YORK.)

FORESTVILLE

George Abbott was born on June 25, 1887, in Forestville, southeast of Buffalo. Abbott is co-author of many successful plays, among them the Pulitzer Prize-winning *Fiorello!* (1960), written with Jerome Weidman.

GARDEN CITY

From 1915 to 1917, **Clare Boothe** attended St. Mary's School in Garden City, on Long Island, and **Christopher Morley** worked in Garden City from 1913 to 1917 for Doubleday Company, the publishing house formerly located here.

GENEVA

Joseph Kirkland, known best for his realistic novel *Zury: The Meanest Man in Spring County* (1887), was born on January 7, 1830, in Geneva, a city on Seneca Lake in western New York. The home he lived in as a child, at 803 South Main Street, is still standing.

GLENORA

Paul Bowles, author of novels dealing with Westerners in the Arab world, where he now lives, spent his childhood summers in this community on the west shore of Seneca Lake, about ten miles from Watkins Glen. The novelist has reported: "The one region in the U.S. which I feel has become a part of me is the country around Seneca Lake (as it was fifty or sixty years ago)."

GLENS FALLS

This city north of Albany was built near a sixty-foot falls in the Hudson River. Cooper's Cave, at the foot of the falls here, was the setting for some of the action in *The Last of the Mohicans* (1826) by **James Fenimore Cooper.** A spiral staircase that once made the cave accessible for the literary visitor has been removed, and now one can only guess from afar at the feelings of Cora and Uncas as they hid with the others to avoid attack.

James Roberts Gilmore, author of a series of Civil War novels carrying the pen name **Edmund Kirke,** spent his last years in Glens Falls, and died on November 16, 1903, in his home at 20 Washington Street, now used by an insurance agency. Gilmore is buried in Pine View Cemetery on Quaker Road.

GLOVERSVILLE

Margaret Widdemer, the poet and novelist, spent the last few years of her life in this city in east-central New York. She lived at 109 First Avenue, and died at Nathan Littauer Hospital on July 14, 1978.

> *He approached the farther end of the cavern, to an outlet, which, like the others, was concealed by blankets, and removing the thick screen, breathed the fresh and reviving air from the cataract. One arm of the river flowed through a deep, narrow ravine, which its current had worn in the soft rock, directly beneath his feet, forming an effectual defence, as he believed, against any danger from that quarter; the water, a few rods above them, plunging, glancing, and sweeping along, in its most violent and broken manner.*
>
> —James Fenimore Cooper, in *The Last of the Mohicans*

Twain grave marker in Elmira.

Arietta Minot Fay, at thirty-seven, still lived in the house in which she, her father and all their known male forbears but the first had been born, a white, Hudson River-bracketed house, much winged and gabled but with a Revolutionary cottage at its core, set in a tiny village, once only a road, on the west shore of the Hudson River, about twenty-five miles from New York.

—Hortense Calisher,
in "Mrs. Fay Dines on Zebra"

Home of Hortense Calisher in Grand View on Hudson, the author's home from 1946 to 1958

GOSHEN

Before the American Revolution, **Michel-Guillaume Jean de Crèvecoeur,** also known as **J. Hector St. John de Crèvecoeur,** settled on a farm near this village in southeastern New York. Crèvecoeur, who was born in France in 1735 and emigrated to Canada and then to New York in 1759, is known best for his *Letters from an American Farmer* (1782), twelve attractive essays based on his experiences in America. A Loyalist, Crèvecoeur fled Goshen when the Revolution broke out. When he returned in 1783, he found that his wife had died, his farm had been destroyed, and his children had been scattered. Crèvecoeur's farm was located midway between Chester and Craigville.

GRAND VIEW ON HUDSON

Maxwell Anderson lived in 1919 in a rented house in this community south of Nyack. **Hortense Calisher** lived at 251 River Road in 1948, when she first began to write professionally. Grand View on Hudson is the setting for her short story "Mrs. Fay Dines on Zebra," which can be read in *The Collected Stories of Hortense Calisher* (1975). The community also figures in her novel *On Keeping Women* (1977).

GREAT NECK

Towns on the north shore of Long Island have long served the wealthy and the well known, and Great Neck is one of the towns most favored. Among authors who have come here are **George M. Cohan,** who lived at 100 Kings Point Road, and **Frederic Dannay,** who with his cousin Manfred B. Lee wrote detective stories under the pseudonym **Ellery Queen.** Dannay lived at 91 Old Mill Road. **Will** and **Ariel Durant,** in May 1928, after *The Story of Philosophy* (1926) was published, took a Colonial-style house at 44 North Drive, still standing. In February 1932, the explicators of the history of ideas moved on to a house at 2 Henry Street, also still standing. In the spring of 1934, the Durants moved once again, to 51 Deepdale Drive, also still standing.

Surely the best known of Great Neck's literary residents of the past are **F. Scott** and **Zelda Fitzgerald,** who moved in October 1922 to a house at 6 Gateway Drive, now private. For the eighteen months that they lived in Great Neck, the Fitzgeralds kept up a social free-for-all, their trademark in the Roaring Twenties. For the many visitors who came to spend weekends at their house, the Fitzgeralds wrote a set of rules, among them: "Visitors are requested not to break down doors in search of liquor, even when authorized

Home of F. Scott Fitzgerald in Great Neck, where Fitzgerald worked on *The Great Gatsby.*

A wafer of a moon was shining over Gatsby's house, making the night fine as before, and surviving the laughter and the sound of his still glowing garden. A sudden emptiness seemed to flow now from the windows and the great doors, endowing with complete isolation the figure of the host, who stood on the porch, his hand up in a formal gesture of farewell.

—F. Scott Fitzgerald,
in *The Great Gatsby*

to do so by the host and hostess." Fitzgerald once estimated that his high living cost him $36,000 a year, and he helped defray his expenses by writing "How to Live on $36,000 a Year," which appeared in the *Saturday Evening Post* on April 5, 1924. Great Neck has a permanent place in American literature as the principal setting used in Fitzgerald's best novel, *The Great Gatsby* (1925), and Fitzgerald worked on it in Great Neck in between grinding out short stories and articles to finance his flamboyant life. The novel remains a monument to, and a condemnation of, the American dream, twentieth-century style. The community of West Egg in the novel was modeled on Kings Point Peninsula here, and Gatsby's home is believed to have been modeled after the Brickman estate, at the end of Kings Point Road.

Other Great Neck residents included **Oscar Hammerstein,** the librettist and lyricist, who lived from the mid-1920s until the mid-1930s in a house at 13 Grace Avenue, no longer standing, and at 40 Shore Drive, in a house still standing, which Hammerstein built in 1926 at a cost of $145,000. **Ring Lardner** moved in 1921 into a three-story house at 325 East Shore Road that Lardner called The Mange. There he entertained such literary friends as F. Scott Fitzgerald, Maxwell Perkins, and Edmund Wilson. Fitzgerald and Lardner became strong friends during Fitzgerald's stay in Great Neck, and it was Fitzgerald who encouraged Lardner to gather a group of his stories for publication in book form, and pressed Maxwell Perkins to publish the book. The result was Lardner's collection *How to Write Short Stories* (1924). Fitzgerald was stunned to find that Lardner had to photograph the stories from the magazines in which they originally appeared because Lardner had not retained manuscript copies. Lardner and Fitzgerald spent many evenings together, and often visited each other's homes. Once, after a number of drinks, they decided that they would like to meet the celebrated novelist **Joseph Conrad,** who then was staying in Great Neck at the home of Frank Nelson Doubleday, Conrad's American publisher. They wandered over to Doubleday's estate and, hoping to attract the novelist's attention, performed a dance on the lawn. They were ejected from the premises by the caretaker before they had a chance to meet Conrad.

Another Great Neck resident in the 1920s was **Grantland Rice,** the sportswriter. A more recent resident was **Herman Wouk,** who lived at 223 Kings Point Road from 1948 to 1951. During that time he wrote the play *The Traitor* (1949); the novel *The Caine Mutiny* (1951), for which he won a Pulitzer Prize; and the novella *The Lomokome Papers* (1956).

GREENPORT

Walt Whitman, during summers in the years before the Civil War, visited his sister Mary and her husband in this town on the northeastern tip of Long Island. It was here, in 1849, that he wrote his "Letters from a Travelling Bachelor," which appeared in the New York *Sunday Dispatch* from October 1849 to January 1850.

HAINES FALLS

Haines Falls, southwest of Catskill, claims the boulder near which Rip Van Winkle is said to have taken his long nap in the story "Rip Van Winkle," by **Washington Irving.** The boulder, once marked with white paint, is on South Mountain, and is accessible to those who relish a hike.

He rubbed his eyes—it was a bright sunny morning. The birds were hopping and twittering among the bushes, and the eagle was wheeling aloft, and breasting the pure mountain breeze. "Surely," thought Rip, "I have not slept here all night."

—Washington Irving,
in "Rip Van Winkle"

HAMBURG

George Abbott lived in this village south of Buffalo and attended Hamburg High School, which was located at the corner of Union and Center streets and now is used as an elementary school. The playwright lived with his family in a house that stood on Main Street and later was moved to a location on South Creek Road. **Clyde B. Davis,** author of the novel *Thudbury* (1952), set in western New York, lived in Hamburg and

The Mange, home of Ring Lardner in Great Neck. While living here from 1921 to 1928, Lardner became a close friend of F. Scott Fitzgerald's.

"When you got a balker to dispose of," said David gravely, "you can't alwus pick an' choose. Fust come, fust served."

—E. N. Westcott, in *David Harum*

. . . it is a repulsive and nasty book. On exhibiting himself, like the silly ostrich, the poet hastens to hide his better, and expose his more indecent parts—as though it were necessary to truth or manliness, to unduly expose human or other nature.

—Editorial, "Leaves of Grass," in *The Long-Islander* of December 10, 1858

When the rifle cracked it shook his soul to a profound depth. Creation rocked and the bear stumbled.

The little man sprang forward with a roar. He scrambled hastily in the bear's track. The splash of red, now dim, threw a faint, timid beam of a kindred shade on the snow. The little man bounded in the air.

"Hit!" he yelled, and ran on. Some hundreds of yards forward he came to a dead bear with his nose in the snow. Blood was oozing slowly from a wound under the shoulder, and the snow about was sprinkled with blood. A mad froth lay in the animal's open mouth, and his limbs were twisted from agony.

The little man yelled again and sprang forward, waving his hat as if he were leading the cheering of thousands. He ran up and kicked the ribs of the bear. Upon his face was the smile of the successful lover.

—Stephen Crane, in *Sullivan County Tales and Sketches*

worked for the Buffalo *Times* from 1931 to 1937. Davis lived at two, possibly more, addresses, including 202 Lake Street and 71 Rosedale Avenue. **Charles Austin Fosdick,** who wrote adventure stories for boys, spent his last years in obscurity in Hamburg, where he died on August 22, 1915.

HARRISON

Maxwell Geismar, the critic, lived for a number of years in Harrison, which lies about twenty miles northeast of New York City. He died at his home on Winfield Avenue on July 24, 1979.

HARTWOOD

Stephen Crane spent winter vacations in the 1890s in the area around Hartwood, in Sullivan County, about forty-five miles west of Newburgh. In 1893 his brother moved to a house on Mill Pond Road, and Crane, in the last years of his brief life, considered this his home, when he was not off traveling. The Hartwood area is the locale for Crane's *Sullivan County Tales and Sketches,* which appeared in newspapers and magazines in 1892, and now can be read in a 1968 edition.

HAWTHORNE

Rose Hawthorne Lathrop, the poet, memoirist, and daughter of Nathaniel Hawthorne, founded Hawthorne's Rosary Hill Home in 1901. This home for the care of patients with incurable cancer is located at 600 Linda Avenue in Hawthorne, about twenty-five miles north of New York City.

HENDERSON

George Wilbur Peck was born on September 28, 1840, in this community near Lake Ontario, southwest of Watertown. Peck originated a phrase that was part of the popular vocabulary for many years: "Peck's bad boy." In Peck's humorous stories, a boy played practical jokes on his father. These stories first appeared in the New York *Sun,* and later were collected in *Peck's Bad Boy and His Pa* (1883) and other volumes.

HERKIMER

The County Courthouse, at 320 North Main Street in Herkimer, southeast of Utica, was the setting in 1906 for the sensational trial of Chester Gillette. Gillette was convicted and sentenced to death for drowning his pregnant girlfriend Grace Brown in Big Moose Lake, north of Herkimer. **Theodore Dreiser** used the tragic story as the basis for his novel *An American Tragedy* (1925). **Jesse Lynch Williams,** the novelist, short-story writer, and playwright, died in Herkimer on September 14, 1929, while visiting friends.

HIGHLAND FALLS

John Burroughs, the naturalist and author, taught in the early 1860s in this village on the Hudson River, about forty-five miles north of New York City. Storm King Mountain, north of Highland Falls, is said to be the setting of the long poem "The Culprit Fay," written by Joseph Rodman Drake in 1816 but not published until 1835, fifteen years after Drake's death.

E[dward] P[ayson] Roe served from about 1866 to 1874 as minister of the Presbyterian Church, in Highland Falls. Roe lived at 12 Mill Street, just north of the church. Both buildings are now privately owned and unmarked, but the United Church of the Highlands, at 140 Main Street, bears a plaque commemorating the novelist. Roe began to write after visiting the scene of the Chicago fire in 1871. *Barriers Burned Away* (1872), his novel based on what he saw, was very popular, as was his second novel, *Opening a Chestnut Burr* (1874). On the strength of these two successes, Roe gave up his ministry, moved to Cornwall-on-Hudson, and devoted the rest of his life to writing.

HILLSDALE

In 1946 **James Agee** bought a farm near Hillsdale, in eastern New York, about twelve miles southeast of Hudson. Agee enjoyed working amid the peaceful countryside, but also needed the noise and energy of New York City as a stimulus for writing. Agee died in New York City on May 16, 1955, and is buried a few miles from his farm here.

HOMER

This village in central New York was the birthplace, on September 24, 1835, of **William Osborn Stoddard,** author of more than one hundred books, mostly stories for boys. Stoddard served as a secretary to Abraham Lincoln, and later wrote of him in such books as *Abraham Lincoln* (1884) and *Lincoln at Work* (1899). His other works included *Little Smoke: A Tale of the Sioux* (1891) and *The Spy of Yorktown* (1903).

Homer was also the home of David Hannum, a banker, storyteller, and sharp-witted businessman whose motto was "Do unto the other fellow what the other fellow 'ud like to do to you—and do it fust." It is believed that **E[dward] N[oyes] Westcott** used Hannum as a model for the title character of his novel *David Harum: A Story of American Life* (1898), and that Homer is the town of Homeville in the novel. Hannum and Westcott's father were good friends and often visited each other here and in Syracuse, where Westcott lived. The two men invested in the promotion of the Cardiff Giant, unearthed in Cardiff, New York, in 1869. As a boy, Edward Noyes Westcott visited Hannum at his home, at 80 South Main Street, and no doubt later used his impressions of Hannum in creating the character David Harum, unlettered but shrewd. Hannum's home, later converted into a thriving restaurant, is now empty and deteriorating.

HOOSICK FALLS

James Fenimore Cooper once said that a man named Shipman was the model for Natty Bumppo, the improbably accurate marksman and scout in a number of Cooper's frontier romances. Cooper may have meant one Nathaniel Shipman, a hunter and trapper who is buried in an unmarked grave in the First Baptist Church cemetery in this village northeast of Troy. Nathaniel Shipman lived near Cooperstown in the early 1800s and was probably known to Cooper. His life had many similarities to that of the fictional Natty Bumppo.

HUNTINGTON

Walt Whitman was born on May 31, 1819, at West Hills, about three miles from the center of Huntington, Long Island. The house in which he was born, 246 Walt Whitman Road, south of Huntington, was built in 1810 by Whitman's father. It is open to visitors daily. Among the Whitman memorabilia collected here are a schoolmaster's desk and secretary used by the poet. Whitman and his family moved to Brooklyn a few days before his fourth birthday, but he spent much of his early manhood on Long Island, teaching school in a number of communities and wandering over the island, collecting impressions he would later form into the poems on which was built his reputation as one of the country's eminent poets. In 1838 Whitman returned to Huntington and started his own weekly newspaper, *The Long-Islander,* in a building at 313 Main Street. Whitman lived in a room over the printing shop and did much of the reporting, writing, printing, and delivery himself. His restless nature was reflected in the erratic publishing schedule of the newspaper—he would miss an issue or two now and then. In 1839 he sold out and went to New York City. The publication thrived, as did Whitman, but its editor had the misfortune, or myopia, to publish an unfavorable review of Whitman's *Leaves of Grass* (original publication 1855), which now, of course, is considered a classic. *The Long-Islander* has survived, and the newspaper is still published from 313 Main Street.

Headquarters of *The Long-Islander* in Huntington, founded by Walt Whitman in 1838. Earliest known photograph of Walt Whitman, taken in 1840.

HYDE PARK

James Kirke Paulding, the author and public official, spent his last years at Placentia, his comfortable estate above Hyde Park, which is just north of Poughkeepsie. Paulding retired here in 1846 after a long and productive career as a writer of stories, comedies, novels, and poems, and government official. His works included the novels *Koningsmarke* (1823), *Westward Ho!* (1832), and *The Puritan and His Daughter* (1849). Paulding died at his home on April 6, 1860. A commemorative marker on Route 9 notes the site of his home but gives the dates of his residence there as 1841 to 1860. A new house has been built on the original foundation of Paulding's home. Paulding's birthplace is not known, but he was born on August 22, 1778, in an area designated in 1697 as Great Nine Partners Patent, adjacent to Hyde Park and then part of Putnam County.

IRVINGTON

Carl Carmer lived from the 1950s until his death in 1976 in this village on the Hudson River, north of New York City. His home, the Octagon House, an eight-sided, eighteen-room mansion on West Clinton Street, is now private and undergoing restoration. When Carmer lived here, the

Birthplace of Walt Whitman near Huntington, and the plaque at the house, which reads:

To mark the birthplace of
WALT WHITMAN.
The Good Gray Poet.
Born May 31, 1819.

> *There was a time, some of the scientists say, when the Hudson, before it reached the highland barrier, wandered off to the southwest through Pennsylvania where it joined the Susquehanna and emptied its waters into Chesapeake Bay. Another stream, whose source was far to the south, gradually by headward erosion moved its beginnings northward until it captured and incorporated the runaway and straightened its course oceanward. That course led through a barrier of ancient mica-laden rocks upreared into the high ridges of the Appalachians. The Hudson worked its way through those ridges, aided by the glaciers that forced their masses southward through its valley, smoothing the northern slopes, carrying boulders away from the southern, leaving behind high-frowning cliffs, digging a channel so deep that its floor is in places nearly a thousand feet below sea level.*
>
> —Carl Carmer, in *The Hudson*

house was filled with books, and Carmer had desks installed in many of the rooms so he could write wherever he was when an idea came to him. The dome and cupola of the house, rising nearly one hundred feet, overlook the river Carmer loved, and wrote of in *The Hudson* (1939). A plaque commemorating Carmer's residence here has been presented for installation in the house when it is opened again. Among Carmer's books are *The Susquehanna* (1955), *My Kind of Country* (1966), and *The Drummer Boy of Vincennes* (1972). **Clarence Day** summered with his family in Irvington when he was a boy. He later wrote of his experiences here in *Life With Father* (1935).

ITHACA

Cornell University, opened in 1868 in this city in central New York, has attracted its share of writers, including **Louis Bromfield,** who studied agriculture here in 1914 and 1915; and **Pearl S. Buck,** who received an M.A. here in 1926. While a student here, she was awarded a prize for her essay "China and the West." **Vladimir Nabokov** was professor of Russian literature at Cornell from 1949 to 1959, during which time his novel *Lolita* (1958), first published in Paris, appeared in the United States. The collection *Nabokov's Dozen,* published in the same year as *Lolita,* contains stories written over many years. In 1959 Nabokov resigned his teaching position to write full time. **Hendrik Willem Van Loon** received a B.A. degree here in 1905 and lectured on European history at Cornell in 1915 and 1916. **Kurt Vonnegut** studied biochemistry at Cornell from 1940 to 1942. **E. B. White** enrolled at Cornell in 1917 but left the next year to serve in World War I. He returned after the war and received his B.A. in 1921. In his senior year he was editor-in-chief of the Cornell *Daily Sun.*

Octagon House, home of Carl Carmer in Irvington. The structure is one of numerous octagonal houses built along the Hudson River in the nineteenth century. Carmer discussed this architectural style in *The Hudson* (1939).

Ithaca is the setting of *Tess of the Storm Country* (1909) by **Grace Miller White,** born here on October 24, 1868. The novel described the conflict between the squatters who lived along the shore of Lake Cayuga and the more respectable people who lived in Ithaca proper. The novelist lived at several places in Ithaca, including 45 South Plain Street and 85 East Seneca Street. Her other novels included *The Shadow of the Sheltering Pines* (1919) and *Susan of the Storm* (1927).

JEFFERSON VALLEY

John Hollander, during the late 1940s, while he attended Columbia University, often visited his parents' home in Jefferson Valley, northeast of Peekskill. It was here that he wrote some of the poems that appeared in his first book, *A Crackling of Thorns* (1958).

> *The tops of the spruces here have always done*
> *Ragged things to the skies arranged behind them*
> *Like slates at twilight; and the morning sun*
> *Has marked out trees and hedgerows,*
> *and defined them*
> *In various greens, until, toward night, they blur*
> *Back into one rough palisade again,*
> *Furred thick with dusk. No wind we know*
> *can stir*
> *This olive blackness that surrounds us when*
> *It becomes the boundary of what we know*
> *By limiting the edge of what we see.*
> *When sunlight shows several spruces in a row,*
> *To know the green of a particular tree*
> *Means disbelief in darkness; and the lack*
> *Of a singular green is what we mean by black.*
>
> —John Hollander, "Jefferson Valley"

KATONAH

The John Jay Homestead, located on Jay Street in this village about thirty miles northeast of New York, was the home from 1801 until his death in 1829 of **John Jay,** first chief justice of the Supreme Court and author of some of the essays published in 1787 and 1788 as *The Federalist,* and signed **Publius. James Fenimore Cooper** was a close friend of Jay's and a frequent visitor to the Jay homestead. It is believed that Jay's stories about the Revolutionary War provided the inspiration for Cooper to write his second novel, *The Spy* (1821), the story of Harvey Birch, a Yankee peddler who acted as an informant for General Washington. The John Jay Homestead is open to visitors on Wednesdays through Saturdays all year.

Clyde Fitch, one of the most successful playwrights of the early 1900s, for a short time owned a home, still standing, on Pea Pond Road (Route 137) opposite Hook Road. **Sinclair Lewis** wrote part of his unsuccessful novel *Mantrap* (1926) while staying in the chauffeur's quarters of a Katonah estate called Loudon Farm.

KINDERHOOK

Washington Irving came to this village southeast of Albany in the spring of 1809, after the death of his fiancée, Matilda Hoffman. Irving stayed at the farm of a family friend, Judge William P. Van Ness, and submerged his sadness in work—tutoring the Van Ness children and writing what was to become *A History of New York* (1809), represented as the work of one **Diedrich Knickerbocker,** Irving's invented Dutch-American scholar. Irving used impressions of life in Kinderhook in writing his story "The Legend of Sleepy Hollow," published in *The Sketch Book* (1820). It is to Kinderhook, therefore, that readers are indebted in part for this classic tale of Ichabod Crane and the headless horseman. The Van Ness home, later called Lindenwald, served as the retirement home of Martin Van Buren, eighth president of the United States, and Irving visited him here. Lindenwald is now undergoing restoration by the National Park Service.

KINGSTON

This city on the west bank of the Hudson River, fifteen miles north of Poughkeepsie, was the birthplace on July 11, 1842, of **Henry Abbey,** a poet popular in the latter half of the nineteenth century. Abbey was born at Rondout, now the eastern district of Kingston, and spent almost his entire life there. He worked in the Rondout bank, and later in his family's feed, flour, and lime business. In Rondout, Abbey also wrote the poetry for which he is remembered, including *Ralph and Other Poems* (1866), *The Poems of Henry Abbey* (1885), and *Phaëthon* (1901). Abbey's home at 11 Linderman Street is not open to the public. **Manuel Komroff,** the novelist, who lived in nearby Woodstock, died in Kingston, at Benedictine Hospital, 105 Mary's Avenue, on December 10, 1974.

LAKE GEORGE

The southern end of Lake George, in northeastern New York, was the site of Fort William Henry, where a bloody massacre took place during the French and Indian War. James Fenimore Cooper wrote of the siege of the fort in *The Last of the Mohicans* (1826). **Edward Eggleston,** the novelist known for his tales of Hoosier life, spent much of his last twenty years—from the early 1880s to 1902—at Owl's Nest, an estate at Joshua's Rock, off Route 9L, near the town of Lake George. The estate, which was owned at that time by Eggleston's daughter and son-in-law, is still intact and includes the house and library built for Eggleston. He worked on two volumes of United States history while at Owl's Nest, and died there on September 2, 1902. His histories were published after his death. A contemporary author, Sloan Wilson, used Lake George as a setting in his novels *A Summer Place* (1958) and *All the Best People* (1970).

LANCASTER

Dorothy Thompson was born on July 9, 1894, in this residential suburb of Buffalo. She went on to become nationally known as a foreign correspondent and author of several books, including *I Saw Hitler* (1932), which established her fallibility as a prognosticator: Thompson predicted that Adolf Hitler would never rise to power.

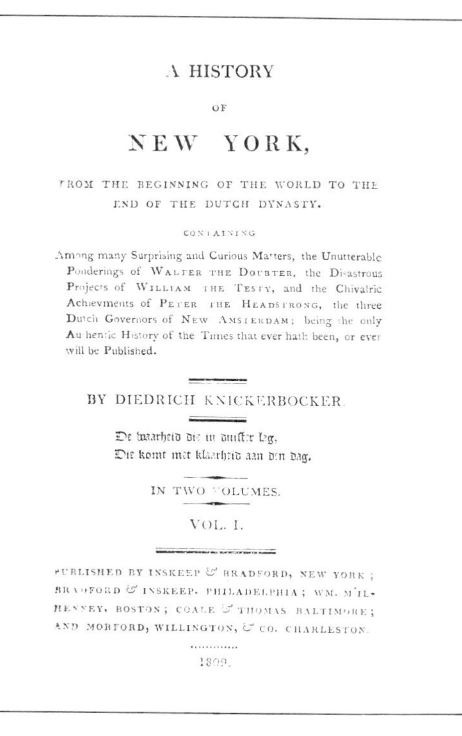

A HISTORY

OF

NEW YORK,

FROM THE BEGINNING OF THE WORLD TO THE END OF THE DUTCH DYNASTY.

CONTAINING

Among many Surprising and Curious Matters, the Unutterable Ponderings of WALTER THE DOUBTER, the Disastrous Projects of WILLIAM THE TESTY, and the Chivalric Achievments of PETER THE HEADSTRONG, the three Dutch Governors of NEW AMSTERDAM; being the only Authentic History of the Times that ever hath been, or ever will be Published.

BY DIEDRICH KNICKERBOCKER.

De waarheid die in duister lag,
Die komt met klaarheid aan den dag.

IN TWO VOLUMES.

VOL. I.

PUBLISHED BY INSKEEP & BRADFORD, NEW YORK; BRADFORD & INSKEEP, PHILADELPHIA; WM. M'ILHENNEY, BOSTON; COALE & THOMAS, BALTIMORE; AND MORFORD, WILLINGTON, & CO. CHARLESTON.

1809.

Title page of Washington Irving's first book, carrying the pen name Diedrich Knickerbocker. The couplet on the title page translates as:
The truth that lay in darkness,
Comes to light in the clearness
of the day.

LARCHMONT

Larchmont, twenty miles northeast of New York City, has had several literary residents. **Edward Albee,** the playwright, lived with his parents at 7 Bay Avenue. **Harry Golden,** the editor and writer, lived in a house on Larchmont Avenue in the late 1920s. **Joyce Kilmer,** the poet, moved with his family to a house at 15 Maple Avenue in October of 1917, just before he entered the U.S. Army. Kilmer died on the Western front on August 1, 1918. **Phyllis McGinley** lived for many years at 12 Hazel Lane, and often wrote of Larchmont in her satirical poetry. **Tad Mosel,** the playwright known principally for his television dramas, lived at 5 Kenmore Road in the 1930s.

LAWRENCE

Louis [Stanton] Auchincloss was born on September 27, 1917, in Lawrence, on the south shore of Long Island. A highly respected attorney and prolific author of novels and short stories, Auchincloss is known as an interpreter of the lives of old-money Eastern families. His works include *Tales of Manhattan* (1967), *The Winthrop Covenant* (1976), and *The Dark Lady* (1977).

LEWISTON

Legend has it that **James Fenimore Cooper** wrote his novel *The Spy* (1821) at Hustler's Tavern, which stood at the corner of Eighth and Center streets in Lewiston, seven miles north of Niagara Falls. What is more, the proprietors of the tavern, Tom and Kate Hustler, are said to have been Cooper's models for Sergeant Hollister and Betty Flanagan in the novel. Cooper observed that Kate Hustler, after mixing a drink, would put the tail feather of a cock into it, and he incorporated this innovation in *The Spy.* Lewiston was on the main transportation route of this region in the early nineteenth century, and many people

> *. . . Betty had the merit of being the inventor of that beverage which is so well known, at the present hour, to all the patriots who make a winter's march between the commercial and political capitals of this great state, and which is distinguished by the name of "cocktail." Elizabeth Flanagan was particularly well qualified, by education and circumstances, to perfect this improvement in liquors, having been literally brought up on its principal ingredient. . . .*
>
> —James Fenimore Cooper, in *The Spy*

I stand as on some mighty eagle's beak,
Eastward the sea absorbing, viewing
(nothing but sea and sky,)
The tossing waves, the foam, the ships
in the distance,
The wild unrest, the snowy, curling
caps—that inbound urge and
urge of waves,
Seeking the shores forever.

—Walt Whitman,
"From Montauk Point"

stopped there on their way to other places. Such a visitor was **Charles Dickens,** who stayed at the Frontier House, at 460 Center Street. The Frontier House is now the site of a lesser establishment, McDonald's Restaurant.

LYONS

Mary Ashley [Van Voorhis] Townsend, a poet and novelist who later lived in, and wrote of, New Orleans, was born on September 24, 1836, in Lyons, which stands between Rochester and Syracuse. Her works included the novel *The Brother Clerks* (1857) and *Down the Bayou and Other Poems* (1882).

MALTA

Katherine Anne Porter lived in Malta for a time at South Hill, her home on Cramer Road, now private. Malta is south of Saratoga Springs.

MAMARONECK

On New Year's Day, 1811, **James Fenimore Cooper** married Susan De Lancey at Heathcote Hall, her father's home on a hill in Mamaroneck that overlooks Long Island Sound. The Coopers lived there for a time after their marriage, before moving nearby to a small cottage of their own, known as Closet Hall, and said to have stood on Tompkins Avenue. Closet Hall, so called by Cooper because of its diminutive proportions, was their home until 1814. Heathcote Hall has been moved from its original site to one at the intersection of Boston Post Road and Fenimore Road, where it functions as a gasoline station, residence, and restaurant, known as the Fenimore Cooper Inn. Cooper's days in Mamaroneck saw him attempting to farm and failing—he had yet to show an interest in writing as a career.

The De Lancey house in Mamaroneck, now in commercial use. James Fenimore Cooper and Susan De Lancey were married in this house in 1811, nine years before Cooper published his first novel.

MASTIC BEACH

Frank O'Hara, the poet and critic, died on July 25, 1966, at Bayview Hospital, in Mastic, Long Island, after being struck by a taxicab on Fire Island.

McCOLLOMS

In 1895 **E. N. Westcott** came to Meacham Lake, near McColloms, a community north of Saranac Lake. Westcott was suffering from the tuberculosis that would kill him within three years, but began writing *David Harum, A Story of American Life,* which appeared posthumously in 1898.

MILTON

Mary Hallock Foote, author of the novel *The Led Horse Claim* (1883) and other novels set in the West, was born on November 19, 1847, in Milton, a town on the Hudson, just downriver from Poughkeepsie.

MONTAUK

Walt Whitman was a frequent visitor to Montauk, the easternmost tip of Long Island. The poet later wrote: "Most people possess an idea (if they think at all about the matter), that Montauk Point is a low stretch of land, poking its barren nose out toward the east, and hailing the sea-wearied mariner, as he approaches our republican shores, with a sort of dry and sterile countenance. Not so is the fact. To its very extreme verge, Montauk is fertile and verdant. The soil is rich, the grass is green and plentiful; the best patches of Indian corn and vegetables I saw last autumn are within gun shot of the salt waves of the Atlantic, being just five degrees east longitude from Washington, and the very extremest terra firma of the good State of New York."

MOUNT KISCO

Philip Barry lived in the 1920s and 1930s at 126 Barker Street, Mount Kisco, thirty-five miles north of New York City. The playwright divided his time between his home in Cannes, France, where he did most of his writing, and this house, which now is used for law offices. **Rebecca Harding Davis,** the novelist, died on September 29, 1910, at Cross Roads Farm, six miles southwest of Mount Kisco. The estate, on Chestnut Ridge Road, was the country home of her son, **Richard Harding Davis,** whose novels, short stories, and newspaper accounts of the major events of the day were very popular in the early twentieth century. Richard Harding Davis died at Cross Roads Farm on April 11, 1916.

In 1927 **Theodore Dreiser** bought a cabin overlooking Croton Lake near Mount Kisco, and set about building a larger house on the property. The new house he named Iroki, Japanese for "beauty." Iroki is now private, and the cabin in which Dreiser first lived still stands on the estate.

MOUNT MORRIS

Jessie B[elle] Rittenhouse, the poet, editor, and critic, and a founder of the Poetry Society of America, was born on December 8, 1869, in this town about thirty-five miles southwest of Rochester. Rittenhouse wrote the volumes *The Door of Dreams* (1918) and *The Lifted Cup* (1921).

MOUNT VERNON

E[lwyn] B[rooks] White, the essayist and humorist, whose writing has for many years graced the pages of *The New Yorker,* was born on July 11, 1899, in this city on the Bronx River, just north of New York City. The house he was born in, at 101 Summit Avenue, is still standing. White attended Lincoln School at 154 East Lincoln Avenue, now the site of a new school, and was graduated in 1917 from Mount Vernon High School, 350 Gramatan Avenue, now used as a middle school.

NEWARK

Charles Jackson grew up in this city about thirty miles southeast of Rochester, and was graduated from Newark High School in 1921. Jackson is best known for his novel *The Lost Weekend* (1944), a grim depiction of several days in the life of an alcoholic.

NEWBURGH

Mary [Hartwell] Catherwood, the novelist and short-story writer, was a teacher in this city on the Hudson River, probably in the early 1870s. She wrote a number of stories for *Wood's Household Magazine* (1867–98), which was founded here in 1867 and later was moved to New York City.

NEW CITY

Maxwell Anderson moved in 1921 to this Rockland County community about twenty-five miles north of New York City. He lived in New City for more than thirty years, first in a two-story farmhouse by a waterfall, on South Mountain Road and, after 1931, in a large house he had built next door to his first home. The playwright produced his best works while living in New City. Among his plays were *Both Your Houses* (1933), for which he won a Pulitzer Prize; *Winterset* (1935); *High Tor* (1937); *Key Largo* (1939); *Candle in the Wind* (1941); and *Anne of the Thousand Days* (1948). Anderson's home at 170 South Mountain Road and his later home next door at 164 South Mountain Road are private and unmarked.

NEW ROCHELLE

This city on Long Island Sound, northeast of New York City, has been the home of a number of writers, including **Robert W. Anderson,** who grew up at 99 Elm Street, still standing, and later wrote his play *The Eden Rose* (1948) there. This house and the city of New Rochelle figure in his play *I Never Sang for My Father* (1968). **Faith Baldwin,** whose many romantic novels have sold more than ten million copies, was born here on October 1, 1893. **Tad Mosel** lived at 323 Oxford Road, still standing, and attended New Rochelle High School from 1938 to 1940.

Probably the best-known writer associated with New Rochelle is **Thomas Paine,** the pamphleteer, philosopher, and author of *Common Sense* (1776); the series of pamphlets known as *The American Crisis* (1776–83); *Rights of Man* (1791–92); and *Age of Reason* (1794–96). For his services to the United States during the Revolution, Paine was given a large estate here by the State of New York. He lived here for a time, but until the end of his life spent most of his time either abroad or at his home in Bordentown, New Jersey. He finally moved to New York City, where he died on June 8, 1809. Paine was buried on his estate in New Rochelle, but William Cobbett, an English essayist and pamphleteer who had attacked Paine during his lifetime, dug up Paine's coffin in 1819 and shipped it to England, intending to make amends and honor Paine with a memorial there. The plan collapsed, and Paine's remains were eventually lost.

Paine's New Rochelle cottage, which he built in 1802 after the original farmhouse burned to the ground, has been moved from its original site to 20 Sicard Avenue and is open to the public, with a modest contribution requested. Nearby is the Huguenot–Thomas Paine Historical Association Museum, 983 North Avenue, which contains such items as Paine's writing kit, death mask, and several of his manuscripts. Also nearby is Paine Park, which contains a one-room schoolhouse and other items of historical interest.

Frederic Remington, best known as an artist and illustrator, owned a studio in New Rochelle at about the turn of the twentieth century. **Margaret Sangster,** a poet, editor, and author of books for children such as *Little Janey* (1855) and *Win-*

> *You will do me the justice to remember that I have always strenuously supported the right of every man to his opinion, however different that opinion may be to mine. He who denies to another this right, makes a slave of himself to his present opinion, because he precludes himself the right of changing it.*
>
> *The most formidable weapon against errors of every kind is reason. I have never used any other, and I trust I never shall.*
>
> —Thomas Paine, in *Age of Reason*

(*left*) Home of Maxwell Anderson at 170 South Mountain Road in New City. While living here, Anderson gained recognition with his second play, *What Price Glory?* (1924), written with Laurence Stallings. (*right*) Home of Thomas Paine in New Rochelle. On Christmas Eve of 1804, Paine was shot at—perhaps by a disgruntled former tenant—while working at his desk. In 1806, he was barred from voting in New Rochelle.

In treating of the early governors of the province, I must caution my readers against confounding them, in point of dignity and power, with those worthy gentlemen who are whimsically denominated governors in this enlightened republic,—a set of unhappy victims of popularity, who are, in fact, the most dependent, hen-pecked beings in the community; doomed to bear the secret goadings and corrections of their own party, and the sneers and revilings of the whole world beside; set up, like geese at Christmas holidays, to be pelted and shot at by every whipster and vagabond in the land.

—Washington Irving,
in *A History of New York*

some Womanhood (1900), was born in New Rochelle on February 22, 1838. **Robert E[mmet] Sherwood,** born here on April 4, 1896, won Pulitzer Prizes for three of his plays: *Idiot's Delight* (1936), *Abe Lincoln in Illinois* (1938), *There Shall Be No Night* (1940); and for *Roosevelt and Hopkins* (1948), a biography. Sherwood was born at 18 Neptune Place, a street that no longer exists. **Augustus Thomas,** author of such plays as *The Witching Hour* (1907) and *As a Man Thinks* (1911), lived in a house, no longer standing, at 11 Thomas Place. The street was named in his honor.

NEW WINDSOR

E[dward] P[ayson] Roe, the minister and author of enormously popular novels, such as *Barriers Burned Away* (1872) and *Opening a Chestnut Burr* (1874), was born on March 7, 1838, in this community just south of Newburgh.

NEW YORK CITY

New York City, the largest single entry of this book, is treated in the following order: Lower Manhattan; Greenwich Village; Lower East Side & East Village; Midtown West; Midtown East; Upper West Side; Upper East Side; Harlem, the Heights, & Upper Manhattan; Schools & Colleges; Publications; Brooklyn; Queens; Staten Island; and The Bronx.

Lower Manhattan

This is the oldest part of Manhattan: first developed, longest inhabited, and therefore much changed since its early days. But even if modern buildings have replaced most of the older structures, the area is still redolent with almost two centuries of literary associations. The first modern hotel in New York was the Astor House, opened in 1836 on Broadway between Vesey and Barclay streets; among the literary figures who stayed or dined here were **Nathaniel Hawthorne, Washington Irving,** and **Edgar Allan Poe.** The nearby Bixby Hotel, at Broadway and Park Row, was a gathering place for writers such as **James Fenimore Cooper, Oliver Wendell Holmes,** and **Ralph Waldo Emerson,** as well as Hawthorne and Irving. A third hotel, the City, at Broadway and Thames Street, was a meeting place for the Bread and Cheese Club, founded about 1822 by Cooper. **William Cullen Bryant** and **Fitz-Greene Halleck** were among the club's members.

One of the most famous names in New York social history is that of the Delmonico brothers, whose first restaurant opened in 1827 at the corner of South William and Beaver streets. The building presently on the site, called Oscar's Delmonico, dates from 1891, and until it closed, the establishment had such eminent visitors as **Theodore Dreiser, William Makepeace Thackeray,** and **Mark Twain.** The Lantern Club, on William Street between John and Fulton streets, was housed in a building said to have been visited by Captain Kidd. Among the club's more reputable visitors were **Stephen Crane, William Dean Howells, Rudyard Kipling, Mark Twain,** and **Booth Tarkington.** Papa Monetta's Tavern, on Mulberry Street, was the meeting place of the Thanatopsis Club, which convened during the early 1920s for poker games. Among the regular players were **Franklin Pierce Adams, Heywood Broun,** and **Alexander Woollcott.**

The present Trinity Church, on Broadway opposite Wall Street, is the fourth church to stand on its site. In its cemetery, a quiet haven in the busy financial district, lie the graves of **William Bradford,** author of the *History of Plimmouth Plantation,* composed between 1630 and 1651 but not published in full until 1856, and **Alexander Hamilton,** principal author of *The Federalist* (1787–88). The grave marked Charlotte Temple is believed to be the resting place of the prototype for the heroine in *Charlotte, A Tale of Truth* (1791), a popular sentimental novel written by **Susanna Rowson.**

Two literary sites no longer standing are Wiley's Bookstore, at 9 Wall Street, and the John Street Theater, on John Street. Wiley's was a favorite meeting place for **James Fenimore Cooper, William Cullen Bryant,** and **Fitz-Greene Halleck,** who gathered here with others in a room they called the "Den." The John Street Theater was New York's first. It opened on December 7, 1767, and saw the first performance of *The Contrast* (1787), by Royall Tyler. *The Contrast* was the second play and the first comedy written by an American. The theater closed down in 1798.

Lower Manhattan is also the birthplace of some of our greatest literary figures. **Philip Freneau,** sometimes called the Poet of the American Revolution, was born on Frankfort Street on January 2, 1752. A classmate and lifelong friend of James Madison, Freneau spent his life in a number of different places, including the West Indies. It was while returning from there in 1778 that Freneau was arrested by British forces in the belief that he was a spy. They released him, but two years later recaptured him and, after a brief trial, imprisoned him for a year. Released in 1780, Freneau lived in New York and Philadelphia, retiring in 1807 to his country home in New Jersey. He died there in 1832, after falling into a bog while walking home late at night.

Ours not to sleep in shady bowers,
When frosts are chilling all the plain,
And nights are cold and long the hours
To check the ardor of the swain,
Who parting from his cheerful fire
All comforts doth forego,
And here and there
And everywhere
Pursues the prowling foe.

—Philip Freneau,
in "The Northern Soldier"

An even more distinguished native of the area is **Washington Irving,** born on April 3, 1786, at 131 William Street, at the corner of Ann Street. A few years after the author's birth, his family moved down the street to number 128, and later in life he lived briefly at 16 Broadway and 53 Greenwich Street. Irving spent much of his life abroad and at Sunnyside, his estate in Tarrytown. The major literary efforts of his years in Lower Manhattan were the satirical miscellanies published as *Salmagundi* (1807–08), which he wrote

In the old mossy groves on the breast of the mountains,
In deep lonely glens where the waters complain,
By the shade of the rock, by the gush of the fountain,
I seek your loved footsteps, but seek them in vain.

Oh, leave not forlorn and forever forsaken,
Your pupil and victim to life and its tears!
But sometimes return, and in mercy awaken
The glories ye showed to his earlier years.

—William Cullen Bryant,
in "I Cannot Forget with What Fervid Devotion"

Portrait of Charles Brockden Brown, who lived in New York City from 1798 to 1801, and here completed such novels as *Arthur Mervyn* (1799–1800) and *Edgar Huntly* (1799). Portrait of Mary Mapes Dodge, author of *Hans Brinker; or, The Silver Skates.*

with his brothers William and Peter and his brother-in-law **James Kirke Paulding,** working often at 17 State Street, William's home; and *A History of New York* (1809). While the city that Irving knew has largely vanished, the author is remembered by a plaque on the New York Insurance Exchange Building, 123 William Street.

The area's third literary native son is **Herman Melville.** The novelist was born at his parents' home, 6 Pearl Street, near the Battery, on August 1, 1819. Allan Melvill, the author's father, had continuing financial problems in his business at 123 Pearl Street and often was forced to borrow money from his wife's parents. In spite of this—or perhaps because of it—he moved his family several times, first to Cortlandt Street in 1824, later to 675 Broadway in 1828, and finally to Bleecker Street. Melvill died in 1832, almost $25,000 in debt, and the family moved shortly thereafter to Mrs. Melvill's parents' home in Albany. The novelist's birthplace has long since been torn down, but the garage now standing on the site is marked with a plaque in honor of the distinguished author of *Moby-Dick* (1851).

Among the many literary figures who have lived here, few have had so strange a residence as **Horatio Alger,** the well-known but now little-read author of inspirational stories for children. After publishing his first *Ragged Dick* stories (1866), Alger was sought out by Charles O'Connor, supervisor of the Newsboys' Lodging House on Fulton Street. They became close friends, and Alger took up residence there as chaplain, benefactor, and spiritual guidance counselor to the troubled young residents. Despite his role here, this was the only real home Alger knew, and the only place where he felt anything remotely resembling happiness.

Maxwell Bodenheim and his wife were found murdered in a furnished room at 97 Third Avenue on February 7, 1954. Bodenheim, a perpetual drifter, had spent his last years living at a number of places in the area. First published in *Poetry* magazine in the 1910s, he produced such volumes of poetry as *The Sardonic Arm* (1923) and *Against This Age* (1929), and novels, including *Crazy Man* (1924), *Sixty Seconds* (1929), and *Naked on Roller Skates* (1931).

An altogether more conventional figure was **Charles Brockden Brown,** America's first professional writer, who came to live with his friend William Dunlap at 15 Park Row in 1798. Brown, the author of such Gothic romances as *Wieland* (1798) and *Arthur Mervyn* (1799), had recently moved to New York City from Philadelphia. **William Cullen Bryant** came to New York City in 1825 to edit the *New York Review and Athenaeum Magazine.* Much of his poetry was published in this magazine, which he edited for nearly fifty years. He lived at first with a family on Chambers Street. His later homes in Lower Manhattan included 67 Varick Street and 88 Canal Street. The novel *The Story of a New York House* (1887), by H. C. Bunner, was written about James Watson's house, part of which is still standing at 7 State Street.

Lydia Maria Child, a novelist and reformer known best for her abolitionist tract *Appeal in Favor of that Class of Americans Called Africans* (1833), lived with her husband at 142 Nassau Street. Another resident was **James Fenimore Cooper,** who lived between 1822 and 1826 at 3 Beach Street. **Mary Mapes Dodge,** the noted children's author and influential editor of *St. Nicholas* magazine, spent part of her childhood at 8 Fourth Avenue, where she met such friends of her father's as Horace Greeley and William Cullen Bryant. They, like John Greenleaf Whittier and Henry Wadsworth Longfellow, were later to write for *St. Nicholas,* which Dodge edited from 1873 until 1905. Of the author's own works, by far the most famous was—and is—*Hans Brinker; or, The Silver Skates* (1865).

Joseph Rodman Drake died at age twenty-five at his home on Park Row below Beekman Street on September 21, 1820. Drake published only one work in his lifetime, a set of satirical verses called the "Croaker Papers," which he wrote in collaboration with his friend Fitz-Greene Halleck. In 1835 his widow collected a group of

Never again might he bask and lie
On that sweet face and moonlight eye,
But in his dreams her form to see,
To clasp her in his revery;
To think upon his virgin bride,
Was worth all heaven, and earth beside.

—Joseph Rodman Drake,
in "The Culprit Fay"

I proceeded, in a considerable degree, at random. At length I reached a spacious building in Fourth Street, which the sign-post showed me to be an inn. I knocked loudly and often at the door. At length a female opened the window of the second story, and, in a tone of peevishness, demanded what I wanted. I told her that I wanted lodging.

"Go, hunt for it somewhere else," said she, "you'll find none here." I began to expostulate; but she shut the window with quickness, and left me to my own reflections.

—Charles Brockden Brown,
in *Arthur Mervyn*

> *. . . there is no human sentiment better than that which unites the Societies of Mutual Admiration. And what would literature or art be without such associations? . . . Who can tell what we owe to the Mutual Admiration Society of which Shakespeare, and Ben Jonson, and Beaumont and Fletcher were members? . . . Was there any great harm in the fact that the Irvings and Paulding wrote in company?*
>
> —Oliver Wendell Holmes, in *The Autocrat of the Breakfast-Table* (1858)

Drake's poems and published them as *The Culprit Fay and Other Poems.* **Fitz-Greene Halleck** was himself a longtime resident here, taking lodgings on East Broadway, two doors below Liberty Street, on moving to New York City in 1811; later he lodged for some years at 25 Franklin Street. The business for which he worked collapsed in 1828, and in 1832 he became confidential secretary to John Jacob Astor. Halleck wrote but little in the last forty years of his life. Perhaps his best-known work is the elegy he wrote for his friend J. R. Drake.

Edward Hoagland, the novelist, has lived at two addresses in the area, 21 Jane Street and, more recently, 463 West Street. He wrote part of his *African Calliope* (1979) at the latter address.

Eugene O'Neill is also associated with Lower Manhattan. O'Neill came on several occasions early in the 1910s to a rooming house on Fulton Street, near the waterfront, which was known as Jimmy the Priest's. His stays here provided material for his plays *Anna Christie* (1921) and *The Hairy Ape* (1922). Jimmy was the model for Johnny-the-Priest in *Anna Christie,* and a ship's foreman named Driscoll was the model for Yank in *The Hairy Ape.* Early in 1912, while O'Neill was here, he attempted suicide and was taken to Bellevue Hospital. Jimmy the Priest's gave O'Neill the setting for *The Iceman Cometh* (1946), but O'Neill is said to have derived the characters who people that play from encounters with clients of the Hell Hole, a saloon and hotel that stood at the corner of Sixth Avenue and Fourth Street, farther uptown.

Thomas Paine was another resident here, staying at various addresses during his last, sad, lonely years. After returning from Europe in 1802, the author of *Common Sense* (1776), *The American Crisis* (1776–83), and *The Rights of Man* (1791–92) found himself ostracized and vilified, and after some time in New Rochelle and in Bordentown, New Jersey, he moved to New York City. Paine stayed with friends at 36 Cedar Street and 85 Church Street, took lodgings for a time on Broome Street, and later moved to 63 Partilion Street, now called Fulton Street. The last period of his life was spent in Greenwich Village, where he died on June 8, 1809.

Also resident here for some time was **James Kirke Paulding,** the novelist, dramatist, poet, and historian, who lived at both 17 Whitehall Street and 43 Vesey Street. Paulding wrote the novels *The Lion of the West* and *The Dutchman's Fireside* (both 1831), which were popular in their day, but the author's greatest claim on the attention of posterity was his contribution to the miscellany *Salmagundi* (1807–08), on which he collaborated with his brothers-in-law: Washington, William, and Peter Irving. **Edgar Allan Poe** also lived and worked in the area, preparing his *Poems* (1831) for publication by Elias Bliss at the latter's bookstore, 111 Broadway, and in 1845 publishing the *Broadway Journal* at 1 Nassau Street. While editing the *Journal,* Poe lived for a time at 195 Broadway. In spite of the popular success of "The Raven," published in the *Mirror* in January 1845, Poe remained as poor as ever, and his home here was in a wretched tenement building.

> *These are the times that try men's souls. The summer soldier and the sunshine patriot will, in this crisis, shrink from the service of their country; but he that stands it* now, *deserves the love and thanks of man and woman. Tyranny, like hell, is not easily conquered; yet we have this consolation with us, that the harder the conflict, the more glorious the triumph. What we obtain too cheap, we esteem too lightly: it is dearness only that gives everything its value. Heaven knows how to put a proper price upon its goods; and it would be strange indeed if so celestial an article as* FREEDOM *should not be highly rated.*
>
> —Thomas Paine, in *The American Crisis*

Another longtime resident was **Seba Smith,** creator of the humorous character Major Jack Downing. Smith lived at 65 Murray Street for much of his time in New York City. He came here in 1839 and during his stay contributed to numerous magazines, edited five others, and published two more. Among the books he published while living here was *May-Day in New York* (1845), republished in the same year as *Jack Downing's Letters.* **Mark Twain** also stayed in the area, renting a room in 1853 in a boarding house on Duane Street while working as a journeyman printer at 97 Cliff Street. The eighteen-year-old Twain wrote to his sister that he liked New York City in spite of the Northern cooking. Later in Twain's life, when he had achieved worldwide fame, he was to return here often, both to live and to visit.

Another eminent author who frequented the area is **Walt Whitman.** The great poet edited various magazines here between 1841 and 1848, including the *Aurora,* at 162 Nassau Street, and *The Evening Tattler,* at 27 Ann Street. Among the places he lived during this period was a boarding house at 12 Center Street. A poet of somewhat lesser stature was **Samuel Woodworth,** who lived at 86 Duane Street and 457 Pearl Street. Woodworth was living at the latter address at the time of his death on December 9, 1842. He is remembered today for his poems "The Old Oaken Bucket," published in *Poems, Odes, Songs* (1818), and "The Hunters of Kentucky," published in *Melodies, Duets, Trios, Songs, and Ballads* (1826).

Greenwich Village

If a list could be compiled of all the authors and artists who have passed through this downtown area of Manhattan, it would include most of the significant figures in our nation's cultural history. Even in the early nineteenth century, when the Village provided some of New York City's most fashionable housing, it was also well known for harboring the city's closest approximation to *la vie bohème.* In recent times the area has been more likely to house stockbrokers than struggling poets, and too much of its nineteenth-century charm has fallen prey to real-estate developers, but Greenwich Village still contains a greater wealth of literary associations than any other district of New York City.

One of the many gathering places here for literary figures is the Cedar Tavern, formerly at 24 University Place, where **Allen Ginsberg, Gregory Corso, Jack Kerouac,** and their friends would gather. When the tavern's original building was torn down, the establishment moved up the street to 82 University Place. The old Brevoort Hotel, formerly on Fifth Avenue just north of Washington Square, attracted literary figures of earlier times, including **Sherwood Anderson, F. Scott Fitzgerald, Sinclair Lewis, H. L. Mencken,** and **Mark Twain.** An establishment that has changed little since its days of literary fame is Chumley's, at 86 Bedford Street, founded as a speakeasy in the days of Prohibition. To this day it bears no sign announcing its presence. In its heyday Chumley's was frequented by such figures as **John Dos Passos, Theodore Dreiser,** and **Edna St. Vincent Millay.** A restaurant is still open in the original quarters today.

The Dauber and Pine Bookshop, at 66 Fifth Avenue, has attracted bibliophiles for many years.

Scheduled for demolition in 1973, Dauber and Pine's was saved by a campaign waged by its dedicated patrons. One of the many historic buildings demolished under the ambitious construction schemes of New York University was a house at 61 Washington Square South that is sometimes referred to as the House of Genius. It included among its tenants in times past **Willa Cather, Stephen Crane, Theodore Dreiser, O. Henry,** and **Frank Norris.** The university's Loeb Center now stands on the site. Still standing nearby is Marta's Restaurant, a favorite in the 1920s of such authors as **William Rose Benét, John Dos Passos,** and **Elinor Wylie,** and more recently of **Eleanor Estes,** the children's author, who has reported that she still eats there.

At 45 Astor Place once stood the offices of the influential *Partisan Review,* first published in 1934, whose noted editors have included **Delmore Schwartz** and whose list of contributors reads like a who's who of distinguished writers. Polly's Restaurant, 137 Macdougal Street, was a favored eating place of the literati. Its patrons in the 1920s included **Max Eastman, Vachel Lindsay,** and **Lincoln Steffens.**

A step away, at both 133 and 139 Macdougal—the first of these was once a stable—stood premises used by the Provincetown Players. This little-theater group, founded in 1915, had many distinguished associates, including **Djuna Barnes, Edna St. Vincent Millay,** and **Eugene O'Neill.** O'Neill's play *Bound East for Cardiff* was on the bill when the Provincetown Players first opened in New York City on November 3, 1916. The Players disbanded about 1929, having produced almost one hundred new plays.

St. Vincent's Hospital, Seventh Avenue and Eleventh Street, has four small connections with our literary past. **Kahlil Gibran,** the Syrian poet, and **Dylan Thomas,** the Welsh poet, died there; and **Denise Levertov,** born in England, worked there. Best of all, **Edna Millay** was given the middle name **St. Vincent** after a relative was treated successfully at the hospital. The San Remo Restaurant, 93 Macdougal Street, no longer operating, was a hangout of Beat writers **William Burroughs, Gregory Corso, Jack Kerouac,** and **Allen Ginsberg.** A more durable gathering place, *sans* food and drink, is the Strand Book Store, at 828 Broadway on the corner of Twelfth Street, a browser's paradise. Also still thriving is the White Horse Tavern, 567 Hudson Street, at which **Norman Mailer, Dylan Thomas,** and numerous other writers have slaked their thirst.

Greenwich Village's greatest landmark and still-favorite outdoor gathering place is Washington Square Park, whose literary associations are too numerous to list. In all likelihood, every author mentioned in the following pages has walked and sat here at some time. Two who came here together were **Robert Louis Stevenson** and **Mark Twain.** During their meeting in 1888, they sat on one of the many benches provided in the park and talked for a while. Of those whose names are closely associated with the square, the two most widely known are probably **Henry James,** one of whose most popular novels (1881) bears the name of the square, and **Edith Wharton,** who mentioned Washington Square in such works as *The Age of Innocence* (1923). An influential presence here for well over a hundred years has been New York University. Its original building stood on the east side of the square until 1894. The university is a major landholder in the area and has instigated much new construction at the expense of the square's once-elegant appearance.

Greenwich Village's two most prominent literary native sons could not be more different. One is **Gregory Corso,** the poet identified with the Beat movement, born on March 26, 1930, at 190 Bleecker Street. Corso's works include *Bomb* (1958) and *Long Live Man* (1962). The other is **Henry James,** born on April 15, 1843, at 2 Washington Place, no longer standing, just off Washington Square. James spent a good part of his childhood in Europe, but he attended school in a small red house on the south side of Waverly Place and spent many hours in his grandmother's house at

The grey and more or less hallowed University building—wasn't it somehow with a desperate bravery, both castellated and gabled?—has vanished from the earth, and vanished with it the two or three adjacent houses of which my birthplace is one.

—Henry James,
in *The American Scene*

Yank—We won't reach Cardiff for a week at least. I'll be buried at sea.
Driscoll—Ssshh! I won't listen to you.
Yank—It's as good a place as any other, I s'pose—only I always wanted to be buried on dry land. But what the hell'll I care—then?

—Eugene O'Neill,
in *Bound East for Cardiff*

White Horse Tavern, a meeting place for writers. The tavern became Dylan Thomas's favorite spot in New York, and he came to know its proprietor and regular patrons well.

The ideal of quiet and genteel retirement, in 1835, was found in Washington Square, where the Doctor built himself a handsome, modern wide-fronted house, with a big balcony before the drawing-room windows, and a flight of white marble steps ascending to a portal which was also faced with white marble. . . . This portion of New York appears to many persons the most delectable. It has a kind of established repose which is not of frequent occurrence in other quarters of the long, shrill city. . . .

—Henry James,
in *Washington Square*

Photograph of James Agee taken in 1941.

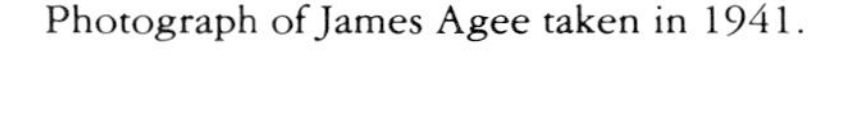

18 Washington Square North, no longer standing. It was this house that was the prototype for the doctor's house in James's *Washington Square* (1881). One of James's later childhood homes was 58 West Fourteenth Street. James also lived or stayed at 11 Fifth Avenue and, in 1911, on his last visit to the United States, at 21 East Eleventh Street. He wrote about his childhood here in *A Small Boy and Others* (1913).

A building on Thirteenth Street between Fifth and Sixth avenues has been home to **Franklin P. Adams,** the columnist and poet known better as **F. P. A.**; to **James Thurber,** the humorist; and to **E. B. White,** the fine essayist identified with *The New Yorker.* **James Agee,** winner of a Pulitzer Prize for his posthumously published novel, *A Death in the Family* (1957), was a longtime resident of Greenwich Village. He lived at 121 Leroy Street, 38 Perry Street, 172 Bleecker Street, and 17 King Street. The King Street address was the author's last: en route from here to his doctor's office, he suffered a fatal heart attack and died on May 16, 1955. Another longtime resident is **Edward Albee,** who wrote *The Zoo Story* (1960) while living at 238 West Fourth Street. He completed the play in three weeks after leaving a job at Western Union. Albee later lived on West Twelfth Street, where he wrote *Who's Afraid of Virginia Woolf?* (1962). Albee had seen the intriguing question written on the mirror of the Ninth Circle, a bar at 139 West Tenth Street, proving perhaps that graffiti are not without redemptive use. Later in the 1960s Albee lived in a carriage house at 50 West Tenth Street.

Edward Albee, who lived in Greenwich Village in the late 1950s and 1960s.

Robert W. Anderson wrote parts of his play *Tea and Sympathy* (1956) while living at 14 West Eleventh Street. He also lived for a time at 14 East Ninth Street. **Sherwood Anderson** lived in 1922 at 12 St. Luke's Place, two doors down from where Theodore Dreiser lived, and also stayed with a friend at 54 Washington Mews. A distinguished novelist of our own day, **Saul Bellow,** lived at 17 Minetta Lane from 1950 until 1952, while he was teaching at New York University. Another contemporary novelist, **Vance Bourjaily,** lived at 49 Grove Street, where he entertained such literary colleagues as **Louis Auchincloss, William Styron,** and **Gore Vidal.**

Randolph Bourne, the essayist, whose writings exerted an enormous influence on much of his generation, was living at 18 West Eighth Street when he died on December 22, 1918, the only known literary victim of the great postwar influenza epidemic. Only thirty-two years old at his death, Bourne had established himself as an important figure with such volumes as *Youth and Life* (1913) and *Education and Living* (1917). Another resident here was **Paul Bowles,** the novelist, who lived for a time at 28 West Tenth Street. **Van Wyck Brooks** lived in 1958–59 at the Hotel Grosvenor, no longer operating, at 35 Fifth Avenue. **William Cullen Bryant** lived in the latter half of the 1820s at various addresses in Greenwich Village, including 147 West Fourth Street, 12 Carmine Street, and a house at Broadway and Fourth Street.

Willa Cather lived at 60 Washington Square South after moving to New York in 1904 to work for *McClure's Magazine.* She moved from here to the Beaux Arts Apartment House, 82 Washington Place, where she lived until 1913, and after that to 5 Bank Street, where she remained until 1927. While living in the Village, Cather wrote many of her novels, including *O Pioneers* (1913), *My Ántonia* (1918), and *One of Ours* (1922), for which she won a Pulitzer Prize. The novelist later lived for five years at the Hotel Grosvenor and there worked on *Shadows on the Rock* (1931). The hotel is no longer standing, but a plaque on the building now occupying the site pays tribute to Cather. A resident of much shorter duration was **James Fenimore Cooper,** who lived at 345 Greenwich Street before sailing for Europe in 1826, and at 145 Bleecker Street after returning in 1833.

From 1919 until 1929, the decade in which **Malcolm Cowley** was writing many of the poems for his volume *Blue Juniata* (1929), he lived at three different addresses here: 107 Bedford Street, 88 West Third Street, and 35 Bank Street. Even more peripatetic a resident was **Hart Crane,** whose verse provides one of the great literary celebrations of New York City. Crane lived in 1917 at 25 East Eleventh Street and 54 West Tenth Street. In 1923, after a period spent away from New York City, the poet lived at 4 Grove Street with his friend **Gorham Munson,** the literary critic, before Crane found an advertising job at the J. Walter Thompson Agency. After he got the job, Crane lived briefly at 6 Minetta Lane and then sublet a place at 45 Grove Street. He left the job within six months and spent the next six months in the country. Crane then returned for a short time to the Grove Street apartment, but soon found another at 15 Van Nest Place, now 79 Charles Street. It was from here, shortly before Easter in 1924, that Crane moved to Brooklyn Heights.

One of the most renowned and enduring residents of the Village was **E. E. Cummings.** The poet moved in 1923 to a house at 4 Patchin Place, a cul-de-sac lined with nineteenth-century brick houses that can be found off West Tenth Street between Sixth and Greenwich avenues. This was his home for nearly forty years—he died on September 3, 1962—and his visitors here included a great number of the most important literary figures of the twentieth century, including **T. S. Eliot** and **Ezra Pound.** Cummings also stayed at one time at 11 Christopher Street. **Peter De Vries** lived at 32 West Eleventh Street early in his career, and it was here that he wrote his first stories published in *The New Yorker.*

Theodore Dreiser liked—possibly had—to move around as much as Hart Crane did, and since he lived more than forty-five years longer than Crane, he had greater opportunity to change addresses. His residences in Greenwich Village include 160 Bleecker Street, now renovated as a luxury apartment house, where he stayed soon after arriving in New York in 1895; 165 West Tenth Street, where he lived from 1914 until 1919 and completed work on *The "Genius"* (1915); 16 St. Luke's Place, scene of a memorable stag party whose guests included F. Scott Fitzgerald and H. L. Mencken, where he lived in 1922–23; 118 West Eleventh Street, his sister Mame's apartment, where he lived in 1924 and worked on *An American Tragedy* (1925); and 116 West Eleventh Street, where he lived in 1937. Dreiser also stayed for a time at Patchin Place, where he first met Van Wyck Brooks.

Another resident here was **Walter D[umaux] Edmonds,** author of such popular historical novels as *Drums Along the Mohawk* (1936). In 1970, well after Edmonds had moved elsewhere, his house at 18 West Eleventh Street was destroyed by a bomb blast. At the time it was in use as a bomb factory operated by members of the Weather Underground. **Kenneth Fearing** lived during the latter part of his life at 311 West Eleventh Street. Fearing's books of verse included *Stranger at Coney Island* (1949) and other works, but his greatest commercial success was the suspense novel *The Big Clock* (1946). Resident here for a time in the late 1930s was **Ford Madox Ford,** the English novelist, who was visited at his apartment at 10 Fifth Avenue by Ezra Pound, among others. **Jean Garrigue,** author of such fine verse collections as *Country Without Maps* (1964), lived for some time in a three-room walkup at 4 Jones Street. **Kahlil Gibran,** one of the greatest moneymakers in the history of poetry, had a home and studio for almost twenty years at 51 West Tenth Street. Gibran was living here at the time of his death in 1931. His enduring work is *The Prophet* (1923).

An architectural landmark and site of literary interest is the row of nineteenth-century brick houses on Hudson Street near St. Luke's Church. The house at 487 Hudson Street was one of the many boyhood homes of **Bret Harte.** Later in life, after his return from California, Harte lived with his sister at 16 Fifth Avenue. Both sites are now marked with plaques honoring the author. Harte is also known to have lived, about 1872, at 713 Broadway. Tradition has it that another eminent short-story writer of the early twentieth century, **O. Henry,** found inspiration for his sentimental story "The Last Leaf" (published in *The Trimmed Lamp and Other Stories,* 1907) in Grove Court, an old passageway between 10 and 12 Grove Street.

William Dean Howells lived for a time at 48 West Ninth Street, his residence at the time of publication of *April Hopes* (1888). A very different writer, **LeRoi Jones,** later called **Imamu Amiri Baraka,** lived for a time at 7 Morton Street. **Joseph Wood Krutch,** the critic and naturalist, has lived at 144 West Twelfth Street and 11 West Eleventh Street. **Stanley Kunitz,** whose first book was *Intellectual Things* (1930), has lived at 157 West Twelfth Street and now makes his home not far away. **Denise Levertov,** another contemporary poet, has lived at 52 Barrow Street and 727 Greenwich Street.

Sinclair Lewis also lived at several addresses here, beginning with 10 Van Nest Place, now 69 Charles Street, where he lived in 1910. Lewis's later addresses include the Hotel Lafayette, on University Place, where he lived in the autumn of 1926 and worked on *Elmer Gantry* (1927), and

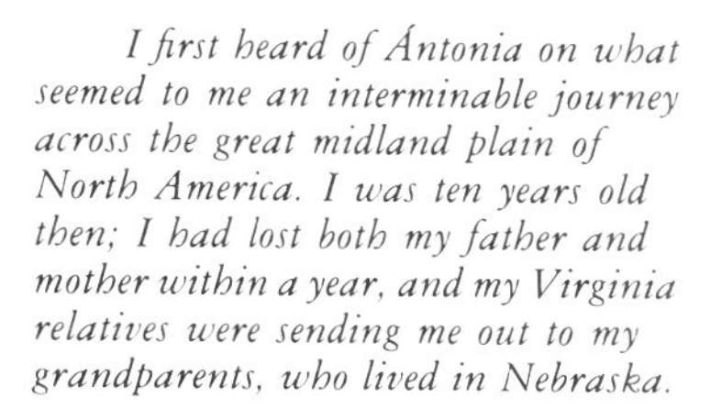

I first heard of Ántonia on what seemed to me an interminable journey across the great midland plain of North America. I was ten years old then; I had lost both my father and mother within a year, and my Virginia relatives were sending me out to my grandparents, who lived in Nebraska.

—Willa Cather,
in the opening lines of
My Ántonia

In a little district west of Washington Square the streets have run crazy and broken themselves into small strips called "places." These "places" make strange angles and curves.

—O. Henry,
in "The Last Leaf"

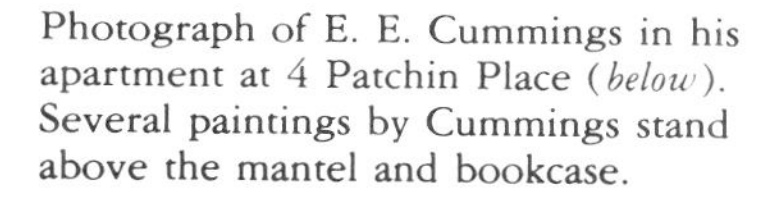

Photograph of E. E. Cummings in his apartment at 4 Patchin Place (*below*). Several paintings by Cummings stand above the mantel and bookcase.

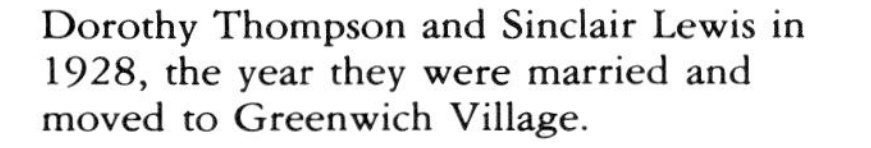

Dorothy Thompson and Sinclair Lewis in 1928, the year they were married and moved to Greenwich Village.

37 West Tenth Street, where he moved in 1928 after his marriage to **Dorothy Thompson.** For the birth of their son, Michael, Lewis and Thompson held a celebratory dinner attended, among others, by H. L. Mencken and George Jean Nathan. A few doors down, at 45 West Tenth Street, was the home of **A. J. Liebling,** the famed columnist, and his wife, **Jean Stafford,** the novelist and short-story writer. They were married in 1959, and until Liebling's death in 1963 divided their time between New York City and East Hampton, Long Island. In his book *Between Meals* (1962), Liebling described his boyhood in New York City. Stafford, who gave up the house here after her husband's death, won a Pulitzer Prize for her *Collected Stories* (1969).

We were very tired,
we were very merry—
We had gone back and forth
all night on the ferry.
It was bare and bright, and smelled
like a stable—
But we looked into a fire,
we leaned across a table,
We lay on a hilltop
underneath the moon;
And the whistles kept blowing, and
the dawn came soon.

—Edna St. Vincent Millay, in "Recuerdo"

A figure of some literary prominence, though not a distinguished writer, was **Mabel Dodge Luhan,** whose Wednesday-evening literary salons at 23 Fifth Avenue during the early decades of the twentieth century attracted such luminaries as Max Eastman, Amy Lowell, Edwin Arlington Robinson, and Lincoln Steffens. **Mary McCarthy** lived for a time at 18 Gay Street, and **Carson McCullers** lived during the early 1940s at 321 West Eleventh Street. **William Starbuck Mayo,** physician and author of such novels as *Never Again* (1873), died at his home at 13 East Eleventh Street on November 22, 1895.

One of the figures most closely associated with the golden age of bohemian Greenwich Village is **Edna St. Vincent Millay.** The poet moved here in 1917, soon after being graduated from Vassar College. Among her addresses in the Village were 25 Charlton Street, 139 Waverly Place, 52 Jane Street, and 75½ Bedford Street. Millay was living in the house on Bedford Street when her second verse collection was published, *A Few Figs From Thistles* (1920). The house is also known for being the narrowest house in the Village; only 9½ feet wide, it was originally built on the site of an alleyway. Millay wrote of the Village at this time as a place where poets and artists were "very, very poor and very, very merry." And Millay, a strikingly handsome woman who was wooed at one time by both Edmund Wilson and John Peale Bishop, was one of the best-known figures in the group of penurious merrymakers. She took herself away from the scene soon after marrying a businessman named Eugen Jan Boissevain in 1923, and spent most of the rest of her life living away from New York City. While still here she helped found the Cherry Lane Theater, at 38 Commerce Street, and acted in productions staged by the Provincetown Players.

Building at 35 West Ninth Street, the last residence of Marianne Moore.

Another resident here was **Henry Miller,** the novelist, who lived in 1917 at 244 Sixth Avenue with his wife Beatrice. **William Vaughn Moody,** the poet and playwright, spent part of the last years of his life at 107 Waverly Place. He married an old friend, Harriet Brainard, in 1909, the year before his death. Moody's collected works, *Poems and Plays,* were published in two volumes in 1912. **Clement Clark Moore,** who wrote the poem "A Visit from St. Nicholas" (1823), was the first warden of St. Luke's Church, on Hudson Street at Grove Street.

Marianne Moore had two homes in Greenwich Village, her stay in them separated by a long residence in Brooklyn. The first was at 14 St. Luke's Place, where she moved with her mother in 1918 and stayed until 1929. During this period Moore worked as librarian at the Hudson Park branch of the New York Public Library, across from her apartment. From 1925 to 1929, Moore was editor of *The Dial,* published at the time from 135 West Thirteenth Street. Among the volumes Moore published in this period were *Poems* (1921) and *Observations* (1929). On her return from Brooklyn in 1965, she took an apartment at 35 West Ninth Street and lived here for the rest of her life, cutting a distinctive figure on the streets in her cape and tricorne hat. After her death on February 5, 1972, the contents of Moore's living room here were transferred to the A. S. W. Rosenbach Foundation in Philadelphia. Another onetime editor of *The Dial,* in the early 1920s, was **Lewis Mumford,** the critic and historian, who lived at the time at 143 West Fourth Street. In the late 1930s Mumford had a home at 393 Bleecker Street.

One of the more enterprising literary residents here was **Anaïs Nin,** the novelist and memoirist. After moving from Europe to New York City in 1940, Nin took a top-floor apartment at 215 West Thirteenth Street. Her novel *Winter of Artifice* had been published in France during the previous year, and she was engaged in looking for an American publisher. Unable to find one, Nin borrowed money—the Gotham Book Mart was one of her benefactors—and bought a hand press. In 1942 she printed five hundred copies of the book herself. Nin managed to sell four hundred copies, which covered her expenses and helped finance her next two books, both collections of short stories, *Under a Glass Bell* (1944) and *Ladders to Fire* (1946). Both books were printed in new premises at 17 East Thirteenth Street, a two-story house where Nin also received such aspiring authors as Allen Ginsberg and Jack Kerouac. **Frank O'Hara,** the poet, a close friend of Ginsberg's, resided at both 90 University Place and 790 Broadway. He was living at the latter address at the time of his death on July 25, 1966.

Eugene O'Neill lived in 1915 in a room he called "the garbage flat" at 38 Washington Square South, and frequented the Golden Swan, a nearby saloon at Sixth Avenue and Fourth Street. In 1916, when he was said to be having an affair with Louise Bryant, the journalist, who had recently married John Reed, O'Neill stayed with her at 42 Washington Square South. O'Neill's relationship with Bryant provided the basis for his Pulitzer Prize-winning play *Strange Interlude* (1928).**Thomas Paine** took up residence in the Village in the last years of his life, a period of poverty and neglect for the Revolutionary War pamphleteer. Paine moved to 293 Bleecker Street in July 1808, when he was in poor health. In May 1809, his health deteriorating further, he was taken to 59 Grove Street, just a short distance away. Paine died there on June 8. The house is marked by a plaque.

Edgar Allan Poe lived at several addresses here in the course of his short, unhappy life. The first was Sixth Avenue and Waverly Place, where he stayed after arriving in New York City in 1837. He then lived at 113½ Carmine Street, which was the author's home through the summer of 1838 but is no longer standing. Among the works Poe worked on here at this time were *The Narrative of Arthur Gordon Pym* (1838), "The Fall of the House of Usher" (1839), and "Ligeia" (1839). Unable to support himself and his family, Poe left New York to live in Philadelphia. When he returned here on April 6, 1844, with his wife, Virginia, and his mother-in-law, Mrs. Maria Clemm, Poe stayed at a boarding house at 130 Greenwich Street, taking working quarters at 4 Ann Street, and was later to move to 18 Amity Street, now West Third Street, where he lived while editing *The Broadway Journal.* He took other lodgings at 85 Amity Street, before moving in 1846 to a cottage in rural Fordham. At one of the many architectural highlights of the Village, the Northern Dispensary at Waverly Place and Christopher Street, Poe was once treated for a head cold.

Katherine Anne Porter lived in the early 1920s at 75 Washington Place, where she worked on some of her first short stories. Another well-known resident of the area was **John Reed,** whose *Ten Days That Shook the World* (1919) provides an eyewitness account of the Russian Revolution. Reed first moved here in 1911 and, with the help of Lincoln Steffens, found a job on *The American Magazine,* but also wrote articles for other publications. In 1913 Reed became editor of *The Masses.* Reed's home here was at 42 Washington Square South, where he lived with some Harvard classmates, including the poet **Alan Seeger.** In 1911, **Lincoln Steffens** himself moved in for a time—his first wife had just died. Reed described his life here in *The Day in Bohemia: or, Life Among the Artists* (1913). Reed spent much of the period from 1914 to 1916 traveling as a reporter in Mexico and Europe. After marrying Louise Bryant in 1917, he went to Russia, where he wrote news accounts of the revolution.

Returning to the United States in April 1918, Reed rented a room at 1 Patchin Place, with the help of an advance from the publisher Boni and Liveright. Reed immediately set about writing his account of the revolution. Working long hours, both at his home and at Polly Holliday's Restaurant, 147 West Fourth Street, Reed is said to have

The light from the icicle bushes threw a patina over all objects, and turned them into bouquets of still flowers kept under a glass bell. The glass bell covered the flowers, the chairs, the whole room, the panoplied beds, the statues, the butlers, all the people living in the house. The glass bell covered the entire house.

—Anaïs Nin,
in "Under a Glass Bell"

Photograph of John Reed taken in 1915, several years before he wrote *Ten Days That Shook the World.*

Old photograph of 293 Bleecker Street, residence of Thomas Paine from 1808 to 1809. The house was demolished in the 1930s.

A visit to Mrs. Manson Mingott was always an amusing episode to the young man. The house in itself was already an historic document, though not, of course, as venerable as certain other old family houses in University Place and lower Fifth Avenue. Those were of the purest 1830, with a grim harmony of cabbage-rose-garlanded carpets, rosewood consoles, round-arched fireplaces with black marble mantels, and immense glazed book-cases of mahogany; whereas old Mrs. Mingott, who had built her house later, had bodily cast out the massive furniture of her prime, and mingled with the Mingott heirlooms the frivolous upholstery of the Second Empire.

—Edith Wharton,
in *The Age of Innocence*

finished the work in a few months. Its publication in 1919 made Reed a celebrity, but he was also considered a troublemaker for his radical activities and was indicted twice, in 1918 and 1919, for sedition. Reed returned that year to the U.S.S.R. to work for the Soviet government. When he came back to New York City as Soviet consul, the United States government protested his appointment, and Reed returned to Moscow. There, in 1920, he died of typhus and was buried beneath the Kremlin wall.

An earlier resident of the Village was **George Ripley,** the editor and critic who was one of the founders of Brook Farm. Ripley lived for a time at 8 University Place after coming to New York in the late 1840s. **Edwin Arlington Robinson,** twice winner of a Pulitzer Prize, lived at several addresses here, including the Judson, later called Judson Hall, 51 Washington Square South, where he stayed in 1906; and a small cottage at 121 Washington Place, where he moved in 1909 and stayed from time to time for several years. This was the period during which Robinson worked in the custom house in downtown New York City. His last home in the Village was a top-floor studio at 28 West Eighth Street. It is believed that Robinson worked on *Tristram* (1927) while living on West Eighth Street and was still resident there when he was awarded a Pulitzer Prize for the work in 1928.

Paul Rosenfeld, co-editor of the well-known publications *The Seven Arts* and *American Caravan* and music critic for *The Dial,* lived at 15 and at 270 West Eleventh Street. An even more eminent resident here for a time was **Delmore Schwartz,** who lived at 775 Greenwich Street. Another writer to spend time here was **Rex Stout,** who moved to 11 West Eighth Street in April 1919. In 1926 Stout had a penthouse apartment at 66 Fifth Avenue, where he entertained such prominent colleagues as John Dos Passos and Robert Sherwood, and later still had a home at 33 Perry Street.

William Styron, who won a Pulitzer Prize for his novel *The Confessions of Nat Turner* (1967), has lived at the University Residence Club, Eleventh Street between Fifth and Sixth avenues, and for a time had a small apartment at 45 Greenwich Avenue. **Ida Tarbell,** the famed muckraking journalist, once lived at The Portsmouth, a residential building at 38–44 West Ninth Street. **Allen Tate,** the noted poet and critic, lived in 1927 in a basement apartment at 27 Bank Street, where he wrote the biography *Stonewall Jackson: The Good Soldier* (1928). **Sara Teasdale,** her life entering what seemed an irreparable decline, in the summer of 1932 rented a room in the high-rise building at 1 Fifth Avenue. Teasdale had been saddened by her divorce from Ernst Filsinger in 1929, and by the gruesome suicide of her intimate friend Vachel Lindsay in December 1931. Teasdale was soon to follow Lindsay. On January 30, 1933, the poet, who was forty-five years old, was found dead in her bathtub from an overdose of sedatives. An even shorter residence in the Village was that of **William Makepeace Thackeray,** the English novelist, who stayed at 604 Houston Street, following his second American lecture tour in 1855–56. **James Thurber,** the brilliant analyst of strategies employed in the war between the sexes, lived at several addresses here: Horatio Street near Ninth Avenue, Thirteenth Street between Fifth and Sixth avenues, and 8 Fifth Avenue. Thurber moved to the third address with his second wife, Helen Wismer, in 1935. In the same year, he published another of his collections of humorous pieces, *The Middle-Aged Man on the Flying Trapeze.*

Mark Twain was yet another distinguished resident of the Village. The great novelist rented a house in 1901 at 14 West Tenth Street and here wrote his article "To the Person Sitting in Darkness," published later that year in *The North American Review.* In 1904, after the death of his wife, Twain returned to the Village to live with his daughter Jean at 21 Fifth Avenue, where he began planning work on his official biography with Albert Bigelow Paine—it was published in 1912, two years after Twain's death.

John Updike lived at 153 West Thirteenth Street while working at *The New Yorker* during 1955–57. After leaving the magazine, Updike published his first novel, *Poorhouse Fair* (1959). **Carl Van Doren,** the Pulitzer Prize-winning biographer and longtime professor of English at Columbia University, lived at both 47 Charlton Street and 123 West Eleventh Street. His brother **Mark Van Doren,** the poet and professor of even longer standing at Columbia, once had a house at 393 Bleecker Street. One of the authors who most brilliantly described an earlier era in the history of the city was **Edith Wharton,** baptized **Edith Newbold Jones** at Grace Church, 802 Broadway, on April 20, 1862. Some twenty years later, in 1881, she was to live in the grand house at 7 Washington Square North. Among Wharton's most notable descriptions of the area is that of her Pulitzer Prize-winning *The Age of Innocence* (1920).

Walt Whitman is believed to have stayed at 113 East Tenth Street when he visited New York City in his later years. **Richard Wilbur,** who won a Pulitzer Prize for *Poems* (1957), lived at 20 Christopher Street in 1942. **Edmund Wilson** lived at several addresses here in the 1920s, when he was an important figure in the area's golden age. Wilson lived at 3 Washington Square North, at 1 University Place, and on Eighth Street

McSorley's Old Ale House, a popular gathering place for artists and writers since the early 1900s.

between Fifth and Sixth avenues. Another literary figure of the same era, **Thomas Wolfe,** lived with Aline Bernstein at 13 East Eighth Street while in the early stages of work on *Look Homeward, Angel* (1929). He later lived with Bernstein at 263 West Eleventh Street, a house he described in *You Can't Go Home Again* (1940), placing it on Twelfth Street. Wolfe also stayed for a time in Room 2220 of the Hotel Albert, on University Place near Eleventh Street. In *Of Time and the River* (1935), Wolfe called this hotel the Leopold.

Alexander Woollcott stayed at 34 West Twelfth Street after being graduated from Hamilton College in 1909, and **Richard Wright** lived at three different places in the Village: 467 Waverly Place, the Beaux Arts Apartment House at 82 Washington Place, and 13 Charles Street.

Lower East Side & East Village

This downtown area, incorporating some of the city's oldest neighborhoods, has been a haunt of literary men and women for generations. The Astor Library on Lafayette Street, now home of the Public Theater, was used by several distinguished figures after opening in 1854. **Washington Irving** did research here for his five-volume *Life of Washington* (1855–59), and **William Makepeace Thackeray,** the English novelist, was known to browse here during his stays in the United States. At Christadora House, a settlement house at 147 Avenue B, members of the Poets' Guild, directed by **Anna Hempstead Branch,** gave lectures and readings from their work. Members at various times included **Robert Frost, Vachel Lindsay, Edwin Markham,** and **Edwin Arlington Robinson.** Colonnade Row, one of the city's architectural treasures, at 428–434 Lafayette Street, across from the former Astor Library, has served as home to both **William Cullen Bryant** and **Washington Irving.** The Cooper Union for the Advancement of Science and Art, at 41 Cooper Square, was opened in 1859 and since then its great hall has provided a speaking platform for noted literary figures, including **Henry Ward Beecher, William Cullen Bryant, William Lloyd Garrison,** and **Mark Twain.** The most famous speech delivered here is that of **Abraham Lincoln,** his first in the East as a presidential candidate, on February 27, 1860.

Lüchow's restaurant, 110 East Fourteenth Street, was a favored eating and drinking place for such men as **Sherwood Anderson, H. L. Mencken, George Jean Nathan, Maxwell Perkins,** and **Thomas Wolfe.** Perkins and Wolfe sometimes dined here while working on the manuscript of Wolfe's *Of Time and the River* (1935). Another well-known dispensary of drink and good cheer is McSorley's Old Ale House, operating since 1854 at 15 East Seventh Street. Among those who enjoyed themselves here are **Brendan Behan, E. E. Cummings,** and **John Reed.** Behan had a favorite spot in the corner near the old pot-bellied stove. A renowned gathering place of an earlier era was Pfaff's Cellar, at 653 Broadway, just above Bleecker Street. This old tavern was headquarters for the so-called Bohemians, who met during the late 1850s. Among the group's members were **Thomas Bailey Aldrich, George Arnold, Fitz-James O'Brien, Bayard Taylor,** and, most prominently, **Walt Whitman.** With the coming of the Civil War the group broke up, and its meeting place is now long gone.

The Lower East Side boasts at least three native sons of literary distinction. **Manuel Komroff,** the editor and novelist, was born at 58 Montgomery Street on September 7, 1890. A longtime New York City resident, Komroff wrote about the city in such novels as *A New York Tempest* (1932). His greatest commercial success, however, came from historical novels, such as *Coronet* (two volumes, 1929). Though not a writer himself, **Maxwell Perkins,** born on September 30, 1884, at the corner of Second Avenue and Fourteenth Street, had a notable influence on our literary history. Joining Charles Scribner's Sons in 1910, Perkins eventually rose to become editor in chief and in his long career worked with such figures as F. Scott Fitzgerald, Ernest Hemingway, Ring Lardner, and Thomas Wolfe. Perkins was the sort of editor all writers dream of but few are lucky enough to have. His work with Thomas Wolfe, in particular, could be considered essential to the author's artistic and commercial success. Another modern literary figure, **Jerome Weidman,** was born on April 4, 1913, at 390 East Fourth Street. Weidman lived here until he was seventeen, attending J.H.S. 64 and P.S. 188. His family later moved to The Bronx. Of his many novels, the best known is his first, *I Can Get It for You Wholesale* (1937).

For almost twenty years, one of the Lower East Side's more prominent residents was **W. H. Auden,** the English-born poet. Auden spent winters from 1953 until 1972 at 77 St. Mark's Place, a house in whose basement Leon Trotsky printed his newspaper *Novy Mir.* Auden entertained numerous friends and visitors here, including **T. S. Eliot,** and was a parishioner of St. Mark's-in-the-Bowery, at 131 East Tenth Street. A much earlier resident nearby was **James Fenimore Cooper,** who lived in 1834 at 6 St. Mark's Place, now the St. Mark's Baths. While living here, Cooper wrote his novel *The Monikins* (1835). **Gregory Corso,** lived for two years of his childhood at 190 Clinton Street. Later in his life he resided for a time at Avenue C and Fifth Street.

Old photograph of Cooper Union. In February 1860, Abraham Lincoln delivered a major address here on the institution of slavery.

Let us have faith that right makes might, and in that faith let us to the end dare to do our duty as we understand it.

—Abraham Lincoln, in his Cooper Union address, February 27, 1860

He loved this old house on Twelfth Street, its red brick walls, its rooms of noble height and spaciousness, its old dark woods and floors that creaked; and in the magic of the moment it seemed to be enriched and given a profound and lonely dignity by all the human beings it had sheltered in its ninety years. The house became like a living presence. Every object seemed to have an animate vitality of its own—walls, rooms, chairs, tables, even a half-wet bath towel hanging from the shower ring above the tub, a coat thrown down upon a chair, and his papers, manuscripts, and books scattered about the room in wild confusion.

—Thomas Wolfe, in *You Can't Go Home Again*

Robert Anderson.

I remember hearing Alexander Woollcott say, reading Proust is like lying in someone else's dirty bath water. And then he'd go into ecstasy about something called Valiant Is the Word for Carrie, *and I knew I'd had enough of the Round Table.*

—Dorothy Parker, quoted in her New York *Times* obituary

LEACH: *I'm hungry. This is my place, why shouldn't I eat? These people never eat. Don't they know it's nutritious? Oh, this boil. Damn this boil. Dream world. Narcotics. I live comfortable. I'm not a Bowery bum. Look at my room. It's clean. Except for the people who come here and call themselves my friends. My friends? Huh! They come here with a little money and expect me to use my hard earned connections to supply them with heroin. And when I take a little for myself they cry, they scream. The bastards.*

—Jack Gelber, in *The Connection*

The Lower East Side was once the center for New York City's thriving burlesque and Yiddish theaters, and among the literary figures who enjoyed them was **E. E. Cummings.** After seeing shows at the National Winter Garden Burlesque Theater, at Second Avenue and Houston Street, Cummings would repair to such nearby establishments as the Café Royale, at the southwest corner of Second Avenue and Twelfth Street, or Moscowitz's Rumanian Broilings, a restaurant on Houston Street near Second Avenue. **John Dos Passos** was another literary patron of Moscowitz's.

Jack Gelber, the contemporary playwright, wrote his best-known play, *The Connection* (1960), while living at 11 Pitt Street. Even better known as a local literary figure is **Allen Ginsberg.** Ginsberg's arrival in the area in 1951 made it a popular district for his fellow Beats. He moved in that year to an apartment at 206 East Seventh Street, which he sometimes shared with **William Burroughs** and **Jack Kerouac,** and later lived at 170 East Second Street. While living at the latter address Ginsberg wrote most of the works published in *Kaddish and Other Poems 1958–1960* (1961), one of his finest volumes. **Harry Golden** lived during childhood at 171 Rivington Street. **Emma Goldman** in 1906 began publishing her magazine *Mother Earth* at 210 East Thirteenth Street. Goldman, the well-known anarchist and essayist, had come to the United States from her native Russia in 1886.

Another literary resident here has been **Edward Hoagland,** the novelist and essayist. Hoagland lived at both 521 East Fifth Street and 261 East Tenth Street while writing his *The Peacock's Tail* (1965). **Leroi Jones,** later known as **Imamu Amiri Baraka,** lived at both 324 East Fourteenth Street and 27 Cooper Square. At the latter address he wrote *Blues People* (1963). **Dwight MacDonald,** at one time editor of *Partisan Review,* lived at 117 East Tenth Street, and **Norman Mailer** once lived in a top-floor apartment at 39 First Avenue.

One of the area's most distinguished residents was **Herman Melville,** who moved in 1847, the year of his marriage to Elizabeth Shaw, to 103 Fourth Avenue, now Park Avenue South. While living here Melville wrote *Mardi* (1849), *Redburn* (1849), and *White Jacket* (1850). It is believed that he met **William Cullen Bryant** and **Washington Irving** while living here, and it is known that he often visited with his friend Evert A. Duyckinck at 20 Clinton Place, now East Eighth Street. Melville left the area in 1850 to live in Pittsfield, Massachusetts.

The Lower East Side was the last home of **Fitz-James O'Brien,** the poet, playwright, and short-story writer. O'Brien was living at 1 Great Jones Street at the outbreak of the Civil War. Soon after the firing on Fort Sumter, he enlisted in the New York National Guard, and died on April 6, 1862, of tetanus that resulted from wounds received in the Battle of Bloomery Gap. O'Brien wrote numerous plays, but today he is best remembered for such short stories as "The Diamond Lens" (1858) and for his active participation in the Bohemian gatherings at Pfaff's Cellar, not far from his Great Jones Street home. A much later resident of the area was **Frank O'Hara.** He lived for about five years in an apartment at 441 East Ninth Street, near Tompkins Square Park. His view across the park of St. Bridget's Church led him to write "Weather Near St. Bridget's Steeples." O'Hara also lived in a loft at 791 Broadway.

Edgar Allan Poe lived for part of his time in New York at 195 Broadway and, possibly, at 47 Bond Street. At the latter address he is said to have worked on "The Bells" (1849). Also resident here, during childhood, was **Robert Charles Sands,** the popular poet and journalist of the early nineteenth century. Sands's home here was at 8 Hester Street. A much shorter stay here was that of **Robert Louis Stevenson,** who had a room at the St. Stevens Hotel on East Eleventh Street during his visit to America in 1888. Longer-term residents were **Richard Henry Stoddard,** the poet and critic, and his wife **Elizabeth Drew Stoddard,** the novelist, who lived during the 1850s or 1860s at Fourth Avenue and Tenth Street. They also lived for a time at Second Avenue and Third Street, where Richard Stoddard wrote his *Songs of Summer* (1857). **Marya Zaturenska** reported that she lived in childhood near the Henry Street Settlement House after she arrived in the United States from Russia in 1910.

Midtown West

The midtown area of Manhattan's West Side houses several of the best-known institutions in our nation's cultural life. As the city's theater district, the area has seen a long list of plays performed by some of the world's finest actors. As a business district, it has been the home of newspapers and magazines that have employed a host of writers. And since writers almost invariably seek out friendly places in which to eat, drink, and spend time with one another, the midtown area has more than its share of clubs, hotels, and restaurants that are identified with the literary life of the city.

One hotel with long-established literary connections is the Algonquin, at 59 West Forty-fourth Street. Its literary patrons past and present form an endless list, of whom **Peter De Vries, F. Scott Fitzgerald, Sinclair Lewis, James Thurber, John Updike,** and **Gore Vidal** are merely a small sample. But the hotel's most famous literary connection is with the Round Table, a group of writers who gathered for lunch in the Algonquin's Oak Room during the 1920s, gaining a reputation for their witty and often devastating conversation. The founding members were **Dorothy Parker, Robert Benchley,** and **Robert Sherwood,** who began to meet here while they were employed by *Vanity Fair* (1868–1935), then located at 19 West Forty-fourth Street. Those who participated in the gatherings came to include **George S. Kaufman,** the renowned *New Yorker* editor **Harold Ross, Russel Crouse, Alexander Woollcott, Donald Ogden Stewart, Heywood Broun,** and **Franklin P. Adams.** While the group continued to meet through much of the 1920s, some of its members in time abandoned it. The Algonquin remains a favorite of literary and publishing figures, and the table at which the famous group used to sit is still used in the Oak Room.

Carnegie Hall, at Seventh Avenue and Fifty-seventh Street, is known best for its concerts, but it has also played host to poetry readings and other literary events. **Mark Twain** spoke here at a Lincoln's Birthday celebration in 1901, and **Edwin**

Markham celebrated his eightieth birthday, in 1932, with a large gathering here at which the poet read "The Man With the Hoe" (1899). The Century Association, founded in 1847 as a literary and artistic club, is now located at 7 West Forty-third Street. **William Cullen Bryant** and **Washington Irving** were among its earlier members, and more recent additions have included **Malcolm Cowley, Alfred Kazin,** and **Archibald MacLeish.**

If there is a hotel that surpasses the Algonquin in the extent of its literary associations, it is the Chelsea, at 222 West Twenty-third Street. This 1884 landmark building has housed distinguished authors and artists since its early days as a cooperative apartment house. American writers who have stayed here include **Nelson Algren,** author of the novel *The Man With the Golden Arm* (1949); **William Burroughs,** author of such experimental novels as *Naked Lunch* (1959); **James T. Farrell,** who was here in the 1950s and wrote of the Chelsea, calling it the Hotel Verve in *Judith and Other Stories* (1974); **William Dean Howells,** the distinguished novelist and man of letters; **Edgar Lee Masters,** a longtime resident who was visited here by, among others, **Theodore Dreiser**—the visit ended a long estrangement between the two; **Mark Twain,** who stayed here in 1888; and **Thomas Wolfe,** who took an eighth-floor suite in 1937. The two writers whose names are associated most closely with the Chelsea are both foreigners who toured America to give readings: the Irish author and playwright **Brendan Behan,** and the Welsh poet **Dylan Thomas.** Thomas was in his room at the Chelsea when he collapsed on November 4, 1953; he died nearby in St. Vincent's Hospital on November 9. He was thirty-nine years old. Brendan Behan and Dylan Thomas, along with Thomas Wolfe, are honored by a plaque at the hotel entrance.

Another midtown establishment favored by the literary is the Coffee House Club, 54 West Forty-fifth Street. Its members have included **George Abbott, Robert W. Anderson, Philip Barry,** and **Thomas Costain.** On February 21, 1972, **W. H. Auden** had his sixty-fifth birthday party here. Not far away, at 234 West Forty-fourth Street, stand the offices of the Dramatists' Guild, formed in 1926 and a part of the Authors' League of America. In the same vicinity is the Gotham Book Mart, 41 West Forty-seventh Street. A literary landmark and browser's delight for over fifty years, this venerable bookstore has seen a long series of eminent visitors. Two of them, **Theodore Dreiser** and **H. L. Mencken,** walked in rather unsteadily one day in 1925 after drinking a good deal at a publisher's party nearby. As the horrified staff looked on, Mencken strode up to a valuable old Bible that was on display, opened it, and wrote on the flyleaf: "If it wasn't for me, Dreiser would be raising chickens in Kansas." The novelist scrawled a suitable reply. Suddenly, the Bible, later acquired for the Lewisohn Collection of Columbia University, was worth a great deal more. The Book Mart, though a longtime champion of the cause of modern literature, tries not to encourage such behavior in even the most distinguished of its visitors.

Robert W. Anderson, Malcolm Cowley, James Gould Cozzens, and **Corliss Lamont** have all been members of the Harvard Club, 27 West Forty-fourth Street. The Lambs Club, down the street a block or two, has been a favored haunt of such figures as **George M. Cohan, Rex Stout,** and **Alexander Woollcott. Frederick Loewe** and **Alan Jay Lerner,** who were to collaborate on such Broadway successes as *Brigadoon* (1947) and *My Fair Lady* (1956), their adaptation of George Bernard Shaw's *Pygmalion* (1912), met here in the 1940s. Formerly located at 128 West Forty-fourth Street, the club now is next door, at number 130.

The main branch of the New York Public Library, at Fifth Avenue and Forty-second Street, is one of the great libraries of the world, and the writers who have used its collections are too numerous to name. Those who have been members of the library staff include **Eleanor Estes,** the children's author, who worked appropriately as children's librarian from 1932 until 1940; and **Marianne Moore,** the distinguished poet, who worked

Photograph of the Algonquin taken in the 1920s, when the Round Table met regularly here.

Those of you who contributed so generously last year to the floating hospital probably have wondered what became of the money. I was speaking on this subject only last week at our up-town branch, and, after the meeting, a dear little old lady, dressed all in lavender, came up on the platform, and, laying her hand on my arm, said: "Mr. So-and-so" (calling me by name) "Mr. So-and-so, what the hell did you do with all the money we gave you last year?" Well, I just laughed and pushed her off the platform, but it has occurred to the committee that perhaps some of you, like that little old lady, would be interested in knowing the disposition of the funds.

—Robert Benchley,
in "The Treasurer's Report"

as an assistant librarian from 1921 until 1925. The library's collection of rare books and manuscripts includes valuable materials from every period of American literary history.

At Fifth Avenue and Fifty-ninth Street stands the Plaza Hotel, another favorite of many literary figures. Those who have stayed here range from **F. Scott** and **Zelda Fitzgerald** to **Georges Simenon,** who has stayed here often; those who have eaten or drunk in the hotel's several bars and restaurants include **Sherwood Anderson, John Dos Passos,** and **J. P. Marquand.** The Royalton Hotel, across the street from the Algonquin at 44 West Forty-fourth Street, has had among its occupants **Robert Benchley,** the humorist, who lived in a two-room suite during the 1930s; and **George Jean Nathan,** the critic and editor, who maintained an apartment here from 1908 until 1958—he died on April 8 in the apartment. Other literary figures who have stayed here include **Ernest Hemingway, Carl Sandburg, William Saroyan,** and **Tennessee Williams.** A famous eating and drinking place is the "21" Club, at 21 West Fifty-second Street, established originally as a speak-easy. **Robert Benchley** was such a regular at "21" that his old table is marked by a wall plaque: "Robert Benchley, His Corner, 1889–1945." Other regulars here have been **George Jean Nathan, Carl Sandburg,** and **Georges Simenon.** Even the men's room at "21" enters literary history, as the arena for a much-publicized disagreement between **John O'Hara** and **Sinclair Lewis,** who happened to meet here. The pair averted fisticuffs only when Lewis turned away hurriedly from the combative O'Hara, who was also twenty years Lewis's junior.

The midtown area was the birthplace of two of our most distinguished literary artists. **Eugene O'Neill** was born on October 16, 1888, in a hotel called Barret House, on the northeast corner of Broadway and Forty-third Street. He spent little time here, and the building was torn down in 1940. In 1928, on leaving his second wife, O'Neill moved to the Hotel Wentworth, at 59 West Forty-sixth Street. He stayed here while attending rehearsals for his plays *Marco Millions* and *Strange Interlude,* both of which had their first performances that year.

Edith Wharton was born **Edith Jones** on January 24, 1862, at her parents' elegant house at 14 West Twenty-third Street, near Madison Square. She passed much of her childhood in trips abroad and in long stays at Newport, and while in New York City moved in a relatively narrow geographical range: no farther south than Washington Square and no farther north than Central Park. Many of the most pleasurable hours of her adolescence were passed in her father's library, on the ground floor of their house. After the death of her father in 1882, Edith's mother moved to a house at 28 West Twenty-fifth Street, from which Edith was married on April 29, 1885, to Teddy Wharton. The wedding ceremony took place at Trinity Chapel, now the Serbian Orthodox Cathedral of St. Sava, at 15 West Twenty-fifth Street, and a select group from the party repaired across the street afterward for a wedding breakfast. Wharton's birthplace, though altered considerably, is still standing.

Many other writers have lived in the midtown area for various periods of time. One was **Sherwood Anderson,** who took a cheap room at 427 West Twenty-second Street in the autumn of 1918. Anderson traveled a good deal over the next few years, finally settling down in 1924 on a farm in Virginia. **S. N. Behrman** lived in an apartment on West Thirty-sixth Street in 1917, after leaving Harvard. The young playwright worked hard and took a graduate degree in English at Columbia, but he had no success with his plays until 1927, when *The Second Man* was produced and became an immediate success. Another young New York-bound writer not long out of Harvard was **Robert Benchley,** the humorist, who came in the autumn of 1912 to the McBurney Branch of the YMCA, at 215 West Twenty-third Street, still standing. Benchley worked at a variety of jobs before achieving success with such volumes as *The Treasurer's Report and Other Aspects of Community Singing* (1930) and *My Ten Years in a Quandary* (1936). **Van Wyck Brooks,** the chronicler of American literature and American writers, lived for a time in one of the row of apartment houses known as London Terrace, on West Twenty-third Street, after returning from Europe about 1908. Brooks made a living by working on dictionaries and other reference books. His first book, *Wine of the Puritans,* appeared in England in 1908, but he was not to publish his second, *America's Coming-of-Age,* until 1915.

Heywood Broun, soon after graduation from Harvard, took a job with the *Telegraph,* published at Eighth Avenue and Fiftieth Street, and there wrote the baseball stories that first won him a wide readership. He found a job with the *Tribune* in 1912, and by 1917 had moved with his wife, Ruth Hale, to a comfortable apartment at Seventh Avenue and Fifty-fifth Street. The couple separated in 1928, and Broun took a penthouse apartment on West Fifty-eighth Street, where he lived until his death on December 18, 1939. A memorial service was held for him at Manhattan Center, still standing, at 311 West Thirty-fourth Street.

Drawing of the dining room of the Plaza Hotel, originally published in a magazine of November 1911.

William Cullen Bryant was widely recognized as one of New York's foremost citizens during the mid-nineteenth century. Early verse efforts, such as "Thanatopsis" (1817) and "The Ages" (1821), had established him as a poet of stature, and through his editorship of the New York *Evening Post* he almost made journalism a fine and noble art. For all his success, however, his spirits were broken by the death, in July 1865, of his wife, Frances. After a trip to Europe he returned to New York, taking a house at 24 West Sixteenth Street, and set about distracting himself by translating *The Iliad* (1870) and *The Odyssey* (1871). He continued writing both prose and verse, producing such melancholy poems as "The Flood of Years" and "A Lifetime" (both 1876); took an interest in his editorial work on the *Evening Post;* and kept himself available for some of the many speeches he was invited to give. It was while returning from a speaking engagement on April 24, 1878, that he fell on the pavement and suffered a fatal concussion. He was eighty-four years old.

Also resident here, in the early 1930s, was **Erskine Caldwell,** who had an apartment on West Fiftieth Street when he wrote his best-selling *Tobacco Road* (1932). The site of Caldwell's apartment is now buried somewhere beneath Rockefeller Center.

Hortense Calisher lived at 24 West Sixteenth Street after her marriage in 1959 to Curtis Harnack. **Malcolm Cowley,** the editor and poet, lived at 360 West Twenty-second Street from 1930 to 1934. **Hart Crane** lived in 1919 in two rooms on the top floor of 24 West Sixteenth Street. This was the building once occupied by **William Cullen Bryant,** but in Crane's time it housed the offices of *The Little Review* (1914–29). Crane's rooms were unheated, and the cold weather of late autumn forced him to move out. An apartment at 165 West Twenty-third Street, decorated with souvenirs gathered from the site of the Battle of Fredericksburg, was the home in 1895 of **Stephen Crane.** The young novelist moved here after publication of his classic novel *The Red Badge of Courage* (1895), and while here wrote about New York City in articles for the New York *Journal.* Five years later, on June 28, 1900, Crane's funeral services were held nearby at Central Metropolitan Temple, Seventh Avenue and Fourteenth Street. The service was sparsely attended, but the small group of mourners included the young **Wallace Stevens.** Long after the death of Crane, scholars found portions of Crane's masterpiece that had been excised from the manuscript. A new edition, published in 1982 by W. W. Norton & Company, reflects the addition of the long-missing material.

Another West Side resident was **E. E. Cummings,** who moved into an apartment at 15 West Fourteenth Street late in 1918, after returning from Army service in Massachusetts. Cummings had entered the Army after serving as a volunteer ambulance driver in France, where he was imprisoned for some months for treasonable correspondence, a charge that eventually was dropped. His prison experience provided the material for his novel *The Enormous Room* (1922), on which Cummings is thought to have done some early work while living on West Fourteenth Street.

Theodore Dreiser lived and worked at a number of addresses on the West Side, beginning with the apartment of his sister and brother-in-law on West Fifteenth Street, where he stayed in 1894. In the summer of 1898 he took an apartment of his own at 232 West Fifteenth Street. After making a partial recovery from the crisis that followed publication of *Sister Carrie* (1900), Dreiser worked from 1905 until 1907 for *Broadway* magazine, published at 7 West Twenty-second Street, and later for the *Bohemian,* published at 40 West Thirty-third Street. In 1927, after *An American Tragedy* (1925) had brought Dreiser considerable wealth, he moved with his companion Helen Richardson to a duplex apartment at the Rodin Studios, 200 West Fifty-seventh Street, where they held a Thursday salon. Some of their guests were **Ford Madox Ford, George Jean Nathan, Alexander Woollcott,** and **Elinor Wylie.** An apartment at 35 or 38 West Fifty-ninth Street, now called Central Park South, was briefly the home of **F. Scott Fitzgerald,** who lived here with his wife **Zelda** while he was finishing *The Beautiful and Damned* (1922). They had their meals sent in from the nearby Plaza Hotel and entertained such guests as **H. L. Mencken, Dorothy Parker, John Peale Bishop,** and **Edmund Wilson.** When the book was finished, the Fitzgeralds sailed for Europe.

Lafcadio Hearn was another resident of the area. He lived at 170 West Fifty-ninth Street, probably in 1889, the year before he set out for Japan, where he died in 1904. While in New York, Hearn is thought to have worked on his travel book *Two Years in the French West Indies* (1890) and on his novel *Youma* (1890), about the Martinique slave rebellion of 1848; he also wrote *Karma* (1890), one of his lesser novels. **O. Henry** spent the last eight years of his life living in Midtown. Released in 1902 from a federal penitentiary where he had served three years and three months for embezzlement, O. Henry made his way to New York and settled first at the Marty Hotel, 47 West Twenty-fourth Street. Living in a small room here, he went out most days and nights to explore the streets of the city, largely the area between Madison Square and Irving Place. What he saw and heard here—in the city he called Baghdad-on-the-Subway—gave him raw material for much of his work. O. Henry himself saw nine volumes of his stories published, including *The Four Million* (1906), *The Trimmed Lamp* (1907), *Heart of the West* (1907), and *Roads of Destiny* (1909). "The Gift of the Magi," perhaps his most widely known story, was published in *The Four Million.* The author moved in 1906 to six sparsely furnished rooms at the Caledonia Hotel, 28 West Twenty-sixth Street. Here he lived for years, until drinking and tuberculosis caught up with him. Taken ill on June 3, 1910, he was brought to the nearby Polyclinic Hospital, where he died two days later. Funeral services were held at the Little Church Around the Corner, 1 East Twenty-ninth Street, and the Caledonia Hotel was later renamed the O. Henry after its celebrated occupant.

William Dean Howells lived briefly in 1865 at 44 West Forty-seventh Street while editing *The Nation,* and on his permanent return to New York in 1891 lived on West Fifty-ninth Street, at a site now occupied by the Hotel St. Moritz. While here, he wrote *The Coast of Bohemia* (1893), *My Literary Passions* (1895), and *Literary Friends and Acquaintances* (1900). In later years he lived at the Bur-

"What is your name?"—"Edward E. Cummings."—"Your second name?"—"E-s-t-l-i-n," I spelled it for him—"How do you say that?"—I didn't understand.—"How do you say your name?"—"Oh," I said; and pronounced it. He explained in French to the mustache that my first name was Edouard, my second "A-s-tay-l-ee-n," and my third "Kay-u-mm-ee-n-gay-s"—and the mustache wrote it all down.

—E. E. Cummings,
in *The Enormous Room*

Photograph of Edith Wharton taken in 1905.

Damon Runyon.

It is along toward 2 o'clock one pleasant AM, *and things are unusually quiet in Mindy's restaurant on Broadway and in fact only two customers besides myself are present when who comes in like a rush of air, hot or cold, but a large soldier crying out in a huge voice as follows:*

"Hello, hello, hello, hello, hello."

Well, when I take a good glaum at him I can see that he is nobody but a personality by the name of West Side Willie who is formerly a ticket speculator on Broadway and when he comes over to me still going hello, hello, hello, hello, hello, I say to him quite severely like this:

"Willie," I say, "you are three hellos over what anybody is entitled to in Mindy's even if there's anybody here which as you can see for yourself is by no means the situation."

—Damon Runyon,
in "Big Boy Blues"

lington, a hotel at 10 West Thirtieth Street; and at 130 West Fifty-seventh Street. **James Weldon Johnson,** the poet and novelist, stayed briefly at the Hotel Marshall, at 127 West Fifty-third Street, on his first visit to New York in 1899. He later headed the Negro Republican Club, at 260 West Fifty-third Street, in association with Theodore Roosevelt's campaign for reelection in 1904. **Imamu Amiri Baraka,** then still known as **Leroi Jones,** moved to 402 West Twentieth Street in 1958, the year in which he founded *Yugen* magazine and Totem Press.

A loft at 149 West Twenty-first Street was home for a short time for **Jack Kerouac,** who finished the first draft of *On the Road* (1957) while living here. Kerouac moved in with Joan Haverty, whom he had recently married, in November 1950. He wrote furiously here, and by the spring of 1951 he had finished writing. Two days later he moved out, abandoning Haverty after six months of marriage. **Denise Levertov** lived for a time at 299 West Fifteenth Street. Levertov, who came to the United States from England in 1948, is the author of many volumes of verse, including *Here and Now* (1957) and *The Jacob's Ladder* (1961). **Sinclair Lewis** moved to the Wyndham Hotel, still standing, at 42 West Fifty-eighth Street, after separating from **Dorothy Thompson** in 1937—they did not divorce until 1942—and another place he sometimes stayed at here was the Dorset Hotel, still standing at 30 West Fifty-fourth Street.

She stoops
to gently dip and deep enough.
Her face resembles
the face of the young actress who played
Miss Annie Sullivan, she who
spelled the word 'water' into the palm
of Helen Keller, opening
the doors of the world.

—Denise Levertov,
in "The Well"

The offices of publisher **Horace Liveright,** in a four-story brownstone at 61 West Forty-eighth Street, are worthy of inclusion in our literary history, because to them came many distinguished authors: for example, it was from here that **H. L. Mencken** and **Theodore Dreiser** sallied forth to the Gotham Book Mart in 1925. Liveright, who founded the Modern Library in 1917 and the publishing firm Boni and Liveright in the following year, both with Albert Boni, played a part in the careers of Dreiser, Ben Hecht, Ernest Hemingway, and Eugene O'Neill, to name but a few. He gave enthusiastic support to contemporary writing, but he was also known for his financial recklessness. When Liveright died in 1933, at 33 West Fifty-first Street, his estate totaled $500, and his publishing firm, which he had officially left in 1930, was in bankruptcy.

Brian Moore, Irish-born but now a Canadian citizen, lived at 150 West Fifteenth Street while writing *An Answer from Limbo* (1962). This novel, like the same author's *I Am Mary Dunne* (1968), is set largely in New York.

The Chelsea district, part of the midtown area, owes its name to a large estate that once occupied the land in the late eighteenth and the early nineteenth centuries. The estate was inherited in 1816 by **Clement Clark Moore,** a Biblical scholar and poet, who gave a large tract of land for establishment of the General Theological Seminary and then subdivided the area around it to form the core of what is known today as Chelsea. The seminary is located at 175 Ninth Avenue, and Moore is honored by a plaque at St. Peter's Church, 346 West Twentieth Street, of which he was designer as well as a founder. For all his importance to this community, Moore is remembered today for a trifle of a poem he wrote for his children at Christmastime of 1822 and which was published in the Troy *Sentinel* early in the following year. Although officially entitled "A Visit from St. Nicholas," the poem is usually referred to by the opening words of its first stanza: " 'Twas the night before Christmas. . . ."

Also resident here for a time was **Frank Norris,** who lived at 10 West Thirty-third Street in 1898. Norris had returned recently from Cuba, where he covered the Spanish-American War for *McClure's Magazine,* and had taken a job with the publishing firm of Doubleday, Page. In addition to working for the firm, he wrote two novels for them, *Blix* and *McTeague*—he had actually been working on *McTeague* since his undergraduate days. Both novels were published in 1899, and Norris remained in New York City for much of the time before his death on October 25, 1902. Another novelist, **John O'Hara,** lived at 470 West Twenty-fourth Street in 1937, when he was completing *Hope of Heaven* (1938). **Dorothy Parker** had a small apartment at 57 West Fifty-seventh Street in 1920, just after leaving *Vanity Fair* in a conflict over the severity of some of her theater reviews. Parker worked on her verse while also writing for *The New Yorker,* and her first book of poetry, *Enough Rope,* was published in 1926. **Edwin Arlington Robinson** lived for most of the period from 1901 to 1906 in a cheap room in Chelsea at 450 West Twenty-third Street, and while here published his second volume of verse, *Captain Craig* (1902).

Damon Runyon absorbed much of the New York City vernacular that enlivens his works while sitting at his favorite table at Lindy's, a once-renowned midtown restaurant that stood on Broadway, just above Times Square, and now exists in a new incarnation at the corner of Fiftieth Street and Avenue of the Americas—Runyon, of course, knew the latter as Sixth Avenue. Runyon, who called the restaurant Mindy's in his stories, spent countless hours in the original Lindy's, drinking numerous cups of coffee and talking with anyone who happened to be within earshot. The colorful people he met and the language he heard went into his classic Broadway stories, collected in *Guys and Dolls* (1932) and other volumes. In the last years of his life, Runyon lived at the Hotel Buckingham, which still operates at 101 West Fifty-seventh Street. He died of cancer on December 10, 1946.

Across the street from the Chelsea Hotel, at 215 West Twenty-third Street, is the McBurney Branch of the YMCA, already mentioned in connection with Robert Benchley. **William Saroyan** also stayed here when he first came to New York in the 1930s, and other onetime residents include **Brendan Behan** and **Edgar Lee Masters.**

Delmore Schwartz, the poet, critic, and teacher, spent the last months of his life in torment at the Columbia Hotel, 70 West Forty-sixth Street. He had moved to New York in January 1966 from Syracuse, where he had an appointment as visiting professor of English, and moved to the Hotel Dixie, at 250 West Forty-third Street. From there he moved to the Columbia, living in complete isolation—none of his friends even knew he was there. When Schwartz went out at all, it was to bars, such as the White Horse in Greenwich Village, or to the New York Public Library, where he would sit in the Main Reading Room, scribbling in his notebook. On July 11, 1966, about 3 AM, he suffered a heart attack while taking his garbage out to the incinerator. He lay on the floor outside Room 406 for over an hour and finally was taken by ambulance to Roosevelt Hospital, 428 West Fifty-ninth Street, where he was pronounced dead at 4:15AM. Schwartz's body lay at the morgue unrecognized and unclaimed for two days; it was only by accident that a reporter scanning the morgue lists happened to see his name and recognize it.

Robert Sherwood lived for a time at 1545 Broadway. **Lincoln Steffens,** the muckraking journalist, moved in 1897 to a large apartment at 341 West Fifty-seventh Street with his first wife, Josephine. Steffens had returned from Europe five years before and was in the midst of a successful career in New York City journalism. **Wallace Stevens,** after marrying Elsie Moll, moved in 1909 to a house at 441 West Twenty-first Street. Stevens lived here until 1916, and while here saw his first poems published in *Poetry* magazine and other publications. Always a hardy and enthusiastic walker, Stevens once walked in a single day from Van Cortlandt Park, in The Bronx, to Greenwich, Connecticut.

Carl Van Vechten, another longtime resident of New York City, had his first home here at 39 West Thirty-ninth Street; **Sinclair Lewis** had a room down the hall. Later, in 1907–08, Van Vechten and his wife lived at the Maison Favre, 528 Seventh Avenue, just behind the Metropolitan Opera House, where the critic and novelist spent so much of his time. Later still, in 1924, he moved to 150 West Fifty-fifth Street, where he entertained such guests as **Theodore Dreiser, W. Somerset Maugham,** and **Paul Robeson.** *Nigger Heaven,* generally considered Van Vechten's best work of fiction, was published in 1926. Van Vechten largely abandoned his writing in 1930 and from then on devoted his energies to photography.

Thomas Wolfe in 1928 rented a room at 27 West Fifteenth Street, still standing, but altered. Although on good terms with his companion, Aline Bernstein, he wanted for the moment to live separately from her. This was all for the best, as in January 1929 he received a contract from Scribner's to publish a work he called *O Lost.* It was the massive novel that would be published that year as *Look Homeward, Angel.* Maxwell Perkins, his editor, wanted a good deal of the novel trimmed away, and Wolfe worked at the job for about three months. As it turned out, he cut only eight pages out of 1100 and had to re-edit in close cooperation with Perkins. **Herman Wouk** lived in 1937 at the Essex House Hotel, still open at 160 Central Park South, while he was working as a scriptwriter for comedian Fred Allen. West Side locations that appear in his books include Rockefeller Center, in *Youngblood Hawke,* (1962); and Radio City Music Hall, 1260 Sixth Avenue, in *Aurora Dawn* (1947). **Richard Wright** lived for a time at 473 West Fortieth Street.

Although it costs you countless agony,
Although you cannot believe it
necessary,
And doubt that the sum is accurate,
Please send me money enough for at
least three weeks.

—Delmore Schwartz,
in "Baudelaire"

Midtown East

From the nineteenth-century townhouses around Gramercy Park to the more recently built-up high-rises of the East Fifties, the fashionable Midtown East district of Manhattan has long housed clubs, restaurants, and other institutions with extensive literary connections. One of them is the former home of the Authors' League of America, at 6 East Thirty-ninth Street. The building was the scene of a 1942 meeting of **Pearl Buck, Russel Crouse, Oscar Hammerstein, J. P. Marquand,** and **Rex Stout** to form the Writers' War Committee. The league now has its offices at 234 West Forty-fourth Street.

Nigger Heaven! That's what Harlem is. We sit in our places in the gallery of this New York theater and watch the white world sitting down below in the good seats of the orchestra. Occasionally they turn their faces up towards us, their hard, cruel faces, to laugh or sneer, but they never beckon.

—Carl Van Vechten,
in *Nigger Heaven*

Bellevue Hospital, at First Avenue and Twenty-seventh Street, has been a temporary stopping place for several writers, including **F. Scott Fitzgerald,** who was brought here after a binge in February 1939. **Sidney Kingsley** investigated the drama of Bellevue as well as other hospitals while doing research for his realistic play *Men In White* (1933). **Malcolm Lowry** spent a few weeks in the Bellevue psychiatric ward after arriving in New York from England in 1939. **David Graham Phillips,** the journalist and novelist, died in Bellevue on January 24, 1911, after being shot near Gramercy Park. **Wallace Thurman,** the bitterly satirical black novelist and playwright, died of tuberculosis here on December 22, 1934.

A more cheerful institution in its day was the Biltmore Hotel, at Madison Avenue and Forty-third Street, still standing but converted for use as an office building. Generations of college students met their dates at the great clock in the Biltmore lobby, and the hotel was popular enough

Old photograph of Lindy's, at its original location on Broadway.

Zelda and F. Scott Fitzgerald.

> *Night could still down the city, absorbing it for all its rhinestone effrontery, but the mornings crept in like applicants for jobs, nuzzling humbly against the masked granite, saying hopefully, "Do you suppose . . . is there anything to be made of me?"*
>
> —Hortense Calisher, in "The Woman Who Was Everybody"

Costello's, a favorite spot of James Thurber, now at 225 East Forty-fourth Street. The bar was located next door when Thurber decorated its walls with his drawings in the winter of 1934-35. After the drawings were mistakenly painted over, Thurber made new ones.

in the 1920s to be chosen as a honeymoon spot by **F. Scott** and **Zelda Fitzgerald.** They came here after their marriage on April 3, 1920, but behaved so boisterously that the management asked them to leave. A more regular and more orderly Biltmore patron was **Peter De Vries,** who came often to the Men's Bar. Another bar, Costello's, at 225 East Forty-fourth Street, was favored by **James Thurber,** the celebrated humorist and cartoonist. Thurber graced the place by drawing murals on the establishment's walls. The Friars Club, now at 57 East Fifty-fifth Street, was a favorite of both **George M. Cohan** and **Ring Lardner.** Lardner is said to have drunk here steadily for sixty hours on one occasion.

Grand Central Terminal, in its more than seventy-five years of existence, has probably seen as many literary travelers as any building in the world. Few, however, have had as dramatic a visit as one by **Thomas Wolfe** on January 26, 1932. Having boarded a train to New Canaan, Connecticut, with his editor, **Maxwell Perkins,** Wolfe suddenly decided—after the train was moving—that he did not want to make the trip after all. He leaped from the train, fell with a crash to the platform, and managed to sever a vein in his arm.

A more sedate and certainly less crowded spot is the Morgan Library, 29 East Thirty-sixth Street, a haven for bibliophiles and art lovers alike. The library holds a continuing series of exhibitions, some drawn from its own vast holdings in printed books, manuscripts, and works of art; some borrowed from other institutions. Two of the many American authors represented in the Morgan collections are Henry David Thoreau and Nathaniel Hawthorne.

Gramercy Park, one of the most enduring areas in New York City, houses The Players, a club still operating at 16 Gramercy Park South. Founded in 1888 by the actor Edwin Booth, The Players has included among its members **Gelett Burgess,** the humorist and author who contributed the words *blurb* and *goop* to the English language, and who lived here for some time; **Richard Harding Davis; Paul Green; Mark Twain,** a charter member who stayed here twice; and **Hendrik Willem Van Loon,** the journalist and historian. Housed next door, at number 15, is the Poetry Society of America, founded in 1910 by **Jessie B. Rittenhouse,** among others. **Robert Frost,** an honorary president, was only one of many distinguished members.

At St. Patrick's Cathedral, Fifth Avenue at Fiftieth Street, were held the funeral services for **Heywood Broun,** who earlier was baptized here after his conversion to Catholicism, **Finley Peter Dunne,** and **George Jean Nathan.** A requiem mass, at which "Over There" was sung, was held for **George M. Cohan** on November 7, 1942. At the rectory, 460 Madison Avenue, **F. Scott Fitzgerald** was married to **Zelda Sayre** on April 3, 1920.

Two blocks away, at 595 Fifth Avenue, stand the offices of Charles Scribner's Sons, the publisher of F. Scott Fitzgerald, Ernest Hemingway, Thomas Wolfe, and many others. In 1937 **Maxwell Perkins** made the error of scheduling meetings back-to-back with two of his authors, **Ernest Hemingway** and **Max Eastman.** Some four years before, Eastman had written a less-than-flattering review of Hemingway's *Death in the Afternoon* (1932). When the pair met, the result, predictable enough, was fisticuffs. Wolfe and Perkins held frequent conferences here, often adjourning for refreshment to the nearby Chatham Hotel, no longer standing, at Vanderbilt Avenue and Forty-eighth Street. Fitzgerald's story "Financing Finnegan" is set partly in Perkins's offices and partly in the offices of Harold Ober, his agent, at 40 East Forty-ninth Street.

One of New York City's most famous hotels, the Waldorf-Astoria at Park Avenue and Fiftieth Street, has played host to a variety of literary figures. **Isaac Asimov** has reported that the hotel's Peacock Alley restaurant is one of his favorites. **Clare Boothe Luce** lived for a time at the hotel. And writers honored by gatherings here include **Ernest Hemingway, Sinclair Lewis,** and **Carl Sandburg.** Sandburg celebrated his eighty-fifth birthday at the Waldorf, an occasion that coincided with publication of his volume *Honey and Salt* (1965). Among those present were **John Steinbeck** and **Mark Van Doren.**

Midtown East is also the birthplace of three authors. **Clarence [Shepard] Day,** the essayist and memoirist, was born on November 18, 1874, when his parents lived at 251 Madison Avenue. The family moved in 1877 to 420 Madison Avenue, and it was this latter neighborhood that Day described in his most popular work, *Life With Father* (1935). **Rachel Field,** the children's author and novelist, was born on September 19, 1894, on East Fortieth Street. Her family moved away when Rachel was less than a year old. Field's novels include *Time Out of Mind* (1935). **Alfred Kreymborg,** born in New York City to poor parents on December 10, 1883, spent much of his youth going to vaudeville and music halls; he also frequented Manhattan's chess clubs, and for some years later in life supported himself as a professional chess player. While his own verse, published in such volumes as *Mushrooms* (1916), is often considered undistinguished, he made valuable contributions to American letters in his capacity as editor of a succession of little magazines.

Authors who have lived here include **Robert W. Anderson,** who has kept city residences on East Fifty-sixth Street and on Sutton Place. At his Sutton Place apartment, the playwright worked on *Getting Up and Going Home* (1978). **Charles Angoff** has lived at 505 Lexington Avenue and 142 East Forty-seventh Street, and **Isaac Asimov** lived at 278 First Avenue in 1948–49. **W. H. Auden** and **Christopher Isherwood** stayed at the George Washington Hotel, still operating at Lexington Avenue and Twenty-third Street, on their arrival here in 1939. In 1947, Isherwood stayed in an apartment at 207 East Fifty-second Street. **Ludwig Bemelmans,** author and illustrator of the well-loved children's book *Madeline* (1938) and its sequels, lived at 20 Gramercy Park and at the National Arts Club, 15 Gramercy Park South, where he died on October 1, 1962.

Robert Benchley lived for a short time with his family at the Gramercy Court, on East Twenty-second Street, and later shared an apartment with **Charles MacArthur** at 536 Madison Avenue. **Van Wyck Brooks** lived for a time at 350 East Fifty-seventh Street, where his neighbor was his Harvard classmate **John Hall Wheelock. Hortense Calisher** lived before her first marriage on East Twenty-third Street and worked as an investigator for the Department of Public Welfare. Her experiences in that job supplied material for her novel *The New Yorker* (1969), and her home on Twenty-third Street is referred to in the story "The Woman Who Was Everybody."

One of the most distinguished literary salons ever held in New York City was that of **Alice** and **Phoebe Cary,** at 52 East Twentieth Street. The Cary sisters, both poets, had come to New York in 1850 from their home in Ohio. While producing now-forgotten works, in such volumes as *Poems of Faith, Hope, and Love* (1868), the sisters held Sunday-evening literary gatherings where guests might include **Horace Greeley, Herman Melville, George Ripley,** the editor and a co-founder of Brook Farm, and **Bayard Taylor,** the poet. Both sisters died in 1871, Alice on February 12, and Phoebe on July 31.

Irvin S. Cobb, author of the humorous volume *To Be Taken Before Sailing* (1930), among others, died on March 10, 1944, at his home at the Hotel Sheraton, now the Sheraton Russell, at Park Avenue and Thirty-seventh Street. **Frank M. Colby,** the essayist and editor, did his last work at 326 East Fifty-seventh Street. The best of his writings were collected in *The Colby Essays* in 1926, the year after his death. **Malcolm Cowley,** known best for critical works such as *Exile's Return* (1934) and *The Literary Situation* (1954), lived during the 1920s at 501 East Fifty-fifth Street. **Hart Crane** stayed with him here briefly, having lived at an earlier date at 135 East Twenty-first Street, Gramercy Park.

Stephen Crane, after moving to New York City about 1890, took modest quarters at 143 East Twenty-third Street, then the location of the Art Students League of New York. Crane lived here for about two years, constantly impoverished and often ill. Occasionally he did some reporting for the *Herald* or the *Tribune,* and he spent much of his time exploring the slums of the nearby Bowery. Yet, while living here, Crane wrote both *Maggie* (1893) and *The Red Badge of Courage* (1895), and the latter novel made him famous. After returning from a trip to the West in 1896, he may have lodged for a time at 33 East Twenty-second Street. Not far away, at 21 East Fifteenth Street, **E. E. Cummings** spent the few months between getting his A.M. at Harvard and leaving, in April 1917, to serve as an ambulance driver in France.

Also resident here for a time from the late 1940s to the 1960s was **James T. Farrell,** who had an apartment at the Beaux Arts, still standing at 310 East Forty-fourth Street. **Clyde Fitch,** who wrote about New York City in such plays as *The City* (1909), lived for a time at 133 East Fortieth Street. **William Goyen,** author of such novels as *The Fair Sister* (1963), lived at 127 East Fifty-seventh Street. **Horace Greeley,** longtime editor of the New York *Tribune,* and a powerful influence in contemporary politics, lived for many years at 25 East Nineteenth Street. **Margaret Halsey** lived at 244 East Forty-eighth Street from 1940 until 1950, the years in which she was working on *Some of My Best Friends Are Soldiers* (1944) and *Color Blind* (1946). Two of the hotels **Ernest Hemingway** stayed at when in New York City were the Sherry Netherland, at 781 Fifth Avenue, and the Barclay, at 111 East Forty-eighth Street. The novelist stayed at the Barclay while doing final revisions on *For Whom the Bell Tolls* (1940), and later that year spent his honeymoon here with **Martha Gellhorn,** the writer who was Heming-

Hortense Calisher.

East Room of the Pierpont Morgan Library. Among the library's extensive holdings are such items as the autograph manuscript of *A Christmas Carol,* by Charles Dickens, and the only complete first edition of Malory's *Morte d'Arthur.*

Last known photograph of O. Henry, taken in 1909, the year before he died. (*far right*) Herman Melville, in a photograph taken late in life.

way's third wife. Among their visitors at the Barclay was **H. G. Wells,** the English novelist, who was in New York City for a lecture tour.

Two other novelists who resided in Midtown East are **John Clellon Holmes,** at 123 and 681 Lexington Avenue, and **William Dean Howells,** who lived at 50 East Fifty-eighth Street, 241 and 330 East Seventeenth Street, and 115 East Sixteenth Street.

A well-known resident of the area was **O. Henry,** who lived from 1903 until 1907 at 55 Irving Place. This home was right in the middle of the author's favorite territory, the small portion of New York City he loved to explore and which gave him many of his story ideas. He enjoyed going to Scheffel Hall, a *Brauhaus* at 130 Third Avenue, now the site of Tuesday's restaurant, and often sat on a bench in nearby Madison Square Park. O. Henry, in a sense himself a literary visitor to New York City, took pleasure in thinking that Washington Irving had once lived in the neighborhood, at 122 East Seventeenth Street—though it now is known that Irving did not in fact live there. After leaving 55 Irving Place, O. Henry stayed briefly in a house on East Twenty-fourth Street between Madison and Fourth avenues. This portion of Fourth Avenue is now known as Park Avenue South.

BILLIE—*You're just not couth!*
BROCK—*I'm just as couth as you are!*

—Garson Kanin,
in *Born Yesterday*

John Treat Irving, the novelist who wrote *The Hawk Chief: A Tale of the Indian Country* (1837), lived at 46 East Twenty-first Street. A regular guest here was the author's uncle, **Washington Irving,** who stayed with him on his trips to New York City from Tarrytown. **Henry James,** before leaving America in 1875 for more permanent residence in Europe, lived for six months at 111 East Twenty-fifth Street. **Garson Kanin,** whose best-known play is *Born Yesterday* (1946), lived for a time at 242 East Forty-ninth Street. **Alfred Kazin,** who has written about his life in New York City in such books as *New York Jew* (1978), has lived at Lexington Avenue and Twenty-fourth Street. Another author who wrote

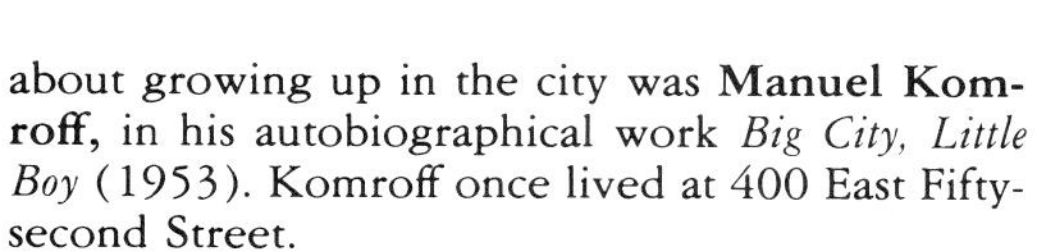

about growing up in the city was **Manuel Komroff,** in his autobiographical work *Big City, Little Boy* (1953). Komroff once lived at 400 East Fifty-second Street.

Stephen Leacock, the English-born Canadian humorist, was married to Beatrix Hamilton in August 1900 at the Little Church Around the Corner, 1 East Twenty-ninth Street. Leacock's books included *Nonsense Novels* (1911). **Max Lerner** lived at 22 Gramercy Park while writing *Public Journal* (1945), and **Sinclair Lewis** stayed at various times at a number of East Side addresses: 309 Fifth Avenue; 66 Park Avenue; the Berkshire Hotel, at Madison Avenue and Fifty-second Street; and the Lombardy Hotel, at 111 East Fifty-sixth Street. Both hotels are still operating. At another hotel, the Roosevelt, at Madison Avenue and Forty-fifth Street, Lewis attended a 1933 dinner honoring a group of fellow Nobel laureates. Irked when photographers began taking pictures of the distinguished group—it included Albert Einstein—Lewis stalked off to his room and had to be coaxed into returning.

I haven't attended a public banquet in five years and I don't see why the hell I should be photographed.

—Sinclair Lewis,
quoted in a 1933 newspaper article

Amy Lowell stayed at both the Hotel Belmont, 541 Lexington Avenue, and the St. Regis, now the St. Regis-Sheraton, Fifth Avenue and Fifty-fifth Street. Whenever she had a new book coming out, she stayed at the St. Regis, renting a suite and holding court for all those whose help she wanted. To create the right atmosphere, Lowell stopped all clocks and hung draperies over the mirrors and windows. A more regular resident of the East Side was **Clare Boothe Luce,** the playwright, who rented a penthouse apartment at 444 East Fifty-second Street after her divorce from her first husband.

Another resident here was **Henry Harland,** who wrote fiction under the name **Sidney Luska.** He lived for a time at 35 Beekman Place but spent his later years in England, where he founded and edited *The Yellow Book,* a quarterly. The fiction he wrote in his later years included *The Cardinal's Snuff-Box* (1900). **Mary McCarthy,** whose best-known novel, *The Group* (1963), is set partly in the East Fifties, lived for a time at 2 Beekman Place. **J. P. Marquand** spent part of his childhood at 51 East Thirtieth Street, and in 1937 took a duplex apartment at 2 Beekman Place that had formerly been occupied by Gertrude Lawrence, the English actress. Farther downtown is the Hotel Bedford, 118 East Fortieth Street, where **Thomas Mann** lived after fleeing Germany in 1938, and where **John Steinbeck** also stayed sometimes.

A longtime resident of East Midtown was **Herman Melville,** who lived first at 150 East Eighteenth Street before moving in 1863 to 104 East Twenty-sixth Street, where he remained until his death. Melville had already published his finest works, but most of the reading public took little notice of him. Retiring from active literary life, he worked from 1868 until 1885 as a customs inspector at the Gansevoort Street wharf and confined his writing primarily to verse. The collections of poetry he published here included *Battle-Pieces and Other Aspects of the War* (1866), *Clarel* (1876), *John Marr and Other Sailors* (1888), and *Timoleon* (1891). In the last five years of his life Melville undertook one more work in prose, the novella *Billy Budd,* which he completed only months before his death and which was not published until 1924. When he died on September 28, 1891, none of the popular newspapers ran his obituary. The site of his last home is occupied now by an office building, 357 Park Avenue South. The delivery entrance carries a plaque recalling Melville's residence.

Another longtime resident of Midtown East was **Alice Duer Miller,** whose most popular work was her book of verse *The White Cliffs of Dover* (1940), a wartime tribute to England and the English people. Miller's other works included the novels *Come Out of the Kitchen* (1916) and *The Beauty and the Bolshevist* (1920); she also published a column in the New York *Tribune* called "Are Women People?" She died at her home here, 450 East Fifty-second Street, on August 22, 1942.

On a summer day in 1930, a young employee in the advertising department of Doubleday and Page, 244 Madison Avenue, jotted down a bit of doggerel instead of working on his copy. He crumpled up the sheet and threw it away, but later retrieved it and sent the verse it contained to *The New Yorker.* It was accepted for publication, and thus was launched the long career of **Ogden Nash,** the first poet to rhyme "Petunia" with "Pennsylvunia." Nash continued to publish in that magazine as well as others, and soon completed his first book, *Free Wheeling* (1931). Although Nash spent most of the rest of his life in Baltimore, he did reside for a time at 333 East Fifty-seventh Street.

John O'Hara started his novel *Appointment in Samarra* (1934) while living at the Pickwick Arms Hotel, still operating at 230 East Fifty-first Street. **Eugene O'Neill** as a child attended St. Stephen's Church, still standing at 149 East Twenty-eighth Street, when his parents were in New York. Later in his life O'Neill stayed at the Barclay Hotel, 111 East Forty-eighth Street. Nearby, at 246 East Forty-ninth Street, was an apartment building in which **Maxwell Perkins** lived during the 1930s.

The Gramercy Park Hotel, 52 Gramercy Park North, was the longtime home of **S. J. Perelman,** who was living here at the time of his death on October 17, 1979. When much of midtown was still a rural area, **Edgar Allan Poe** lived in the Turtle Bay section, perhaps on East Forty-seventh Street, with his wife and mother-in-law. They stayed here a brief time only, moving to even more rural Fordham at the end of May 1846, but while here Poe visited **Horace Greeley** at his home nearby, address unknown. **Ezra Pound** lived for brief periods at 270 Fourth Avenue, now Park Avenue South, and at 24 East Forty-seventh Street.

After the death of her husband in 1932, **Mary Roberts Rinehart** lived in an eighteen-room apartment at 635 Park Avenue. Rinehart continued writing here, even though she had long forgotten the financial troubles that prompted her in 1903 to turn to a literary career. Among the works she wrote after moving here were the story "Tish Marches On" (1937) and the novel *The Wall* (1938). **Edwin Arlington Robinson** often spent his winters at 328 East Forty-second Street and was living here in January 1935 when he entered New York Hospital. Robinson died at the hospital on April 6. **Paul Rosenfeld,** the critic and editor, lived at 77 Irving Place and 20 Gramercy Park. At both these addresses Rosenfeld entertained his circle of distinguished friends, among whom were **Van Wyck Brooks, Hart Crane,** and **Waldo Frank.** Another close friend of Rosenfeld's was **Randolph Bourne,** the essayist, whom Rosenfeld cared for at his Gramercy Park home during the period of Bourne's fatal bout with influenza. Bourne died in Rosenfeld's apartment on December 22, 1918.

Another resident here was **Budd Schulberg,** who lived for a time about 1971 at 340 East Fifty-first Street. **Robert Sherwood** lived for a long time at 25 Sutton Place South. **Cornelia Otis Skinner,** co-author with Emily Kimbrough of *Our Hearts Were Young and Gay* (1942), lived not far away at 22 East Sixtieth Street. A longtime resident of the area was **Leonora Speyer,** who lived at 60 Gramercy Park in her final years. The Pulitzer Prize-winning poet—for *Fiddler's Farewell* (1927)—was living here at her death on February 10, 1956. At 225 East Fifty-seventh Street lived **Frank Morrison Spillane,** known to millions of readers for novels he published under the pen name **Mickey Spillane.** Also resident here for a time was **Rex Stout,** who once had an apartment at 311 East Twenty-ninth Street. **John Steinbeck** lived at 38 Gramercy Park after arriving in New York City in 1925. **Richard Henry Stoddard,** the influential critic and poet, lived for some time at 32 East Fifteenth Street. His home, still standing, houses a small collection of items of literary interest.

The area was also one of the residences of **Edward Streeter,** the popular humorous writer, who lived at 30 Sutton Place. After his death on March 31, 1976, Streeter's funeral services were held at St. Bartholomew's Church, 109 East Fiftieth Street. **Robert Lewis Taylor** lived for a time at 230 East Forty-eighth Street. A sometime resident of the Martha Washington Hotel, still standing at 29 East Twenty-ninth Street, was **Sara Teasdale.** The poet first stayed at the hotel in 1912 while attending the annual dinner of the Poetry Society of America, and returned on several occasions after that. She also stayed, in 1911, in a

To have known him, to have loved him
After loneness long;
And then to be estranged in life,
And neither in the wrong;
And now for death to set his seal—
Ease me, a little ease, my song!

—Herman Melville,
in "Monody," a poem written after the death of Nathaniel Hawthorne on May 19, 1864

I sit in an office at 244
Madison Avenue,
And say to myself You have
a responsible job, havenue?
Why then do you fritter away
your time on this doggerel?
If you have a sore throat
you can cure it by using
a good goggerel.

—Ogden Nash,
in "Spring Comes to Murray Hill"

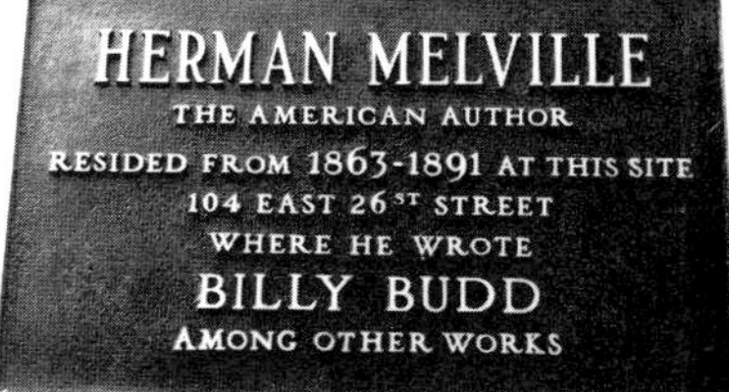

Photograph of Tennessee Williams taken in 1948.

boarding house at 53 Irving Place. **Dorothy Thompson,** the reporter and political commentator, lived at two addresses here, 125 East Sixteenth Street and 237 East Forty-eighth Street. The latter home, in exclusive Turtle Bay Gardens, was Thompson's city home from 1941, when her house in Bronxville burned down, until 1957, when she bought an apartment at 25 East End Avenue. The area was also the home—his first after marrying—of **Carl Van Vechten,** the novelist. He and his wife, Fania, lived for a time at 151 East Nineteenth Street before moving to the West Side. Van Vechten also lived at 151 East Nineteenth Street.

Like so many American writers of his time, **Nathanael West** in 1926 made an extended visit to Paris. When he was forced by a decline in his family's fortunes to return to the United States a year later, West took a job as night manager of the Kenmore Hall Hotel, still standing at 145 East Twenty-third Street. He remained at the job until 1930, entertaining here such friends as **Maxwell Bodenheim, Dashiell Hammett,** and his brother-in-law, **S. J. Perelman.** While on duty during nights when he lacked company, West worked in his office rewriting *The Dream Life of Balso Snell* (1931). In the autumn of 1930, West took a similar job at another hotel, the Sutton Club, also still operating, under the name Sutton East, at 330 East Fifty-sixth Street. Friends continued to visit him here, with some, such as **Erskine Caldwell,** Perelman, and **Edmund Wilson,** taking rooms in the hotel at attractive rates. West said of the hotel: "It was not a place for the successful." Observing the lives of the hotel patrons, West gathered impressions and material that later were to provide some of the basis for *Miss Lonelyhearts* (1939), a novel that West said was written about the people here. The hotel may rightly be considered the origin for what W. H. Auden called West's Disease: "A disease of consciousness which renders it incapable of converting wishes into desires."

A longtime resident of the area was **John Hall Wheelock,** who lived at 350 East Fifty-seventh Street. Wheelock worked for many years at Charles Scribner's Sons, where he succeeded Maxwell Perkins as senior editor. Wheelock combined his work at Scribner's with a career as poet, publishing a number of volumes between 1911 and 1961. His late works, collected in *The Gardener* (1961), were awarded a Bollingen Prize in 1962. Wheelock's neighbor at one time was **Van Wyck Brooks,** a Harvard classmate. The two men and their families, as well as **Allen Tate** and his wife **Caroline Gordon,** celebrated Christmas together in Wheelock's apartment in 1947. Wheelock lived here until his death on March 23, 1978.

Another prominent figure in New York's cultural life since the 1920s, **E. B. White,** has lived at both 239 and 229 East Forty-eighth Street. **Thornton Wilder,** three-time winner of a Pulitzer Prize, lived for four months in 1939 in an apartment at 81 Irving Place. **Tennessee Williams** lived during the late 1940s and early 1950s in a brownstone apartment on Fifty-eighth Street between Lexington and Third avenues. The story goes that Williams arrived home one evening to find **Gore Vidal, Truman Capote,** and a policewoman standing in his living room: the two authors, both feeling the effects of a few drinks, had climbed in through the transom. The policewoman had observed their novel mode of entry and had come in to investigate.

Thomas Wolfe also lived in Midtown East for a time in September 1935, taking an apartment at 865 Third Avenue, two blocks from the home of his editor, Maxwell Perkins. *Of Time and the River* had been published earlier the same year, and Wolfe was taking up much of Perkins's time, calling him at all hours to suggest a trip to some nearby bar, imposing on the family for dinners and lunches, sometimes even camping on one of Perkins's couches. **Alexander Woollcott** once had a penthouse apartment at 450 East Fifty-second Street that shared a balcony with the apartment of **Alice Duer Miller.** When he left the place, Woollcott sold his apartment to Noel Coward. **Elinor Wylie,** the poet, lived from October 1923 to September 1924 at 142 East Eighteenth Street with her husband, **William Rose Benét. Stark Young,** the poet, playwright, novelist, and drama critic, lived in the early 1950s at 329 East Fifty-seventh Street, where he is thought to have worked on his autobiographical study *Pavilions* (1951). Young served as drama critic for *The New Republic* from 1926 until 1947.

Upper West Side

The Ansonia, the neighborhood's great landmark, was built by Stanford White. It looks like a baroque palace from Prague or Munich enlarged a hundred times, with towers, domes, huge swells and bubbles of metal gone green from exposure, iron fretwork and festoons. Black television antennae are densely planted on its round summits. Under the changes of weather it may look like marble or like sea water, black as slate in the fog, white as tufa in sunlight.

—Saul Bellow,
in *Seize the Day*

The literary visitor to the Upper West Side of Manhattan encounters a diversity of interesting sites, including an unsurpassed concentration of homes of Nobel laureates in literature. **Edward Albee,** whose play *The Zoo Story* (1959) is set in Central Park, once worked as a counterman at the luncheonette of the Manhattan Towers Hotel, still operating at 2166 Broadway. **Charles Angoff** wrote short stories, collected in *When I Was a Boy in Boston* (1947), *Adventures in Heaven* (1945), and *Something About My Father and Other People* (1956), while living at the Marie Antoinette, a hotel that stood at Broadway and Sixty-sixth Street. Angoff has also lived at 15 West Seventy-first Street and 75 West Sixty-eighth Street. Not far away, at 12 West Sixty-eighth Street, **Isaac Asimov** had his residence for a time.

Saul Bellow, whose short novel *Seize the Day* (1956) is set largely in the Upper West Side, lived for a time in a white brick house at 333 Riverside Drive—the bathroom provided the only view of the Hudson River from Bellow's apartment. Bellow won a Nobel Prize in literature in 1976, long after leaving New York City. **Konrad Bercovici,** chronicler of gypsy life in such novels as *Ghitza* (1921) and *Singing Winds* (1926), had a home on West Ninety-second Street until neighbors complained about the parties he gave. He moved to 95 Riverside Drive and later to 322 West Seventy-second Street, his home until his death on December 27, 1961.

Heywood Broun lived during early childhood, about 1890, at 140 West Eighty-seventh Street. He was married in 1917 to Ruth Hale at St. Agnes Episcopal Church, which stood on West Ninety-second Street. His best man was **Franklin P. Adams.** When the Brouns's son **Heywood Hale Broun** was an infant, the family moved to a house still standing at 333 West Eighty-fifth Street. One literary visitor, **E. B. White,** noted the untidiness and casual discipline of the household—young Heywood was sometimes to be seen wandering about in his nightclothes late at night. In 1933, the year the Brouns divorced, the couple lived at the Hotel Des Artistes, at 1 West Sixty-seventh Street. Others who have lived in the well-known hotel include **Fanny Hurst,** once called "the sob sister of American literature"; **Margaret Widdemer,** winner of a special Pulitzer award for her verse collection *Old Road to Paradise* (1918); and **Alexander Woolcott,** the acidulous wit and critic who not only inspired the central character of Moss Hart and George S. Kaufman's *The Man Who Came to Dinner* (1939) but toured in the farce, playing that character: Sheridan Whiteside, a meddlesome, demanding, insulting houseguest.

William Burroughs, author of such experimental novels as *Naked Lunch* (1959) and a major figure in the Beat movement, had an apartment on Riverside Drive in the 1940s. Among his visitors here was **Jack Kerouac,** who came to stay with Burroughs in December 1944. **Marc Connelly,** the playwright known best for *The Green Pastures* (1930), lived for some time at the Century Apartments, still standing at 25 Central Park West. **Thomas Costain,** the historical novelist whose works included *The Tontine* (1955), lived at 50 Riverside Drive, and there died of a heart attack on October 9, 1965.

The peripatetic **Theodore Dreiser** lived at several different addresses on the Upper West Side over a period of more than thirty years. At his first home here, 6 West 102nd Street, the novelist began work on *Sister Carrie* (1900). In 1911 he lived at 225 Central Park West, and in 1931 rented a fourteenth-floor suite at the well-known Ansonia Hotel, at 2109 Broadway. Five years later, in December 1936, Dreiser moved with his second wife, Helen, to the Park Plaza Hotel, no longer operating, at 50 West Seventy-seventh Street. **Eleanor Estes,** author of works of juvenile fiction such as *The Sleeping Giant* (1948) and *The Hundred Dresses* (1949), lived for a time at 344 West Seventy-second Street. **Edna Ferber,** who achieved enormous popularity through her first novel, *So Big* (1924), lived at three addresses on the Upper West Side. On arriving in New York, she stayed at the Belleclaire Hotel, now an apartment hotel, at Broadway and Seventy-seventh Street. In 1920–21 she had rooms at the Majestic Hotel, now Majestic Apartments, at 115 Central Park West. In 1923 she sublet an apartment at 50 Central Park West, where she worked on *So Big.*

Ellen Glasgow lived at 1 West Eighty-fifth Street from 1911 to 1916. This was the period in which she worked on her novel *Life and Gabriella* (1916). **Zane Grey,** the dentist who turned Western writer, rented a dental office at 100 West Seventy-fourth Street when he came to New York in 1896. **Oscar Hammerstein** grew up in Manhattan and spent much of his childhood at various addresses here, including the Barnard, at 106 Central Park West; Central Park West and Eighty-seventh Street; the Endicott Hotel, now renovated as an apartment building, on Columbus Avenue between Eighty-first and Eighty-second streets; and the Aylesmere, at the corner of Columbus Avenue and Seventy-sixth Street. **Ben Hecht,** known best today for the plays he wrote with Charles MacArthur, died of a heart attack at his home at 39 West Sixty-seventh Street on April 18, 1964, when he was seventy years old. Hecht is said to have been reading a poem by E. E. Cummings when he was stricken. Memorial services were held for him at Temple Rodeph Sholom, at 7 West Eighty-third Street. One of those who spoke at Hecht's funeral was Menachem Begin, who much later became prime minister of Israel.

Another writer resident on the Upper West Side is **Joseph Heller,** author of *Catch 22* (1961). Heller has lived at 390 West End Avenue, and presently maintains an apartment in the same neighborhood, although he spends much of his time at his home on Long Island. **John Hollander,** the poet and critic, spent part of his childhood on the Upper West Side, and returned to live here for a time as an adult. His childhood addresses were 203 and 124 West Seventy-ninth Street; he lived later at 225 West Eighty-sixth Street and at 88 Central Park West, where he wrote the poems published in *The Night Mirror* (1971), *Tales Told to the Fathers* (1975), and *Reflections on Espionage* (1976). Hollander has reported that he enjoyed eating at the Szechuan Royal, a restaurant still open at 50 West Seventy-second Street.

Alfred Kazin, the critic and memoirist, has lived at various times at 440 West End Avenue, Central Park West and Ninety-seventh Street, and 110 Riverside Drive. **Arthur Kober,** author of the collection of New York sketches called *Thunder Over the Bronx* (1935), lived for some years at 241 Central Park West. **Louis Kronenberger,** known principally for such critical works as *The Cart and the Horse* (1964), lived for a time at 16 West Seventy-seventh Street. **Meyer Levin,** the novelist, lived at 62 West Ninety-first Street for many years. Levin's works included the novel *Compulsion* (1956), which was based on the notorious Leopold-Loeb murder case. Levin spent his summers in Israel, and died in Jerusalem on July 9, 1981.

Sinclair Lewis, the first American to win a Nobel Prize in literature, lived at two addresses on the Upper West Side: 2469 Broadway, which was the home of economist George Soule; and the Eldorado Towers, at 300 Central Park West.

When Caroline Meeber boarded the afternoon train for Chicago, her total outfit consisted of a small trunk, a cheap imitation alligator-skin satchel, a small lunch in a paper box, and a yellow leather snap purse, containing her ticket, a scrap of paper with her sister's address in Van Buren Street, and four dollars in money. It was in August, 1889. She was eighteen years of age, bright, timid, and full of the illusions of ignorance and youth.

—Theodore Dreiser,
the opening lines of *Sister Carrie*

"When you talk to the man upstairs," he said, "I want you to tell Him something for me. Tell Him it ain't right for people to die when they're young. I mean it. Tell Him if they got to die at all, they got to die when they're old. I want you to tell Him that. I don't think he knows it ain't right, because he's supposed to be good and it's been going on for a long, long time. Okay?"

"And don't let anybody up there push you around," the brother advised. "You'll be just as good as anybody else in heaven, even though you are Italian."

"Dress warm," said the mother, who seemed to know.

—Joseph Heller,
in *Catch 22*

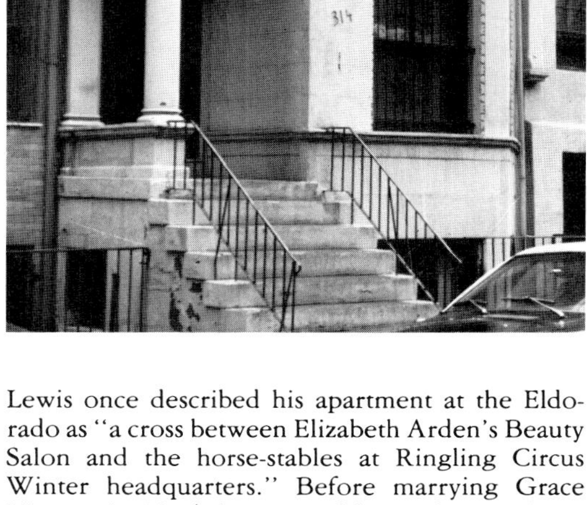

Residence of William Styron at 314 West Eighty-eighth Street. Styron moved to New York in 1947 and began his career as a novelist while living here.

Lewis once described his apartment at the Eldorado as "a cross between Elizabeth Arden's Beauty Salon and the horse-stables at Ringling Circus Winter headquarters." Before marrying Grace Hegger in 1914, he courted her at her mother's home at 345 West Seventieth Street.

The only thing I ever learned in school that did me any good in after life was that if you spit on a pencil eraser, it will erase ink.

—Dorothy Parker, in a remark attributed to her in her New York *Times* obituary

Robert Lowell lived at 15 West Sixty-seventh Street with his second wife, **Elizabeth Hardwick.** It was at their apartment, when Jason and Barbara Epstein were dinner guests and printers were on strike against New York City newspapers, that the *New York Review of Books* was conceived. The *New York Review* began to publish in 1963 and continues to publish from 250 West Fifty-seventh Street.

Malcolm Lowry, author of *Under the Volcano* (1947), lived in 1934–35 in apartments on Central Park West and in the West Seventies, exact addresses unknown. The English novelist's stay in New York City had begun with a period of detention in the psychiatric ward of Bellevue Hospital. **Clare Boothe** was born to wealthy parents in a house on Riverside Drive on April 10, 1903. Another playwright, **Tad Mosel,** was living at 300 Central Park West in 1961, the year in which he won a Pulitzer Prize for *All the Way Home* (1960), a dramatized version of James Agee's novel *A Death in the Family* (1957), for which Agee won a Pulitzer Prize in 1958.

"Drive this foolishness out of your head. The child is yours."

"How can he be mine?" I argued. "He was born seventeen weeks after the wedding."

She told me then that he was premature. I said, "Isn't he a little too premature?" She said, she had had a grandmother who carried just as short a time and she resembled this grandmother of hers as one drop of water does another. She swore to it with such oaths that you would have believed a peasant at the fair if he had used them. To tell the plain truth, I didn't believe her; but when I talked it over next day with the schoolmaster he told me that the very same thing had happened to Adam and Eve. Two they went up to bed and four they descended.

—Isaac Bashevis Singer, in "Gimpel the Fool"

A longtime resident on the Upper West Side has been **Lewis Mumford,** the critic and historian. Born in 1895, Mumford grew up at a succession of addresses: Ninety-seventh Street and Amsterdam Avenue, 230 West Sixty-fifth Street, 200 West 105th Street, 59 West Ninety-third Street, 66 West Eighty-fourth Street, 66 West Ninety-third Street, and 100 West Ninety-fourth Street. He attended Stuyvesant High School, City College of New York, New York University, Columbia University, and the New School for Social Research—without ever receiving a bachelor's degree. Mumford returned to live at 153 West Eighty-second Street from 1948 until 1952.

Another childhood resident was **Anaïs Nin,** the diarist and novelist, who came here from Europe with her mother in 1914, when Nin was eleven. Living in a brownstone at 158 West Seventy-fifth Street, Nin attended public school until she was fifteen. She returned to Europe in 1923 and remained there until the outbreak of World War II. She wrote her first book, *D. H. Lawrence, An Unprofessional Study,* in 1930. **Eugene O'Neill,** the second American to win a Nobel Prize in literature, lived in 1908 at the Lincoln Arcade, at Broadway and Sixty-sixth Street. He shared an apartment there with painters George Bellows and Edward Keefe. A resident of longer tenure was **Dorothy Parker,** who spent most of her childhood living at 57 West Sixty-eighth Street. Parker attended school at the Blessed Sacrament Convent; later, when she got her first job, at *Vogue* magazine, she lived at Broadway and 103rd Street.

One of the Upper West Side's most distinguished and most enduring residents is **Isaac Bashevis Singer,** the Polish-born author who writes his stories in Yiddish. Winner of a Nobel Prize in literature in 1978, Singer has lived on West Eighty-sixth Street for over thirty years. He has reported that he often takes his meals at two local restaurants, the Famous Dairy Restaurant at 222 West Seventy-second Street, and the Four Brothers Restaurant at 2381 Broadway. Approximately six blocks up Broadway, at number 2643, was the residence of **Edmund Clarence Stedman,** the poet and critic. Stedman lived there at the time of his death on January 18, 1908. Stedman's most enduring work was literary criticism and the anthologies he edited, such as *The Poets of America* (1885).

William Styron wrote his first novel, *Lie Down in Darkness* (1951), while living at 314 West Eighty-eighth Street. **Bayard Taylor,** the poet and novelist whose *John Godfrey's Fortunes* (1864) told a tale of the nineteenth-century New York literary world, lived at 31 West Sixty-first Street. Another longtime resident here was **Sara Teasdale,** who moved in September 1917 to the Beresford Apartments at 1 West Eighty-first Street with her husband. It was while living here that Teasdale received a special Pulitzer Prize for her volume *Love Songs* (1917). The couple remained at this address for some years, moving in September 1927 to the San Remo, at Central Park West and Seventy-fourth Street. This was their home together until their divorce in 1929. Both apartment buildings are still standing. **Albert Payson Terhune** was graduated from Columbia College in 1893 and worked in New York City for many years, living during most of that time at 67 Riverside Drive. Terhune's popular stories about collie dogs included *Lad: A Dog* (1919) and *Lad of Sunnybank* (1928). **Dorothy Thompson,** the foreign correspondent and political columnist and in her day among the best-known women in America, lived for a time at 88 Central Park West and at the Alden Hotel, at 225 Central Park West. This was after her divorce in 1942 from **Sinclair Lewis,** whom she had married in 1928.

Another Upper West Side resident was **Richard Walton Tully,** remembered best for his play *The Bird of Paradise* (1912). Tully was living at 50 West Eighty-seventh Street at the time of his death on January 31, 1945. **Louis Untermeyer,** the poet, critic, and anthologist known best as editor of the anthology *Modern American Poetry* (1919), lived for some time at 310 West One Hundredth Street. Among the literary figures he entertained here were **Vachel Lindsay, Robert Frost,** and **Amy Lowell. Carl Van Vechten,** author of *Nigger Heaven* (1926) and other novels, lived at both 101 and 146 Central Park West. **Jerome Weidman,** author of *I Can Get It for You Wholesale* (1937), lived at 160 West Eighty-seventh Street. **Nathanael West,** after his graduation from Brown University in 1924, lived for a time with his parents at their home on Seventy-ninth Street, near Central Park West. The novelist worked on ideas he later incorporated into *The Dream Life of Balso Snell* (1931), his first novel. Years later, after West and his wife died in an automobile accident in December 1940, funeral services were held for them at Riverside Chapel, 180 West Seventy-sixth Street.

W. H. Wright, better known as **S. S. Van Dine,** was the author of the Philo Vance detective novels. Wright had an apartment for some years at 241 Central Park West. After his death here in 1939, Wright was cremated, and his ashes were scattered over New York City from an airplane. **Tennessee Williams** lived in the 1960s in an apartment on the thirty-third floor of a high-rise next door to the famed Dakota apartment building on West Seventy-second Street. **Herman Wouk** lived over many years at many different addresses on the Upper West Side. Among them were 101st Street and West End Avenue; the Normandie, at Riverside Drive and Eighty-fifth Street; 845 and 875 West End Avenue; the Regent Hotel, at Broadway and 104th Street; and West End Avenue and Ninety-first Street. The Upper West Side provided much of the background and setting for Wouk's popular novel *Marjorie Morningstar* (1955).

Today, the most celebrated of all residents of the Upper West Side is one who lived here for only two years. In April 1844, when Broadway was still called Bloomingdale Road, **Edgar Allan Poe** found a cottage at the southwest corner of what is now Eighty-fourth Street and moved in with his wife Virginia and his mother-in-law. Poe was doing freelance work for several newspapers in addition to acting as literary editor of the weekly *New-York Mirror* (1823–60). He also revised and completed "The Raven," which achieved immediate recognition. Some of the imagery of the poem may have come from Poe's house here, which was often buffeted by storms. When weather permitted, Poe enjoyed the view from Mount Tom, in Riverside Park, opposite Eighty-fourth Street. In January 1845 he became part owner of a new publication, the *Broadway Journal;* it failed after a year, and in the spring of 1846 Poe moved away. His house here was replaced by a large building that bears a plaque noting the author's residence. The fireplace that once stood in Poe's cottage—and in which he carved his name—was moved to Philosophy Hall, at Columbia University, but its present location is unknown. In a recent tribute to Poe, a portion of West Eighty-fourth Street was renamed Edgar Allan Poe Street.

Daguerreotype portrait of Edgar Allan Poe, taken in 1848 by Matthew Brady. Also shown is the first page of a manuscript copy of "The Raven." The portrait and the manuscript are at the Free Library of Philadelphia.

The Raven.

Once, upon a midnight dreary, while I pondered, weak and weary,
Over many a quaint and curious volume of forgotten lore —
While I nodded, nearly napping, suddenly there came a tapping,
As of some one gently rapping, rapping at my chamber door.
"'Tis some visiter," I muttered, "tapping at my chamber door —
Only this and nothing more."

Ah, distinctly I remember it was in the bleak December,
And each separate dying ember wrought its ghost upon the floor.
Eagerly I wished the morrow; — vainly I had sought to borrow
From my books surcease of sorrow — sorrow for the lost Lenore —
For the rare and radiant maiden whom the angels name Lenore —
Nameless here for evermore.

And the silken, sad, uncertain rustling of each purple curtain
Thrilled me, filled me with fantastic terrors never felt before;
So that now, to still the beating of my heart, I stood repeating
"'Tis some visiter entreating entrance at my chamber door —
Some late visiter entreating entrance at my chamber door; —
This it is and nothing more."

Presently my soul grew stronger. Hesitating then, no longer,
"Sir," said I, or Madam, truly your forgiveness I implore;
But the fact is I was napping, and so gently you came rapping,
And so faintly you came tapping, tapping at my chamber door
That I scarce was sure I heard you" — here I opened wide the door; —
Darkness there and nothing more.

Deep into that darkness peering, long I stood there, wondering, fearing,
Doubting, dreaming dreams no mortal ever dared to dream before;
But the silence was unbroken, and the stillness gave no token,
And the only word there spoken was the whispered word, "Lenore?"
This I whispered, and an echo murmured back the word "Lenore!" —
Merely this and nothing more.

> *His right name was Frank X. Farrell, and I guess the X stood for "Excuse me." Because he never pulled a play, good or bad, on or off the field, without apologizin' for it.*
>
> *"Alibi Ike" was the name Carey wished on him the first day he reported down South. O' course we all cut out the "Alibi" part of it right away for the fear he would overhear it and bust somebody. But we called him "Ike" right to his face and the rest of it was understood by everybody on the club except Ike himself.*
>
> —Ring Lardner, in "Alibi Ike"

Upper East Side

The Upper East Side, long one of the city's most fashionable areas, is also the location of several major hospitals at which some distinguished literary figures have been treated. **F. Scott Fitzgerald** was brought to Doctor's Hospital, 170 East End Avenue, in 1939 after a terrible drinking binge and a stay in Bellevue. **Ellen Glasgow,** the novelist, was hospitalized for exhaustion at Doctor's Hospital for a month in 1940. **Ring Lardner,** the eminent columnist and short-story writer, stayed here late in 1930 and again in the following year. **Eugene O'Neill** was hospitalized here in 1948 after breaking his arm in a fall.

New York Hospital, at 525 East Sixty-eighth Street, has the dubious distinction of being the place where at least three authors have died. **Charles MacArthur,** known best for his collaboration with Ben Hecht, died here on April 21, 1956. **Edwin Arlington Robinson,** the poet, died here on April 5, 1935. **Robert Sherwood,** a Pulitzer Prize-winner for his anti-war play *Idiot's Delight* (1936), died here following a heart attack on November 14, 1955.

Memorial Sloan-Kettering Cancer Center, formerly Memorial Hospital, at 1275 York Avenue, is the place where **Lynn Riggs,** the poet and playwright, died on June 30, 1954. Riggs is most often remembered for his play *Green Grow the Lilacs* (1931), on which Rodgers and Hammerstein based their 1943 musical *Oklahoma!* Another noted author who died here, on December 10, 1946, was **Damon Runyon,** author of stories about Broadway. Runyon's body was cremated and, according to Broadway legend, his ashes were scattered over New York City.

A more cheerful institution is the bibliophilic Grolier Club, 47 East Sixtieth Street, which mounts exhibitions, makes its extensive library accessible to the public, and publishes works on the history of books and bookmaking.

The Metropolitan Club, at 1 East Sixtieth Street, has been the scene of various literary gatherings. One of them, a dinner on March 19, 1931, in honor of Boris Pilnyak, the Russian novelist, was the occasion of a memorable contretemps between **Sinclair Lewis** and **Theodore Dreiser.** Lewis believed, with good reason, that Dreiser had plagiarized a book written by Dorothy Thompson (Mrs. Lewis) in *Dreiser Looks at Russia* (1928); he may also have heard the rumor, which Dreiser himself spread, that Thompson and Dreiser had had an affair when they were both in Russia. For his part, Dreiser was angry, perhaps also with good reason, that Lewis rather than Dreiser had won the 1930 Nobel Prize. When they met, Dreiser extended congratulations to Lewis, who responded with a Bronx cheer. Later, called on to give a speech, Lewis announced that he did not want to speak "in the presence of one man who has plagiarized 3,000 words from my wife's book on Russia" and sat down. When the dinner was over, Dreiser called Lewis into an anteroom, challenged his accusation, and slapped him on the face. Lewis, making no movement, called Dreiser "a liar and a thief." Dreiser slapped him again, and a friend intervened. When newspapers reported the incident, a fight promoter offered the two novelists the opportunity to slug it out in fifteen rounds at Ebbets Field, erstwhile home of the Brooklyn Dodgers, but Westbrook Pegler suggested they use "ghost-fighters" instead. A more harmonious gathering at the Metropolitan Club was the sixty-seventh birthday party of **Mark Twain,** held in 1902, with such guests as **William Dean Howells** present.

> *Why? Why did a man brought up as I had been, a gentleman born and bred, after so many years of straight conduct, suddenly become a thief? And why did I feel no remorse? It was suggested at the time of my trial that I suffered from megalomania, that I was a kind of sun king of stockbrokers who recognized no distinction between his own accounts and those of his customers, that I strutted up and down between Trinity Church and the East River declaiming: "Wall Street, c'est moi."*
>
> —Louis Auchincloss, in *The Embezzler*

The Metropolitan Club, where Mark Twain's sixty-seventh birthday was celebrated in 1902.

Among the numerous literary residents of the Upper East Side is **Robert W[oodruff] Anderson,** known best for his play *Tea and Sympathy* (1953). Anderson was born in New York City on April 28, 1917, and has lived at two addresses in this area: 1172 Park Avenue and 156 East Seventy-ninth Street, where he worked on his play *Silent Night, Lonely Night* (1959). **Louis Auchincloss,** who has written about the upper crust of New York City society in *Portrait in Brownstone* (1962), *The Embezzler* (1966), and many other novels, has long been a resident here. He lived at 66 East Seventy-ninth Street from 1935 until 1945 and more recently on Park Avenue.

W. H. Auden and **Christopher Isherwood,** shortly after their arrival in the United States in January 1939, moved to an apartment at 237 East Eighty-first Street. During their stay at this address, which lasted a little more than a year, they rewrote the ending of their play *The Ascent of F6* (1936). **Philip Barry,** author of such well-known comedies as *Holiday* (1929) and *The Philadelphia Story* (1939), lived at 510 Park Avenue and died there on December 3, 1949. A requiem mass for the playwright was held at the Church of St. Vincent Ferrer, Lexington Avenue and Sixty-sixth Street. **S. N. Behrman,** author of such polished plays as *No Time for Comedy* (1939), spent his final years at 1185 Park Avenue. Confined to bed for four of those years, he died on September 9, 1973, at the age of eighty. **Stephen Vincent Benét** moved to New York in 1930 and had a home at 220 East Sixty-ninth Street. He moved later to 215 East Sixty-eighth Street, where he died on March 13, 1943, at the age of forty-four. **Clare Boothe,** during the six years of her marriage to George Brokaw, lived in a house at Fifth Avenue and Seventy-ninth Street. **Hortense Calisher,** the short-story writer and novelist, lived for a time at 4 East Ninety-fourth Street.

Truman Capote spent part of his childhood in the home of his mother and stepfather at 1060

Park Avenue. Another resident of Park Avenue was **Willa Cather,** who moved in 1932 to an apartment at 570 Park Avenue. This was the author's New York City home for the rest of her life, and the place where she wrote *Lucy Gayheart* (1935). Cather died here at age seventy on April 25, 1947. **George M. Cohan,** the well-known actor and playwright, lived at 993 Fifth Avenue. Cohan died at his home on November 5, 1942.

Russel Crouse, co-author with Howard Lindsay of such classic Broadway hits as *Arsenic and Old Lace* (1940) and *State of the Union* (1945), lived at 151 East Sixty-first Street, 159 East Seventy-eighth Street, and a five-story townhouse at 141 East Seventy-second Street, his last home. A newspaper reporter for some twenty years, Crouse never abandoned the habit of carrying pocketfuls of change for use in telephone booths. Noticing in his later years that climbing the five flights in his house tired him, he consulted his physician, who recommended a lighter load of change in his coat pockets. **Marcia Davenport,** author of the New York City novel *East Side, West Side* (1947), lived for a time at 1 East End Avenue. **Clarence Day** spent the last years of his life at 139 East End Avenue. He was living here at the time of his death in 1935, within months of the publication of his best-known book, *Life With Father* (1935).

Babette Deutsch, whose volumes of poetry include *Coming of Age* (1959), lived at 124 East Seventy-eighth Street from her birth on September 22, 1895, until her marriage to Avrahm Yarmolinsky in 1921. Deutsch attended the Ethical Culture School and Barnard College, where she began contributing poetry to such periodicals as *The New Republic.* **Edna Ferber** was another longtime resident of New York City. The novelist lived for a time at 791 Park Avenue, but her last home was at 730 Park Avenue, where she died on April 16, 1968, at the age of seventy-one. Not far away, at 785 Park Avenue, **Janet Flanner** spent her final years after returning to the United States from France. Under the pen name **Genêt,** Flanner contributed her famous "Letter from Paris" to *The New Yorker* for many years. She died at Lenox Hill Hospital, Seventy-seventh Street and Park Avenue, on November 7, 1979.

Hamlin Garland lived at 71 East Ninety-second Street from 1916 until 1925, the period in which he wrote his autobiographical volumes *A Son of the Middle Border* (1917) and *A Daughter of the Middle Border* (1921). For the latter, Garland won a Pulitzer Prize in biography.

Two other residents of the Upper East Side have been **William Goyen,** author of such novels as *The House of Breath* (1950), who lived at 225 East Ninety-sixth Street, and **Arthur Guiterman,** whose volumes of poetry include *I Sing the Pioneer* (1926). Guiterman's main residence was in Vermont, but he kept a townhouse at 187 East Sixty-fourth Street.

Oscar Hammerstein, collaborator with Richard Rodgers on some of the best-loved musicals in the history of the American theater, lived for many years in this area, at 1067 Fifth Avenue, at 157 East Sixty-first Street, and at 10 East Sixty-third Street. **Dashiell Hammett** lived for a time at 15 East Sixty-sixth Street and spent his later years in the house owned by **Lillian Hellman** at 63 East Eighty-second Street. Hammett died at Lenox Hill Hospital on January 10, 1961. His funeral services were held at the Frank Campbell Funeral Chapel, 1076 Madison Avenue.

Photograph of Willa Cather taken in 1902, the year before publication of *April Twilights,* her first book and only collection of poetry.

Thomas Heggen, author of the best-selling novel *Mister Roberts* (1946), was found dead in the bathtub of his duplex apartment at 8 East Sixty-second Street on May 19, 1949. Heggen's death, the result of an overdose of sleeping pills, was ruled a suicide. **Ernest Hemingway,** in 1959, two years before his death, bought an apartment at 1 East Sixty-second Street. His principal residence was in Ketchum, Idaho. **John Hersey,** winner of a Pulitzer Prize for his novel *A Bell for Adano* (1944), has lived at several East Side addresses, including 563 Park Avenue, 40 East Sixty-second Street, and 136 East Seventieth Street.

In an earlier time, a distinguished resident of the Upper East Side was **William Dean Howells,** the novelist and critic, who moved to 38 East Seventy-third Street in 1901 and stayed until 1908. While living here Howells wrote the novel *The Kentons* (1902) and completed a critical study, *Heroines of Fiction* (1901). **George S. Kaufman,** collaborator on such familiar plays as *Dinner at Eight* (with Edna Ferber, 1932) and *The Man Who Came to Dinner* (with Moss Hart, 1939), lived for a long time on the Upper East Side. Kaufman's home was at 1035 Park Avenue, and he died here on June 2, 1961. Funeral services were held at the Frank E. Campbell Funeral Chapel. **Emily Kimbrough,** collaborator with Cornelia Otis Skinner on the popular book *Our Hearts Were Young and Gay* (1942), lived for many years at 11 East Seventy-third Street. **Alfred Kreymborg,** whose *Selected Poems* appeared in 1945, learned to play chess at his father's cigar store, at 1667 Third Avenue. Not far away, at 128 East Ninety-fifth Street, once lived **Louis Kronenberger,** the critic and novelist.

Ring Lardner was yet another resident of the Upper East Side, coming to the area in the last years of his life. Suffering from poor health, Lardner lived with his wife first at the Carlyle Hotel, 35 East Seventy-sixth Street, and later, in October 1931, moved to an apartment at 25 East End

> *Lucy did not feel tired, she was throbbing with excitement, and with the feeling of wonder in the air. She put the blinds up high and sat down in a rocking-chair to watch the bewildering, silent descent of the snow, over all the neighbours' houses, the trees and gardens. She was alone on the upper floor. The daylight in her room grew greyer and darker. Lights in the house across the street began to shine softly through the storm. She tried to feel at peace and to breathe more slowly, but every nerve was quivering with a long-forgotten restlessness. How often she had run out on a spring morning, into the orchard, down the street, in pursuit of something she could not see, but knew!*
>
> —Willa Cather,
> in *Lucy Gayheart*

John and Elaine Steinbeck, in a photograph taken about 1956. The Steinbecks lived at 206 East Seventy-second Street from 1951 to 1964. (*far right*) Dorothy Parker.

> *This hotel—the Amazon—was for women only, and they were mostly girls my age with wealthy parents who wanted to be sure their daughters would be living where men couldn't get at them and deceive them; and they were all going to posh secretarial schools like Katy Gibbs, where they had to wear stockings and gloves to class, or they had just graduated from places like Katy Gibbs and were secretaries to executives and junior executives and simply hanging around in New York waiting to get married to some career man or other.*
>
> —Sylvia Plath, in *The Bell Jar*

> *"You haven't said how he was rude."*
> *"His manner. Not his manners. His manner. There's a way of doing all the proper things that's just as bad as if you picked your teeth with the oyster fork. Haughty. He made no effort to be cordial or considerate. He bored the deuce out of all the girls. Made no effort at conversation. He was a New York boor, that's what he was. Trying to put on the dog for the benefit of the country yokels."*
>
> —John O'Hara, in *A Rage to Live*

Avenue. Within two years of this move, Lardner was dead of a heart attack at the age of forty-eight.

Sinclair Lewis and his wife **Dorothy Thompson,** the journalist, in 1931 took an apartment at 21 East Ninetieth Street. The apartment had two living rooms, so Lewis and Thompson could entertain their own friends separately. Lewis worked on his novel *Ann Vickers* (1933) while living here. It was the first book he published after winning the Nobel Prize in 1930. **Walter Lippmann,** one of the foremost American political commentators of the twentieth century, lived on the Upper East Side until he was nearly thirty. His first home was at 121 East Seventy-ninth Street, where he lived until his early teens; his second was at 46 East Eightieth Street. With the exception of four undergraduate years at Harvard, where he was assistant in his final year to George Santayana, Lippmann made his home at this address until 1917, the year of his marriage. By that time Lippmann had already published three books: *A Preface to Politics* (1913), *Drift and Mastery* (1914), and *The Stakes of Diplomacy* (1915). **Carson McCullers** lived here during 1935, when she was struggling to become established in New York. She stayed at the Three Arts Club, which then operated at 240 East Eighty-fifth Street.

Lewis Mumford, the critic and historian, lived in 1942 at 56 East Eighty-seventh Street. This was the period during which he was writing his four-volume series *Technics and Civilization* (1934–51). **John O'Hara** once had a ground-floor apartment at 27 East Seventy-ninth Street and later, from 1945 until 1949, lived at 55 East Eighty-sixth Street, where he worked on his novel *A Rage to Live* (1949). **Eugene O'Neill** and his wife Carlotta rented an apartment at 1095 Park Avenue in July 1931. Their plan was to live in the apartment indefinitely, but they spent only a few months here. O'Neill while here attended rehearsals for *Mourning Becomes Electra,* published in that same year. Many years later, in 1946, the O'Neills lived in an apartment at 35 East Eighty-fourth Street, where he worked on *The Iceman Cometh* (1946).

Dorothy Parker in 1963 returned East from California, after the death of her husband, Alan Campbell, and took a room at the Volney, 23 East Seventy-fourth Street. Parker lived here alone, relatively unproductive and in poor health, until her death on June 7, 1967. The building is no longer used as a hotel, but it is still standing. At another East Side hotel, the Barbizon Hotel for Women, Lexington Avenue and Sixty-third Street, **Sylvia Plath** had a room while she worked as a guest editor at *Mademoiselle* magazine. Plath called the hotel The Amazon in her novel *The Bell Jar* (1963).

John Steinbeck in 1951, the year after he married Elaine Scott, his third wife, moved to a townhouse at 206 East Seventy-second Street. Even though the novelist published *East of Eden* in 1952, his years here were neither his most productive nor his most distinguished. He produced only two works of fiction—a short novel called *The Short Reign of Pippin IV* (1957) and *The Winter of Our Discontent* (1961)—and his reputation suffered something of a decline. In 1962, however, Steinbeck published his immensely popular nonfiction work *Travels with Charley in Search of America,* and in the same year was awarded the Nobel Prize in literature. Steinbeck died six years later, on December 20, 1968.

Another longtime resident of the Upper East Side was **Edward Streeter,** who combined writing with a career in banking. Known best for his novel *Father of the Bride* (1949), Streeter lived at 200 East Sixty-sixth Street. Another writer who combined literary pursuits with another career was **Arthur Cheney Train,** who lived for some years at 113 East Seventy-third Street. Train published his first book while serving as an assistant district attorney in New York City. He was best known for his novels about Ephraim Tutt, a mild-mannered lawyer whose affable exterior concealed an acute and resourceful legal mind. Tutt appeared in such books as *The Adventures of Ephraim Tutt* (1930) and *Mr. Tutt's Case Book* (1937). The latter has found additional use as a textbook for law students.

A distinguished visitor to the Upper East Side

was **Mark Twain,** who enjoyed himself at the new premises of the Lotos Club, 5 East Sixty-sixth Street, in 1893. Twain was also guest of honor at a dinner given at the club in 1900. **Jean Starr Untermeyer,** whose *Love and Need: Collected Poems* appeared in 1940, was living at 235 East Seventy-third Street at the time of her death on July 27, 1970. She had been the wife of **Louis Untermeyer,** also a poet, who accidentally discovered that she wrote poetry and, without informing her, began submitting her poems to magazines in the 1910s. Her first volume, *Growing Pains,* was published in 1918. **Jerome Weidman,** whose novel *I Can Get It for You Wholesale* (1937) was withdrawn briefly from publication during 1938—his publisher found it objectionable—lived at both 1085 Fifth Avenue and 605 East Eighty-fifth Street.

Nathan Wallenstein Weinstein, better known as **Nathanael West,** the novelist, was born on December 17, 1903, at 151 East Eighty-first Street. West, who adopted his pen name after graduation from college, lived here for about four years, moving with his family in 1908 to the more fashionable section of the city known as Harlem. **Edith Wharton** in November 1891 bought a small house on what now is Park Avenue and then was known as Fourth Avenue, near the corner of Seventy-eighth Street: she had just inherited a substantial sum from a distant relative. Wharton soon found the house at 882 Fourth Avenue too small to live in comfortably, and several years later bought the adjoining house, 884 Fourth Avenue. She and her husband, Teddy Wharton, made their home there for a decade, during the relatively few periods when they lived in New York City. While staying at 884 Fourth Avenue, Wharton worked on *The House of Mirth* (1905) and, about New Year's Day of 1905, entertained **Henry James** at home. The house was little used after that, and the novelist in 1910 sold both 882 and 884.

Tennessee Williams lived during the 1960s on the Upper East Side, occupying an apartment at 134 East Sixty-fifth Street. **Edmund Wilson** lived during the late 1940s and the 1950s at various addresses here, including 14 Henderson Place, a cul-de-sac off Eighty-sixth Street between York and East End avenues; 17 East Eighty-fourth Street; 11 East Eighty-seventh Street; 17 East Ninety-seventh Street; and 26 East Eighty-first Street. As one might expect, Wilson did not live for long at any of these residences, regarding them primarily as stopovers on his travels and an easier place to winter in than his principal home on Cape Cod.

P. G. Wodehouse, the creator of Jeeves, Psmith, Bertie Wooster, and other memorable characters in almost one hundred comic novels, lived for a time in a penthouse at 1000 Park Avenue. **Herman Wouk** has been a more persistent resident of the Upper East Side. The bestselling novelist moved to an apartment at 39 East Seventy-second Street in 1951, after the death by drowning of his son Abraham in Cuernavaca, Mexico. While living here he worked on *The Caine Mutiny Court-Martial* (1953), a play based on his novel *The Caine Mutiny* (1951). He also worked on *Marjorie Morningstar* (1955). Wouk and his family moved in 1954 to 116 East Sixty-eighth Street, where the novelist finished *Marjorie Morningstar.*

Harlem, the Heights, & Upper Manhattan

This large area, comprising the northern end of Manhattan Island from 110th Street up, includes several institutions of interest to the literary visitor. The 135th Street branch of the New York Public Library, 103 West 135th Street, has been a gathering place for Harlem writers since it opened in 1905. The building formerly housed the Arthur A. Schomburg collection of historical materials pertaining to the history of black people in America, now in its new home, the Schomburg Center for Research in Black Culture, on Lenox Avenue between 135th and 136th streets. An extension of the 135th Street branch at 104 West 136th Street was named the Countee Cullen Branch, after the distinguished poet who grew up in Harlem and in whose literary education the 135th Street Branch played a central role. In 1945, at the nearby 135th Street YMCA, was founded the Harlem Writers' Workshop, in which **Langston Hughes, Ralph Ellison,** and other important black writers took part.

In the late 1940s the West End Café, still operating at 2911 Broadway, was a favored meeting place of Columbia College students **Allen Ginsberg** and **Jack Kerouac,** and their friend and mentor **William Burroughs.** A slightly younger Columbia student who came here and met these Beats was **John Hollander.**

On a more somber note, the area is also home of a major hospital, the Columbia Presbyterian Medical Center, 622 West 168th Street. At the Harkness Pavilion here died **Heywood Broun,** on December 18, 1939, and **Robert Benchley,** on November 21, 1945.

The area has at least four literary native sons. **Robert W[oodruff] Anderson,** author of such well-known plays as *Tea and Sympathy* (1953) and *I Never Sang for My Father* (1960), was born at 90 Morningside Drive on April 28, 1917. Anderson's family moved to New Rochelle when he

Marjorie loved everything about the El Dorado, even the name. "El Dorado" was perfectly suited to an apartment building on Central Park West. It had a fine foreign sound to it. There were two categories of foreignness in Marjorie's outlook: high foreign, like French restaurants, British riding clothes, and the name El Dorado; and low foreign, like her parents. By moving to the El Dorado on Central Park West her parents had done much, Marjorie believed, to make up for their immigrant origin. She was grateful to them for this, and proud of them.

—Herman Wouk,
in *Marjorie Morningstar*

Schomburg Center for Research in Black Culture.

The apartment was on the top floor—a small living-room, a small dining-room, a small bedroom, and a bath. The living-room was crowded to the doors with a set of tapestried furniture entirely too large for it, so that to move about was to stumble continually over scenes of ladies swinging in the gardens of Versailles.

—F. Scott Fitzgerald, in *The Great Gatsby*

Harlem was like a magnet for the Negro intellectual, pulling him from everywhere.

—Langston Hughes, in *The Big Sea*

Countée Cullen.

was a year old. **James Baldwin,** was born in Harlem on August 2, 1924. Baldwin grew up at various addresses, including one on upper Park Avenue and another at 2171 Fifth Avenue, where he spent his adolescence. His father was a part-time storefront preacher, and young Baldwin spent a good deal of time at the Fireside Pentecostal Assembly Church, Fifth Avenue and 135th Street. He was educated at Frederick Douglass Junior High School and DeWitt Clinton High School, by then situated in The Bronx. Baldwin left Harlem in 1942 to live in New Jersey and later in Europe, but he returned to the United States in the early 1960s, celebrating the publication of *Another Country* (1962) at Small's Paradise, a favorite Harlem restaurant, at Seventh Avenue and 137th Street.

Also born here was **Arthur Miller,** the noted playwright. Miller was born on October 17, 1915, when his family lived on West 111th Street, and was taken as an infant to live at 45 West 110th Street. The fourth of the area's literary native sons is **Richard [Purdy] Wilbur.** He was born at 154 Hart Avenue on March 1, 1921, and lived here until 1923. His first volume of verse, *The Beautiful Changes* (1947), was published in the year he received his A.M. from Harvard. Wilbur has been a professor at Wesleyan University since 1955.

The list of writers who have lived in the area is a long and varied one. **Charles Angoff,** the biographer and novelist, made his home for some time at 614 West 157th Street. There he wrote, among other works, *The Tone of the Twenties* (1966). **John Berryman,** the poet, enjoyed going in his student days at Columbia College to the famed Apollo Theater, still standing but no longer functioning, at 253 West 125th Street. He once commented that he felt more comfortable at the Apollo than in his room in Columbia's Hartley Hall. **Arna Bontemps,** the Louisiana-born writer and editor, lived at 75 St. Nicholas Avenue and 305 Edgecombe Avenue. Among his works were the novels *Black Thunder* (1935) and *Drums at Dusk* (1939). **Heywood Broun** in 1910 stayed with his parents at their home on Claremont Avenue when he worked for the *Morning Telegraph.* This was Broun's first job after leaving Harvard, and the place where he began writing the sports columns that first brought him public prominence.

One of Harlem's most notable literary residents was **Countée Cullen,** born **Countée Porter** on May 30, 1903. Cullen's mother died in 1914, and the eleven-year-old boy was adopted shortly after by Reverence Frederick Cullen, pastor of the Harlem mission of St. Mark's Episcopal Church. Originally operating out of a storefront on St. Nicholas Avenue, the church expanded steadily, and moved in 1924 to a new building at Lenox Avenue and 133rd Street. The Cullen family moved with it, leaving their home at 234 West 131st Street and taking a four-story brownstone at 2190 Seventh Avenue. Cullen meanwhile excelled in his studies at Frederick Douglass Junior High School, DeWitt Clinton High School, New York University (B.A. 1925), and Harvard (A.M. 1926). He married W. E. B. DuBois's daughter Yolande in 1927, having already published his first book of verse, *Color*, in 1925. This was followed by *Copper Sun* (1927), *The Ballad of the Brown Girl* (1927), and *The Black Christ* (1929). Although successful as a poet, Cullen in 1934 began teaching French at his old junior high school and later taught a course in creative writing there. Cullen's health began to deteriorate about 1940, and on January 10, 1946, at the age of forty-two, he died. His funeral, held at his father's church, was attended by over 3,000 people. Cullen had long been a presence at the 135th Street branch library, reading there when young and later meeting with other writers and giving poetry readings.

A Lady I Know

She thinks that even up in heaven
Her class lies late and snores,
While poor black cherubs rise at seven
To do celestial chores.

—Countée Cullen, in "Three Epitaphs"

Theodore Dreiser took an apartment at 439 West 123rd Street in the autumn of 1906, after landing a well-paying job as editor of *Broadway* magazine. He stayed on at the apartment for about four years, a period in which *Sister Carrie* (1900) was reissued, and Dreiser got an even better job as senior editor for the Butterick Publishing Company. When Dreiser moved from 123rd Street, he left behind not only his apartment but his wife Jug, whom he abandoned without even saying goodbye. Two days before that, on October 1, 1910, the novelist had been fired from his job for persisting in his courtship of Thelma Cudlipp, a teenager from Staten Island. About six months later he took an apartment at 3609 Broadway. Jug joined him here in an attempt at reconciliation, but Dreiser used the apartment largely as a mailing address and a place where he could get a good meal. For most of the time he simply avoided his wife, and in June 1912 moved temporarily to 605 West 111th Street, possibly the home of his sister Emma.

Ralph Ellison, known principally for his novel *Invisible Man* (1952), came to New York from Tuskegee Institute in 1936 and lived at the YMCA on West 136th Street. Like many other black writers, he spent time browsing at Lewis Michaux's bookstore, Seventh Avenue and 125th Street. He met **Langston Hughes** at the 135th Street library, where Hughes and Ellison spent a great deal of time. Encouraged to write by Hughes and others, Ellison fairly soon began contributing to various publications. His home from 1945 until 1953, the period in which he wrote his best-known work, was 749 St. Nicholas Avenue, although he slept at times on a bench in St. Nicholas Park.

F. Scott Fitzgerald also came to New York when he was an aspiring young writer, and he too lived uptown. The year was 1919, and Fitzgerald had an apartment at 200 Claremont Avenue while working in advertising for the Barron Collier Agency. He had been here about three months when he sold some short stories to *Smart Set,* edited by H. L. Mencken and George Jean Nathan. Encouraged by his success, Fitzgerald returned to Minnesota to finish *This Side of Paradise,* the novel he had begun writing while he was in the Army. The novel was published in 1920, the following year, and made Fitzgerald a celebrity almost immediately. Tom Buchanan's apartment in *The Great*

Gatsby (1925) may be modeled after the apartment Fitzgerald occupied on Claremont Avenue. **Margaret Halsey** lived at 485 Riverside Drive from 1935 to 1940, here writing *With Malice Toward Some* (1938). **Oscar Hammerstein,** the librettist for such well-known musicals as *Oklahoma!* (1943), lived as a child on both East 116th Street and West 112th Street. Later, after his marriage in 1917, he lived at 509 West 121st Street and at West End Avenue and 122nd Street.

Langston Hughes first came to New York City to study at Columbia University. He stayed only one year, eventually completing a degree at Pennsylvania's Lincoln University in 1929. When Hughes came to Harlem in the 1930s, he lived at various addresses, including 267 West 137th Street and 20 East 127th Street. Among the poet's favorite nightspots here was the Palm Café, where he met the model for his character Jesse B. Simple, the central figure in short pieces collected in *Simple Speaks His Mind* (1950) and other volumes. Hughes was living in Harlem at the time of his death at Polyclinic Hospital on May 22, 1967. Another longtime resident was **James Weldon Johnson,** the poet and novelist, who first achieved distinction writing songs with his brother Rosamond. Johnson lived at several Harlem addresses, including 180 East 135th Street, 187 West 135th Street, and 415 West 148th Street. It was at this last address that Johnson was living at the time of his death in an automobile accident on June 26, 1938. Among the mourners at his funeral, held four days later at Frederick Cullen's Salem Methodist Episcopal Church, 2190 Seventh Avenue, were **Langston Hughes, W. E. B. DuBois,** and **Carl Van Vechten.** In addition to his verse and his prose fiction, Johnson wrote a nonfiction study, *Black Manhattan* (1930), and his autobiography, *Along This Way* (1933).

Some time after changing his name from **LeRoi Jones, Imamu Amiri Baraka** founded the Black Arts Repertory Theater School at 146 West 130th Street. The school is no longer operating. A sometime resident of the area was **Jack Kerouac.** His home here was on 118th Street, between Morningside Drive and Amsterdam Avenue, in the apartment of his future wife, Edie Parker. Kerouac moved into the apartment in 1943, although for much of the couple's life together he divided his time between the apartment here and his mother's house in Queens. Kerouac's friend **William Burroughs,** the novelist, early in 1945 also moved into the apartment along with Parker's friend Joan Vollmer. They too were eventually married. For a time the apartment was something of a literary commune, with other writers—**Allen Ginsberg** was among them—coming to stay from time to time.

Claude McKay, who had the distinction of being the first black writer to have a best seller, lived in Harlem for several years after coming to New York City in 1914 from his native land of Jamaica. For a time he had a room on 131st Street, address unknown, and later lived at 147 West 142nd Street, still standing. McKay dealt with Harlem subjects in several works: his best-selling novel *Home to Harlem* (1928), some of the short stories in *Gingertown* (1932), and the verse collection *Harlem Shadows* (1922). In addition to these and other imaginative works, McKay also wrote an autobiography, *A Long Way from Home* (1937), and a nonfiction study, *Harlem* (1940).

Ann Petry, whose novel *The Street* (1946) is often said to be one of the best about Harlem life, lived here for ten years. Her one recorded address is 2 East 139th Street, where she was living as of 1938. Petry's other novels include *Country Place* (1947) and *The Narrows* (1953). **J. D. Sal-**

Strolling is almost a lost art in New York; at least, in the manner in which it is so generally practiced in Harlem. Strolling in Harlem does not mean merely walking along Lenox or upper Seventh Avenue or One Hundred and Thirty-fifth Street; it means that these streets are places for socializing. One puts on one's best clothes and fares forth to pass the time pleasantly with the friends and acquaintances and, most important of all, the strangers he is sure of meeting.

—James Weldon Johnson,
in *Black Manhattan*

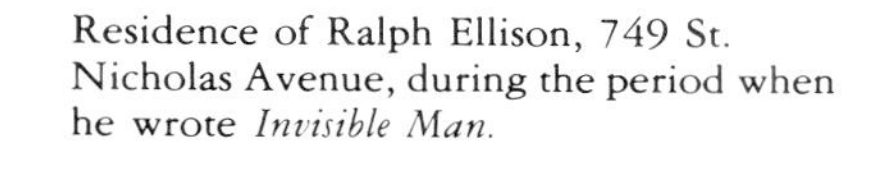
Residence of Ralph Ellison, 749 St. Nicholas Avenue, during the period when he wrote *Invisible Man.*

Abbie kept telling him all the things he could, and could not, do because of The Race. You had to be polite; you had to be punctual; you couldn't wear bright-colored clothes, or loud-colored socks; and even certain food was forbidden. Abbie said that she loved watermelons, but she would just as soon cut off her right arm as go in a store and buy one, because colored people loved watermelons.

—Ann Petry,
in *The Narrows*

White men like ants upon a forage rushed about. Except for the taut hum of their moving, all was silent. Shotguns, revolvers, rope, kerosene, torches. Two high-powered cars with glaring search-lights. They came together. The taut hum rose to a low roar. Then nothing could be heard but the flop of their feet in the thick dust of the road.

—Jean Toomer, in *Cane*

O lovers never barter love
For gold or fertile lands,
For love is meat and love is drink,
And love heeds love's commands.

—Countée Cullen, in *The Ballad of the Brown Girl*

inger, the reclusive author of *The Catcher in the Rye* (1951), spent his childhood at 390 Riverside Drive. The area provides a setting for Salinger's tales of the Glass family, which include *Raise High the Roof Beam, Carpenters* and *Seymour: An Introduction* (both 1963). Not far from this address are two at which **Rex Stout** lived for a time: 8 Morningside Avenue and 364 West 116th Street. **Wallace Thurman,** the prominent black playwright and novelist, had a home at 267 West 137th Street. Although often sharply critical of his colleagues, Thurman made his home a gathering place for writers in the 1920s and early 1930s. Among his short list of major works—he died at age thirty-two—are the play *Harlem* and the novel *The Blacker the Berry* (both 1929).

One of the best-known residents of Morningside Heights was **Lionel Trilling,** the distinguished literary critic. Trilling taught for many years at Columbia University and lived near the campus, at 35 Claremont Avenue. He died here, at age seventy, on November 5, 1975. In addition to a series of influential critical works, beginning with *Matthew Arnold* (1939), Trilling wrote a number of short stories and a novel, *The Middle of the Journey* (1947).

Nathanael West lived at several addresses here during his childhood. His family moved in 1908 to the DePeyster apartments, built by West's father on Seventh Avenue between 119th and 120th streets. They moved later to an apartment in a building on 110th Street near Central Park West, also built by West's father, and later still, during World War I, took a place on Hamilton Terrace, near City College. A prominent Harlem resident was **Walter White,** longtime executive secretary of the NAACP. White's home was at 409 Edgecombe Avenue, in the Sugar Hill area.

Boyhood home of J. D. Salinger, at 390 Riverside Drive.

The three of them talked endlessly. They talked university matters and housing matters and the situation in Washington, they talked child-rearing and usually from there went on to exchange anecdotes of their own childhoods, trying to explain how they had become the people they were today. Again and again they returned to politics, not so much its theory as its gossip.

—Lionel Trilling, in *The Middle of the Journey*

He entertained here such distinguished guests as **Willa Cather, George Gershwin,** and **Sinclair Lewis** and wrote of black America in such novels as *Fire in the Flint* (1924) and *Flight* (1926).

Funeral services for **James A. Wright,** the poet, were held at Riverside Church, on Riverside Drive between 120th and 122nd streets, on March 27, 1980. **Richard Wright** lived at several Harlem addresses after arriving in New York City in 1937, including 139 West 143rd Street and 809 St. Nicholas Avenue. For most of his time in New York City, however, the author of *Native Son* (1940) and *Black Boy* (1945) lived in Brooklyn. After World War II he moved to Paris and spent the rest of his life there.

No discussion of this area of New York City would be complete without mention of the Harlem Renaissance, also called the New Negro Renaissance, during the 1920s and early 30s. The 1920s were a time of rapid social change in Harlem, marked by large-scale immigration and a confluence of different cultures. In a sense, the Renaissance had already begun during World War I, which saw the publication in *The Seven Arts* magazine (1917) of Claude McKay's poem "The Harlem Dancer," but it was not until the 1920s that a movement in black culture was generally perceived. This was a flourishing of all the arts of black America—jazz and the theater as well as literature—and it made Harlem the unofficial black capital of the world.

The prominent figures of the Harlem Renaissance already mentioned include Arna Bontemps, Countée Cullen, Langston Hughes, and Claude McKay; others, though often critical of a movement they mistrusted and therefore on its periphery, included W. E. B. DuBois and Wallace Thurman. Less renowned but still substantial figures were **Jean Toomer,** whose collection *Cane* (1923), set in Georgia and Washington, D.C., was one of the first imaginative treatments of the American black; **Jessie Fauset,** author of the novel *There Is Confusion* (1924); and novelist **Rudolf Fisher,** author of *The Walls of Jericho* (1928). No less important than any of these was **Carl Van Vechten,** the white critic and novelist, whose novel *Nigger Heaven* (1926) dealt sympathetically with the blacks of Harlem. Van Vechten helped promote the careers of black authors: he talked Frank Crowninshield, editor of *Vanity Fair,* into publishing the poetry of Countée Cullen and Langston Hughes, and did much to encourage Alfred Knopf to publish Hughes's first book.

The Harlem Renaissance was at its height in the 1920s, lost much momentum with the beginning of the Depression, and was all but dead by the mid-1930s. With the exception of Langston Hughes, most of its prominent figures were by then dead or inactive, or had turned their attention to other fields. As Langston Hughes put it in *The Big Sea:* "We were no longer in vogue, anyway, we Negroes. Sophisticated New Yorkers turned to Noel Coward. Colored actors began to go hungry, publishers politely rejected new manuscripts, and patrons found other uses for their money." Nonetheless, it may well be that the later black writers—Richard Wright and Ralph Ellison, newcomers to New York in the later 1930s, and native Harlemite James Baldwin—gained something important from the efflorescence of literary activity that had preceded them here.

Schools & Colleges

It is hardly surprising that New York City has some of America's leading educational institutions, both public and private, or that numerous literary figures from all over the country have taught or studied here. The most interesting schools for the literary visitor are City University, Columbia University, and New York University.

The City University of New York has numerous undergraduate and graduate divisions. Its undergraduate divisions include Brooklyn College, Hunter College, Queens College and, principally, City College (CCNY). Among the many literary figures who have taught here are **Joseph Heller; John Hollander,** who taught at Hunter in 1966–67; **Alfred Kazin,** who tutored here while still an undergraduate (B.S.S. 1935), taught English in the summer following his graduation, and returned as a distinguished professor in 1973; **Joseph T. Shipley,** the critic and novelist, who lectured in the graduate division and at Brooklyn College between 1928 and 1938; and **Kurt Vonnegut,** who served as distinguished visiting professor in 1973–74. Students here have included **Paddy Chayefsky** (B.S.S. 1943), the playwright and screenwriter, who studied under Theodore Goodman, a distinguished teacher of the craft of writing; **Arthur Guiterman** (A.B. 1891), author of collections of light verse such as *Song and Laughter* (1924); **Bernard Malamud** (B.A. 1936); **Upton Sinclair,** who entered City College in 1892 at age fourteen and was graduated in 1897—he helped support himself by writing dime novels; and **Jerome Weidman,** who studied here from 1931 until 1933.

The city's oldest surviving educational institution, Columbia University, was founded in 1754 as King's College, named Columbia College in 1784, and renamed Columbia University in 1896. Its principal undergraduate colleges are Columbia College and Barnard College. The many literary figures who have taught here include **Hervey Allen,** the novelist and biographer, who taught English in 1924–25; **W. H. Auden,** who taught at Barnard in 1947 and again in 1956; **William Cullen Bryant,** who became professor of mathematics here in 1825; **Robert P. Tristram Coffin,** Pulitzer Prize-winning poet for his *Strange Holiness* (1935), who lectured here in 1937–38; **Babette Deutsch,** the poet and translator, who taught in the School of General Studies from 1944 until 1971; **Joseph Wood Krutch,** who received his graduate degrees here (M.A. 1916, Ph.D. 1923), taught for a time in the School of Journalism, served as professor of English from 1937 until 1943, and concluded his career here in 1952 as Brander Matthews Professor of Dramatic Literature; **Oliver LaFarge,** Pulitzer Prize-winning novelist for *Laughing Boy* (1929), who was a research associate in anthropology from 1931 until 1933 and taught a course in fiction from 1936 until 1941; and **Budd Schulberg,** the novelist, who taught writing courses and conducted workshops here at various times.

Along with Joseph Wood Krutch, the two literary names most closely associated with the teaching of English at Columbia are those of **Lionel Trilling** and **Mark Van Doren.** Trilling received both undergraduate and graduate degrees at Columbia (Ph.D. 1938), began teaching here in 1931, and remained closely affiliated with the university for the rest of his life. He was professor emeritus at his death in 1975. Apart from critical and historical works such as *Matthew Arnold* (1939) and essay collections such as *The Liberal Imagination* (1950), Trilling published some excellent short stories and a novel, *The Middle of the Journey* (1947). Mark Van Doren also obtained graduate degrees here (Ph.D. 1920), and immediately took a post in the department of English, from which he retired in 1959. Van Doren's students here included John Berryman, Allen Ginsberg, Herbert Gold, Jack Kerouac, and Thomas Merton. A poet of distinction as well as a great teacher, Van Doren won a Pulitzer Prize for his *Collected Poems* (1939). His brother, **Carl Van Doren,** was also a professor of English here, from 1911 until 1930, and won a Pulitzer Prize for his biography *Benjamin Franklin* (1938).

Among the many literary figures who have studied at Barnard or Columbia are **Isaac Asimov,** who received his Ph.D. here in 1948; **John Berryman** (A.B. 1936), who published early works while here and worked on his biography *Stephen Crane* (1950); **Hortense Calisher** (A.B. 1932), who also taught here briefly; **Babette Deutsch** (A.B. 1917); **Max Eastman,** who completed all the requirements for a Ph.D. here without actually taking the degree; and **Allen Ginsberg** (A.B. 1948), the poet, whose career here included serving as editor of the *Columbia Jester,* president of the Philolexian Society, winner of the Woodbury Poetry Prize, member of the debating team, and recipient of a year-long suspension for (a) letting **Jack Kerouac** stay in his room at Livingston Hall and (b) writing obscenities in the grime on his dormitory room window. Ginsberg explained the latter infraction as an attempt to call the cleaning staff's attention to the long-neglected state of his windows.

Others who attended Columbia include **Herbert Gold** (B.A. 1946, M.A. 1948), the novelist; **John Hollander** (A.B. 1952), the poet and critic who is now professor of English at Yale; **Langston Hughes,** another distinguished poet, who spent one unhappy year here; **Jack Kerouac,** who spent the year of 1940 here; **Bernard Malamud,** who received an M.A. in 1942; **Walker Percy,** the novelist and essayist, who received his M.D. degree at the College of Physicians and Surgeons in 1941; **Elmer Rice,** Pulitzer Prize-winner for *Street Scene* (1929), who studied here under **Hatcher Hughes,** also a Pulitzer Prize-winning playwright, for *Hell-Bent for Heaven* (1924); and **Herman Wouk** (A.B. 1934).

The city's third major institution of higher education, New York University, was founded in 1831. Among the literary figures who have taught here are **Charles Angoff,** author of such novels as *In the Morning Light* (1951), who taught from 1958 until 1966; **Saul Bellow,** who taught in the early 1950s; **Ralph Ellison,** who lectured on various subjects; **James Weldon Johnson,** who served as visiting professor of creative literature from 1934 until his death in 1938; **Alfred Kazin,** who was a visiting professor in 1957; **Frederick Prokosch,** author of such novels as *The Conspirators* (1943), who was here in 1936–37; and **Thomas Wolfe,** who taught here on and off from 1924 until 1930, leaving within months of the publication of *Look Homeward, Angel* (1929). The

We are coming we, the young men,
Strong of heart and millions strong;
We shall work where you have trifled,
Cleanse the temple, right the wrong,
Till the land our fathers visioned
Shall be spread before our ken,
We are through with politicians;
Give us Men! Give us Men!

—Arthur Guiterman,
in "Challenge of the Young Men"

One summer a boy had drowned there. From the stoop of his house Rufus had watched as a small group of people crossed Park Avenue, beneath the heavy shadow of the railroad tracks, and come into the sun, one man in the middle, the boy's father, carrying the boy's unbelievably heavy, covered weight. He had never forgotten the bend of the man's shoulders or the stunned angle of his head. A great screaming began from the other end of the block and the boy's mother, her head tied up, wearing her bathrobe, stumbling like a drunken woman, began running toward the silent people.

—James Baldwin,
in *Another Country*

Thomas Wolfe, in a photograph taken in 1935.

Herman Wouk.

novel outraged the citizens of Asheville, North Carlonia, Wolfe's hometown—one letter Wolfe received began with "Sir: You are a son of a bitch"—but it won him a measure of financial security and artistic self-confidence.

Among the writers who have studied at NYU are **Countée Cullen** (A.B. 1925), who while a sophomore won second prize in the Witter Bynner poetry contest, sponsored by the Poetry Society of America, for "Ballad of the Brown Girl," and won first prize the following year; **Joseph Heller** (A.B. 1948); **Lillian Hellman,** the playwright and memoirist, who attended the university from 1922 to 1924 but did not take a degree; **Carson McCullers,** the novelist, who studied here for two semesters; **Delmore Schwartz** (A.B. 1925), who later returned to lecture here; and **Jerome Weidman,** who attended both Washington Square College, undergraduate, and the Law School.

New York City also has several other colleges and universities, less widely known but of literary interest. At the New School for Social Research, some of the professors have been **W. H. Auden, W. E. B. DuBois,** and **Joseph Wood Krutch. Elmer Rice,** the playwright, and **Wallace Stevens,** the poet, both were graduated from New York Law School. **Joseph T. Shipley,** the editor and critic, taught for some time at Yeshiva University, formerly Yeshiva College. **Herman Wouk** also taught at Yeshiva, from 1953 until 1957.

Of New York City's many public high schools, two are worthy of special mention for the literary alumni they claim. DeWitt Clinton High School, once at Tenth Avenue and Fifty-ninth Street and now located in The Bronx, is the high school attended by **James Baldwin, Countée Cullen,** and **Lionel Trilling.** Another student here at the same time as Cullen and Trilling was **Nathanael West.** He left after two years without graduating.

Another New York City school of some literary distinction was Townsend Harris High School, no longer operating, which was located first at Townsend Harris Hall, 138th Street and Amsterdam Avenue, on the CCNY campus, and later at Lexington Avenue and Twenty-third Street, again in a CCNY building. Townsend Harris alumni include **William Gibson, Sidney Kingsley,** and **Herman Wouk.**

Publications

New York City long has been the center of literary publishing in the United States. In its numerous magazines and newspapers have appeared works by distinguished writers from the nineteenth century to the present day. Many of those publications have ceased publishing, but any discussion of New York City's literary heritage would be incomplete if it did not include some of the important magazines published here.

One of the most influential and respected magazines of its day was *The American Mercury,* founded in 1924 by **H. L. Mencken** and **George Jean Nathan.** The magazine had its heyday from its founding until 1933, when Mencken gave up the editorship—Nathan had already taken a minor role by the time he officially left the publication in 1930. Among the many distinguished authors the *Mercury* published were Countée Cullen, William Faulkner, Vachel Lindsay, and Eugene O'Neill, whose *All God's Chillun Got Wings* was first published here, in February 1924. The October 1925 issue carried Mencken's famous attack on William Jennings Bryan.

A somewhat lesser publication was *The Bookman,* a monthly founded in 1895. In its time *The Bookman* carried works by Joseph Conrad, Stephen Crane, Amy Lowell, and others, but it ceased publication in 1933. Longer published was *The Century Illustrated Monthly Magazine,* edited for almost its first thirty years by **Richard Watson Gilder.** Among the most important of the many distinguished works it published were Frank Stockton's still-popular story "The Lady or the Tiger?" (November 1882), Mark Twain's *Huckleberry Finn* (1884), William Dean Howells's *The Rise of Silas Lapham* (1885), Henry James's *The Bostonians* (1886), and Jack London's *The Sea Wolf* (1904).

Collier's, founded as *Once A Week* in 1888 by **Peter Fenelon Collier,** published works by a great many distinguished authors before its demise in 1957. Among them were Willa Cather and O. Henry. Ernest Hemingway and his wife Martha Gellhorn served as war correspondents for the magazine during World War II.

The Cosmopolitan, founded in 1886 by **Paul J. Schlict** and sold by him in 1905 to **William Randolph Hearst,** published numerous works by eminent authors, including Theodore Dreiser, who once interviewed the great baseball player Ty Cobb for the magazine; Henry James; and H. G. Wells, whose *The War of the Worlds* was published here in 1897. This distinguished publication bears no resemblance to the *Cosmopolitan* magazine published today. Another magazine still being published, but in something slightly closer to its original form, is *Esquire,* founded in 1933. Among the contributors to the first issue was Ernest Hemingway, who several years later published in the magazine "The Snows of Kilimanjaro" (1936). Among the prominent contributors of a later era have been Saul Bellow, James Baldwin, and John Updike.

Another respected name in American journalism is that of *Harper's,* founded in 1850 as an offshoot of the successful Harper and Brothers publishing house. The original name of the magazine was *Harper's New Monthly Magazine,* changed later to *Harper's Monthly Magazine,* and finally, in 1925, to its present name. In its early years *Harper's* emphasized serial novels by such eminent English authors as Dickens and Thackeray, but later went on to carry the work of such Americans as Henry James, Sarah Orne Jewett, and Herman Melville. The magazine's "Easy Chair" column has been written at different times by three distinguished figures: George William Curtis, William Dean Howells, and Bernard De Voto.

The weekly *New York Ledger,* founded in 1855, acquired a reputation for procuring the works of eminent authors by paying them, as someone once wrote, "better than they were paid anywhere else." Among those lured by the studied generosity of the magazine were William Cullen Bryant, Charles Dickens, Henry Wadsworth Longfellow, and Harriet Beecher Stowe. The magazine ceased publication in 1903. A very different publication was *The Masses,* founded in 1911 and published until suppressed by the United States government in December 1917. The magazine continued publishing after that date—until

MARTIN. *Lissen, kid . . . when yuh fight, dee idee is tuh win. It don' cut no ice how. An' in gang fightin' remember, take out da tough guys foist. T'ree aw faw a yuh gang up on 'im. Den one a yuh kin git behin' 'im 'an slug 'im. A stockin' fulla sand an' rocks is good fuh dat. An' if ey're lickin' yuh, pull a knife. Give 'em a little stab in ee arm. Ey'll yell like hell an' run.*

—Sidney Kingsley,
in *Dead End*

1918—but it had been fatally banned from the mails. **Max Eastman** was long the editor of *The Masses,* and among its literary contributors were Floyd Dell, John Reed, and Louis Untermeyer.

McClure's Magazine, published from 1893 until its merger with *New Smart Set* in 1929, made a name for itself both in literary and political spheres. It published numerous leading English authors, including Rudyard Kipling, whose *Captains Courageous* (1896–97) was serialized first in its pages, and the then relatively unknown Thomas Hardy. *McClure's* also published "Whistling Dick's Christmas Stocking" (December 1899), when its author, O. Henry, was still in prison. The political impact of *McClure's* came from its publication and encouragement of such pioneering muckrakers as Ray Stannard Baker, Lincoln Steffens, and Ida Tarbell. In 1906 these three authors, accompanied by *McClure's* co-founder **John S. Phillips,** left the magazine to take over command of *The American Magazine.*

The Nation, still publishing though much changed, was founded in 1865 with a first issue that included contributions from Henry James, Sr. and Henry James, Jr. Other early contributors included William Dean Howells and James Russell Lowell; among later contributors were Heywood Broun, who had a column in *The Nation* for many years, and Joseph Wood Krutch, who served as an associate editor. Another distinguished magazine still publishing, but now in Washington, D.C., is *The New Republic,* whose first issue appeared in November 1914. In its early years the magazine had such contributors as Robert Frost, whose well-known poem "The Death of the Hired Man" appeared in the February 1915 issue, and William Faulkner, whose first mature published work, a poem called "L'Après-Midi d'un Faune," appeared in the August 1919 issue. Two of the best-known members of the magazine's editorial staff were Edmund Wilson and Malcolm Cowley.

The magazine most closely identified with New York City is undoubtedly *The New Yorker,* founded in 1925 by **Harold Ross,** its first editor, and still publishing from its original, unassuming headquarters at 25 West Forty-third Street. Since the death of Ross in 1951, the magazine has had only one other editor, **William Shawn.** Well known for its Profiles of the famous and near-famous, its Talk of the Town section, its urbane cartoons, and its nearly impeccable prose, *The New Yorker* has had such distinguished staff members as James Thurber and E. B. White. Just a few of its best-known contributors of the past have been Robert Benchley, Dorothy Parker, S. J. Perelman, and Edmund Wilson. *Here at The New Yorker* (1975), by longtime contributor Brendan Gill, gives an insider's view of the magazine. *St. Nicholas* magazine first appeared in 1873 under the editorship of **Mary Mapes Dodge,** the children's writer known best for her *Hans Brinker; or The Silver Skates* (1865). Dodge, who served as editor until 1905, attracted such eminent contributors as Louisa May Alcott, Rudyard Kipling—his story "Rikki-Tikki-Tavi" first appeared here—and Mark Twain. In the early years of the century *St. Nicholas* solicited contributions from its young readers. Among those who submitted works were several who, under slightly different names, went on to greater literary glory: E. Vincent Millay, Conrad William Faulkner, and Ringold Lardner.

Another magazine that achieved some success in its day was the short-lived *Saturday Press,* founded in 1858 by **Henry Clapp,** a stalwart of the Bohemian group, and by **Edward Howland.** The magazine was published for two years until interrupted, like so much of New York City literary life, by the Civil War. Resuming publication in 1865, it was able to survive only until the following year. In its short life, however, the magazine gave first publication to two of the most celebrated short works in American literature. One appeared in 1859 under the title "A Child's Reminiscence," though it is now known better as "Out of the Cradle Endlessly Rocking," one of Walt Whitman's finest poems. The other, appearing under the title "Jim Smiley and his Jumping Frog" in the issue of November 18, 1865, marked the first time that Mark Twain came to the attention of readers in the East. This story, known today as "The Celebrated Jumping Frog of Calaveras County," won its author the national attention that he was to retain for the rest of his life. The same issue of the *Saturday Press* contained the first published work of Henry Wheeler Shaw, better known by his pen name, Josh Billings. Henry Clapp, publisher of the magazine, did not fare so well as some of his eminent contributors, dying years later in poverty on Blackwell's Island. A fairly notable wit, he had once described Horace Greeley as "a self-made man who worships his creator."

A longtime fixture on the New York literary scene has been the *Saturday Review,* founded in 1924 by an eminent quartet: **William Rose Benét, Henry Seidel Canby, Amy Loveman,** and **Christopher Morley.** Edited successively by Canby, **Bernard De Voto,** and until recently by **Norman Cousins,** the magazine once fought a battle in print with *Poetry* magazine over the decision to give Ezra Pound the 1949 Bollingen Prize. *The Smart Set,* published between 1890 and 1930, flourished particularly from 1914 until 1924, when **H. L. Mencken** and **George Jean Nathan** were co-editors. Among the works they accepted for the magazine was F. Scott Fitzgerald's first commercially published piece of fiction, a short story called "Babes in the Woods" (1919). Other authors promoted by Mencken and Nathan in the pages of the magazine were Theodore Dreiser and Eugene O'Neill.

Vanity Fair, the second magazine of this name published in New York City—the first appeared from 1859 to 1863—was founded in 1868 but achieved its prominence after it was bought in 1913 by **Condé Nast.** Under the editorship of **Frank Crowninshield** from 1914 to 1935, it published the works of such figures as E. E. Cummings and F. Scott Fitzgerald, and among its staff members were Robert Benchley and Dorothy Parker. *Vanity Fair* ceased publication in 1936, when it was absorbed by *Vogue.*

Brooklyn

Brooklyn has attracted literary figures since the early days of the nineteenth century, and claims some of our finest writers among its natives and residents. Among those born in the borough was **Leonie Adams,** whose verse included *High Falcon* (1929) and *This Measure* (1933), born at an unknown address on December 9, 1899. **Hey-**

Crowds of people moved through the street with a dream-like violence. As he looked at their broken hands and torn mouths he was overwhelmed by the desire to help them, and because this desire was sincere, he was happy despite the feeling of guilt which accompanied it.

He saw a man who appeared to be on the verge of death stagger into a movie theater that was showing a picture called Blonde Beauty. *He saw a ragged woman with an enormous goiter pick a love story magazine out of a garbage can and seem very excited by her find.*

—Nathanael West, in *Miss Lonelyhearts*

Floyd Dell.

Photograph of Truman Capote taken in 1979.

Because the man who makes an appearance in the business world, the man who creates personal interest, is the man who gets ahead. Be liked and you will never want. You take me, for instance. I never have to wait in line to see a buyer. "Willy Loman is here!" That's all they have to know, and I go right through.

—Arthur Miller, in *Death of a Salesman*

She looked down into the yard. The tree whose leaf umbrellas had curled around, under and over her fire escape had been cut down because the housewives complained that wash on the lines got entangled in its branches. The landlord had sent two men and they had chopped it down.

But the tree hadn't died . . . it hadn't died.

A new tree had grown from the stump and its trunk had grown along the ground until it reached a place where there were no wash lines above it. Then it had started to grow towards the sky again.

—Betty Smith, in *A Tree Grows in Brooklyn*

wood [Campbell] Broun, who won a wide following for his journalistic commentaries on American political and social affairs, was born in Brooklyn Heights on December 7, 1888. The address of his birthplace is variously given as 55 Clark Street or Pineapple Street. Broun worked first for the New York *Morning Telegraph* and later, from 1912 to 1928, for the *Tribune,* where he wrote his widely read column "It Seems to Me." He also wrote for magazines and published several books, including selections from his journalism and *The Boy Grew Older* (1922), a novel about a journalist. Two years after Broun's death, his son **Heywood Hale Broun,** also a journalist, compiled a *Collected Edition* (1941) of his father's works. **[Harold] Witter Bynner,** author of such volumes of poetry as *Grenstone Poems* (1917) and *A Canticle to Pan* (1920), was born here on August 10, 1881.

Paul Leicester Ford, the distinguished historian and novelist, was born here on March 23, 1865. In addition to using his vast historical knowledge for scholarly editions of Americana, Ford wrote two well-known novels of American life, *The Honorable Peter Stirling* (1894) and *Janice Meredith* (1899). Ford was murdered in 1902 by his brother, who had been disinherited in their father's will. Another Brooklyn-born novelist was **Anna Katharine Green,** born here on November 11, 1846, whose *The Leavenworth Case* (1878) is one of the seminal detective novels and perhaps the first written by a woman. A very different novelist, **Joseph Heller,** author of the war satire *Catch-22* (1961), was born in the Coney Island section on May 1, 1923. He attended P.S. 188, and was graduated from Abraham Lincoln High School in 1941. **John Hollander** was born at the Jewish Hospital and Medical Center on October 28, 1929. Since winning the Yale Younger Poets award in 1958 for *A Crackling of Thorns,* Hollander has published a number of further volumes and a scholarly study of English Renaissance poetry. **Elizabeth Janeway,** the author of *The Walsh Girls* (1943), was born here on October 7, 1913.

Alfred Kazin, distinguished man of letters and author of the important critical work *On Native Grounds* (1942), was born at 234 Sutter Avenue, Brownsville, on June 5, 1915. He attended Franklin K. Lane High School, and later lived at both 91 Pineapple Street and 150 Remsen Street, Brooklyn Heights. He described his boyhood here in the memoir *A Walker in the City* (1951). Another native of Brooklyn is **Arthur Laurents,** author of the well-known play *Home of the Brave* (1946) and the musical *West Side Story* (1957), who was born on July 14, 1918.

Bernard Malamud was born in the Flatbush district on April 26, 1914. Malamud has lived at several locations here, including Albemarle Road and Rogers Avenue, in Flatbush; Ellery Street, in Williamsburg; and Gravesend Avenue, in Sheepshead Bay. He attended P.S. 181 and Erasmus Hall High School. Several of Malamud's novels and short stories have their settings in Brooklyn. *The Assistant* (1957), for example, is laid in a grocery store that may resemble the store Malamud's father ran in Flatbush. Brooklyn also claims among its native sons **S[idney] J[oseph] Perelman.** Perelman's first book was *Dawn Ginsbergh's Revenge* (1929); his later efforts included *Westward Ha!* (1948) and *The Rising Gorge* (1961).

Delmore Schwartz was born in an apartment on Eastern Parkway on December 8, 1913. His family moved to nearby President Street in 1916 and later to Ocean Parkway. The family made its final move in 1921, from Brooklyn to Washington Heights. Schwartz went on to write volumes of poetry such as *In Dreams Begin Responsibilities* (1938) and short-story collections such as *The World Is a Wedding* (1948). For *Summer Knowledge,* his collected poems, Schwartz won the Bollingen Prize for 1960.

Brooklyn's name will probably always be connected with yet another native of the borough, **Betty [Wehner] Smith,** who wrote the popular novel *A Tree Grows in Brooklyn* (1943). Born on December 15, 1904, in the Williamsburg section, Smith was christened at Holy Trinity Church, on Montrose Avenue. She attended P.S. 23, in Greenpoint, where she completed eighth grade. She had no formal schooling after that, but made up for it by reading, as she has said, "all the books" in the library near her home, 81 Devoe Street. Smith left Brooklyn at eighteen and did not begin to write seriously until middle age, when she became a special student at the University of Michigan. She wrote many one-act plays and three novels in addition to her best-known novel, which celebrates the indomitable spirit of the slum-born child who thrives, like the ailanthus, in the harsh environment of city streets.

The list of Brooklyn's literary residents is long and varied. **Isaac Asimov,** probably the most prolific science writer now alive, grew up at various addresses in the borough. His longest residence was at 174 Windsor Place, from 1936 to 1942. Asimov attended school at P.S. 182, P.S. 202, J.H.S. 149, and Boys High School. **W. H. Auden** was one of the central figures in a short-lived experiment—1940 and 1941—in quasi-communal living at 7 Middagh Street, Brooklyn Heights. Auden acted as paterfamilias to a group of writers and artists that included, at one time or another, **Carson McCullers, Christopher Isherwood,** and **Richard Wright.** The household largely dissolved after Auden's departure for Michigan in late 1940, and the entire block was destroyed in 1945 to make way for a new approach to the Brooklyn Bridge. In a basement apartment still standing at 70 Willow Street, a short walk away from Middagh Street, **Truman Capote** lived for some years and wrote two of his most successful works, the novella *Breakfast at Tiffany's* (1958) and the nonfiction chiller *In Cold Blood* (1966).

Theodore Dreiser came to New York City in February 1903 and took a cheap room at 113 Ross Street, in Williamsburg. Dreiser was still in anguish over the difficulties surrounding publication of his novel *Sister Carrie* (1903) and was nearly penniless—he soon found it necessary to take an even cheaper room in the same building. Years later, about the time of publication of his successful work *An American Tragedy* (1925), the novelist returned to Brooklyn and rented an apartment at 1799 Bedford Avenue, near Prospect Park. **Eleanor Estes** lived at 85 State Street while writing her popular children's book *The Moffats* (1941) and at 175 Steuben Street, a faculty house at Pratt Institute, while writing *The Witch Family* (1960) and *The Alley* (1964). The last home of **James Gibbons Huneker,** the critic, novelist, and memoirist, was at 1618 Beverly Road, just south of Prospect Park. Remembered primarily as a music critic of the highest order, Huneker in 1920 published both a novel, *Painted Veils,* and the last volume of his autobiography, *Steeplejack.* He died on February 9, 1921.

Norman Mailer has said that Brooklyn is probably the only "particular place I think of as home," and he has lived or maintained a residence here for much of his life. Mailer's parents moved when he was four to the Eastern Parkway section, where he attended P.S. 161. Mailer was graduated from Boys High School in 1939. Among his addresses in Brooklyn are 128 Willow Street, 102 and 150 Pierrepont Street, and 124 and 142 Columbia Heights, all in Brooklyn Heights. **Edwin Markham,** remembered for his poem "The Man with the Hoe" (1899), lived for a brief time at 545 Third Street after moving East in 1900. A Brooklyn resident for a much longer period was **Arthur Miller,** who won a Pulitzer Prize first for his play *Death of a Salesman* (1949) and a second time for his play *A View from the Bridge* (1955). Miller's family moved here in 1929, when he was fourteen, and their first home was at 1277 Ocean Avenue, their second at 1350 East Third Street. Arthur Miller moved to 62 Montague Terrace in 1940, and later to 18 Schermerhorn Street. In 1944 he lived at 150 Pierrepont Street, where Norman Mailer also lived for a time, and in 1947 bought a house at 31 Grace Court, where Miller wrote *Death of a Salesman.* Miller sold the house to **W. E. B. DuBois** in 1951.

Henry Miller, whose best-known work of fiction is *Tropic of Cancer* (Paris, 1934, New York, 1961), was another longtime Brooklyn resident. Raised until the age of ten at 662 Driggs Avenue, in Williamsburg, Miller later moved with his family to 1063 Decatur Street, in Bushwick. He attended P.S. 85 and what is now Eastern District High School, and later lived at 91 Remsen Street, Brooklyn Heights. *Tropic of Capricorn* (Paris, 1939, New York, 1962) was set partly in Brooklyn. An even more widely known resident of

Photograph of Arthur Miller taken in 1979. (*below left*) Building at 150 Pierrepont Street, where Norman Mailer and Arthur Miller lived at different times. (*below*) 31 Grace Court, where Arthur Miller and W. E. B. DuBois lived at different times.

As for Gil Hodges, in custody of first—
"He'll do it by himself." Now a
specialist—versed

in an extension reach far into the box
seats—
he lengthens up, leans and gloves the
ball. He defeats

expectation by a whisker. The modest
star,
irked by one misplay, is no hero by a
hair;

in a strikeout slaughter when what
could matter more,
he lines a homer to the signboard and
has changed the score.

—Marianne Moore,
in "Hometown Piece for
Messrs. Alston and Reese"

Marianne Moore.

Brooklyn was **Marianne Moore,** a heroically steadfast fan of the Brooklyn Dodgers, who lived for most of her adult life at The Cumberland, 260 Cumberland Avenue, near Fort Greene Park. While living here she published several volumes of verse, including her *Collected Poems* (1951), for which she won a Pulitzer Prize. Moore moved away in 1965, but after her death in 1972, funeral services were held for her at the Lafayette Avenue Presbyterian Church, 85 South Oxford Street.

Former residents of Brooklyn Heights include **Wright Morris, Lewis Mumford,** and **Katherine Anne Porter.** Morris lived for a time at 196 Columbia Heights. Mumford lived at 7 Clinton Street in 1921–22 and at 135 Hicks Street from 1922 to 1924 while he was writing *The Story of Utopias* (1922) and *Sticks and Stones* (1924). Porter finished her collection of stories *Flowering Judas* (1930) while living at 74 Orange Street in 1929. The book, Porter's first, won her immediate critical acclaim. Brooklyn Heights has also been the home of **Norman Rosten,** author of such verse collections as *The Big Road* (1946). Rosten has reported that he "migrated southward" through the borough as a child, and in adult life long maintained a residence at 84 Remsen Street and a writing studio at 20 Remsen Street. Among his literary visitors at his home were **Carl Sandburg** and **Arthur Miller;** among his neighbors, at 88 Remsen Street, was the poet and anthologist **Louis Untermeyer. Irwin Shaw,** who attended grade school in Brooklyn as well as Brooklyn College, wrote a play called *The Gentle People: A Brooklyn Fable* (1939), set in Sheepshead Bay, a section of the borough near Coney Island.

Wallace Stevens attended parochial school briefly at St. Paul's Lutheran Church, at South Fifth and Rodney streets, in Williamsburg. **Richard Henry Stoddard,** the poet and literary figure who played an important role in the intellectual life of New York City late in the nineteenth century, lived for a time at 13 Douglass Street with his wife Elizabeth, also a writer. **Mary Virginia Terhune,** the popular novelist whose works included *A Gallant Fight* (1888), written under the pseudonym **Marion Harland,** lived at 166 Bedford Avenue, in Williamsburg, when her husband Edward was pastor at a nearby church.

Thomas Wolfe was yet another resident of Brooklyn, moving to 40 Verandah Place, in Brooklyn Heights, after returning from Europe in March 1931. In autumn of that year he moved to 111 Columbia Heights, the following summer to 101 Columbia Heights, and about a year later to 5 Montague Terrace, all in Brooklyn Heights. Throughout this period Wolfe worked on *Of Time and the River* (1935), his second published novel, and also wrote short fiction for various magazines to provide himself with an income. He finished the first draft of *Of Time and the River* at Montague Terrace, delivering it to his editor, Maxwell Perkins, just before Christmas of 1933. Still in rough-hewn form, the manuscript was some 700,000 words long. Wolfe kept his apartment here until March 1935, when he gave it up before sailing to Europe: fearful that his novel would be poorly received, Wolfe wanted to be out of the country at the time of publication.

Richard Wright had a somewhat longer residence in Brooklyn. He moved in with his friends Jane and Herbert Newton in 1937 at their home at 343 Grand Avenue, and moved with them in 1938 to 175 Carlton Avenue, where he began work on his novel *Native Son* (1940). Later that year they all moved to 522 Gates Avenue, but after Wright's second marriage, he and his wife Ellen moved to 11 Revere Place in July 1941, and later to 89 Lefferts Place, in Bedford-Stuyvesant, where Wright completed *Black Boy* (1945). Wright often went to nearby Fort Greene Park to work on his manuscripts.

Of all the literary figures associated with Brooklyn, few have lived here longer or made more important places for themselves in the life of the borough than **Henry Ward Beecher,** the clergyman, lecturer, and author, who came to Brooklyn in 1847. Before then he had worked in Boston and in the Midwest, but it was as pastor of Plymouth Church, still standing at Orange and Hicks streets, that he achieved his greatest recognition. Like his father, Lyman Beecher, he delivered brilliant sermons. Like his sisters, Harriet Beecher Stowe and Catharine Beecher, he eloquently opposed slavery and other manifestations of social injustice. And like all of his family he was a writer, contributing essays, in such volumes as *Evolution and Religion* (1885), and a novel, *Norwood* (1868), to our national literature. He suffered greatly from an unsuccessful adultery suit brought against him in 1874, but continued to hold his post at Plymouth Church with the enthusiastic support of the congregation. After Beecher's death on March 8, 1887, a crowd said to number 40,000 turned out for his funeral. Beecher's first home in Brooklyn was at 22 Willow Street, and later he lived at 176 Columbia Heights, 82 and 126 Columbia Street, and, at the end of his life, 124 Hicks Street, now the site of an apartment house. Beecher is buried in Greenwood Cemetery, off Fort Hamilton Parkway.

Another of the authors with whom the borough is closely associated is **Hart Crane,** even though the poet made his home here for only a little more than two years of his short, tormented

Jesus! What a nut he *was! I wondeh what eveh happened to 'im anyway! I wondeh if some one knocked him on duh head, or if he's still wanderin' aroun' in duh subway in duh middle of duh night wit his little map! Duh poor guy! Say, I've got to laugh, at dat, when I t'ink about him! Maybe he's found out by now dat he'll neveh live long enough to know duh whole of Brooklyn. It'd take a guy a lifetime to know Brooklyn t'roo and t'roo. An' even den, yuh wouldn't know it all.*

—Thomas Wolfe,
in "Only the Dead Know
Brooklyn"

life. Crane's first and happiest home in Brooklyn was at 110 Columbia Heights, where he lived from just before Easter of 1924 until the summer of 1925. Crane's apartment here afforded a magnificent view of the East River, the Manhattan skyline, and the Brooklyn Bridge, all of which played central roles in some of the poet's best work. (Not until he had already conceived his masterpiece, *The Bridge* [1930], did Crane realize that John Augustus Roebling, the designer and builder of the Brooklyn Bridge, had once owned 110 Columbia Heights. During construction of the great bridge, Roebling had established his observation post in the same room Crane was to occupy some fifty years later.) The poet returned here for a few weeks in the autumn of 1927, and later lived for about three months at 77 Willow Street, not far away, while working as a file clerk in a Wall Street brokerage house. Crane's final home in Brooklyn was at 130 Columbia Heights, where he lived from the summer of 1929 to the early winter of 1930. After that he visited only occasionally, spending most of his time in restless travel before embarking early in 1931 on his last voyage, on a Guggenheim Fellowship, to Mexico.

Of all the literary figures who have lived in Brooklyn, none is so closely or so prominently identified with the borough as **Walt Whitman,** who moved here in 1823, when he was four years old, and lived here on and off for almost forty years. In that time the great poet lived and worked at a number of addresses. Not all can be verified, and few of his residences are likely to be standing after more than a century—a period in which Brooklyn has been transformed from an independent rural community to a fully developed part of New York City. Whitman's first homes were at 71 Prince Street, where his family lived when they came here from Long Island, and at 251 Adams Street, where they moved in 1826. He attended local schools until the age of ten or eleven, when he became an errand boy in the William Clarke law office, on Fulton Street near Orange Street. A year or two later, leaving home, Whitman went to work as a printer's apprentice on the Long Island *Patriot,* on Fulton Street near Concord Street, and was taken by the publisher to the Dutch Reformed Church on Joralemon Street. Whitman moved around a good deal between 1835 and 1845, living on Long Island as well as in Brooklyn. From 1846 to 1848, he edited the Brooklyn *Eagle,* 30 Fulton Street, for which he wrote political editorials. After leaving the *Eagle* he built a home at 106 Myrtle Street, where he

Photograph of Hart Crane on the roof of the building he lived in, 110 Columbia Heights. In the background is the Brooklyn Bridge, the key symbol in Crane's poem *The Bridge.*

(*far left*) 111 Columbia Heights, where Thomas Wolfe lived from autumn 1931 until the following summer. (*left*) Postcard photograph of statue of Henry Ward Beecher, by Gutzon Borglum, at Plymouth Church.

That window is where I would be most remembered of all: the ships, the harbor, and the skyline of Manhattan, midnight, morning or evening—rain, snow or sun, it is everything from the walls of Jerusalem and Ninevah, and all related in actual contact with the changelessness of the many waters that surround it.

—Hart Crane,
in a letter printed in *Voyager,* by his biographer, John Unterecker

Once I was young and had so much more orientation and could talk with nervous intelligence about everything and with clarity and without as much literary preambling as this; in other words this is the story of an unself-confident man, at the same time of an egomaniac, naturally, facetious won't do—just to start at the beginning and let the truth seep out, that's what I'll do—.

—Jack Kerouac,
in *The Subterraneans*

Jack Kerouac.

operated a printing shop and bookstore, and at the same time, 1848–49, ran his own newspaper, *The Freeman.* His other homes over the next decade include 145 Grand Street, 122 North Portland Avenue, and 91½ Classon Avenue.

Throughout the early 1850s, Whitman worked on the poems that would be published in the first edition of *Leaves of Grass* (1855). He set the poems in type himself at the premises of Andrew and James Rome, 98 Cranberry Street, a site now occupied by part of the Cadman Plaza West. Later in the decade he edited the Brooklyn *Times,* to which he had earlier contributed articles, and during 1861–62 contributed to the Brooklyn *Standard.*

The 1850s were also the period in which Whitman took the ferry from Brooklyn to Manhattan nearly every day, working, enjoying himself, sometimes "riding the whole length of Broadway" on an omnibus. Whitman left Brooklyn in December 1862 for Washington, D.C., and the Civil War front. His brother George had been wounded at the Battle of Fredericksburg, and after visiting him the poet stayed on to work as a nurse in Army hospitals. Whitman returned to Brooklyn in the summer of 1864 to recuperate from illness and exhaustion, walking in quiet country lanes and enjoying the solitude of Coney Island, but Washington was to remain his home until 1873, when Whitman suffered a stroke, and moved to Camden, New Jersey. It was there, on March 26, 1892, that the poet died, and it is there that he was buried, far from the borough of Brooklyn.

Others will enter the gates of the ferry and cross from shore to shore,
Others will watch the run of the flood-tide,
Others will see the shipping of Manhattan north and west, and the heights of Brooklyn to the south and east,
Others will see the islands large and small;
Fifty years hence, others will see them as they cross, the sun half an hour high,
A hundred years hence, or ever so many hundred years hence, others will see them,
Will enjoy the sunset, the pouring-in of the flood-tide, the falling-back to the sea of the ebb-tide.

—Walt Whitman,
in "Crossing Brooklyn Ferry"

Henry Seidel Canby, one of the many distinguished authors to write about Whitman, said that Brooklyn was for Whitman the "port of entry to Paumanok, land of his youth, where he was always happy." Throughout Whitman's life, he retained happy memories of the place, even though he wrote in *Specimen Days* (1882) that "Of the Brooklyn of that time . . . hardly anything remains, except the lines of the old streets." In addition to Brooklyn itself, Whitman loved especially the ferry, which not only took him to Manhattan but afforded a view of "Oceanic current, eddies, underneath—the great tides of humanity also, with ever-shifting movements." It was this ride, of course, that inspired "Crossing Brooklyn Ferry" (1856), one of the poet's finest and best-known works.

Queens

Since the time when Queens was no more than a rural suburb, it has been the home of a number of literary figures. The native sons include **Lewis Mumford,** who was born on October 19, 1895, at 10 Amity Street, in the Flushing section. The house is no longer standing. The Mumford family lived here for a few months and then moved to Manhattan. Mumford later lived at two addresses in Sunnyside: 4112 Goosman Avenue from 1925 to 1927, and 4002 Locust Street from 1927 to 1936. Neither street exists anymore. Mumford's many books include *Sticks and Stones* (1924), *Faith for Living* (1940), and *The City in History* (1961). Another notable writer to come from Queens is **Frederick William Shelton,** the essayist and humorist, who was born in the Jamaica section on May 20, 1815. Shelton's best-known work was *The Trollopiad; or, Travelling Gentlemen in America* (1837), a verse satire.

Among the literary figures who have lived here is **John Berryman.** The poet was here briefly in 1927–28, after the death of his father, and attended P.S. 69. **James Fenimore Cooper** spent summers here in the 1830s, sailing on a sloop every day from Astoria to Manhattan. **Babette Deutsch,** soon after the Sunnyside area was developed, moved here with her husband Avrahm Yarmolinsky. In their nineteen-year stay, the poet published such volumes as *Honey Out of the Rock* (1925), *Fire for the Night* (1930), and *Epistle to Prometheus* (1931) as well as a number of translations. Their neighbors here included **Lewis Mumford,** poet **Horace Gregory,** and Gregory's wife, poet **Marya Zaturenska.**

A longtime resident of Queens was **Jack Kerouac,** one of the major Beat writers, who in 1943 came to live with his mother in her apartment over a drugstore on Cross Bay Boulevard, after being discharged from the Navy on the grounds that he had a schizoid personality. During the following years he traveled extensively, making trips that provided the material for much of his fiction, but he continued to use the Queens apartment as a base. For much of the second half of the 1940s Kerouac lived a fairly secluded life, writing and rewriting the manuscript that eventually was published as *The Town and City* (1950), his first novel. Kerouac is said to have used the table in his mother's kitchen as a workspace. His writing was interrupted briefly in December 1945, when heavy use of amphetamines forced his hospitalization for thrombophlebitis of the leg. Kerouac's mother moved to North Carolina in 1951, and in the following year the novelist took an apartment in a two-story frame house at 94–21 134th Street in Richmond Hill. Here he worked on his later novels of the 1950s, which included *The Subterraneans* (1958), *Dr. Sax* (1959), and *Maggie Cassidy* (1959).

Don Marquis, the newspaperman and humorist well loved for his characters archy (a cockroach) and mehitabel (a cat), lived for some years at 51 Wendover Road, in Forest Hills. Though popular and successful in his own lifetime, he spent his last six years in poverty and sickness, and died at his home in 1937. Two of his best-known works were *The Old Soak* (1921) and *archy and mehitabel* (1927). Another literary resident of Queens was **Thomas Merton,** who was brought here by his parents when he was a year old. The family lived both in Flushing and Douglaston, and Merton later lived in Douglaston before joining the Trappist order in 1941. His address in Douglaston was 50 Rushmore Avenue, but the street numbering system has been changed since then. The bodies of **Nathanael West** and his wife, Eileen McKenney, were brought here after their death in an automobile accident in El Centro, California, on December 22, 1940. The bodies were interred in Mt. Zion Cemetery, Maurice Avenue, in Maspeth. This was only a year after publication of West's novel *The Day of the Locust* (1939). He was thirty-seven years old. **Walt Whitman** lived in Jamaica while working on the Long Island *Democrat* in 1839 and teaching in local country schools in 1840–41.

Staten Island

A number of literary figures have visited or lived in Staten Island. One of these was **James Gould Cozzens,** who won a Pulitzer Prize for his novel *Guard of Honor* (1948). Cozzens spent his boyhood in a house on St. Austin's Place, in West New Brighton, and attended the Staten Island Academy, at 715 Todt Hill Road.

Of those who made their homes on Staten Island, few achieved the local celebrity of **George William Curtis,** the editor and essayist. Curtis in 1854 began writing the "Editor's Easy Chair" for *Harper's Magazine,* a column that gave him enormous influence on public opinion. In 1856 he published *Prue and I,* a collection of magazine sketches which became his most popular book. In that same year, he moved into his spacious home at 234 Bard Avenue, New Brighton. Curtis's move here coincided with his increasing concern with the social and political issues of the day. He not only wrote for magazines but worked actively on campaigns and committees and in 1863 took over the editorship of *Harper's.* Among his guests at his home on Bard Avenue were **Henry David Thoreau** and **James Russell Lowell** and, during the infamous Draft Riots of 1863, Curtis gave shelter to his fellow Abolitionist **Horace Greeley.** Curtis died on August 31, 1892, and was buried in the Moravian Cemetery, Richmond Road, New Dorp. His house is privately owned and not open to visitors, but the Staten Island Institute of Arts and Sciences houses a collection of his papers.

Theodore Dreiser had a short but momentous stay on Staten Island. He rented an apartment at 109 St. Mark's Place, New Brighton, and sometimes stayed with his sister Mame at her home here. In 1909, when Dreiser was thirty-eight, he and his wife Jug paid visits to the home of Mrs. Annie Ericsson Cudlipp in New Brighton; for Dreiser the chief attraction of the visits was Mrs. Cudlipp's seventeen-year-old daughter, Thelma, with whom he fell passionately in love. His open and energetic wooing so alarmed Mrs. Cudlipp that she forbade the novelist to see Thelma and threatened to expose him to his employers, the Butterick Publishing Company. Dreiser ignored her threat, and she carried it out. Dreiser was fired. The episode formed the basis for Dreiser's novel *The "Genius"* (1915), in which Thelma Cudlipp is thinly disguised as Suzanne Dale, and Dreiser as Eugene Witla.

Other short-term visitors included **Maxim Gorky,** the Russian novelist, who stayed in 1906 at the home of an admirer at 37 Howard Avenue, Grymes Hill. Gorky was on a fund-raising trip for the revolutionary cause but found time to work here on his novel *Mother* (1907). **Henry James,** in his childhood during the 1850s, was a guest at New Brighton's elegant Pavilion Hotel, no longer standing. **Herman Melville** stayed with his brother Thomas while the latter served as governor of Sailor's Snug Harbor, a home for retired sailors, from 1867 to 1884. The governor's mansion is no longer standing, and Sailor's Snug Harbor has moved to North Carolina, but the five original Greek Revival buildings still stand and are publicly owned.

Edgar Wilson Nye, better known as **Bill Nye,** and vastly popular in his time for his books and articles, spent some four years in a large house in the Tompkinsville district. The house was torn down in 1920. **Edwin Arlington Robinson,** while working on two plays, *Van Zorn* (1914) and *The Porcupine* (1915), briefly rented the imposing La Tourette house, now the public La Tourette Park Clubhouse, near historic Richmondtown. **Alan Seeger,** whose war poem "I Have a Rendezvous with Death" (1916) has ensured his lasting reputation, spent the first ten years of his life here. He lived in a gabled house on Tompkins Hill and attended the Staten Island Academy.

One of the island's most distinguished literary visitors was **Henry David Thoreau,** who was here from May until October 1843. Thoreau came to Staten Island at the suggestion of Ralph Waldo Emerson, whose brother William was a resident. William's children needed a tutor, and Thoreau decided to take the job. He stayed with the family at their house, called The Snuggery, on what is now called Emerson Hill. Even though New York City provided him with several literary friends and acquaintances, including Horace Greeley and Henry James, Sr., Thoreau was dissatisfied here. He returned to Concord by Thanksgiving, thus ending his longest residence outside his native

Eugene had to escort her to Staten Island and then order the chauffeur to put on speed so as to reach Riverside by four. He was somewhat remorseful, but he argued that his love-life was so long over, in so far as Angela was concerned, that it could not really make so very much difference. Since Suzanne wanted to wait a little time and proceed slowly, it was not going to be as bad for Angela as he had anticipated. He was going to give her a choice of going her own way and leaving him entirely, either now, or after the child was born, giving her the half of his property, stocks, ready money, and anything else that might be divisible, and all the furniture, or staying and tacitly ignoring the whole thing. She would know what he was going to do, to maintain a separate ménage, or secret rendezvous for Suzanne.

—Theodore Dreiser,
in *The "Genius"*

Entrance to Staten Island Institute, former site of Sailor's Snug Harbor. Herman Melville visited Sailor's Snug Harbor during the years when he worked as a customs inspector in New York City.

When a girl leaves her home at eighteen, she does one of two things. Either she falls into saving hands and becomes better, or she rapidly assumes the cosmopolitan standard of virtue and becomes worse. Of an intermediate balance, under the circumstances, there is no possibility. The city has its cunning wiles, no less than the infinitely smaller and more human tempter.

—Theodore Dreiser, in *Sister Carrie*

town. The Snuggery, where **Ralph Waldo Emerson** also visited, burned down in 1855; its stone gate can still be seen today on Douglas Road near Richmond Road.

Phyllis A. Whitney, the novelist, lived at 31 Fort Hill Circle, St. George, for over ten years in the late 1940s and 1950s while teaching at New York University. Staten Island figures in her novel *Step to the Music* (1958). A resident of even longer duration was **William Winter,** drama critic for the *New-York Tribune* (1865–1909) and friend of Walt Whitman and others in the group known as the Bohemians. Winter was author of more than ten books, among them theatrical reminiscences. He lived at 27 Third Avenue, now on the grounds of the Willowbrook State School, and is buried in Silver Mount Cemetery. Of all his works, which included lachrymose funeral poetry and a number of theatrical biographies, the most enduring is his two-volume scholarly study *Shakespeare on the Stage* (1911, 1915). A much later man of the theater, **Paul Zindel,** lived at the now-demolished Horrmann Castle, on Howard Avenue in Grymes Hill. He stayed here in the 1960s while teaching at Tottenville High School. He is best known for his play *The Effect of Gamma Rays on Man-in-the-Moon Marigolds,* for which he won a Pulitzer Prize in 1971.

Of all Staten Island's literary residents, perhaps the most proudly claimed is **Edwin Markham,** who lived at 92 Waters Street, West New Brighton, from 1909 until his death in 1940. Markham's greatest achievements, "The Man with the Hoe" (1899) and *Lincoln, and Other Poems* (1901), were already behind him, but he continued to attract considerable attention. Among the admiring visitors on his sixtieth birthday, April 23, 1912, were the poets **Sara Teasdale, Jessie Rittenhouse,** and **Louis Untermeyer,** and the novelist **John O'Hara.** Markham's eighty-second birthday, in 1930, was proclaimed a holiday on the island, and hundreds of schoolchildren took part in a pageant honoring him. The poet died on March 7, 1940, within days of the death of his dear friend Hamlin Garland. His 15,000-volume library was bequeathed to Wagner College, which houses it today in the Markham Memorial Library. Markham's house is still standing, but it is privately owned and not open to visitors.

Several of the island's literary visitors came here not for the fresh air but for the medical advice of Dr. Samuel Mackenzie Elliott, a renowned eye specialist. Dr. Elliott owned a large tract of land in West New Brighton, and there built some thirty houses. When patients came for treatment, they often would stay either in his own house, 69 Delafield Place, or in one of the other buildings, and among those who did so were the poets **Henry Wadsworth Longfellow** and **James Russell Lowell** and the historian **Francis Parkman.** Lowell came here several times—in 1840 and 1843 seeking treatment for himself, in 1846 for his wife, Maria. Parkman is believed to have spent a good deal of time here after returning from his 1846 journey along the Oregon Trail. Never in good health, Parkman had suffered from his long trip and returned East with weakened eyes and a frail constitution. Nonetheless, while staying here he worked on his classic, *The Oregon Trail* (1849), and on *History of the Conspiracy of Pontiac* (1851). Dr. Elliott's son later recalled that the historian had kept him happily entertained with tales of the West. It was many years later, in 1887, that **Edith Matilda Thomas,** the poet, came to stay with Dr. Elliott, as a guest rather than as a patient. The same year saw publication of *Lyrics and Sonnets,* Thomas's literary debut.

Home of Edwin Markham in West New Brighton. The poet lived in this house for more than thirty years, and died here in 1940.

The Bronx

The Bronx can boast a varied and distinguished literary heritage. **Paddy Chayefsky,** who was born here on January 29, 1923, spent most of his youth in the Riverdale section, and several of his works, including the television play *Marty* (1953) and the play *The Tenth Man* (1959) were set in The Bronx.

In the eastern section of the Bronx called Hunts Point lies Drake Park, named after **Joseph Rodman Drake.** Drake is buried in a small plot here, formerly the family plot of his cousins, the Hunts, who gave their name to the district. The poet died of tuberculosis on September 21, 1820, when he was twenty-five, a coincidence that led some to call him the American Keats. Drake's only book of verse, *The Culprit Fay and Other Poems* (1835), went unpublished until long after his death. Drake's grave marker here bears an inscription from his friend, the poet Fitz-Greene Halleck; Halleck's quatrain on his friend, praised by Poe and others, is a simple but moving tribute:

Green be the turf above thee,
Friend of my better days!
None knew thee but to love thee;
None named thee but to praise.

Drake Park is open to the public and stands at the corner of Hunts Point and Oak Point avenues.

Theodore Dreiser and his wife Jug took a

low-rent apartment at 144th Street and Mott Avenue, in the Mott Haven section, when Dreiser was trying to recover from the crisis that followed publication of his first novel, *Sister Carrie* (1900). One story has it that the publisher, shocked by the theme of the novel, published the work unwillingly and only after Dreiser's insistence that his contract not be breached. The edition was printed badly and went unadvertised. Dreiser was so shaken by the consequent commercial failure that he contemplated suicide. Despite the novel's initial failure, it has come to be regarded as an American classic. In 1981 a new edition of the novel was published, which restored 36,000 words that had been excised from the original text.

Frank Gilroy, whose Pulitzer Prize-winning play *The Subject Was Roses* (1964) is set in the Bronx, grew up at 116 West 176th Street in the Tremont section. **John Hollander,** the poet and critic, attended Bronx High School of Science from 1943 until 1946, and **Howard Nemerov,** the poet and novelist, was graduated from the private Fieldston School in 1937.

Clifford Odets was born in Philadelphia but grew up in the Longfellow Avenue neighborhood of the Morrisania section, attending Morrisania High School from 1921 until 1923. His play *Awake and Sing!* (1935) was set in an apartment on Longwood Avenue, close to Beck Street. **John O'Hara** attended Fordham Preparatory School, on East Fordham Road, for one year in 1920 until he was expelled.

Of all the mansions in the Riverdale section, few have a history as distinguished as that of Wave Hill, at 675 West 252nd Street, at the corner of Sycamore Avenue. **Theodore Roosevelt** and **William Makepeace Thackeray** both lived here briefly, and the place was rented by **Mark Twain** from October 1901 until June 1903. Twain published one volume while living here, the unmemorable *A Double-Barreled Detective Story* (1902); still living in the bitter aftermath of a lecture tour forced on him by bankruptcy, he was not in the period of his greatest creativity. The view of the Hudson that Twain enjoyed here has changed, but visitors may tour the estate. The house is now part of the Wave Hill Center for Environmental Studies.

Jerome Weidman, known for such portraits of New York life as the novel *I Can Get It for You Wholesale* (1937), dramatized by the author in 1962, lived during his childhood at 1075 Tiffany Avenue, near Westchester Avenue, and attended DeWitt Clinton High School, on Mosholu Parkway, from 1927 to 1930. Another prominent novelist, **Herman Wouk,** was born in the South Bronx on May 27, 1915, in a tenement on 167th Street near Southern Boulevard. His family lived at several Bronx addresses, including 974 Aldus Street, 978 Aldus Street, and 1091 Longfellow Avenue, and Wouk attended both P.S. 48 and P.S. 75. The area has decayed dramatically since Wouk lived here, and Wouk has stated that the setting for his work *The City Boy* (1948)—"a predominantly middle-class Jewish Bronx neighborhood" in the 1920s—has vanished completely.

At the northern tip of The Bronx, abutting Van Cortlandt Park, lies Woodlawn Cemetery, where many literary figures are buried: **George M. Cohan** died on November 5, 1942. Although he is remembered as a songwriter principally for his World War I song "Over There"—Cohan won a Congressional Medal of Honor for the song—he also wrote many plays. **Countée Cullen,** who is also buried here, died on January 9, 1946. Cullen first achieved general recognition for his volume of poetry *Color* (1925). **Clarence Day,** also buried at Woodlawn, died on December 28, 1935. **Damon Runyon,** the quintessential New York journalist—he was born, of course, in Manhattan, Kansas—and short-story writer, died on December 10, 1946, and was buried here. **Joseph Pulitzer,** whose endowment of the Pulitzer Prizes in Journalism and Letters ensured that his name will be remembered as long as prizes are awarded for poetry, drama, biography, history, and the novel, was publisher of the New York *World* (1866–1931). He died on October 29, 1911. **Elizabeth Cochrane Seaman,** known better as **Nellie Bly,** the author of sensational newspaper

The building had two entrances. One on Thirty-Eighth Street and one on Broadway. I walked in through the Broadway entrance, slowly, then out the Thirty-Eighth Street side. I did it a few times, maybe five or six. Each time I got the same brisk, excited feeling that the place was full of people moving, working, coining money.

—Jerome Weidman,
in *I Can Get It for You Wholesale*

TIME: *The present . . .*
PLACE: *An apartment in the Bronx, New York City.*

—Clifford Odets,
in the opening of
Awake and Sing!

Wave Hill, home of Mark Twain from 1901 to 1903.

At sea in the old time, the execution by halter of a military sailor was generally from the fore-yard. In the present instance—for special reasons—the main-yard was assigned. Under an arm of that yard the prisoner was presently brought up, the Chaplain attending him. It was noted at the time, and remarked upon afterwards, that in this final scene the good man evinced little or nothing of the perfunctory. Brief speech indeed he had with the condemned one, but the genuine gospel was less on his tongue than in his aspect and manner towards him. The final preparations personal to the latter being speedily brought to an end by two boatswain's-mates, the consummation impended. Billy stood facing aft. At the penultimate moment, his words, his only ones, words wholly unobstructed in the utterance, were these—"God bless Captain Vere!"

—Herman Melville,
in *Billy Budd*

stories, died on January 27, 1922. The greatest of all the literary figures buried at Woodlawn is unquestionably **Herman Melville,** who died on September 28, 1891, only a few months after completing work on the novella *Billy Budd,* which was not published until 1924. It was left to others to give the work its final form. Melville's later years had been spent in isolation and obscurity: when Melville died at his Manhattan home, no New York City newspaper mentioned his passing. Melville is buried here beside his wife, the graves marked by simple stones. Woodlawn is open seven days a week and has entrances at both Webster and Jerome avenues.

Edgar Allan Poe lived in Fordham when that section was still a quiet village in an unspoiled farming district. Poe came here in May 1846 with his wife, Virginia, and Mrs. Maria Clemm, who was both Virginia's mother and the poet's aunt. They moved up from Greenwich Village at the suggestion of Dr. Thomas Holley Chivers, Poe's friend and fellow poet; Chivers thought the country air would benefit Virginia, who was suffering from tuberculosis. Even though they rented a cottage in the middle of the countryside, paying $100 a year, Virginia remained in poor health. This is hardly surprising, since Poe had no regular employment and could barely afford to buy food. His wife died in January 1847, reportedly clutching the cat to her chest to keep herself warm. Poe maintained the cottage after Virginia's death, but traveled a good deal up and down the East Coast. Over the next two and a half years, Poe's mental health was unstable, but he was fit enough to court or become engaged to at least three women. He also composed some of his best poetry: "The Bells," "Annabel Lee," and "Eldorado" (all 1849). These three poems are thought to have been written in whole or in part at the cottage in Fordham. Another composition, "Landor's Cottage" (1849), may have been inspired partly by the cottage.

It was many and many a year ago,
In a kingdom by the sea,
That a maiden there lived whom you may know
By the name of Annabel Lee;—
And this maiden she lived with no other thought
Than to love and be loved by me.

—Edgar Allan Poe,
in "Annabel Lee"

Poe left The Bronx for a trip south in late June 1849. Mrs. Clemm went to stay with friends in Brooklyn, and the cottage was closed up. The author may have intended to return at some point but never did. In Baltimore on October 7, he died at Washington Hospital.

Since Poe's death, his Fordham cottage has had a history of its own. Acquired by the city in 1912, it was taken from its original site at the southeast corner of 194th Street and Kingsbridge Road and twice moved. It stands now at the southeast corner of Grand Concourse and East Kingsbridge Road. Although purchased by New York City some seventy years ago, and presumably placed under public protection, the cottage was looted and vandalized over those years, with some valuable Poe relics—a statue of a raven among them—being removed. Only in the last ten years have the house and its contents been fully restored and protected by the Bronx Historical Society, and today it is open to the public for tours. Among the original furnishings on display in the downstairs rooms are a bed and rocking chair; the upstairs rooms, where Poe had his study, are closed to visitors. In recent years the Bronx Historical Society has celebrated each April the 1846 arrival in New York City of Poe and his young wife. Programs of readings of Poe's work, lectures by Poe scholars, and tours of Poe Cottage mark the days of celebration.

Herman Melville's grave marker, located in section 1–7 of Woodlawn Cemetery. Poe Cottage in The Bronx, as it appeared in 1884.

NIAGARA FALLS

This city on the eastern bank of the Niagara River has drawn many writers to the spectacular beauty of its great falls, including **William Dean Howells,** who wrote of the falls in his autobiographical novel *Their Wedding Journey* (1872). **Jack London** hoboed across the country in 1894 with a couple of thousand unemployed men who had decided to march on Washington to demand work. When the movement faltered, London grew restless and abandoned it, traveling north and east, and eventually arriving at Niagara Falls some time in late June 1894. He found himself immediately overpowered by the beauty and power of the falls and sat watching it for hours. Having no money for a hotel room, he slept in a field outside of town. In the early morning of the next day, London was arrested for vagrancy and shortly thereafter sentenced to thirty days in the Erie County Penitentiary. London wrote of this period in *The Road* (1907).

John O'Hara came to Niagara Falls in the fall of 1923. He was a student at the preparatory school of Niagara University nearby. O'Hara did so well in his studies that he was named class valedictorian in 1924. On the night before graduation, O'Hara got drunk with friends and was still intoxicated when the ceremonies were to begin. O'Hara was relieved of his honors and sent packing. **Abram Joseph Ryan,** the poet, studied for the priesthood at Our Lady of Angels Seminary in the late 1850s. The seminary was the forerunner of Niagara University.

Other visitors to Niagara Falls include **Henry David Thoreau,** who spent five days here in May 1861 and visited the Canadian and American sides of the river as well as Goat Island, and **Frances Trollope,** the English novelist, who spent most of her four-day visit in the spring of 1831 on the Canadian side of the falls. **Mark Twain** wrote of the falls in "Niagara" and in his short story "Extracts from Adam's Diary."

You can descend a staircase here a hundred and fifty feet down, and stand on the edge of the water. After you have done it, you will wonder why you did it; but you will then be too late.

—Mark Twain,
in "Niagara"

NORTHPORT

Writers who have lived in this village on Long Island's north shore, northeast of Huntington, include **Antoine de St. Exupéry,** the French aviator and author of the classic *The Little Prince* (1943), who lived in a house, still standing, on Bevin Road in the nearby community of Asharoken in the winter of 1942–43; and **Jack Kerouac,** who lived at 34 Gilbert Street from 1958 to 1961 and at 7 Judyann Court from 1962 to 1965.

In June 1931 **Eugene O'Neill** moved to a beach house, still standing, at 540 Asharoken Avenue. Here he worked on revisions of his trilogy *Mourning Becomes Electra* (1931), the tragic story of the Mannon family, and made notes for a play he tentatively titled *Nostalgia.* It is believed that this play, which has not been published, became O'Neill's only comedy, *Ah, Wilderness!* (1933).

Booker T. Washington bought a house, still standing, on Cousins Avenue in 1911, but owned it for only a year or two. **Herman Wouk** completed his novel *Aurora Dawn* (1947) and wrote the novel *The City Boy* (1948) while living at 50 Summit Avenue from 1946 to 1948.

NORTH TARRYTOWN

Ruth Herschberger, author of the critically acclaimed collection of poetry *A Way of Happening* (1948), and *Adam's Rib* (1948), a witty study of the status of women in the modern world, was born on January 30, 1917, at 474 Bellwood Avenue in this village north of New York City. She lived here only a short time.

NYACK

Writers associated with this city on the western bank of the Hudson River twenty-five miles north of New York City include **Robert Benchley,** who spent summers here as a boy and young

Eugene O'Neill's Northport home, 540 Asharoken Avenue.

I rode into Niagara Falls in a "side-door Pullman," or, in common parlance, a box car. . . . I arrived in the afternoon and headed straight from the freight train to the falls. Once my eyes were filled with that wonder-vision of down-rushing water, I was lost. . . . Night came on, a beautiful night of moonlight, and I lingered by the falls until after eleven.

—Jack London,
in *The Road*

> *He became aware of the instinct to run away. It suddenly occurred to him that he was hungry. Not merely hungry as one is at supper or breakfast; but a persisting, all-consuming gnawing in his intestines that moved and hurt. He felt that it was not worth staying for. He was too tired. And the oncoming men looked tired. And it seemed to take forever for them to make a contact. But they came like people who couldn't stop themselves, while he himself could not make his feet move to carry him away.*
>
> —Walter D. Edmonds, in *Drums Along the Mohawk*

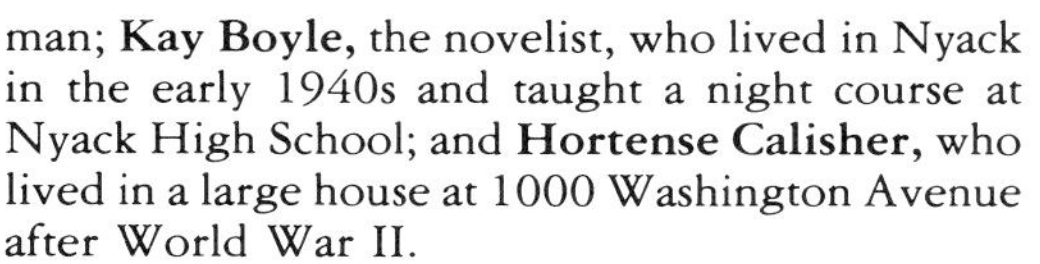

man; **Kay Boyle,** the novelist, who lived in Nyack in the early 1940s and taught a night course at Nyack High School; and **Hortense Calisher,** who lived in a large house at 1000 Washington Avenue after World War II.

Ben Hecht owned a home here for many years until his death on April 18, 1964. His home on Perry Lane was near that of **Charles MacArthur,** the playwright and author with whom Hecht collaborated on such plays as *The Front Page* (1928) and *Twentieth Century* (1933). MacArthur attended Wilson Memorial Academy as a boy. Both Hecht and MacArthur are buried in Oak Hill Cemetery, at 140 North Highland Avenue. Also buried there is **Carson McCullers,** who lived in Nyack in the 1940s and 1950s, and died at Nyack Hospital on September 29, 1967. In September 1944, Carson, her mother, and her sister moved to an apartment at 127 South Broadway. Here she worked on her novel *The Member of the Wedding* (1946). In May 1945, Carson's mother bought a house at 131 South Broadway, and Carson bought the house from her mother six years later. It remained the novelist's home until her death.

Alice Beal Parsons, the novelist and short-story writer, lived in Nyack for nearly thirty years and died, on April 14, 1962, at her home on South Boulevard. Many of her books are set in this area of New York, including the novels *A Lady Who Lost* (1932) and *I Know What I'd Do* (1946). She also wrote of her life in Nyack in the nonfiction collections *The Mountain* (1944) and *The World Around the Mountain* (1947).

OGDENSBURG

This city on the St. Lawrence River in northern New York was described in the novel *The Winds of God* (1941), by Irving Bacheller. Ogdensburg is the birthplace of **William Starbuck Mayo,** a physician and writer, whose experiences in North Africa and Spain were reflected in his novels *Kaloolah; or, Journeyings to the Djébel Kumri* (1849) and *The Berber* (1850). Mayo was born on April 15, 1811, in a house, no longer standing, at 330 Ford Street, and now the site of the Ogdensburg City Hall. Ogdensburg is the home of the Frederic Remington Art Memorial, at 303 Washington Street, which contains a collection of the paintings, sculptures, and personal items of **Frederic Remington.** Remington is best known for his paintings and bronze sculptures of classic scenes of the old West, but he was also the author of a number of books and numerous articles. Among Remington's books were *Pony Tracks* (1895), *Stories of Peace and War* (1899), and *The Way of an Indian* (1906). The Remington museum is open year round.

Carson McCullers's home, 131 South Broadway, Nyack.

OLD BETHPAGE

The school in which **Walt Whitman** taught at Smithtown in 1837–38 stands now in this community in central Long Island, just east of Hicksville. The structure was also used as a cottage by **Edward Everett Tanner III,** who wrote the best-selling novel *Auntie Mame: An Irreverent Escapade in Biography* (1955) under the pen name **Patrick Dennis.** The novel was the basis for two plays by Jerome Lawrence and Robert E. Lee, *Auntie Mame* (1957) and *Mame* (1967).

ORISKANY

The bloody Battle of Oriskany, fought on August 6, 1777, near this village in the Mohawk Valley in central New York, is considered a major battle of the American Revolution. Among the books dealing with this battle and with life in the Mohawk Valley are the novels *Drums Along the Mohawk* (1936) by Walter D[umaux] Edmonds and *In The Valley* (1980) by Harold Frederic.

OSWEGO

Fort Ontario, in this city on the southeastern shore of Lake Ontario, dates from the early eighteenth century, when the British and French fought for control of the region. **Ludwig Bemelmans** was stationed at Fort Ontario during World War I. The fort is open to visitors from mid-April to late October and has information and descriptions of Fort Oswego, where **James Fenimore Cooper,** then a midshipman in the U.S. Navy, was stationed in 1808–09—Cooper never did manage to go to sea. A boulder at West First and Van Buren streets marks the site of Fort Oswego. While in Oswego, Cooper lived in a frame house that stood at 24 West Second Street. Cooper's novel *The Pathfinder* (1840) is set in the Oswego River Valley.

Oswego was the birthplace of two authors of some reputation in their day. **H[enry] C[uyler] Bunner** was born on August 3, 1855, in a house, no longer standing, on West Seneca Street, west of Seventh Street. A poet, author, and editor, Bunner was known best for his short stories, which were collected in *Short Sixes: Stories To Be Read While the Candle Burns* (1891). Bunner wrote of reading Shakespeare in the garden of his grandmother's house in Oswego. The house, at 15 Bronson Street, is now the Delta Kappa fraternity house of the State University of New York at Oswego. **Morgan [Andrew] Robertson,** the short-story writer, was born on September 30, 1861, in a house on West Fifth Street. Among Robertson's collections were *Futility* (1898), *Spun Yarn* (1898), and *Sinful Peck* (1903).

OVID

Thomas Raynesford Lounsbury, the author and teacher, was born on January 1, 1838, in this village in central New York, midway between Seneca and Cayuga lakes. A state historical marker stands at his birthplace, now private, on South Main Street, just north of the village high school. The site of Ovid Academy, which Lounsbury attended in preparation for Yale, is also marked. Lounsbury, associated with Yale for many years, wrote *A History of the English Language* (1879) and a *Life of James Fenimore Cooper* (1882), and edited the complete writings of Charles Dudley Warner (1904).

OWEGO

N[athaniel] P[arker] Willis owned Glenmary, a two-hundred-acre estate on Owego Creek between Owego and Ithaca, in the late 1830s and early 1840s. There he wrote a series of "Letters from Under a Bridge" for the New York *Mirror,* later collected in the volume *A l'Abri, or, the Tent Pitch'd* (1839).

PALISADES

In early 1942 **John Steinbeck** rented a house, no longer standing, at Sneden's Landing, on the western shore of the Hudson River near Palisades. While living here, Steinbeck wrote material for the War Department, and it was here that he learned that the dramatization of his successful book *The Moon Is Down* (1942) was meeting only a lukewarm reception from critics and audiences. "They don't really know what bothered them about the play," he wrote to a friend, "but I do. It was dull. For some reason, probably because of my writing, it didn't come over the footlights."

PALMYRA

This village east of Rochester is the setting for the novel *Canal Town* (1944), by Samuel Hopkins Adams, and the early Mormon settlement near Palmyra is the subject of the historical novel *Children of God* (1939) by Vardis Fisher.

PATCHOGUE

This resort village on Long Island's Great South Bay was the birthplace on October 15, 1880, of **Arthur B[enjamin] Reeve,** author and editor of crime fiction and the creator of the fictional detective Craig Kennedy. **Seba Smith,** a newspaper editor who wrote humorous letters and stories under the name **Major Jack Downing,** is buried in Lakeview Cemetery here, with his wife, **Elizabeth Oakes Smith,** the novelist.

PAWLING

Writers associated with this village in southern New York, near the Connecticut border east of Beacon, include **Sarah Cleghorn,** the poet, who taught in the 1920s at the progressive Manumit School, no longer open. In the summer of 1925, **Hart Crane** spent six weeks at Tory Hill, a farm on South Quaker Hill near Pawling that was owned by Crane's friend, Slater Brown. Here Crane began work on his poems "The Wine Menagerie" and "Passage." At the end of 1925 Crane was invited to live nearby, in a house rented by fellow poet **Allen Tate** and his wife **Caroline Gordon,** the novelist. At first the three writers got along well together, and Crane plunged enthusiastically into reading and preparing early outlines and drafts for his long poem *The Bridge* (1930). Crane's enthusiasm began to get on Tate and Gordon's nerves, particularly because he would often burst in on them and interrupt their own writing. His penchant for heavy drinking in the evenings did not help matters either, so Crane decided he had overstayed his welcome by spring of 1926, and moved out. His friendship with the Tates soon recovered.

I lie on my breast in the grass,
my feet
Lifted boy-fashion, and swinging
free.
The old brown Shakspere in
front of me
And big are my eyes, and my
heart's abeat;
And my whole soul's lost—in
what?—who knows?
Perdita's charms or Perdita's
woes—

—H. C. Bunner,
in "My Shakspere"

Portrait of H. C. Bunner.

Cottage in which Walt Whitman once taught. It stood in Smithtown, but now is in Old Bethpage. Patrick Dennis once lived here.

Phelps was a serene and gracious example of what the small town could be before the coming of the automobile. From end to end the main street was lined by tall elms and maples, with here and there a gnarled locust or a shedding chestnut tree to relieve the monotony. My father's first shingle was hung from the limb of a tree outside his office window.

—Bellamy Partridge, in *Country Lawyer*

Lowell Thomas, the broadcaster and author, lived for many years on his estate on Quaker Hill in Pawling and died there on August 29, 1981. From the estate Thomas made many of the broadcasts for which he was so popular. His many books included *With Lawrence of Arabia* (1924) and *The World of Adventure* (1961). He is buried in the cemetery of Christ Church on Quaker Hill.

PEEKSKILL

Frank Dempster Sherman was born on May 6, 1860, in this city on the Hudson River, north of New York City. The poet grew up in Peekskill and attended Peekskill Military Academy. A park on the former grounds of the academy, on South Division Street at the intersection of Second, Elizabeth, and Academy streets, is named in honor of Sherman. The poet's home from 1871 to about 1887, at 953 Paulding Street, is now private. Among Sherman's books were *Madrigals and Catches* (1887), *Lyrics for a Lute* (1890), and *Lyrics of Joy* (1904).

PHELPS

Bellamy Partridge, born in 1878 in this village about thirty miles southeast of Rochester, is remembered for his best-selling book *Country Lawyer* (1939), about the life of his father. *Country Lawyer* was the first of a series of popular books by Partridge that included *Big Family* (1941) and *Excuse My Dust* (1943). Partridge was born in a house at 127 East Main Street, now private but marked by a sign reading, "Home of the Country Lawyer."

PIERREPONT

Irving Bacheller, known best for his depictions of the people and scenes of upstate New York, was born on September 26, 1859, in this community southeast of Ogdensburg. Among Bacheller's novels were *Eben Holden* (1900) and *D'ri and I* (1901).

PIERREPONT MANOR

Marietta Holley, the humorist and novelist, was buried here, about fifteen miles southwest of Watertown, at Pierrepont Manor Cemetery, on Route 193, west of Route 11. Holley was born and lived most of her life in the community of Ellisburgh, just north of here and near the town of Adams.

PLANDOME

Frances Hodgson Burnett, known primarily for her novels *Little Lord Fauntleroy* (1886) and *The Secret Garden* (1911), spent her last years at Fairseat, her home near the community of Plandome on Long Island's north shore. Burnett died on October 29, 1924, at her home, which is no longer standing, and is buried in the nearby community of Roslyn.

PLEASANTVILLE

This suburban village about thirty miles north of New York City has been the home of several well-known writers. **Hayden Carruth,** the poet, lived during the late 1930s and during part of the 1950s at his parents' home here, 61 Sunnyside Avenue. **Tom Heggen,** author of the best-selling novel *Mr. Roberts* (1946), came here in the summer of 1941 to work at the offices of *Reader's Digest* in nearby Chappaqua. He roomed in a private house at 337 Bedford Road. The novelist left Pleasantville at the end of 1941 but returned to work at *Reader's Digest* after World War II, and lived with his wife in a studio apartment at 220 Mountain Road.

Lillian Hellman in 1940 bought an estate at 600 Hardscrabble Road near Pleasantville and lived there for about twenty years. She and **Dashiell Hammett** bred poodles on the estate, which at one time totaled two hundred acres. Hellman described it in her book of memoirs *Pentimento* (1973). **Marquis James** lived in a house at 182 Broadway in the 1930s and 1940s. Here, in an office and library over his garage, he worked on his two-volume biography of Andrew Jackson. The first volume, *Andrew Jackson, the Border Captain,* was published in 1933. The second, *Andrew Jackson: Portrait of a President* (1937), brought the author a Pulitzer Prize in 1938. This was James's second Pulitzer Prize—the first was for *The Raven, A Biography of Sam Houston* (1930).

Even in those days, 1940, it was one of the last large places in that part of Westchester County. I had seen it on a Tuesday, bought it on Thursday with royalties from The Little Foxes, *knowing and not caring that I didn't have enough money left to buy food for a week. It was called an estate, but the house was so disproportionately modest compared to the great formal nineteenth-century gardens that one was immediately interested in the family who had owned it for a hundred and twenty years. . . . I closed the two guesthouses, decided to forget about the boxwood and rare plants and bridle paths, and as soon as Hammett sold two short stories we painted the house, made a room for me to work in, and fixed up the barn. I wanted to use the land and would not listen to those who warned me against the caked, rock-filled soil.*

—Lillian Hellman, in *Pentimento*

PORT JERVIS

In 1878 the Reverend Jonathan Townley Crane, father of **Stephen Crane,** who was then six years old, became pastor of the Methodist church in Port Jervis, northwest of New York City on the Delaware River. Stephen Crane described his childhood experiences here in his posthumous collection, *Whilomville Stories* (1900).

PORT WASHINGTON

This town on the north shore of Long Island has been the home of several well-known authors, including **William Rose Benét,** who lived in Port Washington before World War I; **Clarence Budington Kelland,** the newspaperman and novelist, who spent many summers here; and **Sinclair Lewis,** who lived in a cottage, still standing, at 20 Vandeventer Avenue in 1914 and 1915. Lewis sold his first story to *The Saturday Evening Post* while living here.

POUGHKEEPSIE

This city on the Hudson River has a distinguished literary history dating back to the nineteenth century. In the 1850s **Henry Wheeler Shaw** settled here and took up the life of an auctioneer and real-estate agent. He wrote for newspapers here and in Massachusetts and, under the name of **Josh Billings,** developed the gift for humor that made him famous. In 1863 he went on the lecture circuit and was so successful that he was able later to move his family to New York City. His home in Poughkeepsie in the 1860s was on Hamilton Street. **Emmet Lavery,** known best for his play about Justice Oliver Wendell Holmes, *The Magnificent Yankee* (1946), was born on November 8, 1902, at 30 North Clover Street. Lavery grew up in Poughkeepsie, and worked on city newspapers from 1921 to 1935 before moving to Hollywood and concentrating on screen and stage plays.

Isaac Mitchell, owner and editor of Poughkeepsie newspapers from 1799 to 1806 and again in 1812, was known for writing one novel, *The Asylum: or, Alonzo and Melissa* (1811), which first appeared in installments in the Poughkeepsie *Political Barometer* from June to October of 1804. In 1811 a plagiarized edition of the book appeared, purporting to be the work of one Daniel Jackson of Plattsburg. A second edition of this pirated novel (1828), taken almost word for word from the newspaper text, became a best seller. No copyright suit was ever instituted by Mitchell or by the newspaper publisher who printed Mitchell's novel. Both had died in 1812, and the stolen version continued to sell well for decades.

Poughkeepsie is the home of Vassar College, whose former students include **Mary McCarthy,** who was graduated in 1933 and used her experiences here in her novel about eight Vassar girls of the class of 1933, *The Group* (1963); and **Edna St. Vincent Millay,** who was graduated from Vassar in 1917.

POUND RIDGE

Sloan Wilson lived in the late 1950s on High Ridge Road in Pound Ridge, about forty miles northeast of New York City, and here wrote his novels *A Summer Place* (1958) and *A Sense of Values* (1960).

PURCHASE

From 1866 until his death in 1912, **Whitelaw Reid,** the editor and diplomat, made his home at Ophir Hall, on Purchase Street. Ophir Hall is now part of Reid Hall and College Chapel at Manhattanville College. Reid succeeded Horace Greeley as editor of the New York *Tribune* in 1872 and held the post until 1905, when he became U.S. ambassador to Great Britain.

QUOGUE

In the summer of 1931, **John O'Hara** rented a beach house on Dune Road, not far from the home of Belle Wylie, whom he was later to marry. Belle Wylie lived with her family at Quogue Street and Shinnecock Avenue in this resort village on the south shore of Long Island. After their marriage, he and Belle spent some time at the Wylie home before buying their own summer cottage on Dune Road. There O'Hara worked steadily on his books, writing every day in the Wylies' living room. At Quogue, during the summer of 1960, O'Hara wrote most of the short stories that appeared in the volume *Assembly* (1961) and worked on the novel *A Rage to Live* (1949), which was completed at his apartment in New York City.

RANDOLPH

Charles Austin Fosdick, who became one of the most popular authors of adventure stories for boys in the second half of the nineteenth century, was born on September 16, 1842, in Randolph, fifty miles south of Buffalo. His family moved to Buffalo when Fosdick was still a child. Under the name **Harry Castlemon,** Fosdick wrote nearly sixty books whose popularity rivaled the success of such contemporaries as Horatio Alger.

REMSENBURG

P[elham] G[renville] Wodehouse, the English novelist who took United States citizenship in 1955, made his principal residence in this town on the South Shore of Long Island from July 1952 until his death on February 14, 1975, at age ninety-three. His home throughout that period was at Basket Neck Lane. The author of almost one hundred comic novels and almost five times as many short stories, motion picture scripts, and song lyrics was made a Commander of the Order of the British Empire on New Year's Day of his last year of life.

The mule is haf hoss and haf Jackass, and then kums tu a full stop, natur diskovering her mistake.

—Josh Billings, in "The Mule"

Home of Sinclair Lewis, 20 Vandeventer Avenue, Port Washington.

She and Mother had talked it over and agreed that if you were in love and engaged to a nice young man you perhaps ought to have relations once to make sure of a happy adjustment. Mother, who was very youthful and modern, knew of some very sad cases within her own circle of friends where the man and the woman just didn't fit down there and ought never to have been married.

—Mary McCarthy, in *The Group*

Don't you ask me, whence this burlesque;
Whence this captious fabrication,
With its huge attempt at satire,
With its effort to be funny,
With its pride in Yankee spirit,
With its love of Yankee firmness,
With its flings at Yankee fashions,
With its slaps at Yankee humbug,
With its hints at Yankee follies,
And its scoffs at Yankee bragging,
With its praise of all that's manly,
All that's honest, all that's noble,
With its bitter hate of meanness,
Hate of pride of affectation.
With its scorn of slavish fawning,
Scorn of snobs, and scorn of flunkies,
Scorn of all who cringe before the
Dirty but "almighty dollar?"
Don't you ask—for I shan't tell you,
Lest you, too, should be a Yankee
And should turn and sue for libel,
Claiming damages—God knows how much.

—Mortimer Neal Thomson,
in *Plu-ri-bus-tah,*
A Song That's By-No-Author

RIGA

Mortimer Neal Thomson [or **Thompson**], born in this community southwest of Rochester on September 2, 1831, was an early American humorist and parodist who wrote under the pen name **Q. K. Philander Doesticks, P.B.** Some of his newspaper letters were collected in *Doesticks, What He Says* (1855). Thomson parodied the poem "Nothing to Wear" by William Allen Butler in an eight-hundred-line poem entitled *Nothing to Say* (1857); and the classic poem *Hiawatha* (1855), by Henry Wadsworth Longfellow, in *Plu-ri-bus-tah, A Song That's-By-No-Author* (1856).

ROCHESTER

This industrial city on the shore of Lake Ontario has a wealth of literary associations, in particular the number of well-known writers who were born here. Among them were **Isabella Alden,** author of many books for children, who was born on November 3, 1841; and **Philip Barry,** the playwright, who was born on June 18, 1896. **Garson Kanin,** the writer, director, producer, and playwright, was born here on November 24, 1912. Among his best-known works is the play *Born Yesterday* (1946). **Henry Francis Keenan,** born here on May 4, 1850, wrote the novel *The Money-Makers* (1885), an anonymously published response to *The Breadwinners* (1884), a novel attacking unions that was written by John Hay. **Charles Warren Stoddard,** born here on August 7, 1843, was a poet and travel writer whose best-known books included *South-Sea Idyls* (1873) and *The Lepers of Molokai* (1885).

Other writers associated with Rochester include **Samuel Hopkins Adams,** the novelist, who attended grade school and high school here in the 1880s. Adams, in childhood, often visited his grandfather's home on South Union Street and there heard the tales he later wrote in his collection *Grandfather Stories* (1955). In 1847 **Frederick Douglass,** the escaped slave who became a leading figure in the Abolitionist movement, established the Abolitionist newspaper *The North Star* here and published it for seventeen years. Douglass's homes in Rochester include a house, no longer standing, at 247 Alexander Street, and a house at 1023 South Avenue, now private. Douglass died on February 20, 1895, and is buried in Mount Hope Cemetery, at 791 Mount Hope Avenue. A statue of Douglass stands in Highland Park, near South Avenue. **Paul Horgan,** the historian and novelist, lived on Windham Street in the 1920s, and attended the Eastman School of Music from 1923 to 1926. Horgan used Rochester and his native city of Buffalo in creating the city of Dorchester for his novel *The Fault of Angels* (1933). **Marjorie Kinnan Rawlings** worked for the Rochester *Journal* in the 1930s. **Mark Twain,** visiting a Rochester book store in 1885 with his friend **George Washington Cable,** came upon a copy of Thomas Malory's *Morte d'Arthur,* which later inspired Twain to write *A Connecticut Yankee in King Arthur's Court* (1889).

Writers associated with the University of Rochester include **George Abbott,** the playwright, who received a B.A. degree here in 1911; **Carl Carmer,** who taught here in 1916–17 and 1919–21; and **William Osborn Stoddard,** the novelist and historian, who received a B.A. here in 1858 and an M.A. in 1861.

ROME

Harold Bell Wright, author of *The Shepherd of the Hills* (1907) and *The Winning of Barbara Worth* (1911), was born on May 4, 1872, in Rome, a city west of Utica.

ROSLYN

William Cullen Bryant lived for many years in this village on the north shore of Long Island. His home, Cedarmere, off Northern Boulevard, is now headquarters of the Nassau County Center for the Fine Arts and is not open to the public, but visitors can tour the estate's 174 acres of trails and gardens. The poet is buried in Roslyn Cemetery, on Northern Boulevard, just east of the entrance to the county center. Bryant Library, on Paper Mill Road, has a collection of Bryant's manuscripts, as well as books by and about him.

William Cullen Bryant's grave marker in Roslyn Cemetery. (*below right*) Cedarmere, his home in Roslyn. The poet moved to Roslyn, then known as Hempstead Harbor, in 1844, and for years divided his time between Roslyn and New York City.

The manuscripts may be seen by appointment.

Also buried in Roslyn Cemetery are **Frances Hodgson Burnett,** who lived near Plandome, just west of here; **Parke Godwin,** the author and editor, who was Bryant's son-in-law; and **Christopher Morley,** the versatile and prolific author, who lived in Roslyn for many years and wrote in a pine cabin he built in the woods behind his house in 1934. The cabin, known as the Knothole, is located in Christopher Morley Park, on Searingtown Road north of the Long Island Expressway, and is open to the public in summer. The work Morley did here includes compiling the 1937 edition of *Bartlett's Familiar Quotations* and writing the popular novel *Kitty Foyle* (1939).

The Knothole, Christopher Morley's cabin in Roslyn. The Latin inscription may be translated as "May you work hard in the library, paradise for you."

ROXBURY

John Burroughs, the naturalist and writer, spent summers from 1910 to 1920 at Woodchuck Lodge, near this community in the Catskill Mountains forty-five miles southwest of Albany. The house, dating to the 1860s, was part of the Burroughs family farm, on which Burroughs was born on April 3, 1837. Toward the end of his life, Burroughs decided that he wanted to spend his summers among the rolling hills and farmland of his youth. He rented and renovated Woodchuck Lodge, made furniture and fixtures for the house, and here continued to write the essays, articles, and books that had made him America's foremost nature writer. He wrote in the open doorway of a nearby barn, or sometimes in a small cabin behind the lodge. Here also he received visitors, including Thomas Edison and Henry Ford, with whom he made several celebrated automobile camping tours. Burroughs died on March 29, 1921, and is buried in a grassy field near Woodchuck Lodge, now known as Memorial Field. Near his grave is the boulder Burroughs referred to as his Boyhood Rock, which bears a bronze plaque in his memory. Woodchuck Lodge, on Burroughs Memorial Road, is a National Historic Landmark and is open to visitors.

RUSHFORD

Philip Wylie, the novelist and short-story writer, is buried here in southwestern New York about twenty-five miles northeast of Olean.

RYE

This city on Long Island Sound, in Westchester County, northeast of New York City, has a number of literary associations. **Edward Albee,** the playwright, who lived in nearby Larchmont as a boy, entered Rye County Day School in 1939 but was expelled in the following year. **Marquis James,** the Pulitzer Prize-winning biographer and historian, moved to Rye in the 1940s, and lived in a converted barn on Stuyvesant Road. James died at his home on November 19, 1955. **John Jay** spent his boyhood on a farm on Boston Post Road in Rye and is buried in the family plot near the United Methodist Church. **J[ohn] P. Marquand** lived here and in Newburyport, Massachusetts, as a boy. In Rye he lived in a fashionable house, believed to be still standing, on the Boston Post Road. **Ogden Nash,** the poet whose humorous verse was known to millions, was born here on August 19, 1902. His birthplace stood near the junction of Route 287 and Boston Post Road.

SAGAPONACK

James Jones lived in this community in eastern Long Island for a time, and died on May 9, 1977, at nearby Southampton Hospital. The novelist is buried in Sagaponack Cemetery.

SAG HARBOR

A number of writers have made their homes in this village on the eastern end of Long Island. **Nelson Algren** lived in Sag Harbor briefly before his death at his home here on May 9, 1981. **Ambrose Bierce** was a frequent visitor to the home here of **Percival Pollard,** the critic and author. **James Fenimore Cooper,** on a visit to Sag Harbor, stayed at the Duke Fordham Inn, no longer standing, in 1819. The site, at the foot of Main Street, is now occupied by a gasoline station and the village post office. Cooper visited Sag Harbor several times between 1819 and 1823, and is said to have worked, while here, on his first novel, *Precaution* (1820).

John Steinbeck made Sag Harbor his summer home from the mid-1950s until his death in 1968. He and his wife Elaine rented in the summer of 1953 and bought a house in 1955. At Sag Harbor Steinbeck worked on his novel *The Short Reign of Pippin IV* (1957), but his house was so small that he was constantly looking for a quiet place to work. He outfitted his station wagon with a work table and, although he got stuck on the beach and had to be towed out the first time he tried this rolling study, he got into the habit of driving to a different place each day and writing in his car. In 1958 Steinbeck built a small studio at the end of Bluff Point. He named it Joyous Garde after a castle of Arthurian legend. Here he worked on his later books. One of them, the novel *The Winter of Our Discontent* (1961), is set in Sag Harbor, called New Baytown in the novel.

Steinbeck was a Western-born writer who moved East; **George Sterling,** the poet, was born in Sag Harbor on December 1, 1869, and moved West as a young man, becoming identified with the literary bohemians of California. His best-known collections of poetry were *The Testimony of the Suns* (1903) and *A Wine of Wizardry* (1909). Sterling is believed to be the model for the character Brissenden in the novel *Martin Eden* (1909) by Sterling's close friend Jack London.

Candy
Is dandy
But liquor
Is quicker.

—Ogden Nash, "Reflection on Ice-Breaking"

"You've got to marry or die to get into it" was our saying about the Front Room. I had quite a vivid picture in my mind of myself getting married in there. Marriage, I thought, was wearing a whole lot of white satin and lace and having a cake with doves in frosting all round it like what I saw in Hanscom's window on Frankford Avenue, while the men got plastered in the kitchen.

Even after Mother died Pop made pathetic attempts to keep the Front Room special. He went in there sometimes to sit in the red velvet easy chair but I don't think he was ever quite easy. It didn't seem right to sit there in shirtsleeves.

—Christopher Morley, in *Kitty Foyle*

In the cold I will rise, I will bathe
In waters of ice; myself
Will shiver and shrive myself,
Alone in the dawn, and anoint
Forehead and feet and hands;
I will shutter the windows from light,
I will place in their sockets the four
Tall candles and set them a-flame
In the gray of the dawn; and myself
Will lay myself straight in my bed,
And draw the sheet up under my chin.

—Adelaide Crapsey
"The Lonely Death"

Robert Louis Stevenson and members of his family, photographed on the porch of their cottage at Saranac Lake. From left to right: Valentine Roch, the Stevensons' maid; Mrs. Baker, owner of the cottage in which Stevenson stayed; Lloyd Osbourne; Fanny Osbourne Stevenson; and Robert Louis Stevenson. (*below*) The cottage where they lived at Saranac Lake.

SARANAC LAKE

Two writers associated with this village on Flower Lake, near Saranac Lake in northern New York, came here seeking treatment for tuberculosis. **Adelaide Crapsey** wrote most of the poems that appeared in her posthumous volume, *Verse* (1915), while being treated here. She died here on October 8, 1914.

Robert Louis Stevenson, seeking relief from a tubercular condition, came to Saranac Lake during the winter of 1887–88. Here he worked on *The Wrong Box* (1888), the novel he wrote in collaboration with his stepson, Lloyd Osbourne;

and *The Master of Ballantrae* (1889), which became one of his best-known novels. The two-story cottage in which Stevenson stayed, at 11 Stevenson Lane, is now known as Robert Louis Stevenson's Memorial Cottage, and is open daily to the public in summer. A modest donation is requested. Among the items on display are Stevenson's writing desk, smoking jacket, and ice skates, as well as a collection of letters and photographs.

SARATOGA SPRINGS

The famous Yaddo artists colony is located in a fifty-five-room mansion off Union Avenue in this city thirty-three miles north of Albany. The list of writers who have visited Yaddo in the last fifty-five years is a virtual who's who of twentieth-century American literature. Among them: **Truman Capote, James T. Farrell, Carson McCullers, Flannery O'Connor, Katherine Anne Porter,** and **William Carlos Williams.** Yaddo was founded in the 1920s as a place for writers and artists to find inspiration and solitude for their work.

Other writers associated with Saratoga Springs include **Edna Ferber,** who stayed in 1937 at the Gideon Putnam Hotel, still standing, and set much of her novel *Saratoga Trunk* (1941) in Saratoga Springs. Marietta Holley wrote of the city in her novel *Samantha at Saratoga* (1887), and Bronson Howard used the city as the setting for his play *Saratoga, or, Pistols for Seven* (1870).

Frank Sullivan, born **Francis John Sullivan** here on September 22, 1892, was graduated from Saratoga Springs High School in 1910, and worked on local newspapers before establishing himself as a journalist and humorist in New York City. He returned to his native city in the 1930s and made it his home until his death on February 19, 1976, at Saratoga Hospital, on Church Street. Sullivan is buried in St. Peter's Cemetery on West Avenue. Among his books were *Broccoli and Old Lace* (1931), perhaps his best-known work; and *The Night the Old Nostalgia Burned Down* (1953). His home at 135 Lincoln Avenue is now private, but visitors can still dine at a restaurant identified as one of his favorites, Siro's Steakhouse, at 168 Lincoln Avenue.

SCARSDALE

From 1920 to 1945 **Robert Benchley,** the theater critic and humorist, owned a home at 2 Lynwood Road in this town in Westchester County, north of New York City. While here, Benchley served as theater critic for *Life* magazine, 1920 to 1929, and for *The New Yorker* magazine, 1929 to 1940. His writing appeared in a number of books, including *My Ten Years in a Quandary* (1936) and *Benchley Beside Himself* (1943). In his later years he became well known as a performer on radio and in films. Benchley died in a New York City hospital on November 21, 1945.

James Fenimore Cooper, who had lived for a time elsewhere in Westchester County and at Cooperstown, moved in 1817 to Angevine Farm, on what now is Mamaroneck Road. The site of Cooper's farm, near Scarsdale Junior High School, is marked. Legend has it that Cooper decided on a career as writer one night in Scarsdale while reading a book aloud to his wife, Susan. He became disgusted with the story and said that he would bet he could write a better book. His wife challenged him to do so, and the result was *Precaution* (1820)—generally regarded as one of his weakest novels—but probably good enough for Cooper to have won his bet. Cooper's daughter Susan was the source of this account, one of the most captivating of all literary anecdotes. What Susan—she was only seven years old when her father wrote *Precaution*—did not know was that Cooper was thirty-one at the time, was fast running through his inheritance, and had no profession. Whatever the truth, Cooper did average a book

Plaque honoring James Fenimore Cooper. It stands off Route 22, near Fenimore Road in Scarsdale. The plaque shows Harvey Birch, the spy in Cooper's novel *The Spy:*

He may be a spy—
He must be one
"but he has a heart
above enmity and
a soul that would
honor a soldier."

Home of Frank Sullivan in Saratoga Springs, where Sullivan was born and lived for most of his adult life.

From behind bedroom doors up and down the hotel corridors could be heard the sounds of gala preparation—excited squeals, the splashing of water, the tinkle of supper trays, the ringing of bells, the hurried steps of waiters and bellboys and chambermaids, the tuning of fiddle and horn. Every gas jet in the great brass chandeliers was flaring; even the crystal chandelier in the parlor, which was lighted only on special occasions. In the garden the daytime geraniums and petunias and alyssum and pansies had vanished in the dusk. In their place bloomed the gaudy orange and scarlet and rose color of the paper lanterns glowing between the trees.

—Edna Ferber,
in *Saratoga Trunk*

I am here in a most wonderful out-of-the-world place, which looks as if it had begun to be built yesterday, and were going to be imperfectly knocked together with a nail or two the day after to-morrow. I am in the worst inn that ever was seen, and outside is a thaw that places the whole country under water.

—Charles Dickens,
writing of Syracuse
in a letter to Charles Fechter,
March 8, 1868

and a half a year for the rest of his life. All this on a dare?

SCHAGHTICOKE

Ann Eliza Bleecker, author of *The History of Maria Kittle* (1793), one of the first American novels, lived in the late 1700s near this community eighteen miles north of Albany. Bleecker based her novel, an account of the kidnaping of a woman by Indians, on an actual incident in 1711 in the settlement of Tomhannick, then a few miles west of present-day Schaghticoke. Bleecker's novel, which appeared posthumously, was the first to depict Indians as villains as well as noble savages, and anticipated the writing of James Fenimore Cooper and others. She also wrote an account of James Yates, a local resident who, in a fit of religious lunacy, murdered his family and killed his livestock in 1781. This account, published in the *New-York Weekly Magazine* in 1796, is believed to have been the inspiration for the novel *Wieland* (1798), by Charles Brockden Brown.

SCHENECTADY

This city northwest of Albany has been the home of at least two well-known authors. **Clyde Fitch,** the playwright, moved with his parents to Schenectady as a child in about 1869 and lived here for ten years. **Kurt Vonnegut** worked in the public relations department of the General Electric Company in Schenectady from 1947 to 1950. While here he wrote short stories and gathered impressions and experiences that he used in his novel *Player Piano* (1952). Vonnegut mentions this period of his life in his essay "Science Fiction," collected in his book *Wampeters, Foma, & Granfalloons* (1974).

Union College, chartered in 1795 in Schenectady, has several literary associations. **Edward Bellamy,** author of the Utopian novel *Looking Backward: 2000–1887* (1888), studied literature for a few months here about 1867. **Robert Traill Spence Lowell,** the clergyman, author, and brother of James Russell Lowell, taught at Union from 1873 to 1879. His book *A Story or Two from an Old Dutch Town* (1878) is based on his experiences in Schenectady. **John Howard Payne,** the playwright and poet, attended Union College for two terms before his father's bankruptcy led him, in 1809, to return to New York City and the profession of acting.

In the Golden Age of Magazines, which wasn't so long ago, inexcusable trash was in such great demand that it led to the invention of the electric typewriter, and incidentally financed my escape from Schenectady. Happy days!

—Kurt Vonnegut,
in "Science Fiction"

SENNETT

Harold Bell Wright, the clergyman and novelist, lived during his childhood in this town twenty miles west of Syracuse. He attended the local school and Presbyterian church. His family was very poor and, after his mother died, the eleven-year-old Wright was sent first to live and work on local farms, and later to live with various relatives.

SMITHTOWN

Walt Whitman taught school in this Long Island community east of Huntington during the fall and winter of 1837–38. The schoolhouse in which he taught has been moved to Old Bethpage.

As the boys skated onward, they saw a number of fine countryseats, all decorated and surrounded according to the Dutchest of Dutch taste, but impressive to look upon, with their great, formal houses, elaborate gardens, square hedges, and wide ditches—some crossed by a bridge, having a gate in the middle to be carefully locked at night. These ditches, everywhere traversing the landscape, had long ago lost their summer film, and now shone under the sunlight like trailing ribbons of glass.

The boys traveled bravely, all the while performing their surprising feat of producing gingerbread from their pockets and causing it to vanish instantly.

—Mary Mapes Dodge,
in *Hans Brinker; or, The Silver Skates*

SOLVAY

Leonard Bacon was born on May 26, 1887, in this village five miles west of Syracuse. Bacon won a Pulitzer Prize for his volume of poetry *Sunderland Capture* (1940).

SOUTHAMPTON

Finley Peter Dunne, the newspaperman and humorist, spent summers here on eastern Long Island's south shore in the 1920s and 1930s. **James Jones** died at Southampton Hospital on May 9, 1977. Southampton College has established a place for writers to work, in memory of John Steinbeck, the novelist and Nobel laureate who once lived in Sag Harbor. **P. G. Wodehouse,** who lived in Remsenburg, died on February 14, 1975, in Southampton Hospital.

SPENCERPORT

John Townsend Trowbridge, one of the most popular novelists of the second half of the nineteenth century, lived until the age of seventeen in this village in Ogden township, about ten miles west of Rochester. Trowbridge was born on his father's farm in Ogden township on September 18, 1827, and grew up in Spencerport in a two-story frame house on Nicholas Road. The author of novels, plays, and poetry was known best for his many popular stories for boys, including *Neighbor Jackwood* (1857) and *Cudjo's Cave* (1864).

STAMFORD

E[dward] Z[ane] C[arroll] Judson, born on March 20, 1823, in this community about fifty miles west southwest of Albany. He was known for his countless dime novels, many of them about the supposed adventures of Buffalo Bill, which Judson wrote under the pen name **Ned Buntline.**

SYRACUSE

This city has a literary heritage dating back to the late 1700s, when Ephraim Webster, the first white settler in this region, established a camp on the shore of the salt marsh here. James Fenimore Cooper is thought to have modeled the sharpshooter Natty Bumppo in part after Webster, who is buried in Onondaga Valley Cemetery, at 1804 Valley Drive. **Charles Dickens** visited Syracuse in March of 1868, during his second American tour, and wrote of his experiences in letters to family and friends back in England. The novelist stayed at the Syracuse House, and gave a reading at the Weiting Opera House. In more recent times, Syracuse was the home of **F. Scott Fitzgerald,** who lived here as a boy from January of 1901 to September of 1903.

Writers born in Syracuse include **Harold McGrath,** a popular and prolific novelist and newspaperman, born on September 4, 1871. McGrath grew up in Syracuse and got his start in journalism in 1890 with the Syracuse *Herald.* He later contributed to newspapers in Albany and Chicago, and wrote for the film industry in its early days. His novels included *Half a Rogue* (1904) and *A Splendid Hazard* (1910). McGrath died at his home here on October 29, 1932. Another writer born in Syracuse was **[Edward] Rod**

[man] **Serling,** the playwright, screenwriter, and author, born on December 25, 1924. Serling is remembered for his television dramas, such as *Patterns* (1955) and *Requiem for a Heavyweight* (1956), which were later published in the book *Patterns* (1957); and for the television series *The Twilight Zone* (1959–64), to which Serling turned when he could not abide censorship of his work by the television networks.

Edward Noyes Westcott, born in Syracuse on September 27, 1846, achieved posthumous literary fame with the publication of his novel *David Harum: A Story of American Life* (1898). The main character of the book is believed to be modeled after two people: Westcott's father, a successful and innovative dentist who served as mayor of Syracuse in 1860; and David Hannum, a friend of the elder Westcott and a shrewd businessman from Homer, about thrity miles south of Syracuse. Westcott worked on his novel in several places, including Meacham Lake, New York, and at his home, no longer standing, in Syracuse at 990 (formerly 826) James Street, where he died on March 31, 1898. (James Street is named for William James, grandfather of Henry James, Jr., and an early developer of the salt industry in Syracuse.) Westcott's novel was rejected by half a dozen publishers before D. Appleton and Company published it in October 1898, after Westcott's death. An almost immediate success, it became one of the best-selling novels of the early twentieth century. Westcott's only other published work was *The Teller* (1901), an unfinished novel.

Edward Noyes Westcott.

Stephen Crane was a student at Syracuse University briefly in 1891. He lived in a boarding house and at his aunt's house before moving into the Delta Upsilon fraternity house, no longer standing, at 426 Ostrum Street, where he worked on the first draft of his novel *Maggie: A Girl of the Streets* (1893). The *Syracuse University Herald* published his story "The King's Favor" while Crane was a student. The present Delta Upsilon fraternity house at 711 Comstock Avenue has a photograph of the house in which Crane lived, as well as a letter of Crane's in which the writer described his memories of life at the university. The university's George Arents Research Library has a collection of Crane material. Other writers associated with Syracuse University include **W. D. Snodgrass,** the poet, who taught here from 1968 to 1977; and **Dorothy Thompson,** who was graduated in 1914 and returned for graduate study in 1937. The university has a collection of her papers, as well as those of Marya Zaturenska, the poet.

TALCOTTVILLE

Edmund Wilson spent summers as a boy in this town in northern New York, the hometown of his mother's family, about thirty-five miles north of Utica. In his later years Wilson divided his time between Talcottville and his home in Wellfleet, Massachusetts.

TANNERSVILLE

Onteora Park, also known as the Onteora Club, just north of Tannersville and about fifteen miles west of Catskill, was an exclusive summer community in the late nineteenth and early twentieth centuries for a number of well-known writers. One of them was **Mary Mapes Dodge,** the author of stories for children. Dodge bought Yarrow Cottage here in 1888 and spent summers at the cottage thereafter. The author of *Hans Brinker; or, The Silver Skates* (1865) died in Tannersville on August 21, 1905, and the children of the community formed the procession for her funeral. **Hamlin Garland** summered at Onteora Park from about 1917 to the 1930s, and there worked on his Middle Border series of books, including the Pulitzer Prize-winning *A Daughter of the Middle Border* (1921).

Other residents included **Laurence Hutton,** the critic and author; **[James] Brander Matthews,** the drama critic and teacher; and **Mark Twain,** who moved with his family to Hutton's cottage, since destroyed, for the summer of 1890. A plaque originally placed on the cottage is mounted now on a boulder at the site of the cottage. This community is private, and permission to visit it must be obtained from the supervisor.

Photograph of Hamlin Garland in front of Grey Ledges, his home near Tannersville. Before buying Grey Ledges, Garland summered in the area in a cabin he called Camp Neshonoc.

I have a rendezvous with Death
At some disputed barricade,
When Spring comes back with rustling
shade
And apple-blossoms fill the air—
I have a rendezvous with Death
When Spring brings back blue days
and fair.

—Alan Seeger
in "I Have a Rendezvous with Death"

TARRYTOWN

Howard Fast lived just before World War II in this community north of New York City and here wrote *Citizen Tom Paine* (1943), one of his works of historical fiction. But it is not the presence of Fast that has made Tarrytown one of America's most important—and most beautiful—literary shrines.

Washington Irving in 1835 bought a small cottage in Tarrytown, calling it his Snuggery, or Wolfert's Roost, after Wolfert Ecker, who built the cottage in the seventeenth century. It was later owned by members of the Van Tassel family, who were immortalized in Irving's classic story "The Legend of Sleepy Hollow" (1820). Under Irving's guidance, his friend George Harvey directed the renovation of the cottage, transforming it into a comfortable and spacious country home, called Sunnyside. On the grounds, Irving built a pond he referred to as "the little Mediterranean," and in 1847 completed the Pagoda, a three-story tower whose rooms were used for guest and servant quarters. Ivy growing on the east wing of the house is said to be from a cutting given to Irving by Sir Walter Scott, the Scottish poet and novelist. Irving served as U.S. minister to Spain from 1842 to 1846, but he spent most of his later years at Sunnyside, where he worked on his books *Astoria* (1836) and *The Adventures of Captain Bonneville, U.S.A.* (1837) and wrote numerous stories and sketches published in the *Knickerbocker Magazine.* Nineteen of these shorter pieces were collected in the volume *Wolfert's Roost and Miscellanies* (1855). At Sunnyside Irving also prepared the Author's Revised Edition of his works (fifteen volumes, 1848–51) and wrote his five-volume *Life of George Washington* (1855–59).

Irving died at Sunnyside on November 28, 1859, and his funeral was held at Christ Episcopal Church on Broadway (Route 9). Irving had served as warden and vestryman at the church. Ivy growing on the church is said to be a cutting from a vine growing at Sunnyside. Irving is buried in Sleepy Hollow Cemetery, adjacent to the Old Dutch Church on Broadway in nearby North Tarrytown, where also are buried **Thomas Beer,** the novelist and biographer; **[Fred] Hayden Carruth,** the editor and humorist; **Whitelaw Reid,** the editor; and **Carl Schurz,** the soldier, statesman, and journalist.

Sunnyside, on West Sunnyside Lane off Broadway, is a National Historic Landmark and is open daily. Admission is charged. Sunnyside has been restored to its appearance in Irving's day, and contains many original furnishings, including the author's desk, his library of nearly 3,000 books, and a portrait of **John Pendleton Kennedy,** the writer, who was a frequent visitor at Sunnyside in Irving's day. At the junction of

Washington Irving, photographed at the entrance to Sunnyside about 1856, three years before his death. This photograph was taken from a stereoscopic slide. (*far right*) Sunnyside as it appears today.

Reader! the Roost still exists. . . . The shade of Wolfert Acker still walks his unquiet rounds at night in the orchard; . . . Mementos of the sojourn of Diedrich Knickerbocker are still cherished at the Roost. His elbow-chair and antique writing-desk maintain their place in the room he occupied, and his old cocked-hat still hangs on a peg against the wall.

—Washington Irving,
in "Wolfert's Roost"

Washington Irving's gravestone in Sleepy Hollow Cemetery and the Old Dutch Church in North Tarrytown.

Broadway and West Sunnyside Lane is a memorial to Irving by Daniel Chester French, featuring a bust of Irving and a relief of Rip Van Winkle.

James Kirke Paulding grew up near Tarrytown, at his family's home, Paulding Manor. He knew Washington Irving well, and Paulding's sister was married to Irving's brother William. **Alan Seeger**, the poet known best for "I Have a Rendezvous with Death" (1916), was graduated from the Hackley School in Tarrytown in 1906. The school is still open and is located at 293 Benedict Avenue. **Mark Twain** from 1902 until 1905

The Marquesas! What strange visions of outlandish things does the very name spirit up! Lovely houris—cannibal banquets—groves of cocoa-nuts—coral reefs—tattooed chiefs—and bamboo temples; sunny valleys planted with bread-fruit trees—carved canoes dancing on the flashing blue waters—savage woodlands guarded by horrible idols—heathenish rites and human sacrifices.

—Herman Melville, in *Typee*

owned land overlooking the Tappan Zee on Tappan Hill at Benedict and Highland avenues. The ten-acre estate, once called Hillcrest, is identified by a historical marker and has been used in recent years as the site of the Tappan Hill restaurant. The restaurant exhibits a portrait of Twain.

TICONDEROGA

Sloan Wilson lived in this village in northeastern New York, at the northern end of Lake George, from the mid-1960s through the late-1970s. While living in a cottage at the Rogers Rock Club here, Wilson wrote his novel *Janus Island* (1967). Later, he lived in a house at 207 Champlain Avenue, where he wrote his autobiographical *What Shall We Wear to This Party? The Man in the Gray Flannel Suit Twenty Years Before and After* (1976), in which he describes in moving detail his life in Ticonderoga and his struggle against depression and insecurity. In the Champlain Avenue house, Wilson also wrote *Small Town* (1978), which incorporates elements of his Ticonderoga experiences, although the names and details of places are changed.

TROY

Troy has several interesting literary associations. The novel *Sunrise to Sunset* (1950) by Samuel Hopkins Adams, set in Troy in the 1830s, described life in Troy's textile mills early in the nineteenth century. A more important association is with the life here of one of America's great novelists: **Herman Melville** lived from the late 1830s until the early 1840s in the community of Lansingburgh, then north of Troy and now part of the city. In 1838 Melville enrolled at Lansingburgh Academy, at the northwest corner of 114th Street and Fourth Avenue, studying engineering and surveying. He was graduated in 1839. The academy building, now undergoing restoration as a library arts center, was near Melville's home at the southeast corner of 114th Street and First Avenue. There are two houses on this lot, both dating to Melville's time, and there is some controversy as to which structure actually was the Melville family's home. The house at Two 114th Street, now headquarters of the Lansingburgh Historical Society and Museum, is identified by a state historical marker as being Melville's home. While living here, Melville contributed three items to the *Democratic Press and Lansingburgh Advertiser:* two pieces under the title "Fragments from a Writing Desk," and one piece of Gothic fiction entitled "The Death Craft." Copies of the newspaper issues in which these appeared are at the Troy Public Library at 100 Second Street.

Unable to find satisfactory work on land, Melville went to sea and later used the experience of this first voyage in his novel *Redburn* (1849). On January 3, 1841, he sailed from Fairhaven, Massachusetts, aboard the *Acushnet,* and began the four-year South Seas adventure that he would use on his return to Lansingburgh in 1844 in writing his novels *Typee* (1846) and *Omoo* (1847). Both these novels, written in Lansingburgh, were successful, but Melville was attacked in the religious press for his negative portrayal of the role of missionaries in the South Seas.

The moon on the breast of the new-fallen snow
Gave the lustre of mid-day to objects below,
When, what to my wondering eyes should appear,
But a miniature sleigh, and eight tiny reindeer,
With a little old driver, so lively and quick,
I knew in a moment it must be St. Nick.
More rapid than eagles his coursers they came,
And he whistled, and shouted, and called them by name;
"Now, Dasher! *now,* Dancer! *now,* Prancer *and* Vixen!
On, Comet! *on* Cupid! *on,* Donder *and* Blitzen!
To the top of the porch! to the top of the wall!
Now dash away! dash away! dash away all!"

—Clement Clarke Moore, in "A Visit from St. Nicholas"

The poem "A Visit from St. Nicholas" by Clement Clarke Moore was first published—without its author's knowledge—in the Troy *Sentinel* on December 23, 1823. Moore, a religious scholar, did not acknowledge authorship of what he considered a trivial poem, and he was not listed as its author until the Troy *Budget* recorded the fact in December 25, 1838. The classic poem was collected in Moore's *Poems* (1844).

TUCKAHOE

Two writers associated with this village north of New York City are **Elizabeth [Chase] Akers,** best known for her poem "Rock Me to Sleep, Mother" (1860), who lived near Tuckahoe in her last years and died here on August 7, 1911; and **Countée Cullen,** who made Tuckahoe his home toward the end of his life, when he taught in New York City. The poet died at Sydenham Hospital in New York City on January 9, 1946.

UTICA

Harold Frederic was born on August 19, 1856, at 324 South Street in this city in the Mohawk Valley, in central New York. The journalist and novelist grew up in Utica and worked as a teenager and young man for the Utica *Observer.* Frederic wrote his first short stories for Utica newspapers before leaving the city. He eventually traveled to Europe and to England, where he wrote all of his ten volumes of fiction, many of them set in the Mohawk Valley. His works included *Seth's Brother's Wife* (1887), *In the Valley* (1890), and *The Damnation of Theron Ware* (1896), considered to be Frederic's best novel. Frederic died on October 19, 1898, and his ashes were buried in the family plot in Forest Hill Cemetery, Utica.

VALHALLA

Margaret Widdemer, the poet and novelist, died on July 14, 1978, and was buried in Kensico Cemetery in Valhalla, a Westchester County community about thirty miles northeast of New York City.

VESTAL

David Ross Locke, the humorist who achieved national fame under the name **Petroleum V[esuvius] Nasby,** was born on September 20, 1833, in this community in south-central New York. Among several collections of his works were *The Nasby Papers* (1864) and *Nasby in Exile* (1882).

WARRENSBURG

In the summer of 1931, **Nathanael West** and a friend, Julian Shapiro, rented a cabin on Viele Pond, near this village northwest of Lake George, in which West worked on his novel *Miss Lonelyhearts* (1933).

WATERFORD

Howard Lindsay, the actor, director, producer, and playwright, was born on March 29, 1889, in Waterford, just north of Albany. Lindsay

is known best for the plays he wrote with Russel Crouse, including *Life with Father* (1939) and the Pulitzer Prize-winning *State of the Union* (1945).

WATERTOWN

Chard Powers Smith was born on November 1, 1894, in this city in northern New York. Smith, a poet, novelist, and biographer, wrote mostly about New England, his home for many years. His books included the verse collection *The Quest of Pan* (1930) and the novel *Turn of the Dial* (1943). He also wrote *The Housatonic* (1946), a volume in the Rivers of America series.

WELLSVILLE

Two best-selling novelists were born in the same house in this village in southwestern New York, about thirty miles east of Olean. **Grace Livingston Hill,** born on April 16, 1865, wrote nearly eighty novels, which altogether sold more than four million copies. Among her books were *A Chautauqua Idyl* (1887), *The Witness* (1917), and *The Enchanted Barn* (1918). Her birthplace at 241 North Main Street, now private and unmarked, was the rectory of the congregational ministry.

This house was also the birthplace of **Charles M[onroe] Sheldon,** born on February 26, 1857. A Congregationalist minister, Sheldon wrote a novel, *In His Steps* (1896), that became one of the all-time best sellers. *In His Steps* described the problems encountered by people in a Midwestern town when they tried to act in accordance with to the question, "What would Jesus do?" The novel sold well from the first, and when a copyright flaw was discovered, competing editions—in foreign languages as well as in English—caused sales to skyrocket. Even though sales reached tens of millions of copies, Sheldon received little or no royalties from most of the unauthorized editions of his work.

WESTFIELD

Charles Austin Fosdick, who wrote novels for boys under the name **Harry Castlemon,** lived in this village on Lake Erie in extreme western New York from 1875 to the early 1900s.

WEST PARK

John Burroughs in 1873 bought a nine-acre farm on the west bank of the Hudson River, across the river from Hyde Park, and a year later began construction of the stone house known as Riverby, which was to serve as his main home until his death in 1921. Burroughs's books and essays on nature brought him such popularity and so many visitors that he decided in 1895 to build Slabsides, a log cabin in the woods about one and a half miles west of Riverby. There he entertained many visitors but also wrote *Walt Whitman: A Study* (1896), *Ways of Nature* (1905), and other works. Among the writer's best-known guests in West Park were Henry Abbey, the poet; Elbert Hubbard; President Theodore Roosevelt; Walt Whitman; and Oscar Wilde. Toward the end of his life Burroughs spent more of his time at Woodchuck Lodge, his retreat near Roxbury, New York, but he maintained his residence in West Park. He died on March 29, 1921, while returning from California, and his funeral service was held in West Park. A marker on Route 9W directs travelers to Slabsides, off Burroughs Drive, open to visitors on the third Saturday in May and the first Saturday in October, at other times by appointment. Riverby, off Route 9W, about one hundred yards from the historical marker, is privately owned but is accessible to students who have obtained permission.

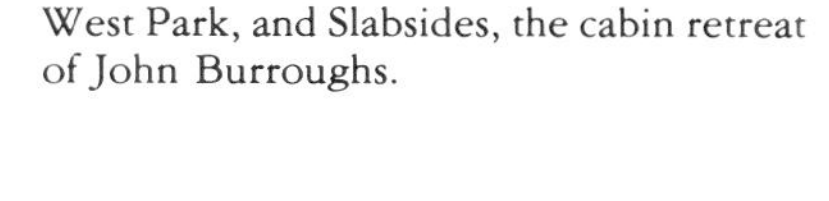
State historical marker off Route 9W in West Park, and Slabsides, the cabin retreat of John Burroughs.

So you see what I mean about Virtue? You have heard—or anyway you will—people talk about evil times or an evil generation. There are no such things. No epoch of history nor generation of human beings either ever was or will be big enough to hold the un-virtue of any given moment, any more than they could contain all the air of any given moment; all they can do is hope to be as little soiled as possible during their passage through it.

—William Faulkner, in *The Reivers*

POEMS

BY

EDGAR A. POE.

TOUT LE MONDE A RAISON.—ROCHEFOUCAULT.

SECOND EDITION.

New York:
PUBLISHED BY ELAM BLISS.
1831.

Title page of the 1831 edition of Edgar Allan Poe's *Poems,* known as the West Point edition.

WEST POINT

The United States Military Academy, located in West Point, fifty miles north of New York City, has a surprisingly large number of literary associations. **George Horatio Derby,** the humorist and author, was graduated from the academy in 1846. **Timothy Dwight,** the clergyman and poet, served as a chaplain here from 1777 until 1779. **William Faulkner** visited West Point in April 1962 and spoke to the cadets in the South Auditorium of Thayer Hall, where his reading of humorous passages from his recently published fictive reminiscence *The Reivers* (1962) gave great pleasure to the cadets. Since the time of Faulkner's brief visit to West Point, the academy library has gathered a major collection of writings by and about Faulkner, including a complete set of Faulkner first editions. **Charles King,** a prolific author of stories of adventure and army life, was graduated from West Point in 1866. Later, when he served as commandant of Michigan Military Academy, King became a guiding force in the life of a young cadet named Edgar Rice Burroughs. Among King's numerous books was the volume *Cadet Days: A Story of West Point* (1894). **James Gates Percival,** the poet, served as assistant surgeon and professor of chemistry here in 1824.

Another poet associated with the academy is **Edgar Allen Poe,** who was a cadet here from July 1830 to February 1831. The rigors of academy routine threatened to keep Poe from working on his poetry. When his foster father, John Allan, refused to send Poe enough money to remain at West Point, Poe neglected his Army duties, and the academy obliged Poe by dismissing him. Yet while he was at the academy, Poe impressed his classmates with his ability to compose satiric verse. Perhaps for this reason, they helped Poe raise a subscription list of 135 cadets, in support of publication of the second edition of his *Poems* (1831). When the copies of this book arrived two months after Poe's departure, many of the cadets must have been bewildered to find that the slim book contained serious poetry. The edition was dedicated by Poe "To the U.S. Corps of Cadets." In the 1960s, at a cost of more than $5,000, the academy purchased a copy of *Poems,* which had sold originally for $1.25. The library keeps *Poems* on display in the West Point Room, adjacent to a memorial to Poe.

Other writers associated with West Point include **Carl Sandburg,** who was a cadet briefly in 1899; and **Upton Sinclair,** who came to the village of West Point in 1897, spent three days on the academy grounds talking with cadets and gathering impressions of the academy. Sinclair later produced a series of dime novels set at the academy, featuring a cadet he named Mark Mallory. The first installment of these largely forgotten works appeared in the magazine *Army and Navy Weekly.* Among Sinclair's Mark Mallory stories, written under the pen name **Lieutenant Frederick Garrison, U.S.A.,** are *Off for West Point* (1903) and *A West Point Treasure* (1903). Sinclair also wrote a similar series set at the U.S. Naval Academy, in Annapolis, Maryland.

Gore Vidal was born in West Point on October 3, 1925. His father was then an instructor in aeronautics at the academy. **Susan Bogert Warner,** who wrote under the name **Elizabeth Wetherell,** lived from 1837 to 1885 on Constitution Island in the Hudson River, opposite West Point. Her best-known work was her first published novel, *The Wide, Wide World* (1850), which described the spiritual growth of an orphaned teenage girl. Her novel *Queechy* (1852) followed similar plot lines, as did her other stories. She wrote several books with her sister, **Anna Bartlett Warner,** a novelist and author of children's books. Susan Bogert Warner died on March 17, 1885, in Highland Falls, near West Point. Anna Bartlett Warner died there on January 22, 1915. Both are buried in the government cemetery at West Point. The Warner home on Constitution Island is open for tours during the summer. Colonel **C[harles] E[rskine] S[cott] Wood,** the soldier, poet, and writer, was graduated from West Point in 1874.

Poe Arch, above the entrance to the West Point Room in the library of the U.S. Military Academy.

WHITE PLAINS

Waldo Frank, the editor and writer, died on January 9, 1967, at the Miller Nursing Home in White Plains, a city north of New York City. **Margaret Halsey,** the novelist and essayist, lived at 113 Ralph Avenue in White Plains from 1950 until 1966. While here she wrote *The Folks at Home* (1952), *This Demi-Paradise: A Westchester Diary* (1960), and *The Pseudo-Ethic: A Speculation on American Politics and Morality* (1963). White Plains also figures in her book *No Laughing Matter: The Autobiography of a WASP* (1977). **Jean Stafford** died on March 26, 1979, at the Burke Rehabilitation Center, at 785 Mamaroneck Avenue.

WHITESBORO

Frances Miriam [Berry] Whitcher, whose humorous sketches were collected in *The Widow Bedott Papers* (1856) and *Widow Spriggins, Mary Elmer, and Other Sketches* (1867), was born on November 1, 1814, in this village near Utica. Her birthplace was a log cabin located just west of the village library, now the Dunham Public Library, at 78 Main Street. Mrs. Whitcher grew up in Whitesboro and spent most of her life here. In her last years she lived at 20 Park Avenue, in a house that is no longer standing and where she died on January 4, 1852. The house had earlier been the home of Hugo White, founder of the village, and a plaque at the site commemorates him. Another literary resident of Whitesboro was **Harold Bell Wright,** the clergyman and novelist, who lived here in the 1870s when he was an infant.

WOODBURY

Walt Whitman taught school during the summer of 1840 in this Long Island community southwest of Huntington. The schoolhouse in which he taught has been restored.

WOODSTOCK

Hart Crane spent the fall and winter of 1923–24 with friends in a rented house here, northwest of Kingston. The house in which they stayed was not far from the home of another poet, English-born **Richard Le Gallienne,** father of Eva Le Gallienne, the well-known actress. Here Crane worked on several poems, including "Recitative" and "Possessions," and among the guests at Thanksgiving dinner that year was John Dos Passos. Other writers associated with Woodstock include **Will** and **Ariel Durant,** who summered here from 1919 to 1928; and **Manuel Komroff,** who had a home here for a number of years before his death in 1974. Komroff is buried in Woodstock Artists' Cemetery, off Rock City Road.

YONKERS

This city, just north of New York City, has a number of noteworthy literary associations. **John Kendrick Bangs,** the humorist, editor, and novelist, was born here on May 27, 1862, and spent much of his life in Yonkers until 1907, when he moved to Maine. Bangs served on the staffs of various magazines before becoming, in 1904, editor of *Puck.* Many of his farcical stories were collected in *Tiddledywink Tales* (1891); *Three Weeks in Politics* (1894), based on his unsuccessful campaign for mayor of Yonkers; and *A House-Boat on the Styx* (1895). **Thomas Beer,** the novelist, short-story writer, and biographer, spent winters at his home at 227 Palisade Avenue for many years before his death in a New York City hotel, on April 18, 1940. Among Beer's books were a biography, *Stephen Crane* (1923); a novel, *Sandoval* (1924); an interpretation of fin-de-siècle America, *The Mauve Decade* (1926); and a collection of short stories, *Mrs. Egg and Other Barbarians* (1933).

Frederick S[wartout] Cozzens owned a summer home, Chestnut Cottage, in Yonkers from 1853 until his death in 1869. The humorist and essayist wrote of his home in *The Sparrowgrass Papers* (1856). **William Makepeace Thackeray** visited Cozzens in Yonkers during the English novelist's second visit to the United States in 1855–56.

Lawrence Ferlinghetti, the poet, was born in Yonkers, probably in 1919 or 1920—the exact date of his birth is not known. **Margaret Halsey,** born here on February 13, 1910, was the author of both fiction and nonfiction works, including *With Malice Toward Some* (1938), a satirical account of her brief stay in England; and *Some of My Best Friends Are Soldiers* (1944), a fictional study of racial intolerance in the United States. Halsey grew up in Yonkers, lived at 130 Lee Avenue, and attended Public School 21 and Yonkers High School. She wrote of her experiences in Yonkers in her book *No Laughing Matter: The Autobiography of a WASP* (1977). **E. D. E. N. Southworth,** a prolific novelist of the second half of the nineteenth century, is said to have lived in, or at least visited, the home of her son, a physician, at 103 Warburton Avenue.

Margaret Halsey, author of *With Malice Toward Some.*

YORKTOWN HEIGHTS

The Wiltwyck School, on Illington Road, near this community about forty miles north of New York City, was established to help troubled boys. The school was the setting for a well-known documentary film, *The Quiet One* (released in 1948), about the struggles of a black youth to deal with his childhood experiences. Narration for the film was written by James Agee. One of the school's best-known former students is **Claude Brown,** author of the best-selling autobiography *Manchild in the Promised Land* (1965). The Wiltwyck School was closed in 1981.

> *When I came out of school that first afternoon, Wiltwyck looked different from the way it had looked the day before. The sun was shining. It was real cold, and everything that wasn't moving around was covered with snow and ice. Some of the trees had little sticks stuck in them, with buckets and tin cans hanging on the end of the little sticks. Boys were sledding on a hill near the four houses that everyone lived in. Some were riding two on a sled, some were fighting over one sled, and some were crashing into trees, but everyone seemed to be having a whole lot of fun. The first thing that had caught my eye when I came out of the school building was the trees with the buckets on them. . . .*
>
> *Mr. Cooper told me they were maple trees and that they were being tapped for the sap to make maple syrup.*
>
> —Claude Brown,
> in *Manchild in the Promised Land*

NEW JERSEY

ALLIANCE

This community in southern New Jersey, just west of Vineland and not marked on current road maps, was founded as a Utopian community by the father of **George Seldes,** the journalist and author of such books as *You Can't Print That* (1929), *Sawdust Caesar* (1932), and *Tell the Truth and Run* (1953). Seldes was born in Alliance on September 10, 1890, according to his birth certificate; on November 16, 1890, according to his family. He attended school in Alliance and in nearby Vineland and lived in a house, no longer standing, on the east side of Henry Avenue, just north of Isaacs Avenue. His brother, **Gilbert [Vivian] Seldes,** was born in Alliance on January 3, 1893. Gilbert Seldes was a critic, scriptwriter, and author of many books, including *The Seven Lively Arts* (1924), a serious examination of motion pictures, comic strips, and other forms of popular art.

Photograph of Stephen Crane taken in January 1896, a few months after publication of *The Red Badge of Courage.*

ALPINE

Harold [Albert] Lamb, author of numerous historical books, including *Genghis Khan* (1927) and *The Plainsman* (1936), was born on September 1892, in this village on the Hudson River just north of New York City.

ARLINGTON

Will Durant, the historian, philosopher, and teacher, moved in 1898 with his family to this town northeast of Newark during his childhood. The Durants lived at 524 Forest Street. After living away from Arlington for a time, Durant returned in the summer of 1910 and taught school here until January 1912.

ASBURY PARK

Stephen Crane, in 1883, when he was eleven years old, moved with his family to this resort city on the Atlantic Ocean. His father, a Methodist minister, had died, and his mother's health was failing. Crane took great care not to upset his mother or give her cause to worry. Thus it happened that Crane failed to tell her that while riding on his pony one day he witnessed the murder of a young woman by her lover. Nor did he tell her that he once saved a companion from drowning in the surf—Crane and his friend had been swimming on Sunday, which would have grieved his religious mother. Crane had not yet begun his writing career, but he must have been greatly influenced by the members of his family: his father had been a writer as well as a minister, his mother wrote for several Methodist newspapers as well as for the New York *Tribune* and the Philadelphia *Press,* and his brother Townley was involved in newspaper work as well. In 1887 Crane left to attend the Hudson River Institute in Claverack, New York, but by 1888 returned to begin his writing career at Townley Crane's newspaper press bureau in Asbury Park. His writing at this time showed some of the qualities that would bring him to prominence, but was not without flaws. As Thomas Beer wrote in his biography of Crane, "No earthly criticism could or ever did make Stephen Crane respect an infinitive."

> *Pleasure seekers arrive by the avalanche. Hotel-proprietors are pelted with hailstorms of trunks and showers of valises. To protect themselves they do not put up umbrellas, not even prices. They merely smile copiously. The lot of the baggageman, however, is not an easy one. He manipulates these various storms and directs them. He is beginning to swear with a greater enthusiasm. It will be a fine season.*
>
> —Stephen Crane, in "Crowding into Asbury Park," New York *Tribune,* July 3, 1892

ATLANTIC CITY

This city is best known as a resort, but it also has a respectable literary history. **Kay Boyle,** the novelist and short-story writer, lived at 210 Seaside Avenue as a girl and, from 1913 to 1915, attended the Friends School, located in the upper story of a building, no longer standing, at 1216 Pacific Avenue. **F. Scott Fitzgerald** saw the premiere of his play *The Vegetable* presented in November 1923 at the Apollo Theater, which stood at 1509–1515 Boardwalk. In the opening-night audience was Fitzgerald's friend Ring Lardner. The play folded before reaching Broadway.

Lewis Mumford lived at 25 St. Catherine's Place in the summer of 1900, and at 108 Folsom Avenue in the spring and summer of 1907. **Morgan Robertson,** whose short stories about the sea were very popular at the beginning of the twentieth century, died in poverty in an Atlantic City hotel room on March 24, 1915. Robertson's stories, which had earned the respect of such authors as Joseph Conrad, had lost their popularity, and Robertson had fallen on hard times. **Leane Zugsmith,** the writer, spent most of her early years in Atlantic City.

AVON-BY-THE-SEA

In the summer of 1891 **Stephen Crane** came to this resort on the Atlantic Ocean, a few miles south of Long Branch, to report on a lecture by **Hamlin Garland.** Crane's article on Garland, which appeared in the New York *Tribune,* so impressed Garland that he asked to meet Crane, and the two writers met several times during Garland's stay here.

BARNEGAT LIGHT

This town at the northern end of Long Beach, a narrow, sandy coastal island about thirty miles north of Atlantic City, is the setting for the novel *The Tides of Barnegat* (1906) by **F. Hopkinson Smith,** the engineer, painter, and author. Smith was involved in the construction of the lighthouse here, now the main attraction at Barnegat Lighthouse State Park.

BELLEVILLE

T[homas] De Witt Talmage, the minister whose popular sermons were collected in such volumes as *Crumbs Swept Up* (1870) and *The Marriage Tie* (1890), served from 1856 to 1859 as minister of the Dutch Reformed Church in this town north of Newark. The church, a National Historic Landmark, stands at 171 Main Street.

BLOOMFIELD

Randolph [Silliman] Bourne, born on May 30, 1886, in this town northwest of Newark, was an essayist and author whose books included *Edu-*

cation and Living (1917) and the posthumous *Untimely Papers* (1919), which reflected Bourne's passionate opposition to American involvement in World War I. Bourne lived at 290 Belleville Avenue, no longer standing, and was graduated from Belleville High School. Deformed by an accident in infancy, Bourne overcame his handicap, and began a promising career in letters before his death in 1918 during the great flu epidemic.

BLOOMSBURY

Louis Adamic, the novelist and author of sociological studies, died here on September 4, 1951, and is buried in a cemetery in this town in northwestern New Jersey, a few miles southeast of Phillipsburg. Adamic lived in the nearby community of Holland.

BORDENTOWN

Thomas Paine, the author and pamphleteer, owned a home in this city just south of Trenton from 1783 until his death in 1809. He lived in a cottage believed to have stood at 2 West Church Street, off Farnsworth Avenue, from 1783 to 1789, when he traveled to France. By the time he returned to the United States in 1802, the publication of his *Age of Reason* (1794–95) had made him a virtual outcast because of its attack on religious institutions and doctrine. Paine spent much of his later life in New Rochelle, New York, and in New York City, where he died on June 8, 1809. A private house in Bordentown, on the site of Paine's cottage, bears a plaque and is identified by a marker on Farnsworth Avenue, but careful research indicates that it is a later structure, not Paine's cottage.

BOUND BROOK

Upton Sinclair died on November 25, 1968, at the Somerset Valley Nursing Home, on Route 22 in this borough about seven miles northwest of New Brunswick. **T[homas] De Witt Talmage** was born in Bound Brook on January 7, 1832. Talmage, a minister and author, served as editor of the *Christian Herald.* Many of his sermons appeared in syndication in newspapers across the country and later were published in book form.

BURLINGTON

James Fenimore Cooper, the first major American novelist, was born on September 15, 1789, in this city midway between Camden and Trenton. His birthplace at 457 High Street is now the headquarters of the Burlington County Historical Society and part of a complex of four buildings of historical significance. The Cooper birthplace is open to visitors on Sundays and Wednesdays. Donations are accepted. On exhibit in the house, the novelist's home for the first fourteen months of his life, are Cooper family furnishings and an elegantly illustrated edition of Cooper's Leather-Stocking Tales.

CAMDEN

This city across the Delaware River from Philadelphia was the birthplace, on December 19, 1858, of **Horace L[ogo] Traubel,** the editor and biographer known best for his five-volume work *With Walt Whitman in Camden* (1906–63), based on his friendship with **Walt Whitman** during the last years of the poet's life. Traubel also served as one of Whitman's literary executors. Traubel was fifteen years old at the time of Whitman's arrival in Camden in 1873, when Traubel's family lived at 509 Arch Street. Whitman, who had recently suffered a stroke, had come to Camden to be with his mother. She had been taken ill while visiting another son, George Whitman, at his home at 322 Stevens Street. Although Whitman had planned to remain in Camden only a short time, he chose, after his mother's death in Camden, to stay on with George and his wife, and in 1874 moved with them to 431 Stevens Street. Whitman wrote his poem "Song of the Redwood Tree" during this period. Whitman lived with his brother and sister-in-law until they decided to move to a farm in Burlington, New Jersey. The

> *I believe in one God, and no more; and I hope for happiness beyond this life.*
>
> *I believe in the equality of man; and I believe that religious duties consist in doing justice, loving mercy, and endeavoring to make our fellow-creatures happy.*
>
> *But, lest it should appear that I believe in many other things in addition to these, I shall, in the progress of this work, declare the things I do not believe. . . .*
>
> *All national institutions of churches, whether Jewish, Christian, or Turkish, appear to me no other than human inventions, set up to terrify and enslave mankind, and monopolize power and profit.*
>
> —Thomas Paine, in *Age of Reason*

(*below left*) House on the site of Thomas Paine's cottage in Bordentown. Although often referred to as Paine's cottage, this building now is considered to be a later structure. (*below*) A postcard view of the birthplace of James Fenimore Cooper, in Burlington. In the photograph, Cooper's house stands to the left of the birthplace of Captain James Lawrence, to whom is attributed the memorable order: "Don't give up the ship."

I claim to be altogether radical—that's my chief stock in trade: take the radicalism out of the Leaves—do you think anything worthwhile would be left?

—Walt Whitman,
quoted by Horace L. Traubel
in *With Walt Whitman in Camden*

Photograph of Walt Whitman, and the house at 330 Mickle Street, Camden, where the poet spent the last decade of his life.

publication in 1882 of Whitman's revised *Leaves of Grass* gave the poet enough money to buy the house at 330, formerly 328, Mickle Street, now a National Historic Landmark. Here, in this simple two-story clapboard house, Whitman spent his declining years working on additions and revisions to a new edition of *Leaves of Grass* (1889) and preparing his final volume of poems and prose, *Good-Bye, My Fancy* (1891). Whitman died at his Mickle Street home on March 26, 1892. The Walt Whitman House is open to visitors.

Whitman is buried in a tomb he designed and had built on a lot in Harleigh Cemetery. The vault, of unpolished Quincy granite, is also the resting place of many members of the poet's family. Rutgers University, whose campus is a short walk from Whitman's house, is the site of an annual two-day Walt Whitman Festival, which includes poetry and essay contests, and features a visit each year by a well-known author.

CAPE MAY POINT

Mary O'Hara, born on July 10, 1885, in this community at the southern tip of New Jersey, was a novelist and composer, and is best remembered for her novels about horses: *My Friend Flicka* (1941), *Thunderhead* (1943), and *The Green Grass of Wyoming* (1946).

CLIFTON

Stephen Crane lived off and on in the early 1890s at his brother's house on Cook Avenue in the community of Lake View, now part of Clifton, a city adjacent to Paterson. During this time Crane was gathering impressions in New York's Bowery district that he incorporated in *Maggie: A Girl of the Streets* (1893), and here, in March 1892, Crane completed the third draft of the story of Maggie Johnson, a girl who was seduced, abandoned, forced into prostitution, and driven to suicide.

Crane family monument in Evergreen Cemetery in Hillside. The author's epitaph is brief:
STEPHEN CRANE
POET—AUTHOR
1871–1900

CONVENT STATION

Frank R. Stockton, after retiring in 1881 from his editorial position on *St. Nicholas,* moved to a gray clapboard house on the south side of Route 24, near this town fifteen miles west of Newark. While living in Convent Station, Stockton saw the publication of many of his best-known books, including the novel *The Casting Away of Mrs. Lecks and Mrs. Aleshine* (1886) and its sequel, *The Dusantes* (1888); and *The Rudder Grangers Abroad* (1891) and *Pomona's Travels* (1894), both sequels to Stockton's novel *Rudder Grange* (1879). Stockton moved to West Virginia in about 1899.

EAST ORANGE

Thomas Caldecott Chubb, born on November 1, 1899, in this city in northeastern New Jersey, was a poet and biographer whose collections of verse included *The White God and Other Poems* (1920) and *Cornucopia: Poems 1919–53* (1953). He also wrote *The Life of Giovanni Boccaccio* (1930) and *Dante and His World* (1967). Other East Orange natives include **Robert Hillyer,** born on June 2, 1895, whose *Collected Verse* (1933) won a Pulitzer Prize; and **Frank Shay,** the editor, writer, publisher, and bookseller, born here on April 8, 1888. Shay established Frank Shay's Traveling Bookshop, helped to found the Provincetown (Massachusetts) Players, and wrote such books as *My Pious Friends and Drunken Companions* (1927).

ELIZABETH

Writers associated with this city just south of Newark include **Donald B[arr] Chidsey,** the novelist and biographer, who was born in Elizabeth on May 14, 1902. Among his many books were the biography *Bonnie Prince Charlie* (1928) and the history *Valley Forge* (1959). **Frank Morrison**

Spillane, known to millions as **Mickey Spillane,** the author of very popular novels, was brought to live in Elizabeth in 1918, and grew up here. He moved to New York after high school, but has returned to attend class reunions. Although Spillane does not use Elizabeth as a setting for his fiction, he has used classmates' names for characters in his books. **Edward Stratemeyer,** creator of the Rover Boys, Tom Swift, and other enormously popular adventure series for boys, was born in Elizabeth on October 4, 1862, and grew up here. While working in a tobacco shop run by his stepbrother, he achieved his first literary success, in 1888, with the sale of a story to a Philadelphia weekly. After 1890, Stratemeyer made his home in Newark.

ENGLEWOOD

Writers associated with this city in northeastern New Jersey include **John Clellon Holmes,** the novelist, who lived at several addresses here as a boy; and **Corliss Lamont,** born in Englewood on March 28, 1902. Lamont's books, such as *You Might Like Socialism* (1939) and *Challenge to McCarthy* (1954), reflect his political views. **Anne Morrow Lindbergh,** the diarist, poet, and author, was born in Englewood on June 22, 1906. She grew up in Englewood, but it was in Mexico, where her father was serving as U.S. ambassador, that she first met her future husband, the aviator Charles A. Lindbergh. The two were married in Englewood on May 27, 1929. A selection of the author's diaries and letters from the years 1922 to 1928 was published as *Bring Me a Unicorn* (1972). Mrs. Lindbergh's other books include *North to the Orient* (1935), an account of her adventures in flying; *Gift from the Sea* (1955), a book of musings on modern life and its problems; and the selections of her diaries and letters, *Hour of Gold, Hour of Lead* (1973) and *The Flower and the Nettle* (1976). The Morrow family estate, at 480 Next Day Hill Drive, now is the home of the Elizabeth Morrow School.

Anne Douglas Sedgwick, born in Englewood on March 28, 1873, was a novelist who spent most of her life in England, where she was taken to live at the age of nine. Among her novels were *Tante* (1911), about a brilliant concert pianist and her relationship with her protegée, who idolized her; and *Adrienne Toner* (1922), about an American girl and her influence on the lives of the members of the English family into which she married.

In 1906 **Upton Sinclair** established Helicon Hall, an experiment in communal living, which lasted only until March of 1907. With the royalties from his novel *The Jungle* (1906), Sinclair bought a former boy's school, which stood near the Palisades overlooking Englewood. About forty people, mostly young writers and their families, sought to create a community based on equality. Among the members was **Sinclair Lewis,** who worked as a janitor there. The noble experiment came to an abrupt end early in the morning of March 7, 1907, when Helicon Hall was destroyed by fire.

FORT LEE

This town in Bergen County, at the western end of the George Washington Bridge, which connects New York and New Jersey, was named after the fort built during the Revolutionary War to keep British troops from sailing up the Hudson River to West Point. **Thomas Dunn English,** the physician and writer best known for his poem "Ben Bolt," lived in Fort Lee from 1856 to 1878. English served for a time as justice of the peace and, from 1863 to 1864, as a state legislator. Of his many plays, only *The Mormons, or Life at Salt Lake* (1858) was published. His home at 511 Main Street is used now for a group of shops.

FREEHOLD

Philip Freneau, spent his last years on a farm about two miles from this borough in east-central New Jersey. On December 19, 1832, the eighty-year-old Freneau, while walking home from Freehold during a blizzard, fell into a bog, broke his leg, and died of exposure. He was buried on a farm he had previously run near the community of Freneau.

FRENCHTOWN

In early 1938 **James Agee** moved to a farm near Frenchtown, a community on the Delaware River about fifteen miles southeast of Phillipsburg. Here he worked on *Let Us Now Praise Famous Men* (1941), his classic account, illustrated by the photographs of Walker Evans, of the lives of Alabama tenant farmers. By May 1938 he had moved to a house on Second Street in Frenchtown, where he brought his manuscript nearly to completion before moving to Brooklyn in the spring of 1939. **Nathanael West** spent six weeks in October and November of 1932 at the Warford House Hotel, where he completed his novel *Miss Lonelyhearts* (1933). West used the Warford House as a setting in his novel *A Cool Million* (1934). **William L. White,** the author and editor, owned a 100-acre farm near Frenchtown. He divided his time between the farm, his home in New York City, and the family home in Emporia, Kansas.

FRENEAU

This community, just south of Matawan in east-central New Jersey, is named after **Philip Freneau,** who lived from the 1790s until about 1818 at Mount Pleasant, his estate here. From Mount Pleasant Freneau edited New Jersey's first newspaper, the *Jersey Chronicle.* In his later years Freneau spent his time farming, editing, writing, and sailing as a sea captain. In 1818 his house was destroyed by fire. Although it was rebuilt, Freneau did not live in it again, preferring to spend his last years on a farm near Freehold, where he died in 1832. Freneau is buried on a tract of his estate now owned by the borough of Matawan. Located on Poet's Drive, off Route 79, Freneau's gravesite is open to visitors at all times. The author's rebuilt house, at 12 Poet's Drive, is private.

GLEN RIDGE

William Hazlett Upson, born on September 26, 1891, in this borough northwest of Newark, was a novelist and short-story writer known especially for his books about Alexander Botts, the

> *He was very, very young and was terribly shy—looked straight ahead and talked in short direct sentences which came out abruptly and clipped. You could not meet his sentences: they were statements of fact, presented with such honest directness; not trying to please, just bare simple answers and statements, not trying to help a conversation along. It was amazing—breath-taking. I could not speak. What kind of boy was this?*
>
> —Anne Morrow Lindbergh, in *Bring Me a Unicorn,* writing of Charles A. Lindbergh

> *She went into the blackness of the final block. The shutters of the tall buildings were closed like grim lips. The structures seemed to have eyes that looked over them, beyond them, at other things. Afar off the lights of the avenues glittered as if from an impossible distance. Streetcar bells jingled with a sound of merriment.*
>
> *At the feet of the tall buildings appeared the deathly black hue of the river. Some hidden factory sent up a yellow glare, that lit for a moment the waters lapping oilily against timbers. The varied sounds of life, made joyous by distance and seeming unapproachableness, came faintly and died away to a silence.*
>
> —Stephen Crane, in *Maggie: A Girl of the Streets*

> *. . . And now, Mrs. Stanley, I have a few small matters to take up with you. Since this corner druggist at my elbow tells me that I shall be confined in this mouldy mortuary for at least another ten days, due entirely to your stupidity and negligence, I shall have to carry on my activities as best I can. I shall require the exclusive use of this room, as well as that drafty sewer which you call the library. I want no one to come in or out while I am in this room.*
>
> —Sheridan Whiteside, speaking in *The Man Who Came to Dinner,* by George S. Kaufman and Moss Hart

tractor salesman whose exploits delighted readers of such volumes as *Alexander Botts, Earthworm Tractors* (1929), *No Rest for Botts* (1951), and *The Best of Botts* (1961). Upson's birthplace at 61 Douglas Road is private, but the Glen Ridge Public Library collects material about Upson.

HACKENSACK

From 1911 to 1913 **F. Scott Fitzgerald** attended the Newman School, a small Catholic preparatory school, no longer in existence, in this city in northeastern New Jersey. While a student here, Fitzgerald met Father Sigourney Webster Fay, a trustee and later headmaster of the school, who became a great influence in the life of the novelist-to-be. Fitzgerald dedicated his novel *This Side of Paradise* (1920) to Father Fay, who was the model for Monsignor Darcy in the novel. The school appears as St. Regis in the novel, and Fitzgerald was writing of his own experiences when he described the main character of the novel, Amory Blaine, in his two-year stay at St. Regis:

> *He went all wrong at the start, was generally considered both conceited and arrogant, and universally detested. He played football intensely, alternating a reckless brilliancy with a tendency to keep as safe from hazard as decency would permit. . . . He was resentful against all those in authority over him, and this, combined with a lazy indifference toward his work, exasperated every master in the school.*

HAMPTON

Glenway Wescott, from the late 1930s to the late 1950s, owned a home, Stone-blossom, near this town in northwestern New Jersey about ten miles east of Phillipsburg. During this time Wescott saw the publication of his novels *The Pilgrim Hawk* (1940), set in Paris in the 1920s; and *Apartment in Athens* (1945), set in Greece during World War II.

HARRISON

A[loysius] M[ichael] Sullivan, the poet whose collections included *This Day and Age* (1944) and *Incident in Silver* (1950), was born on August 9, 1896, in this town on the Passaic River, opposite Newark. The poet's birthplace was on Fourth Street, but the address is unknown. Sullivan also lived for a time in a house at 27 Ogden Street.

HILLSIDE

Stephen Crane is buried in the family plot at Evergreen Cemetery, on Broad Street in this community two miles north of Elizabeth. Crane, who suffered from tuberculosis, died in Germany on June 5, 1900, at the age of twenty-nine.

HOBOKEN

William Heyliger, born in this city on March 22, 1884, wrote such books for boys as the adventure story *High Benton* (1919), which was popular for many years; and *Boys Who Became President* (1932). Heyliger made his home in the early 1900s at 803 Willow Avenue, in a house that is now private.

HOLMDEL

Alexander [Humphreys] Woollcott, the drama critic and essayist, was born on January 19, 1887, near this town in east-central New Jersey, about fifteen miles northeast of Long Branch. Woollcott was born in an eighty-five-room structure that had been used by an experimental agricultural community known as the North American Phalanx, which flourished in the area from 1843 to about 1854. The author's birthplace, located off what is now Route 35 south of Holmdel, was destroyed by fire in 1972. Woollcott, a member of the New York literary group known as the Algonquin Round Table, was an anthologist, a radio essayist, and raconteur. He was also the model for Sheridan Whiteside, the insulter *par excellence* of the play *The Man Who Came to Dinner* (1939), by George S. Kaufman and Moss Hart. Among Woollcott's books were *Enchanted Aisles* (1924), *Two Gentlemen and a Lady* (1928), and the anthology *The Woollcott Reader* (1935).

HOPEWELL

Anne Morrow Lindbergh and her husband Charles A. Lindbergh made their first home as a married couple on a 400-acre estate known as Highfields, on Amwell Road near this borough about fifteen miles north of Trenton. In March 1932, their first child, Charles, Jr., was kidnaped from their house and later found dead. The kidnaping and subsequent trial and execution of Bruno Hauptman, who was convicted of the crime, held the attention of the entire nation for a long time and are still remembered today. The grisly affair spawned its own body of journalistic articles and books. The Lindberghs, devastated by their loss, left Highfields for an extended stay in Europe. Highfields is now used as a rehabilitation center for troubled teenagers.

JERSEY CITY

From 1901 to 1907 **Will Durant** attended St. Peter's College—it was then a combination

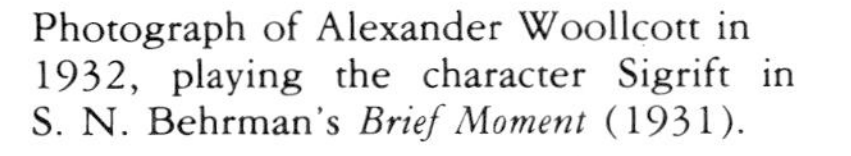

Photograph of Alexander Woollcott in 1932, playing the character Sigrift in S. N. Behrman's *Brief Moment* (1931).

high school and college—still open at 2641 Kennedy Boulevard. While a student here, Durant jeopardized his chances for graduation by writing a letter to a local newspaper in support of the socialist movement. The incident was overlooked, and Durant was permitted to graduate. In 1979 the college conferred on its respected alumnus an honorary degree as well. **Earl Reed Silvers,** author of such books for boys as *Dick Arnold of Raritan College* (1920) and *The Hillsdale High Champions* (1925), was born at 96 Maple Street on February 22, 1891. **Irving Stone,** the biographer and novelist, in the late 1920s directed the Little Theater of the Jewish Community Center, at 604 Bergen Avenue.

KEARNY

Will Durant lived in the late 1890s with his family at 524 Forest Street, in a house still standing in this town north of Newark. Durant attended public school and the school run by St. Cecilia's Church.

KINGSTON

While living in this town about fifteen miles northeast of Trenton, **Francis Fergusson,** the literary critic, wrote several books, including *Dante* (1966) and *Literary Landmarks* (1976).

LAKEWOOD

This community, on Route 9 in east-central New Jersey, became in the late nineteenth century a major resort and vacation center because of its salubrious climate and beautiful setting, as well as its location, just forty-five miles from New York City. One of its famous hotels was the Laurel House, built in 1879 amid thousands of acres of pine woods. Among its guests were **Oliver Wendell Holmes, Sr., Rudyard Kipling,** and **Mark Twain.**

LAMBERTVILLE

James Gould Cozzens, the novelist, lived from 1932 to 1957 on a 124-acre farm near this city about fifteen miles northwest of Trenton. Except for military service during World War II, Cozzens spent most of his time during these years on his farm. Among his highly regarded books are *The Just and the Unjust* (1942), the story of lawyers pursuing their profession in a small town; *Guard of Honor* (1948), Cozzens's Pulitzer Prize-winning novel set at an air base during World War II; and *By Love Possessed* (1957), the story of two turbulent days in the life of a lawyer.

LAWRENCEVILLE

This town a few miles north of Trenton is the home of Lawrenceville School, a preparatory school that is associated with the lives of three well-known writers. **Edward Albee,** the playwright, attended Lawrenceville School in the early 1940s until his failure to attend classes brought about his expulsion. While a student here, Albee wrote a three-act farce entitled *Aliqueen,* never produced professionally.

Owen Johnson, the novelist and short-story writer, attended the school in the 1890s, and it was he who put Lawrenceville on the literary map. On his graduation in 1895, Johnson was allowed to remain at the school for another year so that he could establish the school literary magazine, the *Lit.* Johnson later used the school as the setting for a series of humorous stories about the prep-school adventures of Dink Stover and his friends. These stories, which mention places in Lawrenceville still frequented by students, were collected in such volumes as *The Eternal Boy* (1909), later retitled *The Prodigious Hickey; The Humming Bird* (1910), *Skippy Bedelle* (1923), *The Varmint* (1910), and *The Tennessee Shad* (1911). The Lawrenceville stories, enormously popular for decades, may be read now in an omnibus volume comprising *The Prodigious Hickey, The Varmint,* and *The Tennessee Shad.*

In the 1920s, just as Johnson's stories reached their peak in popularity, **Thornton Wilder** was teaching elementary French at the Lawrenceville School. Wilder taught here until 1925 and, after a period of graduate study at nearby Princeton, returned in 1927 to teach for one more year. Wilder served as assistant master at Davis House, the boys' residence about half a mile from the main campus, on the Lawrenceville Road. At Davis House, now privately owned, Wilder wrote much of his first novel, *The Cabala* (1926), and learned to dislike intensely the Lawrenceville headmaster, whom he used as the model for Breckenridge Lansing in the novel *The Eighth Day* (1967). Wilder wrote of Davis House in his one-act play "The Happy Journey to Trenton and Camden," which was collected in *The Long Christmas Dinner and Other Plays in One Act* (1931).

> *There was about young Stover, when properly washed, a certain air of cherubim that instantly struck the observer; his tousled tow hair had a cathedral tone, his cheek was guileless, and his big blue eyes had an upward cast toward the angels which, as in the present moment when he was industriously transferring a check labeled Baltimore to a trunk bound for Jersey City, was absolutely convincing. But from the limit whence the cherub continueth not, the imp began.*
>
> —Owen Johnson, in *The Varmint*

LONG BRANCH

This city on the Atlantic Ocean, about thirty miles south of New York City, was the birthplace

Davis House, formerly used as a residence hall at Lawrenceville School. Thornton Wilder began his career as a novelist while living here in the 1920s.

Photograph of Bret Harte taken about 1880, when he was serving as U.S. consul at Crefeld, Prussia. Harte summered in Morristown in the 1870s, but lived in Europe after 1878.

THOMAS PAINE
1737 1809
ENGLISH BY BIRTH
FRENCH CITIZEN BY DECREE
AMERICAN BY ADOPTION
Author of
Common Sense
The American Crisis
Rights of Man
The Age of Reason
. . .
YOUR PRESENCE MAY REMIND
CONGRESS (AND THE PEOPLE)
OF YOUR PAST SERVICE
—TO THIS COUNTRY—

GEORGE WASHINGTON

—Main inscription on the Thomas Paine Monument in Morristown

of two writers. **Waldo Frank,** born in Long Branch on August 25, 1899, was a founder and editor of *The Seven Arts,* a monthly magazine that was published for only one year in 1916–17. Frank's books included the novel *The Invaders* (1948) and *The Re-Discovery of America* (1928), *In The American Jungle* (1937), and *Chart for Rough Waters* (1940), nonfiction works dealing with life in America.

Norman Mailer was born in Long Branch on January 31, 1923. Mailer in recent years has concentrated less on writing fiction and more on using factual material to produce novelistic works that transcend fact, but it was a novel and Mailer's first one, at that, that brought him to the front ranks of American authors. *The Naked and the Dead* (1948), a gripping account of an infantry platoon in combat in the Pacific during World War II, was the first success of the post-World War II generation of writers. *The Armies of the Night* (1968) was a third-person account of Mailer's experiences during four days of antiwar demonstrations in Washington, D.C., in October 1967. *The Executioner's Song* (1979) is a novel based on the life of a convicted murderer named Gary Gilmore, whom Mailer interviewed at great length before Gilmore's execution.

LYNDHURST

William Carlos Williams, the poet, novelist, and short-story writer, died on March 4, 1963, at his home in Rutherford, but is buried in Hillside Cemetery, on Rutherford Avenue, in Lyndhurst.

MAHWAH

Joyce Kilmer, the poet, lived in this city in northern New Jersey for a time in the early twentieth century. **Harvey Swados** used his experiences as a metal finisher working on the assembly line at the former Ford Motor Company plant here in writing his novel *On the Line* (1957).

MANASQUAN

Robert Louis Stevenson in early 1888 visited this town on the coast, about fifteen miles south of Long Branch. During this time his wife Fanny was on the West Coast, trying to find a yacht in which she and her husband could sail to the South Seas. Stevenson had brought with him the unfinished manuscript of *The Master of Ballantrae* (1889), but did not work on it much here, preferring instead to spend his days sailing a catboat on the Manasquan River. Charlotte Eaton, the wife of a friend of Stevenson's, wrote of the author's time here in *Stevenson at Manasquan* (1921).

MAPLEWOOD

Margaret E[lizabeth] Sangster, the magazine editor and poet, spent the end of her life in this township west of Newark. Although she had become blind in her later years, she maintained her daily writing schedule by dictating to secretaries. The author died in Maplewood on June 4, 1912. **Agnes Sligh Turnbull** lived here for sixty years. She died at St. Barnabas Medical Center in Livingston, New Jersey, on January 31, 1982. Her first story appeared in *The American Magazine* in 1920. Her last novel was *The Two Bishops* (1980), her first *The Rolling Years* (1936).

METUCHEN

John Ciardi, the poet, lived and wrote for a time at 31 Graham Avenue in this borough west of Perth Amboy.

MILLVILLE

[Lloyd] Logan Pearsall Smith, the essayist and philologist, was born on October 18, 1865, in this city in southwestern New Jersey. His birthplace at 223 North Second Street is now private. Smith was known best for his collections of essays, including *Trivia* (1902, revised edition 1918) and *Afterthoughts* (1931).

MONTCLAIR

Richard Wilbur, the poet, from 1935 to 1938 attended Montclair High School, at 100 Chestnut street in Montclair, a town northwest of Newark.

MORRISTOWN

This town northwest of Newark has a number of noteworthy literary associations. **Bret Harte** lived in Morristown, mostly in summer, for several years before his departure for Europe in 1878. Among the numerous houses he lived in were Fosterfields, known at the time as The Willows, at Route 24 and Kahdena Road, now private; the Watnong House, no longer standing, at Sussex Avenue and Ketch Road; and the Randolph House, at 18 Elm Street, now remodeled for commercial use. Harte also lived in a house at Western Avenue and Anne Street, now an empty lot, and in a boarding house, site unknown, run by Harte's nieces, Anna and Nina Knauft. One of Harte's Morristown residences was the setting for his short story "Thankful Blossom."

Joyce Kilmer taught Latin at Morristown High School during the 1908–09 school year. Thomas Nast, one of this country's best-known political cartoonists, owned from 1873 to 1902 the Victorian house, now private, at 50 MacCulloch Avenue. Among the many visitors to the Nast home, known as Villa Fontana, were **George Washington Cable** and **Mark Twain,** who gave readings in Morristown in November 1884 and stayed at Nast's house. One morning during their visit it was discovered that all the clocks in the house had stopped. Twain had stopped them during the night because their ticking and chiming had kept him awake.

Burnham Park, on Route 24 in Morristown, contains a heroic bronze statute of **Thomas Paine,** the revolutionary pamphleteer whose writings helped to keep the Continental Army together through heartbreaking years of war and deprivation. George Washington twice used Morristown as winter headquarters for his army, and Paine may have stayed in Morristown during Washington's winter encampment here in 1777. Washington, surveying the damage to his army not only from British arms but from dire shortages of food, clothing, and shelter, well understood Paine's words: "These are the times that try men's souls." The statue, said to be the first statue of Paine to be dedicated in this country, bears quotations from a number of well-known revolutionary figures, including Washington, Lafayette, John Adams, and, of course, Thomas Paine.

Other writers associated with Morristown include **Dorothy Parker,** who attended Miss Dana's Seminary for Young Ladies here in 1911 and 1912. The school was housed in a Victorian mansion, no longer standing, at 255 South Street, now the site of a bank. **John Reed,** the journalist, attended the Morristown School from 1904 to 1906 while preparing for Harvard. The frail and sickly Reed participated in sports to build himself up, wrote stories and poems for the school newspaper, *The Morristonian,* and engaged in numerous practical jokes. Once, when visitors were expected at the school, Reed succeeded in scandalizing his schoolmasters and their guests by placing a chamberpot on top of a suit of armor.

NEWARK

Newark counts among its native sons one of this country's finest writers, **Stephen Crane.** The son of a Methodist minister who wrote on controversial religious subjects, Crane was born on November 1, 1871, in the Methodist parsonage at 14 Mulberry Place. Several attempts were made over the years to save Crane's birthplace, but all failed for one reason or another, and the crumbling structure was torn down about 1941 and replaced by a brick-walled, concrete-floored playground that attempted to recall Crane in decorative features. Thirty-six ceramic plaques depicted scenes from the author's stories, Crane's portrait was executed in ceramic, and a bronze tablet identified the site as his birthplace. The work of vandals has taken its toll, and many of the ceramic plaques have been damaged. The park property was recently purchased by a private company, and the area fenced in. Only by peering through the fence can visitors read the plaque bearing the now ironic words:

COULD CRANE REVISIT THIS SCENE
OF HIS BIRTH
HE WOULD REJOICE TO BEHOLD
AT HIS EARLIEST HOME
NO POMP OF MONUMENTAL STONE,
BUT THIS HAPPY AND HUMAN RENDEZVOUS
FOR CHILDREN WHOSE LAUGHTER ILLUMINES
A WORLD HE FOUND DUBIOUS
BUT NOT WITHOUT OCCASION FOR
COURAGE AND FAITH.

Newark's native writers also include **Leslie A. Fiedler,** the critic, who was born here on March 8, 1917, and lived in a house at 60 Wolcott Terrace, over the site of which Route 78 now runs; **Allen Ginsberg,** the poet, born on June 3, 1926; and **[Everett] LeRoi Jones,** the playwright who now uses the name **Imamu Amiri Baraka,** who was born in Newark on October 7, 1934. Baraka lived at 72 Boston Street as a child. In the 1960s, he returned to the city of his birth and became director of a black theater and community center known as Spirit House, at 33 Stirling Avenue, no longer standing. He also became active, as a black leader, in the economic and political life of the city.

Philip Roth, whose novels include *Portnoy's Complaint* (1969) and *The Ghost Writer* (1979), was born in Newark on March 19, 1933. Roth received the National Book Award for his first book, *Goodbye, Columbus* (1959), which included the title novella and five short stories. Roth lived in Newark from 1933 to 1950, first at 81 Summit Avenue and later at 385 Leslie Street.

Dore Schary, the playwright, screenwriter, and film producer, was born on August 31, 1905, in a house at 17 Charlton Street. Schary attended the Thirteenth Avenue Grammar School, at 359 Thirteenth Avenue, from 1910 to 1914; the Morton Street Grammar School, at 75 Morton Street, from 1914 to 1918; and Central High School, at 345 High Street, from 1923 to 1925. While in Newark Schary did newspaper work, and wrote plays, free verse, and sketches. Schary was involved in the production of over 350 motion pictures during his career and wrote more than forty screenplays, including the Academy Award-winning screenplays for *Young Tom Edison, Sunrise at Campobello,* and *Act One,* based on his autobiographical work of that name.

Antoinette [Quinby] Scudder, the poet and playwright, whose works included the collection *Provincetown Sonnets and Other Poems* (1925) and the play *The Grey Studio* (1934), was born in Newark on September 11, 1898, and lived in a house, no longer standing, at 10 Centre Street. **Albert Payson Terhune,** best known for such canine stories as *Lad: A Dog* (1919), was born on December 21, 1872, in the parsonage of his father's church. The Terhune family, including the novelist's mother, **Mary Virginia Terhune,** herself a novelist, lived at 476 High Street and 4 Park Church Place.

Writers who have lived in Newark include **Coningsby Dawson,** the English poet, novelist, and newspaper correspondent, who lived here from some time after 1911 until 1940, when he left for California. From 1920 to 1940 he lived at 533 Mt. Prospect Avenue. Dawson's works written here included the novel *The Coast of Folly* (1924); *The Unknown Soldier* (1929), a novelette dealing with the return to earth of Jesus during

Photograph of Dore Schary, for decades one of Hollywood's prolific screenwriters and producers.

Cockloft Hall is the country residence of the family, or rather the paternal mansion. . . . It is pleasantly situated on the banks of a sweet pastoral stream; not so near town as to invite an inundation of idle acquaintance, who come to lounge away an afternoon, nor so distant as to render it an absolute deed of charity or friendship to perform the journey. It is one of the oldest habitations in the country, and was built by my cousin Christopher's grandfather, who was also mine by the mother's side, in his latter days, to form, as the old gentleman expressed himself, "a snug retreat, where he meant to sit himself down in his old days, and be comfortable for the rest of his life." He was at this time a few years over fourscore . . . .

—Launcelot Langstaff, Esq., in *Salmagundi*

World War I; and the novel *The Moon Through Glass* (1934). **Mary Mapes Dodge,** the editor and author of books for children, moved to Newark in 1858, following the death of her husband. She and her two sons lived in her father's house, which stood at what has since become the junction of Elizabeth and Renner avenues, and is now the site of an apartment building. Mrs. Dodge fixed up a deserted farmhouse near her father's house, and there began to write. **Will Durant** lived in a rented room at 28 Hill Street, no longer standing, in 1912 and 1913, while teaching at the Ferrer School in Newark. He often ate vegetarian meals at Macfadden's Physical Culture Cafeteria, which stood at 54 Clinton Street.

Thomas Dunn English, the physician, public official, and author, lived at 57 State Street from 1878 until his death on April 1, 1902. During his years in Newark, English wrote *The Select Poems of Dr. Thomas Dunn English* (1894) and *Fairy Stories and Wonder Tales* (1897). **Howard R. Garis,** the creator of the children's-story character Uncle Wiggily, worked for the Newark *Evening News* from 1896 to 1908 and 1935 to 1947. His first Uncle Wiggily story was published in the *News* in 1910, and Garis wrote many thousands in the years following. It is believed he coined the term "bedtime stories" to describe these tales for young people. Garis lived at various places in Newark, but only 12 Myrtle Avenue, his home from 1905 to 1920, is known today.

Henry William Herbert, the historical novelist and sports and nature writer, in 1845 built a Tudor cottage he called The Cedars, no longer standing, on property now used by Mount Pleasant Cemetery, 375 Broadway. Herbert, who preferred to be identified as a historical novelist, adopted the pen name **Frank Forester** when he began to write articles and sketches about sports and nature. While he lived in Newark, a collection of these sketches was published as *The Warwick Woodlands* (1845), with a setting in Orange County, New York. His novels *My Shooting Box* (1846) and *The Deerstalkers* (1849), and his *Complete Manual for Young Sportsmen* (1856) were also published during this time. After the failure of a brief marriage, Herbert fell into the last of his periodic depressions and killed himself in New York City on May 17, 1858. He is buried in Mount Pleasant Cemetery.

Washington Irving, as a young man, spent many weekends at Mount Pleasant, the home of Gouverneur Kemble, which stood on the 200 block of Mount Pleasant Avenue, at the northeast corner of Gouverneur Street. Here Irving, his brother **William Irving, James Kirke Paulding,** and others discussed plans for their humorous magazine *Salmagundi; or, The Whim-Whams and Opinions of Launcelot Langstaff, Esq. and Others,* which had twenty issues in 1807 and 1808 and was collected in book form in 1808. Mount Pleasant is referred to in *Salmagundi* as Cockloft Hall, a name devised by Washington Irving.

Edward Stratemeyer, creator and author of the Rover Boys, Tom Swift, and other series of books for boys, lived in Newark from 1890 until his death on May 10, 1930, at his home at 171 North Seventh Street. Stratemeyer organized the Stratemeyer Literary Syndicate, which employed writers to produce books following plots written by Stratemeyer. Stratemeyer himself wrote the Rover Boys series under the name **Arthur M. Winfield,** and the Tom Swift series appeared under the name **Victor Appleton.** After Stratemeyer's death, Tom Swift and other series were carried on by other authors. Stratemeyer, in all, wrote more than 150 books and originated hundreds more.

Louis Untermeyer worked in his father's jewelry factory in Newark from 1902 to 1923, and rose to the position of vice president and general manager before he resigned to devote all his time to literature. During these years a number of Untermeyer's books were published, including his first volume, a twenty-four page collection of poems entitled *The Younger Quire* (1910), a parody of a volume entitled *The Younger Choir,* an anthology of poetry that included works written by Untermeyer. These years also saw publication of other Untermeyer collections, including *First Love* (1911), *These Times* (1917), and *The New Adam* (1920). Untermeyer later wrote of his life in *From Another World* (1939) and *Bygones* (1965).

NEW BRUNSWICK

This city's literary associations derive in part, as one would expect, from the presence of Rutgers University, but New Brunswick was also the birthplace of **Joyce Kilmer,** the poet whose career ended abruptly with his death in battle during World War I. Kilmer was born on December 6, 1886, in a second-floor room of a house at 17 Joyce Kilmer Avenue, formerly Codwise Avenue. The house is now known as the Joyce Kilmer Birthplace. The house, in which the poet spent his early years, is undergoing restoration and will be open to visitors some time in the future. It contains such items as the desk Kilmer used when he was a schoolboy, an oil portrait of Kilmer, and a facsimile of the poem "Trees" written in the poet's handwriting. This best known of Kilmer's poems was inspired, it is said, by a giant oak that stood on the campus of Cook College, a part of Rutgers University. The tree subsequently fell to disease, but a plaque on its massive stump still identifies it. Kilmer attended Rutgers Preparatory School, then located on the Rutgers campus but now at 1345 Easton Avenue in nearby Somerset. While a student at Rutgers Prep, Kilmer edited the school publication, the *Argo,* and contributed poetry under the pseudonym **Malachi Sinclaire.** The school library has original copies of the issues Kilmer edited. After being graduated in 1904, Kilmer enrolled at Rutgers College (now Rutgers University), but transferred to Columbia University in 1906.

Joyce Kilmer photographed at the age of thirty, just before the poet entered the U.S. Army.

KILMER HOUSE
BOYHOOD HOME OF JOYCE KILMER,
WRITER AND POET
FAMOUS FOR POEM "TREES,"
KILLED IN ACTION ON
WESTERN FRONT, 1918

—Marker at Joyce Kilmer Birthplace, New Brunswick

Stephen Longstreet, the novelist and playwright, was brought to live in New Brunswick in 1907, shortly after his birth. He lived in New Brunswick until 1927 in a series of homes, including one, now private, at 19 Prospect Street, and others on French Street, New Street, and Redmond Street. The last house Longstreet lived in, now private, stands at 115 Louis Street. Longstreet used his experiences in New Brunswick for several of his books, including his first novel, *And So Dedicated* (1940), which he wrote under the pen name **Thomas Burton.** New Brunswick also appears in *Nine Lives with Grandfather* (1944) and *Sisters Liked Them Handsome* (1946).

T. DeWitt Talmage, the author of many published sermons, was graduated from the New Brunswick Theological Seminary in 1856. The school is now located at 17 Seminary Place.

Among the numerous writers associated with Rutgers University are **John Ciardi,** the poet, who taught here from 1953 to 1961; **Ralph Ellison,** who has served as visiting professor; and **Philip Van Doren Stern,** the historian and novelist, who was graduated from Rutgers in 1924 and received an honorary degree from the university in 1940. Rutgers University Library maintains collections of the papers of Thomas Paine and of Frances Winwar, the biographer and novelist.

NORTH CALDWELL

Richard Wilbur, the poet, lived from 1923 to 1942 at 9 East Green Brook Road, his family's home in this town southwest of Paterson. The house, built about 1775 and known as the Sandford-Stager House, was bought by Wilbur's father, a noted artist, who lived here until 1971. A marker in front of the house describes its history and notes the residence of Lawrence Wilbur, the poet's father, but does not mention Richard Wilbur. The poet used his experiences in North Caldwell in several of his poems, including "He Was," "Digging for China," and the first part of "Running."

"Far enough down is China," somebody said.
"Dig deep enough and you might see the sky
As clear as at the bottom of a well.
Except it would be real—a different sky.
Then you could burrow down until you came
To China! Oh, it's nothing like New Jersey.
There's people, trees, and houses, and all that,
But much, much different. Nothing looks
the same."

—Richard Wilbur,
in "Digging for China"

NORTH PLAINFIELD

Maxwell Perkins, the noted editor, lived at 95 Mercer Avenue in this town about ten miles north of New Brunswick. He moved here in 1911, a few months after taking a position as advertising manager at Scribner's, and shortly after marrying Louise Saunders. By the time the Perkinses moved to nearby Plainfield in 1916, Perkins had been promoted and had begun to attract attention as a promising young Scribner's editor.

NUTLEY

H. C. Bunner, the poet, editor, and short-story writer, lived in this town north of Newark from 1887 until his death on May 11, 1896. Bunner died at his home, which stood at 119 Whitford Avenue, at the corner of Whitford and Nutley avenues. Bunner commuted to New York City, where he worked as editor of the humor weekly *Puck,* but wrote his short stories and poems at home. Among Bunner's books published during these years were the novel *The Story of a New York House* (1887), the collections of short stories *Short Sixes* (1890) and *Zadoc Pine* (1891), and the collection of poems *Rowen* (1892). The stories in Bunner's collection *Made in France* (1893), adaptations of the tales of Guy de Maupassant, were so good that critics failed to notice that Bunner had included a story of his own.

Another literary resident of Nutley was **Frank R. Stockton,** who lived for a number of years in a house, now renovated, at 203 Walnut Avenue. During these years Stockton held an editorial post with *St. Nicholas.* In Nutley he wrote his first novel for adults, *Rudder Grange* (1879), whose heroine, a woman named Pomona, was modeled after an orphan girl who worked as a maid in the Stockton home. Stockton tried to imagine what the girl would be like when she was grown up, and Pomona was the answer. The book began as a sketch written for *Scribner's Monthly.* The public reaction to it was so warm that Stockton wrote additional sketches, and these were later collected in the novel. It is believed that Stockton wrote "The Lady or the Tiger?" for presentation to a literary society he belonged to, which met in a private house on Grant Avenue. Stockton later sent the tale of the power of love, jealousy, and barbarism over human actions to *Century* magazine, in which it appeared in November 1882. Millions of readers have since wondered at the story's unresolved ending. Some time after Stockton's resignation in 1881 from his job at *St. Nicholas,* he moved to Convent Station, New Jersey.

How often, in her waking hours and in her dreams, had she started in wild horror and covered her face with her hands as she thought of her lover opening the door on the other side of which waited the cruel fangs of the tiger!

But how much oftener had she seen him at the other door! How in her grievous reveries had she gnashed her teeth and torn her hair when she saw his start of rapturous delight as he opened the door of the lady! How her soul had burned in agony when she had seen him rush to meet that woman, with her flushing cheek and sparkling eye of triumph: . . .

Would it not be better for him to die at once, and go to wait for her in the blessed regions of semi-barbaric futurity?

And yet, that awful tiger, those shrieks, that blood!

Her decision had been indicated in an instant, but it had been made after days and nights of anguished deliberation. She had known she would be asked, she had decided what she would answer, and, without the slightest hesitation, she had moved her hand to the right.

—Frank R. Stockton,
in "The Lady or the Tiger?"

PARAMUS

Maxwell Bodenheim, the poet and novelist, died on February 6, 1954, and was buried in Cedar Park Beth-El Cemetery, on Forest Avenue in Paramus in northeastern New Jersey. Also buried there is **Delmore Schwartz,** the poet, who died on July 11, 1966, in New York City. **Robert Molloy,** author of the best-selling novel *Pride's Way* (1945), lived at 249 Monroe Avenue in Paramus from the late 1960s until his death on January 27, 1977. Molloy died at Bergen Pines Hospital, in Paramus.

PASSAIC

Frederick Manfred, the novelist, used this city in northeastern New Jersey as a setting for his novel *The Brother* (1950). In the novel, Thurs, the main character, lived in Passaic and worked in a rubber factory.

PATERSON

Of the several writers associated with this city on the Passaic River, about fifteen miles north of Newark, the one known best is **Stephen Crane,** who as a child lived from the spring of 1876 to the spring of 1878 at 26 Hotel Street, now Hamilton Street. Crane's father, Reverend Jonathan

> *There's a war in Paterson, New Jersey. But it's a curious kind of war. All the violence is the work of one side—the mill owners. Their servants, the police, club unresisting men and women and ride down law-abiding crowds. Their paid mercenaries, the armed detectives, shoot and kill innocent people. Their newspapers, the Paterson Press and the Paterson Call, publish incendiary and crime-inciting appeals to mob violence against the strike leaders . . . control absolutely the police, the press, the courts.*
>
> —John Reed,
> in "War in Paterson"

Townley Crane, was pastor of the Methodist Episcopal Church at Cross (now Cianci) and Elm streets. The building is now St. Michael's Church. In later years Crane lived at his brother's house in Lake View, now a section of the adjacent city of Clifton. Other writers are associated with Paterson. **Allen Ginsberg,** the poet, grew up here and was graduated from high school in Paterson in 1943. **George Middleton,** whose best-known plays included *Cavalier* (with Paul Kester, 1902) and the comedy *Polly with a Past* (with Guy Bolton, 1917), was born in Paterson on October 27, 1880.

John Reed, the radical writer and journalist, came to Paterson in April 1913 to see firsthand the strike of Paterson's silk workers. He wanted to dramatize the strikers' plight and reenact it in New York. Reed was arrested and sentenced to twenty days in the Passaic County Jail for failing to obey a policeman's order to move off a Paterson street. He served four days of his sentence before being bailed out by friends. Returning to New York, Reed wrote the article "War in Paterson," which appeared in the June 1913 issue of *Masses.* He turned to writing and organizing the Paterson Pageant, which was presented in New York's Madison Square Garden on June 7, and included among its cast more than a thousand Paterson workers. The pageant electrified its audience and received warm critical reviews, but failed as a fund-raiser for the strikers. When the numbers were added up, Reed and his friends had spent more on the production than they realized in ticket sales.

Paterson is the setting for the major work of one of this country's most respected poets. William Carlos Williams, who lived most of his life in nearby Rutherford, set his monumental five-volume poetic work *Paterson* (1946–58) in this city.

Grave marker of William Dunlap in St. Peter's churchyard in Perth Amboy. A sharp-eyed visitor has tried to repair the misspelled word "capitol."

PERTH AMBOY

William Dunlap, born in this city south of Newark on February 19, 1766, was a painter, theatrical manager, and playwright. Dunlap, who sometimes is called the father of American drama, was born in a house on High Street, address unknown, and passed the first eleven years of his life in Perth Amboy. In 1777 the family was obliged, because of Royalist sympathies, to seek haven in British-held New York City. It was there that Dunlap spent much of his life and saw the production of thirty-one of his original plays as well as many of his adaptations of French and German works. Among Dunlap's best original works were *The Father* (1789) and *André* (1798). Dunlap died in New York City on September 28, 1839, and is buried with members of his family in the churchyard of St. Peter's Church, on Rector Street, in Perth Amboy. Dunlap was a lifelong member of the church, and a plaque on its north wall notes his accomplishments.

PLAINFIELD

This city in northeastern New Jersey was the birthplace of two distinguished American writers. **Van Wyck Brooks,** the critic and writer who received a Pulitzer Prize for his literary history *The Flowering of New England, 1815–1865* (1936), was born on February 16, 1886, in a house, now private, at 237 East Ninth Street. Brooks grew up in Plainfield and was graduated from Plainfield High School in 1904, where his classmates called him, presciently, Hawthorne the Greater. Years after he left Plainfield, first to attend Harvard University and later, in 1907, to live abroad, Brooks returned to Plainfield to speak at the seventieth anniversary of the founding of the Plainfield Public Library. It was then that he said that "old Plainfield is a marvelous subject if anyone wished to write the story of this town."

William Ellery [Channing] Leonard, the poet, translator, and critic, was born in Plainfield on January 25, 1876, and lived here until 1893. Of his numerous books, perhaps his best known was *The Locomotive-God* (1927), an autobiographical volume dealing with Leonard's search for the cause of the agoraphobia that plagued him. During his years in Plainfield, Leonard experienced two terrifying incidents involving locomotives, the first in June 1878, while waiting on a train platform. Leonard wandered away from his family and, struck with fear at the oncoming rush of the great locomotive, found the path to the safety of his mother's arms blocked by a baggage truck. The second incident occurred in 1885 when, as a student at the Washington School, Leonard was chased home by a band of taunting children, and found his way blocked by a speeding express train.

Maxwell Perkins grew up in Plainfield in the 1880s and 1890s. The son of a New York attorney, Perkins attended the John Leal School for Boys, first located at 483 West Front Street and later at 450 West Seventh Street. Perkins attended Harvard University and then returned to Plainfield off and on, settling down here only after he married Louise Saunders, a Plainfield woman he had known for years. The two were married on December 31, 1910, at the Church of the Holy Cross, 60 Washington Avenue, and, after a honeymoon in New Hampshire, moved into a house at 95 Mercer Avenue in nearby North Plainfield. In 1916 the Perkinses swapped houses with Louise's sister and her husband, who had found their own home at 112 Rockview Avenue too expensive for them to keep. The growing Perkins family remained at the Rockview Avenue home until 1924, when Perkins decided to move to New Canaan, Connecticut. During the years in Plainfield, Perkins began his rise as editor at Scribner's. He worked, for example, with F. Scott Fitzgerald on the manuscripts of *This Side of Paradise* (1920) and *The Beautiful and Damned* (1922), with Ring Lardner on *How to Write Short Stories* (1924), and with J. P. Marquand on the novel *The Unspeakable Gentleman* (1922). Perkins in these early years showed the qualities that helped make him a great editor—openness and sympathy with new writers, and courage to take chances that more conservative editors might have avoided. In his career, these qualities helped Perkins find and help launch writers of future significance, such as Fitzgerald, Hemingway, and Wolfe.

Edmund Clarence Stedman, the poet and critic, lived as a child in the 1830s on his grandfather's farm in Plainfield.

POINT PLEASANT

Writers associated with this coastal resort about fifteen miles south of Long Branch include

Richard Harding Davis, whose family owned a summer house here; and **Hilda Doolittle,** the poet known as **H. D.,** who came here in June 1906 with **Ezra Pound** and **William Carlos Williams** to attend a party at the summer cottage of a college friend, Bob Lamberton. Doolittle came close to death here. While walking into the surf, she was knocked over by the waves and dragged down by the undertow. Williams and Lamberton dragged her from the water unconscious, and revived her back at the cottage.

PRINCETON

Princeton has been the literary center of New Jersey for more than two hundred years, largely because of Princeton University, which was founded as the College of New Jersey and was moved from Newark to Princeton in the 1750s. The college was renamed Princeton University in 1896, long after it had established itself as a major educational institution of the first rank. This prestige and a tradition of welcoming and encouraging scholarly and creative writers have drawn many authors to the university and to the pleasant accommodations available in Princeton. A number of writers who have lived in Princeton have not been associated directly with the university, but there has always existed an indirect connection, through the university's ability to maintain an atmosphere that attracts writers and helps most of them work.

Writers who have lived in Princeton include **W. H. Auden,** who frequently visited friends in Princeton, among them his editor Saxe Commins. Commins summered from 1941 to 1951 in a house on Battle Road Circle, and later in a house at 85 Elm Road. **Louis Bromfield** spent the winter of 1933–34 in the John Gale Hun house, at 18 Hibben Road. He found the lure of nearby New York City difficult to resist and while in Princeton worked only on one short novel, "The Eavesdropper," and revised three others that appeared with it in the volume *Here Today and Gone Tomorrow* (1934).

T[homas] S[tearns] Eliot lived at 14 Alexander Street in 1948, when he was a member of the Institute of Advanced Study—although not officially associated with Princeton University, the institute makes use of university facilities. While in Princeton Eliot worked on his verse play *The Cocktail Party* (1950) in his office at 307 Fuld Hall and at his Alexander Street home. **William Faulkner** in the 1950s made several extended visits to the 85 Elm Road home of Saxe Commins, his editor. There Faulkner worked on his collection of stories *Big Woods* (1955), which included the classic story "The Bear." He also worked on the novels *The Town* (1957) and *The Mansion* (1959). Some scholars believe that Faulkner wrote very little here. Whatever the truth, he wrote the dedication of *Big Woods* in the form of a memo to Commins.

> *Memo to: Saxe Commins*
> *From: Author*
> *To: Editor*
>
> *We never always saw eye to eye*
> *but we were always*
> *looking at the same thing.*
>
> —William Faulkner,
> dedication page of *Big Woods*

Caroline Gordon, the poet and novelist, moved to Princeton with her husband **Allen Tate** in September 1939, when Tate became the first resident fellow in creative writing at Princeton University. The Tates lived in Princeton until June 1942, in a house at 16 Linden Lane, where Gordon completed her novel *The Green Centuries* (1941) and worked on another, *The Women on the Porch* (1944). In 1949 they returned to Princeton and lived until September 1951 in a house at Howe Lane and Nassau Street. There Gordon worked on her novel *The Malefactors* (1956). During the two stays in Princeton, Tate wrote the critical essays collected in *Reason in Madness* (1941) and *The Hovering Fly and Other Essays* (1949), as well as several poems, including "False Nightmare," "The Maimed Man," and "The Swimmers." In 1956 Gordon returned alone to Princeton and moved into the Red House, 145 Ewing Street, which was to be her home until the mid-1970s. Among her later books were *Old Red and Other Stories* (1963) and the novel *The Glory of Hera* (1971).

Dashiell Hammett lived at 90 Cleveland Lane in the mid-1930s. **Lillian Hellman,** who occasionally visited Princeton during this time, has said that Hammett worked in Princeton on a story, possibly "A Man Called Spade," published in the volume *Challenge to the Reader* (1938). Hellman wrote parts of her play *Days to Come* (1936) in Princeton. Other writers who have lived in Princeton include **Sidney Howard,** who rented the John Gale Hun house, 18 Hibben Road, during the winter of 1934–35 and there worked on his play *Ode to Liberty,* which had its premiere at Princeton's McCarter Theatre on December 8, 1934. Here too he worked on *Paths of Glory* (1935), the dramatic adaptation of the novel by Humphrey Cobb. **Helen MacInnes,** a leading author of novels of suspense and intrigue, completed her novel *While Still We Live* (1944) while living at 208 Edgerstoune Road from March 1942 to September 1943.

Princeton's leading literary native son, **John [Angus] McPhee,** was born here on March 8, 1931. McPhee grew up in Princeton, was graduated from Princeton High School in 1948, and was a member of Princeton University's class of 1953. During his senior year at Princeton, McPhee lived in room 236, in 1903 Hall. There and at the university's Firestone Library, he worked on a novel, *Skimmer Burns,* which he submitted as a senior thesis. Bill Bradley, the Princeton basketball star and Rhodes scholar who became a professional basketball player and then

Residence of T. S. Eliot in Princeton.

Old photograph of the cabin in Princeton in which Upton Sinclair wrote *The Jungle.* This picture was published in May 1906, three months after the novel was published in book form.

U.S. senator from New Jersey, was the subject of McPhee's book *A Sense of Where You Are* (1965). Like the Bradley book, much of McPhee's writing appears first in *The New Yorker.* Among his books are *The Pine Barrens* (1968), concerning the nearly 2,000 square miles of pine forest stretching from central to southern New Jersey; *Coming Into the Country* (1977), about Alaska; and *Basin and Range* (1981), dealing with the geologist's view of the planet Earth.

Better than anyone else,
he told the truth about his time,
the first half of the twentieth century.
He was a professional.
He wrote honestly and well.

—Inscription on the gravestone of John O'Hara, Princeton Cemetery

On September 1, 1949, **John O'Hara** moved into a house at 18 College Road West, and one year later into a house, since burned and replaced by a new structure, at 20 College Road West, where he wrote the novel *Ten North Frederick* (1955). On September 1, 1957, O'Hara moved to Linebrook, the mansion, now private, at Pretty Brook Road and Province Line Road, that was to be his home until his death. During his thirteen years at Linebrook, O'Hara wrote many of his books, including the novel *From the Terrace* (1958); the collection *The Horse Knows the Way* (1964); *My Turn* (1966), a collection of letters that originally appeared in *Newsday* and other newspapers; and the novel *The Instrument* (1967). O'Hara died at Linebrook on April 11, 1970. Funeral services were held in the chapel of Princeton University, and the author is buried in the Princeton Cemetery. His tombstone is inscribed with words said to have been written by O'Hara.

Linebrook, John O'Hara's home in Princeton for more than twelve years. He died there in 1970.

Budd Schulberg frequently stayed at the Princeton homes of his editor Saxe Commins in the 1940s and 1950s. At Commins's home on Battle Road Circle, Schulberg worked on his novel *The Harder They Fall* (1947); and at Commins's home at 85 Elm Road, Schulberg worked on *Waterfront* (1955), a novel based on his motion-picture script for *On the Waterfront* (1954). In 1958 Schulberg moved to 343 Jefferson Road, where he worked on his play *The Disenchanted,* based on his 1950 novel. The play, written with Harvey Breit, had its premiere on Broadway on December 3, 1958. **Samuel Shellabarger** was graduated from Princeton University in 1909 and taught there, with interruptions, from 1917 to 1923. In 1928 he returned to Princeton and lived until 1938 at 160 Hodge Road, where he wrote his novel *Tolbecken,* not published until 1956. In 1946 Shellabarger bought a house at 107 Library Place, where he lived until his death on March 21, 1954. There he wrote his novels *Prince of Foxes* (1947), *The King's Cavalier* (1950), and *Lord Vanity* (1953).

Upton Sinclair in May 1903 moved here with his family to make use of Princeton's large Civil War collection in his research for a projected trilogy. The Sinclairs lived in Princeton for a year and a half, at first in a tent three miles north of town. Sinclair then built a small cabin, now remodeled into a guest house, in which he wrote *Manassas* (1904), the first volume of the trilogy. Publication of this novel gave Sinclair the funds to buy a sixty-acre farm at Province Line Road and Ridge View Road. He moved his cabin to the farm and began work on *The Jungle* (1906). **Julian Street** lived at 107 Library Place from 1921 to 1923, and at 86 Stockton Street from 1923 to 1931. While living in Princeton, Street wrote a number of novels and short stories, among them *Rita Coventry* (1922), *Cross-Sections* (1923), and the story "Mr. Bisbee's Princess" (1925).

Writers who have taught at Princeton University include **Kingsley Amis,** the English novelist, who lectured in 1958 and 1959 and lived at 271 Edgerstoune Road. While here he wrote *New Maps of Hell* (1960), a collection of his lectures on science fiction; and worked on the novel *Take a Girl Like You* (1960). He also made notes for his novel *One Fat Englishman* (1963). **Saul Bellow** was a fellow in creative writing in 1952 and 1953. While living at 12 Princeton Avenue, he finished his novel *The Adventures of Augie March* (1953). **John Berryman** taught at Princeton several times between 1943 and 1949. While living at 120 Prospect Avenue, he wrote his poem *Homage to Mistress Bradstreet* (1956), considered by some to be one of his best.

[Maxwell] Struthers Burt was graduated from Princeton in 1904 and taught here from 1907 to 1910. Burt became involved in ranching in Wyoming, but in 1913 returned to Princeton. From 1913 to 1921 he divided his time between his ranches in Wyoming and his Princeton home at 72 Stockton Street. Here he wrote the short story "Each in His Generation" (1920). In 1920

F. Scott Fitzgerald presented Burt with the "first copy" of Fitzgerald's novel *This Side of Paradise*, inscribed: "For Mr. Maxwell Struthers Burt in appreciation of the fact that he does the best short stories in the country." **Richard Eberhart,** the poet, lived in a house on Howe Lane during the winter semester of 1955, and in an apartment at 190 Prospect Avenue during the spring semester of 1956. At the first address he wrote "Thrush Song at Dawn," "Attitudes," and other poems. At the second address he wrote such poems as "Futures," "Anima," "Lucubration," and "Goose Falls." Eberhart was awarded the Bollingen Prize in 1962 after publication of his *Collected Verse Plays*.

Jonathan Edwards served as president of Princeton, then the College of New Jersey, in 1757–58. He arrived in Princeton in January of 1758 but died on March 22 and was buried in Princeton. **Leslie Fiedler** lived at 89 Mercer Street while he was a fellow in creative writing at Princeton in 1956 and 1957. Here he wrote much of *Love and Death in the American Novel* (1960) and *Nude Croquet and Other Stories* (1969). **Randall Jarrell,** a visiting fellow in creative writing in 1951 and 1952, lived at 16 Alexander Street. During that year he wrote his six lectures on W. H. Auden, his novel *Pictures from an Institution* (1954), and several poems, including "The Lonely Man" and "Windows."

Thomas Mann, the German novelist and essayist, lived from October 1938 to March 1941 at 65 Stockton Street. Mann lectured at Princeton and finished his novel *The Beloved Returns, Lotte in Weimar* (1940), as well as his retelling of a legend from India, *The Transposed Heads* (1941). In Princeton he began work on *Joseph the Provider* (1944), the final volume of his *Joseph and His Brothers* tetralogy. **Alfred Noyes,** the English poet, was visiting professor at Princeton from 1914 to 1923, but spent long periods away from the university. While living at 120 Broadmead Street he wrote many poems, including "The Old Meeting House," about Princeton's Quaker Meeting House. **Sean O'Faolain,** the Irish writer, lived at 20 North Stanworth Drive in 1953 and 1954, and there wrote his book of criticism *The Vanishing Hero* (1956). While teaching at Princeton from 1959 to 1961, O'Faolain lived at 66 Stanworth Lane, where he wrote the stories collected in *I Remember! I Remember!* (1962).

> *Its quiet graves were made for peace till Gabriel*
> *blows his horn.*
> *Those wise old elms could hear no cry*
> *Of all that distant agony—*
> *Only the red-winged blackbird, and the rustle*
> *of thick ripe corn.*
>
> —Alfred Noyes,
> in "The Old Meeting House"

Other writers who have taught at Princeton include **Bliss Perry,** who taught here from 1893 to 1900 and was a faculty colleague of Woodrow Wilson's when Wilson taught political economy. Perry wrote of Wilson and of Princeton in his autobiography, *And Gladly Teach* (1935). **Philip Roth** taught at Princeton during 1962–64, living in the first year at 232 Bayard Lane. During the second year, Roth lived in New York City and commuted to classes or to his office, room 126, 1879 Hall. While living in Princeton Roth wrote several short stories and worked on the first draft of his novel *When She Was Good* (1967). **Donald A. Stauffer** was graduated from Princeton in 1923 and received his M.A. in 1927, the year he joined the Princeton faculty. After serving in World War II, Stauffer returned to Princeton to a house at 14 Alexander Street, which was to be his home until his death on August 8, 1952. There he revised the manuscript of his novel *The Saint and the Hunchback* (1946).

Henry Van Dyke, the writer and minister, was graduated from Princeton University in 1873 and from Princeton Theological Seminary in 1877. He was a frequent visitor to Princeton before becoming a member of the faculty in 1899. Van Dyke taught at Princeton until 1923, and lived from 1899 to 1933 at Avalon, his home at 59 Bayard Lane. From 1913 to 1916 he served as U.S. minister to the Netherlands and Luxembourg, and during World War I served as a chaplain in the U.S. Navy. While in Princeton, Van Dyke wrote nearly fifty books—fiction, poetry, and drama—before his death at Avalon on April 10, 1933. Among his many works were a collection of stories, *The Ruling Passion* (1901), and a collection of lectures, *The Spirit of America* (1910).

Princeton University counts among its former students many authors, some of whom contributed to *The Nassau Literary Magazine* or *The Princeton Tiger*. Others wrote for the dramatic productions of the Princeton Triangle Club, and a fair number achieved professional publication while still at Princeton. **John Peale Bishop,** who was graduated in 1917, contributed to *The Nassau Literary Magazine*, and saw his poem "Nassau Inn" published in *Scribner's Magazine* in 1917. While living

> *Outside the New World winters in*
> *grand dark*
> *white air lashing high thro' the virgin*
> *stands*
> *foxes down foxholes sigh,*
> *surely the English heart quails,*
> *stunned.*
>
> —John Berryman,
> in *Homage to Mistress Bradstreet*

Home of Thomas Mann in Princeton. The novelist traveled and lectured throughout the United States during his years here.

F. Scott Fitzgerald.

at Witherspoon Hall, Bishop wrote many of the poems that appeared in the volume *Green Fruit* (1917). Here Bishop became friends with two other Princetonians, **F. Scott Fitzgerald** and **Edmund Wilson. Philip Pendleton Cooke,** poet and author, was graduated from Princeton in 1834. While here Cooke wrote the poems "Song of the Sioux Lovers," "Autumn," and "Historical Ballads, No. 6 Persian: Dhu Nowas" as well as the short prose piece "The Consumptive," all of which appeared in the *Knickerbocker Magazine.*

F. Scott Fitzgerald is probably the best known of Princeton's literary alumni. Fitzgerald failed the entrance examinations twice but succeeded in talking his way past the Princeton admissions committee. He enrolled at the university in September 1913, moving into a boarding house at 15 University Place, called The Morgue by its residents. In his sophomore year, he lived on campus at 107 Patton Hall. As a freshman Fitzgerald wrote the script for the 1914–15 presentation of the Triangle Club, *Fie! Fie! Fi-Fi!* In his sophomore year he wrote the lyrics for Edmund Wilson's Triangle Club script *The Evil Eye,* which was presented during the Club's 1915–16 season. He also made his first contributions to *The Nassau Literary Magazine.* In 1915 he began his junior year at 32 Little Hall, and served briefly on the staff of *The Princeton Tiger,* but became ill and had to leave Princeton. In 1916 he returned as a junior, living at 185 Little Hall. He kept the room into his senior year until, in 1917, he received his commission in the U.S. Army and left for active service. Fitzgerald had already begun work on what would become his first novel, and in January 1918, while on leave, he returned to Princeton and stayed at the Cottage Club. There he continued work on his novel, which was eventually to be published as *This Side of Paradise* (1920). Amory Blaine, the protagonist, had much in common with Fitzgerald, including a stay at Princeton. The unflattering portrait Fitzgerald painted of the university upset many Princetonians. When Fitzgerald visited the Cottage Club in 1920, his behavior was so outrageous that he managed to start a row. Fitzgerald was thrown out of a back window and later was suspended from the club. After Fitzgerald's death in 1940, Princeton University declined to purchase his papers for the trifling sum of $3,750. In 1950 Fitzgerald's daughter, Scottie, presented the papers to the Princeton Library as a gift.

Other writers who attended Princeton include **Philip Freneau,** who was graduated in 1771 from what was then the College of New Jersey. Here the poet collaborated with another student, **H[ugh] H[enry] Brackenridge** (1748–1816), on the poem "The Rising Glory of America," which Brackenridge read at graduation exercises in 1771; and on an early work of fiction, *Father Bombo's Pilgrimage to Mecca,* first published in its entirety in 1975. While a student at Princeton, Freneau also wrote the poems "The History of the Prophet Jonah," "The Village Merchant," and "The Pyramids of Egypt."

Joshua Logan, the playwright, was graduated from Princeton in 1931. As an undergraduate, Logan wrote three plays for the Triangle Club: *Zuider Zee* (1928), *The Golden Dog* (1929), and *The Tiger Smiles* (1930). Logan lived at Holder, Campbell, and Little halls. **Arthur Mizener,** who was graduated from Princeton in 1930, returned in the mid-1940s to conduct research for *The Far Side of Paradise* (1951), his biography of F. Scott Fitzgerald. A Princeton undergraduate who showed neither literary nor scholarly promise was **Eugene O'Neill.** He entered Princeton University in September 1906, raised a little hell, and was sent on his way by the end of the school year.

David Graham Phillips, the newspaperman and novelist, was graduated from Princeton in 1887. **Ernest Poole,** the novelist and playwright, was graduated in 1902. **David Ramsay,** the historian, was graduated from the College of New Jersey in 1765, and later married Frances Witherspoon, daughter of the president of the college. **George R. Stewart,** the teacher and novelist, received his degree from Princeton in 1917.

Booth Tarkington did not manage to get a Princeton degree but made a lasting impression on the university during his two years of study, 1891–93. The novelist and playwright lived in apartment U in the old University Hall, no longer standing, and became a friend of another Princeton student, **Jesse Lynch Williams.** The two founded the Princeton Triangle Club, and Tarkington wrote much of the club's first presentation, *The Hon. Julius Caesar,* and contributed to *The Nassau Literary Magazine* and *The Princeton Tiger.*

James Ramsey Ullman, the novelist and playwright, was graduated from Princeton in 1929. His award-winning senior thesis, *Mad Shelley,* was published by Princeton University Press in 1930. While a graduate student at Princeton in 1945, **Reed Whittemore** lived at 54 Graduate College. There the poet wrote the first two parts of "Reconsiderations" as well as "The Equine Problem in Sophia" and "The Last Resource." All these poems appeared in Whittemore's collection *Heroes and Heroines* (1946).

Thornton Wilder, who taught for several years at the nearby Lawrenceville School, attended graduate school at Princeton in 1925 and 1926. While walking toward town, he got the idea for his Pulitzer Prize-winning novel, *The Bridge of San Luis Rey* (1927). The insight, in Wilder's words, came "actually, on a spot just between the college and the Divinity School." Wilder began to write the novel while at Princeton. His play *Our Town* was first performed at Princeton's McCarter Theatre, on January 22, 1938.

Jesse Lynch Williams was graduated from Princeton in 1892 and received his A.M. degree in 1895. As an undergraduate, he founded, with **Booth Tarkington,** the Princeton Triangle Club. As a graduate student Williams worked on his volume of *Princeton Stories* (1902). From 1900 to 1903 he lived at 70 Washington Road, later the site of the Gateway Club, and served as editor of the *Princeton Alumni Weekly.* In later years Williams visited Princeton from time to time, and in 1928 took a house at 28 Greenholm Street, where Williams worked on his last novel, *She Knew She Was Right* (1930). Williams died in Herkimer, New York, on September 14, 1929. It is thought that his ashes were scattered in Princeton.

Edmund Wilson, who was graduated from Princeton in 1916, contributed to *The Nassau Literary Magazine* and wrote *The Evil Eye,* the musical comedy presented by the Princeton Triangle Club during its 1915–16 season. The lyrics for the musical were written by Wilson's friend **F. Scott Fitzgerald.** Wilson lived at 1B Livingston Hall during

How a Triangle show ever got off was a mystery, but it was a riotous mystery, anyway, whether or not one did enough service to wear a little gold Triangle on his watch-chain. . . . All Triangle shows started by being "something different—not just a regular musical comedy," but when the several authors, the president, the coach and the faculty committee finished with it, there remained just the old reliable Triangle show with the old reliable jokes and the star comedian who got expelled or sick or something just before the trip, and the dark-whiskered man in the pony ballet, who "absolutely won't shave twice a day, doggone it!"

—F. Scott Fitzgerald,
in *This Side of Paradise*

his four years at Princeton. Selections of the journal Wilson kept as a student here were published in his book *A Prelude* (1967).

RAHWAY

Carolyn Wells, the anthologist and author credited with more than 170 books, was born in this city southwest of Elizabeth in the late 1860s or early 1870s. Her works included several series of books for children and about seventy-five mystery and detective stories. Perhaps her best-known book was *A Nonsense Anthology* (1902), a delightful collection of humorous verse.

When that Seint George hadde sleyne ye draggon,
He sate him down furninst a flaggon;
And, wit ye well,
Within a spell
He had a bien plaisaunt jag on.

—Anonymous verse,
collected in *A Nonsense Anthology,*
by Carolyn Wells

RED BANK

Edmund Wilson, the eminent writer and critic, was born on May 8, 1895, in this town near the Atlantic shore, a few miles northwest of Long Branch. Among his books were the volume of criticism *Axel's Castle* (1931); *Travels in Two Democracies* (1936), based on his travels in America and Russia; *The Wound and the Bow* (1941), critical essays; *Memoirs of Hecate County* (1946), a collection of stories; *The Shores of Light* (1952), critical essays; and *Night Thoughts* (1961), a collection of poetry and prose. Wilson attended Princeton University and there became friends with John Peale Bishop and F. Scott Fitzgerald. Wilson collaborated with Bishop on the book *The Undertaker's Garland* (1922), a collection of prose and verse dealing with death and funerals. In 1945 Wilson edited *The Crack-Up,* a collection of Fitzgerald's unpublished notes, journal entries, and letters.

RIEGELSVILLE

Louis Adamic, author of many books, including *The Native's Return* (1934) and *My America* (1938), died at his home here on September 4, 1951 Adamic was found on the couch of his second-floor study, dead from an apparently self-inflicted gunshot wound. His house had been set on fire. There was speculation that Adamic had been killed by Soviet agents, but his death was ruled a suicide. His house, which dates back to the early nineteenth century, is still standing and privately owned. Born in Yugoslavia on March 23, 1899, Adamic came to the United States when he was fourteen. At his death he was at work on a book to be called *The Eagle and the Rock.*

RUTHERFORD

This town in northeastern New Jersey achieved prominence in literary history as the birthplace and lifelong residence of **William Carlos Williams.** Williams, born on September 17, 1883, in a house, no longer standing, on Passaic Avenue, practiced medicine in Rutherford from 1910 to 1952. Williams used Rutherford in the same way that Henry David Thoreau used Concord—as a microcosm from which the careful observer could draw universal truths. Williams grew up in Rutherford, and in 1896 left for a year's study in Europe. After completing high school in New York City, he enrolled at the medical school of the University of Pennsylvania, from which he was graduated in 1906. After serving his internship in New York and completing further study in Europe, Williams returned to Rutherford in 1910 and began his medical practice. It was estimated that between 1910 and 1952 Williams delivered 2,000 babies, most of them in the Rutherford area. In 1912 Williams married Florence Herman—whom he called Flossie—and the couple moved a short time later into the two-story clapboard house, now private, at 9 Ridge Road, which was still their home at the time of Williams's death there on March 4, 1963. Williams used part of the first floor of the house as medical offices and wrote his poetry, especially his work after 1951, in a second-floor study. In this house, over the years, Williams had many literary visitors, including **Maxwell Bodenheim, E. E. Cummings,** and **Ezra Pound.**

Among Williams's many volumes of poetry were *Kora in Hell* (1920), a collection of prose poems; *Complete Collected Poems* (1938); *The Desert Music* (1954); and the Pulitzer Prize-winning *Pictures from Breughel* (1963). He also wrote plays,

Perhaps it was the pure air from the snows before him; perhaps it was the memory that brushed him for a moment of the poem that bade him raise his eyes to the helpful hills. At all events he felt at peace. Then his glance fell upon the bridge, and at that moment a twanging noise filled the air, as when the string of some musical instrument snaps in a disused room, and he saw the bridge divide and fling five gesticulating ants into the valley below.

Anyone else would have said to himself with secret joy: "Within ten minutes myself. . . !" But it was another thought that visited Brother Juniper: "Why did this happen to those *five?"*

—Thornton Wilder,
in *The Bridge of San Luis Rey*

Home of William Carlos Williams in Rutherford.

It's all right to be wise, but you got to watch that too. There's no way to learn it easy. And it brings plenty trouble.

—William Carlos Williams, in the title story from *Life Along the Passaic River*

. . . The barrow he pushed, he did not love. The stones that brutalized his palms, he did not love . . . . A voice within him spoke in wordless language.

The language of worn oppression and the despair of realizing that his life had been left on brick piles. And always, there had been hunger and her bastard, the fear of hunger.

—Pietro Di Donato, in *Christ in Concrete*

collected in *Many Loves* (1961); books of essays, including *The Great American Novel* (1923) and *Selected Essays* (1954); novels, including *A Voyage to Pagany* (1928) and *The Build Up* (1952); short stories, collected in *Life Along the Passaic River* (1938) and other volumes; and his *Autobiography* (1951). Perhaps his best work, certainly his most ambitious, was the five-volume verse epic *Paterson* (1946–58). The William Carlos Williams house is listed in the National Register of Historic Places. Other Rutherford sites associated with the poet include the Rutherford Public Library, on Chestnut Street at Park Place, which contains first editions of the author's books, his Pulitzer Prize diploma, his desk and chair, tape recordings of interviews with Williams, and other items of interest to the literary visitor. The library is open to the public. Williams is buried in Hillside Cemetery on Rutherford Avenue in the adjacent town of Lyndhurst.

SEASIDE PARK

In 1927 **Augusta Huiell Seaman** moved to this community on Island Beach, off the coast of central New Jersey. Seaman used southern New Jersey for a number of her mystery books for young people. Many of her more than forty books were written here, including *The Pine Barrens Mystery* (1937), *The Case of the Calico Crab* (1942), and *The Vanishing Octant Mystery* (1949). Seaman died at home on June 4, 1950.

SHILOH

E[verett] T[itsworth] Tomlinson, born on May 23, 1859, in this community on Route 49 in southwestern New Jersey, was a clergyman, teacher, and author of books for boys. His many books, which included *Boys with Old Hickory* (1898), *Scouting with Daniel Boone* (1914), and *The Story of General Pershing* (1917), sold more than two million copies in his lifetime.

SOMERVILLE

Elinor [Hoyt] Wylie, whose volumes of poetry included *Nets to Catch the Wind* (1921) and *Angels and Earthly Creatures* (1929), was born in this city in central New Jersey on September 7, 1885. Her birthplace on Cliff Street is no longer standing.

SOUTH ORANGE

In 1907 **Will Durant** became a teacher at Seton Hall College in South Orange, west of Newark. In 1908 he entered the Catholic seminary here, but left in January 1910 to resume his teaching post, which he kept until June 1910.

STONE HARBOR

Joseph Hergesheimer, the novelist, short-story writer, and biographer, lived in this community near the southern tip of New Jersey from the mid-1940s until his death. Hergesheimer died in nearby Sea Isle City on April 25, 1954.

SUMMIT

Charles Jackson, best known as the author of *The Lost Weekend* (1944), was born on April 6, 1903, in this city west of Newark. His birthplace at 17 Edgar Place is now private.

TEANECK

Howard Fast lived in the 1950s at 692 Mildred Street, still standing, in this city east of Paterson. Here he wrote the novels *The Passion of Sacco and Vanzetti* (1953), *Silas Timberman* (1954), and *Moses, Prince of Egypt* (1958). **Robert Molloy,** the journalist and novelist, lived in Teaneck from the early 1950s to the mid-1960s, in a house at 560 Palisade Avenue. Molloy's novels include *A Multitude of Sins* (1953), *An Afternoon in March* (1958), *The Reunion* (1959), and *The Other Side of the Hill* (1962).

UNION CITY

This industrial city on the Hudson River in northeastern New Jersey has two notable literary native sons. **Eugène Jolas,** born here in 1894, founded with his wife the literary magazine *transition,* which they published in Paris between 1927 and 1938. The magazine was a forum for avant-garde writers in Europe and America, including Kay Boyle, Hart Crane, James Joyce, Gertrude Stein, and Allen Tate. **Pietro Di Donato,** born on April 13, 1911, in a part of Union City that then was in West Hoboken, used his childhood experiences as the son of an immigrant construction worker as the basis for his powerful and moving novel *Christ in Concrete* (1939). His other books include *Three Circles of Light* (1960), a sequel to the earlier novel; and the biography *Immigrant Saint: The Life of Mother Cabrini* (1960).

UPPER MONTCLAIR

Edward Sylvester Ellis, the prolific author of dime novels and popular histories, lived for many years at 85 Norwood Avenue in this community south of Paterson. Ellis wrote under at least six pen names, had eight publishers, and produced an unknown number of books in his lifetime. He died in Maine on June 20, 1916.

WAYNE

Albert Payson Terhune, the novelist and author of dog stories, lived for many years in

View of Sunnybank, estate of Albert Payson Terhune in Wayne. The property, now a town park, is located on the shore of Pompton Lake.

Wayne township, northwest of Paterson. Terhune spent his boyhood summers at Sunnybank, the estate of his parents, Reverend Edward Payson Terhune and **Mary Virginia Terhune**—his mother wrote under the pen name **Marion Harland.** Sunnybank, overlooking the town of Pompton Lakes, came to be an important part of Albert Payson Terhune's life. He loved its natural beauty and the companionship of the horses and dogs kept on the estate. Terhune turned to writing fiction in order to buy the estate and settle there permanently. This effort, combined with his work for the New York *Evening World,* proved fruitful. In 1912, he bought Sunnybank and, by 1916, was able to leave the *World* and devote his time exclusively to writing books. He found the appropriate medium for his talent when he began to write the dog stories that were collected later in the volume *Lad: A Dog* (1919). Terhune went on to produce many highly popular books about dogs, among them *Bruce* (1920), *Lochinvar Luck* (1923), *Gray Dawn* (1927), and *Lad of Sunnybank* (1928). Not only did Terhune write about dogs, but he had kennels at Sunnybank and raised champion collies, many of whom served as models for the dogs in Terhune's books. In 1928 Terhune was hit by an automobile while he was taking an evening stroll. He never fully recovered from his injuries, and his writing pace slackened. Notwithstanding, in his later years he produced several books, including his autobiography, *To the Best of My Memory* (1930), and *The Book of Sunnybank* (1934). Terhune died at Sunnybank on February 18, 1942, and was buried in the Dutch Reformed Cemetery in Pompton Lakes. Sunnybank, on Route 202 (Terhune Drive) in Wayne, is now known as Terhune Memorial Park and is open to the public. Visitors can see a garden containing plants gathered by Terhune and his wife in their travels around the world, as well as the graves of many of the dogs who lived not only at Sunnybank but in the pages of Terhune's books.

WEST END

Dorothy [Rothschild] Parker was born on August 22, 1893, in West End, which is near Long Branch. She was a poet, dramatist, critic, short-story writer, and trenchant wit, best known for her sardonic verse and eminently quotable remarks. A founder and charter member of the Algonquin Round Table, the New York City literary group, Parker worked on the staffs of *Vogue* and *Vanity Fair* magazines before becoming a steady contributor to *The New Yorker,* from 1925 to 1957. She also wrote for *Esquire* magazine for several years beginning in 1958. Among her many books were the collections of stories *Laments for the Living* (1930) and *Here Lies* (1939); the collection of poems *Death and Taxes* (1931); and the play *Ladies of the Corridor* (1954), written with Arnaud d'Usseau. Parker's volume *Constant Reader* (1970) was a selection of pieces originally published in *The New Yorker.* Although her list of published works was long, it is often the short but brilliant quotations from her verse and articles that readers remember. Among them:

She whose body's young and cool
Has no need of dancing school.

. . .

Scratch a king and find a fool!

Dorothy Parker wrote her own epitaph:

Excuse my dust.

WESTFIELD

Langston Hughes, the poet, lived in this town west of Elizabeth for about nine months in 1929 and 1930, in a house, now private, at 516 Downer Street. While staying at this house, which was the home of friends, Hughes wrote his novel *Not Without Laughter* (1930).

WEST ORANGE

Gertrude Knevels, the author of mystery novels, such as *The Octagon House* (1926) and *The Diamond Rose Mystery* (1928), died on April 7, 1962, in her home at 90 Northfield Avenue in this town north of Newark.

PENNSYLVANIA

ATHENS

Stephen Foster attended Athens Academy in this northern Pennsylvania town in 1840 and 1841. Here he composed his first song, "Tioga Waltz," which took its name from the Indian settlement that formerly stood on the site of present-day Athens.

ATLANTIC

Maxwell Anderson was born near this town in Crawford County, about fifty miles southwest of Erie, on December 15, 1888. Anderson's father was a Baptist minister, and the family moved many times, living not only in Pennsylvania but in Ohio, Iowa, and South Dakota. Among Anderson's best-known plays were *Both Your Houses* (1933), which won a Pulitzer Prize; *Winterset* (1935); and *High Tor* (1937).

BELSANO

Malcolm Cowley was born on August 24, 1898, on a farm on Route 422, near Belsano, which is about fifteen miles north of Johnstown. The Cowley family lived in Pittsburgh and used the farm as a summer home. The farm is now operated as a roadhouse. Cowley achieved a reputation as a poet, editor, literary historian, and translator. His poems were collected in two volumes: *Blue Juniata* (1929) and *Dry Season* (1942). In addition to editing such magazines as *The New Republic,* from 1929 to 1934, and such books as *The Portable Faulkner* (1946), Cowley has published notable works of literary history. His *Exile's Return: A Narrative of Ideas* (1934, revised 1951) was an analysis of American writers of the generation after World War I.

BETHAYRES

Thomas Costain lived here, on the outskirts of Philadelphia, for many of the fourteen years

Spirits that walk beside me in the air—
Having laid by, in your impatience,
The bonds of body and sense—
Tell me how long I must forbear
The ecstasy of going hence
And still submit to wear
The mask of this pretense.

—Elinor Wylie,
in "Absent Thee from Felicity Awhile,"
collected in *Angels and Earthly Creatures*

Men seldom make passes
At girls who wear glasses.

—Dorothy Parker

Stephen Foster in a watercolor portrait made from an ambrotype.

The stranger (Mr. Scratch devil) was . . . "a soft-spoken, dark-dressed stranger" with black boots who drove a handsome buggy, and smiled with many white teeth, "some say were filed to a point."

—Stephen Vincent Benét in "The Devil and Daniel Webster"

Greece sees, unmoved,
God's daughter, born of love,
the beauty of cool feet
and slenderest knees,
could love indeed the maid,
only if she were laid,
white ash amid funereal cypresses.

—H. D., in "Helen"

(1920 to 1934) during which he was associated with *The Saturday Evening Post.* Costain's novels included such historical romances as *The Silver Chalice* (1952) and *The Tontine* (1955).

BETHLEHEM

This industrial city in eastern Pennsylvania has been the home of two noted literary figures, both connected with Lehigh University. **Catherine Drinker Bowen,** known for writing biographies such as *Yankee from Olympus* (1944), which dealt with Justice Oliver Wendell Holmes and his family, came here as a child in 1905, when her father was made president of Lehigh. The house they lived in on the Lehigh campus is still used as the president's residence. **Richard Harding Davis,** author of popular adventure novels, such as *Soldiers of Fortune* (1897), came here in the 1880s to be tutored by his uncle H. Wilson Harding, professor of physics at Lehigh. In 1882 he enrolled at the university but did not take a degree. Davis lived at two different addresses here, his uncle's home at 745 Delaware Avenue and the historic Sun Inn, now the Sun Hotel, at 564 Main Street, whose previous guests had included John Adams, John Hancock, and George Washington. Both buildings are still standing.

Bethlehem was also the birthplace of two distinguished American writers. **Stephen Vincent Benét,** best known for his poem *John Brown's Body* (1928) and the story "The Devil and Daniel Webster" (1937), was born at his aunt's house here on July 22, 1898. Benét's father, an Army colonel, was stationed here over that summer and, when Stephen was only two months old, the family left Bethlehem. Benét's aunt's house, still standing at 827 North Bishopthorpe Street and now used as an apartment building, carries an identifying marker. Bethlehem's other noted literary native was **Hilda Doolittle,** who wrote under the name **H. D.,** the pen name she was given by Ezra Pound. She was born at 118 Church Street on September 10, 1886. The family moved away when Hilda was nine, and the home was later demolished. Probably the finest poet, after Ezra Pound, of the Imagist school, H. D. often wrote on classical themes. Among her books of verse were *Hymen* (1921) and *Hippolytus Temporizes* (1927). Hilda Doolittle died on September 27, 1961, and was buried in Nisky Hill Cemetery, on Church Street, just a few blocks from her birthplace. Also buried in Bethlehem, in "God's Little Acre" off Market Street, is Tschoop, the Mohican Indian, the model for Uncas in *The Last of the Mohicans* (1826), by James Fenimore Cooper.

BRYN MAWR

This residential community near Philadelphia is the home of Bryn Mawr College, long one of the most distinguished colleges in the nation. Many literary notables have lived in Bryn Mawr, some drawn by the college. **W. H. Auden** taught at Bryn Mawr from 1943 to 1945. **Catherine Drinker Bowen,** the biographer, lived here for some years after her marriage to Dr. Thomas McKean Downs in 1939. **Marcia Davenport,** author of biographies such as *Mozart* (1932) and novels such as *East Side, West Side* (1947), attended the Shipley School here in the 1910s. **Hilda Doolittle (H. D.)** attended Bryn Mawr for two years in the early 1900s.

George Kelly, winner of the Pulitzer Prize for his play *Craig's Wife* (1925), died at Bryn Mawr Hospital on June 18, 1974. **Emily Kimbrough,** best known for her collaboration with Cornelia Otis Skinner on *Our Hearts Were Young and Gay* (1942), was graduated from Bryn Mawr in 1921. **Richmond Lattimore,** the poet and translator, has taught at Bryn Mawr from 1935. **Marianne Moore,** the poet, received her BA from Bryn Mawr in 1909. **Owen Wister,** best known for his popular novel *The Virginian* (1902), lived on his estate at 270 South Bryn Mawr Avenue—the house is no longer standing—from 1924 until his death in 1938.

(*below*) Grave marker of H. D. in Nisky Hill Cemetery in Bethlehem. (*right*) God's Little Acre, in Bethlehem. Tschoop's stone is seventh from the bottom in the row to the left of the path.

Elinor Wylie, the poet and novelist, attended the Baldwin School in Bryn Mawr. Another student at the Baldwin School was **Cornelia Otis Skinner,** who moved to Bryn Mawr with her parents in the early 1900s and later attended Bryn Mawr College. While a student at the college she met Emily Kimbrough, with whom she made the European trip that became the subject of their popular book *Our Hearts Were Young and Gay* (1942). Among Skinner's other books were *Tiny Garments* (1932) and *Soap Behind the Ears* (1941).

CARLISLE

Marianne Moore lived from 1896 to 1916 at 343 North Hanover Street in this town, some ten miles southwest of Harrisburg. She moved here with her mother and brother when her mother accepted a job at the Metzger Institute, a girl's school. Moore attended the school herself, from 1899 to 1904, and later attended Bryn Mawr. Following a year at Carlisle Commercial College, she taught arithmetic, typing, stenography, and commercial law at the Carlisle Indian Industrial School. She also worked at her poetry, and her first published works appeared at this time in *Egoist* magazine. Her home on Hanover Street is still standing, but it is privately owned and not open to visitors.

CHADDS FORD

Sidney Lanier, in 1876–77, boarded in Chadds Ford, once a Revolutionary War battleground, ten miles north of Wilmington, Delaware. Lanier and his family stayed with Mrs. Caleb Brinton, probably at Sunnyside, her home west of the Brandywine River. Sunnyside is still standing but is privately owned and not open to visitors.

COLUMBIA

Reginald Wright Kauffman was born on September 8, 1877, in Columbia, a city about twenty-five miles down the Susquehanna River from Harrisburg. He grew up and attended elementary schools here, and his birthplace, at 113 South Second Street, is still standing but privately owned. Kauffman wrote a number of works in various forms, including such novels as *The House of Bondage* (1910) and *The Blood of Kings* (1926). His last novel, *Front Porch* (1933), was set in Columbia. Kauffman died in Roanoke, Virginia, on April 25, 1959, and is buried in Columbia's Mount Bethel Cemetery.

DOYLESTOWN

Several literary figures have lived or spent time in this town north of Philadelphia. **Oscar Hammerstein,** the librettist and lyricist, moved here with his family in 1940 and lived here until his death on August 22, 1960. **James Michener** was brought here from New York City shortly after his birth in 1907 and, living with his adoptive parents, spent his childhood here. **Katherine Anne Porter** completed her story "Old Mortality" and two long stories, *Noon Wine* (1937) and *Pale Horse, Pale Rider* (1939), while staying at Doylestown's Water Wheel Inn in 1936.

Margaret Widdemer, whose collection of verse *Old Road to Paradise* (1918) won a special Pulitzer Prize, was born in Doylestown in 1890. Widdemer's father was rector of St. Paul's Episcopal Church, and the family lived in the rectory, 184 East State Street.

EASTON

At least three literary men have been associated with Lafayette College in Easton, about fifteen miles northeast of Allentown. **Stephen Crane** enrolled at Lafayette in September of 1890 and roomed in 179 East Hall. The novelist and short-story writer stayed only one semester before failing in his studies and transferring to Syracuse University. **Robert Raynolds,** a novelist best known for *Brothers in the West* (1931), *Fortune* (1935), and *The Sinner of Saint Ambrose* (1952), was graduated from Lafayette in 1925. **Theodore Roethke** was at Lafayette from 1931 to 1935, dividing his time between teaching and writing poetry. He also served briefly as director of public relations at the college.

EDGEWORTH

Robinson Jeffers in October 1893 was brought to live in this community, about fifteen

In sleep she knew she was in her bed,
but not the bed she had lain down
in a few hours since, and the room
was not the same but it was a room
she had known somewhere. Her heart
was a stone lying upon her breast
outside of her; her pulses lagged and
paused, and she knew that something
strange was going to happen, even as
the early morning winds were cool
through the lattice, the streaks of light
were dark blue and the whole house
was snoring in its sleep.

—Katherine Anne Porter,
in *Pale Horse, Pale Rider*

(*below left*) Stephen Vincent Benét's birthplace in Bethlehem. (*below*) Water Wheel Inn in Doylestown, where Katherine Anne Porter lived and wrote for a time in 1936.

The world will little note, nor long remember what we say here, but it can never forget what they did here. It is for us the living, rather, to be dedicated here to the unfinished work which they who fought here have thus far so nobly advanced.

—Abraham Lincoln,
at Gettysburg

miles up the Ohio from Pittsburgh. He was six years old at the time, and his family lived in a house called Twin Hollows, 404 Beaver Road, while waiting for a new house to be completed on the same property. Shortly after moving into the new house, the family became dissatisfied with it, and moved back in to Twin Hollows for the rest of their stay here.

ERIE

Charles Erskine Scott Wood was born in this city on Lake Erie on February 20, 1852. Son of the first surgeon-general of the U.S. Navy, Wood received an appointment to West Point in 1870, but retired from the Army in 1884 as a protest against federal mistreatment of the American Indian. From 1884 to 1919, Colonel Wood practiced law in Portland, Oregon, and finally retired to write full time. He wrote several volumes of verse, which he considered his most important work. Among them were *The Poet in the Desert* (1915) and *Maia* (1929). Today, however, Wood is remembered best for his *Heavenly Discourse* (1927), a series of satiric dialogues conducted by such figures as Carry Nation, Voltaire, Thomas Paine, and Joan of Arc.

ERWINNA

S. J. Perelman and **Nathanael West** in 1932 bought a ninety-one-acre farm in this Bucks County community. West stayed here while working on his novel *A Cool Million* (1934), and Perelman and his wife—West's sister Laura—lived here until Laura's death in 1970. Perelman worked on many of his humorous pieces here, including his collection *Acres and Pains* (1947), which makes clear the indignities of country life for the transplanted urbanite.

Lincoln Speech Memorial at the National Cemetery in Gettysburg. The plaques on either side of the Lincoln bust are inscribed with the texts of the Gettysburg Address and the letter inviting the President to make "a few appropriate remarks" at the dedication of the cemetery.

GETTYSBURG

From July 1 to July 3, 1863, the Confederate Army and the Army of the Potomac engaged near this town in one of the bloodiest battles of the Civil War. Almost immediately, the battle began inspiring a wide variety of literary efforts. Poems about the battle or touching on it include "Gettysburg" by Henry Abbey, reported by veterans of the battle to give an accurate description; "John Burns of Gettysburg" by Bret Harte; "Gettysburg" by Edmund Clarence Stedman; "Gettysburg Ode" by Bayard Taylor; and "High Tide at Gettysburg" by Will Henry Thompson. Of the many novels inspired by the conflict, *Long Remember* (1934) by MacKinlay Kantor is usually considered the finest and the most accurate.

But who shall break the guards that wait
Before the awful face of Fate?
The tattered standards of the South
Were shriveled at the cannon's mouth,
And all her hopes were desolate.

—Will Henry Thompson,
in "High Tide at Gettysburg"

Lack of accuracy did not prevent another work from becoming an all-time best seller. *The Perfect Tribute* (1906), by Mary Raymond Shipman Andrews, described apocryphal incidents associated with the writing of President Lincoln's famous speech of November 19, 1863, given in dedication of the national cemetery at Gettysburg. Even today, public libraries have multiple copies of this short work, which tells how Lincoln purportedly scribbled his speech on the back of an envelope.

This speech is undoubtedly the literary masterpiece associated with Gettysburg. Formally entitled "An Address Delivered at the Dedication of the Cemetery at Gettysburg" but best known simply as "Lincoln's Gettysburg Address," this modest gem was delivered in two minutes but, on publication, was recognized as one of the greatest speeches ever delivered on the North American continent, a perception that remains true today. Visitors to the Gettysburg National Cemetery can read Lincoln's speech inscribed on a monument standing where Lincoln once stood.

HARRISBURG

Charles Fenno Hoffman died on June 7, 1884, in Harrisburg's state hospital, where he had been a patient since 1850. During Hoffman's working years, he held editorial positions on the *Knickerbocker Magazine* and the New York *Mirror*, and wrote several novels, the best known being *Greyslaer* (1840). This novel was based in part on a celebrated crime of passion in 1825, known as the Kentucky Tragedy, which involved a revenge murder and the failed double suicide of the killer and his wife. **Conrad Richter** lived near Harrisburg with his family in a 150-year-old farmhouse during the 1920s, during which time he wrote his first published book, *Brothers of No Kin and Other Stories* (1924). Richter won a Pulitzer Prize in 1931 for *The Town* (1950), the third novel in a trilogy.

HAVERFORD

This residential community just northwest of Philadelphia has had several prominent literary residents, most of them connected in some way with Haverford College. **John Dickson Carr,** the mystery writer, was graduated from Haverford in 1928. **Wright Morris** came here in 1944 to take a job as a lecturer at the college, living in nearby Wayne. During his stay here, he completed his novel *The Man Who Was There* (1944). **Frederick Prokosch,** author of such novels as *The Conspirators* (1943), did his undergraduate work at Haverford.

Logan Pearsall Smith attended Haverford from 1881 to 1884, leaving to go to Harvard and, eventually, to take his B.A. at Balliol College, Oxford. Smith's aphoristic wit and polished prose were displayed in such collections as *Trivia* (1902) and *Afterthoughts* (1921). Smith also wrote knowledgeably about the English language in *Words and Idioms* (1925).

Two distinguished authors whose long associations with Haverford began at birth were Catherine Drinker Bowen and Christopher Morley. **Catherine Drinker Bowen** was born here on January 1, 1897. Her family was living at the time on the Haverford College campus, where her father served as college president. In 1905, when her father became president of Lehigh University, the family moved to Bethlehem, Pennsylvania. Bowen returned to live in Haverford later in life and died at her home at 260 Booth Lane on November 2, 1973. She wrote *Yankee from Olympus* (1944), about Justice Oliver Wendell Holmes, and other works of biography. Her home is still standing, but is privately owned and not open to the public.

Christopher Morley, son of a Haverford mathematics professor, was born near the college campus on May 5, 1890. He attended Haverford himself, graduating in 1910. He served as editor of the college literary magazine and, after study at Oxford as a Rhodes Scholar in 1913, worked for various publications, most notably *The Saturday Review of Literature.* Morley's many books included *Parnassus on Wheels* (1917) and the bestselling *Kitty Foyle* (1939), both novels; collections of verse such as *The Old Mandarin* (1947); and *John Mistletoe* (1931), an autobiography. His birthplace has been torn down, but the Haverford college library has a Christopher Morley Memorial Alcove and an extensive collection of the author's books, manuscripts, and memorabilia. The alcove is open to the public during normal library hours.

HOLICONG

George S. Kaufman, the playwright, had a country estate here in Bucks County, not far from New Hope, where his collaborator Moss Hart had his own farm. Barley Sheaf Farm, as the Kaufman estate is known, was a favorite escape from New York: Kaufman stayed here while writing, with Hart, the play *George Washington Slept Here* (1940), a tale of a country house. He also worked here on the dramatization of J. P. Marquand's 1937 Pulitzer Prize novel, *The Late George Apley* (produced in 1946). Barley Sheaf Farm remains in private hands and is not open to visitors.

INDIANA

Agnes Sligh Turnbull, author of many novels, among them *The Rolling Years* (1936), *The Day Must Dawn* (1942), and her best known, *The Bishop's Mantle* (1947), attended Indiana State Teachers College here in 1909–10. Indiana is about twenty-five miles northwest of Johnstown. At the college, Turnbull lived in Sutton Hall, which is open to visitors during normal college hours.

JENKINTOWN

In 1891, when **Ezra Pound** was five years old, his family moved to this small town just north of Philadelphia. Their house, at 417 Walnut Street, is still standing. The Pounds stayed here for a little more than a year before moving in 1892 to the nearby town of Wyncote.

JOHNSTOWN

Conrad Richter began his writing career in this industrial city on the Conemaugh River, working as a reporter for the Johnstown *Journal* in 1910. He later spent two years working for the Johnstown *Leader.*

KENNETT SQUARE

Bayard Taylor was born on January 11, 1825, in this town thirty miles southwest of Philadelphia. Taylor combined his writing career with extensive travel. In addition to covering the California Gold Rush for the New York *Tribune,* he held diplomatic posts in Russia and Germany, explored in Asia and Africa, and served in Commodore Perry's expedition to Japan in 1853. Of Taylor's more than thirty-five books—travel-writing,

Webster's Collegiate Dictionary
. . . describes a farm as "a piece of land leased for cultivation, hence any tract developed for agricultural purposes." I prefer my own definition. A farm is an irregular patch of nettles bounded by short-term notes, containing a fool and his wife who didn't know enough to stay in the city.

. . .

[On moving to the country] I made several important discoveries. The first was that there are no chiggers in an air-cooled movie and that a corner delicatessen at dusk is more exciting than any rainbow.

—S. J. Perelman,
in *Acres and Pains*

Home of Ezra Pound in Jenkintown, now a duplex housing families of obviously independent tastes. The portion numbered 417 Walnut, where the Pound family lived in 1891 and 1892, is at the left in the photograph.

Drawing showing Cedarcroft in 1877, when it was the home of Bayard Taylor.

novels, journalism—his translation of Goethe's *Faust* (1870–71) is considered his finest literary achievement. In 1860 Taylor, then thirty-five, retired to a farm near Kennett Square and built a house he called Cedarcroft. Taylor lived here when not on missions in Europe. Among his later books were the novel *Kennett Square* (1866) and the collection of poems *The Masque of the Gods* (1872). Taylor died in Germany on December 19, 1878, and his body was returned for burial near the town of his birth.

Old parlor stove used by Zane Grey to heat his Lackawaxen study. Just to the right in the photograph is the original painting used for the cover of Grey's novel *The Man of the Forest* (1920).

LACKAWAXEN

In 1904 a New York dentist born **Pearl Zane Gray,** but known later to millions of readers as the Western writer **Zane Grey,** decided to give up his practice. He came to Lackawaxen, in Pike County across the Delaware River from New York State, bought a cabin and five acres of land, and settled down to write. After a trip to the West in 1908, he wrote *The Heritage of the Wind* (1910), the first of his Western novels to be accepted by a publisher. It sold well and was soon followed by the vastly popular *Riders of the Purple Sage* (1912). Grey soon became a wealthy man: in his lifetime, his books sold over thirteen million copies. His home was later opened to the public as the Zane Grey Inn. The inn has since closed, but the room in which he worked is maintained as a museum. Visitors may see such items as the lapboard on which Grey wrote his first ten novels, his Morris chair, and his dentist's drill. On the walls hang paintings used for the dustjackets of his books, and the sign from his dentist's office in New York. Grey died in October 1930 and was buried in Union Cemetery, next door to the Inn. Critic Burton Rascoe once said of Zane Grey: "It is difficult to imagine any writer having less merit . . . yet still maintaining an audience," but among Grey's many admirers was Dwight D. Eisenhower, who considered Grey one of his favorite authors.

LANCASTER

Helen Reimensnyder Martin, author of more than thirty novels about the Mennonites, was born in this city on October 18, 1868. Martin's best-known book was her first, *Tillie: A Mennonite Maid* (1904).

McKEESPORT

Marc Connelly was born on December 13, 1890, in McKeesport, a city about ten miles southeast of Pittsburgh. He probably was born at the old White Hotel on Fifth Avenue, which his father owned and operated, and it was in the hotel that he acquired his love of things theatrical. Buffalo Bill stayed at the hotel when he brought his Wild West show to town, and young Connelly spoke with him. At age nine, Marc with friends staged a number of productions in the Marcus Connelly Opera House, on the second floor of the hotel. Connelly's father died in 1902, but his mother continued to run the hotel until a financial panic forced her to sell in 1908. The family then left McKeesport for Pittsburgh, and Marc eventually went to New York to pursue his career in the theater. While many of his most successful plays were written with collaborators, he alone wrote *The Green Pastures* (1930), for which he won a Pulitzer Prize for drama. This play is generally considered Connelly's finest artistic achievement. The White Hotel has been torn down; on its site today stands the Chestnut Inn.

MARS

Agnes Sligh Turnbull taught high-school English here, about fifteen miles north of Pittsburgh, from 1911 to 1913. The two-room school on Crowe Avenue she taught in is used now as a kindergarten.

MEADVILLE

This town in northwest Pennsylvania is the home of Allegheny College, alma mater of **Ida M. Tarbell,** best known for her exposé *The History of the Standard Oil Company* (1904). Tarbell received her B.A. in 1880 and M.A. in 1883.

MERCERSBURG

Max Eastman, the author and editor whose books include the enormously successful *The Enjoyment of Poetry* (1913), attended Mercersburg Academy, founded in 1836 and still operating. Mercersburg is about fifteen miles southwest of Chambersburg, near the Maryland border. Eastman was graduated in 1900 as class valedictorian, with the highest average attained up to that point in the school's history.

MILTON

Margaret Junkin Preston was born here, about fifteen miles southeast of Williamsport, on May 19, 1820. Preston's husband served on Stonewall Jackson's staff during the Civil War. From what were reported to be Jackson's last words—"Let us cross over the river and rest under the shade of the trees"—she created her best-known poem, "Under the Shade of the Trees." Preston's verse is collected in *Old Songs and New* (1870) and *For Love's Sake* (1886).

What are the thoughts that are stirring
his breast?
What is the mystical vision he sees?
—"Let us pass over the river, and rest
Under the shade of the trees."

—Margaret Junkin Preston,
in "Under the Shade of
the Trees"

NEW ALEXANDRIA

Agnes Sligh Turnbull was born in this village between Pittsburgh and Johnstown on October 14, 1888. She attended New Alexandria Grammar School from 1894 to 1906 and began writing during her childhood. New Alexandria appears in her novel *The Gown of Glory* (1952).

NEW HOPE

This Bucks County town has long been popular with affluent New Yorkers and writers seeking a country retreat. **Moss Hart** maintained a farm on Aquetong Road and paid frequent visits to his collaborator George S. Kaufman in nearby Holicong. **Stanley Kunitz,** the poet, lived for a time in an eighteenth-century stone house near Bowman's Hill Tower. **Budd Schulberg** lived in a stone house on Old York Road in the late 1940s and early 1950s. While living here, Schulberg wrote his novels *The Harder They Fall* (1947), about the underworld's role in prizefighting, and *The Disenchanted* (1950), a recounting of the last year in the life of a once-great novelist whose misfortunes bring to mind F. Scott Fitzgerald.

NEWTOWN

James Michener taught from 1933 to 1936 at the George School, still operating in this town, about twenty-two miles northeast of Philadelphia.

OXFORD

Joseph Henry Taylor, the trapper, editor, and author, died in Washburn, North Dakota, on April 9, 1908. He was buried near the graves of his parents in the Friends Persuasion Churchyard here in southwestern Pennsylvania.

PATTON

Conrad Richter, in the early 1900s, edited the *Weekly Courier* in Patton, which is about fifteen miles northwest of Altoona.

PERKASIE

Pearl S. Buck made her home from 1937 to 1972 in Perkasie, which lies between Philadelphia and Allentown. During this period she wrote prolifically and, in 1938, was awarded the Nobel Prize for literature. Among the books she wrote here were *Dragon Seed* (1942), *What America Means to Me* (1943), *Peony* (1948), *My Several Worlds* (1954), *Imperial Woman* (1956), and *The Living Reed* (1963). The novelist and her husband raised their children here—five of them were adopted—and after her husband's death in 1960, Buck adopted and raised four more. Pearl Buck died on March 6, 1973, and was buried here. Today her home, Green Hills Farm, at 520 Hilltown

Photograph of Agnes Sligh Turnbull taken on her ninetieth birthday, October 14, 1978.

Road, also called Dublin Road, is headquarters for the Pearl S. Buck Foundation, which she established in 1964 to aid children of mixed American and Asian parentage. The main house contains original furnishings and memorabilia, including the Chinese desk at which Buck wrote her most famous novel, *The Good Earth* (1931), one of many she set in China, and for which she won a Pulitzer Prize. The farm is a National Historic Site, open to the public.

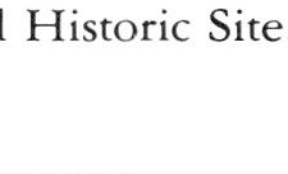

PHILADELPHIA

Established in the 1680s by William Penn, Philadelphia soon became an important port and commercial center, and by the time of the American Revolution was perhaps the major center for publishing in the Colonies. It was publishing that drew young Benjamin Franklin to Philadelphia, and years later it was Franklin and his influence that drew Thomas Paine here. Philadelphia became a center of magazine publishing, and among the many magazines founded or published here, three stand out as powerful influences in the growth of popular American literature. *Godey's Lady's Book* (1830–98), founded here, was one of the most popular magazines in America in the years before the Civil War. *The Ladies' Home Journal,* under the editorship of **Edward Bok,** reached an enormous readership. Bok introduced departments to the magazine that dealt with such subjects as child care, gardening, home design, and decoration. He and his publisher, Cyrus Curtis, banned the advertising of patent medicines and published fiction by such writers as Bret Harte, William Dean Howells, Rudyard Kipling, and Mark Twain. *The Saturday Evening Post,* founded in Philadelphia in 1821, reached its peak under the editorship of **George Horace Lorimer,** the magazine's editor from 1899 to 1936. Lorimer solicited contributions from virtually every well-known writer of the early twentieth century. He lived in Wyncote, a Philadelphia suburb, until his death on October 22, 1937, and is buried in Laurel Hill Cemetery, 3822 Ridge Avenue in Philadelphia.

Philadelphia is also noteworthy for the number of its native sons and daughters who became prominent writers, including **Louisa May Alcott,** who is discussed below, and **William Bartram,** the botanist and travel writer, born on February 9, 1739. Bartram's father, **John Bartram,** was himself a botanist and author and is considered America's first botanist. The elder Bartram's home at Fifty-fourth Street and Lindbergh Avenue is open to visitors daily. A donation is requested. The property adjacent to the Bartram House was used as this country's first botanical garden.

Donald Barthelme, the novelist and short-story writer, was born in Philadelphia on April 7, 1931; and **Charles Brockden Brown,** considered to be America's first professional novelist, was born in Philadelphia on January 17, 1771, place unknown. When young, Brown lived in a house on Chancellor Street between Second and Third streets. He attended the Friends School, located on Fourth Street below Chestnut Street, and later served for a time as assistant master of the school and then as headmaster in 1794–95. Brown also lived in a house, no longer standing, at 117 Second Street. After living in New York City for several years, and failing—despite the fact that he completed all his major works there—to establish himself in that city's literary world, Brown returned to Philadelphia. He initially worked with his brothers in a business at 23 North Water Street, but soon left to found a monthly, the *Literary Magazine and American Register,* which he edited from 1803 to 1807. In 1807 he became

Gravestone of Pearl S. Buck, on the grounds of her home in Perkasie, bearing an oriental rendering of her name. A sign on Route 313 directs travelers to the author's home (*below*).

editor of the *American Register, or General Repository of History, Politics and Science* and continued in that post until his death. In his last years Brown worked as a businessman and journalist, wrote a number of books now generally considered to be hack-work, and lived in a house that stood at 74 (later 124) Eleventh Street. He died there on February 22, 1810, and is buried in the churchyard of the Arch Street Friends Meeting House, at 304 Arch Street. Philadelphia was the setting for one of Brown's best novels, *Arthur Mervyn* (1799–1800), which covered the period of Philadelphia's yellow fever epidemic of 1793.

Richard Harding Davis, the novelist, short-story writer, and war correspondent, was born on April 18, 1864, in a house at 1429 Girard Avenue. In 1865 the Davis family moved to 1817 North Twelfth Street. There Davis's mother, **Rebecca Harding Davis,** wrote two of her novels, *Waiting for the Verdict* (1868) and *Dallas Galbraith* (1868). In 1869 the family moved to 230 South Twenty-first Street, when Davis's father became managing editor of the Philadelphia *Inquirer.* He served later on the staff of the Philadelphia *Public Ledger,* from 1889 until his death in 1904. Rebecca Harding Davis wrote her later fiction in the house on Twenty-first Street, including *John Andross* (1874), *Silhouettes of American Life* (1892), *Kent Hampden* (1892), and *Doctor Warrick's Daughters* (1896). Richard Harding Davis no doubt owed much of his own development to the influence of his parents. He began his journalistic career in Philadelphia by working as a reporter for the Philadelphia *Record* and for the *Press.* After the success of his short story "Gallegher: A Newspaper Story" (1890), in which a street-wise copyboy at the Philadelphia *Press* helps solve a crime and, through extraordinary efforts, gets the story of the culprit's arrest to the newspaper in time to beat the competition, Davis moved to New York City and began his steady rise to the upper ranks of American journalism. He died in 1916 at his home near Mount Kisco, New York, and his ashes were interred in the family plot in Leverington Cemetery, at Ridge Avenue and Conarroe Avenue, in Philadelphia.

Other writers born in Philadelphia include **Miriam A[llen] de Ford,** a versatile writer known especially in her time for her mystery stories, who was born here on August 21, 1888; **Ignatius Donnelly,** the historian, novelist, and politician, who was born here on November 3, 1831; and **Thomas Dunn English,** the physician and author, who was born here on June 29, 1819. English, best known as author of "Ben Bolt," a popular poem that was set to music, grew up in Philadelphia, received his medical degree from the University of Pennsylvania Medical School, and was admitted to the Philadelphia Bar in 1843.

James Hall, who was born here on August 19, 1793, went on to found, edit, and largely write the *Illinois Monthly Magazine,* an early literary periodical—the first of its kind published west of Ohio. Hall also recorded in several books the legends and folklore of the then rapidly advancing Western frontier. His work included *Letters from the West* (1828) and *Legends of the West* (1832). **Henry George,** author of the classic economic treatise *Progress and Poverty* (1879), was born on September 2, 1839, in a house at 413 South Tenth Street. His birthplace is open to visitors.

Photograph of Edward Bok, editor of the *Ladies' Home Journal* for thirty years. Bok had an enormous influence on American literature in the late nineteenth and early twentieth centuries.

Don't you remember sweet Alice, Ben Bolt,—
Sweet Alice whose hair was so brown,
Who wept with delight when you gave her a smile,
And trembled with fear at your frown?
In the old church-yard in the valley, Ben Bolt,
In a corner obscure and alone,
They have fitted a slab of the granite so gray.
And Alice lies under the stone.

—Thomas Dunn English, in "Ben Bolt"

Also natives of Philadelphia were **Joseph Hergesheimer,** author of such novels as *The Three Black Pennys* (1917), set in part in Philadelphia, and *Java Head* (1919), was born on February 15, 1880; **Thomas A[llibone] Janvier,** the journalist and fiction writer, born on July 16, 1849; and **George Kelly,** believed to have been born in 1887. Kelly, an actor and playwright, used Philadelphia as a setting for his plays *The Torch Bearers* (1922) and *The Show-Off* (1924), and won a Pulitzer Prize for his play *Craig's Wife* (1925). **Bayard Kendrick,** creator of the blind detective Duncan

Philadelphia home of John and William Bartram. Linnaeus once referred to John Bartram as "the greatest natural botanist in the world."

I have but to close my eyes to see the house in which I lived in my youth. It stood in the city of Penn, back from the low bluff of Dock Creek, near to Walnut Street. The garden stretched down to the water, and before the door were still left on either side two great hemlock-spruces, which must have been part of the noble woods under which the first settlers found shelter. . . . The house was of black and red brick, and double; that is, with two windows on each side of a white Doric doorway, having something portly about it. I use the word as Doctor Johnson defines it: a house of port, with a look of sufficiency, and, too, of ready hospitality. . . .

—S. Weir Mitchell, in *Hugh Wynne: Free Quaker*

Maclain, who appeared in *Blind Man's Bluff* (1943) and other novels, and first president of Mystery Writers of America, Inc., was born here on April 8, 1894.

S[ilas] Weir Mitchell, born on February 15, 1829, achieved eminence as a physician and reputation as a novelist. He lived in childhood at 238 South Eighth Street; at 1226 Walnut Street, no longer standing, from 1861 to 1866; and at 1524 Walnut Street, also no longer standing, from 1873 until his death on January 4, 1914. At his last home, Mitchell wrote his novels, short stories, and collections of verse. Among his works were the novels *In War Time* (1885), *Hugh Wynne: Free Quaker* (1897), and *The Red City* (1907), the last two set in Philadelphia during the Revolutionary War. Philadelphia's native-born authors also include **Clifford Odets,** the playwright known especially for *Waiting for Lefty* (1935), born on July 18, 1906; and **Robert E. Spiller,** the editor and literary historian, born on November 13, 1896, in a house at Forty-second and Pine streets. Spiller lived from 1897 to 1921 at 14409 Pine Street, and there wrote *The American in England During the First Half-Century of Independence* (1926). Spiller was graduated from the University of Pennsylvania with a B.A. in 1917, received his M.A. there in 1921, and his Ph.D. in 1924. He taught at the university from 1919 to 1921 and 1945 to 1967.

Frank R. [Francis Richard] Stockton, the novelist and short-story writer, was born in Philadelphia, address unknown, on April 5, 1834. Stockton grew up in Philadelphia, where he learned illustrating and engraving, his first career, but soon turned to editing and writing. While living here, he wrote a pamphlet, *A Northern Voice for the Dissolution of the United States of North America* (1860). Stockton left Philadelphia in the mid-1860s. He died in Washington D.C., on April 20, 1902, and was buried in Philadelphia's Woodlands Cemetery, Thirty-ninth Street and Woodland Avenue. The final native Philadelphian worthy of mention is **Kate Douglas Wiggin,** author of *Rebecca of Sunnybrook Farm* (1903) and other books for children. Wiggin was born here on September 28, 1856.

Photograph of South Tenth Street taken in 1958. Henry George's birthplace is the house with a bay window in its second story.

Writers have been attracted to live and work in Philadelphia since Colonial times. **A. Bronson Alcott** in the 1830s established a school at 222 South Eighth Street and operated it for about a year before moving the school to 5425 Germantown Avenue in Germantown, now part of Philadelphia. In the Germantown house, no longer standing, was born Alcott's daughter, **Louisa May Alcott,** author of *Little Women* (1868), on November 29, 1832.

Robert Montgomery Bird, the dramatist and novelist, spent most of his adult life in Philadelphia. Bird was graduated in 1827 from the University of Pennsylvania Medical School. In 1828 he moved to a house at 247 South Tenth Street, where he began to write plays for a competition sponsored by the actor, Edwin Forrest, himself a native of Philadelphia. Of the four plays Bird wrote for the competition, two are believed to have been completed here: *The Gladiator* (1831) and *Pelopidas* (never produced). Bird also lived at 268 South Eleventh Street, where he worked on his novels *Calavar* (1834), *The Infidel* (1835), and *The Hawks of Hawk Hollow* (1835), and at 140 North Twelfth Street, where he is believed to have written his best-known novel, *Nick of the Woods; or, The Jibbenainosay* (1837). In his later years Bird taught at the Pennsylvania Medical College, wrote for the Philadelphia daily newspaper *The North American,* and lived in a house on Filbert Street, between Ninth and Tenth streets. Bird died there on January 23, 1854, and is buried in Laurel Hill Cemetery, at Lehigh Avenue and Thirty-fourth Street.

Edward Bok, the editor and journalist who won a Pulitzer Prize for his autobiography, *The Americanization of Edward Bok* (1920), served as editor of *The Ladies' Home Journal* from 1889 to 1919. Bok lived in Philadelphia until about 1895, first in a house, no longer standing, at 1502 Walnut Street; then at 128 South Nineteenth Street; and later in a succession of Philadelphia hotels, before buying a home in Merion. **Joseph Dennie** lived in Philadelphia from 1799 until his death here on January 7, 1812. He founded and edited *The Port Folio* (1801–27), a weekly publication that contained both political writing and literature. Dennie was a leading figure in Philadelphia literary circles until his death. He is buried in the churchyard of St. Peter's Episcopal Church, 313 Pine Street.

John Dickinson, the Penman of the Revolution, spent much of his adult life in Philadelphia. He came here from Delaware in 1750 to study law, and thereafter made Philadelphia his home, with extended stays in Delaware, until 1785, when he returned to Delaware to live. While living in Philadelphia, Dickinson wrote his influential *Letters From a Farmer in Pennsylvania to the Inhabitants of the British Colonies* (1768), first published in the Philadelphia *Pennsylvania Chronicle and Universal Advertiser* (1767–68). Dickinson helped draft the Articles of Confederation and, while serving as president of Pennsylvania's Supreme Council from 1782 to 1785, lived in a house at the southeast corner of Sixth and Market streets.

Theodore Dreiser, suffering from writer's block and a nervous condition—largely a reaction to the failure of his novel *Sister Carrie*—came to Philadelphia in July 1902, and took a room at 3225 Ridge Avenue. Unable to work on *Jennie Gerhardt,* his novel-in-progress, Dreiser wrote several articles for local periodicals and the sketch "Christmas in the Tenements." In February 1903, Dreiser left for New York City, but Philadelphia was the setting for *The Financier* (1912). This novel was based on the Philadelphia years of Charles Tyson Yerkes, Jr., a financial operator and transportation magnate, on whom Dreiser modeled Frank Cowperwood, the main character of the novel.

Philip [Morin] Freneau lived in Philadelphia in the 1780s and worked for a time in the Philadelphia post office. He contributed to *The United States Magazine,* which was edited by his Princeton classmate, H. H. Brackenridge. He also contributed to and helped edit *The Freeman's Journal; or, the North American Intelligencer.* The publisher of this journal, Francis Bailey, later published *The Poems of Philip Freneau* (1786) and *The Miscellaneous Works of Mr. Philip Freneau Containing His Essays, and Additional Poems* (1788). After spending several years at sea, and then working in New York City, Freneau returned to Philadelphia in 1791 and became translating clerk of the Department of State. In October of that year, he published the first issue of *The National Gazette* (1791–93), which was supported by Thomas Jefferson in its attacks against the Federalists. In 1793, after the *Gazette* suspended publication and Jefferson had retired as secretary of state, Freneau left Philadelphia.

Sarah Josepha Hale, the editor and writer, served as editor of *Godey's Lady's Book* from 1837 to 1877. Hale lived in Philadelphia from 1841, when the magazine was moved to Philadelphia from Boston, until her death on April 30, 1879. She died at her home, 1413 Locust Street, and was buried in Laurel Hill Cemetery. A popular writer of poetry, plays, and articles as well as an extremely successful editor, Mrs. Hale is best remembered for her poem "Mary's Lamb" (1830).

George Lippard, the newspaperman and prolific novelist, spent much of his short life—he died at the age of 31—in Philadelphia, the setting of his best-known work, *The Monks of Monk Hall* (1844). This novel, enlarged and published in 1845 as *The Quaker City,* was an account of corruption in Philadelphia, and is considered this country's first muckraking novel. Lippard dramatized it for presentation at Philadelphia's Chestnut Street Theatre, but the performance was banned by the mayor, who feared that the play would provoke a riot. Lippard lived in childhood in a house, no longer standing, at the southwest corner of Sixth and Callowhill streets. As a young man he wrote for the daily newspaper *Spirit of the Times,* then located at the northwest corner of Third and Chestnut streets. He moved in 1847 to the house at 965 North Sixth Street in which Edgar Allan Poe three years earlier had spent his last day in Philadelphia. In the 1840s and early 1850s, Lippard was drawn away from Philadelphia on extensive lecture tours, and lived in Cleveland for a time, but he returned to Philadelphia in 1853. In 1854 he moved to 1509 Lawrence Street, where he died on February 9, 1854. His grave in Odd Fellows Cemetery, 3111 West Lehigh Avenue, is marked by a monument placed there by the Brotherhood of the Union, later called the Brotherhood of America, a fraternal organization that Lippard had founded at his North Sixth Street home.

James Russell Lowell lived for several months in early 1845 in a house, no longer standing, at the northeast corner of Fourth and Arch streets. He wrote for *Graham's Magazine,* then located at the southwest corner of Third and Chestnut streets, and for the Abolitionist publication *The Pennsylvania Freeman* (1838–54), located on North Fifth Street. In 1889 **Ezra Pound,** then three years old, was brought by his parents to live in Philadelphia. The Pounds lived at 208 South Forty-third Street until 1891, then moved to Wyncote, where Pound grew up. In 1901 he entered the University of Pennsylvania, but transferred in 1903 to Hamilton College, in Clinton, New York. Pound returned in 1905, completed graduate studies at the university, and received his M.A. in 1906. During part of his college career, Pound's parents lived at 502 South Front Street, and it is believed that he stayed there during school holidays.

Gilbert Seldes, the critic and author, worked for the Philadelphia *Evening Ledger* from 1914 to 1916. Years later, from 1959 until 1963, he served as professor and dean of the Annenberg School of Communications of the University of Pennsylvania and lived at 3416 Sansom Street. **Logan Pearsall Smith,** the critic and essayist, grew up in Germantown in the 1870s and 1880s. **Walt Whitman** was a frequent visitor to the Smith household. Smith was fond of books from childhood on, and it must have come as a surprise to his family when he chose not to follow in the footsteps of his grandfather and uncle as librarian of Philadelphia. He also chose not to take over his father's manufacturing business; he left the United States for England in 1888, and made his home there for the rest of his life.

John Augustus Stone, a prominent actor and playwright of the early 1800s, performed on Philadelphia stages in the 1820s and 1830s, and is said to have lived in Philadelphia toward the end of his life. The playwright suffered periods of mental instability and, on May 28 or 29, 1834, committed suicide by jumping into the Schuylkill River. His grave, in Mount Moriah Cemetery, at Kingsessing Avenue and Cemetery Road, is marked by a monument erected by Edwin Forrest, the actor, who achieved fame and fortune as the lead in Stone's play *Metamora; or, the Last of the Wampanoags* (unpublished, produced 1829). Philadelphia has the distinction of being the place of publication of Stone's only known surviving play, *Tancred; or, the Siege of Antioch* (1827). Another play by Stone, *The Ancient Briton,* was first produced at Philadelphia's Arch Street Theatre in March 1833. Stone acted at several Philadelphia theaters during his career, including the Walnut Street Theatre at Ninth and Walnut streets, which has been restored as a performing-arts center.

T. De Witt Talmage, the author of popular sermons, served as pastor of the Second Dutch Reformed Church here from 1862 to 1869. **William Tappan Thompson,** the humorist and editor, grew up in Philadelphia in the 1810s and 1820s, and worked for a time for the Philadelphia

Sarah Josepha Hale, in an engraving made from a painting by Thomas Buchanan Read (1822–72). Read was a painter and poet, known widely for his poem "Sheridan's Ride."

I am a FARMER, settled after a variety of fortunes, near the banks, of the river Delaware, *in the province of* Pennsylvania. *I received a liberal education, and have been engaged in the busy scenes of life: But am now convinced, that a man may be as happy without bustle, as with it.*

—John Dickinson, the opening sentence of *Letters From a Farmer in Pennsylvania, to the Inhabitants of the British Colonies*

Things lived on each other—that was it. Lobsters lived on squids and other things. What lived on lobsters? Men, of course! Sure, that was it! And what lived on men? he asked himself. Was it other men?

—Theodore Dreiser, in *The Financier*

But an attempt to fix a standard of any particular class of people is highly absurd: As a friend of mine once observed, it is like fixing a light house on a floating island. It is an attempt to fix *that which is in itself* variable; *at least it must be variable so long as it is supposed that a local practice has no standard but a* local practice; *that is, no standard but* itself. *While this doctrine is believed, it will be impossible for a nation to follow as fast as the standard changes—for if the gentlemen at court constitute a standard, they are above it themselves, and their practice must shift with their passions and their whims.*

—Noah Webster, in *Dissertations on the English Language*

Daily Chronicle. **Noah Webster** lectured in Philadelphia in the 1780s and became a friend of Benjamin Franklin's. Webster lived in Philadelphia for ten months in 1786–87 and taught in an Episcopal school here. During these months Webster talked with Franklin about Franklin's system of phonetic spelling. Franklin's ideas so impressed Webster that he drew on them in writing his "Essay on a Reformed Mode of Spelling," which was published in *Dissertations on the English Language* (1789).

John Greenleaf Whittier lived in Philadelphia in the late 1830s and served as editor of *The Pennsylvania Freeman.* Whittier edited the newspaper first from offices at 223 (later 619) Arch Street, then in May 1838 from offices in the newly opened Pennsylvania Hall, on North Sixth Street. Shortly after the move, a pro-slavery mob attacked and burned the building. **Alexander Woollcott** lived in Germantown in the early 1900s, and was graduated from Philadelphia's Central Public High School in 1904.

Philadelphia also has associations with a number of writers who attended school here, worked here, or used the city as a subject or setting for their writing. Philip Barry used Philadelphia as the setting for his play *The Philadelphia Story* (1939). **Thomas B. Costain,** the journalist and author, was associate editor of *The Saturday Evening Post* from 1920 to 1934. **Emily Dickinson** visited Philadelphia in March 1855—the visit was long thought to have taken place in 1854—while en route to Washington, D.C., where her father was serving in Congress. It was in Philadelphia that she is said to have met Rev. Charles Wadsworth, pastor of the Arch Street Presbyterian Church, who was to be her friend—Dickinson was then twenty-three—for the rest of her life. Although their relationship remains somewhat obscure to biographers, there is speculation that Dickinson fell in love with the married clergyman, and that her years of seclusion at her home in Amherst, Massachusetts, had their root in this impossible love.

Washington Irving served as editor of Philadelphia's monthly *The Analectic Magazine* (1813–21) in 1813 and 1814. Although Irving did not live in Philadelphia at this time, he did visit Rebecca Gratz, who lived in a house, no longer standing, at 110 (later 317) Chestnut Street. Irving described Rebecca Gratz in a letter to his friend Sir Walter Scott, who used Irving's description as the basis for Rebecca in the novel *Ivanhoe* (1820). Miss Gratz died in 1869 and was buried in the Cemetery of Mikveh Israel Congregation, at Spruce and Darien streets.

Edgar Lee Masters died in a nursing home in the Philadelphia suburb of Melrose Park on March 5, 1950. **Carson McCullers** saw her play *The Member of the Wedding* produced for the first time on June 5, 1950, at the Empire Theatre in Philadelphia. The Philip H. and A. S. W. Rosenbach Foundation, 2010 Delancey Place, maintains the literary papers of Marianne Moore. The foundation also maintains a Marianne Moore Room, a faithful reconstruction of the living room of the poet's home in Greenwich Village. The Rosenbach Foundation is open to visitors, and a modest admission fee is charged. Scholars may visit by appointment.

John O'Hara in the fall of 1937 completed the last pages of his novel *Hope of Heaven* (1938) at the Benjamin Franklin Hotel, Ninth and Chestnut streets. **Horace Traubel,** best known for his friendship with Walt Whitman during the last years of Whitman's life in nearby Camden, New Jersey, founded and edited two publications in Philadelphia: the *Conservator* (1890–1919) and *The Artsman* (1903–07). **Mark Twain** worked as a journeyman printer here in 1853 and 1854, first for the Philadelphia *Inquirer* and then for the *North American.* About 1851 Twain had written two anecdotes for *The Saturday Evening Post,* for which he received no payment. Twain was later quoted by Albert Bigelow Paine, his biographer, as saying: "Seeing them in print was a joy which rather exceeded anything in that line I have ever experienced since." **Leon Uris,** the best-selling novelist, attended John Bartram High School in Philadelphia, but left school at age seventeen to serve in the U.S. Marine Corps during World War II.

Walt Whitman did not live in Philadelphia, but in the 1870s and 1880s was a frequent visitor from his home in Camden and a well known figure along Market Street. The Walt Whitman Bridge, which spans the Delaware River, connects the cities of Camden and Philadelphia, serving to remind the toll-paying motorist of the former presence of one of the country's great poets. A replica of the statue of Whitman by Jo Davidson marks the Philadelphia access to the bridge.

Philadelphia's position as a major American city made it a favorite stopping point for English literary visitors in the nineteenth century. **Charles Dickens** stayed at the United States Hotel, then on Chestnut Street near Fourth Street, in March 1842, during his first visit to the United States. Dickens remarked a number of things he admired about the city in his book *American Notes* (1842), but strongly condemned the practice of solitary confinement then in use at the Eastern Penitentiary, at Fairmount and Corinthian avenues. Dickens also met **Edgar Allan Poe,** then living in Philadelphia. One might imagine that the two authors discussed literature, but they did not: their concern was international copyrighting, and Poe asked for help in finding English publishers for his works. Dickens promised to help, but proved unable to interest English publishers.

Fanny [Frances Anne] Kemble, the English actress and author, came to Philadelphia for the first time in 1832, when she met Pierce Butler. They were married on June 7, 1834, at Christ Church, on Second Street near Market Street. The Butlers lived in Philadelphia at Butler Mansion, at the northwest corner of Eighth and Chestnut streets, and also at a farm known as Butler Place, about six miles from Philadelphia on Old York Road, but no longer in existence. Kemble's *Journal* (1835) records her acting tour of 1832–34. In addition to his property in Philadelphia, Pierce Butler owned land and slaves in Georgia, and slavery was anathema to Kemble. The question of slave ownership was one of the major differences of opinion that led to the couple's separation in 1841, and divorce in 1848. After 1841, Fanny Kemble divided her time between the United States and Europe. She lived for a time in a house at 1812 Rittenhouse Square before moving to a cottage called York Farm, across the road from Butler Place. She lived there for several years and

SANDY. *No, they can't. Not till they get their story.*
TRACY. *Story? What Story?*
SANDY. *The Philadelphia story.*
MARGARET. *And what on earth's that?*
SANDY. *Well, it seems Kidd has had Connor and Imbrie and a couple of others down here for two months doing the town: I mean writing it up. It's to come out in three parts in the Autumn. "Industrial Philadelphia," "Historical Philadelphia"—and then the third—*
TRACY. *I'm going to be sick.*
SANDY. *Yes, dear, "Fashionable Philadelphia."*
TRACY. I am *sick.*

—Philip Barry, in *The Philadelphia Story*

had many visitors, among them her grandson, **Owen Wister,** who one day would become a distinguished literary Philadelphian.

William Makepeace Thackeray visited Philadelphia during his American tours of 1852–53 and 1855–56, and **Frances Trollope** visited the city in June 1830. Mrs. Trollope described her two-week stay here in *Domestic Manners of the Americans* (1832).

Of all Philadelphia's literary figures, four are most closely identified with the city: Benjamin Franklin, Thomas Paine, Edgar Allan Poe, and Owen Wister. **Benjamin Franklin** came to Philadelphia from Boston in 1723. The young Franklin established himself as a printer, and in 1729 purchased and became editor of *The Universal Instructor in All Arts and Sciences and Pennsylvania Gazette.* The weekly newspaper had been founded the year before by Samuel Keimer, who had begun publishing it to thwart Franklin's own plan for starting a newspaper. Keimer soon ran into debt, and Franklin bought the newspaper, renamed it the *Pennsylvania Gazette,* and edited it until 1766. Franklin published his first *Poor Richard* almanac in Philadelphia in 1732. This almanac, purportedly compiled by one Richard Saunders of Philadelphia, was issued annually by Franklin until 1757. It not only contained practical information but also such maxims as "The rotten apple spoils his companions," "Early to bed and early to rise makes a man healthy, wealthy, and wise," and "God helps them that help themselves." Franklin, through industry, practicality, and shrewd business sense, became a leading figure in Philadelphia. Aside from his printing and publishing interests, Franklin founded The Junto, a group of twelve Philadelphia workingmen who met once a week for approximately forty years to discuss topics of mutual interest or importance to the community. It was The Junto, at Franklin's urging, that formed Philadelphia's first subscription library.

Franklin's interests were remarkably varied. He served as a member of the Pennsylvania Assembly and later of the Continental Congress, experimented with electricity, helped to found Pennsylvania Hospital, organized Pennsylvania's first militia and Philadelphia's first volunteer fire company, was elected first postmaster general by Congress, and produced a stream of essays, letters, and pamphlets on subjects ranging from the function and importance of lightning rods to his "Advice to a Young Man on the Choice of a Mistress." Perhaps his greatest contribution to Philadelphia, and to the country-to-be, was his founding of the American Philosophical Society in 1743. The society was formed to promote the advancement of knowledge in the Colonies. In the 1780s its members built the brick structure at Fifth and Chestnut streets, next to Philadelphia's Independence Hall. The building is still used for meetings of the society, but is not open to visitors. Independence Hall, which is open to visitors, was the site of the signing, on August 2, 1776, of the Declaration of Independence. Franklin, one of the signers, was also one of the five men who drafted the document. The Declaration of Independence is considered to have been largely the work of Thomas Jefferson, but Franklin made several changes of phrase in Jefferson's text.

The houses in which Franklin lived during his Philadelphia years—he also spent periods in England and in France—no longer stand, but visi-

Fanny Kemble, grandmother of Owen Wister, and one of nineteenth-century Philadelphia's most colorful residents.

Marianne Moore Room, with the living-room furniture of the poet's last home, at the Rosenbach Foundation in Philadelphia. The Foundation holdings include Moore's literary papers.

My life closed twice before its close;
It yet remains to see
If Immortality unveil
A third event to me,

So huge, so hopeless to conceive,
As these that twice befell.
Parting is all we know of heaven,
And all we need of hell.

—Emily Dickinson,
"My Life Closed Twice
Before Its Close"

tors can see the foundation of Franklin's last home, Franklin Court, on Market Street between Third and Fourth streets. Franklin Court, now managed by the National Park Service, is part of Independence National Historical Park. Franklin lived there in 1775 and 1776, and from 1785 until his death. At Franklin Court he turned his attention to his last and best-known work, his *Autobiography,* never completed. It was published in 1868 for the first time as Franklin had written it—earlier editions in English were retranslations of French translations. Benjamin Franklin died at Franklin Court on April 17, 1790, and was buried in the churchyard of Christ Church, at Second and Market streets.

The Body of
B Franklin Printer,
(Like the Cover of an old Book
Its Contents torn out
And stript of its Lettering & Gilding)
Lies here, Food for Worms.
But the Work shall not be lost;
For it will, (as he believ'd) appear once more,
In a new and more elegant Edition
Revised and corrected,
By the Author.

—Benjamin Franklin,
his own epitaph, written in 1728,
at the beginning of his career

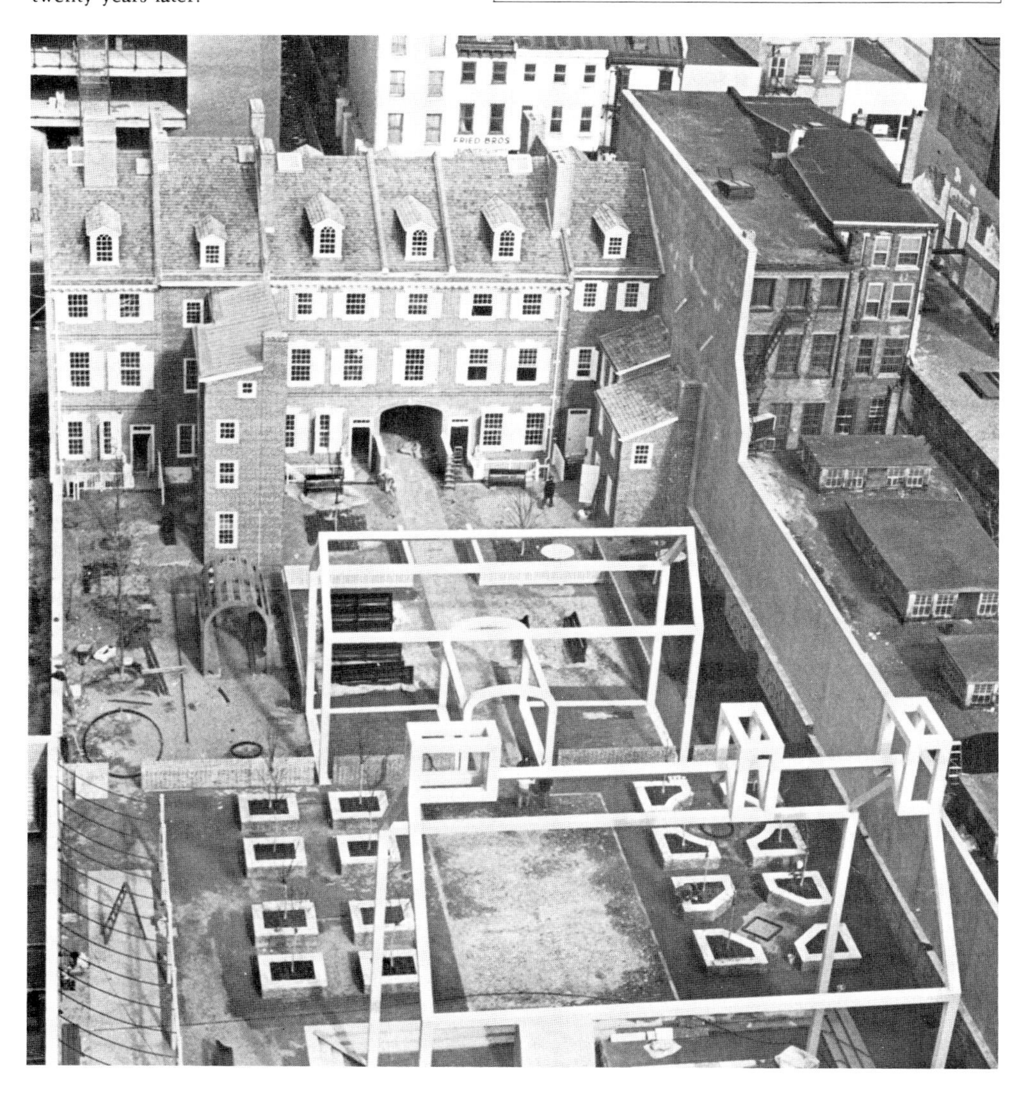

Franklin Court, site of Benjamin Franklin's last home in Philadelphia, as it appears today. The ghost framing in the foreground suggests how Franklin's home probably looked. Franklin built the house in the 1760s, and died there in 1790. The house was torn down some twenty years later.

Thomas Paine arrived in Philadelphia in November 1774, and made the city his home during the Revolutionary War. Through the assistance of Benjamin Franklin, Paine became editor of the *Pennsylvania Magazine* in February 1775. Paine held the position until late 1775 and, during that time, met a number of leaders of the growing movement for Colonial independence. Among them was Benjamin Rush, who suggested to Paine that he write an essay on independence: the result was *Common Sense,* published in Philadelphia in January 1776. The pamphlet, bought and read by thousands of people throughout the Colonies, established Paine as a major political figure. In 1777 he was appointed secretary of the Committee of Foreign Affairs by Congress, and in 1781 he traveled to France with a delegation seeking aid from the French government. During the Revolutionary War, Paine lived and worked in Philadelphia, when he was not encamped with General Washington's army, or traveling on government business. In Philadelphia Paine wrote a number of the sixteen pamphlets known as *The American Crisis* (1776–83)—Paine said he wrote thirteen of them. With the war won, Paine, virtually unpaid for his services to Congress, in 1783 was given a farm in New Rochelle, New York, by that state's legislature in recognition of his contribution to American independence. Paine spent some of his time there, or at his home in Bordentown, New Jersey, but also lived in Philadelphia.

Peacetime is not always easy for revolutionaries, and Paine found increased resistance to his ideas among politicians and intellectuals. In 1787, after failing to find financial support for his new project—the construction of a wrought-iron bridge of his own design—Paine left the United States for Europe. Upon his return in 1802, Paine found himself virtually a pariah because of the publication of his volume *Age of Reason* (1794–95) and his radical political activity. Finding little friendship in the city that had once been his home, he visited Philadelphia only occasionally in the years before his death in New York City in 1809.

Edgar Allan Poe moved to Philadelphia in the summer of 1838 with his wife, Virginia, and her mother, Mrs. Maria Clemm. Poe made Philadelphia his home for six years, the most productive of his life. It was during this time that he attained national recognition as a writer and editor. The Poes and Mrs. Clemm stayed first in a boarding house at Twelfth and Arch streets, but soon moved to another boarding house at the northeast corner of Fourth and Arch streets. In September 1838, the Poe household moved to a small house on Sixteenth Street near Locust. Poe began to contribute to the monthly *Burton's Gentleman's Magazine* and soon became its editor. The magazine offices stood at Lodge and Dock streets. Poe contributed some of his best stories to this publication, notably "The Fall of the House of Usher," "William Wilson," "Morella," and "The Conversation of Eiros and Charmion." It is believed that the Poe family stayed at their Sixteenth Street home until late 1839, or early 1840, when they moved to a row house on Coates Street and Fairmount Avenue. This house, overlooking the Schuylkill River, was later identified as standing at 2502 Fairmount Avenue, but was demolished during construction of the Benjamin Franklin Parkway. It is believed that this three-story brick house was the one in which Poe wrote many of the reviews and tales that appeared in *Graham's Magazine.* Poe served as its literary editor from April 1841 to May 1842. He had left *Burton's Gentleman's Magazine* in 1840 and tried to establish his own journal, which he tentatively called *Penn Magazine.* In October 1840, *Burton's* was bought by George R. Graham, who renamed the journal *Graham's Magazine.* Poe began to contribute to it. In February 1841, two months before he became the magazine's editor, Poe wrote a review of Charles Dickens's novel *Barnaby Rudge,* then appearing serially, and predicted the novel's resolution. When told of Poe's prediction, Dickens is said to have remarked: "The man must be the devil." *Barnaby Rudge* introduced to Poe the symbol of the raven, which he developed in his classic poem.

Poe worked for *Graham's Magazine* both at his Coates Street home and at the magazine's offices at Third and Chestnut streets, and circulation rose from 5,000 to 40,000. Before becoming editor, Poe had submitted to the magazine one of his best stories, "The Man of the Crowd." As editor, he contributed the classic story "The Murders in the Rue Morgue." But Poe's growing recognition and enormous literary productivity were not destined to last. In January 1842 Virginia, then only twenty years old, burst a blood vessel in her throat while singing at a party. This marked the beginning of the decline in her health and also marked the end of Poe's rising literary fortunes, as the sensitive man began to neglect his work in direct proportion to Virginia's worsening condition. Graham was finally forced to release Poe from his position at the magazine, and the last issue for which Poe served as editor appeared in May 1842.

Some time after leaving *Graham's Magazine,* Poe moved his family to a three-story brick house at 530 North Seventh Street, which was to be their home until shortly before their departure from Philadelphia in April 1844. In this house, the only Poe residence in Philadelphia still standing, Poe continued his efforts to establish his own magazine, now to be called *The Stylus.* His labors on the projected magazine went unfulfilled, but his writing met with greater success. He won a $100 prize from the Philadelphia *Dollar Newspaper* (then located in the same building as *Graham's Magazine,* but one floor below) for his story "The Gold-Bug." Poe also saw publication of "The Tell-Tale Heart" and "The Black Cat." These stories, all published in 1843, are thought to have been written a good deal earlier. It is believed that while living at North Seventh Street Poe wrote "A Tale of the Ragged Mountains," "The Balloon Hoax," and the poem "Eulalie—A Song," as well as an early draft of his best-known poem, "The Raven." He may also have written here the stories "The Oblong Box" and "The Premature Burial," and the poem "Dream-Land."

The Poe House at 530 North Seventh Street has been restored and is now operated as a museum by the National Park Service. The literary visitor may see photographs, facsimiles of Poe manuscripts, and first editions. Original Poe material, including manuscripts of "The Murders in the Rue Morgue" and "The Raven," are kept by the Free Library of Philadelphia on Logan Square, Twentieth Street and Benjamin Franklin Parkway. The Rare Book Department of the Library has many literary treasures, including items relating to Charles Dickens, Munro Leaf, and Beatrix Potter. It is open to visitors.

Society in every state is a blessing, but government, even in its best state, is but a necessary evil; in its worst state, an intolerable one.

—Thomas Paine, in *Common Sense*

Portrait of Thomas Paine executed by John Wesley Jarvis (1781–1839) and now on exhibition at the National Gallery of Art in Washington, D.C.

I dwelt alone
In a world of moan,
And my soul was a stagnant tide,
Till the fair and gentle Eulalie became my blushing bride—
Till the yellow-haired young Eulalie became my smiling bride.

—Edgar Allan Poe, in "Eulalie—A Song"

Last and only remaining Philadelphia home of Edgar Allan Poe, 530 North Seventh Street, the building on the left in this photograph. Poe lived in this house from about 1842 to 1844.

Owen Wister in his prime. Wister began his career as a writer in Philadelphia in the fall of 1891.

. . . Trampas spoke. "Your bet, you son-of-a-____."
The Virginian's pistol came out, and his hand lay on the table, holding it unaimed. . . .
"When you call me that, smile!"

—Owen Wister, in *The Virginian*

Owen Wister, the novelist, short-story writer, and biographer, known best for his novel *The Virginian* (1902), was born in Germantown on July 14, 1860. His family then lived in a house, variously reported as being at 5103 or 5203–05 Germantown Avenue, apparently still standing, but the Wisters soon moved to a house at 5253 Germantown Avenue. When Owen Wister was ten years old, his family moved to the house called Butler Place, about six miles from Philadelphia, where Wister's grandfather, Pierce Butler, and his grandmother, **Fanny Kemble,** had lived together for a time. Wister grew up at Butler Place, no longer in existence, and attended Germantown Academy for a time, before going off to school in Concord, New Hampshire. During vacations he visited his grandmother, Fanny Kemble, then living in a cottage across from Butler Place. After graduation from Harvard in 1882, Wister sought a career in music, but settled for work as a bank clerk in New York City. A physical collapse in 1885 brought him back to Philadelphia. **S. Weir Mitchell,** the family physician and a well-known writer—he was the author of *Hugh Wynne: Free Quaker* (1897)—gave him advice worthy of a Horace Greeley editorial—Go West, young man. Wister followed Mitchell's advice and was well enough by the fall of 1885 to enter Harvard Law School, from which he was graduated in 1888. He was admitted to the Pennsylvania bar in 1889, but his heart was in the West, not the law, and he spent each summer away from Philadelphia—out West, in the land that had attracted his Harvard classmate Theodore Roosevelt. Wister, like Roosevelt, admired the kind of life that even then was rapidly fading. The ranges, once wide open and free, were largely fenced in, the wildlife disappearing under the hunter's gun, and the cowpokes and gunslingers rapidly becoming the stuff of history books and Western novels.

One evening, while engaged in discussion at the Philadelphia Club, at Thirteenth and Walnut streets, it occurred to Wister that he should capture that vision of the West in fiction, just as Roosevelt had tried to capture it in factual books. Legend has it that Wister set to work immediately, writing that evening in the library of the Philadelphia Club his first Western story, "Hank's Woman." From then on, Wister's attention was drawn to the writing of books, but he continued to work at the law. He was with the firm of Robert Ralston and Francis Rawle, at 402 Walnut Street, now the site of the General Accident Assurance Company building, and later, from 1892 to 1909, in Francis Rawle's law office at 328 Chestnut Street. Wister's own offices, from 1909 to 1928, were at Broad Street and South Penn Square, 1421 Chestnut Street, and finally at the southeast corner of Broad and Chestnut streets.

In 1889 Wister married Mary Channing Wister, his second cousin, and the couple moved to a house at 913 Pine Street, where they lived until 1913, before moving back to Butler Place. In 1925–26 Wister lived at 1112 Spruce Street, and then moved to Bryn Mawr, which remained his home until his death on July 21, 1938, at his summer home in Saunderstown, Rhode Island. He is buried in Laurel Hill Cemetery. Among Wister's best-known works were the novels *Lin McLean* (1898); *Philosophy Four* (1903), based on his Harvard experiences; and *Lady Baltimore* (1906), set in Charleston, South Carolina. He also wrote *Ulysses S. Grant* (1900); *The Pentecost of Calamity* (1915), dealing with the crisis of Europe just before this country's entry into World War I; and *Roosevelt, the Story of a Friendship, 1880–1919* (1930). His major work, however, was *The Virginian* (1902), which surely romanticized life in the West and helped establish the cowboy and gunfighter as folk heroes.

No account of Philadelphia's literary heritage would be complete without mention of the contribution of its educational and historical institutions. The University of Pennsylvania counts among its former students **Erskine Caldwell; Zane Grey,** who studied dentistry here and was graduated in 1896; **James Gates Percival,** the poet, who studied here in 1818–19; **Charles Dudley Warner,** the author and editor, who was graduated from the University of Pennsylvania Law School in 1858; and **William Carlos Williams,** the poet and physician, who was graduated from the University of Pennsylvania Medical School in 1906. **W. E. B. Du Bois** worked for the university in 1896 and 1897, conducting a study of Philadelphia blacks. **Lewis Mumford** taught at the university in 1950–51, from 1952 to 1956, and in 1960–61. He lived from 1952 to 1955 at the Hotel Drake, 1512 Spruce Street, and from 1959 to 1961 at 3414 Sansom Street. **Philip Roth** was a visiting writer here from 1967 to 1977. The University of Pennsylvania Library maintains collections of the papers of Theodore Dreiser, Howard Fast, George Seldes, and Walt Whitman.

The Drexel University Library has literary manuscripts of Charles Dickens and William Makepeace Thackeray, and the Historical Society of Pennsylvania, 1300 Locust Street, has a copy of *Poor Richard,* as well as other examples of Benjamin Franklin's writing. Dickens is recalled once more in Philadelphia by a statue of the English novelist that stands in Clark Park, at Forty-third Street and Baltimore Avenue. The statue, which was erected in 1901, is said to be the only one of Dickens anywhere in the world.

PINE GROVE

Conrad Richter was born on October 30, 1890, at 33 Mifflin Street, in Pine Grove, which is about thirty miles northwest of Reading. Richter's family had deep roots in the town. His great-grandfather, a tavern keeper, helped name it. His grandfather was pastor of St. John's Lutheran Church, at 222 South Tulpehocken Street (original building torn down). Richter's father, also drawn to the religious life, moved the family to nearby Selinsgrove in 1899 so he could pursue theological training. For a time in the 1940s Richter rented a house, still standing, at 179 South Tulpehocken Street. In 1950, at age sixty, the novelist returned here with his wife. They made their permanent residence at 11 Maple Street, where Richter died in 1968. Among the works Richter wrote here were *The Waters of Kronos* (1960) and *A Simple Honorable Man* (1962), novels set in Unionville, a town based on Pine Grove, and peopled with members of Richter's own family. Richter is buried in St. John's Lutheran Cemetery, at Wood and Walter streets. Both his homes here are privately owned, but visitors can see numerous trees that were planted throughout the town by the Conrad Richter Memorial Fund. The novelist is further commemorated by a plaque outside the Mifflin Street house, placed there by his publisher and literary agent.

PITTSBURGH

Pennsylvania's second largest city, spreading over the area where the Allegheny and Monongahela rivers meet to form the Ohio, claims a number of literary residents, natives, and visitors. **Willa Cather** lived here from 1895 to 1906, teaching at Allegheny High School, working for the *Home Monthly* magazine, and writing her first two books—*April Twilights* (1903) and *The Troll Gardens* (1905). **Marc Connelly** came here with his mother in 1908, working at odd jobs before becoming a reporter for the *Sun* and later the *Dispatch,* and living for a time at the Lancaster Apartments, 314 McKee Place, still standing. **Malcolm Cowley** attended Peabody High School on East Liberty Street from 1911 to 1915—he later described his period of study here as "the time of my life."

Marcia Davenport lived in the 1920s with her husband in a house at 507 Shady Avenue, no longer standing. **Theodore Dreiser** worked as a reporter for the *Dispatch* in 1894, living for a time in a cheap room on Wylie Avenue. **Arthur Guiterman,** the poet and playwright, died at Pittsburgh's Presbyterian Hospital on January 11, 1943. He had come to Pittsburgh to deliver a lecture on the subject of "brave laughter." **Richard Realf,** best remembered for his volume of poetry *Guesses at the Beautiful* (1852), worked for the *Commercial* from 1872 to 1877. **George Seldes,** writer of many exposés, such as *Iron, Blood, and Profits* (1934), about the munitions industry, and *Witch Hunt: The Techniques of Redbaiting* (1941), worked for the *Leader* from 1909 to 1911 and for the *Post* from 1913 to 1916.

Of the numerous authors born in Pittsburgh, several came from Allegheny, formerly a separate town but now part of the city's North Side. **Margaret Deland,** born on February 28, 1857, set several of her novels, such as *The Awakening of Helena Richie* (1906), in this area. Mrs. Deland grew up in the town of Manchester, on the Ohio River near Allegheny and now part of Pittsburgh. Manchester was the model for Old Chester, the fictional town of several of her books, including *Old Chester Tales* (1898). **George Washington Harris,** born here on March 20, 1814, was best known for his humorous writings under the name of **Sut Lovingood. Robinson Jeffers,** born here on January 10, 1887, lived in a house at 723 (possibly 318) Ridge Avenue, which has since been torn down. **Gertrude Stein** was also born on the North Side, on February 3, 1874. Stein's birthplace was torn down long ago, and even the street in which it stood has disappeared.

Elizabeth Cochrane Seaman, who was to achieve fame and wealth as the sensationalist New York journalist **Nellie Bly,** was born on May 5, 1867, in what once was a community known as Cochrans Mills and now is part of the Pittsburgh metropolitan area. She began working on newspapers at an early age and, in the 1880s, went to work for the New York *World,* for which she did most of her famous reporting. In one exploit she had herself committed to New York's infamous asylum on Blackwell's Island, where she stayed for ten days and gathered material for a shocking exposé. Published first in the *World* and later in her book *Ten Days in a Mad House* (1887),

He thought of all he had once known and loved . . . the red roofs and green trees, the life and talk and tender thought that went on under them; the brave brick schoolhouse and its white belfry; the shining railroad and its yellow and brown station; his grandfather's church that skilled hands had put together of stone; the mill that ground the town's staff of life and the shirt factory that had covered the men's backs; the old blacksmith shop which, the last time he saw it, served as an automobile body shop; the gas service station where the old tannery had stood; and his family's frame house where his mother long ago with her bare white hands had thrown a blazing oil lamp out the window.

—Conrad Richter,
in *The Waters of Kronos*

Birthplace of Conrad Richter in Pine Grove, and the plaque mounted at the corner of the house.

Lute belonged on Lantenengo Street, and she as his wife belonged on Lantenengo Street. And not only as his wife. Her family had been in Gibbsville a lot longer than the great majority of the people who lived on Lantenengo Street. She was a Doane, and Grandfather Doane had been a drummer boy in the Mexican War and had a Congressional Medal of Honor from the Civil War.

—John O'Hara,
in *Appointment in Samarra*

I had a dream de udder night,
When eb'rything was still;
I thought I saw Susanna,
A coming down de hill;
De buckwheat-cake was in her mouth,
De tear was in her eye;
Says I, I'm coming from de South,
Susanna, don't you cry.

—Stephen Foster,
in "Oh! Susanna"

the exposé led to reforms for the suffering inmates. Bly became an international celebrity in 1889, when she undertook an around-the-world trip to beat the record of Phileas Fogg in Jules Verne's romance *Around the World in Eighty Days* (1873). Bly accomplished the feat in 72 days, 6 hours, 11 minutes, and 14 seconds. She recounted this adventure in *Nellie Bly's Book: Around the World in Seventy-Two Days* (1890).

Writers born in Pittsburgh proper include **Hervey Allen,** born on December 8, 1889, and known for historical novels such as *Anthony Adverse* (1933) and for biographies such as *Israfel: The Life and Times of Edgar Allan Poe* (1926). **George S. Kaufman,** born here on November 16, 1889, collaborated on many Broadway hits, such as the Pulitzer Prize-winning *Of Thee I Sing* (1932) and *The Man Who Came to Dinner* (1939). **Mary Roberts Rinehart,** born here on August 12, 1876, attended the Pittsburgh Training School for Nurses and lived in this area for many years.

Of all the literary men and women who have lived in Pittsburgh, perhaps none is so famous as **Stephen Foster,** born on July 4, 1826, in a white frame cottage in Lawrenceville, now part of the city proper. With the exception of the four years he spent in Cincinnati, 1846 to 1850, Foster lived in Pittsburgh or its environs from birth until 1860, when he moved to New York City. After leaving The White Cottage, as the family home was known, the Fosters took a house on Water Street, no longer standing. They moved later to Allegheny, and it was to Allegheny that Foster returned from Cincinnati in 1850. His parents had sent him away to learn the practical art of bookkeeping, but he preferred writing songs. When his *Songs of the Sable Harmonists* (1848), which included "Oh! Susanna," won him some popularity, his parents agreed to let him return home and become a professional writer. He soon became wealthy and increasingly popular, but his move to New York in 1860 marked the beginning of an unexplained decline. He drank too much, did hackwork unworthy of his talents and, in 1864, died a pauper. Pittsburgh commemorates Foster in the Stephen Collins Foster Memorial Building, at Forbes Street and Bigelow Boulevard, which houses a collection of memorabilia.

(*below*) Monument to Stephen Foster on the University of Pittsburgh campus. (*right*) Birthplace of John O'Hara at 123 Mahantongo Street in Pottsville.

PLOWVILLE

John Updike, while in his teens, lived on a farm near here, about five miles southwest of Reading.

POTTSTOWN

After graduation from Swarthmore College in 1929, **James Michener** taught at the Hill School in Pottstown, which is about fifteen miles southeast of Reading. He left in 1931 to travel in Europe on a grant from Swarthmore. Among Hill School's other literary graduates are **John Dickson Carr,** the mystery writer, who was graduated in 1924, and **Edmund Wilson,** the eminent man of letters, who was graduated in 1912. Wilson served as an editor of the school literary magazine, the *Record.*

POTTSVILLE

Pottsville's most distinguished native son is **John O'Hara,** born on January 31, 1905, at 123 Mahantongo Street, now used by the Continental Hair Stylists. (Mahantongo appears as Lantenengo in O'Hara's work.) O'Hara, one of eight children, had to leave school in 1924, after his father's death. He worked as a reporter for the Pottsville *Journal* before moving to New York City to pursue his career. Pottsville provided a model for O'Hara's fictional Gibbsville, which is the setting for several of his works, including his first novel, *Appointment in Samarra* (1934); the stories collected in *The Doctor's Son and Other Stories* (1935); and the novel *Ten North Frederick* (1955). In such works O'Hara chronicled the public and private lives of the citizens of Gibbsville. While in Pottsville, O'Hara also lived at 606 Mahantongo Street, now used as an apartment building.

Conrad Richter, who lived in nearby Pine Grove, died at Pottsville Hospital on October 30, 1968.

READING

Conrad Richter lived in the Pennside area of Reading from 1915 to 1922. During that time he launched his own magazine, *Junior Magazine Book,* for which he wrote and edited every word and acted as advertising salesman. The magazine failed after one year of publication, and Richter moved on to other things.

Reading's primary claim to literary fame rests on other shoulders. Reading was the birthplace of **Wallace Stevens,** one of America's leading twentieth-century poets. Stevens was born on October 2, 1879, at 323 North Fifth Street. He attended a private kindergarten at Walnut and Sixth streets, three blocks from his home and, later, the parochial school attached to St. John's Evangelical Lutheran Church, on Walnut Street. After graduation from Reading High School in 1896, he went on to Harvard, New York University Law School, and a long career with the Hartford

(Connecticut) Accident and Indemnity Company. At the same time that he worked at his business career, Stevens wrote verse steadily. He did not publish until 1915; his first book, *Harmonium* (1923), appeared when he was forty-three years old. Despite the fact that *Harmonium* sold only a few hundred copies before being remaindered, the poet's reputation grew from then on. His later books included *Notes Toward a Supreme Fiction* (1942); *The Necessary Angel* (1951), a book of essays; and *Collected Poems* (1954), for which Stevens was awarded a Pulitzer Prize. Stevens kept separate his business life and his life as a poet, so separate that many of his business colleagues knew him for years without being aware of his literary importance. Stevens's birthplace on North Fifth Street is not open to visitors, but a plaque outside the house pays tribute to him.

RIDGEWAY

Katherine Mayo, whose news reporting was considered among the best of its kind in the early twentieth century—before the term "investigative journalism" had been invented—was born in Ridgeway in northwestern Pennsylvania on January 24, 1867. Her birthplace stood on the site now occupied by the Centennial School. Among Mayo's books were *Justice to All* (1917), which detailed the work of the Pennsylvania State Police, and *Mother India* (1927), an account of the practice of child marriage in India.

ROSEMONT

Elinor Wylie lived with her family from 1887 to 1897 in this suburb of Philadelphia, attending a private school in nearby Bryn Mawr.

SCHUYLKILL HAVEN

Elsie Singmaster was born on August 29, 1879, in the parsonage of St. Matthew's Lutheran Church, 23 Dock Street, in this town five miles south of Pottsville. Singmaster spent most of her life in Pennsylvania Dutch country, and many of her books were set there. Among her novels were *The Loving Heart* (1937) and *A High Wind Rising* (1942). St. Matthew's parsonage, which was designed by Elsie's father, Reverend John Alden Singmaster, is still in use today.

SCRANTON

Charles MacArthur was born in Scranton on November 5, 1895. MacArthur worked for some years as a newspaperman and, in the 1920s, was a regular figure at New York City's Algonquin Round Table. His best plays were those he wrote in collaboration with Ben Hecht. Two that still are produced today are *The Front Page* (1928), which stands as the classic romanticization of American newspaper life, and *Twentieth Century* (1932), a satire on Hollywood.

SELINSGROVE

In 1899, when **Conrad Richter** was nine years old, his father moved the family from Pine Grove to this town in Snyder County, on the Susquehanna River. The family had a house at High and Sassafras streets. During subsequent years, Richter's father studied for the ministry at Susquehanna College, and his son attended grade school at Susquehanna Academy. Father and son were graduated in 1904.

SEWICKLEY

This residential town twelve miles northwest of Pittsburgh has been home to several twentieth-century writers. **Robinson Jeffers** was not quite two years old in 1888, when his family moved to a three-story brick house, still standing at 44 Thorn Street. They lived here until 1893. **Mary Roberts Rinehart** lived in a house at 1314 Linden in the early 1900s, when she was establishing her career as a writer of mysteries with such successful books as *The Circular Staircase* (1908). The main house is no longer standing, but the carriage house has been turned into a residential property. **George R. Stewart,** a prolific writer of fiction and nonfiction, was born here on May 31, 1895.

This is the story of how a middle-aged spinster lost her mind, deserted her domestic gods in the city, took a furnished house for the summer out of town, and found herself involved in one of those mysterious crimes that keep our newspapers and detective agencies happy and prosperous.

—Mary Roberts Rinehart, in *The Circular Staircase*

Plaque marking the birthplace of Wallace Stevens in Reading (*shown below the plaque*).

"What, Wally a poet!"

—One of Stevens's business friends, on being told of his poetry-writing

Birthplace marker of Ida M. Tarbell.

Among Stewart's works were the novels *Doctor's Oral* (1939) and *The Years of the City* (1955), and such nonfiction works as *Pickett's Charge* (1959) and *Donner Pass and Those Who Crossed It* (1960).

SHILLINGTON

After his birth at nearby Reading Hospital—Reading is only five miles from Shillington—on March 18, 1932, **John Updike** was brought by his parents to their home in Shillington. Updike lived here until he was thirteen. The boyhood home of the novelist, at 117 Philadelphia Street, is a private residence and not open to visitors.

SWARTHMORE

This residential suburb of Philadelphia is the home of Swarthmore College, one of the finest liberal-arts colleges in America. **W. H. Auden** taught here from 1942 to 1945 on a Guggenheim fellowship. He lived at various addresses, including 16 Oberlin Avenue, and often took his lunch at the Dew Drop Inn, now the Village Restaurant, at 407 Dartmouth Avenue. **James Michener** was graduated from Swarthmore in 1929.

TITUSVILLE

Ida M. Tarbell, the famous muckraker, spent her childhood here, in the region of northwest Pennsylvania that is identified with the early growth of the oil industry. She remained here until after her graduation from Titusville High School in 1875, and her years in the oil area provided perfect background when she did research for *The History of the Standard Oil Company* (1904), her best known and most important book.

TREMONT

Conrad Richter was graduated in 1906 from Tremont High School on Clay Street, still standing in this town north of Pine Grove, where young Conrad was born. His family had come to Tremont in 1904, when his father, a Lutheran minister, took his first parish here. Richter drew on his memories of his father's parish and on his father's character when writing the novel *A Simple Honorable Man* (1962).

TYRONE

Dan Wickenden, author of novels such as *The Wayfarers* (1945) and *The Dry Season* (1950), was born here, fifteen miles northeast of Altoona, on March 24, 1913.

UNIONTOWN

John Dickson Carr, whose mystery novels included *The Bride of Newgate* (1950) and *The Dead Man's Knock* (1958), was born on November 30, 1906, in Uniontown, not far from the West Virginia border. His father served in Congress from 1913 to 1915 and later returned here to become the town's postmaster. Carr was so prolific that he wrote some of his 120 novels under the names **Carter Dickson** and **Carr Dickson.** Carr's birthplace now is said to be used as the Fayette Bank Building.

UPPER DARBY

Hilda Doolittle, the poet known as **H. D.,** lived in this district of Philadelphia in the early 1900s, while her father was head of the Flower Observatory. The observatory, owned by the University of Pennsylvania, was located along the West Chester Pike, and the Doolittles lived in a house on the grounds. Among H. D.'s visitors here were William Carlos Williams and Ezra Pound: the trio used to walk together near the observatory, and Williams later recalled grape hyacinths growing along the edge of the road. The observatory is no longer standing, its site occupied by the St. Lawrence Church and School.

WASHINGTON

Rebecca [Blaine] Harding Davis, author of novels such as *Margaret Howth* (1862) and *Waiting for the Verdict* (1868), was born here on June 24, 1831, and lived here for a few years. Washington is about twenty-five miles southwest of Pittsburgh. Mrs. Davis's work reflected deep concern over racial discrimination and political corruption. She

Boyhood home of John Updike in Shillington. Photograph of the first producing oil well in the United States, drilled in Titusville in 1859 by Edwin Laurentine Drake (1819–80). Ida M. Tarbell, who lived in Titusville, wrote *The History of the Standard Oil Company,* a classic work of muckraking journalism.

was the mother of Richard Harding Davis, the journalist and author.

WATTSBURG

Ida M. Tarbell was born on November 5, 1857, in her grandfather's log cabin on a farm near Wattsburg, in Erie County. Her family moved to Iowa when she was young, but later returned to nearby Titusville. Her grandfather's cabin is no longer standing, but a marker on Route 8, about two miles east of the site, commemorates her birth.

WAYNE

This community just outside Philadelphia is home to Valley Forge Military Academy, attended by **Edward Albee** and **J. D. Salinger.** While lecturing at nearby Haverford College in the 1940s, **Wright Morris** lived here at Eagle Road and Beachtree Lane.

WEST CHESTER

Joseph Hergesheimer lived for many years in West Chester, which is about twenty miles west of Philadelphia. The novelist's home was a prerevolutionary farmhouse called Dower House, which Hergesheimer described in his autobiographical *From an Old House* (1925). He died on April 25, 1954, and is buried in West Chester's Oakland Cemetery. Another resident here, though for a much shorter period, was **Sidney Lanier.** The poet lived in a boarding house run by a Mrs. Thompson.

WILKES-BARRE

Erskine Caldwell came to this city on the Susquehanna River as a teenager, in late 1923, and stayed here until the next year, when he began his studies at the University of Pennsylvania. While in Wilkes-Barre, Caldwell lived at 40 North Main Street and worked in a five-and-ten-cent store and in a restaurant in the railroad station.

WILKINSBURG

W[illiam] D[eWitt] Snodgrass was born in this residential town east of Pittsburgh on January 5, 1926. Snodgrass's first book, *Heart's Needle* (1959), won a Pulitzer Prize for poetry; his later volumes of poems include *After Experience* (1968) and *Six Troubadour Songs* (1977).

WYALUSING

Philip Van Doren Stern was born in Wyalusing, in Bradford County, northeastern Pennsylvania, on September 10, 1900. Stern wrote many books, both fiction and nonfiction; among his novels were *The Man Who Killed Lincoln* (1939) and *The Drums of Morning* (1942). One of his noteworthy achievements was his work as general manager of "Editions for the Armed Services" during World War II. This publishing program provided over 120 million paperback books to U.S. servicemen, and is credited by some as having paved the way for the explosive growth of paperback book publishing after the war.

WYNCOTE

Christopher Morley came to live here about 1917, when he was working nearby, in Philadelphia, for *The Ladies' Home Journal.* During his stay in Wyncote, Morley also wrote his first book, the novel *Parnassus on Wheels* (1917). Another resident, from age six to twenty-two, was **Ezra Pound,** whose parents brought him here about 1891. Their home at 166 Fernbrook Avenue is still standing but privately owned. Young Ezra attended several different day schools until he was twelve. He then became a boarding student at Cheltenham Military Academy, about a mile from home. Later, when he was studying at the University of Pennsylvania, a friend and fellow student, **William Carlos Williams,** came to stay with him and his family at the Fernbrook Avenue house.

. . . a very old stone house . . . empty of ornament, with severely whitewashed walls and walnut furniture; once filled by a constant stream of people, the Dower House is now mostly a tranquil place of books.

—Joseph Hergesheimer, writing in 1942

DELAWARE

ARDEN

Writers associated with this community near Wilmington include **Ross Santee,** the novelist, who lived in a house, now private, at 1914 Sherwood Road; and **Upton Sinclair,** who came to live in Arden in the spring of 1910. Sinclair and his wife Meta first made their home in three tents before moving the following spring to a new two-story cottage, still standing at 2321 Woodland Lane. It is said that during this period Meta Sinclair had an affair with **Harry Kemp,** a poet who stayed with the Sinclairs for a time at the invitation of the novelist. In August 1911, Mrs. Sinclair left with Kemp for a hideaway in New York. Kemp wrote about this sidelight on literary history in

Every woman hungers for a real mate. Upton Sinclair is an essential monogamist.

—Meta Sinclair

Home of Upton Sinclair in Arden, where he lived when he filed suit for divorce from Meta Sinclair.

Woodburn, now the Governor's residence in Dover, and once a setting for Townsend's novel *The Entailed Hat.*

his book *Tramping on Life* (1922). Sinclair, in *The Brass Check* (1919), cited the news reports of the affair as an example of journalistic irresponsibility. In 1911, Sinclair managed to get himself arrested for "gaming on the Sabbath," after becoming involved in a local dispute. He and ten others were found guilty of the charge: on a Sunday morning, Sinclair had played tennis, and the others had played baseball, on the village green. All refused to pay their fines, and spent the night in the New Castle County Workhouse.

CLAYMONT

Anne Parrish came to Delaware as a young girl, living for several years in a painted brick house off Route 13, near this community northeast of Wilmington. She wrote many stories here, some in collaboration with her brother Dillwyn, and the area around Claymont served as the setting for her best-selling novel *The Perennial Bachelor* (1925).

DOVER

The Hall of Records Archives Bureau in Dover, the capital of Delaware, contains many of the writings of **John Dickinson,** Penman of the Revolution, who lived just southeast of here in the 1740s. Woodburn, a Georgian-style brick mansion at King's Highway and South Pennsylvania Avenue, was used by George Alfred Townsend as the setting for a scene in his novel *The Entailed Hat* (1884). Woodburn was also used as a station on the Underground Railroad in the 1800s. Today it is the official residence of the Governor of Delaware and is open to visitors on Saturdays.

GEORGETOWN

This town in southern Delaware was the birthplace, on January 30, 1841, of **George Alfred Townsend.** In his best-known novel, *The Entailed Hat* (1884), Townsend used the story of Patty Cannon, a notorious woman who kidnaped free blacks and sold them into slavery. She was also suspected of committing numerous murders. Arrested and indicted for murder in 1829, Patty died in jail in Georgetown and was buried in the jail yard. Townsend's other books included the collection of stories *Tales of the Chesapeake* (1880) and the novel *Katy of Catoctin* (1887). He also published numerous articles in national journals under the pen name **Gath.**

KITTS HUMMOCK

The John Dickinson Mansion, just east of Route 113, off Kitts Hummock Road, near this community on Delaware Bay, was the boyhood home from 1740 to 1750 of **John Dickinson,** Penman of the Revolution. Dickinson made Philadelphia his home after 1750, but returned to Kitts Hummock in later years for visits, notably for extended stays in 1776 and 1781. After the mansion was gutted by fire in 1804, Dickinson restored it to something less than its original elegance—

John Dickinson Mansion near Kitts Hummock. The main house, on the right in the photograph, was originally a three-story structure.

> *Lily ate fudge, read "Richard Carvel" and "My Lady Peggy Comes to Town," wishing that she had lived in the time of brocade and powdered hair, saw "If I Were King" and fell mildly in love with Gibson men, Christy men, and C. Allen Gilbert men, made bead chains and got her threads in terrible knots, and took up pyrography, burning crooked fleurs-de-lis and* Art Nouveau *water-lilies on wooden picture frames and glove boxes.*
>
> —Anne Parrish, in *The Perennial Bachelor*

he intended it to be used as a tenant's house—but never rebuilt the third floor. Dickinson, a major figure in American politics, wrote his *Letters from a Farmer in Pennsylvania to the Inhabitants of the British Colonies* (1768) in an attempt to use reason and conciliation in avoiding the coming conflict. Although Dickinson refused to sign the Declaration of Independence, he did take up arms in the fight for independence. Dickinson also helped frame the Constitution, and his letters in support of it, written under the name **Fabius,** were instrumental in its ratification. The Dickinson Mansion, which once looked out over a 1,300-acre plantation, was one of the finest structures of its day. It is now open to the public free of charge. Visitors are guided through its various restored rooms, where they may see such Dickinson memorabilia as his cradle, a small desk, his books, and samples of his writing, as well as period furniture and furnishings. A beautiful garden in front of the mansion evokes the elegance of the estate in Dickinson's time.

MILFORD

This city in central Delaware was the birthplace in 1798 of **John Lofland,** known as The Milford Bard. Like his friend Edgar Allan Poe, Lofland knew too well both alcohol and reduced circumstances and, also like Poe, died in 1849. Unlike Poe, however, Lofland has faded into literary obscurity, although his poems and stories inspired by and depicting Delaware settings have made him a major figure in Delaware's literary history. The Towers, a large house at 101 N.W. Front Street, stands on the site of Lofland's birth. This house, painted light blue, features a large turret, stained-glass windows, and ornate exterior trim. It is used as a private home except for an extension to the building, which at present houses a beauty salon.

NEW CASTLE

Among the many historic structures standing in this city in northern Delaware is the Booth House, 216 Delaware Street, which was the birthplace of **Robert Montgomery Bird,** on February 5, 1806. The first floor of this brick building now houses a tavern and an antique shop, and upper floors are used for apartments. Bird wrote four plays that were bought by actor Edwin Forrest for $1,000 apiece. Forrest made a fortune with them but refused to pay Bird anything more, despite his agreement to do so if the plays proved successful. Enraged at Forrest and thwarted in his attempts to publish the plays, Bird turned to writing novels, at which he had great success, this time financial as well as literary. Among his novels were *Hawks of Hawk Hollow* (1835) and the best-selling *Nick of the Woods* (1837), an early study of dual personality, set against a pioneer background.

WILMINGTON

Authors born in Delaware's largest city include **Henry Seidel Canby,** born on September 6, 1878, a founder and the first editor of *The Saturday Review of Literature.* His book *The Age of Confidence* (1934) depicted town life, particularly of Wilmington, in the 1890s. **J[ohn] P[hillips] Marquand,** a Pulitzer Prize-winning novelist, was born on November 10, 1893, in a house still standing at 1305 Pennsylvania Avenue. **Howard Pyle** was born here on March 5, 1853. Pyle, author and illustrator of such children's books as *The Merry Adventures of Robin Hood* (1883), *Men of Iron* (1892), and *Howard Pyle's Book of Pirates* (1921), lived here for many years and ran an art school. His students included N. C. Wyeth and Maxfield Parrish.

Among the writers who have lived in Wilmington, the best known is **F. Scott Fitzgerald** who, with his wife Zelda, lived in the city's Edgemoor section, in northeastern Wilmington, from 1927 to 1929. On the advice of Fitzgerald's editor, Maxwell Perkins, the Fitzgeralds leased Ellerslie, a spacious mansion built in the 1840s, in which they were visited by Perkins and other literary figures, such as **John Dos Passos, Ernest Hemingway, Thornton Wilder,** and **Edmund Wilson.** It was here that Wilder first met Wilson and began a friendship that was to last for nearly thirty years. Wilson wrote of a party here in "A Weekend at Ellerslie," collected in his book *The Shores of Light, A Literary Chronicle of the Twenties and Thirties* (1952). **Zelda Fitzgerald** described Ellerslie in " 'Show Mr. and Mrs. F. to Number ——,' " first published in *Esquire* in 1934, and Fitzgerald wrote of the move to Ellerslie in "My Lost City." Ellerslie was torn down in 1970 to make way for an office and laboratory for the Du Pont Company's pigments plant.

Other authors associated with this city include **John Dickinson,** who died in Wilmington on February 14, 1808, and is buried in the churchyard of the Wilmington Friends Meeting House, off West Street between Fourth and Fifth streets. **John Lofland** used the area around Wilmington's Brandywine Creek as the inspiration and setting for many of his poems and tales. Lofland, who died in 1849, is buried in the churchyard of St. Andrew's Protestant Episcopal Church, at the corner of Eighth and Shipley streets.

> *I had never yet visited the Fitzgeralds in the house just outside Wilmington that they had taken on their return from Europe at the end of 1927, but rumors had reached me of festivities on a more elaborate scale than their old weekends at Westport or Great Neck. Dos Passos had attended a party that Scott had given for his thirtieth birthday, which he described to me as "a regular wake"—Scott had been lamenting the passing of youth ever since his twenty-first birthday, and he had apparently commemorated his twenties with veritable funeral games.*
>
> —Edmund Wilson,
> in "A Weekend at Ellerslie"

Brandywine Creek in Wilmington, inspiration for poems and tales by John Lofland, who is buried in Wilmington.

The average schoolmaster is and always must be essentially an ass, for how can one imagine an intelligent man engaging is so puerile an avocation?

—H. L. Mencken, in *Prejudices,* Third Series

In 1911, after being found guilty in Arden of "gaming on the Sabbath," **Upton Sinclair** and ten others refused to pay their fines, and spent a night in the New Castle County Workhouse, about three miles from the center of Wilmington. Upon their release the next day, Sinclair and the others related to a score of reporters the dreadful conditions they had seen. The ensuing investigation prompted reforms at the workhouse.

MARYLAND

ANNAPOLIS

Among the literary figures associated with this port and capital city of Maryland is **Hervey Allen,** the novelist known best for *Anthony Adverse* (1933). Allen attended the U.S. Naval Academy here for two years in 1909–11, and later became a member of the board of governors of St. John's College, the distinguished institution founded in Annapolis in 1696.

James M[allahan] Cain, the novelist whose great success was *The Postman Always Rings Twice* (1934), was born in Annapolis on July 1, 1892. **Winston Churchill,** the historical novelist, was graduated from the Naval Academy in 1894. Churchill used Maryland as the background for his first, best-selling historical romance, *Richard Carvel* (1899). **Thorne Smith** was born in Annapolis in 1892, probably at the Naval Academy, where his father, Commodore James Thorne Smith, was quartered. Smith is remembered for creating the character Cosmo Topper in *Topper* (1926) and *Topper Takes a Trip* (1932). Cosmo Topper, an inhibited banker, was constantly bedeviled by the activities of a pair of ghosts who appeared and disappeared unexpectedly. Eventually, Topper himself died and acquired the same capability to materialize without warning.

Lionel Carvel, Esq., of Carvel Hall, in the county of Queen Anne, was no inconsiderable man in his Lordship's province of Maryland, and indeed he was not unknown in the colonial capitals from Williamsburg to Boston. When his ships arrived out, in May or June, they made a goodly showing at the wharves, and his captains were ever shrewd men of judgment who sniffed a Frenchman on the horizon, so that none of the Carvel tobacco ever went, in that way, to gladden a Gallic heart. Mr. Carvel's acres were both rich and broad, and his house wide for the stranger who might seek its shelter, as with God's help so it shall ever be. It has yet to be said of the Carvels that their guests are hurried away, or that one, by reason of his worldly goods or position, shall be more welcome than another.

—Winston Churchill, in *Richard Carvel*

BALTIMORE

Among America's great institutions are Baltimore's Johns Hopkins University and Johns Hopkins Hospital, and both have literary associations. **F. Scott Fitzgerald,** while living near Baltimore in the summer of 1934, was treated at Johns Hopkins Hospital for excessive drinking. **Frank Slaughter** was a student at the Johns Hopkins Medical School, where he received an M.D. degree in 1930. He used his experiences as a physician in many of his popular novels, which carry such titles as *Spencer Brade, M.D.* (1942) and *Air Surgeon* (1943). **Gertrude Stein** also studied medicine here, after graduation from Radcliffe in 1897—in her first book, *Three Lives* (1909), Bridgepoint is possibly a fictionalized version of Baltimore. **Thomas Wolfe** was treated at Johns Hopkins for pneumonia and brain infection in the summer of 1938. Wolfe died here on September 15, two weeks before his thirty-eighth birthday.

Writers who have lived in Baltimore include **James M. Cain,** who worked in 1917–18 for the Baltimore *American,* now the *News-American,* and in 1919–23 for the Baltimore *Sun.* **John Dos Passos** for many years divided his time between his house in Westmoreland, Virginia, and a succession of Baltimore apartments: 552 Chateau Avenue, 3911 Canterbury Road, 1821 Sulgrave Avenue, and 3B Hamill Road, Cross Keys Village. The Hamill Road apartment is believed to be the home where he was living at the time of his death, on September 28, 1970. **Frederick Douglass** lived as a house slave in the homes of Hugh Auld on Aliceanna Street, on Philpot Street, and on Fells (now Thames) Street—all in the Fells Point district.

F. Scott Fitzgerald, although born in Minnesota, always pointed to his Maryland connections, especially to Francis Scott Key, the distant relative after whom he was named. Fitzgerald lived in Baltimore off and on during the 1930s, moving here in February 1932 when his wife, Zelda, became a patient at Johns Hopkins' Phipps Psychiatric Clinic. At first he stayed at the Rennert Hotel, no longer standing; later, after returning in November 1933 from La Paix in nearby Towson, he took an apartment at 1307 Park Avenue, where he usually stayed while Zelda was confined in various psychiatric clinics. Throughout this period Fitzgerald continued to work, both at 1307 Park Avenue and at the Cambridge Arms, 1 East 34 Street, where he moved in September 1935, writing "Pasting It Together," "Handle with Care," and "The Crack-Up."

Lizette Woodworth Reese, the poet, who taught at Western High School here, died here on December 17, 1935. **John Banister Tabb,** the poet, was ordained a Roman Catholic priest here in 1884 in St. Mary's Seminary. **Hendrick Willem Van Loon,** the historian and journalist, worked for the Baltimore *Sun* in 1923–24.

Among Baltimore's native sons are **Daniel Henderson,** born on May 27, 1880, and **John Beauchamp Jones,** born on March 6, 1810. Henderson's books of verse included *Life's Minstrel* (1919) and *Frontiers* (1933). Jones was better known. Based on his experiences in the Confederate War Department, Jones wrote his highly regarded *A Rebel War Clerk's Diary* (1866). His most popular work was *Wild Western Scenes* (1841), which sold over 100,000 copies.

John Pendleton Kennedy was born here on October 25, 1795. Kennedy's career spanned politics and literature; he served as state representative and member of congress and became secretary of the Navy in Millard Fillmore's administration. Kennedy wrote several books of political satire under the pen name **Mark Littleton.** His novel *Rob of the Bowl* (1838) was set in St. Marys City, the capital of Maryland until 1694. While in Congress, Kennedy satirized Jacksonian democracy in *Quodlibet* (1840). A well-known literary figure in his day, Kennedy was a member of Baltimore's Delphian Club—Francis Scott Key was also a member—and a judge in a literary competition in which Edgar Allan Poe won first prize for his short story "Ms. Found in a Bottle." When **William Makepeace Thackeray** was on a lecture tour of America in 1851–56, Kennedy met him and provided information that the English novelist used in *The Virginians.* Kennedy died on August 18, 1870, and is buried here in Green Mount Cemetery, at Greenmount and Oliver streets.

Baltimore natives also include **Munro Leaf,** known especially for his *Ferdinand the Bull* (1936), born on December 4, 1905. **Walter Lord,** best known for the novel *A Night to Remember,* was born here on October 8, 1917. **Frank O'Hara,** the poet and novelist, was born here on June 27, 1926. **James Ryder Randall** was born here on January 1, 1839. He is known principally for his poem "My Maryland" (1861), which was later set to the tune of "O Tannenbaum," and became an unofficial anthem of the Confederacy.

Karl Shapiro, the poet, was born here on November 10, 1913. He studied at Johns Hopkins in 1937–39 and taught there in 1947. **Upton Sinclair** was born here on September 20, 1878. **F. Hopkinson Smith,** best known in his time for his short novel *Colonel Carter of Cartersville* (1891), was born here on October 23, 1838. **Leon Uris,** the best-selling author of *Exodus* (1958) and *QB VII* (1970), was born here on August 3, 1924.

Malcolm Braly, author of *On the Yard* (1967), died in a traffic accident in Baltimore on April 7, 1980.

In discussing literary Baltimore, we most often refer to Sidney Lanier, H. L. Mencken, and Edgar Allan Poe. **Sidney Lanier,** the greatest flutist of his day, came here in December 1873 as first flutist for the Peabody Symphony Concerts, and stayed until just before his death in Lynn, North Carolina, on September 7, 1881. He lived on Lexington Street, near the Peabody Institute and near Johns Hopkins, where he lectured on English literature from 1879 until his death. These lectures provided the basis for Lanier's famous prose work *The Science of English Verse* (1880), and it was during these last two years of his life that he wrote some of his most notable poetry, including "The Revenge of Hamish" and "The Song of the Chattahoochee." Lanier is buried in Green Mount Cemetery, where John Wilkes Booth is also buried. Lanier's gravestone is inscribed "I am lit with the Sun," a line from his last poem, "Sunrise." He had written this poem in Baltimore in December 1880 while prostrate with the tubercular fever from which he died less than a year later. Johns Hopkins' Evergreen House at 4545 North Charles Street has a collection of Lanier's books, music, and furniture. The university's Milton S. Eisenhower Library has a collection of his manuscripts.

Bust of Sidney Lanier, at Johns Hopkins University in Baltimore. The poet lectured at the university for three semesters before his death in 1881.

H[enry] L[ouis] Mencken, the Sage of Baltimore, was born on September 12, 1880, in a house on the south side of Lexington Street, near Fremont. At the age of two he moved with his family to the brick row house at 1524 Hollins Street, where he lived, except for the five years of his marriage, until his death on January 29, 1956. After graduating in 1898 from the Baltimore Polytechnic Institute—at the top of his class, though he said he had no "taste for mechanics"—Mencken became a police reporter for the Baltimore *Morning Herald* in 1899. His first book was a collection of poems, *Ventures into Verse* (1903). In 1906 he joined the Baltimore *Sun,* for which he continued to write for most of the rest of his life. In 1906 he became associated with the magazine *Smart Set,* editing it with George Jean Nathan from 1914 to 1923. In the following year he and Nathan founded *The American Mercury.* Mencken left the magazine in 1933. Mencken always considered himself a journalist. He achieved fame through his coverage of the Scopes trial in 1925 and, through his magazines, became known as a literary bad boy, especially because of his vitriolic attacks on the "booboisie" and on religion, government, and the established culture.

Mencken's best-selling work and his greatest achievement, however, was a pioneering survey of the development of English in the United States, *The American Language* (1919), which he expanded in three subsequent editions and supplemented with two additional volumes in 1945 and 1948. Mencken's output was vast: he wrote plays, essays, criticism, travel books, autobiographical volumes, and a dictionary of quotations. On August 27, 1930, Mencken married Sara Powell Haardt in the Episcopal church at North and Warwick avenues. They lived at 704 Cathedral Street until she died on May 31, 1935. Mencken suffered a severe stroke in 1948, which ended his literary career. After his death, his ashes were placed in a grave next to his wife's grave in Loudon Park Cemetery in southwest Baltimore. Mencken's home at 1524 Hollins Street is open to the public on two Saturdays and Sundays each month. The Enoch Pratt Library at 400 Cathedral Street has a Mencken Room containing memorabilia, including his desk and typewriter.

Edgar Allan Poe is the most famous literary figure associated with Baltimore. His first stay here began in May 1829, when his foster father, John Allan, a wealthy exporter living in Richmond, Virginia, sent him to live with Maria Clemm, a younger sister of Poe's father, at a house on Eastern Avenue, between High and Exeter streets, that no longer stands. Poe's second book, *Al Aaraff, Tamerlane, and Minor Poems* (1829), was published at that time by the Baltimore firm of Hatch and Dunning. At the end of the year, he set off for West Point—his foster father had secured an Academy appointment for him—but returned in March 1831 to Mrs. Clemm's house, where he began writing the stories that made him famous. "Metzengerstein," "A Descent into the Maelström," and "Ms. Found in a Bottle" were probably written here.

The eating room in all these houses is the living room in winter. It has a round table in the centre covered with a decorated woolen cloth, that has soaked in the grease of many dinners, for though it should be always taken off, it is easier to spread the cloth upon it than change it for the blanket deadener that one owns. The upholstered chairs are dark and worn, and dirty. The carpet has grown dingy with the food that's fallen from the table, the dirt that's scraped from off the shoes, and the dust that settles with the ages. The sombre greenish colored paper on the walls has been smoked a dismal dirty grey, and all pervading is the smell of soup made out of onions and fat chunks of meat.

—Gertrude Stein,
in *Three Lives*

About an hour ago, I made bold to trust myself among a group of the crew. They paid me no manner of attention, and, although I stood in the very midst of them all, seemed utterly unconscious of my presence. Like the one I had at first seen in the hold, they all bore about them the marks of a hoary old age. Their knees trembled with infirmity; their shoulders were bent double with decrepitude; their shriveled skins rattled in the wind; their voices were low, tremulous, and broken; their eyes glistened with the rheum of years; and their gray hairs streamed terribly in the tempest.

—Edgar Allan Poe,
in "Ms. Found in a Bottle"

In 1833 the Clemm family moved to 203 Amity Street, where Poe is said to have written the stories "Berenice," "Morella," "King Pest," and "Hans Pfaal." While living here, Poe won a $50 prize from the Baltimore *Saturday Visitor* for "Ms. Found in a Bottle." The prize was awarded at the Latrobe House on Mulberry Street. With the help of John Pendleton Kennedy, Poe took a job on the *Southern Literary Messenger,* becoming its editor in 1835. In that same year Poe left Baltimore with his aunt and his cousin, Virginia Clemm. It is believed by some that Poe and Virginia, who was only thirteen years old, were married secretly on September 22, 1835, at St. Paul's Episcopal Church, at Charles and Saratoga streets.

Poe's death in Baltimore in 1849 is still surrounded by mystery. He had spent the summer in Richmond, Virginia, and traveled by boat to Baltimore, stopping off in Norfolk, Virginia, to give a lecture. Poe arrived in Baltimore on September 28; on October 3 he was found semiconscious in front of Cooth's and Sergeant's Tavern on Lombard Street. He was taken to Washington Hospital, now known as the Church Home and Infirmary, at Broadway and Fairmount Avenue, where he died during the morning of October 7. His last words are said to have been: "Lord, help my poor soul."

Poe is buried in Westminster Church Cemetery on West Fayette Street. The house at 203 Amity Street, containing period furniture of the type used by Poe and the Clemms, is maintained as a museum by the Edgar Allan Poe Society of Baltimore. A bronze statue of the author stands at Twenty-ninth Street, between Maryland Avenue and Oak Street, near Johns Hopkins. The Poe Room at Enoch Pratt Free Library, 400 Cathedral Street, is private and has only a few items related to Poe.

Francis Scott Key wrote "The Star-Spangled Banner" in Baltimore. On the night of September 13–14, 1814, Key was detained on a British warship at the mouth of the Patapsco River in Chesapeake Bay, where he watched the bombardment of Fort McHenry. In the morning, when he saw the U.S. flag still over the fort, he was inspired to write his famous lines. That same day, after being released by the British, Key completed the song at the Fountain Inn, which stood at the corner of Light and Redwood streets, later the site of the Southern Hotel. It was published a few days later in the Baltimore *American.* Sung to a British drinking song, it became popular at once. President Woodrow Wilson made "The Star-Spangled Banner" the national anthem in 1916, a designation confirmed by Congress in 1931.

The autographed manuscript is on display at the Maryland Historical Society, 201 West Monument Avenue. Key died on January 11, 1843, at his daughter's home on Mt. Vernon Place, now the Mt. Vernon Place Methodist Church, and his body was kept for a time in the family vault in St. Paul's Cemetery before burial in Frederick, Maryland. A monument to Key stands at Eutaw Place and Lanvale Street in Baltimore.

BERLIN

Charles Heber Clark was born in this Eastern Shore town on July 11, 1847. He was a newspaperman in Philadelphia who, under the name **Max Adeler,** wrote a collection of farcical sketches called *Out of the Hurly-Burly* (1874) and several other works of fiction.

Statue of Edgar Allan Poe, by Moses Ezekiel, in Baltimore. In 1930, an admirer of Poe's work noticed that the monument bore a misquotation of a line from "The Raven": "Dreaming dreams no mortals ever dared to dream before." The mistake was the "s" added to "mortal." The sharp-eyed reader, Edmund Fontaine, announced his intention to remove the letter on May 30, and the police were ready for him. But Fontaine outwitted them. Equipped with chisel and hammer, he had set things right on May 29. (*far right*) Poe's home at 203 Amity Street, Baltimore.

BETHESDA

The remains of **F. Scott** and **Zelda Fitzgerald** were brought to Tichnor's Funeral Home in this suburb of Washington, D.C., prior to burial—Scott in 1940, Zelda in 1947—in nearby Rockville. In 1943 **Jack Kerouac** spent some months as a psychiatric patient at the Bethesda Naval Hospital.

CATONSVILLE

Catherine Drinker Bowen, the biographer, in the early 1900s attended St. Timothy's Church Day School, 200 Ingleside Avenue, in Catonsville, just southwest of Baltimore. **John Banister Tabb** taught at St. Charles College here from 1889 until his death on November 19, 1909, and published a number of volumes of verse. His *Poems* (1894) established his reputation. Two other volumes were *Later Lyrics* (1902) and *Quips and Quiddities* (1907).

CHEVY CHASE

Robert Benchley and his family lived from January to May 1918 in a rented apartment here, when Benchley was working as a press representative in nearby Washington, D.C.

COLLEGE PARK

Katherine Anne Porter lived here at 6100 Westchester Park Drive for a little over ten years before the spring of 1980. Porter died in a nursing home in Silver Spring on September 18 of that year.

DAVIDSONVILLE

In 1785, at the age of twenty-six, **Mason Locke Weems** served as rector of All Hallows Church in this town between Annapolis and Washington, D.C. The church, built in 1710 and completely restored in 1940, is located at the junction of All Hallows Road and South River Club Road.

EASTON

Frederick Douglass spent most of his youth near this county seat on the Eastern Shore, near Chesapeake Bay. He lived for two years at Wye House, five miles north of here. After one of his attempts to escape from slavery, he spent several days in the county jail on Federal Street.

ELKTON

Martha [Farquharson] Finley, author of the immensely popular Elsie Dinsmore stories, visited here in 1867 and subsequently bought a house at 259 East Main Street, where she lived until her death on January 30, 1909. It was here that she wrote her seven-volume *Mildred* series (1878–94). The house, now used by the Gee Funeral Home, is not open to visitors.

ELLICOTT CITY

When **H. L. Mencken** was a boy, he spent two summers with his father and brother in this western suburb of Baltimore. They stayed at The Vineyards, 3611 Church Road. There are two houses on the property and, from Mencken's description, it is impossible to tell which he stayed in. The Vineyards now is private and not open to visitors.

"Shoot, if you must, this old grey head
But spare your country's flag," she said.

—John Greenleaf Whittier, in "Barbara Frietchie"

Old drawing of the home in Frederick of Barbara Frietchie, heroine of John Greenleaf Whittier's poem.

FREDERICK

In this city, about forty-five miles northwest of Washington, D.C., lived Barbara Frietchie who, at age ninety-six, supposedly waved a Union flag in defiance at Confederate General Thomas Jonathan (Stonewall) Jackson's troops while they rode past her house on September 6, 1862. John Greenleaf Whittier, who celebrated the event in his poem "Barbara Frietchie" (1863), acknowledged that the real facts are unclear. Though he based his poem on a story told him by novelist E. D. E. N. Southworth, Whittier later said it was a younger woman, Mary Quantrell, who had actually waved the flag. It is Barbara Frietchie, however, who is commemorated in Frederick.

Her homesite is now submerged in Carroll Creek, but a reproduction of her house stands at 154 West Patrick Street. It is open daily. Mrs. Frietchie is buried in Mt. Olivet Cemetery on South Harbor Street, which also contains the grave of **Francis Scott Key,** who was born near here. Key practiced law in Frederick from 1801–02 with Roger Brooke Taney, who later became Key's brother-in-law and, later still, chief justice of the U.S. Supreme Court. Taney's house, at 123 South Bentz Street, is now a museum owned by the Francis Scott Key Memorial Foundation. It contains some Francis Scott Key memorabilia, and for eleven months of the year is open by appointment. During the Bell and History Tour in May, it keeps regular hours. The Tourist Center at Second and Rosemont avenues will supply more information.

Natalie S[avage] Carlson attended Frederick Visitation Academy from 1910 to 1914. The academy was the setting for her novel *Luvvy and the Girls* (1971).

GARRETT PARK

Munro Leaf, author of the classic children's book *Ferdinand the Bull* (1936), lived at 11121 Rokeby Avenue in this suburb of Washington, D.C., at the time of his death on December 21, 1976. His wife still lives in the house and is happy to show it to visitors who make an appointment.

Katherine Anne Porter, who lived in College Park from the late 1960s until shortly before her death in 1980.

The inn at Haggerstown was one of the most comfortable I ever entered. It was there that we became fully aware we had left Western America behind us. Instead of being scolded, as we literally were at Cincinnati, for asking for a private sitting room, we here had two, without asking at all. . . . The charges were in no respect higher than at Cincinnati.

—Frances Trollope, in *Domestic Manners of the Americans*

HAGERSTOWN

Abram Joseph Ryan, known as the Poet of the Confederacy and the Poet of the Lost Cause, was born here on February 5, 1838. Father Ryan, a Roman Catholic priest, was baptized at Hagerstown's St. Mary's Church.

Hagerstown, near the Pennsylvania border and an important stop on the old National Road, chief route west in the first half of the 1800s, was visited in 1830 by English author **Frances Trollope** (1780–1863). She had come to America in 1827, landing in New Orleans and traveling to Cincinnati, to open a store, which failed. She then returned to England, but only after making a tour of the country by riverboat and stage coach. Trollope had a keen eye for scenery. Of the approach to Hagerstown, she wrote: ". . . world of mountains rising around you in every direction, and in every form; savage, vast and wild. . . . Through kalmias, rhododendrons, azaleas, vines and roses; sheltered from every blast that blows by vast masses of various colored rocks. . . ." This town provided material for her caustic *Domestic Manners of the Americans* (1832), which launched her career in England as a novelist and travel writer.

HERRING BAY

"I cannot tell a lie. I did it with my little hatchet."

—Mason Locke Weems, in *The Life of Washington the Great*

Mason Locke Weems, clergyman, biographer, and bookseller, was born on October 11, 1759, at Marshes Seat, a house near Herring Bay, about fifteen miles south of Annapolis. Educated in England, Parson Weems was one of the first two Anglican clergymen to be ordained in America after an oath of allegiance to the English crown was no longer required. He had a keen interest in making money, which he pursued for thirty-one years by traveling up and down the Atlantic coast in a wagon, fiddling, preaching, and selling books. To his credit, Parson Weems had much to do with keeping literacy alive in rural areas. His best-known book was *A History of the Life and Death, and Virtues and Exploits, of General George Washington* (1800), for the most part completed and awaiting publication before Washington's death in 1799—an early example of instant book publishing. The biography went through more than eighty-four editions in thirty years. Weems relied primarily on legend and imagination in writing the book. For example, in the fifth edition, *The Life of Washington the Great, enriched with a number of very curious anecdotes* (1806), Weems introduced the apocryphal cherry tree. His other biographical studies included *Life of Benjamin Franklin* (1815) and *Life of William Penn* (1822), both fanciful, and his many moral tracts included *God's Revenge Against Murder* (1807) and *The Drunkard's Looking Glass* (1823).

Birthplace of Francis Scott Key in Keysville.

HYATTSVILLE

James M. Cain, known best for *The Postman Always Rings Twice* (1934) and *Mildred Pierce* (1941), moved from California to this community north of Washington D.C. in 1950. He lived at 6707 Forty-fourth Avenue for many years, until his death there on October 27, 1977. In his last years he was in ill health and was cared for by neighbors, but he still managed to produce such books as *The Magician's Wife* (1965), *Rainbow's End* (1975), and *The Institute* (1976).

KEYSVILLE

Francis Scott Key was born on August 1, 1779, supposedly at Terra Rubra, an estate on the Keysville-Bruceville road, off Route 194. Key spent a good deal of time here, especially in summer, and in 1813 bought 700 acres of the Terra Rubra estate, which he farmed until his death on January 11, 1843. In 1856 a severe storm damaged the house irreparably. With wood from the old house, a new house was built that still stands on or near the original site. A monument to Key was erected on the grounds in 1915, but the property is privately owned and not open to visitors.

LEXINGTON PARK

Dashiell Hammett, the detective novelist, was born here on May 27, 1894. Lexington Park is just north of Point Lookout. Hammett's birthplace, on unmarked Sanner's Pond Road, off Great Mills Road, is privately owned and not open to visitors.

OXFORD

Hervey Allen bought Bonfield Manor, a large farm near Oxford, on the Eastern shore of Chesapeake Bay, with the royalties from his best-selling historical romance *Anthony Adverse* (1933). Believing that everyone should be self-sufficient, Allen worked Bonfield Manor as a farm.

POINT LOOKOUT

During the Civil War, the Union Army prison here at the mouth of the Potomac River affected the lives of two poets. **Sidney Lanier,** a volunteer in the Confederate Army, was captured late in 1864 and transferred here from Fort Monroe, Virginia. During his four-month imprisonment he contracted tuberculosis, from which he eventually died. Lanier managed to write while here, producing poems that included "The Palm and the Pine" and "Spring Greeting." One of Lanier's fellow prisoners was **John Banister Tabb.** Captured in

1864 and sent to Point Lookout, he met Lanier, who urged him to write poetry.

RELAY

George Bronson Howard was born in this town nine miles southwest of Baltimore on January 7, 1884. Howard worked for several newspapers, including the Baltimore *American,* the New York *Herald,* and the San Francisco *Chronicle;* he also served as a war correspondent during the Russo-Japanese War. Howard's plays included *The Snobs* (1911) and, with Eric Howard, *The Alien* (1927). His best-known novel was *God's Man* (1915).

RELIANCE

This hamlet on the Maryland-Delaware state line was the site of Joe Johnson's Tavern, which is also known as Patty Cannon's Tavern. Patty Cannon was a murderer and a kidnaper of blacks who were taken to be sold as slaves. The house now on the site, at the junction of Routes 392 and 577 and Delaware Route 20, is open to visitors curious to see a house that stands where the tavern once stood. George Alfred Townsend, of Georgetown, Delaware, used the infamous Patty Cannon in his novel *The Entailed Hat* (1884).

ROCKVILLE

F. Scott Fitzgerald came to this city just north of Washington, D.C., in January 1931, to attend the burial of his father in the cemetery of St. Mary's Catholic Church, 520 Viers Mill Road. A scene based on the funeral appears in *Tender Is the Night* (1934). Fitzgerald later said he would like to be buried in the same cemetery, but permission was denied after Fitzgerald's death because Fitzgerald had renounced Catholicism, some of his books had been proscribed, and he had not received last rites. In December 1940 he was buried instead at Rockville Union Cemetery. In 1948, **Zelda Fitzgerald** was buried beside her husband. The Fitzgeralds were reinterred finally at St. Mary's in 1975.

SILVER SPRING

Katherine Anne Porter died on September 18, 1980, at the Carriage Hill Nursing Home here.

TOWSON

In the summer of 1932, **F. Scott** and **Zelda Fitzgerald** moved into La Paix, an eighteen-room house built and owned by their friend Bayard Turnbull, in this northern suburb of Baltimore. Scott, his income severely reduced by the effects of the Depression, hoped that La Paix would provide a haven in which he could finish writing *Tender Is the Night* (1934). He also worked on some of the short pieces collected later in *The Crack-Up* (1945). Zelda was receiving outpatient treatment at the Phipps Psychiatric Clinic, part of the Johns Hopkins Hospital. The cost of her treatment was only one of the burdens on Fitzgerald, who counted on *Tender Is the Night* to solve his financial problems.

Zelda's physical health bloomed at La Paix, but she made little progress in her battle against mental illness. In June 1933, while trying to burn some old clothes in an upstairs fireplace, she started a fire that ruined several rooms. By October, a draft of *Tender Is the Night* was complete, and in November Scott and Zelda moved into Baltimore. By January 1934, Zelda was back in Towson—at the Sheppard and Enoch Pratt Hospital, a psychiatric clinic next door to La Paix. She was released in March, but returned in May. By that time, *Tender Is the Night* had been published. Its sale of 13,000 in the first printing was a great disappointment to Fitzgerald. La Paix was torn

Marker at the site of Patty Cannon's Tavern in Reliance, just across the border from Delaware. A notorious kidnaper and murderer, she led a gang that terrorized parts of Maryland and Delaware in the early 1800s.

La Paix, the Towson home of F. Scott Fitzgerald in 1932 and 1933, shown in a photograph taken shortly before it was razed. (*below*) Photograph of the graves of Scott and Zelda Fitzgerald in Rockville, with the closing sentence of *The Great Gatsby* inscribed: "So we beat on, boats against the current, borne back ceaselessly into the past."

Capitola at first was delighted and half-incredulous at the great change in her fortunes. The spacious and comfortable mansion of which she found herself the little mistress; the high rank of the veteran officer who claimed her as his ward and niece; the abundance, regularity and respectability of her new life; the leisure, the privacy, the attendance of servants, were all so entirely different from anything to which she had previously been accustomed, that there were times when she doubted its reality and distrusted her own identity or her sanity.

—E. D. E. N. Southworth, in *The Hidden Hand*

Inebriate of air am I,
And debauchee of dew,
Reeling, through endless summer days,
From inns of molten blue.

—Emily Dickinson, in "I Taste a Liquor Never Brewed"

Old photograph of Prospect Cottage, once the home of E. D. E. N. Southworth in Washington, D.C.

down in 1961 to make room for St. Joseph Hospital, which is located at 7620 York Road. The remains of the house shared by the Fitzgeralds lie buried under a parking lot near the hospital laundry.

Towson is also the site of Goucher College, where **H. L. Mencken** in the spring of 1923 delivered a lecture on "How to Catch a Husband." While in Towson he met Sara Powell Haardt, whom he married in Baltimore on August 27, 1930.

TUCKAHOE

Frederick Douglass was born in a slave cabin on a farm near here, on what is now Route 303, three miles west of Denton in the center of the Eastern Shore. His birthdate was long thought to be late 1817, but recent evidence suggests that February 1818 is closer to correct. He lived here until he was six. The cabin, no longer standing, was located in the woods near Tuckahoe Creek. Several miles away, on Route 328 at the Tuckahoe Bridge, the Maryland Historical Society has erected a monument commemorating Douglass's birth and paying tribute to his achievements.

WAVERLY

Lizette Woodworth Reese, described by H. L. Mencken as the outstanding lyric poet of her generation, was born on January 9, 1856, in this village, now part of Baltimore. She began teaching at seventeen, ending her career forty-eight years later in the nearby Baltimore city schools. In 1923 teachers, students, and alumni presented a bronze plaque to Western High, Baltimore, with lines from her "Tears," which Louis Untermeyer has called "one of the most famous sonnets written by any American." Reese's first book of verse, *A Branch of May,* was published in 1887; her last was *Pastures* in 1933. She also published two biographical works, *The New York Road* (1931) and *A Victorian Village* (1929), in which she provided recollections of the Civil War, including the funeral procession of Abraham Lincoln passing through Baltimore. She is buried in St. John's Episcopal Church in Waverly.

WEVERTON

This hamlet just east of Harpers Ferry, West Virginia, was the site of the Savage family home. **Natalie S[avage] Carlson** lived there for many years, and used the Savage home as the setting for her novel *The Half Sisters* (1970).

WASHINGTON, D.C.

Writers born in our nation's capital include **Edward [Franklin] Albee,** winner of a Pulitzer Prize for his play *A Delicate Balance* (1967). Albee was born on March 12, 1928, and was adopted at the age of two weeks by Reed Albee, the wealthy theater owner. **Channing Pollock,** the drama critic and playwright, was born in Washington on March 4, 1880, when his family was living at 821 Thirteenth Street NW. Pollock worked as drama critic for both the Washington *Post* and Washington *Times* before leaving the city in 1900. **Marjorie Kinnan Rawlings,** winner of a Pulitzer Prize for her novel *The Yearling* (1938), was born in Washington on August 8, 1896, and was graduated from Western High School. **Samuel Shellabarger,** the historian and author of such novels as *Captain from Castile* (1945), was born here on May 18, 1888. **Frank G[ill] Slaughter,** the physician and author of many novels dealing with the medical profession, was born here on February 25, 1908.

E[mma] D[orothy] E[liza] N[evitte] Southworth, one of the most popular and prolific novelists of the nineteenth century, was born on December 26, 1819, in a house, no longer standing, at 224 North Capitol Street NW, and grew up in Washington. In 1840 she married and moved away from Washington but returned in 1843, after her marriage failed. In 1851 Southworth moved to Prospect Cottage, her home at 3600 Prospect Street in Georgetown, now a fashionable section of northwest Washington. Southworth there wrote many of her sentimental novels, among them *The Hidden Hand* (1859), *Ishmael* (1863), and *Self-Raised* (1876); and there she died on June 30 1899. She is buried in Oak Hill Cemetery, at Thirtieth and R streets NW. As a footnote to literary history, it is recalled that Southworth related to John Greenleaf Whittier the story on which he based his poem "Barbara Frietchie." Also buried in Oak Hill Cemetery is **John Howard Payne,** the playwright and actor, now remembered for his sentimental lyrics for "Home, Sweet Home." Also born in Washington were **Leonora Speyer,** the poet, on November 7, 1872; and **Jean Toomer,** author of the classic collection of stories *Cane* (1923), on December 26, 1894.

Not unexpectedly, various writers have served in the U.S. Congress. **Ignatius Donnelly** represented Minnesota from 1863 to 1869. **Thomas Dunn English** represented Pennsylvania from 1891 to 1895. **John Pendleton Kennedy** represented Maryland from 1838 to 1844. **Clare**

Boothe Luce represented Connecticut from 1943 to 1947. The final name in this group of literary members of Congress is that of **Joseph Pulitzer,** the newspaper publisher, who served briefly in 1885 and 1886 but discovered that he could not simultaneously serve both Mammon and his constituents. He resigned from Congress on his thirty-ninth birthday, April 10, 1886, and devoted full attention from then on to the business of running the St. Louis *Post and Dispatch* and the New York *World.*

Writers who have lived in Washington include **Henry Adams,** grandson of John Quincy Adams, who came here in 1878. In 1884 Adams had built for him a house, no longer standing, at 1603 H Street NW. The suicide of Adams's wife in 1885 left him grief-stricken, and in the following year he went off to Japan. On his return to Washington, Adams completed his monumental *History of the United States of America During the Administrations of Jefferson and Madison* (9 volumes, 1889–91). Adams then set out on a trip to the South Seas. For the rest of his life Adams lived part of each year in Washington and spent the rest traveling. His later years were highly productive: his work in that period included two books of permanent value, *Mont-Saint-Michel and Chartres* (privately printed 1904; published 1913) and *The Education of Henry Adams* (privately printed 1907; published 1918). Adams died at his Washington home on March 27, 1918, and he was buried beside his wife in Rock Creek Cemetery, on Rock Creek Church Road NW. Their grave is marked by a statue by Augustus Saint-Gaudens, commissioned by Adams as a memorial to his wife.

Louisa May Alcott served as a volunteer nurse from December 1862 to January 1863 at the Union Hotel, used as a hospital, which stood at the northeast corner of Thirtieth and M streets NW. Here she became severely ill, and the calomel treatment administered to her damaged her health permanently. Alcott's *Hospital Sketches* (1863) was based on her letters to her family from Washington. **Joel Barlow,** the poet and statesman, owned a home called Kalorama, off Kalorama Road, from about 1805 to his death in 1812. Barlow was appointed minister to France, and he died in Poland on December 24, 1812.

Ambrose Bierce worked as Washington correspondent for the San Francisco *Examiner* in 1896—he had stopped writing short stories by this time—and from 1900 to 1906. One place in which Bierce is known to have lived in Washington is 603 Fifteenth Street NW. **John Burroughs,** the naturalist and author, lived at 1332 U Street NW from 1867 to 1873, while working for the U.S. Treasury Department. It was in Washington, on Pennsylvania Avenue, that Burroughs first met Walt Whitman. **Jonathan Daniels,** the newspaperman and author, attended St. Albans School, Wisconsin and Massachusetts avenues NW, from 1913 to 1921, while his father served as secretary of the Navy. **Emily Dickinson** stayed with her family for a time in 1855 at the Willard Hotel, while her father was serving as a congressman. The Willard, on Pennsylvania Avenue at Fourteenth and F streets, is still standing and undergoing restoration.

John Dos Passos lived in childhood at 1201 Nineteenth Street NW and spent school vacations here from 1907 to 1911, when he was a student at Choate School in Wallingford, Connecticut. Dos Passos's parents did not marry until 1912, so Dos Passos divided his time between his mother's home here, and his father's home in Virginia. Dos Passos inherited the house after his mother's death in 1915, and he sold it in 1917.

Frederick Douglass, the abolitionist editor and author, lived from 1871 to 1877 at 318 A Street NE, now used by the African Art Museum and open to the public. Douglass lived from 1877 until his death on February 20, 1895, at Cedar Hill, a mansion at 1411 W Street SE, now administered by the National Park Service and open to visitors. During his residence in Washington, Douglass served in several official capacities, including recorder of deeds for the District of Columbia, and minister to Haiti. **Lloyd C. Douglas,** the clergyman and novelist, served from 1909 to 1911 as minister of the Luther Place Memorial Church, 1226 Vermont Avenue NW.

Allen Drury, author of the best-selling novel *Advise and Consent* (1959), worked as a newspaperman in Washington from 1943 to 1959. For *Advise and Consent,* a tale of Washington political intrigue, Drury won a Pulitzer Prize. **Eleanor Estes** wrote her children's book *A Little Oven* (1955) while living at 3017 Cambridge Place. **Hinton Rowan Helper,** an influential Southern writer whose books denounced the presence of blacks in America, lived in Washington from 1890 until his death by suicide on March 8, 1909, in a lodging house at 628 Pennsylvania Avenue NW.

Langston Hughes worked in the early 1920s as a busboy at the Wardman Park Hotel, now the Sheraton Park Hotel, at 2660 Woodley Road NW. One evening he left a sheaf of his poetry at the dining table of one of the hotel guests, **Vachel Lindsay.** Lindsay brought attention to Hughes's work by reading three of his poems before an audience in the little theater of the hotel. **Sinclair Lewis** lived at several addresses here in

There they were! "our brave boys," as the papers justly call them, for cowards could hardly have been so riddled with shot and shell, so torn and shattered, nor have borne suffering for which we have no name, with an uncomplaining fortitude, which made one glad to cherish each as a brother. In they came, some on stretchers, some in men's arms, some feebly staggering along propped on rude crutches, and one lay stark and still with covered face, as a comrade gave his name to be recorded before they carried him away to the dead house.

—Louisa May Alcott,
in *Hospital Sketches*

Adams memorial in Rock Creek Cemetery.

America has always taken tragedy lightly. Too busy to stop the activity of their twenty-million-horsepower society, Americans ignore tragic motives that would have overshadowed the Middle Ages; and the world learns to regard assassination as a form of hysteria, and death as neurosis, to be treated by a rest-cure. Three hideous political murders [Lincoln, Garfield, McKinley], *that would have fattened the Eumenides with horror, have thrown scarcely a shadow on the White House.*

—Henry Adams,
in *The Education of Henry Adams*

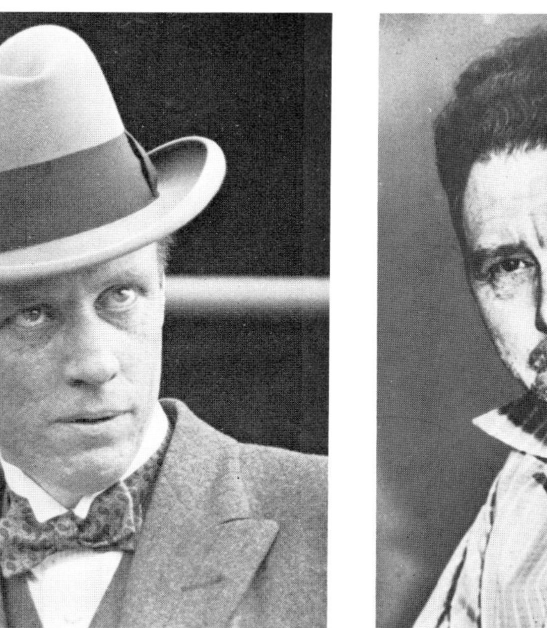

Sinclair Lewis (*right*), in a photograph taken in the 1920s. Ezra Pound (*far right*), in a photograph taken in Rome in 1942.

1919 and 1920, including 1814 Sixteenth Street NW, 1639 Nineteenth Street NW, and Corcoran Street NW. He also rented an office at 1127 Seventh Street NW, where he worked on *Main Street* (1920). In the fall of 1926, intending to stay a year, Lewis returned to Washington and rented a house at 3028 Q Street NW. He also rented a room at the Hotel Lafayette to use as an office and there worked on the manuscript of *Elmer Gantry* (1927). Lewis's marriage was in trouble, and he did not stay the year. By November 1926, he had left for New York City. **Ruth McKenney** lived at 5019 Reno Road NW in the 1940s.

Joaquin Miller lived in the early 1880s in a log cabin on Meridian Hill, on Sixteenth Street near Crescent Street NW. Miller built the cabin himself and decorated it on the outside with bear skins, elk horns, bows, and arrows. It soon became a tourist attraction, and Miller thrived on the attention until a man bought property adjacent to his and built a treehouse. People soon began to stop and ask directions to the man who lived in the tree. This humiliated Miller, and in 1886 he sold out and moved to California. Miller's cabin, now restored, was moved in 1912 to Rock Creek Park, on Beach Drive. For the occasion Miller wrote a poem entitled "To My Log Cabin Lovers."

Other literary residents of Washington include **Katherine Anne Porter,** who lived at 3601 Forty-ninth Street NW in the 1960s; and **Ezra Pound,** whose stay at St. Elizabeth's Hospital, in Congress Heights, from 1945 to 1958 was anything but voluntary. Pound, who lived in Europe during World War II, was arrested at the end of the war and charged with treason for making broadcasts over Italian radio in support of fascism.

The cabin Joaquin Miller built on Meridian Hill, in Washington. It now stands in Rock Creek Park.

Dear, loyal lovers, neighbors mine,
What word of mine or deed or sign,
Can compensate what ye have done.
This housing in your hearts my home,
My lowly old log cabin home,
Aye, dear the friends and memories
Of London, Dresden, storied Rome,
The Arctic, the Antipodes,
But dearer far than all of these,
Your holding of my hearth and home,
My lordly, kingly, cabin home. . . .
Yea, many hands have been most fair,
Yea, many trumps of fame and faith
Mine ears have heard both here and there
That said as only true love saith,
But nothing ever seemed so dear,
As this, your brave log cabin cheer.

—Joaquin Miller,
"To My Log Cabin Lovers"

After several months of being held prisoner in a cage—for some time his only protection from the sun was a sheet of tar paper rolled over the top of the cage each morning—Pound was flown to Washington for medical examination and trial. At Gallinger Hospital, now D.C. General Hospital, at Nineteenth Street and Massachusetts Avenue SE, Pound was found to be insane and unfit to stand trial. In late December 1945, Pound was sent to St. Elizabeth's. It was not until 1958 that appeals from many American writers met with success: the charges against Pound were dropped, and the poet was released.

Mary Roberts Rinehart lived from 1920 to 1932 at 2419 Massachusetts Avenue. **Harriet Prescott Spofford** came to live in Washington after her marriage in 1865, and here wrote many of the stories for which she was known. Her novel about the city, *Old Washington,* was published in 1906. **Gore Vidal** lived in childhood at the home of his maternal grandfather, Senator Thomas Gore, of Oklahoma. The house, located at 4500 Broad Branch Road NW and now used for the Malaysian embassy, figures in Vidal's story "A Moment of Green Laurel," which appeared in the collection *A Thirsty Evil* (1956). Vidal also wrote of the house in his novel *Washington, D.C.* (1967). Vidal has lived also at the Wardman Park Hotel, now the Sheraton Park Hotel, at 2660 Woodley Road NW; and at Tilden Gardens, an apartment house at 3000 Tilden Street NW.

Walt Whitman came to Washington in late December 1862 to find and care for his brother, who had been wounded in a Civil War battle. Moved by the suffering of the wounded soldiers, Whitman decided to stay on in Washington and work in the hospitals. What Whitman had thought would be a visit of about ten days turned out to last eleven years. Whitman at first found lodgings at 394 L Street NW, near Fourteenth Street, and took a part-time job as a copyist in the Paymaster's office, at the corner of Fifteenth and F streets NW. He worked there in 1863 and 1864, and moved in 1865 to 468 N Street NW, near Twelfth Street, and took a job as a clerk for the Department of the Interior. His job there lasted only until James Harlan, secretary of Interior, discovered that Whitman was *the* Walt Whitman, author of *Leaves of Grass,* a work Harlan judged to be morally offensive, and fired the poet. Whitman remained in Washington until 1873, when a stroke forced him to move to Camden, New Jersey, to live with his brother. While living in Washington, Whitman wrote the poems collected in *Drum-Taps* (1865) and *Sequel to Drum-Taps* (1965–66), the latter containing the poems "When Lilacs Last in the Dooryard Bloom'd" and "O Captain! My Captain!" In Washington, too, Whitman first met **John Burroughs** and was a frequent visitor at the naturalist's Washington home. It was Burroughs who gave Whitman the idea of the mockingbird as the central image in the poem "Out of the Cradle Endlessly Rocking."

Writers who have lived in Washington also include **Herman Wouk,** who lived from 1964 to 1966 at 2220 Massachusetts Avenue NW and from 1966 to 1968 at 2230 S Street NW. While living in Washington, Wouk wrote most of his novel *The Winds of War* (1971).

Elinor [Hoyt] Wylie, the poet and novelist, moved with her family to 1205 Vermont Avenue NW about 1897, when she was twelve years old. Daughter of a prominent family, Wylie was educated in Washington schools, and in 1906 married Philip Hichborn, a young Washington lawyer and son of a rear admiral. The marriage was an unhappy one—Hichborn was subject to severe depression and violent rages—and soon Elinor was attracted to Horace Wylie, a married man fifteen years her senior. In December 1910 the lovers left for England. In 1912 Hichborn committed suicide, and three years later Elinor and Horace returned to the United States and married. After a time they returned to Washington, but finding themselves outcasts among the city's social elite, they left Washington in 1919. **Shelby Foote** was playwright in residence at the Arena Stage, Sixth and M streets SW, in 1963 and 1964; and **Bret Harte,** who with Mark Twain wrote the play "Ah Sin," first produced at the National Theater, at 1321 E Street NW on May 7, 1877. **Imamu Amiri Baraka,** born **LeRoi Jones,** was graduated from Howard University in Washington in 1954. **Norman Mailer** used his experiences during a four-day antiwar protest here in 1967 for his book *The Armies of the Night* (1968), for which he received a Pulitzer Prize.

Writers associated with Washington also include **Carl Sandburg,** who addressed a joint session of Congress on February 12, 1959, the 150th anniversary of Abraham Lincoln's birth; **Frank R. Stockton,** who died on April 20, 1902, in a private home at 2129 P Street NW; **William Makepeace Thackeray,** who visited here in 1853 during his first American tour; and **Frances Trollope,** who visited Washington in 1830. **Mark Twain** served as secretary to Senator William M. Stewart of Nevada during the winter of 1867–68, and was a frequent visitor to the city during his writing career. While serving as Stewart's secretary, Twain stayed at the Willard Hotel.

Among Washington sites and institutions of special interest to literary visitors is, of course, the Library of Congress, on First Street between East Capitol Street and Independence Avenue SE. The largest library in the world, it has employed numerous writers as consultants over the years, including **Conrad Aiken,** who wrote about half of his autobiographical novel *Ushant* (1952) here while serving as a Fellow in American Letters at the library from 1947 to 1954. Other writers who have worked here include **James Dickey, Paul Laurence Dunbar, Robert Frost, Jeremy Ingalls, Randall Jarrell, Stanley Kunitz, Clare Boothe Luce, Robert Lowell, Archibald MacLeish,** who served as Librarian of Congress from 1939 to 1944, **Karl Shapiro,** and **Allen Tate.** The Library of Congress has collections of the papers of many authors, including John Ciardi, Stephen Foster, Oliver Wendell Holmes, Sinclair Lewis, Bernard Malamud, Ayn Rand, and Walt Whitman. Just across First Street from the library, at 209 East Capitol Street, is the eminent Folger Shakespeare Library, established by Henry Clay Folger to house his collection of Shakespeare material.

Visitors to the White House should bear in mind that its heritage is literary as well as political. **Louis Adamic** dined with Franklin D. Roosevelt and Winston Churchill, then prime minister of Great Britain, in January 1942, and used his experiences here in *Dinner at the White House* (1946).

And thenceforward all summer in the sound of the sea,
And at night under the full of the moon in calmer weather.
Over the hoarse surging of the sea,
Or flitting from brier to brier by day,
I saw, I heard at intervals the remaining one, the he-bird,
The solitary guest from Alabama.

Blow! blow! blow!
Blow up sea-winds along Paumonok's shore;
I wait and I wait till you blow my mate to me.

—Walt Whitman, in "Out of the Cradle Endlessly Rocking"

The atmosphere of the oval room known as the President's Study—sometimes as the Lincoln Study—did not derive so much from the mixture of fine old period pieces and chintz-covered armchairs, the too-numerous paintings and prints occupying much of the wall space, and the cheerful, light-reflecting green-and-yellow curtains, as from its proportions and historic associations. I had heard the rumor (since printed in a slightly different version by Mrs. Roosevelt) that sometimes late at night when everything is quiet some of the White House residents imagine hearing steps like those of a very tall man thoughtfully pacing the carpeted floor. . . .

—Louis Adamic, in *Dinner at the White House*

Out of the dusk a shadow,
Then a spark;
Out of the cloud a silence,
Then a lark;
Out of the heart a rapture,
Then a pain;
Out of the dead, cold ashes,
Life again.

—John B. Tabb,
in "Evolution"

Let us cross over the river and rest under the shade of the trees.

—Dying words of
Stonewall Jackson

He had given the finishing touches to his [inaugural] *address that very morning. None knew so well as he what consequences would surely follow any blunder in tone or mistake in declaration. He looked worn and pale and anxious, but from the first to the last his voice rang out clear, firm, unhesitating, resonant with faith and courage, while its every tremor and modulation seemed to vouch for his sincerity. He was making his last appeal for peace and his last solemn protest against needless bloodshed.*

—William Osborn Stoddard,
in *Abraham Lincoln*

Churchill sued Adamic successfully for libel because of a footnote in the book suggesting that Churchill had been bailed out of bankruptcy by British bankers who were creditors of the Greek government.

Abraham Lincoln has his unique place in literary history as author of the Gettysburg Address, one of the finest pieces ever written by an American, and as the subject of countless novels, plays, poems, short stories, and biographies. **Robert E. Sherwood,** from 1940 to 1945 an unofficial member of Franklin D. Roosevelt's White House staff, wrote the Pulitzer Prize-winning history *Roosevelt and Hopkins: An Intimate History* (1948). **William Osborn Stoddard** served as secretary to Abraham Lincoln from 1861 to 1864, and wrote of his experiences and of Lincoln in *Abraham Lincoln* (1884), *Inside the White House in War Times* (1890), *The Table Talk of Lincoln* (1894), and *Lincoln at Work* (1899).

Among the many statues to be found in Washington are several that commemorate literary figures. Benjamin Franklin is remembered in two statues: one stands at the intersection of Pennsylvania Avenue and Tenth Street NW, the other in the Senate Chamber of the Capitol Building. Henry Wadsworth Longfellow is represented at the intersection of M Street with Rhode Island and Connecticut avenues NW, and a statue of Lew Wallace stands at the Capitol Building. Last and most worthy of mention here, of course, is the Lincoln Memorial, in Potomac Park. This beautiful structure, whose focus is an awesome statue of the president, provides the literary visitor with the opportunity to read and ponder some of Lincoln's most memorable statements.

VIRGINIA

ABINGDON

This town in the southwestern corner of Virginia enters literary history as the home of the Barter Theatre. This unique institution was established in the 1930s, when unemployed actors were willing to practice their craft in exchange for food, and playwrights were given food in lieu of royalty payments. **Maxwell Anderson, Noel Coward, Robert Sherwood,** and **Thornton Wilder** were among the playwrights so nourished. Actors who received early training here included Hume Cronyn, Patricia Neal, and Fritz Weaver. Now designated the State Theatre of Virginia, the Barter Theatre is located just off Route 81.

ALEXANDRIA

This city on the Potomac River just south of Washington, D.C., which boasts innumerable historical associations, has little to offer the literary visitor. Gadsby's Tavern Museum, at 134 North Royal Street, is a restoration of the tavern that served as a center of political and social activity in the late 1700s and early 1800s. Practically every political leader involved in the founding of the United States visited the Georgian-style tavern, or stayed at the hotel that was added in 1792, but only **Francis Scott Key,** one of its patrons, can be considered a literary figure. The museum is open daily except for Monday, and an admission fee is charged.

AMELIA COURT HOUSE

Father **John B[anister] Tabb,** whose *Poems* (1894) had nearly a score of editions, was born on March 22, 1845, at The Forest, a house that stood about seven miles north of Amelia Court House, on Route 609.

APPOMATTOX

On April 9, 1865, Robert E. Lee surrendered the Confederate troops under his command to Ulysses S. Grant. The surrender took place in the parlor of a private home near Old Appomattox Courthouse, midway between Roanoke and Richmond, bringing to an end the bloodiest war in American history. The Civil War has been the subject of thousands of histories, biographies, novels, plays, and poems, and Lee's surrender at Appomattox figures in many of them. *A Stillness at Appomattox* (1953), by Bruce Catton, is a brilliant account of the final year of the war. Herman Melville wrote of the war's ending in "The Surrender at Appomattox." The site of the surrender is part of Appomattox Courthouse National Historical Park, on Route 24, three miles northeast of the town of Appomattox.

As billows upon billows roll,
On victory victory breaks;
Ere yet seven days from Richmond's fall
And crowning triumph wakes
The loud joy-gun, whose thunders run
By sea-shore, streams, and lakes.
The hope and great event agree
In the sword that Grant received from Lee.

—Herman Melville,
in "The Surrender at Appomattox"

ARLINGTON

Arlington National Cemetery is the burial place of those who have served their country, among them **Hervey Allen, Ludwig Bemelmans,** and **Dashiell Hammett.**

BIG STONE GAP

The town of Big Stone Gap, in the Cumberland Mountains in southwestern Virginia, is host each July and August to a musical adaptation of a novel about mountain feuding and passion, *The Trail of the Lonesome Pine* (1908), by **John Fox, Jr.** Fox lived in a two-story frame house here, at 117 Shawnee Avenue. The model for the character June Tolliver in the novel was a girl named June Morris, who boarded in a brick house on Route 613, now known as the June Tolliver House.

BOYCE

John Esten Cooke wrote most of his later books in Boyce, in northern Virginia, a few miles from the west bank of the Shenandoah River. He lived in a house known as The Briars, to which he came shortly after the Civil War and where he lived until his death. It was the Civil War and the periods before and after it that served as the subject for a number of his books, both fiction and nonfiction. His works included *Wearing of the Gray* (1867), a collection of essays; *The Heir of Gaymount* (1870), a novel; *Virginia* (1883), a history of the state; and *My Lady Pokahontas* (1885), a novel. Cooke died at The Briars on September 27, 1886.

BUCHANAN

Mary Johnston, the author of historical romances, many of them set in Virginia, was born on November 21, 1870, in Buchanan, about twenty-five miles northeast of Roanoke. Among her most popular works was the novel *To Have and To Hold* (1900), set in the Virginia of the 1600s.

CHANCELLOR

This town, west of Fredericksburg, was known as Chancellorsville when, in early May of 1863, Union and Confederate forces met in one of the pivotal battles of the Civil War. Although Robert E. Lee led the Confederate army to victory here, his success was followed only by a greater defeat at Gettysburg two months later. But it was at Chancellorsville that Stonewall Jackson received the wounds—from the fire of his own troops—that led to his death on May 10, 1863. The Battle of Chancellorsville is the scene of action in *The Red Badgé of Courage* (1895) by Stephen Crane, although this great novel of a young soldier caught in the fury of battle never mentions the actual battle by name. The battle is described in other novels, including *The Long Roll* (1911), by Mary Johnston, and *The Southerner* (1913), by Thomas Dixon. Stonewall Jackson's dying words inspired the title of the novel *Across the River and into the Trees* (1950) by Ernest Hemingway, as well as several poems, including "Under the Shade of the Trees," by Margaret Junkin Preston, a poet of the Confederacy whose sister was Jackson's wife.

CHARLOTTE COURT HOUSE

This town, the seat of Charlotte County in south central Virginia, was the home of **Mary Virginia [Hawes] Terhune** from 1856 to 1859. The novelist, who wrote under the pen name **Marion Harland,** lived with her husband, Rev. Edward Payson Terhune, in the manse of the Village Presbyterian Church. The old manse, where she wrote several of her early books, is still in use. Mary Terhune was the mother of **Albert Payson Terhune,** author of *Lad: A Dog* (1919) and other stories about collies. Together they wrote *Dr. Dale—A Story Without a Moral* (1900).

CHARLOTTESVILLE

This city in central Virginia, the seat of Albemarle County and the home of the University of Virginia, has an abundance of literary associations. Among the writers who lived in Charlottesville are **Erskine Caldwell,** who attended the University of Virginia here in 1920 and 1921; **Thomas H. Dickinson,** playwright, author, and editor, who died in Charlottesville on June 12, 1961; and **Willard Huntington Wright,** born here on June 18, 1888. Wright established himself as a literary critic and editor and then, in the 1920s, began to write detective novels under the name **S. S. Van Dine.** He was the creator of the popular fictional detective Philo Vance.

The University of Virginia has attracted many writers since its founding. Writers have come here to study, to teach, and to confer about their craft. In 1931 the university was host to a Southern writers' conference that included thirty authors, among them **Sherwood Anderson, James Branch Cabell, William Faulkner, Ellen Glasgow,** and **DuBose Heyward.** Faulkner returned to the university to serve as lecturer and writer in residence from 1957 to 1960.

One of the university's first students, and perhaps its best known, was **Edgar Allan Poe,** who enrolled in February 1826, one year after the school opened. Poe was an excellent student, but he ran up large gambling debts in a self-defeating attempt to win enough money to remain in school, and it was for lack of funds that Poe was eventually forced to leave in December 1826, at the close of the school session. Poe was under threat of arrest because he could not pay his debts. His room, at Number 13, West Range, is maintained by the Raven Society as a memorial to the poet's residence here.

Other writers associated with the university include **Karl Shapiro,** the poet, who studied here in 1932 and 1933; **John Reuben Thompson,** editor of the *Southern Literary Messenger,* who was graduated in 1845; and **William Peterfield Trent,** a founder and editor of the *Sewanee Review,* who received an M.A. degree in 1884. The Virginia Authors Collection of the University of Virginia's Alderman Library contains collections of the letters and papers of numerous writers, including Sherwood Anderson, James Branch Cabell, John Esten Cooke, Clifford Dowdey, John Fox,

The level sheets of flame developed great clouds of smoke that tumbled and tossed in the mild wind near the ground for a moment, and then rolled through the ranks as through a gate. The clouds were tinged an earth-like yellow in the sunrays and in the shadow were a sorry blue. The flag was sometimes eaten and lost in this mass of vapor, but more often it projected, sun-touched, resplendent.

. . . To the youth it was an onslaught of redoubtable dragons. He became like the man who lost his legs at the approach of the red and green monster. He waited in a sort of horrified, listening attitude. He seemed to shut his eyes and wait to be gobbled.

—Stephen Crane,
in *The Red Badge of Courage*

Edgar Allan Poe's room at the University of Virginia in Charlottesville, in a photograph taken about 1930.

James Whitcomb Riley.

Jr., Ellen Glasgow, Mary Johnston, John Pendleton Kennedy, and Thomas Nelson Page.

CHINCOTEAGUE ISLAND

This island off the Delmarva Peninsula is the home of the small horses known as the Chincoteague ponies, which used to run free over Chincoteague Island. The horses are believed to be descended from horses abandoned on the islands by early colonists. Marguerite Henry brought them to wide attention with the publication of her book *Misty of Chincoteague* (1947) and other books for children.

CISMONT

Near this town just east of Charlottesville is Castle Hill, the home for many years of **Amélie Rives,** whose married name was **Princess Troubetzkoy:** Russian Prince Pierre Troubetzkoy, a portrait painter, was Rives's second husband. Her best-known work was her novella *The Quick or the Dead?* (1888). This psychological study portrays a widow in love with her cousin, who strongly resembles her dead husband. As a child, Rives was brought to live at Castle Hill, her grandfather's home near the junction of routes 231 and 640. Castle Hill is actually two houses joined together—the first was constructed in the 1760s and the second in 1825. Amélie Rives, after a productive career as novelist and playwright, spent many of her later years in Castle Hill and died here on June 15, 1945. The house is not open to visitors.

COURTLAND

This town in southeastern Virginia was called Jerusalem at the time when Nat Turner led the bloodiest slave uprising in American history. Turner, a slave, was owned by Joseph Travis, who operated a small plantation about twelve miles southwest of Jerusalem. Turner, believing that he was called on by God to deliver the slaves to freedom, instigated a revolt in August 1831 that lasted barely two days but claimed the lives of nearly sixty whites and an unknown number of blacks. Turner escaped capture until October 30. Tried and convicted of insurrection, he was hanged on November 11. William Styron used the confession dictated by Turner while in prison in Courtland as the basis for *The Confessions of Nat Turner* (1967), the novel that won a Pulitzer Prize for Styron.

Bel Air, home of Mason Locke Weems near Dumfries.

DENNISVILLE

Mary Virginia [Hawes] Terhune, born here on December 21, 1830, wrote under the pen name **Marion Harland.** She concentrated at first on writing novels, including *Alone* (1854) and *Sunnybank* (1866), but her phenomenal success with *Common Sense in the Household* (1871) led her to write books and syndicated columns on cooking and home economics. She lived until the age of thirteen in this community just south of Amelia Court House, on Route 38.

DUMFRIES

The Weems-Botts Museum, at 300 Duke Street, in this town southwest of Alexandria, is a memorial to **Mason Locke Weems,** an Episcopal clergyman best known for his biography of George Washington, and to Benjamin Botts, a prominent attorney. Weems came to live in the area in the 1790s and, in 1798, bought the house in which the museum is located. Botts bought the property from Weems in 1802. It is believed that Weems used the building as a bookstore and as a resting place during his travels up and down the East Coast to sell books. Weems's experience as a bookseller may have given him insight into the tastes of the reading public and inspired him to write his now legendary biography of George Washington. The project saw light first as an eighty-page booklet mixing fact and fantasy and designed to be instructional, entertaining, uplifting and, above all, profitable. The fifth edition of the book (1806) introduced the apocryphal story of Washington and the cherry tree, and Weems could afford to move to Bel Air, a Federal-style brick house off Route 640, not far from Dumfries. Bel Air remained his home for the rest of his life, although he still spent much of his time on the road, selling bibles and biographies of Washington as well as moralistic tracts. When he was not out selling books, he was writing them. Among his books were *Life of Doctor Benjamin Franklin* (1815), and *Life of William Penn* (1822). Weems died on May 23, 1825, and was buried on his estate here. Bel Air is private and not open to visitors.

FALLS CHURCH

Cherry Hill Farm, once owned by Judge Joseph Riley and his wife Mary, in Falls Church, northwest of Alexandria, was visited by their nephew **James Whitcomb Riley,** who wrote of the farm in his poem "Out to Old Aunt Mary's." The farm, fully restored, is now a public park. **James Thurber** lived in Falls Church as a boy in the summer of 1901. His father had taken a post as secretary to an Ohio congressman and moved the family to a house at 319 Maple Avenue, now the site of a group of town houses known as James Thurber Court. A game of William Tell, played with his brothers in the backyard of their home, proved fateful for Thurber. He was hit in the left eye with an arrow, and the local doctor advised against removal of the eye. By the time the eye was removed, Thurber's condition had so deteriorated that he eventually lost sight in his remaining eye.

FREDERICKSBURG

Near this city, southwest of Alexandria, stands Chatham, a Georgian-style mansion that was impressed into service as a Federal hospital and military headquarters during the Battle of Fredericksburg in December 1862. **Walt Whitman** came here at that time to care for his brother George, who had been wounded in battle. The plight of the wounded soldiers helped Whitman make up his mind to work in Union hospitals for the duration of the war. Chatham, on the north bank of the Rappahannock River, overlooking Fredericksburg, is now part of Fredericksburg-Spotsylvania National Park. It contains two rooms of museum exhibits, and one exhibit is devoted to Whitman.

GORE

Willa [Sibert] Cather was born near this community at the northern tip of Virginia on December 7, 1873. Her birthplace, on the north side of Route 50, which links Gore with the city of Winchester, is listed in the National Register of Historic Places and bears a state historical marker. When Cather was very young she moved with her family to another house, Willow Shade, about a mile away on Route 50. The house, now known as the William Cather-Willa Cather House, is located just east of the Back Creek bridge, on the north side of the road. Cather lived at Willow Shade for eight years before moving to Nebraska, where she grew up. The novelist used her impressions of Gore in her novel *Sapphira and the Slave Girl* (1940).

HAMPTON

This city at the mouth of Chesapeake Bay stands on the site of one of the earliest English settlements in America. Its citizens were harassed by pirates in the 1700s, its streets resounded to the march of British boots during the War of 1812, and it was almost totally destroyed by its own citizens in 1861, in the face of Federal occupation. Just south of Hampton, on a spit of land known as Old Point Comfort, stands Fort Monroe, now a National Historic Landmark, where **Sidney Lanier** was held prisoner late in the Civil War.

Edgar Allan Poe, who had enlisted in the U.S. Army as E. A. Perry, served at Fort Monroe from December 1828 to April 1829. The fort is associated with the founding of another Hampton landmark, Hampton Institute, on Queen Street. At the end of the Civil War, a wave of freed slaves, perhaps having no other place to go, came to Fort Monroe. In April 1868 a school opened in the barracks of the old Hampton Hospital, with the goal of helping America's newest citizens learn to read and write. Two years later it became the Hampton Normal and Industrial Institute. **Booker T. Washington** arrived at the Institute in 1872 and was graduated with honors three years later at the age of nineteen. Washington's autobiography, *Up from Slavery* (1901), covers his years here. **Thomas Wolfe** worked at another Hampton landmark, Langley Field, in June 1918. He used his experiences there in the novel *Look Homeward, Angel* (1929) and in the short story "The Face of the War," which was collected in the volume *From Death to Morning* (1935).

JAMESTOWN ISLAND

This island just south of Williamsburg is the site of James Towne, capital of Virginia Colony from 1607 to 1699. The original settlement is virtually gone now—destroyed by man or by the tides—but the Jamestown Church, dating to the seventeenth century, survives and contains memorials to Pocahontas, John Rolfe, and John Smith. The Jamestown Church also has a memorial to **George Sandys.** While living in the colony, Sandys translated the last ten books of Ovid's *Metamorphoses.* These were published, with the first five books, which Sandys had translated previously, as *Ovid's Metamorphoses Englished by G.S.* (1626).

Pocahontas, the Indian girl who is said to have converted to Christianity, saved Capt. John Smith's life, and married John Rolfe, figures in numerous poems, tales, and novels. Among them are *My Lady Pokahontas* (1885), the novel by John Esten Cooke, and *The Bridge* (1930), the poem by Hart Crane. The Virginia Colony is the setting for the novel *To Have and To Hold* (1900), by Mary Johnston.

LEESBURG

James Dickey, the poet, is said to have rented the Harrison-Trone house at 47 North King

Wasn't it pleasant, O brother mine,
In those old days of the lost sunshine
Of youth—when the Saturday's chores were through,
And the "Sunday's wood" in the kitchen, too,
And we went visiting, "me and you,"
Out to Old Aunt Mary's?—

—James Whitcomb Riley, in "Out to Old Aunt Mary's"

Willow Shade, childhood home in Gore of Willa Cather. The house, built in 1858 by her grandfather, William Cather, served during the Civil War as headquarters for Union Army officers.

Is life so dear, or peace so sweet, as to be purchased at the price of chains and slavery? Forbid it, almighty God! I know not what course others may take; but as for me . . . give me liberty or give me death.

—Patrick Henry,
in a speech on March 23, 1775

Street, in this town northwest of Washington, D.C., while working on his novel *Deliverance* (1970).

LYNCHBURG

This city is the home of Randolph-Macon Women's College, from which **Pearl Buck** was graduated in 1914. While here, she wrote for the school newspaper, served as class president, and received two literary prizes. After graduation she taught at the college for a semester. Another student at Randolph-Macon was **Josephine Haxton,** who writes novels and short stories under the name **Ellen Douglas.**

McLEAN

Gore Vidal lived for a time at Merrywood, in this suburb of Washington, D.C. The slave-cabin and river-edge scenes in Vidal's third novel, *The City and the Pillar* (1948), are based on Merrywood in the 1930s, and the river edge appears again at the end of his novel *Kalki* (1978).

MARION

Sherwood Anderson moved to this town in 1924 and became editor of both of the town's newspapers: the Republican *Smyth County News,* and the Marion *Democrat,* both of which are still publishing. To give each of the newspapers its own editorial slant, Anderson had the town sheriff write editorials for one newspaper, and the town postmaster write editorials for the other. Anderson lived over the printshop here in Marion for a time but later built a home near Trout Dale, about ten miles to the southeast. Anderson is buried in Round Hill Cemetery in Marion.

MARTINSVILLE

Vein of Iron (1935), a novel of rural life in Virginia by Ellen Glasgow, is set partially in the community once known as Ironside, northwest of Martinsville. The "vein of iron" is the strength of the protagonist, who held her family together during the years of the Depression.

MONTPELIER

Off Route 33 in Montpelier, about twenty miles northwest of Richmond, is Sycamore Tavern, built in the late 1700s. In the early 1900s, **Thomas Nelson Page** gave the small frame building to the county for use as a library. Page was born on April 23, 1853, at Oakland, his family's plantation home located north of Montpelier. He experienced life on an antebellum Southern plantation and then lived through the war that changed forever the economic system essential to plantation life. Oakland survived the Civil War but was destroyed by fire in 1898. It was rebuilt later and was the home of Page's brother. Page died at Oakland on November 1, 1922. Many of Page's later stories and novels were based on his experiences here.

Sherwood Anderson.

NEWPORT NEWS

William Styron, author of such novels as *Lie Down in Darkness* (1951), *Set This House on Fire* (1960), and the Pulitzer prize-winning novel *The Confessions of Nat Turner* (1967), was born in Newport News on June 11, 1925. The city appears as Port Warwick in *Lie Down in Darkness.* **Thomas Wolfe** worked as a munitions checker on the government docks in Newport News in the summer of 1918, and later used his experiences in the novel *Look Homeward, Angel* (1929) and in the short story "The Face of the War." Wolfe lived in a furnished room near the docks, probably no longer to be seen because nearly all of the houses standing in 1918 have been torn down. The house still standing at 114 Twenty-seventh Street gives the visitor an idea of what the houses looked like in Wolfe's day.

NORFOLK

Writers associated with this city include **A[mos] Bronson Alcott,** the distinguished educator and philosopher, who came here as a young man in 1818 to peddle small items to people in and near Norfolk. He returned here again in 1819 and 1821, still a peddler. **Augustin Daly,** best known for his dramatic adaptations of European plays, attended school in Norfolk in the 1840s. **F. Scott Fitzgerald** visited his aunt Eliza Delihant and his cousin Cecilia Delihant at their home at 4107 Gasnold Avenue, now a private residence. Cecilia Delihant was the model for the character Clara in Fitzgerald's novel *This Side of Paradise* (1920). **Joseph Heller,** author of *Catch-22* (1961), worked as a blacksmith's helper at the Norfolk Navy Yard before joining the U.S. Army Air Force in October 1942. **Karl Shapiro,** the poet, attended Maury High School, on Shirley Avenue, in 1929 and 1930, serving as joke editor of the *Maury News.*

PETERSBURG

[Nathaniel] Beverley Tucker, best known in his time for his novels *George Balcombe* (1836) and *The Partisan Leader* (1836), was born on September 6, 1784, at Matoax, his family's home. Matoax was built in 1770 on a knoll overlooking the Appomattox River. The original house was destroyed by fire, and a new house was built on its site in 1853. The site is now part of the campus of Virginia State College. Tucker's novel *George Balcombe* greatly impressed one of his contemporaries, Edgar Allan Poe.

RICHMOND

This city, the capital of Virginia and former capital of the Confederacy, is also the literary capital of the state. In addition to an abundance of historical associations, Richmond has a literary heritage so extensive and colorful that it is well worth a visit, especially to Hollywood Cemetery, where many of Richmond's literary figures are buried. Some of Richmond's literary sites no longer exist, but we must remember that the city was evacuated and burned by its citizens in the face of Union occupation at the end of the Civil War.

Civil War Richmond was the setting for the melodrama *Secret Service* (1895), written by William Gillette, the actor and playwright. This play, which concerned a Southern girl and her love for a Union spy, was one of Gillette's most popular dramas. The career of **Patrick Henry,** who was a principal contributor to the Bill of Rights, is associated with St. John's Church, on East Broad Street between North Twenty-fourth and North Twenty-fifth streets, the site, in March 1775, of the Second Virginia Convention. There Patrick Henry is reported to have delivered one of his noteworthy orations, his defiant statement on liberty. John Esten Cooke, in his romance *The Virginia Comedians* (1854), dealt with Henry's life. In the churchyard is the grave of Elizabeth Arnold Poe, mother of Edgar Allan Poe, whose strong association with Richmond is described below. Richmond figures in the novel *The Virginians* (1859), by the English novelist **William Makepeace Thackeray,** who visited the city during his American tour of 1852–53, and again on his tour of 1855–56.

Authors born in Richmond include **Clifford Dowdey,** the novelist and historian, who was born on January 23, 1904, in a house, now private, at 406 North Twenty-fourth Street. He lived during childhood at 2504 Kensington Avenue and attended John Marshall High School. After living elsewhere for a number of years, Dowdey returned to his boyhood home at 2504 Kensington Avenue, where he was to live for the rest of his life. He worked as a reporter for the Richmond *News Leader,* in 1925 and 1926, and did research for his nonfiction books, beginning with *Experiment in Rebellion* (1946). Richmond figures in his novel *Bugles Blow No More* (1937). Dowdey died on May 30, 1979. **Amélie Rives,** was born on August 23, 1863, in a house, no longer standing, near the corner of Eighth and East Grace streets. One of Rives's romantic novels with a Virginia setting was *Virginia of Virginia* (1888).

John Reuben Thompson, owner and editor of *Southern Literary Messenger,* published in Richmond from 1847 to 1860, was born on October 23, 1823, in a house, no longer standing, at the corner of Fourteenth and Main streets. Thompson lived at 108 Mayo (now Ballard) Street from 1839 to 1858, at Franklin Street between Eighth and Ninth streets in 1858–59, and at 802 East Leigh Street from 1859 to 1864. Thompson died on April 30, 1873, and is buried in Hollywood Cemetery, at 412 South Cherry Street. Thompson was a friend of Edgar Allan Poe, who had edited the *Messenger* ten years before Thompson became its editor. Thompson's lecture on the poet, *The Genius and Character of Edgar Allan Poe,* was not published until 1929, nine years after the posthumous publication of Thompson's *Collected Poems* (1920). *Southern Literary Messenger,* in its final years, became a spirited defender of the Confederate cause, which led ultimately to the demise of the journal in 1864. **William Peterfield Trent,** a founder and editor of *Sewanee Review,* was born on November 10, 1862, in a house, no longer standing, at the corner of Seventh and East Marshall streets.

Writers who have resided in Richmond include **Mary Johnston,** the novelist, who lived from 1902 to about 1905 at 113 East Grace Street and, until 1911 or 1912, at 110 East Franklin Street. She is buried in Hollywood Cemetery. **Thomas Nelson Page** lived in Richmond from 1874 to 1893. None of the houses in which he lived has a marker noting his residence. Two of the houses, at 404 North Twelfth Street and 631 East Main Street, are no longer standing. His home at 609 East Main Street, where he lived when his story "Marse Chan" (1884) appeared in *Century* magazine, is now a real-estate office. His home at 722 East Franklin Street, where he lived when his collection of stories *In Ole Virginia* (1887) was published, is gone. Page is known to have lived at 814 West Grace Street, now the site of a movie theater, when *Two Little Confederates* (1888), his story for children, was published. He also lived at 1101 Grove Avenue, a site now occupied by an apartment building.

James Branch Cabell, long a resident of Richmond, shown in a photograph taken about 1950.

Mary Virginia [Hawes] Terhune lived at 506 East Leigh Street in the 1840s. By the age of fourteen she was a contributor to the local newspaper, and by sixteen she had written her first novel, *Alone,* which was published in 1854 under her pen name, **Marion Harland.** In 1856 she married Reverend Edward Payson Terhune and went with him to Charlotte Court House, Virginia.

Of the many writers who have lived in Richmond, three have become most closely associated with the city: James Branch Cabell, Ellen Glasgow, and Edgar Allan Poe. Of them, only Poe was not born in Richmond.

James Branch Cabell was born on April 14, 1879, in a house at 101 East Franklin Street. The land on which it stood is now occupied by the Richmond Public Library, which displays a bust of Cabell and a plaque noting the novelist's birth. Cabell worked for the Richmond *Times* in 1898 and, after a brief stint as a reporter for the New York *Herald,* returned to Richmond to work as a reporter for the Richmond *News.* In 1902 he turned to freelance writing, but his books did not sell, so in 1911 he went to work for his uncle's coal company in West Virginia. In 1913 he married Priscilla Bradley Shepherd. The couple lived at her home in nearby Dumbarton Grange before moving back to Richmond. For a long time Cabell lived at 3201 Monument Avenue, where he wrote many of his nearly fifty books, working directly on a typewriter and declining to make carbon copies of his manuscripts. Among Cabell's best-known

"As I felt the sickening sweep of the descent, I had instinctively tightened my hold upon the barrel, and closed my eyes. For some seconds I dared not open them—while I expected instant destruction, and wondered that I was not already in my death-struggles with the water. But moment after moment elapsed. I still lived. The sense of falling had ceased; and the motion of the vessel seemed much as it had been before, while in the belt of the foam, with the exception that she now lay more along. I took courage and looked once again upon the scene."

—Edgar Allan Poe,
in "A Descent into the Maelström"

works are *The Cream of the Jest* (1917), *Jurgen* (1919), and *The High Place* (1923), all part of a series of tales set in the imaginary medieval kingdom of Poictesme. It was the publication of *Jurgen,* and its subsequent suppression on the grounds of obscenity, that established Cabell's reputation. His books had lost much of their popularity by the time of his death in Richmond on May 6, 1958, but editions of his books today are introducing a new generation of readers to his writing. Cabell is buried in Hollywood Cemetery.

Ellen Glasgow was born on April 22, 1873, in a house, no longer standing, at 101 East Cary Street. The author of such novels as *Barren Ground* (1925), *The Romantic Comedians* (1926), *Vein of Iron* (1935), and the Pulitzer Prize-winning novel *In This Our Life* (1941) lived for many years in a Greek Revival house at 1 West Main Street. Here, in an upstairs study, she wrote all of her books with the exception of the novel *Life and Gabriella* (1916). Over the years she was visited by many writers, among them Hamlin Garland, H. L. Mencken, Burton Rascoe, Marjorie Kinnan Rawlings, and Gertrude Stein. Glasgow died here on November 21, 1945, and is buried in Hollywood Cemetery.

Edgar Poe, a child two years old in 1811, became an orphan when his mother, Elizabeth Arnold Poe, died in Richmond. He was taken in by the family of John Allan, a Richmond merchant, and although Poe was never formally adopted into the family, he became known as **Edgar Allan Poe.** Except for the five years from 1815 to 1820, when Poe and the Allans lived in England and Scotland, the poet lived in Richmond from 1811 to 1826, and returned to the city several times in later years.

The houses in which Poe lived here are no longer standing. The Allans lived at 12 North Fourteenth Street before moving to Great Britain in 1815. On their return in 1820 they moved into a house at Fifth and Clay streets, on land now occupied by the Richmond Coliseum, and for a time lived in the home of Charles Ellis—Allan's partner in the firm of Ellis and Allan—at the southwest corner of Franklin and Second streets, where the Richmond Public Library now stands. The death of a relative made John Allan one of the richest men in the country, and in 1825 the family moved to a large house on the southeast corner of Fifth and Main streets. This was the last home Poe shared with the Allans and, in his writings, he described some of its furnishings: tapestries, the agate lamp that stood on the table at the door to his room, and the like.

In February 1826 Poe entered the University of Virginia. Despite Allan's wealth, Poe's foster father refused to aid him financially, and the poet was forced to leave the university after one session. Poe then spent two years in the Army. A brief period at West Point was brought to a halt, at least in part, by John Allan's lack of affection and support. Allan wanted Poe to turn his attention to business or to law, but the poet knew what he wanted to do with his life. He managed to have his poems published in book form, and his poetry and stories began to appear in newspapers and magazines. In 1835 Poe returned to Richmond, took lodgings in a house on the southeast corner of Eleventh and Bank streets, and worked for the *Southern Literary Messenger* (1834–64), which was founded in Richmond by Thomas W. White at the southeast corner of Main and Fif-

The Ellen Glasgow House on West Main Street, Richmond. The novelist came here with her family when she was in her teens, and remained for most of her life. (*below*) Photograph of Ellen Glasgow.

The Allan House, at Fifth and Main streets, in Richmond, where Edgar Allan Poe once lived. The drawing was made by Harry Fenn from a photograph taken shortly before the house was torn down.

teenth streets. Poe worked for the *Messenger* for only a short time before he broke with White over an incident involving Poe's drinking. Poe traveled to Baltimore and lived with his aunt, Maria Clemm, and her daughter, Virginia. It is believed that while there he secretly married Virginia Clemm. He returned to Richmond, patched up the rift with White, and became editor of the *Messenger.* Mrs. Clemm and Virginia later joined him in Richmond and, on May 16, 1836, he married—remarried?—his cousin here. The Poes and Mrs. Clemm remained in Richmond until early 1837, when Poe again broke with the *Messenger.* During his association with the *Messenger,* Poe helped to increase its subscription list sevenfold, to more than 3,500. The journal published a number of his tales, including "Berenice," "Morella," and "Lionizing," and reprinted such works as the story "The MS. Found in a Bottle" and the poem "Israfel," but Poe's greatest contribution to the *Messenger* was his unsparing and often brilliant literary criticism. Poe brought to American letters a demanding form of criticism it had lacked until then, and he soon became known for his ability to launch devastating literary attacks.

Poe returned to Richmond during the summer of 1849 and took lodgings in the old Swan Tavern, at the northeast corner of Ninth and Broad streets, where Richmond City Hall now stands. During this last summer of his life, Poe revisited many of the scenes of his youth and paid calls on friends old and new. Although still suffering from acute poverty and very ill, Poe had reason to feel hopeful about his future. His public readings in Richmond were being widely reported, he had a growing circle of friends who were looking out for his welfare, and he had decided to return to live in Richmond permanently. At the offices of the Richmond *Examiner* he worked on his poetry, completing the final version of his best-known poem, "The Raven." Poe gave his last public lecture at the Exchange Hotel, which stood at the southeast corner of Fourteenth and Franklin streets. He visited Elmira [Royster] Shelton, to whom he had been engaged briefly in 1825–26. Her home at 2407 East Grace Street is now the headquarters of the Historic Richmond Foundation, Inc., and is open to visitors. Poe's engagement to Elmira Royster had been broken through the efforts of her parents, who had intercepted letters between the two—Poe was then attending the University of Virginia—causing each to feel abandoned by the other. After many years, Poe and Mrs. Shelton, by then a widow, renewed the romance that had been thwarted so many years earlier. Poe may have proposed to her here, for the two once again became engaged to be married, on October 17. The marriage never took place: Poe died in Baltimore on October 7, 1849.

Mrs. Shelton's house is one of two existing homes Poe is known to have visited. The other is Talavera, 2315 West Grace Street, now headquarters of Help Encourage Landmark Preservation (H.E.L.P.) Inc., which is in process of remodeling the house. Talavera was the home of **Susan Archer [Talley] Weiss,** a poet and literary patron whom Poe visited on September 25, 1849, just before he left Richmond for Baltimore, en route to Philadelphia and New York. Mrs. Weiss recalled later that Poe was reluctant to say goodbye to his friends: "We were standing in the portico, and after going a few steps he paused, turned, and again lifted his hat in a last adieu. At that moment a brilliant meteor appeared in the sky directly over his head, and vanished. . . ."

If I could dwell
Where Israfel
Hath dwelt, and he were I,
He might not sing so wildly well
A mortal melody,
While a bolder note than this might swell
From my lyre within the sky.

—Edgar Allan Poe,
in "Israfel"

Helen, thy beauty is to me
Like those Nicéan barks of yore,
That gently, o'er a perfumed sea,
The weary, way-worn wanderer bore
To his own native shore.

On desperate seas long wont to roam,
Thy hyacinth hair, thy classic face,
Thy Naiad airs have brought me home
To the glory that was Greece,
And the grandeur that was Rome.

Lo! in yon brilliant window-niche
How statue-like I see thee stand,
The agate lamp within thy hand!
Ah, Psyche, from the regions which
Are Holy-Land!

—Edgar Allan Poe,
"To Helen"

Visitors to Richmond may find interest in two other sites associated with Poe. The poet attended the Monumental Church, at 1224 East Broad Street, when he was a boy. The grave of Mrs. Jane Stith Stanard, in Shockoe Hill Cemetery, may also be of interest. This woman is said to have been the inspiration for Poe's poem "To Helen" (1831). Poe had fallen in love with her when he was a boy, and was devastated by her death in 1824. The Poe legend has it that he often visited her grave at night and wept with grief at his loss.

The main Richmond attraction for those who admire Poe is a building that has no known connection with the poet's life in this city. The Edgar Allan Poe Museum, 1914-16 East Main Street, occupies five buildings that open on enclosed gardens. The main building is the Old Stone House, which dates to the 1700s. Poe may have visited the house as part of an honorary color guard for the Marquis de Lafayette, who was entertained here in 1824. The other buildings have been added to the museum to house a growing collection of Poe material. Here visitors may view a scale model of Richmond as it was in Poe's day, and examine some of the poet's personal items, including his trunk, his walking stick, his wife's mirror, and her trinket box. The museum has portraits of Poe, his wife, his mother, his foster parents, and others. The Poe Museum is open daily, and an admission fee is charged.

ROCKY MOUNT

The Booker T. Washington National Monument stands sixteen miles northeast of Rocky Mount, which is south of Roanoke. The visitor will see here a restoration of the plantation on which **Booker T. Washington** was born on April 5, 1856. The author of the classic autobiography *Up from Slavery* (1901) spent his childhood here in slavery. At the end of the Civil War, he and his family moved to Malden, West Virginia. The original buildings of the plantation no longer exist, but several have been reconstructed, including the small cabin in which Washington was born. Nearby, on Route 122, is a marker commemorating Washington's birth.

Ripshin Farm, Sherwood Anderson's home in Trout Dale, in a photograph taken about 1947.

SURRY

Bacon's Castle, a brick mansion in Surry, built in the seventeenth century, was visited by **Sidney Lanier** frequently during the Civil War. Bacon's Castle is a National Historic Landmark, private and not open to visitors. Surry is about thirty-five miles east of Petersburg.

TROUT DALE

In the summer of 1925 **Sherwood Anderson** and his third wife stayed in this town in southwestern Virginia, near the North Carolina and Tennessee borders. Here Anderson worked on *Tar, A Midwest Childhood* (1926), a fictional work based on the author's life. He also explored the area around Trout Dale. Anderson liked the region and the people who lived here, and soon decided to buy a farm on Ripshin Creek, naming it Ripshin Farm. There he built a house of native stone and wood, and worked on his last books in a cabin near the main house, which is now a National Historic Landmark. Ripshin Farm, near the junction of Routes 603 and 732, is open to visitors by arrangement with Mrs. Sherwood Anderson, the author's fourth wife, who lives in nearby Marion.

WARM SPRINGS

This mountain resort community, seventy-five miles west of Charlottesville, was the home for many years of **Mary Johnston.** The novelist built her home, Three Hills, about 1912 and lived here for the rest of her life. Among her books written during these years were *Croatan* (1923), *The Slave Ship* (1924), and *The Great Valley* (1926). She died at Three Hills on May 9, 1936. Three Hills now is used for apartments.

WARRENTON

Mary H. Eastman was born in 1818 in Warrenton, thirty-five miles west of Alexandria. She was the author of *Dacotah, or Life and Legends of the Sioux Around Fort Snelling* (1849), said to have been a source of inspiration for *Hiawatha* (1855), by Henry Wadsworth Longfellow.

WESTMORELAND

In Westmoreland, on the southern bank of the Potomac River, near where the river flows into Chesapeake Bay, is Spence's Point, once the home of **John Dos Passos.** Dos Passos spent part of his boyhood on his father's farm in Westmoreland County. Later he inherited part of the estate and settled here permanently in the late 1940s. His home on Sandy Point Neck is a National Historic Landmark, private and not accessible to the public.

WILLIAMSBURG

This city, settled in 1633 and the capital of Virginia from 1699 to 1780, is the site of one of the best-known and largest colonial restorations. The Williamsburg Historic District comprises nearly five hundred restored or reconstructed buildings, dating from the seventeenth and eighteenth centuries. Although colonial Williamsburg is not thought of primarily as a literary site, it does have some connections with historical figures known, at least in part, for their writings or speeches, notably Patrick Henry and Thomas Jefferson.

A number of authors have written about Williamsburg and its residents. John Esten Cooke called it Martinsburg in his novel *Leather Stocking and Silk* (1854). The play *The Common Glory* (1947), by Paul Green, presented here annually until 1976, was first performed in Williamsburg. The lakeside amphitheater in which it was presented is now neglected. Walter Havighurst wrote of Alexander Spotswood, governor of Virginia in the early 1700s, in *Alexander Spotswood: Portrait of a Governor* (1967). Elswyth Thane [Beebe] has written a series of novels about the generations of an American family in Williamsburg from early times to the twentieth century. The series includes *Dawn's Early Light* (1943), *Yankee Stranger* (1944), *Kissing Kin* (1948), and *Homing* (1951).

Williamsburg is also the home of William and Mary College, founded in 1693. Writers associated with the college include **James Branch Cabell,** who received a B.A. here in 1898. Cabell met **Ellen Glasgow,** a fellow Richmond native, in Williamsburg on the porch of the Colonial Inn, no longer standing, but ten years were to pass before the writers became close friends. **Carl Sandburg** read his poem "The Long Shadow of Lincoln" (1945), considered one of the best of his later poems, as the Phi Beta Kappa poem at William and Mary in December 1944. **St. George Tucker,** the lawyer, poet, and editor, studied law in Williamsburg in the 1770s, and later became a professor of law at William and Mary. His works included *A Dissertation on Slavery* (1796), which proposed a plan for gradual emancipation of blacks, and the *Probationary Odes of Jonathan Pindar* (1796), a collection of political satires. His son, **[Nathaniel] Beverley Tucker,** the novelist, also taught at William and Mary. The Tucker House in Williamsburg is now open to the public. The Tucker-Coleman Room at the Earl Gregg Swem Library, William and Mary College, contains a collection of St. George Tucker's books and his son's personal items. The library also contains other literary collections, including papers of Thomas Nelson Page.

WINCHESTER

This city in northern Virginia is associated with several writers. **Natalie S[avage] Carlson,** who was born here on October 3, 1906, is best known for her books for children, including *The Talking Cat and Other Stories of French Canada* (1952) and *Alphonse, That Bearded One* (1954). **John Esten Cooke,** the historian and novelist, was born near Winchester on November 3, 1830. Cooke wrote about Virginia in a number of his best-known novels, including *The Virginia Comedians* (1854), *Surry of Eagle's Nest* (1866), and *My Lady Pokahontas* (1885). His nonfiction included *Life of Stonewall Jackson* (1863) and a history, *Virginia* (1883).

WEST VIRGINIA

BUCKHANNON

The Pearl Buck manuscripts, property of the Pearl S. Buck Birthplace Foundation, Hillsboro, West Virginia, are now stored at West Virginia Wesleyan College on College Avenue in Buckhannon. They are to be transferred to Hillsboro when adequate storage space can be arranged for them there.

CHARLES TOWN

John Peale Bishop was born in Charles Town on May 21, 1892, and lived on Congress Street, in a house still standing but unmarked and not open to visitors. Bishop grew up in Charles Town and attended the local elementary school. His books included *The Undertaker's Garland* (1922), a collection of poetry and prose he wrote with his friend **Edmund Wilson,** whom he met at Princeton University. Bishop succeeded Wilson as editor of *Vanity Fair,* and then lived in Paris for several years, where he came to know other American expatriates, including Ezra Pound, E. E. Cummings, and Archibald MacLeish. At Princeton Bishop also became acquainted with F. Scott Fitzgerald, who is said to have modeled Tom D'Invilliers in *This Side of Paradise* (1920) after Bishop. Bishop's only novel was *Act of Darkness* (1935), which he based on recollections of his boyhood in Jefferson County, of which Charles Town is the county seat. The novel upset many old-timers in town who saw real people in its pages, and the author was told not to bother to come home again. Bishop's *Collected Essays* and *Collected Poems* both appeared posthumously in 1948.

White pine, yellow pine,
The first man fearing the forest
Felled trees, afraid of shadow,
His own shade in the shadow of pinewoods.

Slash pine, loblolly,
The second man wore tarheels,
Slashed pine, gashed pine,
The silent land changed to a sea-charge.

Short leaf, long leaf,
The third man had aching pockets.
Mill town, lumber mill,
And buzzards sailed the piney barrens.

Cut pine, burnt pine,
The fourth man's eyes burned in starvation.
Bone-back cattle, razor-back hogs
Achieve the seedling, end the pinewoods.

—John Peale Bishop,
in "Southern Pines"

Amory's companion proved to be none other than "that awful highbrow, Thomas Parke D'Invilliers," who signed the passionate love-poems in the Lit. *He was, perhaps, nineteen, with stooped shoulders, pale blue eyes, and, as Amory could tell from his general appearance, without much conception of social competition and such phenomena of absorbing interest. Still, he liked books, and it seemed forever since Amory had met anyone who did. . . .*

—F. Scott Fitzgerald,
in *This Side of Paradise*

Frank R. Stockton, in an engraving made from a drawing by John White Alexander (1856–1915).

CHEAT MOUNTAIN

Ambrose Bierce, nineteen years old in 1861, reenlisted in the Union army and was sent to Cheat Mountain in central West Virginia. Here he participated in the Battle of Cheat Mountain in September 1861. The area provided background for one of his best stories, "A Horseman in the Sky."

CLARKSBURG

Melville Davisson Post, the writer of many mystery and detective stories, was living at the time of his death in a house he called The Chalet. The house is of Swiss design and stands eleven miles south of Clarksburg, at Lost Creek. Post died in Clarksburg on January 23, 1930, after a fall from a horse. The Chalet is not open to visitors.

GRAFTON

Ambrose Bierce arrived in Grafton, twelve miles south of Fairmont, in June 1861 as a member of the Ninth Indiana Infantry. Here he participated in the Battle of Girard Hill, distinguishing himself by carrying a wounded soldier to safety under fire. The campaign over, Bierce returned to Indiana in late July.

HARPER'S FERRY

John Brown's Body (1928), the Pulitzer Prize-winning epic poem by Stephen Vincent Benét, about the causes and course of the Civil War, described Brown's raid of October 16, 1859, on the Federal arsenal in Harper's Ferry. Brown was captured while seeking guns to use in Abolitionist raids. Tried in the county courthouse in nearby Charles Town, and convicted of treason, he was hanged in an open field outside Charles Town on December 2, 1859. The Twelfth Massachusetts Regiment first popularized the marching song "John Brown's Body" (origin uncertain), and it later was adopted generally by Union troops: "John Brown's body lies a-mouldering in the grave. His soul is marching on!" The words were sung to an old Methodist hymn, with the memorable chorus "Glory, Glory, Hallelujah!" The John Brown of the song is not thought to be the John Brown of Harper's Ferry or of Stephen Vincent Benét's poem, but probably a Sergeant Brown of the Twelfth Massachusetts. Many people of the time found the words of the marching song offensive, and Julia Ward Howe's "Battle Hymn of the Republic," also written in 1861 and sung to the same tune, soon came to be more widely known: "Mine eyes have seen the glory of the coming of the Lord: He is trampling out the vintage where the grapes of wrath are stored. . . ." Harper's Ferry is currently being restored by the National Park Service to look as it did at the time of Brown's raid. Interpretive progr
and automobile tours, and living-h
strations are planned.

Frank Stockton, the novelist
writer, lived from 1899 until his de
Court, a large nineteenth-century
of brick painted white. The house
and now is owned by the Claymon
mont Court, as yet unmarked as S
lies off Route 340 between Harp
Charles Town. Perhaps it goes unn
a more important figure, George Wa
once owned the property. It is belie
more, that it was Washington's asso
the property that enabled the house t
struction during the Civil War. Whi
lived here, he wrote the novels *Kate Bon*
and *The Captain's Toll-Gate* (1903) in ad
the story "John Gayther's Garden." Having
the winter of 1901–02 in New York City, St
ton was on his way home to Claymont Court wh
he died, on April 21, 1902, in Washington, D.C.

Birthplace of Pearl Buck in Hillsboro. The house was built in 1858 by her maternal grandparents.

HILLSBORO

Pearl [Comfort Sydenstricker] Buck was born on June 26, 1892, in Hillsboro, a tiny mountain hamlet south of Marlinton. When she was three months old, her missionary parents completed their furlough in Hillsboro and returned to China. Buck did not return to the United States until she was ready for college. Buck, whose books have been translated into more than fifty foreign languages, is thus far the only woman to win both a Pulitzer Prize, for *The Good Earth* (1931), and a Nobel Prize for Literature, in 1938. Over the years, she came back to Hillsboro for short visits several times but did not live here again. Hillsboro has become a tourist center because of Buck's birthplace. Buck described the house she was born in, and where her mother grew up, in the book *The Exile* (1936), a biography she wrote of her parents:

> *Around the white house that had been her childhood home there were wide spaces, and about the doorway a flower garden and a flagged path leading to the square porch that had vines covering its open sides, and wooden seats in the green shadows. A big white door faced one then, thrown wide in all seasons but winter, and there was a brass knocker on it and above it a fan-shaped glass window.*

The Pearl Buck Birthplace Foundation Museum has been restoring the house, which is on Route

219 just northeast of Hillsboro. It is open daily, and an admission fee is charged. A collection of manuscripts and correspondence will be displayed when the restoration is complete.

HINTON

In early February 1902, **Theodore Dreiser** and his wife stayed in Hinton while he worked on *Jennie Gerhardt* (1911), the first of his books that proved successful. Dreiser had come to Hinton to improve his health at the curative springs, but he became more rather than less nervous and after a few weeks decided to move on to Lynchburg, Virginia.

MALDEN

After the Emancipation Proclamation in 1865, **Booker T. Washington,** then nine years old, came from Hale's Ford, Virginia, with his mother, to live with Washington's stepfather in a log cabin near the center of Malden. Malden is on Route 64, a few miles south of Charleston. Young Washington worked in the salt furnaces and later in a coal mine, and during this period completed his elementary-school education in night school. It is known that Washington worked in 1871–72 as a houseboy for Mrs. Viola Ruffner, a mine owner's wife. He left in 1872 for Hampton Normal and Agricultural Institute in Virginia. When he returned to Malden in 1875, he taught school for two years. A monument to Washington, formerly located on Route 60 near Malden, has been moved to the back lawn of the state capitol in Charleston.

ROMINES MILLS

Melville Davisson Post was born in Romines Mills on April 19, 1871. Romines Mills lay south of Clarksburg, very near Lost Creek, where Post was living when he died. Post, who wrote mystery and detective stories and novels, was known best for his character Uncle Abner, a Virginia squire who turned to detective work in order to uphold justice. The squire appeared, for example, in *Uncle Abner: Master of Mysteries* (1918), a collection of tales that proved popular in its day. Post had been a lawyer for several years, and he first attracted attention and considerable criticism by writing about an unscrupulous attorney in *The Strange Cases of Randolph Mason* (1896). His last book was *The Silent Witness* (1930).

WHEELING

An ancestor of the Western writer **Zane Grey** was Colonel Ebenezer Zane, who in 1770, with his brothers, founded Wheeling at the mouth of Wheeling Creek on the Ohio River. Within their settlement they built a stockade on a hill at the top of Main Street. It was called Fort Fincastle, but was known after 1776 as Fort Henry, named for Patrick Henry. Zane Grey's novel *Betty Zane* (1904) commemorated Elizabeth Zane, a sister of Ebenezer who, on September 1, 1777, when the fort was besieged by Indians loyal to the British, carried under fire a keg of gunpowder from a house to the fort sixty yards away. A plaque on the fort, which is at Main and Eleventh streets, reminds us of Elizabeth Zane's heroism.

Frances Trollope saw Wheeling in 1830 while on the visit to the United States that she was to record in *Domestic Manners of the Americans* (1832). The Englishwoman, who later turned to writing novels, found in Wheeling "little of beauty to distinguish it, except the ever lovely Ohio [River]." From Wheeling she traveled to Baltimore and Washington by coach along the Great National, or Cumberland, Road, across the Allegheny Mountains to Cumberland, Maryland. "The whole of this mountain region," she wrote, "through ninety miles of which the road passes, is a garden. The almost incredible variety of plants, and the lavish profusion of their growth, produce an effect perfectly enchanting." With Trollope's book in hand, the literary pilgrim today can enjoy the same trip from Wheeling to Cumberland on Route 40.

The country was wooded everywhere except at the bottom of the valley to the northward, where there was a small natural meadow, through which flowed a stream scarcely visible from the valley's rim. This open ground looked hardly larger than an ordinary door-yard, but was really several acres in extent. Its green was more vivid than that of the inclosing forest. Away beyond it rose a line of giant cliffs similar to those upon which we are supposed to stand in our survey of the savage scene, and through which the road had somehow made its climb to the summit. The configuration of the valley, indeed, was such that from this point of observation it seemed entirely shut in, and one could but have wondered how the road which found a way out of it had found a way into it, and whence came and whither went the waters of the stream that parted the meadow more than a thousand feet below.

—Ambrose Bierce,
in "A Horseman in the Sky"

Booker T. Washington (*below*) and Washington standing in front of his cabin in Malden (*left*). The identity of the woman standing at the door is unknown.

OHIO
ILLINOIS
INDIANA
WEST VIRGINIA
Ohio R.
LOUISVILLE
FRANKFORT
LEXINGTON
VIRGINIA
MISSOURI
KENTUCKY
RALEIGH
WINSTON-SALEM
KNOXVILLE
NASHVILLE
MEMPHIS
NORTH CAROLINA
ARKANSAS
CHARLOTTE
TENNESSEE
FT. SMITH
SOUTH CAROLINA
LITTLE ROCK
Mississippi R.
ATLANTA
COLUMBIA
BIRMINGHAM
AUGUSTA
CHARLESTON
JACKSON
MISSISSIPPI
MACON
ALABAMA
SAVANNAH
SHREVEPORT
MONTGOMERY
GEORGIA
JACKSONVILLE
ATLANTIC OCEAN
MOBILE
PENSACOLA
TALLAHASSEE
LOUISIANA
BATON ROUGE
FLORIDA
NEW ORLEANS
TAMPA
GULF OF MEXICO
MIAMI

SOUTHEASTERN STATES

NORTH CAROLINA

ASHEVILLE

Numerous literary figures are associated with this city on a plateau in the Blue Ridge Mountains, among them **Olive [Tilford] Dargan,** the novelist and dramatist, who died in an Asheville nursing home on January 22, 1968. Several writers came here in hope of regaining lost health. One was **F. Scott Fitzgerald,** sent here in May 1935 to be treated for tuberculosis. Fitzgerald stayed in Rooms 441–443 of the Grove Park Inn, still open at 290 Macon Avenue. He returned the following summer to be near his wife, **Zelda Fitzgerald,** then a patient at Asheville's Highland Hospital, 49 Zillicoa Street. In spite of a cheering visit from novelist Marjorie Kinnan Rawlings, Fitzgerald found the summer a time of almost unbroken gloom. He saw little of Zelda, since he had to employ a nurse to tend to his own health and drinking habits, and spent much of his time lying in bed with the blinds drawn. Zelda returned to the hospital here several times, her last stay commencing in autumn of 1947. On March 10, 1948, a fire broke out in the hospital kitchen and rapidly consumed the building. Trapped in the third story, Zelda perished in the fire along with eight other women.

Sidney Lanier passed through Asheville in May 1881. He was in desperate search of a climate

Grove Park Inn in Asheville. F. Scott Fitzgerald stayed here in 1935 and again in 1936. During the second visit Fitzgerald met Marjorie Kinnan Rawlings, who described him later as acting "nervous as a cat."

The last voyage, the longest, the best.

—*Look Homeward, Angel*

Death bent to touch his chosen son with mercy, love and pity, and put the seal of honor on him when he died.

—*The Web and the Rock*

that would help his fight against tuberculosis. With the help of his brother Clifford, Lanier pitched a tent three miles north of the city, on the grounds of Richmond Hill, the estate of Richmond Pearson, probably a distant relative. The poet was joined there in July by his wife and youngest son. They stayed about two months, during which time Lanier occasionally played his flute with local musicians. In August they moved on, looking for a still milder climate in which to spend the autumn. The Pearson estate is still standing but is expected to be torn down to make way for condominiums.

A third writer to visit Asheville for reasons of health was **Ring Lardner,** who came here in 1924. He stayed at the Grove Park Inn, the same hotel Fitzgerald was to stay in ten years later.

Asheville is the birthplace of **Ward Greene,** born on December 23, 1892, in Victoria, formerly a suburb but now part of the city. Greene worked for King Features Syndicate, becoming an executive editor in 1921. Among his novels were *Weep No More* (1932) and *Death in the Deep South* (1936).

By far the most famous of Asheville's literary men is **Thomas Wolfe,** who was born on October 3, 1900, and lived in Asheville until he reached the age of sixteen. Wolfe's birthplace, at 91 or 92 Woodfin Street, is now the parking lot of the local YMCA. A marker on nearby College Street notes the site of the house. Wolfe lived there with his family until 1906, when his mother bought a boarding house, Old Kentucky Home, at 48 Spruce Street. His first novel, *Look Homeward, Angel* (1929), is set primarily in Altamont, Catawba, the fictional name Wolfe gave Asheville, North Carolina. The protagonist in the novel, Eugene Gant, lives in Dixieland, a boarding house run by his mother, Eliza Gant.

Wolfe's portrayal of the townspeople of Altamont so enraged the residents of Asheville that Wolfe's work was banned by the local Pack Memorial Public Library—when **F. Scott Fitzgerald** came here in 1935 and found none of Wolfe's books in the library, he bought two copies of *Look Homeward, Angel* and donated them to the library. After leaving Asheville, Wolfe returned several times on visits and, when he died in 1938, memorial services were held at the First Presbyterian Church, 40 Church Street. Wolfe was buried in the family plot in Riverside Cemetery, 53 Birch Street, and his grave is marked with a stone bearing inscriptions from his works (*shown at left*).

Asheville today pays ample tribute to its greatest native son. The family home on Spruce Street is run as the Thomas Wolfe Memorial, open year-round. Many of the Wolfe family's possessions are displayed, and one room of the house is furnished with part of the contents of Wolfe's New York City apartment. Best of all, perhaps, the Pack Memorial Public Library, on Pack Square, which once banned Wolfe's works, now houses a Thomas Wolfe Collection, which includes every known publication by or pertaining to the author.

Riverside Cemetery, where Wolfe is buried, also contains the grave of **William Sidney Porter,** better known as **O. Henry,** who came to Asheville for rest cures during the last three years of his life, after he married Sara Lindsay Coleman, a native of Asheville. Porter died in New York City on June 5, 1910.

Old Kentucky Home in Asheville, home of Thomas Wolfe from 1906, when he was six years old, until he left for college.

BANNER ELK

Marjorie Kinnan Rawlings in 1936 began writing *The Yearling* (1938) while living in a mountain cabin in Banner Elk, a town in what is now the Cherokee National Forest, in the northern part of the state.

BATH

This lovely town on the Pamlico River, the oldest town in the state, was a headquarters of Edward Teach, the notorious English pirate known better as Blackbeard, but Bath has a more respectable claim to fame. It is the site of North Carolina's first public library, started in 1700 with 176 books brought from England by the Reverend Thomas Bray. The town also has an interesting literary association. An epitaph on a grave in St. Thomas Episcopal Church here, the oldest church in the state, is used with only minor changes as the epitaph for Suzanne Ravenal in the novel *Show Boat* (1926), by **Edna Ferber.** In 1925 Ferber, whose novel *So Big* had just won a Pulitzer Prize for its author, came to Bath while gathering material for *Show Boat.* She visited the James Adams Floating Palace Theater here—"the only show boat experience I ever had"—and stayed at Marsh House, formerly the governor's mansion. Marsh House still stands, but the Floating Palace Theater is gone.

BLACK MOUNTAIN

This town near Asheville was the home from 1933 to 1957 of Black Mountain College, a liberal arts college that promoted experimentation in literature and the other arts. Among those who taught or studied here were poets **Charles Olson, Edward Dorn, Robert Duncan, Joel Oppenheimer,** and **Robert Creeley.** Creeley was graduated in 1954, taught here later, and edited the influential *Black Mountain Review,* which was published from 1954 to 1957.

BLOWING ROCK

While on vacation stays in this resort town about one hundred miles northeast of Asheville, **John Hersey** wrote his novels *A Bell for Adano* (1944) and *Hiroshima* (1946).

BOONE

This town northeast of Asheville, in the highest reaches of the Appalachians, is the home of the Daniel Boone Theater, where the outdoor drama *Horn in the West* (1952), by Kermit Hunter, is performed regularly each summer between late June and Labor Day.

CARY

Cary, a city west of Raleigh, was the birthplace of **Walter Hines Page,** born on April 15, 1855. His home, on Route 64 and half a block from Academy Street, is now a private residence. Page lived here until 1871. Frustrated by the persistence of what he called the South's three ghosts—the Confederate dead, religious orthodoxy, and white supremacy—he moved north and achieved a reputation as editor of several noted publications, including *The Atlantic Monthly,* from 1896 to 1899. His only novel was *The Southerner* (1909), but he helped Thomas Dixon and Charles Waddell Chesnutt find publication for their works of fiction.

CHAPEL HILL

This city is prominent as the principal home of the University of North Carolina, America's first chartered state university (1789) and one of its finest. One of the many literary North Carolinians who studied here was **Jonathan Daniels,** who went on to an outstanding career in journalism after he was graduated in 1921 and received an M.A. from the university in 1922. His books include *Tar Heels: A Portrait of North Carolina* (1941). **Shelby Foote** attended the university from 1935 to 1937. His books about the South include the novel *Shiloh* (1952).

Paul Green, who was graduated in 1921, won a Pulitzer Prize for his play *In Abraham's Bosom* (1927). Green taught at the university from 1923 to 1944 and was living in Chapel Hill, on Old Lystra Road, at the time of his death on May 4, 1981. Green was named dramatist laureate of North Carolina in 1979, and is commemorated by the Paul Green Theater at the university. Green is buried in the old Chapel Hill Cemetery. **Frances Gray Patton,** who was graduated in the mid-1920s, is best known for her collections of stories about North Carolina life, such as *The Finer Things of Life* (1951). She has also taught here. **Robert Ruark,** who was graduated in 1935, has written many novels, including *Poor No More* (1959), which tells the story of a poor North Carolina boy who grew up to become a ruthless tycoon.

Thomas Wolfe, who was graduated in 1920, used the name Pulpit Hill for Chapel Hill in *Look Homeward, Angel* (1929). Today the university's Louis Round Wilson Library has a large collection of material pertaining to Wolfe, including papers, books, correspondence, manuscripts, and memorabilia. Several of the writers who attended the university, including Paul Green and Thomas Wolfe, participated in the Carolina Playmakers, a theater group that was formed at the university in 1918 by Frederick H. Koch (1877–1944). The Playmakers, whose productions were performed in a converted nineteenth-century building still standing on campus, have made a major contribution to regional and folk drama. Another writer associated with the Playmakers was **Betty Smith,** who wrote her best-known work, the novel *A Tree Grows in Brooklyn* (1943), while she was living in Chapel Hill. She lived first at 502 North Street and later at 315 Rosemary Street, in houses still standing. The house on Rosemary Street, officially called the Michle-Mangum-Smith House, is sometimes referred to as The Betty Smith House. Listed in the National Register of Historic Places, it is a private residence today.

Yet another writer associated with Chapel Hill was **Wilbur Daniel Steele,** who lived here for a time. Steele was the author of such volumes as *Tower of Sand* (1929), a collection of short stories, and *Taboo* (1925), a novel. **James Street** lived in Chapel Hill from 1946 until his death on September 28, 1954. Here he wrote several of his novels, including *Tomorrow We Reap* (1949), *The High Calling* (1951), and *Goodbye, My Lady* (1954). Chapel Hill was the scene in a time gone by of an untoward incident. When **Langston Hughes,** the poet, was here to give readings, he stayed in a room in the home of some white students. The students later were evicted for showing hospitality to a black man.

CHARLOTTE

This city, the largest in the Carolinas, has been home to **Erskine Caldwell,** best-selling author of *Tobacco Road* (1932) and many other books, who lived here as a child, and to **Carson McCullers,** who lived here from September 1937 until the following spring. The two buildings in which McCullers and her husband lived, 311 East Boulevard and 806 Central Avenue, are still standing. At the latter address McCullers wrote the first six chapters of her first novel, *The Heart Is a Lonely Hunter* (1940).

Look out, Death; I am coming.
Art thou not glad? what talks we'll have,
What memories of old battles.
Come, bring the bowl, Death.
I am thirsty.

—Sidney Lanier,
in one of his last poems (untitled)

Windy Oaks Farm, home of Paul Green on Old Lystra Road about four miles south of Chapel Hill. Green lived here from 1965 until his death in 1981.

Harry Golden, publisher of *The Carolina Israelite,* as he appeared in 1967.

The contemporary literary figure most closely identified with Charlotte is the essayist **Harry Golden,** who lived here for many years and died here on October 2, 1981. Humorous sketches from his newspaper, *The Carolina Israelite,* were collected in the best-selling volumes *Only in America* (1958), *For 2¢ Plain* (1959), and *Enjoy, Enjoy!* (1960). Golden came to Charlotte in 1939 and worked for the Charlotte *Observer* until 1941, when he began to publish *The Carolina Israelite.* His first address in Charlotte was 1219 Elizabeth Avenue. Golden was living at 1701 East 8th Street at the time of his death.

CHEROKEE

Unto These Hills, a play written by Kermit Hunter in 1950, about the betrayal by white settlers of the Cherokee, is presented on summer nights in this small town about forty miles west of Asheville.

DURHAM

This city is the home of Duke University, alma mater of such novelists as **Frank G[ill] Slaughter,** who was graduated in 1926, and **William Styron,** who was graduated in 1947. The Rare Book division of the university library contains the noted Trent Collection of Walt Whitman material, which includes letters, manuscripts, and notes.

Ellen Glasgow came to Durham in 1930 to receive the honorary degree of Doctor of Laws. By mistake she was made Doctor of Letters. To compound the mix-up, contrary to the novelist's specific instructions, photographers insisted on taking her picture during the ceremony. As a result she kept her back to the audience for much of the time. But not all academic celebrations go wrong at Duke. On June 9, 1915, when Duke was still known as Trinity College, **Owen Wister** delivered the graduation speech, which was well received and was published later in an expanded version as *The Pentecost of Calamity* (1915).

EDENTON

This old town on Albemarle Bay and its environs are the literary domain of **Inglis Fletcher,** whose twelve-novel Carolina Series tells the story of the state from 1585 to 1789. Although Fletcher was born in Illinois and lived for a long time in San Francisco, she became deeply interested in North Carolina in 1935 while doing genealogical research on her own family. From that research came *Raleigh's Eden* (1940), the first novel of the Carolina Series. In 1944 Fletcher and her husband came to live at Bandon Plantation, north of Edenton off Route 32, in the area now known as Arrowhead Beach. The novelist lived here for the rest of her life, dying in an Edenton nursing home on May 30, 1969. Bandon Plantation's main house burned down some years ago, but the original kitchen and smokehouse are still there. A third original building, built as a schoolhouse, has been moved to the grounds of the James Iredell House in Edenton, and is open to the public.

FAYETTEVILLE

Carson McCullers and her husband moved in the spring of 1938 to Fayetteville, about sixty miles southwest of Raleigh. Their first home here was on Rowan Street, possibly number 102, now the site of a gasoline station. They moved later to an apartment at 119 North Cool Spring Street, in the oldest building in Fayetteville, which had once been the Cool Spring Tavern. Later still, after a stay in McCullers' hometown of Columbus, Georgia, they returned to Fayetteville and lived at 1104 Clark Street. A building still stands on the site, but it may not be the same one they lived in. When McCullers moved here, she was still working on her first novel, *The Heart Is a Lonely Hunter* (1940), which was to bring her great criti-

Carl Sandburg's study at Connemara Farm in Flat Rock. It was Sandburg's habit to begin writing late at night and not stop until early morning.

cal acclaim. After finishing that, she went on to write *Reflections in a Golden Eye* (1941), a novel set in an Army camp in the South.

FLAT ROCK

On Little River Road here, twenty-five miles south of Asheville, on Route 25, stands Connemara Farm, a large estate built as a summer home about 1838 by C. G. Memminger, later the first secretary of the treasury of the Confederacy. **Carl Sandburg** bought and restored the farm in 1945 and moved here with his family, a prize herd of goats, and a 10,000-volume library. Connemara was to be Sandburg's home for the rest of his life. While living here he wrote numerous poems; *Remembrance Rock* (1948), his only novel; and *Always the Young Strangers* (1953), a memoir of his youth. In addition to working at his writing, the poet entertained guests, paid visits to the goats—tended largely by Sandburg's wife, Paula—and took long walks over Connemara's 240 acres. After his death here on July 22, 1967, funeral services were held for Sandburg at the nearby St. John of the Wilderness Episcopal Church, one of the writer's favorite spots in the area. Connemara Farm is owned and run now by the National Park Service as the Carl Sandburg Home National Historic Site, open year-round to visitors.

Shadows fall blue on the mountains.
Mountains fall gray to the rivers.
Rivers fall winding to the sea.
Oldest of all the blue creep,
the gray crawl of the sea
And only shadows falling older than the sea.

—Carl Sandburg,
in "Shadows Fall Blue on the Mountains,"
collected in *Honey and Salt* (1963)

FLETCHER

Rugby Grange, the home of George Westfeldt, a Swedish diplomat, near this small town south of Asheville, was a stopping-off point in September 1881 for **Sidney Lanier,** his wife, and youngest son. Lanier, who was searching for a climate that might help cure his life-threatening tuberculosis, stayed in Fletcher for a few days. He gave a piano recital here—it turned out to be his last—and developed a deep affection for his host. A few weeks later, when Lanier was dying, he said: "All my life I have searched for the father of my soul, but never have I found him until now, when I greet him in the person of George Gustav Westfeldt." And he asked that Westfeldt be sent a copy of "Sunrise," Lanier's last poem, "that he may know how entirely we are one in thought."

Less than a mile down the road from Rugby Grange is Buck Shoals, once the home of **Edgar Wilson Nye,** the humorist better known as **Bill Nye,** who built the house and moved into it with his family in 1891. Nye, who won international celebrity through collections such as *Bill Nye and Boomerang* (1881), lived here until his death in 1896. He is buried in the churchyard of Calvary Episcopal Church, a few miles away on Route 25. Buck Shoals is still privately owned.

And ever my heart through the night
shall with knowledge abide thee,
And ever by day shall my spirit, as
one that hath tried thee,
Labor, at leisure, in art,—
till yonder beside thee
My soul shall float, friend Sun,
The day being done.

—Sidney Lanier,
in "Sunrise"

GREENSBORO

Randall Jarrell taught at the Women's College of the University of North Carolina here from 1947 until his death in 1965. While living here Jarrell wrote many of his most important poems and translations, among them *The Woman at the Washington Zoo* (1960) and *The Lost World* (1965).

Greensboro is the birthplace of one of America's best-loved writers of the short story: **William Sydney Porter,** known better as **O. Henry.** Born at Worth Place, a plantation near Greensboro, on

Postcard view of Connemara.

September 11, 1862, William was taken to live at 440 Market Street after his mother's death in 1865. He attended Lena Porter's School, worked in W. C. Porter's Drugstore, and went to church at the First Presbyterian Church, originally located at 130 Summit Avenue. The church building, now the Greensboro Historical Society Museum, houses an extensive O. Henry collection. Among the exhibits are reconstructions of Lena Porter's School and the drugstore where Porter worked, a collection of memorabilia, and the Eli Oettinger collection of O. Henry manuscripts, pictures, and letters. The museum is open to the public. The Greensboro Public Library honors its most famous native son with an O. Henry collection of portraits, first editions, and other rare books.

The O. Henry Memorial Award for high achievement in short-story writing, established in 1918 by the Society of Arts and Sciences, has been given five times to another Greensboro native: **Wilbur Daniel Steele,** born here on March 17, 1886. In addition to his numerous short stories, Steele has also written novels, such as *That Girl from Memphis* (1945).

Monument in Oakdale Cemetery in Hendersonville, perhaps the best-known funerary sculpture associated with American literature.

HENDERSONVILLE

F. Scott Fitzgerald, in February 1935, came to stay in Hendersonville, on Route 26, southeast of Asheville. In deep financial trouble, he tried to earn money by writing short stories while staying at the Skylands Hotel. Later that year, in November, he returned to the hotel for another brief stay. Now called the Skyland Motor Inn, the hotel is still standing at its original address, 538 North Main Street. A more regular visitor here was **DuBose Heyward,** who spent summers at a house called Dawn Hill, on Price Road off Old Kanuga Road. Heyward is best known for his novel *Porgy* (1925), which served as the basis for George Gershwin's opera *Porgy and Bess* (1935). Dawn Hill is privately owned and not open to visitors. Perhaps the most famous of Hendersonville's literary figures is likely to be there for many decades to come: the funerary monument over the grave of Margaret E. Johnson, in Oakdale Cemetery on Sixth Avenue. **Thomas Wolfe** first saw the monument, which is in the form of an angel, when it was standing on the porch of Wolfe's Monument Shop, his father's business establishment in Asheville. Later it provided the central image—and the title—for Wolfe's first and most celebrated work, *Look Homeward, Angel* (1929).

LILLINGTON

Paul Green was born on March 17, 1894, in a house on Route 1 in Lillington, about twenty-five miles south of Raleigh. Green went to school in nearby Buies Creek. After graduation he saved enough money to enter the University of North Carolina, Chapel Hill, but his studies were interrupted by World War I. North Carolina is the setting for many of Green's plays, including *The Field God* (1927) and the Pulitzer Prize-winning *In Abraham's Bosom* (1927).

LYNN

Sidney Lanier spent the spring and summer of 1881 going from place to place in what turned out to be a losing race against his fatal tuberculosis. He stopped for brief periods in nearby Asheville and Fletcher while looking for a climate that would ease his suffering, but Lynn is where the search ended. Lanier died here on September 7, 1881, in the old Wilcox House on Route 108. The red-brick house is privately owned but can be seen easily from the road. Lanier also stayed at the nearby Mimosa Inn, a nineteenth-century building now used as a private residence.

MARION

Sinclair Lewis in 1929 came to this city, about thirty-five miles east of Asheville, to report on a strike against the Marion Manufacturing Company and the Clinchfield Mills, two local textile factories. His reports, written for the Scripps-Howard newspapers, were also published as *Cheap and Contented Labor: The Picture of a Southern Mill Town* (1929). Lewis is said to have stayed at the Mariana Hotel, later renamed the James Hotel, at 2 North Main Street. The building, still standing, now houses a bookstore and a barber shop.

MOCKSVILLE

Hinton Rowan Helper was born on December 27, 1829, near Mocksville, about twenty miles southwest of Winston-Salem, on Route 64, and spent most of his childhood in Mocksville before going to California in 1849 to join the gold rush. His three years there led to his writing *Land of Gold: Reality Versus Fiction* (1855) after his return to North Carolina. But he is remembered best in literary history for his tract *The Impending Crisis of the South: How To Meet It* (1857), which created a furor by arguing that slavery should be abolished because of its harmful economic consequences. In the North the book was adopted as a basic abolitionist text; in the South it was reviled and banned as insurrectionary. These reactions are ironic in light of Helper's passionate racism, which he freely expressed after the Civil War in several books, beginning with *Nojoque: A Quest for a Continent* (1867).

MORGANTON

H. Allen Smith, author of *Low Man on a Totem Pole* (1941), is brought to mind for devotees of humorous writing by the totem pole that stands in the garden of Smith's daughter's house at 310 Shore Drive in Morganton, about fifty miles east of Asheville. The daughter, Mrs. Don Van Nappen, welcomes guests who wish to see the totem pole, but asks that they make an appointment.

PINEHURST

J[ohn] P[hillips] Marquand, in the last ten years of his life, came often to this town, about sixty miles southwest of Raleigh. He liked to play golf at the Pinehurst Country Club and, in the 1950s, bought a house here, address unknown, in which he spent winters.

PLYMOUTH

Augustin Daly was born in this town near Albemarle Sound on July 20, 1838. Most of the numerous plays Daly wrote and produced were adapted from French or German originals. Of his own works, *Horizon* (1871) and *Divorce* (1871) are usually considered the finest. Daly also had considerable success as a theatrical producer.

RALEIGH

In 1887, years before **Thomas Dixon** wrote the novels that made him famous in his own day, he served as minister of the Second Baptist Church, at the corner of Hargett and Person streets in Raleigh, capital city of North Carolina. During Dixon's pastorate the church was given its present name of Tabernacle Baptist Church. The novelist lived for many years in Boston, but he returned to Raleigh in 1937 to a house at 1507 Hillsborough Street, a site now occupied by the Velvet Cloak Motor Hotel. Dixon's best-known novel was *The Clansman* (1905), which he later adapted for the stage and for motion pictures. The film version, entitled *The Birth of a Nation* (1915) and directed by D. W. Griffith, is considered a classic. Another literary resident here, from 1875 to 1889, was **Albion W[inegar] Tourgée.** His reconstructionist ideas—carried out as a Federal pension agent and judge, and expressed in such novels as *A Fool's Errand* (1879) and *Bricks Without Straw* (1880)—made him vastly unpopular with the people of his adopted state.

Two of North Carolina's most distinguished twentieth-century authors were born in Raleigh a few years apart. **Jonathan Daniels,** a newspaperman, political commentator, and biographer, was born at 125 East South Street on April 26, 1902, and also lived at 1614 Iredell Drive and 1540 Caswell Street. The three houses are still standing but privately owned. It was at the Caswell Street address that Daniels wrote *Clash of Angels* (1930), his only novel. After editing the Raleigh *News and Observer* (1933–42) and working in Washington as an assistant to Franklin D. Roosevelt (1942–45), Daniels returned to Raleigh, where he wrote *The Man of Independence* (1950), a biography of Harry Truman. **Frances Gray Patton** was born on March 19, 1906, at 530 North Blount Street, still standing. She attended the University of North Carolina, Chapel Hill, where she joined the Carolina Playmakers, and later taught creative writing both at Chapel Hill and at Greensboro. Patton is best known for such volumes of short stories as *The Finer Things of Life* (1951) and *Good Morning, Miss Dove* (1954).

ROCKY MOUNT

In the summer of 1951, **Jack Kerouac** and his mother came to this city about sixty miles east of Durham, to visit Kerouac's sister at her house near town. By the end of the year, Kerouac's mother had bought a home in the area, and Kerouac paid her a number of visits in the following years. He wrote the novel *Visions of Gerard* (1963) while staying here in 1956, and wrote about another visit to the town in the novel *The Dharma Bums* (1958). Mrs. Kerouac's little white house is still standing, as is the pine forest in which Kerouac used to practice meditation. The house and the forest can be seen at West Mount Crossroads on West Mount Drive, north of Rocky Mount off Route 301-A.

SHELBY

Thomas Dixon was born in Shelby, about forty miles west of Charlotte, on January 11, 1864. Dixon was known best for his novel *The Clansman* (1905), which was later made into the motion picture *The Birth of a Nation* (1915).

TRYON

Margaret Culkin Banning, the novelist and short-story writer, died at Friendly Hills, her home in Tryon, on January 4, 1982. Banning completed her last published novel, *Such Interesting People* (1979), while living at Friendly Hills, and was at work on a novel before she died.

F. Scott Fitzgerald, in 1935, a time of despair for him, came to this city to be near his wife, **Zelda,** who was hospitalized in Asheville. Fitzgerald was in financial difficulties, and the reception of *Tender Is the Night* (1934) had not lived up to his expectations. While living in Tryon, Fitzgerald worked on short stories and on the autobiographical essay "The Crack-Up," published in 1936 in *Esquire.* Fitzgerald returned to Tryon once more, in January 1937, while waiting to be offered a screenwriting job in Hollywood. He ate often at Misseldines Drug Store here and wrote a paean to it, making the institution one of the few drug stores so honored:

> *Oh Misseldines, dear Misseldines,*
> *A dive we'll ne'er forget,*
> *The taste of its banana splits*
> *Is on our tonsils yet.*

The Oak Hall Hotel at 201 Chestnut Street, where Fitzgerald stayed, has been demolished.

Another writer to visit Tryon was **William Gillette,** whose most popular work in his own

Hinton Rowan Helper as he appeared about 1900. Born in Mocksville, Helper was a popular and controversial writer in the years just before the Civil War.

Jonathan Daniels, newspaper editor and author.

> *He worked for hours and finally dislodged the monster, and it went bounding down the mountainside. . . . The big boulder crashed through a livery stable, shot down the main street, went through the First National Bank and finally came to rest in the rear of that institution . . . when Bob's uncle arrived on the run. He shoved everyone aside and approached the rock, which he scrutinized carefully on all sides. Finally he straightened up and said: "Nope. No moss."*
>
> —H. Allen Smith,
> in *Low Man on a Totem Pole*

Catfish Row, in which Porgy lived, was not a row at all, but a great brick structure that lifted its three stories about the three sides of a court. The fourth side was partly closed by a high wall, surmounted by jagged edges of broken glass set firmly in old lime plaster, and pierced in its center by a wide entrance-way. . . . Within the high-ceilinged rooms, with their battered colonial mantels and broken decorations of Adam designs in plaster, governors had come and gone, and ambassadors of kings had schemed and danced. Now before the gaping entrance lay only a narrow, cobbled street, and beyond, a tumbled wharf used by negro fishermen. Only the bay remained unchanged.

—DuBose Heyward, in *Porgy*

day was *Sherlock Holmes* (1899), a dramatization of stories by Arthur Conan Doyle. Gillette stopped here in 1889, intending to go on to Asheville to recuperate from an illness. He decided to stay in Tryon and had a house built in a nearby pine forest. He boarded the house up in 1910 and never returned. After he sold the house in 1925, it was converted into the Thousand Pines Inn. The building is no longer open as a hotel, but its present owner, Selina Lewis, will show it to interested visitors.

DuBose Heyward died here on June 16, 1940, while on a visit. The novelist had been staying in a house at 606 Glenwalden Circle, which is still privately owned. **Sidney Lanier** died in nearby Lynn in 1881. The Lanier Library Association here, founded in 1890, received its first gift of books from Lanier's widow.

WASHINGTON

Washington, on Route 17, near the Pamlico River, was the birthplace of **William Churchill De Mille,** a playwright, born at 127 Bridge Street on July 25, 1878. De Mille collaborated on a number of plays with his better-known brother, Cecil B. De Mille, the motion-picture director, and with David Belasco, the producer. Of his own works, the best known was *Strongheart* (1905).

WILMINGTON

Inglis Fletcher stayed at Clarendon, a large estate about eight miles south of this deep-water port on the Cape Fear River, while writing her novel *Lusty Wind for Carolina* (1944). She is buried in the Wilmington National Cemetery here. **Thomas Godfrey,** author of *The Prince of Parthia* (1759), the first play written by a native American to be acted professionally, in 1767, is buried in the churchyard of St. James Church, at Market and Sixth streets. The original grave marker was lost during the nineteenth century but has been replaced.

Home of William Gillette in Tryon. The playwright and actor achieved his greatest successes while living here.

SOUTH CAROLINA

BEAUFORT

This resort city on the South Carolina coast has several literary associations. **Samuel Hopkins Adams,** the novelist and magazine writer, wintered here regularly at 127 Nort Hermitage Drive, where he died on November 16, 1958. The house is not open to the public. Marshlands, a house at 501 Pinckney Street, figured in the novel *A Sea Island Lady* (1939), by Francis Griswold, who wrote frequently about the Carolina low country. **Mason Locke Weems,** the biographer, died in Beaufort on May 23, 1825, some say at the John Cross Tavern, which still stands at 807–813 Bay Street. Weems was buried first in the churchyard of St. Helena's Episcopal Church in Beaufort but later was reburied at his home in Bel Air, near Dumfries, Virginia.

BRADLEY

Erskine Caldwell lived as a child in this small town situated on the edge of Sumter National Forest. The novelist's father, a Presbyterian minister, often changed parishes, so Bradley was only one of several towns young Caldwell lived in.

CAMDEN

Camden has a special place in United States history. It was here, on August 16, 1780, that the Revolutionary forces were defeated by troops under command of General Cornwallis. For a time thereafter, it seemed as if the entire South had been lost to the British. The Battle of Camden, Camden itself, and the bitter partisan conflict in South Carolina between loyalists and patriots were depicted in *The Partisan* (1835), by William Gilmore Simms. This was one of several connected narratives Simms wrote about the Revolutionary War in the South.

Mary Boykin Chesnut and her husband James, a Confederate general, lived over the years before and after the Civil War in a number of Camden houses. She is known for *A Diary from Dixie,* not published until 1905 and enlarged in 1949. It was republished in 1981. The work is a moving account of the South during the War. Frogden, one of the houses the Chesnuts lived in, was built in about 1848 on Union Street, where it still stands. Another of their houses, Kamsehatka, on Kirkwood Lane, where they lived from 1853 to 1860, was renovated in 1938. Mulberry Plantation, on what is now Route S21, was the Chesnut country home. Sarsfield, in the 100 block of Chesnut Street, is the last of the houses built by the Chesnuts, and they lived in it from 1873 to 1886. The Chesnut homes are not open to the public.

CHARLESTON

Citizens of this handsome port city take justifiable pride in the restoration that has been done on many old homes, churches, and other buildings. Walking tours and short trips by car can take one to many historic sites, a number of which have literary interest. Boone Hall, perhaps America's most photographed plantation, entered literary history in 1939 as the main location of the motion picture based on *Gone with the Wind,* by Margaret Mitchell. Boone Hall, on Route 17 North, is open to the public. The Charleston Library Society, founded in 1748, is a private research institution with interest for readers of literature. The society, which is located at 164 King Street and is open to the public, includes among its holdings many rare books by important Charleston authors, such as DuBose Heyward, William Gilmore Simms, and Henry Timrod. The South Carolina Historical Society, at 100 Meeting Street, maintains an extensive collection of valuable literary research material, including papers of DuBose Heyward, William Gilmore Simms, and Henry Timrod.

The Citadel, a military college operating here since 1842, is the setting of *The Lords of Discipline* (1980), a novel by Pat Conroy. The Dock Street Theatre, at 135 Church Street, is located on the site of the first building in America devoted exclusively to drama. The original building, erected in 1736, was replaced in 1809 by a hotel and then restored as a theater in the 1930s.

DuBose Heyward was born at 38 Savage Street on August 31, 1885. As a young man, Heyward worked at the Charleston docks, among black laborers, an experience he later drew on in the novel *Porgy* (1925) and in the dramatization of the novel that he wrote with his wife Dorothy in 1927. *Porgy* was further adapted in succeeding years, becoming one of the best-known American operas, *Porgy and Bess* (1935), with music by George Gershwin. The model for the character Porgy, a crippled beggar, is believed to have lived at 89–91 Church Street. Heyward wrote *Porgy* while living at 76 Church Street; he later lived at 96 Church Street and 24 South Battery. The three homes are standing but are not open to the public. Funeral services for Heyward were held at St. Philip's Protestant Episcopal Church, 142 Church Street, on June 18, 1940, and were followed by burial in the graveyard next to the church, near the grave of John C. Calhoun, the nineteenth-century South Carolina senator and orator.

William Gilmore Simms, one of the South's great writers, was born in Charleston on April 17, 1806. At first ostracized by the city's literary establishment because he lacked an aristocratic background, Simms ultimately won the highest local social honor—membership in the St. Cecilia Society. He first gained recognition as a poet, having published several volumes of verse by 1832, when his best-known poem, *Atalantis,* appeared. From 1828 to 1832, Simms was an editor and part owner of Charleston's *City Gazette,* which failed. He then turned to writing romances—he is known as the Southern Cooper—as well as histories and biographies, which were collected in ten volumes in 1882. *The Yemassee* (1835), in its day perhaps Simms's most widely read novel, *The Partisan* (1835), and *Beauchampe* (1842) were representative of his fiction. *Nathaniel Green* (1849) was one of his successful biographies. Simms lived from 1869 to his death on June 11, 1870, at 13 Society Street in a building that still stands and is marked with a plaque. Other Simms sites here are his grave, in Magnolia Cemetery, twelve miles northwest of the city, on Route 61, and a bust in White Point Park, at the Battery.

Portrait of William Gilmore Simms, which appeared first in a collection of his poems that was published in 1853.

Home of DuBose Heyward at 76 Church Street in Charleston. The house bears a plaque noting its designation as a national historic landmark.

Bust of Henry Timrod in Washington Park, at Broad and Meeting streets in Charleston. Plaques on the pedestal bear inscriptions by and about the poet.

Henry Timrod, called the Laureate of the Confederacy, was born in Charleston on December 8, 1829. His father owned The Sign of the Golden Ledger, a bookbinding shop then at 105 East Bay Street. While attending Coates School here, he met Paul Hamilton Hayne who, with Timrod and William Gilmore Simms, was to become the core of a literary group frequenting Russell's Bookstore at 285 King Street, now a jewelry store. The three writers were also instrumental in publishing *Russell's Magazine* (1857–60), named after the bookstore. Timrod became a correspondent for the Charleston *Mercury* (1851–64) and, in 1864, editor of Columbia's *South Carolinian.* During the Civil War, Charleston was under almost constant siege by Union naval forces from 1862 to 1865. Gilmore, in his "Charleston," wrote in April of 1863:

> *. . . In the temple of the Fates*
> *God has inscribed her doom:*
> *And, all untroubled in her faith, she waits*
> *The triumph or the tomb.*

Timrod's poem "At Magnolia Cemetery" was read for the 1866 Memorial Day services at Magnolia Cemetery. The poem has been read annually since then as part of the city's celebration of Confederate Memorial Day. A monument to Timrod Charleston's Washington Park bears the words of "At Magnolia Cemetery."

> *Sleep sweetly in your humble graves,*
> *Sleep, martyrs of a fallen cause;*
> *Though yet no marble column craves*
> *The pilgrim here to pause.*
>
> *In seeds of laurel in the earth*
> *The blossom of your fame is blown,*
> *And somewhere, waiting for its birth,*
> *The shaft is in the stone!*
>
> —Henry Timrod, in "At Magnolia Cemetery"

Charleston has other literary sons and daughters. **Octavus Roy Cohen,** born here on June 26, 1891, attended Porter Military Academy, now the Porter Gaud School, on Albemarle Road. His works, many of them about Southern blacks, included *Polished Ebony* (1919) and *Florian Slappey Goes Abroad* (1928). **Paul Hamilton Hayne,** known for his *Sonnets and Other Poems* (1857) and *Legends and Lyrics* (1872), was born here on January 1, 1830. Hayne was a friend and admirer of Henry Timrod's, and Timrod was little known as a poet until 1873, six years after his death, when Hayne collected Timrod's later and best lyrics in *Poems* (1873).

> *Let the only walls the foe shall scale*
> *Be ramparts of the dead!*
>
> —Paul Hamilton Hayne, in "Vicksburg"

Edwin Clifford Holland was born here in 1794, date unknown, and died here on September 12, 1824. He is buried at Archdale, one of several church cemeteries on Archdale Street. In his brief career, he produced a number of significant works, including *Odes, Naval Songs, and Other Occasional Poems* (1813) and a dramatization, in 1818, of Lord Byron's *Corsair* (1814). **Jane McElheney,** who wrote the novel *Only a Woman's Heart* (1866), was born in Charleston in 1836, date unknown. **Robert Molloy** was born here on January 9, 1906, at 147 Tradd Street, in a building still standing but not open to visitors. His novel *Pride's Way* (1945) and his nonfiction work *Charleston: A Gracious Heritage* (1947) were written about his native city.

Josephine [Lyons Scott] Pinckney, who wrote about Charleston in her novel *Three O'Clock Dinner* (1945), was born here on January 25, 1895, and later lived at 36 Chalmers Street in a house no longer standing. Pinckney was the first student to attend Ashley Hall, at 172 Rutledge Avenue. She is buried in Magnolia Cemetery. **Frank L[ebby] Stanton,** in his time called Poet Laureate of Georgia, was born in Charleston on February 22, 1857. His poem "Mighty Lak' a Rose" (1901) was set to music and became a popular song.

Before the Civil War, Charleston was the cultural center of the South, especially noteworthy as the home of one of the country's leading literary reviews, *The Southern Review* (1828–32). Its editors and contributors included **Stephen Elliott, Hugh Swinton Legaré,** and **William Gilmore Simms.** It was later revived as *The Southern Quarterly Review* (1842–57)—somewhat less literary and more political—and once more became well known, chiefly through its advocacy of free trade and states' rights. Simms, once more, was a frequent contributor.

Charleston has always had many literary residents and visitors. **Hervey Allen** moved here about 1918 and, in the early 1920s, taught English at the Porter Military Academy and at Charleston High School. Allen and **DuBose Heyward** collaborated on a volume of poems, *Carolina Chansons* (1922), and together organized the Poetry Society of South Carolina, a group that sponsored lectures and readings by local authors and by writers from across the country.

William Ioor, who was best known for his play *The Battle of Eutaw Springs* (1807), lived and worked here in the late eighteenth and early nineteenth centuries. **Owen Wister,** while wintering in Charleston in 1901–02, completed his novel *The Virginian* (1902) and gathered material for *Lady Baltimore* (1906), a novel set in the fictional Kings Port, a city modeled on Charleston. **Ralph Waldo Emerson,** seeking to recover from a chest ailment, arrived in Charleston by sea aboard the *Clematis* on December 7, 1826. This was his first major trip outside New England, and he is reported to have stayed at a boarding house on East Bay Street. **F. Scott** and **Zelda Fitzgerald** vacationed here during the first week of September in 1937.

Ellen Glasgow came here during the winter of 1890 to attend her first ball, the famous St. Cecilia Ball, which now is held at Hibernian Hall, 105 Meeting Street. The St. Cecilia Society has sponsored this social event almost every year since 1819. **Joseph Pulitzer** came to Charleston in 1907 to arrange for construction of his yacht *Lib-*

erty. It was on this ship that the publisher died on October 29, 1911. **Carl Sandburg** on several occasions gave readings of his poems in Charleston, at meetings of the Poetry Society of South Carolina. **William Makepeace Thackeray** gave a series of talks at Hibernian Hall in the spring of 1853, during his first tour of America. The English novelist received the large sum of $665 for his appearances.

CLINTON

The romantic novel *Horse-Shoe Robinson* (1835), by John Pendleton Kennedy, was set in western South Carolina, near Musgrove's Mill, the site of a Revolutionary War battle and ten miles north of Clinton, off Route 56. In real life, Musgrove's daughter Mary was devastated by the death of the Revolutionary soldier she loved, but eventually married another man and raised a family. She is buried near the mill. In *Horse-Shoe Robinson,* Mildred Lindsay, in love with a patriot, Arthur Butler, was the daughter of a Tory who wanted Mildred to marry a British spy. She married Butler secretly and was protected by Horse-Shoe Robinson, a local blacksmith, while Butler was a prisoner of war. The spy was eventually executed, and the couple reunited.

COLUMBIA

South Carolina's state capital and largest city has some interesting literary and historic buildings. The Chesnut Cottage, at 1718 Hampton Street, was the home of **Mary Boykin Chesnut** and her husband, Gen. James Chesnut, from July 1864 until February 1865, when General Sherman's Union army shelled Columbia. The house is closed to the public but carries a marker: "Temporary wartime home of General and Mrs. James Chesnut. Here they entertained Jefferson Davis, President C.S.A., and his staff, October 5, 1864. President Davis addressed the citizens of Columbia from the front steps of the cottage." The Chesnuts also stayed for extended periods during the Civil War at the Hampton-Preston Mansion, 1615 Blanding Street. This house is open to the public. Some of Mary Chesnut's letters and memorabilia and a noted collection of the works of William Gilmore Simms are housed in the South Caroliniana Library of the University of South Carolina, which has the distinction of being the oldest American educational institution supported by state funds.

Other writers who have lived in Columbia include **Frederick William Thomas** and **Henry Timrod.** Thomas, author of the novel *Clinton Bradshaw; or, The Adventures of a Lawyer* (1835) and remembered as a friend of Edgar Allan Poe's, worked here on the *South Carolinian,* a journal. Timrod lived at 1108 Henderson Street from 1864 until his death on October 7, 1867. The house is no longer standing, but one can visit his grave in the Trinity Episcopal Churchyard, at 1100 Sumter Street. The churchyard is just across from the South Carolina State House, which displays a portrait of Henry Timrod. Another portrait of Timrod hangs in the Reference Department of the Columbia Library, 1400 Sumter Street.

DAUFUSKIE ISLAND

Joel Chandler Harris gathered some of the material for his Uncle Remus stories on visits to Daufuskie Island, near Hilton Head Island and twelve miles east of Harris's home in Savannah, Georgia.

DORCHESTER

William Ioor was born in the mid-eighteenth century in Dorchester, which is about thirty miles northwest of Charleston. He moved later to Charleston, where his plays were produced.

EUTAW SPRINGS

This town, about forty-five miles northwest of Charleston, was the scene of the Battle of Eutaw Springs, September 8, 1781, a turning point in

Sweetest li'l' feller, everybody knows;
Dunno what to call him, but he's
mighty lak' a rose;
Lookin' at his mammy wid eyes so
shiny blue
Mek' you think that Heav'n is comin'
clost ter you.

—Frank L. Stanton,
in "Mighty Lak' a Rose"

(*below left*) Hibernian Hall in Charleston, where William Makepeace Thackeray delivered three lectures in 1843. They were so successful that he gave a second series of lectures here ten days later. (*below*) Chesnut Cottage, the home of Mary Boykin Chesnut toward the end of the Civil War.

the Southern campaign of the American Revolution. The battle was described by Philip Freneau in the poem "To the Memory of the Brave Americans" (1786) and by William Ioor in the play *The Battle of Eutaw Springs* (1807), as well as by William Gilmore Simms in the novel *Eutaw* (1856).

> *At Eutaw Springs the valiant died:*
> *Their limbs with dust are covered o'er;*
> *Weep on, ye springs, your tearful tide;*
> *How many heroes are no more!*
>
> *If in this wreck of ruin they*
> *Can yet be thought to claim a tear,*
> *O smite thy gentle breast, and say*
> *The friends of freedom slumber here!*
>
> —Philip Freneau,
> in "To the Memory of the Brave Americans"

FLORENCE

A park and literary shrine named for **Henry Timrod** is located at Timrod Drive and South Coit Street in the industrial city of Florence, in northeastern South Carolina. The poet taught school in Florence in 1859, in a one-room schoolhouse that now stands in the park. A marker reads: "Within this building he taught among others, 'Katie,' later to become his wife." An obelisk nearby commemorates Timrod.

Eight miles east of Florence, at Mars Bluff, is Carolina Hall. This two-story mansion served as the model for a plantation that was constructed in Hollywood for the filming of *Carolina* (1934). The motion picture was based on a play by Paul Green, *The House of Connelly* (1931), which portrayed decadent Southern planters of the early twentieth century.

FOLLY ISLAND

DuBose Heyward lived for a time at 706 West Ashley Street on Folly Island, eight miles south of Charleston. The writer found the house a good place to work during his collaboration with Ira Gershwin on the libretto for George Gershwin's opera *Porgy and Bess* (1935). The unmarked house is not open to the public.

Photograph taken in 1865 of Woodlands, about a year after the plantation was razed by General Sherman's troops.

FORT MOTTE

Julia Peterkin, author of the novel *Scarlet Sister Mary* (1928), for which she won a Pulitzer Prize, lived near here on Lang Syne Plantation, Route 26, about thirty-five miles southeast of Columbia. Her house, in which she lived from 1903 until her death, is unmarked and not open to visitors. After her death in a hospital in nearby Orangeburg, on August 10, 1961, the author was buried in Peterkin Cemetery, in Fort Motte.

GEORGETOWN

Joseph Blyth Allston, known as a poet only for one poem, "Stack Arms," which he wrote while he was a Civil War prisoner at Fort Delaware, was born in 1833 near the port city of Georgetown.

> *"Stack Arms!" In faltering accents, slow*
> *And sad, it creeps from tongue to tongue.*
> *A broken, murmuring wail of woe,*
> *From manly hearts by anguish wrung,*
> *Like victims of a midnight dream,*
> *We move, we know not how nor why,*
> *For life and hope but phantoms seem,*
> *And it would be relief—to die!*
>
> —Joseph Blyth Allston,
> in "Stack Arms"

Anya Seton, in her historical romance *My Theodosia* (1941), wrote about Theodosia Allston, who was the daughter of Aaron Burr and wife of Joseph Allston, governor of South Carolina from 1812 to 1814. The Allston home, called The Oaks, stood fifteen miles northeast of Georgetown. The novel told of the tragic disappearance of Mrs. Allston at sea in 1812 as she was on her way by ship from Georgetown to New York to visit her father, who also figured in the novel. Near The Oaks, off Route 17, is Brookgreen Gardens, the model for Blue Brook Plantation in the novel *Scarlet Sister Mary* (1928), by Julia Peterkin. This outdoor museum is open to the public. Admission is charged.

GREENVILLE

This textile and university center in western South Carolina was the home in his final years of **John Dickson Carr,** known for his mystery novels, including *The Bride of Newgate* (1950) and *The Dead Man's Knock* (1958). He died here on February 27, 1977.

HILTON HEAD ISLAND

Jonathan Daniels lived here from 1970 until his death on November 6, 1981. The editor and biographer came to the island to spend his retirement years, but he established *The Island Packet* newspaper and wrote columns for it. Hilton Head Island is at the southern tip of South Carolina.

McCLELLANVILLE

Archibald [Hamilton] Rutledge, known in his time as Poet Laureate of South Carolina, was born at Hampton Plantation, near McClellanville, on October 23, 1883. He lived on the plantation for much of his life and was buried in the cemetery on the grounds. The Greek Revival house in which he lived is located eight miles north of McClellanville, which is in eastern South Carolina on the Intercoastal Highway. The house, of yellow pine and cypress, was built in 1730. The plantation, which is open to visitors on Sundays, was treated in Rutledge's prose work *Home by the River* (1941).

MIDWAY

During winters from 1836 to 1870, **William Gilmore Simms** lived at Woodlands, a plantation on Route 78 near Midway, northwest of Charleston. Simms, while living at Woodlands, would spend mornings at work on his writing, and afternoons and evenings entertaining his many houseguests. In 1865, General William Tecumseh Sherman's troops destroyed Simms's 10,000-volume library as well as most of the house—and with it, his fortune. Only one wing was left standing, and this has been renovated as a private home, not open to the public.

ORANGEBURG

Jonathan Daniels, author of the biographies *The Man of Independence* (1950) and *Stonewall Jackson* (1959), reported that he frequented Berry's restaurant in Orangeburg, at 450 John C. Calhoun Drive, SE. **Julia Peterkin,** the novelist, who lived in nearby Fort Motte, died at the Orangeburg Regional Hospital on August 10, 1961.

PROSPERITY

Erskine Caldwell spent part of his childhood in this small town thirty-five miles northwest of Columbia. From 1906 to 1911, he lived with his parents in the parsonage of the Associated Reform Presbyterian Church, where his father was pastor. The parsonage, on McNary Street, Route 391, and the church building, neither of which is now used for religious purposes, are unmarked and not open to the public. Caldwell returned to Prosperity in 1970, long after he had become known as a writer of best-selling fiction, to serve as chief speaker at the South Carolina Tricentennial celebration.

SULLIVAN'S ISLAND

While serving in the U.S. Army, **Edgar Allan Poe** was stationed for over a year at Fort Moultrie on this island off the coast of South Carolina, near Charleston. During his stay, which began on November 18, 1827, Poe gathered impressions and experiences that he used in "The Gold Bug" (1843) and in "The Balloon Hoax" (1844), as well as in his celebrated poem "Israfel," which was included in *Poems* (1831). As a result the island boasts a Gold Bug Avenue among its various street names. Fort Sumter National Monument, which includes Fort Moultrie, is open to the public. Admission is free.

WILLINGTON

Willington Academy, founded in 1801 in this western South Carolina town, appeared in the semi-autobiographical novel *Master William Mitten* (1864), by **Augustus Baldwin Longstreet.** Longstreet, while a student at the academy, built a cabin for himself on the academy grounds. No buildings of the academy still stand, but a marker on the site relates some of the history of the school.

This island is a very singular one. It consists of little else than the sea sand, and is about three miles long. Its breadth at no point exceeds a quarter of a mile. It is separated from the mainland by a scarcely perceptible creek, oozing its way through a wilderness of reeds and slime, a favorite resort of the marsh-hen. The vegetation, as might be supposed, is scant, or at least dwarfish. No trees of any magnitude are to be seen.

—Edgar Allan Poe,
in "The Gold Bug"

GEORGIA

ANDERSONVILLE

The Andersonville National Cemetery, fifty miles south of Macon, on Route 49, was the site of Camp Sumter, the infamous Confederate prison that held as many as 32,000 prisoners at one time in about twenty-seven acres, and in which died more than 12,000 of the 50,000 Union soldiers who entered its gates during the Civil War. The wretched life of prisoners housed there was described by **MacKinlay Kantor** in his novel *Andersonville* (1955), which won a Pulitzer Prize. Today this historical site includes the graves of the Union soldiers, original Civil War buildings, and other installations. The cemetery is open daily throughout the year.

ATHENS

This city, sixty miles northeast of Atlanta, is the home of the University of Georgia, founded in 1805. Literary men who have studied here include **Francis Robert Goulding,** a clergyman and novelist known best in his time for his boys' novel *Young Marooners* (1852); **Paul Hamilton Hayne,** author of several volumes of verse; **Charles Henry Smith [Bill Arp],** author of celebrated satirical letters which were collected in such volumes as *Bill Arp, So-Called* (1866); and **Henry Timrod,** sometimes called the Laureate of the Confederacy. Timrod immersed himself here in the study of classical poetry, a discipline that was to have a profound effect on his poems, including "Ethnogenesis" (1861) and "Ode Sung at the Occasion of Decorating the Graves of the Confederate Dead" (1867). Poor health and financial problems forced him to leave the university before he had completed his education.

South of the North, yet North of the South, lies the city of a hundred hills, peering out from the shadows of the past into the promise of the future.

—W. E. B. Du Bois,
in "The Wings of Atlanta,"
The Souls of Black Folk

ATLANTA

The capital of Georgia, and one of the great cities of the South, was the home for many years of **W[illiam] E[dward] B[urghardt] Du Bois.** Du Bois taught economics and history at Atlanta University from 1897 to 1910 and directed an important annual lecture series on the problems of American blacks. While living here he also wrote *The Souls of Black Folk* (1903), perhaps his single most important book. Du Bois left Atlanta in 1910 to edit *The Crisis,* the magazine of the newly formed National Association for the Advancement of Colored People (NAACP), but returned to Atlanta University in 1934 as professor

More than 30,000 men, crowded upon twenty-seven acres of land, with little or no shelter from the intense heat of a Southern summer, or from the rain and from the dew of night . . . with little or no attention to hygiene, with festering masses of filth at the very doors of their rude dens and tents, with the greater portion of the banks of the stream flowing through the stockade a filthy quagmire of human excrements alive with working maggots. . . .

—a description of the prison,
quoted in *Andersonville,*
by MacKinlay Kantor.

"I don't keer w'at you do wid me, Brer Fox," sezee, "so you don't fling me in dat briar-patch. Roas' me, Brer Fox," sezee, "but don't fling me in dat briar-patch," sezee.

—Uncle Remus, speaking of Brer Rabbit in "The Briar Patch."

of sociology, serving for ten years in that post.

Another long-time resident of Atlanta was **Joel Chandler Harris,** author of the Uncle Remus stories, who moved here in 1876 from Savannah to escape an epidemic of yellow fever. Already an established newspaperman, Harris soon found a job on the Atlanta *Constitution;* he was to stay with the paper for twenty-four years, the period in which his stories enjoyed huge success. The first was published in the *Constitution* in 1879. Harris and his family had several homes during their early years in Atlanta: initially in the Kimball House, a hotel; in the Harrison home in nearby Decatur; and in a five-room frame house, no longer standing, at 201 Whitehall Street. They later moved to a house at 1050 Gordon Street, SW, which was to be Harris's home for the rest of his life. One day he observed a wren building her nest in his mailbox; leaving it undisturbed, he was able to watch the bird laying eggs and raising her young, and from that time on the house was called The Wren's Nest, by which name it still is known. While living here, Harris wrote more than twenty-five books—not only those about the Uncle Remus characters but others, such as *Gabriel Tolliver: A Story of Reconstruction* (1902), on more realistic themes. The Wren's Nest is open to visitors under the auspices of the Joel Chandler Harris Memorial Association. Among its memorabilia are the round library table at which Harris did much of his writing; his favorite rocking chair and his old-fashioned typewriter; books by himself and by friends; and photographs of the author with Theodore Roosevelt, Andrew Carnegie, and other visitors. The Wren's Nest is open every day, and an annual memorial service is held here on December 9, Harris's birthday. Harris died on July 3, 1908, and is buried in Westview Cemetery, at 1679 Westview Drive, SW.

The Wren's Nest, home of Joel Chandler Harris in Atlanta from 1881 until his death on July 3, 1908. In this old photograph, the creator of Uncle Remus stands in front of the house.

James Weldon Johnson was also a resident of Atlanta. He attended the preparatory school of Atlanta University in 1887–88 and took a degree at the university in 1894. Among his works were a book of sermons in verse, *God's Trombones* (1927), and a novel, *The Autobiography of an Ex-Colored Man* (1912).

Perhaps Atlanta's most famous native literary figure is **Margaret Mitchell,** author of the vastly popular *Gone with the Wind* (1936). Mitchell was born on November 8, 1900, in a two-story cottage (no longer standing) at 296 Cain Street. When she was two years old, her family moved to a house at Jackson Street and Highland Avenue. They moved again a year later, this time to a larger house at 179 Jackson Street, now the National Linen Supply Company Warehouse. There they stayed until about 1912, when they moved to 1401 Peachtree Street, still standing but privately owned. In the early 1920s, Mitchell worked as a reporter for the Atlanta *Journal* Sunday magazine, in which capacity she once interviewed actor Rudolph Valentino. In 1926, while living at 979 Crescent Avenue, in a house (no longer standing) that she called "The Dump," Mitchell began to do research for *Gone with the Wind,* her only novel. She worked almost every day at the Atlanta Public Library, and the papers she accumulated cluttered her quarters: at one point some of them were used to prop up a sofa. In 1932 she moved to 4 Seventeenth Street NE. Here she remained until 1936, and here she completed her novel.

Mitchell worked on the book for ten years—and, by some reports, submitted it to many publishers who rejected it—before sending her manuscript to Macmillan. Quick to see its potential, the publisher launched an extensive advertising campaign while the author revised her manuscript and corrected proofs over the next year. *Gone with the Wind* was published in 1936 and sold fifty thousand copies in one day and well over a million copies (in hardback, of course) in its first year. It won a Pulitzer Prize for its author in 1937, and in 1939 was made into a motion picture, one of the most popular of all time. In that year, Mitchell and her husband, John R. Marsh, moved to 1268 Piedmont Avenue, but she was to enjoy her success for only one more decade: on August 12, 1949, Margaret Mitchell was struck by a car while crossing Atlanta's Peachtree Street with her husband. She died of her injuries on August 16 and was buried in Oakland Cemetery, 248 Oakland Avenue, SE.

Frank L[ebby] Stanton, the poet, spent his most productive years in Atlanta. Stanton joined the staff of the Atlanta *Constitution* in 1889, and for nearly forty years the newspaper published his column "Just from Georgia." Stanton wrote several collections of poetry, including *Songs of the Soil* (1892), *Songs from Dixie* (1900), and *Up from Georgia* (1902). In 1925 he was named poet laureate of Georgia. Stanton lived in Atlanta at 645 Highland Avenue, and later at 1421 Fairview Road NE, where he died on January 7, 1927.

Today Atlanta pays tribute not only to Margaret Mitchell and Frank L. Stanton but to other Southern writers as well. The Atlanta Historical Society, at 3099 Andrews Drive, NW, has memorabilia of Mitchell, Erskine Caldwell, Joel Chandler Harris, Sidney Lanier, Flannery O'Connor, Mark Twain, and others. The society is open to the public. The Atlanta Public Library, 10 Pryor Street, SW, has a Margaret Mitchell Memorial—

a collection of books on the South—and assorted Mitchell memorabilia, including photographs. The State Archives, at 330 Capitol Avenue, SE, displays items relating to a number of Southern authors and is open to the public. The cyclorama in Atlanta's Grant Park has on display an enormous painting of the Battle of Atlanta (1864), which is described in *Gone with the Wind,* and lectures on the battle are given regularly here. The Old Lamppost, a survivor of the battle itself, burns today at the northeast corner of Whitehall and Alabama streets. A hole in its base, torn by a shell, serves as a reminder of the terrible battle.

AUGUSTA

This old city, not far from the South Carolina border, has been home to a number of notable Southern writers. **Paul Hamilton Hayne,** the poet, editor, and biographer, retired near here after the Civil War, living out the rest of his life in poverty and ill health. **James Ryder Randall,** best known for his lyric "Maryland, My Maryland" (1861), worked for several newspapers here. Father **Abram Joseph Ryan,** author of such popular Confederate poems as "The Conquered Banner," worked here briefly for *The Pacificator* and the *Banner of the South.* **William Tappan Thompson,** author of such enormously successful books as *Major Jones's Courtship* (1843), lived here from 1838 to 1842 while founding and editing the Augusta *Mirror,* Georgia's first purely literary magazine. Augusta's most distinguished literary native son is **Augustus Baldwin Longstreet,** born on September 22, 1790. His best-known book was *Georgia Scenes, Characters, and Incidents* (1835), humorous sketches of life on the Southern frontier.

Richard Henry Wilde was also an Augustan. His lyric fragment entitled "Stanzas," but intended for "The Lament of the Captive," an epic treatment of the Seminole War, caused an extraordinary controversy in the poet's lifetime. The poem was first published in 1819, without Wilde's consent, and was variously said to be the work of one or another contemporary or of the Greek poet Alcaeus. Not until 1843 did Wilde claim the work as his own, but in the meantime it had been set to music by various composers, including Sidney Lanier, and was widely known by its first line, "My life is like the summer rose." Verdery Cottage, Wilde's home in Augusta, still stands at 2229 Pickens Road, near Augusta College; it is privately owned and not open to visitors. Wilde is buried, along with Hayne and Randall, in the Poets' Corner of Augusta's Magnolia Cemetery, 702 Third Street. Hayne, Randall, Ryan, and Sidney Lanier are also commemorated by the Poets' Monument in the center of Green Street, between Seventh and Eighth streets.

Poets' Monument in Augusta.

My life is like the prints, which feet
Have left on Tampa's desert strand;
Soon as the rising tide shall beat,
All trace will vanish from the sand;
Yet, as if grieving to efface
All vestige of the human race,
On that lone shore loud moans the sea—
But none, alas! shall moan for me!

—Richard Henry Wilde,
in "The Lament of the Captive"

Margaret Mitchell in 1936, the year in which *Gone with the Wind* was published.

The world became an inferno of noise and flame and trembling earth as one explosion followed another in ear-splitting succession. Torrents of sparks shot to the sky and descended slowly, lazily, through blood-colored clouds of smoke.

—Margaret Mitchell,
in *Gone with the Wind.*

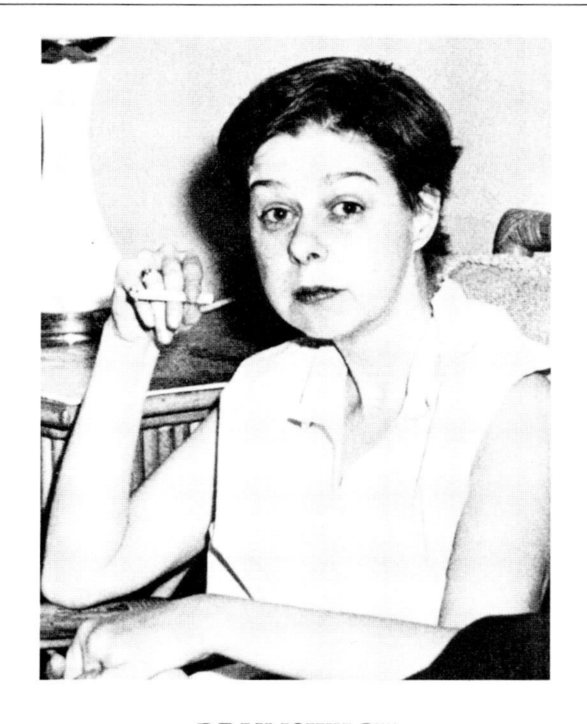

Photograph of Carson McCullers taken some years after the novelist achieved recognition.

BRUNSWICK

The marshes near this coastal town were the inspiration for the poem "The Marshes of Glynn" (1878), by **Sidney Lanier.** The poet used to come to Brunswick to visit his sister at her home on Albany Street. He is said to have conceived the poem while sitting under what now is called Lanier's Oak, near the Saint Simons Causeway. North of Brunswick, on Saint Simons Island, once stood a plantation that was owned in the early nineteenth century by Pierce Butler, husband of **Fanny Kemble,** the English actress and diarist. The journal she kept during her stay on the plantation and at nearby Butler Island was published in England in 1863 and soon after in the United States. Kemble's strong feelings against the practice of slavery are apparent in the pages of her *Journal of a Residence on a Georgia Plantation in 1838–1839* (republished 1961).

> *While I was sitting by the window, Abraham, our cook, went by with some most revolting-looking "raw material" (part, I think, of the interior of the monstrous drumfish of which I have told you). I asked him, with considerable disgust, what he was going to do with it; he replied: "Oh! we colored people eat it, missis."*
> *Said I: "Why do you say we colored people?"*
> *"Because, missis, white people won't touch what we too glad of."*
>
> —Fanny Kemble, in *Journal of a Residence on a Georgia Plantation in 1838–1839*

CARTERSVILLE

Charles Henry Smith [Bill Arp] in about 1877 moved to a farm near the town of Cartersville, approximately thirty-five miles northwest of Atlanta. Here he continued writing the humorous-satirical letters, begun in 1861, that made his reputation. He died here on August 24, 1903, and was buried in nearby Rome.

COLUMBUS

Two miles south of Columbus, a town on the Alabama border, stood Torch Hill, the home of **Francis Ticknor** from 1857 until his death in 1874. The house was torn down some years ago, but a plaque has been placed near the site. Ticknor was a physician as well as a poet, and one of his most memorable poems, "Little Giffen" (1867), took form as a result of his experiences tending wounded Confederate soldiers at a Columbus hospital. He and his wife took young Giffen, one of his patients, to stay with them for over six months; when Giffen left for the front, the Ticknors never heard from him again. Years later, when Ticknor himself took sick, he died after returning to his medical duties before he had recovered fully. Ticknor is buried in Linwood Cemetery in Columbus.

> *"Johnston pressed at the front," they say;—*
> *Little Giffen was up and away!*
> *A tear, his first, as he bade good-by,*
> *Dimmed the glint of his steel-blue eye.*
> *"I'll write, if spared!" There was news of fight,*
> *But none of Giffen—he did not write!*
>
> —Francis Ticknor, in "Little Giffen

Columbus has a firmer position in literary history as the birthplace of two women novelists, Augusta Jane Evans [Wilson] and Carson McCullers. **Augusta Jane Evans,** author of the popular romance *St. Elmo* and other works of the same genre, was born on May 8, 1835, in her maternal grandparents' home at the corner of Wildwood Avenue and Garrard Street. The house has been torn down, but a marker has been placed on the site. *St. Elmo,* set partly in Columbus, was so popular in its day for the moralistic story it told of a woman who wrote moralistic novels that towns, streets, hotels, and even children were named for the novel.

It is through the other woman novelist born here, **Carson McCullers [Lula Carson Smith],** that Columbus achieved its lasting place in the history of American literature. McCullers was born here on February 19, 1917, at 423 Thirteenth Street. The house, though still standing, is privately owned and not open to visitors. The same is true of three other houses she lived in here: Macon Road, number unknown, where she lived in 1924; 2417 Wynnton Road, in 1925; and 1519 Starke Avenue, from 1926 until about 1933. McCullers described the Starke Avenue house in *The Member of the Wedding* (1946). The schools she attended in Columbus were the Sixteenth Street School, the Wynnton School, and Columbus High School, from which she was graduated in 1933. She moved soon thereafter to New York City, returning for the summer of 1935 to work on the Columbus *Ledger,* and for a longer stay in 1936, when she became seriously ill. It was while convalescing in Columbus that McCullers began writing her first novel, *The Heart Is a Lonely Hunter* (1940), which was to win critical acclaim. On a later visit, beginning in the autumn of 1939, she began work on a novel she called *The Bride and Her Brother* and on a story, "A Tree, A Rock, A Cloud." None of the Columbus houses McCullers lived in is marked in any way, but the Bradley Memorial Library here has an extensive collection of writings about and by the novelist.

DECATUR

Robert Frost visited Agnes Scott College in Decatur, just east of Atlanta, twenty times between 1935 and 1962, and the college library today has a Robert Frost Memorial Room, containing first editions, special editions, anthologies, translations, letters, and critical material pertaining to the poet.

EATONTON

Joel Chandler Harris, author of the well-loved Uncle Remus stories, among other works, was born on December 9, 1848, in this town about forty miles northeast of Macon. The house in which he probably was born, at the corner of West Marion and West Lafayette streets, just west of the courthouse, has a marker outside but is closed to the public. The courthouse pays tribute to Harris with a statue of one of his most famous creations, Brer Rabbit. Nearby, on Jefferson Street, a plaque refers to Harris as the "most distinguished son of Putnam County and beloved of all the world." Still more tribute to Harris is found south of the town center on Route 441, at the Uncle Remus Museum. The museum, in Turner Park, is housed in buildings constructed from two slave cabins that were part of the home of Joseph Sidney Turner, the "little boy" in the Uncle Remus tales. Displays and memorabilia can be seen here daily. Also nearby is Turnwold, the plantation where Harris was an apprentice printer from 1862 to 1866, working on *The Countryman,* a magazine published by the scholar Joseph Addison Turner. Turner encouraged the young printer's literary talents by giving the boy free access to his large library. It was at Turnwold, in the slave quarters, that Harris first heard bits of African animal lore; transformed, these became his famous Uncle Remus stories.

FAYETTEVILLE

Carson McCullers and her husband Reeves lived in this town, just south of Atlanta, in a brick house, address unknown. The Margaret Mitchell Library, named for **Margaret Mitchell,** who visited Fayetteville about 1936, when *Gone with the Wind* was published, contains a number of books donated through Mitchell's efforts, and a very good Civil War reference section. The Fife House, at 140 West Lanier Street, is the former home of Fayetteville Academy, the school attended by Mitchell's heroine, Scarlett O'Hara.

FORSYTH

Joel Chandler Harris, author of the Uncle Remus stories, lived here, twenty-two miles northwest of Macon, from 1867 to 1870, while working for the Monroe *Advertiser.*

FORTVILLE

Francis Ticknor was born in this small town, not far from Macon, on November 13, 1822. Ticknor was known best for his poetry of the Civil War, much of it collected in his *Poems* (1879).

JONESBORO

Sidney Lanier was visiting the Kell house in Sunnyside, about twelve miles south of Jonesboro, a city near Atlanta, when he wrote "Corn" in August of 1874. "Corn" was the first poem by Lanier to attract national attention. The Kell house is still standing today but is privately owned and closed to visitors.

LAWRENCEVILLE

Charles Henry Smith [Bill Arp] was born in Lawrenceville, twenty-five miles northeast of Atlanta, on June 15, 1826. Smith's work took the form of hopelessly misspelled letters written by one Bill Arp. The letters were published for thirty years in the Atlanta *Constitution* and collected in a number of books, the first of which was *Bill Arp So-Called* (1866). The house Smith was born in no longer stands, but visitors may see—from the outside only—one house he lived in, at 341 West Crogan Street, Route 29. Gwinnett Court House Square, where his father ran a general store, has a plaque dedicated to Smith.

MACON

Harry Stillwell Edwards was born in Macon on April 23, 1855, and lived here most of his life. He wrote *Sons and Fathers* (1896) as well

Thou lustrous stalk, that ne'er mayst
walk nor talk,
Still shalt thou type the poet-soul
sublime
That leads the vanward of his timid
time
And sings up cowards with
commanding rhyme—

—Sidney Lanier,
in "Corn"

Kingfisher Cabin, writing retreat of Harry Stillwell Edwards in Macon. Edwards built the cabin in 1929 and here wrote his weekly column for the Atlanta *Journal.*

I tell you, my friend, we are the poorest peepul on the face of the yearth—but we are poor and proud. We made a bully fite . . . and the whole Amerikan nation ought to feel proud of it.

—Bill Arp,
in a letter to Artemus Ward,
September 1, 1865

Ambrotype portrait of Sidney Lanier at the age of fifteen.

as other novels while working as a newspaper editor and owner. He died here on October 22, 1938, and was buried in Rose Hill Cemetery. **Joel Chandler Harris** worked briefly in 1866 as a typesetter for the Macon *Telegraph* in its office at 94 Cherry Street. The building, its address changed to 596 Cherry Street, is now occupied by the Pearle Vision Center; the *Telegraph* continues to publish at 120 Broadway. **Laurence Stallings,** best known as co-author (with Maxwell Anderson) of the World War I play *What Price Glory?* (1924), was also born here, on November 24, 1894. His birthplace and the area around it have since been cleared to make room for Route 75.

Macon's most distinguished literary native son is **Sidney Lanier,** who was born here on February 3, 1842. His birthplace, on Second Street near Pine Street, has been torn down to make way for a hospital and medical offices, but the home of his grandparents, at 935 High Street, has survived. Lanier and his family lived here for some years, in what now is known as the Sidney Lanier Cottage. The cottage can be visited by arrangement with the Middle Georgia Historical Society, also located at 935 High Street. Lanier came from an intellectual, music-loving family that encouraged artistic pursuits, and in his grandfather's home Lanier's great passions for music and poetry had ample opportunity to flourish. Lanier attended Bibb County Male Academy, no longer standing, on Academy Square. After his graduation from college in 1860, Lanier volunteered for service in the Civil War. He was captured by Federal troops in 1864 and imprisoned for five months at Point Lookout, Maryland. Ill with tuberculosis at war's end, Lanier walked back to Macon, a twenty-dollar gold piece in his pocket and his beloved flute in his hand. He soon left to work in Alabama but returned in 1867 to Macon. In that year his only novel, *Tiger-Lilies,* was published, and he married Mary Day, on December 19. Christ Episcopal Church, at 538 Walnut Street, where they were married, is the oldest church in Macon. From 1871 to 1873, the couple lived in a home, no longer standing, on Orange Street between Georgia Avenue and Bond Street.

Lanier is honored in Macon not only by the Sidney Lanier Cottage but by a park and a high school that bear his name. Washington Memorial Library, on Washington Avenue at College Street, houses a marble bust of the poet by Gutzon Borglum, a manuscript and first edition of Lanier's poems, one of his flutes, and a copy of the invitation to his wedding.

MILLEDGEVILLE

In 1856 **Sidney Lanier** came to Milledgeville, thirty miles northeast of Macon, to study at Oglethorpe University. After taking a year off to work as a postal clerk in Macon, he was graduated in 1860 at the top of his class. Since then

Sidney Lanier Cottage in Macon. A marker identifies the cottage as Lanier's birthplace but this claim is disputed by some scholars, who place the poet's birth in an earlier Lanier family home.

the university has moved to a site near Atlanta, and most of the buildings have been torn down. The sole survivor is Thalian Hall, a three-story brick building in which Lanier lived as a student. The building, now part of a hospital, is open to visitors who want to see Lanier's dormitory room.

Another important writer who resided in Milledgeville was **Flannery O'Connor.** She moved here after her father became incurably ill with lupus erythematosus, the same disease that was eventually to take her life as well. The house O'Connor lived in, her family home, is still standing at 311 West Greene Street but is not open to visitors. O'Connor was graduated from Peabody High School here and, in 1945, from Georgia State College for Women, in Milledgeville. In 1951 she moved to Andalusia Farm, 500 acres on Eatonton Road three miles north of town. The farm, still in private ownership, looks today much as it did when the novelist died at Milledgeville Hospital on August 3, 1964.

Only thirty-nine years old when she died, Flannery O'Connor had already established herself as a major contemporary writer through her novels, such as *Wise Blood* (1952), and her short stories, collected in *A Good Man Is Hard to Find* (1955). These works, along with the novel *The Violent Bear It Away* (1960) and a second book of short stories, *Everything That Rises Must Converge* (1965), were completed at Andalusia. *The Complete Stories* (1971) contains all known stories of this outstanding Southern writer, whose often grotesque tales are widely admired. The Ina Dillard Russell Library at O'Connor's alma mater, Georgia State College for Women, now called Georgia College, has a Flannery O'Connor Collection, which consists of newspaper clippings, magazines, manuscripts, photographs, critical writings, and memorabilia. A Flannery O'Connor Memorial Room, furnished in Victorian decor, with a carpet reminiscent of the fantail of a peacock, was opened in the college library in 1973.

MORELAND

Biographers give either Coweta County or White Oak, Georgia, as the birthplace of novelist **Erskine Caldwell** on December 17, 1903. But Caldwell once wrote: "I was born in the country many miles from railroad or post office. The place where I was born was so remote it had no name. . . . The nearest landmark was a church several miles away." He has reported recently that he was born in Moreland, southeast of Atlanta. Caldwell's books—the best known are *Tobacco Road* (1932) and *God's Little Acre* (1933)—have made him one of the most popular authors of this century: by 1962 his books, thirty-eight in all, had sold over twenty-five million copies, and many of them still sell well today.

PHILOMATH

The Bartram Trail Library and Museum in this small town in northeastern Georgia houses a collection of books and maps relating to botanist **William Bartram,** usually considered America's first native-born botanist, who traveled through Georgia in 1765. Bartram was the father of **John Bartram,** the botanist and travel writer. The museum is open daily.

POWELTON

Richard Malcolm Johnston, educator and humorist, was born on March 8, 1822, at **Oak Grove,** his family's plantation near the tiny village of Powelton in central Georgia. The town of Dukesborough in Johnston's *Dukesborough Tales* (1871), an expanded version of his earlier *Georgia Sketches* (1864), was based on Powelton.

ROME

Charles Henry Smith [Bill Arp] practiced law in this city from 1851 to 1877, except during the Civil War. In 1861 he launched his Bill Arp letters, the first of which was published in the Rome *Confederacy.* After the war he returned to serve as state senator and as mayor of Rome. Smith is buried in Oak Hill Cemetery.

ROSWELL

Francis Robert Goulding, who was best known for his novel *Young Marooners* (1852), died on August 22, 1881, in Roswell, now part of the Atlanta metropolitan area. He was buried in the old Presbyterian Cemetery on Route 19.

SAVANNAH

Among the literary residents of this beautiful city, the oldest in Georgia, have been **Joel Chandler Harris,** who was associate editor of the Savannah *Morning News* from 1870 to 1876 and who met and married Esther LaRose here in 1873, and poet **Ogden Nash,** who lived as a child in the Low House at 329 Abercorn Street. The Low House, now a hotel known as the Colonial Dames House, saw such eminent visitors as **William Makepeace Thackeray,** in 1853 and 1856, and General **Robert E. Lee,** in 1870. The city has not always been hospitable to literary visitors: **Langston Hughes** was once accosted by a Savannah policeman for buying a newspaper in the waiting room for whites in the railroad station. The Pirates' House at 20 East Broad Street, now a restaurant, is traditionally considered to have been the model for the Admiral Benbow, the inn that appears in *Treasure Island* (1883), by Robert Louis Stevenson.

> *I never saw in my life a more dreadful looking figure. He stopped a little from the inn, and, raising his voice in an odd sing-song, addressed the air in front of him:-*
>
> *"Will any kind friend inform a poor blind man, who has lost the precious sight of his eyes in the gracious defense of his native country, England, and God bless King George! where or in what part of this country he may now be?"*
>
> *"You are at the 'Admiral Benbow,' Black Hill Cove, my good man," said I.*
>
> —Robert Louis Stevenson, in *Treasure Island*

The Pirates' House in Savannah. Built in 1754, the inn was once a seamen's tavern and, by tradition, a meeting place for pirates.

> *The long street, in the moonlight, was like a deep river, at the bottom of which they walked, making scattered, thin sounds on the stones, and listening intently to the whisperings of elms and palmettos. And their house, when at last they stopped before it, how strange it was! The moonlight, falling through the two tall swaying oaks, cast a moving pattern of shadow and light all over its face.*
>
> —Conrad Aiken, in "Strange Moonlight."

Savannah's greatest contributions to literature have come in the twentieth century. **Conrad Aiken** was born at 503 Whitaker Street, then a boarding house, on August 5, 1889. The house is still standing but is privately owned. Aiken's family later moved into a home at 228 East Oglethorpe Avenue in which they lived for ten years. His father, a physician, had offices in the basement and the family—in the great American tradition—lived upstairs. Aiken's short story "Strange Moonlight," in *Collected Short Stories of Conrad Aiken* (1960), described this house, which was also mentioned in the autobiographical *Ushant* (1952) and in "The Coming Forth by Day of Osiris Jones" (1931). Other Savannah landmarks mentioned in "Strange Moonlight" include the Oglethorpe Avenue fire station, the Colonial Cemetery on Oglethorpe Avenue, and the Habersham Street police station. Aiken lived most of his life in New England but in 1962 bought a house at 230 Oglethorpe Avenue, next door to his childhood home, and spent much time there until his death in Savannah on August 17, 1963. Plans have been made to install a plaque in honor of the novelist and poet in the space between his two Oglethorpe Avenue homes. He is buried in the magnificent and historic Bonaventure Cemetery on Bonaventure Road, about two miles from the downtown area.

Flannery O'Connor was born at 207 East Charlton Street on March 25, 1925. The house, an old brownstone, has been converted into apartments, but a plaque on the building commemorates the author's birth. Also standing are two schools the novelist attended: Sacred Heart School, at 209 East Thirty-eighth Street, and St. Vincent's Academy, at 207 East Liberty Street.

SEA ISLAND

Eugene O'Neill and his wife Carlotta moved in 1931 into Casa Genotta, a beach cottage they built on this lovely resort island off Georgia's Atlantic coast. The money to build the cottage came from the success of *Mourning Becomes Electra* (1931), and during O'Neill's five years here the playwright continued, as always, to work with enormous diligence, seldom venturing into town. Occasionally he would write for twenty-four hours at a stretch. He and Carlotta did entertain visitors here, however: **Sherwood Anderson, Somerset Maugham,** and **George Jean Nathan,** among others, but for most of the time O'Neill worked. While here he wrote much of *Days Without End* (1934) and his only comedy, *Ah, Wilderness!* (1933). He also outlined his projected cycle of eleven plays, finishing the first draft of *A Touch of the Poet* (1958) and creating dialogue for *More Stately Mansions* (1964). Casa Genotta is privately owned now and not open to visitors.

WASHINGTON

Thomas Holley Chivers was born on October 18, 1809, at Digby Manor, his father's plantation near Washington. The manor is no longer standing, but a marker has been placed near the site. Chivers also lived briefly at 501 South Alexander Street in Washington itself, but the building that stands there today bears no resemblance to Chivers's home. The poet, whose works included *The Lost Pleiad* (1845) and *Eonchs of Ruby* (1851), is remembered today primarily as a friend and correspondent of Edgar Allan Poe, who called him "at the same time one of the best and one of the worst poets in America."

WRENS

This small town on Highway 1, southwest of Augusta, was one of the many boyhood homes of **Erskine Caldwell.** The church in which the novelist's father preached is still in use. The Caldwells lived in the manse of the Associate Reformed Presbyterian Church, at 401 King Street, which is now used for Sunday School classes and other church functions and is known as the Presbyterian Church House. Caldwell lived in Wrens for about a year and was graduated from Wrens Institute in 1920. It was by riding around this rural area

Casa Genotta, home of Eugene O'Neill on Sea Island, looking out upon the Atlantic Ocean. O'Neill worked in a second-floor study built to resemble a ship's cabin. (*below right*) Birthplace of Conrad Aiken in Savannah. All three of Aiken's homes in Savannah are still standing but closed to visitors.

with his father that Caldwell learned so much about the tenant farmers who later peopled his works of fiction.

FLORIDA

BRADENTON BEACH

Joseph Hayes, while living at Bradenton Beach, on the Gulf coast of Florida, wrote his novel *The Desperate Hours* (1954), a suspense thriller in which three convicts terrorize a family. Hayes later dramatized the book and wrote a screenplay for the motion picture of the same title. Bradenton Beach was the landing place on May 25, 1539, of the Spanish explorer Hernando De Soto, commemorated by a National Memorial Park.

DAYTONA BEACH

Robert Lewis Taylor, the prolific popular biographer, lived in Daytona Beach at the Bellair Apartments in 1971–72 and at Peck Plaza in 1975–76.

DUNEDIN

While living in Dunedin, on the Gulf coast of Florida, **Erskine Caldwell** wrote the novels *The Weather Shelter* (1969), *The Earnshaw Neighborhood* (1971), and *Annette* (1973).

GAINESVILLE

The University of Florida library, in Gainesville, maintains a collection of manuscripts, photographs, and memorabilia, as well as the personal library, of Marjorie Kinnan Rawlings.

HAWTHORNE

Marjorie Kinnan Rawlings from 1928 to 1941 lived in Cross Creek, a backwoods town near Hawthorne, which itself is about ten miles southeast of Gainesville. Her property consisted originally of a weatherbeaten house and a seventy-two-acre orange grove. Rawlings used Cross Creek characters and themes in writing *The Yearling* (1938), the novel for which she won a Pulitzer Prize. After her autobiographical work *Cross Creek* (1942) was published, an old friend charged Rawlings with libel, but the charge was ultimately reduced to invasion of privacy. Rawlings considered selling the Cross Creek land because of the unpleasantness, but after her neighbors refused to testify against her, and she decided to stay on. After marrying Norton S. Baskin in 1941, Rawlings made her home in St. Augustine but retained her Cross Creek home as a place where she could concentrate on her work. As her readership grew, the demands on her time increased, making it difficult for her to work, even in Cross Creek. Rawlings retreated to Banner Elk, North Carolina, and there wrote much of her later work, but she is buried in Antioch Cemetery, a few miles east of Cross Creek, on Route 325. Her epitaph reads: "Through her writings she endeared herself to the people of the world." Her Cross Creek home now is the Marjorie Kinnan Rawlings State Historic Site, on Route 325, open daily. A small admission fee is charged.

HIBISCUS ISLAND

Sholem Asch, the novelist whose works dealt with Jewish life and with New Testament themes, lived in the 1950s at 121 Hibiscus Avenue, on Hibiscus Island, southeast of Miami, in Biscayne Bay. Among Asch's later novels was a trilogy: *The Nazarene* (1939), *The Apostle* (1943), and *The Prophet* (1955). **Damon Runyon,** wanting a place in which to write and relax, built a large, comfortable white villa, Las Melaleuccas, on Hibiscus Island in the early 1930s. Unfortunately, central heating was not one of the amenities provided, and the house proved less than comfortable in winter. In 1935 Runyon here completed the farce *A Slight Case of Murder* (1935), written with Howard Lindsay.

ISLAMORADA

Robert Lewis Taylor lived in a rented house in Islamorada on Upper Matecumbe Key in 1959, when he was gathering material for *A Journey to Matecumbe* (1961).

JACKSONVILLE

This city in northern Florida has several literary associations. **Stephen Crane** spent some time here in the late 1890s, staying at the St. James Hotel, thought to have been located on Duval Street, between Laura and Hogan streets, until it burned down in 1901. On a voyage to Cuba in 1896, Crane was shipwrecked off the Florida

> *A soft rain fell in the night. The April morning that followed was clear and luminous. The young corn lifted pointed leaves and was inches higher. The cow-peas in the field beyond were breaking the ground. The sugar-cane was needle-points of greenness against the tawny earth. It was strange, Jody thought, whenever he had been away from the clearing, and came home again, he noticed things that he had never noticed before, but that had been there all the time.*
>
> —Marjorie Kinnan Rawlings, in *The Yearling*

Front porch of the home of Marjorie Kinnan Rawlings in Cross Creek. When weather permitted, the novelist would entertain guests here.

Lift ev'ry voice and sing
Till earth and heaven ring,
Ring with the harmonies of Liberty;
Let our rejoicing rise
High as the list'ning skies,
Let it resound loud as the rolling seas;
Sing a song full of the faith that the dark past has taught us,
Sing a song full of the hope that the present has brought us;
Facing the rising sun
Of our new day begun,
Let us march on till victory is won.

—James Weldon Johnson, in *Lift Every Voice and Sing*

coast, an experience he was able to use in the masterful story "The Open Boat" (1898).

George Dillon, the poet, was born on November 12, 1906, in Jacksonville, where his family had a house on the northwest corner of Boulevard and Silver streets. Dillon won a Pulitzer Prize for his collection of poems *The Flowering Stone* (1931). **James Weldon Johnson** was born on June 17, 1871, in a house on the northwest corner of Lee and Houston streets. He attended Stanton Central Grammar School, graduating from eighth grade in 1887. When Johnson returned from college, he became principal of his former school and restructured it as a high school for blacks. Johnson, who later became a lawyer, was the first black admitted to the Florida bar after Reconstruction. His books included a novel, *The Autobiography of an Ex-Colored Man* (1912), and a collection of poems, *God's Trombones* (1927). He also wrote the popular black anthem, *Lift Every Voice and Sing* (1900).

Vachel Lindsay spent some time in Jacksonville in 1906 before embarking on a tramping trip through Florida, during which he earned room and board by giving poetry readings. **Sidney Lanier** in 1875 was commissioned by a railroad line to write a Florida guidebook—Lanier was visiting Jacksonville at the time. The book was published in the next year as *Florida, Its Scenery, Climate and History.* **Frank G. Slaughter,** known for his novels on medical and Biblical themes, lived in several places in Jacksonville. He wrote many of his works, including *Sangaree* (1948) and *The Road to Bithynia* (1951), while living at 3202 Garibaldi Avenue, now 4275 Garibaldi Avenue. He wrote *That None Should Die* (1941), which has a Florida setting, while living at 2210 St. Johns Avenue.

JASPER

Lillian Smith was born on December 12, 1897, in Jasper, a small town in northern Florida near the Georgia border. An incident in her early life here made a deep impression on her: her parents took in and cared for a homeless girl, but when they found that the girl was the offspring of a white and a black, she was sent away. When Lillian Smith grew up, she earned her living by doing social work, and what she saw in that work was reflected in her later writing, especially in her best-known novel, *Strange Fruit* (1944). The novel, which dealt with the unhappy life of a mulatto woman, was a popular and critical success.

KEY WEST

This southernmost city of the United States can be thought of as Hemingway country, but other writers are also associated with Key West, beginning with **John James Audubon,** who lived and worked here in 1832. His residence, now known as Audubon House, stands at Green and Whitehead streets. Audubon's complete folio of *Birds of America* (1827–38) is on display for visitors daily for a small admission charge. **Stephen Crane** is said to have stayed in 1898 either at the Duval House, then in the 100 block of Duval Street, or at the Key West Hotel.

Ernest Hemingway and his second wife, Pauline, moved to Key West in 1928, living first in a rented apartment across from the present post office. Hemingway worked here on *A Farewell to Arms* (1929). Later, the Hemingways rented a home at the intersection of United and Whitehead

Ernest Hemingway home at 907 Whitehead Street in Key West. Now a tourist attraction, the handsome structure housed Hemingway and his second wife, Pauline, for many years, and it was here that Hemingway wrote many of his best-known works, including *For Whom the Bell Tolls.*

Now the brave actions of a coward are very valuable in psychological novels and are always extremely valuable to the man who performs them, but they are not valuable to the public who, season in and season out, pay to see a bullfighter. All they do is give that bullfighter a seeming value which he does not have.

—Ernest Hemingway, in *Death in the Afternoon*

streets, where Hemingway was unable to write. He had broken his right arm on a hunting trip in Montana. On April 29, 1931, the Hemingways bought a rambling old Spanish colonial structure at 907 Whitehead Street, built with rock quarried from the grounds. Since the carriage house was to be Hemingway's studio, it was renovated first. The kitchen was then redone: counters and appliances were raised to enable Hemingway to work comfortably, and a marble countertop was installed so he could clean the many fish he caught. Once, while Hemingway was away, his wife spent $20,000 installing a swimming pool; when Hemingway found out how much the pool had cost, he pressed a penny into the wet cement, protesting that the pool had taken his last cent. The Hemingways also added gardens and walkways to the grounds, the writer himself planting many trees and shrubs that are still there today.

One of Hemingway's favorite haunts in Key West was Sloppy Joe's, 201 Duval Street. In light of the prodigious drinking Hemingway did here, he considered himself a silent partner in the bar. Even now, if Hemingway's home is a literary shrine, Sloppy Joe's must be considered a watering place for anyone interested in Hemingway: when Tennessee Williams visited Sloppy Joe's in 1941, he found many of Hemingway's former cronies still in place. Among the books Hemingway completed in Key West are *A Farewell to Arms* (1929), *Death in the Afternoon* (1932), *Green Hills of Africa* (1935), *To Have and Have Not* (1937), and *For Whom the Bell Tolls* (1940). Here he also wrote his play *The Fifth Column* (1938) and two of the best of his many stories, "The Snows of Kilimanjaro" (1936) and "The Short Happy Life of Francis Macomber" (1936). Hemingway moved to Cuba in December of 1939, but Pauline and their two sons remained in Key West. After his death, his sons and his last wife, Mary, sold the property. The present owners lived in the house until 1964, when they turned it into a tourist attraction, the Ernest Hemingway Home and Museum. Visitors to this National Historic Landmark can see Hemingway's books, his workroom, and a mounted wildebeest taken by him on his 1933 African safari, as well as international furnishings, clippings, and photographs. Many of the nearly fifty cats living on the property are descendants of cats that lived here in Hemingway's time—and like their ancestors, many have extra toes. The museum is open daily for an admission fee.

Other writers have visited and worked in Key West. **John Dos Passos,** visiting Hemingway in 1929, stayed at the Overseas Hotel, on Fleming Street, later destroyed by fire. **Carson McCullers,** while on vacation here in 1955 with Tennessee Williams, worked on her play *The Square Root of Wonderful* (1958) and her novel *Clock Without Hands* (1961). **Mark Twain,** traveling to New York in 1867 on the steamship *San Francisco,* was detained at Key West for a short time while a store of medicine was taken aboard to combat an outbreak of Asiatic cholera.

Tennessee Williams, on a visit here in 1941, stayed at the Tradewinds, a mansion converted to a boarding house. In a small shack behind the house, Williams began rewriting *Battle of Angels* (1940), which closed in Boston. It was later produced successfully as *Orpheus Descending* (1957). In 1946 Williams returned to Key West with his grandfather. They took a two-room suite at the top of the Hotel La Concha, where Williams completed *A Streetcar Named Desire* (1947), for which he was awarded a Pulitzer Prize. In his *Memoirs* (1975) Williams noted that he had done his best work in Key West. The Monroe County Public Library at 700 Fleming Street has a commemorative plaque to Williams, and Florida Key Community College has a Tennessee Williams Fine Arts Building.

Old photograph of Sloppy Joe's Bar in Key West. Still open, the establishment looks much as it did when Ernest Hemingway was one of its many regular patrons.

. . . there, ahead, all he could see,
as wide as all the world, great, high,
and unbelievably white in the sun, was
the square top of Kilimanjaro. And
then he knew that there was where
he was going.

—Ernest Hemingway,
in "The Snows of Kilimanjaro"

LAKE WALES

Edward Bok, following his grandmother's dictum: "Make you the world a bit better or more beautiful because you have lived in it," in 1922 established the Bok Mountain Lake Sanctuary and Singing Tower, three miles north of Lake Wales, off Route 27. Now known as Bok Tower Gardens, it is open daily and charges admission. Bok, whose autobiography, *The Americanization of Edward Bok* (1920), was a Pulitzer Prize-winner, was editor of *Ladies' Home Journal* from 1889 to 1920. Bok died at Mountain Lake on January 9, 1930, within sight of the sanctuary he built.

LEESBURG

Bayard Kendrick wrote *The Flames of Time* (1948), a historical novel about Florida, while living at Leesburg, a fishing resort northwest of Orlando.

LONG KEY

Zane Grey came to this Florida key several times to fish and write. On one visit he wrote all of his novel *Code of the West* (1934)—in five weeks.

MADISON

Shirley Anne Grau wrote of Madison in her Pulitzer Prize-winning novel, *The Keepers of the House* (1964).

STELLA. Your face and your fingers are disgustingly greasy. Go and wash up and then help me clear the table. *(He hurls a plate to the floor.)*
STANLEY. That's how I'll clear the table! *(He seizes her arm.)* Don't ever talk that way to me! "Pig—Polack—disgusting—vulgar—greasy!"—them kind of words have been on your tongue and your sister's too much around here! What do you two think you are? A pair of queens? Remember what Huey Long said—"Every Man Is a King!" And I am the king around here, so don't forget it! *(He hurls a cup and saucer to the floor.)* My place is cleared! You want me to clear your places?

—Tennessee Williams,
in *A Streetcar Named Desire*

Mom is an American creation. Her elaboration was necessary because she was launched as Cinderella. Past generations of men have accorded to their mothers, as a rule, only such honors as they earned by meritorious action in their individual daily lives. Filial duty *was recognized by many sorts of civilizations and loyalty to it has been highly regarded among most peoples. But I cannot think, offhand, of any civilization except ours in which an entire division of living men has been used, during wartime, or at any time, to spell out the word "mom" on a drill field, or to perform any equivalent act.*

—Philip Wylie,
in *Generation of Vipers*

MANDARIN

Harriet Beecher Stowe and her husband spent their winters from 1867 to 1884 in this town on the St. John's River, twelve miles south of Jacksonville. Their intention was to help their son regain his health and conduct philanthropy among the blacks, but during their stay Mrs. Stowe wrote a collection of sketches about the area, *Palmetto Leaves* (1873). The cottage the Stowes stayed in is no longer standing, but a marker commemorates their long residence.

MARIANNA

Caroline Lee Hentz lived with one of her grown daughters in Marianna, a city between Pensacola and Tallahassee, for much of the time between 1849 and her death in 1856. Her work during this period included the novels *Linda* (1850), *Rena* (1851), and *Robert Graham* (1855). Hentz died here on February 11, 1856.

MIAMI

Hervey Allen, author of the enormously successful novel *Anthony Adverse* (1933), lived on The Glades Estate in the Coconut Grove section of Miami. He was completing work on a series of historical novels when he died here on December 28, 1949. **Upton Sinclair** wrote the play *The Machine* here in 1909, and **Irving Stone,** living in a rented room, wrote his play *Truly Valiant* (1934), which closed after one night.

Philip Wylie, an overnight sensation for his attack on the American mother in *Generation of Vipers* (1942), lived in South Miami from 1928 until his death. Wylie died at Baptist Hospital in Miami on October 25, 1971, and a memorial service was held for him at Brockway Hall, University of Miami.

OCALA

Bayard Kendrick, the mystery-story writer who created the blind detective Duncan Maclain, died in a nursing home here on March 22, 1977. Two of his Maclain novels were *The Last Express* (1937) and *Blind Man's Bluff* (1943).

PALM BEACH

John Hersey wrote his first book, *Men on Bataan* (1942), while vacationing in Palm Beach. **Henry James** was a guest at the Breakers Hotel here in 1905, while he was on tour. **Edwin O'Connor** based his novel *The Oracle* (1951) on his experiences as a radio announcer in Palm Beach and elsewhere.

PENSACOLA

John Greenleaf Whittier wrote an angry poem, "The Branded Hand," praising Jonathan Walker, a white carpenter convicted in the 1840s by a Pensacola court of helping slaves escape to freedom. Walker was sentenced to jail and was branded "SS" to show that he was a slave stealer. After his release, his book *Trial and Imprisonment of Jonathan Walker* was published in Boston, where it came to Whittier's attention.

ST. AUGUSTINE

St. Augustine, the oldest permanent city in the United States, sighted first by Spanish colonists on August 28, 1565, has celebrated its long history each year by presenting the musical drama *Cross and Sword* (1964), by Paul Green. The play is performed at the St. Augustine Amphitheater from late June to early September. *Cross and Sword,* which recounts the violent early history of the city, is regarded as the official state play of Florida.

Marjorie Kinnan Rawlings, after her marriage to Norton S. Baskin in 1941, divided her time between her home in Cross Creek, near Hawthorne, and the Castle Warden Hotel in St. Augustine, where Baskin worked as manager. She also maintained a small cottage outside town, where she was at work on *The Sojourner* (1953) when she took ill. The novelist died at Flagler Hospital in St. Augustine on December 14, 1953.

William Cullen Bryant, in 1850, wrote *Letters of a Traveller* about this area. **Henry James** once stayed at the Hotel Ponce de Leon here, and **Sinclair Lewis** and **James Branch Cabell** met at this same hotel in 1941. **Sara Teasdale,** while on vacation in January 1910, stayed at the Valencia Hotel here.

ST. PETERSBURG

George Washington Cable came to this city in 1924 with his third wife, Hannah Cowing, to spend the winter. He died here on February 25, 1925. **F. Scott Fitzgerald** revised *Tender Is the Night* (1934) while staying at the Don-Ce-Sar Hotel here in the early 1930s. **Jack Kerouac,** who lived in St. Petersburg after 1965, died here on October 21, 1969. St. Petersburg is more happily remembered as the locale for "The Golden Honeymoon" (1922), one of the best-known stories by Ring Lardner.

SANIBEL ISLAND

In 1936, while **Edna St. Vincent Millay** was staying at the Palms Hotel on Sanibel Island, off southwest Florida, the manuscript for her dramatic narrative *Conversation at Midnight* (1937) was destroyed by fire. The poet had been gathering shells

What lips my lips have kissed, and where, and why,
I have forgotten, and what arms have lain
Under my head till morning; but the rain
Is full of ghosts tonight, that tap and sigh
Upon the glass and listen for reply;
And in my heart there stirs a quiet pain
For unremembered lads that not again
Will turn to me at midnight with a cry.

Thus in the winter stands the lonely tree,
Nor knows what birds have vanished one by one,
Yet knows its boughs more silent than before:
I cannot say what loves have come and gone;
I only know that summer sang in me
A little while, that in me sings no more.

—Edna St. Vincent Millay,
"What Lips My Lips Have Kissed"

on the beach when she happened to glance back at the hotel and see it engulfed in flames.

Robert Lewis Taylor from 1972 to 1974 owned a condominium at 3-B Sunset South, where he and his family spent the winter months. The Taylors sought quiet and seclusion, but were dismayed at the island's tourism and development. The building they lived in is private and not accessible to the public.

SARASOTA

Joseph Hayes lived here when he wrote *Missing and Presumed Dead* (1977), a novel about Sarasota. **MacKinlay Kantor,** author of such novels as *The Romance of Rosy Ridge* (1937) and *Happy Land* (1942), but known best for *Andersonville* (1955), built a home in Sarasota in 1937. It was his principal residence until his death here on October 11, 1977. Kantor wrote many novels and stories after moving to Sarasota, among them *Andersonville,* but none had the success of that account of the Confederate prison camp in Georgia.

Thorne Smith, creator of the character Cosmo Topper, died in Sarasota of a heart attack on June 21, 1934, while he was on vacation. *Topper* (1926), *The Bishop's Jaegers* (1932), and *Topper Takes a Trip* (1932) were among his most popular works.

Robert Lewis Taylor in 1955 bought a home at Lido Beach, near Sarasota. He wanted to be near what were then the winter quarters of Ringling Brothers and Barnum & Bailey Circus so he could write profiles of the performers. In 1956 Taylor moved to a larger home on Longboat Key.

SEBRING

Sebring's literary association rests, as far as is known, on one writer, a long-time resident: **Rex Beach** was living at 2107 Northeast Lakeview Drive here when he took his life on December 7, 1949. Beach, author of novels on the Klondike, including *The Spoilers* (1906), had been fighting cancer of the throat. Over the years, Beach had taken part in the Sebring Fire Department's annual minstrel show and served as a director of Highlands Hammock State Park, on Route 634, west of Sebring. Lake Jackson, nearby, was renamed Lake Rex Beach during the novelist's lifetime, but the name Jackson was restored after Beach died. In the cornerstone of the Girl Scout log house at 442 South Eucalyptus Street are several of Beach's novels, and a tree on the property carries a bronze marker bearing the author's name.

TAMPA

John Berryman moved to Tampa with his family in 1924, when he was ten years old. **Erskine Caldwell** also lived in Tampa as a child. **Sidney Lanier** had rooms at the Orange Grove Hotel, 806 Madison Street, when he wrote numerous poems, including "The Waving of the Corn" (1884) and "Tampa Robins" (1877). **Alec Waugh,** the English novelist, died at Tampa General Hospital on September 3, 1981. Waugh, known best for his novel *Island in the Sun* (1955), lived at 717 Bungalow Terrace from late 1980 until his death.

TARPON SPRINGS

Lois Lenski, author and illustrator of *Strawberry Girl* (1946), *I Like Winter* (1950), *Houseboat Girl* (1957), and other children's books, was a resident of Tarpon Springs at her death, on September 11, 1974. She was eighty years old when she died.

WEST PALM BEACH

Edwin O'Connor, who won a Pulitzer Prize for his novel *The Edge of Sadness* (1961), worked at radio station WJNO in West Palm Beach in the early 1940s. **Maurice Thompson,** whose most successful work was his historical romance *Alice of Old Vincennes* (1900), during 1867 conducted a survey of plant, animal, and bird life near West Palm Beach, at Lake Okeechobee.

WINTER PARK

Rex Beach, the novelist, is buried beside his wife on the Rollins College campus in Winter Park. Beach was living in nearby Sebring at the time of his death. **Winston Churchill,** whose novels included *Richard Carvel* (1899), *The Crisis* (1901), and *The Crossing* (1904), spent several winters at Research Studio in Winter Park, where he died of a heart attack on March 12, 1947. **Constance Fenimore Woolson,** grandniece of James Fenimore Cooper, was herself a novelist and poet. She lived here late in her life, during the 1870s, and several of her sketches and poems were written about Florida: "The Acklawah," "The Ancient City" [St. Augustine], and "The Voyage on an Unknown River." Some of her stories of Southern life were collected in *Rodman the Keeper: Southern Sketches* (1880), and her novel *East Angels* (1886) was set in St. Augustine. Woolson died in England in 1894. The Constance Fenimore Woolson House on the campus of Rollins College is open to visitors.

I'll south with the sun, and keep my clime;
My wing is king of the summer-time;
My breast to the sun his torch shall hold;
And I'll call down through the green and gold
Time, take thy scythe, reap bliss for me,
Bestir thee under the orange-tree."

—Sidney Lanier,
in "Tampa Robins"

The Constance Fenimore Woolson House in Winter Park. On display for visitors are etchings and paintings by the novelist and poet, and first editions of her books.

Back to where the roses rest
Round a Shrine of holy name,
(Yes—they knew me when I came)
More of peace and less of fame
Suit my restless heart the best.

—Father Abram Joseph Ryan,
in "St. Mary's"

Monument honoring Father Abram Joseph Ryan, the poet, in Father Ryan Park in downtown Mobile.

ALABAMA

BIRMINGHAM

One of the most popular writers associated with this city was **Octavus Roy Cohen,** newspaperman, short-story writer, and novelist, and a member of a literary group known as The Loafers, which thrived in Birmingham in the 1920s. Many of Cohen's stories were set in Birmingham, the most popular of them humorous sketches about a group of blacks with the unlikely names of Florian Slappey, Epic Peters, and Pullman Porter.

Mary Johnston came here from Virginia in 1886, when she was sixteen. The frail girl was educated mostly at home, but became strong enough in later years to travel to New York City for extended visits, accompanying her father, a former Confederate officer turned railroad president. Except for these trips, Johnston lived in Birmingham until 1902, when she decided to return to Richmond, Virginia. The house in which she lived in Birmingham, on Seventh Avenue North, between Twenty-second and Twenty-third streets, is no longer standing, but a later residence still exists at 1120 North Twenty-fifth Street. There she worked on her very popular historical romance *To Have and To Hold* (1900), a retelling of the history of the Virginia colony.

Walker Percy was born in Birmingham on May 28, 1916. The birthplace of the novelist and essayist, at 1042 South Twenty-fourth Street, has been replaced by apartment buildings. His father committed suicide when Percy was eleven years old. After his mother died two years later, Percy and his two younger brothers went to live with relatives in Greenville, Mississippi.

The hue of the sunset lingered in cloud and water, and in the pale heavens above the rose and purple shone the evening star. The cloudlike ship at which we had gazed was gone into the distance and the twilight; we saw her no more. Broad between its blackening shores stretched the James, mirroring the bloom in the west, the silver star, the lights upon the Esperance that lay between us and the town. Aboard her the mariners were singing, and their song of the sea floated over the water to us, sweetly and like a love song. We passed the ship unhailed, and glided on to the haven where we would be. The singing behind us died away, but the song in our hearts kept on. All things die not: while the soul lives, love lives: the song may be now gay, now plaintive, but it is deathless.

—Mary Johnston,
in *To Have and To Hold*

COURTLAND

The Lost Virgin of the South (1833), by Don Pedro Casender, the first novel published in Alabama, appeared here, under the imprint of one M. Smith. Courtland is on Route 72 in northwestern Alabama.

FAIRHOPE

In the 1890s a group of people from Des Moines, Iowa, founded a single-tax colony in Fairhope, on Mobile Bay, and attempted to put into effect the economic ideas of Henry George, the author of *Progress and Poverty* (1879). In 1918, when **Sherwood Anderson** lived in Fairhope, he was inspired by the community's resident artists and the surrounding scenery to produce paintings of his own. Another visitor to Fairhope was **Upton Sinclair,** who lived here during the winter of 1909–10.

FLORENCE

Caroline Lee Hentz lived from 1834 to 1843 in Florence, a city in northwestern Alabama, where her husband ran a school for girls. Hentz wrote poems and stories that appeared in the popular magazines of her day. Florence is the setting for a trilogy by T. S. Stribling that chronicled the middle-class Vaiden family from before the Civil War to the twentieth century. The trilogy comprised *The Forge* (1931); *The Store* (1932), a Pulitzer Prize-winning novel; and *Unfinished Cathedral* (1934).

LAFAYETTE

James Still, novelist, short-story writer, and poet, was born on July 16, 1906, at Double Creek, near this city in east-central Alabama. Still is best known for his stories about life in Kentucky, where he has lived for many years.

LOCKHART

William Edward March Campbell, who wrote under the pen name **William March,** was born on September 18, 1893, in this community just north of the Florida border. March grew up in the South and served with distinction in World War I. He later worked as an executive of a steamship company, and lived in Europe for a time. Much of March's fiction reflected his childhood experiences in Alabama and Florida. Among his works with an Alabama setting was the novel *The Looking-Glass* (1943), which depicted the lives of psychologically disturbed people. One of his best-known books was *Company K* (1933), a group of stories about the experiences of the members of an infantry company during World War I.

MARION

John Trotwood Moore, a poet, novelist, and short-story writer, was born on August 26, 1858, in this city northwest of Selma. Born John Moore, he took the name Trotwood from a character in *David Copperfield.* Elmcrest, Moore's birthplace, was built in 1851 and later was remodeled by Judson College for use as an infirmary.

MOBILE

Writers born in this port city at the head of Mobile Bay include **Mary Raymond [Shipman]**

Andrews, born probably in the 1860s. The best-known work of this novelist and short-story writer is *The Perfect Tribute* (1906), which first was published in *Scribner's Magazine* and later in book form, when it became a best seller. *The Perfect Tribute* retells one of the most enduring, albeit apocryphal, stories of Abraham Lincoln's life: his use of a stubby pencil and brown wrapping paper to write the Gettysburg Address while he was en route to the dedication of a national cemetery at the Civil War battle site.

William March lived in Mobile on and off for a number of years, and here wrote the novel *The Bad Seed* (1954). Maxwell Anderson successfully dramatized the novel. Most of the apartments March lived in have been torn down, and none is marked.

Father **Abram Joseph Ryan,** known as the poet of the Confederacy and the poet of the Lost Cause, served at the Cathedral of the Immaculate Conception, occupying the 300 block of Dauphin Street, from 1870 to 1877, and lived at the Bishop's house, 307 Conti Street, now identified by a historical marker. In 1877 Ryan became pastor of St. Mary's Church, which stood on the southwest corner of Lafayette Street and Old Shell Road. A second church was built near the site in 1927, and a third, on the site of Father Ryan's church, replaced the 1927 church in the early 1950s. Ryan is buried in the Roman Catholic cemetery at 1700 Stone Avenue. A bronze statue of Ryan stands in Ryan Park, at the junction of Springhill Avenue and Saint Francis and Scott streets.

Augusta Jane Evans spent most of her life in or near Mobile, arriving here with her family in 1849. When she was seventeen, she wrote her first novel, *Inez, a Tale of the Alamo* (1855). From 1859 to 1868, she lived at The Cottage, 2564 Springhill Avenue, her only Mobile residence still standing. There is a marker in front of the house, now closed to visitors. Among the moralistic novels she published during her years here were *Beulah* (1859) and the best-selling *St. Elmo* (1867), the story of a man who renounced his life of dissipation when he fell under the influence of a woman who wrote moralistic novels. In December 1868, Evans married a Mobile businessman, Lorenzo Madison Wilson, and moved with him to his home just outside the city, where she continued to write novels, including *Vashti* (1869) and *At the Mercy of Tiberius* (1887). When her husband died, in 1891, she moved back to Mobile and lived until her death on May 9, 1909, in a house at 930 Government Street. She is buried in Magnolia Cemetery, on Virginia Street.

William Russell Smith, known as the father of Alabama literature, is also associated with Mobile. He is credited with writing the first play produced in Alabama, *Aaron Burr, a Tragedy,* performed here in 1837, and the first book of poetry by an Alabama poet, *College Musings, or Twigs from Parnassus* (1833). He also published Alabama's first literary magazine, *The Bachellor's Button,* issued here and in Tuscaloosa in 1837.

MONROEVILLE

Truman Capote spent much of his first ten years in this town, the seat of Monroe County in southwestern Alabama. He lived at the home of a relative, Miss Sook Faulk, on whom Capote later based a character in *The Thanksgiving Visitor* (1969). Capote's next-door neighbor was **[Nelle] Harper Lee,** born in Monroeville on April 28, 1926, and brought up here. Capote used her as the model for Idabel Thompkins in his first novel, *Other Voices, Other Rooms* (1948). Lee, in turn, used Capote as the model for the boy Dill in her Pulitzer Prize-winning novel *To Kill a Mockingbird* (1960). The action of the novel is placed in May-

Maycomb was an old town, but it was a tired old town when I first knew it. In rainy weather the streets turned to red slop, grass grew on the sidewalks, the courthouse sagged in the square. Somehow, it was hotter then: a black dog suffered on a summer's day; bony mules hitched to Hoover carts flicked flies in the sweltering shade of the live oaks on the square.

—Harper Lee,
in *To Kill a Mockingbird*

Augusta Jane Evans, the nineteenth-century novelist, lived in the Mobile area for sixty years, for nine of them in the Georgia Cottage in Mobile.

Being in love, she concluded, is simply a presentation of our pasts to another individual, mostly packages so unwieldy that we can no longer manage the loosened strings alone. Looking for love is like asking for a new point of departure, she thought, another chance in life.

—Zelda Fitzgerald, in *Save Me the Waltz*

comb, her fictional name for Monroeville. The homes in which Lee and Capote lived are gone now—Lee's has been replaced by Garrett's Dairy Dream—but the tree mentioned in her novel as the one in which Boo Radley hid gifts for Scout and Jem Finch may still be standing behind the Monroeville Elementary School.

MONTGOMERY

The best-known Montgomery-born writer is actually known more for whom she married than for what she wrote. **Zelda Sayre,** author of the autobiographical novel *Save Me the Waltz* (1932), under the name **Zelda Fitzgerald,** was born on July 24, 1900, in a house on South Street. The Sayre family moved in 1906, living in three different homes before settling, in 1907, at 6 Pleasant Avenue, where Zelda Sayre grew up. This house, no longer standing, made a lasting impression on her, and her memories of it found expression in many of her paintings as well as in her writing. She drew on her years in Montgomery for *Save Me the Waltz* and for her unfinished novel *Caesar's Things.* She attended Chilton Elementary School, later called Sayre Street School, at 500 Sayre Street, and Sidney Lanier High School, 1756 South Court Street. She first met Lieut. **F. Scott Fitzgerald,** then stationed at nearby Camp Sheridan, at a dance at the Montgomery Country Club in July 1918. The club, now extensively remodeled, faces Fairview Avenue. Fitzgerald was transferred in October 1918 to Camp Mills, on Long Island, New York. His unit was ordered back to Alabama in November, and Fitzgerald again was stationed at Camp Sheridan, this time until February 1919, when he was released from active duty and left for New York City. There, he and Zelda were married on April 3, 1920. They returned to visit her family in Montgomery several times in the years following, and lived in Montgomery in the fall of 1931, first at the Greystone Hotel, now headquarters of the First Southern Federal Savings and Loan Company; later, at the Jefferson Davis Hotel, now the site of a bank. They finally moved to a house at 919 Felder Avenue, now used for apartments. After Fitzgerald's death in 1940, Zelda lived with her mother in a house at 322 Sayre Street, but had to be taken from time to time to Highland Hospital in Asheville, North Carolina, for psychiatric treatment. In 1942 the Montgomery Museum of Fine Arts held an exhibit of Zelda Fitzgerald's paintings. Today, the museum houses a memorial to Zelda: a glass case containing such memorabilia as a scarf, a golden handbag, and a copy of *Save Me the Waltz.*

Other writers associated with Montgomery include **Shirley Ann Grau,** the novelist, who attended Booth School here; **Johnson Jones Hooper,** who founded and edited the *Mail* here from 1850 to 1861; and **Sidney Lanier,** who came to Montgomery in 1866 to work as a bookkeeper and clerk at the Exchange Hotel, no longer standing. While here, Lanier worked on *Tiger Lilies* (1867), his first book and only novel. The library of the Alabama Archives in Montgomery maintains a collection of the letters of American writers, including poets William Cullen Bryant, Sidney Lanier, Henry Wadsworth Longfellow, and Edwin Markham.

Ivy Green main house, childhood home of Helen Keller in Tuscumbia and now a museum dedicated to her. On display are Keller's library of books in Braille, her Braille typewriter, and other possessions of the author.

PRATTVILLE

Sidney Lanier served in 1867–68 as principal of Prattville Academy, on Washington Street, in this city northwest of Montgomery. He boarded at the Mims Hotel, 145 Chestnut Street, now a run-down apartment building. While here, Lanier wrote a number of poems, including "The Ship of Earth" and "Tyranny." Prattville Primary School now stands on the site of Prattville Academy.

SELMA

Just before the end of the Civil War, **Ambrose Bierce** resigned his commission in the U.S. Army to take a post here with the U.S. Treasury Department. His job involved confiscation of Confederate property, which meant seizure of nearly everything that had not been destroyed in the war. Bierce found the assignment distasteful and left in 1866 to join General Hazen's surveying expedition to the West.

SHAWMUT

James Still lived here for a time, in east-central Alabama near the Georgia border.

TUSCALOOSA

Authors associated with this city, home of the University of Alabama, include **Caroline Lee Hentz,** the novelist and playwright, who lived here from 1843 to 1845; **William March,** the novelist and short-story writer, who lived in Tuscaloosa for a time and is buried here; and **William Russell Smith,** who lived in Tuscaloosa and practiced law here. Smith issued *The Bachellor's Button,* Alabama's first literary magazine, here and in Mobile in 1837. After serving in Congress from 1851 to 1857, he returned to live in Tuscaloosa, where he conducted a law practice from 1870 to 1879 and served briefly as president of the University of Alabama. Smith died on February 26, 1896, and is buried in Tuscaloosa.

Writers associated with the University of Alabama include **Carl Carmer,** who came to teach here in 1921. He remained in Alabama for six years and traveled extensively, gathering local tales and legends he recounted in his book *Stars Fell on Alabama* (1934). In the 1930s the University of Alabama became a center of literary activity when **Hudson Strode** began teaching his course in the writing of fiction. From the late 1930s to Strode's retirement in 1963, his students published scores of novels and more than one hundred short stories. Among Strode's students were **Borden Deal,** a novelist and short-story writer who lived in Tuscaloosa for many years, and his wife, **Babs H. Deal,** a novelist. The Deals first met while she was a student of Strode's. Strode lived at 49 Cherokee Road for many years before his death on September 22, 1976.

The University of Alabama's Amelia Gayle Gorgas Library has special attraction for those interested in Southern literature. The library houses collections of rare and first editions of many authors, including William March, Lafcadio Hearn, Robinson Jeffers, and James Street. It also maintains a vast collection of manuscripts, including papers of Zelda [Sayre] Fitzgerald, poet Samuel Mintern Peck, Hudson Strode, and Augusta Jane Evans.

TUSCUMBIA

Helen Keller was born on June 27, 1880, in this city on the Tennessee River in northwestern Alabama. Her birthplace and childhood home, Ivy Green, at 300 West North Common, is now a ten-acre shrine that includes the house in which she was born and the cottage that was used as a schoolhouse for her after her teacher, Anne Sullivan, arrived in Tuscumbia in 1887. Also on the grounds is the water pump at which Miss Keller learned her first word, "water," and a memorial garden, which was dedicated to Miss Keller's memory in 1971. At the age of two, Helen Keller was stricken with scarlet fever and, as a result, lost her sight, hearing, and speech. With Miss Sullivan's help, Keller learned to associate objects with words represented by hand movements. From this small beginning, she eventually learned to read, write, and speak. She left Tuscumbia with Miss Sullivan to continue her studies and, while a student at Radcliffe College, wrote her first book, *The Story of My Life* (1902). Her other books included *The World I Live In* (1908), *Out of the Dark* (1913), and *Let Us Have Faith* (1940). Each summer at Ivy Green a local theater group performs the play *The Miracle Worker* (1960), by William Gibson, which tells the inspiring story of Helen and her great teacher. The Helen Keller Public Library, 509 North Main Street, maintains a collection of Helen Keller's books.

TUSKEGEE

Caroline Lee Hentz lived here from 1845 to 1848, while she worked on *Aunt Polly's Scrap Bag* (1846), one of her best-known novels. Tuskegee, in eastern Alabama, is known principally, however, as the home of Tuskegee Institute, founded in 1881 by **Booker T. Washington** as a school for blacks, with special emphasis on practical knowledge. The first classes were held in a building next to the Butler Chapel A.M.E. Zion Church, which was used as an assembly hall. Both structures stood on the site of the present two-story brick church and school, and a granite monument there commemorates the founding of the institute, since incorporated into the Tuskegee Institute National Historic Site. Among sites of interest is Washington's home from 1899 to 1915, The Oaks, on Old Montgomery Road. Its bricks were made by Tuskegee Institute students and teachers. While here, Washington published his landmark autobiography, *Up from Slavery* (1901). He also lived in another house, now moved to 303 Franklin Road and not open to visitors. Washington died on November 14, 1915, and was buried in the Tuskegee Institute Cemetery. Also buried there is the institute's best-known teacher, George Washington Carver. A replica of Washington's birthplace is located in Tuskegee National Forest, off Route 29 east of Tuskegee.

Writers associated with Tuskegee Institute include **Paul Laurence Dunbar,** the poet and novelist, who in 1902 wrote "The Tuskegee Song," which is still the school song; and **Ralph Ellison,** author of *Invisible Man* (1952), who studied music here from 1933 to 1936. Ellison's novel presents a devastating picture of the life of a sensitive student at an all-black college. **Claude McKay,** the poet and novelist, studied at Tuskegee for one year after emigrating from Jamaica in 1912.

My experience is that there is something in human nature which always makes an individual recognize and reward merit, no matter under what color of skin merit is found.

—Booker T. Washington, in *Up from Slavery*

MISSISSIPPI

BANNER

The novel *Losing Battles* (1970), by Eudora Welty, is set near the northern Mississippi hamlet of Banner, about twenty miles southeast of Oxford on Route 9 W. In the novel, the Vaughn family—

The Oaks in Tuskegee. Built in 1899 by the students of Tuskegee Institute, The Oaks reflects Booker T. Washington's belief that students should learn practical skills to advance their lives.

> *Due to an accident years ago, she had the mental age of a child of ten. But anyone on earth, meeting her for the first time, would have found this incredible. Mrs. Johnson had managed in many tactful ways to explain her daughter to young men without wounding them. She could even keep them from feeling too sorry for herself. "Every mother in some way wants a little girl who never grows up. Taken in that light, I do often feel fortunate. She is remarkably sweet, you see, and I find a great deal of satisfaction." She did not foresee any such necessity with an Italian out principally to sell everything for the gentlemens. No, he could not offer them anything else. No, he certainly could not pay the check. He had been very kind . . . very kind . . . yes, yes, very kind. . . .*
>
> —Elizabeth Spencer, in *The Light in the Piazza*

three generations of rural Mississippians—gathers for two days of reunion and story-telling:

> *Now there was family everywhere, front gallery, and back, tracking in and out of the company room, filling the bedrooms and kitchen, breasting the passage. The passageway itself was creaking; sometimes it swayed under the step and sometimes it seemed to tremble of itself, as the suspension bridge over the river at Banner had the reputation of doing. With chairs, beds, windowsills, steps, boxes, kegs, and buckets all taken up and little room left on the floor, they overflowed into the yard, and the men squatted down in the shade. Over in the pasture a baseball game had started up. The girls had the swing.*

BILOXI

This port city on the Gulf of Mexico has had several literary visitors. **F. Scott** and **Zelda Fitzgerald** in 1932 spent their "second honeymoon" at the "biggest hotel" in the city, probably the Edgewater Gulf Hotel, now razed for the Edgewater Shopping Plaza, which stands on Route 90. Father **Abram Joseph Ryan,** the poet, came here in October 1881 for reasons of health and lived at 1428 West Beach Boulevard, in a house known now as the Father Ryan House. **Thomas Wolfe** visited Biloxi in January 1937 to see a college friend, Garland Porter. Wolfe thoroughly enjoyed the fried chicken and biscuits prepared Southern style. When Porter expressed admiration for *Of Time and the River* (1935) and quoted relevant passages, Wolfe is said to have replied, "Damned if I don't believe you've read the book. Very few people have ever waded through it."

CARROLLTON

Elizabeth Spencer, author of a number of novels, among them *Fire in the Morning* (1948), *This Crooked Way* (1952), and *The Light in the Piazza* (1960), was born on July 19, 1921, at 1010 College Street, a private house that is unmarked. Valley Hill and Cemetery Road are Carrollton sites that figure in her books. Tarsus, the fictional town in *Fire in the Morning,* typifies a central Mississippi county seat like Carrollton. The Merrill Building Museum here exhibits Spencer's published works and family memorabilia. Carrollton is on Route 82, fifteen miles east of Greenwood.

COLUMBUS

Thomas Lanier Williams, who later took the name **Tennessee Williams** and went on to become one of America's foremost playwrights, twice winning a Pulitzer Prize, was born in Columbus on March 26, 1911, and was baptized at St. Paul's Episcopal Church. Williams lived for three years in Columbus, at the church rectory, home of his grandfather, Reverend Walter Dakin, at 318 Second Avenue. Williams adopted the name "Tennessee" when he was a young man: "the Williamses had fought the Indians for Tennessee and I had already discovered that the life of a young writer was going to be something similar to the defense of a stockade against a band of savages."

COMO

Stark Young, the poet, playwright, and critic, was born on October 11, 1881, on Oak Street in Como, which stands between Routes 310 and 51, northwest of Oxford. His birthplace is private, but a marker in the front yard identifies the house. A house in Young's novel *Heaven Trees* (1926) is believed to be the Oak Street house. Another novel by Young, *So Red the Rose* (1934), also set in Mississippi, has been recognized as outstanding in its portrayal of the South. Young died in Fairfield, Connecticut, but was buried in Friendship Cemetery, off Route 55, north of Como.

Rectory of St. Paul's Episcopal Church in Columbus, birthplace and childhood home of Tennessee Williams. The playwright, who visited here in the 1950s, had no clear memories of the three years he lived here but is said to have described St. Paul's, where he was baptized, as "one of the handsomest churches I have ever seen."

FALKNER

This village in northeastern Mississippi was named for William Faulkner's great-grandfather, Colonel **William C. Falkner,** a Confederate officer in the Civil War and author of *The White Rose of Memphis* (1880). Falkner stands on Route 15, north of Ripley, on the eastern edge of Holly Springs National Forest.

FOOTE

This village, about twenty-five miles south of Greenville, between Route 1 and Lake Washington, enters literary history through its association with the early life of **Shelby Foote,** the historian and novelist, whose grandfather had a home here. The thirty-room brick house, called Mt. Holly, appears in Foote's novel *Tournament* (1949) as Solitaire, and the character Hugh Bart in the same work is modeled on Foote's grandfather.

GREENVILLE

This city, which once stood on the bank of the Mississippi River and now—thanks to the U.S. Army Corps of Engineers—stands near Lake Ferguson, is rich in literary associations. **[William] Hodding Carter, Jr.,** came in 1936 to Greenville, where he was to achieve national reputation as publisher of the *Delta Democrat-Times* and, in 1946, win a Pulitzer Prize for his editorials advocating religious and racial tolerance. His newspaper, the result of several consolidations, is published today at 988 North Broadway, but was published until the 1960s at 201 Main Street, at the foot of the levee that protects Greenville from the waters of the Mississippi. This former address is reflected in the title of Carter's autobiography, *Where Main Street Meets the River* (1953). Carter did not confine himself to newspaper work. The many books he wrote in Greenville included two nonfiction works, *Southern Legacy* (1950) and *Gulf Coast Country* (1951), and the novels *The Winds of Fear* (1944) and *So the Heffners Left McComb* (1965), which dealt with the racial problems of his region. Carter died on April 4, 1972, at his home in Greenville and is buried in Greenville Cemetery, 1000 South Main Street.

David Cohn, who wrote *Where I Was Born and Raised* (1948) and other works, and who is said to have originated the phrase "the Mississippi Delta begins in the lobby of the Peabody Hotel in Memphis, and ends on Catfish Row in Vicksburg," was born on September 20, 1898, in Greenville and attended Greenville High School. He wrote an earlier version of *Where I Was Born and Raised,* which he called *God Shakes Creation* (1935), at the Greenville home of William Alexander Percy.

Shelby Foote, the historian and novelist, was born in Greenville on November 17, 1916, and lived at Allen Courts, 222A South Washington Avenue, apartments later remodeled and renumbered as 212 South Washington. Foote's works of history deal primarily with the South and especially with the Civil War, for example, *Shiloh* (1952). While attending Greenville High School, Foote worked at the *Delta Star,* a newspaper later incorporated into the *Delta Democrat-Times.* Foote left Greenville after high school, but wrote about the city in *Jordan County* (1954), a collection of stories set in Mississippi in a town called Bristol, which is taken to be Greenville.

William Alexander Percy, the lawyer and poet, was born in Greenville on May 14, 1885, and lived here until his death, except for the years he spent in university studies, foreign travel, and military service. Percy lived at 601 Percy Street, in a house that served as a literary and cultural center until Percy's death, and it was here—in a house that Hodding Carter once described as "never locked"—that he wrote his poetry, including the volumes *In April Once, and Other Poems* (1920) and *Selected Poems* (1930), and his autobiography, *Lanterns on the Levee* (1941). The Percy house was demolished in 1969, well after Percy's death on January 21, 1942, at King's Daughters

I saw one man come over, running sort of straddle-legged, and just as he cleared the rim I saw the front of his coat jump where the shots came through. He was running down the slope, stone dead already, the way a deer will do when it's shot after picking up speed. This man kept going for nearly fifty yards downhill before his legs stopped pumping and he crashed into the ground on his stomach. I could see his face as he ran, and there was no doubt about it, no doubt at all: he was dead and I could see it in his face.

—Shelby Foote, in *Shiloh*

Home of Sherwood Bonner in Holly Springs. The short-story writer died in this house on July 23, 1883.

The pit of death was giving up its secrets. The hoist was busy, and cage-load after cage-load came up, with bodies dead and bodies living and bodies only to be classified after machines had pumped air into them for a while. Hal stood in the rain and watched the crowd and thought that he had never witnessed a scene so compelling to pity and terror.

—Upton Sinclair,
in *King Coal*

Hospital here. The William Alexander Percy Memorial Library, 341 Main Street, maintains collections of his memorabilia and those of Hodding Carter. William Alexander Percy is buried in Greenville Cemetery.

Walker Percy, author of the novels *The Moviegoer* (1961) and *The Second Coming* (1980), spent his formative years in Greenville. Percy's father died when Percy was eleven years old, and his mother died two years later. Walker Percy's father had been first cousin to William Alexander Percy, who took the orphaned boy in and adopted him. In an article written for *Saturday Review* in 1973, Walker Percy described his adoptive father as "the most extraordinary man I have ever known." At the Percy Street home, Walker Percy met poets, politicians, preachers, and musicians, and he was introduced to music, poetry, and the works of Shakespeare. As a schoolboy, Percy was unsuccessful in selling his writing to magazines, but he had great success in selling his sonnets to classmates who needed them for their English classes. Price—50 cents a sonnet.

GULFPORT

Upton Sinclair wrote *King Coal* (1917) in 1915 while staying at Ashton Hall, the Gulfport summer home of his second wife, Marg Kimbrough Sinclair. The Confederate Inn Motel occupies the site today.

Birthplace of Eudora Welty in Jackson. The house has been considerably altered since Welty's day.

HERMANVILLE

Maxwell Bodenheim was born on May 26, 1893, in Hermanville, on Route 18 south of Vicksburg. Bodenheim went on to become a well-known but tragic literary figure in New York City. He wrote *Minna and Myself* (1918) and many other volumes of verse as well as several novels and plays. Bodenheim, who had a serious drinking problem, was murdered in New York City on February 6, 1954.

HOLLY SPRINGS

Katharine Sherwood Bonner McDowell, who wrote under the name **Sherwood Bonner,** was born on February 26, 1849, at her family's plantation, a mile and one-half northwest of Holly Springs, a town on Route 7 north of Oxford. She later lived in Holly Springs, in a house once known as the Bonner House and later called Cedarhurst, which stands at 411 Salem Street. During a stay in Boston in the early 1870s, Bonner worked as personal secretary to Henry Wadsworth Longfellow but soon began to produce her own sketches, which were collected in *Dialect Tales* (1883) and *Suwanee River Tales* (1884). She also wrote the novel *Like Unto Like* (1878), covering the Civil War and Reconstruction periods. Cedarhurst, a Gothic mansion that was built in 1858, is sometimes open to visitors. Bonner died on July 23, 1883, and was buried in a now-unmarked grave in the Bonner family plot in Hill Crest Cemetery in Holly Springs.

Joseph Holt Ingraham, a writer of historical romances, came to Holly Springs in September 1858 as rector of Christ Church and head of St. Thomas Hall, the diocesan boys' school. While here he wrote *The Pillar of Fire* (1859) and *The Throne of David* (1960), romances with religious themes. Ingraham died in Holly Springs in December 1860, when he accidentally shot himself while standing in the vestry of his church.

JACKSON

The capital of Mississippi holds considerable attraction for visitors interested in the history of the South and offers a special treat for the literary visitor who wishes to walk where a favorite author has walked and see some of the things that author has seen. Two writers associated with Jackson—much different in outlook and background but equally concerned with literary quality—have made national reputations. One of them, Eudora Welty, has lived most of her life in Jackson, and her stories and novels based on Mississippi people and places attract an ever-widening circle of admirers. The other, Richard Wright, spent a few of his formative years in Jackson and here—fiercely resentful toward the intolerable racial attitudes he saw about him—began to write during the few hours he could spare from schooling and odd jobs.

Eudora Welty was born to a socially prominent Jackson family on April 13, 1909, in a house at 741 North Congress Street. She attended Central High School, then located at 259 North West Street, and studied for two years at Jackson's Mississippi State College for Women, now Mississippi University for Women. She went on to further study at the University of Wisconsin and at the

Columbia University School of Business, in New York City. She wanted to remain in New York but was unable to find a job—it was the early 1930s, and the Great Depression was in full bloom. Welty returned to Jackson and, with her family's help, found work at a local radio station, WJDX. She worked later, from 1933 to 1936, as a publicity writer for the Jackson bureau of the WPA. Her job was to travel through the state to gather information on the lives of Mississippians and write articles for publication in various newspapers. This experience provided invaluable background for the stories she wrote after hours—stories that went unpublished until 1936, when one of them was accepted by *Manuscript,* a little magazine. With this story, "Death of a Traveling Salesman," Welty's literary career was launched. She soon began to publish regularly, her work appearing most notably in *The Southern Review.* After 1941, the year of publication of her first collection of stories, *A Curtain of Green,* Welty was able to work exclusively as a writer of fiction. *The Robber Bridegroom* (1942), Welty's first novel, was followed by a second collection of short stories, *The Wide Net* (1943). She continued to write stories and novels. For one novel, *The Optimist's Daughter* (1973), she was awarded a Pulitzer Prize. Readers now can savor the full flavor of Welty's pictures of Southern life in *The Collected Stories of Eudora Welty* (1980). The Mississippi Department of Archives and History in Jackson maintains the Eudora Welty Collection of papers and manuscripts.

The other prominent writer associated with Jackson is **Richard Wright,** a victim of poverty and racial injustice, who came to Jackson in 1920, when he was twelve years old, to live with his grandparents at 1107 Lynch Street, now the address of the College Inn. Wright stayed in Jackson only a short time, leaving in 1925 to go first to Memphis, then to Chicago and finally, after World War II, to Paris, France, where he lived the life of an expatriate author until his death in a hospital there on November 28, 1960. His ashes—along with those of a copy of *Black Boy*—are in Père Lachaise Cemetery in Paris. Wright's novel *Native Son* (1940) describes the violent life of a black in Chicago, and *Black Boy* (1945), his autobiography, reflects the experiences of a black child growing up in the deep South. While a student at the Smith-Robertson and Jim Hill schools in Jackson, Wright began to write. His first published story, "The Voodoo of Hell's Half-Acre," appeared in the *Southern Register,* a local newspaper published for blacks. Wright was then in the eighth grade. To support himself while he was in school, Wright worked at various jobs, including one as delivery boy for the W. J. Farley clothing store, another as ticket collector at the Alamo Theater. While working as a bellboy at the Edward House, he carried liquor to prostitutes who used the hotel, and it was this source of income, plus money he made by scalping theater tickets, that enabled him to save enough money to leave Jackson.

LELAND

William Alexander Percy often visited the Percy Place, his father's family's plantation on the west bank of Deer Creek here. His mother's family had a plantation called Camelia, which stood several miles upstream of the Percy Place. Leland is a few miles west of Greenville, where Percy made his home.

LUMBERTON

James [Howell] Street was born on October 15, 1903, in Lumberton, a city on Route 11, about fifty miles north of Gulfport. Street wrote about the South in several novels, including *Look Away!* (1936) and *Tomorrow We Reap* (1949). His works of nonfiction included *The Civil War* (1953) and *The Revolutionary War* (1954).

MACON

Ben Ames Williams, a popular novelist in his day, was born on March 7, 1889, in Macon, on Route 45, about thirty miles south of Columbus. His birthplace was his grandmother's home, which stood about a block away from the county courthouse until the house was destroyed by fire. Williams grew up in Jackson, Ohio, and spent most of his adult life in Maine, the principal locale for several of his novels.

NATCHEZ

Richard [Nathaniel] Wright, author of *Native Son* (1940), was born in the village of Roxie, about twenty miles east of Natchez, on September 4, 1908, and lived at 20 Woodlawn Street in Natchez from 1911 to 1913, when his family moved to Memphis, Tennessee. After some years there, Richard Wright moved to Jackson, Mississippi, to live with his grandparents.

Josephine Haxton, who writes under the name **Ellen Douglas,** was born in Natchez on July 12, 1921. Her mother had returned to her hometown from Arkansas to give birth. Douglas's works include the novels *A Family's Affairs* (1961), *Where the Dreams Cross* (1968), and *Apostles of Light* (1973). Natchez appeared in the novel *So Red the Rose* (1934), by Stark Young, and one of the sites described was The Towers, 801 Myrtle Avenue.

NEW ALBANY

This city, about forty miles east of Oxford, was the birthplace, on September 25, 1897, of the great American novelist, Nobel laureate **William [Cuthbert] Faulkner,** original name Falkner. He was born at 204 Cleveland Street, which now is the site of the parsonage of the New Albany Presbyterian Church. A bronze marker at the church lot identifies the site, and a plaque at the Union County Fair Grounds nearby identifies William Faulkner Park. Some readers of Faulkner see New Albany, along with Oxford, Ripley, and Holly Springs, as sites within Yoknapatawpha County, the most famous of all literary geographic subdivisions of the South.

OXFORD

The county seat of Lafayette County, Mississippi, is Oxford, and the county seat of Yoknapatawpha County, Mississippi, is Jefferson. Both are **William Faulkner** country. Visitors to Oxford may see for themselves the county courthouse that is the hub of Yoknapatawpha, the beautiful old

> *Ashamed, shrugging a little, and then shivering, he took his bags and went out. The cold of the air seemed to lift him bodily. The moon was in the sky.*
>
> *On the slope he began to run, he could not help it. Just as he reached the road, where his car seemed to sit in the moonlight like a boat, his heart began to give off tremendous explosions like a rifle, bang bang bang.*
>
> *He sank in fright onto the road, his bags falling about him. He felt as if all this had happened before. He covered his heart with both hands to keep anyone from hearing the noise it made.*
>
> *But nobody heard it.*
>
> —Eudora Welty, in "Death of a Traveling Salesman"

> *There was no day for him now, and there was no night; there was but a long stretch of time, a long stretch of time that was very short; and then—the end. Toward no one in the world did he feel any fear now, for he knew that fear was useless; and toward no one in the world did he feel any hate now, for he knew that hate would not help him.*
>
> *Though they carried him from one police station to another, though they threatened him, persuaded him, bullied him, and stormed at him, he steadfastly refused to speak. Most of the time he sat with bowed head, staring at the floor; or he lay full length upon his stomach, his face buried in the crook of an elbow, just as he lay now upon a cot with the pale yellow sunshine of a February sky falling obliquely upon him through the cold steel bars of the Eleventh Street Police Station.*
>
> —Richard Wright, in *Native Son*

William Faulkner receiving the Nobel Prize from King Gustav VI of Sweden on December 10, 1950. In his acceptance speech Faulkner referred to his work as an attempt at making "out of the material of the human spirit something which was not there before."

homes, and the streets of this Southern city where Faulkner is part of a living legend. Readers may also stay at home and participate in the life of Yoknapatawpha by tracing the lives and motivations of Faulkner's fictional Sartoris family and of the Compsons, Benbows, McCaslins, and Snopeses through fourteen novels: *Sartoris* (1929), *The Sound and the Fury* (1929), *As I Lay Dying* (1930), *Sanctuary* (1931), *Light in August* (1931), *Absalom, Absalom!* (1936), *The Unvanquished* (1938), *The Hamlet* (1940), *Intruder in the Dust* (1948), *Knight's Gambit* (1949), *Requiem for a Nun* (1951), *The Town* (1957), *The Mansion* (1960), and *The Reivers* (1962), and through the collection of stories called *Go Down, Moses* (1942).

But the literary visitor to Oxford itself does not go unrewarded. Oxford has become one of the country's principal literary shrines, if not the country's leading literary growth industry. The University of Mississippi periodically plays host to Faulkner scholars, there is an annual Oxford Pilgrimage for tourists who wish to see the city's fine homes, and local stores trade on their association with Faulkner's life in Oxford. Faulkner lived here for most of his life after reaching the age of four, first in a house on what now is called South Lamar Street and later, after 1930, in the house at 719 Garfield Avenue that Faulkner called Rowan Oak. The name came from Scottish legend, which ascribes to the rowan tree the property of being able to ward off evil spirits and afford refuge and privacy to those seeking its protection. Rowan Oak stands off Old Taylor Road, in a grove of tall trees on a thirty-one-acre tract. It now is owned

Rowan Oak, William Faulkner's home in Oxford. Faulkner wrote many of his books in this spacious antebellum mansion.

by the University of Mississippi, whose campus is adjacent to the property. When the university is in session, Ole Miss graduate students conduct tours of Rowan Oak for visitors. One can see Faulkner's study, with its Underwood portable typewriter, an ashtray made from an artillery shell casing, and a bottle of horse liniment—all just as Faulkner left them. But the most interesting feature of the study is the outline for *A Fable*, written in Faulkner's own hand. For *A Fable* (1954), Faulkner received a Pulitzer Prize. Outside the house are the garden that Faulkner designed and the smokehouse in which he cured bacon, ham, and sausage. Throughout the city of Oxford, the visitor can see stores frequented by Faulkner, and can meet people who knew and will talk about him.

Faulkner attended the University of Mississippi in 1919–20 and worked on *The Mississippian*, the student newspaper. He wrote poetry as well—his collection *The Marble Faun* was published in 1924—before going off to New York City to work in a bookstore and then, after another stay in Oxford, on to New Orleans. It was during his stay in New Orleans that Faulkner met Sherwood Anderson, who encouraged Faulkner in his writing, helping him to arrange publication of his first novel, *Soldier's Pay* (1926).

The Mississippi Collection of the University of Mississippi Library includes much Faulkner material, Faulkner's Nobel Prize, and a portrait of the novelist. Faulkner's work did not sell well during much of his lifetime—in 1944, for example, only one of the nine books he had produced by then was still in print—and he was chronically in debt. After Faulkner died on July 6, 1962, in an Oxford hospital, a copy of his unpublished one-act play *Marionettes*, written while Faulkner was a student at the University of Mississippi, brought $34,000 when it was sold at auction. Faulkner is buried in St. Peter's Cemetery, Jefferson Avenue and North Sixteenth Street.

Faulkner's younger brother, **John Faulkner**, also spent most of his life in Oxford. He came here with his family in 1902, when he was one year old. He was known—by no means as well as his brother—for his novels and short stories, and for his painting. His works included the novels *Men Working* (1941) and *Dollar Cotton* (1942), and his last book was a memoir, *My Brother Bill*, published in 1963. John Faulkner lived in Memory House, at 406 University Avenue. Memory House is open to visitors during the Oxford Pilgrimage, which is held each year late in the month of April. John Faulkner died in Memphis, Tennessee, on March 28, 1963, and was buried in St. Peter's Cemetery.

Another writer associated with Oxford was **Stark Young**, the playwright, novelist, and longtime drama critic for *The New Republic*, who lived at 418 University Avenue and returned to Oxford every summer for many years to visit members of his family. The house he lived in is now the residence of the pastor of First Presbyterian Church. Young is believed to have written *So Red the Rose* (1934), his novel about the Civil War, while staying at Cedar Hill Farm, which stood about three miles outside Oxford. A character in *So Red the Rose* hid from Federal raiders at Shadow Lawn, a mansion still standing at Eleventh and Fillmore streets.

PASCAGOULA

William Faulkner, while visiting friends during 1926 in this city in the southeast corner of Mississippi, began work on *Mosquitoes* (1927), a satirical novel set in New Orleans, and completed work on *Mayday*, an allegorical novella. The latter work was never published, but Faulkner printed it himself in a manuscript volume and presented it to a friend. Faulkner also used "Mayday" as a working title for his first novel, *Soldier's Pay*, which was published in 1926. While in Pascagoula, Faulkner spent much of his spare time with neighborhood children, including a girl named Helen Baird, to whom he dedicated *Mosquitoes*.

The Longfellow House, a resort on Beach Drive in Pascagoula, is said to be the place where **Henry Wadsworth Longfellow** wrote "The Building of the Ship" in 1850.

PORT GIBSON

Irwin Russell, a poet who wrote in black dialect, was born on June 3, 1853, at 411 Jackson Street in this small city south of Vicksburg. Russell traveled extensively in Texas and through the Mississippi River valley. He died in 1879 in Port Gibson, of yellow fever. His work was published, with the assistance of Joel Chandler Harris, in the volume *Poems by Irwin Russell* (1888), and reissued in enlarged form in 1917 as *Christmas-Night in the Quarters*, which took its title from Russell's best-known poem. The Irwin Russell Memorial at Southeast College and Coffee streets, now the site of Port Gibson's city hall, displays Russell's manuscripts and memorabilia.

RIPLEY

John [Wesley Thompson] Faulkner, the novelist and younger brother of William Faulkner, was born on September 24, 1961, in Ripley, a town on Route 15, northeast of Oxford. Faulkner lived here for only a year, but his great-grandfather, Col. **William Clark Falkner**, himself an author and the prototype of Col. Sartoris in William Faulkner's novels, lived in the 300 block of North Main Street in Ripley from 1840 until his death.

But above all, the courthouse the center, the focus, the hub; sitting looming in the center of the county's circumference like a single cloud in its ring of horizon: laying its vast shadow to the uttermost rim of the horizon; musing, brooding, symbolic and ponderable, tall as cloud, solid as rock, repository and guardian of the aspirations and hopes.

—William Faulkner,
in *Requiem for a Nun*

Sometimes it seemed like Bill used up all the words he knew in one sentence. It was said that he didn't use enough periods. One day Bill said that the next book he published was going to have a full page of periods inserted in the back with a note that if anyone felt Bill had used too few periods they were free to take as many as they wished from the extra page and put in their own.

—John Faulkner,
in *My Brother Bill*

Home of Stark Young in Oxford. Young taught at the University of Mississippi from 1904 to 1907, when he left the state. Young summered in Oxford for many years until his death in 1963.

Colonel Falkner was shot to death in November 1889 in the courthouse square, by a man with whom he had been feuding. The colonel's house was torn down in 1937, but a statue of him marks his grave in the family burial plot in Ripley Cemetery. Falkner's work included *The White Rose of Memphis* (1880), a tremendous best-seller in its day, and *Rapid Ramblings in Europe* (1884). William Faulkner, in *Sartoris* (1929), supplied an accurate description of his great-grandfather's burial monument when he wrote of Colonel Sartoris's burial monument:

> *He stood on a stone pedestal, in his frock coat and bareheaded, one leg slightly advanced and one hand resting lightly on the stone pylon beside him. His head was lifted a little in that gesture of haughty pride which repeated itself generation after generation with a fateful fidelity, his back to the world and his carven eyes gazing out across the valley where his railroad ran, and the blue changeless hills beyond, and beyond that, the ramparts of infinity itself.*

ROXIE

Richard Wright, author of *Native Son* (1940), was born on September 4, 1908, on a farm in Roxie, Route 84, about twenty miles east of Natchez. Roxie, then a village of two hundred, has grown somewhat and now has a population of six hundred. Wright's parents farmed here until the work proved too much for them. His mother took the three-year-old Wright and her other children to live in Natchez.

LOUISIANA

ALEXANDRIA

Arna Bontemps was born in this central Louisiana city, northwest of Baton Rouge, on October 13, 1902. His novels about blacks included *Black Thunder* (1936), which described a slave revolt in Virginia in 1800, and *God Sends Sunday* (1931), about a jockey. The latter novel was dramatized by Bontemps and Countée Cullen as *St. Louis Woman* (1946). John Clellon Holmes described Alexandria in *Get Home Free* (1964), his novel about the marriage of a bohemian couple.

BATON ROUGE

This city, the capital of Louisiana, has many literary associations. **Robert Lowell** and his wife **Jean Stafford** lived at 1106 Chimes Street in 1941, when Lowell taught at Louisiana State University. Stafford did secretarial work for *Southern Review,* a literary quarterly published by the Louisiana State University Press between July 1935 and April 1942. Among the notable contributors to *Southern Review* were **W. H. Auden, Katherine Anne Porter, John Crowe Ransom, Wallace Stevens, Eudora Welty,** and **Yvor Winters,** and two of its editors were **Cleanth Brooks** and **Robert Penn Warren.** Warren lived at Park Drive (Southdowns), RD3 from 1936 to 1939, when he moved to Hammond Road. Warren's character Willie Stark in the novel *All the King's Men* (1946), which won a Pulitzer Prize, calls to mind Huey P. Long, governor of Louisiana from 1928 to 1931.

Gene Stratton-Porter, the novelist and nature writer, lived for a time in Baton Rouge, and **Katherine Anne Porter** lived here in the late 1930s for a time, at 1050 Government Street. **Lyle Chambers Saxon** was born here on September 4, 1891. He was known as a schoolboy prankster who delighted in riding through the city streets on a one-eyed white pony. In 1912, after Saxon was graduated from Louisiana State University, he reviewed vaudeville shows for the Baton Rouge *State Times,* 329 Florida Boulevard, then and now the city's major evening newspaper. Among Saxon's books on Louisiana were *Father Mississippi* (1927) and *Old Louisiana* (1929). Saxon is buried in Magnolia Cemetery, 300–20 Nineteenth Street. **Mark Twain** visited Baton Rouge briefly in the 1880s.

CLOUTIERVILLE

Kate [O'Flaherty] Chopin, novelist and short-story writer, in 1880 came with her six children and her husband to live at his Cloutierville plantation. Here she wrote the stories "Bayou Folk" and "A Night in Acadie." After her husband died in 1882, Chopin supervised the cotton plantation for two years before returning to her native St. Louis. The Kate Chopin Home, now restored as the Bayou Folk Museum, is open to the public. Admission is charged. Two generations earlier, **Harriet Beecher Stowe** visited Robert McAlpin, a relative, at the Old Chopin Plantation, later known as Little Eva Plantation, in the Cane River Valley, southwest of Cloutierville on Route 1. It is believed that McAlpin was the prototype for Simon Legree in *Uncle Tom's Cabin* (1852).

GRAND ISLE

Kate Chopin used this island, off the southern coast of Louisiana, as a setting for her masterpiece, the novel *The Awakening* (1899). In 1884 **Lafcadio Hearn** visited Grand Isle, staying at Krantz's Hotel and beginning work on his novel *Chita: A Memory of Last Island* (1889). In *Chita* Hearn described the great tidal wave of 1856, which killed hundreds of people vacationing on Isle Dernière, off the Louisiana coast.

HAMMOND

William Hodding Carter, Jr. was born here on February 3, 1907, on Happy Woods Road, the last house on the right. The house is not open to visitors. Carter was brought up on a 300-acre farm north of Hammond. Many of his early experiences, such as seeing the body of a black woman hanging from a tree, are recounted in his autobiographical works, *Where Main Street Meets the River* (1953) and *First Person Rural* (1963). In 1932 Carter bought a broken-down press and began publishing the Hammond *Daily Courier,* actively opposing the Huey Long political machine. Fol-

> *There was the same scene every Saturday at Foché's! A scene to have aroused the guardians of the peace in a locality where such commodities abound. And all on account of the mammoth pot of gumbo that bubbled, bubbled, bubbled out in the open air. Foché in shirt-sleeves, fat, red and enraged, swore and reviled, and stormed at old black Douté for her extravagance. He called her every kind of a name of every animal that suggested itself to his lurid imagination. And every fresh invective that he fired at her she hurled it back at him while into the pot went the chickens and the panfuls of minced ham, and the fistfuls of onion and sage and piment rouge and piment vert. If he wanted her to cook for pigs he had only to say so. She knew how to cook for pigs and she knew how to cook for people of les Avoyelles.*
>
> —Kate Chopin,
> in "A Night in Acadie"

Funerary monument to Colonel William C. Falkner in Ripley Cemetery. Falkner's novel *The White Rose of Memphis* sold 160,000 copies in about twenty-five years.

lowing Huey Long's death in 1935, Carter sold the newspaper and ran unsuccessfully for a seat in the Louisiana legislature.

LACOMBE

François Dominique Rouquette, born on January 2, 1810, in Lacombe, which is north of New Orleans, across Lake Pontchartrain and in the heart of Choctaw country, often extolled the scenic bayous of Louisiana in his poetry. François and his brother Adrien, who became a Catholic priest, spent much of their youth among the Indians before going to France for their education. Adrien, who established several Indian missions in St. Timothy Parish, was revered by the Choctaw. The brothers wrote a number of books, chiefly in French—both were well known in French literary circles. François was the author of *Les Meschacébéennes* (1839) and *Fleurs d'Amérique* (1856), volumes of poetry. Visitors can find the Rouquette Monument, a russet granite memorial with a stone cross, in the old cemetery off Route 190.

MARKSVILLE

Ruth McEnery was born in Marksville, which is southeast of Alexandria, on May 21, 1849. Her works included *A Golden Wedding and Other Tales* (1893) and *In Simpkinsville: Character Tales* (1897).

METAIRIE

Shirley Ann Grau wrote her novel *The House on Coliseum Street* (1961), describing the loveless experiences of a young girl, in the town of Metairie, near New Orleans.

NATCHITOCHES

Natchitoches, founded in 1714, is the oldest town in Louisiana and has splendid examples of plantation architecture. While **Lyle Chambers Saxon** lived near here in a house called Yucca, on Melrose Plantation, he wrote several books, including his novel *Children of Strangers* (1937), set on what Saxon called Yucca Plantation and dealing with the lives of the plantation mulattoes, descendants of the French settlers. Yucca itself, which was used as a hospital during the Civil War and as a home for indigent former slaves after the war, was built about 1795 by Louis Metoyer, a freeman who was the son of a slave mother and a French father. Melrose Plantation was owned in the 1930s by Mrs. Cammie Garrett Henry, a woman who was interested in the arts. It now is owned by a Natchitoches historical association. Melrose comprises six acres of land and eight buildings, of which only Yucca and a small cabin are fully restored, but all the buildings are open to visitors on Saturday and Sunday afternoons in summer, at other times by appointment. Some of the other writers who stayed at Yucca when Mrs. Henry owned Melrose were **Roark Bradford, Gwen Bristow, Rachel Field,** and **Rose Franken,** all of whom did some of their writing there.

NEW IBERIA

West of the city of New Iberia and off Route 14 is Bob Acres Plantation, whose spacious, confortable white frame main house was the home in 1870 of Joseph Jefferson, the actor famous for his portrayal of Rip Van Winkle, Washington Irving's character. Several plaques mark the property, but only a few of the groves of trees and extensive fruit orchards remain. The property, a National Historic Landmark, has been private but is expected to be open to the public in the future.

NEW ORLEANS

New Orleans, the literary capital of Louisiana, has been home and working place for so many

> *Fog covered Cane River. The plantation was drowned in a milky cloud that lay along the face of the earth. A negro man in faded blue overalls stood on the river bank holding a silver watch between his forefinger and thumb, watching the last few minutes of sleeping-time as they ticked away. When the hands of the watch pointed to five o'clock, he sighed and began to ring the plantation bell. Only the rope was visible to him, a rope stretching up into the mist, and it seemed that the voice of the bell should be muffled in the fog; but the sound was silver clear as it rolled out over the invisible fields.*
>
> —Lyle Saxon, in *Children of Strangers*

Home of Kate Chopin in Cloutierville, built in the early 1800s. The marker in front identifies the house today as the Bayou Folk Museum.

Artist's 1907 pencil sketch of the home of Thomas Bailey Aldrich in New Orleans. Aldrich lived here for a time in the 1840s.

prominent authors that, in the French Quarter particularly, a visitor can scarcely avoid coming upon a literary site. All the established restaurants in the Quarter—particularly Antoine's, Arnaud's, Galatoire's, and the venerable Morning Call Coffee Shop—have fed such important writers as **Sherwood Anderson, William Faulkner, Eudora Welty, Tennessee Williams,** and **Thomas Wolfe.** *Double Dealer Magazine,* at 204 Common Street, which began publishing here in 1921, included Anderson, Faulkner, Ernest Hemingway, and Thornton Wilder among its contributors.

The richness of literary life in New Orleans is so great that it may be best to begin with those authors for whom New Orleans was their native city. **George Washington Cable** was born here on October 12, 1844. When Cable was fourteen, his father died and young Cable became the support of his family. His first attempt at writing professionally was a submittal for the "Drop Shot" column in the New Orleans *Picayune.* He became a reporter for the newspaper in 1869 and was asked to do drama criticism, a task forbidden by his religious views. He left and became an accountant and correspondence clerk for A. C. Black and Company, a firm of cotton factors. Cable married Louise S. Bartlett on December 7, 1869, and began writing stories in his spare time. When Black and Company was dissolved, Cable resumed his career as a writer, using New Orleans sites often in his work. For example, Mme. Délicieuse, a character in his story "Sieur George" (1871), lived at 253 Royal Street. The protagonist of his novel *Madame Delphine* (1881) lived at 294 Barracks Street, near Royal Street, and at the LeMonnier House, 640 Royal Street, known also as Sieur George's House, considered a skyscraper in 1876 because it was a four-story building. Cable and his wife lived at 1313 Eighth Street, in a house that was occupied in 1884–85 by **Joaquin Miller,** the poet, then on a newspaper assignment in New Orleans to cover the Cotton Exposition.

George Washington Cable, photographed in the 1880s.

In a much later time, **Truman Capote** was born in New Orleans, on September 30, 1924, and it was here he wrote "My Side of the Matter" for *Story* magazine in 1941 and sections of his first published novel, *Other Voices, Other Rooms* (1947), which became an enormous success. **Shirley Ann Grau,** born here on July 8, 1930, lived at 921 Chartres Street, attending Ursuline Academy at 2635 State Street. Before 1955, when she was graduated from Newcomb College, Tulane University, Grau began to write her books. The first was *The Black Prince and Other Stories* (1955). Her novel *The House on Coliseum Street* (1961) is set in New Orleans. Grau won a Pulitzer Prize for her novel *The Keepers of the House* (1964), which deals with the troubled lives of a Southern family.

Lillian Hellman, the playwright, was born in New Orleans on June 20, 1905, at 1718 Prytania Street. Hellman recalled her life in New Orleans in her memoir *An Unfinished Woman* (1969). She moved to New York City when she was five years old, but attended New Orleans schools and often returned for long visits in later years. Her successful play *Toys in the Attic* (1960) is set here. **Harnett [Thomas] Kane** was born here on November 8, 1910, and lived at 5919 Freret Street. Kane was editor of the Warren Easton High School newspaper, *Old Gold and Purple,* and was a member of the school's debating team. At Tulane University Kane was editor of *Hullabaloo,* the student newspaper, in addition to working afternoons for the New Orleans *Item.* The Louisiana political scandals of the 1930s furnished material for his *Louisiana Hayride: American Rehearsal for Dictatorship* (1941), a well-received account of the political career of Huey Long.

Moses Koenigsberg, who founded the Newspaper Feature Syndicate in 1913 and King Features Syndicate in 1916, was born in New Orleans on April 16, 1878. **Miriam Leslie,** the editor and feminist, was born here in 1836. Her works included *Rents in Our Robes* (1888) and *Are Men Gay Deceivers?* (1893). Joaquin Miller, who met Leslie when she traveled to California, modeled the heroine of his novel *The One Fair Woman* (1876) on Leslie.

Adrien Emmanuel Rouquette was born here on February 13, 1813. Rouquette usually wrote in French and in Creole dialects, but his best-known work, a volume of verse entitled *Wild Flowers* (1848), was written in English. Father Rouquette was the first native-born Roman Catholic priest to be ordained after Louisiana was purchased by the United States. He died in New Orleans on July 15, 1887. **Robert Tallant,** the folklorist, was born here on April 20, 1909, and was graduated from Warren Easton High School. After spending many years in other work, Tallant decided to become a writer and worked in the early 1940s for the Louisiana Writers' Project, under Lyle Chambers Saxon. Tallant, who lived for a time at 3324 Carondelet Street, wrote *Voodoo in New Orleans* (1946) and other works of folklore that had New Orleans as their settings, and the novel *Mrs. Candy and Saturday Night* (1947), also set in New Orleans.

The literary associations of New Orleans go well beyond a recital of authors who were born here. **Thomas Bailey Aldrich,** who lived here as a child, set part of *The Story of a Bad Boy* (1870) in New Orleans. Hervey Allen, in writing *Anthony Adverse* (1933), relied heavily on the journal of a European businessman, Vincent Nolte, who stayed for a time at 708 Toulouse Street. The building at that address appears also in the novel *The Crossing* (1904) by the American author Winston Churchill. **Sherwood Anderson** married Elizabeth, his third wife, in 1923; they lived at first at 715 Governor Nicholls Street, moving later to the third floor of the old Pontalba Building on Jackson Square. Anderson's memorable novel *Dark Laughter* (1925) is set partially in New Orleans. **Ambrose Bierce** often journeyed to New Orleans from Selma, Alabama, to visit his former Army commanding officer, who lived here.

Roark [Whitney Wickliffe] Bradford, author of *Ol' Man Adam an' His Chillun* (1928), which Marc Connelly adapted as the enormously successful play *The Green Pastures* (1930), lived in the old Pontalba Building while working as night editor and, later, as Sunday editor of the New Orleans *Times-Picayune.* On November 13, 1948, Bradford died in New Orleans of a disease contracted in French West Africa. **William Hodding Carter, Jr.** worked as a reporter for the New Orleans *Item-Tribune* from 1929 until he became night manager for the United Press in 1930. **Kate Chopin** in her novel *The Awakening* (1899) used Esplanade Street as a setting for the home of her protagonist. In 1895, **Stephen Crane** arrived in New Orleans, spending one night at a boarding house before registering at the Hotel Royal, where he stayed until Mardi Gras. New Orleans is one of the settings in *Big River, Big Man* (1959) and *Virgo Descending* (1961), novels by **Thomas Duncan.** The novelist often takes meals at The Famous Door, 339 Bourbon Street, and at Arnaud's, 813 Bienville Street.

A frequent visitor to New Orleans was Colonel **William C. Falkner,** author of the best seller *The White Rose of Memphis* (1880). Falkner, a great-grandfather of William Faulkner, had joined the Tippah Guards, who assembled at Vicksburg and New Orleans for service in the Mexican War. Falkner became the prototype for Col. Sartoris, founder of the Sartoris family of Yoknapatawpha County in Faulkner's novels. **William Faulkner** himself came to New Orleans in about 1925 to write his first novel, *Soldier's Pay* (1926), with guidance from Sherwood Anderson. *Double Dealer Magazine* had carried Faulkner's first published poem in 1922. While in the city, Faulkner lived first at 626 Pirate's Alley, in 1924–25, and later at the LeMonnier House. **F. Scott Fitzgerald,** read galley proofs for *This Side of Paradise* (1920) while living in a boarding house at 2900 Prytania Street. Fitzgerald also began to write a novel to be called *The Demon Lover,* but abandoned the

After my loaf of bread, I went looking for a bottle of soda pop and discovered, for the first time, the whorehouse section around Bourbon Street. The women were ranged in the doorways of the cribs, making the first early evening offers to sailors, who were the only men in the streets. I wanted to stick around and see how things like that worked, but the second or third time I circled the block, one of the girls called out to me. I couldn't understand the words, but the voice was angry enough to make me run toward the French Market.

—Lillian Hellman,
in *An Unfinished Woman*

(*left*) The LeMonnier House in New Orleans. While living here, William Faulkner enjoyed the companionship of Sherwood Anderson, Roark Bradford, Lyle Saxon, and other writers. F. Scott Fitzgerald lived at 2900 Prytania Street (*below*) in January and February of 1920.

Travelling south from New Orleans to the Islands, you pass through a strange land into a strange sea, by various winding waterways. You can journey to the Gulf by lugger if you please; but the trip may be made much more rapidly and agreeably on some one of those light, narrow steamers, built especially for bayou-travel, which usually receive passengers at a point not far from the foot of old Saint-Louis Street, hard by the sugar-landing, where there is ever a pushing and flocking of steam-craft—all striving for place to rest their white breasts against the levée, side by side,—like great weary swans.

—Lafcadio Hearn, in *Chita*

project after many false starts and interruptions in order to turn to writing more readily realizable and salable short stories. "May Day," published in *The Smart Set,* may have found its inspiration in Fitzgerald's abortive novel-writing experience in New Orleans.

Stephen Foster and his wife, on Foster's only trip South, came to this city aboard a ship for the 1852 Mardi Gras. The trip is said to have inspired Foster to write "Massa's in de Cold, Cold Ground" (1852).

A woman whose pen name, **Dorothy Dix,** became a household name, **Elizabeth Meriwether Gilmer,** lived here at 6334 Prytania Street. Mrs. Gilmer, whose husband was an invalid for more than thirty years, supported both of them by working for the New Orleans *Picayune,* at first earning $5 a week as a women's-page columnist—her first piece was called "Sunday Salad"—but later writing a column of advice for the lovelorn. The column, soon syndicated, was reported at one time to have a readership of more than sixty million. From her columns, in which she probed the hearts of her readers, Dorothy Dix was able to create widely read volumes, such as *How to Win and Hold a Husband* (1939).

Another writer whose career received a boost in New Orleans was **Joel Chandler Harris,** creator of Uncle Remus, who found work during his teen years as secretary to the editor of *The Crescent Monthly,* then published at 82 Baronne Street. Harris met Mark Twain and Lafcadio Hearn while working in New Orleans. **Lafcadio Hearn** lived in many places in New Orleans, among them 228 (now 813) Baronne Street; 105 (now 516) Bourbon Street; 68 Gasquet (now Cleveland) Street; 39 and 1458 Constance Street; and 278 Canal Street. At one time Hearn invested money in the Five Cent Restaurant—sometimes called the Hard Times Restaurant—at 160 Dryades Street. He described his first impressions of the city in the Cincinnati *Commercial,* using the pen name **Ozias Midwinter.** In the late 1870s he revitalized the colorless *Daily City Item,* at 39 Natchez Alley, and then was hired by the *Times Democrat.* Hearn was a book collector, and his favorite New Orleans sources were Fournier's Book Shop, on Royal Street near Toulouse Street, and Armand Hawkin's Book Store, at 196½ Canal Street. Hearn completed his novel *Chita: A Memory of Last Island* (1889) while living in New Orleans.

Marquis James, who won a Pulitzer Prize for *The Raven: A Biography of Sam Houston* (1929), worked as a newspaperman in New Orleans. He met and married Bessie Rowland while they both worked for the *Item.* The New Orleans *Picayune,* which appears so often in the lives of the city's literary figures, had its beginning in 1837, when **G. W. Kendall** and **Francis A. Lumsden,** both printers, launched a four-page sheet called *The Picayune.* This modest newspaper became the forerunner of modern newspapers in one sense by devoting space to war reporting. **Frances Parkinson Keyes** in 1944 bought the Beauregard house, 1113 Chartres Street, as a winter home, and there worked on *Dinner at Antoine's* (1948), a successful murder mystery.

Oliver La Farge is another author who contributed to New Orleans' considerable stake in American literary history. La Farge completed *Laughing Boy* (1929) here. This novel about life among the Navajo Indians won a Pulitzer Prize in the following year. **Sinclair Lewis** and his wife Grace stayed on Rampart Street for a few days in 1917; much later, in 1939, Lewis and Marcella Powers rented an apartment for a few months at 1536 Nashville Avenue. Here Lewis wrote his story "Carry Your Own Suitcase," which appeared in *This Week* on February 25, 1939.

Stephen Longstreet is yet another literary figure associated with New Orleans. He wrote of the city in *The Real Jazz Old and New* (1956),

(*below*) Old photograph of house at 516 Bourbon Street, Lafcadio Hearn's residence for a time in the 1880s. It stood almost across the street from the French Opera House, which Hearn visited frequently. (*right*) 622 St. Ann Street. While staying here in December 1946 Malcolm Lowry spent three days checking page proofs for *Under the Volcano.*

one of his historical books. Longstreet enjoyed relaxing at Commanders Palace, a restaurant at 1403 Washington Avenue. Before **Malcolm Lowry** and his wife left for Haiti in December 1946, they stayed at a boarding house at 622 St. Ann Street, off Chartres Street. The novelist and his wife worked here on page proofs for Lowry's masterpiece, *Under the Volcano* (1947). **William March,** whose real name was **William Edward March Campbell** and whose last novel was *The Bad Seed* (1954), the macabre story of an eight-year-old girl guilty of murder, died at his home here on May 15, 1954. *The Bad Seed* was dramatized by Maxwell Anderson in 1955.

William Sydney Porter—O. Henry—went into hiding on Bienville Street in 1896 before fleeing to Honduras to avoid indictment for embezzlement in Austin, Texas. While the length of his stay in New Orleans is not known, Porter was able to take temporary jobs on the New Orleans *Delta* and the *Picayune.* He also managed to assimilate some of the Creole dialect he used in his New Orleans stories, which can be read in *Roads of Destiny* (1909). The *Delta* has a further claim to literary fame: it was the first newspaper to print "Maryland, My Maryland" (April 26, 1861), the lyric by **James Ryder Randall.**

[Thomas] Mayne Reid, Irish-born author of adventure tales such as *Afloat in the Forest* (1866), arrived in the port of New Orleans in 1840. Father **Abram Joseph Ryan,** although living in Mobile, Alabama, was editor-in-chief of the *Star,* a Catholic weekly then published in New Orleans. **Lyle Chambers Saxon** arrived in New Orleans in the 1930s, staying first at the St. Charles Hotel, now at 2203 St. Charles Avenue. He later bought a charming three-story house with a patio at 536 Royal Street. Still later he restored a home at 432-34 Madison Street, which still stands. Saxon served as editor of *The New Orleans City Guide* (1938) and of the *Louisiana State Guide* (1941) and wrote feature articles for the *Item* and *Times-Picayune.* After Saxon's death on April 9, 1946, at Baptist Hospital, 2700 Napoleon Avenue, his books, pictures, and manuscripts were donated to the Tulane University Library to form the Lyle Saxon Memorial Collection.

Harold Sinclair, a poet and musician, has written of the city in *Port of New Orleans* (1942) and *Music Out of Dixie* (1952). **Charles Testut,** a New Orleans physician and writer, wrote *Portraits Littéraires de la Nouvelle Orléans* (1850), in which he discussed the city's leading writers of his day. **William Makepeace Thackeray,** the English novelist, greatly admired nineteenth-century New Orleans. During his trip to America, Thackeray stopped at the St. Charles Hotel, where he is said to have been mistaken for a Kentucky farmer because he wore a broad-brimmed hat.

Edward Larocque Tinker was introduced to the charms of New Orleans as a result of his marriage to Frances McKee, a native of the city. He wrote of the city in several books, including *Gombo: The Creole Dialect of Louisiana* (1934), *Creole City* (1953), and four novellas collected as *Old New Orleans* (1930), the last-named in collaboration with his wife. Another writer who described the city was **Mary Ashley Townsend.** Using the pen name **Xariffa,** among others, she often contributed articles to the New Orleans *Delta Crescent* and to the *Picayune. Xariffa's Poems* (1870) and

Architect's drawing of 722 Toulouse Street, where Tennessee Williams lived in 1940.

the novel *The Brother Clerks* (1857) were among Townsend's works. **Frances Trollope,** another English novelist who visited this city, arrived after a seven-week sea voyage to rest for a few days before continuing on to Ohio. In *Domestic Manners of the Americans* (1832), Trollope attacked slavery and other practices of American life.

Mark Twain, still calling himself Samuel Clemens, and hoping to travel to the Amazon, left Cincinnati in 1857 aboard the riverboat *Paul Jones* as an assistant to the pilot, Horace Bixby. Bixby was immortalized in Twain's *Life on the Mississippi* (1874), a book that gives the modern reader a sense of New Orleans in Twain's time. It was during this period that another riverboat pilot, Isaiah Sellers, a man of literary bent, was sending reports to the *Picayune,* under the name Mark Twain. After Sellers learned that Clemens had parodied Sellers's articles for the New Orleans *True Delta,* the pilot, deeply hurt, never wrote for the newspapers again. After Sellers's death, Clemens regretted his plagiarism but took the name Mark Twain for himself.

Yet another literary immortal was a resident for a time in New Orleans. **Walt Whitman,** with his brother, took quarters at the Fremont Hotel, across the street from the St. Charles Hotel, on February 25, 1848. Whitman was employed to write editorials for the New Orleans *Crescent,* but quit after a few months because of differences with the owners of the newspaper. The *Picayune* later invited Whitman to submit reminiscences from the diary he kept while here.

Tennessee Williams, whose play *A Streetcar Named Desire* (1947) not only won a Pulitzer prize for Williams but turned public transportation in New Orleans into a tourist attraction, arrived in New Orleans in September 1940. While here, he completed work on *I Rise in Flame, Cried the Phoenix* (1951) and continued work on *Stairs to the Roof,* a play that was produced only once, in 1947. Williams lived at 722 Toulouse Street and supported himself by working as cashier at a local restaurant. When he returned to New Orleans much later, in 1946, he stayed at the Hotel Pontchartrain, now located at 2031 St. Charles Street. After a short while, he moved to a furnished apartment on Orleans Street and, in 1947, took an apartment near the corner of St. Peter and Royal streets. At one time Williams also leased a small

The old French part of New Orleans—anciently the Spanish part—bears no resemblance to the American end of the city: the American end which lies beyond the intervening brick business center. The houses are massed in blocks; are austerely plain and dignified; uniform of pattern, with here and there a departure from it with pleasant effect; all are plastered on the outside, and nearly all have long, iron-railed verandas running along the several stories. Their chief beauty is the deep, warm, varicolored stain with which time and the weather have enriched the plaster. It harmonizes with all the surroundings, and has as natural a look of belonging there as has the flush upon sunset clouds. This charming decoration cannot be successfully imitated; neither is it to be found elsewhere in America.

—Mark Twain,
in *Life on the Mississippi*

Beautiful is the land, with its prairies
and the forest of fruit-trees;
Under the feet a garden of flowers, and
the bluest of heavens
Bending above, and resting its dome
on the walls of the forest.
They who dwell there have named it
the Eden of Louisiana.

—Henry Wadsworth Longfellow,
in *Evangeline*

pink house with white shutters on Dauphine Street. **Thomas Wolfe,** on his 1937 visit to New Orleans, stayed at the Roosevelt Hotel, now called the Fairmont, at 123 Baronne Street.

The most recent—and poignant—literary association of New Orleans worthy of citation is with *A Confederacy of Dunces* (1980), written by John Kennedy Toole. The novel, set in New Orleans, is reported to have been completed by Toole years earlier, but the author could not find a publisher for his work. Toole is reported to have committed suicide in 1969, whereupon his mother took up the search for a publisher and, with the help of Walker Percy, finally arranged for publication. *A Confederacy of Dunces* proved to be a best seller, and Toole was posthumously awarded a Pulitzer Prize for the work in 1981.

NEW ROADS

James Ryder Randall wrote the lyric "Maryland, My Maryland" on April 26, 1861, on the present site of Poydras College, northwest of Baton Rouge. The Randall Oak, off Route 1, marks the spot.

PLAQUEMINE

Truman Capote, after his parents were divorced, lived for a time with his aunt in Plaquemine, a small city southwest of Baton Rouge.

PRAIRIEVILLE

Robert Penn Warren first came to this town in 1942, when he began teaching at Louisiana State University.

ST. FRANCISVILLE

John James Audubon in 1821 came to St. Francisville to work at Oakley, a plantation on Route 965. His job was to teach art to the daughter of the plantation owner. Captivated by the scenery along the Mississippi River northwest of Baton Rouge, he sent for his family. Audubon, with his wife, also taught at the private school at Beech Woods, a nearby plantation, and while here produced thirty-two of his famous paintings of birds. Audubon Memorial State Park here is open to visitors.

ST. MARTINVILLE

George Washington Cable referred to St. Martinville in his *Strange True Stories of Louisiana* (1889), and recounted an Acadian tale in *Bonaventure* (1887). The most important literary sites in St. Martinville do indeed concern the exiled Acadians, who began to arrive here from Canada about 1765 and whose descendants are the Cajuns of today. The Evangeline Oak at East Port Street and Bayou Teche marks the place where many of them landed at the end of their long wanderings. Legend has it that the oak was the meeting place of Emmeline Labiche and Louis Arceneaux—the legendary Evangeline and Gabriel of *Evangeline* (1847), the narrative poem by **Henry Wadsworth Longfellow** that immortalized the Acadians. Under the tree, which carries a plaque, are benches from which one can look out over the bayou, with its water hyacinths and moss-draped oaks. In town, on Main Street, is St. Martin de Tours Church, whose old cemetery exhibits the Evangeline Monument, for which Dolores del Rio posed. Miss del Rio played the role of Evangeline in an early motion picture based on the legend. Another literary site is the Evangeline Museum, at the foot of Teche Bridge. About a mile north of St. Martinville, on Route 31, is Longfellow-Evangeline State Park. The park includes a three-story Acadian cottage, built in 1765 and believed to have been the home of Louis Arceneaux.

The Evangeline Monument in St. Martinville. (*right*) Evangeline Oak in St. Martinville.

Constructed originally without windows, the cottage has handhewn cypress timbers fastened with pegs and is enclosed by adobe and brick walls. Nearby is the centuries-old Gabriel Oak, beneath which one can sit and reflect on the tale of the famous lovers so long separated by war and finally reunited in death.

SHREVEPORT

Roark Bradford, while living in New Orleans, often visited his Little Bee Bend Plantation near Shreveport, which is on the Red River, close to the Texas border.

ARKANSAS

CARLISLE

During the 1870s **Opie Read,** the humorist and novelist, founded, with H. C. Warner, the short-lived newspaper *The Prairie Flower* in Carlisle, which is about thirty miles east of Little Rock.

CLOVER BEND

For nearly thirty years before World War I, **Alice French,** who wrote short stories and novels under the name **Octave Thanet,** spent her winters in the community of Clover Bend in northeastern Arkansas, about twenty-five miles northwest of Jonesboro. Here Thanet absorbed material and background for much of her work, including the novels *Expiation* (1890), *Otto the Knight* (1893), and *By Inheritance* (1910). *By Inheritance* was a study of black life after the Civil War. Thanet's home, Thanford, a three-story white frame house overlooking the Black River, is located on Route 228 near Clover Bend. The house has been moved several hundred yards from its original site, and is somewhat altered since Thanet's residence. It is listed in the National Register of Historic Places.

CONWAY

Opie Read was editor of a newspaper in this city north of Little Rock in the late 1870s.

ELAINE

Richard Wright moved in the summer of 1916 to the community of Elaine, near the Mississippi River and about twenty-five miles southwest of Helena, with his mother, brother, and aunt. Wright's uncle, Silas Hoskins, owned a saloon that catered to blacks employed at local sawmills. While living here with his uncle, Wright found for the first time in his life that he could have all he wanted to eat. He also enjoyed going on buggy rides around town with his brother and uncle. This relatively idyllic life lasted only a short time. Hoskins was shot to death by whites jealous of his success in business, and the terrified Wright family fled to West Helena. Wright wrote of his childhood experiences here in his autobiographical work *Black Boy* (1945).

Thanford, home of Alice French in Clover Bend. The novelist built the house in 1897, after an earlier house had been destroyed by fire, and spent winters here until 1909.

EUREKA SPRINGS

Writers associated with this Ozark town in northwestern Arkansas, well known for its mineral springs, include **Glenn Ward Dresbach,** who lived at Bon Repos, his two-story white-stucco cottage on the town's east side, from 1940 until his death on June 27, 1968. Dresbach wrote many of his poems in the downstairs studio of this house, now privately owned. His home and the area around Eureka Springs provided inspiration for his poems, including "Nameplate for Our House" and "Mountain Air—the Ozarks."

Carry Nation is said to have written her autobiography, *The Use and Need of the Life of Carry Nation* (1904), in Eureka Springs. In 1909 the temperance advocate bought land thirty-five miles east of Eureka Springs and began building Highland Farm Home as a retirement home. She repaired to Highland Farm Home several times toward the end of her life. Nation's followers built Hatchet Hall, a frame building on Steel Street in Eureka Springs, to house an academy for prohibitionists. The project failed when Nation died, and no classes were ever held there. Hatchet Hall is now open to visitors. Carry Nation gave her last temperance lecture in Eureka Springs, on January 13, 1911. After her talk she staggered from

This was as close as white terror had ever come to me and my mind reeled. Why had we not fought back, I asked my mother, and the fear that was in her made her slap me into silence.

—Richard Wright, in *Black Boy.*

Hatchet Hall in Eureka Springs, as it appeared about forty years ago. By at least one account, Carry Nation lived at Hatchet Hall for a time and conducted a day school next door.

Kisses upon a hill-top
Where rhododendrons ran aflame
Against a wall of glittering leaves.
Cool through the twilight moved your
fresh bright body,
Bringing me offerings of joy.
Beware of love's late fire;
From its black traces you will never
part;
Your will has changed to water, your
best flame to destroy.

—John Gould Fletcher,
in "Autobiography"

the rostrum and collapsed. She was taken to a hospital in Leavenworth, Kansas, where she died five months later.

Vance Randolph lived in Eureka Springs in the 1950s and early 1960s. Much of Randolph's writing concerned the people and folklore of the Ozarks. Among his books were *Ozark Mountain Folks* (1932) and *The Talking Turtle, and Other Ozark Folktales* (1957).

FAYETTEVILLE

This city is the home of the University of Arkansas, whose library houses a collection of the papers of **Glenn Ward Dresbach.** Dresbach is buried here, in the National Cemetery on Route 71 South. **Charles J. Finger** moved to Fayetteville in 1920, one year before he received the Newbery Medal for his *Tales from Silver Lands* (1924), a children's book. His later books included *Ozark Fantasia* (1927) and *Courageous Companions* (1929). Finger died on January 7, 1941, at Gayeta Lodge, his home on Finger Road. This house is private. Finger is buried in a cemetery in Farmington, a few miles west of Fayetteville.

Charles Morrow Wilson, born on June 16, 1905, on a farm near Fayetteville, attended Fayetteville schools and was graduated, in 1926, from the University of Arkansas. After following a newspaper career elsewhere, Wilson returned in 1952 to live on a farm near Fayetteville, where he stayed for several years. Among his books with an Arkansas background were *Acres of Sky* (1930), *Backwoods America* (1934), and *Man's Reach* (1944), a biographical novel about Archibald Yell, a Fayetteville man who served as governor of Arkansas and U.S. representative in the 1840s.

FORT SMITH

Katharine Susan Anthony lived in Fort Smith during her childhood. She was born on November 27, 1877, about thirty miles east of Fort Smith, in the hamlet of Roseville. Her literary work included several biographies, among them *Margaret Fuller* (1920), *Queen Elizabeth* (1929), and *Louisa May Alcott* (1938). *The Adventures of Captain Bonneville, U.S.A.* (1837), by Washington Irving, was written about the career of Benjamin Louis Eulalie Bonneville, who lived at 3215 North O Street. Bonneville served as commander of the military fort from which the city derives its name, and retired here after leaving the U.S. Army in 1871.

HOPE

Josephine Haxton, who wrote novels under the pen name **Ellen Douglas,** lived as a girl in this city in southwest Arkansas.

LITTLE ROCK

Writers associated with Little Rock include **John Gould Fletcher,** born here on January 3, 1886. Fletcher lived in what is now known as the Pike-Fletcher-Terry Mansion, 411 East Seventh Street, the subject of his poem "The Ghosts of an Old House." After he was graduated from Harvard in 1907, Fletcher lived in Massachusetts and in England. He returned to Arkansas in the 1930s and became associated with the Agrarians, a Southern literary coterie that included Robert Penn Warren, John Crowe Ransom, and Allen Tate. In 1936 Fletcher moved into Johnswood, the house on Route 10 that remained his home until he died in 1950, eleven years after winning a Pulitzer Prize for his *Selected Poems* (1938). On May 10, 1950, Fletcher drowned in a pond near his home. He was buried in Mount Holly Cemetery at West Twelfth Street and Broadway. **Charlie May [Simon] Fletcher,** wife of John Gould Fletcher, was a long-time resident of Little Rock. Author of twenty-seven books, including the biography *Faith Has Need of All the Truth* (1974), whose subject was Teilhard de Chardin, the French religious philosopher. Mrs. Fletcher died in Little Rock on March 22, 1977. **Don L. Lee,** the poet and writer who uses his pen name **Haki R. Madhubuti,** was born in Little Rock on February 23, 1942. His collections of poetry include *Think Black* (1967) and *Don't Cry, Scream* (1969).

The Pike-Fletcher-Terry Mansion on East Seventh Street was built in 1840 for **Albert Pike,** the soldier, explorer, lawyer, and writer, whose poem "The Widowed Heart" anticipated Edgar Allan Poe's "The Raven" (1845) by two years, and bears some resemblance to the better-known work. Pike, who is credited with writing the words to the song "Dixie," lived in Little Rock until shortly after the Civil War. His poems were collected and published in three volumes between 1900 and 1916.

Opie Read served as city editor of the *Daily Arkansas Gazette* in Little Rock from 1878 to 1881. In 1883 Read and a partner founded *The Arkansas Traveler,* a lively publication that quickly became one of the most popular weeklies in the country and established Read's national reputation as a humorist. In 1887 success forced Read to move the offices of his newspaper to Chicago. **James Street** in 1927 worked as state editor of the *Daily Arkansas Gazette.* The material he gathered in Lit-

Pike-Fletcher-Terry Mansion in Little Rock, home of both Albert Pike and John Gould Fletcher.

tle Rock and in other Southern cities was reflected in his book *Look Away! A Dixie Notebook* (1936).

PIGGOTT

Ernest Hemingway in late 1928 brought his second wife, Pauline Pfeiffer Hemingway, and their son to stay at the home of his wife's family at 1021 West Cherry Street while he traveled in the West. During the few days Hemingway stayed in this town in northeastern Arkansas, he worked at his writing in a remodeled carriage house behind the main house. Here Hemingway is said to have worked on the first draft of *A Farewell to Arms* (1929), which Hemingway dedicated to Pauline's uncle, Gus Pfeiffer, a resident of Piggott. Both the main house and carriage house are still standing but are private.

PINE BLUFF

C[larence] P[endleton] Lee set his novel *The Unwilling Journey* (1940) in this city on the Arkansas River, about forty miles southeast of Little Rock. Opie Read used a plantation near Pine Bluff as a setting for his novel *An Arkansas Planter* (1896).

VARNER

C[larence] P[endleton] Lee was born on May 26, 1913, in this small community about thirty miles southeast of Pine Bluff on Route 65. Lee is the author of the novels *Athenian Adventure* (1957) and *The Unwilling Journey* (1960).

WASHINGTON

Ruth [McEnery] Stuart moved from New Orleans to this city in southwestern Arkansas in 1879, when she married a local cotton planter. Until her husband's death in 1883, when the novelist returned to New Orleans, the Stuarts lived in a house that stood near the Baptist Church. Stuart wrote of Arkansas places and people in such books as the novel *Sonny* (1896) and the collection *In Simpkinsville: Character Tales* (1897).

WEST HELENA

Richard Wright lived during his childhood in this city in eastern Arkansas, first for several weeks in 1917, then for a longer period after a stay of a few months in Jackson, Mississippi. While here, Wright's family rented half of a two-family house. The landlady, a prostitute, lived and worked in the other half. Wright, spurred on by his playmates, once climbed on a chair in his room to peer through a crack in the wall, hoping to find out for himself just what his friends had been gossiping about. He lost his balance and fell, and the noise alerted the landlady to his snooping. She gave him a severe scolding and later threatened to throw the family out for scaring her customers away. Wright was a student in West Helena in November 1918, when World War I ended. The jubilation of that passing moment, causing blacks and whites to laugh and sing together, contrasted sharply with the ever-increasing dread of racial terror that had come to dominate Wright's perception of his life in Arkansas.

TENNESSEE

ATOKA

Erskine Caldwell lived during childhood in Atoka, twenty-five miles north of Memphis. He attended middle school here for one year.

BEERSHEBA SPRINGS

Mary [Noailles] Murfree, the short-story writer and novelist, spent summers at her family's home, Crag-wylde, in Beersheba Springs, thirty-five miles north of Chattanooga. She wrote of her experiences here in her first story collection, *In the Tennessee Mountains* (1884).

CADE'S COVE

The Prophet of the Great Smoky Mountains (1885), a novel by Mary Murfree, was set in the Cade's Cove area, twenty-five miles west of Sugarlands, now the Great Smoky Mountain Park, south of Maryville in eastern Tennessee.

CAMPBELLSVILLE

Donald [Grady] Davidson, the poet and historian, was born here on August 18, 1893. His works included *The Tennessee* (1946–48), a two-volume history of the Tennessee River and Tennessee River Valley, and *The Tall Men* (1927), a long poem in blank verse. Davidson was one of the founders of *The Fugitive* magazine (1922–25), published at Vanderbilt University.

CHATTANOOGA

Ambrose Bierce was sent in June 1864 to direct the advance of the Army of the Cumberland at Kenesau Mountain. After suffering a head wound, Bierce spent a month recovering at a base hospital in Chattanooga. **Jeannette [Augustus] Marks,** who is remembered as one of the great teachers at Mt. Holyoke College, in Massachusetts, was born here on August 16, 1875. Her plays included *The Merry Merry Cuckoo* (1927), and she also wrote poetry and biography.

CHICKAMAUGA

The short story "Chickamauga," by **Ambrose Bierce** is thought to have been based on his ride through the brush at the Battle of Chickamauga on September 20, 1863.

CLARKSVILLE

Evelyn Scott, the poet and novelist, was born in Clarksville, forty-five miles northwest of Nashville, on January 17, 1893. Her works included the experimental novel *The Wave* (1929) and *Background in Tennessee* (1937), an autobiography. **Allen Tate** and his wife, the novelist **Caroline Gordon,** did some of their writing on a farm near Clarksville, and **Robert Penn Warren** attended Clarksville High School in the 1920s.

Thou art gone from me forever;
—I have lost thee, Isadore!
And desolate and lonely I shall be
forever more:
Our children hold me, Darling,
or I to God should pray
To let me cast the burthen of this long,
dark life away,
And see thy face in Heaven, Isadore!

—Albert Pike,
in "The Widowed Heart"

The water gleamed with dashes of red, and red, too, were many of the stones protruding above the surface. But that was blood; the less desperately wounded had stained them in crossing.

—Ambrose Bierce,
in "Chickamauga"

Ambrose Bierce.

Opie Read, the popular and prolific novelist. Read spent part of his boyhood in Gallatin and did his early newspaper work in Tennessee, Arkansas, and Kentucky.

CLIFTON

T[homas] S[igismund] Stribling was born on March 4, 1881, in Clifton, east of Jackson. He wrote primarily about the South and won a Pulitzer Prize for his novel *The Store* (1932). His novels *Teeftallow* (1926) and *Bright Metal* (1928) were laid in Tennessee.

COLUMBIA

John Trotwood Moore, the poet and novelist, lived after 1885 on a farm outside Columbia, where he raised horses and continued to write. His novel *A Summer Hymnal* (1901), written here, was set in Tennessee, as were many of his poems and stories, collected in *Songs and Stories from Tennessee* (1897).

DAYTON

The celebrated Scopes trial of 1925, with Clarence Darrow opposing William Jennings Bryan in a test of the right of a teacher to expound evolution in his classes, was conducted in Dayton at the courthouse on Market Street between Second and Third avenues. Robert E. Lee and Jerome Lawrence wrote the play *Inherit the Wind* (1955) about the trial. The Scopes trial also was responsible for bringing **Irving Stone** to Dayton, to do research for *Clarence Darrow for the Defense* (1941).

GALLATIN

Opie Read, the humorist, worked as a reporter for the *Examiner,* a newspaper published in Gallatin, twenty-six miles east of Nashville.

GREENEVILLE

The Davy Crockett Birthplace Park, nine miles east of Greeneville off Route 11E, in the community of Limestone, is open daily, with free admission. The park has a replica of the log cabin in which Crockett was born. Although **Davy Crockett** may not have written his autobiography, *A Narrative of the Life of David Crockett* (1834), the book is a major contribution to the folklore of the American frontier.

HALLS

Roark Bradford, author of *Ol' Man Adam an' His Chillun* (1928), was born on August 21, 1896, in Nankipoo, a settlement west of Halls. Halls lies on Route 51, just south of Dyersburg.

JELLICO

John Fox Jr. set his story "A Mountain Europa" in Jellico, a town on the Tennessee-Kentucky border. Uncle Belly in Fox's *The Trail of the Lonesome Pine* (1908) was patterned after William Beans, a resident of Jellico in Fox's time.

JOHNTOWN

The Cave (1959), a novel by Robert Penn Warren, is set in this small town between Knoxville and Nashville.

> *Nick Pappy, much to the astonishment of Johntown, wore a diamond ring on his right hand, but the hand looked so awe-inspiringly competent that no citizen of Johntown ever expressed the astonishment. Nick had the best car in Kobeck County, a Cadillac, new at that, or at least new by the standards of Johntown, Tennessee, and his idea of the Good Life was to take Sunday afternoon off, drape his bulk in his expensive hound-tooth tweed jacket, which he had bought in Nashville, clamp a cigar in his left jaw, establish his wife beside him, with her drugstore blond hair advertising itself, and him, to the hicks, and cruise contemptuously up hill and down dale over half the natural beauty and historicity of the Volunteer State.*
>
> —Robert Penn Warren, in *The Cave*

KNOXVILLE

James [Rufus] Agee was born on November 27, 1909, at the Knoxville General Hospital, still standing at 991 Cleveland and Dameron avenues, but no longer a hospital. Agee was six years old when his father died in an automobile accident. His family was living at that time on Highland Avenue, in a neighborhood of old houses in poor repair. Agee attended school in Knoxville for a time, including one year, 1924–25, at Knoxville High School. Agee set his novels *The Morning Watch* (1954) and *A Death in the Family* (1957) in the Cumberland Mountain region where he was raised, and readers of the latter novel will see the parallel with Agee's own life.

Critic **Joseph Wood Krutch** was born on November 25, 1893, probably at the family home on West Main Street, now a dentist's office. He was graduated from Knoxville High School in 1911.

MADISONVILLE

Charles William Todd, author of Tennessee's first novel, *Woodville* (1832), lived for a time in Madisonville, which is south of Oak Ridge. He edited *The Tennessean* here for a short time in 1833, but things did not go well for Todd in Madisonville. In 1833 he was divorced by his wife, and in 1834 he was indicted by the county grand jury on a charge of "assault with an intent to kill and murder in the first degree." He left town after the indictment and never was heard of again.

MEMPHIS

Memphis is rich in literary associations, including the novel *If Beale Street Could Talk* (1974), by James Baldwin, which is, of course, set in Memphis. **William March Campbell,** who wrote novels and short stories under the name of **William March,** worked for a time as vice president of operations at the Waterman Steamship Company in Memphis. Colonel **William C. Falkner,** William Faulkner's great-grandfather, ran the Federal blockade during the Civil War, and his novel *The White Rose of Memphis* (1881) described the city in the nineteenth century. **John Faulkner,** the novelist and brother of William Faulkner, worked as manager of Mid South Airways at the old Municipal Airport at Airways and Brooks, now part of the international airport. He was living in Whitehaven, a section of Memphis, at the time of his death on March 28, 1963, in a Memphis hospital.

Shelby Foote wrote the first volume of *The Civil War* (1958) while living at 697 Arkansas Street. He wrote the second (1963) while living at 507 Yates Road. The latter home is still standing but private. Justine's, at 919 Coward Place, which Foote has described as his favorite Memphis restaurant, is still open, and some of the Memphis sites mentioned in Foote's novel *September, September* (1978) are Handy Park, Peabody Hotel, and the Rivermont Holiday Inn.

In June 1858, **Mark Twain** was involved in a tragic incident in Memphis. Twain and his brother Henry Clemens had been en route to New Orleans aboard the steamship *Pennsylvania,* when Twain fell into an argument with the pilot and left the ship. Henry remained aboard and, near Memphis, the ship exploded and sank. Henry, severely wounded, was taken to a makeshift hospital in Memphis, at Main and Madison streets, where he was treated by a young physician. At Twain's urging, morphine was administered to ease Henry's pain. Henry fell into a deep sleep and died soon after. Twain later wrote that he considered himself responsible for his brother's death, feeling that he should have remained aboard the *Pennsylvania* and been with his brother during the accident. Twain also felt that he had made a mistake in urging the use of morphine, since the physician had said he was uncertain of the proper amount to be injected.

Richard Wright moved to Memphis in 1911, when he was three years old, living with his family in California Flats, south of Blues Street. He entered school in Memphis, attending Howard Institute, at 476 St. Paul Street, no longer standing, but left when his family moved on to Jackson, Mississippi. Later, when Wright was seventeen, he returned to Memphis, living at 570 Beale Street and working as a dishwasher at the Lyle Drugstore, 190-4 South Main Street, which now is Bert's Men and Boys' Shop. In order to be able to borrow books from the public library, Wright borrowed the library card of a white friend and wrote an explanatory note for himself addressed to the librarian: "Dear Madam: Will you please let this nigger boy have some books by H. L. Mencken?"

MURFREESBORO

Andrew Nelson Lytle, the biographer and novelist, was born on December 26, 1902, in Murfreesboro in central Tennessee. His books include the novels *The Long Night* (1936) and *The Velvet Horn* (1957). Lytle served as managing editor of *Sewanee Review* in 1942–43.

Mary [Noailles] Murfree, the short-story writer and novelist, who used the pen name **Charles Egbert Craddock,** was born on January 24, 1850, at Grantlands, her family's estate near Murfreesboro. She lived there until she was six years old. In 1856 her family moved to Nashville, and it was there that Murfree grew up. When she was four years old, Murfree contracted "slow fever," thought now to have been polio, and was left partially paralyzed and somewhat lame. Grantlands was destroyed during the Battle of Murfreesboro, but Murfree's father built New Grantlands on the estate about 1871, and Murfree returned to Murfreesboro and began to write there. After her father's death, Murfree lived with her sister Fannie in a nearby cottage, and it was there, on July 31, 1922, that Murfree died, leaving behind a substantial body of work, including the stories collected as *In the Tennessee Mountains* (1884) and several novels, among which were *The Prophet of the Great Smoky Mountains* (1885) and *The Frontiersmen* (1904).

NASHVILLE

Nashville has many interesting literary associations, in part due to the presence of Vanderbilt University, which has long attracted writers to its faculty. Among the most distinguished are John Crowe Ransom, Donald Davidson, Allen Tate, Robert Penn Warren, Jesse Stuart, Cleanth Brooks, and Randall Jarrell. **Donald Davidson,** a faculty member at the university, wrote many poems about the South and about Tennessee in particular, including *The Tall Men* (1927) and "On a Replica of the Parthenon in Nashville."

The Parthenon itself is worthy of attention by the literary visitor. This building at West End Avenue and Twenty-fifth Avenue North provides exhibits of art for visitors, but its interest for readers focuses on the east portico and one of the figures in its pediment. **Randall Jarrell,** born in Nashville on May 6, 1914, grew up there. He lived at 2007 Nineteenth Avenue South in 1935, still standing, and at 2524 Westwood Street from 1937 to 1940, thought now to have been razed. When Jarrell was about twelve years old, he posed for the sculptor who was at work on the Parthenon

> *A few minutes before ten, the phone rang. Mary hurried to quiet it: "Hello?"*
>
> *The voice was a man's, wiry and faint, a country voice. It was asking a question, but she could not hear it clearly.*
>
> *"Hello?" she asked again. "Will you please talk a little louder? I can't hear. . . . I said I can't hear you! Will you talk a little louder please? Thank you."*
>
> *Now, straining and impatient, she could hear, though the voice seemed still to come from a great distance.*
>
> *"Is this Miz Jay Follet?"*
>
> *"Yes; what is it?" for there was a silence; "yes, this is she."*
>
> *After further silence, the voice said, "There's been a slight—your husband has been in an accident."*
>
> —James Agee,
> in *A Death in the Family*

The Parthenon in Nashville. Originally built in 1897 as a temporary structure for the Tennessee Centennial Exposition, the Parthenon was rebuilt in the 1920s. This structure is of special literary interest because Randall Jarrell posed for one of the figures in the pediment.

Etching of Mary Noailles Murfree, who took her pen name Charles Egbert Craddock from the main character of one of her first short stories.

pediment. Thus, the literary visitor viewing Ganymede among the heroic figures can recall the once youthful Jarrell. Jarrell attended the still-functioning Hume-Fogg Technical Institute and Vocational High School, at 700 Broadway, and contributed articles to the school paper, *The Echo.* In his school yearbook for 1931, Jarrell's classmates wrote of him: "Appearance: Hopefully hopeless; Always Seen: In the clouds and below; Distinguished By: His white tennis shoes; Greatest Ambition: To be appreciated." Jarrell went on to attend Vanderbilt University, where he received a B.S. degree in 1935 and an M.A. in 1939. His volumes of poetry included *Blood for a Stranger* (1942), *Little Friend, Little Friend* (1945), and *The Seven-League Crutches* (1951).

Mary Murfree lived during childhood at 120 Ninth Avenue South, now a parking lot, and at Vauxhall Place. Murfree was tutored in French at Nashville Female Academy. After their father died in 1922, Murfree and her sisters bought a house at 216 North Spring Street, and Murfree died in Murfreesboro. Frank Norris once said that there was never any material to write about in Nashville. To prove him wrong, O. Henry wrote "A Municipal Report," published in *Hampton's Magazine* in 1909 and later collected in one of O. Henry's volumes of stories, *Strictly Business* (1910).

John Crowe Ransom was a member of the class of 1903 at the Bowen Secondary School, 1309 Broad Street, no longer standing. He went on to edit *The Fugitive* at 2202 Eighteenth Avenue South. The publication is defunct, but the building is still there. Ransom was a 1909 graduate of Vanderbilt University and taught there from 1914 to 1937. The collections of poetry he wrote while there included *Poems About God* (1919) and *Chills and Fever* (1924). **Opie Read,** the humorist, was born on December 22, 1852, at Fifth Avenue South, now the Donaho Nursery. Near his home is an Opie Read Well, said to hold especially delicious water.

Jesse Stuart, a graduate of Vanderbilt, wrote an autobiographical work as a term paper. It was published as *Beyond Dark Hills* (1938). **Wilkins Tannehill** in 1825–26 was mayor of Nashville, where he also edited *The Daily Orthopolitan* and later, without success, started the Nashville *Herald.* Tannehill lived at Belle View on Franklin Road, a house that was torn down in 1960. **Allen Tate** helped to establish the Fugitives, a literary group that included Donald Davidson, John Crowe Ransom, and Robert Penn Warren. The group published *The Fugitive* from 1922 until 1925: the list of contributors to a single issue of this journal, that of February-March 1923, gives some idea of the quality of the journal. It included Allen Tate, Louis Untermeyer, Donald Davidson, John Crowe Ransom, John Gould Fletcher, and Witter Bynner.

NEW MARKET

Frances Hodgson lived in New Market, in northeastern Tennessee, from 1875 until she married Dr. S. M. Burnett and moved to Washington, D.C. The Hodgson-Burnett House, a log house, was where she wrote *Surly Tim and Other Stories* (1877) and *Little Lord Fauntleroy* (1886).

PEGRAM

Horace McCoy, who was best known for the novel *They Shoot Horses, Don't They?* (1935), was born on April 14, 1897, in Pegram, on Route 70, southwest of Nashville. This novel, about the Depression phenomenon of dance marathons, still has power to shock the reader. Other novels by McCoy included *I Should Have Stayed Home* (1938) and *Kiss Tomorrow Good-Bye* (1948).

PULASKI

John Crowe Ransom was born on April 30, 1888, in Pulaski, a city on Route 64, south of Columbia. He had a distinguished career as a poet and as a teacher of literature at Vanderbilt University, in Nashville, and at Kenyon College, in Gambier, Ohio, where in 1939 he founded and edited *Kenyon Review.*

ROGERSVILLE

Davy Crockett, the frontiersman folk hero, was born in Rogersville, northwest of Johnson City, on August 17, 1786. He was the subject of various biographies and children's books and the putative author of an autobiography, *A Narrative of the Life of David Crockett, of the State of Tennessee* (1834), and several other works.

RUGBY

Thomas Hughes, the English author of *Tom Brown's School Days* (1857) and *Tom Brown at Oxford* (1861), founded the Rugby colony in 1879. Many of the original buildings of the failed community remain, although most of them are private and can be seen only from the outside. All the homes, including Kingston Lisle, which Hughes lived in while on visits to Rugby, are open daily from March to November for an admission fee. The Thomas Hughes Free Public Library, built here in 1882, has 7,000 volumes—all dating back to before 1900—which form one of the best collections of Victorian literature in the United States.

Kingston Lisle, home of Thomas Hughes in Rugby, on Route 52 about seventy-five miles northwest of Knoxville. Hughes stayed here during his annual visits to Rugby Colony.

RUTHERFORD

Davy Crockett moved in 1823 to western Tennessee and lived until 1835 on the Rutherford Fork of the Obion River. A reproduction of the cabin he lived in was built in 1956 off Route 45W on the Rutherford High School grounds. A rocking chair Crockett had built, as well as other nineteenth-century items, are displayed daily for an admission fee.

SEWANEE

James Agee lived with his mother in Saint Andrews, a community near Sewanee, from the summer of 1918 until 1925, except during 1924, when he attended Knoxville High School. Agee's mother rented a cottage in Saint Andrews and found the community so attractive that she decided to make the family home there. Agee attended St. Andrew's School, where he established a friendship with Father Flye, a history teacher, who remained his confidant until Agee's death. Their correspondence was published as *Letters to Father Flye* (1962).

Henry Sydnor Harrison, author of the novel *Queed* (1911), was born on February 12, 1880, on University Avenue in the Nauts House, a private home adjoining the University of the South campus in Sewanee. Harrison lived here until he was two years old. **William Alexander Percy** was a student at the turn of the century at the University of the South. He described his years here in his autobiography, *Lanterns on the Levee* (1941). Percy's family had a home in Sewanee called Brinkwood, and he often returned here in later years.

William Peterfield Trent, a faculty member of the University of the South, was founder and editor of *The Sewanee Review* in 1892, the same year in which his *Life of William Gilmore Simms* was published. *Sewanee Review* has been published ever since, making it the oldest literary journal in the United States. Its editors have included Allen Tate and Andrew Lytle.

SHILOH

The town of Shiloh stands on the west bank of the Tennessee River, about forty miles southeast of Jackson, and here on April 6 and 7, 1862, was fought one of the bloodiest battles of the Civil War: casualties on each side were estimated at 10,000—and the outcome of the battle is still in dispute to this day. Inevitably, the Battle of Shiloh has attracted writers of fiction. Shelby Foote, for one, wrote a novel entitled *Shiloh* (1952), and Andrew Lytle wrote the novel *The Long Night* (1936), which gives a vivid account of the battle.

It is also interesting to note that **Lew Wallace,** author of *Ben Hur* (1880), commanded a Union division at Shiloh, but even more interesting for readers was the presence of **Ambrose Bierce** in Union ranks during the battle. One of Bierce's best stories, "An Occurrence at Owl Creek Bridge," relates an incident that some believe took place at Shiloh, where there is a modern, concrete structure named Owl Creek Bridge. Why, one may wonder, did Bierce identify the bridge in his story as "a railroad bridge in northern Alabama"? We cannot say for sure, but we do note that the Alabama state line is a few miles below Shiloh. What is important is that in his story Bierce created a minor masterpiece, which intrigues readers even today, as they share the last thoughts of a condemned man on the gallows.

TRENTON

Peter [Hillsman] Taylor, playwright and short-story writer, was born in Trenton, a town north of Jackson, on January 8, 1917. His works, which depict middle-class people in today's South, include the collections *A Long Fourth* (1948) and *Happy Families Are All Alike* (1959).

WOODSTOCK

The novel *Penhally* (1930) by Caroline Gordon, depicting four generations of a Southern family, is set in Woodstock, a town situated just north of Memphis.

Firm and upright he walked for one so old,
Thrice-pondered; and I dare not prophesy
What age must bring me; for I look round bold
And seek my enemies out; and leave untold
The sideway watery dog-glances I
Send fawning on you, thinking you will not scold.

—John Crowe Ransom, in "Old Man Pondered"

KENTUCKY

ASHLAND

Jean Thomas, author of *Devil's Ditties* (1931) and *The Traipsin' Woman* (1933), was born here on November 14, 1881. Her home, Wee House in the Wood, with a gateway and entrance modeled after Anne Boleyn's estate in Yorkshire, England, has been coverted into a museum displaying Jean Thomas memorabilia. **Dalton Trumbo,** the novelist and screenwriter, known particularly for his anti-war novel *Johnny Got His Gun* (1939), was imprisoned in Ashland in 1950 after he refused to testify about his political beliefs before a Congressional investigating committee. Ashland also has historical interest: Mary Todd lived at 574 West Main Street from childhood until she married Abraham Lincoln. For a time the building served as a rooming house, with a grocery store on the first floor; now it is open to visitors by appointment.

AUGUSTA

Stuart Walker, the playwright who established the touring Portmanteau Theatre, was born on March 4, 1888, in this city on the Ohio River, southeast of Cincinnati, Ohio.

BARDSTOWN

Stephen Collins Foster, the composer, wrote "My Old Kentucky Home" about a mansion on Federal Hill in this city southwest of Lexington. Now the area is known as My Old Kentucky Home State Park and is open daily to visitors.

BEREA

Berea College, noted for its history of service to southern Appalachia, maintains a **James Still** collection in its library. Located on the edge of the Cumberland Mountains, in Madison County, Berea claims Still as a local author. His work often deals with the lives of mountain families.

He stands at the gate of his own home. All is as he left it, and all bright and beautiful in the morning sunshine. He must have traveled the entire night. As he pushes open the gate and passes up the wide white walk, he sees a flutter of female garments; his wife, looking fresh and cool and sweet, steps down from the veranda to meet him. At the bottom of the steps she stands waiting, with a smile of ineffable joy, an attitude of matchless grace and dignity. Ah, how beautiful she is! He springs forward with extended arms. As he is about to clasp her he feels a stunning blow upon the back of the neck; a blinding white light blazes all about him with a sound like the shock of a cannon—then all is darkness and silence!

Peyton Farquhar was dead; his body, with a broken neck, swung gently from side to side beneath the timbers of the Owl Creek bridge.

—Ambrose Bierce, in "An Occurrence at Owl Creek Bridge"

Then suddenly I saw a steeple that shone like silver in the moonlight, growing into sight from behind a rounded knoll. The tires sang on the empty road, and, breathless, I looked at the monastery that was revealed before me as we came over the rise. At the end of an avenue of trees was a big rectangular block of buildings, all dark, with a church crowned by a tower and a steeple and a cross: and the steeple was as bright as platinum and the whole place was as quiet as midnight and lost in the all-absorbing silence and solitude of the fields.

—Thomas Merton,
in *The Seven Storey Mountain,* on first coming to the Abbey of Our Lady of Gethsemani

FRANKFORT

When Governor-elect William Goebel, of Kentucky, was shot in front of the old State House on Broadway on January 30, 1900, **Irvin S. Cobb,** then a young reporter, helped carry him to a doctor in the Capital Hotel. Cobb wired his reports on the assassination and the ensuing near-civil war to the Louisville *Post* and the New York *Sun.* At that time Cobb shared a room in Miss Eliza Overton's apartments on the second floor of the old Farmer's Bank Building on Main Street. *The Kentuckians* (1898), a novelette by **John Fox, Jr.,** used the old capitol, the old Governor's mansion, the Capital Hotel, and the Hanna House at the head of Bridge Street in various scenes. While Fox was working on *The Kentuckians* here, he also gathered local material for *The Little Shepherd of Kingdom Come* (1903), one of his most successful novels, which describes many Frankfort sites. Fox lived in Frankfort during 1895 and 1896, sometimes sharing a room with **Robert Burns Wilson,** a painter and poet, at Miss Ludie Ware's home on the corner of Ann and Clinton streets. It was a line in Wilson's poem "Battle Song" (1898) that supplied the Spanish-American War slogan, "Remember the Maine."

FRANKLIN

Opie Read, the humorist and novelist, author of *A Kentucky Colonel* (1890), worked in 1873 on the Franklin *Kentucky Patriot.* His salary was fifty cents a week plus room and board. Franklin lies southwest of Bowling Green, near the Tennessee border.

FULLERTON

Jesse Stuart, the novelist, while living here was principal of the McKell High School and worked also as an editorial writer for a county newspaper.

FULTON

[Arthur] Burton Rascoe, the literary critic, was born here in southwestern Kentucky on October 22, 1892. He attended Carr Institute School, now called Carr Elementary School, on West State Line. Rascoe's works include *Titans of Literature* (1932), *Theodore Dreiser* (1925), and the autobiographical *Before I Forget* (1937).

GETHSEMANE

Thomas Merton entered the Trappist Abbey of Our Lady of Gethsemani in 1941, when he was twenty-six. Ordained here as Father M. Louis, he remained in the abbey for most of the rest of his life, receiving many visitors who sought his guidance. Merton died in a tragic accident in Bangkok, Thailand, on December 10, 1968, while attending a congress of Roman Catholic monks. Merton's autobiography, *The Seven Storey Mountain* (1948), written here, became a best seller, and it is for this inspirational book that he is best remembered. Merton wrote many other books of similar character while at the abbey, including *The Ascent to the Truth* (1951), *No Man Is an Island* (1955), and *The New Man* (1962). While living here, Merton received permission to build a small private retreat at the monastery. There he found the solitude he needed to write his books. Merton's hermitage now is used as a retreat for those monks who wish to spend one week of the year in solitude. It is not accessible to visitors. Gethsemane is on Route 52, south of Bardstown.

GREENSBURG

Lincoln's law partner **William Henry Herndon,** born in this central Kentucky city on December 25, 1818, lived on South Main Street. The building is still standing and is marked with a plaque. With Jesse W. Weik, Herndon wrote *Herndon's Lincoln: The True Story of a Great Life*

Thomas Merton's hermitage at the Abbey of Gethsemani. Completed in 1961, the structure had neither kitchen nor bathroom and was heated only by a fireplace.

(1889), which supplied fresh material on Lincoln's early days. The book was controversial because of its revelations about Lincoln's personal life.

GUTHRIE

Robert Penn Warren, author of the Pulitzer Prize-winning novel *All the King's Men* (1946), was born in Guthrie on April 24, 1905. He attended Guthrie public schools from 1911 to 1920. Kentucky's rolling hills figure in several of Warren's poems and novels, including the novel *Night Rider* (1938), which dealt with the fight between tobacco growers and manufacturers at the turn of the twentieth century. Guthrie is a small tobacco town southeast of Hopkinsville, on the Tennessee border.

HAZARD

This mining town in southeastern Kentucky figured in *Youngblood Hawke* (1962), by Herman Wouk. Hazard was called Hovey in the novel.

HENDERSON

John James Audubon lived from 1810 to 1819 in a house at the corner of Main and Second streets, near the general store he owned. In 1818 George and Georgiana Keats, brother and sister-in-law of John Keats, the English poet, stayed here in the same house with Audubon. The two men quarreled after George Keats lost his money on an investment in Ohio River cargo recommended by Audubon. The Keatses left Audubon's house and went to Louisville. North of Henderson, on Route 41, is Audubon Memorial State Park. The John James Audubon Memorial Museum in Henderson is open to the public. Admission is charged.

HINDMAN

James Still, who lived here from 1932 to 1939 and from 1951 to 1961, served for a time as librarian for Hindman Settlement School, still standing, at the mouth of Troublesome Creek. His books, including a volume of poetry, *Hounds on the Mountain* (1937), and a volume of short stories, *Way Down Yonder on Troublesome Creek* (1974), reflect his deep understanding of the hill people of southeastern Kentucky. Wolfpen Creek, in *Sporty Creek* (1977), and Dead Mare Branch and Little Carr Creek, in *River of Earth* (1940), are near Hindman, which is on Route 80, in the foothills of the Cumberland Mountains.

HODGENVILLE

Abraham Lincoln was born on February 12, 1809, on the old Rock Spring farm, three miles south of Hodgenville, which is on Route 31E, forty-five miles north of Louisville. The log cabin on view here in the Abraham Lincoln National Historic Park, open daily except Christmas day, is believed to be the house in which he was born. The Lincolns lived here until about 1811. Mark Twain, Ida Tarbell, and others formed the Lincoln Farm Association about 1900 in order to preserve the site and memorialize Lincoln. Lincoln also lived at the Knob Creek Farm on Route 31E, about ten miles from Hodgenville, from 1811 to 1816. The farm is open daily from April 1 to November 1. It was here that Lincoln had his first schooling.

Log cabin believed to be the birthplace of Abraham Lincoln. Lincoln's father paid $200 for the cabin and three hundred acres in 1807.

The floor was packed-down dirt. One door, swung on leather hinges, let them in and out. One small window gave a lookout on the weather, the rain or snow, sun and trees, and the play of the rolling prairie and low hills. A stick-clay chimney carried the fire smoke up and away.

—Carl Sandburg, in *Abraham Lincoln: The Prairie Years*

Near Hodgenville, on May 30, 1851, was born **Albery A[llson] Whitman,** whose poetry, the most ambitious attempted by a black in the nineteenth century, dealt with the lives of blacks and Indians. Whitman's volumes included *Not a Man and Yet a Man* (1877), *The Rape of Florida* (1884), and *The Octoroon* (1900).

LEXINGTON

Lexington, probably best known for its thoroughbred horses, has often been called the Athens of the West. Its notable educational and cultural institutions include Transylvania University, founded in 1780, and the Public Library, founded in 1795, both the first of their kind west of the Alleghenies. It should also be noted that the first newspaper published west of the Alleghenies, the *Kentucky Gazette,* was established here in 1787.

Lexington's literary figures include **James Lane Allen,** born here on December 21, 1849. Allen used Bluegrass settings in many of his novels, including *The Kentucky Cardinal* (1894) and *The Choir Invisible* (1897). He was born on Parker's Mill Road, about five miles from Lexington, in what was known as the Old Poindexter Place, later razed by fire. Allen became principal of an academy here in 1878 and opened a private school for boys in the old Masonic Temple in 1882. He is buried in the Lexington Cemetery.

Gertrude Atherton, the novelist, was sent to Sayre Institute here because of her frail health. The building is now used as a school. The novel *Big River, Big Man* (1959), by Thomas Duncan, is set partially in Lexington. **A. B. Guthrie** moved here from Montana in 1926 and worked as a cub reporter and executive editor for the Lexington *Leader,* still publishing on Midland Avenue. Some of his residences were 251 Rand Avenue, in 1928; 211 Forest Park Road, in 1937; and 246 Tahoma Road, in 1945. Guthrie's novel *The Big Sky* (1947) told of the adventures of a Kentucky boy who headed West in the 1830s, seeking the freedom of a trapper's life. **Elizabeth Hardwick,** the novelist and essayist, was born here on July 27, 1916. She was graduated from Henry Clay High School and from the University of Kentucky, which is

When the vengeance wakes, when the battle breaks,
And the ships sweep out to sea;
When the foe is neared, when the decks are cleared,
And the colors floating free;
When the squadrons meet, when it's fleet to fleet
And front to front with Spain,
From ship to ship, from lip to lip,
Pass on the quick refrain,
"Remember, remember the Maine!"

—Robert Burns Wilson, in "Battle Song"

The Wiggses lived in the Cabbage Patch. It was not a real cabbage patch, but a queer neighborhood, where ramshackle cottages played hop-scotch over the railroad tracks. There were no streets, so when the new house was built the owner faced it any way his fancy prompted. Mr. Bagby's grocery, it is true, conformed to convention, and presented a solid front to the railroad track. . . .

—Alice Hegan Rice,
in *Mrs. Wiggs of the Cabbage Patch*

located here. The university library has a collection of the works of James Still.

LOUISVILLE

Louisville is the largest city of Kentucky and the richest in literary associations. It also boasts the oldest city university in the United States, the University of Louisville, founded in 1798. A few years later, about 1818, George and Georgiana Keats arrived in Louisville, where they prospered and had eight children. George's brother, John Keats, the English poet, expressing the hope that one of the children would become "the first American poet," prophesied:

Little child
O' the western wild
Bard art thou completely!—
Sweetly with dumb endeavour—
A Poet now or never!
Little Child
O' the western wild
A Poet now or never!

George died of tuberculosis on December 24, 1841, and was buried, with other members of his family, in historic Cave Hill Cemetery. Keats's prophecy was unfulfilled.

But there was a Keats of Kentucky: **Madison Cawein,** who earned that accolade by producing more than thirty volumes of verse, was born in Louisville on March 23, 1865. Cawein, whose works included *Lyrics and Idylls* (1890) and *Vale of Tempe* (1905), was graduated from Louisville Male High School in 1881. His verse, sometimes lyrical, fell often into sentimentality. Yet lines describing a lynching, from "The Man Hunt" (1905), give an idea of the reality he could present:

A rope, a prayer, and an oak-tree near.
And a score of hands to swing him clear.
A grim black thing for the setting sun
And the moon and stars to look upon.

Another Louisville native was **John Patrick Goggan,** who wrote under the name **John Patrick,** born on May 27, 1905. Patrick first gained notice for his play *The Hasty Heart* (1945), a comedy set in a hospital in Burma during World War II. His most successful play was *The Teahouse of the August Moon* (1953), a comedy that won a Pulitzer Prize; it was based on a novel by Vern Sneider.

Rupert Sargent Holland, author of such novels for young people as *Blackbeard's Island* (1916) and *A Race for a Fortune* (1931), was born in Louisville on October 15, 1878. **George Horace Lorimer,** editor of the *Saturday Evening Post* from 1899 to 1936, was born here on October 6, 1868. Lorimer's own writings included *Letters from a Self-Made Merchant to His Son* (1902) and *Old Gorgon Graham* (1904). The character Gorgon Graham was based on Philip Danforth Armour, the Chicago meatpacker, for whom Lorimer once worked.

Elizabeth Robins was born here on August 6, 1865. The novelist used the pseudonym **C. E. Raimond** for *Below the Salt* (1896) and other early works, but later reverted to her own name for such novels as *The Magnetic North* (1904) and *My Little Sister* (1913). **Charles Hanson Towne,** the poet and novelist who also worked for *The Smart Set, McClure's Magazine,* and *Harper's Bazaar,* was born here on February 2, 1877. **Leane Zugsmith,** author of the novels *All Victories Are Alike* (1929), *Goodbye and Tomorrow* (1931), and *A Time to Remember* (1936), was born here on January 31, 1903.

Other writers associated with Louisville include **Irvin S. Cobb,** who wrote the "Sour Mash" column for the Louisville *Evening Post* from 1898 to 1901. **Shirley Ann Grau** wrote the novel *Evidence of Love* (1977) while living in Louisville, and Louisville is the setting for the novels *Voices on the River* (1964) and *River to the West* (1970), by **Walter Havighurst.** Joseph Hayes set his novel *Winners Circle* (1980) in Louisville, including scenes at Churchill Downs.

William Vaughn Moody, the playwright and poet, studied drawing and painting here at the Pritchett Institute of Design. In the early 1920s, **Marjorie Kinnan Rawlings** wrote advertising copy and special articles for the Louisville *Courier-Journal.* **Alice Hegan [Caldwell] Rice** lived in Louisville during most of her life. She set her popular children's story *Mrs. Wiggs of the Cabbage Patch* (1901) in the area west of Central Park here, between South Fourth Street and Magnolia Avenue, along the railroad tracks.

Abram Joseph Ryan, the Catholic priest known as the Tom Moore of Dixie and as the Poet of the Lost Cause, retired in 1886 to St. Boniface Friary, a Franciscan monastery here, a few weeks before his death. His principal works of literature were *Father Ryan's Poems* (1879) and a book of devotions, *A Crown for Our Queen* (1882). Ryan died here on April 22, 1886. A bronze plaque on the front wall of the church marks the site of his death.

MURRAY

Cleanth Brooks, who was born in this city twenty-two miles southeast of Mayfield, on October 16, 1906, is known best for *Understanding Poetry* (1938) and *Understanding Fiction* (1943), written in collaboration with Robert Penn Warren. **Felix Holt,** author of *The Gabriel Horn* (1951), which was adapted by A. B. Guthrie for the motion picture *The Kentuckian,* was born here in 1898. The Jesse Stuart Suite at Murray State University Library here houses the papers of Jesse Stuart.

PADUCAH

Irvin S. Cobb, known as the Sage of Paducah, was born on June 23, 1876, at 321 South Third Street. The site of Cobb's birth was marked with a sidewalk tablet, but the tablet was torn out when Paducah's new post office was built. Cobb wrote over three hundred humorous stories about Paducah newspaper life, the Paducah courthouse, and many local people such as Judge Priest, one of his favorite characters. A beauty parlor and barber shop, a cigar, and even a Kentucky mint julep have been named after Cobb. He is also commemorated by the Irvin Cobb building, a former hotel at Sixth Street and Broadway, now renovated for senior citizen's housing. Cobb was editor of the Paducah *Daily News* and managing editor of the *News-Democrat* from 1901 to 1904. Cobb died on March 10, 1944, and was buried in Oak Grove Cemetery in Paducah.

Grave marker of Irvin S. Cobb in Oak Grove Cemetery, 1613 Park Avenue, Paducah. A red line on the road directs visitors to the grave site.

PARIS

John Fox, Jr. was born near here, in Stony Point, in 1863, exact date unknown. He understood the life and customs of the mountaineers, and captured their dialect in such sentimental best sellers as *The Little Shepherd of Kingdom Come* (1903) and *The Trail of the Lonesome Pine* (1908). He is buried in Paris, France, where he died on July 8, 1919.

PERRYVILLE

Elizabeth Madox Roberts was born in 1886 in this city ten miles west of Danville. Her birthplace was at the top of the hill on East Third Street, in an unnumbered house since destroyed. After winning several prizes for her poetry, including the John Reed Memorial Prize, Roberts turned to fiction. Her first novel, *The Time of Man* (1926), about Kentucky sharecroppers, established her position near the top among novelists of the post-World War I period. Roberts was best known for her pastoral novel *My Heart and My Flesh* (1927) and her historical novel *The Great Meadow* (1930), about pioneer life in Kentucky. In her novels Roberts referred to the area around Perryville as Pigeon Roost.

PINEVILLE

In early November of 1931, while **Theodore Dreiser** was in Pineville, a summer resort north of Middlesboro, he stayed at the Continental Hotel, at the corner of Walnut and Virginia, now a parking lot. Pineville lies on the Cumberland River and is surrounded by coal mines. Dreiser, a member of a volunteer committee that included John Dos Passos, was investigating conditions in the coal mines. A local judge wrote an article accusing Dreiser of being a communist intent on stirring up trouble among the miners. Dreiser was also charged with adultery by a grand jury, which set him free on $200 bail. Dreiser avoided prosecution by not returning to Kentucky.

RICHMOND

John Fox, Jr. wrote in his novel *Crittenden* (1900) about Woodlawn, a Georgian brick house in Richmond that was occupied by armies of the North and the South during the Civil War. Richmond is south of Lexington in central Kentucky, at the junction of Routes 227 and 52.

RIVERTON

Jesse [Hilton] Stuart, one of Kentucky's best-known writers, was born on August 8, 1907, in W-Hollow, about five miles from Riverton, which is near Load, in Greenup County. Greenup County lies in the heart of mountain country, in the extreme northeast corner of the state. While living in W-Hollow, Stuart has taught school—he was superintendent of schools in Greenup, the county seat, from 1941 to 1943—and has written his many novels and short stories, nearly all of them about this region of Kentucky. He first gained recognition with a collection of short stories, *Head O' W-Hollow* (1936), and followed his success with more short stories and an autobiographical work, *Beyond Dark Hills* (1938). Then came his first novel, *Trees of Heaven* (1940); *Taps for Private Tussie* (1943), an extremely popular humorous novel about the behavior of a mountain family that suddenly comes into money; and *The Thread That Runs So True* (1949), an autobiographical novel. These were followed by more than twenty books, novels, collections of stories, and reminiscences. Greenup County is called Greenwood in Stuart's fiction. A courthouse plaque in Greenup, provided by Stuart's friends and former pupils, includes the following verse:

By your own soul's law learn to live,
And if men thwart you, take no heed,
If men hate you, have no care;
Sing your song, dream your dream,
Hope your hope, and pray your prayer.

Jesse Stuart in 1970. The novelist has lived in the Riverton area nearly all his life.

SCOTTSVILLE

Opie Read, after learning the newspaperman's trade in Franklin, worked as a reporter on the Scottsville *Argus,* in the hills of Allen County in southern Kentucky. Read did not establish himself as a humorist until after he moved on to Arkansas in the 1880s.

SHELBYVILLE

Alice Hegan [Caldwell] Rice, who was to achieve great success as an author of children's stories, particularly *Mrs. Wiggs of the Cabbage Patch* (1901), was born on January 11, 1870, in this tobacco-market city west of Frankfort.

SPRINGFIELD

Elizabeth Madox Roberts, who was born in nearby Perryville, lived on North Walnut Street in Springfield, about twenty-five miles west of Danville. Roberts's last published work was *Not by Strange Gods* (1941), a collection of stories mainly about Kentucky women. Roberts died on March 31, 1941, in Orlando, Florida, her winter home for many years, but she was buried on Cemetery Hill in Springfield.

TILFORD

Olive [Tilford] Dargan was born in 1869 in this village on Route 105, off Route 62, southwest of Elizabethtown. She attended public schools here until she was ten. Dargan's works included a collection of lyric poetry, *Lute and Furrow* (1922); and the novel *Call Home the Heart* (1932).

WINCHESTER

[John Orley] Allen Tate, the poet and critic, was born on November 19, 1899, in this town twenty miles east of Lexington. He received his early education at home. Among Tate's works were *Mr. Pope and Other Poems* (1928) and *The Mediterranean and Other Poems* (1936). Tate was married for a time to **Caroline Gordon,** also a native of Kentucky, who was born on October 6, 1895, in Todd County. Her novel *Penhally* (1931) followed a Kentucky family through four generations. Tate and Gordon wrote *The House of Fiction* (1950), for many years a standard work for students of literature.

Furl that Banner, for 'tis weary;
Round its staff 'tis drooping dreary;
Furl it, fold it—it is best;
For there's not a man to wave it,
And there's not a sword to save it,
And there's not one left to lave it
In the blood which heroes gave it;
And its foes now scorn and brave it;
Furl it, hide it—let it rest!

—Abram Joseph Ryan,
in "The Conquered Banner"

SASKATCHEWAN
MANITOBA
C A N A D A
ONTARIO
MONTANA
NORTH DAKOTA
MINNESOTA
LAKE SUPERIOR
BISMARCK
DULUTH
MINNEAPOLIS
MICHIGAN
WYOMING
SOUTH DAKOTA
PIERRE
RAPID CITY
WISCONSIN
LAKE MICHIGAN
LAKE HURON
LANSING
ST. PAUL
SIOUX FALLS
MADISON
MILWAUKEE
Missouri R.
IOWA
NEBRASKA
DES MOINES
OMAHA
CHICAGO
DETROIT
LAKE ERIE
COLORADO
LINCOLN
COUNCIL BLUFFS
INDIANA
CLEVELAND
OHIO
ILLINOIS
SPRINGFIELD
KANSAS CITY
JEFFERSON CITY
ST. LOUIS
INDIANAPOLIS
COLUMBUS
CINCINNATI
TOPEKA
KANSAS
WICHITA
Ohio R.
MISSOURI
SPRINGFIELD
KENTUCKY
NEW MEXICO
OKLAHOMA
OKLAHOMA CITY
AMARILLO
ARKANSAS
TENNESSEE
Mississippi R.
FORT WORTH
DALLAS
EL PASO
TEXAS
AUSTIN
HOUSTON
LOUISIANA
MEXICO
GULF OF MEXICO

MIDWEST & GREAT PLAINS STATES

OHIO

AKRON

Ruth McKenney, best known as the author of *My Sister Eileen* (1938), worked for the Akron *Beacon-Journal* in 1932. She returned to Akron in 1936 to collect material for *Industrial Valley* (1939), her study of labor conditions in Akron in the early 1930s.

ANDOVER

Clarence Darrow, one of the best-known trial lawyers of his day, moved to this village in the northeastern corner of Ohio in 1880. Darrow lived here for several years and it was here, well before he began to write books, that he conducted his first law practice.

ASHTABULA

Clarence Darrow moved to the city of Ashtabula, in northeastern Ohio, in 1883 and practiced law for four years. **William Dean Howells,** born in Ohio, lived in Ashtabula briefly as a boy in the early 1850s.

BEREA

Raymond Moley was born on September 27, 1886, in Berea, a city just southwest of Cleveland. In 1906 he received a Ph.B. degree from Baldwin-Wallace College at 275 Eastland Road. Moley was founder and editor of the political journal *Today,* and became editor of *Newsweek* when it merged with *Today* in 1936. His books included *After Seven Years* (1939), about his experiences as a member of Franklin D. Roosevelt's administration.

CAMDEN

Sherwood Anderson, best known for *Winesburg, Ohio* (1919), his collection of stories describing the frustrations of life in a small Ohio town, was born in Camden on September 13, 1876. Camden, west of Dayton, was then a quiet farming community. The Anderson family lived here until about 1880, when Sherwood was four years old. Sherwood Anderson's birthplace at 142 South Lafayette Street is private but marked. The building at 29 North Main Street, now private, where Anderson's father operated a harness shop, looks much as it did in Anderson's day. The Eleanor I. Jones Archives of the Prebie County District Library, at 104 South Main Street, has considerable information on Anderson. Anderson's *Tar, A Midwest Childhood* (1926) was a fictional account of his life.

William Dean Howells.

You might have painted that picture,
I might have written that song;
Not ours, but another's the triumph,
'Tis done and well done—so 'long!

—Edith Matilda Thomas,
in "Rank-and-File"

DICTIONARY, *n. A malevolent literary device for cramping the growth of a language and making it hard and inelastic. This dictionary, however, is a most useful work.*

PROOF-READER, *n. A malefactor who atones for making your writing nonsense by permitting the compositor to make it unintelligible.*

—Ambrose Bierce,
in *The Devil's Dictionary*

CEDARVILLE

Whitelaw Reid, newspaperman, historian, and diplomat, was born on October 27, 1837, at his family's home at 2587 Conley Road, about three miles from Cedarville, northeast of Xenia. Reid grew up and attended school in Xenia. After his graduation from college, he returned home and taught school for two years before buying and editing the Xenia *News.* Reid's home is privately owned and is being restored by its owners. It is open for tours conducted by the local historical society and upon request.

CHAGRIN FALLS

Hart Crane spent Christmas of 1930 with his father and stepmother in this town southeast of Cleveland. The poet's father ran Crane's Canary Cottage, a restaurant at 87 West Street, in a building now used to house several shops and businesses. The exterior of the building looks much as it did in 1930. A photograph of Crane standing outside of this building is reproduced on the dustjacket of *Voyager, A Life of Hart Crane* (1969), the standard biography by John Unterecker.

CHATHAM

Edith Matilda Thomas, whose poems in the classical style were collected in such volumes as *The Inverted Torch* (1890) and *The Flower From the Ashes* (1915), was born on August 12, 1854, in Chatham, about twenty miles west of Akron.

CHESTER

Ambrose [Gwinett] Bierce was born on June 24, 1842, in a small settlement on Horse Cave Creek southeast of the community of Chester in Meigs County, southeastern Ohio. Bierce lived here until about age nine, when his family moved to Indiana. Bierce had a powerful influence on American literature in the late nineteenth century and especially on writers on the West Coast. His unsparing critical style earned him the title "Bitter Bierce." His best-known works included the collections *Tales of Soldiers and Civilians* (1891) and *Can Such Things Be?* (1893), and the still-read *The Devil's Dictionary* (1911), an excellent example of Bierce's cynical and often brilliant wit.

CHILLICOTHE

Martha Farquharson Finley, born in this city in southern Ohio on April 26, 1828, wrote approximately one hundred novels for girls, and is best remembered for her Elsie Dinsmore series, comprising twenty-eight volumes published between 1867 and 1905. **Charles Fletcher Lummis,** from 1882 to 1884, was editor of the Scioto *Gazette,* forerunner of the current Chillicothe *Gazette.* In Lummis's time, the newspaper was located at 54 West Main Street, in a building that is no longer standing. The *Gazette* is now published at 50 West Main Street.

Burton E. Stevenson was born in Chillicothe on November 9, 1872, and spent most of his life here. Stevenson worked for several newspapers in Chillicothe before becoming, in 1899, head of the Chillicothe and Ross County Public Library at 140–146 South Paint Street. Stevenson held that post until 1957 and during that time wrote nearly forty books. He edited numerous others, including *The Home Book of Verse* (1912) and *The Home Book of Quotations* (1934). The library has a portrait of Stevenson as well as a collection of his books. Stevenson lived in a large frame house at 46 Highland Avenue, where he died on May 13, 1962.

CINCINNATI

In 1790 the small community of Losantiville, on the Ohio River at the mouth of the Licking River, was given the name Cincinnati. The introduction of steamboat travel on the Ohio and Mississippi rivers in the early 1800s turned Cincinnati

Birthplace of Whitelaw Reid shown in a drawing first published in 1874 in the *Combination Atlas Map of Greene County, Ohio.* The house was remodeled shortly after the drawing appeared and now looks quite different.

into a boom town and cultural outpost at the edge of the western frontier. Cincinnati was still growing rapidly by 1835, when the *Western Messenger,* a religious and literary monthly, began publishing. This journal, published here and in Louisville, Kentucky, until 1841, was edited in 1838 and 1839 by **William Henry Channing,** who was pastor of the Unitarian Church in Cincinnati during those years. The *Western Messenger* published works by many well-known writers, including Ralph Waldo Emerson, Margaret Fuller, and Francis Parkman.

Cincinnati has been the subject or locale of numerous books, essays, and travel notes by both English and American writers. **Charles Dickens** traveled through Cincinnati in 1842, and soon wrote of the city in one of the more benevolent passages of his book *American Notes* (1842). A century later Walter Havighurst wrote about the city in *Voices on the River* (1942) and again in *River to the West* (1970), which traces the history of the Ohio River. H. L. Mencken wrote of Cincinnati in "A Neglected Anniversary," his classic tongue-in-cheek essay on the introduction of the bathtub to America, first published in the New York *Evening Mail* on December 28, 1917. According to Mencken, the first bathtub was installed in 1842 in "a large house with Doric pillars, standing near what is now the corner of Monastery and Oregon streets." People took Mencken's imaginings as facts, despite his repeated protests, and the essay even entrapped unwary compilers of reference books.

Writers born in or near Cincinnati include **Alice Cary,** born on April 26, 1820, and her sister **Phoebe Cary,** born on September 4, 1824. Both poets were born in a farmhouse near the site on which was to stand their own home, now known as Cary Cottage, at 7000 Hamilton Avenue in the Cincinnati suburb of North College Hill. Cary Cottage, now a museum, is open to visitors on Sundays free of charge. **Louis Kronenberger,** born in Cincinnati on December 9, 1904, contributed to American letters as an editor, critic, historian, and novelist. He also taught theater arts at Brandeis University and served as its librarian for many years. Among his books were the novel *A Month of Sundays* (1961) and *The Republic of Letters* (1955), a critical literary history. Kronenberger's home at 3442 Reading Road is no longer standing, but Hughes High School, from which he was graduated in 1921, still operates at 2515 Clifton Avenue.

Writers who have lived in Cincinnati include **Kay Boyle,** the novelist and short-story writer, who studied at the Cincinnati Conservatory of Music, on Highland Avenue at Oak Street, from 1916 to 1919. She worked in the city in 1919 and 1920. **Mary Catherwood,** the novelist and short-story writer, lived and wrote in Cincinnati in the late 1870s. **Irvin S. Cobb** worked for the Cincinnati *Post* in 1902 and 1903. The newspaper was then located on the site of the present Cincinnati Convention Center. In 1910 **Russel Crouse,** the playwright and producer, worked as a reporter for the Cincinnati *Commercial-Tribune,* then located at Fourth and Race streets but no longer publishing. In 1917 Crouse worked as reporter for the Cincinnati *Post.* **Stephen Foster** came to Cincinnati in 1846 to work as a bookkeeper at his brother's store. While living in the city, Foster saw publication of his song "Open Thy Lattice, Love." After working in Cincinnati for several years, Foster returned to his parents' home in Pennsylvania, convinced that his future was in music, not business.

Lafcadio Hearn lived in Cincinnati from 1869 to 1877; from 1869 to 1871 he stayed at Henry Watkins's print shop at 230 Walnut Street, and from 1872 to 1874 he lived at 215 Plum Street and worked for the Cincinnati *Enquirer,* then located at 247 Vine Street. In 1874 Hearn married Mattie Foley, a woman of partially black ancestry. He broke the law in doing so, and lost his job because of the marriage. The Hearns settled at 114 Longworth Street, and Hearn found a position with the Cincinnati *Commercial-Tribune* in 1875. In 1877 the newspaper sent Hearn to New Orleans to write a series of articles, and then fired him when he failed to do so. Hearn decided not to return to Cincinnati.

Cincinnati is a beautiful city; cheerful, thriving, and animated. I have not often seen a place that commends itself so favourably and pleasantly to a stranger at the first glance as this does: with its clean houses of red and white, its well-paved roads, and footways of bright tile.

—Charles Dickens, in *American Notes*

Lafcadio Hearn.

Home of Alice and Phoebe Cary in Cincinnati. The house, built by their father in 1840, was made of bricks manufactured on the Cary farm.

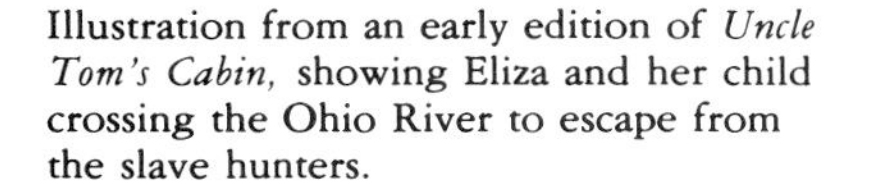
Illustration from an early edition of *Uncle Tom's Cabin,* showing Eliza and her child crossing the Ohio River to escape from the slave hunters.

Other writers associated with Cincinnati include **Sinclair Lewis,** who stayed at the Queen City Club, at 331 East Fourth Street, while on a promotional tour of nearby cities for his novel *Main Street* (1920). Lewis also worked on *Babbitt* (1922) here, and toured the city for background material for the novel. **Joaquin Miller,** the Poet of the Sierras, and his friend Harr Wagner visited Cincinnati for several weeks in the late 1890s during a cross-country lecture tour. They were the guests of Rev. E. R. Wagner, who lived in the parsonage of St. Paul's English Lutheran Church, at Cook and Draper streets. While here, Miller gave a reading at the University of Cincinnati.

> *The bravest battle that ever was fought;*
> *Shall I tell you where and when?*
> *On the maps of the world you will find it not;*
> *It was fought by the mothers of men.*
>
> —Joaquin Miller,
> in "The Bravest Battle"

The Beecher family home in Cincinnati, where Harriet Beecher lived until her marriage to Calvin Stowe.

David Graham Phillips worked as a correspondent and reporter for the Cincinnati *Times-Star* in 1887, and for the Cincinnati *Commercial Gazette* from 1888 to 1890. It was the Cincinnati *Gazette* that hired **Whitelaw Reid** as a correspondent during the Civil War.

Perhaps the best known of the writers associated with Cincinnati are Harriet Beecher Stowe, Frances Trollope, the English novelist and travel writer, and Mark Twain. In 1832, **Harriet Beecher** moved here with her family when her father, Lyman Beecher, was made president of Lane Theological Seminary, which stood at 2820 Gilbert Avenue, now the site of a Cadillac dealership. The Beechers lived in a house at 2950 Gilbert Avenue, which, until recently, was a museum administered by the Ohio Historical Society. Harriet Beecher worked as a teacher in Cincinnati, and during this time published *An Elementary Geography* (1835). In 1836 she married Calvin Stowe, a teacher at her father's school, and lived in Cincinnati with him until 1850. During these years, Mrs. Stowe witnessed the growing storm over slavery that in a few years would engulf the nation. Across the Ohio River was the slave state of Kentucky, from which the character Eliza in Mrs. Stowe's novel *Uncle Tom's Cabin* (1852) fled across the ice floes on the river to safety in Ohio. While in Cincinnati, Mrs. Stowe is said to have written the tales and sketches that appeared in her collection *The Mayflower* (1855).

Frances Trollope arrived in Cincinnati in February 1828. She, her children, and her servants stayed at a hotel briefly before renting rooms, and in May moved to larger quarters in Mohawk, then a suburb of Cincinnati. The highlight of Mrs. Trollope's stay in Cincinnati, until she left in March 1830, was the construction of her Bazaar, a four-story brick building intended to house a center for trade, business, and culture. The building, topped by a twenty-four-foot dome crowned with a Turkish crescent, and located on Third Street opposite the current Western and Southern Life Insurance parking garage, was doomed from the beginning. When Mrs. Trollope could not pay the men who were building it, they walked off the job. She asked her husband, who had remained behind in England, to buy and ship fine items to sell in the Bazaar, but he sent junk. She fell ill, the building was auctioned off, and the family was evicted from the cottage in Mohawk. By the time she left Cincinnati, Mrs. Trollope had determined to see the rest of the young country and write a book about her experiences in it. She described her adventures in Cincinnati in that book, *Domestic Manners of the Americans* (1832). When her son **Anthony Trollope,** the distinguished English novelist, visited Cincinnati in the 1850s, the Bazaar, known by the city's residents as Trollope's Folly, was being used as a medical institute of sorts. It was used for a variety of other purposes over the years, but was eventually torn down.

During the winter of 1856–57, **Mark Twain,** still calling himself Samuel Clemens, worked as a journeyman printer for a company at 167 Walnut Street and lived in a boarding house at 76 Walnut Street. Twain later wrote in his *Autobiography* (1924) that while living at the boarding house, he met a man named Macfarlane, whose philosophy of evolution and man greatly influenced him. Macfarlane's philosophy, as expounded by Twain, so closely reflects Twain's own later views of man and the universe, that it is questionable whether Macfarlane ever existed.

CLEVELAND

Cleveland, Ohio's largest city, has an impressive literary heritage. Writers born in this city include **Charles W. Chesnutt,** born on June 20, 1858, a pioneer in black literature and author of *The Conjure Woman* (1899) and *The House Behind the Cedars* (1900); and **Avery Hopwood,** a successful playwright in the early 1900s, born on May 28, 1882. Cleveland was the birthplace of another playwright, **Jerome Lawrence,** on July 14, 1915. Lawrence lived at 8822 Esterbrook Avenue as a child, and later at 12313 Edmonton Avenue. He attended Patrick Henry Junior High School and Glenville High School. While in school he wrote plays, interviews, and a newspaper humor column. **Alfred Henry Lewis,** best known for his stories about life in the Southwest, was born in Cleveland in 1858. **Sarah Chauncey Woolsey,** author of the series of books for girls that began with publication of *What Katy Did* (1872), was born in Cleveland on Janurary 29, 1835.

Among the writers associated with Cleveland is **Sherwood Anderson,** who suffered a nervous breakdown in late November 1911 in the city of Elyria, southwest of Cleveland. After several days of wandering, Anderson, then thirty-five years old, appeared at a drug store at 152nd Street and Aspinwall Avenue, dazed and apparently suffering from exposure. He was taken to Huron Road Hospital. This breakdown marked a turning point in Anderson's life, forcing him to choose between a career as businessman and the life of a writer.

Hart Crane in 1908 came to his grandparents' spacious home at 1709 East 115th Street, no longer standing, and lived there until he was seventeen years old. Crane lived in a tower room of the mansion, and some have noted that towers figure prominently in the poet's work. While in Cleveland, Crane attended East High School on Eighty-second Street and published his first poem, "C33," in the Greenwich Village publication *Bruno's Weekly* in September 1916. By the end of 1916, Crane acted on his resolve to move to New York City, but he returned to Cleveland in 1918 and remained here for several more years. **Charles B. Driscoll,** newspaperman and writer, was associate editor of the Cleveland *Press,* at 901 Lakeside Avenue, in 1924.

Herbert Gold lived in Cleveland in the early 1950s. While here he lectured at Case Western Reserve University, reported for the Cleveland *Press,* and worked on his novel *The Prospect Before Us* (1954), which is set in Cleveland, on Prospect Street. Other Cleveland sites that figure in Gold's writing are the Central Market at East Fourth and Bolivar, University Circle, and Quincy Avenue.

Walter Havighurst wrote of Cleveland in his book *The Long Ships Passing* (1942). **Langston Hughes** lived here in the early 1900s and attended Central High School. He wrote his first short story and poems for the school magazine. **Ruth McKenney** moved to Cleveland with her family when she was eight years old. She grew up in the city, worked nights in a print shop in her early teens, and was graduated from Shaw High School in 1928. **George Jean Nathan,** the critic and editor, moved here with his family in 1886, when he was four years old. He wrote his first long article, "Love: A Scientific Analysis," here at the age of sixteen. Nathan attended Cleveland High School in the late 1890s. **Opie Read,** the newspaperman and humorist, was an editor of the Cleveland *Leader* in 1881.

Conrad Richter worked as a private secretary for two years here. In September 1913, while living in Cleveland, Richter had his first work of fiction accepted for publication. Richter also wrote children's stories here for *John Martin's Book,* and published his own magazine for children, the *Junior Magazine Book,* for which he wrote nearly everything, including the advertisements. **Edward Rowland Sill,** the poet and essayist, died in a Cleveland hospital on February 27, 1887. **Rex Stout** came to live in Cleveland in 1907, after his discharge from the Navy. He worked here in several capacities, including window dresser at the May Company department store, clerk in a law office on Euclid Avenue, and salesman in a tobacco store. **Constance Fenimore Woolson,** the novelist and short-story writer, lived in Cleveland from shortly after her birth in 1840 to 1869. She used Cleveland as a setting for her book *The Old Store House* (1873), which she published under the name **Anne March. Philip Wylie** lived in Cleveland early in his life, after his family moved here in 1904.

> *We quitted Cincinnati the beginning of March, 1830, and I believe there was not one of our party who did not experience a sensation of pleasure in leaving it. We had seen again and again all the queer varieties of its little world; had amused ourselves with its consequence, its taste, and its* ton, *till they had ceased to be amusing. . . . The only regret was, that we had ever entered it; for we had wasted health, time, and money there.*
>
> —Frances Trollope, in *Domestic Manners of the Americans*

CLYDE

Sherwood Anderson moved to this village in northern Ohio with his family in 1884 and remained here until 1896. His family lived in a house still standing at 129 Spring Avenue, and later moved to a house in a run-down section of Clyde known as Piety Hill. Anderson's father, once the owner of a harness shop, started here as a harness maker in another man's shop, but his fondness for liquor and long stories made it difficult to keep the family coffers supplied. The elder Anderson eventually started a painting company and employed his sons to paint while he, more often than not, regaled them with yarns about the Civil War. Sherwood Anderson attended school regularly until about the age of eleven, after which he worked at different jobs to help his family, including stints as newsboy and stable groom. He worked with such intensity that he earned the nickname Jobby. Anderson later used his recollections of his youth in Clyde in several somewhat autobiographical books, and it is believed that Clyde was his model for the fictional town whose stories are told in *Winesburg, Ohio* (1919).

Snapshot of Sherwood Anderson's home on Spring Avenue in Clyde. The Clyde Public Library, a short walk away at 222 West Buckeye Street, maintains a large collection of books by and about Anderson.

COLUMBUS

Ohio's capital city has been the home and workplace of an impressive list of writers. Some have lived here by choice, and others seemed brought to Columbus by fate. In the latter group are the native-born writers, including **Ketti Frings,** author of the Pulitzer prize-winning play *Look Homeward, Angel* (1958), based on the novel by Thomas Wolfe; and **Donald Ogden Stewart,** the screenwriter and author, born on November 30, 1894.

Always a deep sleeper, slow to arouse . . . I was at first unconscious of what had happened when the iron cot rolled me onto the floor and toppled over on me. . . . The racket, however, instantly awakened my mother, in the next room, who came to the immediate conclusion that her worst dread was realized: the big wooden bed upstairs had fallen on father.

—James Thurber, in "The Night the Bed Fell"

Perhaps the writer most closely associated with Columbus is **James [Grover] Thurber**—lovers of humor say he put Columbus on the literary map with "The Night the Bed Fell." Thurber was born on December 8, 1894, in a house at 251 Parsons Avenue. In 1898 the Thurber family moved to a three-story brick house at 921 South Champion Avenue, where they lived until 1901, when the family moved to Washington, D.C. The three years here were among the happiest of Thurber's life. In 1903 the Thurbers returned to Columbus, to live in a boarding house known as the Park Hotel, before moving in 1905 to Thurber's grandfather's home, a two-story brick mansion on Bryden Road, where they remained until 1907. Thurber was never happy in his grandfather's house, and from 1905 to 1910 spent much time at the home of the practical nurse who had brought him into the world, "Aunt" Margery Albright, who lived at 185 South Fifth Street.

Thurber attended the Sullivant School, in a three-story brick building. One of his classmates was Donald Ogden Stewart. At Sullivant an older, tougher student named Floyd became bodyguard to the frail Thurber when he found out that Thurber knew how to pronounce "Duquesne." In 1907, the Thurber family moved to a house on South Seventeenth Street, the first of a series of houses in which Thurber lived in the next twenty years. He attended Douglas Junior High School and East High School. At the latter he wrote his first published story, "The Third Bullet," which appeared in the school magazine *The X-Rays* in May 1913. Thurber missed a chance to become editor in chief of *The X-Rays* when his mother, knowing her son would be tapped for the job, asked that he not be appointed because he might strain his already poor eyesight. Thurber did become senior class president and was graduated with honors in 1913. He enrolled at Ohio State University in the fall and studied there on and off until he eventually was graduated in 1919. He began his newspaper career while in college and wrote for the Columbus *Dispatch* from 1920 to 1927.

We reached Columbus . . . and stayed there . . . having excellent apartments in a very large unfinished hotel called the Neil House, which were richly fitted with a polished wood of the black walnut, and opened on a handsome portico and stone verandah, like rooms in some Italian mansion.

—Charles Dickens, in *American Notes*

Other writers associated with Columbus include **Charles Dickens,** who stayed at the Neil House, one of the most elegant hotels of its day, during his tour of America in 1842, and later wrote of his stay in his book *American Notes* (1842). The Neil House made its way into American literature in the opening scenes of the novel *Jennie Gerhardt* (1911), in which Theodore Dreiser told the tragic story of a self-sacrificing woman. In 1890, **Zane Grey,** then about eighteen years old, lived with his family in a house still standing at 108 Lexington.

In the fall of 1851, **William Dean Howells** moved to Columbus with his family and, at the age of thirteen, took a job as compositor for the *Ohio State Journal,* no longer publishing. Howells worked his way up to news editor for the newspaper. While in Columbus, he also worked as correspondent for the Cincinnati *Gazette* and the Cleveland *Herald.* It was while Howells was working as a newspaperman that he and John J. Piatt, a reporter for the *Journal,* published *Poems of Two Friends* (1860). In 1860 Howells wrote a biography of Abraham Lincoln, which helped Lincoln in his campaign for the presidency. As a reward for his campaign work, Howells was appointed U.S. consul in Venice. Howells described his life in Columbus in his autobiographical book *Years of My Youth* (1916). **Ruth McKenney** worked for the Columbus *Dispatch,* at 34 South Third Street, and for the *Ohio State Lantern,* at 242 West Eighteenth Street. This was in the early 1930s, when she was studying at Ohio State University.

William Sydney Porter, soon to become known as **O. Henry,** had an association with Columbus that was not of his choosing. Convicted of embezzlement of bank funds in Texas, Porter entered the Ohio State Penitentiary, at Spring Street and Neil Avenue, in April 1898, to serve a five-year sentence. During his three years and three months in the penitentiary—his sentence was reduced for good behavior—Porter worked as prison pharmacist. He also wrote a number of stories here, including "Whistling Dick's Christmas Stocking," "A Chapparal Christmas Gift," "A Fog in Santone," and "The Enchanted Kiss." Best of all for admirers of O. Henry's work, Porter met a safe-cracker here, one Jimmy Connors, who became the model for the character Jimmy Valentine in the short story "A Retrieved Reformation." The story in turn provided the basis for the hit play *Alias Jimmy Valentine* (1909) by Paul Armstrong. Ohio State Penitentiary has yet another foothold in the O. Henry saga. Porter, while here, experimented with a number of pen names, finally settling on the name O. Henry. The source of the name, and the reason for Porter's preference for it over others he tried, remain a mystery, but the penitentiary honors the name today by calling its athletic field the O. Henry Field.

A far less restrictive Columbus institution is Ohio State University, where **Harlan Hatcher,** who received a Ph.D. here in 1928, was a teacher and administrator until 1951. Hatcher was state director of the Ohio Federal Writers' Project and edited *The Ohio Guide* (1940). Hatcher's other books included *The Buckeye Country: A Pageant of Ohio* (1940) and *The Western Reserve: The Story of New Connecticut in Ohio* (1949). **Jerome Lawrence** attended Ohio State from 1933 to 1937, and served as a visiting professor in 1969. Columbus figures in Lawrence's play *Jabberwock* (1974), written with Robert E[dwin] Lee. Lawrence and Lee collaborated on several plays, the most notable being *Inherit the Wind* (1955). Ohio State University has a Lawrence and Lee collection.

CUYAHOGA FALLS

Edward Rowland Sill, the poet and essayist, after the death of his parents in the early 1850s came to live with his uncle in this city in northeastern Ohio. His uncle's home then stood on Front Street, between Portage Trail and Broad Boulevard. Sill, who published commercially only one book of poems in his lifetime, *The Hermitage and Other Poems* (1868), grew up in Cuyahoga Falls, and made the city his permanent home after his retirement from the University of California in 1882. He died here four years later and was buried in Oakwood Cemetery, on Sixth Street. A junior high school in the city is named in his honor.

Let the noisy crowd go by:
In thy lonely watch on high,
Far from the chattering tongues of men,
Sitting above their call or ken,
Free from links of manner and form
Thou shalt learn of the winged storm—
God shall speak to thee out of the sky.

—Edward Rowland Sill, in "Solitude"

DAYTON

Writers associated with Dayton include **Irving Babbitt,** critic, scholar, and leader of the

New Humanism movement, who was born here on August 2, 1865; **Hart Crane,** the poet, who lived with his family in Dayton briefly; and **Paul Laurence Dunbar,** who was born in Dayton on June 17, 1872, in a house that apparently is no longer standing. Dunbar's home from 1872 to 1906, a two-story brick house at 219 North Summit Street, is now a museum open to visitors. Dunbar was graduated from Steele High School, at Main Street and Monument Avenue, where he edited the student magazine and wrote the class song for graduation exercises in 1891. Later, while working as an elevator operator, Dunbar himself published a book of his poems, *Oak and Ivy* (1893), and sold copies to people who rode his elevator. Dunbar's ability to capture black speech and humor, while reflecting the realities of black life, makes his verse readable even today. In 1896 he received national attention when William Dean Howells wrote a full-page review of Dunbar's second collection, *Majors and Minors* (1895), for *Harper's Weekly.* After this exposure to a larger audience, Dunbar spent much time traveling and lecturing in the United States and abroad. He returned to Dayton in about 1900 with an advanced case of tuberculosis, to spend his last years in his home on North Summit Street, where he died on February 9, 1906. Among his books were *Lyrics of Lowly Life* (1896), *Lyrics of Love and Laughter* (1903), and four novels, including *The Sport of the Gods* (1902). Dunbar is buried in Woodland Cemetery, on Woodland Avenue.

William Dean Howells, who helped Dunbar achieve national recognition, lived briefly in Dayton as a boy. He learned to set type on his father's newspaper, the Dayton *Transcript.*

Lay me down beneaf de willers in de grass,
Whah de branch'll go a singin' as it pass,
An' when I'se a-layin' low,
I kin heah it as it go,
Sayin' "Sleep, mah honey, tak' yo' res' at las'."

—Inscription on the gravestone of
Paul Laurence Dunbar

DELAWARE

This city north of Columbus is the home of Ohio Wesleyan University. **Walter Havighurst** was a student here from 1919 to 1921. The University's Bayley Collection contains—among many other items of interest—a rare photograph of the Camden, New Jersey, home of Walt Whitman. **Paul Kester,** best known in his time for his dramatic adaptations from novels and foreign-language plays, was born in Delaware on November 2, 1870. Kester's birthplace at 113 West Winter Street is now an apartment house.

ELYRIA

Sherwood Anderson in 1907 founded a mail-order paint company on Lodi Street in this city southwest of Cleveland. The building that housed the paint company no longer exists, but Anderson's apartment at 401 Second Street and his home at 229 Seventh Street are still in use. While living in Elyria, Anderson divided his time between his business career and his writing. The strain of this regimen slowly wore him down until,

House in Dayton where Paul Laurence Dunbar lived. Dunbar became well known for his verse in the black dialect.

Step by step I am assendin the ladder of fame; step by step I am climbin to a proud eminence.

—Petroleum V. Nasby,
in "The Great Presidential Excursion"

Oh, singing was only in Heaven
Ere Lucifer's melody came,
But when Lucifer's harp-strings grew loud in their sighing,
When he called up the dragons by name—
The song was the sorrow of sorrows,
The song was the Hope of Despair,
Or the smile of a warrior falling—
A prayer and a curse and a prayer—

—Vachel Lindsay,
in "The Last Song of Lucifer"

in November 1911, he suffered a nervous breakdown. On the fateful day, he is said to have remarked to his secretary, "My feet are cold and wet. I have been walking too long on the bed of a river." He left the paint factory after saying those words and disappeared. Four days later he reappeared in Cleveland, dazed and in poor physical condition. Anderson tried afterward to make it seem as if the breakdown had been a stunt he had thought up to gain attention, but the incident proved a signal to him that he must abandon the businessman's life and become a writer.

Robert E[dwin] Lee, who collaborated with Jerome Lawrence on such plays as *Inherit the Wind* (1955) and *The Gang's All Here* (1959), was born in Elyria on October 15, 1918. His home at 137 Pasadena Avenue, believed to be his birthplace, is now private.

FARMDALE

Clarence Darrow was born on April 18, 1857, in this community near Kinsman in the northeastern corner of Ohio. His birthplace, a one-and-a-half-story white frame cottage at 5788 Mayburn-Barclay Road, is marked with a plaque. Darrow lived here until the age of seven.

FINDLAY

Russel Crouse, born in this city forty miles south of Toledo on February 20, 1893, collaborated with Howard Lindsay on such plays as *Life with Father* (1939) and the Pulitzer Prize-winning play *State of the Union* (1945). Crouse's birthplace at 324 West Hardin Street is believed to be a private residence. On March 21, 1861, **David Ross Locke** published, in the Findlay *Jeffersonian,* the first of his many satirical sketches written under the name **Petroleum V[esuvius] Nasby.** Locke was editor of the *Jeffersonian* from 1861 to 1865, the Civil War years. Perhaps the most faithful reader of Nasby's sketches was Abraham Lincoln, who treated his guests in the White House to readings of favorite passages from these humorous and sometimes brilliant commentaries on affairs in the war-torn country. Lincoln is reported to have said: "I am going to write to Petroleum to come down here, and I intend to tell him if he will communicate his talent to me, I will swap places with him." Nasby's letters were published in *The Nasby Papers* (1864) and other volumes.

FREMONT

The Rutherford B. Hayes Library, at 1337 Hayes Avenue in this city southeast of Toledo, has photographs, letters, and manuscripts of David Ross Locke [Petroleum V. Nasby].

GAMBIER

This town fifty miles northeast of Columbus is the home of Kenyon College, from which the literary quarterly *Kenyon Review* has been issued, with only a nine-year interruption, since 1939. The *Review* was edited at the start by **John Crowe Ransom,** who taught at the college from 1937 to 1958. Ransom lived for a time in a building owned by the college and now used as a craft center. In the 1950s Ransom built a house behind the college-owned house. This later house is private. Among the original members of the quarterly's advisory board were Allen Tate, Mark Van Doren, and Robert Penn Warren. Under Ransom, the *Review* became established as a major literary publication, and having one's work appear in its pages has been a sign of accomplishment ever since. *Kenyon Review* ceased publication in 1970 but was revived in 1979. Ransom died in Gambier on July 3, 1974, and was buried in Kenyon College Cemetery. The Kenyon administration building is named Ransom Hall in his honor.

Other writers associated with Kenyon College are **Randall Jarrell,** who taught here from 1937 to 1939, and **Robert Lowell,** who was graduated in 1940. Both poets were attracted to Kenyon College by the presence of Ransom on the faculty, as were other writers. **Peter Hillsman Taylor,** the novelist and short-story writer, was graduated from Kenyon in 1940 and later taught at the college. **Yvor Winters,** the poet, taught at Kenyon from 1948 to 1950, and **James Arlington Wright,** whose *Collected Poems* (1971) won a Pulitzer Prize, received a B.A. degree from Kenyon in 1952.

Ransom Hall at Kenyon College in Gambier. The building, formerly the Alumni Library, was dedicated to John Crowe Ransom in 1964.

GARRETTSVILLE

[Harold] Hart Crane, best known for his long poem *The Bridge* (1930), was born on July 21, 1899, in a house at 10688 Freedom Street, now private. Crane's grandfather, a prominent community and business leader in Garrettsville, lived in the house next door, now private. Crane and his family lived in Garrettsville until late in 1903. In 1978, long after Hart Crane's death at sea on April 27, 1932, a rose granite marker at High and Maple streets was dedicated to the memory of the dead poet. An inscription on the side of his father's gravestone in Park Cemetery, on Maple Street, reads:

Harold Hart Crane
1899–1932
Lost at Sea

GENEVA

Edward S. Ellis, born on April 11, 1840, in this city in northeastern Ohio, was one of the most successful dime novelists writing in the nineteenth century, his works appearing under seven different pen names. One of his novels, *Seth Jones; or, The Captives of the Frontier* (1860), is said to have sold 450,000 copies when it was first issued. Ellis also wrote a six-volume *History of the United States* (1896).

GRANVILLE

Mary Catherwood, a successful writer of historical romances, attended Granville Female College in this city northeast of Columbus and was graduated in 1868 after completing a four-year course in three years. After graduation she went on to teach here and elsewhere.

GREENVILLE

Lowell [Jackson] Thomas, the well-known travel writer and news commentator, was born on April 6, 1892, in the community of Woodington, northwest of Greenville, in western Ohio.

HAMILTON

William Dean Howells spent his childhood in this city just north of Cincinnati. His family moved here in the 1890s, when Howells was two years old, and left when Howells was ten. Howell's father was editor of the Hamilton *Intelligencer,* and the family lived in three different houses in Hamilton, none believed to have survived. A plaque honoring Howells is mounted on a stone at the northeast corner of the High Street bridge. Howells wrote of his boyhood in Hamilton in *A Boy's Town* (1890). Hamilton was the birthplace, on October 19, 1889, of **Fannie Hurst,** the novelist. Hurst was taken to St. Louis by her parents shortly after her birth, and it was there that she was raised. **Robert McCloskey,** a writer of children's books, was born on September 15, 1914, at 552 Franklin Street. He grew up in Hamilton and was graduated from Hamilton High School in 1932. The Lane Public Library at 300 North Third Street has some of McCloskey's papers. McCloskey, who was also an artist, received his first important art commission when he was hired in the 1930s to do bas-reliefs for the city's municipal building.

HEBRON

Mary Catherwood, after her parents died in the late 1850s, came to live with her maternal grandparents in this community east of Columbus. She attended the local public schools and had an early poem, "Willetta," published in the school newspaper. She used her experiences in Hebron in writing her novel *Craque-O'-Doom* (1881).

HIRAM

While studying at Hiram College in northeastern Ohio from 1897 to 1900, **Vachel Lindsay** produced little more than some illustrations for college annuals. He left without a degree to continue his education in Chicago, not returning until 1931. By then his poetry was nationally recognized, and Hiram College awarded him an honorary degree. The Vachel Lindsay Reading Room of the Teachout-Price Memorial Library at Hiram contains an original copy of Lindsay's poem "The Last Song of Lucifer," Lindsay's painting "Lucifer," an original copy of the tribute "In Memory of Vachel Lindsay," by Sara Teasdale, and several framed copies of Lindsay's poems. **Harold Bell Wright,** the novelist, like Lindsay, attended Hiram College but did not complete his studies. A student for two years, beginning in 1894, Wright contracted pneumonia, injured his eyesight, and was forced to abandon his studies.

KELLEY'S ISLAND

This island in Lake Erie near Sandusky Bay appears as Hazard Island in the novel *Signature of Time* (1949) by Walter Havighurst.

KINSMAN

The Octagon House at 8405 Main Street in Kinsman, in northeastern Ohio, was the home of **Clarence Darrow** from the mid-1860s until the late 1870s. Darrow, seven years old, had moved here with his family from the nearby community of Farmdale. Darrow grew up in Kinsman, studying at Kinsman Academy and attending United Presbyterian Church, on Church Street. He worked in his father's furniture and cabinet shop, which stood next door to the family home. Darrow's parents were free-thinking individuals who had something of a reputation as town characters. Their choice of an eight-sided home may have reflected their unconventionality. Darrow used his experiences in Kinsman in his novel *Farmington* (1904), and wrote of his childhood and youth in *The Story of My Life* (1932). The Octagon House, now private, is being restored. It is listed in the National Register of Historic Places, and a sign marks it as Darrow's boyhood home. The

Clarence Darrow, the great trial lawyer, who spent part of his childhood in Kinsman.

Vachel Lindsay, top row on the left, photographed with other students on the steps of Bowler Hall, the oldest dormitory of Hiram College. Olive Catherine Lindsay, the poet's sister, appears in the top row, extreme right.

My mind goes back to Kinsman because I lived there in childhood, and to me it was once the centre of the world, and however far I have roamed since then it has never fully lost that place in the storehouse of miscellaneous memories gathered along the path of life.

—Clarence Darrow,
in *The Story of My Life*

building that housed his father's shop is now used for apartments. In Darrow's later years, he returned to visit his boyhood town and enjoyed sitting in the front parlor or on the porch of the Colonial Inn, now the site of the Kinsman Clinic, and debating with guests and townspeople.

Other literary residents of Kinsman have included **Edmond Hamilton,** best known for his science fiction, and his wife, **Leigh Brackett,** a screenwriter and author.

LAKEWOOD

Herbert Gold, the novelist and essayist, has reported that he was born in this suburban city five miles west of Cleveland on March 9, 1924; other sources state that he was born in Cleveland or East Cleveland. At the time of Gold's birth, his family lived in Lakewood at 1548 Glenmont Road. Gold attended Lakewood High School at 14100 Franklin Avenue and lived, after his marriage, at 1242 Cove Avenue.

LANCASTER

In 1905 **Lloyd C. Douglas,** later a highly successful novelist, became minister of the First English Lutheran Church, at 220 North Columbus Street, in this city southeast of Columbus. Douglas lived at the church parsonage, at 1023 North Columbus Street.

LIVERPOOL

Charles Bertrand Lewis, once known for his humorous sketches, collected in such volumes as *Brother Gardener's Lime Kiln Club* (1882), was born in this town south of Cleveland on February 15, 1842.

LUCAS

Malabar Farm, on Bromfield Road, Route 1, near Lucas in north-central Ohio, was the home of **Louis Bromfield** from 1940 until his death in 1956. The novelist, born in nearby Mansfield, bought four run-down farms in 1939 and began systematic rejuvenation of the soil. His intention was to build a model farm, which he named for India's Malabar coast, the setting of Bromfield's highly successful novel *The Rains Came* (1937). Bromfield also expanded an existing farmhouse, making it into a thirty-two-room mansion known as the Big House. Here, in a book-lined writing room, Bromfield worked on his later books. He wrote on a card table that stood behind a massive desk of his own design. Bromfield did not like the desk—it was too high—and found the card table more practical, though less elegant. He wrote about his farm in many of his later books, including *Pleasant Valley* (1945), *Malabar Farm* (1948), and *From My Experience* (1955). Bromfield turned Malabar Farm into an agricultural showcase, and many people visited him at his spacious and elegant home. Among them were two motion-picture stars, Lauren Bacall and Humphrey Bogart, who were married in the Malabar rose garden and spent their honeymoon at the farm. Malabar Farm, now run by the state of Ohio, is listed in the National Register of Historic Places, and is open to the public. An admission fee is charged for guided tours of the house.

The Big House on Louis Bromfield's Malabar Farm. The house, kept as it was during the author's lifetime, contains many antiques and works of art as well as a very good natural resources library.

LURAY

Mary [Hartwell] Catherwood was born on December 16, 1847, near this town east of Columbus. Her birthplace was a farmhouse located just west of the Ohio Canal. Her family lived in Luray until the 1850s. Catherwood later wrote in her stories of Buckeye Lake, near here, and of the area's pigeon swamps, but she was best known for her historical romances, such as *The Romance of Dollard* (1889) and *The Story of Tonty* (1890).

MANSFIELD

Louis Bromfield was born in this city on December 27, 1896. His birthplace at 323 West Third Avenue is now private. Bromfield used his memories of life in and around Mansfield in his novels *The Green Bay Tree* (1924), *Possession* (1925), the Pulitzer Prize-winning novel *Early Autumn* (1926), and *The Farm* (1933). The setting of *The Farm* was modeled after his grandparents' farm, which was located just north of Mansfield, on Route 30. The Mansfield campus of Ohio State University maintains a collection of Bromfield's writings, letters, and manuscripts.

MARIETTA

Wilbur Schramm, best known for his books on education and mass communications and for his collections of stories, was born on August 5, 1907, in this city on the Ohio River. His family's home at 602 Cutler Street is now private. Schramm's stories, written in the vein of the American tall tale, were collected in such books as *Windwagon Smith and Other Stories* (1947).

MARTINS FERRY

This city stands on the west bank of the Ohio River, opposite Wheeling, West Virginia. The writer most readers associate with Martins Ferry is **William Dean Howells,** born here on March 1, 1837, in a small brick house, no longer standing, on Second Street between Hickory and Walnut streets. Howells began his apprenticeship in American letters early in life by working on the various newspapers his father published in small towns in Ohio—the family moved often during the 1840s and 1850s—and later on larger newspapers. Howells eventually became editor of *The Atlantic Monthly* and then author of the "Editor's Easy Chair" column in *Harper's Magazine.* This activity made Howells one of the country's foremost literary figures. The book for which he is best known today is *The Rise of Silas Lapham* (1885), a classic novel depicting the life of a self-made man. This work helped prepare the way for the realism of such authors as Stephen Crane, Hamlin Garland, and Frank Norris. Although Howells wrote of his childhood and youth in Ohio in such volumes as *A Boy's Town* (1890) and *Years of My Youth* (1916), other places figure more prominently in these works, since Howells lived in Martins Ferry for only three years.

James Arlington Wright, winner of a Pulitzer Prize in 1972 for his *Collected Poems* (1971), was born in Martins Ferry on December 13, 1927. At Martins Ferry High School, where Wright was enrolled in a vocational training program, he was encouraged by perceptive teachers to study English and Russian literature. Many of Wright's poems mention people and sites in and near Martins Ferry.

MAUMEE

Theodore Dreiser lived in Maumee, southwest of Toledo, for some months during 1899 as a guest of Arthur Henry, editor of the Toledo *Blade.* It was Henry who encouraged Dreiser to turn his talents from newspaper work to fiction. Dreiser here wrote his short story "The Shining Slave Makers."

MINERVA

Ralph Hodgson, the English lyric poet, emigrated to the United States and lived for more than twenty years on his farm in Minerva, about fifteen miles southeast of Canton. His best-known poems, among them "Eve" and "Time, You Old Gypsy Man," were published many years before he moved here, but Hodgson wrote and illustrated a number of books of poetry that were published in Minerva. Hodgson died at his home here on November 3, 1962, at the age of ninety-one.

NILES

This city in northeastern Ohio was the birthplace, on December 3, 1911, of **Kenneth Patchen,** who wrote such volumes of poetry as *Before the Brave* (1936), *Selected Poems* (1947), and *Because It Is* (1960).

NORWICH

The National Road–Zane Grey Museum, on Route 40 in Norwich, east of Zanesville, contains exhibits and items related to the development of the National Road: at one time this road was the primary route between East and West, stretching from Cumberland, Maryland, to Vandalia, Illinois. The museum also contains a collection of items relating to Zane Grey, author of three historical novels dealing with the settlement of Ohio early in our nation's history, about the time of the building of the National Road. Grey's great-grandfather Ebenezer Zane blazed the first public trail into Ohio, and it was on Zane's land that the city of Zanesville was built. Today Route 40, a national highway, traces the route of the original National Road. The museum's collection of materials on Zane Grey includes items from the novelist's days as a college student and as a dentist, his hunting and fishing equipment, and some of his early manuscripts. Also here is a reconstruction of Grey's study in Altadena, California, including a lifelike figure of Grey in his Morris chair, at work on a manuscript. Operated by the Ohio Historical Society, the National Road–Zane Grey Museum is open to the public.

OBERLIN

This city in northern Ohio is the home of Oberlin College, whose former students include **Sinclair Lewis,** who studied here in 1902–03, and **Thornton Wilder,** who attended Oberlin from 1915 to 1917.

OXFORD

Jeremy Ingalls, the poet, has reported that she lived in Oxford at various times between 1941 and 1947, but this town in southwestern Ohio finds its way into literary history primarily because it is the home of Miami University. **Walter Havighurst** began teaching at Miami in 1928 and was a member of the faculty until 1969. While living at 21 University Avenue in the 1930s, Havighurst wrote the novels *Pier 17* (1935) and *The Quiet*

Eve, with her basket, was
Deep in the bells and grass
Wading in bells and grass
Up to her knees,
Picking a dish of sweet
Berries and plums to eat,
Down in the bells and grass
Under the trees.

—Ralph Hodgson,
in "Eve"

Zane Grey, who was born in Zanesville, is well represented in the collections of the National Road-Zane Grey Museum in Norwich.

A pole stuck up out of the wagon like a ship's mast, and on it a square of canvas turned sideways to catch the quartering wind.

A little man in blue denim was riding on the wagon seat. . . . When he hopped down from the wagon he walked with a sailor's roll and sway. . . .

"Ahoy!" he said out of the silence. "Think I'll drop anchor and come ashore for a bit of refreshment."

—Wilbur Schramm,
in "Windwagon Smith"

Shore (1937), as well as *The Upper Mississippi* (1937), a volume in the Rivers of America series. The Walter Havighurst Special Collections Library is part of the university's King Library, which also contains material relating to **William Holmes McGuffey,** who taught at Miami University from 1826 to 1836.

McGuffey, in 1828, bought a house and property on East Spring Street, now on the campus of the university, and by 1833 had completed construction of a six-room brick addition to the original frame house, joining both buildings to form one spacious house. Here McGuffey assembled the first four of his six famous *Eclectic Readers,* which were published between 1836 and 1857. These books, combining McGuffey's own writing with selections from the works of many other writers, sold in the millions in their time and were standard school books for many years. His home, added to and modified by subsequent owners, now houses the McGuffey Museum of Miami University. The museum has a collection of children's textbooks published prior to 1900 as well as copies of the famous McGuffey readers, spellers, and primers. The museum also has a collection of manuscripts, letters, and diaries relating to McGuffey. A National Historic Landmark, the museum is open to the general public.

Miami University also has a collection of books, art material, and memorabilia of **Whitelaw Reid.** Reid, who went on to become editor of the New York *World,* attended the university long after McGuffey had become a household name. It is reported that Reid, while admiring a portrait of McGuffey at the home of McGuffey's son-in-law, Dr. Andrew D. Hepburn, said, in astonishment, "McGuffey had a daughter! Why, I always thought he was a reader." Reid was graduated from Miami University in 1856, at the age of nineteen. **[Frederic] Ridgely Torrence,** the poet and playwright, studied at Miami University from 1893 to 1895 and taught here in 1920 and 1921. He also was resident in creative writing in 1941.

McGuffey Museum of Miami University in Oxford. The house was the home in the 1830s of William Holmes McGuffey, author of the famous series of readers.

> *At first, she loved nought else but flowers,*
> *And then—she loved the rose;*
> *And then—herself alone; and then—*
> *She knew not what, but now—she knows.*
>
> —Ridgely Torrence,
> in "House of a Hundred Lights"

PORTAGE

Don Carlos Seitz, a prolific author and poet, was known best for his biography of *Joseph Pulitzer* (1924). Seitz, who was born in Portage, in northwestern Ohio, on October 24, 1862, worked as a newspaperman throughout his career and was business manager of the New York *World* for twenty-five years. His poetry was collected in *The Buccaneers* (1912) and *Farm Voices* (1918), and his biographies included *Artemus Ward* (1919), *Horace Greeley* (1926), and *James Gordon Bennett* (1928).

RAVENNA

William Tappan Thompson was born near Ravenna, a city in northeastern Ohio, on August 31, 1812. Thompson was known best for his humorous stories about Georgia, where he spent much of his life. Many of Thompson's stories featured Major Jones, a Southern character, whose adventures were collected in such volumes as *Major Jones's Courtship* (1843) and *Major Jones's Sketches of Travel* (1848).

ST. MARY'S

Jim Tully, the novelist, was born on June 3, 1891, in the western Ohio community of Glynwood, some miles from St. Mary's. The son of Irish immigrants, Tully overcame the handicaps of life in an orphanage and early poverty, and learned to read and write. He followed a series of occupations, including hobo, prize fighter, and farmer, before establishing himself as a writer with the help of H. L. Mencken. Among Tully's books was *Beggars of Life* (1924), an account of his early life, which Maxwell Anderson dramatized in *Outside Looking In* (1928).

SHANDON

Albert Shaw, who served as editor of *Review of Reviews* for the entire time the monthly was published, was born on July 23, 1857, in Shandon, a town in southwestern Ohio. *Review of Reviews,* published from 1891 until 1937, featured articles reprinted from other magazines and journals. It ceased publication when it was merged into *Literary Digest,* which itself fell victim to reduced readership following the failure of its straw poll to predict the actual winner of the 1936 presidential election.

SPRINGFIELD

Writers associated with Springfield, the home of Wittenberg University, include **Sherwood Anderson** who, in 1899 and 1900, attended Wittenberg Academy, the preparatory school for what was then Wittenberg College. Anderson and his brother Karl lived at The Oaks, a boardinghouse, now private, at 153 Wittenberg Avenue, where artists and writers often congregated. Anderson's speech at graduation exercises in the spring of 1900 so impressed Harry Simmons, advertising manager of *Woman's Home Companion,* that he offered Anderson a job with the magazine in Chicago. Anderson accepted the offer and did not continue at Wittenberg. Wittenberg Academy no longer exists, but its building, which now is called Recitation Hall, is used for offices by Wittenberg University.

Lloyd C. Douglas received a B.A. degree from Wittenberg College in 1900, an M.A. in 1903, and an honorary doctorate in 1945, after he achieved popularity as a novelist. **Lois Lenski,** author of many books for children, was born in Springfield on October 14, 1893. Her home at 422 Cedar Street is still standing. Warder Public Library, 13 East High Street, has a collection of Lois Lenski material, including manuscripts, photographs, and first editions of her books. **Herbert Quick** was editor, from 1909 to 1916, of the magazine *Farm and Fireside,* published in Springfield by P. P. Mast and Company, predecessor of Crowell-Collier Publishing Company.

STEUBENVILLE

Tad Mosel was born in this steel city on the Ohio River on May 1, 1922, and lived as a child first at 1116 Oregon Avenue and later at 303 South Bend Boulevard. Mosel received a Pulitzer Prize in 1961 for his play *All the Way Home* (1960), adapted from James Agee's novel *A Death in the Family* (1957), itself a Pulitzer Prize-winner.

TOLEDO

Writers associated with this port city at the western tip of Lake Erie include **Russel Crouse,** the playwright, who attended Toledo public schools in the early 1900s; and **David Ross Locke,** who became editor of the Toledo *Blade* in 1865. Locke left Toledo in 1871 to be managing editor of the New York *Evening Mail,* but soon returned to Ohio. He published his humorous and satirical sketches under the name of **Petroleum V. Nasby** in the *Blade* until December 1887. Locke was elected alderman from Toledo's third ward, and held that post at the time of his death, in Toledo, on February 15, 1888. The offices of the Toledo *Blade* are now at 541 Superior Street, and the lobby of the building has been designated by the Society of Professional Journalists, Sigma Delta Chi, as the David Ross Locke Historical Site in Journalism. A plaque honors Locke's contribution to journalism. **Kenneth Rexroth,** as a child, lived in Toledo with his family during and shortly after World War I. The Rexroths lived first at 2552 Lawrence Avenue and later at 2404 Detroit Avenue.

VERNON

Clarence Darrow worked in Vernon as a teacher after the financial panic of 1873 interrupted his education. Darrow lived in nearby Kinsman while teaching in Vernon. During his three years as a teacher, he liberalized the school's rules—he ended the practice of corporal punishment and lengthened the lunch hours and recesses—and began to study law. He later wrote in his autobiography, *The Story of My Life* (1932): "No matter when I go back to my old home I am sure to meet some of the thinning group whom I tried to make happy even if I could not make them wise." The school in which Darrow taught was destroyed by fire in 1914.

WARREN

Writers associated with this city in northeastern Ohio include **Earl Derr Biggers,** born here on August 26, 1884, and best known as the creator of fictional detective Charlie Chan. Biggers's family lived successively at 3 East Clinton, 105 Main, and 309 (now 670) Mercer streets. Only the last of these houses still stands. Biggers attended Warren High School, now the Elm Road Elementary School.

Hart Crane moved to Warren with his parents in late 1903 and lived here until 1908, when family problems and his mother's nervous breakdown caused the family to break up. Here the young Crane gathered impressions he later worked into his long poem *The Bridge* (1930). In that poem, Crane mentioned his father's cannery at the southwest corner of Franklin and South Pine streets, and Central Grammar School on Harmon Street. Crane also described life in his family's homes: first at 249 High Street, and then, in 1908, on North Park Avenue.

Another literary native of Warren was **[Robert] Forrest Wilson,** born on January 20, 1883, who won a Pulitzer Prize for *Crusader in Crinoline* (1941), his biography of Harriet Beecher Stowe.

WIILBERFORCE

This village three miles northeast of Xenia is the home of Wilberforce University, where **W. E. B. DuBois,** the educator and author, taught from 1894 to 1896.

WILLIAMSFIELD

Albion W[inegar] Tourgée, born on May 2, 1838, in this town north of Youngstown, fought for the Union in the Civil War. He later moved to North Carolina, where he became a strong supporter of radical Reconstruction. Tourgée's popular novels, including *A Fool's Errand* (1879) and *Bricks Without Straw* (1880), were based on his experiences in the postwar South.

WOOSTER

Dorothy [Hartzell Kuhns] Heyward was born on June 6 (7?), 1890, in Wooster, a city west of Canton. She collaborated with her husband, DuBose Heyward, on several plays, including *Porgy* (1927)—based on the novel *Porgy* (1925), written by DuBose Heyward—and *Mamba's Daughters* (1939). Dorothy Heyward also wrote the novels *Three-a-Day* (1930) and *The Pulitzer Prize Murders* (1932).

XENIA

Writers associated with this city include **William Dean Howells,** who moved in October

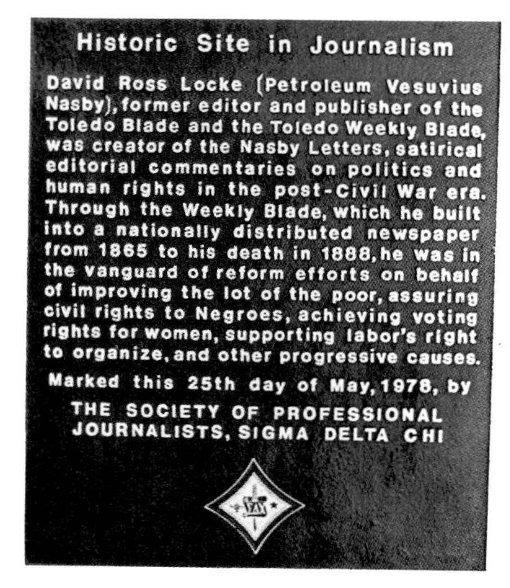

Plaque honoring David Ross Locke in the lobby of the offices of the Toledo *Blade.*

> *The skripter sez, in substance:*
> *Ther was a certin man who hed 2 suns. The yungist hed a taist for that branch of agricultooral persoots known ez soowin wild oats, so he askt the old man for his sheer in the estait. He got it, turnd it into greenbax, and went off. He commenst livin high—bording at big hotels, and keepin trottin hosses, and playin bilyards, and sich. In about a year he run thro his pile, and wuz dead broak. Then his credit playd out, and he wuz in a tight place for his daily bred. The idee struck him that he had better put for hum, wich he did.*
>
> —Petroleum V. Nasby, recounting the parable of the prodigal son

There were cracks in the shingles, through which we could see the stars, when there were stars, and, which, when the first snows came, let the flakes sift in upon the floor.

—William Dean Howells, in *Years of My Youth*

1850 with his family to a dilapidated cabin on the Little Miami River, just southwest of Xenia. Here his father operated a gristmill and a sawmill for a time before moving the family to Columbus. Howells described his life here in *My Year in a Log Cabin* (1893) and in *Years of My Youth* (1916). In the 1840s and 1850s, during the same period that Howells lived near here, **Whitelaw Reid** was growing up at his family's home near Cedarville, a few miles northeast of Xenia. Reid attended Xenia Academy, which stood on East Church Street, and in 1853 left to attend Miami University in Oxford, Ohio. He returned to Xenia in 1858 and became editor of the Xenia *News.* Reid gained national attention in 1860, when he supported Abraham Lincoln in the Republican presidential primary against Ohio's favorite son, Salmon P. Chase.

[Frederic] Ridgely Torrence, the poet and playwright, was born in Xenia on February 27, 1874. Torrence's poetry was praised by such contemporaries as the English poet A. E. Housman and by Edwin Arlington Robinson. Robert Frost, in dedicating his poem "A Passing Glimpse," wrote "To Ridgely Torrence, On Last Looking Into His 'Hesperides.' " Among Torrence's best-known volumes of verse were *Hesperides* (1925) and *Poems* (1941, enlarged edition 1952). His play "Granny Maumee," published in *Plays for a Negro Theater* (1917), was described by Max J. Herzberg as ". . . perhaps the first serious drama about Negroes by an American writer, and the first to be acted by a Negro cast." Torrence's birthplace at 208 North King Street was damaged severely by a tornado in 1974 but has been restored. Torrence is buried in Woodland Cemetery in Xenia.

Birthplace of Ridgely Torrence in Xenia.

YELLOW SPRINGS

This village in southwestern Ohio, about eight miles south of Springfield, is the home of Antioch College. Writers associated with Antioch include **Rod Serling,** the author, playwright, and television writer, who was graduated from Antioch in 1950; **Ridgely Torrence,** who was resident in creative writing here in 1938; and **Hendrik Willem Van Loon,** who was professor of history at Antioch in 1922 and 1923.

YOUNGSTOWN

Writers associated with this industrial city include **Clarence Darrow,** whose first job was in a Youngstown law office in 1878, and **Harry Kemp,** born here on December 15, 1883. Kemp grew up in Youngstown and in the state of Kansas, and through his poetry and his travels earned the sobriquet "the tramp poet." Among his books were the poetry collection *Chanteys and Ballads* (1920) and the novel *Tramping on Life* (1922). It is believed that Kemp's boyhood home at 956 Poland Avenue is no longer standing.

ZANESVILLE

This city in southeastern Ohio was named after Ebenezer Zane, an early pioneer in Ohio and the great-grandfather of **Zane Grey.** The Western novelist was born **Pearl Zane Gray**—he later dropped his first name and changed the spelling of his last name—on January 31, 1875, in a seven-room frame house at 363 (now 705) Convers Avenue. Grey attended the Eighth Ward School and Moore Grammar School. It is said that he wrote his first story in a cave behind his home. **Elizabeth Robins,** who wrote under the name **C. E. Raimond,** attended the Putnam Female Seminary, in a building on Woodlawn Avenue that is no longer standing. She lived in the Stone House, her grandmother's home at 115 Jefferson Street, which was built in 1809 in the hope, unrealized, that the building would be used as the Ohio capitol—Zanesville was the capital of the state from 1810 to 1812. Stone House later was used for the Putnam Female Seminary and was a station on the Underground Railroad. Elizabeth Robins wrote of this historic house in her novel *The Open Question* (1898). Another student of the Putnam Female Seminary was **Jean Starr Untermeyer,** who was born on May 13, 1886, in a house, no longer standing, at 221 North Seventh Street. Among her collections of poetry were *Love and Need: Collected Poems* (1940) and *Private Collection* (1965).

INDIANA

ANDERSON

As a young man **James Whitcomb Riley** worked for two summers in the 1870s with a group of house, barn, and sign painters known as the Graphic Company, which was based in this city northeast of Indianapolis. He later worked as reporter for the Anderson *Democrat.* Riley's residence here was the home of the Ethell family, 501 West Eighth Street, now the site of a building called Riley Place. A marker on this building commemorates his stay as well as his visit to Anderson on Riley Day in June 1913. The Grand Theatre, where Riley gave readings, is also marked with a plaque.

AUBURN

Will Cuppy, the critic and humorist, was born in Anderson on August 25, 1884. Cuppy's books included *How to Tell Your Friends from the Apes* (1931), *How to Become Extinct* (1941), and the best-selling *The Decline and Fall of Practically Everybody* (1950).

AURORA

Elmer Davis, whose best-known works included *A History of the New York Times* (1921) and the novel *Giant-Killer* (1928), was born on January 13, 1890, in this city on the Ohio River in southeast Indiana. His home, at 422 Sunnyside Avenue, is private.

BLOOMINGTON

This city owes its literary associations to the presence of Indiana University, which was built here in 1820, initially as a seminary. **Theodore Dreiser** studied here in 1888–89 and used his Bloomington experiences in his autobiography, *Dawn* (1931). **John Hollander,** the poet and critic, attended the university from 1952 to 1954 and taught here in 1953–54. Hollander frequented the Stardust, a tavern now called the Bluebird, at 216 North Walnut Street; and the Regulator, known as the Reg, a restaurant at 319 North Walnut Street. **Ernie Pyle,** the World War II combat correspondent, studied at the university from 1919 to 1922 but left school six months before graduation.

Ross Lockridge, the novelist, received a B.A. here in 1935 and an M.A. in 1939. He also taught here from 1936 to 1939. The author of the best-selling novel *Raintree County* (1948) lived in Bloomington for many years, and he created Raintree County as a fictional county in Indiana. From 1924 to 1937, Lockridge lived with his family in a house on South High Street. After his marriage to Vernice Baker in July 1937, the couple lived at 612 South Park. Lockridge's final residence was a house at 817 South Stull Avenue, where he lived from January 1948 until his death, reportedly a suicide, on March 6, 1948. A funeral for Lockridge was held the next day at the First Methodist Church, and he was buried in Rose Hill Cemetery.

Indiana University's Lilly Library contains portraits of James Whitcomb Riley as well as black-and-white illustrations by [John] Will[iam] Vawter, Riley's friend and illustrator. The library also houses an Upton Sinclair Archive, containing manuscripts by Upton Sinclair as well as clippings, photographs, and editions of all his works.

BROOK

Hazelden Farm, two miles east of this town in northwest Indiana, was the home of **George Ade** from 1915 until his death on May 16, 1944. Ade, a newspaperman and satirical writer, was the author of *Fables in Slang* (1900) and other works portraying country characters, including the plays *The College Widow* (1900) and *Just Out of College* (1905). As an indication of Ade's popularity, it is worth noting that he left an estate valued at more than $400,000. Today, Hazelden Farm comprises various buildings bearing the Ade name, including the George Ade Memorial Hospital and the George Ade Hazelden Country Club. The George Ade Hazelden Home, restored by the George Ade Memorial Association, is open by appointment only.

BROOKVILLE

This town, sixty miles southeast of Indianapolis, was the birthplace on April 10, 1827, of **Lew[is] Wallace,** author of *The Fair God* (1873)

Alexander III of Macedonia was born in 356 BC, on the sixth day of the month of Lous. He is known as Alexander the Great because he killed more people of more different kinds than any other man of his time. He did this in order to impress Greek culture upon them. Alexander was not strictly a Greek and he was not cultured, but that was his story, and who am I to deny it.

—Will Cuppy, in *The Decline and Fall of Practically Everybody*

Always advise a Friend to do that which you are sure he is not going to do. Then, if his Venture fails, you will receive credit for having warned him. If it succeeds, he will be happy in the Opportunity to tell you that you were Dead Wrong.

—George Ade, in "The Fable of Uncle Silas and the Matrimonial Game"

Home of George Ade in Brook, his permanent residence after 1915. Ade built the house in 1904, four years after publication of *Fables in Slang.* George Ade (*below*) celebrating his seventieth birthday, February 9, 1936, at his winter home in Miami Beach, Florida.

Shock treatments. Why bother with insulin, metrazol or electricity? Long ago they lowered insane persons into snake pits; they thought that an experience that might drive a sane person out of his wits might send an insane person back into sanity. By design or by accident, she couldn't know, a more modern "they" had given V. Cunningham a far more drastic shock treatment now than Dr. Kik had been able to manage with his clamps and wedges and assistants.

—Mary Jane Ward, in *The Snake Pit*

and the highly successful *Ben-Hur* (1880). A bronze plaque marks the site of Wallace's birthplace and childhood home, 306 High Street, no longer standing.

COLUMBIA CITY

Lloyd C[assell] Douglas, author of *Magnificent Obsession* (1929), *The Robe* (1942), and many other novels, was born here, in northwestern Indiana, on August 27, 1877. The novelist's birthplace, at 401 North Main Street, is no longer standing, but the site is marked.

COVINGTON

This city on the Wabash, seat of Fountain County, was the home of **Lew Wallace** from 1850 until 1853, while he served as county prosecuting attorney. His home, where he is said to have written the novel *The Fair God* (1873), was located at Eighth and Crockett streets.

CRAWFORDSVILLE

Authors associated with this city northwest of Indianapolis include **Meredith Nicholson.** The novelist-diplomat was born in Crawfordsville on December 9, 1866, and lived until age six at 205 South Walnut Street. The city named an elementary school in his honor. **[James] Maurice Thompson,** poet, novelist, and editor, first came to Crawfordsville in 1868, and made his home here until his death. It was here that he wrote his most successful historical novel, *Alice of Old Vincennes* (1900). Thompson's early residences included houses on Smartsburg Road and at 908 East Main Street. His home from 1892 until his death on February 15, 1901, was located on Pine Street, between East Pike Street and East Wabash Avenue, and is no longer standing. Thompson is buried in Oak Hill Cemetery in Crawfordsville.

Lew Wallace came to Crawfordsville in 1853 and made it his home, with interruptions for military and government service, until his death on February 15, 1905, four years to the day after the death of his neighbor, Maurice Thompson. Wallace's home on East Wabash Avenue, between Wallace and Elston avenues, was a couple of blocks away from Thompson's. Part of Wallace's home was later incorporated into a new brick house now standing here. Wallace practiced law from an office still standing at 201 East Main Street. The General Lew Wallace Study, at Pike Street and Wallace Avenue, was built in 1896 on the grounds of Wallace's estate. A statue near the study marks the site of the Ben Hur Beech, where Wallace is said to have written much of *Ben-Hur* (1880).

Crawfordsville is also the home of Wabash College, where poet **Ezra Pound** taught briefly in 1907.

Lew Wallace study in Crawfordsville. The statue of Wallace, a facsimile of one in the Hall of Fame in Washington, D.C., marks the site of the Ben Hur Beech.

DECATUR

Gene[va] Stratton-Porter lived in this city, southeast of Fort Wayne, for three years before moving with her husband to Geneva, Indiana, in 1895. The house in which she lived here is no longer standing, but in the Decatur courthouse yard is a monument to the novelist that was erected with money contributed by Adams County schoolchildren.

ELKHART

Ambrose Bierce moved at age seventeen to this city in northern Indiana, where he lived with his family in a house, now greatly altered, at 518 West Franklin Street. Bierce, who hated life in Elkhart, was one of the earliest volunteers for duty in the Civil War. **Kenneth Rexroth,** the poet, lived with his family at several addresses here during his childhood in the early 1900s. Among them were a house on Franklin Street and half of a double house, now private, at 714 West Marion Street.

EVANSVILLE

Theodore Dreiser lived in Evansville with his family in the early 1880s, in a brick cottage at 1413 East Franklin Street. **Jean Garrigue,** the poet, was born **Gertrude Louise Garrigus** here on December 8, 1914, in a two-story frame house at 26 Madison Avenue. Her family later lived at two other houses on South Third Street. Garrigue's volumes of poetry included *The Ego and the Centaur* (1947) and *Country Without Maps* (1964).

FAIRFIELD

[James] Maurice Thompson was born on September 9, 1844, in this community in southeastern Indiana, northeast of Brookville. Thompson achieved prominence initially as a poet and as editor of the New York *Independent* from 1889 to 1901, but his greatest success came with the publication of his historical novel *Alice of Old Vincennes* (1900), set in Indiana during George Rogers Clark's expedition of 1779. Thompson's family lived in Fairfield for only one year before moving to Missouri. The Thompson homestead was a small one-story structure, still standing in 1970.

FAIRMONT

Mary Jane Ward, known best for her novel *The Snake Pit* (1946), describing a young woman's terrifying experiences in a mental hospital, was born on August 27, 1905, in Fairmont, which lies midway between Fort Wayne and Indianapolis.

FORT WAYNE

This city earns its place in literary history not through the presence of an author, but as the home of a native American literary subject: Fort Wayne was the home of Johnny Appleseed in the 1830s and 1840s. Born John (or Jonathan) Chapman, the renowned orchardist spent many years planting and tending fruit trees in advance of the wave of settlers moving west. He became the subject of tall tales and legends by virtue of his skill in woodcraft and the help he gave frontiersmen. Vachel Lindsay wrote the poem "In Praise of Johnny Appleseed" about this symbol of America's creative restlessness. At the time of Chapman's death in 1845, he had a log cabin here on eleven cleared acres and was preparing to build a barn. It is said that he caught pneumonia during a fifteen-mile walk to repair an orchard fence. He died at the Worth family home on the St. Joseph River and was interred in a private burial ground nearby. A memorial to Johnny Appleseed can be seen in Swinney Park, off Jefferson Avenue.

GENEVA

This community in eastern Indiana, about thirty-five miles south of Fort Wayne off Route 27, was the home of **Gene[va] Stratton-Porter** from the mid-1890s to 1913. She and her husband built a log house they called Limberlost Cabin. It stood at the edge of Limberlost Swamp. The Limberlost area provided the setting for a number of other books Stratton-Porter wrote here, works of fiction and nonfiction. Among them were *Freckles* (1904), *A Girl of the Limberlost* (1909), and *Laddie* (1913). When draining of Limberlost Swamp began, in 1913, the Porters moved to Rome City. Limberlost Cabin, at 200 East Sixth Street, is operated today as a state memorial and is open year round.

GOSHEN

On June 28, 1911, **Ring Lardner** married Ellis Abbott at her family's home at 313 East Lincoln Avenue in Goshen. The building, slightly altered, is now an apartment house.

GREENFIELD

James Whitcomb Riley was born in Greenfield, east of Indianapolis, on October 7, 1849. His birthplace and boyhood home, at 250 West Main Street, is open to the public. When Riley was born, Greenfield was a village of 300, and the Riley home was a two-room log cabin. This cabin became the kitchen of the larger home his father built here a year after Riley's birth. Riley was an indifferent student, less interested in school than in the sights and sounds of Greenfield, and he soon began to submit poems to local newspapers, among them the Greenfield *Times,* which Riley edited a few years later.

One of Riley's favorite haunts as a boy was the Hancock County Court House, at the intersection of Main and State streets. Today a statue of Riley stands in front of the court house: it was

Birthplace of James Whitcomb Riley, 250 West Main Street, in Greenfield.

In that town, in those days, all the women who wore silk or velvet knew all the other women who wore silk or velvet, and when there was a new purchase of sealskin, sick people were got to windows to see it go by. Trotters were out, in the winter afternoons, racing light sleighs on National Avenue and Tennessee Street; everybody recognized both the trotters and the drivers; and again knew them as well on summer evenings, when slim buggies whizzed by in renewals of the snow-time rivalry. For that matter, everybody knew everybody else's family horse-and-carriage, could identify such a silhouette half a mile down the street, and thereby was sure who was going to market, or to a reception, or coming home from office or store to noon dinner or evening supper.

—Booth Tarkington, in *The Magnificent Ambersons*

paid for by contributions from schoolchildren. Another of Riley's favorite places was the old swimmin' hole, subject of his poem of that title, and now a main attraction of Greenfield's Riley Park. Riley's poem "Little Orphant Annie" and his story "Where Is Mary Alice Smith?" were both inspired by a real person, an orphan who had lived with the family when Riley was a boy.

INDIANAPOLIS

Writers associated with this city include two of Indiana's best-known men of letters: **James Whitcomb Riley** and **Booth Tarkington.** Riley came here in 1877 to work for the Indianapolis *Journal* and remained with the newspaper until 1885. A number of his well-known rustic poems first appeared in the *Journal,* and the newspaper's business manager helped Riley arrange publication of *The Old Swimmin'-Hole and 'Leven More Poems* (1883), which included "When the Frost Is on the Punkin." The immediate success of the volume marked the real beginning of Riley's long career as a popular poet. From 1893 to his death in 1916, Riley lived in the home of friends at 528 Lockerbie Street, a Victorian brick house now designated a National Historic Landmark. The James Whitcomb Riley Lockerbie Street Home is open to visitors. For a small admission charge, one may see the rooms in which Riley entertained literary figures of his day, the desk on which he wrote, and the original copy of his poem "Out to Old Aunt Mary's." Riley died here on July 22, 1916, and was buried in Crown Hill Cemetery, 3402 Boulevard Place.

[Newton] Booth Tarkington was born on July 19, 1869, at 520 North Meridian Street, in what was then the city's best residential section. This house has been destroyed, and so have Tarkington's later homes on New York Street and at 570 North Delaware Street. It is not known whether his home at 1100 North Pennsylvania Street, where he lived from the late 1870s to 1924, is still standing. Tarkington's best-known books were published while he lived on North Pennsylvania Street, among them *Penrod* (1914); *The Magnificent Ambersons* (1918), for which Tarkington won a Pulitzer Prize; and *Alice Adams* (1921), for which he won a second Pulitzer Prize. From 1924 until his death, Tarkington traveled from time to time in Europe and spent summers in Maine, but his principal residence was 4270 Meridian Street, Indianapolis, and it was there he died on May 19, 1946. He was buried in Crown Hill Cemetery. Tarkington was still active as a writer at the time of his death, at work on the novel *The Show Piece,* which was published in 1947. Tarkington's last home is reported now to be headquarters of the International Travel Films Company.

Indianapolis was the birthplace of several other writers. **Janet Flanner** was born here on March 13, 1892. She worked for a time as a motion-picture critic for the Indianapolis *Star,* but became known principally as a contributor to *The New Yorker,* from its founding in 1925 until her death in New York on November 7, 1978. Her "Letter From Paris," written under the name **Genêt,** reported on cultural, political, and economic developments. Flanner's books included *The Cubical City* (1926), a novel; and *Men and Monuments* (1957), about modern French artists.

Joseph Hayes was born here on August 2, 1918, and later set his novel *The Desperate Hours* (1954) in Indianapolis. The novel became the basis for a successful play and motion picture. **Charles Major,** who wrote the novel *When Knighthood Was in Flower* (1898) under the pen name **Edwin Caskoden,** was born in Indianapolis on July 25, 1856. **Kurt Vonnegut [Jr.]** was born here on November 11, 1922, and grew up here. At Shortridge High School, he was editor of *Echo,* the school newspaper. His many novels include *Cat's Cradle* (1963) and *Slaughterhouse Five* (1969).

(*below*) This statue of James Whitcomb Riley stands in front of the Hancock County Courthouse, at 40 East Main Street, about half a mile east of Riley's birthplace. (*below right*) The Old Swimmin' Hole, Riley Park, Greenfield.

Writers associated with Indianapolis include **Eleanor Atkinson,** the novelist, who was graduated in the 1880s from Indianapolis Normal Training School; and **Ambrose Bierce,** who was stationed in 1861 at Camp Morton here. A plaque at 2011 North Meridian Street commemorates the camp, which later was used as a prison for Confederate soldiers. **Mary [Hartwell] Catherwood** lived in Indianapolis with her husband from 1879 to 1882, when she worked as a drama critic and wrote three book-length serials, among them *Craque-O'-Doom* (1881).

Jean Garrigue, the poet, attended Shortridge High School a few years before Vonnegut did and also served as editor of *Echo,* then a daily and now a weekly. **[Frank Mc]Kin[ney] Hubbard** lived in Indianapolis and worked for many years as a columnist for the Indianapolis *News,* becoming known for humorous sketches of Indiana rustics. He died on December 26, 1930, at his home, 5437 North Meridian Street. Hubbard is buried here in Crown Hill Cemetery, as is **Meredith Nicholson.** Known best in his time for his novels set in Indiana, especially Indianapolis, Nicholson first came to this city in childhood in 1872 and spent much of his adult life here. His homes included a house at 1322 North New Jersey Street, and The House of a Thousand Candles, a house Nicholson built at 1500 North Delaware Street. Here he lived from 1904 to 1920, and wrote *The House of a Thousand Candles* (1905), perhaps his most successful novel, as well as *The Poet* (1914), a fictionalized biography of James Whitcomb Riley. He also lived at 1321 North Meridian Street, from 1921 to 1923; at Golden Hill, now private, 3736 Spring Hollow Road, from 1924 to 1929; and at 5417 North Meridian Street, his home at the time of his death on December 21, 1947.

Lew Wallace came to live in Indianapolis in 1837, when his father became governor of the state. Wallace reported on the Indiana House of Representatives for the Indianapolis *Daily Journal,* and studied law in his father's office. He was licensed to practice law in 1849, the year in which his family moved to Covington.

KENTLAND

George Ade, remembered particularly for his *Fables in Slang* (1899), was born on February 9, 1866, in this town northwest of Lafayette, near the Illinois border. His birthplace, at 317 East Graham Street, has been replaced by a store. A longtime resident of nearby Brook, Indiana, Ade died there on May 16, 1944, and was buried in Fairlawn Cemetery, one mile south of Kentland.

James Whitcomb Riley memorial at Crown Hill Cemetery in Indianapolis.

KOKOMO

In 1877, **James Whitcomb Riley,** then working for the Anderson *Democrat,* had a poem published in the Kokomo *Dispatch.* "Leonainie" was written in the style of Edgar Allan Poe and represented by Riley as a long-lost work of Poe's. Riley was trying to prove that a famous poet's name was the only requirement for ensuring success of a poem. "Leonainie" received a great deal of attention, and the hoax was exposed. As a result, Riley lost his job at the *Democrat.* The incident showed one aspect of Riley's stylistic ability, but he was often troubled in later years by memories of his literary prank.

LAFAYETTE

After graduation from Purdue University in 1887, **George Ade** studied law here for a time. He soon turned to writing advertising copy for a patent-medicine company, and then worked as a newspaper reporter, first for the Lafayette *Morning News* and later for the *Evening Call,* which was located at 47 South Fourth Street. Ade lived at the Stockton House, which stood at 132 South Street, and often visited the Grand Opera House, which stood on Columbia Street. Ade moved to Chicago in 1890.

(*below left*) 4270 Meridian Street in Indianapolis, last home of Booth Tarkington. The novelist died here on May 19, 1946. (*below*) The House of a Thousand Candles, home of Meredith Nicholson at 1500 North Delaware Street in Indianapolis. The house is named for Nicholson's popular novel, which he wrote here in his third-floor study.

Bronze plaque on Route 27 in Liberty, commemorating the birth of Joaquin Miller. The property on which Miller was born is across the road from the plaque.

George Barr McCutcheon was born near Lafayette on July 26, 1866, and moved to this city during childhood. While attending school here, he wrote some of his early stories. As a student during 1882–83 at Purdue University, where he became acquainted with George Ade, McCutcheon was a college stringer for the Lafayette *Journal.* He eventually left the university to work full time for the *Journal.* In 1893 he became city editor of the Lafayette *Daily Courier* and remained with the *Courier* until 1901. His first literary success came in that year with publication of the romance *Graustark,* for which he was paid $5000. *Graustark* proved a bonanza for his publishers, who later agreed to pay him royalties on sales of the novel. On the heels of this success, McCutcheon wrote the equally successful novel *Brewster's Millions* (1902). McCutcheon died on October 23, 1928, and his ashes were buried in a cemetery in Lafayette.

LAGRO

Gene[va] Stratton-Porter was born on August 17, 1863, in this community northeast of Wabash. Hopewell Farm, her birthplace and childhood home, is not open to visitors, and a new house stands on the site of the farmhouse in which she was born. Hopewell Church, standing at the junction of Routes 300N and 500E, was built on a portion of her family's acreage. The church, which has a collection of Stratton-Porter memorabilia, holds open house each year on or about August 14. Across from the church is a memorial to the novelist, and the church, which is used as a community house, is open to visitors by arrangement with the sexton.

> *Linguists have gone pretty far in discerning inside its symmetrical structure the nature of the persons who fabricated language. But they have had nothing to say as to whether language is predominately a man-made tool. My guess would be yes, and particularly so when it comes to sex. And now liberated woman proves her liberation by declaring, "What a man can say, I can say." But does it convey* her *experience? That doesn't seem to matter to her.*
>
> *If you were born middle class and went to college, you can use the once-forbidden vocabulary as a sign that you've come a long way, baby.*
>
> *But if you were born down among the wild ones, the renters, the ranters, the sharecroppers, the men who went off to the factories; if you lived among the tobacco fields, the corn rows, the woods lots, the stands of sassafras and hackberry, those words prove just one thing: you haven't moved an inch or learned a thing. You're still out there in the haymow with the hired man or riding home in the spring wagon with the Poor House handyman.*
>
> —Jessamyn West,
> in *The Life I Really Lived*

LIBERTY

Joaquin Miller, the Poet of the Sierras, was born **Cincinnatus Hiner** (or **Heine**) **Miller** on November 10, 1841, on a small farm off Route 27 near this town in east-central Indiana, about fifteen miles south of Richmond. His best-known work was *Songs of the Sierras* (1871). Nothing remains of his birthplace, but he is recalled by Joaquin Miller Road, which leads into the property, and by a boulder half a mile west on Route 27, which bears a plaque commemorating his birth.

MADISON

This city in southeastern Indiana was the model for the town of Sutherland in the novel by **David Graham Phillips,** *Susan Lennox: Her Fall and Rise* (1917). Phillips was born on October 31, 1867, in a two-story frame house that stood at the northwest corner of East and High streets. His family later lived in a house on Main Street, next to the National Branch Bank, as well as in rooms over the bank, where Phillips's father worked for many years. An incident on Main Street was the inspiration for *Susan Lennox,* a long-suppressed, posthumous novel considered to be Phillips's best work. One day Phillips saw a young woman sitting alone in a wagon. A look of revulsion crossed the beautiful girl's face as an elderly farmer climbed into the wagon and silently drove away with the girl. Phillips in 1908 built on this incident to create his powerful novel about a woman whose sin was to be born out of wedlock.

MISHAWAKA

Rose Hartwick Thorpe, best known for her ballad "Curfew Must Not Ring Tonight" (1867), was born in this town in northern Indiana on July 18, 1850. Her birthplace and childhood home, at 225 South Union Street, has been replaced by a church. **Ruth McKenney,** author of *My Sister Eileen* (1938), was born here on November 18, 1911.

MOORESVILLE

James Whitcomb Riley in childhood often visited the home of relatives named Albinus and Sarah Marine at 229 East Main Street, now a private residence. It is believed that he later gave a public reading at the old Academy Building in this town, southwest of Indianapolis. The academy building is being restored by the Morgan County Historical Society.

MUNCIE

Emily Kimbrough, collaborator with Cornelia Otis Skinner on *Our Hearts Were Young and Gay* (1942), was born on October 23, 1899, in this city in east-central Indiana. Her birthplace, at 715 East Washington Street, is not open to visitors, and her grandfather's house at 615 East Washington Street is now used as local headquarters of United Way.

NEW ALBANY

William Vaughn Moody, son of a steamboat captain, spent his early years, 1870 to 1885, in this city on the Ohio River. His family's home at 411 West Market Street is no longer standing, but a historical marker indicates its former location, taking note of the literary contributions of the poet and playwright.

NEW CASTLE

This city, twenty miles south of Muncie, was the model for the city of Freehaven in the novel *Raintree County* (1948), by Ross Lockridge.

NOBLESVILLE

Rex [Todhunter] Stout, creator of gourmet detective Nero Wolfe, was born in this city, northeast of Indianapolis, on December 1, 1886. Stout's birthplace and home until age two, 1151 Cherry Street, is not open to visitors.

NORTH MANCHESTER

Lloyd C. Douglas had his first ministry, from 1903 to 1905, at the Zion Evangelical Lutheran Church, 113 West Main Street. Before his marriage in 1904, Douglas lived at the Sheller Hotel, 202 Walnut Street. It was during his stay in this town in northeastern Indiana that Douglas wrote—and paid for the printing of—his first book, *More Than a Prophet*, an unsuccessful venture that was to keep him in debt for many years.

NORTH VERNON

Jessamyn West was born on July 18, 1902, near this town in southeastern Indiana. Her birthplace was located on the Greensburg Road, three miles north of here. West's fiction dealing with Quaker life includes *The Friendly Persuasion* (1956), a series of sketches of Quaker life in Indiana; and the novel *The Life I Really Lived* (1979), which carries the reader along with a novelist as she journeys from a small Kentucky town to Los Angeles and on to Hawaii.

OAKFORD

Mary [Hartwell] Catherwood lived in this town, southeast of Kokomo, in the 1870s. After her marriage to James Steele Catherwood in 1877, the couple lived in rooms over the railway station. Here she first met James Whitcomb Riley, who was living in Kokomo, and they began to collaborate on a work to be called *The Whittleford Letters*. The book was never completed. The Catherwoods later lived in a two-story brick structure on the west edge of Oakford, known as the W. H. Thompson House, and Riley spent several days with them at the house. It is said that Riley wrote his poem "Prior to Miss Belle's Appearance" while staying with the Catherwoods.

PENDLETON

Lloyd [Downs] Lewis was born on May 2, 1891, on a farm, now private, near this town northeast of Indianapolis. Lewis spent his boyhood years here and was graduated from the local high school. Known best for his biography of General William Tecumseh Sherman, Lewis is buried in the Friends Burying Ground at Fall Creek Meeting, on Route 28, two miles east of Pendleton.

ROME CITY

Cabin in Wildflower Woods, a stone-and-log structure on Sylvan Lake off Route 9, southeast of Rome City, was built by **Gene[va] Stratton-Porter** and her husband in 1913 and was their home from 1914 to 1920. Among the books she wrote here were the novels *Michael O'Halloran* (1915), *A Daughter of the Land* (1918), and *Her Father's Daughter* (1921). The cabin contains many original furnishings and other Porter memorabilia and is now a state memorial, open year round for a small admission fee. Admission to the grounds is free.

SOUTH BEND

Among the writers associated with this city in northern Indiana are **Ring Lardner,** who worked for the now-defunct South Bend *Times* in 1906; and **Kenneth Rexroth,** the poet, who was born here on December 22, 1905, in a house at 828 Park Avenue, now private.

Edwin O'Connor was graduated in 1939 from the University of Notre Dame near South Bend. While an undergraduate, he worked for the campus radio station and wrote a short story that was published in *Script*, the university's literary magazine. Years later, O'Connor dedicated his Pulitzer Prize-winning novel, *The Edge of Sadness* (1961), to his English professor at Notre Dame, Frank O'Malley.

SPENCER

William Vaughn Moody was born on July 8, 1869, in this town about fourteen miles northwest of Bloomington. The birthplace of the poet and playwright stood on Washington Street.

STRAUGHN

This town in east central Indiana, served by Route 40, was the model for the town of Waycross in *Raintree County* (1948), by Ross Lockridge.

SULLIVAN

In 1878, **Theodore Dreiser** moved with his mother and two siblings to Vincennes, about twenty-eight miles to the south, but they later returned to Sullivan, and Dreiser lived here until he was fourteen. Dreiser remembered collecting coal along the railroad tracks to heat the family house, no longer standing, on Depot Street.

TERRE HAUTE

Theodore Dreiser was born on August 27, 1871, in this city and spent his early years here. His birthplace, believed to have been at 525 South Ninth Street, is no longer standing. Dreiser's family, plagued by poverty, had five different addresses during the years Dreiser spent here as a child. One home was at Sixth and Mulberry streets. A two-story brick house at First and Farrington streets was the birthplace, on April

What makes you come here *fer, Mister,*
So much to our *house?—Say?*
Come to see our big sister!—
An' Charley he says 'at you kissed her
An' he ketched you, th'uther day!—

—James Whitcomb Riley,
in "Prior to Miss Belle's Appearance"

Was it for this our fathers kept the law?
This crown shall crown their struggle and their ruth?
Are we the mighty eagle nation Milton saw
Mewing its mighty youth,
Soon to possess the mountain winds of truth,
And be a swift familiar of the sun
Where aye before God's face his trumpets run?

—William Vaughn Moody,
in "An Ode in Time of Hesitation"

Snapshot of birthplace of Paul Dresser in Terre Haute. Dresser's brother, Theodore Dreiser, never lived in this house.

"Want to be a school-master, do you? You? Well, what would you *do in Flat Crick deestrick,* I'd *like to know? Why, the boys have driv off the last two, and licked the one afore them like blazes. . . . They'd pitch you out of doors, sonny, neck and heels, afore Christmas."*

—Edward Eggleston, in *The Hoosier Schoolmaster*

21, 1875, of Dreiser's brother **Paul Dresser** (who changed the spelling of his name). Dresser, called the King of Tin Pan Alley, is credited with writing Indiana's state song, "On the Banks of the Wabash Far Away" (1899). Theodore Dreiser's biographer, W. A. Swanberg, said that Dreiser wrote the first verse and the chorus of the song. Dresser's birthplace is now a museum.

Other authors associated with Terre Haute include **Sinclair Lewis,** who visited Eugene V. Debs, the labor leader, at Debs's home, 451 North Eighth Street, in 1926; **Irving Stone,** who stayed here in 1945–46 to collect material for his biographical novel about Debs, *Adversary in the House* (1947); and **Booth Tarkington,** who often visited at the home of his grandmother Booth here.

TREATY

This town, southeast of Wabash, was the birthplace on May 3, 1853, of **E[dgar] W[atson] Howe,** known as the Sage of Potato Hill. He wrote the novel *The Story of a Country Town* (1883) about the narrowness of life in the Midwest.

VEVAY

Edward Eggleston, author of the novels *The Hoosier Schoolmaster* (1871) and *The Hoosier Schoolboy* (1883), was born on December 10, 1837, in this Ohio River town in southeastern Indiana. His birthplace, a two-story brick house at 306 West Main Street, is now a private home, open to visitors by request as well as during the town's Swiss Alpine Festival, held in August each year. This house was also the birthplace of Eggleston's brother, **George Cary Eggleston,** born on November 26, 1839, whose experiences as a teacher in adjacent Jefferson County provided the basis for *The Hoosier Schoolmaster.* George, who served in the Confederate Army, later wrote for the New York *Evening Post* and New York *World.* His books included a memoir, *A Rebel's Recollections* (1874).

Postcard photograph of birthplace of Edward and George Cary Eggleston in Vevay, where the brothers lived until 1850.

VINCENNES

Theodore Dreiser lived here for a time as a small boy in the 1870s. After the Dreiser family split up because of poverty, Dreiser's mother brought her three youngest children to live here in an apartment over a firehouse. When it turned out that they were, in fact, living over a brothel, Mrs. Dreiser moved her family once again.

Other writers associated with Indiana's oldest town, on the eastern bank of the Wabash, include Walter Havighurst, whose books *George Rogers Clark* (1952) and *Proud Prisoner* (1964) used this locale; and Maurice Thompson, who used the life of Alice Tarleton Roussillon, member of a pioneering Vincennes family, in his best-known work, the historical romance *Alice of Old Vincennes* (1900). Two different sites have been marked as the possible site of the Roussillon home: the southwest corner of Second and Barnett streets, and—Thompson's choice—the west end of DuBois Street, on the south side of the street. Thompson's immortality was ensured many years ago, when the athletes at Lincoln High School in Vincennes were given the nickname "the Alices," in honor of Thompson's novel.

WABASH

Gene[va] Stratton-Porter moved here in 1874 with her family, from their farm at Lagro. Four months later, her mother died and Gene, transplanted from a rural home to this city in northern Indiana, became a rebellious student at the old Miami School. The family lived at several different addresses here, but stayed longest in a house at 54 Elm Street. In 1886 Gene Stratton married Charles Darwin Porter at her sister's home, 112 Hill Street, where the Wabash Christian Church now stands. Her novel *The White Flag* (1923) was set in Wabash, which she called Ashwater in the novel.

WARSAW

As a boy, **Ambrose Bierce** lived with his family on an eighty-acre farm at Walnut Creek, about three miles south of this city. About 1857, when Bierce was fifteen, he moved to Warsaw and worked as a printer's devil for the *Northern Indianan.* A second author of reputation, **Theodore Dreiser,** moved here with his family in 1884 and studied at the West End School. His teacher, Miss May Calvert, was one of the first people to recognize something special about the shy boy. Dreiser later acknowledged her inspiration and guidance in his autobiographical book *A Hoosier Holiday* (1916).

WEST LAFAYETTE

This city in west-central Indiana is the home of Purdue University. Writers associated with Purdue include **George Ade,** who was graduated with a B.S. degree in 1887 and worked for a number of years on newspapers in nearby Lafayette. Ade served as a trustee of the university from 1908 to 1915, and promoted the Ross-Ade Stadium here in 1923–24 with manufacturer David E. Ross. **George Barr McCutcheon,** who became a close friend of Ade's, studied here in 1882–83. McCutcheon left his studies to work as a reporter for a Lafayette newspaper. His brother, **John T. McCutcheon,** a writer and artist, also became a friend of Ade's and later worked with him in Chicago. **Booth Tarkington** studied here in 1891 and became a friend of Ade's and of the McCutcheon brothers.

ILLINOIS

ADDISON

Rosamond [Neal] du Jardin, novelist, short-story writer, and poet, lived from the late 1950s until her death on March 31, 1963, on Glen Ellyn Road in the village of Addison, about twenty miles west of Chicago. While living here, she wrote *Wedding in the Family* (1958) and *Double Wedding* (1959).

ALTON

A train wreck in 1893 near Alton, a city north of St. Louis on the Mississippi River, gave a boost to the career of **Theodore Dreiser,** who was then twenty-two years old. Dreiser reported on the accident for the St. Louis *Globe-Democrat* in a front-page article for which he received a bonus and a raise.

Alton is known to historians as the site, on what now is Broadway, of the seventh and final debate between Abraham Lincoln and Stephen A. Douglas. Near Broadway, at the end of Monument Avenue, is the Alton City Cemetery, where one may visit the grave of **Elijah P. Lovejoy,** the abolitionist editor murdered by a mob on November 7, 1837, for his editorials against slavery in the Alton *Observer.* His memorial, a statue of Victory atop a ninety-three-foot column, bears Lovejoy's proud pronouncement:

I have sworn eternal opposition to slavery
and by the blessings of God I will never
turn my face.

ARCOLA

This city, about thirty-five miles southeast of Decatur, was the birthplace, on December 24, 1880, of **John Gruelle,** a cartoonist and writer known for his children's stories featuring Raggedy Ann and Andy. His family lived in a house on South Locust Street when Gruelle was a child. Gruelle's books included *Raggedy Ann* (1918) and *Raggedy Andy* (1920).

AURORA

Writers associated with this city, some thirty-five miles west of Chicago, include **Frederick Irving Anderson,** a short-story writer, who was born here on November 14, 1877. He attended West High School, on Oak Avenue, where the Mary A. Todd Alternative School now stands. He worked in 1895 and 1896 as a reporter for the now-defunct Aurora *Daily News.*

Elizabeth Frances Corbett, author of more than forty-five books, including the popular series of stories for girls centering around the Graper family, was born here on September 30, 1887, and lived at 69 South Broadway. Corbett's best-known character was Mrs. Meigs, the central figure in several of her novels. **Vernon L[ouis] Parrington** was born here on August 3, 1871. He won a Pulitzer Prize in 1928 for the first two volumes of his influential three-volume examination of American ideas and their effects on our literature, *Main Currents in American Thought* (1927–30).

BARRY

Floyd Dell was born in this city, about twenty-five miles southeast of Quincy, on June 28, 1887, at Blair's Boarding House, no longer standing. He lived in several homes here, but the only one still standing is at 888 Mortimer Street, now private. Dell worked on various newspapers in the Chicago area before moving to New York to write for the radical periodicals *Masses* and *The Liberator.* His best-known works included the novels *Moon-Calf* (1920) and *The Briary-Bush* (1921), the play *Little Accident* (1928), and his autobiographical *Homecoming* (1933). The Barry Public Library, on Bainbridge Street, was an early influence on Dell, who used Barry locales in his writing. Local residents can point out many of these sites, and interested readers should visit the library, which is reported to be collecting material on Dell.

Now the yard was hard-packed clay without a blade of grass, there was not a bush or flower on the place, it was naked and bare; the gray little weatherbeaten house seemed not to have been painted in the thirty years since I was there as a child; only the trees remained, thirty years taller than they had been before, giving some shady refuge from the blazing summer sun.

—Floyd Dell,
in *Homecoming*

BLOOMINGTON

Writers who have lived in this city in north-central Illinois include **Rachel Crothers,** born here on December 12, 1878. Her girlhood home at 414 East Jefferson Street is no longer standing. Among Crothers's plays were *Nice People* (1921), *As Husbands Go* (1931), and her best-known drama, *Susan and God* (1937). **Walter Havighurst,** novelist and historian, lived as a boy at 704 East Grove Street in the early 1900s. **Elbert Hubbard,** a writer and publisher of inspirational essays, was born in Bloomington on June 19, 1856. Hubbard is remembered primarily for his essay *A Message to Garcia* (1899), which related an incident in the Spanish-American War. This account of extraordinary persistence by an American officer caught the attention of the public, and more than 40,000,000 copies of the essay were distributed.

Harold Sinclair, author of the novel *The Horse Soldiers* (1956), came to Bloomington as a child in the early 1900s and lived at 509 North Roosevelt Street. His books *American Years* (1938), *The Years of Growth* (1940), and *Years of Illusion* (1941) depicted the development of a town in Illinois from 1830 to the eve of World War I. Sinclair spent most of his life in Bloomington and died here on May 24, 1966. He was buried in Park Hill Cemetery in Bloomington.

CARBONDALE

This city in southern Illinois, home of Southern Illinois University, was the birthplace, on September 24, 1912, of **Robert Lewis Taylor,** known best for his Pulitzer Prize-winning novel *The Travels of Jamie McPheeters* (1958). Taylor grew up in Carbondale and attended University Grammar School and University High School

Memorial to Elijah P. Lovejoy on a hilltop in Alton City Cemetery. The statue of Victory can be seen for miles around.

The other sister was Different.
She began as Mary, then changed to Marie, and her Finish was Mae.
From earliest youth she lacked Industry and Application.
She was short on Intellect but long on Shape.

—George Ade,
in "The Fable of Sister Mae, Who Did as Well as Could Be Expected"

here. He also studied at Southern Illinois University in 1929. Taylor's books include *W. C. Fields: His Follies and Fortunes* (1949) and the novel *A Journey to Matecumbe* (1961). The University maintains a collection of Taylor's papers, as well as those of Kay Boyle and H. Allen Smith.

CARLINVILLE

Mary Austin, poet, novelist, and playwright, was born on September 9, 1868, in this city about forty miles southwest of Springfield. Her birthplace, at 511 East First South Street, is now greatly altered and is attached to the rear of a funeral home. Other houses in which she lived with her family have also been extensively remodeled. They include a house at the end of South Plum Street, on the west side, where her family lived from 1871 to 1879; a house at 610 East Second Street, her home from 1879 to 1884; and a house at 328 Johnson Street, her home from 1884 to 1888. This last residence is three blocks from Blackburn College, from which she was graduated in 1888. After graduation she moved with her family to California, and her works reflected her interest in California and New Mexico.

CHAMPAIGN

When the Illinois Central Railroad, in the 1850s, built its depot not in Urbana but in the prairie two miles to the west, a small community formed around the depot, taking the name West Urbana. In 1860, the community was incorporated as Champaign. Thus **William Osborn Stoddard,** who was associate editor of the West Urbana *Central Illinois Gazette,* is considered a Champaign literary figure. Stoddard served as secretary to Abraham Lincoln and wrote a biography of Lincoln and several other books about him, as well as many books for boys. The best-known was *Little Smoke: A Tale of the Sioux* (1891).

CHICAGO

Chicago has been the birthplace of an extraordinary number of well-known writers, including **Franklin P[ierce] Adams,** the newspaper columnist, born on November 15, 1881. Adams grew up in Chicago, was graduated from the Armour Institute of Technology in 1899, and wrote his first column for the Chicago *Journal* in 1903. Before working for the *Journal,* Adams spent some time as an insurance solicitor and once called at the home of a prospective client named **George Ade.** Adams was so impressed by the fact that Ade was breakfasting on strawberries—they were out of season and the time was 11 AM—that Adams decided he would become a writer like Ade. Adams is remembered best for his column "The Conning Tower," which he signed **F.P.A.**

Margaret Barnes, who won a Pulitzer Prize for her novel *Years of Grace* (1930), was born in Chicago on April 8, 1886. **Raymond Chandler,** creator of Philip Marlowe, the private eye, was born here on July 23, 1888. **James Gould Cozzens,** author of *The Just and the Unjust* (1942) and several other novels, was born here on August 19, 1903. **Peter De Vries,** the novelist and short-story writer, was born in Chicago on February 27, 1910, and grew up here. His family lived at several addresses in Chicago, including 7315 South Aberdeen Street, 7147 South Peoria Street, and 7357 South Sangamon Street. **John [Roderigo] Dos Passos,** author of the trilogy *U.S.A.* (1938), was born in Chicago on January 14, 1896.

There was a time when Archey Road was purely Irish. But the Huns, turned back from the Adriatic and the stock-yards and overrunning Archey Road, have nearly exhausted the original population,—not driven them out as they drove out less vigorous races, with thick clubs and short spears, but edged them out with the more biting weapons of modern civilization,—overworked and under-eaten them into more languid surroundings remote from the tanks of the gas-house and the blast furnaces of the rolling-mill.

—Finley Peter Dunne,
in the preface to *Mr. Dooley in Peace and in War*

But that five dollars a day
paid to good, clean American workmen
who didn't drink or smoke cigarettes or read or think,
and who didn't commit adultery
and whose wives didn't take in boarders,
made America once more the Yukon of the sweated workers of the world;
made all the tin lizzies and the automotive age, and incidentally,
made Henry Ford the automobileer,
the admirer of Edison, the birdlover,
the great American of his time.

—John Dos Passos,
in *U.S.A.*

Finley Peter Dunne, the newspaperman and humorist, was born on July 10, 1867, at his family's home on Adams Street, across the street from St. Patrick's Church. Dunne was graduated from West Division High School in 1884 and began writing for Chicago newspapers. In late 1892 he took a job with the Chicago *Post* and there wrote his first essay featuring Mr. Dooley, the Irish philosopher and bartender. Dooley, whose fictional tavern was located on Archey Road in Chicago's Irish neighborhood, was modeled after a bartender of Dunne's acquaintance. Dooley was widely known at the turn of the twentieth century for his philosophical and humorous remarks about the people and events of his day. Dunne's essays about Dooley were collected in several books, including *Mr. Dooley in Peace and in War* (1898) and *What Mr. Dooley Says* (1899).

James T[homas] Farrell, born on February 27, 1904, at 269 West Twenty-second Street, now a hospital, used Chicago as the setting for his Studs Lonigan trilogy: *Young Lonigan* (1932), *The Young Manhood of Studs Lonigan* (1934), and *Judgment Day* (1935). Farrell lived at several addresses in Chicago, including Twenty-fifth Street and La Salle Avenue, the 4900 block of Indiana Avenue, and 5816 South Park Avenue. Farrell attended St. Cyril High School, since rebuilt and now named Mount Carmel High School, at 6410 South Dante Street. He also studied at De Paul University in 1924 and at the University of Chicago for eight quarters in the late 1920s. The broad area in which Farrell set his trilogy is a section of Chicago bounded roughly by Garfield Boulevard on the north, Washington Park on the east, Sixty-first Street on the south, and State Street on the west, but Lonigan's own turf was the area between Fifty-seventh and Fifty-eighth streets from Wabash to Calumet avenues. Many of the sites mentioned in the novels are located here, and readers can still follow Lonigan through his youth and untimely death, right where he lived in Chicago.

John Gunther, the journalist and author, was born in Chicago on August 30, 1901, and **Albert Halper,** the novelist and short-story writer, was born in Chicago on August 3, 1904, and lived here until 1928. Halper was graduated in 1921 from Marshall High School, 3250 West Adams Street, and attended Northwestern University in Chicago from 1922 to 1924. Many of Halper's stories and novels derive from his Chicago days. For example, his novel *The Chute* (1937) was based upon his experience as an order filler in a Chicago mail-order house. **Lorraine Hansberry** was born here on May 19, 1930, and wrote the play *A Raisin in the Sun* (1959) about a black family living in Chicago. **Hutchins Hapgood,** the newspaperman and novelist, was born in Chicago on May 21, 1869.

Meyer Levin was born on October 8, 1905, in the West Side's Nineteenth Ward, later the home and battlefield of Chicago's gangsters. He attended the University of Chicago and began his newspaper career with the Chicago *Daily News.* Levin's novel *Citizens* (1940) dealt with the killing of ten steelworkers in Chicago on Memorial Day in 1937.

Harriet Monroe, born in Chicago on December 23, 1860, was a poet and dramatist but is best remembered as the founder in 1912 of *Poetry: A Magazine of Verse,* the influential magazine that has published the work of many poets of the first rank, including Hart Crane, T. S. Eliot, Vachel Lindsay, Ezra Pound, and Carl Sandburg. Monroe was editor of the magazine until her death on September 26, 1936.

Willard Motley was born in Chicago on July 14, 1912, and used the city as a setting for his novels *Knock on Any Door* (1947), *We Fished All Night* (1951), and *Let No Man Write My Epitaph* (1958). **[Benjamin] Frank[lin] Norris,** born here on March 5, 1870, used Chicago as the setting for his posthumously published novel, *The Pit* (1903), which dealt with grain speculation. **Ernest Poole,** author of the Pulitzer Prize-winning novel *His Family* (1917), was born in Chicago on January 23, 1880, and **Lew Sarett,** whose collections of poetry included *Slow Smoke* (1925) and *Wings Against the Moon* (1931), was born in Chicago on May 16, 1888.

Chicago was also the birthplace, on February 23, 1904, of **William L[awrence] Shirer,** the journalist and author of *Berlin Diary* (1940) and *The Rise and Fall of the Third Reich* (1960). **Harold Sinclair,** author of such novels as *Journey Home* (1936) and *The Horse Soldiers* (1956), was born here on May 8, 1907, and **Cornelia Otis Skinner,** author with Emily Kimbrough of *Our Hearts Were Young and Gay* (1942), was born here on May 30, 1901. **Julian Street,** the newspaperman and novelist, was born in Chicago on April 12, 1879. **B[erick] Traven [Torsvan],** born here on March 5, 1890, wrote *The Death Ship* (1934), *The Treasure of the Sierra Madre* (1935), and other novels.

Edward Wagenknecht, the author and anthologist, was born on March 28, 1900, in a house at 882, later 1819, South California Avenue, which was his home until 1915. Wagenknecht attended Chicago schools and received undergraduate and graduate degrees from the University of Chicago, where he taught from 1923 to 1925. Wagenknecht returned to Chicago to teach from 1943 to 1947 at the Illinois Institute of Technology. During that time he lived at 1721 Chancellor Street and there worked on a number of books, including *The Fireside Book of Christmas Stories* (1945), *When I Was a Child* (1946), *Abraham Lincoln: His Life, Work, and Character* (1947), and *Cavalcade of the American Novel* (1952). Chicago was also the birthplace, on October 17, 1900, of **[Arthur] Yvor Winters,** the critic and poet, author of *Edwin Arlington Robinson* (1946), *Collected Poems* (1952), and *The Poetry of W. B. Yeats* (1960).

Writers who have lived in Chicago include **George Ade,** who came here in 1890 to take a job with the Chicago *Morning News,* which was soon after renamed the *Record.* At the *Record* Ade wrote his first fable in slang, a moralistic tale in which he used colloquialisms and capitalization of key words to achieve humorous effect. This fable, later titled "The Fable of Sister Mae, Who Did as Well as Could Be Expected," appeared in the *Record* on September 17, 1897. The fable was well received, and Ade wrote more for his column. A collection, *Fables in Slang,* was issued in 1897 and sold 69,000 copies during its first year. By the time he left the *Record,* Ade was a nationally known humorist, and his subsequent volumes of fables were very popular. **Nelson Algren** lived in Chicago for many years and used the city as the setting for *The Man With the Golden Arm* (1949), a novel that deals with the problems of drug addiction.

Sherwood Anderson came to live in Chicago in late 1896, staying at 708 Washington Boulevard and working in a warehouse, where his job was rolling kegs of apples. Anderson left Chicago in the spring of 1898 to return home to Clyde, Ohio, where he joined the local militia, then mobilizing to fight in the Spanish-American War. Anderson returned to Chicago in 1900 to take a job in the Chicago office of *The Woman's Home Companion.* He left in a short time to work as a copywriter for an advertising agency and rose rapidly in that career. In 1906 he left Chicago to become head of a company in Cleveland and did not return to Chicago until 1912, after his nervous collapse and abandonment of his own company in Elyria, Ohio. Back in Chicago he turned once again to copywriting, and now met a number of other writers, including **Floyd Dell, Theodore Dreiser,** and **Carl Sandburg.** Anderson became a member of the group of writers and artists known as the

> *Studs couldn't stay in one place, and he kept walking up and down Indiana Avenue, wishing that the guys would come around. As he passed Young Horn Buckford and some punk he didn't know, Young Horn said hello to him. He gruffed a reply. He heard Young Horn say, as he walked on: "You know who that is? That's STUD LONIGAN. He's the champ fighter of the block."*
>
> *Studs laughed to himself, proud.*
>
> —James T. Farrell, in *Young Lonigan*

James T. Farrell in his middle years. The novelist used Chicago's South Side as the setting for his Studs Lonigan trilogy.

I am an American, Chicago born—Chicago, that somber city—and go at things as I have taught myself, free-style, and will make the record in my own way: first to knock, first admitted; sometimes an innocent knock, sometimes a not so innocent. But a man's character is his fate, says Heraclitus, and in the end there isn't any way to disguise the nature of the knocks by acoustical work on the door or by gloving the knuckles.

—Saul Bellow,
in *The Adventures of Augie March*

Chicago Renaissance, who for a short time in the 1910s made the city an important literary center. In his room at 735 Cass Street, Anderson worked on his novels and short stories. Cass Street no longer exists. During this period Anderson saw the publication of his first novel, *Windy McPherson's Son* (1916), which he had written largely in Elyria, and which found a publisher mainly through the efforts of Floyd Dell. Also during this time Anderson first published sketches that would be collected in his masterwork, *Winesburg, Ohio* (1919).

Saul Bellow, Nobel laureate in literature in 1976, came to live in Chicago in 1924 at the age of nine. He attended the University of Chicago—he has taught there since 1963—and was graduated from Northwestern University in 1937. Bellow has used Chicago as a setting for some of his books, notably *The Adventures of Augie March* (1953) and *Humboldt's Gift* (1975), for which Bellow won a Pulitzer Prize.

Edgar Rice Burroughs lived in childhood at 646 Washington Boulevard where he made friends with the Alvin Hulbert family, who lived at 194, later 2005, Park Avenue. The Hulberts had several daughters, one of whom Burroughs married in 1900. Burroughs and his wife, Emma, lived at several addresses in Chicago—Burroughs was to move many times during his life—including the Park Avenue house from 1904 to 1908. During 1907 and 1908, Burroughs worked for Sears, Roebuck and Company. Although he had a promising career ahead of him at Sears, Burroughs quit his job to try to establish his own business. Several attempts at entrepreneurship failed, and Burroughs was father of a growing family with no income. In 1911 a friend lent him office space at Market and Monroe streets, and Burroughs set up a business that sold pencil sharpeners. While waiting for his salesmen to return from their rounds, Burroughs began to write a story. After the business collapsed, he took a job with his brother at 222 West Kinzie Street and continued to write his story there. The result was his first published story, "Under the Moons of Mars," printed in book form as *A Princess of Mars* (1917). But before the story's publication, Burroughs had to take a job with *System* magazine, a business publication, at Wabash and Madison avenues. Despite his dismal business record, Burroughs was hired to give advice to businessmen who wrote to the magazine seeking help. The sale of his first story to *All-Story Magazine,* which published it in 1912, made him decide to devote himself to writing. It was in Chicago that Burroughs wrote his best-known novel, *Tarzan of the Apes* (1914). He moved to California in 1913, but returned to Chicago in 1914 and soon after bought a house in Oak Park, a suburb of Chicago.

Mary Catherwood lived in Chicago from 1899 until her death on December 26, 1902, at her home at 4852 W. Washington Boulevard. In Chicago Catherwood wrote her novel *Lazarre* (1901). **Clarence Darrow** moved to Chicago in the 1880s and lived here for many years. In the early 1900s Darrow formed a law partnership here with **Edgar Lee Masters.** Darrow's Chicago home addresses included 4219 South Vincennes Avenue, 1321 Michigan Avenue, and the Hunter Building, an apartment house at 1537 East Sixtieth Street, now a vacant lot. Darrow lived at the Hunter Building for thirty years and died there on March 13, 1938. His ashes were taken to Jackson Park and cast into Lake Michigan.

Floyd Dell lived in Chicago in the 1910s. It was at his house at Stony Island Avenue and Fifty-seventh Street that he introduced Sherwood Anderson to Carl Sandburg and other literary figures. Dell came to Chicago in the 1900s to work as a newspaperman. In 1909 he became assistant literary editor of the Chicago *Evening Post,* and editor in 1911. Dell helped make the newspaper's Friday *Literary Review* one of the foremost newspaper supplements in the country. It drew attention to the literary movement here that came to be known as the Chicago Renaissance. In 1914 Dell left Chicago to become associate editor of the *Masses* in New York City.

Theodore Dreiser first came to Chicago in 1884, when he was twelve. His family moved to an apartment in a house at West Madison and Throop streets. The Dreisers stayed in Chicago for only a few months, but Dreiser returned in the summer of 1887. He took a room on West Madison Street and found a job as a dishwasher. In the fall his family was back in Chicago, and Dreiser moved with them to an apartment on Ogden Avenue near Robey Street, and later to 61 Flournoy Street. During this time Dreiser worked at and was fired from a series of jobs. In the summer of 1889 he escaped from the tedium of menial work when a teacher he had studied with in Warsaw, Indiana, financed a year of study for Dreiser at Indiana University. In 1890 he returned to the family apartment on Flournoy Street and went back to work. The death of Dreiser's mother split up the family, and Theodore in 1891 moved with a brother and sister to an apartment on Taylor

Home of Mary Catherwood in Chicago. The novelist spent the last three years of her life here.

Street. He drifted from job to job, including one as bill collector for a store at 65 East Lake Street. In 1892 Dreiser found his first newspaper job, at the Chicago *Globe,* which then was located on Fifth Avenue. He covered the Democratic National Convention for the *Globe* and obtained the news scoop that Grover Cleveland would be renominated by the Democratic Party. In November of 1892, Dreiser left Chicago to take a job with the St. Louis *Globe-Democrat.* He returned to Chicago in 1912 and 1913 to conduct research on the life of Charles Tyson Yerkes, the financier and public-utilities magnate who was the model for Frank Cowperwood in Dreiser's trilogy: *The Financier* (1912), *The Titan* (1914), and *The Stoic* (1947). While in Chicago Dreiser met many of the writers then living here, including Floyd Dell and Edgar Lee Masters. Dreiser used Chicago as a setting not only for the Cowperwood trilogy but also for *Sister Carrie* (1900).

Edna Ferber worked for the Chicago *Tribune* in 1920 and 1921 and lived at 5414 East View Park and at the Windermere Hotel. During this time Ferber worked on her novel *So Big* (1924), for which she won a Pulitzer Prize. **Eugene Field** moved to Chicago in 1883 and became a columnist for the Chicago *Morning News,* later renamed the *Record.* He wrote his column "Sharps and Flats" for this newspaper until his death in 1895. In Buena Park, a suburb of Chicago, Field built a rambling house he called Sabine Farm, no longer standing, and in which he died on November 4, 1895. Field's years in Chicago were his most productive, and here he wrote the two poems for which he is remembered: "Wynken, Blynken and Nod" and "Little Boy Blue." A monument to Field was constructed in 1922 in Chicago's Lincoln Park. **Hamlin Garland** lived in Chicago on and off from 1893 to 1915 and wrote of the city in his semiautobiographical novel *Rose of Dutcher's Coolly* (1895).

Ben Hecht lived in Chicago and worked for city newspapers from 1910 to 1925. A leading member of Chicago's literary circle, Hecht became well known for his newspaper column "1001 Afternoons in Chicago" in the Chicago *Daily News.* Some of the sketches published in the column were collected in *1001 Afternoons in Chicago* (1922). In 1923 Hecht founded the Chicago *Literary Times* and served as its editor until it ceased publication two years later. During his Chicago years he wrote several books, including the novel *Erik Dorn* (1921), which was influenced by a year he spent as head of the Berlin office of the *Daily News;* the novel *Fantazius Mallare* (1922), which was sequestered by the government as "obscene literature"; and the collection *Tales of Chicago Streets* (1924). It was the play *The Front Page* (1928), written in collaboration with Charles MacArthur, that brought Hecht national recognition, but this tough melodrama of Chicago newspaper life—star reporter Hildy Johnson was always at odds with Walter Burns, his managing editor—was not written until several years after Hecht left Chicago.

Ernest Hemingway lived in a boarding house on Division Street for a time in 1919 while editing the house publication of the Cooperative Society of America. **Robert Herrick,** the novelist and short-story writer, lived in Chicago while teaching at the University of Chicago from 1893 to 1923. During these thirty years Herrick wrote most of his books, including the novels *The Web of Life* (1900) and *The Common Lot* (1904), both set in Chicago; the short novel *The Master of the Inn* (1908); and the novel *One Woman's Life* (1913). He also used Chicago as a setting for his novel *Chimes* (1926), which dealt with the University of Chicago. **Ruth Herschberger** lived at 3415 West Sixty-fourth Street from 1919 to 1938, and studied at the University of Chicago from 1935 to 1938.

Monument to Eugene Field in Lincoln Park, Chicago. The base of the monument displays stanzas from "Wynken, Blynken, and Nod" and "The Sugar-Plum Tree."

Langston Hughes worked on his autobiography *The Big Sea* (1940) while staying at the Hotel Grand in Chicago in 1939. **MacKinlay Kantor** came to Chicago in 1925 and worked at several jobs while contributing to the Chicago *Tribune.* Chicago was the setting for Kantor's first novel, *Diversey* (1928), which described the city during the heyday of gangsterism. **Emily Kimbrough** attended school in Chicago, and from 1921 to 1923 was editor of *Fashions of the Hour,* a magazine published by the Marshall Field department store. **Ring Lardner** came to Chicago for the first time in the early 1900s and worked at various jobs before leaving the city after a few years. He returned in 1908 to cover sports for the Chicago *Examiner.* By 1909 he was working for the Chicago *Tribune* and remained with the newspaper until November 1910, when he left for St. Louis. Lardner returned to Chicago in late 1911 and worked for the Chicago *American,* then the *Examiner,* and in 1913 once again for the *Tribune.* Lardner stayed with the *Tribune* until June 1919.

Tom Lea, the artist and novelist, lived in Chicago from 1924 to 1933, attending the Art Institute of Chicago, on Michigan Avenue at Adams Street, from 1924 to 1926. **Sinclair Lewis** lived in Chicago in March and April 1916, first at 2147 Washington Boulevard and then at the Brewster Hotel, on Diversey Parkway. During this time he worked on his novel *The Job* (1917). Lewis returned to Chicago for September and October of 1933 and stayed at the Sherry Hotel, near the apartment of **Lloyd Lewis,** with whom he worked on the play *Jayhawker* (1935). Lloyd Lewis worked for the Chicago *Daily News* from 1930 to 1945 and was managing editor and a

Presently a branch of the filthy, arrogant, self-sufficient little Chicago River came into view, with its mass of sputtering tugs, its black, oily water, its tall, red, brown, and green grain-elevators, its immense black coal-pockets and yellowish-brown lumber-yards.

Here was life; he saw it at a flash. Here was a seething city in the making.

—Theodore Dreiser,
in *The Titan*

WALTER—*Duffy! Listen! I want you to send a wire to the Chief of Police of La Porte, Indiana. . . . That's right. . . . Tell him to meet the twelve-forty out of Chicago. . . . New York Central . . . and arrest Hildy Johnson and bring him back here. . . . Wire him full description. . . . The son of a bitch stole my watch!*

The curtain falls.

—Ben Hecht and
Charles MacArthur,
in *The Front Page*

Further encouragement revealed that in Mr. Kaplan's literary Valhalla the "most famous tree American wriders" were Jeck Laundon, Valt Viterman, and the author of "Hawk L. Barry-Feen," one Mock-tvain. Mr. Kaplan took pains to point out that he did not mention Relfvaldo Amerson because "He is a poyet, an' I'm talkink abot wriders."

—Leonard Q. Ross, in *The Education of H*y*m*a*n K*a*p*l*a*n*

columnist for the Chicago *Sun-Times* from 1945 until his death on April 21, 1949.

Claude McKay, the poet and novelist, spent the last five years of his life in Chicago, doing research for the National Catholic Youth Organization and writing poetry. McKay died in a Chicago hospital on May 22, 1948. **Edgar Lee Masters** lived in Chicago from the 1890s to 1920, when he abandoned the practice of law to devote his energies to writing. He lived at 2128 S. Michigan Avenue, no longer standing, Groveland Avenue just north of Thirty-first Street, and the Tudor Building, near the corner of Forty-second Street and Ellis Avenue, no longer standing. **Wright Morris** lived at 218 Menominee Street in the 1920s. The novelist's books include *The Works of Love* (1952) and *The Field of Vision* (1956).

James Farl Powers, the short-story writer, lived in Chicago in the 1930s and 1940s, and studied at the Chicago campus of Northwestern University from 1938 to 1940. Powers worked as a clerk in several Chicago stores and also on the Chicago Historical Records Survey, a WPA project, but it was while working in a Chicago bookstore that Powers saw publication of his first story: "Lions, Harts, Leaping Does" appeared in 1943 in the literary quarterly *Accent,* published from the campus of the University of Illinois.

Burton Rascoe, the editor and columnist, was book and drama critic for the Chicago *Tribune* from 1912 to 1920. In one of his book reviews, Rascoe was less than flattering in remarks he made about Mary Baker Eddy's writing. Many wealthy and influential Christian Scientists then lived in Chicago, and when Rascoe got to the *Tribune* offices the day his review was published, he found that he was out of a job. He went to work as manager of the Chicago bureau of the Newspaper Enterprise Association, but left Chicago for New York in 1921.

In 1887 **Opie Read** moved his humorous literary weekly, *The Arkansas Traveler,* from Little Rock, Arkansas, to Chicago. Here Read lived for many years and wrote most of his popular novels, among them *Len Gansett* (1888), *The Jucklins* (1895), *My Young Master* (1896), and *An Arkansas Planter* (1896). He also wrote the autobiographical books *I Remember* (1930) and *Mark Twain and I* (1940). Read lived for many years at 5000 Harper Avenue, the home of a friend, and there he died on November 2, 1939.

Leo Rosten was brought to Chicago from his native Poland in about 1911, when he was three years old. Rosten grew up in Chicago, was graduated from the University of Chicago in 1930, and received a Ph.D. degree from the university in 1937. This was the same year in which was published his comic masterpiece *The Education of H*Y*M*A*N K*A*P*L*A*N,* issued under the pseudonym **Leonard Q. Ross. Philip Roth** attended the University of Chicago in 1954 and 1955 and taught there from 1956 to 1958. While living at Fifty-sixth Street and University Avenue, Roth wrote his collection *Goodbye, Columbus* (1959), consisting of five stories and the title novella.

Carl Sandburg was a leading figure in the Chicago Renaissance of the 1910s. Sandburg came to Chicago in 1912 to work for the *Daily Socialist,* but a year later became associate editor of *System,* the same magazine for which Edgar Rice Burroughs worked in 1912. Sandburg also worked for the Chicago *Day Book,* located at 500 South Peoria Street, until it ceased publication in 1917. During this period he found the city an inspiration for his poetry, especially the poems that later were published as *Chicago Poems* (1916). In the fall of 1913 Sandburg took a number of these poems to **Harriet Monroe,** founder of the literary magazine *Poetry,* which then was published at 543 Cass Street, no longer in existence. The meeting resulted in acceptance of nine of his poems for the March 1914 issue of *Poetry.* Among them was "Chicago," Sandburg's memorable celebration of the city. From 1912 to 1914 Sandburg lived at 4646 North Hermitage Avenue. In 1914 he moved to the Chicago suburb of Maywood and soon to Elmhurst, but he maintained his ties with Chicago. In 1917 he worked for three weeks for the Chicago *American* and then took a job with the Chicago *Daily News,* for which he worked until he moved on to Harbert, Michigan, in 1928.

Karl Shapiro attended public schools in Chicago in the 1920s. From 1950 to 1955, he served

Home of Carl Sandburg at 4646 North Hermitage Avenue in Chicago. While living here Sandburg saw publication of "Chicago" in *Poetry* magazine.

Hog Butcher for the World,
Tool Maker, Stacker of Wheat,
Player with Railroads and the Nation's
Freight Handler;
Stormy, husky, brawling,
City of the Big Shoulders:
They tell me you are wicked and I believe them,
for I have seen your painted women
under the gas lamps luring the farm boys.

—Carl Sandburg, in "Chicago"

as editor of *Poetry,* and from 1966 to 1968 taught at the University of Illinois. **Bert Leston Taylor** worked for the Chicago *Journal* from 1899 to 1901, and then went to work for the Chicago *Tribune.* There he launched his column, "A Line o' Type or Two," but left Chicago in 1903. In 1909 he returned to the *Tribune,* revived his column, and continued it until shortly before his death on March 19, 1921. Taylor's column, which he signed with the initials **B. L. T.,** became one of the most popular and respected newspaper columns in the country. Taylor published several books of material from his column, including *Line-o'-Type Lyrics* (1902), *A Line-o'-Verse or Two* (1911), *Motley Measures* (1913), *A Penny Whistle* (1921), and *The So-Called Human Race* (1922). **Sara Teasdale** spent much time in Chicago in the early twentieth century, and **Dorothy Thompson** lived for a time with an aunt in Chicago. Here Thompson attended high school and the Lewis Institute before enrolling at Syracuse University in 1910.

Richard Wright came to Chicago in 1927, at the age of nineteen. Chicago remained his home for ten years, and here he worked at a series of jobs, including porter, postal clerk, hospital worker, and insurance salesman. During this period Wright fashioned his experiences into stories, and began the struggle to become a writer. Life was hard for Wright and his family—at one point in the early 1930s he was unemployed and went on relief—and it was about this time that Wright joined the Communist Party. In 1935 he found a job with the Federal Writers' Project in Chicago. He then became publicity agent for the Federal Negro Theatre. Wright's literary ambitions and political interests were all the while coming closer together. They finally merged in the spring of 1937, when Wright left Chicago to become Harlem editor of the *Daily Worker* in New York City. His Chicago addresses included 4831 South Vincennes Avenue, 4804 South St. Lawrence Avenue, and 3743 South Indiana Avenue. The South St. Lawrence site is now a vacant lot. Wright used his experiences in Chicago in the novel *Native Son* (1940), set here and based in part on the life of Robert Nixon, a Chicago black, called Bigger Thomas in the novel. Nixon was electrocuted in 1938 for murdering a white woman.

Other black writers are associated with Chicago. **Paul Laurence Dunbar,** the poet, worked at the Haiti Building at the Chicago World's Fair of 1893. His superior in that job was **Frederick Douglass,** who had served as U.S. minister to Haiti from 1889 to 1891. Here Dunbar met **James Weldon Johnson,** who had come to Chicago to see the fair.

Zona Gale, the novelist and short-story writer, died in a Chicago hospital on December 27, 1938. **William Dean Howells** based his novel *A Hazard of New Fortunes* (1890) in part on the Haymarket Riot in Chicago and its aftermath. The riot occurred during a mass demonstration by striking workers on May 4, 1886, when a bomb exploded, killing a number of people. **William Inge** came to Chicago in 1945 to see a production of *The Glass Menagerie* (1944), by Tennessee Williams. Inge was so moved by the experience that he decided to become a playwright. After **Upton Sinclair** took part in an investigation of conditions in the Chicago stockyards in 1904, he wrote *The Jungle* (1906). The novel was instrumental in establishing federal pure-food laws.

Chicago institutions of special interest to literary travelers include the Newberry Library, at 60 West Walton Street, which has the papers of Sherwood Anderson and Malcolm Cowley; and the Art Institute of Chicago, at South Michigan Avenue and East Adams Street, which includes among its former students **Vachel Lindsay, Kenneth Rexroth,** and **Ross Santee.**

Many writers, in addition to those already cited, have been associated with the University of Chicago. **Vardis Fisher** received his M.A. here in 1922 and Ph.D. in 1925. **Janet Flanner** attended the university in 1912 and 1913. **Jean Garrigue** was graduated from the university in 1937. **Jeremy Ingalls** studied here in 1938 and 1940 and from 1945 to 1947. She lived on and off from 1941 to 1947 in an apartment house that stood on Kimbark Avenue, a few blocks from the university. **William Vaughn Moody** taught at the University of Chicago from 1895 to 1899 and 1901 to 1907. **James Purdy,** a student at the university in 1945 and 1946, set his novel *Eustace Chisholm and the Works* (1967) in Chicago of the Depression years.

Vincent Sheean studied at the University of Chicago for three and a half years, before the death of his mother caused him to leave school and go to work for the Chicago *Daily News.* After a couple of weeks he was fired. Legend has it that Sheean walked directly from the editor's office to the train station, and boarded a train for New York City. **Carl Van Vechten** was graduated from the University of Chicago in 1903. **Kurt Vonnegut** attended the university from 1945 to 1947. In 1946 he worked as a police reporter for the Chicago City News Bureau. Vonnegut received an M.A. degree from the university in 1971. **Glenway Wescott** studied here from 1917 to 1919, and **Thornton Wilder** taught here from 1930 to 1936. Wilder would teach for six months of each year and spend the rest of the year writing

Richard Wright in 1946, a year after publication of *Black Boy,* recounting his childhood and youth in the South.

Home of Richard Wright at 4831 South Vincennes Avenue in Chicago. Wright moved here in 1929, eleven years before the appearance of his celebrated novel *Native Son.*

He was in the street now, being dragged over snow. His feet were up in the air, grasped by strong hands.

"Kill 'im!"

"Lynch 'im!"

"That black sonofabitch!"

They let go of his feet; he was in the snow, lying flat on his back. Round him surged a sea of noise. He opened his eyes a little and saw an array of faces, white and looming.

"Kill that black ape!"

—Richard Wright,
in *Native Son*

Thornton Wilder and his sister Isobel in 1935. At the time this photograph was taken, Wilder was living at 6020 Drexel Avenue in Chicago. (*far right*) Old photograph of Emerson Hough, who wrote a number of novels set in the West.

and lecturing around the country. For the first four years he had an apartment on campus, and then an apartment at 6020 Drexel Avenue. While here Wilder published of *The Long Christmas Dinner and Other Plays in One Act* (1931) and the novel *Heaven's My Destination* (1935).

DECATUR

Walter Havighurst, the novelist and historian, lived during his high-school years at 710 West Main Street. He was graduated from Decatur High School in 1919. Eight miles southwest of Decatur is the Lincoln Trail Homestead State Park, site of the first Illinois home of Abraham Lincoln and family. Lincoln and his father and stepmother, who had moved from Indiana in 1830, lived here for one year.

ELGIN

Emerson Hough is said to have written *The Covered Wagon* (1922), his most popular novel, in this city northwest of Chicago. Although Hough was based in Chicago, he spent most of his life as a wanderer, camping and living an outdoor life throughout the West.

ELMHURST

In the early 1940s, **Rosamond [Neal] du Jardin** lived at 194 May Street in this suburb west of Chicago. **Carl Sandburg** lived in Elmhurst from 1919 to 1928, and Happiness House, his home at 331 South York Street, was one of the oldest buildings in what was then a somewhat rural area. While here, Sandburg worked for the Chicago *Day Book* and the Chicago *Daily News* and wrote many poems, as well as *Rootabaga Stories* (1922). He also wrote most of his two-volume *Abraham Lincoln: The Prairie Years* (1926), which formed the first part of his massive six-volume Lincoln biography. The 1920s brought a great surge of people to the Chicago area, and Elmhurst lost much of its rural atmosphere. As a result, the Sandburgs decided to move to Harbert, Michigan, and their Happiness House gave way to a parking lot.

He got up and began to shave. It was his custom while shaving to prop up before him a ten-cent copy of King Lear *for memorization. His teacher at college had once remarked that* King Lear *was the greatest work in English literature, and the Encyclopaedia Britannica seemed to be of the same opinion. Brush had read the play ten times without discovering a trace of talent in it, and was greatly worried about the matter. He persevered, however, and was engaged in committing the whole work to memory.*

—Thornton Wilder, in *Heaven's My Destination*

EVANSTON

Writers associated with this city north of Chicago include **Walter Kerr,** the drama critic, born on July 8, 1913, in a house still standing at 2142 Asbury; **Ring Lardner,** who lived here for a time in 1917 and 1918 while working for the Chicago *Tribune;* **Edward Wagenknecht,** who lived at 1721 Chancellor Street from 1943 to 1947, compiling seven anthologies during that time and working on his literary history, *Cavalcade of the American Novel* (1952); and **Mary Jane Ward,** who moved here at age ten with her family and lived in Evanston for most of her life. Her homes included 1406 Chicago Avenue, 2720 Woodbine Avenue, and 1144 Asbury Avenue. Ward is known best for *The Snake Pit* (1946), her shocking account of a young woman confined in a mental hospital.

Northwestern University, founded here in 1851, had only four students when it opened in 1855; the current enrollment exceeds 15,000. Writers associated with Northwestern include **Bernard De Voto,** who taught here in the 1920s and married one of his students, Helen Avis MacVicar; **Tom Heggen,** author of the best-selling novel *Mister Roberts* (1946), who received midshipman training here in 1942; and **Lew Sarett,** who taught here from 1920 to 1953. Sarett's books included *The Box of God* (1922), *Slow Smoke* (1925), *Wings Against the Moon* (1931), and *The Collected Poems of Lew Sarett* (1941), the last with an introduction by his friend Carl Sandburg. The Northwestern University library houses a collection of the fiction manuscripts of Anaïs Nin, the fiction writer and diarist.

FAIRLAND

Rosamond [Neal] du Jardin was born on July 22, 1902, in Fairland, southeast of Urbana.

When she was two, her father moved the family to Chicago, where she grew up. Mrs. du Jardin wrote many short stories and novels. Her novels for adults included *All Is Not Gold* (1935) and *Tomorrow Will Be Fair* (1946). She also wrote such novels for teenagers as *Practically Seventeen* (1949) and *Wedding in the Family* (1958).

GALESBURG

Of the authors associated with this city in northwest Illinois, the best known is **Carl Sandburg,** born on January 6, 1878, in a three-room frame house at 331 East Third Street. The walls of the house were of rough vertical siding, and cracks between the boards were sealed with newspaper and paste. After the Sandburgs moved to another house in Galesburg, a carpenter added clapboards to the exterior, plastered the interior walls, and built an addition to the house. The Carl Sandburg Birthplace has since been restored to its improved condition. It now contains period furnishings as well as Sandburg family items, photographs, and the family Bible. Now designated the Lincoln Room, the addition to the house displays a portrait of Lincoln by N. C. Wyeth, a collection of Sandburg's autographed books, and the typewriter Sandburg used in his early work. The ashes of Sandburg and his wife are interred here under Remembrance Rock, a red granite boulder named for Sandburg's only novel, written in 1948. The Carl Sandburg Birthplace is open free of charge.

I am the people—the mob—the crowd—the mass.
Do you know that all the great work of the world is done through me?
I am the workingman, the inventor, the maker of the world's food and clothes.

—Carl Sandburg,
in "I Am the People, the Mob"

Sandburg lived with his family at other homes in Galesburg, including a house on South Street; and one at 806–810 Berrien Street, now private. After completing the eighth grade, in 1891, he left school to work as a milkman and as a porter in a barbershop at the Union Hotel. He left Galesburg in 1897 on a hobo tour of the West, and later volunteered for the Spanish-American War. He returned here to study at Lombard College (later absorbed by Knox College), where he met Professor Philip Green Wright, who encouraged Sandburg in his writing and published Sandburg's first volume of poetry, *In Reckless Ecstasy* (1904), from a press in the basement of his home.

Writers associated with Knox College here include **Eugene Field,** a student at the college in 1869, and **George Fitch,** who was graduated with the class of 1897. Fitch wrote stories about the fictional Siwash College (presumably Knox College) which later were collected in *The Big Strike at Siwash* (1909) and *At Good Old Siwash* (1911). Abraham Lincoln in 1858 debated Stephen A. Douglas outside Old Main, a building at Knox. **Edgar Lee Masters** studied at a preparatory school at Knox in 1889–90. **Emerson Hough,** the novelist, is buried in Galesburg's Hope Cemetery.

GALVA

George Fitch, known best for his stories about Siwash College, was born in this town northeast of Galesburg on June 5, 1877. Fitch worked on newspapers in Iowa and Illinois after his graduation from Knox College.

GLENCOE

Archibald MacLeish, born in this city north of Chicago on May 7, 1892, has received three Pulitzer prizes for his works: the epic poem *Conquistador* (1932), *Collected Poems 1917–52* (1952), and the verse drama *J. B.* (1958). His birthplace, at 459 Longwood Avenue, is no longer standing.

This is a man with an old face, always old . . .
There was pathos in his face and in his eyes,
And early weariness; and sometimes tears in his eyes,
Which he let slip unconsciously on his cheek,
Or brushed away with an unconcerned hand.
There were tears for human suffering, or for a glance
Into the vast futility of life,
Which he had seen from the first, being old
When he was born.

—Edgar Lee Masters,
in an unpublished poem
about Clarence Darrow

Birthplace of Carl Sandburg. The Pulitzer Prize-winning poet and novelist lived in this three-room cottage in Galesburg until he was three.

I stand again in the field
Where first my father broke
The prairie sod, and the sweating team
Passed with the creak of yoke
And straining tug, and steam
Arose from the furrows that rolled,
Tough and straight and black, under
the morning's gold.

—Glenn Ward Dresbach,
in "The Field"

For it's always fair weather
When good fellows get together. . . .

—Richard Hovey
in his long poem "Spring,"
which includes "A Stein Song"

Grave of Mary Catherwood in Floral Hill Cemetery in Hoopeston. The novelist lived in Hoopeston for several years in the 1880s.

MacLeish died on April 20, 1982, at Massachusetts General Hospital, in Boston. At the time of his death, he lived in Conway, Massachusetts.

GLEN ELLYN

Rosamond [Neal] du Jardin made her home in this suburb west of Chicago from the mid-1940s to the late 1950s. Her residences here included 621 Prairie Avenue and 670 Kenilworth Avenue. She is buried in Forest Hill Cemetery.

GLENVIEW

Edward Wagenknecht served as minister of the Congregational Church in this village north of Chicago from 1918 to 1920.

GODFREY

This community, just north of Alton on Route 67, is the home of Monticello College, founded in 1835 as the Monticello College and Preparatory School for Girls. **Zoë Akins,** who won a Pulitzer Prize in 1935 for her dramatization of Edith Wharton's novel *The Old Maid,* studied here. **Lucy Larcom,** taught here from 1849 to 1852. Larcom, by then, was well known as an abolitionist and contributor to *The Lowell Offering,* published by workers in the textile mills that flourished in Lowell, Massachusetts.

HOOPESTON

Mary [Hartwell] Catherwood, the novelist, first met her future husband, James Steele Catherwood, while visiting her aunts in this city twenty-five miles north of Danville, in eastern Illinois. She and her husband came to live here after his retirement in the early 1880s, remaining until 1889. After her death in 1902, she was buried in Floral Hill Cemetery in Hoopeston.

HOPE

This community west of Danville, in eastern Illinois, was the birthplace of the distinguished Van Doren brothers. **Carl Van Doren,** born on September 10, 1885, received a Pulitzer Prize for his biography *Benjamin Franklin* (1938). His other books included the biography *Sinclair Lewis* (1933) and *What Is American Literature?* (1935). He also served as managing editor of the three-volume *Cambridge History of American Literature* (1917–21). **Mark Van Doren,** born on June 13, 1894, won a Pulitzer Prize for his *Collected Poems* (1939). His books included the biographies *Henry David Thoreau* (1916) and *Nathaniel Hawthorne* (1949), the collections of poetry *Spring Thunder* (1924) and *New Poems* (1948), and *Autobiography* (1958).

HUDSON

The site of the boyhood home of **Elbert Hubbard** in this village north of Bloomington is now marked by a boulder and plaque. A similar boulder and a plaque a few blocks away mark the birthplace of **Melville E. Stone,** born on August 22, 1848. He was a founder of the Chicago *Daily News* and first general manager of the Associated Press.

JACKSONVILLE

J[ames] F[arl] Powers was born here, thirty miles west of Springfield, on July 8, 1917. Powers attended grade school here and has used Jacksonville in his writing. His books include *Morte D'Urban* (1962), *The Presence of Grace* (1956), and *Look How the Fish Live* (1975).

KENILWORTH

Eugene Field is buried here with his wife in the cloister of the parish churchyard at the Church of the Holy Comforter, 333 Warwick Road. Best known for "Little Boy Blue" (1889) and other children's verse, Field was buried first in Chicago's Graceland Cemetery. His body remained there for many years before being reinterred in Kenilworth in 1926. The walls within the cloister exhibit such memorabilia as Field's wedding ring, copies of "Little Boy Blue," and the Lord's Prayer copied in Field's handwriting.

The little toy dog is covered with dust,
But sturdy and stanch he stands;
And the little toy soldier is red with rust,
And his musket moulds in his hands.
Time was when the little toy dog was new,
And the soldier was passing fair,
And that was the time when our Little Boy Blue
Kissed them and put them there.

—Eugene Field,
in "Little Boy Blue"

LANARK

Glenn Ward Dresbach was born on September 9, 1889, on his grandfather's farm at 775 South Beede Road, near this town in northwest Illinois. His father later moved the family into Lanark, where they lived at 604 East Pearl Street, and Dresbach attended school here until 1908, leaving to enter the University of Wisconsin. Many of his poems were inspired by his experiences on his grandfather's farm and in the town of his youth. Among the poems drawn from his memory of farm life were "The Field," "The Drouth," and "The Last Corn Shock." A cider press at the corner of Franklin and Will streets in Lanark was the inspiration for "Cider Press." Dresbach was said at one time to be the most financially successful poet writing in the English language. His books included *The Road to Everywhere* (1916), *The Wind in the Cedars* (1930), and *Collected Poems 1914–1948* (1949).

LEBANON

Charles Dickens passed through this city twenty miles east of East St. Louis during his American tour of 1842. The English novelist and his party stopped on April 12 at the Mermaid House, 114 East St. Louis Street, praising it for its "cleanliness and comfort." Another famous guest, Abraham Lincoln, is claimed for the Mermaid House. The Lebanon Historical Society bought the Mermaid House in 1964 and has spent several years restoring the site, now listed in the National Register of Historic Places.

LEWISTOWN

Edgar Lee Masters lived as a boy in this city near the Spoon River, about thirty-seven miles southwest of Peoria. In 1880 the Masters family occupied a house, still standing, at 306 North Adams Street. In 1883 the family moved to a house at the corner of South Main and East Avenue D. This house is now used as a residence and insurance office, but is marked by a plaque commemorating Masters. It is believed that Masters used both Lewistown and Petersburg, Illinois, as models for the town in his collection of verse epitaphs, *Spoon River Anthology* (1915), which discloses the secret lives of the people buried in a town cemetery. Lewistown sites that appear in the work include a monument in Oak Hill Cemetery to one William Cullen Bryant (not the poet), who died at age twenty-four while hunting ducks on nearby Thompson's Lake. In *Spoon River*, Bryant is called Percy Bysshe Shelley. Masters used the Ross Mansion, 409 East Milton Avenue, as the model for the McNeely mansion in *Spoon River*. Masters also mentioned Major Newton Walker, who served in the Illinois legislature with Abraham Lincoln. Walker entertained Lincoln several times at his home at 1127 North Main Street.

LIBERTYVILLE

Lloyd Lewis, whose best-known books were *Sherman, Fighting Prophet* (1932) and *Captain Sam Grant* (1950), made his home during the 1940s in this village north of Chicago. Lewis's home on Little Saint Mary's Road, where he died on April 22, 1949, was located on the bank of the Des Plaines River. The funeral of the newspaperman and author was attended by many literary friends, including Marc Connelly and Carl Sandburg.

LINCOLN

This city, thirty miles northeast of Springfield, was named for Abraham Lincoln who, on hearing his name proposed for the new town, commented that he "never knew of anything named Lincoln that amounted to much." Literary figures associated with the city include **Langston Hughes,** who lived here for a brief time as a boy; and **William Keepers Maxwell,** born on August 16, 1908, in a house, now private, at 455 Eighth Street. Maxwell's novels included *The Folded Leaf* (1945), *Time Will Darken It* (1948), which is set in Illinois, and *The Chateau* (1961). **Carl Sandburg** was listed as associate editor of the Lincoln *Herald* from 1954 until his death in 1967.

MAYWOOD

This village on the Des Plaines River, twelve miles west of Chicago, was the home of **Carl Sandburg** from 1914 to 1919. Sandburg, working as a roving reporter for the Chicago *Daily News*, lived at 616 South Eighth Avenue.

McLEANSBORO

This city in southeastern Illinois is the birthplace of **H[arry] Allen Smith,** born on December 19, 1907. The humorist wrote entertaining and popular books, such as *Low Man on a Totem Pole* (1941), *People Named Smith* (1950), and *How to Write Without Knowing Nothing* (1961). In *Lo, the Former Egyptian* (1947), Smith wrote that neither he nor his relatives knew exactly where he was born. This should come as no surprise to those who know and admire Smith's brand of humor. Residences on North Hancock Street and on East Main Street were both possible sites. Smith eventually decided that the North Hancock Street house was his birthplace, and he posed for photographs in front of it in 1946.

NORMAL

Richard Hovey was born on May 4, 1864, in this town just north of Bloomington, in a house at 202 West Mulberry Street, now identified by a marker. Hovey collaborated with his friend Bliss Carman on several well-received volumes of poetry, beginning with *Songs from Vagabondia* (1894). He also translated eight plays by Maurice Maeterlinck. Hovey's other work included a posthumous collection of poems, *To the End of the Trail* (1908).

OAK PARK

Edgar Rice Burroughs, the creator of Tarzan—Lord Greystoke—lived from 1910 to 1919 in this residential community ten miles west of Chicago. Burroughs lived in several homes, including those at 821 South Scoville Avenue, 414 Augusta Street, 325 North Oak Park, and 700 Linden Avenue. At his office, 1020 North Boulevard, Burroughs pursued his literary career and, during World War I, wrote patriotic articles and recruited men for the Illinois Reserve Militia. Among the books Burroughs published during his Oak Park years were *Tarzan of the Apes* (1914), *A Princess of Mars* (1917), and *The Gods of Mars* (1918).

Other, no less memorable writers are also

Percy Bysshe Shelley
. . . At Thompson's Lake the trigger
of my gun
Caught in the side of the boat
And a great hole was shot through
my heart.
Over me a fond father erected this
marble shaft,
On which stands the figure of a woman
Carved by an Italian artist . . . .

—Edgar Lee Masters,
in *Spoon River Anthology*

Recent photograph of Mermaid House in Lebanon. Visitors to this old house included Charles Dickens and Abraham Lincoln.

A walk of half a mile through a thick grove of oaks brought us out upon a lovely, grassy knoll, which rose two hundred feet or more above the Rock River, and from which a pleasing view of the valley opened to the north as well as to the south. The camp . . . was in fact a sylvan settlement of city dwellers—a colony of artists, writers and teachers out for a summer vacation.

—Hamlin Garland,
in *A Daughter of the Middle Border*

associated with Oak Park. **Kenneth Fearing,** the poet and novelist, was born on July 28, 1902, at 720 North East Street and grew up in Oak Park. Fearing's work included *Poems* (1935) and *The Big Clock* (1946). **Ernest Hemingway** was born on July 21, 1899, in a house that is still standing at 339 (formerly 439) North Oak Park Avenue. He and his family also lived in houses at 161 North Grove Street and 600 North Kenilworth Avenue. Hemingway attended Oak Park and River Forest Township High School, where he contributed news columns and short stories to the school publication. **Edward Wagenknecht,** the writer and anthologist, also attended Oak Park High School, from 1915 to 1917, and lived from 1915 to 1925 at 934 Wenonah Avenue.

OREGON

The area surrounding this city in northern Illinois was an early inspiration for **[Sarah] Margaret Fuller,** the famous nineteenth-century feminist. It was she who named Eagle's Nest, a hill sloping down to the Rock River. Below is Ganymede's Spring, also named by her and now marked as the place where she wrote "Ganymede to His Eagle" (1843). In Rock River itself is Margaret Fuller Island. In 1898, sculptor Lorado Taft and others founded Eagle's Nest Artists Colony, on a hill overlooking the island, as a place in which writers and artists could live and work in tranquility. Writers associated with the colony include **Hamlin Garland** and **Harriet Monroe,** the poet and editor. Garland courted Lorado Taft's sister Zulime here, and later wrote about the courtship in *A Daughter of the Middle Border* (1921). The site of the colony is used now as an outdoor education area by Northern Illinois University.

Harry Leon Wilson was born in **Oregon** on May 1, 1867. Wilson's best-known works included the novels *Ruggles of Red Gap* (1915) and *Merton of the Movies* (1922) as well as a play, *The Man from Home* (1907), on which he collaborated with Booth Tarkington. *Merton of the Movies* was dramatized in 1922 by Marc Connelly.

Birthplace and childhood home of Ernest Hemingway in Oak Park.

OTTAWA

[Arthur] Johnston McCulley, author of more than sixty novels as well as plays, screenplays, and radio dramas, was born on February 2, 1883, in this city in north central Illinois. McCulley was the creator of the romantic swordsman Zorro, who appeared in *The Mark of Zorro* (1922), *Zorro Rides Again* (1931), and other novels by McCulley.

PALOS PARK

Sherwood Anderson lived in this city, ten miles southwest of Chicago, while working on his novel *Poor White* (1920). Anderson's home, at 12410 Ridge Avenue, is private.

PANA

This city, about thirty-five miles southwest of Decatur, is the birthplace of **[James] Vincent Sheean,** born on December 5, 1899. He was educated at local schools and the University of Chicago. Sheean, the quintessential foreign correspondent, reported on such momentous events as the Fascist march on Rome, the Spanish and Chinese civil wars, and the assassination of Mahatma Gandhi. Perhaps the most interesting of his many books was *Personal History* (1935), in which he described his intellectual development during the decade after World War I.

PEORIA

George Fitch, creator of Siwash College, lived here in the early 1900s. He served as editor of the Peoria *Herald-Transcript* from 1905 to 1911.

PETERSBURG

The boyhood home of **Edgar Lee Masters,** at Eighth and Jackson streets, is now the Edgar Lee Masters Memorial Museum, open daily during summer. This was his home during the eight years in which his father served as state's attorney for Menard County. After the family moved to Lewistown in 1880, Edgar returned each summer to his grandparents' farm outside of Petersburg. It is believed that Masters used this city, which is twenty miles northwest of Springfield, and Lewistown as models for the fictional town of his

Petersburg is my heart's home. There
I knew at first earth's sun and air;
Still I can see the hills around it,
The people that walked its business square.

—Edgar Lee Masters,
in "Petersburg"

Spoon River Anthology (1941). His poem "Petersburg" was published in *Illinois Poems* (1941). Masters is buried in Oakland Cemetery, just southwest of Petersburg. Also buried there are **Vachel Lindsay,** who died in 1931, and Ann Rutledge, who died in 1835. Lines by Masters that echo Ann Rutledge's place in the Lincoln legend appear on her gravestone:

Out of me unworthy and unknown
The vibrations of deathless music;
"With malice toward none, with charity for all."

QUINCY

This city on the Mississippi River in west-central Illinois was the model for the fictional Riverton in the novel *Eagle on the Coin* (1950), by R. V. Cassill. In Washington Park is a monument commemorating the sixth debate between Abraham Lincoln and Stephen A. Douglas, held here on October 13, 1858. **J[ames] F[arl] Powers** lived at 1658½ Jersey Street as a boy and attended Quincy Academy, now defunct.

RIVERSIDE

Ring Lardner built a house at 150 Herrick Road in this Chicago suburb during the fall of 1913. The experience gave Lardner material for the stories he later collected in the volume *Own Your Own Home* (1919). The Lardners lived in this house, now private, from 1914 to 1917, when Lardner was having his first successes as a writer.

ROBINSON

James Jones, known particularly for his best-selling novel *From Here to Eternity* (1951), was born on November 6, 1921, in this city in southeastern Illinois. *From Here to Eternity* dealt with U.S. Army life in Hawaii at the outbreak of World War II. Jones served in the Pacific during the war.

ROCK ISLAND

Cornelia Meigs was born in Rock Island on December 6, 1884. Her literary output included history, biography, short stories, and books for children. Among her children's books was her noteworthy biography of Louisa May Alcott, *Invincible Louisa* (1933). Her other books included *The Trade Wind* (1927), *Railroad West* (1937), and *The Violent Men* (1949).

ROSCOE

Edward Wagenknecht lived from 1921 to 1923 in this community north of Rockford, serving as minister of the Roscoe Congregational Church, now at 10780 Third Street.

SHARPSBURG

John G. Neihardt, Poet Laureate of Nebraska from 1921 until his death in 1973, was born on January 8, 1881, near this town southeast of Springfield. Neihardt's major work was the epic poem *A Cycle of the West* (1949), which included five long poems, conceived as a unified work but first published individually. These poems described the exploration and settlement of the frontier, combining historical fact with Neihardt's own observations of the land and knowledge of Indian life and lore. His other books included *The Splendid Wayfaring* (1920) and *Collected Poems* (1926).

SPRINGFIELD

Walter Havighurst spent his late youth in the capital city of Illinois. His family's home was on South Fifth Street, across from that of Vachel Lindsay. Havighurst later wrote of Springfield in *The Heartland* (1962), a history of the settlement of the states of Illinois, Indiana, and Ohio. Abraham Lincoln lived in Springfield from 1837 to 1861. The only home Lincoln ever owned—he lived in it from 1844 to 1861—stands at Eighth and Jackson streets and is open to visitors. At the entrance to Lincoln's tomb, at the end of Monument Avenue in Oak Ridge Cemetery, is a bronze bust of Lincoln by Gutzon Borglum.

[Nicholas] Vachel Lindsay was born in Springfield on November 10, 1879, in a two-and-a-half-story frame house at 603 South Fifth Street, once owned by Lincoln's brother-in-law. It was in 1861, in this house, that a farewell reception was held for Lincoln, who was leaving Springfield to serve as President of the United States. Lindsay, growing up surrounded by the Lincoln legend, later wrote in the poem "Abraham Lincoln Walks at Midnight" (1914):

It is portentous, and a thing of state
That here at midnight, in our little town
A mourning figure walks, and will not rest,
Near the old court-house pacing up and down. . . .

Lindsay's *The Golden Book of Springfield* (1920) evokes memories of earlier days in the town's history. Lindsay traveled widely after his graduation

They got me all most drove crazy
& if any body ever says to you
build a house bust them in the jaw.

—Ring Lardner,
in *Own Your Own Home*

Sketch by Virginia Stuart Brown of Vachel Lindsay's birthplace in Springfield. The poet lived in this house on and off throughout his life and died here in 1931.

i do not see why men
should be so proud
insects have the more
ancient lineage
according to the scientists
insects were insects
when man was only
a burbling whatisit

—Don Marquis,
in *archy and mehitabel*

from Springfield High School but returned often to his home on South Fifth Street, and it was there, on December 5, 1931, that he committed suicide. The Vachel Lindsay House is open to visitors, and there is a small charge for admission. The house contains more than sixty of Lindsay's art works as well as a collection of his letters, manuscripts, and other memorabilia. Among the numerous Springfield sites in Lindsay's poetry is the first cabin built in Springfield, standing at what are now Klein and Jefferson streets. Lindsay wrote of this cabin in his poem "On the Building of Springfield." Vachel Lindsay Bridge across Lake Springfield, southeast of the city, reminds visitors of Springfield's most illustrious literary son.

We must have many Lincoln-hearted men.
A city is not builded in a day.
And they must do their work,
and come and go,
While countless generations pass away.

—Vachel Lindsay,
in "On the Building of Springfield"

STERLING

Writers associated with this city in northwestern Illinois include **Odell Shepard,** born here on July 22, 1884, author of the Pulitzer Prize-winning biography *Pedlar's Progress, The Life of Bronson Alcott* (1937); and **Jesse Lynch Williams,** born here on August 17, 1871, who wrote the first play to be awarded a Pulitzer Prize, *Why Marry?* (1917), based on his novel *And So They Were Married* (1914).

STREATOR

Clarence E. Mulford was born on February 3, 1883, in this city fifty miles northeast of Peoria. His home, at 311 West Bridge, is now used for apartments. Mulford introduced Hopalong Cassidy in his first book, *Bar-20* (1907), and wrote twenty-seven more novels featuring Hopalong. Mulford's books sold in the millions, but Hopalong Cassidy today is known primarily as a creature of motion pictures and television, made to conform more with children's notions of the straight-shooting cowboy hero than with the character created by Mulford.

The Little Brick House in Vandalia. Portraits of James Hall, the editor and pioneer folklorist, are among the items on display here.

URBANA

Writers associated with this university city in east-central Illinois include **Frank Crane,** born here on May 12, 1861, whose syndicated newspaper columns had an enormous readership at the turn of the century; and **James Garfield Randall,** an authority on Abraham Lincoln and the Civil War era, who lived at 1101 West Oregon with his wife **Ruth Painter Randall,** author of the best selling *Mary Lincoln: Biography of a Marriage* (1953). Professor Randall was visited often here by **Carl Sandburg.** Mrs. Randall later lived at 1506 South Race Street. She died on January 22, 1971, and was buried in Mt. Hope Cemetery in nearby Champaign. Also associated with Urbana are **Carl** and **Mark Van Doren,** who moved here in 1900 to a house, no longer standing, at 712 West Oregon Street.

Writers associated with the University of Illinois, founded here in 1867, include Professor Randall, who taught here from 1920 to 1949; **Lew Sarett,** the poet, who taught here from 1912 to 1920; **James Still,** who received a B.S. degree here in 1931; **Carl Van Doren,** who was graduated in 1907 and taught here in 1907–08; and **Mark Van Doren,** class of 1914, who had a poem published in *The Smart Set* while he was a student here. The University of Illinois library houses most of the papers of Carl Sandburg.

VANDALIA

James Hall lived here in a house, no longer standing, at Third and Madison streets from 1829 to 1833. An early chronicler of frontier lore, Hall founded *The Illinois Monthly Magazine* here in 1830 and edited it until 1832, writing most of its contents himself. Hall's books included *Letters from the West* (1828), *Western Souvenirs* (1829), and *Legends of the West* (1832). The collection *Seven Stories* (1975) is the first modern edition of Hall's writing. The Little Brick House, 621 St. Clair Street, is a museum of frontier life and the life of Abraham Lincoln. Its James Hall Library contains an oil portrait and pencil sketch of Hall. A bronze marker in the Vandalia cemetery, dedicated to Hall's wife and child, recognizes Hall's accomplishments.

WALNUT

Don[ald] Marquis was born on July 29, 1878, in Walnut, which is on Route 92, about forty miles north of Peoria. Marquis became one of the best-known columnists in New York City and established, in 1912, the celebrated "Sun Dial" column in the New York *Sun.* Marquis is remembered today as the creator of archy and mehitabel. Archy, a poet reincarnated as a cockroach, used Marquis's typewriter to turn out accounts in free verse of his adventures with mehitabel, a cat. Marquis collected the pieces in *archy and mehitabel* (1927) and several other volumes. The free-verse accounts were set in lower-case type, because archy was not strong enough to depress the typewriter shift key. Marquis's many other books included *Poems and Portraits* (1922), *The Revolt of the Oyster* (1922), and *Chapters for the Orthodox* (1934).

MICHIGAN

ANN ARBOR

This city, thirty-five miles west of Detroit, is the home of the University of Michigan, one of America's great state universities. Numerous literary figures have spent their college days here —as just one example, **Arthur Miller** received the A.B. degree in 1938—or taught at the university on visiting professorships. One writer among the many who studied here is worthy of special mention because she was an irregular student: **Betty Smith,** author of *A Tree Grows in Brooklyn* (1943), attended the university as a special student from 1927 to 1930 and first became interested in writing while she was here.Twenty-three years old and mother of two young children, Smith arranged to take classes that did not conflict with her children's schedules. As it happened, many of the courses that fitted into her busy life were in writing and literature, and she managed as well to find time to work for the Detroit *Free Press* and for the Newspaper Enterprise Association.

Three of the most prominent literary figures who have taught on the Michigan faculty are **Robert Frost; Donald Hall,** the poet and critic, a member of the department of English since 1957; and **Allan Seager,** the novelist, who taught in the English department from 1934 until 1966, two years before his death. Seager died of cancer at St. Joseph Mercy Hospital on May 10, 1968; the hospital was located then at 326 North Ingalls Street but has since been moved to 5301 East Huron River Drive.

Another novelist who lived in Ann Arbor, **Lloyd C. Douglas,** was pastor from 1915 to 1921 of the First Congregational Church, still located at 608 East William Street. This was long before Douglas began writing the novels, such as *Magnificent Obsession* (1929), that would make him one of the most popular authors of his day.

ATWOOD

This small town south of Charlevoix, near Grand Traverse Bay, is the birthplace of **Rex Beach,** born on September 1, 1877. Beach wrote novels set in Florida, New York City, and elsewhere, but he is remembered best for his stories of life in the Klondike, such as *The Spoilers* (1906) and *The Iron Trail* (1913).

BATTLE CREEK

This town twenty miles east of Kalamazoo is the home of the Battle Creek Sanitarium, founded in 1866 to promote good health through good diet, and operating to this day at 197 North Washington Street. **Upton Sinclair** came here in October 1907, a year after publication of his novel *The Jungle* (1906), which portrayed conditions in the stockyards of Chicago. His wife, Meta Sinclair, was a patient at the sanitarium at the time. Other notable literary figures who have visited the sanitarium include **John Burroughs, George Bernard Shaw,** and **Booker T. Washington.**

Kenneth Rexroth, the poet, was a resident of Battle Creek for three years, 1912 to 1915, during his childhood. With his parents he lived first at 93 Garrison Avenue and later at 123 North McCamly. Rexroth attended public school in what now is the McKinley School.

BENZONIA

When **Bruce Catton** was three years old, in 1902, his family moved from Petoskey to this northern Michigan town. They left for a time to live in Boyne City, but returned here to stay in 1906, when Catton's father became headmaster of the Benzonia Academy. Bruce Catton attended that school, and continued to make his home here in later years. One of his favorite spots in town was the Civil War monument in Benzonia Cemetery, built by veterans in honor of their dead comrades. Catton, of course, wrote extensively of the Civil War in *A Stillness at Appomattox* (1953), a Pulitzer Prize-winner, and in other books. He wrote of his boyhood in Benzonia in *Waiting for the Morning Train* (1972).

BOYNE CITY

In the summer of 1919, while staying in nearby Petoskey, **Ernest Hemingway** and three friends amused themselves by shooting out the streetlights here on their way home from a fishing trip.

CHARLEVOIX

In this town on Lake Michigan, near Grand Traverse Bay, once stood Altasand, a large house owned by the family of **Sara Teasdale.** The family spent many summers there before the death of the poet's father in 1921. The house no longer stands, but the barn is now used as a garage.

DEARBORN

The Greenfield Village–Henry Ford Museum at Village Road, between Southfield Freeway and Oakwood Boulevard, includes the courthouse in which Abraham Lincoln practiced law and the chair in which he sat at Ford's Theatre in Washington, D.C., on the night of his assassination; the Stephen Foster Memorial Home; and the house Robert Frost lived in when he was teaching at the University of Michigan.

DETROIT

This city, the center of America's automotive industry, was the birthplace of **Nelson Algren.** The novelist, whose best-known works are *The Man with the Golden Arm* (1949) and *A Walk on the Wild Side* (1956), was born here on March

> *Early youth is a baffling time. The present moment is nice but it does not last. Living in it is like waiting in a junction town for the morning limited; the junction may be interesting but some day you will have to leave it and you do not know where the limited will take you. . . . In this respect early youth is exactly like old age; it is a time of waiting before a big trip to an unknown destination. The chief difference is that youth waits for the morning limited and old age waits for the night train.*
>
> —Bruce Catton,
> in *Waiting for the Morning Train*

Snapshot of Chickaming Goat Farm, home of Carl Sandburg in Harbert. The success of *Abraham Lincoln: The Prairie Years,* published in 1926, enabled Sandburg to leave Chicago for a more comfortable home.

28, 1909. **Bronson Howard,** born here on October 7, 1842, portrayed the upper classes in such plays as *The Banker's Daughter* (1878). **Charles A. Lindbergh,** who collaborated on *We* (1927), the story of his history-making transatlantic flight, and was sole author of his autobiography, *The Spirit of St. Louis* (1953), was born here on February 4, 1902. Writers who lived here for brief periods include **Herbert Gold,** who wrote the novel *The Man Who Was Not With It* (1956) while teaching at Wayne State University from 1954 to 1956; and **Margaret Halsey,** the novelist and satirist, who lived at 201 East Kirby Street from 1966 to 1968.

Two writers who lived here for somewhat longer periods were **Edgar Guest,** the popular poet, and **Clarence Budington Kelland,** a newspaperman and novelist. Guest arrived in Detroit with his family in 1901, when he was nine. Four years later he left school to work as a copyboy on the Detroit *Free Press.* Progressing to editor, in 1905 he took over a weekly verse column; it was popular enough to be turned into a daily "Breakfast Table Chat." Guest wrote the column for the rest of his life, producing a poem a day for decades. All his collections of verse, from *A Heap O' Livin'* (1916) to *Between You and Me* (1938), were enormously successful. Guest died in Detroit on August 5, 1959.

Kelland was educated in private schools here and received a degree from Detroit College of Law in 1902. From 1904 to 1909, he worked for the Detroit *News* in that newspaper's offices, now destroyed, at 65 Shelby Street. After leaving the *News* he edited *American Boy,* whose offices were in the Majestic Building, also destroyed, at the corner of Michigan and Woodward avenues. It was during Kelland's time at *American Boy* that he wrote *Mark Tidd* (1913), the first of his series of novels for boys.

W. D. Snodgrass, the poet, taught at Wayne State University from 1959 to 1967. While living here Snodgrass was awarded the Pulitzer Prize for his collection of poems *Heart's Needle* (1959).

Mary Catherwood, author of the collection *Mackinac and Lake Stories* (1899).

FRANKFORT

Bruce Catton spent summers in a house here, on the shores of Crystal Lake near Lake Michigan, in the last twenty years of his life. The road leading up to the house is called Glory Road, after *Glory Road* (1952), one volume in Catton's Civil War trilogy. The author's sister Ruth now lives in the house; it is not open to visitors.

GRAND RAPIDS

Stewart Edward White was born in this city in the southwestern part of the state on March 12, 1873. The site of his birthplace is now occupied by the Grand Rapids Museum. Most of White's books were set in the western states, but *The Blazed Trail* (1902), sometimes considered his finest work, was a story of the lumber camps of northern Michigan. Grand Rapids is the home of Calvin College, alma mater of two novelists, **Peter De Vries** and **Frederick Manfred.** The college is located at 3201 Burton Street, S.E.

HARBERT

The Chickaming Goat Farm here, sixty miles east of Chicago, on sand dunes overlooking Lake Michigan, was for many years the country retreat of **Carl Sandburg.** Sandburg's family often stayed here without him while the poet was on lecture tours, but he lived here much of the time between 1932 and 1945, and it was here that he wrote *The War Years* (1939), the second part of his massive Pulitzer Prize-winning biography of Abraham Lincoln.

HORTON BAY

Ernest Hemingway, who spent summers in his youth at his family's cottage in nearby Petoskey, often visited this small bayside town. On September 3, 1921, in the local Methodist Church, Hemingway was married to Hadley Richardson. Hemingway used the small community as the setting for his short story "Up in Michigan," referring to it as Horton's Bay. He also described the town's general store, which still operates and has changed little since Hemingway's time.

HUDSON

Will Carleton, the poet and short-story writer, was born on October 21, 1845, in this small town ten miles from the Ohio border. The house in which Carleton was born, at 1499 Carleton Road, is privately owned. A commemorative monument stands in the front yard. The local school, where Carleton was a student and later a teacher, is also standing. The town library has a collection of Carleton's works and a portrait in its small museum. His books included *Farm Ballads* (1873) and *City Ballads* (1885). *Farm Ballads* contained his best-known poem, "Over the Hill to the Poor-House."

KALAMAZOO

This city twenty miles west of Battle Creek is the birthplace of **Edna Ferber,** born on August 15, 1885. Her home, at 825 South Park Street, was torn down over sixty years ago, and two residential buildings now stand on the site, renumbered as 901 South Park Street. Ferber's most popular novels were *So Big* (1924), *Show Boat* (1926), and *Cimarron* (1930).

LANSING

Ray Stannard Baker, the essayist and historian, was born in this city in south-central Michigan on April 17, 1870. His home, on Walnut Street, near the old Lewis Cass building, has since been demolished. Under his own name Baker wrote several books on President Woodrow Wilson, whose friend and sometime assistant he was. Under the pseudonym **David Grayson,** he wrote seven volumes of essays, the most popular of which was *Adventures in Contentment* (1907).

LITCHFIELD

Rose Hartwick Thorpe, the poet and author of books for children, grew up in the 1850s and 1860s in Litchfield, about thirty miles southeast of Battle Creek in southern Michigan. By the time she was eleven years of age, Rose Hartwick had begun to write poetry, and she was still a student at Litchfield High School, from which she was graduated in 1868, when she wrote her best-known poem, "Curfew Must Not Ring Tonight." This poem, based on a story Rose Hartwick read in *Peterson's Magazine,* recounted the tale of a young woman who saves the life of her lover by clinging to the clapper of the local church bell so that it cannot toll curfew, the time set for her lover's execution.

MACKINAC ISLAND

This small island in Lake Huron has long been popular with visitors. **Mary [Hartwell] Catherwood,** the historical novelist, spent her summers here after 1890; and **Peter De Vries** visited Mackinac Island while traveling in Michigan with his parents. **Walter Havighurst,** while staying at the Hotel Windemere on Mackinac Island, wrote *Three Flags at the Straits* (1966), a history of the forts that commanded the Straits of Mackinac.

MACKINAW CITY

Henry David Thoreau and Horace Mann, Jr. came to this city at the Straits of Mackinac, the meeting point of Lakes Huron and Michigan, in June 1861. They were on their way to Minnesota, where Thoreau hoped to repair damage to his health from a bout of bronchitis he had suffered several years earlier. Mann was active during their stay here at the Mackinaw House, but Thoreau is said to have spent most of the time sitting by the fire.

MANISTEE

This small town on the shore of Lake Michigan, fifty miles south of Traverse Bay, has played host to at least two twentieth-century literary men. **Ross Lockridge** and his family rented a small cottage at 101 Lake Shore Drive in August 1945, when Lockridge was revising his novel *Raintree County* (1948). During that autumn he went to Boston to confer with his publisher, Houghton Mifflin, but his family stayed on in Manistee and he rejoined them early in 1946. Today the cottage where they stayed is privately owned and not open to the public. Twenty years earlier, in August 1926, **Stuart P[ratt] Sherman,** the editor and critic, vacationed here with his wife. On August 20, their canoe capsized while they were out on the lake. Swimming to shore, Sherman suffered a heart attack and, despite all attempts to revive him, died on the beach. He was forty-four at the time of his death. For two years co-editor of the *Cambridge History of American Literature* (1917–21), Sherman also was editor of the literary supplement of the New York *Herald-Tribune* and professor of English at the University of Illinois.

MONROE

This city off Route 75, between Toledo and Detroit, was the birthplace of **Elizabeth Bacon,** the wife of General George Armstrong Custer. Born on April 8, 1842, she lived with her family in a house at 126 South Monroe Street. That house has been moved to Seventh and Cass streets, but some of its original features have been changed and it is not open to visitors. At the original site now stands the Monroe County Historical Museum. A plaque on the south wall is dedicated to Custer. After his death, his widow wrote several books, including *Boots and Saddles* (1885) and *Tenting on the Plains* (1887).

MUSKEGON

J[ohn] F[rederick] Nims was born in Muskegon, thirty-five miles northwest of Grand Rapids, on November 20, 1913. Nims served as an editor of the magazine *Poetry* for a time in the 1940s and has taught at Notre Dame and Illinois universities. Among his collections of verse are *The Iron Pastoral* (1947) and *Knowledge of the Evening* (1960).

NILES

Ring[old] Lardner was born in Niles, ten miles north of South Bend, Indiana, on March 6, 1885. His family's house, at 519 Bond Street, is now privately owned and not open to visitors,

Over the hill to the poor-house—
I can't quite make it clear!
Over the hill to the poor-house—
it seems so horrid queer!
Many a step I've taken a-toilin'
to and fro,
But this is a sort of journey I never
thought to go.

—Will Carleton,
in "Over the Hill to the Poor-House"

Birthplace of Ring Lardner in Niles. A marker at the house reads in part: "Sportswriter, humorist, sardonic observer of the American scene . . . began his career writing sketches of sporting events for the Niles Sun and later worked for papers in Chicago and New York, where he wrote a popular syndicated column."

". . . I grew up in and around a beautiful greenhouse owned by my father and uncle; I hated high school and Michigan. . . ."

—Theodore Roethke, in a biographical statement supplied for *Twentieth Century Authors*

although it does have a plaque outside. Lardner is further commemorated by the Ring Lardner Junior High School; by the Fort St. Joseph Museum, off Route 60 near Niles, which has some memorabilia; and by the Trinity Episcopal Church, which has a piece of wood—once part of the church organ—on which the youthful Lardner carved his initials. Lardner is reported to have said of Niles: ". . . we didn't have no telephones and neither did anybody else and those was the happiest days of my life."

PETOSKEY

This town on Michigan's Little Traverse Bay is the birthplace of **Bruce Catton.** The historian was born at 430 Pearl Street on October 9, 1899. The house, in which Catton lived until he was three, is still privately owned and not open to visitors, nor are there any markers outside. The Little Traverse Historical Society here has copies of all of Catton's books and the manuscripts of *Grant Takes Command* (1969) and *Waiting for the Morning Train* (1972), as well as family photographs and other memorabilia. The society's museum is open to the public during the summer months. The Petoskey Public Library at 451 East Mitchell Street has a bronzed stone and plaque dedicated to Catton.

Another writer born in the same year as Catton had even more extensive connections with Petoskey. **Ernest Hemingway** came here for the first time when he was only a year old: his parents had recently built a summer house, Windemere, on Walloon Lake, nine miles south of the town. Ernest was to spend seventeen summers here, and additional time in Petoskey itself. Much of what he knew about nature and outdoor life was learned here and was reflected in his Nick Adams stories. In October 1919, back from World War I with massive leg wounds, he rented a room at 602 State Street in Petoskey, and in December of that year gave a talk on his war experiences at the public library. Two years later, in September 1921, he returned to Windemere for his honeymoon with Hadley Richardson. Windemere today is occupied by Hemingway's sister Madelaine. Although registered as a National Historic Landmark, the house is not open to visitors and its official plaque is not displayed. The Emmanuel Episcopal Church, at 1020 East Mitchell Street, has a stained glass window dedicated to Hemingway, and the public library displays a fine collection of his works. The Little Traverse Historical Museum has a number of family photographs of Hemingway and autographed copies of three of his books.

PORTLAND

Clarence Budington Kelland, the novelist, was born here, twenty miles north of Lansing, on July 11, 1881. Kelland's mother operated a millinery shop, and the family lived here until Kelland was ten years old. The family's first home, on West Bridge Street, was burned down in a firefighting training program, and the site is now occupied by a park. The Kellands also lived at 290 Lincoln Street, in a house that is privately owned and has no marker. Kelland returned here twice, first in 1929, when he was already known well as the author of *Mark Tidd* (1913) and other novels, and again in 1944. His *Scattergood Baines* (1921), which recounted the experiences of a shrewd Yankee affecting ingenuousness, was Kelland's most memorable work.

SAGINAW

Theodore Roethke was born here, ninety miles northwest of Detroit, on May 25, 1908. The house, at 1805 Gratiot Street, is privately owned and not open to the public. Roethke's father and uncle owned a large greenhouse, which may have

Madelaine Hemingway Miller, holding the plaque designating Windemere (*right*) as a National Historic Landmark. Mrs. Miller has stated: "I accepted the plaque with the understanding that I would not display it outdoors. Possible thievery or vandalism has to be considered."

provided much of the horticultural imagery of Roethke's poetry. For *The Waking* (1954), Roethke won a Pulitzer Prize, and for *Words for the Wind* (1958), a Bollingen Prize. After the poet's death on August 1, 1963, in the state of Washington, he was cremated and his ashes brought back to Saginaw for burial in Oakwood Cemetery.

SENEY

From this isolated town in the Upper Peninsula, **Ernest Hemingway** and two friends set out in the summer of 1919 for a fishing trip. Their trip provided Hemingway with the background for his story "Big Two-Hearted River," published in the collection *In Our Time* (1925). The Big Two-Hearted River is a real river, but Hemingway and his friends are believed to have fished another river in the area.

SPRING LAKE

Edgar Lee Masters often came to this popular holiday area, west of Grand Rapids, while he was practicing law in Chicago. On the lakefront is a cement bench on which Masters is said to have sat while working on his masterpiece, *Spoon River Anthology* (1915). In 1917 he bought a large farm at 18067 Fruitport Road near here; the house has been extensively remodeled and is now privately owned, but the current owners will show it to visitors who make an appointment. Masters also wrote *Toward the Gulf* (1918), another collection of poetry, while living here.

To scratch dirt over scandal for money,
And exhume it to the winds for revenge,
Or to sell papers,
Crushing reputations, or bodies, if need be,
To win at any cost, save your own life.
—Edgar Lee Masters,
in *Spoon River Anthology*

WHITEHALL

Frank R[amsay] Adams died on October 8, 1963, at 5746 Murray Road here, on the southwestern fringe of the Manistee National Forest. Adams's novels included *Gunsight Ranch* (1939) and *Arizona Feud* (1941), but he is remembered particularly for his lyrics to the popular song "I Wonder Who's Kissing Her Now."

WISCONSIN

APPLETON

Edna Ferber and her family came to Appleton about 1898, when her father opened a store at 772 College Avenue, now renumbered 120 College Avenue. The Ferber store was replaced eventually by a Gimbel's department store. The family lived first at 319 Drew Street, across from the city park, and later in a house still standing at 216 North Street. Ferber was graduated in 1903 from Ryan High School, no longer standing; when her father went blind she found a job as a reporter—Appleton's first female reporter, in fact—on the Appleton *Daily Crescent;* her salary was $3 a week. Ferber was later to say that her experience on the *Crescent,* which once stood at 758, now 134, College Avenue, gave her material for the *Oklahoma Wigwam,* the fictional newspaper she created for her novel *Cimarron* (1930). But the job on the *Crescent* lasted only a year, and Ferber went off to Milwaukee, where once more she worked as a reporter, working so hard that by 1909 she was back in Appleton to recover from a nervous collapse. For this reason it was in Appleton that Ferber wrote—at 216 North Street—her first novel, *Dawn O'Hara* (1911), and her first short story, "The Homely Heroine." Another Appleton figure is **Walter Havighurst,** who was born here, possibly on John Street, on November 28, 1901. His family lived at the time at 502, now 804, South Street, and later at 783, now 539, Lawe Street. By 1910 the Havighursts had left Appleton and Wisconsin, but Walter Havighurst did not forget his native state. He wrote lovingly of it in *The Winds of Spring* (1940).

Carl Sandburg and his wife Lillian, shortly after their marriage in Milwaukee in June 1908, took rooms in Appleton in a house at 353 Second Avenue, now 1037 East Wisconsin Avenue, where they lived until April 1909. During their stay in Appleton, Sandburg worked as a district organizer for the Social Democratic Party, often riding on the campaign train—dubbed the Red Special by political opponents—that was used by Eugene V. Debs, the Socialist candidate for president.

Winter settled like a siege around the scattered habitations. The cold crept into the cabins; the windows glazed over. Many people boarded their houses tight, preferring closeness and darkness to the bitter prying wind. In the storerooms the food froze solid; milk, butter, cheese, bread, all turned to solid lumps and the potatoes blackened slowly with frost. Night and day the wind cried through the naked trees, the wolves grew bold with hunger and came almost to the cabin door. In every house the life grew tight about the red embers in the stove's grating. Fire was a magnet. It became a strict radius that pulled stronger as one left it. And the mind shrank from the cold horizons, from the naked forest, even from the preempted acres where the snow lay drifted. People scraped the frost from the windows and wondered whether spring would ever find this land again.
—Walter Havighurst,
in *The Winds of Spring*

BEAVER DAM

Carl Sandburg and his wife lived here for a time in 1909, after leaving Appleton. Beaver Dam is eighty-five miles northeast of Madison.

Home of Edna Ferber on North Street in Appleton.

Summer passes all too swiftly for an active boy of eleven. Rascal and I often went fishing below the dam in the river at Indian Ford. Sometimes I cast for bass and pickerel among the water lilies. More often I fished for big, fighting silver catfish in my favorite hole below a pleasant sand bar in the river. Rascal meanwhile fished in the shallows along the edge of the bar, often seizing a crayfish—those little monsters that look so much like small fresh-water lobsters. These he washed and ate, tail first, with obvious delight.

—Sterling North,
in *Rascal*

CATO

Thorstein [Bunde] Veblen, remembered for his influential work *The Theory of the Leisure Class* (1899) and for his phrase "conspicuous consumption," which remains to this day a favorite term of derogation among social and economic critics, was born in Cato on July 30, 1857. Cato is a town just west of Manitowoc.

DOWNSVILLE

The Caddie Woodlawn Memorial Park, at the junction of Route 25 and County Trunk Y, eleven miles south of Menomonie, was named after *Caddie Woodlawn* (1935), a children's book by **Carol Ryrie Brink.** The book was based on stories told to Brink by her grandmother, Caroline Caddie Woodhouse, who came to Wisconsin in 1857. Their original home has been preserved by the Dunn County Historical Society and moved to its present site in the park, where it is open to the public daily without charge.

EAU CLAIRE

The Paul Bunyan Camp at Carson Park on Half Moon Lake in the western Wisconsin city of Eau Claire is free and open daily. It has a cook shack, bunkhouse, stable, and blacksmith shop built to resemble the structures found in old lumber camps. For literary visitors, the park recalls the legendary American lumberjack, hero of the tall tales popular in timber country.

EDGERTON

Sterling North, the newspaperman and novelist, was born on November 4, 1906, in Edgerton, which is twenty-three miles southeast of Madison. His birthplace was his family's farmhouse on R.D.2, overlooking Lake Koshkonong. The farmhouse is still standing but private. North also lived at 409 West Rollin Street, and this house is also still standing but private and unmarked. North spent his youth in and around Edgerton where, as he once wrote, he was "blessed by a richly rewarding childhood in a far-off time." In a number of his books, Edgerton appeared as Brailsford Junction. North's work included the novel *So Dear to My Heart* (1947) and *Rascal* (1963), a popular story of a pet raccoon that he himself had kept when he was a boy.

Home of Carol Ryrie Brink in Caddie Woodlawn Memorial Park, Downsville.

GENOA

William Ellery Leonard based his play *Red Bird* (1923) on an 1827 Indian raid that took place near Genoa, a town about fifteen miles south of La Crosse. The Winnebago found themselves displaced from their mines by a band of white miners and tried, in a bloody encounter, to drive the miners away. After several miles of pursuit, Chief Red Bird and several of his warriors surrendered to U.S. troops rather than subject their people to open warfare. Red Bird died in prison a few months later, and the other members of the Winnebago attack party were eventually pardoned. A plaque on Route 35 marks the site of the raid.

HURLEY

This logging center on Route 51 in northern Wisconsin, near the Michigan border, was the setting for the novel *Come and Get It* (1936), by Edna Ferber. The novelist used Hurley because of its reputation for licentiousness and violent crime. Lolla, one of the principal characters in the book, was found behind Hurley's Dew Drop Inn, her head split by an ax. At last report, the Dew Drop Inn was functioning as Trolla's Meat Market.

IRON RIVER

Henry Wadsworth Longfellow used the countryside near Iron River, on Lake Superior, in northwestern Wisconsin as the scene of his poem *Hiawatha* (1855). It is interesting to note that Hiawatha in all probability never visited this region, since he belonged to the Iroquois, who were bitter enemies of the Ojibway, the inhabitants of this part of Wisconsin.

JOHNSTOWN CENTER

Ella Wheeler Wilcox, the poet and novelist, was born here on November 5, 1850. Of the more than forty volumes of verse that she wrote, her best-remembered poem is "Solitude," which begins:

Laugh and the world laughs with you,
Weep and you weep alone.

KEWASKUM

Glenway Wescott was born on April 11, 1901, in Kewaskum, north of Milwaukee. He used Wisconsin as a setting for some of his novels, including his first, *The Apple of the Eye* (1924), which he began writing in Kewaskum, and *The Grandmothers* (1927). A collection of his short stories, *Goodbye Wisconsin* (1928), also contained Wisconsin material. Wescott grew up in Kewaskum and lived for a time in his uncle's house after leaving home because of differences with his father.

LA CROSSE

George Wilbur Peck, creator of the mischievous lad who entered American lore as Peck's Bad Boy, lived at Ninth and King streets in the city of La Crosse, which lies on the Mississippi River, in southeastern Wisconsin. Peck was editor and one of the owners of the La Crosse *Evening Democrat* in 1871–72. Two years later, he founded his own newspaper, the *Sun,* which he published from a building at Second and Main streets. The newspaper moved later to offices in the Opera Block, a building at the corner of Fourth and Main streets, where the newspaper was published until 1878, when Peck moved the *Sun* to Milwaukee. Peck began to write his stories for the *Sun,* and they later were collected in several volumes, including *Peck's Bad Boy and His Pa* (1883).

MADISON

Sarah N[orthcliffe] Cleghorn, the poet, lived during her childhood at Fairfields, a small farm near Madison. She was a year old in 1877 when her family moved to Fairfields. **Zona Gale** studied at the University of Wisconsin in Madison, receiving a B.A. degree in 1895 and an M.A. degree in 1899. While Gale was an undergraduate, her work was published in *The Aegis,* the university's literary magazine, and she sold her first short story to the *Evening Wisconsin,* in Milwaukee. Gale, who received a Pulitzer Prize for her novel *Miss Lulu Bett* (1920), was granted an honorary doctorate by the university in 1929.

Perhaps the most interesting literary figure on the faculty of the university in time past was **William Ellery Leonard,** who taught in Madison from 1906 until his death in 1944. Leonard fell in love with Charlotte Freeman, a resident of Madison, and married her on June 23, 1909. The marriage ended in tragedy on May 6, 1911, when Charlotte committed suicide. Leonard suffered a mental breakdown and, for the rest of his life, a phobia made it increasingly difficult for him to leave his house. He wrote a series of sonnets describing his love affair and attempting to explain the reasons for his wife's death. The sonnets were collected in *Two Lives* (privately printed in 1922 and reprinted in 1925).

Leonard also wrote a remarkable autobiography, *The Locomotive-God* (1927), which may have been the first popular work to use psychoanalytic terminology to explain behavior. After Leonard's death on May 2, 1944, it was reported that he had in recent years "ventured only once out of a five-block radius beyond his apartment and classroom. This adventure from his 'phobia prison' consisted of a trip to a downtown motion-picture theater." Among Leonard's students in his years at the university were at least two who would one day win a Pulitzer Prize: **Marjorie Kinnan Rawlings** for the novel *The Yearling* (1938), and **Marya Zaturenska** for the book of poetry *Cold Morning Sky* (1937).

A curious Madison faculty note is also worthy of mention: **Sinclair Lewis** was engaged to teach a creative-writing course at the university during the 1940–41 semester. In the middle of the first term, Lewis decided that he had taught his class all he could. He left his class and Madison. During his short stay, the novelist lived at 1712 Summit Avenue, in a house still standing but private.

William Ellery Leonard, poet and professor of English at the University of Wisconsin.

Frederic Prokosch was born here on May 17, 1908, probably at his family home, 1619 Jefferson Street, now located within the university grounds. Nearly all of Prokosch's novels reflect his worldwide travels. They include *The Asiatics* (1935); *The Skies of Europe* (1941); *The Conspirators* (1943), set in Lisbon during World War II; and *Storm and Echo* (1948), about Africa. **[John] Herbert Quick,** author of *Vandemark's Folly* (1921), was associate editor of La Follette's *Weekly Magazine* from December 1908 to July 1909. The magazine was published from 1909 to 1929 at 119 West Main Street, now reported to be the home of The New Paradise Lounge Tavern. **Mark Schorer** completed his first novel, *A House Too Old* (1935), while he was a student at the University of Wisconsin in the 1930s.

Thornton Wilder was born on April 17, 1897, in a rented flat at 140 Langdon Street, no longer standing. Wilder won three Pulitzer Prizes for his work: *The Bridge of San Luis Rey* (1927), *Our Town* (1938), and *The Skin of Our Teeth* (1942). On October 5, 1940, Wilder gave the 100th Anniversary Lecture at Madison's First Congregational Church, at University Avenue and Breese Terrace.

Among the other literary figures who have studied or taught at the University of Wisconsin is **Joyce Carol Oates,** who received an M.A. degree in 1961. She lived at 625 North Frances Street, where **Eudora Welty** had lived years before. Welty received a B.A. degree in 1929. The building at 625 North Frances Street is still standing. **R[obert] W[ooster] Stallman** studied at the university—earning three degrees, including a Ph.D.—and taught here from 1939 to 1942. **Wallace Stegner** and **Lionel Trilling** also taught at the university.

MARINETTE

Edna Ferber, while working as a reporter for the Milwaukee *Journal,* came up to Marinette,

> *The song of the axe was a magic song, bringing quick and marvelous change. For first, with the land cleared and houses built, the prairie ordered itself, and then, on the bare, turned earth, grain grew. But, as the song called to the earth, so it called too, to men, and more came, less afraid, less powerful, for with the wildness changing into peace, it needed less courage to come, less motive, and less power to build. But all of them, all new, watched with awe how the earth worked, watched the changes of the seasons on the face of the prairie, on the sides of the hills, on the wall of the forest; and into this change their changes fitted themselves.*
>
> —Mark Schorer, in *A House Too Old*

As one speaks of History, and more specifically Memory, my symbol of this is the house on Jefferson Street, that stood for whatever I learned during childhood and youth. Through its associations, both conscious and unconscious, the house is a signature of my existence—other signs are in my verse, my translations, my critical writings—and all these are my very life, however it may take form. Today, the visible being of the house is gone. It belongs to Memory.

—Horace Gregory, in *The House on Jefferson Street*

just across the border from the town of Menominee, Michigan, to interview an old-time logger. This trip later provided material for her novel of early Wisconsin logging days, *Come and Get It* (1935).

MILWAUKEE

Local history has it that the first practical typewriter was perfected in Milwaukee in 1869, the site of the invention being 318 State Street. With that much interest in improving the legibility of manuscript, it is not surprising that the city has many literary associations. **Edna Ferber** worked as a reporter at the Milwaukee *Journal* for a number of years early in the 1900s. The *Journal* was published at 182–184 Fourth Street, now 734 North Fourth Street. Ferber lived during that period at a number of rooming houses, including Kahlo's, across the street from the courthouse square, which she used as background for her first novel, *Dawn O'Hara* (1911). She also lived at the Avon, 444 Cass Street, now 760 Cass Street. The Avon, known more generally as Ma Haley's, was a colorful place, which Ferber described as sinister and having something of a whorehouse atmosphere. Two other places where Ferber stayed were the Walther League Lutheran Girls' Home and a rooming house at 638 Astor Street, now 1220 North Astor Street. Ferber arrived in Milwaukee shortly after leaving high school, and her autobiographical works, especially *A Peculiar Treasure* (1938), described her experiences in Milwaukee as she made friends and established a career as a newspaper reporter. One of her favorite hangouts, beloved of other newspaper reporters as well, was Martini's Bakery and Coffee House, at 443 East Water Street, which has since moved to larger quarters. Ferber worked so hard at the *Journal* that she was forced, for reasons of health, to return in 1910 to Appleton, where she moved in with her mother to recuperate. The Edna Ferber Room at the University of Wisconsin, Milwaukee, exhibits some of Ferber's personal effects.

Zona Gale also worked for a time for the Milwaukee *Journal* and, before her job at the *Journal*, for the *Evening Wisconsin*, on Milwaukee Street at the northeast corner of Michigan Street. Gale lived at various places during this period, including 70 Prospect Street, now 1344 North Prospect Street, in 1897; 3023 Mt. Vernon Avenue, now 3027 Mt. Vernon Avenue, in 1898; and 107 Thirty-first Street, now North Thirty-first Street, in 1899 and 1900.

Horace Gregory, the poet, was born on April 10, 1898, at 587 First Avenue, now probably 1737 North Sixth Street, and attended the German-English Academy, on Broadway. He also attended, from 1913 to 1916, the Milwaukee School of Fine and Applied Arts, at the corner of Downer Avenue and Kendall Boulevard. Gregory lived at 717 Jefferson Street, which he recalled later in his book of memoirs *The House on Jefferson Street.* After leaving Milwaukee, Gregory studied at the University of Wisconsin, and it was there that he met and married **Marya Zaturenska,** the poet.

Milwaukee also figures in the novel *The Long Ships Passing* (1942), by Walter Havighurst. **Edith J. R. Isaacs,** editor of *Theatre Arts Monthly* (1924–46) and one of the developers of theater in America, was born on March 27, 1878, at 638, now 1220, Astor Street. **Charles King,** an army officer in the Civil War and author of popular adventure books, died in Milwaukee on March 18, 1933. Among his novels were *The Colonel's Daughter* (1883) and *Under Fire* (1894). **James Russell Lowell,** who had become widely known as the author of the first series of *Biglow Papers* (1848), written in opposition to the Mexican War, lectured in 1855 on the English poets at Young's Hall, at the corner of Wisconsin and Main streets. Main Street is now Broadway, and Young's Hall has disappeared. **George Wilbur Peck** brought his newspaper, the *Sun,* to 416, now 306, Milwaukee Street in 1878. His popular book *Peck's Bad Boy and His Pa* was published in 1883, launching a series of books that sold hundreds of thousands of copies. Peck was elected mayor of Milwaukee, and served as governor of Wisconsin from 1890 to 1894. He lived at 814, now 484, Marshall Street in 1878; 206 Biddle Street, now Kilbourn Street, in 1880; 140, now 1534, Prospect Street in 1884; the Plankinton House in 1888; and 195, now 1627, Prospect Street in 1889.

Carl Sandburg was an organizer for the Wisconsin Social Democratic Party in 1907, when he met his future wife, Lillian Steichen. They were married on June 15, 1908, at party headquarters in Milwaukee, and went off to live for a time in Appleton before returning to Milwaukee. After working from 1910 to 1912 as personal secretary to Emil Seidel, Milwaukee's first socialist mayor, Sandburg became a reporter and editorial writer, working for the *News,* the *Sentinel,* the *Journal,* and finally *The Leader,* all of Milwaukee. He also served as city editor of the Milwaukee *Social Democratic Herald.* When the Sandburgs came to Milwaukee to live, they stayed first in Wauwatosa, then a suburb of Milwaukee, in a plain, "almost shabby" house with no carpets and little furniture. After a year they moved, in 1911, to 907 Eighteenth Street, now 2469 North Eighteenth Street, and then to 934 Cambridge Street, now 3324 North Cambridge Street, in 1912. Their first child, Margaret, was born at the North Eighteenth Street house, which still stands.

R[obert] W[ooster] Stallman was born in Milwaukee on September 11, 1911. The Stallman family is reported to have lived in 1911 at 337½ Twenty-sixth Street, now 1125 North Twenty-sixth Street, and in 1912 at 2408 Chestnut Street, now 2406 West Juneau Street. Both residences are private. Stallman attended West Division High School from 1925 to 1929. The old building was demolished in 1959, but a new school stands at 2300 West Highland Avenue.

We put new wine in old bottles, but also we use new bottles to hold our old wine. For, consider the name of our main street: is this Main or Clark or Cook or Grand Street, according to the register of the main streets or towns? Instead, for its half-mile of village life, the Plank Road, macadamized and arc-lighted, is called Daphne Street. Daphne Street! I love to wonder why. Did our dear Doctor June's father name it when he set the five hundred elms and oaks which glorify us? Or did Daphne herself take this way on the day of her flight, so that when they came to draught the town, they recognized that it was *Daphne Street, and so were spared the trouble of naming it? Or did the Future anonymously toss us back the suggestion, thrifty of some day of her own when she might remember us and say,* "Daphne Street!"

—Zona Gale, in *Friendship Village*

PEPIN

Laura Ingalls Wilder, author of the autobiographical series of Little House books for children, was born on February 7, 1867, in a log cabin near Pepin. Pepin lies close to the mouth of the Chippewa River on Lake Pepin, a natural swelling of the Mississippi River. Her many books, among them *Little House in the Big Woods* (1932), *Little House on the Prairie* (1935), *On the Banks of Plum Creek* (1937), *The Long Winter* (1940), and *Little Town on the Prairie* (1941), became the basis for a popular television program called *Little House* *on the Prairie.* A restored cabin of the period stands at its site on the bluffs north of Pepin.

PORTAGE

The town of Portage played a vital role in the early history of the United States as the site of the historic portage between the Fox and Wisconsin rivers, enabling early traders to travel from Lake Michigan to the Mississippi River. Jonathan Carver (1710–1780) wrote of the portage site in his *Travels Through the Interior Part of North America* (1778), a work that was published in more than thirty editions, but the literary history of Portage did not end with Carver's work.

Zona Gale was born at 605 De Witt Street on August 26, 1874, and lived in Portage for all but fourteen years of her life. She was educated in the local schools and at the University of Wisconsin, completing her M.A. degree in 1899. After working on newspapers in Milwaukee and New York City, Gale returned to Portage in 1904, living at 506 West Edgewater Street, in a home she had previously had built for her mother. The building now is the home of the Women's Civil League, to whom Gale donated the building. Gale also lived in a house at the corner of West Franklin Street and MacFarlane Road, and this house she donated to the Portage Free Public Library. Friendship Village and the town of Burrage, which appear in many of Gale's stories, were based on the town of Portage, although there is an actual village called Friendship, some thirty miles northwest of Portage. The interested reader can find Portage in Gale's *Friendship Village* (1908), *Yellow Gentians and Blue* (1927), and *Bridal Pond* (1930). Gale dramatized her novel *Miss Lulu Bett* (1920), a study of the bleakness of small-town life, and was awarded a Pulitzer Prize for drama in 1921. She died in a Chicago hospital on December 27, 1938, but was buried in Portage, in Silver Lake Cemetery.

Frederick Jackson Turner, the eminent historian, was born in Portage on November 14, 1861. Among his most important works were *The Frontier in American History* (1920) and *The Significance of Sections in American History* (1932), for which Jackson was awarded a Pulitzer Prize.

RACINE

Ben Hecht was twelve years old when his family moved in 1906 from New York to Racine. His father established the Ladies' Garment Manufacturing Company at 1221 Chestnut Street, and his mother operated a dress shop at 505 Main Street. Hecht recalled learning acrobatics from the circus people he met at Mrs. Frances A. Costello's rooming house, 838 Lake Avenue, where the Hecht family lived for some time: Mrs. Costello was the widow of P. T. Barnum's partner. The Hechts later moved to an apartment house on Eighth Street and College Avenue, no longer standing. After graduation from Racine High School, Hecht became a newspaper reporter in Chicago. In Hecht's erotic novella *Fantazius Mallare* (1922), some characters were named for Hecht's high-school friends; Hecht, his co-author, and the publisher were sued successfully and had to pay $2,000 in damages.

RIVER FALLS

Edgar Wilson Nye, later to be known as **Bill Nye,** came to River Falls about 1852 with his family. His father wanted to farm in the rich

I will premise generally that I hate this business of lecturing. To be received at a bad inn by a solemn committee, in a room with a stove that smokes but not exhilarates, to have three cold fish-tails laid in your hand to shake, to be carried to a cold lecture-room, to read a cold lecture to a cold audience, to be carried back to your smoke-side, paid, and the three fish-tails again—well, it is not delightful exactly.

—James Russell Lowell,
in a letter dated April 9, 1855,
from Wisconsin

Replica of the cabin near Pepin in which Laura Ingalls Wilder was born. (*below*) A sign marking the park in Pepin that was named for her.

Snapshot of August Derleth at work in his house, Place of Hawks, in Sauk City.

valley of the St. Croix River, near the confluence of the St. Croix and Mississippi rivers. Nye was only two at the time. By the time Nye was graduated from the local academy he had made up his mind about a career for himself: "I decided that I would be a miller, with flour on my clothes. . . . One day the proprietor came upstairs and discovered me in a brown study, whereupon he cursed me in a subdued Presbyterian way, abbreviated my salary from $26 per month to $18 and reduced me to the ranks." It was not until about 1883 that Nye found his true calling as a spinner of yarns, which later were collected in such works as *Bill Nye and Boomerang* (1881).

SAUK CITY

August Derleth was born on February 24, 1909, in Sauk City, a town of about 2,000 people. Derleth wrote about his hometown in his *Sac Pacific Saga,* a series of books that included eight novels, among them *Bright Journey* (1940) and *The Hills Stand Watch* (1960). Derleth died in Sauk City on July 4, 1971, and was buried in the Catholic cemetery on Lueders Road. **Mark Schorer,** the novelist and biographer, who was born here on May 17, 1908, began writing short stories for publication in 1933. Many of his early stories and his first novel, *A House Too Old* (1935), use Sauk City as a background.

WEST SALEM

Hamlin Garland, Wisconsin's best-known native-born author, was born in West Salem on September 4, 1860, in a squatter's log cabin, no longer standing but thought to have stood near the county hospital. Many years later, in 1893, after making a name for himself as a writer, Garland returned to West Salem, bought a cabin, and began to enlarge it. This became the Garland Home, a National Historic Landmark, at 357 West Garland Street. It is open to the public, and an admission fee is charged. The house, which Garland sold in 1938, has broad wings and a long row of kitchen and sheds. When fire destroyed the west wing of the lower floor of the house, Garland's study and many of his possessions, including one of his manuscripts—possibly *The Captain of the Gray-Horse Troop* (1902)—were destroyed. Garland at once set about having the house restored to the exact specifications of the original. Garland died on March 4, 1940, in California, but his ashes are buried in Neshonoc Cemetery, one mile from the Garland Home on Route 108. The plaque at the family plot reads:

Hamlin Garland
1860–1940
'A Son of the Middle Border'
is buried here
with his wife and
pioneer parents.

Garland described West Salem in *A Son of the Middle Border* (1917) and *A Daughter of the Middle Border* (1921)—the second book won a Pulitzer Prize. The West Salem Public Library, 155 South Leonard Street, has a Hamlin Garland collection.

West of the city of West Salem is Green's Coulee, where Garland's family had a farm of 160 acres from 1861 until 1869. Green's Coulee, short

Hamlin Garland Homestead on West Garland Street in West Salem, and the living room of the house. An outside chimney Garland added has been removed, but the house is otherwise much the same as it was in the novelist's time.

and narrow, contained five farms, the middle one belonging to the Garlands. Readers of Garland's *A Son of the Middle Border* can visit Green's Coulee by going west from West Salem on Route 16, turning right at Route 157 until the first stop sign. A right turn there takes one to Green's Coulee. Garland's mother's family lived about one mile to the west of West Salem, off County Trunk M, at Gill's Coulee. Dutcher's Coolly, in Garland's novel *Rose of Dutcher's Coolly* (1895), was actually Gill's Coulee. The novel told the story of a motherless farmgirl with literary ambitions who sought escape from the monotony of farm life.

MINNESOTA

AUSTIN

Richard Eberhart was born on April 5, 1904, at 811 North Kenwood Avenue in this town, thirty-five miles southwest of Rochester. The poet's house is still standing, but the street is now known as Fourth Street, N.W. Eberhart's poetry, for which he won a Bollingen Prize in 1962, may be read in such volumes as *Collected Poems* (1960) and *Collected Verse Plays* (1962).

BEMIDJI

This town in Minnesota's northern lake country stakes its claim to literary immortality by insisting that the legend of Paul Bunyan originated here. Bunyan, the hero of lumberjacks, logged even when the weather was cold enough to cause usswords to freeze in the air. Huge figures of Bunyan and Babe, his blue ox, stand near Lake Bemidji, and the information center at Third Street and Bemidji Avenue displays various exhibits relating to Bunyan. Louis Untermeyer, in *The Wonderful Adventures of Paul Bunyan* (1945), retold the Bunyan legend in verse.

BENSON

Martha Ostenso, born in Norway, was twelve years old and living in this small town in western Minnesota when she wrote stories for the junior page of the Minneapolis *Journal*, receiving eighty cents for a column. Her novel *O River, Remember!* (1943) described life in the Red River Valley, along the border of Minnesota and North Dakota.

BRAINERD

Martha Ostenso once lived on Route 6 in Brainerd, sixty miles north of St. Cloud. She also lived at Gull Lake, a community ten miles northwest of Brainerd.

BUFFALO

Margaret Culkin Banning, the novelist known for her depictions of the problems of contemporary living, was born on March 18, 1891, in the town of Buffalo, about thirty miles northwest of Minneapolis. Her novels included *The Women of the Family* (1926) and *Too Young to Marry* (1938).

DELLWOOD

F. Scott Fitzgerald often visited this summer resort town on White Bear Lake, just north of St. Paul. During his school days he spent summer weekends at the cottage of a friend, Cecil Read, at 20 Peninsula Road. This cottage has since been moved and added to a nearby building. Fitzgerald frequented the White Bear Yacht Club, at 56 Dellwood Avenue, and in 1913 directed and acted in a production of his Civil War drama, *The Coward.* When another actor discovered a live bullet in a prop gun, Fitzgerald covered for him by adlibbing until the actor could run outside and fire the gun into the lake. The next year Fitzgerald saved the production of his play *Assorted Spirits* by jumping onstage during a lighting failure and delivering a monologue until the lights came on again. In 1921 Scott and Zelda Fitzgerald leased the Mackey J. Thompson Cottage, 14 Route 96, on White Bear Lake. Fitzgerald wanted to find the seclusion he needed to work on his writing, but he was unsuccessful. One cold night he forgot to tend the cottage furnace; the water pipes burst, and the cottage owner asked the Fitzgeralds to leave. In the summer of 1922, the Fitzgeralds were in Dellwood once more, staying at the White Bear Yacht Club while Scott worked on a play and a collection of stories. One of the stories, "Winter Dreams," captured the atmosphere of the club. The club is still open, but the original building burned in 1937 and has been replaced.

> *In the fall when the days became crisp and gray, and the long Minnesota winter shut down like the white lid of a box, Dexter's skis moved over the snow that hid the fairways of the golf course. At these times the country gave him a feeling of profound melancholy—it offended him that the links should lie in enforced fallowness, haunted by ragged sparrows for the long season. . . . In April the winter ceased abruptly. The snow ran down into Black Bear Lake scarcely tarrying for the early golfers to brave the season with red and black balls. Without elation, without an interval of moist glory, the cold was gone.*
>
> —F. Scott Fitzgerald, in "Winter Dreams"

DULUTH

Margaret Culkin Banning grew up in the Hunter's Park neighborhood of Duluth. As an adult she also lived in Duluth, writing in an office that overlooked Lake Michigan. She also served as trustee of the Duluth Public Library, now a strikingly modern structure on the Fifth Avenue Mall. **Walter Havighurst** described Duluth in *The Long Ships Passing* (1942), a history of shipping on the Great Lakes. He had seen Duluth when he worked as a deckhand aboard a freighter. **Sinclair Lewis** moved to Duluth in 1944, staying at first at the Hotel Duluth, 231 East Superior Street, then renting a large house at 2601 East Second Street, which he bought in 1945. It was here he wrote the novel *Cass Timberlane* (1945). His home on Second Street is owned now by the Holy Rosary Convent and is not open to the public.

Sinclair Lewis at the door of his home in Duluth in 1945. Though Lewis liked the city, he sold the house a year after he bought it, and returned to the East.

EXCELSIOR

Sinclair Lewis worked during 1942 on his novel *Gideon Planish* (1943) in a house near Excelsior, on Lake Minnetonka and just west of Minneapolis.

FERGUS FALLS

Herbert Krause was born on a farm near here on May 25, 1905, and lived in the area until 1938, writing about farm life in western Minnesota. *Wind Without Rain* (1939) was his first novel,

O. E. Rölvaag as a student at Royal Frederick University in Norway in 1906.

and his later work included *The Oxcart Trail* (1954) and *The Builder and the Stone* (1958).

HASTINGS

Ignatius Donnelly, author of the novel *Atlantis: The Antediluvian World* (1882), had a home near this Mississippi River town, twenty miles southeast of St. Paul. He had abandoned his Philadelphia law practice to found a new city, to be called Nininger, Minnesota. In his pamphlet *Nininger City* (1856), he described the abortive attempt. Donnelly later in life turned to politics, serving as lieutenant-governor of Minnesota from 1859 to 1863 and as congressman from 1863 to 1869.

ISABELLA

Sections of the book of essays *Red Wolves and Black Bears* (1976), by Edward Hoagland, are set in this small northeastern Minnesota town located in Superior National Forest.

LAKE CITY

[Fred] Hayden Carruth, author and editor, was born on October 31, 1862, on a farm in Mount Pleasant, near the town of Lake City, which is fifteen miles southeast of Red Wing. Carruth worked for *The Woman's Home Companion* from 1905 to 1932, writing the column "The Postscript." He also wrote many humorous sketches, which were collected in such volumes as *Mr. Milo Bush and Other Worthies* (1899) and *The Adventures of Jones* (1903). *Track's End* (1911), a boy's adventure book, was well known in its time.

Statue of Hiawatha standing in Minnehaha Park in Minneapolis, recalling Longfellow's "The Song of Hiawatha."

The towers of Zenith aspired above the morning mist; austere towers of steel and cement and limestone, sturdy as cliffs and delicate as silver rods. They were neither citadels nor churches, but frankly and beautifully office-buildings.

—Sinclair Lewis,
in *Babbitt*

LITTLE FALLS

Charles A[ugustus] Lindbergh, author of *The Spirit of St. Louis* (1953), the Pulitzer Prize-winning account of his 1927 transatlantic flight, spent summers at his family's home in Little Falls, in central Minnesota, until he went to college. He also lived here in winter during World War I. The Lindbergh house, on Lindbergh Drive, is a National Historic Landmark, open to the public. An Interpretive Center nearby contains memorabilia and audio narratives of three generations of Lindberghs.

MADISON

Robert Bly, author of *The Light Around the Body* (1967) and *Sleepers Joining Hands* (1972), was born here on December 23, 1926. The poet lived as an adult at the Odin House in Madison and in a home he maintained near this western Minnesota community.

MANKATO

In the summer of 1919 **Sinclair Lewis,** with his wife Grace and their child, lived at 315 Broad Street in the town of Mankato, sixty-five miles southwest of Minneapolis. Lewis worked here on his novel *Main Street* (1920), in which the town is mentioned.

MARCELL

O[le] E[dvart] Rölvaag wrote much of *Giants in the Earth* (1927), the first book of his trilogy about Norwegian immigrants who farmed the Northwest prairie, while he lived in a cabin on Big Island Lake, near Marcell. Marcell is a town on Route 38, about twenty-five miles north of Grand Rapids.

MINNEAPOLIS

Although Henry Wadsworth Longfellow never visited Minnesota, he left his mark on Minneapolis. Travel accounts and pictures of Minnehaha Falls, in Minneapolis, inspired a number of passages in Longfellow's "The Song of Hiawatha" (1855). The falls and a statue of Hiawatha are in Minnehaha Park in southeastern Minneapolis, and nearby are Lake Hiawatha, Hiawatha Avenue, and Minnehaha Parkway. In June 1861, **Henry David Thoreau** and Horace Mann, Jr. stayed at Mrs. Hamilton's boarding house, between Lake Calhoun and Lake Harriet. The lakes, now part of western Minneapolis, were then surrounded by woods rich in plants and animals of much interest to Thoreau and Mann—although he was only seventeen, Mann was already a proficient field naturalist. The pair spent a month in the area, but Thoreau's failing health forced them to return to Concord, and Thoreau died not long after.

Many of Minneapolis's literary figures have some connection with the University of Minnesota, which is located east of the downtown area on both banks of the Mississippi River. **Saul Bellow** taught at the university several times during the 1940s and 1950s. **John Berryman** lived at 33 Arthur Avenue, SE, and taught at the university from 1954 until 1972. During these productive years, he wrote his Pulitzer Prize-winning *77 Dream Songs* (1964) and *His Toy, His Dream, His*

Rest (1968). Berryman died in Minneapolis on January 7, 1972, by jumping off a Mississippi River bridge. The university maintains a collection of the papers of Natalie S. Carlson, the author of books for young people. **Richard Eberhart** was a student here in 1922 and 1923.

F. Scott Fitzgerald, who had gone off to study at Princeton University, attended a Psi U dance here in February 1916, dressing as a woman and bringing a male friend as his escort, an incident that was reported in the local newspapers. **James Gray,** critic and novelist, was born in Minneapolis on June 30, 1899. He lived in this city for his first thirteen years, and returned later to attend the university and, from 1948 to 1957, to serve as professor of English. **Tom Heggen,** known for the novel *Mister Roberts* (1946), was a student here from 1939 to 1942 and lived with his parents at 4621 Beard Avenue South. In his last college year, Heggen worked as a copy chief and wrote a column for the student newspaper, the Minnesota *Daily.* From 1942 until his death in 1949, Heggen often returned to Minneapolis to visit his parents.

In 1918 and part of 1919 **Sinclair Lewis** and his family lived at 1801 James Avenue South, and Lewis rented an office on Harmon Place, where he wrote. In May 1919 the Lewises lived at the Hotel Maryland, 1346 La Salle Avenue, while they searched for a house in Mankato. In 1942 Lewis was again in Minneapolis, living at 1500 Mount Curve, teaching writing at the university, and working at his office in Folwell Hall, University Avenue and Fifteenth Avenue, SE. In May 1944, Lewis once again was in the city, this time visiting friends, researching *Cass Timberlane,* and staying at the Hotel Nicollet at Nicollet and Washington avenues.

Max Shulman was a student here. His humorous *Barefoot Boy with Cheek* (1943) has several references to university sites, including the Ski-U-Mah Room in the Coffman Student Union. **Allen Tate** taught English at the university from 1951 to 1968, occupying an office at 219 Folwell Hall and living at 2019 Irving Avenue South. His important publications of this period included *Collected Essays* (1959) and *Poems* (1960). **Robert Penn Warren** taught English here from 1942 to 1950. During this period he wrote his best-known novel, *All the King's Men* (1946), for which he won a Pulitzer Prize.

Other literary figures are connected with Minneapolis. **Gladys Hasty Carroll** wrote the novels *As the Earth Turns* (1933), *A Few Foolish Ones* (1935), and *Neighbor to the Sky* (1937) while living with her husband and son at several addresses here: 3210 Girard Avenue South; 3948 First Avenue South; 2724 West River Road; and 5038 Fremont Avenue South. **Sarah N[orthcliff] Cleghorn,** author of *The Turnpike Lady* (1907), lived as a child in the 1880s with her family at 1706 Laurel Avenue. **Ignatius Donnelly** died here on January 1, 1901, while visiting at his father-in-law's house.

In 1918, when **Mary McCarthy** was five years old, she and her brothers were brought here to be placed in the care of their stern aunt and uncle. The children's parents had died in the great flu epidemic of that year. Mary attended St. Stephen's School, where she won a state prize for her essay "The Irish in American History." McCarthy has described these school years in her autobiography, *Memories of a Catholic Girlhood* (1957). **Frederick [Feikema] Manfred,** author of the novel *The Golden Bowl* (1944), worked as a reporter for the Minneapolis *Journal* but was dismissed in 1939, partly because he helped organize the American Newspaper Guild. **Martha Ostenso** once lived in St. Louis Park, a western suburb of Minneapolis.

MORTON

In 1861 **Henry David Thoreau** and Horace Mann, Jr., traveling by riverboat up the Minnesota River, arrived on June 20 at Redwood, the Lower Sioux Indian Agency trading post near Morton. They saw a ceremonial dance and witnessed treaty payments to the Indians by agents of the federal government. Thoreau was impressed with Chief Little Crow, who in the next year was to lead a major Sioux uprising in this area. In a letter dated June 25, 1861, Thoreau wrote

> *A regular council was held with the Indians, who had come in on their ponies, and speeches were made on both sides thro' an interpreter, quite in the described mode; the Indians, as usual, having the advantage in point of truth and earnestness, and therefore of eloquence.*

NEW ULM

Wanda Gág, author and illustrator of many children's books, including *Millions of Cats* (1928), was born here on March 11, 1893, at 512 Third Street, twenty-five miles northwest of Mankato. Although the home has been remodeled and does not resemble the original structure, it is open to visitors. Some of her stories were set in New Ulm: at Turner Hall, on State and First South streets;

Wanda Gág.

Home of Wanda Gág at 226 North Washington Street, shown in a photograph taken about 1930. Gág spent most of her childhood and young adult years here.

There were immense exhibits of grain, livestock and farming machinery; there were horse races and automobile races and, lately, aeroplanes that really left the ground; there was a tumultuous Midway with Coney Island thrillers to whirl you through space, and a whining, tinkling hoochie-coochie show. As a compromise between the serious and the trivial, a grand exhibition of fireworks, culminating in a representation of the Battle of Gettysburg, took place in the Grand Concourse every night.

—F. Scott Fitzgerald,
in "A Night at the Fair"

at the Hermann Monument, at Center and Monument streets; and in "Goosetown," south of Center Street and near the Minnesota River. The New Ulm Public Library, at 27 North Broadway, displays some of her illustrations. The Historical Museum, at the same address, owns many of her childhood drawings and other memorabilia.

NORTHFIELD

This town, home of St. Olaf and Carleton colleges, about thirty-five miles south of St. Paul, has attracted a number of literary figures. **Robert Bly** attended St. Olaf from 1946 to 1947. **Herbert Krause** was once a student of the English department of St. Olaf. The most prominent author connected with St. Olaf was **O[le] E[dvart] Rölvaag,** a student from 1901 to 1905, and a teacher from 1906 to 1931. Rölvaag established a Norwegian church service on the school radio station and served as chairman of the Norwegian literature department. The house at 311 Manitou Street in which Rölvaag lived during most of his faculty years is still standing but not open to visitors. One can see a bronze bust and a portrait of the writer in the reference room of the Rölvaag Memorial Library, a building that was dedicated to him in 1942 and now holds his papers. **Thorstein Veblen,** who studied Hume, Kant, and other social philosophers at Carleton College, was graduated in the class of 1880.

Home of F. Scott Fitzgerald at 599 Summit Avenue in St. Paul.

OLIVIA

Kathleen Winsor, remembered for her titillating novel *Forever Amber* (1944), was born on October 16, 1919, in this southwestern Minnesota town. Her birthplace on Sixth Street is still standing but is not open to visitors.

PEQUOT LAKES

Sinclair Lewis spent the summer of 1926 in a cottage east of here, working on his novel *Elmer Gantry* (1927). The cottage stood on a large tract of land that included a mile of shore front on Lake Pelican.

RED WING

On June 23, 1861, **Henry David Thoreau** and Horace Mann, Jr., on their way home to Concord, Massachusetts, stopped at Red Wing. They stayed at the former Metropolitan House on Main Street, near the boat landing. While here, Thoreau and Mann went swimming in the Mississippi, climbed nearby Red Wing Bluff, and visited the grave of Chief Red Wing, for whom the town was named.

ROCHESTER

Ernest Hemingway was treated twice at St. Mary's Hospital here for depression and anxiety.

ST. PAUL

F. Scott Fitzgerald was born on September 24, 1896, in a three-story brick house at 481 Laurel Avenue, still standing but not open to the public. In 1898 the Fitzgeralds moved away but by 1908 were back. Over the next decade, the family lived at the following addresses, all fairly close to one another: 294 Laurel Avenue; 514, 509, and 499 Holly Avenue; and 593 and 599 Summit Avenue. Only the building at 599 Summit Avenue has a plaque marking it as an F. Scott Fitzgerald house. From 1908 to 1911 Fitzgerald attended St. Paul Academy, still at 1712 Randolph Avenue. His first published story, "The Mystery of the Raymond Mortgage" (1909), appeared in his school magazine, *The St. Paul Academy Now & Then.* Fitzgerald also wrote four plays for the local Elizabethan Dramatic Club. One of them, a melodramatic western entitled *The Girl from Lazy J,* was produced in 1911. He went away to school in 1911, thereafter returning to St. Paul only for brief periods. During one of his visits, in January 1915, he met Ginevra King at the Town and Country Club, 2279 Marshall Avenue. She was his first love and the inspiration for Isabelle in *This Side of Paradise* (1920).

In July 1917, Fitzgerald went to Fort Snelling, just south of St. Paul, to take the qualifying examinations to become a U.S. Army officer, and he received a commission on October 26. In 1919 Fitzgerald moved into the third floor of his parents' home, then at 599 Summit Avenue, to recover from a drinking binge and to revise *This Side of Paradise.* In October 1921, Fitzgerald and his pregnant wife, Zelda, stayed at the Commodore Hotel, 79 Western Avenue, while they waited for the birth of their child. During the next six months, they rented a house at 626 Goodrich Avenue, where Fitzgerald wrote his stories "The Popular Girl" (1922) and "The Diamond as Big as the Ritz" (1922), and worked on *The Beautiful and Damned* (1922). After spending part of a day writing in a rented room in downtown St. Paul, he would stop in at the Kilmarnock Book Shop at 84 East Fourth Street, a meeting place for writers. One of the store owners, Tom Boyd, was literary editor of the St. Paul *Daily News*. The store, no longer standing, has been replaced by the Minnesota Building, which occupies the entire city block. In August 1922, the Fitzgeralds again stayed at the Commodore Hotel after being ousted from the White Bear Yacht Club in Dellwood. They enjoyed the hotel's formal dining room and popular rooftop garden, where they could socialize, dance, and listen to bands. The Fitzgeralds also frequented the University Club, still at Ramsey Hill and Summit Avenue.

Two other literary figures of some reputation were born in St. Paul. **Kay Boyle,** author of many short stories, including the collection *Wedding Day* (1930), and many novels, including *Plagued By the Nightingale* (1931), was born here on February 19, 1902. **Max Shulman** was born in St. Paul on March 14, 1919. He is known best for his humorous novel *Barefoot Boy with Cheek* (1943).

A number of other writers have chosen St. Paul as a place to live and work. **Charles B[enedict] Driscoll,** author of *Doubloons, The Story of Buried Treasure* (1930), was a reporter on the former *Daily News* here, at 94 East Fourth Street, in the block now occupied by the Minnesota Building. While living in St. Paul, Driscoll owned a home at 1626 Dayton Avenue, not open to the public. **James Gray** worked from 1920 to 1946 as a drama critic and book reviewer for the St. Paul *Pioneer Press & Dispatch,* still at 55 East Fourth Street. Gray lived at 1554 Lincoln Avenue, in a house that is still standing but not open to the public. The regional novels *Shoulder the Sky* (1935) and *Wake and Remember* (1936) and a collection of reviews, *On Second Thought* (1946), are a few of the books Gray wrote here.

Sinclair Lewis visited St. Paul several times. In 1902 he took his entrance examinations for Yale University at the Clarendon Hotel here. During the winter of 1917–18 Lewis, with his wife and child, lived in a lavish house at 516 Summit Avenue. A few lines from the *Pioneer Press & Dispatch* of January 18, 1918, give the flavor of the lives they led that winter: "A group of society folk will assemble tomorrow evening at the home of Mr. and Mrs. Sinclair Lewis . . . to hear Danny Reed, producing director of the Little Theater Association, read the dramatization of Mr. Lewis's play, 'Hobohemia,' which recently appeared in *The Saturday Evening Post.*" During short visits to St. Paul in 1943 and 1947, Lewis did research at the Minnesota Historical Society, 690 Cedar Street. He used some of this research in *The God-Seeker* (1949), his historical novel about Minnesota.

SAUK CENTRE

For readers, this central Minnesota town will always be known principally as the birthplace of **[Harry] Sinclair Lewis** and as the model for the town of Gopher Prairie in his novel *Main Street* (1920). Lewis was born on February 7, 1885, at 811 Third Avenue, now Sinclair Lewis Avenue. When he was four, the family moved across the street to 812 Third Avenue, now known as the Sinclair Lewis Boyhood Home. His birthplace is not open to the public, but his boyhood home is open daily for a small admission charge. Lewis's stepmother entertained the Gradatim Women's Club in the living room here, and some believe that this group was the basis for the Thanatopsis Club known to readers of *Main Street.* Lewis lived for seventeen years in Sauk Centre, until he went away to college. One of his haunts was the Bryant Library, now known as the Sauk Centre Public Library and marked by a plaque commemorating Lewis. The Palmer House, at Main Street and Sinclair Lewis Avenue, where Lewis worked as a night clerk in 1902, has been restored to turn-of-the-century decor. Lewis was on duty from 6 PM to 6 AM and earned room and board plus $5 a week. Lewis came back to Sauk Centre in 1916 with his wife Grace to visit friends and to take notes for *Main Street.* They rented a room above Rowe's Hardware Store, now the Main Street Drug Store, also at Main Street and Sinclair Lewis Avenue. Over the years he made a number of visits to his home town, and after he died in Rome, Italy, on January 10, 1951, his body was brought back and buried in Greenwood Cemetery, on Sinclair

Street marker in Sauk Centre.

(*below left*) Birthplace of Sinclair Lewis in Sauk Centre, at 811 Sinclair Lewis Avenue. (*below*) The Lewis family on the front porch of 812 Sinclair Lewis Avenue, photographed about 1897. Sinclair Lewis, then about twelve years old, appears at the far right. A historical marker on the front lawn of the house reads in part: "he described [Minnesota] as 'the newest empire of the world . . . a land of dairy herds and exquisite lakes, of new automobiles and tarpaper shanties and silos like red towers, of clumsy speech and a hope that is boundless.' "

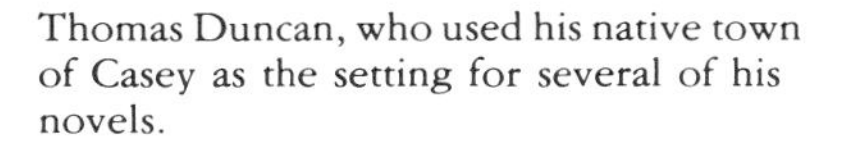

Thomas Duncan, who used his native town of Casey as the setting for several of his novels.

Lewis Avenue. On the real Main Street, at Route 94, is the Visitor Center, an interpretive museum of Lewis's work that exhibits autographed first editions and other memorabilia. The museum is open daily and admission is free.

SHAKOPEE

Eleanor Gates was born here, just southwest of Minneapolis, on September 26, 1875. Her best-known works were the novel *Poor Little Rich Girl* (1913), which later was dramatized, and the play *We Are Seven* (1913).

WALNUT GROVE

Laura Ingalls Wilder lived as a child with her family in the 1870s in a dugout house on a farm near the southwestern Minnesota community of Walnut Grove. The house was described in her novel *On the Banks of Plum Creek* (1937). The homesite is marked, and a small museum in town describes Wilder's early life here.

IOWA

It's a wonderful short course in human nature, being a doctor's son in a town of eight hundred. Excitement always ran high.

—Thomas Duncan,
quoted in *Current Biography* (1947)

ALBIA

[Charles] Badger Clark was born on January 1, 1883, in a Methodist parsonage in Albia, a coal-mining town on Route 34, in southern Iowa. He went on to write verse on cowboy and Western themes, which was collected in volumes such as *Sun and Saddle Leather* (1915) and *Sky Lines and Wood Smoke* (1935). **James [Floyd] Stevens** was born here on November 15, 1892. Stevens wrote primarily about the Northwest and the West. His first novel, *Paul Bunyan* (1925), was representative of the former, and his short-story collection *Homer in the Sagebrush* (1928) dealt with life in the West.

ANITA

Stuart [Pratt] Sherman, author of *On Contemporary Literature* (1917), *Americans* (1922), and *The Genius of America* (1923), was born on October 1, 1881, in this town on Route 83, northeast of Atlantic. His father owned a drug store and eighty acres of land. When Sherman was one year old, his family moved on to Rolfe, Iowa, where he grew up.

ATLANTIC

Paul Corey, the novelist and editor, lived during childhood at 101 Birch Street and attended Atlantic High School from 1918 to 1921. Atlantic is on Route 6, east of Council Bluffs.

BOONE

Bess Streeter Aldrich taught school in Boone in 1902–03, living at 321 Story Street, which now is the site of a gasoline station. Boone is just west of Ames.

BURLINGTON

Hartzell Spence, the novelist, worked part time as a reporter for the Burlington *Hawk-Eye* while attending high school in Burlington, a city in southeastern Iowa, on the Mississippi River. **Elswyth [Mrs. William Beebe] Thane** was born here in 1900 and went on to write novels such as *Riders of the Wind* (1925) and *Kissing Kin* (1948).

CASEY

Thomas Duncan, the novelist, was born in Casey on August 15, 1905, and was graduated from the local public schools in 1922. Duncan, who was a doctor's son, set type for *The Vindicator,* a newspaper published in Casey by his uncle. He also wrote for the high-school newspaper. Casey, on Route 6, west of Des Moines, appeared as Sioux Creek and Litchfield in Duncan's novels *Gus the Great* (1947), *The Labyrinth* (1967), and *The Sky and Tomorrow* (1974).

CEDAR FALLS

Bess Streeter Aldrich, in her novel *Song of Years* (1939), created the city of Sturgis Falls, closely resembling Cedar Falls, the city in which she was born, on February 17, 1881. The novel spans the growth of this city, which is on Route 20, adjacent to Waterloo, from a settlement of a few cabins to a thriving community. Aldrich was graduated from Cedar Falls High School and State Teacher's College. Her work included the novels *A Lantern in Her Hand* (1928), the story of a pioneer woman, and *White Bird Flying* (1931), the story of the children of the pioneers. **R[onald] V[erlin] Cassill** was born here on May 17, 1919. Cassill dealt with controversial themes in his many novels, which included *Eagle on the Coin* (1950) and *The President* (1964). **Ruth Suckow,** the novelist and short-story writer, who died on January 23, 1960, in Claremont, California, is buried in Greenwood Cemetery, in Cedar Falls.

CEDAR RAPIDS

Cedar Rapids, a manufacturing center that has been home for several writers, has a Shakespeare

Garden located in Ellis Park, where visitors may see all the species of flowers grown in William Shakespeare's own garden.

Paul Engle was born in Cedar Rapids on October 12, 1908. He lived at 1602 Fifth Avenue S.E. and worked for the East End Pharmacy, still standing at 394 Sixteenth Street, while he attended Washington High School. As a boy, Engle delivered the Cedar Rapids *Gazette,* a newspaper still publishing at 500 Third Avenue S.E. His novel *Always the Land* (1941) was set in the Squaw and Indian Creek areas northwest of the city. **MacKinlay Kantor,** the novelist, for a short time attended Washington High School and later worked as a reporter for the Cedar Rapids *Republican.* After he was laid off by the *Republican* in 1927, Kantor moved back to Webster City to write full time, vowing never again to work at a regular job.

William L. Shirer lived with his grandparents at 811 Second Avenue from 1913 to 1925. He studied at Jackson Elementary School and Washington High School, and attended the First Presbyterian Church. Shirer was graduated in 1925 from Coe College, and *Cosmos,* the college newspaper, gave him his first byline. He also worked as a sportswriter for the Cedar Rapids *Republican.* Immediately after graduation from college, Shirer borrowed money from an uncle and from the president of Coe College to travel to Europe by cattleboat and spend the summer. The summer lasted fifteen years: Shirer returned to the United States in 1940, a celebrated foreign correspondent and soon to become known for his best seller *Berlin Diary* (1940).

Carl Van Vechten was born here on June 17, 1880, and lived at 253 Second Avenue from 1884 to 1892, and at 845 Second Avenue from 1895 to 1903, the latter site reported now to be occupied by Denny's Muffler Center. As a student, Van Vechten often attended performances at Green's Opera House, no longer standing. Green's, a railway stopover for touring Broadway shows, provided the young Van Vechten with the opportunity to develop the critical ability he later used in writing for the New York press. Maple Valley, in Van Vechten's novel *The Tattooed Countess* (1924), is thought to be Cedar Rapids. Van Vechten's other novels included *Nigger Heaven* (1926) and *Spider Boy* (1928).

CLARION

Hartzell Spence, whose novels *One Foot in Heaven* (1940) and *Get Thou Behind Me* (1942) were based on his memories of Iowa, was born on February 15, 1908, in Clarion, which is on Route 3, in north-central Iowa.

COLFAX

James Norman Hall, author with Charles Nordhoff of the novel *Mutiny on the Bounty* (1932), was born on April 22, 1887, in this city on Route 6, east of Des Moines. He attended public schools and lived at 416 East Howard Street. The house is not open to visitors.

COUNCIL BLUFFS

Thomas Beer was born on November 22, 1889, in this Mormon settlement on the Missouri River in southwestern Iowa. His best-known works were *Stephen Crane: A Study in American Letters* (1923) and *The Mauve Decade* (1926), which dealt with late nineteenth-century America. His other works included a novel, *The Road to Heaven* (1928), and a collection of short stories, *Mrs. Egg and Other Barbarians* (1933).

DAVENPORT

Davenport, a large city on the Mississippi River, has several literary associations. **George Cram Cook,** a novelist and playwright known best as one of the organizers of the Provincetown Players in Provincetown, Massachusetts, was born here on October 7, 1873, and attended Griswold College, a military prep school in Davenport. Cook wrote the novel *The Chasm* (1911) and, with his third wife, Susan Glaspell, two one-act plays, *Suppressed Desires* (1914) and *Tickless Time* (1919). Cook's great-grandfather was one of Davenport's first settlers, in 1836, and several buildings have been named for him, including Cook Memorial Library and Cook Memorial Church.

> *Gone are the three, those sisters rare*
> *With wonder-lips and eyes ashine.*
> *One was wise and one was fair,*
> *And one was mine.*
>
> *Ye mourners, weave for the sleeping hair*
> *Of only two your ivy vine.*
> *For one was wise and one was fair,*
> *But one was mine.*
>
> —Arthur Davison Ficke,
> in "The Three Sisters"

Arthur Davison Ficke, a poet best known for *Sonnets of a Portrait Painter* (1914), was born in this city on November 10, 1883. Ficke wrote plays and a novel in addition to many volumes of poetry, and he had the gift of satire. In 1916 he and Witter Bynner wrote *Spectra,* a brilliant hoax that caricatured the new poetry. For the book Ficke used the name **Anne Knish,** and Bynner called himself **Emanuel Morgan.**

Alice French, pen name **Octave Thanet,** moved with her family to Davenport in 1856, when she was six. Thanet's novel *The Man of the Hour* (1905) concerned a struggle in Fairport, a fictionalized Davenport, and her collection *Stories of a Western Town* (1893) was also about Davenport. The novelist died in Davenport on January 9, 1934, when she was eighty-three. **Susan Glaspell,** a founder of the Provincetown Playhouse with her husband, George Cram Cook, was born in Davenport on July 1, 1876, and grew up here. After high-school graduation, she became a reporter for a local newspaper at a salary of three dollars a week. Glaspell attended Drake University, in Des Moines, and covered state government news for the Des Moines *Daily News* until she returned to Davenport to write short stories for magazines. She often used Freeport, representing Davenport, as the locale for stories she wrote for *Harper's* and *The Youth's Companion.* Glaspell won a Pulitzer Prize for her novel *Alison's House* (1930), about the effect of a poet's life on the life of her family.

> *"And another thing," Jeremiah went on, "Sturgis Falls is the county-seat . . . has all the court records. . . ."*
>
> *Sarah gave her preliminary sniff. " 'All the court records,' " she mimicked. "A little black book bought at Dubuque. Has two items in it, they say—and one of* them's *the seventy cents they paid for the book. Court-house is a little two-by-four room in the loft over Mullarky's store where you couldn't stand up straight if you tried. . . ."*
>
> —Bess Streeter Aldrich,
> in *Song of Years*

Ruth Suckow, Famed Iowa Author,
Established Herself As A Writer
While Living In Earlville 1920–1926.
She Earned A Living
By Operating An Apiary.
She Lived In A White House
On This Site 1925–26.

—From a stone marker on Ruth Suckow's home in Earlville

. . . most of the afternoon, he did not attempt to talk with the rest of us. He rocked and smoked, and at times he murmured with a sigh, "Ach, ja, ja . . . dot was all so long ago!"

—Ruth Suckow, in a newspaper article on her German grandfather

The novel *Your Life Lies Before You* (1935), by **Harry Hansen,** opens in Davenport, where Hansen lived. He began to write for newspapers, working successively on his high school newspaper, the Davenport *Republican,* and the Davenport *Times.* His *Midwest Portraits* (1923) was a study of Davenport writers. **Charles Edward Russell,** who became publisher of two of William Randolph Hearst's newspapers, the *American* and the *Examiner,* was born in Davenport on September 25, 1860. Russell won a Pulitzer Prize for his biography *The American Orchestra and Theodore Thomas* (1927).

DES MOINES

Thomas Duncan lived from 1922 to 1942 at 1050 Thirty-third Street, in Des Moines. While he was a student at Drake University, Duncan wrote articles for the Des Moines *Register* and the Des Moines *Tribune.* After graduation, he worked as a reporter for the *Tribune* and book reviewer for the *Register.* He also wrote scripts for radio station KRNT and taught English at the Des Moines College of Pharmacy. In his novels, Duncan wrote about the Des Moines area, sometimes calling it Tamarack.

Susan Glaspell, after graduation from Drake University in 1899, worked for two years as a political reporter for the Des Moines *Daily News* before returning to Davenport. **William Haines,** the novelist known best for *High Tension* (1938) and *Command Decision* (1947), was born on September 17, 1908, at 2824 Grand Avenue, now an apartment building. **MacKinlay Kantor** during childhood lived in Des Moines for a time with his mother. Later, when the Kantors were living in Webster City, MacKinlay won a $50 prize from the Des Moines *Register* for his first published short story, "Sheridan Rhodes." The story was published later in Kantor's *Author's Choice* (1944).

Cottage of Ruth Suckow in Earlville. While living here in 1925 and 1926 the author completed much of her novel *The Odyssey of a Nice Girl* and the collection *Iowa Interiors.*

DOON

Frederick Feikema, who changed his name in 1951 to **Frederick Manfred,** was born on January 6, 1912, on a farm in Doon, north of Sioux City. After attending Doon Christian School and the Calvinistic Hull Academy, he worked as a newspaperman and then turned to writing fiction. Doon and its surrounding country are called Siouxland in Manfred's novels, which included *The Brother* (1950) and *Conquering Horse* (1959).

DUBUQUE

Richard Bissell, who collaborated with George Abbott on the musical comedy *The Pajama Game* (1954), was born in Dubuque on June 27, 1913. The musical was adapted from Bissell's novel *7½ Cents* (1952). Bissell also collaborated with his wife Marian Bissell and Abe Burrows in writing the musical comedy *Say Darling* (1959). Dubuque is the model for the Mississippi River town in Bissell's novel *Goodbye Ava* (1960).

EARLVILLE

Ruth Suckow lived in a cottage in Earlville, west of Dubuque, while she was writing the novels *Country People* (1924) and *The Odyssey of a Nice Girl* (1925). Her cottage is no longer standing, but the site has now been turned into Ruth Suckow Park. The site of the cottage is marked with a granite slab bearing an inscription and a drawing of the house. The Orchard Apiary Site, where she kept bees for seven years, is on the edge of town. The Earlville–Ruth Suckow Memorial Library, at Elizabeth and Northern avenues, has a collection of the novelist's books, pictures, and memorabilia, including manuscript, a honey label, her typewriter, and a letter she received from H. L. Mencken.

FORT DODGE

Thomas Heggen, known best for his novel *Mister Roberts* (1946), was born on December 23, 1919, in this northern Iowa town, west of Dubuque, and attended Fort Dodge High School, now North Junior High School, located at 1015 Fifth Avenue N. *Mister Roberts* sold over one million copies in its original edition. Heggen lived at various places in Fort Dodge: 1909 Seventh Avenue N., 1285 Eighth Avenue N., 1615 Elmhurst, and 629 Forest Avenue. The last house is still standing but is not open to visitors.

GRAND JUNCTION

Frank Luther Mott edited the Grand Junction *Globe* from 1914 until 1917, when he left this town west of Ames to study at Columbia University in New York City.

GRANGER

Thomas Duncan found inspiration in an abandoned circus winter quarters in Granger, north of Des Moines, for *Gus the Great* (1947), his novel about circus life.

GRINNELL

While living in this college town south of Marshalltown from 1942 to 1944, **Thomas Duncan** worked on *Gus the Great* (1947). He also taught and was public-relations director at Grinnell College. One of Grinnell High School's most successful graduates was **Ruth Suckow,** author of *The Odyssey of a Nice Girl* (1925), a novel recounting the struggle of an Iowa girl to grow free of the repression of her family.

GRUNDY CENTER

Grundy Center, a town on Route 6, west of Des Moines, may be thought of as **[John] Herbert Quick** country, even though Quick was born miles away, near Steamboat Rock, Iowa. His vivid accounts of prairie and pioneer life in *Vandemark's Folly* (1922) and *The Hawkeye* (1923) brought Iowa history to life for readers. Quick was seventeen when he began teaching school, a career he followed for twenty years. The desk he used as a teacher, the children's lunch buckets, old maps, and a waterbucket may be seen by visitors to Orion Park, in Grundy Center, where a country schoolhouse stands as a tribute to Quick.

HAWARDEN

Ruth Suckow was born on August 6, 1892, in the Congregational Church parsonage in Hawarden, on Route 10, in northwestern Iowa. The parsonage was moved later to 1022 Central Avenue, where it stands today. Suckow lived in many places in Iowa, and her fiction employed Iowa themes and settings. Her collection of stories *Iowa Interiors* (1962) and her memoir *Some Others and Myself* (1952) are representative contributions of this Iowa author.

IOWA CITY

The University of Iowa, founded in Iowa City in 1847, has a history rich in literary associations. Perhaps the major contributor to this literary heritage has been the Writers Workshop at the University of Iowa, which for years has drawn to the campus many successful writers, who teach and share their experience, and far more writers-to-be, who seek to learn their craft. The Writers Workshop has included among its faculty such distinguished writers as **John Berryman,** who taught here in the spring of 1954, and **Paul Engle,** who taught poetry and fiction writing in the 1950s and lived during that period at 724 Bayard Street. One product of Engle's years here was the volume *Midland* (1960), a collection of poetry and short fiction by his students. Other writers associated with the program include **Flannery O'Connor,** who studied under Engle; and **Philip Roth,** who wrote his first novel, *Letting Go* (1962), while teaching here from 1960 to 1962. **Harvey Swados,** the novelist and short-story writer, taught at the workshop in 1956–57, and **Kurt Vonnegut** lectured here during 1965–67.

The University of Iowa counts among its former students **Robert Bly,** a native of Madison, Minnesota, who studied here in 1955–56; and **R. V. Cassill,** the novelist and short-story writer, who was graduated from the university in 1939, received his M.A. here in 1947, and taught at the Writers Workshop from 1948 to 1952 and 1960 to 1966. **Emerson Hough** was graduated from the University of Iowa in 1880, and **W. D. Snodgrass,** the poet, received his B.A. in 1949, his M.A. in 1951, and his M.F.A. in 1953. Snodgrass composed some of his early poems while he was a student here. **Tennessee Williams** completed his college education at the University of Iowa and was graduated in 1938. He paid his college expenses by working as a waiter at the Iowa State Hospital.

One of the university's most distinguished teachers was **Frank Luther Mott,** who taught American literature and short-story writing from 1921 to 1927, when he became head of the School of Journalism. Mott retained that position until 1942, and during that time wrote the first three volumes of the authoritative five-volume work *A History of American Magazines* (1930–68). Mott was awarded a Pulitzer Prize in 1939 for the second and third volumes of this scholarly and colorful study.

The University of Iowa library maintains a collection of the papers of Paul Corey, the novelist, who was graduated from the university in 1925, and a collection of papers and manuscripts of Thomas Duncan.

Photograph of Frank Luther Mott taken in the late 1930s, when Mott was dean of the University of Iowa School of Journalism.

Boyhood home of Herbert Quick near Grundy Center. Quick lived here from 1869 to about 1880, attended the town's district school, and received his teaching certificate from a normal school in Grundy Center.

"Pigs is pigs," he declared firmly. "Guinea-pigs or dago pigs or Irish pigs is all the same to the Interurban Express Company an' to Mike Flannery. Th' nationality of the pig creates no differentiality in the rate, Misther Morehouse! 'Twould be the same was they Dutch pigs or Rooshun pigs. Mike Flannery," he added, "is here to tind to the expriss business and not to hould conversation wid dago pigs in sivinteen languages fer to discover be they Chinese or Tipperary by birth an' nativity."

—Ellis Parker Butler, in "Pigs Is Pigs"

KEOKUK

Carl Sandburg recalled in *Always the Young Strangers* (1953), a memoir of his early years, that he once spent a day in this city in the extreme northeastern tip of Iowa. The poet, on a five-month wandering tour of the West, took work on July 4, 1897, as a short-order cook in a lunch counter at the end of Main Street, probably at 28 South Water Street. He worked only the one day, leaving with sandwiches to feed him on the next leg of his journey.

Mark Twain came to Keokuk more than once. During 1855–56 he lived at the Ivens House, a building then standing at First and Johnson streets, and worked as a printer's assistant with his brother Orion at the Ben Franklin Book and Job Printing Office, 56 Main Street. The building is no longer there, and the building now at the site is numbered 202 Main Street. Twain and a fellow employee are understood to have annoyed the members of a music class meeting on the floor below by playing tenpins with empty wine bottles. While Twain lived in Keokuk, he wrote letters under the pseudonym **Thomas Jefferson Snodgrass** that were published by the Keokuk *Evening Post.* He was paid five dollars a letter, the first payments Twain ever received for his writing. The story has it that Twain soon became tired of being a printer's assistant in Keokuk and, finding fifty dollars in the street, decided to go off on an expedition to the Amazon. Keokuk saw Twain again in 1882, when he returned to gather material for his autobiographical *Life on the Mississippi* (1883), and he came back again in 1885, on tour with **George Washington Cable,** the novelist. On that trip Twain lectured at the Keokuk Opera House, at Seventh and Blondeau streets. In July 1886 he returned once more, to visit his mother, who then was living at 628 High Street, in a house that still stands today.

KEOSAUQUA

Phil[ip Duffield] Stong, the novelist, was born on January 27, 1899, in Keosauqua, a town on Route 1, in southeastern Iowa. Many of Stong's novels dealt with Iowa themes, including *Stranger's Return* (1933), *Village Tale* (1934), and his best-known work *State Fair* (1932). *State Fair* was Stong's first book, and it became known to millions as a motion picture. Stong worked in Hollywood for many years as a writer and had made enough money by 1933 to repurchase Linwood Farm, the onetime homestead of the Stong family in Keosauqua, where Stong had spent much of his early life. He operated the farm but continued to live elsewhere and, on April 26, 1957, died in Washington, Connecticut, where he had a home. Stong is buried in Keosauqua.

Only Abel felt the dews which came toward morning and saw the trees faintly begin to silhouette themselves against a dawning gray light. A little later he could see the sunrise from the side door of the truck. . . . Only Abel saw the freshly gilded dome of the Capitol suddenly shine out over Des Moines, and the newly whitewashed Fair Buildings which promised them carnival from the near side of the town. At the gates he halted. . . . Mrs. Frake started up.

"Wake up, children," she cried, "Wake up! We're here! We're at the Fair!"

—Phil Stong, in *State Fair*

LAKE MILLS

Wallace Stegner was born on February 18, 1909, on a farm in this town on Route 69 in northwestern Iowa. The farmhouse is still standing but is private. Stegner became a well-known author, whose many works included *Remembering Laughter* (1937), a novella set in Iowa, and *The Big Rock Candy Mountain* (1943), a novel about a man in the Far West who went from place to place with his wife and children in a fruitless search for riches. Stegner won a Pulitzer Prize for his novel *Angle of Repose* (1971), which depicted the lives of four generations of a family spanning the late nineteenth century and recent times.

MARNE

Paul [Frederick] Corey, the novelist, was born on July 8, 1903, on a farm six miles from this village on Route 83, east of Council Bluffs. Corey is the author of a trilogy with a Midwestern setting: *Three Miles Square* (1939), *The Road Returns* (1940), and *County Seat* (1941). He has reported that most people living in the northwest corner of Cass County "would know where the Corey farm, or 'old Corey' farm is."

MARSHALLTOWN

Bess Streeter Aldrich, the novelist, taught school for three years in Marshalltown, a city on Route 14, northwest of Des Moines. **Carl Glick,** a playwright and author of books for children, was born here on September 11, 1890. Among his plays were *It Isn't Done* (1928) and *The Laughing Buddha* (1937).

Merle Miller grew up in Marshalltown. While attending high school, he worked at Wee Dug Inn, his father's hamburger stand, and also wrote a weekly column for his high-school paper. Miller has written of his early years in a frank book about his life, *On Being Different* (1972).

MASON CITY

[John] Herbert Quick, who wrote novels about pioneer life, came to Mason City, on Route 18 in northcentral Iowa, in 1881 to teach at Garfield School, 318 Sixth Street S.E. Quick's novel *Aladdin and Company* (1904) was set in the city of Lattimore, which is taken to be an amalgam of Sioux City and Mason City.

MONTOUR

Merle Miller, whose best-known book is *Plain Speaking, An Oral Biography of Harry S. Truman* (1974), was born on May 17, 1919, in Montour, a small town on Route 135, northwest of Des Moines. His wartime experience as editor of *Yank* magazine, 1941–45, helped launch his writing career, and he has produced many books since then, including *We Dropped the A-Bomb* (1946), *Reunion* (1954), and *Only You, Dick Daring* (1964).

MUSCATINE

Ellis Parker Butler, the humorist remembered for a single story, the enormously popular "Pigs Is Pigs" (1905), was born on December 5, 1869, in Muscatine, a city on Route 92, west of Davenport. His home, at 607 West Third Street, is still standing but private. "Pigs Is Pigs" is a classic account of a disagreement between a private citizen and a bureaucratic express agent over a shipment of a pair of guinea pigs. While the men argue over the rates that apply to the shipment, the guinea pigs reproduce and reproduce and reproduce.

Muscatine is also associated with the life of **Mark Twain,** who frequently visited his mother and his brother Orion Clemens in Muscatine in the 1850s. They lived at 109 Walnut Street, in a one-room house. The house was torn down in the 1940s and is reported now to be the site of a car wash. Twain worked in Muscatine from late 1853 to early 1855 as a reporter for the *Journal,* a newspaper owned partially by Orion Clemens.

NEWTON

Emerson Hough, who wrote historical Western romances, was born on June 27, 1857, in Newton, a city on Route 6, east of Des Moines. His family lived in a house that stood on First Street N. When Hough was eighteen, his family moved to a stucco house at 423 East Seventh Street N., still standing but privately owned. A commemorative marker stands in the front yard. Hough's works included *The Covered Wagon* (1922), *North of 36* (1923), and *Mother of Gold,* published in 1924, the year after Hough died.

OSAGE

This town on Route 9 northeast of Mason City is associated with the early life of **Hamlin Garland** who, as a boy of ten, plowed the virgin land near here. A marker reading "Boyhood Home of Hamlin Garland," on Route 218, directs visitors to Garland's father's farm of a century ago. The property lies four miles northeast of Osage. Garland's father built the house that still is in use on the property, and trees planted by Garland and his father over one hundred years ago can still be seen and enjoyed. Garland studied at a country school just north of the farm and at Cedar Valley Seminary in Osage, graduating in 1881. One school building, called Old Central, survives on the Lincoln School grounds and serves the Mitchell County Historical Society. Visitors are welcome on Saturday and Sunday afternoons, but readers can experience Garland's life in Osage at any time in Garland's *Boy Life on the Prairie* (1899).

OTTUMWA

This city on Route 34, in southeastern Iowa, is associated with the lives of two novelists. **Edna Ferber** lived here for a short time in her childhood. Her parents operated Ferber's Bazaar, a general store that stood on Main Street, and the family lived on Wapello Street, at the foot of a steep hill. Ferber's autobiographical books, *A Peculiar Treasure* (1939) and *A Kind of Magic* (1963), give the reader an account of her life here.

Honoré Willsie Morrow, who wrote of the West in her early novels, such as *The Heart of the Desert* (1913) and *Still Jim* (1915), was born in Ottumwa in 1880. In addition to her fiction, Morrow wrote biography. Her most ambitious work was a trilogy on the life of Abraham Lincoln, published in 1935 as *Great Captain.* The individual volumes were *Forever Free* (1927), *With Malice Toward None* (1928), and *The Last Full Measure* (1930).

SIGOURNEY

Sigourney, on Route 92, about twenty-five miles northeast of Ottumwa, is named for Lydia Huntley Sigourney, a Connecticut poet whose work was popular in the first half of the nineteenth century. The poet was so pleased that the county seat of Keokuk County was to have her name that she presented the town with fifty books for its library and arranged for the planting of trees that stand around the Sigourney town square.

SIOUX CITY

Josephine [Frey] Herbst, the novelist, was born on March 5, 1897, in Sioux City, which stands at the point where Iowa, Nebraska, and South Dakota meet. Her parents' home, at 427 Center Street, is still standing but privately owned. Herbst became known for her trilogy, which traced one American family from the Gilded Age to the years of the Great Depression: *Pity Is Not Enough* (1933), *The Executioner Awaits* (1934), and *Rope of Gold* (1939). She died in New York City

> *Every change of work brought joy, like a release from prison. From the seeding, corn-planting seemed very desirable; but when the hoes had clicked behind their heels for a couple of days the boys longed for breaking or fence-building. Burning brush seemed glorious sport until they had tried it, and found it very hot and disagreeable, after all. The fact is, they considered any continuous labor an infringement of their right to liberty and the pursuit of knowledge.*
>
> —Hamlin Garland, in *Boy Life on the Prairie*

(*below left*) Home of Ellis Parker Butler in Muscatine. It is not known whether the author was born here, but he did live in the house until about 1900. (*below*) Old snapshot of the farm in Marne where Paul Corey was born and raised.

> *. . . the little, straggling, treeless village of Monterey Centre, now, in summer embowered in verdure; the board sidewalks, each built according to the taste and fancy of the proprietor in the absence of any ordinance, wide, narrow, high or low, with wide gaps here and there across which the pedestrian made his way over earth paths in dry and through puddles of mud in wet weather. . . .*
>
> —Herbert Quick, in *The Hawkeye*

on January 28, 1969, and was cremated. Her ashes were buried on March 24, 1969, in Graceland Cemetery, 2701 South Lakeport, Sioux City.

[John] Herbert Quick, the novelist, served as mayor of Sioux City in 1898 and 1899. He lived in a house, still standing but private, at the northeast corner of Sixth Avenue and South Newton Avenue, then called St. Louis Avenue. A boulder monument and bronze plaque in South Ravine Park commemorate the novelist.

STEAMBOAT ROCK

[John] Herbert Quick was born on a farm near Steamboat Rock, a village on Route 118, about forty miles west of Waterloo, on October 23, 1861. His family moved to Grundy Center when he was quite young. Quick later wrote of Steamboat Rock in his books, calling it Monterey Centre and Monterey. His books also mentioned Root's Hotel, which is still standing here but is divided into apartments.

THORNBURG

Ross Santee, the Western writer who illustrated his own books, was born on August 16, 1888, in Thornburg, a village on Route 22, about forty miles north of Ottumwa. Santee described his childhood experiences in Iowa in the book *Dog Days* (1955), but his best-known work was *Cowboy* (1928).

TIPTON

Bess Streeter Aldrich lived in a house one mile west of Tipton after her marriage in 1907. The house is still standing but is unmarked and private. **Frank Luther Mott** learned to set type at the Tipton *Advertiser,* a newspaper his father edited in this city on Route 130, northwest of Davenport. Mott's father edited newspapers in various Iowa cities, including Audubon and Marengo. With his father, Mott worked as co-editor of the Marengo *Republican* from 1907 to 1914 and then as sole editor of the Grand Junction *Globe.* Mott described his early life in Tipton in *Time Enough: Essays in Autobiography* (1962).

Birthplace of MacKinlay Kantor in Webster City. Kantor wrote his first published short story and first published novel while living in Webster City.

TRAER

Margaret Wilson, winner of a Pulitzer Prize for her first novel, *The Able McLaughlins* (1923), was born on January 16, 1882, in Traer, a town on Route 63, about twenty-five miles south of Waterloo. Wilson was the granddaughter of a Scottish immigrant, and she dealt in *The Able McLaughlins* with the difficult lives of Scottish-American pioneers in Iowa. Wilson married in 1923 and took up permanent residence in England. Her other novels included a second book about the McLaughlins, *Law and the McLaughlins* (1936).

WATERLOO

Sinclair Lewis worked for a few months during 1908 as writer, critic, and proofreader for the Waterloo *Daily Courier,* still publishing in this central Iowa city as the Waterloo *Courier.*

WEBSTER CITY

MacKinlay Kantor, known best for his novel *Andersonville* (1955), was born on February 4, 1904, in this town twenty miles west of Fort Dodge. He lived at 1627 Willson Avenue, the house in which he was born. The house is still standing but is privately owned. When Kantor was a child, his mother worked as an editor for the Webster City *News,* which was published during the 1920s at 546 Second Street. Kantor worked as a reporter for the *News* while attending Webster City High School. While he was a student, he wrote a story that won a $50 prize in a contest sponsored by the Des Moines *Register.* It appeared in the issue of February 26, 1922, and was his first published piece of fiction. He was to wait more than five years to sell his next story. In 1927, while Kantor was working on his first published novel, *Diversey* (1928), he and his wife lived at his grandfather's house, 1718 Willson Avenue—Kantor had also lived in that same house during part of his childhood. *Diversey* took only three months to write, and it was the first book published by the newly formed firm of Coward, McCann. Kantor died in Sarasota, Florida, on October 11, 1977, and was cremated there. His ashes were buried in a Webster City cemetery. The Kendall Young Library, 1201 Willson Avenue, maintains files and scrapbooks on Kantor, as well as autographed copies of all of Kantor's novels.

WHAT CHEER

What Cheer, a town on Route 21, east of Des Moines, was the birthplace of **Frank Luther Mott,** on April 4, 1886. Mott, who had a distinguished career as a teacher of journalism and became the foremost historian of American journalism, had his first taste of newspaper life in What Cheer. As he was to do so many other times, Mott worked for his father, who was editor of many Iowa newspapers during his lifetime. The young boy's job was folding copies of the What Cheer *Patriot.*

MISSOURI

BELTON

This small town about fifteen miles south of Kansas City was the home for some years of **Carry Nation,** the hatchet-wielding temperance agitator. Born Carry Amelia Moore, she moved here with her parents in 1867, when she was twenty. Shortly afterward she married Charles Gloyd, an incurable alcoholic who made their marriage a "literal Hell." She soon left him and supported herself by teaching in nearby Holden until she was fired, supposedly over a dispute about the proper pronunciation of the letter *A.* At her parents' home again, and penniless, she prayed to God to send her a husband. Within weeks she met and married a lawyer and preacher named David Nation, and they spent most of their twenty-four miserable years together in Texas, where Carry supported the family by running hotels. In the late 1890s, when they lived in Kansas, Carry began having visions that impelled her to launch the antisaloon activities for which she is still known. The year 1904 saw the publication of her only book, *The Use and Need of the Life of Carry A. Nation,* and saloon-busting and speech-making took her as far from home as New York. In 1911, her mind increasingly strained by the illness that caused her visions, she broke down while making a public speech. She spent the last five months of her life in a Leavenworth, Kansas, hospital and was buried with her parents in the Belton cemetery, her grave marked by a simple granite shaft.

BOONVILLE

Will Rogers entered Kemper Military Academy in this town, twenty miles west of Columbia, in January 1897. His schoolboy pranks are something of a Kemper legend. He wore his uniform untidily, carried a lariat at all times, and frequently asked his fellow students to "stoop over, run down the hall, and beller like a calf." The great humorist left after only two terms, abandoning formal studies to become—as might be expected—a cowboy.

BRANSON

The area around this town in the Ozarks, about thirty-five miles south of Springfield, provided the setting for the novel *The Shepherd of the Hills* (1907), by **Harold Bell Wright.** Wright, who had come to the mountains to regain his health, started the book here, and its success helped him decide to leave the ministry and become a full-time writer. He wrote more than a dozen books about the wide-open spaces of the Southwest, the most popular of which was *The Winning of Barbara Worth* (1911), a novel said to have sold more than a million and a half copies. Seven miles west of Branson is the Shepherd of the Hills Farm, Wright's home for a time after 1896. The farm is open daily until early November, free of charge. A fee is charged for a tour of the grounds, and for an outdoor pageant held on the grounds each evening.

BUTLER

Robert A. Heinlein was born in this town near the Kansas border on July 7, 1907. His family moved to Kansas City in 1908 but later returned to Butler. The house in which the science-fiction novelist was born, at 308 West Pine, near the Presbyterian Church, is still standing but privately owned and not open to visitors. Heinlein's works have been published in twenty-eight languages, and among his most popular novels are *The Green Hills of Earth* (1951), *Starship Troopers* (1959), and *Stranger in a Strange Land* (1961).

Well, I got all my feet through but one.

—Will Rogers,
commenting on his ability to perform rope tricks

CLAYTON

Tennessee Williams and his family lived during the 1960s in a two-story Georgian house on Pershing Street in this city just west of St. Louis.

I—I have done what I could.

—Carry Nation
in her last public speech,
Jan. 13, 1911

COLUMBIA

Columbia, in the central part of the state, is the home of the University of Missouri and Stephens College, both of which have attracted literary figures as students, teachers, and visitors. **William Inge,** author of such popular plays as the Pulitzer Prize-winning *Picnic* (1953) and *Bus Stop* (1955), taught English and drama at Stephens College from 1938 to 1943. **Frank Luther Mott,** best known for his four-volume *A History of American Magazines* (1930–1957), was dean of the university's school of journalism from 1942 until 1951. **[John] Herbert Quick,** best known for such novels about his native Iowa as *Vandemark's Folly* (1921) and *The Hawkeye* (1923), died at University Hospital on May 10, 1925, after giving a lecture to the Journalism Week Book Banquet.

Tennessee Williams entered the University of Missouri in 1929, living for a time in the Alpha Tau Omega fraternity house, but left after two years due to lack of funds. He then took a job in a shoe company—as Tom, the brother in Williams's great play *The Glass Menagerie* (1945), was to do. **John G. Neihardt** died at age ninety-two in the home of his daughter in Columbia, on November 3, 1973. Neihardt's works included *A Cycle of the West* (1915–1941), a five-part epic poem, and *Eagle Voice* (1953), a novel.

AMANDA: *Most young men find adventure in their careers.*

TOM: *Then most young men are not employed in a warehouse.*

—Tennessee Williams,
in *The Glass Menagerie*

DANVILLE

Theodore Dreiser here courted Jug White, the woman he later married, at her parents' home on the outskirts of this town, about five miles south of Montgomery City in the eastern part of the state. The house is no longer standing.

DONIPHAN

Olive Tilford Dargan, the poet and dramatist, moved with her family to this city in Ripley County in about 1880, when Olive was a child. Her parents, both teachers, opened a school here but moved on three years later to Arkansas.

FLORIDA

John Clemens and his wife Jane moved in 1835 to this small town on the Salt River about

It was a heavenly place for a boy, that farm of my uncle John's. The house was a double log one, with a spacious floor (roofed in) connecting it with the kitchen. In the summer the table was set in the middle of that shady and breezy floor, and the sumptuous meals—well, it makes me cry just to think of them.

—Mark Twain,
in *Autobiography*

Well, when Tom and me got to the edge of the hill-top, we looked away down into the village and could see the three or four lights twinkling . . . and the stars over us was sparkling ever so fine; and down by the village was the river, a whole mile broad, and awful still and grand.

—Mark Twain,
in *Huckleberry Finn*

thirty miles southwest of Hannibal. On November 30 of that year, Jane gave birth to **Samuel Langhorne Clemens,** their second son, who was to become, as **Mark Twain,** one of America's best-loved and most distinguished authors. His family stayed in Florida for only four years, moving in 1839 to Hannibal, but young Sam spent summers at the Florida home of his uncle James Quarle. He returned again in 1861, during his brief and disastrous tenure in the Confederate Army—described in *The Private History of a Campaign that Failed* (1885)—and throughout his life he remembered the town with deep affection. Florida, in turn, pays abundant tribute to its most eminent native son. The site of his birthplace is marked with a monument, and the house itself has been moved to the Mark Twain Memorial Shrine in Mark Twain State Park, on the south fork of the Salt River. Among the exhibitions in the shrine, which is open to visitors throughout the year, are a manuscript of *The Adventures of Tom Sawyer* (1876), some of Twain's possessions, and first editions of all the author's works.

HANNIBAL

This town on the Mississippi might have come to be remembered largely for its importance as a river port had it not been for the arrival here, in 1839, of small-time businessman John Clemens and his family. Clemens never achieved much success in business, and his activities here had small impact on the town. It was in the writings of Clemens's younger son Samuel, known throughout the world as **Mark Twain,** that Hannibal achieved lasting and universal fame. Twain used the place—a "little white town drowsing in the sunlight"—as the model for the town of St. Petersburg in *The Adventures of Tom Sawyer* (1876) and *The Adventures of Huckleberry Finn* (1884), his two most famous works. Hannibal also provided much of the background of *Life on the Mississippi* (1883) and *The Tragedy of Pudd'nhead Wilson* (1894). And it was in Hannibal that young Sam fell in love with the Mississippi, its great riverboats, its spectacle and drama, its promise of unknown worlds. When Sam and his family moved here, they lived first in the Pavey Hotel, no longer standing, and then in what is now known as the Mark Twain House, at 208 Hill Street. When John Clemens's business suffered a serious reversal, the family let that house and moved into an apartment in the Virginia house, at the corner of Hill and Main streets, to return later to the Hill Street house. Sam attended school in a log house, no longer standing, on Main Street but may have cared for school even less than did Huck Finn, who found that "the longer I went . . . the easier it got to be." Whatever his opinion of the place, Sam did not stay long: after his father died in 1847, when Sam was only twelve, the young boy had to leave school and go to work as a printer's apprentice. He spent two years at the Hannibal *Courier,* leaving to join the *Journal,* which his older brother, Orion, had recently bought. Sam added writing to the more typical responsibilities of a printer, contributing pseudonymous letters—sometimes followed by replies he also wrote—squibs, and verse. While working for the *Journal* Sam, still only sixteen years old, saw his story "The Dandy

Birthplace of Mark Twain, now located in the enclosed Mark Twain Birthplace Shrine, on Route 107 just south of Florida, Missouri.

(*far left*) Building in Hannibal in which Sam Clemens's father maintained a law office and where young Sam one night came upon the corpse of a murder victim and fled through a window: "I was not scared," he later wrote, "but I was considerably agitated." (*left*) The Pilaster House, also known as Grant's Drug Store, in Hannibal. The Clemens family lived here from late 1846 until after the death of Sam Clemens's father in 1847.

Frightening the Squatter" (1852) published in *The Carpet-Bagger,* a Boston magazine. It was his first published work of fiction.

Sam left town in 1853 to become a journeyman printer and, though he returned many times to his hometown, he never again lived in Hannibal for long. In his greatest works, however, Twain showed that Hannibal had never lost its importance for him. Places in and around the town, still to be seen today, appeared frequently in the stories of Huck Finn and Tom Sawyer. Holliday's Hill, the scene of so much adventure, became Cardiff Hill in Twain's imagination. The hill is still the home of the cave in which Injun Joe met his lonely and anguished death, and where Tom and Becky Thatcher became lost. Twenty years after moving away, Twain could still describe the cave with all the excitement it produced in him as a boy, and under its new name of Mark Twain's Cave, it still is visited today. A few miles down river from Hannibal is the island on which Tom and his gang held their meetings, and where Huck and Jim hid from their various pursuers. Called Glasscock's Island in the nineteenth century—Jackson's Island in Twain's books—it is now known as Pearl Island, or Tom Sawyer's Island.

Of the author's many famous characters, several were based on his Hannibal contemporaries. The prototype for Pap in Huckleberry Finn was Jimmy Finn, town drunkard and "a monument of rags and dirt," who "slept with the hogs in an abandoned tanyard." Becky Thatcher had her origin in Laura Hawkins, Twain's childhood sweetheart. And Huck Finn, whom T. S. Eliot called "one of the permanent symbolic figures of literature," was drawn from a drunkard's son named Tom Blankenship. The Blankenship family were considered lowlifes in Hannibal, and Sam Clemens continually sought their company, despite his parents' admonitions. Tom Blankenship

Mark Twain Museum (*standing left*) and the Mark Twain Boyhood Home in Hannibal. On a return visit to his boyhood home in 1902, Twain remarked: "It all seems so small to me. . . . I suppose if I should come back ten years from now it would be the size of a bird-house."

Cardiff Hill, beyond the village and above it, was green with vegetation, and it lay just far enough away to seem a delectable land, dreamy, reposeful, and inviting.

—Mark Twain,
in *Tom Sawyer*

Three miles below St. Petersburg . . . there was a long, narrow, wooded island, with a shallow bar at the head of it, and this offered well as a rendezvous. It was not inhabited; it lay far over towards the farther shore, abreast a dense and almost wholly unpeopled forest.

—Mark Twain,
in *Tom Sawyer*

The mouth of the cave was high up the hillside, an opening shaped like the letter A. Its massive oaken door stood unbarred. Within was a small chamber, chilly as an icehouse, and walled up by Nature with solid limestone that was dewy with a cold sweat. It was romantic and mysterious to stand here in the deep gloom and look out upon the green valley shining in the sun.

—Mark Twain,
in *Tom Sawyer*

was his favorite: he and Sam used catcalls to summon one another, just as Tom Sawyer and Huck Finn did, and they heroically resisted all efforts to "sivilize" them. Twain later said that Tom Blankenship was "ignorant, unwashed, insufficiently fed; but he had as good a heart as ever any boy had. . . . He was the only really independent person—boy or man—in the community, and by consequence he was tranquilly and continuously happy, and was envied by all the rest of us." The site of the Blankenships' ramshackle house on Hill Street is marked, though the house itself is no longer standing. A statue of Tom Sawyer and Huck Finn stands at the intersection of North and Main streets. The Mark Twain House, at 208 Hill Street, is open daily, as is the Mark Twain Cave, one mile south on Route 79, which can be seen by guided tour. Hannibal itself has changed considerably, but its citizens are as proud as ever of living in the town once lived in by one of America's greatest writers.

HUMANSVILLE

Zoë Akins was born in this town northwest of Springfield on October 30, 1886. Akins is best known for her Pulitzer Prize-winning dramatization in 1935 of *The Old Maid* (1924), by Edith Wharton, and for original plays such as *The Greeks Had a Word for It* (1930). She also wrote the screenplay for the 1936 motion-picture version of the novel *Show Boat* (1926), by Edna Ferber.

JOPLIN

[James] Langston Hughes was born on February 1, 1902, in this city some 140 miles south of Kansas City. Among Hughes's books of poems were *Shakespeare in Harlem* (1942) and *Montage of a Dream Deferred* (1951).

Huck Finn and Tom Sawyer in Hannibal. The statue stands at the foot of Cardiff Hill, about half a block from the site of the home of Tom Blankenship, the model for Huck Finn.

I set still and listened. Directly I could just barely hear a "me-yow! me-yow!" *down there. That was good! Says I,* "me-yow! me-yow!" *as soft as I could, and then I put out the light and scrambled out of the window onto the shed. Then I slipped down to the ground and crawled in amongst the trees, and sure enough there was Tom Sawyer waiting for me.*

—Mark Twain,
in *Huckleberry Finn*

KANSAS CITY

The second largest city of Missouri has many literary associations. **Russel Crouse,** best known for his theatrical collaborations with Howard Lindsay, worked as a sports columnist and reporter for the Kansas City *Star* from 1911 to 1916, still published today from Eighteenth Street and Grand Avenue. **Eugene Field,** best remembered for his children's poems, such as "Dutch Lullaby," known better as "Wynken, Blynken and Nod," and "Little Boy Blue," worked on the Kansas City *Times* during 1880–81. **Robert A. Heinlein,** the science-fiction novelist, lived here during most of his childhood and was graduated from Central High School in 1924. **Ernest Hemingway** worked as a reporter on the Kansas City *Star* in 1917–18.

Sinclair Lewis, staying in a two-room suite at the Hotel Ambassador, did research here for *Elmer Gantry* (1927). While in Kansas City, Lewis was awarded, and declined to accept, the 1926 Pulitzer Prize for fiction, for his novel *Arrowsmith.* **Richard Lockridge,** best known for the Mr. and Mrs. North mystery novels he wrote with his wife Frances Lockridge, lived in 1905 in a house at 4223 Windsor, still standing but privately owned. Lockridge returned to Kansas City in 1922 to work as a reporter. **Irving Stone,** best known for his fictionalized biographies, such as *Lust for Life* (1934), about Vincent van Gogh, came here in 1943 with his wife to do research on the lives of Jessie Benton Frémont and John Charles Frémont for *Immortal Wife* (1944). **Harold Bell Wright,** author of such bestsellers as *The Winning of Barbara Worth* (1911), served as pastor at the Forest Avenue Christian Church from 1903 to 1905. The church, then located at the corner of Forest Avenue and Sixteenth Street, was razed in 1939.

Kansas City's native sons include **Evan [S.] Connell, Jr.,** who was born here on August 17, 1924. Connell is the author of such works as *The Anatomy Lesson and Other Stories* (1957) and the novel *Mrs. Bridge* (1959), an ironic depiction of the life of a suburban matron. **Courtney Ryley Cooper,** born on October 31, 1886, attended public schools in Kansas City but ran away when he was sixteen to become a clown in a small circus. He worked his way up in the circus world, serving as Buffalo Bill Cody's press agent and later as press agent for Ringling Brothers Barnum and Bailey. In addition to writing on circus life in such volumes as *Memoirs of Buffalo Bill* (1920) and *Under the Big Top* (1923), Cooper wrote extensively about crime in the underworld in such books as *Here's to Crime* (1937).

KIDDER

Paul Armstrong was born on April 25, 1869, in this town about thirty miles east of St. Joseph. His best-known play, *Alias Jimmy Valentine* (1909), inspired a long succession of imitators in the popular genre of gangster drama.

KIRKWOOD

This suburb of St. Louis boasts two distinguished literary native daughters: **Josephine Johnson** and **Marianne Moore.** Johnson, born on June 20, 1910, lived here and attended school until she was twelve, when she moved with her family to a 100-acre farm. Since completing her education at Washington University in St. Louis, she has lived most of her life in the countryside here and has written extensively on rural subjects. Her *Now in November* (1934), a novel about farm life, won a Pulitzer Prize. Her other works include *Winter Orchard and Other Stories* (1935) and the novels *Jordanstown* (1937) and *Wildwood* (1946). Marianne Moore was born on November 15, 1887, at the home of her grandfather, the Reverend John R. Warner, who raised her. Warner served at the time as pastor of Kirkwood's Presbyterian Church. One of the most distinguished poets of her generation, Moore saw *Poems* (1921), her first book, published without her knowledge by two friends, poets Hilda Doolittle (H. D.) and Robert McAlmon. Moore's later volumes included *Observations* (1924), *The Pangolin and Other Verse* (1936), and *What Are Years?* (1941). Her *Collected Poems* (1951), with an introduction by fellow Missouri native T. S. Eliot, won a Pulitzer Prize.

LANCASTER

Rupert Hughes was born in this city in the northeastern corner of the state on January 31, 1872. Not long after the author's birth, the Hughes family moved to Iowa, but he returned later to attend boarding school in St. Charles, which is near St. Louis. Hughes wrote over two dozen plays and books of stories, but he was known best for his three-volume biography of George Washington, published in 1926–30, which treated its subject as a great man, while doing away with many of the Washington myths and legends. Hughes is also credited with having originated the cruelly ironic line, "Her face was her chaperone."

LEBANON

Harold Bell Wright, author of popular novels set in the Southwest, served from 1905 to 1907 as pastor at the First Christian Church in Lebanon, which is about seventy-five miles southwest of Jefferson City.

LEXINGTON

Edwin Milton Royle was born in this city, about thirty miles east of Independence, on March 2, 1862. Royle's play *The Squaw Man* (1906) was adapted for motion pictures and became the first professional assignment of a young director named Cecil B. De Mille.

MANSFIELD

About a mile east of Mansfield, due east of Springfield, lies Rocky Ridge Farm, home at one time of **Laura Ingalls Wilder** and **Rose Wilder Lane,** who was Laura Wilder's daughter. Laura Wilder achieved enormous popularity through the series of eight children's books she wrote about her prairie childhood, beginning with *Little House in the Big Woods* (1932) and concluding with *These Happy Golden Years* (1943). By far her best-known work today is *Little House on the Prairie* (1935), the third novel in the series. Wilder died here on February 10, 1957, and was buried in Mansfield Cemetery. Her home, on Route 60, is now open to visitors. Admission is charged. Mrs. Wilder's daughter, Rose Wilder Lane, was known best for vivid portrayals of life in the Ozarks, such as *Hill-Billy* (1926) and *Old Home Town* (1935). A hard-line conservative, she announced in 1944 that she would cease writing in order not to pay income taxes, which she regarded as unjustifiable government interference in individual affairs. Rose Wilder Lane is buried next to her mother and father in Mansfield Cemetery. She and her mother are both commemorated by a plaque on nearby Route 5, and there is a Laura Ingalls Wilder Library in Mansfield.

MARSHALL

Emory Holloway was born in this city, about seventy miles east of Kansas City, on March 16, 1885. For *Whitman: An Interpretation in Narrative* (1926), Holloway won the 1927 Pulitzer Prize for biography.

MARYVILLE

Dale Carnegie was born in this small town in northwest Missouri on November 24, 1888. Though more a lecturer and businessman than an author, Carnegie wrote an enormously successful self-help book, *How to Win Friends and Influence People* (1936). It has sold millions of copies and has been translated into thirty-one languages. **Homer Croy** was born on a farm some miles from Maryville on March 11, 1883. Among Croy's novels were *Jesse James Was My Neighbor* (1949) and *He Hanged Them High* (1952); his works of nonfiction included *Wheels West, the Story of the Donner Party* (1955).

PALMYRA

John Clemens, father of **Mark Twain,** rode in February 1847 to this town about ten miles northwest of Hannibal to be sworn in as a clerk in Surrogate Court. Returning home in rain and sleet, he contracted the pneumonia that caused his death a month later. Clemens's death forced his son Sam to leave school, and the boy became apprenticed as a printer at the Hannibal *Courier.*

ST. JOSEPH

This city on the Missouri River was for several years the home of **Eugene Field,** the poet and journalist, who returned here from a European tour in the fall of 1873, and on October 16 married Julia Comstock, a native of the city. She and Field had become engaged two years before, when Julia was only fourteen. After their honeymoon the couple returned here, Field to his job on the St. Joseph *Gazette,* a morning newspaper. One of the homes the Fields had here was in a building now called the American Apartments, still standing but unmarked, at 425 North Eleventh Street. Also standing today is Christ Episcopal Church, 201 North Seventh Street, in which the couple were married. One of Field's many popular poems, "Lover's Lane, Saint Jo," is about St. Joseph's Rochester

> *What happens to a dream deferred?*
>
> *Does it dry up*
> *like a raisin in the sun?*
> *Or fester like a sore—*
> *And then run?*
> *Does it stink like rotten meat?*
> *Or crust and sugar over—*
> *like a syrupy sweet?*
>
> *Maybe it just sags*
> *like a heavy load.*
>
> Or does it explode?
>
> —Langston Hughes,
> "A Dream Deferred"

Strephon kissed me in the spring,
Robin in the fall,
But Colin only looked at me
And never kissed at all.

Strephon's kiss was lost in jest,
Robin's lost in play,
But the kiss in Colin's eyes
Haunts me night and day.

—Sara Teasdale,
"The Look"

Road, which runs north and northeast from North Eighteenth Street.

St. Joseph was the birthplace of novelist **Richard Lockridge,** born on September 26, 1898, and best known as the co-author—with his wife **Frances Lockridge,** who died on February 18, 1963—of novels about a pair of amateur detectives who managed somehow to solve crimes. Lockridge's birthplace has not been identified with certainty, but it may be a house at 712 South Eleventh Street, still standing but privately owned. In addition to their novels about Mr. and Mrs. North, the Lockridges wrote a number of other books, including *Cats and People* (1950) and *The Faceless Adversary* (1956).

ST. LOUIS

As the largest city in Missouri and one of the great centers of the Midwest ever since frontier days, St. Louis has long attracted literary visitors and residents. While on a lecture tour in 1971, **W. H. Auden** celebrated his sixty-fourth birthday here at the Jefferson Hotel, drinking champagne and doing crossword puzzles with his friend Orlan Fox. **William Cowper Brann** wrote editorials for the St. Louis *Globe-Democrat* in the late 1870s, long before 1891, when he founded *Iconoclast,* a magazine that he published first in Austin and later in Waco, Texas. **Charles Dickens** visited St. Louis on his American tour of 1842 and wrote about the city in his *American Notes for General Circulation* (1842).

Theodore Dreiser worked for the *Globe-Democrat* and the *Republic* here between 1892 and 1894, living first at 708 Pine Street and later in rented rooms at Tenth and Walnut. It is said that **Langston Hughes** wrote one of his best-known poems, "The Negro Speaks of Rivers" (1922), while on a train going past St. Louis. **Fannie Hurst,** who was brought here by her parents when she was only a few weeks old, spent the first twenty years of her life in St. Louis, living in a house at 5641 Cates Avenue. She was graduated from Central High School and, in 1909, from Washington University.

William Inge, the playwright, served as drama critic for the St. Louis *Star-Times* from 1943 to 1946, and taught at Washington University from 1946 to 1949, when he lived at 1213 North Seventh Street. **Howard Nemerov,** author of such volumes of poetry as *The Next Door of the Dream* (1964), has taught at Washington University since 1969. **Joseph Pulitzer** lived here from 1865 to 1883 and founded, by merger, the St. Louis *Post-Dispatch.* **Odell Shepard,** winner of a Pulitzer Prize for *Pedlar's Progress, The Life of Bronson Alcott* (1937), taught English at Smith Academy here for a year early in his career. **Robert Lewis Taylor,** author of *The Travels of Jamie McPheeters* (1958), spent three years working for the St. Louis *Post-Dispatch.*

Mark Twain came here many times during his long life, delivering his famous "Sandwich Islands" lecture in 1867 at Mercantile Hall, now the Mercantile Library, and helping, in 1902, to dedicate the St. Louis World's Fair. **Walt Whitman,** while on a trip to the Rocky Mountains, visited his brother Jefferson here in the autumn of 1879. **Tennessee Williams** moved here with his family when he was about twelve years old, living in modest apartments, first at 4633 Westminster Place and then at 5 South Taylor, which later inspired the setting for *The Glass Menagerie* (1944). **Thomas Wolfe** came to St. Louis when his mother moved here in 1904—Tom was four years old. She opened a boarding house, The North Carolina, at 5095 Fairmont (now Cates) Avenue. Wolfe used some details of the house in *Look Homeward, Angel* (1929) and in his short story "The Lost Boy," published posthumously in 1941 in *The Hills Beyond.*

Maya Angelou, the poet whose first volume of verse was *I Know Why the Caged Bird Sings* (1970), was born here on April 4, 1928. **Sally Benson,** born here on September 3, 1900, is best known for her collection of short stories *Meet Me in St. Louis* (1942). She grew up in a house at 5135 Kensington Avenue. **William S. Burroughs,** born here on February 5, 1914, is known for his experimental novels, including *The Naked Lunch* (1959), published in the United States in 1962 as *Naked Lunch,* and *The Soft Machine* (1961). **Kate [O'Flaherty] Chopin,** born here on February 8, 1851, achieved a still-growing reputation with her novel *The Awakening* (1899). Publication of the work brought so much criticism down on her because of its treatment of mental disturbance and adultery that she stopped writing. Chopin lived in St. Louis all her life except for the two years she spent in Louisiana with her husband.

Winston Churchill, born here on November 10, 1871, was the author of such historical novels as *Richard Carvel* (1899) and *The Crisis* (1901). *The Crisis* is set in St. Louis. **Eugene Field** was born on September 2, 1850. His birthplace, at 634 Broadway, is now a museum that houses manuscripts, books, furniture, and assorted memorabilia. There is some uncertainty as to whether this was, in fact, Field's birthplace. Field's brother Roswell identified an earlier family residence on Collins Street as the actual site of Field's birth, but Field himself identified the house on Broadway. Whatever the truth, 634 Broadway was dedicated as Eugene Field's birthplace on June 6, 1902. The guest of honor at the ceremony was none other than Mark Twain. **Alice Corbin Henderson,** born in 1881, was co-founder of *Poetry* magazine in 1912. Her poetry collections included *Red Earth* (1920) and *The Sun Turns West* (1933). **Augustus Thomas,** born on January 8, 1857, wrote of the American experience in such plays as *Alabama* (1891) and *In Mizzoura* (1893). He spent the first thirty years of his life here before moving to New York in 1888.

Two poets born in St. Louis conclude the entry for this city: Sara Teasdale and T. S. Eliot. **Sara Teasdale** was born on August 8, 1884, in a house at 3668 Linden Boulevard, no longer standing. She was educated at home and then attended Hosmer Hall, a private school, from which she was graduated in 1903. Her family's wealth enabled her to travel frequently to Europe and to New York, where she met Vachel Lindsay. He fell desperately in love with her and courted her for some years, the courtship finally ending in 1914, when Sara married a St. Louis businessman named Ernest Filsinger. The couple moved into the Arthur Hotel in early 1915 and, later that year, into the Usona Hotel. Sara's first book, *Sonnets to Duse and Other Poems,* had been published

Eugene Field House in St. Louis, where Field lived from 1850 until his mother's death in 1856.

in 1907, with two other volumes following in 1911 and 1915. *Love Songs* (1917) won her a Pulitzer Prize for poetry and brought her public acclaim. In her personal life, however, Sara Teasdale was tormented. Never fully satisfied with her marriage, she divorced her husband in 1929 and moved to New York City shortly afterward. Her rejected suitor, Vachel Lindsay, committed suicide in 1931, and Sara became deeply despondent. On January 29, 1933, she died from what was reported to be an overdose of sleeping medicine.

Teasdale's poetry is neglected today, but the works of another St. Louis native, 1948 Nobel laureate **T[homas] S[tearns] Eliot,** are counted among the great achievements in English and American literature. T. S. Eliot was born on September 26, 1888, in his father's home, no longer standing, at 2635 Locust Street. This was his home for some sixteen years, years in which he attended Smith Academy and began to build the wide knowledge that underlies his works of poetry and criticism. In the early 1900s Eliot moved with his family to a house, still standing, at 4446 Westminster Place. Early and late, his poetry showed influences from his childhood in St. Louis. The title of "The Love Song of J. Alfred Prufrock" (1915) recalled the name of a furniture store at 1104 Olive Street, and the city itself can be seen as providing some of the "half-deserted streets" of the poem's setting. More than twenty-five years later, in "The Dry Salvages" (1941, collected in 1943 as one of the *Four Quartets*), Eliot revived his childhood memories of the Mississippi River, on whose banks St. Louis was founded. And long after Eliot had renounced his American citizenship—he became a British subject in 1927—he said to a St. Louis audience:

> *I think I was fortunate to have been born here, rather than in Boston, or New York, or London.*

SPRINGFIELD

Marquis James was born in this city in southwest Missouri on August 29, 1891. His family was living at the time in the Metropolitan Hotel, since razed. James won a Pulitzer Prize for *The Raven, A Biography of Sam Houston* (1929).

NORTH DAKOTA

BISMARCK

This city's literary history is closely connected with the history of one of its newspapers, the Bismarck *Tribune.* **James W. Foley,** who was to be named poet laureate of North Dakota, lived in Bismarck as a boy and attended school here. He began working for the *Tribune* in 1892 and remained with the newspaper for many years. He later left North Dakota, but on February 8, 1924, the state celebrated his fiftieth birthday in all its public schools, and his birthday has been celebrated in the schools every year since then. Among Foley's numerous books were *Complete Verses* (1911), *Tales of the Trails* (1914), and *The Voices of Song* (1916). The Bismarck *Tribune* is itself honored today for a story of heroism known to nearly every American as Custer's Last Stand. The founder of the *Tribune,* Clement A. Lounsberry, telegraphed, on July 5, 1876, to the New York *Herald*—and the world—the first news of General George Armstrong Custer's defeat by Sitting Bull in the Battle of Little Bighorn, in southeast Montana. A marker commemorating Lounsberry's telegram stands near Fifth Street, a short distance from the station platform of the Burlington Northern depot.

> *I bathed in the Euphrates when dawns were young.*
> *I built my hut near the Congo and it lulled me to sleep.*
> *I looked upon the Nile and raised the pyramids above it.*
> *I heard the singing of the Mississippi when Abe Lincoln went down to New Orleans, and I've seen its muddy bosom turn all golden in the sunset.*
>
> —Langston Hughes, in "The Negro Speaks of Rivers"

1917 photograph of T. S. Eliot's study in his home at 4446 Westminster Place in St. Louis. When this photograph was taken, Eliot was living in England and had just published his first collection of poems, *Prufrock and Other Observations.*

(*right*) Maltese Cross Cabin, in which Theodore Roosevelt lived in 1884 and early 1885. The cabin was considered very comfortable in its time: it had a floor and three rooms. (*far right*) Log cabin built by Joseph Henry Taylor in Painted Woods. It now stands in Washburn.

FORT YATES

Hamlin Garland and his brother came to Fort Yates, about fifty miles south of Bismarck, in July 1897, when Fort Yates was headquarters of the Indian Agency for North and South Dakota. Garland wanted to talk with the Sioux Indians of the Standing Rock Reservation. He interviewed many people, including Rain-in-the-Face—one of the leaders of the Sioux resistance to encroachment on Indian land—and, after several weeks, continued on his Western tour, visiting other Indian agencies as well as the Custer battlefield site in Montana. He used the information gathered on this trip for several books, including his novel *The Captain of the Gray-Horse Troop* (1902) and the volume *The Book of the American Indian* (1923).

GRAND FORKS

This city on North Dakota's eastern border is the home of the University of North Dakota. **Maxwell Anderson** received a B.A. degree here in 1911. In the same year he married Margaret Haskett, whom he had met while studying at the university.

Let me get out in the Hills again,
I and myself alone,
Out through the wind and the lash of rain
To find what we really own.
Under the stars while a campfire dies,
Let me sit and look myself in the eyes.

—Badger Clark,
in "I and Myself"

JAMESTOWN

Maxwell Anderson moved to this city in southeastern North Dakota in 1907. The playwright studied at Jamestown High School, from which he was graduated in 1908. On March 22 of the same year, **Louis L'Amour,** at present North Dakota's most widely read native author, was born in Jamestown. L'Amour grew up here and attended local grade and high schools but, restless for a wider world, left Jamestown when he was fifteen. L'Amour is the author of scores of highly respected and very popular Western novels, noted for their historical accuracy as well as their literary merit.

The Sons of Mary seldom bother, for
they have inherited that good
part;
But the Sons of Martha favour their
Mother of the careful soul and
the troubled heart.
And because she lost her temper once,
and because she was rude to the
Lord her Guest,
Her Sons must wait upon Mary's Sons,
world without end, reprieve, or
rest.

—Rudyard Kipling,
in "The Sons of Martha"

MEDORA

This village in western North Dakota is most closely associated with **Theodore Roosevelt,** who owned a cattle ranch near here in the late 1800s. Roosevelt first visited this region in September 1883. He stayed at the Maltese Cross Ranch, went hunting with the men who ran it, and one day decided impulsively to invest in its cattle operations. Roosevelt later returned to the ranch and eventually bought his own spread, known as the Elkhorn Ranch. Roosevelt, while living here, wrote the biography *Thomas Hart Benton* (1887). He wrote at lightning pace, when he was not hunting or occupied in supervision of his cattle operations. Roosevelt used his hunting experiences for the book *Hunting Trips of a Ranchman* (1885), which he worked on at the Maltese Cross Ranch, and for a series of six articles for *The Century Illustrated Monthly Magazine* that later were published as *Ranch Life and the Hunting Trail* (1888). He wrote these articles and others to help finance work on his four-volume history *The Winning of the West* (1889–96). Roosevelt, who at one time considered abandoning politics to live as a writer and rancher, returned to New York and by 1898 had sold out his ranching interests.

The cabin in which Roosevelt stayed at the Maltese Cross Ranch has been moved to Medora, and now stands behind the visitors center of the south unit of the Theodore Roosevelt National Park, just north of Medora. Guided tours of the cabin are given daily during the summer, and visitors can enter the cabin unattended during the rest of the year. The site of Roosevelt's Elkhorn Ranch, about thirty-five miles north of Medora, is accessible only by a rough road that may be closed because of weather conditions. Visitors must also ford the Little Missouri River, and it is advisable to ask at the park for exact directions before setting out. One of the ranches operating at the time of Roosevelt's visit was the Custer Trail Ranch, about five miles from Medora and located on the site of an encampment used by Custer's Seventh Cavalry Regiment while on its way to the Little Bighorn in 1876. **Ernest Thompson Seton** visited the Custer Trail Ranch in September 1897. During the month he spent there, he gathered material he was later to use in his book *Lives of the Hunted* (1901).

MINNEWAUKAN

From 1911 to 1913, **Maxwell Anderson** taught in the high school of this city on Route 19, in northeastern North Dakota.

WASHBURN

Visitors to this city, about thirty-five miles northwest of Bismarck, may be surprised to find, on Route 83, a marker bearing lines from the poem "The Sons of Martha," written by Rudyard Kipling in 1907. The marker was donated to the city by Harry F. McLean, a well-known construction engineer who was born in Bismarck. McLean may have seen in the poem ample tribute to those who serve others through hard manual labor in dangerous settings. The marker bears the legend "In Loving Memory of Those Who Worked and Died Here." Not far from this marker is the Joseph Henry Taylor Cabin Historic Site. In the 1870s, **Joseph Henry Taylor** built this cabin in a small community called Painted Woods, about fifteen miles southeast of Washburn. Taylor became the settlement's first postmaster—he used a hole cut into a tree as the first post office—and later moved to Washburn, where he became editor of the Washburn *Leader*. In Washburn, Taylor composed and printed several of his books in his own shop. His books included *Frontier and Indian Life* (1889), printed in Pottstown, Pennsylvania; and *Kaleidoscopic Lives* (1902) and *Beavers and Their Ways* (1904). These last two books Taylor composed and printed in Washburn. Taylor died in his shop on April 9, 1908. His cabin, open to the public by appointment, was moved from Painted Woods to Washburn in 1932.

SOUTH DAKOTA

ABERDEEN

L[yman] Frank Baum, best known for *The Wonderful Wizard of Oz* (1900), moved to this northeastern South Dakota town in the 1880s and ran a small general store, Baum's Bazaar, until it failed. From 1888 to 1891 he served as editor of the weekly *The Aberdeen Saturday Pioneer,* but his humorous and imaginative columns, collected in *Our Landlady* (1941), did not prevent this enterprise from failing as well. **Hamlin Garland,** who lived in nearby Ordway and visited Aberdeen on many occasions, wrote of the former Lacey Drug Store, which stood on the corner of Main Street and First Avenue, in his collection of stories *Main-Travelled Roads* (1891).

Aberdeen has a further claim to literary fame. The home of Henry Wadsworth Longfellow in Cambridge, Massachusetts, has been reproduced here, in design and furnishings, as a residence on North Main Street and Twelfth Avenue.

CANTON

O[le] E[dvart] Rölvaag lived in this community, about twenty miles southeast of Sioux Falls, from 1898 to 1901, attending Augustana Academy, a Lutheran preparatory school whose buildings are used now as a drug and alcohol rehabilitation center.

CUSTER

[Charles] Badger Clark, poet laureate of South Dakota, lived near Custer from 1925 to 1957. His home, called The Badger Hole, still stands on Route 16A, just east of Legion Lake. The four-room cabin, which Clark himself built and which contains such memorabilia as his collection of boots and his library, is open by appointment free of charge. Nearby, at Legion Lake, is a marker for the cabin.

DEADWOOD

Badger Clark attended public school in the 1890s in this Black Hills gold-rush town, now known for its Wild West tourist attractions. **Mary Hallock Foote,** author of stories of Western life, including the novel *The Led-Horse Claim* (1883), lived in Deadwood for a time in the 1880s.

DE SMET

De Smet is a shrine for the many readers of the novels of **Laura Ingalls Wilder.** She and her family came to live in this community about thirty miles east of Huron in 1879, spending their first winter in the Surveyor's House, which is still standing at First Street and Olivet Avenue and which was described in Wilder's novel *By the Shores of Silver Lake* (1939). For a short time they lived at Second and Main streets, in the Ingalls Store Building, no longer standing. Then they moved just southeast of town to a homestead whose buildings are no longer standing but are marked by a plaque mounted on a boulder. Some of the nearby cottonwood trees were planted by Pa Ingalls. Each summer, on selected dates, a Long Winter Pageant is held on land near the Ingalls Homestead. Admission is charged.

On August 25, 1885, Laura married Almanzo Wilder, and they lived until 1889 at a homestead

> *"I read in the paper," said our landlady, as she finished sewing a button on the doctor's overcoat and bit off the thread, as a woman will, "that the great air ship is goin' ter be a success, after all."*
>
> *"Yes," said the Colonel, by way of reply, "it looks as though they will be able to make it work after they've rectified a few mistakes."*
>
> *"If it does," said Tom, "the railroads will be ruined."*
>
> —L. Frank Baum, in *Our Landlady*

(*below left*) Surveyor's House, Laura Ingalls Wilder's first home in De Smet. (*below*) The Ingalls family in an 1891 photograph. Laura is seen standing behind her father.

Sign marking the home of Laura Ingalls Wilder and birthplace of her daughter, Rose Wilder Lane.

site about one mile north of town on Route 25. A marker near the site indicates that the Wilder's daughter, **Rose Wilder Lane,** was born there on December 5, 1887. Lane became a well-known novelist and short-story writer in her own right, describing South Dakota pioneer life in her novels *Free Land* (1938) and *Let The Hurricane Roar* (1933). In 1887 Laura's father built a house on Third Street near Loftus Avenue. Here Pa, Ma, and Mary Ingalls spent the rest of their lives. Laura and Almanzo Wilder's next home was located on a tree claim a mile and a half north of the Wilder homestead. Although the house has burned down, some of the trees planted by Laura and Almanzo are still standing, and a highway marker indicates the site. In 1892 the Wilders bought a house in town at Fourth Street and Loftus Avenue, which was described in Laura's diary, *On the Way Home* (not published until 1962). They lived here until 1894, when they sold the house and moved to Mansfield, Missouri. Although this was their last home in De Smet, the Wilders actually left De Smet from the Ingalls home on Third Street.

Other sites connected with Laura Ingalls Wilder include Brewster School, one of the schools in which Laura taught and which is now located behind the Surveyor's House; the Loftus store, on Main Street, where she worked; the Original Congregational Church, at Second Street and Loftus Avenue, which Pa Ingalls helped build and which the Ingalls family attended; the cemetery, on Prairie Avenue just south of town, where are buried Laura's parents, other relatives, and many residents of De Smet who figured in Wilder's books; and the City Library, at First and Main streets, which houses many Ingalls family possessions. Tours to most of these sites are available, and an admission fee is charged. See PEPIN, WISCONSIN.

ELK POINT

Twenty-year-old **O. E. Rölvaag** came from Norway in 1890 to work with his uncle as a hired hand on the S. S. Eidem farm, eight miles from this southeastern South Dakota town. The farm is still in the Eidem family, although the original house no longer exists. In his partially autobiographical work *Letters from America* (1912), Rölvaag referred to Elk Point as Clarkfield and Mr. Eidem as Boss.

HOT SPRINGS

Badger Clark lived with his parents here from about 1910 to 1925 and was buried in a

(*above left*) Laura Ingalls Wilder; (*above right*) Rose Wilder Lane. (*right*) The Third Street home of Pa and Ma Ingalls.

"This country's going to be covered with trees," Pa said. "Don't forget that Uncle Sam's tending to that. There's a tree claim on every section, and settlers have got to plant ten acres of trees on every tree claim. In four or five years you'll see trees every way you look."

—Laura Ingalls Wilder, in *By the Shores of Silver Lake*

Hot Springs cemetery in 1957, after his death in Rapid City. He described the unique Western atmosphere of the Black Hills in two collections of poems, *Sun and Saddle Leather* (1915) and *Grass Grown Trails* (1917).

HURON

Badger Clark lived from 1893 to 1898 in this east-central South Dakota city, at the Methodist Parsonage, no longer standing, that adjoined the Church at Fourth and Kansas S.E., where Clark's father was minister. **Charles Partlow,** known as **Charles "Chic" Sale,** was born here on August 25, 1885, and lived for eleven years at 643 Illinois S.W., still standing but not open to the public. He was the author of *The Specialist* (1929), a humorous best seller dealing with the construction of outhouses.

KEYSTONE

Stewart Edward White lived in this Black Hills mining community and used the area as a setting for his novels *The Westerners* (1901) and *The Claim Jumpers* (1901). He had to leave town because local prospectors became enraged when they read unflattering stories about themselves in the novels.

MITCHELL

Badger Clark lived from 1885 to 1893 in the town of Mitchell, off Route 90 in eastern South Dakota, and attended Wooden West Side School. He went on to study at Dakota Wesleyan University, on South Sanborn and McCabe streets, which his father—a local minister—helped build and which, in 1923, awarded Badger Clark an honorary degree.

Josephine Herbst set her novel *Rope of Gold* (1939) in the area around Mitchell. The novel dealt with the Depression years and the Dust Bowl. The Friends of the Middle Border Museum, founded at Dakota Wesleyan University in 1939, contains manuscripts and galley proofs of authors connected with South Dakota, including L. Frank Baum, Hamlin Garland, Frederick Manfred, John G. Neihardt, and O. E. Rölvaag. The museum, at 1311 South Duff, is open to the public, and an admission fee is charged.

ORDWAY

Hamlin Garland stayed with his parents in 1881 and again in 1887 and 1889 on their homestead about two miles above this northern South Dakota town. Some of their neighbors and the town were described in *Main-Travelled Roads* (1891) and *A Son of the Middle Border* (1917). In 1883 Hamlin worked in his father's grocery store, thirty miles west of Ordway, while building a cabin on his own land claim, about six miles farther west of the store.

RAPID CITY

Badger Clark died on September 27, 1957, at Rapid City Hospital, in the largest city in the Black Hills.

SHADEHILL

Just west of Shadehill, in northwestern South Dakota, and half a mile south of the forks of the Grand River, is a monument to Hugh Glass, a nineteenth-century figure who is the subject of several literary works. John G[neisenau] Neihardt's epic poem "The Song of Hugh Glass" (1915) described how Glass, in 1823, after being mauled by a bear and abandoned by his companions, crawled several miles to Fort Kiowa, vowing he would track down the men who had left him to die. Glass did regain his strength and did hunt the men, only to forgive them when he finally caught up with them.

SIOUX FALLS

Herbert Krause, author of the novels *Wind Without Rain* (1939) and *The Thresher* (1945), lived in Sioux Falls and taught at Augustana College, on South Summit Avenue, from 1939 until his death on September 23, 1976.

YANKTON

Laura Ingalls Wilder mentioned the town of Yankton, in southeastern South Dakota, in an entry in her diary dated July 23, 1894. The diary was published as *On the Way Home* (1962):

> *We reached Yankton at 4 o'clock. Drove by the insane asylum. The buildings look nice and they stand in the middle of a large farm of acres and acres of corn and potatoes. Manly wanted to stop and go through the asylum but I could not bear to, so we did not. We passed by the Yankton College, the buildings are very nice.*

> *The frail shanty, cowering close, quivered in the wind like a frightened hare. The powdery snow appeared to drive directly through the solid boards, and each hour the mercury slowly sank. . . . This may be taken as a turning point in my career, for this experience permanently chilled my enthusiasm for pioneering the plain.*
>
> —Hamlin Garland,
> in *A Son of the Middle Border*

NEBRASKA

BANCROFT

This northern Nebraska community, at the southern edge of the Omaha Indian Reservation, was named in honor of a historian, George Bancroft, but its literary significance rests mainly on its association with the career of **John G. Neihardt.** Neihardt came to Bancroft about 1899 to work among the Indians and eventually became an authority on Indian life and culture. He was co-owner and editor of the Bancroft *Blade* from 1903 to 1905, before he began to write poetry and fiction. From 1908 until 1920, Neihardt lived in a house on Pennsylvania Street. During most of this period, he maintained a study in a one-story frame building at the northwest corner of Washington and Grove streets. There, in 1912, he began work on five epic poems relating the exploration and settlement of the West. They were finally collected in *A Cycle of the West* (1949). Neihardt was named poet laureate of Nebraska in 1921 and retained the honor until his death in 1973. His study, at Washington and Grove, stands near the grounds of the John G. Neihardt Center.

John G. Neihardt, Nebraska's poet laureate from 1921 to 1973.

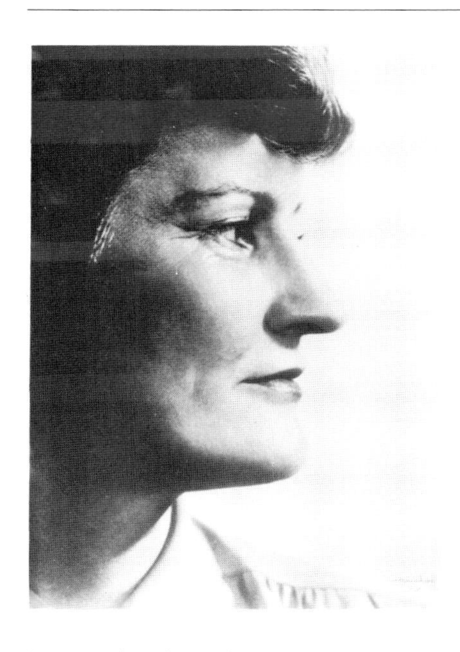

Mari Sandoz, the Nebraska novelist.

Completed in 1976, the center houses a large collection of materials relating to Neihardt, as well as a research library containing copies of Neihardt's books. Also on the grounds is the Sioux Prayer Garden. The center is open to the public.

CENTRAL CITY

Wright Morris was born on January 6, 1910, in this city, formerly called Lone Tree, in east-central Nebraska. Central City appears in Morris's novels as Lone Tree. Among his works with Nebraska settings are *The Home Place* (1948), with photographs and text by Morris, and the novel *Ceremony in Lone Tree* (1960).

ELLSWORTH

Mari Sandoz was born, grew up, and was buried near this northwestern Nebraska community. One account has it that Sandoz was born in 1901 in an unpainted shack, no longer standing, near the Sandoz Bridge, which spans the Niobrara River. Many of the places associated with her life here are gone now, but a sod house she lived in can be seen eighteen miles north of Ellsworth, off Route 27. Twenty-four miles north of Ellsworth, off Route 27, stands the home of Jules Sandoz, the novelist's father, whose efforts to make his living as a frontier farmer were recounted in *Old Jules* (1935). A few miles from her father's house is a schoolhouse in which she taught. Another Sandoz work about Nebraska is *Slogum House* (1935), a novel set in the late nineteenth century. The novelist died on March 10, 1966. Her grave, stands less than a mile west of her father's house and is accessible to visitors.

ELMWOOD

Bess Streeter Aldrich made her home from 1909 to 1946 in this community east of Lincoln, and it was here that she wrote novels of pioneer life, such as *The Rim of the Prairie* (1925), *Lantern in Her Hand* (1928), and *White Bird Flying* (1931). The rim of the prairie in the title of Aldrich's 1925 novel is a ridge of ground east of Elmwood. Aldrich's home is across from the town park, one block east of the main street, and is private. Within the park stands a marker honoring the novelist, and she is recalled also in the Bess Streeter Aldrich Library Foundation of Elmwood, created in 1978. Aldrich is buried in Elmwood Cemetery.

Fairview, William Jennings Bryan's home in Lincoln.

HASTINGS

Helen [Josephine] Ferris was born in Hastings, south of Grand Island, on November 19, 1890, and lived first at 315 North Saunders and later, until 1896, at 732 North Lexington, in a building no longer standing. Ferris wrote many books for children, including *When I Was a Girl* (1930).

LINCOLN

Lincoln has been home for many literary figures. **Bess Streeter Aldrich** lived in a house, still standing, at 1000 South Fifty-second Street here, from 1946 until her death on August 3, 1954. **William Jennings Bryan** lived at 1625 D Street from 1887 to 1902, and at Fairview, a red-brick mansion on Sumner Street, off Forty-eighth Street, from 1902 into the early 1920s. A statue of Bryan stands opposite Fairview, now a National Historic Landmark. The great orator founded his weekly, *The Commoner,* in Lincoln and served as its editor from 1901 until 1923.

Willa Cather studied at the preparatory school of the University of Nebraska in Lincoln during 1890–91 and enrolled at the university in the following school year. Her first published story appeared in the university magazine, *The Hesperian.* Early in her college career, Cather began writing for Lincoln newspapers, and had several years' experience by the time she was graduated, in 1895. Her stay at the university is commemorated in the Willa Cather Garden, located on campus near Love Library, and by a commemorative plaque embedded in the sidewalk nearby. During her stay in Lincoln, Cather lived in many places, her residence for the greatest length of time being a building, no longer standing, at 1029 L Street. In the 1890s she worked as drama critic, columnist, feature writer, and special correspondent for the Nebraska *State Journal,* now published as the Lincoln *Evening Journal.* A plaque in the lobby of the *Journal* office, at 926 P Street, commemorates the beginning of Cather's professional writing career with the newspaper. It was at the *State Journal* that Cather met Stephen Crane, during his brief stay in Lincoln in February 1895.

Helen Ferris, the writer and editor of children's books, lived in Lincoln from 1896 to 1912, and **Dorothy Canfield Fisher,** the novelist, lived here from 1891 to 1895. Her father, James Canfield, served during those years as chancellor of the University of Nebraska. **John G. Neihardt** lived at 5835 Vine Street here during his last years, before his death in 1973 in Columbia, Missouri, on a visit to his daughter's home. A bust of Neihardt, Nebraska's poet laureate from 1921 to 1973, is on view at the Hall of Fame in the State Capitol. Also on view there is a bust of **Mari**

Sandoz, who read proof for the Lincoln *Star* and the Nebraska *State Journal* while working on *Old Jules* (1935), the biography of her pioneering father. In 1934 and 1935, Sandoz was director of research for the Nebraska State Historical Society, at Fifteenth and R streets, and was associate editor of *Nebraska History Magazine.* During her time in Lincoln, she lived at Shurtleff Arms Apartments, 645 South Seventeenth Street.

Karl Shapiro taught at the University of Nebraska from 1956 to 1966 and served from 1956 to 1963 as editor of the literary magazine *The Prairie Schooner,* still published from offices on the university campus. During Shapiro's tenure as editor, this respected publication was known as *The Schooner.* Shapiro lived at 940 Eldon Drive and wrote a number of books here, including a volume of poetry, *Poems of a Jew* (1958), and a volume of criticism, *The Bourgeois Poet* (1964).

Writers born in Lincoln include **Mignon [Good] Eberhart,** the novelist, born on July 6, 1899, probably in a house on University Place. Her novels, the first of which was *The Patient in Room 18* (1929), often had nurses as central characters, and Sarah Keate was the detective-heroine in a number of them. **Ernest K. Gann,** who was born in Lincoln on October 13, 1910, has written a number of novels, including *Blaze of Noon* (1946) and *The High and the Mighty* (1952).

NORTH PLATTE

Just north of this city, near the confluence of the Platte and North Platte rivers, is the Buffalo Bill Ranch State Historic Park, site of Scout's Rest Ranch, once owned by **William F. (Buffalo Bill) Cody.** Cody wrote several books, including *The Life of Hon. William F. Cody* (1879) and *True Tales of the Plains* (1908), but he was more important as a subject of hundreds of dime novels. A less colorful figure, but of literary quality, is also associated with North Platte. Thomas Duncan wrote of the North Platte area in his novel about circus life, *Gus the Great* (1947).

OMAHA

Omaha writers include **Colin Clements,** born on February 25, 1894, and best known for *Harriet* (1943), a play about Harriet Beecher Stowe, which Clements wrote with his wife, **Florence Ryerson. Carl Jonas** was born here on May 22, 1913. His novels include *Jefferson Selleck* (1951) and *Riley McCullough* (1954). The Jonas family lived at 106 South Thirty-first Avenue when Jonas was born. He grew up in Omaha, attended the public schools and, after college and a year of travel, returned in 1938 to work for one year for the Omaha *World Herald.* He also lived in Omaha in his later years and taught at Omaha University. In 1975 he donated his home at 14323 Edith Marie Avenue to the Fontenelle Forest Nature Center, which uses it as the Jonas Interpretive Center, open to visitors.

Authors associated with Omaha include **William Jennings Bryan,** who served as editor of the Omaha *World Herald* from 1894 to 1896; and **Hamlin Garland,** who came here in 1891 while on a tour of the Midwest, gathering material for articles for a monthly magazine, *The Arena.* Garland's experiences here were reflected in his novel *A Spoil of Office* (1892). **Wright Morris** lived here during his childhood, shortly after World War I. **Gilbert Patten** visited Omaha in 1889 long enough to pick up some local color he used in a series of dime-novel Westerns he wrote under the pen name **William "Wyoming Will" West Wilder.** Later, as Burt L. Standish, Patten became widely known for his enormously successful series of Frank Merriwell novels.

PONCA

Virgil Geddes was born on May 14, 1897, on a farm near this community in the northeast corner of Nebraska. His works included the plays *The Earth Between* (1929), *The Stable and the Grove* (1930), and the trilogy *Native Ground* (1932), all set in the Midwest.

RED CLOUD

It takes only one author of stature to establish a town in the first ranks of literary America. **Willa Cather** spent her childhood in Red Cloud, after moving with her family from Virginia in 1883 to her grandfather's homestead near this small city. The contrast between the wooded, rolling land she knew in Virginia and the flat prairie that was to be her home impressed her deeply. She later wrote: "The land was open range and there was almost no fencing. As we drove further and further out into the country, I felt a good deal as if we had come to the end of everything—it was a kind of erasure of personality." But Cather in time came to understand, respect, and love the Midwest, using it again and again as the setting for some of her best stories and novels.

In 1884 the family moved to a house in Red Cloud, at 245 Cedar Street, which is now the site of the Willa Cather Historical Center, a National

Much usually of moment to Jules went unnoticed that summer and fall. The new ranches deep in the hills were infested with road agents and hide-outs. Shooting, knifing, and disappearances became common as sand lizards and wind. The Northwestern was held up, the mail sacks found slit and rifled in a blowout. The divorces of the community were offset by shotgun weddings. A neighbor on the south table hanged himself in his well curbing and dangled unnoticed for two days in plain sight of the road, so nearly did his head resemble a windlass swinging.

—Mari Sandoz, in *Old Jules*

Home of Willa Cather on Cedar Street in Red Cloud, now the Willa Cather Historical Center. On display are letters, first editions, and Cather memorabilia.

Grave of Anna Pavelka, model for Willa Cather's Ántonia Shimerda. Cather called the heroine of *My Ántonia* "a rich mine of life, like the founders of early races."

Historic Landmark and open to visitors. Cather described this house, which was her home until 1890, in her novel *The Song of the Lark* (1915) and in the stories "Old Mrs. Harris" and "The Best Years." In 1904, Cather's family moved to a house at Sixth and Seward, now private. Although the novelist never lived in this house, she often visited it to see her family. Red Cloud has various fictional names in Cather's work: Moonstone in *The Song of the Lark* (1915), Black Hawk in *My Ántonia* (1918), and Sweetwater in *A Lost Lady* (1923).

Red Cloud also supplied characters for Cather's works. Silas Garber, a Red Cloud resident and governor of Nebraska from 1874 to 1876, was the model for Captain Forrester in *A Lost Lady*. Garber, in 1889, built the Farmers and Merchants Bank, a brick and stone structure on Webster Street. The building is the home today of the Willa Cather Pioneer Memorial and Educational Foundation. The foundation's museum and archives contain a great deal of Cather material, and the foundation conducts tours of Red Cloud and of Webster County, noting in each tour the many sites that figure in Cather's books. Among the sites of interest are the gravesite of Anna Pavelka, the model for Ántonia Shimerda, the unforgettable Bohemian girl in *My Ántonia;* St. Juliana's Church, where Anna married John Pavelka, the model for Anton Rosicky in the story "Neighbour Rosicky," which was collected in the volume *Obscure Destinies* (1932); and the Burlington Railroad Depot, described in *My Ántonia* as well as in several Cather stories. Five miles south of Red Cloud, on Route 281, is the Willa Cather Memorial Prairie, 610 acres of grassland set aside in memory of the writer who has given us our most vivid account of early life on the Nebraska prairie.

Seen from a balloon, Moonstone would have looked like a Noah's Ark town set out in the sand and lightly shaded by grey-green tamarisks and cottonwoods. A few people were trying to make soft maples grow in their turfed lawns, but the fashion of planting incongruous trees from the North Atlantic States had not become general then, and the frail, brightly painted desert town was shaded by the light-reflecting, wind-loving trees of the desert, whose roots are always seeking water and whose leaves are always talking about it, making the sound of rain.

—Willa Cather,
in *The Song of the Lark*

TECUMSEH

Eugene Manlove Rhodes was born in Tecumseh, in the southeastern corner of the state, on January 19, 1869. His family homesteaded in Nebraska and Kansas before moving on to New Mexico in 1881. Rhodes worked as a cowboy for many years before writing the books for which he was called the novelist of the cattle kingdom. Among his works were *Good Men and True* (1910), *Bransford in Arcadia* (1914), and *The Proud Sheriff* (1935).

UNADILLA

Clyde Brion Davis was born on May 22, 1894, in Unadilla, a community southeast of Lincoln. He gained wide experience working on newspapers across the country and used this background in *The Great American Novel* (1938), perhaps Davis's best novel. He also wrote *Nebraska Coast* (1939), *Temper the Wind* (1948), and an autobiographical work, *The Age of Indiscretion* (1950).

WAYNE

John G. Neihardt moved with his family in 1891, when he was a boy, to Wayne, a city in northeast Nebraska. He attended Nebraska Normal College here and received a diploma in 1897, when he was sixteen. He then taught for two years in country schools and worked on his first book, *Divine Enchantment* (1900). A granite and bronze monument in the Wayne city park honors Neihardt, and a dormitory at Wayne State College here bears his name.

KANSAS

ATCHISON

E[dgar] W[atson] Howe, known as the Sage of Potato Hill, made his home in Atchison for many years, living in a two-story brick house at 1117 North Third Street. From 1877 to 1911, Howe edited and published the Atchison *Daily Globe* from a two-story brick building at 123 South Fifth Street. After relinquishing management of the *Globe* to his son, Howe concentrated his energies on *E. W. Howe's Monthly,* which he published until 1933. Author of twenty-eight books, Howe achieved his greatest success with his first work, *The Story of a Country Town* (1883). This melodramatic novel, dealing with the narrowness of Midwestern life, was rejected for publication by all the publishers he approached. Howe finally published the work himself from the *Globe* offices. It received critical praise from William Dean Howells and Mark Twain, and eventually went through more than fifty editions. Howe died at his home on October 3, 1937, and was buried in Atchison's Mount Vernon Cemetery.

CLIFTON

Robert [Menzies] McAlmon, an American expatriate poet and author published mainly in France, was born on March 6, 1896, in this community northwest of Manhattan. His several volumes of poetry included *The Portrait of a Generation* (1926), which established him as a spokesman for the "lost generation" of American writers. McAlmon was known best for his *Village: As It Happened Through a Fifteen Year Period* (1924). This series of sketches about a fictional, repressive town called Westworth has been compared with Sherwood Anderson's *Winesburg, Ohio.*

EL DORADO

William Allen White is most closely associated with the city of Emporia, about fifty-five miles to the northeast, but he grew up in El Dorado. White's parents moved the family here in 1869, the year after White's birth in Emporia. After studying at the College of Emporia and the University of Kansas, White returned here to run the El Dorado *Republican.* He also worked at several other newspapers before settling permanently in Emporia.

EMPORIA

William Allen White, known as the Sage of Emporia, was born on February 10, 1868, in this city in east central Kansas. Raised in El Dorado, White returned to study at the College of Emporia from 1884 to 1886 and, after further study and employment on several newspapers, borrowed $3,000 and bought the Emporia *Gazette* on June 1, 1895. The smartly dressed young editor contrasted with his rural neighbors not only in appearance but in political views: he was an outspoken opponent of the Populists and a supporter of Republicanism, "independent manhood," and the free-market system. On Saturday, August 13, 1896, White was surrounded on the street by a group of farmers who thought him either a fool or a tool of the wealthy. White stormed to the *Gazette* office at 517 Merchant Street and wrote the editorial "What's the Matter With Kansas?" This caustic attack on the Populists was widely reprinted, and White was transformed into a national figure. The recognition he received also helped the sales of *The Real Issue* (1896), the first of his many books.

His best-known editorials were "Mary White" (May 17, 1921), written after the death of his daughter, and "To an Anxious Friend" (July 22, 1922), which won a Pulitzer Prize. White received a second Pulitzer Prize for his posthumous *The Autobiography of William Allen White* (1946), which was completed by his son. During his career, White spent his mornings at the *Gazette* and worked during the rest of the day on articles and books at his home at 917 Exchange Street, a three-story house he bought in 1900, a few months before the birth of his son. White died at home on January 29, 1944, and was buried in the family plot at the Maplewood-Memorial Lawn Cemetery on Prairie Street. Numerous sites in Emporia have been named after him, including the William Allen White Library at Emporia State University, the William Allen White Memorial Drive, a civic auditorium, and an elementary school. A bust of White stands in Peter Pan Park. The land for the park was deeded to the city by the White family. On the monument is inscribed White's editorial tribute to his daughter.

White's son, **William L[indsay] White,** born in Emporia on June 17, 1900, had established himself as an award-winning author and journalist by the time of his father's death. His first book, *What People Said* (1938), was based on a 1933 Kansas bond scandal. His award-winning 1940 radio broadcast from the Mannerheim Line in Finland inspired Robert Sherwood to write the play *There Shall Be No Night* (1940). William L. White succeeded his father as editor of the *Gazette* in 1944. Among his best-known books was *They Were Expendable* (1942), the story of a U.S. boat squadron in the Pacific in World War II. White died on July 26, 1973, and was buried in Maplewood-Memorial Lawn Cemetery. His grave, like the graves of his father and sister, is marked by a marble composing stone once used by the *Gazette.* His widow lives in the family home at 927 Exchange Street, a National Historic Landmark, still private.

FORT SCOTT

Eugene Fitch Ware first came to live in Fort Scott, in southeastern Kansas, in 1867. His residence for many years, a two-story frame house, still stands at 202 South Eddy Street. Before a large fireplace in his study, Ware wrote articles for literary, historical, and legal journals of the day, as well as poetry published under the name **Ironquill.** Ware died on July 1, 1911, and was buried in the National Cemetery on East National Avenue. Among his best-known books was *The Rhymes of Ironquill* (1908).

GARNETT

Edgar Lee Masters was born on August 23, 1868, in this city in eastern Kansas. His birthplace at Second and Oak streets is no longer standing. His family moved to Illinois when Masters was one year old.

GIRARD

E[manuel] Haldeman-Julius was publisher of a socialist weekly in Girard, in the southeastern corner of Kansas, in 1919, when he created a series of inexpensive paperbound volumes known even today—in antiquarian bookstores—as the Little Blue Books. In thirty years the series sold more than 300 million copies, mostly by mail order. The list of 2,000 titles included self-help books, literary classics, and works of such contemporary writers as Will Durant, Bertrand Russell, and Upton Sinclair. Haldeman-Julius also wrote a number of books, including *The First Hundred Million*

What's the matter with Kansas?
We all know; yet here we are at it again. We have an old mossback Jacksonian who snorts and howls because there is a bathtub in the State House; we are running that old jay for Governor. We have another shabby, wild-eyed, rattle-brained fanatic who has said openly in a dozen speeches that "the rights of the user are paramount to the rights of the owner"; we are running him for Chief Justice, so that capital will come tumbling over itself to get into the state. We have raked the old ash heap of failure in the state and found an old human hoop skirt who has failed as a businessman, who has failed as an editor, who has failed as a preacher, and we are going to run him for Congressman-at-Large.

—William Allen White, in *"What's the Matter With Kansas?"*

Home of William Allen White in Emporia. White moved here about four years after gaining national attention with his editorial "What's the Matter With Kansas?" William Lindsay White also lived here.

The village of Holcomb stands on the high wheat plains of western Kansas, a lonesome area that other Kansans call "out there." Some seventy miles east of the Colorado border, the countryside, with its hard blue skies and desert-clear air, has an atmosphere that is rather more Far West than Middle West. The local accent is barbed with a prairie twang, a ranch-hand nasalness, and the men, many of them, wear narrow frontier trousers, Stetsons, and high-heeled boots with pointed toes. The land is flat, and the views are awesomely extensive; horses, herds of cattle, a white cluster of grain elevators rising as gracefully as Greek temples are visible long before a traveler reaches them.

—Truman Capote,
in *In Cold Blood*

(1928), an account of his success as a publisher. He died at his home here on July 31, 1951.

HOLCOMB

This town in southwestern Kansas entered American literary history in the aftermath of a gruesome crime. **Truman Capote** came here in November 1959 to investigate the murder of a farmer and his immediate family. Capote interviewed many people who had known the family and spent countless hours with the two men charged with the murders. His research provided the source material for his book *In Cold Blood* (1966), and Capote's novelistic approach to his true-to-life account made the work an enormous success.

INDEPENDENCE

William Inge, known best for his Pulitzer Prize-winning play *Picnic* (1953), was born here, in southeastern Kansas, on May 3, 1913, at 504 North Ninth Street, in a house that is still standing.

LEAVENWORTH

F. Scott Fitzgerald in November 1917 came to Fort Leavenworth, about three miles north of this city, to train as a U.S. Army officer. Fitzgerald's hopes of seeing combat in World War I were not realized, and he spent much of his time here working in the officer's club on a manuscript that was later published as *This Side of Paradise* (1920). In this, his first novel, Fitzgerald portrayed the life of the generation that came to maturity in the Jazz Age.

MANHATTAN

[Alfred] Damon Runyon was born in this city on October 4, 1880. (The marker in front of his birthplace at 400 Osage Street states that he was born in 1884.) His family name was Runyan, but in later years Runyon adopted the new spelling, which was the result of a typographic error. Runyon's father published the Manhattan *Enterprise.* The newspaper failed when Damon was two years old, and the family moved from Manhattan, eventually settling in 1887 in Pueblo, Colorado, where Runyon grew up.

Snapshot of Damon Runyon's birthplace in Manhattan. Runyon worked as a newspaperman before turning to writing short stories.

PITTSBURG

Harold Bell Wright served from 1898 to 1903 as pastor of the Forest Avenue Church in this city in the southeastern corner of Kansas. Here he wrote his first novel, *That Printer of Udell's* (1903), which incorporated much autobiographical material and included Wright's opinions and observations of the church.

SCRANTON

Frederic Wakeman, author of *The Hucksters* (1946), a satirical novel that attacked the advertising business, was born on December 26, 1909, in this city south of Topeka. His birthplace is on Route 56, opposite the Scranton Grade School, which Wakeman attended.

SEVERY

Walter Stanley Campbell was born on August 15, 1887, in this city east of Wichita. Campbell wrote biographies and poetry and, under the pen name **Stanley Vestal,** a number of historical works about the Old West.

TOPEKA

Writers associated with the state capital of Kansas include **Gwendolyn Brooks.** The poet and novelist, whose poetic narrative about a black woman's life, *Annie Allen* (1949), won a Pulitzer Prize, was born here on June 7, 1917. Her birthplace at 1311 North Kansas Avenue is no longer standing. It is believed that **Hal George Evarts** was born on August 24, 1887, in a house, no longer standing, at 213 West Third Street. Evarts wrote stories set in the West. Among his best-known works were *The Passing of the Old West* (1921) and *Tumbleweeds* (1923).

Rex Stout, creator of detective Nero Wolfe, lived in Topeka as a boy. In 1896 his family moved to Bellview, then three miles from the center of Topeka. In 1899 the family moved into Topeka, to a house on Quincy Street. In 1902 the Stouts moved to 900 Madison, and a few years later to 423 East Eighth Street. Stout worked at Crawford's Opera House after completing high school in 1903.

WAKARUSA

Rex Stout lived from 1888 to 1896 with his family on a forty-acre farm just outside this community south of Topeka.

WICHITA

Authors associated with Wichita include **Charles B. Driscoll,** born on October 19, 1885, on a farm just to the south of the city. Driscoll attended school in Wichita, and in 1912 was graduated from Friends University, located at 2100 University Avenue. He worked for a time as newsboy for the Wichita *Beacon,* which later published his syndicated column "The World and All."

William Inge, the playwright, worked as a news announcer for radio station KFH in the 1930s. KFH now operates from the seventh floor of a building at 209 East William. **Paul Wellman,** author of works of history and historical novels, attended high school in Wichita and was graduated in 1918 from Fairmount College, now the University of Wichita. Wellman worked as a newspaperman in Wichita for many years, on the *Beacon* and on the *Evening Eagle.*

OKLAHOMA

CLAREMORE

Lynn Riggs, best known for his play *Green Grow the Lilacs* (1931), which later was transformed into the world-famous musical comedy *Oklahoma!* (1943), by Richard Rodgers and Oscar Hammerstein, was born on a farm near Claremore on August 31, 1899. At that time this region was in Indian Territory. Riggs, educated in the public schools of Claremore, a town northeast of Tulsa, captured the vitality of Oklahoma life and speech in his plays, especially in *The Cherokee Night* (1936) and *Russet Mantle* (1936). *The Cherokee Night* dealt with the defeat of the Cherokee spirit resulting from occupation of Indian land by U.S. forces. The Thunderbird Library in Claremore maintains a Lynn Riggs Collection.

Will[iam Penn Adair] Rogers was born on a ranch near Oologah, which is about fifteen miles north of Claremore, on November 4, 1879. On July 4, 1899, he entered a steer-roping contest in Claremore and won first prize. A lariat became a standard prop in the thousands of public appearances Rogers made from then on. Just west of Route 66 here is Will Rogers Memorial Building, financed by the state in 1937. The rambling ranch-style building has four main foyers built around a rotunda that displays a statue of the cowboy humorist. Saddles and trappings from all over the world are displayed in the east foyer. Rogers's personal effects, including a battered typewriter and his Russian government credentials, can be found in the north and west sections. In the south foyer are thirteen dioramas depicting Rogers' life. Rogers, who had died in 1935, was reinterred in a crypt in the building on May 22, 1944, and his wife and son were buried next to him. As one of Rogers's friends has put it, "I feel his spirit in the wind as it hits my face." Another visitor to the Memorial was overheard saying, "Claremore and Will Rogers, they belong together, part of the prairie; and no matter how far a man goes; at the end he likes to come back to his roots." While Will Rogers wrote several books, he is remembered best for his witty one-liners and for his part in the apotheosis of the American Cowboy. Well after his death, a selection of his writings was published as *The Autobiography of Will Rogers.* The editor was Donald Day.

ENID

Marquis James came to Enid in 1901, when he was ten, and here attended school and learned the printing trade. When he was fourteen, James wrote for the Enid weekly *Events*—one other distinguished writer, **Russel Crouse,** also wrote for *Events* in his day. After high school, James attended Oklahoma Christian University, now Phillips University, in Enid, but he left school for what he thought was the more enjoyable life of a wandering newspaperman.

Paul I[selin] Wellman, the historian and author of historical romances, was born in Enid on October 14, 1898, but moved with his family to Portuguese West Africa when he was only six

Statue of Will Rogers in the foyer of the Will Rogers Memorial, in Claremore. The base of the statue bears the humorist's best-known saying: "I never met a man I didn't like." (*below left*) View of the Rogers Memorial.

I am an invisible man. No, I am not a spook like those who haunted Edgar Allan Poe; nor am I one of your Hollywood-movie ectoplasms. I am a man of substance, of flesh and bone, fiber and liquids—and I might even be said to possess a mind.

—Ralph Ellison,
the opening sentences of
Invisible Man

Ralph Ellison, author of *Invisible Man.*

months old. Wellman's histories included *Death on the Prairie* (1934) and *Death in the Desert* (1935). His novels included *The Walls of Jericho* (1947) and *The Comancheros* (1952).

LAWTON

John Allyn Smith, the father of **John Berryman,** sometimes brought his young son to National Guard maneuvers at Fort Sill, near this city. The boy's family lived in McAlester at the time. **Carl Carmer** served, at the end of World War I, as a first lieutenant and instructor at Fort Sill.

MCALESTER

John Berryman was born on October 25, 1914, in this eastern Oklahoma city on Route 1. He had already become widely known for his long poem *Homage to Mistress Bradstreet* (1956), when he won a Pulitzer Prize for *77 Dream Songs* (1964). Berryman, whose original name was **John Allyn Smith,** was twelve when his father shot himself. He was legally adopted when he was a teenager and took his adoptive family's surname. Berryman wrote later about his father's death, and in 1972 the poet himself died a suicide.

NORMAN

Lynn Riggs entered the University of Oklahoma at Norman, south of Oklahoma City, in 1920. In the summer of 1922 he wrote his first play, *Cuckoo,* but it was never published.

OKLAHOMA CITY

Ralph [Waldo] Ellison, author of the widely admired novel *Invisible Man* (1952), was born on March 1, 1914, at 407 Northeast First Street, in a house no longer standing. His father died when Ellison was three, leaving Ellison's mother to support the family by working as a domestic. Ellison attended Douglass Elementary and High School, worked as a newsboy on the corner of Main Street and Broadway, and shined shoes at Grand Street and Broadway and on Robinson Street between Main and Grand streets. These were just the first of a long series of jobs for Ellison. For free meals and tips, he worked as a bread-and-butter boy at the Oklahoma Club. He also worked as a waiter at the Skirvin Hotel, Huckins Hotel, Chamber of Commerce, University Club, and Oklahoma City Golf and Country Club at Nichols Hills. He worked at the Randolph Drug Store, still operating at 331 North East Second Street; he washed cars at a gasoline station on Classen Boulevard; and operated an elevator at the Lewinsohn Clothing Store on West Main Street. Withal, Ellison managed to spend considerable time at the Negro Library in the Slaughter Building, on Second Street N.E., and it was this effort that enabled him to make his way at age twenty-one to Tuskegee Institute, in Alabama. Ellison is remembered in Oklahoma City through the Ralph Ellison Branch Library, at 2000 North East Twenty-third Street, where a stainless-steel and bronze bust of the novelist is displayed.

Edna Ferber is believed to have conducted research for her novel *Cimarron* (1930) at the State Historical Library here. **Tom Heggen** and his family moved in 1935 from Iowa to a bungalow at 1014 North West Twenty-second Street. Heggen, then sixteen, completed his secondary education at Classen High School and entered Oklahoma City University. While at the university, he worked on the weekly newspaper, *The Campus,* writing a "Pi-Lines" column dealing with campus personalities. His "Like a Dry Twig," in the unofficial university magazine, *Gold Brick,* was his first published fiction. Heggen's best-known work was *Mister Roberts* (1946), a novel about the crew of a Navy cargo ship in the Pacific during World War II.

PERKINS

There is said to be a marker placed at a country school near here that commemorates the visit of **Washington Irving** and his hunting party in October 1832. About a mile and a half northwest of the school, in an alfalfa field, is the actual place of Irving's campsite, where he spent a single night. The literary detective should first locate Wild Horse Creek, next to which camp was made. Perkins is fifteen miles south of Stillwater on Route 33, next to the Cimarron River.

Our encampment was in a spacious grove of lofty oaks and walnuts, free from underwood, on the border of a brook. . . . [It was] a good neighborhood for game, as the reports of rifles in various directions speedily gave notice. One of our hunters soon returned with the meat of a doe, tied up in the skin, and slung across his shoulders. Another brought a fat buck across his horse. Two other deer were brought in, and a number of turkeys. All the game was thrown down in front of the Captain's fire, to be portioned out among the various messes. . . .

—Washington Irving,
writing about the Camp of the Wild Horse, in *A Tour on the Prairies.*

SEMINOLE

Burton Rascoe came to Seminole after graduation from high school in nearby Shawnee. He left soon to attend the University of Chicago for two years and then worked in Chicago for the *Tribune.* Fired for making an irreverent remark about Mary Baker Eddy, Rascoe returned to Seminole to his father's cotton plantation, about a mile down old Route 99. The town librarian can direct visitors to the site of the house, which has since been moved. Local historians are trying to determine if Rascoe's home still exists. The family farm was adjacent to the property of one Billy Bowlegs, a descendant of a Seminole chief. Bowlegs was shot one night, his wife was beaten and raped, and his son was savagely beaten. Mrs. Bowlegs was able to crawl toward the Rascoe farm, where she died in the arms of Rascoe's mother. The Bowlegs's son gave testimony at the trial that followed, and the murderers were hanged.

SHAWNEE

While **Burton Rascoe** was in Shawnee, in central Oklahoma, probably about 1910, he lived at 244 South Oklahoma Avenue and worked for the Shawnee *Herald,* at 129 North Broadway, no longer publishing. Meanwhile, the Rascoe family bought a farm that yielded oil and made his father wealthy overnight. At one time, Rascoe was an assistant at the Carnegie Library here, and he earned extra money as a ghostwriter for ladies of the Hawthorne Club and the Round Table Club who enjoyed reading—but not writing—papers.

TULSA

Although there is a Sidney Lanier School at 1727 South Harvard in Tulsa, Lanier is not known to have lived here. William A. Owens, who has never lived here either, set his novel *Walking on Borrowed Land* (1954) in Tulsa, referring to Pleasant Valley and Happy Hollow, which probably are in the Greenwood section of Tulsa. **Lynn Riggs,** who did live in Tulsa for a time, worked as a reporter for the Tulsa *Oil and Gas Journal,* still publishing at 1421 South Sheridan Road.

WATONGA

Edna Ferber is reported to have stayed at the Thompson-Benton-Ferguson House, 521 North Weigel Street, in Watonga, northwest of Oklahoma City, while she was conducting research for *Cimarron* (1930). Now called the Ferguson Museum, the building is open to the public. The character Sabra in *Cimarron* was based on Mrs. Ferguson. Ferber, on the same trip, visited the Watonga Republican Building, 123 East Main Street, a building that still looks the same as it did in the 1920s but now houses an insurance and real-estate company.

> *Down the street toward her came a galloping cowboy in sombrero and chaps and six-shooters. Sabra was used to such as he. . . . She realized, in a flash of pure terror, that he was making straight for her. She stood, petrified. He . . . kissed her full on the lips, released her, leaped on his horse, and was off with a bloodcurdling yelp and a clatter and a whirl of dust.*
>
> —Edna Ferber, in *Cimarron*

TEXAS

ALICE

J[ames] Frank Dobie, the author of books on Southwestern history and folklore, attended high school in Alice, a city forty miles west of Corpus Christi. His family's ranch was situated about forty miles north of Alice, in Live Oak County.

ALPINE

J. Frank Dobie served as principal of the Alpine Public School in 1910–11 and is thought to have lived across from the school, in a boarding house at 601 North Seventh Street. The house still stands but the second floor, on which Dobie lived, was destroyed by fire. The old school building was eventually torn down and replaced at the same site by the present Alpine Elementary School, 200 West Avenue A. While living in Alpine, Dobie made the acquaintance of John Young, a real-estate dealer who had led a colorful life in the Southwest. Dobie wrote of Young in *A Vaquero of the Brush Country* (1929). Dobie described his Alpine days in *Some Part of Myself* (1967).

H. Allen Smith, the humorist, came to Alpine in 1967 to live in a Spanish-style house he helped design, at 1108 Loop Road, where he and his wife remained until 1976. Smith died in San Francisco on February 24, 1976, but his ashes were brought to his home in Alpine and scattered in the cactus garden on the east hillside. The house still stands and can be visited if telephone arrangements are made with the present owners. Several of Smith's thirty-seven books were written here, including *The View from Chivo* (1971), *Low Man Rides Again* (1973), *Return of the Virginian* (1974), and his last one, *The Life and Legend of Gene Fowler* (1977). Smith did most of his writing at home, but he also wrote and did considerable research at the Bryan Wildenthal Memorial Library, Sul Ross State University, in Alpine; at Terlingua; and at Big Bend National Park, which is near Terlingua. Most of Smith's literary effects are kept at the University of Southern Illinois, but his private library and some papers are at Sul Ross State University.

AUSTIN

William Sydney Porter, known as **O. Henry,** lived for many years in Austin, the capital of Texas. His stay here bridged—sometimes sadly—the years in which Porter began to emerge as a man of letters. In 1884, when Porter was twenty-two, he came to Austin to visit Richard Hall, a man he had worked for in Cotulla, Texas. Porter stayed on to take a job for a few months at Harrell's Drug Store, in the Driskill Hotel, 117 East Seventh Street. From 1884 until 1887, Porter clerked at Morley Brothers Drug Store, now the Grove Drug Store, at 209 East Sixth Street, and kept books for Maddox Brothers and Anderson, a real-estate firm. The bookkeeping job is generally considered Porter's first real employment. He stayed with Maddox for two years, earning $100 a month and acquiring skills that he would later find useful and dangerously tempting. When his friend Richard Hall was elected land commissioner of Texas, Porter learned drafting in three months and went to work at the Land Office, Eleventh and Brazos streets, which now is a museum for the Daughters of the Confederacy and the Daughters of the [Texas] Republic. Porter worked at the Land Office until 1891, when an election was won by a political opponent of Hall's. Porter then took a job as a teller—a fateful appointment—at the First National Bank, Sixth and Congress streets. The bank building now is occupied by a five-and-ten-cent store.

Porter was married on July 1, 1887, at the home of Reverend R. K. Smoot, 1316 West Sixth Street. Porter had courted Athol Estes at the James Baird Smith House, now a dentist's office, at 502 West Thirteenth Street, where she often visited

> *My pupils varied. Shy, bold, genial, generous-natured, they were for the most part as intelligent as university professors. I asked Georgia if she had "gone over" an assigned book. "Twice," she replied. Then, too honest to deceive, she, laughing and blushing, explained that she had "gone over" the book by stepping over it.*
>
> —J. Frank Dobie, in *Some Part of Myself*

As Bob Buckley, according to the mad code of bravery that his sensitive conscience imposed upon his cowardly nerves, abandoned his guns and closed in upon his enemy, the old, inevitable nausea of abject fear wrung him. . . . The hot June day turned to moist November. And still he advanced. . . .

The distance between the two men slowly lessened. The Mexican stood, immovable, waiting.

—O. Henry,
in "An Afternoon Miracle"

and where Porter once carved her name in a limestone window ledge. The carving now is on view at the O. Henry Museum, 409 East Fifth Street. In 1893 Porter, his wife, and their daughter moved into a cottage located at 308 East Fourth Street and lived there for three years. The cottage eventually was moved to 409 East Fifth Street, where since 1934 it has been the O. Henry Museum, open daily. In 1894 Porter, eager to get away from his dull job at the bank, became a publisher in his spare time. For $250 he bought a monthly magazine called *Iconoclast* and its press, and began to publish a weekly humor magazine he named *The Rolling Stone.* Porter managed to keep his weekly alive from April 14, 1894, to April 27, 1895, publishing at Tenth and Congress streets, now the site of a Trailway Bus Station. It was in the pages of *The Rolling Stone* that Porter first tested his skills as a writer—among other things, he conducted a readers' question-and-answer column, for which he wrote the questions as well as the answers—and from this time on he never doubted that writing was the most suitable profession for him. Porter's year as a publisher is recognized by a plaque at the bus station.

Unfortunately, *The Rolling Stone* proved a financial failure, and Porter was fired from the bank because of shortages in his accounts. Porter went to Houston and worked for eight months for the Houston *Post,* but on February 10, 1896, was indicted for embezzlement of more than $6,000 on the complaint of the First National Bank. Porter sent his daughter and his wife, who was now dying of tuberculosis, to live with his wife's mother in Austin, and he fled to New Orleans, a port of embarkation for Honduras. In the following year, at the urging of his wife, Porter returned to Austin from Honduras, where he had gone to avoid prosecution. Athol Porter died on July 25, 1897, and Porter went on trial in February 1898. Convicted of the federal offense of embezzling funds from a national bank, Porter was sent to the penitentiary in Columbus, Ohio, to begin serving a sentence of five years. Porter had sold his first story to a national publication in December 1897, when the S. S. McClure Company bought "The Miracle of Lava Canyon," later rewritten as "An Afternoon Miracle," but the story did not appear in print until after Porter went to the penitentiary. Devoting himself to writing while in prison, Porter was able to sell many of his stories. He began to call himself O. Henry, and it was under this name, of course, that he became known to the literary world.

Photograph of O. Henry working as a teller, about 1892, in the First National Bank in Austin. It was for embezzling funds from this bank that O. Henry was convicted and sent to jail.

It was in Austin, in 1891, that **William Cowper Brann** first published *Iconoclast,* the monthly magazine O. Henry later bought, but *Iconoclast* did not achieve its lasting reputation until Brann reestablished it in Waco, Texas.

Next to O. Henry, the most important author associated with Austin is **J. Frank Dobie,** who lived at 702 Park Place, in a building that is still standing. Dobie liked to socialize at the Town and Gown Club, which still meets in the Headliner Club in Austin. He wrote a number of books about the history of Texas and the folklore of the Southwest, and also taught at the University of Texas, in Austin, from 1914 to 1947. Dobie's books included *A Vaquero of the Brush Country* (1929), *On the Open Range* (1931), *Guide to Life and Literature of the Southwest* (1943), *Tales of Old-Time Texas* (1955), and *Cow People* (1964). Dobie died at his home on September 18, 1964, and funeral services were held for him in the Hogg Memorial Auditorium of the University of Texas. He was buried in the Texas State Cemetery, 901 Navasota Street, and his gravestone bears the inscription "Story Teller of the Southwest."

The University of Texas Library has manuscripts, letters, and papers of many American authors. The collection of the Humanities Research Center includes papers of E. E. Cummings, T. S. Eliot, William Faulkner, William Goyen, Ernest Hemingway, Tom Lea, H. L. Mencken, Marianne Moore, and Ellery Queen—who is, of course, Manfred B. Lee and Frederic Dannay. Dannay, who was once a visiting professor at the university, donated his collection of detective stories to the Humanities Research Center. The Lamar Library at the university has collections of works by Eugene O'Neill, Edgar Allan Poe, and Ezra Pound, among others.

BLOSSOM

William A. Owens was born on November 2, 1905, in Pin Hook, a hamlet that once stood near Blossom, which is about fifteen miles east of Paris. Owens frequently uses Texas locales in his novels: *Fever in the Earth* (1958) has scenes from Beaumont and neighboring areas, and *Look to the River* (1963) and *This Stubborn Soil* (1966) both portray Blossom.

CHILDRESS

Vaida Montgomery, the poet and editor, was born in Childress on August 28, 1888. With her husband, **Whitney Maxwell Montgomery,** she founded the poetry magazine *The Kaleidoscope* in 1929. Childress is on Route 287 in the Texas Panhandle.

CLARKSVILLE

William Humphrey was born on June 18, 1924, in Clarksville, on Route 82 west of Texarkana. His first book was *The Last Husband and Other Stories* (1953); his first novel, *Home from the Hill* (1958), became a best seller.

COLLEGE STATION

William A. Owens was on the faculty of Texas A&M University here from 1937 to 1947. At first he lived at 314 Ayrshire, later at the YMCA. Owens became a member of the faculty of Columbia University in 1947, but served as writer in residence at A&M in 1976. The university library maintains special collections of the papers of Owens and J. Frank Dobie.

COTULLA

William Sydney Porter [O. Henry] first came to Texas in 1882, when he was twenty years old. Porter's mother and grandmother had died of tuberculosis, and he was advised to go to the arid country of Texas to clear up his persistent cough. Porter spent two years as a cowboy, working for Richard Hall at the Dull Brothers Ranch east of Cotulla, and regained his health. Every few days he would ride to Fort Ewell, fourteen miles away, to post his letters, and about once a month he and the other cowboys would ride into Cotulla, spending the night at the Boutrell Home, down by the tracks of the International and Great Northern Railway. The original buildings of the ranch no longer exist, and there are no plaques or markers.

DALLAS

Grace Noll Crowell, whose collections of poetry included *White Fire* (1925) and *Flames in the Wind* (1930), lived in Dallas. She died here on March 31, 1969, and was buried in Sparkman-Hillcrest Memorial Park. **Whitney Maxwell Montgomery,** poet and, with his wife Vaida Montgomery, co-founder of *The Kaleidoscope,* later *The Kaleidograph,* a national poetry magazine, lived at 624 North Vernon Avenue. He is buried in Laurel Land Memorial Park in Dallas. **William A. Owens** worked as a clerk at Butler Brothers Department Store and lived in a rented room at 4914 Santa Fe Avenue. The house has since been torn down. In 1931–32 Owens, then a student at Southern Methodist University, lived at 5807 Reiger Avenue, which is still standing.

EL PASO

Glen [Ward] Dresbach, the poet, once worked as an auditor and credit manager for the Peyton Packing Company in El Paso. **Tom Lea** was born at Hotel Dieu Hospital on July 11, 1907. During his childhood Lea lived at 1400 East Nevada Street. He was graduated from the Lamar School in 1920 and from the El Paso High School in 1924. Much later, while living at 1520 Raynolds Boulevard, he wrote the novels *The Brave Bulls* (1949), which was very popular, and *The Wonderful Country* (1952). In the latter work, El Paso was called Puerto. Lea is also known for his two-volume history *The King Ranch* (1957) and for his paintings, which are widely exhibited. He illustrates his own books.

EUREKA

Whitney Montgomery was born in Eureka on September 14, 1877. Eureka lies south of Corsicana, on Route 287. Collections of Montgomery's poetry included *Corn Silk and Cotton Blossoms* (1928) and *Brown Fields and Bright Lights* (1930).

GALVESTON

Jean Laffite and his band of pirates and smugglers, following their departure from Louisiana waters, came to Galveston Bay in about 1817 to resume their trade. Looked upon as a hero, Laffite quickly became a national legend and fair game for writers of dime novels and popular romances. Joseph Holt Ingraham seems to have scored first with his romance *Lafitte; or, The Pirate of the Gulf* (1836). Hervey Allen kept the Laffite tradition alive almost one hundred years later, in 1933, when he included the pirate in *Anthony Adverse.*

In the 1880s **William Cowper Brann** worked in Galveston on the *Evening Tribune,* before moving on to Houston and then to Austin, where he initially established his magazine *Iconoclast.* **Stephen Crane** stayed for several days in March 1895 at the Hotel Tremont, at Tremont and Church streets. The hotel was razed in 1928, and on the site now stands the W. L. Moody and Company Bank, which was built there in 1952.

William A. Owens, the novelist. Owens lived near Blossom much of the time until early manhood.

Home of Tom Lea on East Nevada Street in El Paso.

> *James was this mysterious, wandering boy. . . . He was a wild country boy brought to live in the city of Houston when his parents moved there from a little town down the road south. He said he wanted to be a cowboy, but it was too late for that; still, he wore boots and spurs. He hated the city, the schools, played away almost daily.*
>
> —William Goyen, in "The Faces of Blood Kindred"

Crane reported that Galveston resembled a New England town more than a Texas one.

Laura Krey was born in Galveston on December 18, 1890, and grew up on a cotton plantation on the nearby Brazos River. Many years later, she wrote about the real life of a cotton plantation in her novel *And Tell of Time* (1938), which unfavorably depicted the Northerners who came to the South during Reconstruction. Krey wrote about Spanish Texas and the days of the Texas Republic in *On the Long Tide* (1940). As a child Krey used to visit a cousin in Galveston at 2709 Broadway, no longer standing. The Rosenberg Library, 2310 Sealy Avenue, has a collection of Ellery Queen books, papers, and manuscripts.

HOUSTON

Donald Barthelme, the fiction writer, grew up in Houston, at 819 Harold Avenue. His father was a professor of architecture at the University of Houston. **William Cowper Brann** worked in the 1880s as an editorial writer for the Houston *Post.* **Allen Drury,** the author of the Pulitzer Prize-winning novel *Advise and Consent* (1959), was born in Houston on September 2, 1918. **Edna Ferber,** while gathering material for *Giant* (1952), stayed at the old Rice Hotel, 518 Main Street, at Texas Avenue. The hotel has since been converted into an apartment building.

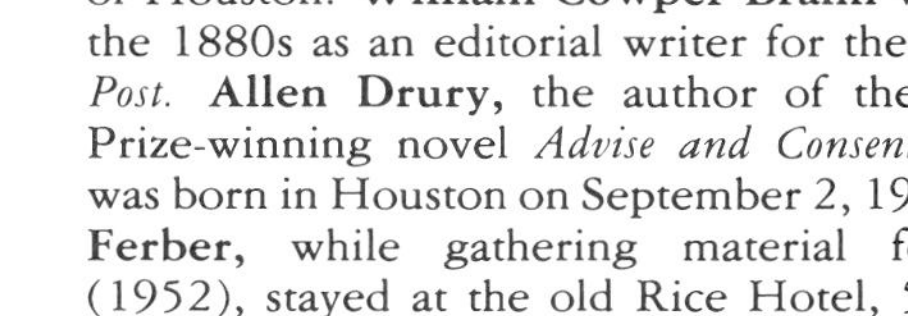

William Goyen lived from 1923 to 1940 at 614 Merrill Street. Here he wrote the short story "The White Rooster." Merrill Street itself was depicted in "Zamour, or A Tale of Inheritance," which appeared in his collection of stories *The Faces of Blood Kindred* (1960), and Houston, Texas, was called Rose, Texas, in his novel *Come, the Restorer* (1974). Goyen attended Houston public schools from the third grade, and was graduated in 1932 from Sam Houston High School, 9400 Irvington Avenue. He attended Rice University here, receiving a B.A. in 1936 and an M.A. in 1939. There is now a collection of his papers and manuscripts at the Fondren Library, Woodson Research Center.

Albert Joseph Guérard was born at 5218 Austin Street on November 2, 1914. Guérard has written a number of important critical studies, including *Joseph Conrad* (1947) and *André Gide* (1951). His novels include *The Past Must Alter* (1937), *The Hunted* (1944), *The Bystander* (1958), and *The Exiles* (1963). **William Sydney Porter [O. Henry]** was a columnist for the Houston *Post* for eight months in 1895–96, before he was indicted in Austin for bank embezzlement. **Dixon Wecter,** author of the novel *When Johnny Comes Marching Home* (1944), was born in Houston on January 12, 1906.

INDIAN CREEK

Katherine Anne Porter, one of our finest short-story writers, was born in Indian Creek, in central Texas south of Brownwood, on May 15, 1890. When she was three, her family moved to Kyle. It is interesting to note that her father was a first cousin of William Sydney Porter.

KINGSVILLE

King Ranch, near Kingsville, about thirty-five miles southwest of Corpus Christi, is not usually thought of as a literary site, but **Richard Harding Davis,** the journalist, did visit the ranch in 1892, and **Edna Ferber,** conducting research for her novel *Giant* (1952), also stayed at the ranch.

KYLE

Katherine Anne Porter, in 1892, a year after her mother's death, was taken to live at her grandmother Porter's home at 508 West Center Street in Kyle, about twenty miles southwest of Austin. She spent the next nine years there, along with her father and her sister. Aunt Cat, Katherine's grandmother, was recalled in a 1944 reminiscence of Porter's entitled "Portrait: Old South," which appears in *The Collected Essays and Occasional Writings of Katherine Anne Porter* (1970):

> *She believed it was her duty to be a stern methodical disciplinarian, and made a point of training us as she had been trained even to forbidding us to cross our knees, or to touch the back of our chair when we sat, or to speak until we were spoken to: love's labors lost utterly, for she had brought up a houseful of the worst spoiled children in seven counties, and started in again hopefully with a long series of motherless grandchildren—for the daughters of that after-war generation did not survive so well as their mothers, they died young in great numbers, leaving young husbands and children—who were to be the worst spoiled of any.*

McALLEN

Thomas Duncan lived in 1959 and 1960 in McAllen, a city west of Brownsville, while revising his novel *Virgo Descending* (1961).

MATAGORDA

Charles A. Siringo was born on February 17, 1855, near Matagorda, which is on the Gulf

Home of Katherine Anne Porter in Kyle.

coast. Siringo was known best for *A Texas Cowboy, or Fifteen Years on the Hurricane Deck of a Spanish Pony* (1885).

RIO HONDO

Nelson Algren, who once was stranded at an abandoned gas station in Rio Hondo, north of Brownsville, wrote of his experience in his first published short story, "So Help Me" (1933).

SAN ANTONIO

One of the great American shrines and a lodestone for romantic writers is the Alamo, once a mission, in downtown San Antonio. As Stephen Crane once wrote, "Statistics show that 69,710 writers of the state of Texas have begun at The Alamo." On February 23, 1836, a Mexican force of four thousand troops, commanded by Gen. Antonio Lopez de Santa Anna, demanded surrender of the mission building, which had been made into a fort by its defenders. The Texans, numbering fewer than two hundred, refused, but the Alamo fell on March 6. Among the novels written about the battle are *Inez, A Tale of the Alamo* (1855), by Augusta Evans Wilson; *Remember the Alamo* (1888) by Amelia E[dith] Barr, an Englishwoman who lived in Texas for a few years after the days of the Republic; and *The Wine of San Lorenzo* (1945), by Herbert Gorman.

Stephen Crane, on tour after his success with *The Red Badge of Courage,* visited San Antonio in mid-March 1895. Crane stayed at the Mahncke Hotel, on the northeast corner of Houston and St. Mary streets. The Mahncke was replaced by the Gunter Hotel, still open at 205 East Houston Street. **H. L. Davis,** the poet and novelist, died at Nix Memorial Hospital here on October 31, 1960.

Sidney Lanier was suffering from tuberculosis when he came to San Antonio, staying from November 1872 to April 1873. Here he decided to concentrate his career on music and poetry instead of the law. There is a Sidney Lanier High School in San Antonio, and the Menger Hotel has Lanier tablets and exhibits. In 1894–95, **William Sydney Porter** was a frequent visitor to San Antonio, and often stopped at a small house at 827 South Presa Street, later renumbered 903 South Presa Street. Porter's association with the house is not clear, but it is known as the O. Henry House. In 1960 it was moved to 520 Lone Star Boulevard and is open daily, admission free. San Antonio places often appeared in O. Henry's stories. In "A Fog in Santone," he mentioned the Johnson Street bridge, which crosses the San Antonio River and is now named the O. Henry Bridge. He referred to the Menger Bar and Hotel, still open at 204 Alamo Plaza, in many of his stories. "Seats of the Haughty," "Hygeia at the Solito," "The Higher Abdication," "The Missing Chord," and "The Enchanted Kiss" all have San Antonio locales.

Katherine Anne Porter, who once said she spent her childhood in San Antonio, did live near enough to the city to visit it frequently. She also spent two years at Our Lady of the Lake Convent in the city, and was a day student at the Thomas School for Young Ladies on West End, now Woodlawn Lake, which must have been near her home. She often picnicked at The Alamo, still a ruin in her time. When she was thirteen, Katherine Anne Porter moved with her family to New Orleans.

TRINITY

William Goyen, born in Trinity on April 24, 1915, lived here through his second year of schooling. His childhood home was the setting for *The House of Breath* (1950), his novel about a boy growing up in a small town in Texas. Trinity is about twenty miles northeast of Huntsville, and near the Trinity River.

WACO

William Cowper Brann in Waco reestablished his previously unsuccessful magazine *Iconoclast* in 1894, this time calling it *Brann's Iconoclast.* The monthly soon developed a circulation of nearly 50,000, making it one of the most popular magazines then published. *Brann's Iconoclast* was anything but moderate in its editorial content and tone, idolizing womanhood, attacking blacks, and questioning religion. When Brann learned that a Brazilian girl training at Waco's Baylor University to become a missionary was pregnant, he began to attack the university in print. His assaults on the university lasted more than a year, and tension mounted. On April 1, 1898, a man named T. E. Davis, whose daughter was a student at Baylor, shot Brann in the back as they passed on the street. Although wounded, Brann turned and shot Davis. Both men died of their wounds. Brann's funeral procession was two miles long, and two bands played. Brann was buried in Oakwood Cemetery, where his marble monument originally bore a lamp inscribed with the word "Truth" and Brann's likeness but, within a few weeks, someone had shot away a piece of the marble profile. *Brann's Iconoclast* lasted for some years after its owner's death, but it had already passed its heyday.

Madison Cooper wrote his two-volume novel *Sironia, Texas* (1952), which described life in small Texan towns, while living at 1801 Austin Avenue. **C[harles] Wright Mills,** a popular author and sociologist known widely for *White Collar* (1951) and *The Power Elite* (1956), was born at 1409 Barron Street on August 28, 1916.

Frances Winwar's original copy of her biography *The Immortal Lovers: Elizabeth Barrett and Robert Browning* (1950) is in the cornerstone of the Armstrong Browning Library, Baylor University, located at Eighth and Speight streets. The library maintains an extensive collection of Robert and Elizabeth Barrett Browning manuscripts, first editions of their books, more than 2,000 of their letters, and other memorabilia. The literary visitor may see Browning's desk, which stood originally in his London study, and a portrait of the poet done by his son. Stained-glass windows depict many of Browning's poems and Elizabeth Barrett Browning's sonnets. Dr. A. J. Armstrong, once chairman of the Baylor English Department, collected the Browning material and encouraged the study of poetry at Baylor by bringing to the campus such figures as **Vachel Lindsay, Amy Lowell, Robert Frost,** and **Carl Sandburg.**

> *The purchaser of the morphine wanders into the fog, and at length finds himself upon a little iron bridge, one of the score or more in the heart of the city, under which the small tortuous river flows. He leans upon the rail and gasps, for here the mist has concentrated. . . . The iron bridge guys rattle to the strain of his cough. . . .*
>
> —O. Henry,
> in "A Fog in Santone"

O. Henry House in San Antonio, in which are displayed some of O. Henry's personal possessions.

KAUAI
NIIHAU
HONOLULU
OAHU
MOLOKAI
LANAI
MAUI
HAWAII
HAWAII
BRITISH COLUMBIA
ALBERTA
SASKATCHEWAN
CANADA
SEATTLE
OLYMPIA
SPOKANE
WASHINGTON
Columbia R.
PORTLAND
SALEM
OREGON
CASCADES
MISSOULA
Milk R.
Missouri R.
HELENA
MONTANA
N. DAK.
S. DAK.
IDAHO
BOISE
Snake R.
ROCKY MOUNTAINS
WYOMING
CASPER
CHEYENNE
Green R.
NEB.
PACIFIC OCEAN
SIERRA NEVADA
NEVADA
RENO
CARSON CITY
SACRAMENTO
SAN FRANCISCO
CALIFORNIA
SALT LAKE CITY
OGDEN
UTAH
DENVER
COLORADO SPRINGS
PUEBLO
COLORADO
KAN.
LAS VEGAS
Colorado R.
OKLA.
LOS ANGELES
ARIZONA
SANTA FE
ALBUQUERQUE
PHOENIX
SAN DIEGO
Gila R.
TUCSON
NEW MEXICO
TEXAS
Rio Grande R.
MEXICO
UNION OF SOVIET SOCIALIST REPUBLICS
Yukon R.
FAIRBANKS
ANCHORAGE
CANADA
JUNEAU
ALASKA

MOUNTAIN STATES & FAR WEST

MONTANA

BILLINGS

Ernest Hemingway, after an automobile accident late in 1930, was hospitalized for two months at St. Vincent's Hospital, 1233 North Thirtieth Street, in Billings. Suffering from a bad fracture, Hemingway was confined to bed and became increasingly restless and irritable. To distract himself, he talked with another patient, a small-time Mexican gambler, and with a nun, Sister Florence, and he listened for many hours to a radio. His experiences in this hospital led Hemingway to write "The Gambler, the Nun, and the Radio." This story, published originally in *Scribner's Magazine* in May 1933 with the title "Give Us a Prescription, Doctor," has been anthologized many times.

BOZEMAN

Harold Brainerd Hersey was born here on March 2, 1893. He wrote of Montana in much of his poetry, some of which was collected in *Gestures in Ivory* (1919) and *Bubble and Squeak* (1927).

CHOTEAU

A[lfred] B[ertram] Guthrie was raised in this town about fifty miles west of Great Falls. His father was a teacher and principal at the local high school, from which Guthrie was graduated in 1919. Guthrie is known best for his novels *The Big Sky* (1947) and *The Way West* (1949). His novel *Three Thousand Hills* (1956) described Montana cattlemen in the 1880s.

COOKE CITY

When **Joaquin Miller** wrote *The Illustrated History of Montana* (1894), he predicted that Cooke City, would one day be the scene of a great gold strike. The prophecy is as yet unfulfilled.

GREAT FALLS

Archie Lynn Joscelyn, who was born in Great Falls on July 25, 1899, used many pen names, including **Al Cody, Lynn Westland, A. A. Archer,** and **Evelyn McKenna,** for his many short stories and Western novels. His Westerns included *Hell for Leather* (1951), *The Thundering Hills* (1952), and *The Sundowners* (1956).

Wallace Stegner, the novelist, lived in this city in west-central Montana in 1920–21, attending eighth grade in the Largent School, at 915

Wallace Stegner.

> *When he had been in Montana for less than a month and things were going very poorly indeed, he stumbled on his great discovery. He had lost his way when riding in the hills, and after a day without food he began to grow hungry. As he was without his rifle, he was forced to pursue a squirrel, and in the course of the pursuit he noticed that it was carrying something shiny in its mouth. Just before it vanished into its hole—for Providence did not intend that this squirrel should alleviate his hunger—it dropped its burden. . . . In ten seconds he had completely lost his appetite and gained one hundred thousand dollars. The squirrel . . . had made him a present of a large and perfect diamond.*
>
> —F. Scott Fitzgerald, in "The Diamond as Big as the Ritz"

First Avenue South. The school now is used for adult education classes in addition to serving as an Indian educational resource center.

HINSDALE

Lee Floren was born in this small town in the northeastern portion of Montana on March 22, 1910, and was raised on his father's ranch in nearby Milk River Valley. Floren used many pen names for his novels, including **Brett Austin, Wade Hamilton, Lew Smith,** and **Claudia Hall.** Floren's first novel, *The Long S* (1945), as well as much of his other work, was set in Milk River country.

JUDITH GAP

Will James, the Western writer, claimed in his autobiography, *Lone Cowboy: My Life Story* (1930), that he was born on June 6, 1892, "close to the sod of the Judith Basin country" in central Montana. Although James did spend much of his life in Montana, he appears to have been born in St. Nazaire de Acton, Quebec, and christened **Joseph Ernest Nephtali Dufault.** This discrepancy is characteristic of the Will James story as he told it.

LAME DEER

Hamlin Garland came in 1897 to Lame Deer, a town about one hundred miles east of Billings, from the Standing Rock Reservation, in North Dakota. Garland had previously visited the town of Crow Agency, on the Crow Indian Reservation in Montana. The research he did at the reservations helped him with his novel *The Captain of the Gray-Horse Troop* (1902), which described the abuse of Indians by frontiersmen, and *The Book of the American Indian* (1923).

MISSOULA

While teaching at the University of Montana from 1941 to 1964, **Leslie Fiedler** lived in a number of houses in Missoula, including 102 McLeod Avenue, currently the Unitarian Universalist Fellowship Center. The university's annual writers conference here has attracted well-known scholars and authors, including **William Faulkner, Allan Nevins,** and **Walter Prescott Webb.**

PARK CITY

While **Ernest Hemingway** and **John Dos Passos,** late in 1930, were driving to Billings from northern Wyoming, where they had been hunting, their car went off the road. The accident, which occurred between Park City and Laurel, about twenty miles east of Billings, fractured Hemingway's arm severely enough to require him to be hospitalized in Billings for two months.

PRYOR

Will James lived from 1927 until his death in Hollywood on September 3, 1942, on the Rocking R Ranch, located on the Crow Indian Reservation, about thirty-five miles south of Billings, between Pryor and Fort Smith. Early in his stay at the ranch, James wrote his autobiography, *Lone Cowboy: My Life Story* (1930). After his death, James's body was cremated, and his ashes were spread over the Rocking R Ranch.

WHITE SULPHUR SPRINGS

F. Scott Fitzgerald spent the summer of 1915 with his friend "Sap" Donahoe on a ranch near White Sulphur Springs, about ninety miles north of Bozeman. This visit later was exploited as background for Fitzgerald's "The Diamond as Big as the Ritz" (1922).

IDAHO

BOISE

Vardis Fisher moved to Boise from his father's farm near Ririe in 1935, when the novelist became Idaho director of the Federal Writers'

Snapshot of Will James at the Rocking R Ranch, his home near Pryor.

Project of the WPA, which was responsible for initiating the American Guide series. Because Fisher was persistent and hardworking, his *Idaho Guide* was the first volume published, in 1937, and writers for other state guides were instructed to use Fisher's work as their model.

Mary Hallock Foote and her husband, a civil engineer, settled in this area in 1884 and lived in the Boise Valley for the next twelve years. Initially, they rented a home across the Boise River from a state park immediately below the Lucky Peak Dam. Later, economic necessity forced them to move to temporary quarters nearby, and Mrs. Foote supported the couple with her novels and short stories. They were able eventually to build a house of their own in Boise, only to have to move later to a house southwest of the city. Mrs. Foote's novels included *The Chosen Valley* (1892), *The Prodigal* (1900), and *Ground-swell* (1919). Wallace Stegner used the Boise area as part of the setting for his Pulitzer Prize-winning novel, *Angle of Repose* (1971).

Mary Hallock Foote settled near Boise one year after publication of her novel *The Led Horse Claim* (1883).

CHESTERFIELD

Frank Chester Robertson, the prolific novelist and short-story writer, lived in Chesterfield, east of Pocatello and south of the Blackfoot River, from 1901 to 1924, homesteading from 1914 to 1921. He attended Chesterfield Grade School but never progressed beyond the eighth grade. When his farm failed, Robertson turned to writing Westerns. Among his books, many of which were made into motion pictures, were *Foreman of the Forty-Bar* (1925), *Fall of Buffalo Horn* (1928), and *The Hidden Cabin* (1929). His autobiography, *Ram in the Thicket* (1950), described his early years in Idaho.

HAGERMAN

In Hagerman, in the Thousand Springs Valley, fifteen miles southwest of Gooding, **Vardis Fisher** built himself a stone house. The prolific novelist was still living here when he died on July 9, 1968, in a hospital in nearby Twin Falls.

HAILEY

Ernest Hemingway, who was converted to Catholicism when he married his second wife, was a friend of Father O'Connor, then pastor of the Catholic Church in Hailey, a city one hundred miles east of Boise. Hemingway at one time contributed the cost of a new roof for the church and, in 1958, at the request of Father O'Connor, Hemingway spoke in the Parish House of the church to a group of teenagers. This was only the second time the novelist had done any public speaking.

Ezra Pound, the most illustrious of Idaho's literary native sons, was born here, in what was then a frontier town, on October 30, 1885. His birthplace, a white frame house at 314 Second Avenue South, bears a plaque that reads:

> *Ezra Pound, the poet, was born in this house on the 30th of October, 1885.*
> *"I have beaten out my exile."*
> *From "The Rest," in* Personae *(1962).*

Although Pound left Idaho—in a blizzard—when he was fifteen months old, he was to write in *Indiscretions* (1923) that he believed his parents, and therefore he himself, had been "enriched" by their "Hailey experience." Pound's works include *Personae* (1926), *ABC of Reading* (1934), *The Pisan Cantos* (1954), and *Love Poems of Ancient Egypt* (1962).

KETCHUM

Lillian Ross, on assignment in 1947 to write a profile of **Ernest Hemingway** for *The New Yorker,* visited Ketchum briefly to interview the novelist. Her profile, reprinted in *Reporting* (1964), is considered one of the best the magazine has yet published. Hemingway had come to Ketchum, a mining village one mile from Sun Valley, in 1947. He was determined to lose weight and bring down his blood pressure, and he succeeded. In 1958, Hemingway and his fourth wife, Mary Welsh, rented a cabin near the Edelweiss Motel here. In the following year, they bought an ultra-modern concrete house, with seventeen acres, on the northern border of Ketchum. Mrs. Hemingway still summers there, and the house is occasionally open to the public for house tours. The house, which has no street address, stands on a hillside near the Wood River and is not without attraction for the morbidly curious. During the last weeks of his life, Hemingway was working here on a novel about his life in Paris in the 1920s. He had recently been discharged from the Mayo Clinic after two months of treatment for hypertension—he had also been hospitalized in Sun Valley. News accounts had it that on July 2, 1961, Hemingway, an avid hunter, was cleaning his gun when it went off. The death certificate left blank the cause of death, but the death was later ruled to be self-inflicted. A similar fate, with the same gun, had befallen his father, Dr. Clarence H. Hemingway, in 1928. A memorial to Hemingway stands outside Ketchum, overlooking Trail Creek. The inscription is from a eulogy Heming-

> *[Hemingway] was standing on the hard-packed snow, in dry cold of ten degrees below zero, wearing bedroom slippers, no socks, Western trousers with an Indian belt that had a silver buckle, and a lightweight Western-style sports shirt open at the collar. . . . He looked rugged and burly and eager and friendly and kind.*
>
> —Lillian Ross, in *Reporting*

> *Now—I've finally gathered me a little scope of range like I've always hankered for—A place away from lanes, and in the heart of a wide-open cow and horse country—only a hundred miles from where I was born—I have my ponies, cattle, corrals and all to my taste—There's hundreds of wild horses around, thousands of cattle from neighboring outfits—timber—big creeks with trout in 'em—plenty of grass on both sides, and on the ridges where riders fog down off of to drop in and say hello or rest and feed up while on their way from one cow camp to another—I'm at home—*
>
> —Will James, in *Lone Cowboy: My Life Story*

When at last we were down, out of the sky and the mountains, we found ourselves on a river bottomland, with marshes and bogs smelling of sulphur, with a mountainside of jungle looming above us, with the furious river before us, with all the wild things everywhere—beaver, otter, mink, muskrat; deer and elk and bear; badger and coyote and wolf and mountain lion—and more birds than I have ever seen anywhere since that day.

—Vardis Fisher,
in "Hometown Revisited,"
collected in *Thomas Wolfe as I Knew Him and Other Essays*

way delivered in 1939 at the funeral in Ketchum of his friend Gene Van Guilder, accidentally killed by a shotgun blast.

Carlos Baker, in *Ernest Hemingway Selected Letters 1917–1961,* cited now-ironic lines Hemingway wrote in 1926: "The world is so tough and can do so many things to us and break us in so many ways that it seems as though it were cheating when it uses accidents or disease . . . But all you can do about hell is last through it. If you can last through it. And you have to. Or at least I always will I guess."

LORENZO

Vardis Fisher was born on March 31, 1895, in Lorenzo, a village known in Fisher's time as Annis. Fisher's father, the first settler in the upper Snake River Valley, had been sent here by the Church of Jesus Christ of Latter-Day Saints. Fisher was known best for his four novels about the life of his alter ego, transparently named Vridar Hunter: *In Tragic Life* (1932), *Passions Spin the Plot* (1934), *We Are Betrayed* (1935), and *No Villain Need Be* (1936).

MOSCOW

Frank Chester Robertson, the Western novelist and story writer, was born on January 21, 1890, in this city north of Lewiston. Among Robertson's many works were *Back to the West* (1935), *Boomerang Jail* (1947), and *Idaho Range* (1951).

PARMA

Edgar Rice Burroughs, the novelist who created the Tarzan books, moved to Parma, west of Boise, with his wife in 1903. They joined Burroughs's brother, Harry, who was running a dredge on the Snake River. Here Burroughs wrote, on the backs of old letters, his first story, "Minidoka 937th Earl of One Mile Series M." The next year, Burroughs ran for the office of town trustee and won by one vote. Unfortunately, his brother's dredging company failed in that same month, so Burroughs and his wife soon left Parma to seek their livelihoods elsewhere.

POCATELLO

Edgar Rice Burroughs ran a book and stationery store here in the late 1890s.

RIRIE

Vardis Fisher returned to his father's farm near this town northeast of Idaho Falls in 1931. He was to live in Idaho for the rest of his life. Fisher had been teaching at various universities in the East and in Idaho, but began writing full time here. Although several of the novels he wrote were published, the income they produced was inadequate, and he accepted a job as director of the Federal Writers' Project Idaho State Guide. He wrote much later of the area south of Idaho's Snake River in the essay "Hometown Revisited."

Ernest Hemingway holding trout he caught while on a fishing trip to Sun Valley in 1939. Hemingway memorial between Ketchum and Sun Valley.

SUN VALLEY

Ernest Hemingway first came to Sun Valley in 1903, when it was just beginning to become a ski resort. He was drawn here by the game-bird hunting and the skiing. Many years later, with his third wife, Martha Gellhorn, he moved into Suite 206 at the Sun Valley Lodge. In the mornings, he worked on *For Whom the Bell Tolls* (1940), his novel about the Spanish Civil War, and in the afternoons he hunted. When his wife left on a magazine assignment, Hemingway stayed on, naming the suite Hemingstein's Mixed Vicing and Dicing Establishment. Hemingway came again to Sun Valley in 1941, with his three sons. He wanted to take them on an antelope-hunting expedition. Staying once more at the Sun Valley Lodge, Hemingway found time to enjoy old and new Hollywood friends, among them Gary Cooper, Barbara Stanwyck, and Howard Hawks. In 1961, the year of Hemingway's death, he was back at Sun Valley, this time as a patient at the Moritz Community Hospital. Hemingway's association with Sun Valley is reflected in the memorial dedicated to him on July 21, 1966: a bronze head on a pedestal, placed about half a mile east of town, in a small grove of alder, willow, and red birch.

WYOMING

CASPER

Owen Wister, the Philadelphian best known for his vastly popular novel *The Virginian* (1902), visited this city in Natrona County on the North Platte River on June 12, 1891. Casper was much smaller then than it is now, and Wister reported being unimpressed with the place. The Goose Egg Ranch house, a stone building on the banks of the Platte River, about fourteen miles west of Casper, figured in Wister's novel. Built in the late 1870s, the house by the 1930s was abandoned and falling to ruin.

CHEYENNE

George Abbott, the playwright who collaborated on such Broadway hits as *A Tree Grows in Brooklyn* (1951) and the Pulitzer Prize-winning *Fiorello* (1959), attended public schools in Wyoming's capital from the fifth to the eighth grade. **Willard Motley,** author of such naturalistic novels as *Knock on Any Door* (1947), once spent thirty days in jail here on a charge of vagrancy. **Owen Wister,** the novelist, visited Cheyenne in the summer of 1894 while collecting material for his books. Robert Burns, the Scottish poet, who never visited the city, is honored by a statue that stands at the intersection of Carey and Randall avenues.

DOUGLAS

Owen Wister visited this town in Converse County on the North Platte River on June 11, 1891, and wrote to his mother, Fanny Kemble Wister, daughter of the English actress Fanny Kemble, from the Valley House, a local hotel, no longer standing. Wister described Douglas as a town that "though laid out at right angles with wide streets, is a hasty litter of flat board houses standing at all angles, with the unreal look of stage scenery."

GRANITE

Mary O'Hara, the novelist, owned a ranch near Granite, formerly known as Granite Canyon and located southeast of Laramie. Here she wrote her novels *My Friend Flicka* (1941), *Thunderhead* (1943), and *The Green Grass of Wyoming* (1946). The novelist lived here until 1948.

JACKSON

Struthers Burt, the novelist and newspaperman, first came to Jackson Hole, northwest of Jackson, a town south of Grand Teton National Park, in 1908, and spent most of his summers here until his death in 1954. He considered himself a citizen of Wyoming and wrote about the state in the nonfiction work *Powder River: Let 'Er Buck* (1938). His wife, **Katharine Newlin Burt,** was also a writer, and used Western settings in such novels as *The Branding Iron* (1919) and *When Beggars Choose* (1937).

KAYCEE

Owen Wister, in the summer of 1885, visited the TTT Ranch, not far south of this town, which lies sixty miles northwest of Casper. Wister was a house guest of the Tisdale brothers, who owned the ranch, and he often jotted notes during his visit. The notes were later used in Wister's novel *The Virginian.* Although the main ranch house burned down some years ago, the present owner welcomes interested visitors.

The entire state is in reality a mountain and its ranges are merely the peaks of the mountain. One hour you are traveling through hot plains, the next you are in the cool recesses of incredible hills. Thousands of people cross southern Wyoming every year convinced it is a semidesert state. They do not know that to the north and south of them . . . all around them . . . are green valleys and greener forests and luminous uplands.

—Struthers Burt,
in *Powder River: Let 'Er Buck*

Photograph of monument to Robert Burns in Cheyenne taken during dedication ceremonies on November 11, 1929.

A company incorporated itself and started a paper of which I took charge. The paper was published in the loft of a livery stable. This is the reason they called it a stock company. You could come up the stairs into the office or you could twist the tail of the iron grey mule and take the elevator.

—Bill Nye,
commenting on the founding of the Laramie *Boomerang*

LARAMIE

Edgar Wilson Nye, the humorist known better by the name of his fictional character **Bill Nye,** in 1876 came to live in this city about fifty miles west of Cheyenne. Although never graduated from law school—he was never graduated from any school, for that matter—Nye was admitted to the bar, practiced law, and served as justice of the peace. He also served at other times as postmaster, reporter for the Laramie *Daily Sentinel,* superintendent of schools, member of the territorial legislature, and United States Commissioner. Fame came to him after he founded the Laramie *Boomerang*—named for his mule—in 1881 and began writing his humorous sketches under the name of Bill Nye. Among the collections of his sketches were *Bill Nye and Boomerang* (1881) and *Baled Hay: A Drier Book Than Walt Whitman's "Leaves of Grass"* (1884). The library of the University of Wyoming here has Owen Wister's diaries in its special collections.

MEDICINE BOW

Owen Wister came to this small town about sixty miles northwest of Laramie in 1885, when he was twenty. He used the town as part of the setting of *The Virginian* (1902), and Medicine Bow has the distinction of being the only real town mentioned by name in the novel. A cabin he built near Moose has been moved to Medicine Bow, where it occupies a site just opposite the Virginian Hotel—there is some disagreement about whether it is the original Wister home, but many ranchers in the area claim that it is. The Virginian Hotel has a suite and a dining room named after the novelist, and a monument to him was erected in town in 1939.

Snapshot of Owen Wister Monument in Medicine Bow. The plaque reads:
"When you say that, smile."
OWEN WISTER
whose writings acquainted the nation with pioneer Wyoming ranch life, made Medicine Bow the beginning of his most popular novel, "The Virginian."

MOOSE

Owen Wister built a cabin in this tiny town in Grand Teton National Park in the summer of 1912. He and his family stayed here for about two weeks that summer, but Wister's wife died the following winter, and the family never returned. The cabin was moved to Medicine Bow, where it stands in a city park.

RIVERSIDE

Owen Wister is said to have accompanied the Tisdale brothers on trips to Riverside when they came to collect their mail. Unfortunately, Kaycee, where the Tisdales lived, is about 150 miles from Riverside, the latter town standing about one hundred miles west of Laramie. Be that as it may, the post office was on the Coable and Parker Ranch. Visitors picked up their mail from a sack by the door and could, if they were hungry, settle down and cook themselves a meal there. Wister was so fascinated by the place that he returned here years later and rented a cabin, where he wrote his novel *Lin Maclean* (1898).

WOLF

Mary Roberts Rinehart, the novelist, had a summer home for a time in this small town near the Bighorn National Forest.

COLORADO

ASPEN

Luke Short, whose real name was **Fred Glidden,** wrote many of his Westerns in his downtown Aspen office and in the yard of his home here. **Carl Jonas,** known best for his novel *Jefferson Selleck* (1951), lived in Aspen for a time after 1946. **J. P. Marquand,** after his second marriage in 1937, lived in Aspen for a while. He found the mountains not at all to his taste, and returned to the East. **Harold Ross,** the first editor of *The New Yorker* magazine, was born on November 6, 1892, in a house at 601 West Bleeker Street. **Damon Runyon** worked as a bellhop in Aspen, and **Leon Uris** has lived here.

Hunter S. Thompson, who gained wide attention with his novel *Fear and Loathing In Las Vegas* (1971), has lived for many years near Aspen as well as in San Francisco. *Hell's Angels: A Strange and Terrible Saga of the Outlaw Motorcycle Gangs* (1966) established his reputation as a counter-culture reporter. A sometime reporter for *The National Observer,* Thompson now writes for *Rolling Stone.*

BOULDER

Thomas Duncan lived in the resort and university city of Boulder in 1944 and 1945 and worked on his novel *Gus the Great* (1947) and some short fiction. **Allen Ginsberg,** with others, established the Jack Kerouac School of Disembodied Poetics, Naropa Institute, at 1111 Pearl Street. Ginsberg has served as co-director and has taught there during the summer months. **Jean Stafford,** novelist and short-story writer, grew up in Boulder, where she attended the University Hill School and the State Preparatory School, now the Boulder High School, at 1604 Arapahoe Street. She earned an M.A. degree from the University of Colorado in 1936. Stafford here met Robert Lowell, whom she married in 1940. They were introduced by their mutual friend Ford Madox Ford, the English novelist and critic. Stafford bequeathed her manuscripts, letters, and typewriter to the University of Colorado. Included in the bequest were English compositions she wrote while in school and articles she wrote for *The Owl,* her high school newspaper.

Betty MacDonald, who wrote *The Egg and I* (1945), a humorous novel that sold more than a million copies, was born in Boulder on March 26, 1908. Her name at birth was **Anne Elizabeth Campbell Bard,** and she used the name by which we know her only after her second marriage, to a man named MacDonald. The novelist lived only briefly in Boulder. Her father, a mining engineer, had to move with his work, but when Betty MacDonald was nine, her family settled in Seattle, Washington. It was her experience in running a chicken farm in Washington during the early 1940s that led to the novel that established her career as a writer.

Thomas Wolfe, during the summer of 1935, spent twelve days in Boulder as visiting novelist at the Writer's Conference of the University of Colorado. Afterward he traveled through Colorado and was inspired to write "The Hound of Darkness," a description of nighttime in America. This work was published as "A Prologue to America" in the February 1, 1938, issue of *Vogue.* Wolfe also used much of this same material in his novels *The Web and the Rock* (1939) and *You Can't Go Home Again* (1940).

CANON CITY

Joaquin Miller served for a time as judge, mayor, and minister in the early days of Canon City, a mining town thirty-five miles southwest of Colorado Springs. Canon City stands at the mouth of Royal Gorge, called the Grand Canyon of the Arkansas River. Miller wanted to change the name of the town to Oreodelphia (City of the Mountains), but the miners protested that they could not even pronounce it.

COLORADO SPRINGS

Helen Hunt Jackson in the 1870s came to this residential resort city for her health, staying at the Colorado Springs Hotel. In 1875 she married her second husband, William S. Jackson, and moved with him to 228 East Kiowa Street. The house is no longer standing, and the site now is occupied by the Colorado Springs Police Department. Three rooms of the original house, with authentic furnishings, have been reconstructed as the Pioneers' Museum, 25 East Kiowa Street, open to the public. Donations are accepted. Four more rooms have been added since the museum moved to a new site at 215 South Tejon Street.

Jackson died in 1885 and was buried on Cheyenne Mountain, at a site above Seven Falls in South Cheyenne Canyon. It was an area she liked to visit, and her gravesite is still marked by a cairn. In 1891, Jackson's remains were removed to Evergreen Cemetery at 1001 South Hancock Avenue, where her tombstone inscription reads: "Helen, Wife of William S. Jackson, Died August 12, 1885." On the foot marker is "Emigravit" (she migrated), the title of one of her best-known poems. (While visiting Cheyenne Mountain, the literary traveler may be interested in seeing the Shrine of the Sun Memorial, which includes a 100-foot tower dedicated to the memory of Will Rogers and a bust of the cowboy-humorist.) *Ramona* (1885) was Jackson's best-known novel. A romance about the mission Indians of southern California, *Ramona* did much to change the negative attitudes of Americans toward Indians. An earlier nonfiction work by Jackson, *A Century of Dishonor* (1881), recounted the government's long history of injustice to Indians.

> *Who knows what myriad colonies there are*
> *Of fairest fields, and rich, undreamed-of gains*
> *Thick planted in the distant shining plains*
> *Which we call sky because they lie so far?*
> *Oh, write of me, not "Died in bitter pains,"*
> *But "Emigrated to another star!"*
>
> —Helen Hunt Jackson,
> in "Emigravit"

Charles Kingsley, the English author still read in such novels as *Hypatia* (1853) and *Westward Ho!* (1855) and the delightful children's book *The Water Babies* (1863), toured Canada and the United States in 1874. While in Colorado Springs, he came down with pleurisy, and soon returned to England, where he died on January 23, 1875. **William Vaughn Moody,** the playwright and poet, lived in Colorado Springs for a short time. He had come here in poor health to take advantage of the mountain climate, but died of a brain tumor on October 17, 1910, at his home at 5 West View Place. Moody was only forty-one when he died. His home is still standing but not open to the public. Moody was widely known for

> *Ramona looked anxiously at her. "I have never disobeyed you, Señora," she said, "but this is different from all other things; you are not my mother. I have promised to marry Alessandro."*
>
> *The girl's gentleness deceived the Señora.*
>
> *"No," she said icily, "I am not your mother; but I stand in a mother's place to you. You were my sister's adopted child, and she gave you to me. You cannot marry without my permission, and I forbid you ever to speak again of marrying this Indian."*
>
> —Helen Hunt Jackson,
> in *Ramona*

(*below*) Exhibit of Helen Hunt Jackson's books and furnishings at the Pioneers' Museum, now located at 215 South Tejon Street. (*below left*) Original gravesite of the novelist on Cheyenne Mountain southeast of Colorado Springs. So many visitors carried rocks away that Mrs. Jackson's body was reinterred in town.

Wynken, Blynken and Nod one night
Sailed off in a wooden shoe—
Sailed on a river of crystal light
Into a sea of dew.

—Eugene Field,
in "Dutch Lullaby,"
often called "Wynken, Blynken and Nod"

And despite the apparent eagerness to fill his own money-drawers and build his own imperial dream, I am naïve enough to believe that Bonfils had an actual, deep love for the region which made him a journalistic czar. The many sides of his character offered puzzling testimony as to his actual feelings on any subject.

—Gene Fowler,
in *Timberline*

his immensely popular play *The Great Divide,* produced in 1906 under the title *A Sabine Woman,* and produced and published in 1909 under the new title. Generations of students knew Moody and [Robert Morss] Lovett as the authors of what once was the standard textbook on English literature.

Scattering wide or blown in ranks,
Yellow and white and brown,
Boats and boats from the fishing banks
Come home to Gloucester town.
There is cash to purse and spend,
There are wives to be embraced,
Hearts to borrow and hearts to lend,
And hearts to take and keep to the end,—
O little sails, make haste!

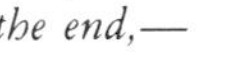

—William Vaughn Moody,
in "Gloucester Moors"

Anne Parrish, best known for her novel *The Perennial Bachelor* (1925), was born in her family's home at 209 North Weber Avenue on November 12, 1888. The Junior Achievement Building stands on the site today. Parrish was educated in private schools in Colorado Springs and elsewhere. **Damon Runyon** once roomed in a South Nevada Avenue boarding house. In 1901 he worked as a reporter for the Colorado Springs *Gazette,* 15 Pikes Peak Avenue, in a building no longer standing. The newspaper continues to publish as the Colorado Springs *Gazette Telegraph.* Runyon was known to treat friends to draughts of redeye, a potent decoction he bought at Mr. Farringer's Drug Store on Pikes Peak Avenue. **Gladys [Bagg] Taber,** the author of many books about life in Southbury, Connecticut, was born in Colorado Springs on April 12, 1899.

CREEDE

Richard Harding Davis recalled in *The West from a Car Window* (1892), a collection of his articles, a sidetrip he made from Denver to Creede in 1890 in order to view the silver mines. In that year, silver was discovered in Colorado and Creede was founded. By 1893, Creede had 8,000 inhabitants, but shortly thereafter, when the price of silver dropped, the mines closed and the population declined. Today Creede, on Route 149 in southern Colorado, has fewer than 700 inhabitants. **Carl Sandburg** is said to have worked as a dishwasher at the Creede Hotel and Bar in 1897, while on his hobo trip. The establishment is still open to the public.

CRIPPLE CREEK

Julian Street, a newspaperman and novelist, once submitted such a thorough report to *Collier's Weekly* about the Cripple Creek red light district, whose heart was Myers Avenue, that the local residents began to call the thoroughfare Julian Street. His travel book *Abroad at Home* (1915) included material about Cripple Creek, which is southwest of Colorado Springs. **Lowell Thomas,** the newscaster and writer of travel books, was raised in the gold-mining camps near Cripple Creek, and later worked on the Cripple Creek newspaper.

DENVER

Among the many writers who lived and worked in Denver are Mary Coyle Chase and Eugene Field. **Mary Coyle Chase,** who won a Pulitzer Prize for the play *Harvey* (1944), was born here on February 25, 1907, when her family was living at 532 West Fourth Avenue. Chase was graduated from the West Denver High School in 1922 and later attended the University of Denver and the University of Colorado at Boulder. *Harvey,* a comedy about an alcoholic whose constant companion is a six-foot white rabbit, has been widely produced. Among Chase's other plays were *Mrs. McThing* (1952), *Bernadine* (1953), and *Midgie Purvis* (1961). Mrs. Chase died in Denver on October 20, 1981.

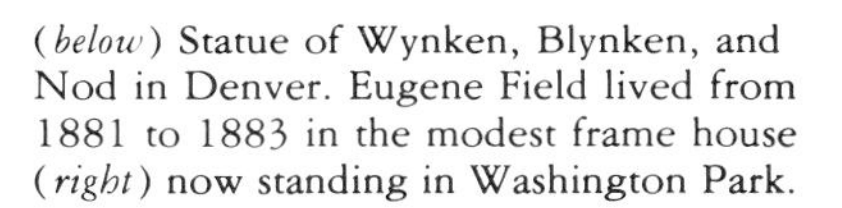

(*below*) Statue of Wynken, Blynken, and Nod in Denver. Eugene Field lived from 1881 to 1883 in the modest frame house (*right*) now standing in Washington Park.

In 1881, when **Eugene Field** became managing editor of the Denver *Tribune,* he rented a house at 315 West Colfax Avenue, where he lived for two years. Field wrote the column "Odds and Ends" for the *Tribune,* providing some poetry as well as satiric tidbits about prominent city residents. In 1930 the Colfax Avenue home was donated to Denver and became the Eugene Field Memorial House, with the stipulation that it be moved to a city park. It has stood since 1930 at 715 South Franklin Street in Washington Park, and is marked with a plaque. The Denver Public Library, now at 1357 Broadway, which exhibits several volumes of Field's work, used the site as a branch library from 1930 to 1970, but the building today is headquarters for various civic organizations. The Eugene Field Branch Library, opened in 1970 at 815 South University Boulevard and East Ohio Avenue, houses Field's collected works. A Wynken, Blynken and Nod statue, commemorating Field's famous characters, was added to the northwest section of Washington Park, off Downing Street at Exposition Avenue.

Gene Fowler, the son of a patternmaker named Devlan, who lived at 34 South Twelfth Street, was born in Denver on March 8, 1890. Fowler was later adopted by his stepfather, Frank G. Fowler. During the time Fowler worked under publisher Frederick G. Bonfils and editor Harry H. Tammen of the Denver *Post,* its circulation increased from 6,000 to 150,000. Fowler recalled his days of flamboyant journalism in *Timberline: A Story of Bonfils and Tammen* (1938).

After **Horace Greeley** heeded John Babson Soule's advice to "Go West, Young Man!" he found he did not care for the brawls and gunfights common among early Denver cowboys, and precipitately arranged for a stagecoach to carry him farther west. Greeley made his trip West for the New York *Tribune* in the summer of 1859, contributing articles for the *Tribune* that later were collected in *An Overland Journey* (1860). His stay in Denver, only two weeks in June, enabled him to visit the gold digging sites in the area and observe the lives of the miners and other local types:

> *Prone to deep drinking, soured in temper, always armed, bristling at a word, ready with the rifle, revolver, or Bowie-knife, they give law and set fashions which, in a country where the regular administration of justice is yet a matter of prophecy, it seems difficult to overrule or disregard. I apprehend that there have been, during my two weeks' sojourn, more brawls, more fights, more pistol-shots with criminal intent in this log city of one hundred and fifty dwellings, not three-fourths completed nor two-thirds inhabited, nor one-third fit to be, than in any community of no greater numbers on earth.*

Jack Kerouac visited Denver in 1947 on his first trip to the West, and two years later rented a small house in the foothills outside Denver. **Alexander K. McClure,** on his way to Salt Lake and the Northwest, was detained in Denver by Indian uprisings for a month in 1867. McClure recounted his experiences in *Three Thousand Miles Through the Rocky Mountains* (1869). **Katherine Anne Porter** lived for a time at 1510 York Street, in a house still standing and much enlarged.

Damon Runyon lived in several places in Denver, including 1420 Logan Street in 1907, now the Emerson House Community Center; 1458 Court Place in 1908, no longer standing; and the Denver Athletic Club at 1325–1335 Glenarm Place, still standing. Runyon worked for the Denver *News, Times, Republic,* and *Post.* The *Times* and the *Republic* are no longer publishing. While working for the *Post,* Runyon wrote two volumes of verse, *Tents of Trouble* (1911) and *Rhymes of the Firing Line* (1912).

Carl Sandburg, on his 1897 hobo trip through the West, which he described in *Always the Young Strangers* (1953), stopped in Denver for two weeks, working as a dishwasher at the Windsor, a first-class hotel. He was paid $1.50 plus room and board for the two weeks. **Upton Sinclair,** who came here in 1914, later wrote *The Brass Check, A Study of American Journalism* (1919), in which he criticized Denver newspapers for their biased coverage of striking miners and the Ludlow massacre. His novel *King Coal* (1917) dealt with the plight of the Colorado coal miners, and the locale of his novel *Mountain City* (1930) is thought to be Denver.

Wilbur Daniel Steele, who was raised in Denver, lived at 2161 South Josephine Street while attending the University of Denver, where his father was a professor of Biblical literature. **Irving Stone** stopped at the Brown Palace Hotel, at Seventeenth and Tremont streets, while he researched *Men to Match My Mountains* (1956). **Ruth Suckow,** after being graduated from the University of Denver in 1917, worked as an assistant in the English department at the university for a year. She later worked for a Denver map company and as an assistant, in 1919, at Delia Weston's Apiary. The novelist and short-story writer lived at 3817 Vrain Street in 1917, 1981 South York Street in 1918, and 134 Pearl Street in 1919. **Bayard Taylor,** in *Colorado: A Summer Trip* (1867), summarized his travel experiences in the state, which included a stay in Denver.

> ELWOOD. *I am trying to be factual. I then introduced him to Harvey.*
> WILSON. *To who?*
> KELLY. *A white rabbit. Six feet tall.*
> WILSON. *Six feet!*
> ELWOOD. *Six feet and a half!*
> WILSON. *Okay—fool around with him, and the Doctor is probably some place bleedin' to death in a ditch.*
> ELWOOD. *If those were his plans for the evening he did not tell me.*
>
> —Mary Coyle Chase, in *Harvey*

Entrance to the Emerson House in Denver. While living here in 1907, Damon Runyon worked as a reporter.

Jed was going to the city! Going to have his first ride in one of those transcontinental "flyers" which had treated him so contemptuously all through his childhood. Going to stand on the rear platform of an elegant maroon-colored coach, while the engine gave two long and two short blasts of the whistle, and whirled him past the little yellow station of Alamito, and the lonely cabin from which he had gazed during the first twelve years of his life. . . . Nowadays rural America learns to know great cities upon the moving picture screen. But nothing can lessen the thrill of the country-bred boy who steps for the first time onto the crowded sidewalk, and gazes up at the towers of office-buildings, and beholds the evidence of wealth and pride and power.

—Upton Sinclair,
in *Mountain City*

Lowell Thomas worked as part-time reporter for the *Rocky Mountain News* and the Denver *Times,* as well as night desk clerk for a downtown hotel, while studying at the University of Denver. In 1911 Thomas met Gene Fowler, who worked on the rival Denver *Post.* They shared stories and sometimes a scoop. **Artemus Ward,** the newspaperman and humorist, recalled his 1864 Denver trip in *Artemus Ward: His Travels* (1865). **Oscar Wilde,** on a lecture trip through the United States, stayed in April 1882 at the Windsor Hotel in Denver. The press had been ridiculing Wilde, so Eugene Field, dressed in vintage Wilde fashion—a long wig, broad hat, and fur-trimmed overcoat—drove alone in Wilde's carriage along the streets of the city. The crowds, not in the least fooled, cheered and booed. Wilde's prescient comment: "What a splendid advertisement for my lecture!" **Thomas Wolfe** spent a week at the Brown Palace Hotel in May 1938.

DOVE CREEK

Zane Grey, while writing one of the most popular Westerns of all time, *Riders of the Purple Sage* (1912), lived in the tiny town of Dove Creek, near the Utah border and about thirty miles northwest of Cortez.

ESTES PARK

William Allen White, the crusading Kansas journalist, had a summer home about six miles west of the village of Estes Park, in Rocky Mountain National Park. Although the White complex, consisting of cabins, studio, and other buildings, is not open to the public, a history walk sponsored by the park takes visitors to one of the cabins. From mid-June to mid-August, the walk is conducted one or more times a week.

Main cabin of the William Allen White summer home in Rocky Mountain National Park, near Estes Park. The cabin has been restored so that it looks as it did in the early 1940s.

GLENWOOD SPRINGS

Damon Runyon, in order to gain experience in writing for newspapers, worked for a time as a reporter for the *In-It Daily,* probably the only newspaper ever named for a swimming pool. It appears that Glenwood Springs, now a city on Route 70 in the west central part of the state, looked on its outdoor swimming pool as its center of civic pride.

GOLD HILL

Thomas Duncan has reported that he and his wife, whose pen name is **Carolyn Thomas,** worked during 1945 on the novel *Gus the Great* (1947) while they were living in a mountain cabin in this ghost town seven miles northwest of Boulder.

GRAND JUNCTION

Dalton Trumbo lived, when he was a child, in Grand Junction at 117 South Ninth Street in 1909, 502 Chipeta Avenue in 1910, and 1124 Gunnison Avenue from 1912 to 1925. All these residences are unmarked and private. Trumbo attended Hawthorne Elementary School here and Grand Junction High School. While at school he worked for the Grand Junction *Daily Sentinel,* 634 Main Street. The Sentinel Building is now a group of shops called Sentinel Square. Trumbo's novel *Eclipse* (1935) was set in Shale City, which is believed to be Grand Junction.

LA JUNTA

Ken [Elton] Kesey was born on September 17, 1935, in the plains city of La Junta, sixty miles east of Pueblo. Kesey is known best for his first novel, *One Flew Over the Cuckoo's Nest* (1962).

LEADVILLE

Mary Hallock Foote lived in 1879 at 216 West Eighth Street in Leadville, a mining city of central Colorado. The house is no longer standing. Her life in Leadville was reflected in *The Led-Horse Claim: A Romance of a Mining Camp* (1883), her first novel. It may be difficult to picture **Oscar Wilde** in a Colorado mining town but that, in April 1882, is just where he was. He visited the Matchless Mine in Leadville. Wilde had been warned that Leadville was a tough mining town, but he was able to keep the attention of his lecture audience even while he read passages from the autobiography of Benvenuto Cellini—in translation—and described James McNeill Whistler's paintings.

MESA VERDE

Willa Cather in 1915 viewed the cliff dwellings of Mesa Verde, near Cortez in southwestern Colorado. Her impressions of the cliff dwellings can be read in the Tom Outland section of *The Professor's House* (1925).

MONTROSE

Dalton Trumbo, author of one of the best-known American antiwar novels, *Johnny Got His Gun* (1939), was born on December 9, 1905, at 233 Main Street, in a room over the Montrose Public Library. The building in which Trumbo was born is no longer standing.

PUEBLO

Damon Runyon was raised in Pueblo and attended the three-story brick Hinsdale School, at Seventh Street and Grand Avenue. When Runyon was thirteen, his father published two of Damon's stories in the Pueblo *Adviser,* a commercial weekly. Runyon later gained invaluable writing experience under Col. W. B. McKinney of the *Evening Press,* in South Pueblo. He worked also for the Pueblo *Chieftain,* then located in the 100 block of East Fourth Street, now at 825 West Sixth. Runyon's home addresses included the Fifth Avenue Hotel at 503 Sante Fe Avenue in 1897; and 702 Summit (now Albany) Avenue in 1900–02. The house on Summit has since been torn down to make way for the freeway. Runyon drew attention to the 1903 strikes here by the United Mine Workers with his poem "One Chance Men," about coal-mine inspectors. The Pueblo Baseball Park has a Runyon plaque.

TRINIDAD

John Reed and **Max Eastman,** working as journalists in 1914, investigated conditions in the coal mines of Ludlow, a town of a few hundred that stood just north of Trinidad, near the New Mexico border. National Guard troops machine-gunned the miners and burned Ludlow on April 20, because the miners were striking. The incident became known as the Ludlow massacre. Reed wrote about the incident in "The Colorado War," which was published in July 1914 in *Metropolitan Magazine:*

> *It is a long story, but I shall tell it in full, because it is not a personal story, but a story of eleven thousand miners with their wives and children, living in slavery in lonely mountain fortresses, making a desperate fight for the rights of human beings, and crushed back into their slave-pens by all the agencies of capitalist repression.*
>
> *I had been to Colorado, and knew intimately the conditions. Now the strike was on, and the miners and their families living in tent-colonies had been raided, beaten, shot up by gun-men. Finally a couple of machine-guns had been turned loose on them, their tent-colony at Ludlow had been burned, and three women and fourteen children had been suffocated to death.*

VICTOR

Lowell Thomas lived in Victor, just west of Colorado Springs, when he was a child. He delivered newspapers when he was in the sixth grade, by his report to earn enough money to buy a burro. Later he sorted ore in the mines and heard tales of prospecting for Klondike gold that instilled in him the desire to travel to Alaska. In 1911, when Thomas was nineteen, he worked as a reporter for the Victor *Daily Record.* Six months later, he became editor of the Victor *Daily News.* His memories of Victor are recalled in *Book of the High Mountains* (1964) and *Good Evening Everybody* (1976).

VIRGINIA DALE

Artemus Ward stopped here briefly on a trip to California in 1868–69. He reported: "Virginia Dale is a pretty spot, as it ought to be with such a pretty name; but I treated with no little scorn the advice of a hunter I met there, who told me to give up 'literatoor,' form a matrimonial alliance with some squaws, and 'settle down thar.' " Ward made ample use of his Western visit in the many enormously popular lectures he gave in the United States and England.

NEW MEXICO

ALBUQUERQUE

Robert Creeley, the poet, attended the University of New Mexico, and taught here in the 1960s and early 1970s. **Harvey Fergusson** was born at 1801 Central Avenue N.W., now the Manzano School, on January 28, 1890. His *Blood of the Conquerors* (1921), dealing with the Spanish landowners, is considered the first realistic novel about modern New Mexico. Several of his other novels, including *Grant of Kingdom* (1950), trace the historical development of the state.

Paul Horgan, the novelist, lived at 120 South Walter Street from age twelve to fourteen, and then at 409 North Twelfth Street until 1922, when he was nineteen. Horgan attended local public schools and, in 1921 and 1922, while he was a cadet at the New Mexico Military Institute in Roswell, worked for the Albuquerque *Morning Journal,* then published at 310 West Gold Street. Years later, while Horgan lived in Roswell, he wrote about Albuquerque in *The Common Heart* (1942).

Conrad Richter and his wife moved here in 1928 staying, at least until 1930, at 720 West New York Avenue, probably no longer standing. In later years Richter lived also at 1617 Las Lomas Road, and in the early 1950s at 1421 Las Lomas Road.

CIMARRON

Will James, the Western writer and illustrator, worked in the early 1920s on the Springer C S Ranch here. Cimarron, which is northeast of Santa Fe, in the eastern foothills of the Sangre de Cristo Mountains, was the site of a famous frontier hotel, the Maxwell House. Near Cimarron today is Philmont National Boy Scout Ranch, the site of the Seton Memorial Museum and Library. Ernest Thompson Seton, a naturalist born in Canada, wrote many books on wildlife and was closely associated with the Boy Scout movement.

LOS ALAMOS

The Bandelier National Monument, twelve miles south of Los Alamos, was the site of a sixteenth-century community of cliff dwellers. A museum, trails, and walks now at Bandelier explain the significance of the ruins. Bandelier takes its

> *It all hung together, seemed to have a kind of composition: pale little houses of stone nestling close to one another, perched on top of each other, with flat roofs, narrow windows, straight walls, and in the middle of the group, a round tower. . . . There was something symmetrical and powerful about the swell of the masonry. The tower was the fine thing that held all the jumble of houses together and made them mean something. It was red in colour, even on that grey day. In sunlight it was the colour of winter oak-leaves. A fringe of cedars grew along the edge of the cavern, like a garden. They were the only living things. Such silence and stillness and repose—immortal repose. That village sat looking down into the canyon with the calmness of eternity.*
>
> —Willa Cather,
> in *The Professor's House*

> *When his car got into the sandhills and followed the winding road, Peter felt transported, as if he were lost in a desert whole continents away. This was because the hills rose sharply, hiding the view of the town below, and alternated with one another so clearly, as if eroded blocks of earth had been set down in a maze pattern something like immense cog teeth. When the rain happened in torrents on the mesa, this winding place became a canyon boiling with pale-brown water that rushed upon the streets of the town and fanned out for blocks.*
>
> —Paul Horgan,
> in *The Common Heart*

One afternoon in the autumn of 1851 a solitary horseman, followed by a pack-mule, was pushing through an arid stretch of country somewhere in Central New Mexico. He had lost his way, and was trying to get back to the trail, with only his compass and his sense of direction for guides. The difficulty was that the country in which he found himself was so featureless—or rather, that it was crowded with features, all exactly alike. As far as he could see, on every side, the landscape was heaped up into monotonous red sand-hills, not much larger than haycocks, and very much the shape of haycocks. One could not have believed that in the number of square miles a man is able to sweep with the eye there could be so many uniform red hills.

—Willa Cather,
in *Death Comes for the Archbishop*

name from **Adolph Bandelier** (1840–1914), an explorer and pioneer ethnologist who studied the ancient people of the area and wrote the novel *The Delight Makers* (1890), which dealt with the lives of the early Pueblos and Navajos.

POJOAQUE

Thomas Duncan, the novelist, lived in this town north of Santa Fe at 907 Garcia Street, still standing, while he completed *Big River, Big Man* (1959) and *Virgo Descending* (1961).

ROSWELL

Paul Horgan attended Roswell's New Mexico Military Institute, now at North Main Street and College Boulevard, as a cadet from 1919 to 1923. He returned to the school to work as librarian from 1926 to 1942. During part of this time, he lived at 1313 North Pennsylvania Avenue, where he wrote the novels *Figures in a Landscape* (1940) and *The Common Heart* (1942), as well as stories that were published in leading magazines. Horgan moved later to ½ Park Road, where he wrote *The Devil in the Desert* (1952) and *Great River, The Rio Grande in North American History* (1954), which won a Pulitzer Prize. The library at the military institute has been renamed The Horgan Library.

SANDOVAL

Richard Wilbur, the poet, in 1952–53 lived in Sandoval, which is northwest of Albuquerque.

SAN PATRICIO

Paul Horgan lived in 1946–47 at the Sentinel Ranch in San Patricio, west of Roswell.

SANTA FE

The long history of Santa Fe, which was founded in 1609–10 by Spaniards and thus is the oldest of all U.S. capital cities, is marked by what often were bloody conflicts between the Pueblo Indians and French and Spanish missionaries. We are not surprised, therefore, that writers who used Santa Fe as a locale have gone beyond the natural beauty of the area in finding their subjects. **Mary Austin** lived in Santa Fe from 1918 until her death in 1934. From 1924, she made her home at 439 Camino del Monte Sol. The house Austin called La Casa Querida (Beloved House) is now a private art gallery. One of her visitors was **Sinclair Lewis,** who met Austin at her home in February 1926. Her first work, a book of sketches entitled *The Land of Little Rain* (1903), described the beauty of life in the Western desert. Austin died in Santa Fe on August 13, 1934, and her ashes were buried in a rock crevice on Mt. Picacho, east of the city. **Adolph Bandelier** and his first wife, Josephine, lived in a large adobe house at 352 East DeVargas Street from 1882 to 1892, while he studied the pueblos and other early settlements. **Witter Bynner,** whose poems about the Pueblo Indians were collected in *Indian Earth* (1929), lived at 342 East Buena Vista Street. The building now is used as private student housing for St. John's College.

Santa Fe is the setting for one of America's great works of fiction, *Death Comes for the Archbishop* (1927), by Willa Cather, who had previously written of the people of the area around Santa Fe in her short novel *My Mortal Enemy* (1926). *Death Comes for the Archbishop* celebrates the tireless and dedicated work of missionaries in the American Southwest. The novel is based on the lives of Bishop Jean Baptiste Lamy and his vicar-general, Father Joseph Machebeuf, called Bishop Jean Latour and Father Joseph Vaillant in the novel. The plot concerns Latour's struggle to have a cathedral built in Santa Fe. A statue of Bishop Lamy stands in front of St. Francis Cathedral, 131 Cathedral Place, in Santa Fe.

The cathedral is mentioned also in the work of **Paul Horgan,** who lived at the Arroyo Chamiso Apartments while writing the novel *A Distant Trumpet* (1960). Horgan's *Centuries of Santa Fe*

(*below*) Statue of Archbishop Lamy standing in front of St. Francis Cathedral in Santa Fe. (*right*) Home of Mabel Dodge Luhan in Taos. Mrs. Luhan, known for her literary salons, wrote of D. H. Lawrence and other writers in her four-volume autobiography, *Intimate Memories* (1933–37).

(1956) gave historical accounts of the city. **Oliver La Farge** came to Santa Fe in 1946 and lived at 647 Old Santa Fe Trail, an unmarked private building that can be seen by appointment. Several of La Farge's books, including the novel *Laughing Boy* (1929), for which he won a Pulitzer Prize, and the novel *The Enemy Gods* (1937) described the lives of the Navajo Indians. La Farge also wrote *Santa Fe: The Autobiography of a Southwestern Town* (1959). The Santa Fe Public Library has named a branch library for him.

D. H. Lawrence and Frieda Lawrence stayed in 1924 at the De Vargas Hotel, 210 Don Gaspar Avenue, still operating today. While here they visited with **Witter Bynner,** who later was to write *Journey with Genius* (1951), about D. H. Lawrence. **Tom Lea,** the novelist, lived during the mid-1930s at the Rancho San Sebastian, about five miles east of Santa Fe. **Lynn Riggs,** the poet and playwright, was in Santa Fe briefly in 1923 and returned in 1939, when he took up residence at 770 Acequia Madie. **Ernest Thompson Seton** lived from 1930 until his death in 1946 at a ranch six miles south of Santa Fe, off Route 84. Seton Village, the site of the ranch, is now a registered National Historic Landmark. His daughter, **Anya Seton,** herself a novelist, made use of a visit here, of memories of her grandmother, and of advice from the novelist Mary Austin to shape the plot of *Turquoise* (1946), her romantic novel about nineteenth-century America.

While **Lew Wallace** was governor of the New Mexico Territory, he wrote part of *Ben-Hur* (1880). As governor, he lived in the Palace of the Governors, on the Plaza, which is now a museum, open to the public. The Palace of the Governors figured prominently in Paul Horgan's novel *From The Royal City* (1936).

Interior view of the D. H. Lawrence shrine at Kiowa Ranch in Taos. The author's ashes are contained in the solid block of concrete bearing his initials.

> *The steady motion of excitement was slowed then, in the last of the day, by the rocks and the piñons, by the reflection of the sky in the pool where flat, vague silhouettes of horses stooped to drink. The voices of many people, the twinkling of fires continued the motif, joining the time of quiet with elation past and to come; a little feeling of expectation in Laughing Boy's chest, a joyful emptiness, part hunger and part excitement.*
>
> —Oliver La Farge, in *Laughing Boy*

TAOS

Taos for many years has attracted writers and artists interested in the heritage of early Indian culture in the magnificent setting of the Sangre de Cristo Mountains. **Willa Cather** came here in 1915 and again in 1916. **Walter Van Tilburg Clark,** author of *The Ox-Bow Incident* (1940), lived here for a time. **William Goyen,** the novelist, also lived here, for varying lengths of time, between 1945 and 1957. He used a Taos setting for his romance *In a Farther Country* (1955), calling Taos El Prado.

Mabel Dodge Luhan, who moved to New Mexico in 1918, attracted many famous writers to her Taos mansion, including **Aldous Huxley, Robinson Jeffers, D. H. Lawrence,** and **Thomas Wolfe.** She described her life here in *Lorenzo in Taos* (1932), which included an account of her friendship with Lawrence. Her house is owned now by a nonprofit corporation called Las Palomas de Taos, and workshops on art and history of the Southwest are held there.

Even if no other writer had ever lived or worked in Taos, its attraction as a literary shrine would be ensured by the presence here of **D[avid] H[erbert] Lawrence,** whose stay was measured in months, yet has provided source material for numerous biographical and scholarly studies and has established Taos as a place of ever-growing interest for travelers with a literary bent. The great English novelist and his wife, Frieda, had come to New Mexico in the autumn of 1922 at the invitation of Mabel Dodge Luhan. After a brief stay in town, the Lawrences moved into a cabin on Lobo Mountain, about two miles below Kiowa Ranch, owned by Mrs. Luhan. Kiowa Ranch, which is fifteen miles north of Taos, on Route 3, became their principal residence in the Taos area. It eventually was given to Frieda Lawrence by Mrs. Luhan and is now owned by the University of New Mexico. Road markers direct visitors to Kiowa.

The first three months the Lawrences spent at Kiowa proved to be a time of such tension—

Cabin in which D. H. Lawrence lived at Kiowa Ranch. Near it stands a reconstructed horno, an Indian adobe oven similar to the one Lawrence used.

> *Lou and Rico had a curious exhausting effect on one another: neither knew why. They were fond of one another. Some inscrutable bond held them together. But it was a strange vibration of the nerves, rather than of the blood. A nervous attachment, rather than a sexual love. A curious tension of will, rather than a spontaneous passion. Each was curiously under the domination of the other. They were a pair—they had to be together. Yet quite soon they shrank from one another. This attachment of the will and the nerves was destructive. As soon as one felt strong, the other felt ill. As soon as the ill one recovered strength, down went the one who had been well.*
>
> —D. H. Lawrence, in *St. Mawr*

principally because of Mrs. Luhan's attentions to Lawrence—that the Lawrences moved on to Del Monte, a ranch seventeen miles away and also owned by Mrs. Luhan. At Del Monte, Lawrence completed a volume of poetry, published as *Birds, Beasts and Flowers* (1923).

The Lawrences interrupted their stay in Taos after six months, in March 1923, to go off to Mexico. They were to visit Mexico one more time, in 1924, before returning to Kiowa Ranch for the last time together. During his time at Kiowa, Lawrence wrote poems, essays, and two novellas, *The Princess* (1924) and *St. Mawr* (1925). Kiowa Ranch today is similar in appearance to the Kiowa of Lawrence's time, despite the fact that some of the ranch outbuildings have been torn down, so there is considerable interest for the visitor who wishes to walk the paths that Lawrence walked and see the vistas Lawrence saw:

> *I . . . stood, in the fierce proud silence of the Rockies, on their foot-hills, to look far over the desert to the blue mountains away in Arizona, blue as chalcedony, with the sage-brush desert sweeping grey-blue in between, dotted with tiny cube-crystals of houses, the vast amphitheatre of lofty, indomitable desert, sweeping around to the ponderous Sangre de Cristo mountains on the east, and coming up flush at the pine-dotted foot-hills of the Rockies! What splendour! Only the tawny eagle could really sail out into the splendour of it all.*

The association of Lawrence with Kiowa did not end with his departure, in 1925. Lawrence died in Vence, in southern France, on March 2, 1930, and was initially buried there. In April 1935, Frieda Lawrence was able to return to Kiowa with an urn that held the novelist's ashes, which she intended to reinter at the ranch. Her journey to Kiowa was beset with problems: the urn was left mistakenly in the New York customs shed, then was left behind on the train platform in Lamy, New Mexico, where friends had gathered to meet the widow, and was left behind once again at a friend's house in Santa Fe. Each time the urn was retrieved, and Lawrence's ashes were finally brought to Kiowa Ranch. Frieda Lawrence had planned an elaborate ceremony for the interment—with Indian dancers in performance—but this arrangement was abandoned. The ashes were mixed in what is described as a ton of concrete and placed in a chapel on the ranch.

WASHINGTON

BELLEVUE

Nard [Maynard Benedict] Jones, author of many works of fiction and nonfiction about Washington, lived in his later years in this eastern suburb of Seattle and died here on September 3, 1972. His novel *Swift Flows the River* (1940) concerned pioneer settlers on the Columbia River, and *Evergreen Land: A Portrait of the State of Washington* (1947) depicted the geographic and cultural variety of the state. **Mary McCarthy,** the novelist, attended Forest Ridge Convent, now Forest Ridge School, at 4800 139th Street, S.E. Even though she became disillusioned with Christianity here, she continued, on her grandfather's orders, to attend church every Sunday.

GOLDENDALE

Frederic Homer Balch, known best for his novel *The Bridge of the Gods* (1890), spent part of his childhood, from 1871 to 1880, in this south-central Washington community. He developed a strong interest in Indian traditions and listened eagerly to elderly Indians telling tales about the history of their tribe.

LYLE

Frederic Homer Balch moved in 1880 to a log farmhouse here, about thirty-five miles southwest of Goldendale and near the Columbia River. Here Balch wrote his first novel, *Wallulah* (unpublished), about Indians of the area. To support himself he worked ten hours a day doing construction work on the Oregon Railway and Navigation line, then under construction on the south shore of the river. While living here Balch became deeply religious. He gave up writing for a time and burned the manuscript of *Wallulah,* which he saw as being full of "agnostic thoughts." In June 1891 Balch died and was buried in a local cemetery, only a few yards from the grave of Genevra Whitcomb, the girl he had loved but never married. The cemetery now is known as Balch Cemetery.

Gravestone of Frederic Homer Balch in Balch Cemetery, Lyle. The novelist, who hoped "to make Oregon as famous as Scott made Scotland," died at the age of twenty-nine, his hopes unrealized.

MARBLEMOUNT

Jack Kerouac worked for eight weeks during the summer of 1956 as a fire-watcher in the Cascades, not far from this town and about one hundred miles northeast of Seattle. From his wooden structure, high on Desolation Peak, he could see the surrounding peaks, such as Mt. Baker and Mt. Hozomeen. He had plenty of time to write, read, think, and meditate, but he found the life too lonely. Kerouac used this experience in *The Dharma Bums* (1958) and *Desolation Angels* (1965).

NEAH BAY

Archie Binns at age eighteen spent nine months on a lightship offshore from this northwestern coastal town south of Vancouver Island, Canada. Binns described his experiences in his novel *Lightship* (1934).

ONALASKA

Robert Cantwell, author of the novel *Laugh and Lie Down* (1931), which described life in a Northwest lumber-mill city, lived as a child in Onalaska, about seventy miles southwest of Tacoma. Onalaska bears a marked resemblance to the town described in his novel.

PORT LUDLOW

Archie Binns was born on July 30, 1899, in this town northwest of Seattle, on Puget Sound,

Old headquarters of the Seattle *Times*, at Fourth and Olive streets, where E. B. White worked in 1922 and 1923. Photograph of Eugene O'Neill and his wife Carlotta Monterey, taken at their home in Seattle in 1936, when O'Neill won the Nobel Prize.

the body of water about which he wrote a comprehensive history, *The Sea in the Forest* (1953). His novel *The Laurels Are Cut Down* (1937) provided a picture of what it was like for a boy to grow up in this area early in the twentieth century.

SEATTLE

Seattle has had its share of authors attracted either as visitors or as residents, as well as a number of native literary sons and daughters. **Sinclair Lewis** and his wife Grace arrived here on September 12, 1916, staying for about a month at the Hotel Chelsea, now the Chelsea Court Apartments, at 620 West Olympic Place. Lewis lectured to journalism students at the University of Washington, did some writing, and explored the city with notebook in hand, showing special interest in the activity down at the Seattle docks. While here he looked up **Anna Louise Strong,** the poet and journalist, whom he had met before and who at the time was the sole woman member of the Seattle school board. Soon she was to begin working for the Seattle *Union Record,* a union newspaper. Strong recorded her Seattle experiences in her memoir *I Change Worlds* (1935).

Eugene O'Neill and his wife Carlotta Monterey in November 1936 rented a home on Magnolia Bluff, overlooking Puget Sound, where he hoped to do a lot of writing. Within a few days, on November 12, O'Neill received word that he had been awarded the Nobel Prize for Literature. The excitement and publicity surrounding the award made it impossible for him to get much work done, so the O'Neills left the city in December to find another place to live and work.

Just out of college, **E[lwyn] B[rooks] White** lived here in 1922 and 1923, working as a reporter for the Seattle *Times,* then located at Fourth and Olive streets and now located at North Fairview and John streets. The newspaper paid White $40 a week and gave him his own column, but White was discharged in June 1923 during a general layoff. James Thurber has said that White left the job because he had become disgusted with the city editor. According to Thurber, White once wrote a story about a man who exclaimed, "My God, it's her!" upon finding his wife's body at the city morgue, and the city editor changed the quote to "My God, it is she."

Thomas Wolfe arrived in town on July 2, 1938, staying at the New Washington Hotel, now the Josephinum Residence for the Aged, at 1902 Second Avenue. He was at that time on a long tour from Portland south to lower California, east to the Grand Canyon, north to Glacier Park, and back west to Seattle. He wrote about this motor tour in *A Western Journal* (published posthumously in 1951). On a boat trip from Seattle to Vancouver and back, Wolfe took sick and was taken to Firlands Sanitarium, north of the city in Richmond Highlands. On August 6 he was transferred to the Providence Hospital on Seventeenth Avenue for X rays and further treatment. He also stayed for a short time at the Spring Apartment Hotel, now the Kennedy Hotel, 1100 Fifth Avenue, before leaving the city on September 6 on the train *Olympian.* He died within ten days, of tuberculosis of the brain.

Many of Seattle's literary figures have been connected with the University of Washington, either as teachers or students. **John Berryman,** the poet, lectured here in 1950; and from 1950 to 1956, **Archie Binns** taught creative writing. **Robert [Emmett] Cantwell** attended classes at the university "for one barren and miserable year," 1924–25. **Richard Eberhart** was poet in residence in 1952–53. **Glenn Hughes** became director of the Drama School in 1930, having made a name for himself with his play *Happiness for Six* (1928) and his critical study *The Story of the Theater* (1928). **Robinson Jeffers** studied forestry here for a year, living at 4215 North East Brooklyn Avenue. **Stanley J. Kunitz** was poet in residence in 1955–56. **Vernon Louis Parrington,** winner of the Pulitzer Prize in 1928 for the first two volumes of his three-volume landmark study *Main Currents in American Thought* (1927–30), taught English here from 1907 until his death on June 16, 1929.

Every night at 8 the lookouts on all the different mountaintops in the Mount Baker National Forest have a bull session over their radios—I have my own Packmaster set and turn it on, and listen. It's a big event in the loneliness.

—Jack Kerouac,
in *Desolation Angels*

I was eleven years old, a seventh-grader, when I was first shown into the big study hall in Forest Ridge Convent and issued my soap dish, my veil, and my napkin ring. The sound of the French words awed me, the luster of the wide moire ribbons cutting, military-wise, across young bosoms, the curtained beds in the dormitories, the soft step of the girls, the curtsies to the floor, the white hands of the music master (a Swedish baron in spats), the cricket played in the playground, the wooden rattle of the surveillante's *clapper.*

—Mary McCarthy,
in *Memories of a Catholic Girlhood*

I was not allowed to go out with boys, but one night the captain of the track team drove up to my house in his roadster and honked for me to join him. My grandfather flashed on the front porch lights and thundered at him to go away, and that was the end of my conquest.

—Mary McCarthy,
in *Memories of a Catholic Girlhood*

Theodore Roethke taught English at the university from 1947 to 1963. It was during this time that he won a Pulitzer Prize for his volume of poetry *The Waking* (1953) and a Bollingen Prize for yet another volume, *Words for the Wind* (1958). He lived at 3802 East John Street and had a summer house on Bainbridge Island in Puget Sound. His frequent leaves of absence from his teaching duties, ascribed to manic-depressive seizures, were defended by the chairman of the English Department, Robert Heilman, citing Roethke's "inestimable service to the University." On August 1, 1963, Roethke was swimming in a neighbor's pool on Bainbridge Island when he suffered a heart attack. Although neighbors quickly pulled him out of the water and volunteer firemen tried to revive him, Roethke died. Funeral services were held at the Episcopal Church of the Epiphany, Thirty-eighth Avenue and East Denny Way.

The list of prominent authors who taught at the University of Washington goes on: **Irving Stone,** known for such biographical novels as *The Agony and the Ecstasy* (1961), taught creative writing here in 1961. **Edward [Charles] Wagenknecht** taught here from 1925 to 1943 and received a Ph.D. from the university in 1932. During the years he lived in Seattle, he published many of his important works: *Values in Literature* (1928), *Utopia Americana* (1929), *A Guide to Bernard Shaw* (1929), *The Man Charles Dickens* (1929), and *Mark Twain, The Man and His Work* (1935). For his first eleven years in Seattle, he lived at 5637 Twentieth Avenue; from 1936 to 1943, at 4703 Thackeray Place. Besides teaching at the university, Wagenknecht served from 1935 to 1940 as literary editor of the Seattle *Post-Intelligencer,* at Sixth and Wall streets, and from 1940 to 1943 as minister of the Friends Memorial Church. **William Carlos Williams,** in the summer of 1948 and again in the fall of 1950, lectured and gave readings of his poetry at the university. **James Arlington Wright,** the poet, attended graduate school here, earning an M.A. in 1954 and a Ph.D. in 1959. Wright was a student of Theodore Roethke's.

Seattle's literary native sons and daughters include **Frederick Faust,** who was born on May 29, 1892, at 1229 Seventh Avenue, in a building no longer standing. Under his own name he wrote two volumes of poetry, *The Village Street and Other Poems* (1922) and *Dionysus in Hades* (1931). It was under the pen name **Max Brand,** however, that Faust was best known. Indeed, he was called king of the pulp writers, specializing in Western novels. Brand's best-known work was *Destry Rides Again* (1930). **Nard [Maynard Benedict] Jones** was born in Seattle on April 12, 1904, and used the city as a setting in his novel *The Petlands* (1931).

Seattle would have ample literary interest even if only one writer, **Mary McCarthy,** had been born there. The novelist was born on June 21, 1912, at 747 East Sixteenth Avenue, in a house still standing but not open to the public. She lived for a time in nearby Bellevue but returned to Seattle at age eleven, an orphan, to live with her grandfather, Harold Preston, a Seattle lawyer. She attended Garfield High School, at 400 East Twenty-third Avenue, for one year, 1925–26, but her grades were poor—by her own account she spent too much time dreaming about certain school athletes—so the next year saw her enrolled in a boarding school in Tacoma.

Another literary native of Seattle was **Audrey [May] Wurdemann,** the youngest poet to win the Pulitzer Prize—she was twenty-three. Wurdemann was born on January 1, 1911, at 218 North Fortieth Avenue, in a home still standing but not open to the public. She did not attend elementary school, but at age eleven entered high school, St. Nicholas School for Girls. She went on to the University of Washington, graduating with honors in 1932. Her first collection of poems, *The House of Silk,* appeared in 1926. The second, *Bright Ambush,* won a Pulitzer Prize after it appeared in 1934. Other collections followed soon afterward: *The Seven Sins* (1935), *Splendor in the Grass* (1936), and *The Testament of Love* (1938).

Seattle has further claim on the attention of readers. Edna Ferber set her historical novel *Great Son* (1945) here. The Seattle waterfront was one of the settings for *Pier 17* (1935), the first published novel by **Walter Havighurst.** *Pier 17* was based on the novelist's memories of the waterfront strikes of 1921–22, when he was a crew member on several ships, including a lumber schooner that worked in Puget Sound. **Martha Ostenso,** known especially for the novel *Wild Geese* (1925), died in Seattle on November 24, 1963, after she had come here with her husband, Douglas Durkin, to visit his sons. Ostenso's later novels included *Milk Route* (1948), *The Sunset Tree* (1949), and *A Man Had Tall Sons* (1958). **James [Floyd] Stevens** lived here in the 1940s, working as public-relations representative for the West Coast Lumbermen's Association. Many of his works, including the tall tales collected in *Paul Bunyan* (1925) and the novel *Big Jim Turner* (1948), dealt with lumbering and the lore of the lumberjack. **Yvor Winters,** the poet and critic, lived with his family here as a teenager from 1912 to 1914. His *Collected Poems* (republished in 1960) won a Bollingen Prize in 1961.

There was too much of everything. But not for Vaughan Melendy. Himself of heroic stature, he fitted well into the gorgeous and spectacular setting that was the city of Seattle. Towering and snow-capped like the mountains that ringed the city, he seemed a part of it—as indeed he was. Born into this gargantuan northwest region of towering forests, limitless waters, vast mountains, fertile valleys, he himself blended into the lavish picture and was one with it. He loved it, he understood it. Breathing deep of its pine and salt air, a heady draught, he digested it like the benevolent giant he was.

—Edna Ferber,
in *Great Son*

SEQUIM

At the time of his death, on June 28, 1971, **Archie Binns,** the novelist, was a resident of this town on Puget Sound, about fifty miles northwest of Seattle.

SPOKANE

Stoddard King, known for his humorous verse and for his column "Facetious Fragments," published in the Spokane *Spokesman-Review* from 1916 to 1933, lived at 447 West Twenty-fourth Avenue until his death on June 13, 1933. Much of his newspaper verse was republished in *Listen to the Mocking Bird* (1928) and in his other collections. The *Spokesman-Review* is still published each morning at 927 West Riverside Avenue.

You may deprive him of his pen and ink,
And still he'll write with fluency and taste;
But he will totter on starvation's brink!
If you remove his scissors and his paste!

—Stoddard King,
in "Epigram on a Hack Writer"

Vachel Lindsay moved here in July 1924, staying in Room 1129 of the Davenport Hotel, 807 West Sprague Avenue. Lindsay's hotel bills were secretly subsidized by a local businessman who admired the poet. Lindsay enjoyed the hotel and gave at least one successful poetry recital in it, but he had complaints over some aspects of the establishment. One of them was the ballroom orchestra. His poem "The Jazz of This Hotel" (1926) elaborated his complaint. While here, Lindsay was befriended by Stoddard King, who made a place in his newspaper column for some of Lindsay's new poems, including "Butterfly Hieroglyphics," "Virginia," and "Nancy Hanks" (all 1926). On May 19, 1925, Lindsay married a twenty-three-year-old high-school teacher, Elizabeth Conner, who became the poet's personal secretary and manager, handling his correspondence and setting up speaking engagements. They lived at first in the Davenport Hotel but, in the spring of 1926, rented a small apartment on a hill behind Cliff Park. Finally, in the fall of 1926, they moved to a larger apartment, at 2318 West Pacific Avenue. The extra space was essential to accommodate their two children, Susan and Nicholas, both born in Spokane, but Lindsay still needed space in which to write. He once again rented his room at the Davenport, and it was there that he wrote two volumes of verse in 1926: *Going-to-the Stars* and *The Candle in the Cabin*. Lindsay began to dislike Spokane after a time and yearned to return to Springfield, Illinois, his hometown. He and his family moved there for good in April 1929.

Why do I curse the jazz of this hotel?
I like the slower tom-toms of the sea;
I like the slower tom-toms of the thunder;
I like the more deliberate dancing knee
Of outdoor love, of outdoor talk and wonder.
I like the slower deeper violin
Of the wind across the fields of Indian corn;
I like the far more ancient violoncello
Of whittling loafers telling stories mellow
Down at the village grocery in the sun;
I like the slower bells that ring for Church
Across the Indiana landscape old.
Therefore I curse the jazz of this hotel
That seems so hot, but is so hard and cold.

—Vachel Lindsay,
in "The Jazz of This Hotel"

SPRINGFIELD

Theodore Dreiser, a year before his death, married Helen Richardson on June 13, 1944, in this small lumber town, about thirty-five miles east of Portland. His wife picked Springfield because it was a stop for Dreiser's New York-Portland train and required couples to wait only four days after applying for a marriage license. The ceremony was performed at 43 Russell Street, in a building no longer standing, by a justice of the peace who had no idea that the groom was a well-known writer. The newlyweds stayed at the Sampson Hotel, which overlooked the Columbia River.

Davenport Hotel in Spokane, and plaque noting the residence here of Vachel Lindsay in the 1920s. Poetry readings honoring Lindsay have been held in the hotel.

VACHEL LINDSAY, THE POET, LIVED HERE
AND WROTE HIS POEMS IN ROOM 1129
FROM JULY 1924 TO JANUARY 1929.
EASTERN WASHINGTON STATE HISTORICAL SOCIETY

The attitude of the Spokane Gentry, who are all millionaires, or pretend to be, is that if I be a good boy all my days, maybe I can be a columnist on the evening paper, or maybe a special writer on the morning paper in the far, far future. They have not the remotest notion they are insolent.

—Vachel Lindsay,
in a letter quoted by Edgar Lee Masters in *Vachel Lindsay: A Poet in America* (1935)

I first heard of Marcus Whitman more than thirty years ago. It was a day in spring and in the time of youth. With other freshmen at Whitman College I climbed a brown hill near Walla Walla, Washington, and looked down upon a quiet valley. I learned that this hill once marked for the weary emigrant the "Place of the Rye Grass"—Waiilatpu was how the Indians said it. Below us was a farm, and somewhere under its fertile earth lay the ruins of a dream begun on December 10, 1836, by two missionaries, Dr. Marcus Whitman and his wife Narcissa. Their mission stood almost in the center of a mysterious, magnetic foreign land called Oregon.

—Nard Jones,
in *The Great Command*

TACOMA

Mary McCarthy from 1926 to 1929 attended an Episcopal boarding school, Annie Wright Seminary, now Annie Wright School, at 827 North Tacoma Avenue, in this city about thirty miles south of Seattle. Here the future author became enthusiastic about her Latin classes, the Latin teacher, and the Latin club, which put on an original play, *Marcus Tullius,* with Mary McCarthy in the cast. She recalled these experiences in *Memories of a Catholic Girlhood* (1957). **James Stevens** lived here from 1924 to 1929, writing and sending articles to *American Mercury* in Baltimore. During this time Stevens published his collection of tall stories, *Paul Bunyan* (1925), and two semi-autobiographical novels, *Brawnyman* (1926) and *Mattock* (1927).

VADER

Robert [Emmett] Cantwell was born on January 31, 1908, in this community, then known as Little Falls, on the Cowlitz River in southwestern Washington. Cantwell became known for his proletarian novels, such as *The Land of Plenty* (1934), which dealt with strikes in the lumber mills.

WAITSBURG

Genevieve Taggard, the poet, was born on November 28, 1894, on an apple farm near Waitsburg, twenty miles north of Walla Walla. Until Taggard was a teenager, she lived with her farming parents at the home of her grandparents, 401 West Sixth Street, no longer standing.

WALLA WALLA

Both **Nard Jones** and **Stoddard King,** a writer of humorous verse, attended Whitman College, at 345 Boyer Street, in this southeastern Washington town. Jones received a B.A. in 1926, and King an M.A. in 1927. During Jones's school years, 1922 to 1926, he worked as a reporter for the local newspaper. After college he wrote the novel *Wheat Women* (1933), set in the countryside surrounding Walla Walla. The history of Walla Walla began with the founding of an Indian mission by the Whitmans, nineteenth-century pioneers and the subject of Jones's biography *The Great Command: Marcus and Narcissa Whitman and the Oregon Country Pioneers* (1959) and of the novel *We Must March* (1925), by Honoré [Willsie] Morrow. The Whitman Mission, seven miles west of town, is now a National Historic Site, open daily. There is no admission charge.

WOODINVILLE

David Wagoner, author of a collection of poems based on the lore and legends of Northwest Coast and Plateau Indians, *Who Shall Be the Sun?* (1978), lived for a time at 12235 Woodinville Drive in this northeastern suburb of Seattle.

OREGON

ALBANY

Samuel L. Simpson, the poet, practiced law in 1867–68 here in western Oregon, about ten miles northeast of Corvallis. Simpson lived and worked with J. Quinn Thornton, whose office was located where Albany's J. C. Penney building now stands. On April 18, 1868, just a few days after Simpson dissolved his partnership with Thornton, the Albany *States Rights Democrat* published what

Joaquin Miller as he appeared a few years before his death in 1913. The cabin in Canyon City in which Miller lived in the late 1860s.

was to become Simpson's best-known poem, "Beautiful Willamette." It brought Simpson recognition that his later poems never matched: Simpson once remarked that the success of this early poem "exercised a sort of tyranny over me." His poetry was collected in the posthumous volume *The Gold-Gated West* (1910).

ANTELOPE

H[arold] L[enoir] Davis, who won a Pulitzer Prize for his novel *Honey in the Horn* (1935), lived in the early 1900s in this small town southeast of The Dalles, a city in northern Oregon. Davis worked on a ranch, set type for the Antelope *Herald* and, at the age of fifteen, played for the town baseball team, the Antelope Maroons. In 1917, Davis revived the defunct *Herald* and ran it until 1921. Antelope, said to be a ghost town now, is on Route 218 east of Route 197.

ASTORIA

This city at the mouth of the Columbia River, in the northwest corner of Oregon, owes much of its small claim on literary history to the vanity of John Jacob Astor, the nineteenth-century tycoon for whom the city was named, and whose philanthropy led to the founding of the great New York Public Library, far across the continent. Working in Astor's home in New York City, Washington Irving put together—for a fee—an adulatory account of Astor's fur trading in the Northwest that was published as *Astoria* (1836). His collaborator in this glorification of the Astor commercial empire was his nephew, Pierre Munro Irving, who later became Irving's literary executor and wrote the first biography of Irving, *The Life and Letters of Washington Irving* (four volumes, 1862–64). A century later, John Jennings wrote the novel *River to the West* (1948) about Astor and his fur trading in the Northwest, and Archie Binns set his novel of the 1890s, *You Rolling River* (1947), in scenic Columbia River country.

BEND

James Stevens worked from 1919 to 1923 as a logger, mostly along the Columbia River, and it is said that he wrote his first story, about Paul Bunyan, in an Oregon logging camp. He lived for a time in this city in west-central Oregon before April 1924, when he began writing for *American Mercury* and moved to Tacoma, Washington. Stevens's books included *Paul Bunyan* (1925) and *The Saginaw Paul Bunyan* (1932).

CANYON CITY

In the 1860s, **Joaquin Miller** brought his wife and child from Eugene to this town about fifty-five miles north of Burns in east-central Oregon. Miller served from 1866 to 1870 as the first judge of Grant County, whose seat is Canyon City, and lived during that time in a four-room cabin that now stands on the grounds of the Grant County Historical Museum. It was while living in Canyon City that Miller's first two volumes of poetry were published, *Specimens* (1868) and *Joaquin et al.* (1869). Miller also helped found the Canyon City Literary Society and planted the town's first orchard. Visitors to Miller's cabin may see photographs of the poet and his family, pioneer artifacts, and other memorabilia. The cabin is open to visitors during limited hours, and admission is charged.

I know a grassy slope above the sea,
The utmost limit of the westmost land.
In savage, gnarl'd, and antique majesty
The great trees belt about the place, and stand
In guard, with mailèd limb and lifted hand,
Against the cold approaching civic pride.

—Joaquin Miller,
in "By the Sun-down Seas"

When Miller returned to visit Canyon City in 1907, he was appalled at the condition of the road linking it with the city of Burns. He wrote a letter to the judges and commissioners of Grant and Harney counties, entitled "A Royal Highway of the World," in which he protested the lack of maintenance on the road. Route 395 now follows the path of the original road and is known as the Joaquin Miller Highway.

CASCADE LOCKS

Frederic Homer Balch wrote of this section of the Columbia River, some forty-five miles west of the city called The Dalles, in his novel *The Bridge of the Gods* (1890), making use of the Indian legend that a natural stone bridge once spanned the Columbia River at Cascade Locks. According to the legend, the god Tyhee Sahale destroyed the bridge in a rage after his two sons had argued over a beautiful young woman. The sons and the woman died in the collapse of the bridge, forming the cascades, a stretch of treacherous rapids. The three were reborn as Mount Hood, Mount Adams, and Mount Saint Helens. It is, of course, the last of these mountains that resumed volcanic activity in 1980. The Cascade Locks were built in 1896 to make this section of the Columbia safer for navigation and, where the legendary bridge is said to have stood, the river is spanned by a toll bridge named the Bridge of the Gods. Over the years, *The Bridge of the Gods* has continued to be read: a 1965 copy of the work indicates that the novel was then in its thirty-third impression.

Other writers have written of the cascades and the Columbia. Washington Irving wrote of explorer Robert Stuart's travels here in 1812, and Nard Jones wrote of the Columbia in his novel *Swift Flows the River* (1940).

COBURG

Joaquin Miller, the poet, first came to Oregon in childhood, arriving by covered wagon in 1852 with his family. The site of his father's homestead in Coburg, just north of Eugene, has been marked by Lane County authorities.

CORVALLIS

Literary residents of this city in northwest Oregon, about thirty miles south of Salem, include

From the Cascades' frozen gorges,
Leaping like a child at play,
Winding, widening through the valley,
Bright Willamette glides away;
Onward ever,
Lovely River,
Softly calling to the sea,
Time, that scars us,
Maims and mars us,
Leaves no track or trench on thee.

—Samuel L. Simpson,
in "Beautiful Willamette"

Over the valley Mount Adams towered, wrapped in dusky cloud; and from Mount Hood streamed intermittent bursts of smoke and gleams of fire that grew plainer as the twilight fell. Louder, as the hush of evening deepened, came the sullen roar from the crater of Mount Hood. Below the crater, the ice-fields that had glistened in unbroken whiteness the previous day were now furrowed with wide black streaks, from which the vapor of melting snow and burning lava ascended in dense wreaths. Men wiser than these ignorant savages would have said that some terrible convulsion was at hand.

—Frederic Homer Balch,
in *The Bridge of the Gods*

Bernard Malamud, who lived in Corvallis for twelve years. Malamud set his novel *A New Life* in an Oregon college town.

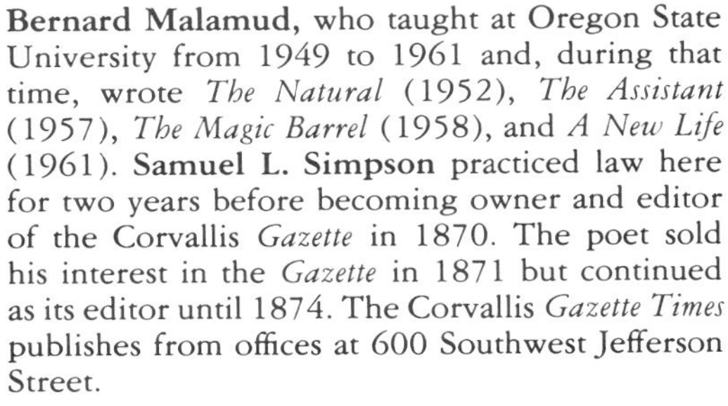

Bernard Malamud, who taught at Oregon State University from 1949 to 1961 and, during that time, wrote *The Natural* (1952), *The Assistant* (1957), *The Magic Barrel* (1958), and *A New Life* (1961). **Samuel L. Simpson** practiced law here for two years before becoming owner and editor of the Corvallis *Gazette* in 1870. The poet sold his interest in the *Gazette* in 1871 but continued as its editor until 1874. The Corvallis *Gazette Times* publishes from offices at 600 Southwest Jefferson Street.

Though a seeker since my birth,
Here is all I've learned on earth,
This is the gist of what I know:
Give advice and buy a foe.
Random truths are all I find
Stuck like burrs about my mind.
Salve a blister. Burn a letter.
Do not wash a cashmere sweater.
Tell a tale but seldom twice.
Give a stone before advice.

—Phyllis McGinley,
in "A Garland of Precepts"

THE DALLES

This city on the Columbia River was named by early French settlers, who thought the stone banks of the river resembled the flagstones, *les dalles,* used on village streets in France. The romance of the name The Dalles carries into the city's single claim to literary fame. Genevra Whitcomb, the girl who inspired the novel by **Frederic Homer Balch,** *Genevieve, A Tale of Oregon* (1932), was an eighteen-year-old in 1886, the year she died here. Genevra was a student at The Dalles Academy, now known as St. Mary's Academy. Although Balch came to know it only after her death, Genevra was the great love in his life. In a cruel twist of fate, Balch, who had ended their relationship to become a Congregational minister—this part of the story is in dispute—presided over her funeral. Balch himself died five years later, and he and Genevra Whitcomb are buried in the same cemetery, near Lyle, Washington, about ten miles northwest of The Dalles.

The prairie! the mighty, rolling, and seemingly boundless prairie! With what singular emotions I beheld it for the first time! I could compare it to nothing but a vast sea, changed suddenly to earth, with all its heaving billows. Thousands upon thousands of acres lay spread before me like a map, bounded by nothing but the deep blue sky. What a magnificent sight!

—Emerson Bennett,
in *The Prairie Flower*

EUGENE

Ken Kesey, the novelist, attended the University of Oregon here in the 1950s. A member of the wrestling team and actor in numerous plays presented on campus, Kesey was graduated in 1957 with a B.A. degree. The University of Oregon claims another former student who went on to write novels and short stories: **Edison Marshall** attended classes here from 1913 to 1915. The university library houses a collection of the papers of Philip Van Doren Stern, the historian and novelist. **Joaquin Miller** in the late 1850s attended Columbia College, a school that came into being here in 1856 but closed its doors a few years later. Miller, an excellent student, wrote the valedictory poem for the 1859 commencement exercises. Three years later, Miller bought the *Democratic-Register,* a newspaper sympathetic toward the Confederate states. He renamed it the Eugene *City Review* and served as its editor, meanwhile studying law. One contributor to the newspaper was Minnie Myrtle Dyer, a poet whom Miller later married. Denied the use of the mails for his newspaper in 1863—by then he called it the *Democratic Review*—Miller sold out and moved to Canyon City.

FLORENCE

George Melvin Miller, a brother of Joaquin Miller, was one of the developers of this coastal town, about fifty miles west of Eugene, but Florence has another, less tenuous claim to literary reputation. **Ken Kesey** worked at a logging camp near here and used his experiences as the basis for his novel *Sometimes a Great Notion* (1964), considered by some to be one of the best novels produced by an Oregon author. Kesey's book, which describes life in a small town on the Oregon coast, was the basis for a 1971 motion picture bearing the same title.

GRANTS PASS

Zane Grey, the Western writer, owned a cabin at Winkle Bar on the Rogue River, near this city in southwestern Oregon. Grey first came here on a fishing trip in September 1924, and returned often in the years before his death, in 1939. Among Grey's writings were "Rocky Riffle on the Rogue," first serialized as "Tales of Freshwater Fishing" in *Field and Stream* in 1926, and "Shooting the Rogue," which was serialized in *Country Gentleman* in the same year. His novel *Rogue River Feud* was published in 1948.

HOOD RIVER

Frederic Homer Balch came in 1886 to this city on the Columbia River northeast of The Dalles. He was pastor of the congregational Church in Hood River and served other settlements in the area as well. Balch lived in a house, still standing, at 3896 Barrett Road. Fascinated since childhood by the culture and history of the Indians of Oregon, Balch was determined to become a writer and "make Oregon as famous as Scott made Scotland; to make the Cascades as widely known as the Highlands . . . to make the splendid scenery of the Willamette the background for romance full of passion and grandeur. . . ." Yet, when he decided to enter the ministry, Balch destroyed the manuscript of a novel he had written—its working title was *Wallulah*—and, according to some literary historians, parted from his sweetheart, Genevra Whitcomb. When the young woman died, Balch came to the realization, too late, that she was his one great love. He also realized that he still wanted to write, so here, in Hood River, Balch began to work on the novel

Genevieve, A Tale of Oregon (1932), to be dedicated to Genevra Whitcomb. He set *Genevieve* aside in 1887 to write the only book he published in his lifetime, *The Bridge of the Gods* (1890). The novel is still read by those who are interested in Oregon and in Indian legend.

In 1889, Balch left Hood River to attend theological school in Oakland, California, but returned to Hood River in March 1891. In a few months he was to die in Portland, Oregon, only twenty-nine years old. In the few years before his death Balch had planned several more books he did not live to write, but *Genevieve* was published in 1932—more than forty years after the death of its author—and the following year, 1933, saw publication of his only other book, *Memaloose, Three Poems and Two Prose Sketches,* in a privately printed limited edition.

Congregational Church four miles south of Hood River, where Frederic Homer Balch preached in the 1880s. Balch's two novels date from his residence in Hood River.

LEBANON

This city east of Corvallis in northwest Oregon was the birthplace, on December 14, 1861, of **Frederic Homer Balch.** Although his exact birthplace is not marked, there is a plaque to Balch's memory in the park near the Lebanon City Library, 626 Second Street.

MEDFORD

Edison Marshall, the novelist and short-story writer, lived in this city in southwest Oregon for many years in the early 1900s. Marshall was known for his adventure stories, such as *The Voice of the Pack* (1920) and *The White Brigand* (1937).

ONTARIO

Phyllis McGinley was born on March 21, 1905, in this town on the Oregon-Idaho border, northwest of Boise. A prolific writer of light verse and children's books, she received a Pulitzer Prize in 1961 for her collection *Times Three* (1960). Her many other books of verse include *Pocketful of Wry* (1940) and *Husbands Are Difficult* (1941). *Province of the Heart* (1959) is representative of her amusing verse essays on life in the suburbs.

OREGON CITY

Of the authors associated with this city, the best known is **Edwin [Charles] Markham,** who was born here on April 23, 1852. Neither his birthplace nor the other house in which his family also lived still stands, but the poet is not forgotten. A marker near Eleventh and Main streets and one outside the Oregon City Public Library, at 606 John Adams Street, commemorate the poet's birth here. Markham lived in Oregon City until he was five years old. He wrote little about Oregon, but his birth here led to his designation as poet laureate of Oregon.

Several other authors are associated with Oregon City: **Eva Emery Dye,** a historical novelist, lived for many years in a frame house at the northeast corner of Ninth and Jefferson streets. Her books included *McLoughlin and Old Oregon* (1900), *Stories of Oregon* (1900), and *McDonald of Oregon* (1906). **Ella [Rhoads] Higginson** lived here as a girl in the late 1800s and, at the age of fourteen, had her first poem published, in a local newspaper. In addition to such collections of poetry as *The Snow Pearls* (1897) and *The Vanishing Races and Other Poems* (1911), Mrs. Higginson wrote a travel book, *Alaska, the Great Country* (1908).

Sidney Walter Moss, an Oregon settler in the 1840s, became a wealthy and active member of the community and something of a local character. Proprietor of the Main Street Hotel, located first at the south end of Main Street and later at the southwest corner of Main and Third streets, Moss is said to have called his guests to dinner by walking up Main Street ringing a cow bell. Moss died in 1901 and was buried in Mountain View Cemetery on Hilda Street. He merits mention in a discussion of Oregon's literary history because he is said to have written much of *The Prairie Flower* (1849), the first novel published in Oregon. **Emerson Bennett,** who wrote many novels and short stories, is usually considered the author of *The Prairie Flower.*

When the lid of the coffin was removed, and I looked upon her dead face with the little heap of salt on the still lips, it all came back, and I knew that I was looking for the last time upon the face of the only girl I would ever love.

—Frederic Homer Balch,
describing the funeral of
Genevra Whitcomb

PORTLAND

The largest city in Oregon and, perhaps, its literary capital, claims only one well-known author as a native, but many who have lived here. **Ella [Rhoads] Higginson,** the poet, lived here in her childhood during the 1870s with her family, first on a farm at Risley's Landing, about three miles south of the city of Milwaukie on Portland's southern boundary, and later in a white colonial house at the northeast corner of Lincoln and Fourth streets. **Stewart H. Holbrook** lived from the late 1940s until his death in 1964 in a house at 2670 Lovejoy Street. Born in Vermont, Holbrook worked as a logger in Oregon before turning to writing. He became associate editor of *The Lumber News* in Portland in 1923, and editor from 1926 to 1934, when he turned to freelance writing. He was also a feature writer for the Portland *Oregonian* from 1930 to 1937. His numerous books included *Ethan Allen* (1940), *Lost Men of American History* (1946), and *Dreamers of the American Dream*

Snapshot of marker outside the Oregon City Public Library, honoring Edwin Markham.

Photograph of Samuel L. Simpson used as the frontispiece of the poet's posthumous collection *The Gold-Gated West.*

> *I went back to Petrograd riding on the front seat of an auto truck, driven by a workman and filled with Red Guards. We had no kerosene, so our lights were not burning. The road was crowded with the proletarian army going home, and new reserves pouring out to take their places. Immense trucks like ours, columns of artillery, wagons, loomed up in the night, without lights, as we were. We hurtled furiously on, wrenched right and left to avoid collisions that seemed inevitable, scraping wheels, followed by the epithets of pedestrians.*
>
> *Across the horizon spread the glittering lights of the capital, immeasurably more splendid by night than by day, like a dike of jewels heaped on the barren plain.*
>
> *The old workman who drove held the wheel in one hand, while with the other he swept the far-gleaming capital in an exultant gesture.*
>
> *"Mine!" he cried, his face all alight. "All mine now! My Petrograd!"*
>
> —John Reed,
> in *Ten Days That Shook the World*

(1957). A frequent contributor to *American Heritage, American Mercury, Reader's Digest,* and *The Saturday Evening Post,* Holbrook died in Portland on September 3, 1964.

John Reed, a native son, was born in Portland on October 20, 1887, at his grandmother's elegant estate, Cedar Hill, which stood at the top of B Street, highest point in Portland. Reed was christened at Trinity Episcopal Church, N.W. Nineteenth Avenue and N.W. Everett Street. While still a boy, he decided to become a writer and, at Cedar Hill, began to write a "Comic History of the United States." It was during this time that he moved with his parents to a house in Portland's West End, and the family later moved once more, to an apartment hotel known as The Hill. Reed attended Portland Academy, where he wrote for the school newspaper. Reed left Portland when he was sixteen to attend preparatory school and, later, Harvard University.

Reed settled finally in New York City but returned to Portland several times as an adult. During one visit, he composed the long poem *The Day in Bohemia, or Life Among the Artists* (1913, privately printed), which described his life in Greenwich Village. In Portland also he first met Louise Bryant, whom he later married. Reed wrote many books, but his best-known work is *Ten Days That Shook the World* (1919), an eyewitness account of the Russian Revolution that was influential in its time. A later edition of this work carried an introduction by Lenin.

Samuel L. Simpson spent his last days at the St. Charles Hotel, at the corner of S.W. Front and S.W. Morrison streets. Simpson, whose greatest success came at the beginning of his career, produced an admirable body of poetry, but none of his poems ever received the recognition accorded "Beautiful Willamette." In the grips of alcoholism, he struggled on, writing in "The Gorge of Avernus" the lines

> *"I have banished the spectre of sorrow,*
> *And conquered the dragon of drink. . . ."*

while asking in his poem "Quo Me, Bacche?":

> *"So, Bacchus, whither dost thou bear*
> *What still is left of me,—*
> *Down to the valley of Despair,*
> *Down to the wailing sea?"*

Simpson fell one day while on a walk and was brought to his room at the St. Charles, where he died on June 14, 1899. He was buried in Lone Fir Cemetery on S.E. Twentieth Avenue. His poems were collected in the posthumous volume *The Gold-Gated West* (1910).

Other writers who lived in Portland include **Frances Fuller Victor.** The poet and historian spent the last four years of her life in near poverty. She lived first in a house at 624 Salem Street for a year, before moving to a boardinghouse at 501 Yamhill Street, where she died on November 14, 1902. Her best-known historical work was her two-volume *History of Oregon* (1886, 1888), written for *History of the Pacific States,* a series edited by Hubert Howe Bancroft. Victor's collection *Poems* appeared in 1900.

C[harles] E[rskine] S[cott] Wood, a poet and lawyer, lived in Portland from the mid-1880s until 1919, when he moved to San Francisco. Wood served as an Army officer in several Indian wars, resigning in disgust at the brutal treatment accorded Indians. As a lawyer, Wood fought for those with little power or privilege, but also represented some of the most powerful corporations on the West Coast. He was an early influence on John Reed, a playmate of Wood's son. Like Wood, Reed came to love adventure but, more importantly, learned to support unpopular individuals and movements. Colonel Wood's favorite of all his own literary efforts was the long poem *The Poet in the Desert* (1915), a colloquy between Truth and a Poet, first published in Portland, as was much of his early writing. His other books written in Portland included *A Book of Tales, Being Some Myths of the North American Indians* (1901), *A Masque of Love* (1904), *Maia* (1918), and *Circe* (1919), but his best-known work, written in California, was *Heavenly Discourse* (1927).

Portland is associated with still more authors. **Frederic Homer Balch** used nearby Sauvie Island as the scene for the main action of his novel *The Bridge of the Gods* (1890), and it was here, at Good Samaritan Hospital, that Balch died on June 3, 1891, at the age of twenty-nine. **Rudyard Kipling** came here in the summer of 1889 on a fishing excursion and later wrote of his experiences in *From Sea to Sea* (1899). The first two books written by Joaquin Miller, *Specimens* (1868) and *Joaquin et al.* (1869), were first published here.

Reed College, a first-rate institution founded in Portland in 1911, has no connection with John Reed, who wrote many articles for *The Masses,* a magazine condemned by political conservatives—*The Masses* was actually suppressed in 1918—but the school for many years was the victim of an insidious misconception. Its first president was William T. Foster, whose name was easily confused with that of William Z. Foster, a labor leader who led a violent strike and later became head of the Communist Party in the United States. As a result, the story was abroad that John Reed had founded Reed College, and that William Foster trained Communists there.

A Portland site with an actual literary association is Mt. Tabor Park, in which stands a statue of Harvey W. Scott. Scott's many articles on Oregon history were collected in the six-volume *History of the Oregon Country* (1924). The statue is not far from the park entrance, at Sixty-ninth Avenue and S.E. Yamhill Street.

PORT ORFORD

This town on the southwest coast of Oregon was the home of Minnie Myrtle Dyer, a poet, who was the daughter of a local judge. **Joaquin Miller** came to visit here in 1862 and, within several days of their first meeting, they were married. Some critics have seen her influence in Miller's writing but, whatever the validity of this claim, the marriage ended in divorce in 1870.

Jack London is said to have worked on his novel *The Valley of the Moon* (1913) at the Knapp Hotel here. Built in the 1860s, the hotel, no longer standing, served as an unofficial lighthouse. The owner would light a lamp in the hotel window to warn vessels away from the dangerous coast. In later years, rooms in the hotel were named after famous persons who stayed here, among them Jack London.

SALEM

Samuel L. Simpson, the poet, was graduated in 1865 from Willamette University in Salem. He served in 1865–66 as editor of the *Oregon Statesman,* now published at 280 Church Street N.E.

SPRINGFIELD

Ken Kesey, the novelist, lived in his teens in this city five miles east of Eugene. Springfield High School, where Kesey was voted the student most likely to succeed, has been replaced by a new building at the original site, 902 North Tenth Street. His family's home during Kesey's years in high school stood at 120 West Q Street.

YONCALLA

H[arold] L[enoir] Davis was born in this town, about thirty-five miles south southwest of Eugene, on October 18, 1896. Davis was known best for his realistic novel of pioneer life in Oregon, *Honey in the Horn* (1935), which won a Pulitzer Prize in 1936. His other books included *Beulah Land* (1949), *Winds of Morning* (1951), and *Distant Music* (1957).

CALIFORNIA

ALAMEDA

In the early 1880s, John and Flora London, the stepfather and mother of **Jack London,** farmed twenty acres here. Alameda was then an island farming community separated by a narrow strip of water from the city of Oakland. Jack London attended his first school here, the West End School. It was while living on his parents' farm that he had his first bout with whiskey, which he later described in *John Barleycorn* (1913). On January 12, 1883, London's seventh birthday, the family moved to a farm near Colma, south of San Francisco. The Alameda land the Londons farmed became in later years part of the site occupied by N. Clark & Sons Pottery, the firm that made the Spanish tiles for the roof of Wolf House, London's home in Glen Ellen.

> *She is a nun, withdrawing behind her veil;*
> *Gray, mysterious, meditative, unapproachable.*
> *Her body is tawny with the eagerness of the Sun*
> *And her eyes are pools which shine in deep canyons.*
> *She is a beautiful swart woman*
> *With opals at her throat,*
> *Rubies at her wrists*
> *And topaz about her ankles.*
> *Her breasts are like the evening and the day stars.*
>
> *She sits upon her throne of light, proud and silent,*
> *Indifferent to wooers.*
>
> —C. E. S. Wood,
> in *The Poet in the Desert*

ALTADENA

Zane Grey, the writer of Western novels, lived in the 1920s and 1930s in this suburb of Pasadena, occupying a three-story house at 396 East Mariposa Street. From 1928 to his death in 1939, Grey used as his office a two-story wing he had added to the house. In the upper story he wrote his books; in the lower story he kept his large collection of trophies and awards. Grey died of a heart attack in this house on October 23, 1939. The house is now private, and the owner reports that he is assembling a private collection of material on Zane Grey.

ANAHEIM

Henryk Sienkiewicz, the Polish author of the novel *Quo Vadis?* (1895), came to this city southeast of Los Angeles in 1876, well before 1905, when he won a Nobel Prize for Literature. He and a fellow countryman were looking for a site for a utopian colony organized by a group that included the renowned actress Helen Modjeska and her husband Count Charles Chlapowski. They selected a ranch just east of Anaheim, at what is now State College Boulevard and Center Street. The members of the colony knew little about farming, so the enterprise did not last long. Another problem was that many of the members became homesick because they could not speak English and the residents of Anaheim could not speak Polish. Madame Modjeska returned to the stage to raise money for the colony, but the ranch was soon sold. Sienkiewicz moved to Los Angeles in 1878 and returned to Poland a year later. While in Anaheim he wrote a series of letters for the Polish newspaper *Gazeta Polska* as well as some sketches. He later wrote of his experiences in his book *Portrait of America* (published in English in 1959). The cabin in which he lived is no longer standing, but the Mother Colony House here, a frame structure built in 1857 and often mistaken for the site of the Polish colony, bears a plaque noting Sienkiewicz's short and unfulfilling stay in Anaheim.

ANGELS CAMP

In the 1850s Angels Camp was the center of quartz mining in Calaveras County, southeast of Sacramento, but its claim to literary fame rests on its association with the writings of **Bret Harte** and **Mark Twain.** Harte became acquainted with this mining community during his travels in the area in the 1850s. He later used it as the setting for his stories "The Bell Ringer of Angels" and "Mrs. Skaggs's Husbands" as well as for several poems, including "The Spelling Bee at Angels." In later years the town established Bret Harte Union High School in his honor. The school also paid homage to Mark Twain, naming its football

> *Waltz in, waltz in, ye little kids, and gather round my knee,*
> *And drop them books and first pot-hooks, and hear a yarn from me.*
> *I kin not sling a fairy tale of Jinny's fierce and wild,*
> *For I hold it is unchristian to deceive a simple child;*
> *But as from school yer driftin' by, I thowt ye'd like to hear*
> *Of a "Spelling Bee" at Angels that we organized last year.*
>
> —Bret Harte,
> in "The Spelling Bee at Angels"

Drawing by J. A. Wilmot of the Angels Hotel Building in Angels Camp, as it appeared in the 1970s.

team the Jumping Frogs. Twain visited Angels Camp during the winter of 1864–65, and it is said that here he first heard of a legendary jumping-frog contest that became Twain's "The Notorious Jumping Frog of Calaveras County." According to legend, Twain heard the story from Ben (or Ross) Coon, bartender at the Angels Hotel, a two-story stuccoed stone building on the corner of Main Street and Birds Way. After returning to San Francisco, Twain wrote his version of the tale as "Jim Smiley and His Jumping Frog," and it appeared first in the New York *Saturday Press* on November 18, 1865. Angels Camp recognizes its literary debt by acting as host to an annual Jumping Frog Jubilee.

APPLE VALLEY

Thomas Duncan owned a house from 1949 to 1968 at 13819 Quinnault Road in Apple Valley, a town northeast of San Bernardino. It was here that he wrote the novel *The Labyrinth* (1967) and began work on *Big River, Big Man* (1959) and *The Sky and Tomorrow* (1974).

AVALON

Writers associated with this resort city on Santa Catalina Island include **Zane Grey,** who had a summer home on the southern tip of the island; and **Gene[va] Stratton-Porter,** the novelist, who lived in Avalon briefly in the early 1920s.

AZUSA

Azusa Pacific College, eighteen miles northeast of Los Angeles, maintains a collection of the research material used by **Irving Stone** for his fictional biographies *Love Is Eternal* (1954) and *Those Who Love* (1965).

BAKERSFIELD

In the early 1940s **Allen Drury,** known best for his Pulitzer Prize-winning novel, *Advise and Consent* (1959), was county editor of the Bakersfield *Californian,* a newspaper still published in this city. Bakersfield is the setting for *Love's Old Sweet Song* (1941), a farce by William Saroyan.

BEL AIR

Gene[va] Stratton-Porter lived in this suburb of Los Angeles in the early 1920s. The novelist was killed on December 6, 1924, in an automobile accident in Los Angeles.

BELMONT

Jack London worked for three months in 1897 in the laundry of Belmont Academy, a school located on the grounds of the present Immaculate Heart of Mary School, 1000 Alameda De La Pulgas. London worked six days a week, and his hours were so long that on his one day off he was too tired to read any of the trunkful of books he kept in his room. He later used his unhappy experiences here in his autobiographical novel *Martin Eden* (1909). **Frank Norris,** the novelist, attended Belmont Academy in 1885.

BENICIA

This town just north of Oakland, and the capital of California in 1853 and 1854, has a tragic claim to literary fame as well as two happier ones. Gertrude Atherton based her novel *Rezanov* (1906) on the life of Doña María Concepción Argüello, a teacher at St. Catherine's Seminary from 1854 to 1857. Doña María had fallen in love with a Russian, Count Rezanov, during his visit to San Francisco, where she lived with her family. He had come to California to purchase supplies for the Russian settlement in Sitka, Alaska. Rezanov returned to Russia to secure the Czar's permission to marry, but died before he could return to California. After waiting in vain for news of him, Doña María finally joined the Dominican Sisterhood. After her death in 1857, she was buried in St. Dominic's Cemetery, located on Route 21. Bret Harte also wrote of the tragic figure in one of

He was dead. His soul seemed dead. He was a beast, a work beast. He saw no beauty in the sunshine sifting down through the green leaves, nor did the azure vault of the sky whisper as of old and hint of cosmic vastness and secrets trembling to disclosure. Life was intolerably dull and stupid, and its taste was bad in his mouth. . . .

—Jack London,
in *Martin Eden*

Father had an excellent library and any book was available to his small son. Which ones Stephen had been browsing in goodness only knows, for he appeared at the lunch table from the library one day, turned to Father, and said gravely: "In the event of William's dying without issue, what do I do?"

"We'll talk about this later," replied his startled parent.

—Laura Benét,
in *When William Rose, Stephen Vincent and I Were Young*

his poems, "Concepcion de Arguello."

Stephen Vincent Benét spent part of his childhood in Benicia in the early 1900s. His father was stationed at the military arsenal here from 1905 to 1911. Benét's brother and sister, **William Rose Benét** and **Laura Benét,** came here during their vacations from college. Laura Benét wrote of their experiences here, especially those of young Stephen, in her book *When William Rose, Stephen Vincent and I Were Young* (1976). **Wilson Mizner,** born here on May 19, 1876, led a colorful life before turning to writing plays and reminiscences about the famous people he knew. His best-known works were the plays *Alias Jimmy Valentine* (with Paul Armstrong, 1909) and *The Only Law* (with Bronson Howard, 1909).

BERKELEY

This city is named in honor of **George Berkeley,** the Irish-born Anglican bishop, who traveled to the New World in 1729 to found a college in Bermuda, and whose poetical words "Westward the course of Empire takes its way" inspired the town's trustees to name their community after him. Berkeley, of course, is the home of the University of California. Among the numerous writers associated with the university were **Frank Norris,** who studied here from 1890 to 1894 but left before earning his degree; and **Mark Schorer,** who taught at the university from 1945 to 1973 and made his home at 68 Tamalpais Road. During his residence in Berkeley, Schorer published a number of his best-known books, including the biographies *William Blake* (1946), which took ten years to write; *Sinclair Lewis: An American Life* (1961); and *D. H. Lawrence* (1968). **Louis Simpson** taught at the university from 1959 to 1967 and while here won a Pulitzer Prize for his collection of poetry *At the End of the Open Road* (1963).

Lincoln Steffens, a student here from 1885 to 1889, changed from a rowdy youth, involved in such escapades as smashing furniture in the president's house and stealing chickens from a professor's coop—the professor later invited the culprit to Sunday chicken dinner—into a thoughtful young man who served as commandant of the university's cadet corps. He worked as a drill instructor at another school in Berkeley during his last two years at the university, and then launched his career in journalism. **Irving Stone** received a B.A. degree from the university in 1923 and later wrote about his undergraduate years in his novel *Pageant of Youth* (1933). Stone lived in Berkeley for several years after graduation and taught at the university from 1924 to 1926. In 1937, while living in Berkeley, Stone and his wife Jean edited the letters of Vincent Van Gogh to his brother Theo. The letters were published as *Dear Theo: The Autobiography of Vincent van Gogh* (1937).

Writers who lived in Berkeley include **Ambrose Bierce,** who had a comfortable apartment here in the 1890s; **Allen Ginsberg,** who lived in the 1950s at 1624 Milvia Street; **Jack Kerouac,** who visited Ginsberg in Berkeley and later lived briefly in a house, no longer standing, at 1943 Berkeley Way. **Richmond Lattimore,** the poet and critic, was graduated from high school in Berkeley in the 1920s, and **Thornton Wilder** completed his last two years of high school here.

BEVERLY GLEN

Henry Miller lived in 1942 in a three-room cabin about a mile from the UCLA campus. The cabin, at 1212 Beverly Glen, stood in the hills north of Bel Air. While here, Miller began to paint, and a gallery in Hollywood offered to give a showing of his paintings.

BEVERLY HILLS

Many writers, some of them drawn to California by the chance to write for motion pictures, have lived in this posh Hollywood suburb. In early 1935 **Edgar Rice Burroughs** moved with Florence, his second wife, to a house, still standing, at 806 North Rodeo Drive. In August of 1939 the Burroughs family moved to a house, also still standing, at 716 North Rexford Drive, where they lived for eight months. Burroughs wrote a number of books while living in Beverly Hills, including *Back to the Stone Age* (1937), *Tarzan the Magnificent* (1939), *Carson of Venus* (1939), and *Synthetic Men of Mars* (1940). **Erskine Caldwell** lived in Beverly Hills, while working as a Hollywood scriptwriter, three different times during the 1930s and 1940s. Caldwell met **Theodore Dreiser** for the first and only time at a Beverly Hills restaurant, where they were luncheon guests of the motion-picture agent who represented both of them. **Thomas Duncan** has reported that one of his favorite spots in Beverly Hills is Dave Chasen's Restaurant, at 9039 Beverly Boulevard, frequented by many writers and motion-picture people.

Rachel Field spent the last four years of her life at 714 North Camden Drive. The poet and novelist, whose last published work was *And Now Tomorrow* (1942), died on March 15, 1942, in a Los Angeles hospital after an operation. **Oscar Hammerstein,** the librettist, lived in the mid-1930s in a French provincial house, still standing, at 1106 Benedict Canyon Drive. John O'Hara used Beverly Hills as a setting in his novel *Hope of Heaven* (1938). In the late 1960s and early 1970s, **Budd Schulberg,** the novelist and short-story writer, lived at 1666 North Beverly Drive. **Upton Sinclair,** finding that his political activities had made him a celebrity, could find no peace at his home in Pasadena, so he and his wife spent part of the years 1933 to 1936 living in Beverly Hills, at 614 North Arden Drive, where they were able to avoid well-wishers and cranks alike.

In 1946 **Irving Stone** and his wife built a house at 717 North Maple Drive, where they made their home until they left for Europe in the 1950s. On their return in 1958, they rented a house at 301 North Alpine, where Stone wrote *The Agony and the Ecstasy* (1961), his novel based on the life of Michelangelo. Stone has reported that Ah Fong's at 424 North Beverly Drive and The Bistro at 246 Cañon Drive North are two of his favorite Beverly Hills restaurants.

Other writers who have lived in Beverly Hills include **Robert Lewis Taylor,** who rented a house here while gathering material for *W. C. Fields, His Follies and Fortunes* (1949); **James Thurber,** who lived in a bungalow during the summer of 1939 while collaborating with Elliot Nugent on *The Male Animal* (1940); and **Dalton Trumbo,** the novelist and screenwriter, who lived at 121 South Beverly Drive in the 1940s.

> *There is a fascination in places that hold our past in safekeeping. We are drawn to them, often against our will. For the past is a shadow grown greater than its substance, and shadows have power to mock and betray us to the end of our days. I knew it yesterday in that hour I spent in the storeroom's dusty chillness, half dreading, half courting the pangs which each well-remembered object brought.*
>
> —Rachel Field, in *And Now Tomorrow*

Photograph of Frank Norris taken about the turn of the century by Arnold Genthe, the well-known photographer.

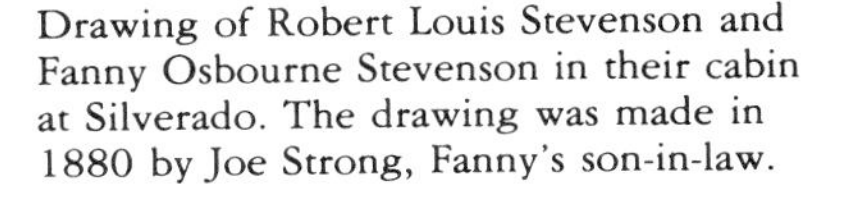

Drawing of Robert Louis Stevenson and Fanny Osbourne Stevenson in their cabin at Silverado. The drawing was made in 1880 by Joe Strong, Fanny's son-in-law.

BIG SUR

Jack Kerouac used this coastal community and the area around it as the setting for his novel *Big Sur* (1962). From 1941 to 1960 **Henry Miller** made his home in a rugged and beautiful section of California's coast near the village of Big Sur. Among the works he completed during these years were *The Rosy Crucifixion,* comprising *Sexus* (1949), *Plexus* (1953), and *Nexus* (1960); and *Big Sur and the Oranges of Hieronymus Bosch* (1958), which discussed his life in California.

BISHOP

Mary Austin, the novelist and short-story writer, taught from 1895 to 1897 at Bishop Academy in this town in east-central California.

I never saw a Purple Cow;
I never Hope to See One;
But I can Tell you, Anyhow,
I'd rather See than Be One.

—Gelett Burgess,
"The Purple Cow"

CALISTOGA

After their marriage in May 1880, Scottish novelist **Robert Louis Stevenson** and Fanny Osbourne traveled with Mrs. Osbourne's son Lloyd to this community northeast of Santa Rosa in the belief that they would have no trouble finding a place to live. Their romantic notion was to come upon an abandoned house near the worked-out mines in the area, and they eventually settled in a three-story cabin at the entrance to the Silverado Mine, some miles northwest of Calistoga, on the slope of Mount St. Helena. They stayed on until July, while Fanny nursed the ailing Stevenson back to a degree of health and Stevenson made notes for his book *The Silverado Squatters* (1883), which recounts this happy period in his life. The area around the mine is now Robert Louis Stevenson State Park, and a monument in the form of a book atop a pillar marks the site of the cabin in which they lived.

CARMEL

From the early 1900s this community on the wooded coast of central California has been known as a refuge for writers and artists. The first writer to move to Carmel was a poet, **George Sterling,** who left his job in his uncle's San Francisco insurance office to build a cabin at what is now the southeast corner of Tenth and Torres streets. Sterling's works included *Thirty-five Sonnets* (1917) and *Selected Poems* (1923). His home soon became a gathering place for the many writers and artists who moved to Carmel. Among the writers who gathered at Sterling's home was **Mary Austin,** known for novels and other works on California and Indian themes. She acquired ten wooded lots near Third and Lincoln streets, on which she first built a small cottage and then Wickiup, a structure resembling a treehouse. There she completed her novel *Isidro* (1905), a romance set in California during the time of Mexican rule. Austin later moved out of Wickiup, taking a house nearby that now is a private residence, and storms destroyed Wickiup in 1914. Austin traveled in Europe during the years 1907 to 1912, and returned to Carmel only three times before selling her house in 1924 and settling in New Mexico.

Van Wyck Brooks lived in Carmel several times during his career as biographer and critic. After marrying Eleanor Kenyon Stimson here in 1911, he stayed on to write *The Ordeal of Mark Twain* (1920) and *The Pilgrimage of Henry James* (1925). **Gelett Burgess,** the humorist, lived in Carmel for two years before his death on September 18, 1951, at the Community Hospital of the Monterey Peninsula, on W. R. Holman Highway. **Holman Day,** the novelist, poet, and screenwriter, lived in Carmel for a time before moving to Hollywood in 1922. **Shirley Ann Grau** wrote her novel *The Hard Blue Sky* (1958) while living here. **Langston Hughes,** while living in Carmel, wrote a number of short stories that later were collected in the volumes *The Ways of White Folks* (1934) and *Laughing to Keep from Crying* (1952).

Carmel and the Monterey Peninsula area were the settings for a number of the works of **Robinson Jeffers.** Jeffers came to Carmel after one of his uncles had died, in 1914, and left the poet an inheritance substantial enough to permit him to concentrate on writing. Jeffers and his wife Una bought property on Carmel Point in 1919 and began building Tor House, their magnificent stone home at 26304 Ocean View Drive. Jeffers supervised all the work he did not actually do himself. Using native granite boulders rolled up from the beach just below the building site, Jeffers did much of the stonework in the main house and adjacent forty-foot tower during his forty-three-year residence here. He wrote nearly all his later books and poems in the stone tower,

Here were new idols again to praise him;
I made them alive; but when they looked up at the face before they had seen it they were drunken and fell down.
I have seen and not fallen, I am stronger than the idols,
But my tongue is stone how could I speak him? My blood in my veins is seawater how could it catch fire?

—Robinson Jeffers,
in *The Women at Point Sur*

which is connected to the main house by stone walls that form a courtyard. He lived a life of seclusion, interrupted only rarely by visits of a few close friends and by travels in the West and in the British Isles. Among the many works Jeffers produced here were *Roan Stallion, Tamar, and Other Poems* (1925), *The Women at Point Sur* (1927), *Solstice* (1935), and *Medea* (1946). Jeffers died on January 20, 1962, at Tor House, which is occupied by his son and family, and has recently been purchased by the Tor House Foundation. It will one day be open to visitors.

Sinclair Lewis came to Carmel in January 1909 and took a position as secretary to two authors, Alice MacGowan and Grace MacGowan Cooke, now forgotten. While living here, Lewis first met William Rose Benét, who traveled from his home in Benicia to meet Lewis. After leaving Carmel, Lewis returned once more, in 1933, and stayed at the Del Monte Lodge. **Jack London** and his wife Charmian visited Carmel in the early 1900s, but decided to settle in Glen Ellen rather than Carmel. George Sterling and James Hopper, who were close friends of London's, appear thinly disguised as the characters Mark Hall and Jim Hazard in London's novel *The Valley of the Moon* (1913), which described the lives of London's friends in Carmel. Sterling, who died a suicide in San Francisco on November 17, 1926, is said also to have been the prototype for Brissenden, a poet in *Martin Eden* (1909). In the novel Brissenden met the same fate as Sterling.

Other writers associated with Carmel include **Georges Simenon,** the Belgian novelist, who completed several of his novels featuring Inspector Maigret while staying here in 1949 and 1950; **Upton Sinclair,** who spent several months here in the winter of 1908–09; and **Lincoln Steffens,** who from 1927 until his death on August 9, 1936, made his home on San Antonio Avenue. This house, the second one southeast of Ocean Avenue, is still standing. Steffens's *Autobiography* (1931), still read, recounts much of the political life of his time, in which he played an active part. **Harry Leon Wilson,** author of *Ruggles of Red Gap* (1915) and of *Merton of the Movies* (1922), later dramatized by Marc Connelly and George S. Kaufman, lived in Carmel Highlands from 1912 until his death on June 28, 1939.

CLAREMONT

Ruth Suckow, from 1952 until her death on January 23, 1960, lived in a house, now private, at 420 West Eighth Street. While Suckow was living here, she wrote *The John Wood Case* (1959), a novel about the effects of an embezzlement on the friends of the embezzler. Claremont, about thirty miles east of Los Angeles, is the home of Pomona College. **Wright Morris,** a student at the college from 1930 to 1933, used it as the setting for *The Huge Season* (1954), in which Pomona appears as Colton College. **Willard Huntington Wright** who, under the name **S. S. Van Dine,** created the fictional detective Philo Vance, studied at the college in 1903–04.

At the hotel he hurried up to Brissenden's room, and hurried down again. The room was empty. All luggage was gone.

"Did Mr. Brissenden leave any address?" he asked the clerk, who looked at him curiously for a moment.

"Haven't you heard?" he asked.

Martin shook his head.

"Why, the papers were full of it. He was found dead in bed. Suicide. Shot himself through the head."

—Jack London,
in *Martin Eden*

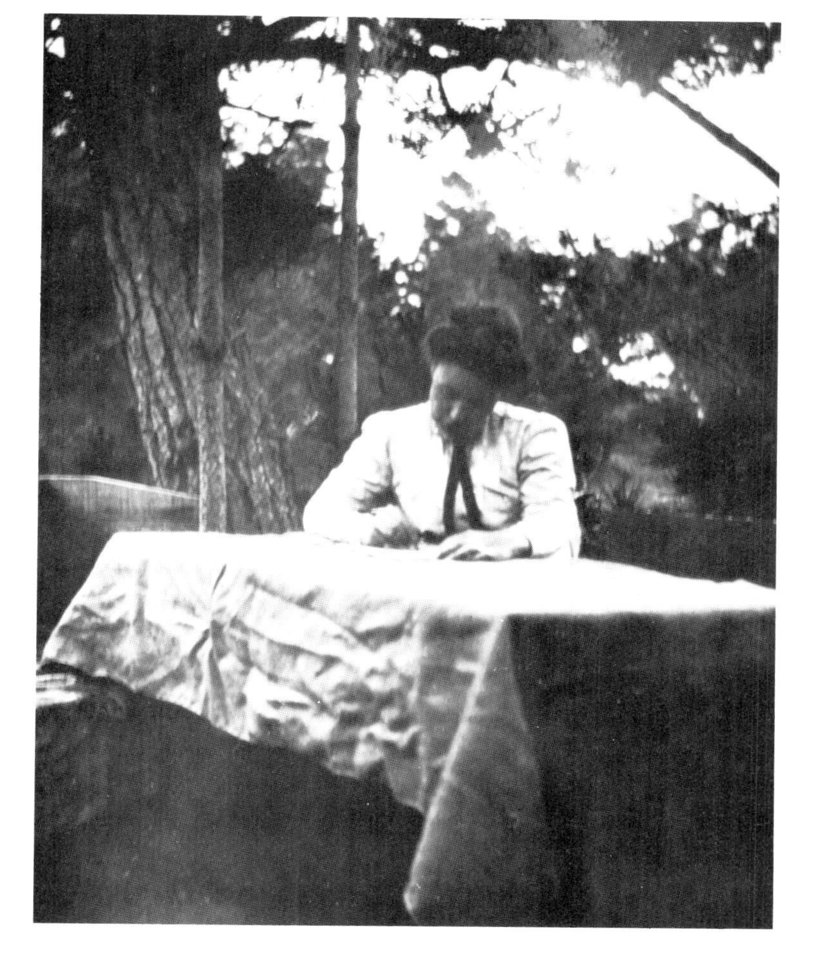

Mary Austin at work in her wickiup at Carmel in the early 1900s. Tower at Tor House, home of Robinson Jeffers in Carmel. The poet built the tower himself.

> *San Juan is the only romantic spot on the coast. The country here for several miles is high table-land, running boldly to the shore, and breaking off in a steep cliff, at the foot of which the waters of the Pacific are constantly dashing. For several miles, the water washes the very base of the hill, or breaks upon ledges and fragments of rocks which run out into the sea.*
>
> —Richard Henry Dana, Jr., in *Two Years Before the Mast*

COLFAX

Frank Norris, ill with tropical fever, returned to San Francisco in the spring of 1896 from South Africa, where he had gone to write travel sketches. He eventually came to Colfax, about forty miles northeast of Sacramento, to work at the Big Dipper Mine, where he completed *McTeague* (1899), his naturalistic novel about life in San Francisco.

COLMA

On January 12, 1883, the seventh birthday of **Jack London,** he and his family moved from Alameda to a seventy-five-acre ranch near the small community of Colma, just south of San Francisco. The two years the Londons spent here were unhappy ones for Jack, broken only by occasional outings he enjoyed with his stepfather, John London, on the shore of San Francisco Bay, where they would gather clams or admire the natural beauty of the area. Since 1901 Colma has been the burying ground for the dead of San Francisco, due to an ordinance prohibiting burial within the city and county of San Francisco, and much of Colma's area is taken up by cemeteries. Among those buried here are **Ina Coolbrith,** the poet; **William Randolph Hearst,** and **Richard Realf,** the poet, all of them buried in Cypress Lawn Cemetery.

COLOMA

In the 1870s **Edwin Markham,** the poet, lived in this village northeast of Sacramento and taught in a one-room schoolhouse here.

CORONA

For a time in the 1940s **Upton Sinclair** and his wife lived in a cottage in the hills near this city southwest of San Bernardino. They returned in the 1950s after Sinclair's wife had suffered a heart attack. Sinclair characterized their home, in *Autobiography of Upton Sinclair* (1962), as "the most comfortable cottage we ever had."

CORONADO

From September 1913 to January 1914, **Edgar Rice Burroughs** rented a house at 550 A Avenue in this community across the bay from San Diego. **Henry James,** on a tour of the United States in 1905, stayed at the Hotel Del Coronado here. It is a massive resort hotel sporting numerous gables, turrets, and towers, and now is listed in the National Register of Historic Places.

COVELO

Percy Marks, born on September 9, 1891, in this town in northern California, first came to public attention as a writer in 1924, on publication of his novel of college life, *The Plastic Age,* which caused a stir by virtue of its depiction of life in the jazz age. His next novel, *Martha* (1925), about a half-Indian girl, used Covelo as a setting. Marks's birthplace has been replaced by the local bank building.

DANA POINT

This village, on a rocky stretch of coast overlooking the Pacific midway between Los Angeles and San Diego, is named for **Richard Henry Dana, Jr.,** as is Dana Point Harbor, over which rise steep cliffs. Dana visited this part of California when he was a seaman aboard the ship *Pilgrim,* a hide trader out of Boston. When the ship anchored off the point in 1835 to gather hides from nearby San Juan Capistrano Mission, Dana found time to explore the coast and made notes he would later use in writing *Two Years Before the Mast* (1840).

Tao House, home of Eugene O'Neill in Danville. O'Neill wrote his last plays during the seven years he lived in this house.

DANVILLE

Tao House, at 1000 Kuss Road in the mountains near Oakland, was the home of **Eugene O'Neill** and his wife from 1937 to 1944. O'Neill and Carlotta Monterey built Tao House, whose Chinese name means "the right way of life," and from it they could view the San Ramon Valley and mist-shrouded Mount Diablo. O'Neill worked five to eight hours each day, when his health permitted, in his second-floor study, completing some of his best-known plays, among them: *The Iceman Cometh* (1946) and *Long Day's Journey Into Night,* which was published posthumously in 1956. When debilitating illness put an end to O'Neill's career as a playwright, he burned all his unfinished manuscripts and sold Tao House. Today it is owned by the Eugene O'Neill Foundation, Tao House. Listed in the National Register of Historic Places, Tao House is not yet open to the public, but the foundation expects to open it to visitors in the future.

EL CENTRO

Nathanael West and his wife, Eileen McKenney West, were killed on December 22, 1940, when their car was involved in a collision at the intersection of Routes 111 and 180, a few miles east of this city on the Mexican border. The novelist and his wife had been on their way home from a hunting trip, and West apparently failed to obey a stop sign at the intersection. West, who was known to let his attention wander while driving, may have been preoccupied with thoughts of his friend F. Scott Fitzgerald, who had died the day before.

ENCINO

Writers associated with this community northwest of Los Angeles include **Edgar Rice Burroughs,** who lived at 5465 Zelzah Avenue from December 1945 until his death on March 19, 1950. Although Burroughs suffered increasing pain from heart trouble in the last years of his life, he continued to work on his popular stories: about fifteen were in process at the time of his death. From October 1938 to May 1940, **F. Scott Fitzgerald** lived in the guest house at Belly Acres, Edward Everett Horton's estate at 5521 Amestoy Avenue, now Edward Everett Horton Lane. Horton's estate has since been demolished. While here, Fitzgerald wrote a number of short stories and worked on the draft of his unfinished novel, *The Last Tycoon,* which was published posthumously in 1941. **Irving Stone** lived from 1936 to 1946 at 16751 Oak View Drive. Here he wrote *Clarence Darrow for the Defense* (1941), *They Also Ran* (1943), and *Immortal Wife* (1944).

ESCONDIDO

Harold Bell Wright spent the last ten years of his life at Quiet Hills Farm, his home near this city north of San Diego. The novelist sold the farm and moved to San Diego in 1944, a month before his death.

FIDDLETOWN

This quiet community in the foothills of the Sierra Nevada Mountains east of Sacramento took its name from its residents' delight in fiddle music. Bret Harte used the mining community that thrived here in the 1850s as the setting for his story "An Episode of Fiddletown." Fiddletown is designated a California Historical Landmark, and many of its buildings, dating back to the 1850s, are representative of the architecture of mining towns of the period.

FRESNO

For readers and playgoers, Fresno is Saroyan, and Saroyan is Fresno. **William Saroyan** was born on August 31, 1908, in a house, no longer standing, at 484 I Street. His father died when Saroyan was three, and the boy was sent to live in an orphanage in San Jose. He remained there for five years, when his widowed mother finally could afford to bring her large family together again in Fresno. Saroyan grew up from then on in the city that would figure prominently in many of his works, living first at 2226 San Benito Avenue and then at 3204 El Monte Way, the latter home still standing. Although he early showed a bent for learning and tried to read his way through the Fresno Public Library, Saroyan left school in 1921 and worked at a succession of jobs. He began to send stories to publishers when he was sixteen years old, but it was not until he was twenty-five that his first story, "The Broken Wheel" (1933), appeared in print, in an Armenian magazine called *Hairenik.* Saroyan came to the attention of a wider audience when the story was collected in *Best Short Stories of 1934,* edited by Edward J. O'Brien. In Saroyan's later years he divided his time between

Mrs. Tretherick faced quickly about. Standing in the doorway was a little girl of six or seven. . . .

"Whose child are *you?" demanded Mrs. Tretherick still more coldly, to keep down a rising fear.*

"Why, yours," said the little creature with a laugh.

—Bret Harte,
in "An Episode of Fiddletown"

Photograph of Fiddletown as it appeared about fifty years ago. From 1878 to 1932, Fiddletown, known to readers through Bret Harte's story "An Episode of Fiddletown," was called Oleta.

Passport photograph of William Saroyan taken in 1978, about two years before the author's death.

Do not pay any attention to the rules other people make. Forget Edgar Allan Poe and O. Henry and write the kind of stories you feel like writing. Learn to typewrite, so you can turn out stories as fast as Zane Grey.

—William Saroyan, in the preface to *The Daring Young Man on the Flying Trapeze*

Fresno and San Francisco, but he maintained close ties until his death with the city of his birth and its Armenian-American working people. Saroyan's last address in Fresno was 2729 West Griffith Way, his home at the time of his death. Among Saroyan's works with Fresno settings are his story collection *The Daring Young Man on the Flying Trapeze* (1934); the Pulitzer Prize-winning play *The Time of Your Life* (1939)—Saroyan declined the award—and another story collection, *My Name Is Aram* (1940). Saroyan died on May 18, 1981, at the Veterans Administration Hospital in Fresno. The Associated Press has reported that Saroyan called five days before his death to leave a final message:

> *Everybody has got to die, but I have always believed an exception would be made in my case. Now what?*

Armine von Tempski, known best for her autobiographical account of her youth in Hawaii, *Born in Paradise* (1940), died in Fresno on December 2, 1943.

FULLERTON

Jessamyn West, the novelist and short-story writer, was graduated in 1919 from Fullerton Union High School, 201 East Chapman Avenue, and then spent a year at Whittier College before returning to attend Fullerton Junior College for two years.

GILROY

Frank Norris owned a ranch near this city, southwest of San Jose. The novelist had hoped to live and write on his ranch, but his sudden death on October 25, 1902, after an appendectomy in San Francisco, ended this dream. His cabin, ten miles west of Gilroy off Route 152, is private and listed in the National Register of Historic Places.

GLENDALE

James M. Cain used Glendale, a suburb of Los Angeles, as the setting for his novel *Mildred Pierce* (1941). In 1919 **Theodore Dreiser** and his wife Helen bought a small white cottage near Pioneer and Columbus streets in Glendale. They lived there until 1923, and Dreiser used a corner of the kitchen to work on his collection of poems *Moods* (1926), his autobiographical *A Book About Myself* (1922, republished in 1931 as *Newspaper Days*), and his essays on New York City, *The Color of a Great City* (1923). He also began work here on his highly acclaimed *An American Tragedy* (1925). Adjacent to the cottage was a corner lot on which Dreiser grew zinnias that were the envy of the neighborhood. Much later, from December 1938 to May 1939, the Dreisers lived in an apartment at 253-A West Lorraine, no longer standing, where Dreiser worked on the unfinished philosophical book that he intended to call *The Formulae Called Life* or *The Formula Called Man.* Following Dreiser's death in Hollywood on December 28, 1945, funeral services were held for him at Forest Lawn Church of the Recessional, and he was buried in Forest Lawn Memorial Park, 1712 South Glendale Avenue. The bronze tablet marking his grave bears lines from his poem "The Road I Came." Also buried in Forest Lawn is **Johnston McCulley,** the novelist and playwright, who died in Hollywood on November 23, 1958.

GLEN ELLEN

The Jack London State Historic Park, just west of this community in the wine-growing region northwest of Sonoma, preserves nearly fifty acres of 1300-acre Beauty Ranch, once owned by **Jack London.** The novelist was attracted to Glen Ellen in the early 1900s and rented a cottage on the grounds of Wake Robin Lodge, the home of Ninetta Eames, who had interviewed him for a magazine article in 1900. Following the failure of his marriage to Bess Maddern, London married Mrs. Eames's niece, Charmian Kittredge. In 1905 he bought 130 acres of neglected farmland and began to build the ranch of his dreams. He wanted to create a self-sufficient farm community that would incorporate all modern techniques of agriculture and animal husbandry. He interrupted his work on Beauty Ranch almost immediately to begin building the ketch *Snark,* intended for a seven-year voyage around the world. The voyage, begun in April 1907, was abandoned in the South Pacific twenty-seven months later when London, his health shattered, had to leave the *Snark.* He returned to Glen Ellen in 1909 to find himself nearly bankrupt and his reputation as a writer in jeopardy. In the next few years, he reestablished his position in American letters and devoted more and more energy to developing Beauty Ranch.

In 1910 London began to build Wolf House at Beauty Ranch. The twenty-three-room mansion was constructed of native lava boulders and redwood. Intended to last for centuries, Wolf House was nearly complete by August 1913, when it was suddenly destroyed by fire. Surveying the destruction of his greatest dream, London is said to have remarked: "I would rather be the man whose house was burned than the man who burned it." The fire may have been set deliberately, but this has never been proved. London intended to rebuild Wolf House but never did. He and Charmian lived on in a small cottage London had built in 1911, and London remodeled and added to the cottage while turning out books and stories to pay his ever-mounting debt: in addition to buying adjacent properties when they became available, London renovated existing buildings, built new cement-block silos, built a modern piggery, built a horse barn, and imported the finest livestock. He planted nearly 140,000 eucalyptus trees as an investment, never realized, and kept an ever-increasing number of people on his payroll. Meanwhile, he also managed to travel to Mexico and to Hawaii, and to sail aboard his yawl *Roamer.* His interest in the land can be seen in several of his later books, including *Burning Daylight* (1910), *The Valley of the Moon* (1913), and *The Little Lady of the Big House* (1916).

In London's final years his health failed rapidly, and he began to rely on drugs to relieve his severe pain. On November 22, 1916, he died at the cottage on his beloved Beauty Ranch after injecting himself with a fatal dose of morphine. His ashes were buried on the ranch, beneath a maroon lava boulder standing near the graves of two pioneer children. In 1919 Charmian built The House of Happy Walls on the ranch. It is a spacious stone structure that served as her home until her death in 1955. It now is a museum containing much of the specially designed furniture that had been intended for use in Wolf House, London's desk and Dictaphone, and other items relating to his life and travels. Along a ranch trai! nearby are many of the buildings built or renovated by London, the ruins of Wolf House, and London's gravesite. The cottage in which London lived is not yet open to the public, but Jack London State Historic Park is open daily.

GRASS VALLEY

Wallace Stegner used Grass Valley, a mining town forty-five miles west of Lake Tahoe, as a setting in his Pulitzer Prize-winning novel, *Angle of Repose* (1971). **Mary Hallock Foote,** author of novels and stories of Western life, lived in Grass Valley from 1895 to 1929. Her husband was hired in 1895 to build a power house at the North Star Mine on Lafayette Hill, and he stayed on to work as manager of the mine until 1913. From 1895 to 1906 the Footes lived in the North Star Mine Cottage, on Old Auburn Road, about two miles south of the city limits. During these years Mrs. Foote wrote *The Little Fig Tree Stories* (1899), *The Prodigal* (1900), and *The Desert and the Sown* (1902). From 1906 to 1929, the Footes lived at North Star House, also on Old Auburn Road and now owned by the North Star Conference Center, a religious organization. The house is being considered for designation as a State Historical Landmark because of its architecture. Mrs. Foote's literary output at the cottage included *The Royal Americans* (1910), *The Valley Road* (1915), and *Edith Bonham* (1917). Mary Hallock Foote died on June 25, 1938, and was buried in Greenwood Cemetery, on Marysville Highway.

HEMET

In the 1920s **Jessamyn West** taught in a one-room schoolhouse near this city in southeast California, east of Los Angeles.

HIGHLAND PARK

While **Robinson Jeffers** attended Occidental College in nearby Los Angeles in the early 1900s, his family lived in Highland Park, in a two-story frame house, now private, at 346 Avenue 57. **Dalton Trumbo** owned a home at 6231 Annan Trail from the mid-1950s to the mid-1960s. Trumbo was living here when he won an Academy Award for his script for the motion picture *The Brave One.* Trumbo had written the script under the name **Robert Rich,** because he was the victim of blacklisting by the motion-picture industry.

HOLLISTER

At the turn of the century **Frank Norris** visited Rancho Santa Anita, which stood about five miles south of this city in west central California. Here Norris worked on his novel *The Octopus* (1901), a realistic study of California wheat growers. The only remaining buildings of Rancho Santa Anita, which was part of Rancho Ana y Quien Sabe, are a few small sheds off Santa Anita Road. Rancho Ana y Quien Sabe, the largest ranch in the area, is mentioned in *The Octopus.*

HOLLYWOOD

Officially a district of the city of Los Angeles, Hollywood has a unique literary heritage that qualifies it for independent consideration by literary travelers. Hollywood has drawn many writers to its studios and corporate offices, to work as screenwriters, actors, and corporate executives, or simply to visit. Many have found the glamour industry a fit subject for their works, and others have written of Hollywood and the greater Los Angeles area without using the movie business as their prime subject.

Among the writers associated with Hollywood are **Maya Angelou,** who performed in the 1966 production of Jean Anouilh's play *Medea* with the Theatre of Being in Hollywood, and wrote the screenplay for *Georgia, Georgia* (1972), the first original screenplay by a black woman to be produced as a motion picture. **S[amuel] N[athaniel] Behrman** worked as a screenwriter here, as did **Ludwig Bemelmans,** who based his satirical novel *Dirty Eddie* (1947) on his Hollywood experiences. Other Hollywood screenwriters included **Robert Benchley, Konrad Bercovici,** and **Erskine Caldwell.** Caldwell first met **William Saroyan** in Hollywood, at the Stanley Rose Bookshop, on Hollywood Boulevard, a popular gathering place for writers. **Raymond Chandler** worked as a screenwriter here, and used Hollywood as a setting in his novel *The Little Sister* (1949), which featured the fictional detective Philip Marlowe. Marlowe's office was located in Hollywood.

Paddy Chayefsky worked in Hollywood as a screenwriter, as did **Russel Crouse** and **William**

Oh, space!
Change!
Toward which we run
So gladly,
Or from which we retreat
In terror—
Yet that promises to bear us
In itself
Forever.

Oh, what is this
That knows the road I came?

—Theodore Dreiser,
in "The Road I Came"

House of Happy Walls at Glen Ellen, built by Charmian London in 1919, three years after Jack London's death. The building houses a large collection of London memorabilia and is one of the main points of interest in Jack London State Historic Park.

Meanwhile, vast armies mount themselves, the world revolves, the traveller clutches his breast. From out the unyielding contradictions of labor stolen from men, the march to the endless war forces its pace. Perhaps, as the millions will be lost, others will be created, and I shall discover brothers where I thought none existed.

But for the present the storm approaches its thunderhead, and it is apparent that the boat drifts ever closer to shore. So the blind will lead the blind, and the deaf shout warnings to one another until their voices are lost.

—Norman Mailer, in *Barbary Shore*

Faulkner. Faulkner worked in Hollywood in the 1930s and 1940s, on such films as *Gunga Din, To Have and Have Not, The Big Sleep,* and *Mildred Pierce.* Faulkner did not like Hollywood and was always eager to finish his assignments so he could return to Oxford, Mississippi. One story, probably apocryphal, has it that at a script meeting Faulkner asked his boss, "Do you mind if I work on this at home?" The studio man was surprised later to find that Faulkner had taken the script home to Oxford.

Other writers who worked in Hollywood include **Paul Green, Ben Hecht, Lillian Hellman, Charles Jackson, Garson Kanin, MacKinlay Kantor,** and **Emily Kimbrough.** Kimbrough came to Hollywood with **Cornelia Otis Skinner** to work on the screenplay of their book *Our Hearts Were Young and Gay* (1942), but left the production before it was completed, taking legal action to prevent their names from being used on the film. Kimbrough wrote of their experiences in *We Followed Our Hearts to Hollywood* (1943).

Arthur Kober, the playwright, writer, and producer, worked as a screenwriter here, as did **Horace McCoy,** who used Hollywood as the setting for his novels *They Shoot Horses, Don't They?* (1935) and *I Should Have Stayed Home* (1938). The former, considered one of the best novels about Hollywood, deals only peripherally with the movie industry. This story of two young people trapped in the endlessly dreary marathon-dance circuit is an exploration of the despair resulting from failed human dreams and aspirations.

While working as a scriptwriter in Hollywood, **Norman Mailer** began to write his novel *Barbary Shore* (1951). **J. P. Marquand** worked on the screenplay of the 1941 film *H. M. Pulham, Esq.,* which was based on his 1941 novel of the same title. **Wilson Mizner,** the playwright, confidence man, and wit, came to Hollywood in the 1920s and became, at least nominally, a writer for Warner Brothers. According to Alva Johnston in *The Legendary Mizners* (1953), Mizner slept his way through most of his story conferences, allowing himself to be wakened on occasion to supply ideas or dialogue. Mizner died here on April 3, 1933.

Ogden Nash worked on the play *One Touch of Venus* (1943) with **Kurt Weill** and **S. J. Perelman** while employed in Hollywood, and **Clifford Odets** worked as a screenwriter and director here for many years. **John O'Hara** wrote screenplays here, and used Hollywood as a setting for his novel *Hope of Heaven* (1938). Other authors who have worked in Hollywood include **Ayn Rand, Conrad Richter, Lynn Riggs, Damon Runyon, William Saroyan,** and **Robert E. Sherwood. Harry Leon Wilson,** who lived for many years in Carmel, California, satirized Hollywood in his novel *Merton of the Movies* (1922), which is considered one of his best books.

Hollywood has been the home of a number of writers, including **Edgar Rice Burroughs,** who lived at 7933 Hillside Avenue for a time in 1934, and stayed at Sunset Plaza apartments in June 1937. **James M[allahan] Cain** lived in Hollywood in the 1930s and 1940s. During this period Cain worked on scripts for motion pictures and wrote his best-known novels, including *The Postman Always Rings Twice* (1934), *Mildred Pierce* (1941), and *Double Indemnity* (collected in the omnibus volume *Three of a Kind,* 1943). Cain's addresses included 2966 Belden Drive, 1714 North Ivar Avenue, and 8523 Sunset Boulevard, the last a business address.

Theodore Dreiser moved in May 1939 to a small apartment at 1426 North Hayworth Avenue. Here Dreiser worked on a book he tentatively titled *Is America Worth Saving?,* which was published in 1941 as *America Is Worth Saving.* In the fall of 1940 Dreiser succeeded in selling the motion-picture rights to his novel *Sister Carrie* (1900), and moved to a Spanish-style house at 1015 North Kings Road. There he lived with his long-time companion Helen Richardson, whom he married in 1944, and completed his novels *The Bulwark* (1946) and *The Stoic* (1947). Dreiser died at his North Kings Road home on December 28, 1945, and was buried in Forest Lawn Cemetery in Glendale.

1931 photograph of Theodore Dreiser.

F. Scott Fitzgerald lived in Hollywood for several short periods in the 1920s and 1930s. In early 1927 he and his wife, **Zelda Fitzgerald,** stayed at the Ambassador Hotel, and during this time visited Pickfair, the home of one of Hollywood's greatest stars, Mary Pickford. Zelda used her experiences here in her story " 'Show Mr. and Mrs. F. to Number ——.' " In late 1931 Scott Fitzgerald returned to Hollywood alone, staying at the Hotel Christie. During this trip Fitzgerald met Irving Thalberg, the movie producer, who became the model for the character Monroe Stahr in Fitzgerald's last, unfinished novel, *The Last Tycoon* (1941). In 1937 Fitzgerald went once more to Hollywood and took a bungalow at the Garden of Allah Hotel and Villas, no longer in existence, at 8152 Sunset Boulevard. The Garden of Allah, at one time the home of the actress Alla Nazimova, was a favorite temporary residence for writers, and housed at one time or another such literary figures as **Robert Benchley, John O'Hara, Dorothy Parker,** and **Herman Wouk,** who used it as a setting in his novel *Youngblood Hawke* (1962). It was here, at Robert Benchley's bungalow, that Scott Fitzgerald met Sheilah Graham, who was to be Fitzgerald's companion in the last years of his life, when he was dogged by ill health. In April 1940 he and Graham took apartments at 1403 North Laurel Avenue—his apartment was one floor above hers—and Fitzgerald worked on his last two memorable pieces of fiction, the series of Pat Hobby stories he wrote for *Esquire,* and *The Last Tycoon.* Fitzgerald had gone to work for the movie industry, and used his experiences as a scriptwriter in the stories and the novel. The seventeen Pat Hobby stories dealt with the life of an over-the-hill screenwriter, perhaps the sort of person Fitzgerald foresaw that he would become if his talent failed him. The novel, an ambitious study of life in Hollywood, supported Fitzgerald's belief that he was still a great writer: Edmund Wilson, who prepared *The Last Tycoon* for publication, termed it "far and away the best novel we have had about Hollywood." Fitzgerald did not live to complete the novel. Heart disease had forced him to move downstairs to Sheilah Graham's apartment—his doctor advised against climbing the stairs—and in that apartment Fitzgerald died on December 21, 1940.

Hamlin Garland lived at 2405 De Mille Drive from the early 1930s until his death on March 4, 1940. His home stood in McLaughlin

Park, across the street from Cecil B. DeMille's mansion. While there, Garland completed his books *Afternoon Neighbors* (1934), *Forty Years of Psychic Research* (1937), and *The Mystery of the Buried Crosses* (1939).

Other writers who have lived in Hollywood include **Zane Grey,** who rented a house here in 1918; and **William Inge,** the playwright, who came to Hollywood in the 1960s after winning an Academy Award for the scenario of *Splendor in the Grass* (1961). Inge died, an apparent suicide at his home on Oriole Drive in Hollywood Hills on June 10, 1973. **Christopher Isherwood** moved to 7136 Sycamore Trail in 1939, and lived in 1943 at the Vedanta Center, on Igar Avenue, where he was visited by **Tennessee Williams. Will James,** the Western writer, spent winters in Hollywood and in San Francisco after 1927, and lived in Hollywood for a year before his death on September 3, 1942, at Presbyterian Hospital.

Sinclair Lewis moved to Hollywood in 1943, living at the Chateau Marmont on Sunset Boulevard. Lewis worked on screenplays with Dore Schary at Schary's home. **Anita Loos,** the actress, screenwriter, playwright, and novelist, lived in Hollywood and worked for D. W. Griffith's American Biograph Company. She used her Hollywood experiences for the satirical novels *A Mouse Is Born* (1951) and *No Mother to Guide Her* (1961) and for *Kiss Hollywood Good-by* (1974). Her book *Cast of Thousands* (1977) is a lavishly illustrated memoir of the Hollywood figures she knew. **Johnston McCulley,** the prolific novelist, playwright, and screenwriter, owned a home at 6533 Hollywood Boulevard until his death at Presbyterian Hospital on November 23, 1958. McCulley was buried in Forest Lawn Cemetery in Glendale.

Dorothy Parker moved to Hollywood in the 1930s and lived here for a number of years, working on motion-picture scenarios, short stories, and book reviews. **S. J. Perelman** lived in the 1930s at 5734 Cazaux Drive, still standing. He worked as a scriptwriter, notably on the Marx Brothers motion pictures. Perelman wrote of his years in Hollywood in his book *The Last Laugh* (1981). **Dore Schary** lived in Hollywood for many years and worked as a screenwriter, playwright, producer, and studio head for M-G-M and RKO studios. While living in Hollywood, Schary wrote a number of plays, as well as thirty-five screenplays.

Budd Schulberg grew up in Hollywood and worked as a screenwriter here from 1937 to 1940. Schulberg used his Hollywood experiences in his first novel, *What Makes Sammy Run?* (1941), the story of Sammy Glick, a relentlessly ambitious young man from New York's Lower East Side who rises to the top of Hollywood's movie industry. Schulberg once worked with **F. Scott Fitzgerald** on a movie script when Fitzgerald was near the end of his remarkable literary career and Schulberg's career was just beginning. Parallels have often been perceived between Fitzgerald's Hollywood experiences and those of Manley Halliday, the aging writer in Schulberg's novel *The Disenchanted* (1951). Schulberg has stated that the character was not modeled on Fitzgerald alone.

Robert W. Service, the poet, wrote his autobiography *Ploughman of the Moon* (1945) in Hollywood, and **Irving Stone** wrote and then destroyed an autobiographical novel while living at 500A Manola Way. **Jessamyn West** in Hollywood wrote *To See the Dream* (1957), an account of the filming of her novel *The Friendly Persuasion* (1945). She later wrote two works of fiction set in California, *Cress Delehanty* (1953) and *South of the Angels* (1960).

The Hollywood novel that comes to mind most readily for many readers is apt to be *The Day of the Locust* (1939), by **Nathanael West.** West first came to Hollywood in the 1930s and worked as a scriptwriter for Columbia Pictures. He lived at the home of his brother-in-law, **S. J. Perelman**—Perelman had married West's sister Laura—at 5734 Cazaux Boulevard. West returned to the East Coast in the fall of 1933, but was back in Hollywood by the summer of 1935. He took rooms at the Pa-Va-Sed (or Parva Sed) Apartments, on North Ivar Street above Hollywood Boulevard, not far from Stanley Rose's bookstore, where West received encouragement for his writing from such authors as **William Faulkner, F. Scott Fitzgerald, John O'Hara,** and **William Saroyan.** At the Pa-Va-Sed, West wrote an early draft of *The Day of the Locust,* which he tentatively titled *The Cheated.* This novel, dealing with the emptiness and illusion of modern life, with the despair and boredom of the lives of people who feel they have been cheated, sold fewer than two thousand copies when it was published, but it has since come to be regarded by many as a minor classic of twentieth-century American literature.

West used a number of Hollywood places as settings for the novel. For example, the Pa-Va-Sed Apartments was the model for the apartment hotel in which lived the main character, Tod Hackett. West also lived on Alta Loma Terrace, address unknown; at 6614 Cahuenga Terrace; on North Stanley Street; and at 2225 Canyon Terrace. In the late 1930s West achieved some success as a

> *The first time I saw him he couldn't have been much more than sixteen years old, a little ferret of a kid, sharp and quick. Sammy Glick. Used to run copy for me. Always ran. Always looked thirsty.*
>
> *"Good morning, Mr. Manheim," he said to me the first time we met. "I'm the new office boy, but I ain't going to be an office boy long."*
>
> *"Don't say ain't," I said, "or you'll be an office boy forever."*
>
> *"Thanks, Mr. Manheim," he said, "that's why I took this job, so I can be around writers and learn all about grammar and how to act right."*
>
> —Budd Schulberg, in *What Makes Sammy Run?*

1403 Laurel Avenue, Hollywood, where F. Scott Fitzgerald worked on *The Last Tycoon* until his death in 1940.

Vanzetti walked to the chair and sat down. Then he spoke—words which he had made the subject of much thought. "I wish to tell you that I am innocent and never committed any crime, but sometimes some sin. I thank you for everything you have done for me. I am innocent of all crime, not only of this one, but of all. I am an innocent man."

The guards, well trained, went on with their work, paying no attention to eloquence. . . . "I wish to forgive some people for what they are now doing to me."

—Upton Sinclair, in *Boston*

screenwriter, working on *Advice to the Lovelorn,* based on his novel *Miss Lonelyhearts* (1933), and on other scripts. In 1939, shortly after publication of *The Day of the Locust,* West married Eileen McKenney, and the couple settled into a comfortable house at 12706 Magnolia Boulevard in December 1940. Among the people who visited West here were **Scott Fitzgerald** and Sheilah Graham, **Ring Lardner, Jr.,** and **Elliot Paul.** The Wests lived here only briefly. On December 22, 1940, while returning from a hunting trip to Mexico, they died in an automobile accident near El Centro, California. Their deaths came one day after the death of Scott Fitzgerald, and four days before the New York City opening of the play *My Sister Eileen,* based on the humorous sketches by Eileen's sister, Ruth McKenney.

INDEPENDENCE

Mary Austin lived in the early 1900s in this town in east-central California. She wrote of this area in her collection of sketches *The Land of Little Rain* (1903). The State Historical Marker in front of her house at 253 West Market Street includes an excerpt from this book.

INDIO

Zane Grey, at the time of his death in 1939, owned a small ranch eighteen miles south of this city on Route 86 in southern California.

LA JOLLA

Carol Ryrie Brink, author of books for children, died in La Jolla on August 15, 1981. **Raymond Chandler** spent his last years in this suburb of San Diego. He died on March 26, 1959, in his home at 6925 Neptune Place, still a private residence. Chandler's later books included the novel *The Long Goodbye* (1954), one of his most popular stories of the career of Philip Marlowe, a hard-boiled private detective. **Harold Bell Wright,** who had recently moved to San Diego, died in a La Jolla hospital on May 24, 1944. He was seventy-two years old.

LIVERMORE

In the mid-1880s young **Jack London** moved with his parents to a ranch of about seventy acres near this city, off what is now Alden Lane. London's father here made his final attempt to become an independent farmer. Although it seemed for a time that he would succeed, it became clear by 1886 that he had failed, so in that year the family moved back to Oakland. Life at the ranch had been lonely for the young London, but it was here that he read the books that kindled his lifelong passion for adventure and romance: *Signa* (1875), by the English novelist Ouida, pen name of Louise de La Ramée; and *The Legends of the Alhambra* (1832), by Washington Irving. **Howard Pease,** author of adventure stories, lived at 1358 Ora Avenue, Livermore, in the 1960s.

LONG BEACH

James Hilton, the English novelist, lived during the last ten years of his life at 235 Argonne Avenue, in this city southeast of Los Angeles. In a workroom over the garage, Hilton wrote in longhand, sitting at a large oak desk once owned by a popular character actor, Sir C. Aubrey Smith. Hilton's later novels included *Nothing So Strange* (1947) and *Time and Time Again* (1953). Hilton died on December 20, 1954, at Seaside Hospital, now El Cerrito Health Services Center, at 1401 Chestnut Avenue. He was buried in Sunnyside Mortuary and Memorial Park at 4725 Cherry Avenue. **Upton Sinclair** lived on Alamitos Bay in Long Beach during the late 1920s and early 1930s, first at 10 Fifty-eighth Place and later at 43 Fifty-seventh Place. While here, Sinclair wrote his novels *Oil!* (1927) and *Boston* (1928). *Boston* recounted the cause célèbre of the early twentieth century, the Sacco-Vanzetti case.

James Hilton's home in Long Beach.

LOS ANGELES

Louis Adamic worked, first as a secretary and later as a journalist, in Los Angeles early in the twentieth century. Adamic's investigation of the circumstances of the bombing of the Los Angeles *Times* building in 1910 inspired him to write *Dynamite* (1931), a history of violence in the American labor movement. **Zoë Akins,** winner of a Pulitzer Prize for her play *The Old Maid* (1935), which she adapted from the novella written by Edith Wharton in 1924, lived for some years at 324 South Ardmore Avenue, where she died on October 29, 1958.

Edgar Rice Burroughs, who first came to Los Angeles in 1916, lived in at least eight homes here during the following twenty years. From September 1916 to June 1917, he was at 355 South Hoover Street. In February 1919, Burroughs and his family lived first at the Alvarado Hotel, then in a rented house at 1729 North Wilton Avenue, before moving to Tarzana, the family's new ranch—the acreage later was incorporated into the city of Tarzana. In 1924 the Burroughs family returned to Los Angeles, living first at 544 South Gramercy Place, and then at 674 South New Hampshire Avenue. In July 1926 the peripatetic family returned to Tarzana. In April 1934 Burroughs separated from his wife but did not abandon his proclivity for changing residences. He stayed for a short time in a house at 2029 Pinehurst Road and by May 1936, married for the second time, was in a house at 2315 North Vermont Avenue, but again only for a short stay. Burroughs moved once again early in 1937, this time to the Chateau Elysée, owned by William Randolph Hearst, at 5930 Franklin Avenue, where his tenure extended only to June 1937. It is known that he worked at Chateau Elysée on the story "Tarzan and the Elephant Men," which was published in *Tarzan the Magnificent* (1939).

Raymond Chandler lived in Los Angeles and several of its neighboring communities for many years. He used Southern California and Los Angeles as settings for a number of his detective stories and novels, including *The Big Sleep* (1939), *Farewell, My Lovely* (1940), *The High Window* (1942), and *The Lady in the Lake* (1943). These novels, featuring Philip Marlowe, are known for their hard-hitting style and realistic depiction of the seamier side of Los Angeles life. It was in 1912 that Chandler first came to Los Angeles from England, where he had spent his youth. After several years of work as an accountant and bookkeeper—from time to time he made attempts at writing—Chandler left to join the Canadian army and fight in World War I. He returned about 1919, working for the Los Angeles *Daily Express* for just six weeks and then drifting through other jobs before beginning to rise in the early 1920s through the executive ranks of a Los Angeles oil syndicate. In February 1924 Chandler married Cissy Pascal, eighteen years his senior, and the couple took up residence at 2863 Leeward Avenue. The Chandlers moved frequently during the next twenty years. Their Los Angeles addresses included 2315 West Twelfth Street and 1024 South Highland Avenue. Chandler, who was not happy as a businessman, began to express his dissatisfaction by drinking heavily and womanizing. He also began to neglect his job, and in 1932 was fired.

Chandler turned to writing in order to earn his livelihood. His reading of pulp detective magazines, which he was accustomed to buying during business trips up and down the California coast, led him to view detective fiction as a genre in which he could learn the trade. During the next five years he wrote and sold a number of stories, which appeared in such magazines as *Black Mask* and *Dime Detective Magazine,* and during this period moved frequently, living at one time at 1637 Redesdale Avenue. Chandler became a member of a group of magazine and motion-picture writers known as The Fictioneers, whose monthly meetings were held at Steven's Nikobob Café, at the corner of Ninth Street and Western Avenue. In 1938 Chandler started work on his first novel, *The Big Sleep,* which he based largely on material from two earlier stories, "Killer in the Rain" (1935) and "The Curtain" (1936). *The Big Sleep* (1939) met with encouraging critical response and had good sales, and Chandler followed it with *Farewell, My Lovely* (1940). In 1942 he moved to 12216 Shetland Lane, in the Brentwood section of Los Angeles, where he worked on revisions of his third novel, *The High Window* (1942). Chandler then moved out of Los Angeles for a time and completed his novel *The Lady in the Lake* (1943).

While Chandler's novels brought him a respectable literary reputation, they produced only small income, and Chandler was drawn inevitably into working for the motion-picture industry. In Hollywood he was assigned to write the screenplay for *Double Indemnity* (1944), the film made of James M. Cain's novel of 1943. He achieved recognition as well as substantial income for his work as a screenwriter but soon began to hate Hollywood. He moved to 6520 Drexel Avenue in Los Angeles, where he wrote the screenplay for *The Blue Dahlia* (1945). Chandler soon enough decided that the pay he received for his motion-picture work was not worth the restrictions and compromises he was forced to endure, so he moved in 1946 to La Jolla and resumed work on his novels.

Jerome Chodorov lived in Los Angeles in the 1940s and 1950s and wrote many motion-picture screenplays as well as, with Joseph Fields, several plays, including *My Sister Eileen* (1940), *Junior Miss* (1941), *Anniversary Waltz* (1954), and *The Ponder Heart* (1957). While it has been said that nobody was ever born in California, Los Angeles was the birthplace, on July 6, 1913, of **Eleanor Clark,** the novelist and essayist whose works include the novels *The Bitter Box* (1946) and *Baldur's Gate* (1971), and the collection of essays *Rome and a Villa* (1952, expanded edition 1975). Clark is married to Robert Penn Warren. **Gregory Corso** worked in 1950 as a file clerk in the morgue of the Los Angeles *Examiner.* His volumes of poetry include *The Vestal lady of Brattle* (1955) and *The Happy Birthday of Death* (1960).

Lloyd C. Douglas served from 1926 to 1929 as pastor of the Los Angeles First Congregational Church, now listed at 540 South Commonwealth Avenue. During this time he conceived and wrote *Magnificent Obsession* (1929). Douglas left Los Angeles in 1929, but returned in the late 1930s to live at 214 St. Pierre Road in the Bel Air section. Douglas lived there until about 1944, during which time he wrote his novels *Invitation to Live* (1940) and the enormously popular *The Robe*

"Who is this Hemingway person at all?"

"A guy that keeps saying the same thing over and over until you begin to believe it must be good."

"That must take a hell of a long time," the big man said.

—Raymond Chandler,
in *Farewell, My Lovely*

Photograph of Raymond Chandler taken in 1954, five years before his death in La Jolla.

I didn't see him [Philip Marlow] again for a month. When I did it was five o'clock in the morning and just beginning to get light. The persistent ringing of the doorbell yanked me out of bed. I plowed down the hall and across the living room and opened up. He stood there looking as if he hadn't slept for a week. He had a light topcoat on with the collar turned up and he seemed to be shivering. A dark felt hat was pulled down over his eyes.

He had a gun in his hand.

—Raymond Chandler,
in *The Long Goodbye*

> *So I had to stay there and read the* Hollywood Reporter *and* Variety *and try to get my mind off the sound of the dynamo or the generator or whatever it was that made that sound. That sound never let up, and if you let yourself listen to it it had the effect of the dentist's drill, or the bastinado. That sound is in every studio that I've ever worked in, and I never have been able to determine just what it is. Some say it's a dynamo; some say it's the ventilating system; others say it's just water in the pipe-lines. Whatever it is, it's always near the writers' offices.*
>
> —John O'Hara,
> in *Hope of Heaven*

(1942). Douglas spent his last years in Las Vegas, Nevada, but died in Los Angeles at Good Samaritan Hospital on February 13, 1951. He was buried in Los Angeles.

Theodore Dreiser moved to Los Angeles in late 1919 with Helen Richardson. The couple lived at first in a house on Alvarado Street but made a number of moves in the Los Angeles area in the next three years. During this period Dreiser completed his autobiographical *A Book About Myself* (1922) and began to plan his novel *An American Tragedy* (1925). **Eleanor Estes** wrote the juvenile novel *Ginger Pye,* for which she won the Newbery Medal in 1951, while living at 510 Harvard Boulevard. **Langston Hughes** founded The New Negro Theater in Los Angeles in 1939. The first production of the company was Hughes's play *Don't You Want to Be Free?* In its opening season the company also performed two skits by Hughes, "Em-Fuehrer Jones" and "Limitations of Life."

Will and **Ariel Durant,** the historians and philosophers, lived for many years at 5608 Briarcliff Road in Los Angeles. Ariel Durant died there on October 25, 1981. Will Durant, who was a patient at Cedars-Sinai Hospital at the time of his wife's death, died at the hospital two weeks later, on November 7, 1981.

Christopher Isherwood, the playwright and writer of fiction, lived in Los Angeles for a year in the early 1950s. **Robinson Jeffers** also lived in Los Angeles for a short time, occupying a house with his family at 1623 Shatto Street. Jeffers was graduated in 1905 from Occidental College, in Los Angeles, and studied for three years at the UCLA Medical School. **Jerome Lawrence,** the playwright, worked from 1937 to 1939 as a continuity editor for radio station KMPC, at 5858 West Sunset Boulevard, and was a graduate student at UCLA in 1938–39. **Vachel Lindsay** visited Los Angeles in September 1912, during his hobo tour of the Southwest. It was at about this time that Lindsay wrote one of his most popular poems, "General William Booth Enters into Heaven."

Photograph of Lloyd C. Douglas in the study of his home in Bel Air in 1944.

> *Booth died blind and still by faith he trod,*
> *Eyes still dazzled by the ways of God.*
> *Booth led boldly, and he looked the chief,*
> *Eagle countenance in sharp relief,*
> *Beard a-flying, air of high command*
> *Unabated in that holy land.*
>
> —Vachel Lindsay,
> in "General William Booth Enters into Heaven"

Ross Lockridge came to Los Angeles late in 1947 to discuss with motion-picture studio executives the possibility of adapting his novel *Raintree County* (1947) for the screen. The studio heads were not enthusiastic about the project, and Lockridge left after three weeks. In 1958, ten years after Lockridge's death, a movie based on the novel was produced and released. **Stephen Longstreet,** the novelist and playwright, was staff lecturer for the Los Angeles Art Association in 1954 and served also as a trustee of the association. Longstreet taught at UCLA in 1955 and in 1958–59, and was a literary critic for the Los Angeles *Daily News.* Los Angeles was the setting for Longstreet's novel *The Beach House* (1952).

Malcolm Lowry came to Los Angeles from Mexico in the summer of 1939. His father, who lived in England, hired an attorney to exercise some control over Lowry's financial affairs. The attorney checked Lowry into the Hotel Normandie and tried to bring some order into the novelist's chaotic life. While here, Lowry hired a typist to redo the second draft of *Under the Volcano* (1947) and worked on a novella, *The Last Address*—this work was incorporated after Lowry's death into *Lunar Caustic* (1968). In Los Angeles Lowry met Margerie Bonner, the former Hollywood starlet he later married. At the time they met, Lowry was involved in divorce proceedings with his first wife, Jan Gabrial. Lowry and Bonner fell in love at once but had little time to pursue the matter, for Lowry's father's attorney decided that Canada would be a better place for Lowry to live. He drove him off to Vancouver about the end of July 1939.

One of the best-known literary figures in Los Angeles during the late nineteenth and early twentieth centuries was **Charles Fletcher Lummis,** the author, editor, and poet, who came to Los Angeles in February 1885 after walking from Cincinnati through the Southwest, a journey of more than 3,500 miles. Lummis had agreed to supply weekly accounts of his walk to the Los Angeles *Times* in return for a job when he arrived. On February 1, 1885, he walked into Los Angeles in less than prime condition after the 143-day hike—he had broken an arm in a fall from a cliff and set the broken bone himself—and on the next day became the first city editor of the Los Angeles *Times.*

Lummis remained with the newspaper until 1887, and during that time began his lifelong study of the ethnological, historical, and archaeological backgrounds of California and the American Southwest. In 1894 Lummis founded and edited *Land of Sunshine,* forerunner of *Out West* magazine, and in 1897 began construction of El Alisal (The Sycamore), his home at 200 East Avenue 43. This stone house, built by Lummis himself as recreation from his literary activity, is now the Charles Fletcher Lummis Home State Historical Monument, open to visitors. Here he wrote many books dealing with the Southwest, including *A Tramp Across the Continent* (1892), an account of his walking trip; *The Land of Poco Tiempo* (1893); *The Spanish Pioneers* (1893); *Spanish Songs of Old California* (1923); and *Mesa, Cañon and Pueblo* (1925). Lummis served from 1905 until 1910 as librarian of the Los Angeles Public Library and helped found the Southwest Museum, located at 234 Museum Drive. Lummis died at his home, and his ashes were interred in a wall at El Alisal, behind a plaque that commemorates his life and accomplishments:

CHARLES F. LUMMIS
Mar. 1, 1859—Nov. 25, 1928
He founded the Southwest Museum
He built this house
He saved four Missions
He studied and recorded Spain in America
He tried to do his share

Other writers are associated with Los Angeles. **Edwin Markham,** the poet, died at his home in New York City on March 7, 1940, but he was buried beside his wife in Calvary Cemetery, 4201 Whittier Boulevard, in Los Angeles. **Brian Moore** lived on Look Out Mountain Drive from 1965 through 1967 and there wrote his novel *I Am Mary Dunne* (1968). Moore used Los Angeles, Malibu Beach, and Hollywood as settings for his novel *Fergus* (1970). **Wright Morris** lived in Los Angeles while writing his novel *My Uncle Dudley* (1942). **Anaïs Nin,** the author and diarist, was living in the Silver Lake district of the city at the time of her death on January 14, 1977, at Cedars-Sinai Medical Center, 8700 Beverly Boulevard. The manuscripts of Nin's diaries are stored in the Rare Book Room of the UCLA Research Library. **Clifford Odets** for many years before his death divided his time between New York City and the Los Angeles area. He died in Los Angeles on August 14, 1963. At his death, Odets was living on Beverly Hills Drive, in Beverly Hills. He had just completed work on the musical version of his successful play *Golden Boy* (1937). The musical was produced in the year following the playwright's death.

John O'Hara in 1936 lived at 10735 Ohio Avenue in West Los Angeles and there worked on *Hope of Heaven* (1938). **Lynn Riggs,** the playwright, came to Los Angeles in 1919, working as a theater extra and reading proof for the Los Angeles *Times.* During this time, Riggs saw publication in the *Times* of his first poem, which the editor ungenerously entitled "Spasm." **Budd Schulberg,** son of a Hollywood studio executive, was graduated from Los Angeles High School in 1931. He wrote stories for his school newspaper and served as editor. In the 1960s, when Schulberg was enjoying the successes of several of his novels and screenplays, he owned a home at 8484 Grand View Drive as well as a home in Mexico. **Robert W. Service,** whose poems celebrated the Yukon and the Klondike, lived in Los Angeles for a time in the mid-1890s. From 1895 until 1901, he traveled widely along the Pacific Coast, working for short spells as bank clerk, logger, fruit picker, and gardener, this last position in the employ of a landscaped brothel. **Stuart P. Sherman,** the literary critic, lived in childhood during the 1880s at several Los Angeles addresses, including 107 North Figueroa Street.

Henryk Sienkiewicz, the Polish novelist who won the Nobel Prize in literature in 1905, lived in Los Angeles for a time in the late 1870s and there wrote four of his Hania stories. **John Steinbeck** early in 1930 lived at 2741 El Roble Drive, in Eagle Rock, which now is part of Los Angeles. It was there that he completed the manuscript of his novel *To a God Unknown* (1933). The cottage in which he lived has been remodeled, and the neighborhood long ago lost the rural atmosphere Steinbeck admired. **Irving Stone** lived in 1919 with his family at Thirty-first Street and Vermont Avenue. He attended Manual Arts High School, on Vermont Avenue, where he wrote for the school newspaper. Stone was graduated in February 1920 and went on to the University of Southern California. **Gene Stratton-Porter,** the novelist and nature writer, lived in West Los Angeles for a time during the early 1920s. She died in Los Angeles on December 6, 1924, in an automobile accident.

Dalton Trumbo lived in Los Angeles during the 1920s and 1930s, first in a house off Fifty-fifth Street and later in a house on Cahuenga Boulevard. While attending school, Trumbo worked in a bakery at the corner of Second Street and Beaudry Avenue. He attended classes at the University of Southern California from 1928 until 1930, and became associate editor of the Hollywood *Spectator* in 1933. During this period Trumbo sold several stories to magazines, including *The Saturday Evening Post* and *Liberty.* Toward the end of his life Trumbo made his home at 8710 St. Ives Drive, where he died on September 10, 1976. **Nathanael West** lived in Hollywood during the 1930s and visited Los Angeles frequently while he was writing his novel *The Day of the Locust* (1939). One of the places West visited was Aimée

Sketch of Charles Fletcher Lummis, published originally in the April 1896 edition of *Munsey's Magazine.*

Home of Charles Fletcher Lummis in Los Angeles. Lummis played a major role in preserving the missions and historical sites of the Old Southwest, and in recording its folklore and cultural heritage.

John Steinbeck.

Semple McPherson's Angelus Temple, main church of the International Church of the Foursquare Gospel, 1100 Glendale Boulevard. The impressions he gathered there were used later in *The Day of the Locust.* **Tennessee Williams** worked as a chicken plucker on a ranch just outside Los Angeles in 1939.

Yvor Winters, the poet, winner of a Bollingen Prize in 1961, lived in Los Angeles from about 1904 until 1914. **Willard Huntington Wright [S. S. Van Dine]** the critic and detective novelist—he invented the character Philo Vance, a master detective—worked as literary editor of the Los Angeles *Times* from 1907 until 1913. Wright was at the *Times* on a day in 1910 when a powerful bomb was set off there, killing twenty-one people, among them the person at the desk next to Wright's. Wright escaped the blast—he had gone home early to nurse a headache.

No matter how one leans
One yet fears not to know.
God knows what all this means!
The mortal mind is slow.

—Yvor Winters,
in "A Song in Passing"

LOS GATOS

John Steinbeck and his wife Carol lived from 1936 to 1938 in a white house on Greenwood Lane, two miles from Los Gatos. The house, standing on a hill shaded by oak, manzanita, and madrone, is surrounded by a wooden fence Steinbeck built, and the gate still bears the sign he put there: "Arroyo del Ajo" (Garlic Gulch). Steinbeck moved to this house while it was still being built, and completed his novella *Of Mice and Men* (1937), which he had begun to write when he lived in Pacific Grove. In 1938 the four sections of *The Red Pony* were published in *The Long Valley* (1938), and in the fall of that year the Steinbecks moved to a new home on Brush Road, where Steinbeck completed *The Grapes of Wrath* (1940) and began work on the screenplays for *Of Mice and Men* and *The Grapes of Wrath.* About this time, his marriage breaking up, Steinbeck returned to Pacific Grove.

Colonel C[harles] E[rskine] S[cott] Wood lived in Los Gatos for many years. The poet died here on January 22, 1944, and his ashes were scattered over the grounds of his estate.

Billy dropped the knife. Both of his arms plunged into the terrible ragged hole and dragged out a big, white, dripping bundle. His teeth tore a hole in the covering. A little black head appeared through the tear, and little slick, wet ears. A gurgling breath was drawn, and then another. Billy shucked off the sac and found his knife and cut the string. For a moment he held the little black colt in his arms and looked at it. And then he walked slowly over and laid it in the straw at Jody's feet.

—John Steinbeck,
in *The Red Pony*

MALIBU

Edgar Rice Burroughs lived from 1932 to 1934 at 90 Malibu La Costa Beach, in a house he never liked even though it was located in an ideal spot near the ocean. It was here that Burroughs decided to terminate his marriage with his first wife, Emma. He moved to Los Angeles in 1934, but Emma and the children stayed on in Malibu. **F. Scott Fitzgerald** rented a bungalow in Malibu Beach in the summer of 1938. **Jerome Lawrence,** while living at 18106 Pacific Coast Highway, collaborated with Robert E. Lee on *Inherit the Wind* (1955), the play about the Scopes trial in Tennessee in 1925, in which Clarence Darrow and William Jennings Bryan debated the propriety of teaching evolution in Tennessee schools.

MANHATTAN BEACH

In the early 1900s **Robinson Jeffers** moved with his family from Highland Park to a house on Third Street, one block from the ocean, and later to other homes in Manhattan Beach, now part of Los Angeles.

MELONES

Bret Harte came to California in the 1850s and lived for a time with a mining crew near this community on the Stanislaus River, about forty miles northeast of Modesto. Melones, previously known as Slumgullion, appears in Harte's "A Lonely Ride," the story of a night stagecoach ride from Wingdam (the town of Murphys) to Slumgullion (Melones).

MENLO PARK

Ken Kesey took a job in 1960 as a psychiatric aide at the Veterans Administration Hospital in this city south of San Francisco. Kesey lived at 9 Perry Lane, now Perry Avenue. The novelist had worked earlier at the Veterans Administration Hospital in nearby Palo Alto and had taken part in experiments on the effects of hallucinogenic drugs. Kesey used his experiences in Palo Alto and Menlo Park for his macabre comic novel *One Flew Over the Cuckoo's Nest* (1962).

MILL VALLEY

Holman Day moved to this city, northwest of San Francisco, just three months before his death on February 19, 1935. In the summer of 1947 **Jack Kerouac** visited a friend, Henri Cru, who lived in a one-room cabin here. Kerouac worked as a night watchman during his brief stay. When the novelist returned to Mill Valley in 1956, it was to visit another friend, the poet **Gary Snyder. Howard Pease,** the short-story writer, who lived for a time at 130 Linden Lane, died in Mill Valley at the home of friends on March 17, 1974.

MONROVIA

In the fall of 1942 **Upton Sinclair** and his wife Craig moved to a large house at 464 North Myrtle Avenue in this city northeast of Los Angeles. Sinclair moved his manuscripts, books, and papers here from his Pasadena home in four specially built boxes said to be as large as small houses. Sinclair later turned this massive collection over to Indiana University's Lilly Library. During his residence in Monrovia, from 1942 to 1966, Sinclair worked in the garage, which he had turned into an office. Sinclair wrote most of his later books here, including many of his eleven Lanny Budd novels. While living here, Sinclair was awarded a Pulitzer Prize for *Dragon's Teeth* (1942), the third volume in that series. While here he also wrote *The Autobiography of Upton Sinclair* (1962). The Sinclair home is now designated a National Historic Landmark.

MONTECITO

Charles B. Nordhoff died on April 11, 1947, at his home in this community, southeast of Santa Barbara. **Kenneth Rexroth** died in his home here on June 6, 1982.

MONTEREY

Writers associated with Monterey include **Josh Billings,** the humorist, who died on October 14, 1885, in a Monterey hotel while in the midst of a lecture trip; **Richard Henry Dana, Jr.,** who came here as a seaman aboard the *Pilgrim* in the 1830s, and later described Monterey in *Two Years Before the Mast* (1840); and **Robinson Jeffers,** who used the area around Monterey as the setting for a number of poems, including his allegorical narrative "Roan Stallion" and his free-verse narrative "Tamar." **John Steinbeck** in 1944 bought the Soto House at 460 Pierce Street, now used for offices of the Monterey Institute of International Studies. Among Steinbeck's novels with Monterey settings are *Tortilla Flat* (1935) and *Cannery Row* (1945). The latter dealt with the lives of people living and working in the row of fish canneries that once thrived between David and Reeside avenues and now are all but gone.

Robert Louis Stevenson stopped in Monterey briefly in August 1879 before traveling to the Santa Lucia Mountains. He returned after several weeks and took up residence at 536 Houston Street, in a house that now is part of Monterey State Historic Park and known as the Stevenson House. Stevenson lived here until December 1879, writing articles for $2 a week for the Monterey *Californian,* no longer publishing. He also worked on drafts of *The Amateur Emigrant* (first published in its entirety in the twenty-seven-volume Edinburgh Edition of Stevenson's works, 1894–98) and *Prince Otto* (1885). He also wrote the story "The Pavilion on the Links," which appeared in *New Arabian Nights* (1882), as well as several essays, including one on Monterey entitled "The Old Pacific Capital." It is said that the forests, beaches, and other natural features of the Monterey area served Stevenson as models for settings in *Treasure Island* (1883). Stevenson had come to Monterey to be near Mrs. Fanny Osbourne, who was living in another house here and whom Stevenson intended to marry. Also staying with Mrs. Osbourne was her son Lloyd, who later described Stevenson's condition during this time in *An Intimate Portrait of R.L.S.* (1924): "I was old enough to appreciate how poor he was, and it tore at my boyish heart that he should take his meals at a grubby little restaurant with men in their shirt-sleeves, and have so bare and miserable a room in the old adobe house on the hill." That grubby little restaurant was run by Jules Simoneau, who befriended the starving writer and provided him with good food and many hours of conversation. Stevenson later sent an inscribed first edition of each of his books to Simoneau, and these were given eventually to the city of Monterey. Stevenson House is one of ten buildings located in Monterey State Historic Park and is open daily. The house contains a collection of items relating to Stevenson and his writing.

Portrait of Josh Billings made originally for *Harper's Magazine.*

Other writers associated with Monterey include **Charles Warren Stoddard,** the poet and author, who died here on April 23, 1909, and was buried in the Catholic Cemetery, at Fremont and Via Mirada. Stoddard's former colleagues at Catholic University, in Washington, D.C., supplied the monument that marks his grave. **Thomas B[angs] Thorpe** came west with General Zachary Taylor in 1846 and used his experiences for several books, including *Our Army at Monterey* (1847).

Robert Louis Stevenson House in Monterey. Stevenson described his quarters as "great airy rooms with five windows opening on a balcony."

The road from Wingdam to Slumgullion knew no other banditti than the regularly licensed hotelkeepers; lunatics had not yet reached such depth of imbecility as to ride of their own free will in California stages; . . .

—Bret Harte,
in "A Lonely Ride"

Days departing linger and sigh:
Stars come soon to the quiet sky;
Buried voices, intimate, strange,
Cry to body and soul of change;
Beauty, eternal fugitive,
Seeks the home that we cannot give.

—George Sterling,
in "The Last Days"

MONTROSE

John Steinbeck and his wife lived for a short time in 1930 at 2527 Hermosa Avenue in this community, northwest of Los Angeles.

MOUNT SHASTA CITY

Anita Loos was born on April 26, 1893, in this city, then a resort town known as Sisson. Her father, editor of the Sisson *Mascot,* moved the family to San Francisco when Loos was four years old. Among her best-known works are the satirical novel *Gentlemen Prefer Blondes* (1925) and the play *Gigi* (1952), which Loos based on a novel by Colette. The area around Mount Shasta City was the setting of the novel *Shadows of Shasta* (1891), by Joaquin Miller. In the novel, Miller protested the displacement of the Indians from their native lands to government reservations. This area was also the setting for *The Shoulder of Shasta* (1895), a novel by Bram Stoker.

MURPHYS

This town in the mining district northwest of Sonora is believed to be the town of Wingdam, which appears in such stories of Bret Harte as "A Night in Wingdam" and "A Lonely Ride." The Mitchler Hotel on Route 4 is said to be the hotel described in "A Night in Wingdam."

NEVADA CITY

Richard Walton Tully was born on May 7, 1877, in this city about forty-five miles west of Lake Tahoe. The birthplace of the playwright and producer, at 449 Broad Street, now private, was his family's home from 1870 to 1879.

NEW ALMADEN

In the 1870s **Mary Hallock Foote** lived in this mining town southwest of San Jose while her husband, a civil engineer, worked at the New Almaden Mine. Here Mrs. Foote wrote and illustrated her first article, "A California Mining Camp," which was published in *Scribner's Monthly.*

OAKLAND

This city—part of a metropolitan area that includes Alameda, Berkeley, and San Francisco—has a colorful and unique literary history reflecting the influence of pioneer and aesthete, the activity of Oakland's busy streets and the solitude of its nearby hills, the wildness of its waterfront and the discipline of its schools. **Frederic Homer Balch** saw the publication of his novel *The Bridge of the Gods* (1890) while attending Pacific Theological Seminary from 1889 to 1891. The seminary, then located at 463 Plymouth in Oakland, is now at 1798 Scenic Avenue in Berkeley. **Ambrose Bierce,** whose career was closely associated with San Francisco, moved to an apartment in Oakland during the winter of 1886–87. In March 1887 he received a visit from **William Randolph Hearst,** who had recently been expelled from Harvard and, just a few days before visiting Bierce, had become publisher of the San Francisco *Examiner.* Hearst had big plans for the faltering newspaper and asked Bierce to write for it. Bierce, who had severed his ties with the *Wasp,* a San Francisco weekly, needed the job and readily accepted. Bierce wrote of this turn in his career—he was to remain with the Hearst publishing empire for many years—in "A Thumbnail Sketch," published in Bierce's *Collected Works* (1909–12).

Oakland was the birthplace on January 7, 1919, of **Robert Duncan,** the poet whose collections include *Caesar's Gate* (1955) and *Selected Poems* (1960). **Bret Harte** lived in the 1850s in a house, no longer standing, at Fifth and Clay streets. The house was the home of Harte's stepfather, Col. Andrew Williams, who was a model for Harte's fictional character Colonel Starbottle. Harte called Oakland Encival in several stories, including "The Devotion of Henriquez" and "Chu Chu." Today Harte is honored with a Bret Harte Boardwalk, a group of stores and shops in the 500 block of Fifth Street.

Sidney [Coe] Howard, born in Oakland on June 26, 1891, was the author of a number of popular and critically acclaimed plays, including the Pulitzer Prize-winning drama *They Knew What They Wanted* (1924), set in California's Napa Valley. **Edwin Markham** became principal of the Tompkins Observation School here in 1890. It is said that while here Markham wrote his best-known poem, "The Man With the Hoe."

Other writers associated with Oakland include **Frank Norris,** who lived for a year in the mid-1880s with his family in a house on the shore of Lake Merritt, which now lies within Oakland but then was just outside the city limits. Norris died on October 25, 1902, in San Francisco, and was buried in Mountain View Cemetery, 5000 Piedmont Avenue in Oakland. **Richard Realf,** the English-born poet, came to California in the summer of 1878. He was in bad health, and his attempts to secure a job as a clerk at the United States Mint in San Francisco were unsuccessful—he was, however, hired as a laborer. On October 28, 1878, Realf took poison and died at the Winsor Hotel, which stood on Washington Street near the corner of Ninth Street. He was buried in Lone Mountain Cemetery, later called Laurel Hill Cemetery, in Oakland, but was reinterred in Cypress Lawn Cemetery in Colma when Laurel Hill was closed. *Poems by Richard Realf, Poet, Soldier, Workman* appeared in 1898, twenty years after his death.

William Saroyan spent part of his childhood in Oakland, separated from his family in Fresno because his widowed mother could not afford to keep her family together. Saroyan lived from 1910 until 1916 at the Fred Finch Orphanage, at 3800 Coolidge Avenue, and attended the Sequoia School. **Mark Schorer,** the critic and novelist, died on August 11, 1977, at Kaiser Hospital in Oakland. **George Sterling,** the poet, worked as a clerk in the real-estate office of his uncle, Frank C. Havens, about the turn of the twentieth century. Havens's office was located in San Francisco, and Sterling lived at Havens's house on Oakland Avenue near Vernal Avenue, in what is now the city of Piedmont. Sterling hated his job and was doubtless overjoyed when a gift from an aunt enabled him to quit. He moved with his wife Carrie to Carmel in 1908 to enjoy the bohemian life and write poetry.

Robert Louis Stevenson was brought to

He worked mechanically. When a small bobbin ran out, he used his left hand for a brake, stopping the large bobbin and at the same time, with thumb and forefinger, catching a flying end of twine. Also, at the same time, with his right hand, he caught up the loose twine-end of a small bobbin. These various acts with both hands were performed simultaneously and swiftly. Then there would come a flash of his hands as he looped the weaver's knot and released the bobbin. There was nothing difficult about weaver's knots. He once boasted that he could tie them in his sleep. And for that matter, he sometimes did, toiling centuries long in a single night at tying an endless succession of weaver's knots.

—Jack London,
in "The Apostate"

(*far left*) Robert Louis Stevenson. (*left*) Photograph of Jack London taken early in his career.

Oakland in 1880 after his physical collapse in his rented room in San Francisco. His fiancée, Fanny Van de Grift Osbourne, moved Stevenson to the Tubbs Hotel, which stood on East Twelfth Street between Fourth and Fifth avenues. There he suffered a relapse. Under the care of an Oakland physician, Dr. William Bamford, Stevenson regained some of his strength and was moved to Mrs. Osbourne's home at 554 East Eighteenth Street. Mrs. Osbourne's house no longer stands. Dr. Bamford's house at 1235 East Fifteenth Street, which Stevenson may have visited, is still standing but greatly altered. By May 1880 Stevenson was well enough to marry Mrs. Osbourne—she had been divorced from her first husband in January of that year. After the wedding in San Francisco, the newlyweds—along with Mrs. Osbourne's son Lloyd—went to live in Calistoga. Stevenson returned to Oakland in 1888 to prepare for his departure for the South Seas aboard the schooner *Casco.* Tradition has it that Stevenson, while supervising the loading of supplies, often stopped by Johnny Heinold's First and Last Chance Saloon, now located in Jack London Square, on the waterfront at the foot of Broadway. Some like to think that **Jack London,** then about twelve years old, was in the saloon during one of Stevenson's visits, and that this putative meeting was the origin of London's own desire to sail to the South Seas. Regardless of whether the two actually met, Stevenson had the ninety-five-foot yacht towed out past the Golden Gate on June 28, 1888, to start the six-year adventure that would end with his death in Samoa in 1894.

Charles Warren Stoddard, the poet and writer, studied in 1863–64 at the College of California, which was moved to Berkeley in the early 1870s and became the University of California. **Thornton Wilder,** as a student at Berkeley High School in the early twentieth century, often visited Oakland's Ye Liberty Theater, in a building that still stands on Broadway between Fourteenth and Fifteenth streets.

Two important writers associated with Oakland are Jack London and Joaquin Miller. **Jack London** first came to live in Oakland when he was a boy. His parents, John and Flora London, moved to Oakland in the 1880s after they had failed in several attempts to establish themselves as independent farmers. Oakland was to be Jack London's home until the early 1900s, by which time he had become a successful writer. Oakland was perhaps the greatest influence on London's life and writing, not only because of the length of London's residence here but because of the variety of experiences the city offered him. Here, on the Oakland waterfront, he found the adventure that he sought to counterbalance the enormous pressures of duty to his impoverished family. But he also found stimulation in the books of the Oakland Public Library, whose head librarian, **Ina Coolbrith,** the poet and friend of such writers as Bret Harte, Joaquin Miller, and Charles Warren Stoddard, helped London find his way to the world of books.

The excitement of the waterfront and the mystery of books helped draw London's attention away from the hard realities of his life. While a child—small for his age, and by nature, shy—he sold newspapers before and after school to help pay family expenses. Being a small boy in a rough city was hard on London, and he frequently had to defend himself against larger and tougher youths: he later used these experiences for his story "Moon Face." After he had grown a bit, London worked in a cannery, putting in ten or twelve hours a day for ten cents an hour. Several years later he worked in a jute mill here and earned the same wage. London also worked at shoveling coal in an electric power plant in Oakland, hoping to become an electrical engineer by starting at the bottom and working up. He quit after learning that he had been hired to replace two grown men. London's experiences as a laborer in the jute mill were the basis for his story "The Apostate."

After a time, London sickened of the life of a laborer, bought a sloop, and sailed the bay as an oyster pirate. He and his friends would raid the commercial oyster beds at night and race to

His gaze dropped from the clouds to the bay beneath. The sea breeze was dying down with the day, and off Fort Point a fishing boat was creeping into port before the last light breeze. A little beyond, a tug was sending up a twisted pillar of smoke as it towed a three-masted schooner to sea. His eyes wandered over toward the Marin County shore. The line where land and water met was already in darkness, and long shadows were creeping up the hills toward Mount Tamalpais, which was sharply silhouetted against the western sky.

Oh, if he, Joe Bronson, were only on that fishing boat and sailing in with a deep-sea catch! Or if he were on that schooner, heading out into the sunset, into the world! That was life, that was living, doing something and being something in the world.

—Jack London,
in *The Cruise of the Dazzler*

Jack London signature slab: "Jack London miner author Jan 27 1898," taken from the Yukon cabin in which London spent the winter of 1897–98.

the Oakland docks to sell their catch. When they were not raiding, they were in the waterfront saloons, drinking, telling stories, planning future raids, brawling, and having a good time. London was smart enough to realize that if he remained an oyster pirate he would probably end up in prison or dead, so he went over to the other side, joining the Fish Patrol, whose mission was to bring to justice the oyster pirates. London's experiences as Prince of the Oyster Pirates made him an especially able officer of the law and saw him through many dangerous encounters. When the excitement and adventure of the Fish Patrol wore off, London abandoned that line of work and set off on new adventures. Years later he used his experiences as a pirate and as a member of the Fish Patrol in such books as *The Cruise of the Dazzler* (1902), *Tales of the Fish Patrol* (1905), and *John Barleycorn* (1913).

Oakland was the starting point for other adventures: London's months in 1893 as a seaman aboard the sealing schooner *Sophie Sutherland,* which gave him material for his novel *The Sea Wolf* (1904); his hobo expedition across the country in 1894, from which London drew the material for his book *The Road* (1907); and his trip to the Yukon in 1897–98, the source of the novels and stories for which London is best remembered. Oakland was also the birthplace of London's lifelong interest in socialism. As a student at Oakland High School, and later at the University of California at Berkeley, London was looked on by students and faculty as an unusual student, not only because of his somewhat rough appearance, but because of his varied experiences and wide reading. London's study of the works of Darwin, Marx, Nietzsche, and Spencer, combined with his experiences as a laborer, formed the basis of his energetic, if somewhat contradictory, socialism. London even went so far as to allow himself to be arrested for addressing a gathering at what is now City Hall Plaza. Partly on the basis of his notoriety as "the boy socialist" and his popularity as a writer, London ran for mayor on the Socialist ticket in 1901 and 1905, but was not elected.

In Oakland Jack London frequented Johnny Heinold's Saloon, also called First and Last Chance Saloon—it offered passengers on the Oakland–San Francisco ferry the chance to quench their thirst before and after a trip. Heinold, the figure on the right in the photograph, taken about 1890, helped London when he was struggling to become a writer.

London's years in Oakland may have been the basis for the contradictions in his view of life. Poverty and hard work suggested to him that only the strong could force their way to the top, and reading encouraged his belief that capitalism was a corrupt system that had to be overthrown and that the capitalist society would use violent means to protect itself. Thus it is in Oakland that one finds the origins not only of Wolf Larsen, the superman captain in the novel *The Sea Wolf,* but also of Ernest Everhard, the protagonist of the novel *The Iron Heel* (1907), who sought the overthrow of the capitalist oligarchy.

Of the many houses in which London lived in Oakland, only three still stand: 1639 Twenty-second Avenue, 1645 Twenty-fifth Avenue, and 1914 Foothill Boulevard. But there are other Oakland sites associated with London. Jack London Village is a commercial plaza on the Oakland waterfront off Alice Street. It bears virtually no resemblance to the waterfront London knew. Two blocks to the west, at the foot of Broadway, is Jack London Square, which features a reconstruction of the cabin in which London lived while in the Yukon. Nearby is Johnny Heinold's First and Last Chance Saloon, London's favorite watering hole during his Oakland days. According to London in his book *John Barleycorn* (1913), it was Johnny Heinold who encouraged London's love of books and learning, gave London his *Webster's* unabridged dictionary to study, and later lent him the money to enroll at the University of California. London never forgot the countless kindnesses shown him by the cigar-chewing saloonkeeper, and after London had become a famous writer, he stopped by the saloon many times to have a drink with his old friend. London's lasting tribute to Heinold can be found in the pages of *John Barleycorn.*

One of Jack London's early literary acquaintances was **Joaquin Miller,** the bearded and buckskinned poet and writer known as the Poet of the Sierras. In 1886 Miller bought a tract of land in the foothills just outside Oakland. Here, at the Hights, as he called his estate, Miller built a house called the Abbey, so named in honor of his wife Abbie. In this house, Miller wrote many of his poems, including "Columbus," and he completed work on *The Building of the City Beautiful* (1893) and *The Complete Poetical Works of Joaquin Miller* (1897). At the Hights, Miller continued to live the colorful life that had made him a well-known figure in America and in Europe. He planted many trees on his property and in the Oakland area, and in 1886 helped organize the first Arbor Day in California. On his property Miller constructed monuments to Robert Browning, General John C. Frémont, and Moses. He also built his own funeral pyre. Among the many writers who visited Miller at the Hights were Ambrose Bierce, Ina Coolbrith, Jack London, Edwin Markham, and George Sterling. Miller died on February 17, 1913. His body was cremated—but, because of local law, not on the pyre he had built for himself—and his ashes were cast over the grounds of the Hights. Miller's house, the Abbey, is standing and marked, and visitors can still enjoy the natural beauty of Miller's estate, now incorporated into Joaquin Miller Park, off Joaquin Miller Road.

> *Behind him lay the gray Azores,*
> *Behind the Gates of Hercules;*
> *Before him not the ghost of shores,*
> *Before him only shoreless seas.*
> *The good mate said: "Now must we pray,*
> *For lo! the very stars are gone.*
> *Brave Adm'r'l, speak; what shall I say?"*
> *"Why, say 'Sail on! sail on! and on!'"*
>
> —Joaquin Miller,
> in "Columbus"

OCEANSIDE

From 1946 to 1960, **Ben Hecht** owned a beachfront home, now private, at 2035 South Pacific Street in this city north of San Diego. During these years the novelist and dramatist divided his time between Oceanside and his home in Nyack, New York.

OXNARD

Erle Stanley Gardner, who created lawyer Perry Mason, was admitted to the California Bar in 1911 and soon after began to practice law here from offices at 423 B Street. For the five years he worked as a lawyer in this city west of Los Angeles, Gardner lived at 100 C Street, in a building that is still standing.

PACIFIC GROVE

Gelett Burgess, the humorist, who died in Carmel on September 18, 1951, was buried at Little Chapel By-The-Sea, 65 Asilomar Avenue, Pacific Grove.

John Steinbeck and his wife Carol moved in 1930 to a small cottage, built and owned by Steinbeck's father, at 147 Eleventh Street. This was to be the couple's home in Pacific Grove until 1936. The struggling writer held several temporary jobs during this period, tended a garden next to a small pond near the house, caught fish in Monterey Bay, and scrimped to buy paper and ink. Here he met Ed Ricketts, owner of Pacific Biological Laboratories, who rekindled Steinbeck's early interest in marine biology. (Steinbeck's 1945 novel *Cannery Row* describes the Western Biological Laboratory, owned by a sympathetic character called Doc.) By the time Steinbeck left Pacific Grove in 1936, he had completed work on his novel *To a God Unknown* (1933) and on his first two successful novels, *Tortilla Flat* (1935) and *In Dubious Battle* (1936). In 1941 Steinbeck returned to Pacific Grove and bought a small house at 425 Eardley Street. His first marriage had failed, and he would shortly move East with Gwyndolyn Conger, who became his second wife. While in Pacific Grove, Steinbeck conducted research at Ed Rickett's laboratory for *The Sea of Cortez* (1941), the book they wrote together. In September 1948, after the failure of Steinbeck's second marriage, he returned once more to Pacific Grove, where he worked on the script for the motion picture *Viva Zapata!* Here, in 1949, he first met Elaine Scott, whom he later married.

PACIFIC PALISADES

Writers associated with this community west of Los Angeles include **Raymond Chandler,** the detective-story writer, who lived here in the early 1940s; and **Henry Miller,** who lived in a home at 444 Ocampo Drive from 1963 until his death on June 7, 1980. **Will Rogers** lived from 1928 until his death in 1935 on a ranch at 14253 Sunset Boulevard, now Will Rogers State Historic Park. Rogers was a motion-picture star and radio personality as well as a syndicated columnist. At the height of his career, he and aviator Wiley Post died together in an airplane crash in Alaska. The accident, on August 15, 1935, made newspaper headlines everywhere. Rogers's ranch is still maintained as it was when Rogers died. It includes a main house, stables, corrals, riding ring, polo field, and riding trails designed by Rogers and is open daily. An admission fee includes an audio tour of the ranch.

[Edward] Rod[man] Serling, best known as a writer of plays for television, lived at 1490 Monaco Drive from the late 1950s until his death on June 28, 1975.

PALM SPRINGS

From October 1935 to May 1936, **Edgar Rice Burroughs** and his wife Florence lived in

> *More than once, in the brief days of my struggle for an education, I went to Johnny Heinold to borrow money. . . . In distress, when a man has no other place to turn, when he hasn't the slightest bit of security which a savage-hearted pawnbroker would consider, he can go to some saloon keeper he knows. Gratitude is inherently human. When the man so helped has money again, depend upon it that a portion will be spent across the bar of the saloon keeper who befriended him.*
>
> —Jack London,
> in *John Barleycorn*

The Abbey, home of Joaquin Miller in Oakland. The poet lived here for more than twenty-five years, until his death in 1913.

a rented house on Arenas Road in this city. **Moss Hart,** the playwright and librettist, died on December 20, 1961, just three weeks after moving into a new home in Palm Springs.

PALO ALTO

An early resident of this city was **Isabella Alden,** author of more than one hundred and twenty books, many of them for children, who lived in Palo Alto for more than thirty years. At first a winter resident, she later lived in Palo Alto year-round, and died here on August 5, 1930. **Kenneth Patchen,** the poet, died on January 8, 1972, at his home, now private, at 2340 Sierra Court.

Adjacent to Palo Alto is the spacious campus of Stanford University. **Maxwell Anderson,** the playwright, taught English at Stanford while studying for his M.A. degree, which he received in 1914. In 1936, just after Anderson's verse tragedy *Winterset* (1935) was produced, the university established the Maxwell Anderson Award for verse drama. **Van Wyck Brooks,** an instructor at Stanford from 1911 to 1913, wrote the biography *John Addington Symonds* (1914) while living in Palo Alto. **Allen Drury,** the novelist, was graduated from Stanford in 1939. **John Steinbeck** studied at Stanford intermittently between 1919 and 1925. He majored in English but took courses that appealed to him instead of courses required for a degree. While here, Steinbeck published a short story and a sketch in the Stanford *Spectator*. **Yvor Winters** taught at Stanford from 1928 until his retirement in 1966. The poet, who made his home at 143 West Portola Avenue, died in Palo Alto on January 25, 1968. Among his many books were *Edward Arlington Robinson* (1946), *Collected Poems* (1952), and *The Poetry of W. B. Yeats* (1960).

> *I remember when I came to the end*
> *of all the Talmud said, and the commentaries,*
> *then I was fifty years old—and it was time*
> *to ask what I had learned. I asked this question*
> *and gave myself the answer. In all the Talmud*
> *there was nothing to find but the names of things,*
> *set down that we might call them by those names*
> *and walk without fear among things known.*
>
> —Maxwell Anderson,
> in *Winterset*

Stanford University maintains collections of the papers of Ambrose Bierce; Bernard De Voto; Allen Drury; and Wallace Stegner, the author of the novel *The Big Rock Candy Mountain* (1943) and the Pulitzer Prize-winning novel *Angle of Repose* (1971), who has taught at the university for many years.

PASADENA

This city northeast of Los Angeles was the home of the Pasadena Playhouse, at 39 South El Molino Avenue, from 1925 to its closing in 1969 one of the best-known theaters in the United States. It attained wide recognition for its 1928 production of *Lazarus Laughed,* by Eugene O'Neill. In the same year, the Playhouse opened its College of Theatre Arts, which gave early training to a number of actors and directors who became well known. The Pasadena Playhouse also ran a laboratory theater, in which the works of new writers were produced, and the Playhouse had the distinction of being the first theater to produce all of Shakespeare's history plays in chronological order. In 1937 a State of California resolution gave the Playhouse the title "State Theater of California," and in 1946 the Playhouse presented the world première of *In Savoy; or, "Yes" Is for a Very Young Man,* by Gertrude Stein. The Pasadena Playhouse in 1969 was forced to close for financial reasons, but the City of Pasadena has bought the building since then and plans to reopen it as a theater and communications center.

Writers associated with Pasadena include **Zoë Akins,** the dramatist, who moved in 1929 to Green Fountains, her home at 2207 Brigden Road. **Earl Derr Biggers,** the playwright and novelist, lived at 2000 East California Boulevard, in a house no longer standing, for eight years before his death on April 5, 1933, at Pasadena Hospital, now Huntington Memorial Hospital, at 100 Congress Street. Biggers is best known for his wise and quotable detective Charlie Chan who, with his number one son, solved perplexing crimes in such thrillers as *Charlie Chan Carries On* (1930) and *Keeper of the Keys* (1932).

In late 1927 **Hart Crane,** working as secretary to millionaire Herbert A. Wise, lived in Wise's villa, no longer standing, at 2160 Mar Vista. The poet kept the job until March 1928, and then moved to his mother's bungalow, also no longer standing, at 1803¼ North Highland Avenue. **F[rancis] O[tto] Matthiessen,** the literary critic, born here on February 10, 1902, spent some of his early years in Pasadena. His family's home was at 100 North El Molino. Matthiessen's many works of criticism included *American Renaissance* (1941) and *Henry James: The Major Phase* (1944). He was editor of *The Oxford Book of American Verse* (1950). **Frederick Jackson Turner,** the historian, spent the last years of his life in Pasadena, and served as a research associate at the Huntington Library in nearby San Marino. Turner died on March 14, 1932, at home, 26 Oak Knoll Gardens.

Of all the writers who have lived in Pasadena, perhaps the best known is **Upton Sinclair,** who moved in 1916 with his wife Craig to a house at 1050 North Hudson, and in 1917 to a house at 1497 Sunset Street. From 1919 until their move to Monrovia in the fall of 1942, the couple lived in a third house at 1513 Sunset Street. Over the years the Sinclairs bought seven adjacent lots to add to their property. They also moved four houses onto the property and had them attached to the original house to produce a single, massive home. In the 1930s Sinclair found himself something of a celebrity because of his political beliefs, so he and his wife sometimes moved to temporary homes to avoid the many people who wanted to visit them. None of Sinclair's Pasadena homes is still standing. Sinclair worked on various books during his stay in Pasadena: his tract on journalistic practices, *The Brass Check* (1919); the novels *Oil!* (1927) and *Boston* (1928); and his autobiographical *American Outpost* (1932). In 1934 Sinclair ran unsuccessfully for governor on the EPIC (End Poverty In California) ticket.

> *Oaths rolled from his [Wolf Larsen's] lips in a continuous stream. And they were not namby-pamby oaths, or mere expressions of indecency. Each word was a blasphemy, and there were many words. They crisped and crackled like electric sparks. I had never heard anything like it in my life, nor could I have conceived it possible. With a turn for literary expression myself, and a penchant for forcible figures and phrases, I appreciated, as no other listener, I dare say, the peculiar vividness and strength and absolute blasphemy of his metaphors.*
>
> —Jack London,
> in *The Sea-Wolf*

PIEDMONT

In February 1902 **Jack London** and his wife Bess moved to a house in Piedmont at 56 Bayo Vista Avenue, no longer standing, overlooking San Francisco Bay. Piedmont now is a suburb of Oakland. From here, London traveled to England to gather material for his book *The People of the Abyss* (1903). Among the books London worked on while in Piedmont are two of his best works: *The Call of the Wild* (1903) and *The Sea-Wolf* (1904). In 1903, his marriage to Bess having failed, London moved to Oakland. He later built another home in Piedmont for Bess and their two daughters.

PLACERVILLE

Edwin Markham lived in this city, east of Sacramento, for seven years beginning in 1879, when he was selected County Superintendent of Schools. The Edwin Markham Elementary School in Placerville is named in his honor.

PORTERVILLE

Allen Drury, the novelist, lived during childhood at 200 Carmelita Way in this city, about forty-five miles north of Bakersfield.

REDLANDS

In 1907 **Harold Bell Wright** lived at 314 Buena Vista while serving as minister of the First Christian Church (Disciples of Christ) in this city southeast of San Bernardino. He resigned after a year to devote his time to writing. It is said that Wright returned to Redlands while writing *The Winning of Barbara Worth* (1911), a novel that was to sell a million and a half copies in twenty-five years. Wright stayed at the home of W. F. Holt, who became the model for one of the characters in the novel. Redlands is the setting for Wright's novel *The Eyes of the World* (1914).

SACRAMENTO

Lincoln Steffens moved in 1870 with his parents to Sacramento, the capital of California. The family lived in larger and larger homes over the years. Finally, in 1888, Steffens's father bought a Victorian mansion at the corner of Sixteenth and H streets, which served as the family home until 1903, when the elder Steffens sold it to the State of California for use as the governor's mansion, a function it fulfilled until 1967. Ironically, later in his life, Steffens frequently returned to this house to plead on behalf of individuals and causes. The former governor's mansion is now listed in the National Register of Historic Places, but the highways now surrounding it have robbed the mansion of its original grandeur.

The area along the Sacramento River west of Route 5 is designated as the Old Sacramento Historic District, the largest restoration of buildings of the Gold Rush Era on the West Coast. Here stands the reconstructed building that once housed the Sacramento *Union,* the newspaper that sent **Mark Twain** to Hawaii in 1866. At that time the newspaper was located at 49 and 51 Third Street, between J and K streets, now a vacant lot. Twain used the material he gathered during his Hawaii trip for a seriocomic lecture that was well received across the country, and later wrote about his trip in *Roughing It* (1872).

ST. HELENA

In the late 1800s **Ambrose Bierce** and his family moved to a small white cottage at 1515 Main Street in this city in the Napa Valley. This was Bierce's home for several years, but he spent much of his time at Angwin's Camp, about seven miles away.

The Silverado Museum, at 1490 Library Lane, houses a major collection of items relating to Robert Louis Stevenson who, with his wife and stepson, lived for several months in the summer of 1880 at the Silverado Mine on Mount St. Helena, about eighteen miles northwest of St. Helena. The museum's holdings include many of Stevenson's books, manuscript notes for *The Master of Ballantrae,* letters, and paintings, as well as Stevenson's toy soldiers, wedding ring, desk, and other items.

SALINAS

Jackson Gregory was born on March 12, 1882, in this city, the seat of Monterey County. Gregory's more than thirty popular Westerns and detective novels included *The Splendid Outlaw* (1932) and *Hermit of Thunder King* (1945).

The author most closely associated with Salinas, however, is **John [Ernst] Steinbeck,** born on February 27, 1902, in a large two-story house at 132 Central Avenue. Steinbeck grew up in Salinas, attended school here, worked on ranches all along the Salinas Valley, and became familiar with the land and the people he would later celebrate

Home of Lincoln Steffens in Sacramento.

It was a deluge of a winter in the Salinas Valley, wet and wonderful. The rains fell gently and soaked in and did not freshet. The feed was deep in January, and in February the hills were fat with grass and the coats of the cattle looked tight and sleek. In March the soft rains continued, and each storm waited courteously until its predecessor sank beneath the ground. Then warmth flooded the valley and the earth burst into bloom—yellow and blue and gold.

—John Steinbeck,
in *East of Eden*

in his books. As a child he began to read extensively, and it was during these years that he first read *Crime and Punishment, Paradise Lost,* Malory's *Morte d'Arthur,* and other classics whose universal themes were to play an important role in shaping Steinbeck's novels and stories. At Salinas High School, Steinbeck wrote for the school newspaper and was senior class president. He left after graduation in 1919 to attend Stanford University and later to travel to New York, but he returned to Salinas a number of times before his marriage to Carol Henning in 1930.

Steinbeck's ashes were buried in the Hamilton family plot—Steinbeck's mother was a Hamilton—in Garden of Memories Memorial Park, 768 Abbott Street. His birthplace at 132 Central Avenue is now used as a luncheon restaurant, and the front bedroom in which Steinbeck was born is now the restaurant's reception room. Tours of the house are available by arrangement with the Valley Guild, which owns the house. The John Steinbeck Library, 110 San Luis Street, maintains a collection of 30,000 items relating to the Nobel laureate, including letters, photographs, first editions, and a collection of taped interviews with people who had known Steinbeck and the locales used in his books.

SAN DIEGO

Writers associated with San Diego include **Edgar Rice Burroughs,** who lived at 4036 Third Street for several months early in 1914; Raymond Chandler, who used the San Diego waterfront as a setting for his mystery novel *Playback* (1958); **Richard Henry Dana, Jr.,** who worked in a hide shack at Point Loma—where the Naval Fuel Depot now stands—when he served as a seaman on a hide-gathering expedition in the 1830s; and **George Horatio Derby,** an engineer with the U.S. Army in San Diego in 1853. J. J. Ames, editor of the San Diego *Herald,* asked Derby to run the newspaper while Ames was out of town. Derby's injection of humor into the sedate journal was well received, and he began a rapid rise to national reputation as a humorist. The *Herald* was originally located at Fourth and California streets. Derby lived at the Pendleton House, now a historical site, located on Harney Street between Juan Street and San Diego Avenue in San Diego's Old Town section.

Carl Glick, known for his books on China, taught at California Western University in San Diego from 1955 to 1961 and lived at 4674 Point Loma Avenue for several years before his death on March 7, 1971. He is buried in Fort Rosecrans National Cemetery on Cabrillo Memorial Drive. The cemetery dates from the 1850s, as does the rebuilt Chapel of the Immaculate Conception, on Conde Street west of San Diego Avenue, described in the novel *Ramona* (1884), by Helen Hunt Jackson. In the novel Ramona married Alessandro in this adobe church. Gaspara, who performed the marriage, was modeled after Father Antonio Ubach, a priest who served at the church for many years. Originally located on San Diego Avenue, Immaculae Conception was replaced by a new structure in 1916 and, in 1937, was rebuilt on its present site.

Henry Miller first met anarchist Emma Goldman in San Diego, shortly after World War I. He later described the event for *Twentieth Century Authors* (1942), by Stanley Kunitz and Howard Haycraft, as ". . . the most important encounter of my life. . . . She opened up the whole world of European culture for me and gave a new impetus to my life, as well as direction." **Rose Hartwick Thorpe,** the poet, died on July 19, 1939, at the home of her son-in-law at 3361 Fourth Avenue, now a doctor's office. The ashes of **Harold Bell Wright** are buried in Greenwood Memorial Park.

Photograph of Chapel of the Immaculate Conception in San Diego, as it appeared in 1876.

SAN FRANCISCO

Writers born in San Francisco include **Gertrude [Franklin Horn] Atherton,** the novelist, who was born on October 30, 1857, in a house on Rincon Hill near South Park, an avenue connecting Second and Third streets. The area was originally developed as a community for the wealthy, and Atherton wrote of it in her novel *A Daughter of the Vine* (1899). Toward the end of her life, Atherton returned to live in San Francisco at 2280 Green Street. She continued to write her daily thousand words until shortly before her death on June 14, 1948, at Stanford Hospital, on Webster Street. Atherton described her childhood and youth in her autobiography, *The Adventures of a Novelist* (1932).

David Belasco, the playwright and theatrical producer who wrote, among other plays, *The Girl of the Golden West* (1905) and *Madame Butterfly* (with John Luther Long in 1900), was born in San Francisco on July 25, 1854. Belasco spent part of his childhood and youth in San Francsico, and began to work in the theater here. In 1874 he was stage manager at Maguire's Opera House, which originally stood on Washington Street on the east side of Portsmouth Square, but after a fire in 1873 moved to new quarters in the Bush Street district of the city. In 1876 Belasco became assistant manager of the Baldwin Theater, which stood on the east corner of Powell and Market streets. There he acted with such noted performers as Edwin Booth, and in 1878 became the theater's sole manager.

Barnaby Conrad, the painter and writer known best for his works dealing with bullfighting, was born in San Francisco on March 27, 1922; and **Charles Caldwell Dobie,** the novelist, short-story writer, and playwright, was born here on March 15, 1881. Dobie's books included *San Francisco: A Pageant* (1933), *San Francisco Tales* (1935), and *San Francisco's Chinatown* (1936). **Martin Flavin,** who won a Pulitzer Prize for his novel *Journey in the Dark* (1943), was born here on November 2, 1883. **Kathryn Forbes,** born **Kathryn Anderson** here on March 20, 1909, wrote the novel *Mama's Bank Account* (1943), set in San Francisco and adapted by John Van Druten for the stage as *I Remember Mama* (1944). Forbes also wrote the novel *Transfer Point* (1947), also set in San Francisco.

Robert Frost, who celebrated New England life, was born in San Francisco on March 26, 1874, and lived here until about the age of eleven. San Francisco honors him with a Robert Frost Plaza, at the corner of California and Market streets.

Shirley Jackson, author of the collection of stories *The Lottery* (1949) and the novels *Hangsaman* (1951) and *The Haunting of Hill House* (1959), was born in San Francisco on December 14, 1919. **Peter B[ernard] Kyne,** creator of Cappy Ricks, the tough old sea captain with a heart of gold, was born here on October 12, 1880. Kyne wrote twenty-five novels, including *Three Godfathers* (1913) and *The Valley of the Giants* (1918), but he was known more for the half a hundred short stories he wrote about Cappy Ricks's adventures. Alden P. Ricks appeared in such volumes as *Cappy Ricks* (1916) and *Cappy Ricks Comes Back* (1934). Kyne lived toward the end of his life at 2351 Bay Street. He died on November 25, 1957, at the Veteran's Administration Hospital in San Francisco.

Jack London was born in San Francisco on January 12, 1876, the son of Flora Wellman and one William Chaney, an itinerant astrologer and writer. Flora Wellman and Chaney had been living together at 122 First Avenue when Flora became pregnant. She begged Chaney to marry her but he denied paternity and refused. Flora Wellman made an unsuccessful and melodramatic attempt to shoot herself. The local newspapers picked up the story that Chaney had turned her out of their home and that she had refused to abort the pregnancy. Chaney left for Oregon, and Flora was taken to the home of a friend at 615 Third Street, where Jack London was born. The infant did not receive his surname until his mother married John London in San Francisco on September 7, 1876. The full name given was **John Griffith London.** Soon after the marriage the London family moved across San Francisco Bay to live on a series of small farms before settling in Oakland.

Thirty years after Jack London's birth, his birthplace was destroyed, along with most of San Francisco, in the earthquake and fire of April 18, 1906. London, by then living in Glen Ellen and one of his country's most popular writers, rushed to the burning city to record the disaster in an article, "The Story of an Eye-Witness" (1906). London used this experience in describing the burning of Chicago in his novel *The Iron Heel* (1908). The site of London's birthplace, at Third and Brannan streets, now is marked by a plaque.

Jack London's birthplace in San Francisco was destroyed in the great fire of 1906. This commemorative plaque stands just to the left of the main entrance of the Wells Fargo Bank, Third and Brannan streets.

Other writers born in San Francisco include **Kathleen [Thompson] Norris,** the prolific novelist, on July 16, 1880; and **[Joseph] Lincoln Steffens,** on April 6, 1866. Steffens was born in San Francisco's Mission district and lived here until 1870. His exposés of corruption in government and business, collected in *The Shame of the Cities* (1904) and other volumes, made him the premier muckraker of his time. After Steffens's death in Carmel on August 9, 1936, funeral services were held at Cypress Lawn Cemetery, in Colma. **Irving Stone,** the biographer and novelist, was born **Irving Tennenbaum** on July 14, 1903, in a house on Stockton Street north of Washington Square. Stone grew up in San Francisco and lived at a number of addresses, including 158 Downey Street, 2449 Lake Street, and 882 Haight Street. Stone was graduated in 1919 from Lowell High School, then on Hayes Street but now located at 1101 Eucalyptus Drive. **Alice B[abette] Toklas,** who was to become the friend and secretary of Gertrude Stein for many years, was born on April 30, 1877, in a house at 922 O'Farrell Street. She later lived with her family at 2300 California Street.

Writers who have lived or worked in San Francisco include **Maxwell Anderson,** the playwright, who taught English at Polytechnic High School here from 1914 to 1917, and worked for the San Francisco *Bulletin* and the *Chronicle* before moving to New York City in 1918. During his time in San Francisco, Anderson also worked toward his M.A. degree at Stanford University.

Ambrose Bierce came to San Francisco in 1866 after serving in a military expedition led by Gen. William B. Hazen. Bierce, who had distinguished himself in the Civil War and risen to the rank of brevet major, found that if he wanted to rejoin the peacetime Army he would have to accept the rank of second lieutenant. Bierce chose instead to take a job with the U.S. Mint in San Francisco, living at first in a furnished room. Soon he moved to the Russ House, a hotel that stood on Montgomery Street between Pine and Bush streets. In 1868 Bierce began contributing to the San Francisco *News-Letter* and in December of that year became managing editor of the weekly. Bierce wrote much of the "Town Crier" page, and under his guidance the *News-Letter* began to attract national attention. During this time Bierce met such writers as **Ina Coolbrith, Bret Harte, Charles Warren Stoddard,** and **Mark Twain.**

The business man has failed in politics as he has failed in citizenship. Why? Because politics is business. That's what's the matter with everything,—art, literature, religion, journalism, law, medicine,—they're all business, and all—as you see them.

—Lincoln Steffens,
in *The Shame of the Cities*

I passed out of the house. Day was trying to dawn through the smoke-pall. A sickly light was creeping over the face of things. Once only the sun broke through the smoke-pall, blood-red and showing quarter its usual size. The smoke-pall itself, viewed from beneath, was a rose-color that pulsed and fluttered with lavender shades. Then it turned to mauve and yellow and dun. There was no sun. And so dawned the second day on stricken San Francisco.

—Jack London,
in "The Story of an Eye-Witness"

If California ever becomes a prosperous country, this bay will be the center of its prosperity.

—Richard Henry Dana, Jr., writing of San Francisco Bay in *Two Years Before the Mast*

"Nine years ago, w'en that big brute killed the woman who loved him better than she did me!—me who had followed 'er from San Francisco, where 'e won her at draw poker! . . ."

—Ambrose Bierce, in "The Haunted Valley"

Harte published Bierce's first story, "The Haunted Valley," in the *Overland Monthly.* As Bierce's literary stature grew, so did his social position: on December 25, 1871, he married Ellen Day, known as Mollie, who was the daughter of a well-to-do mining engineer and mine owner. In March 1872 the couple left for England. They returned to San Francisco in 1875 and moved to a house on Harrison Street. In 1877 Bierce became associate editor of the *Argonaut,* where he conducted his "Prattle" column until 1879. In that year Bierce left the city on a mining expedition to the Black Hills of Dakota Territory, now South Dakota.

The mining venture failed, and Bierce returned to San Francisco in late 1880, soon after becoming editor of *Wasp,* a satirical weekly. He revived his "Prattle" column and in it published the first definitions that would be collected later in *The Cynic's Wordbook* (1906), which was republished in 1911 as *The Devil's Dictionary.* Bierce wrote against vice, stupidity, corruption, and literary pretentiousness with such vigor that he was often called Bitter Bierce. He remained with the *Wasp* until 1885, although by that time he had moved out of the city. In March 1887 he returned to San Francisco, this time to work for the San Francisco *Examiner,* newly purchased by George Hearst, a wealthy and politically ambitious miner, for his son, **William Randolph Hearst.** Bierce revived his "Prattle" column for the *Examiner,* then located at 10 Montgomery Street, and continued it in the newspaper until 1896. During his years with the Hearst organization, Bierce wrote the stories collected in *Tales of Soldiers and Civilians* (1891) and *Can Such Things Be?* (1893), often considered among the best of his writings.

Although Bierce's vicious attacks in print against practically everybody in the fields of art and literature left him with few friends, he was respected enough to be endured as a member of the Bohemian Club. Bierce was a charter member of this society for San Francisco newspapermen. The Bohemian Club originally met for Sunday breakfast at the home of one James Bowman, who worked for the San Francisco *Chronicle.* In the late 1870s it moved to 430 Pine Street, and by the mid-1930s was located at 624 Taylor Street, its present address. Members of the club have included **Allen Drury, Henry George, Bret Harte, Oliver Wendell Holmes, Jack London, George Sterling, Charles Warren Stoddard, Irving Stone, Mark Twain,** and **Herman Wouk.** In 1872 the club held its first annual High Jinks, which evolved into a summer encampment and festival in a private grove north of San Francisco. Since then each High Jinks has featured a play written and performed by the members. The Bohemian Club now includes among its members not only writers and artists but also millionaires, politicians, and Pentagon officials.

Entrance of the Bohemian Club, 624 Taylor Street, San Francisco. The ivy-covered building takes up half a city block.

Early one June morning in 1872 I murdered my father—an act which made a deep impression on me at the time. This was before my marriage, while I was living with my parents in Wisconsin. My father and I were in the library of our home, dividing the proceeds of a burglary which we had committed that night.

—Ambrose Bierce, in "An Imperfect Conflagration"

Gelett Burgess lived in the 1890s in the Russian Hill section of San Francisco. In 1894 Burgess became associate editor of *The Wave* and wrote for it until 1901. In 1895 he and some friends founded the short-lived literary journal *The Lark* (1895–97), whose first issue carried Burgess's best-known poem, "The Purple Cow." In 1897 Burgess moved to New York City, taking with him a reputation as writer of light verse. In later years Burgess felt that the notoriety of his humorous quatrain kept others from appreciating his more serious work. He summarized this feeling in yet another quatrain, entitled "Cinq Ans Après":

Ah, yes, I wrote the "Purple Cow"—
I'm sorry, now, I Wrote it!
But I can Tell you, Anyhow,
I'll Kill you if you Quote it!

Ina [Donna] Coolbrith, the poet, came to live in San Francisco with her family in the mid-1860s. Here she became a teacher and, in 1868, co-editor with **Bret Harte** of the newly founded *Overland Monthly.* Coolbrith was librarian of the Oakland Public Library from 1874 to 1893, librarian of the Mercantile Library in San Francisco from 1897 to 1899, and librarian of the Bohemian Club—and the club's only woman member—from 1899 to 1906. She was also, with Bret Harte and Charles Warren Stoddard, a member of what was known as the Golden Gate Trinity, considered the pinnacle of San Francisco's literary world. Coolbrith was an early influence on Joaquin Miller, raised Miller's daughter Cali-Shasta, and as librarian in Oakland helped Jack London, then just a youth, to find books to read. It is thought that her influence on London helped him turn to the life of a writer. London, who acknowledged his debt to her, is thought to have used Coolbrith as a model for Mrs. Mortimer in his novel *The Valley of the Moon* (1913). Coolbrith was poet laureate of California from 1915 until her death on February 29, 1928. For many years she lived on Russian Hill in a house near what is now called Ina Coolbrith Park. Russian Hill for decades was the home of many San Francisco artists and writers, including **Mary Austin, Ambrose Bierce, Gelett Burgess, Will Irwin, Peter B. Kyne,** and **George Sterling.**

Richard Henry Dana, Jr. came to San Francisco during the winter of 1835–36 as a seaman aboard the ship *Alert.* Dana described San Francisco in *Two Years Before the Mast* (1840). **Miriam A[llen] deFord,** the poet, short-story writer, novelist, and biographer, made her home in San Francisco from the mid-1930s until her death on March 22, 1975. For a time she lived at 415 Stockton Street, and in the late 1930s moved to the Ambassador Hotel, at 55 Mason Street, where she lived and worked until her death. Mrs. deFord wrote many books, mostly biographies and translations, for the E. Haldeman-Julius series of Little Blue Books, and her poems and stories have appeared in more than fifty anthologies. She also wrote such books as the biography *They Were San Franciscans* (1941); the novel *Shaken With the Wind* (1942); the collection of poetry *Penultimates* (1962); the story collection *The Theme is Murder* (1967); and a book of science fiction, *Elsewhere, Elsewhen, Elsehow* (1971).

Lawrence Ferlinghetti for many years has been one of San Francisco's leading literary figures. His books of poetry include *A Coney Island of the Mind* (1958), *Starting from San Francisco* (1961), and *Who Are We Now?* (1976). Ferlinghetti was co-founder in 1953 of City Lights Bookshop, at 261 Columbus Avenue, which became a gathering place for writers, artists, and literary travelers. **Erle Stanley Gardner** was president of Consolidated Sales Company in San Francisco from 1916 to 1920. The automotive supply company did so well at first that other businesses hired Gardner as a management consultant. When a buyers' strike in 1920 found the company overexpanded, the business failed. Gardner, still more than a decade away from his writing career, returned to the practice of law.

While living in San Francisco in the 1870s, **Henry George** wrote his economic tract *Progress and Poverty* (1879). It was while riding in the hills outside of San Francisco that George came upon the central idea of his book: progress produces great wealth for a few, but it also produces economic hardship for many. No publisher would take George's book, so he published an edition of 500 copies himself. Soon enough he found a publisher, who issued the book in 1880, the same year in which George moved to New York City.

Allen Ginsberg, one of the Beat poets, lived in San Francisco in the 1950s. City Lights Bookshop published Ginsberg's volume *Howl and Other Poems* (1956). A year later Lawrence Ferlinghetti was arrested on obscenity charges stemming from publication of the volume by City Lights. After a colorful trial Ferlinghetti was acquitted. Ginsberg's *Kaddish and Other Poems* (1961) also was published by City Lights. Ginsberg has been associated since the 1950s with the literary life of San Francisco.

Dashiell Hammett lived and worked in San Francisco from 1921 until 1930, completing four of his five novels here, *Red Harvest* (1929), *The Dain Curse* (1929), *The Maltese Falcon* (1930), and *The Glass Key* (1931), as well as about one hundred short stories. Hammett, considered the founder of the hardboiled school of detective fiction, often used San Francisco as a setting, notably in *The Dain Curse* and *The Maltese Falcon*. Hammett worked on the latter while living at 891 Post Street. For a time he worked as advertising manager for a jeweler at 856 Market Street, and as a detective for the Pinkerton detective agency, whose office was located in the Flood Building, at 870 Market Street. This was the location of the fictional Continental Detective Agency that appeared in a number of Hammett's tales. In Burritt Alley is a plaque marking the site of the murder of Miles Archer, the partner of Hammett's fictional detective Sam Spade in *The Maltese Falcon*. Spade ate at John's Grill, at 63 Ellis Street, which now features a Dashiell Hammett room containing a collection of Hammett memorabilia.

Bret Harte came to San Francisco in 1860 and took a job as compositor for *The Golden Era*, a weekly newspaper noted for its sketches, fiction, and poetry. Harte soon became a member of its editorial staff, and his story of the indomitable M'liss was first published in *The Golden Era*. In the same building in which *The Golden Era* was published, the *Morning Call* had its offices. This was the newspaper for which **Mark Twain** worked. Twain and Harte became friends, and Twain contributed articles to *The Golden Era*. In 1861 Harte became a clerk in the surveyor-general's office, and in 1863 took a job with the United States Mint, for which he worked for six years. These jobs did not slow his literary production. After Charles Henry Webb founded *The Californian* in 1864, Harte contributed most of his writing for the next several years to this literary journal and

> *Spade did not look at the pistol. He raised his arms and, leaning back in his chair, intertwined the fingers of his two hands behind his head. His eyes, holding no particular expression, remained focused on Cairo's dark face.*
>
> *Cairo coughed a little apologetic cough and smiled nervously with lips that had lost some of their redness. His dark eyes were humid and bashful and very earnest. "I intend to search your offices, Mr. Spade. I warn you that if you attempt to prevent me I shall certainly shoot you."*
>
> *"Go ahead." Spade's voice was as empty of expression as his face.*
>
> —Dashiell Hammett, in *The Maltese Falcon*

John's Grill in San Francisco, one of the numerous San Francisco sites visited by Sam Spade, Dashiell Hammett's celebrated fictional detective. Old postcard view of City Lights Bookshop, one of San Francisco's best-known literary sites.

Joaquin Miller (*left*), George Sterling, and Charles Warren Stoddard (*right*), in a photograph taken in 1905.

served as its editor from time to time. In 1868 he became editor of the *Overland Monthly,* and it was with this publication that Harte achieved a national reputation. The *Overland Monthly* printed Harte's stories "The Luck of Roaring Camp" and "The Outcasts of Poker Flat." Harte, suddenly a celebrity, went East in 1871.

Other writers who have lived or worked in San Francisco include **Will James,** who wintered here and in Hollywood toward the end of his life; and **Jack Kerouac,** who lived here off and on in the early 1950s. Kerouac in 1952 revised his novel *On the Road* (1957) while living at the home on Russian Hill of **Neal Cassady,** the poet. Kerouac visited San Francisco again in 1954 and worked on an unpublished collection of poetry, *San Francisco Blues,* at the Cameo Hotel, which was located on Third Street. **Sinclair Lewis** came to San Francisco in the summer of 1909. He took rooms at 1525 Scott Street and found a job in the Alameda County office of the San Francisco *Evening Bulletin,* but he was fired after two months. Lewis then worked for the Associated Press from December 1909 to February 1910. Lewis wrote, somewhat inaccurately, of his San Francisco experiences in an autobiographical sketch, "You Get Around So Much" (1947).

> *NICK: I run the lousiest dive in Frisco, and a guy arrives and makes me stock up with champagne. The whores come in and holler at me that they're ladies. Talent comes in and begs me for a chance to show itself. Even society people come here once in a while. I don't know what for. Maybe it's liquor. Maybe it's the location. Maybe it's the crazy personality of the joint. The old honky-tonk. Maybe they can't feel at home anywhere else.*
>
> —William Saroyan, in *The Time of Your Life*

Frank Norris came to San Francisco in 1885, when he was fifteen years old. His family took a house at 1822 Sacramento Street, and Norris entered Belmont Academy in Belmont, south of San Francisco. His studies were interrupted when his family took an extended trip to Europe. On their return in 1889, Norris attended the University of California. In 1895 he traveled to South Africa as a correspondent for the San Francisco *Chronicle.* Returning here in 1896, Norris joined the staff of *The Wave,* but by 1898 had left San Francisco once again, this time for New York City. In 1902 Norris returned to California and made plans to live on a ranch near Gilroy, but he died in San Francisco on October 25, 1902, at Mount Zion Hospital, following an operation for appendicitis. Norris had been living at 1921 Broderick Street. San Francisco was the setting for Norris's novel *McTeague* (1899), the story of an unlicensed dentist on Polk Street and his fall from comfort and prosperity.

Kenneth Patchen, the poet, lived in San Francisco for a time in the 1950s. **Howard Pease,** who wrote many adventure stories for young readers, lived in the 1970s at 801 Sutter Street. Pease's books set in San Francisco included *Long Wharf* (1939) and *Mystery on Telegraph Hill* (1961). **Richard Realf,** the English-born poet, spent the last year of his life working at the United States Mint in San Francisco and trying to avoid one of his many wives who, unlike the other women Realf married and left, resisted abadonment. Realf, in despair, committed suicide in an Oakland hotel on October 28, 1878. He was buried first in Lone Mountain Cemetery, in Oakland, and when that cemetery was closed, in Cypress Lawn Cemetery, in Colma.

Kenneth Rexroth lived at 250 Scott Street in the 1960s and worked as a columnist for the San Francisco *Examiner* from 1960 to 1968. **William Saroyan** first came to San Francisco in the 1920s and drifted from job to job while gathering experience he later used in his short stories, plays, and novels. Saroyan lived at 2378 Sutter Street from 1928 to 1932, and at 348 Carl Street from 1932 to 1939. San Francisco was the setting for his collection of stories *My Name Is Aram* (1940) and his novel *The Human Comedy* (1942). His play *The Time of Your Life* (1939), for which Saroyan was awarded a Pulitzer Prize that he declined, was set in a San Francisco waterfront bar peopled by a diversity of interesting eccentrics.

Gertrude Stein early in the 1890s lived for about a year in a house on Turk Street. **John Steinbeck** moved in 1928 to 2953 Jackson Street to be near Carol Henning, who later became Steinbeck's first wife. Here Steinbeck worked on his second novel, *To a God Unknown* (1933). Steinbeck also lived at 2441 Fillmore Street. **George Sterling** lived in San Francisco toward the end of his life and died a suicide in his room at the Bohemian Club on November 17, 1926. From the beginning of the twentieth century, Sterling had been a colorful, bohemian presence in San Francisco's literary scene and, for some time, a protégé of Ambrose Bierce. Sterling's later life, however, was marked by despair: his estranged wife, Carrie, committed suicide in 1918, and the bohemian life he loved disappeared forever. Sterling is commemorated by a simple monument—a park bench at the junction of Hyde, Greenwich, and Lombard streets. The bench, inlaid with colorful tiles, bears a plaque honoring the poet.

Robert Louis Stevenson first came to San Francisco in 1879. Stevenson, twenty-nine years old, had come to the United States from his native Scotland to be near Mrs. Fanny Van de Grift Osbourne, with whom he had fallen in love. She was then living in Oakland. Stevenson stayed at first in Monterey, but in December 1879 moved to San Francisco. He lived first in a boarding house on Montgomery Street and then at the Gailhard Hotel, on Pine Street, before moving in late December to a second-floor room of a house at 608 Bush Street. Here Stevenson spent the bitter winter of 1879–80, struggling against the cold and the hunger his poverty obliged him to endure. Reduced to one full meal a day, Stevenson ate at a little French restaurant at 425 Bush Street.

In his room up the street from the restaurant, Stevenson began work on an account of his early years, *From Jest to Earnest,* of which twenty-three manuscript pages survive. He also worked on an account of his trip across the Atlantic, *The Amateur Emigrant,* (published in the Edinburgh Edition of Stevenson's work, 1894), and wrote a number of poems that appeared in the volume *Underwoods* (1887). Stevenson's landlady, Mary Carson, watched as the frail and intense young writer slowly lost his battle with hunger. To her credit, she would provide Stevenson with a little food with his tea, but his ordeal did not abate. Stevenson took daily walking tours of the city, occasionally with Fanny Osbourne, whose divorce was in process. On one of these walks, up Telegraph Hill, Stevenson met **Charles Warren Stoddard.** It is said that Stevenson stopped to rest in front of Stoddard's house, and Stoddard asked him in. Stevenson noticed a copy of Stoddard's *South-Sea Idyls* (1873) in the house, and at the end of the visit Stoddard let him have a copy of the book as well as a copy of Melville's South Seas novel, *Omoo.* Thus it may have been that this encounter led eventually to Stevenson's later voyage to Samoa, where he spent his last years and where he died. While Stevenson was in San Francisco, the two writers visited back and forth and Stevenson, going through Stoddard's collection of autographs of famous men one day, suggested that Stoddard write a series of essays about the men whose signatures were in the collection. The result was Stoddard's volume *Exits and Entrances* (1903).

It was Stoddard who took Stevenson to the Bohemian Club, then located at 430 Pine Street, and there Stevenson spent many pleasant hours. By March 1880 other matters occupied Stevenson: Robbie Carson, his landlady's son, developed pneumonia. Stevenson took it on himself to watch over him; the child recovered but Stevenson collapsed. Fanny Osbourne and Mrs. Carson helped move Stevenson to Oakland, where he could be cared for better. By May he was well, and on May 19 Stevenson and Fanny Osbourne were married in San Francisco at the home of Rev. William Anderson Scott, at 521 Post Street. Toward the end of the summer of 1880, the Stevensons rented rooms at 7 Montgomery Avenue but prepared to travel to Scotland that same year. Stevenson returned to San Francisco once again in June 1888, while preparing to sail for the South Seas aboard the yacht *Casco.* He stayed at the Occidental Hotel, and was visited there by his former landlady, Mrs. Carson. On June 28 he sailed from San Francisco Bay for the South Seas.

Stevenson's days in San Francisco were over, but he did not forget the city: San Francisco was the setting for Stevenson's novel *The Wrecker* (1892), which he wrote with his stepson, **Lloyd Osbourne.** The character Mr. Speedy in this novel was modeled after Mary Carson's husband, and Speedy's house was modeled after 608 Bush Street. San Francisco also figured in Stevenson's novel *The Ebb-Tide* (1894), also written with Lloyd Osbourne. In 1897, three years after Stevenson's death, a monument to Stevenson was erected in Portsmouth Square in San Francisco. At the dedication, Mrs. Carson placed a wreath at the base of the monument.

Mark Twain came to San Francisco in May 1864 and became a reporter for the San Francisco *Morning Call.* During this time Twain also wrote letters for the Virginia City, Nevada, *Territorial Enterprise.* Here he met **Bret Harte,** who began to advise Twain in matters of literary style, and here began Twain's conversion from newspaperman to author. Twain had tried unsuccessfully to get the *Call* to publish his articles exposing official corruption in the city but, after he left the *Call,* the *Territorial Enterprise* published a series of Twain's letters that angered city politicians and police. This animosity was a major reason for Twain's departure for the gold fields in the Tuolumne Hills, in Calaveras County. There he heard a story that he wrote as "Jim Smiley and His Jumping Frog" when he returned to San Francisco. It appeared in the New York *Saturday Press* on November 18, 1865, and created a stir across the country.

On October 2, 1866, Twain delivered his first lecture at Maguire's Academy of Music, which stood on Pine Street near Montgomery Street. Two months later he made arrangements with the San Francisco *Alta California* to write letters about his excursion to the Holy Land aboard the steamship *Quaker City.* These letters helped Twain gain a national audience and formed the basis for his book *The Innocents Abroad* (1869). Twain wrote of his experiences in San Francisco and much of the Far West in his next book, *Roughing It* (1872).

Kate Douglas Wiggin lived at 1107 Pine Street in the later 1880s and established in San Francisco the first free kindergarten on the Pacific Coast. The kindergarten, on Silver Street, was so successful that a second school was needed to handle the overflow. To raise funds for the second kindergarten, Wiggin wrote her book *The Birds' Christmas Carol* (1887).

Chinatown by a thousand eccentricities drew and held me; I could never have enough of its ambiguous, interracial atmosphere, as of a vitalised museum; never wonder enough at its outlandish, necromantic-looking vegetables set forth to sell in commonplace American shop-windows, its temple doors open and the scent of the joss-stick streaming forth on the American air, its kites of Oriental fashion hanging fouled in Western telegraph-wires, its flights of paper prayers which the trade-wind hunts and dissipates along Western gutters.

—Robert Louis Stevenson
and Lloyd Osbourne,
in *The Wrecker*

Memorial to Robert Louis Stevenson in Portsmouth Square, San Francisco.

> *The world which lay on the surface of his mind was like the world he saw and believed in every day in San Francisco. It was richer than the average world, but not entirely beyond the reach of most people with ambition. And so advertising ideas conceived in its image and stated with optimism through a beautifully brushed set of smiling white teeth were likely to be the golden answer, flattering to the purchaser, who was starving to spend his money and only needed to be told how. . . .*
>
> —Paul Horgan, in *Give Me Possession*

Other writers associated with San Francisco include Howard Fast, who used the city as a setting for his novel *The Immigrants* (1977). Paul Horgan wrote of San Francisco in the novel *Give Me Possession* (1957). **Helen Hunt Jackson** died in San Francisco on August 12, 1885. Francis Scott Key is commemorated with a monument in Golden Gate Park. Stephen Longstreet wrote of San Francisco in the history *The Wilder Shore* (1968) and in the novel *The General* (1974). **Edwin Markham** saw first publication of his poem "The Man With the Hoe" in the San Francisco *Examiner* on January 15, 1899. **Joaquin Miller** in the mid-1880s worked as associate editor of *The Golden Era,* then located at 420 Montgomery Street. **H. Allen Smith,** the humorist, died in his sleep at the Clift Hotel, 495 Geary Street, on February 24, 1976.

SAN JACINTO

In 1882, while visiting this city southeast of San Bernadino to gather information for *Report on the Conditions and Needs of the Mission Indians* (1883), **Helen Hunt Jackson** visited the Cahuilla Indian Reservation here. This visit gave her much of the material she used later in her novel *Ramona* (1884). The model for Ramona was a Cahuilla Indian named Ramona Lubo, whose husband Juan Diego was the model for Alessandro, Ramona's husband, even though the lives of the real-life couple bore only slight resemblance to the lives of their fictional counterparts. Ramona Lubo and Juan Diego are buried in a cemetery on the reservation, not accessible to visitors.

SAN JOSE

Frank Ernest Hill, the poet, was born in San Jose on August 29, 1888. Hill's family then lived at 86 Fox Street and moved later to a house at 1350 Sherman Street. **Jack Kerouac** stayed here with his friend **Neal Cassady** for two months in 1952 and for a time during the winter of 1954.

From the 1870s to the 1890s, **Edwin Markham** lived at his mother's house at 432 South Eighth Street, but stayed for extended periods at a number of different places during those years. Markham first came to San Jose when the State Normal School, at which he was a student, was moved here from San Francisco. He was graduated from the school, now San Jose State University, in 1872 and began a career in education, working first as a teacher and later as a school principal and county superintendent of schools. Meanwhile, he wrote poetry and over many years produced fragments of a poem inspired by Jean-François Millet's painting "The Man with a Hoe." In 1898 Markham, then principal of a school in Oakland, was asked to compose a poem for recitation at a San Francisco literary gathering. He is said to have returned to his mother's house in San Jose to complete the poem. The poem Markham read, "The Man with the Hoe," was greeted by near silence, but one of the guests, an editor of the San Francisco *Examiner,* was impressed enough to publish it in the January 15, 1899, issue of the *Examiner.*

> *Bowed by the weight of centuries he leans*
> *Upon his hoe and gazes on the ground,*
> *The emptiness of ages in his face,*
> *And on his back the burden of the world.*
>
> —Edwin Markham, in "The Man with the Hoe"

The poem was widely reprinted, and Markham became an international figure. He soon moved to the East Coast, where he spent the remainder of his life. The home in San Jose eventually was acquired by San Jose State University for use as a health facility and was moved some distance from its original site. One room is reserved for meetings of the Edwin Markham Poetry Society and maintained as a memorial to Markham. The original site of Markham's house, 432 South Eighth Street, is marked by a boulder and plaque. Nearby, at

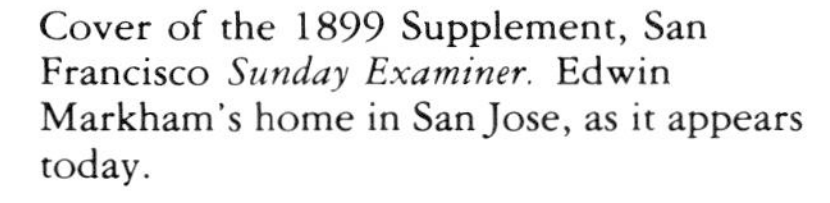

Cover of the 1899 Supplement, San Francisco *Sunday Examiner.* Edwin Markham's home in San Jose, as it appears today.

the university's Tower Hall, a small brass plaque bears the words of "Outwitted," Markham's shortest poem:

He drew a cricle that shut me out—
Heretic, rebel, a thing to flout.
But Love and I had the wit to win:
We drew a circle that took him in.

SAN MARINO

The Henry E. Huntington Library, 1151 Oxford Road, is known worldwide as a center for research in English and American history and literature. Huntington, a nephew of railroad magnate Collis P. Huntington, collected 100,000 books and one million manuscripts, as well as numerous works of art, and housed them in two large buildings on his 550-acre estate. In 1919 this collection became the nucleus of the library, which has expanded over the years to include about six million items. Among the many writers whose letters and manuscripts are housed here are Conrad Aiken, Zoë Akins, Ambrose Bierce, Benjamin Franklin, Nathaniel Hawthorne, Jack London, Wallace Stevens, and Henry David Thoreau. The library complex also includes an art gallery and botanical gardens.

SAN MATEO

In the fall of 1881 **Lincoln Steffens** enrolled at St. Matthew's Hall, a military school in San Mateo. He became captain of cadets, served as editor of the school's literary magazine, and had some early poetry published in his hometown newspaper, the Sacramento *Record-Union.* He was also frequently in trouble because of his numerous pranks. After reading *Tom Brown's School Days* (1857), by Thomas Hughes, Steffens introduced fagging to St. Matthew's in imitation of the English practice, but its existence was discovered and halted. Another time, for bringing a keg of beer onto the school grounds, Steffens spent twenty-two days in the guardhouse. This respite from school routine gave him a chance to read, and he became interested in the works of Herbert Spencer and Charles Darwin. One of the books he read in the guardhouse supplied the theme for his graduation speech from the military school: a denunciation of war.

SAN RAFAEL

After **Ambrose Bierce,** master of the suspense story, married Mary Ellen Day in December 1871, they moved north of San Francisco to San Rafael, where they lived until March 1872. The couple returned to San Rafael after several years and lived here while Bierce worked, from 1877 to 1879, as an associate editor of the San Francisco *Argonaut.* During these years Bierce made the daily trip to his office by ferry. It appears that the only Bierce house still standing in San Rafael is the two-story brown-shingled house at 814 E Street, now private.

SAN SIMEON

Literary visitors may want to stop in at the Hearst San Simeon State Historic Monument, off Route 1, to see the former home of **William Randolph Hearst.** Hearst named his 123-acre estate, located amid 240,000 acres of Hearst-owned land, La Cuesta Encantada (The Enchanted Hill). Here he built his one-hundred-room castle, La Casa Grande, and surrounded it with beautiful gardens, spacious guest houses, and fine sculpture. Among its facilities is Hearst's private library, from which he directed his publishing empire. La Casa Grande, begun in 1919, was still uncompleted at Hearst's death in 1951. Today it is open to tourists for an admission fee.

SANTA BARBARA

Among the first writers to visit this city northwest of Los Angeles was **Richard Henry Dana, Jr.,** who traveled here in the mid-1830s. **Wallace Stegner,** while living here, wrote the text for *One Nation* (1945), a photographic study of racial and ethnic intolerance in America. **Bradford Torrey,** author of many books on birds and nature and editor of the works of Henry David Thoreau (fourteen volumes, 1906), lived in solitude in a small cabin here toward the end of his life, and died here on October 7, 1912.

Kate Douglas [Smith] Wiggin, known for her children's books, came to Santa Barbara as a young woman in 1873 with her family. They had moved in hope that her stepfather's health would be helped by the gentle climate. When he died three years later, Kate wrote her first story to help the family through its financial crisis. She sold the three-part story to *St. Nicholas* magazine for $150. In 1881 she married Samuel Bradley Wiggin. The couple lived here and in San Francisco until 1884.

SANTA MONICA

Writers who have lived in this city include **Robert W. Anderson,** who wrote the play *The Days Between* (1965) and the motion picture *The Sand Pebbles* (1966) while living at 42 Haldemann Road; **Raymond Chandler,** who lived in an apartment on San Vincente Boulevard in 1940; and **Paul Green,** the playwright, who lived here while writing motion-picture scripts. **Christopher Isherwood** has lived in Santa Monica for many years. He came here in the early 1940s. The numerous former addresses of the playwright include 137 Entrada Drive, 165 Mabey Road, 333 East Rustic Road, and 434 Sycamore Road. **Katherine Anne Porter,** the novelist, lived at 843 Sixth Street in the late 1940s. **Tennessee Williams** lived in a two-room apartment on Ocean Boulevard while working on motion-picture scripts in the early 1940s. During this time he also worked on *The Glass Menagerie* (1944), the play that was to become his first success.

SONOMA

From 1947 to 1962, **Paul Corey** lived at 255 Cavedale Road, where he wrote a number of stories for national magazines as well as the book *Homemade Homes* (1951).

SOUTH PASADENA

Wynyate, the three-story Victorian house at 851 Lyndon Street, was a gathering place for writers in the 1880s. The house was owned by the

Tennessee Williams. The playwright came to California to start work on a screenplay. When his outline was rejected, Williams began to rework his idea into what eventually became *The Glass Menagerie.*

Seven of them sat against the rough log walls, silent, motionless, and the room being small, not very far from the table. By extending an arm any one of them could have touched the eighth man, who lay on the table, face upward, partly covered by a sheet, his arms at his sides. He was dead.

—Ambrose Bierce,
in "The Damned Thing"

first mayor of South Pasadena. His wife, a writer named Margaret Collier Graham, conducted a salon for literary people, including **Mary Austin, Charles Fletcher Lummis,** and **John Muir.** In 1889, Muir planted a lemon tree that still grows on the grounds near the house. Wynyate takes its name from the Welsh word for vineyard. Now privately owned, the house is threatened by a planned freeway extension.

SPRING VALLEY

The Bancroft Ranch House, on Bancroft Drive, off Route 94 in this community just east of San Diego, was the home of **Hubert Howe Bancroft** from 1885 to 1918. Bancroft had come to California in 1852 and worked at mining briefly before turning to bookselling and publishing. Over the years he collected every book, pamphlet, and document he could find that related to the Pacific Coast. In 1869 he hired twenty writers and a large staff of researchers to assemble the thirty-nine volumes of Bancroft's *History of the Pacific States of North America,* which appeared between 1882 and 1890. There has been much controversy over how much of this work Bancroft actually wrote, but there is no doubt that he was the moving force behind it. In 1900 he presented his massive collection of books and manuscripts to the University of California at Berkeley. Bancroft continued to live at his Spring Valley home until his death on March 2, 1918. The Bancroft Ranch House now is listed both as a California Historical Landmark and as a National Historical Landmark.

STOCKTON

Howard Pease, author of many adventure stories, was born here on September 6, 1894. His home at 112 Miner Avenue is no longer standing. The site is occupied now by the Stockton Civic Auditorium. Pease wrote his first story, "Turn Back, Never!" when he was in the sixth grade at El Dorado School. The opening scenes of his book *Thunderbolt House* (1944) are set in Stockton, and Pease is buried in Stockton Rural Cemetery, on Cemetery Lane.

TARZANA

In 1919 **Edgar Rice Burroughs** bought Mil Flores, the 540-acre estate of Gen. Harrison Gray Otis, founder and publisher of the Los Angeles *Times.* Burroughs renamed the estate Tarzana and, after attempting unsuccessfully to turn it into a working ranch, subdivided about fifty acres of his land into lots to launch the community now known as Tarzana. The community grew slowly, and it was not until 1928 that it adopted the name Tarzana, not until 1930 that it obtained its own post office. In 1924 Burroughs sold the main house and 120 acres to a country club and moved to Los Angeles. He returned to Tarzana in 1926, building a small house at 5046 Mecca Avenue. The country club failed, and Burroughs took back the main house after foreclosure proceedings. In 1927 he built offices for his enterprise, Edgar Rice Burroughs, Inc., at 18354 Ventura Boulevard. His ashes are buried at the base of a tree that stands in front of one of the buildings here.

> *Men—motes in the sunshine—perished, were shot down in the very noon of life, hearts were broken, little children started in life lamentably handicapped; young girls were brought to a life of shame; old women died in the heart of life for lack of food. In that little, isolated group of human insects, misery, death, and anguish spun like a wheel of fire.*
>
> *But the* Wheat *remained. Untouched, unassailable, undefiled, that mighty world-force, that nourisher of nations, wrapped in Nirvanic calm, indifferent to the human swarm, gigantic, resistless, moved onward in its appointed grooves.*
>
> —Frank Norris, in *The Octopus*

TEMECULA

Erle Stanley Gardner owned Rancho Del Paisano, a 400-acre ranch in Temecula, about forty miles south of Riverside. Gardner first came to Temecula in the fall of 1937 on a camping trip, and eventually decided to make his home in the area. At first he wanted only enough land on which to park his trailer. In the intervening years Gardner built nearly thirty buildings on his ranch—everything from guest houses to a special building for his papers. Here he lived for many years, working on his books in the morning and spending the rest of the day with his many visitors. He remained at Rancho Del Paisano until his death in 1970.

TORRANCE

Frank G. Slaughter wrote several books, including *In a Dark Garden* (1946) and *Medicine for Moderns* (1948), while working during the 1940s in a hospital in this city southwest of Los Angeles.

TULARE

Allen Drury lived at 237 L Street, Tulare, in 1940 and 1941, when he was editor of the Tulare *Bee.* The novel *The Octopus* (1901), by Frank Norris, was set in the fictional city of Bonneville, the name Norris used for Tulare.

TUTTLETOWN

Nestled in the Tuolumne Hills, about six miles west of Sonora, Tuttletown in the 1850s was a mining and trading center for prospectors drawn to the area by the chance to find gold. One mile west of Tuttletown is Jackass Hill, where **Mark Twain** lived for several months during the winter of 1864–65 in a cabin built by Dick Stoker. Stoker became the model for Dick Baker in Twain's *Roughing It* (1872), and Stoker's cat Tom Quartz was immortalized in that book. Twain had been invited by a friend, Jim Gillis, to spend some time at the cabin with him and Stoker. Twain had made enemies in San Francisco by writing about police corruption for the Virginia City, Nevada, *Territorial Enterprise* and decided that it was a good time to vacation in the hills for a while. With Gillis and Stoker as companions, he gathered the material that would later appear in such stories as "The Notorious Jumping Frog of Calaveras County," "The Californian's Tale," and "What Stumped the Blue Jays." As the years passed, Tuttletown virtually disappeared, and by the 1930s all that remained were the store where Twain traded and the Tuttletown Hotel where, it is said, **Bret Harte** once stayed. Dick Stoker's cabin was destroyed, but a replica was built and dedicated on the site in 1922.

UKIAH

In 1900 **Percy Marks** moved with his parents to this city north of Santa Rosa. The family lived at 612 West Smith Street, and Marks attended local schools. Marks wrote about Ukiah in *No Steeper Wall* (1940), his novel about the life of an upper-class Bostonian on a California ranch.

VACAVILLE

Edwin Markham and his mother lived in the 1860s on a ranch in the Suisun Valley, near Vacaville, southwest of Sacramento. After studying at a nearby school, sixteen-year-old Markham wanted to continue his education in Vacaville at a preparatory school run by the Methodist Church. His mother demanded that he remain on the ranch and work. Markham left, wandering from place to place until the rift between him and his mother was mended. It was agreed that he could continue his education at Ulatis Academy, then located near Callen Street, but he would have to pay the expenses himself. Markham paid his school costs partly from $900 in gold coins he found buried on the ranch. Local legend has it that the bandit Black Bart once kidnapped Markham and tried to persuade him to join Bart's gang of outlaws. After Markham respectfully declined the invitation, Bart hid the coins on the property and took off for the hills.

VAN NUYS

Irving Stone wrote *Sailor on Horseback* (1938), his biographical novel about Jack London, while living in a rented house at 13914 Davana Terrace in Van Nuys, northwest of Los Angeles.

VENTURA

Erle Stanley Gardner was a trial lawyer in private practice in Ventura in the 1920s and 1930s. His firm changed its name several times, as members came and went, and changed locations several times as well, but the only office site still standing is the building at 494 East Main, now the American Commercial Bank Building. From 1916 to 1934, Gardner lived in five different houses. Again, only one is still standing, the house at 2420 Foster Avenue, where he lived from 1931 to 1934. During this time Gardner saw publication of his first book featuring Perry Mason, *The Case of the Velvet Claws* (1933). Gardner followed this with many novels about Mason's courtroom adventures, including *The Case of the Lucky Legs* (1933) and *The Case of the Curious Bride* (1934).

VISTA

William Gilbert Patten who, as **Burt L. Standish,** wrote more than two hundred dime novels featuring a wholesome college athlete named Frank Merriwell, died on January 16, 1945, at his son's home on Terracino Way in Vista, north of San Diego. The manuscript of Patten's autobiography, found much later, was published as *Frank Merriwell's Father* (1964).

WHITTIER

Maxwell Anderson became head of the Whittier College English department here in 1917. The playwright was asked to resign after he defended a Whittier student for resisting the draft of World War I. **Jessamyn West** was graduated from Whittier in 1923, and the college maintains a collection of her papers.

YORBA LINDA

Jessamyn West was six years old when her family moved to California. Her father built a two-story house on a hill in this city southeast of Los Angeles, and it is believed that West attended Yorba Linda's first schoolhouse, a one-room building on Olinda Street used in later years for offices of the Yorba Linda Water Company. This building, dating to 1911, is being considered for restoration as a one-room country schoolhouse. West also attended, and in 1915 was graduated from, a school, no longer standing, on School Street.

People that come to me don't come to me because they like the looks of my eyes, or the way my office is furnished, or because they've known me at a club. They come to me because they need me. They come to me because they want to hire me for what I can do.

She looked up at him then. "Just what is it that you do, Mr. Mason?" she asked.

He snapped out two words at her. "I fight!"

—Erle Stanley Gardner, in *The Case of the Velvet Claws*

Replica of the Mark Twain cabin on Jackass Hill near Tuttletown.

By and by, an old friend of mine, a miner, came down from one of the decayed mining camps of Tuolumne, California, and I went back with him. We lived in a small cabin on a verdant hillside, and there were not five other cabins in view over the wide expanse of hill and forest.

—Mark Twain, in *Roughing It*

[Carson City] was a "wooden" town; its population two thousand souls. The main street consisted of four or five blocks of little white frame stores which were too high to sit down on, but not too high for various other purposes; in fact, hardly high enough. They were packed close together, side by side, as if room were scarce in that mighty plain. The sidewalk was of boards that were more or less loose and inclined to rattle when walked upon.

—Mark Twain,
in *Roughing It*

YUBA CITY

In the early 1930s **Jessamyn West** lived with her husband in Yuba City, north of Sacramento, in an apartment house, no longer standing, at 156 B Street.

NEVADA

AURORA

Mark Twain, in the fall of 1861 and still known as Samuel Clemens, visited the hilly mining community of Aurora, south of Hawthorne and near the California border. Friends had acquired mining claims, and Twain himself bought some claims in partnership with his brother Orion Clemens, who then was living in Carson City. In February 1862, Twain returned to Aurora for a longer stay, this time to dig in his own mines and to supervise the work of hired hands. He soon ran out of money and had to rely on Orion to send him enough to keep the unprofitable mines going. He himself worked as a laborer in a quartz mill to earn $10 a week plus board and also to learn the art of refining so that he could apply his knowledge when his own mines began producing. As it turned out, this was never to be, but in that summer Twain wrote several letters that appeared, under the name **Josh,** in the Virginia City *Territorial Enterprise.* The editors liked Twain's writing so much that by July he was working for the *Territorial Enterprise* at $25 a week, and his days in Aurora were at an end.

Home of Orion Clemens in Carson City. The original frame construction has been stuccoed.

CARSON CITY

Mark Twain and his brother Orion Clemens, who had been appointed secretary of the Nevada Territory, arrived in Carson City, thirty miles south of Reno, on August 14, 1861, after an overland stage journey. They stayed first at the Ormsby Hotel, on the south side of South Carson Street between Second and Third streets, no longer standing, and then at a boardinghouse run by a Mrs. Murphy, who was to appear in literature as the character Mrs. O'Flannigan in *Roughing It* (1872). Twain served as his brother's unpaid personal secretary for a while but was out of town too much on trips to Lake Tahoe, where he searched for timber claims, and to Unionville and Aurora, where his objective was a lucrative mining claim. Twain moved away from Carson City by the fall of 1862 but often was sent back by the *Territorial Enterprise* to cover sessions of the territorial legislature and the Constitutional Convention of 1863. A letter datelined Carson City, February 2, 1863, published by the *Territorial Enterprise* was the first writing that Samuel Clemens signed with the name Mark Twain. Orion Clemens in 1864 purchased a house, still standing and bearing a marker, at the northwest corner of North Division and Spear streets, where Twain sometimes stayed when he was visiting his brother. Twain left Nevada in 1864 but made several visits in 1866 and again in 1868 to give public lectures on his foreign travel.

Will James, the author and illustrator of many Western stories, including *Cowboys, North and South* (1924) and *The Drifting Cowboy* (1925), was also associated with Carson City, but in a less creditable way. In 1915–16, James served time in a jail two miles east of the city after being convicted of rustling cattle.

ELKO

Ernest Hemingway, his wife Mary, and their friend A. E. Hotchner—who later was to write *Papa Hemingway: A Personal Memoir* (1966)—driving from Ketchum, Idaho, to Key West, Florida, stopped overnight on March 16, 1959, in this northeastern Nevada town on Route 80. They stayed at the Stockman's Hotel, the largest casino-hotel in town and still operating at 340 Commercial Street.

LAS VEGAS

Joan Didion set some of her novel *Play It as It Lays* (1971) at the Sands Hotel, 3355 Las Vegas Boulevard South, and part of *The Book of Common Prayer* (1977) at Caesar's Palace, 3570 Las Vegas Boulevard South, thus ensuring a place in literary history—as well as American folklore—for two of the largest casino-hotel establishments in the city that is called the gambling capital of the world. **Lloyd C. Douglas,** author of the New Testament historical novels *The Robe* (1942) and *The Big Fisherman* (1949), and a Lutheran clergyman, is also associated with Las Vegas. Douglas moved here in 1944, after the death of his wife, to live with his daughter at 721 East Charleston

Boulevard, now the site of medical offices. He was still resident here when he died, on February 13, 1951.

Ernest Hemingway, on the same 1959 automobile trip that took him through Elko, stayed for two days at the Sands Hotel. Hemingway is not remembered at the gaming tables, but he did hold court in the hotel cocktail lounge and gambling casino, discoursing on prize fighting, the Battle of the Bulge, and American literature. **Mario Puzo,** author of the best-selling novel *The Godfather* (1969), set his novel *Fools Die* (1978) in Las Vegas, and *The Godfather* itself used Las Vegas for some of its gory scenes. The main characters in *Fools Die* were guests at the Hotel Xanadu, a fictional composite of several of the better-known Las Vegas establishments. Puzo visited the city more than once to conduct research for his novel and deepen his understanding of the psychology of the gambler. **Frank Chester Robertson,** the author of about a thousand Western stories and novels, including *Rawhide* (1961) and *A Man Called Paladin* (1964), died of a heart attack in Las Vegas on July 29, 1969.

RENO

One might expect that many authors would have visited Reno for the required stay before a divorce back in the days when Reno was popular among those looking for termination of a marriage, but only two appear to have visited the city for that reason. **Sherwood Anderson** was one. Anderson was here from February 1923 to April 1924, staying first at the Overland Hotel and then in an apartment at 33 East Liberty Street, while he waited for his divorce from his second wife, Tennessee Mitchell. The dramatic colors of the desert inspired Anderson to paint, and while here he also wrote. He worked on his collection of stories about horse racing, *Horses and Men* (1923), and on his autobiography, *A Story Teller's Story* (1924).

Among those who came to Reno for purposes other than securing a divorce was young **Walter Van Tilburg Clark.** Only eight years old in 1917, Clark came with his family to Reno because his father, Walter Ernest Clark, had been appointed president of the University of Nevada, located in the northern section of the city. Walter Van Tilburg Clark attended Reno public schools and then the university, receiving a B.A. in 1931 and an M.A. in 1932. He later was to use his knowledge of Reno in his novel about the city, *The City of Trembling Leaves* (1945), the title reflecting the characteristic of the poplars prevalent in the area. Clark's other novels also had Nevada settings. *The Ox-Bow Incident* (1940) dealt with a hanging in 1885 cattle country, which Clark called Bridger's Wells. *The Track of the Cat* (1949) was set on the eastern slopes of the Sierra Mountains. Clark became writer in residence at the University of Nevada in 1962, and he died in Reno on November 11, 1971. Services were held for him at the Walton Funeral Home here.

Langston Hughes lived in a boardinghouse for blacks in Reno in late 1934 and there wrote several of his stories, including "On the Road" (1935) and "Slice Him Down" (1936). His story "Mailbox for the Dead," unpublished, was also written here. Inspiration for the story has been attributed to a curious incident in Hughes's life. He had seen the gate of a rural cemetery outside Reno that appeared in silhouette to resemble a mailbox. While writing the first draft of the story in November of 1934, Hughes had persistent thoughts of his father. Soon afterward, he learned that his father had just died.

Will James married Alice Conradt, a native of Reno, on July 7, 1920. Two years later, the couple took a studio apartment on a ranch just west of the city, where they lived for some time. The account of Reno's place in literary lore closes—for now—with the second of Reno's divorce-seekers. **J. P. Marquand,** author of the Pulitzer Prize-winning novel *The Late George Apley* (1937), lived for a few months at the Riverside Hotel, 17 South Virginia Avenue, while he waited for his divorce decree.

SILVER SPRINGS

A washed-out road near Silver Springs brought this city, about forty miles northwest of Carson City, into literature. **Mark Twain,** on his way to Carson City from Unionville in January of 1862, spent nine days at Honey Lake Smith's stage station on the Carson River. Like the other passengers and teamsters stranded here, Twain took amusement in playing endless games of chance. Unlike the others, Twain reported his experiences in a letter that was published in the *Territorial Enterprise.* It read in part:

> *The whole place was crowded with teamsters, and we wore out every deck of cards in the place, and then had no amusement but to scrape up a handful of vermin off the floor or the beds, and "shuffle" them, and bet on odd or even. Even this poor excuse for a game broke up in a row at last when it was discovered that Colonel Onstein kept a "cold deck" down the back of his neck!*

STEAMBOAT SPRINGS

Mark Twain in August 1863 visited Steamboat Springs, a spa south of Reno on Route 395, staying at the Steamboat Springs Hotel, no longer standing. Along with the other guests, he took advantage of the salubrious effects of the waters.

UNIONVILLE

Mark Twain in December 1861 came to Unionville, a town off Route 50, about 120 miles northeast of Reno. He was searching for the fortune he thought he could make in the Humboldt mining district. With three other men, Twain hauled some eighteen hundred pounds of provisions and mining equipment from Carson City and built a crude shack in Unionville for shelter during their prospecting. For reasons uncertain, Twain stayed less than two weeks, leaving before any serious prospecting could be accomplished.

VIRGINIA CITY

Charles Farrar Browne, the humorist and lecturer known better as **Artemus Ward,** spent the latter part of December 1864 in Virginia City,

> *On the luckiest day of Jordan Hawley's life he betrayed his three best friends. But yet unknowing, he wandered through the dice pit of the huge gambling casino in the Hotel Xanadu, wondering what game to try next. Still early afternoon, he was a ten-thousand-dollar winner. But he was tired of the glittering red dice skittering across green felt.*
>
> —Mario Puzo, in *Fools Die*

> *The north-east quarter of Reno, with the ranching valley on the east of it and the yellow hills with a few old mines on the north, is drawn out of the influence of the university and Peavine into the vortex of the race track. . . . It was in this quarter that Tim Hazard lived when he was a boy, on the street right next to the track, so that he got to see a good many horse races and rodeos, and even circuses that set up their tents outside the fence. He lived in a square, white-board house with a shallow porch with a dirt walk and three big poplars in front of it.*
>
> —Walter Van Tilburg Clark, in *The City of Trembling Leaves*

A crowd of miners and men of all trades, and of no trade, had collected for the evening at a popular North C Street saloon. The bar was well lined with persons engaged in imbibing their favorite stimulants, while most of the seats along the walls of the place were occupied by men who had taken their drinks and were sufficiently exhilarated to find a fair degree of comfort in hearing themselves talk.

—Dan de Quille,
in "Big Bill Scuffles, or The Fight at Foster's Bar," *The Nevada Monthly,* March, 1880

about twenty-five miles southeast of Reno. Browne made his headquarters at the editorial offices of the *Territorial Enterprise,* still standing on C Street. **Walter Van Tilburg Clark** for many years before his death in 1971 lived in an old frame house at the corner of G and Taylor streets.

Mark Twain lived from September 1862 to May 1864 in Virginia City, working as a reporter for the *Territorial Enterprise.* A plaque on the newspaper building commemorates his stay in town. Twain reported local news, wrote some unsigned editorials, and periodically went to Carson City to report on Nevada territorial affairs. He was known to fabricate news about Virginia City when real news items were scarce, and he could often be found in saloons or in a brewery on D Street, with an informal group of habitués who gathered to talk and drink. Twain was obliged to leave Nevada in 1864, when he violated a new statute by challenging a rival newspaperman to a duel. Twain went to California, still working for the *Territorial Enterprise,* but as its San Francisco correspondent. He returned to Virginia City once more, on October 30, 1866, to lecture about his recent travels across the Pacific at Pipers Opera House, still operating and open to the public at North B Street and Union Street.

From one spring the boiling water is ejected a foot or more by the infernal force at work below, and in the vicinity of all of them one can hear a constant rumbling and surging, somewhat resembling the noises peculiar to a steamboat in motion—hence the name.

—Mark Twain,
in a letter dated August 23, 1863, in the *Territorial Enterprise*

William Wright, who wrote under the name **Dan de Quille,** worked from 1862 to 1893 as city editor of the *Territorial Enterprise.* He was a friend of Mark Twain's in Virginia City, and Twain later wrote an introduction to de Quille's *History of the Big Bonanza* (1877), a rich source of information and anecdotes about mining the Comstock Lode. **Bret Harte,** who worked as editor of the San Francisco *Overland Monthly* from 1868 to 1871, stayed in Virginia City for a short time in 1868. The visit inspired him to write a poem, "The Stage-Driver's Story" (1868), about Geiger Grade, a steep, winding road that still can be traversed. A more recent author, **Louis L'Amour,** set one of his best-selling Western novels, *Comstock Lode* (1981), in the Virginia City area.

WASHOE CITY

Will James and his wife lived from 1923 to 1927 in a cabin built for them about five miles from Washoe City, a town in western Nevada. James's property, some five acres with a view of Washoe Lake, is now privately owned.

UTAH

CEDAR CITY

About thirty-five miles southwest of this town in the southwestern corner of Utah, in September 1857, a wagon train of emigrants from Arkansas and Missouri was attacked. After a siege of several days, the pioneers were massacred by a group of Indians and Mormon settlers. The Mountain Meadows Massacre, as it came to be called, was the inspiration for one of the stories in the book *The Star Rover* (1915), a series of related stories by Jack London. Mark Twain included an account of the massacre in his book *Roughing It* (1872). The massacre site is in Dixie National Forest, off Route 18, north of the community of Central.

GREEN RIVER

Zane Grey visited this community in east-central Utah in the fall of 1929. He traveled into the untamed badlands area known as the San Rafael Swell, and visited Robber's Roost, the outlaw hideout of Butch Cassidy and his gang the Wild Bunch, who operated in this area in the 1890s. The hideout was the inspiration for Grey's novel *Robber's Roost* (1932).

KANAB

Zane Grey is said to have visited this city in southern Utah, a few miles north of the Arizona border, while on a visit to the Southwest early in his writing career. During his stay in the area, he collected impressions that he incorporated into his novel *The Heritage of the Desert* (1910) and also into his best-known novel, *Riders of the Purple Sage* (1912). This trip to Arizona and Utah turned Grey to the writing of Western novels and was

Composing room of the *Territorial Enterprise* in Virginia City in a photograph taken about 1881. The upright desk against the far wall is said to be the one at which Mark Twain worked. Standing near the desk is W. E. Sutherland, who came to Nevada in 1875 and worked as a printer for the *Territorial Enterprise.* Sutherland wrote *The Wonders of Nevada* (1878), thought to be the first guidebook to Nevada.

Artemus Ward (*left*) and Frank Chester Robertson (*right*). Brigham Young University has a collection of Robertson's papers.

a strong factor in establishing Grey as a leading author in the field. It is believed that the locale of both these novels is southeastern Utah.

OGDEN

This city about thirty-five miles north of Salt Lake City was the birthplace on January 11, 1897, of **Bernard [Augustine] De Voto,** the author and editor. He grew up in Ogden and on his grandparents' farm in what is now Uintah, just southeast of Ogden. He was graduated in 1914 from Ogden High School, at the southwest corner of Monroe Avenue and Twenty-fifth Street. A newer structure that is used as a junior high school stands on the site. After serving in the U.S. Army during World War I and studying at the University of Utah and at Harvard, De Voto returned in 1920 to his family's home at 2561 Monroe Avenue, no longer standing. There he suffered the first of a series of nervous collapses that were to plague him throughout his life. From November 1921 to June 1922, De Voto taught history at North Junior High School, no longer standing, at 1420 Washington Boulevard. During this time he also worked on an unpublished novel he entitled *Cock Crow.* In 1922 De Voto left to teach at Northwestern University. He visited Ogden in later years, but made his home in the East. He used Ogden as a model for the city of Windsor in *The Crooked Mile* (1924), and for Windsor Springs City in *The House of Sun-Goes-Down* (1928).

Another resident of Ogden was **Phyllis McGinley,** the poet, who moved here with her family about 1920. She attended Sacred Heart Academy and Ogden High School. After studying at the University of Utah in the 1920s, she returned to teach in Ogden for one year.

SALT LAKE CITY

This city, the capital of Utah and seat of the Mormon Church, is also the literary center of Utah. Writers associated with the city include **Charles Farrar Browne,** better known as **Artemus Ward,** the humorist, who visited the city during his tour of the West in 1863–64, after publication of *Artemus Ward, His Book* (1862). The Mormons subsequently became the subject of his humorous lecture "Artemus Ward Among the Mormons," which he delivered in the United States and abroad from 1864 until his death in 1867.

Edgar Rice Burroughs lived with his wife in rooms at 111 North First West Street, no longer standing, from May to October of 1904. During that time he worked as a depot policeman for the Oregon Railroad Company in Salt Lake City. Sir Arthur Conan Doyle, creator of Sherlock Holmes, used the area around Salt Lake City as a locale in his story *A Study in Scarlet* (1887), the first story in which the famous English detective appeared. Salt Lake City is the setting of the play *The Mormons, or, Life at Salt Lake* (1858), by Thomas Dunn English, the writer, physician, and public official. This was the only one of the author's numerous plays to be published. **Frank Chester Robertson,** a prolific author of novels, short stories, and articles about the West, lived in Salt Lake City in the early 1930s. He made his home at 1045 Blaine Avenue and later at 136 Lincoln Avenue. Among the books Robertson published while living here were *The Mormon Trail* (1931), *Fight for River Range* (1932), and *Song of the Leather* (1933).

Harold Ross, for many years editor of *The New Yorker,* moved with his family to Salt Lake City in 1896, when he was four years old. Ross grew up in the city and completed two years of high school before working for the Salt Lake City *Tribune* from 1904 to 1906. **Mark Twain** visited Salt Lake City in the summer of 1861, while en route to Nevada with his brother Orion. He stayed at the Salt Lake House, which stood on what is now Main Street, between First and Second South streets. Twain wrote of the city in his book *Roughing It* (1872).

Salt Lake City was the birthplace of two au-

It was the Geiger Grade, a mile and a half from the Summit:
Black as your hat was the night, and never a star in the heavens,
Thundering down the grade, the gravel and stones we sent flying
Over the precipice side—a thousand feet plumb to the bottom.

—Bret Harte,
in "The Stage-Driver's Story"

I have no politics. Nary a one. I'm not in the bizniss. If I was I spose I should holler versiffrusly in the streets at nite and go home to Betsy Jane smellen of coal ile and gin, in the mornin. I should go to the Poles arly. I should stay there all day. I should see to it that my nabers was thar. I should git carriages to take the kripples, the infirm and the indignant thar. I should be on guard agin frauds and sich. I should be on the look out for the infamus lise of the enemy, got up jest be4 elecshun for perlitical effeck.

—Artemus Ward,
in *Artemus Ward, His Book*

thors. **Whit Burnett,** born here on August 14, 1899, was an editor and short-story writer, and founder with his wife, Martha Foley, of *Story* magazine. **Wallace Thurman,** born in Salt Lake City on August 16, 1902, was an editor, novelist, and playwright whose promising career came to an end with his death at the age of thirty-two. His works included the novels *The Blacker the Berry* (1929) and *Infants of the Spring* (1932), and the play *Harlem* (1929), written in collaboration with W. J. Rapp.

Salt Lake City is also the seat of the University of Utah. **Vardis Fisher,** the novelist, received a B.A. from the university in 1920, and taught here from 1925 to 1928. Fisher's novel *Children of God* (1939) concerned the Mormon settlement of Utah. While attending the University of Utah in the 1920s, **Phyllis McGinley** had several poems published in magazines. **Wallace Stegner** attended East High School, 840 South 1300 East Street, before studying at the University of Utah, from which he was graduated in 1930. He taught at the university from 1934 to 1937. While in Salt Lake City, Stegner wrote his first book, the novella *Remembering Laughter* (1937). Salt Lake City is the setting for Stegner's novel *Recapitulation* (1979).

Oak Creek Canyon, just south of Flagstaff, where Zane Grey once built a cabin.

SPRINGVILLE

Frank Chester Robertson lived for more than thirty years, from the mid-1930s to the late 1960s, in this city in north-central Utah. Most of his more than 150 books and thousands of stories and articles about the West were published while he lived in Springville. Among them was his autobiography, *A Ram in the Thicket* (1950).

VERNAL

Paul I[selin] Wellman, the novelist and historian, lived in childhood in this city in eastern Utah in the early 1900s. Here he became friends with old cowboys, who told him stories of life in the West during the Indian wars. Here, too, Wellman witnessed the last major uprising of the Ute Indians, who in 1907 left their reservation near Vernal and began a trek toward the Black Hills. Wellman later worked as a cowboy for a time, and studied the folklore and history of the Old West, the subject of many of his books, which included *Death on the Prairie* (1934) and *Glory, God, and Gold* (1954).

ARIZONA

BISBEE

The First Bisbee Poetry Festival, with funding by the National Endowment of the Arts, was held in August 1979 at the Old Sacred Heart Church in Bisbee, ninety miles south of Tucson. **Lawrence Ferlinghetti** was among those who gave poetry readings. **Allen Ginsberg** appeared at the Bisbee Poetry Festival in the following year.

BUCKEYE

Upton Sinclair lived at 108 North Seventh Street in Buckeye, which is about thirty miles west of Phoenix. While here, he wrote *The Cup of Fury* (1956). To keep out the curious, Sinclair built a six-foot wall around his small house, which now is privately owned and closed to visitors.

FLAGSTAFF

Zane Grey gathered background material in Flagstaff for many of his books, including *The Last of the Plainsmen* (1908). In 1907, when he was on a tour of the West with Colonel C. J. "Buffalo" Jones, Grey stayed first at the Commercial Hotel, 14 East Santa Fe Street (burned down in 1975). He later moved to the Monte Vista, 100 North San Francisco Street, then the city's finest hotel and still standing now as a hotel and restaurant. A daughter of the night clerk at that time recalled later that she once ate bear meat from a kill made during one of Grey's hunting trips. Still later, Grey stayed at the Weatherford Hotel, Suite 2748, 23 North Leroux Street, which has been restored and is open for business. Slightly to the south of Flagstaff, on Route 89A, Grey built a cabin and cultivated a garden. The site is now a picnic area, and the fruit trees there are said to have been planted by Grey himself. Grey also stayed at Mayhew's Lodge, on Route 89A, which is not open to the public. It was about the Oak Tree Canyon area just south of Flagstaff that Grey wrote in his *Call of the Canyon* (1924), and Al Doyle, a stockman portrayed in several of Grey's stories, once lived in the 0–100 block of Birch Avenue near Beaver Street, Flagstaff, now the site of the Bank of Northern Arizona. Doyle's barns and corrals occupied the tract on which now stands Flagstaff's City Hall.

GLOBE

Discouraged by life in New York City, **Ross Santee** made his first trip out West in the spring of 1915 to visit his mother, then living in Globe, seventy miles east of Phoenix. Santee stayed at the Kinney House and worked at the Old Dominion Smelter and Mine and as a cowboy. He then served in the U.S. Army during World War I, returning later to work again as a cowboy. After establishing himself as a writer and artist, Santee divided his time between Arizona and New York, living in Arizona in the town of Payson. Santee died in a Globe hospital on June 28, 1965.

GRAND CANYON

In 1908 **Zane Grey** hunted mountain lions for zoos with guide "Uncle Jim" Owen. The area of northern Arizona and southern Utah, which Grey often visited, provided background for many of his stories. Grey sometimes stayed in Owen's cabin, which was on Route 67, four miles north of the Grand Canyon National Park entrance to the North Rim. He also stayed at the El Tovar Hotel in Grand Canyon.

John Burroughs, the naturalist and writer, stopped at the El Tovar in 1909 and 1911. Burroughs later told of a woman he met who "thought that they had built the canyon too near the hotel." Just west of the village of Grand Canyon is a plaque to Major **John Wesley Powell,** the first white man to explore the Grand Canyon. He wrote a vivid report of his two trips down the river in *Exploration of the Colorado River of the West and its Tributaries* (1875; revised as *Canyons of the Colorado,* 1895). Wallace Stegner wrote an excellent account of Powell's explorations in *Beyond the Hundredth Meridian* (1954).

PARADISE VALLEY

Erskine Caldwell has reported that he lived in this suburb of Phoenix during the early 1940s and wrote some of *Georgia Boy* (1943) here.

PAYSON

The cabin retreat of **Zane Grey** near the rugged southern edge of the Mogollon plateau (sometimes called the Mogollon Mountains) northeast of Payson is open to the public for an admission charge. Grey used the white clapboard cabin as a hunting lodge and studio. When Grey left Arizona in 1929, the building fell into disrepair, but it has been restored, and some original family furnishings and photographs are displayed. At the cabin a marker reads, "An Ohio born dentist Zane Grey spent many years under the Mogollon Rim writing 'To the Last Man' and a dozen other Westerns with Arizona settings and characters. His prolific writings popularized the American cowboy as a taciturn, romantic figure." To reach the site from Payson, the visitor should travel nineteen miles east on Route 260, then turn left and go four miles to the Zane Grey cabin and State Fish Hatchery.

Ross Santee owned the Bar F Bar and the Cross S ranches near Payson for a long time before his death in 1965 in Globe. He worked on his books and relaxed by doing ranch chores.

PHOENIX

Erskine Caldwell once lived on North Central Avenue in Phoenix, the largest city in Arizona. **Raymond Moley** lived at 109 East Palm Lane, in a house still standing. He wrote *Lessons in Democracy* (1917) and edited *Today,* staying on as an editor after *Today* merged with *Newsweek* in 1936. **Rosemary Taylor,** born here on May 8, 1899, wrote *Chicken Every Sunday* (1943) and *Ridin' the Rainbow* (1944), recalling her early experiences in Arizona. *Chicken Every Sunday* was also dramatized successfully and made into a motion picture.

PRESCOTT

The Sharlot Hall Museum, 415 West Gurley Street, is open to the public, with free admission. It offers a historical perspective on the Arizona Territory. The museum was named after **Sharlot Hall,** the Arizona poet and short-story writer, who lived in the Old Governor's Mansion for many years. She came to Arizona with her family in 1881, worked as an editor for *Out West* magazine in 1906–07, and wrote a book of verse, *Cactus and Pine* (1911). Hall died on April 9, 1943, at the Pioneer's Home in Prescott and was buried in Pioneer's Cemetery.

No beggar she in the mighty hall where
her bay-crowned sisters wait,
No empty-handed pleader for the right
of a free-born state,
No child, with a child's insistence,
demanding a gilded toy,
But a fair-browed, queenly woman,
strong to create or destroy—
Wise for the need of the sons she
has bred in the school where
weaklings fail,
Where cunning is less than manhood,
and deeds, not words, avail—
With the high, unswerving purpose
that measures and overcomes,
And the faith in the Farthest Vision
that builded her hard-won homes.

—Sharlot Hall,
in "Arizona"

Restored home of Zane Grey near Payson.

Photograph of Erskine Caldwell taken in 1974.

It was in the Klondike I found myself. There nobody talks. Everybody thinks. You get your true perspective. I got mine.

—Jackson London, in an autobiographic sketch written for the Macmillan company

Preacher Hawshaw was always coming to our house and trying to make my old man promise to go to church on Sunday, but Pa always had a good excuse for not going, usually saying Ida, our sugar mule, had the colic and that he couldn't afford to leave her all alone until she got well, or that Mr. Jess Johnson's hogs were running wild and that he had to stay at home to keep them from rooting up our garden, and so I thought they were arguing about the same thing as they always did.

—Erskine Caldwell, in *Georgia Boy*

SCOTTSDALE

Clarence Budington Kelland was stranded here in 1937, when his automobile broke down during a transcontinental trip. Kelland, known best for his stories collected in *Scattergood Baines* (1921), stayed in Phoenix for a while and then decided to buy the Broken Diamond K Ranch, 5320 Casa Blanca Road. He used the Spanish-style ranch home, on what now is Sixty-sixth Street, north of Chapparal Road, as his winter home and writing studio. Kelland and his wife also owned the Scattergood Date Gardens, and Kelland at one time was also vice president and director of Phoenix Newspapers, Inc., publishers of the *Arizona Republic* and the Phoenix *Gazette*. Kelland died at his home on February 18, 1964.

TOMBSTONE

Badger Clark, because of frail health, came to Tombstone when he was a young man. He had left his job as reporter for a newspaper in Lead, South Dakota, to live and work on a ranch twelve miles north of Tombstone. There, Clark began to write cowboy verses. He sent them to his mother, and she sent them on to a magazine called *Pacific Monthly*, which published them. After working as a cowboy for four years, Clark recovered his health and decided to become a writer.

TUCSON

Erskine Caldwell worked on *Georgia Boy* (1943) and many other books while living in the 1940s and 1950s at the El Encanto Estates. Caldwell has reported that his visitors during that time included S. J. Perelman and Sherwood Anderson. **Jeremy Ingalls** visited Tucson twice in 1943 and 1944. Her experiences with the Hopi and Navajo Indians here figure in her long poem *Tahl* (1945). A motion picture was made in 1940 of the Clarence Budington Kelland novel *Arizona* (1939), using sites at the Old Tucson village, Tucson Mountain Park.

While **Georges Simenon** lived in Tucson in the late 1940s, he wrote the novels *La jument perdue* (1947), *Les vacances de Maigret* (1947), and *La neige était sale* (1948). **Ruth Suckow,** the novelist and short-story writer, lived in Tucson from 1947 to 1951, working on her book of stories and memoirs, *Some Others and Myself* (1952). **Sara Teasdale** lived in a small cottage at the corner of Mountain Avenue and Speedway during 1908 and 1909. Teasdale wrote "Day's Ending" (1921) here. **Thornton Wilder,** exhausted from long anticipation of the opening of his play *Our Town* (1938), fled to Tucson in March 1938 to work on *The Merchant of Yonkers* (1938), which he later revised as *The Matchmaker* (1954). Still later, through the work of others, the play became the enormously successful musical *Hello, Dolly!* (1963). **Harold Bell Wright,** suffering from tuberculosis, worked for a time at the Cross Anchor Ranch near Tucson to improve his health.

TUMACACORI

Georges Simenon, Belgian author of the popular detective novels featuring Inspector Maigret, lived for a time in this town north of Nogales.

WALNUT CANYON NATIONAL MONUMENT

Willa Cather visited Walnut Canyon National Monument in 1912 and modeled Panther Canyon in the *Song of the Lark* (1915) on some of the locales near the park. Walnut Canyon National Monument, twelve miles east of Flagstaff, is open daily.

WICKENBURG

J. B. Priestley, the British novelist, stayed in Wickenburg in 1935–36, completing a novel about London. While here, he also assembled notes for his autobiography, *Midnight on the Desert* (1937). Wickenburg is on Route 93, northwest of Phoenix.

ALASKA

BARROW

Will Rogers, the cowboy philosopher, died on August 15, 1935, in an airplane crash near Point Barrow, the northernmost extent of the United States. He and his pilot, Wiley Post, were on the Fairbanks–Point Barrow leg of an Alaskan tour. Post was one of the foremost pilots of his day. Rogers, who was fascinated by air travel, had flown with him often.

CHILCOOT PASS

In 1897, this rugged mountain pass, just north of Skagway in southern Alaska, was the first great barrier for thousands of people who hoped to make their fortunes in the Yukon, prospecting for gold. Among the early adventurers who scaled the Chilcoot was **Jack London,** who arrived here in August 1897. All travelers into the Canadian interior were required to carry 1,000 pounds of food in addition to their mining and camping equipment, and it was not uncommon for a prospector's supplies to total one ton. London found the cost of hiring Indian porters excessive and, like many others, he was obliged to carry his supplies himself, 100 to 200 pounds at a time, often beating the Indian porters in races up the pass. His ability to beat the professionals at their own game remained one of London's fondest memories of his year of prospecting. Though he never found any gold, his experiences during the Klondike Gold Rush gave him the basis for numerous books and stories, from which he did earn a fortune, albeit a small one.

Joaquin Miller, the poet, then working as correspondent of the Hearst newspaper syndicate, crossed the Chilcoot Pass in 1897. Like London, Miller found beauty, excitement, and adventure in the north, but—again like London—no fortune in gold.

ESTER

This town ten miles west of Fairbanks off Route 3 was founded in the early 1900s, when gold was discovered along the Ester Creek. About 15,000 people were working claims here by 1908, and among the businesses that opened to serve their needs was the Malemute Saloon, the setting for "The Shooting of Dan McGrew," one of the best-known poems of Robert W. Service, which was published in *The Spell of the Yukon* (1907). Today the Malemute Saloon is operated by Don Pearson, who bought the mining camp in 1958, turned its mess hall and bunkhouse into the Cripple Creek Resort, and restored the saloon as faithfully as possible to its appearance in the early 1900s. Two family shows are performed each night at the Malemute Saloon, and a featured attraction is Pearson's rendition of Service's poems.

A bunch of boys were whooping it up
in the Malamute saloon;
The kid that handles the music-box
was hitting a jag-time tune;
Back of the bar, in a solo game, sat
Dangerous Dan McGrew,
And watching his luck was his light-o'-love, the lady that's known as Lou.

—Robert W. Service,
in "The Shooting of
Dan McGrew"

Snapshot of Malemute Saloon, near Cripple Creek, in Ester.

FAIRBANKS

There is a place named Baranof in Alaska, but some see a strong similarity between Fairbanks and the Baranof **Edna Ferber** wrote about in her novel *Ice Palace.* Ferber visited Fairbanks and other Alaskan cities in the 1950s, and the setting of her novel is believed to be a composite of several places. *Ice Palace* was published in the spring of 1958, and the favorable impression it made on American readers is said to have helped Alaska win statehood a few months later.

NOME

This city in northwestern Alaska was the scene in 1900 of one of the most audacious claim-jumpings in history. Alexander McKenzie, aided by a judge named Arthur H. Noyes, attempted to rob miners of the richest claims in the Nome district. The scheme worked for a time but the miners, usually fiercely independent, finally got together and summoned federal authorities, who put a stop to the theft. **Rex Beach,** who had spent some time in Nome, used the McKenzie-Noyes scandal as the foundation for his novel *The Spoilers* (1906). He also wrote several articles about the affair for *Booklovers Magazine,* under the title "The Looting of Alaska, or The True Story of a Robbery by Law" (1906).

Wilson Mizner, the playwright, lived in Nome from 1899 to 1902. Here he made and lost fortunes through means illegal as well as legal. Unlike Beach, Mizner was not sympathetic to the

Photograph of Dawson, Yukon Territory, taken about 1898. Gold strikes in 1896 led to the celebrated Klondike gold rush. Among the Klondikers were such literary figures as Jack London, Rex Beach, Wilson Mizner, Robert W. Service, and Joaquin Miller.

There was a man of the Island of Hawaii, whom I shall call Keawe; for the truth is, he still lives, and his name must be kept secret; but the place of his birth was not far from Honaunau, where the bones of Keawe the Great lie hidden in a cave. This man was poor, brave, and active; he could read and write like a schoolmaster; he was a first-rate mariner besides, sailed for some time in the island steamers, and steered a whaleboat on the Hamakua coast.

—Robert Louis Stevenson, in "The Bottle Imp"

See Naples and die but see Hawaii and live.

—Jack London

I have spoken of the outside view—but we had an inside one, too. That was the yawning dead crater, into which we now and then tumbled rocks, half as large as a barrel, from our perch, and saw them go careering down the almost perpendicular sides, bounding three hundred feet at a jump; kicking up dust-clouds wherever they struck; diminishing to our view as they sped farther into distance; growing invisible, finally, and only betraying their course by faint little puffs of dust; and coming to a halt at last in the bottom of the abyss, two thousand five hundred feet down from where they had started! It was magnificent sport. We wore ourselves out at it.

—Mark Twain, in *Roughing It,* writing of Haleakala crater

plight of the miners affected by the McKenzie-Noyes affair. His biographer, Alva Johnston, reported that Mizner once remarked:

> *"The truth is that most of the fellows up there were the worst sissies on earth. I was in court when two hundred of them were robbed of their claims by a crooked judge and a set of thieving politicians. Did they string up the judge, as the forty-niners would have done? Did they tear the politicians limb from limb? No. They just sat there crying into their beards. Then they slunk back to their cabins and had to be treated with smelling salts."*

RAMPART

Rex Beach lived for a time in Rampart, a settlement on the Yukon River northwest of Fairbanks. An important trading post during the Gold Rush, Rampart provided background for Beach's novel *The Barrier* (1907). The cabin in which the novelist lived has been preserved.

SEWARD

Rockwell Kent, the artist and writer, lived in 1918–19 with his son in a cabin on Fox (Renard) Island, in Resurrection Bay, near Seward. Kent described his carefree days in Alaska in his book *Wilderness, A Journal of Quiet Adventure in Alaska* (1920).

SKAGWAY

This city, below the Chilcoot Pass and at the head of the Lynn Canal, was an early stop for prospectors on their way to the gold-rich Klondike and Yukon. Just to the west of Skagway stood the tent city of Dyea, where thousands of the hopeful began the task of portaging their supplies up the heartbreaking Chilcoot Trail to Lake Lindeman and the Canadian interior. A second route into Canada, more difficult and more dangerous than the Chilcoot, began in Skagway and wound its way up White Pass, northeast of the city, to Lake Bennett and on to the Yukon. By 1900 a railroad traversed this second route, connecting Skagway with the mining district, and making the trip much easier for those who could afford the rates. Visitors can still ride this narrow-gauge railroad train from Skagway to Whitehorse, in the Yukon.

In the early days of the Gold Rush, Skagway was run by a band of crooks controlled by a con man known as Soapy Smith. **Wilson Mizner** came to know Smith well when Mizner came to Skagway to manage affairs for the Alaska Commercial Company. He also learned a few shady practices from Smith that he later put to use in Nome. Smith, who died in a shootout in 1898, was immortalized by Mizner in a number of his Alaska anecdotes. Mizner resigned his post in Skagway to return to San Francisco and organize his own expedition into the Yukon. When he returned to Skagway, he was accompanied by a dance-hall girl named Rena, who proved to be a valuable and able traveling companion.

Nowadays Skagway is host, several times a week during the summer months, to a "Days of '98" show, which recalls some of the more colorful moments of the Gold Rush. Among the features is a dramatization of Robert W. Service's poem "The Shooting of Dan McGrew" (1907). The show is performed in Eagle's Hall, on Broadway.

HAWAII

HAWAII ISLAND

Most Westerners today know the Hawaiian Islands primarily through the enormously popular book *Hawaii* (1959), by **James Michener,** but these islands attracted writers long before Michener came here, and many of them wrote of this exotic land. **May Sarton,** the poet and novelist, spent three months on Hawaii Island in 1957, interviewing people whose backgrounds reflected the ethnic diversity of Hawaii. In South Kona, on Hawaii Island, she met an American-Japanese family she later wrote of in an article entitled "Sukiyaki on the Kona Coast," which was published in *The Reporter* on June 27, 1957. **Robert Louis Stevenson,** on a visit to the Hawaiian Islands in 1888, also spent time in South Kona and later wrote the story "The Bottle Imp" based on his experiences there.

Mark Twain, in a trip to the Hawaiian Islands in the 1860s—he later referred to Hawaii as "the first paradise that man ever invaded"—crossed the erupting volcano at Kilauea, twenty-three miles east of Hilo. Twain's biographer, Albert Pigelow Payne, wrote that Twain raced "across the burning lava floor, jumping wide and bottomless crevices, when a misstep would have meant death."

KAUAI ISLAND

Lumahi Beach, between the cities of Hanalei and Haena, on the island of Kauai, was the setting used for Bloody Mary's Beach in the motion picture *South Pacific,* which was based on the musical comedy by Richard Rodgers and Oscar Hammerstein, in turn derived from *Tales of the South Pacific* (1948), the series of sketches for which James Michener won a Pulitzer Prize.

MAUI ISLAND

Jack London and his wife Charmian visited the Haleakala Ranch on Maui in 1907. When London asked Louis von Tempsky, the ranch manager, for permission to visit, von Tempsky replied that anyone who could write *The Call of the Wild* was welcome. The Londons, von Tempsky, and von Tempsky's teenage daughters, Gwen and Armine, started a climb up Maleakala Mountain to a cabin at 6,500 feet. They continued on a cinder path to the crater at the top—described notably by **Mark Twain** in *Roughing It* (1872)—and camped at the crater for two days. The return trip was made through Kaupo Gap, a break in the southern rim of the crater. During this trip down the mountain, London read some of the juvenile works of **Armine von Tempski**—she spelled her name this way in her published works—but did not care for the writing. Nevertheless, he gave her encouragement, and Armine von Tempski went on to become a prolific novelist, highly regarded in Hawaii. Her novels included *Hula* (1927), *Lava* (1930), and *Hawaiian Harvest* (1933). In her autobiography, *Born in Paradise* (1940), von Tempski wrote of her experiences with the Londons.

The Jack London party went on to Hana, then an isolated area in the rain forest of eastern Maui. It was dusk when they reached the ditchtender's home in Keanae Valley, near the peninsula on which Mormon missionaries had converted many islanders. Haleakala Volcano National Park is near the sites that London and Twain visited, and the literary visitor may view Kaupo Trail and the sixty-five-foot Bottomless Pit in the center of the crater from Puv Ulvala Visitor's Center at the peak of Haleakala.

Charmian London returned to Haleakala Ranch after her husband's death in 1916, and later wrote an account of her days there in *Our Hawaii* (1922).

Herman Melville arrived on Maui on May 2, 1843, aboard the whaler *Charles and Henry.* Melville was unable to find work ashore and stayed only sixteen days before departing for Honolulu. Prior to coming to Maui, where he stayed primarily in Lahaina, now a city, Melville had visited a valley in the Marquesas, where the Taipi—Melville spelled this Typee—tribe lived. He was there for several weeks and was able later on to use his experience in writing his first novel, *Typee* (1846). Melville is reported to have stayed in Lahaina at the Full Gospel Church, on Front Street, when he needed a place in which to sober up after a binge.

MOLOKAI ISLAND

This island was known through the years as the site of a leper colony, founded by the government of the islands in 1860. **Jack London** and Charmian visited the lepers at Kalaupapa during the Fourth of July celebration of 1907. He was able to photograph the lepers' colorful parade and festivities, which he described in "The Lepers of Molokai" for *The Woman's Home Companion.* The literary visitor can follow a dangerous path over the Kalaupapa cliffs, since called the Jack London Trail because it was used by London and his wife. The trail is not marked, but townspeople can direct visitors to it.

Robert Louis Stevenson in 1889 visited the leper colony, which stands on the northern coast of the Makanalua peninsula. It is believed that Stevenson was attracted to the place because he had read an account of the colony by Charles Warren Stoddard. Stevenson arrived a month after the death of Joseph Damien de Veuster, the humanitarian Belgian priest known as Father Damien, who had devoted himself to the lepers. Father Damien died as a result of the leprosy he had contracted here. Stevenson visited for a week with the leper girls and, after his return to Honolulu, sent a piano to the leper settlement for their use. In the following year, Stevenson published a defense of Father Damien, whose work had been slandered by a minister.

Charles Warren Stoddard, who lived in Hawaii from 1881 until 1884, described his own visit to Molokai in his *Diary of a Visit to Molokai in 1884* (published posthumously in 1933) and wrote about Father Damien and his work in *The Lepers of Molokai* (1885). Stoddard was the first writer to call attention to the work of Father Damien.

Armine von Tempski, photographed in 1908, a year after she met Jack London. Her initial attempts at writing were severely criticized by London, but von Tempski went on to write many popular novels, and her autobiography, *Born in Paradise,* became a best seller.

OAHU ISLAND

Henry Adams, on a trip to the South Seas in 1890, interviewed Kalakaua, the last of the Hawaiian kings. Adams was impressed with King Kalakaua and included accounts of him in various pieces he wrote about his trip. **Rupert Brooke,** the English poet, came to Oahu in 1913, staying at Honolulu, the capital of Hawaii and principal city of Oahu. Brooke had been on a tour of the United States and took the side voyage to Hawaii, where he wrote the sonnet "Waikiki":

Warm perfumes like a breath from vine and tree
Drift down the darkness. Plangent, hidden from eyes,
Somewhere an eukaleli *thrills and cries*
And stabs with pain the night's brown savagery;
And dark scents whisper; and dim waves creep to me,
Gleam like a woman's hair, stretch out, and rise;
And new stars burn into the ancient skies,
Over the murmurous soft Hawaiian Sea.

And I recall, lose, grasp, forget again,
And still remember, a tale I have heard, or known,
An empty tale, of idleness and pain,
Of two that loved—or did not love—and one
Whose perplexed heart did evil, foolishly,
A long while since, and by some other sea.

Edgar Rice Burroughs, the creator of Tarzan, visited Honolulu several times between 1935 and 1945. He and his family lived at 2623 Halelena Place for a few months in 1940 and moved later

> *These celebrated warriors appear to inspire the other islanders with unspeakable terrors. Their very name is a frightful one; for the word "Typee" in the Marquesan dialect signifies a lover of human flesh. It is rather singular that the title should have been bestowed on them exclusively, inasmuch as the natives of all this group are irreclaimable cannibals. The name may, perhaps, have been given to denote the peculiar ferocity of this clan, and to convey a special stigma along with it.*
>
> —Herman Melville, in *Typee*

They passed Miss Thompson on the road. The doctor took off his hat, and she gave him a "Good morning, doc," in a loud, cheerful voice. She was dressed as on the day before, in a white frock, and her shiny white boots with their high heels, her fat legs bulging over the tops of them, were strange things on that exotic scene.

"I don't think she's very suitably dressed, I must say," said Mrs. Macphail. "She looks extremely common to me."

—Somerset Maugham, in "Rain"

to the Niumalu Hotel, at 2005 Kalia Road, now the site of the Hilton Hawaiian Village. While living here, Burroughs wrote in an office at 1298 Kapiolani Boulevard, at the corner of Piikoi Street.

Padraic Colum, the Irish poet and folklorist, early in the 1920s was invited by the Hawaiian legislature to study local mythology and folk traditions. He wrote many children's stories based on his studies, which were published as *At the Gateways of the Day* (1924) and *The Bright Islands* (1925). Two centuries earlier, in 1777, Captain James Cook had discovered Hawaii for the Western world and, on a return voyage in 1779, met his death on Oahu at the hands of islanders. The literary visitor may trace Cook's voyage of discovery in the journal kept by Cook and Lieutenant James King: *A Voyage to the Pacific Ocean,* which was republished in the twentieth century. **James Jackson Jarves** came to Oahu from Boston in 1837 to found and edit the *Polynesian,* the first newspaper published in the Hawaiian Islands, and stayed until 1848. Jarves, who was also a novelist, is credited with writing the first Hawaiian novel, *Kiana* (1857). He also wrote a fictionalized autobiography, *Why and What Am I* (1857), about his experiences in Hawaii.

Another novel, written by another writer almost a century later and set largely in Hawaii, had an entirely different tone. **James Jones** served as an enlisted man in the U.S. Army before and during World War II and was stationed at the time of the Japanese attack in 1941 at Schofield Barracks, Pearl Harbor. In 1942 Jones was transferred to Guadalcanal. The novel for which Jones is remembered best is *From Here to Eternity* (1951). It was set in Hawaii at the time of the attack on Pearl Harbor, and the air raid figures prominently in the climactic scene of the novel.

Jack London and Charmian, waiting for their yacht *Snark* to be repaired off Honolulu in 1907, stayed for a time at a small cottage they rented in Pearl Locks, near Pearl Harbor. London was extremely busy during those months, completing many stories, among them "Flush of Gold" and "To Build a Fire." The Londons also stayed during 1907 at a four-room tent-house at the Seaside Hotel, later called the Outrigger Club. London spent a good deal of time with Fred Church, manager of the Seaside, reminiscing about their Klondike days: London's friendship with Church dated from their first meeting in Dawson City. While in Honolulu, the Londons stayed also with **Lorrin Andrews Thurston,** the publisher of the Honolulu *Advertiser,* and at Thurston's home London started work on his novel *Martin Eden* (1909). London enjoyed Waikiki Beach, learning to surf and becoming friendly with Duke Kahanamoku, the Hawaiian who went on to become an Olympic swimming champion. But Jack continued his practice of writing a thousand words every morning. A few years later, in 1915, the Londons were guests at Waikiki of the Hawaiian queen, Liliuokalani. Since London was working on two novels, *Jerry of the Islands* (1917) and *Michael, Brother of Jerry* (1917), he saw visitors only in the afternoon. The Londons left Hawaii in July 1915, returning once more in December and departing for the last time in July 1916. Jack London, despite his strong desire to return, did not live to see his beloved islands again.

J. P. Marquand lived in Honolulu in the 1930s and here conceived the character Mr. Moto, the Japanese detective who appeared in "Lunch at Honolulu" and other short stories. **[William] Somerset Maugham,** the British novelist, passed through Honolulu en route to Russia while working as a British secret agent during World War I. Maugham is reported to have remarked that Iwilei, an area just north of Sand Island, was "a desolate, unattractive place." Iwilei at that time was a red-light district. When Maugham sailed from Honolulu, one of the prostitutes of Iwilei sailed aboard the same ship. She is believed to have provided Maugham with the character Sadie Thompson in Maugham's memorable story "Rain." In another of Maugham's stories, "Honolulu," he referred to the city as "the meeting place of East and West."

Herman Melville came to Honolulu in May 1843 and was accused of deserting his ship. To raise money for the trip home, he worked as a pinsetter in a bowling alley, and for six weeks as clerk-bookkeeper in the Isaac Montgomery store here. On August 17, 1843, Melville signed aboard the U.S. warship *United States* as an ordinary seaman.

James Michener lived at 260 Lewers Road in Honolulu for several years, beginning in 1949. Toward the end of his stay, he had an apartment on the eighth floor of the Ainahu Apartments, address not known. Even though Michener returned to the mainland in June 1956, he maintained an address in Honolulu—he has reported it as being on Pupukea Cliffs—until 1960. While living in Hawaii, Michener saw publication of several of his books: *Return to Paradise* (1951), *The Voice of Asia* (1951), *The Bridges at Toko-ri* (1953), and *Sayonara* (1954). His massive chronicle *Hawaii* (1959) tells the story of Hawaii from prehistory until its emergence as a state of the United States.

Joaquin Miller came to Hawaii in 1894 as

Hobron's Cottage at Pearl Locks near Pearl Harbor. In 1907 Jack London lived and worked at this cottage, almost certainly no longer standing.

a correspondent for *Overland Monthly.* Some of Miller's poems were published in local newspapers, and he covered an armed insurrection here in January 1895. He also spent a social evening with Sanford Dole, president of the newly formed Hawaiian Republic, but it became known that Miller was sharing a Honolulu hotel room with Alice Oliver, not Miller's wife but obviously pregnant. Miller was advised to leave the islands, and he did. Alice Oliver stayed behind to give birth before returning to California.

> *I tell you, my boy, the man who has not seen the Sandwich Islands, in this one great ocean's warm heart, has not seen the world.*
>
> —Joaquin Miller

Lewis Mumford and his family stayed in a rented home on Kalakaua Avenue in 1938 while Mumford's wife was recovering from pneumonia at Queens Hospital, Vineyard Street. Mumford had come to Honolulu to prepare a report on the planning of the city's parks. **Charles B. Nordhoff** and **James Norman Hall,** some years earlier, in the 1920s, had come to Hawaii to write a series of articles for *Harper's Magazine.* The collaborators later achieved lasting reputation as authors of *Mutiny on the Bounty* (1932). **Ernie [Ernest Taylor] Pyle,** who died in 1945 on Iwo Jima from machine-gun bullet wounds suffered during that bloody battle of World War II, is buried in the National Cemetery of the Pacific, in Punchbowl Crater near Honolulu. His war dispatches were collected in such volumes as *Here Is Your War* (1943), *Brave Men* (1944), and *Last Chapter* (1946).

Robert Louis Stevenson may have spent some time in a small building known as The Grass House, now standing in the Manoa Valley area, 3016 Oahu Avenue, where the house is cared for by the Salvation Army. Stevenson completed *The Master of Ballantrae* (1889) while living near Diamond Head, in Waikiki. In addition to the house he and his family occupied, Stevenson had a shack away from the main house, and it was there that he worked on *The Master of Ballantrae* as well as on a lesser-known work, *The Wrong Box* (1888). The Grass House may have been the shack in which Stevenson worked, but this is not certain. The Stevensons' first dinner in Waikiki was a memorable one at the old Royal Hawaiian Hotel: Stevenson was the guest of Kalakaua, last of the Hawaiian kings. The king is said to have drunk five bottles of champagne and two bottles of brandy in four hours. When he dined with Stevenson for a second time some days later, he demonstrated that his drinking habits had not changed appreciably.

Charles Warren Stoddard visited Oahu for the first time in 1864, when he was twenty-one. Stoddard recalled his visits to the Hawaiian Islands in sketches he later collected as *South Sea Idyls* (1873). While Stoddard lived in Hawaii, he wrote *A Troubled Heart* (1885), about his conversion to Catholicism. He also wrote about the islands in *Hawaiian Life* (1894) and *The Sea of Tranquil Delights* (1904). It is not known whether Stoddard and Robert Louis Stevenson met while they were on Oahu, but Stevenson once observed: "There are but two writers who have touched the South Seas with genius, both Americans: Melville and Charles Warren Stoddard."

Genevieve Taggard lived on Oahu beginning in 1896, when she was two years of age, until she was twenty. Her parents were teachers, and Taggard grew up with the children of Chinese, Japanese, and Hawaiian plantation laborers. Taggard had little contact with *haole* (white settlers) until she was sent to the Punahou School, at Punahou and Clement streets in Honolulu. She left high school when her father became ill, and went to work as a substitute teacher at a plantation school. Taggard's story "Hiawatha in Hawaii" related an incident in her work at the school, and some of her poems, collected in *Hawaiian Hilltop* (1923), dealt with the islands.

Robert Lewis Taylor worked in the 1930s as a reporter for the Honolulu *Advertiser*—its publisher was **Lorrin Andrews Thurston,** who wrote *Memoirs of the Hawaiian Revolution* (1936) and *Writings* (1936). Thurston was born in Honolulu on July 31, 1858.

Mark Twain arrived on March 18, 1866, in Oahu, in what then were called the Sandwich Islands, where he met a group of Americans, including Anson Burlingame, the U.S. minister to China, who encouraged Twain—Twain was thirty-one years old and had not yet begun to publish widely—to refine his writing style. Twain left the islands on July 19, 1866, but wrote about his Hawaiian experiences in the Sacramento, California, *Union* as well as in *Roughing It* (1872). When Twain returned to Hawaii in 1895, he was unable to go ashore because cholera had broken out in the islands. The last author associated with Oahu is Herman Wouk, who used Hawaii and Pearl Harbor prominently in *The Winds of War* (1971), *The Caine Mutiny* (1951), and *War and Remembrance* (1978).

Then Maui rose and climbed at night
The mountain. Dim and deep
Within the crater's bowl he saw
The sprawling sun asleep.

He looped his ropes, the mighty man,
He whirled his sisal cords;
They whispered like a hurricane
And cut the air like swords.

Up sprang the spider. Maui hurled
His lasso after him.
The spider fled. Great Maui stood
Firm on the mountain-rim.

—Genevieve Taggard, in "Solar Myth"

Robert Louis Stevenson Grass Hut, a tourist attraction in Honolulu. Stevenson's association with the hut is not clear.

ACKNOWLEDGMENTS

We wish to express our gratitude to the many talented and hard-working people who assisted in the development of THE OXFORD ILLUSTRATED LITERARY GUIDE TO THE UNITED STATES. *We owe special thanks to the men and women who researched or helped in the writing of the manuscript. They are listed first. Others to whom we are particularly grateful performed photo research or made trips to literary sites to take photographs for the volume. They are listed second. Finally, we acknowledge the indispensable help of the many authors, librarians, and local historians who supplied basic information, verified leads developed in our research, or corrected misinformation that had crept into our files. The support of the members of this last group, from all fifty states, enriched the entire project: in most cases they supplied much-needed local detail that would otherwise have been nearly impossible for us to collect. To each of these many hundreds of supporters we are indebted, and we thank them for their generous assistance.*

Lee Ackerman, Loretta Annese, Marianne Bunch, Christopher Carruth, Hayden Carruth, Mary Egner, Norma S. Ehrlich, Richard Ehrlich, Rosalie B. Ehrlich, Tamara Glenny, Margaret Huffman Graff, Christine Hewitt, Susan Horton, Anne Marie Kennedy, Joan Lizzio, Perry Morse, Alison Porter, Susan Steckbeck Scheffel, David H. Scott, Janet Shepard, Cheryl Silberstein, Ellen Silberstein, Richard Stukey, and Leslie Tierney.

Marvin Beinema, Marion Bodine, Ilene Cherna, Gérard Cochepin, Joseph Lizzio, Larry R. Nicodemo, Ken Riehl, and Patricia Vestal.

George Abbott; *Patricia K. Abe*, Harvard University Press, Cambridge MA; *J. Richard Abell*, Public Library, Cincinnati OH; *John R. Abram*, Public Library, Newark NJ; *Peter Abramov III*, Sayville Library, Sayville NY; *Bruce Adams*, Public Library, Schenectady NY; *E. Ruth Adams*, Public Library, Westport CT; *Fran Adams*, Payson Library, Malibu CA; *Mrs. Reuben H. Adams*, Dallas County Historical Commission, Dallas TX; *Helen G. Ahearn*, Public Library, Derby CT; *Mrs. Conrad Aiken*, Savannah GA; Albany Library, Albany OR; *Betty Albsmeyer*, Public Library, Quincy IL; *Mrs. Edward Allen*, Cazenovia NY; *Janice Allen*, Paine Memorial Free Library, Willsboro NY; *Mrs. V. P. Allbert*, Kansas State Historical Society, Topeka KS; Allegheny Regional Library, Pittsburgh PA; *Mrs. Grace Allison*, County Historical Society, Warren OH; *T. F. Altmann*, Public Library, Milwaukee WI; Alton Telegraph, Alton IL; Amenia Free Library, Amenia NY; *Alan H. Anderson*, New City NY; *Barbara Anderson*, County Library, San Bernadino CA; *Eleanor Anderson*, Marion VA; *Barbara Andrews*, Nantucket Atheneum, Nantucket MA; *Marjorie Andrews*, Public Library, Lafayette AL; *Elizabeth Anesi*, Public Library, Portland MI; *Charles Angoff*, Ann Arbor Library, Ann Arbor MI; *John R. Angell*, Free Library of Philadelphia, Philadelphia PA; *Carolyn Anthony*, Baltimore County Library, Towson MD; *Linda E. Arbaugh*, Burlington County Library, Bordentown NJ; *Elizabeth Armand*, Salem Library, Salem MA; *Katie Armitage*, Elizabeth M. Watkins Community Museum, Lawrence KS; Armstrong Library, Natchez MS; *Mrs. Lloyd R. Arnold*, Hailey ID; *Margaret J. Arnold*, Free Library, Wellesley MA; Artz Library, Frederick MD; *Judith Ashlock*, Presbyterian Church Library, La Crescenta CA; *Isaac Asimov*; *Sallie O. Athern*, Public Library, Montezuma IN; *Barbara Athey*, Taylor Memorial Public Library, Cuyahoga Falls OH; *J. M. Atkinson*, Jack London State Historic Park, Sonoma CA; Audubon State Park, Henderson KY; Aurora Library, Aurora IL; *Betty Austin*, Public Library, Farmington ME; *Roxanna Austin*, Athens Regional Library, Athens GA; *Eleanor Ayoub*, New Dorp Regional Library, Staten Island NY;

Katherine Babbitt, Highland Falls NY; *Betty Babicz*, Public Library, Bloomfield NJ; *Mrs. Bennie Baham*, Southeastern Louisiana University Library, Hammond LA; *Marjorie Bahlke*, Houghton Mifflin Company, Boston MA; *Alevina D. Bailey*, Public Library, Millville NJ; *Billie Bailey*, Grout Museum of History and Science, Waterloo IA; *Rosalind Bailey*, City Library, Manchester NH; *Bruce D. Bajema*, Free Library, San Rafael CA; *Grace Baker*, Public Library, Galesburg IL; *Maggie Baker*, Public Library, Fort Dodge IA; *Pearl Baker*, Green River UT; *Scott Baker*, Public Library, Troy NY; *Tracey Baker*, Minnesota Historical Society, St. Paul MN; County Library, Baker OR; *Jean D. Baldwin*, Free Library, Guilford CT; *Lockett Ford Ballard*, Jr., Litchfield Historical Society, Litchfield CT; Bandelier National Monument, Los Alamos NM; *Frances Barba*, Public Library, Old Saybrook CT; *Barbara Barber*, Public Library, Topeka KS; *Dee S. Barber*, Seton Village, Santa Fe NM; *Sylvia P. Barber*, Free Library, Salisbury NH; *Morgan J. Barclay*, Public Library, Toledo OH; *Rev. Bailey B. Barnes*, St. Peter's Episcopal Church, Perth Amboy NJ; *David C. Barnes*, Whitman Mission National Historic Site, Walla Walla WA; *Margarie Barnes*, Public Library, Tipton IA; *Choice Bartlett*, Public Library, Brook IN; *Jean Bartlett*, Office of the Village Historian, Bronxville NY; *Mrs. J. Floyd Barton*, Memorial Library, Kent CT; *Richard L. Barton*, Brunswick ME; *William H. Barton*, Wyoming State Archives, Cheyenne WY; *Nancy Batton*, Harriette Person Memorial Library, Port Gibson MS; *Evelyn C. Bauerle*, Public Library, Pottstown PA; Bay City Library, Bay City TX; *J. Lyle Bayless, Jr.*, Live Oak Gardens, Ltd., Jefferson Island LA; *J. Beamguard*, Public Library, Tampa FL; *Charles E. Beard*, Ina Dillard Russell Library, Milledgeville GA; *Patricia A. Beaton*, Public Library, Lynn MA; County Library, Beaufort SC; Beaufort, Hyde, Martin Regional Library, Washington NC; *Gerald Becham*, Ina Dillard Russell Library, Milledgeville GA; *Elswyth Thane Beebe*; *Cecil Beeson*, Public Library, Hartford City IN; *Marvin Beinema*, Bethlehem PA; *Marie Belcinski*, Public Library, Seymour CT; *Natalie Bell*, Perrot Memorial Library, Old Greenwich CT; *Saul Bellow*; *Robert J. Belvin*, Free Library, Geneva NY; *Rowland Bennett*, Maplewood Memorial Library, Maplewood NJ; Bennington Library, Bennington VT; *Isabel Benson*, Public Library, Stockton CA; *William L. Bergeron*, Public Library, Cranston RI; *Sister Hildemar Berkemeier*, Christ the King Seminary Library, East Aurora NY; Berlin Library, Berlin MD; *Bruce R. Berney*, Public Library, Astoria OR; *James M. Berry*, Malabar Farm State Park, Lucas OH; *Denise B. Bethel*, Poe Museum, Richmond VA; *George R. Beyer*, Pennsylvania Historical and Museum Commission, Harrisburg PA; *Elizabeth Bezera*, Public Library, Fayetteville NC; *Gladys S. Bickford*, Center Harbor NH; *Judy Bieber*, Public Library, Oregon City OR; *Ann E. Billesbach*, Willa Cather Historical Center, Red Cloud NE; *Florence C. Bingham*, Silsby Free Public Library, Charlestown NH; Public Library Information Services, Binghamton NY; Bishop Museum, Honolulu HI; *Hazel Blackstone*, City Library, Ellsworth ME; *Marilyn Blackwell*, Vermont Historical Society, Montpelier VT; *Anne Blatt*, Public Library, Elmsford NY; *Patricia S. Blalock*, Public Library, Selma AL; *Phyllis Blodget*, Public Library, Lynn MA; Blodgett Memorial Library, Fishkill NY; Bloomington Library, Bloomington IL; *Robert Bly*; *Hilda Bohem*, University of California Library, Los Angeles CA; *Sam J. Boldrick*, Public Library System, Miami FL; *Gladys E. Bolhouse*, Historical Society, Newport RI; *Carbilene G. Bolin*, Public Library, Fulton KY; *Ruth Bolin*, Finkelstein Memorial Library, Spring Valley NY; *Charles K. Boll*, Free Public Library, Elizabeth NJ; *George S. Bolster*, Saratoga Springs NY; *Sarah Davis Bolz*, O. Henry Museum, Austin TX; *Shirley W. Boone*, Chapin Memorial Library, Myrtle Beach SC; *Mary Hard Bort*, Manchester Historical Society, Manchester VT; *Howard H. Borthwick*, Montgomery County Department of Public Libraries, Rockville MD; *Lucille Bousquet*, Public Library, Bridgewater MA; *Janet Bowen*, Public Library, Anderson IN; *Paul Bowles*; *Persis Boyesen*, Public Library, Ogdensburg NY; *Hope H. Boykin*, Historic Camden, Camden SC; *Lucile Boykin*, Public Library, Dallas TX; *Kay Boyle*; Mayor *Tom Bradley*, Los Angeles CA; *Linda Brammer*, Public Library, Elkton MD; *Dorothy Brandon*, Frothingham Free Library, Fonda NY; *Eugene A. Bratek*, Rutgers Preparatory School, Somerset NJ; *Susannah W. Breaden*, University of Utah Library, Salt Lake City UT; *Marguerite Breland*, Noxubee County Library, Macon MS; *Dorothy Brennan*, Free Public Library, Fort Lee NJ; *Patricia Brewer*, Public Library, Middletown OH; Public Library, Brewster NY; *Roger D. Bridges*, Illinois State Historical Library, Springfield IL; *Juanita S. Brightwell*, Lake Blackshear Regional Library, Americus GA; *Harry Brinton*, Public Library, Jacksonville FL; *Warren F. Broderick*, Lansingburgh NY; *Mrs. Warren J. Broderick*, Lansingburgh NY; *Delia S. Bronson*, Middlebury CT; Bronx County Historical Society, Bronx NY; Public Library, Bronxville NY; *Valerie Brooker*, Santa Fe Library, Santa Fe NM; *Kenneth Browand*, Public Library, Lansing MI; *Betty R. Brown*, Public Library, Rock Falls IL; *Eleanor Logan Brown*, Clarence Darrow Octagon House, Kinsman OH; *Helen Brown*, Halle Memorial Library, Pound Ridge NY; *Janet D. Brown*, Dudley-Tucker Library, Raymond NH; *Marie Brown*, Rogers Memorial Library, Southampton NY; *Mary A. Brown*, Rye Free Reading Room, Rye NY; *Richard E. Brown*, Public Library, Berkeley CA; *Stephen Brown*, Cedarville College Library, Cedarville OH; *Betty Bruce*, Public Library, Key West FL; *Mrs. K. R. Bruner*, Public Library, Greenwood Lake NY; *Jeannette M. Brush*, Public Library, Phoenix AZ; *Mary Bryant*, Historical Society, Kennebunkport ME; *Barbara Bublitz*, Public Library, Muscatine IA; *Doris Buchheit*, Plumb Memorial Library, Shelton CT; *Clifford M. Buck*, Salt Point NY; Bucks County Free Library, Doylestown PA; *Donna M. Bugieda*, Philadelphia Convention and Visitors Bureau, Philadelphia PA; *Cheryl Bugnone*, Kinsman Library, Kinsman OH; *Linda Bunyan*, Reed Memorial Library, Ravenna OH; *Marion Burch*, Public Library, Newton IA; *Larry E. Burgess*, Smiley Public Library, Redlands CA; *Mrs. Harry Burk*, Overbrook Library, Scranton KS; *Bettie Burke*, Mount San Jacinto College Library, San Jacinto CA; *Betty Burnett*, Public Library, Lexington KY; *Eugene H. Burrell*, Public Library, Elkhart IN; *Hubert Burroughs*, Tarzana CA; *Marci Burstedt*, Public Library, Challis ID; *Mary Burtschi*, Little Brick House, Vandalia IL; *Samuel Busey*, Public Library, Urbana IL; *Francis Butler*, Haynes Memorial Library, Alexandria NH; *Joan Butler*, Public Library, New London CT; *Natalie S. Butler*, Farmington Public Library, Farmington ME; Byers Memorial Library, Monongahela PA; *Thomas E. Byrne*, Elmira NY;

Cabinet Press, Inc., Milford NH; *Mary Cabral*, Free Public Library, New Bedford MA; *Frank Caddy*, Greenfield Village & Henry Ford Museum, Dearborn MI; *Helen H. Cahill*, Horace Greeley Museum, East Poultney VT; *Herbert Cahoon*, Morgan Library, New York NY; Public Library, Cairo NY; *Erskine Caldwell*; Calhoun County Library, Saint Matthews SC; *Hortense Calisher*; *Lawrence Calkins*, Elko County Library, Elko NV; *Virginia Callahan*, Public Library, Leavenworth KS; *Elizabeth Calloway*, Gardiner ME; *Georgie Ruth Geddes Calmer*, St. Louis MO; *Dennis Camp*, Vachel Lindsay Home, Springfield IL; *Dr. Hilbert H. Campbell*, Virginia Polytechnic Institute and State University, Blacksburg VA; *Marjorie H. Campbell*, Public Library, Weston MA; *Victoria Campbell*, Thayer Public Library, Braintree MA; Canfield Memorial Library, Arlington VT; *Sharon Cantrall*, Mendocino County Library, Ukiah CA; *Charlotte Capers*, Mississippi Department of Archives and History, Jackson MS; *Gioconda Capitolo*, Public Library, Salt Lake City UT; *Marie T. Capps*, United States Military Academy Library, West Point NY; *Jerry Carbone*, Brooks Memorial Library, Brattleboro VT; *Margaret Cardwell*, Public Library, Kokomo IN; *Anne R. Carey*, Chester County District Library Center, West Chester PA; *Esther D. Carey*, Torrington Library, Torrington CT; *Jacalyn Carfagno*, Arkansas Historic Preservation Program Little Rock AR; *Natalie Savage Carlson*; *Thomas W. Carneal*, Nodaway County Historical Society, Maryville MO; Carnegie Free Library, McKeesport PA; *Thomas Carney*, Public Library, Cedar Rapids IA; *Eric J. Carpenter*, State University of New York at Buffalo, Buffalo NY; *Thelma Carpenter*, Public Library, Oregon IL; *William H. Carpenter*, Division of Parks and Recreation, Concord NH; *Evelyn Carrell*, Public Library, Lake City MN; *Gladys Hasty Carroll*; *Robert G. Carroon*, Milwaukee County Historical Society, Milwaukee WI; *John E. Carter*, State Historical Society, Lincoln NE; *Merton M. Carter*, Little Traverse Regional Historical Society, Petoskey MI; *Jerry C. Cashion*, North Carolina Department of Cultural Resources, Raleigh NC; *Nina Caspari*, Kern County Library System, Bakersfield CA; *Mrs. Joseph N. Cathcart*, New Orleans LA; *Roger Catherwood*, Austin MN; Catskill Public Library, Catskill NY; *William D. Cecil*, Lewiston Historical Society, Lewiston NY; *Dale W. Cerkins*, City-County Library, San Luis Obispo CA; *Freda Chabot*, Brooks Memorial Library, Brattleboro VT; *Sue Chamberlain*, Sharlot Hall Historical Society, Prescott AZ; *Richard Chamberlin*, Stabley Library, Indiana University, Indiana PA; *Elizabeth Chapman*, Atlanta GA; *Florence Chapman*, Boothbay Harbor Memorial Library, Boothbay Harbor ME; Charleston County Library, Charleston SC; *Bettie L. Chase*, Waitsburg Historical Society, Waitsburg WA; *Vera O. Chase*, General Electric Company, Schenectady NY; *Sarah Chatham*, Gulfport-Harrison County Library, Gulfport MS; *Margaret S. Cheney*, Mark Twain Memorial, Hartford CT; Chicago Association of Commerce and Industry, Chicago IL; *Emmett D. Chisum*, American Heritage Center, University of Wyoming, Laramie WY; *Jerome Chodorov*; *Cliff Choquette*, Public Library, Chelmsford MA; *Roger Christian*, Public Library, Mobile AL; *Claire Christiansen*, Public Library, Springfield OR; *Virginia W. Christy*, Williams Public Library, Woodstock VT; *Judith Ciampoli*, Missouri Historical Society, St. Louis MO; *John Ciardi*; Cincinnati and Hamilton County Library, Cincinnati OH; *Douglas Cisney*, Public Library, Nahant MA; *Katherine T. Citizen*, Yorba Linda Library District, Yorba Linda CA; *Anna L. Cladek*, Free Public Library, Perth Amboy NJ; *Irma Clark*, Free Public Library, Tyringham MA; *Sarah B. Clark*, Richland County Public Library, Columbia SC; *W. Dean Clark*, Alberta Culture Resources, Edmonton, Alberta, Canada; *William Clark*, Santa Fe NM; Clifton Springs Library, Clifton Springs NY; *Mrs. Robert E. Clise*, Historical Society and Museum, Geneva NY; *Dr. Jon P. Cobes*, Ohio State University, Mansfield OH; Cohoes Public Library, Cohoes NY; *Doris Colbath*, Village Library, Raymond ME; *Virginia Colby*, Cornish Historical Society, Cornish NH; *Helen Colchin*, Public Library, Fort Wayne IN; *Margie L. Cole*, Inyo County Free Library, Independence CA; *Pauline Cole*, Public Library, West Haven CT; *Earle E. Coleman*, Mudd Manuscript Library, Princeton University, Princeton NJ; *George Collins*, Boston Globe Newspaper Library, Dorchester MA; *Marion Collins*, Indianapolis-Marion County Library, Indianapolis IN; *Sue Collins*, Public Library, Stillwater MN; Columbia County Historical Society, Kinderhook NY; Columbia Library, Columbia PA; *Grace Comes*, Town

Library, Lancaster MA; *Myrtle Comes*, Public Library, Jersey City NJ; *Dorothy Connolly*, Public Library, Salem MA; *Teri Conrad*, Public Library, Salt Lake City UT; *Marian F. Conway*, Grass Valley CA; *Dorothy H. Cook*, Fryeburg Women's Library, Fryeburg ME; *Henry G. Cook*, Government of the Northwest Territories, Yellowknife, Canada; *Jean M. Cook*, Yukon Archives, Whitehorse, Yukon; *Lurline Cook*, Helen Keller Property Board, Tuscumbia AL; *Sterling Cook*, Miami University Art Museum, Oxford OH; *William R. Cook*, Onondaga County Public Library, Syracuse NY; *Bertha Cookinham*, City Library, Santa Clara CA; *Clarice M. Cooks*, Public Library, San Mateo CA; *Mrs. Robert Cooley*, Wolcott Library, Litchfield CT; *Carol Coon*, Public Library, San Francisco CA; *Bryna Coonin*, Asheville-Buncombe Library System, Asheville NC; *JoAnn Cooper*, Public Library, Inc., Winchester KY; *Louise Field Cooper*, Town Library, Woodbridge CT; *Rita Cooper*, Public Library, Independence KS; *Ruby Cooper*, Public Library, Watonga OK; *Paul Corey*; *Denise Corless*, Public Library, Brockton MA; *Ilene J. Cornwell*, Tennessee Historical Commission, Nashville TN; *Gregory Corso*; Corvallis Library, Corvallis OR; *Shirley Costa*, Hotchkiss Library Association, Sharon CT; *M. J. Costenink*, Public Library, Passaic NJ; *Alice R. Cotten*, University of North Carolina Library, Chapel Hill NC; *Judy Countryman*, Fairbanks North Star Borough Public Library, Fairbanks AK; *Kay Courtnage*, Public Library, Great Falls MT; *Tommy Covington*, Library, Ripley MS; *Dorothy S. Cowan*, Ashtabula County District Library, Ashtabula OH; *Malcolm Cowley*; *Ann Crandall*, The Oxford Eagle, Oxford MS; *Robert Creeley*; *Dorothy Crews*, Augusta Regional Library, Augusta GA; *Therese Critchlow*, Princeton Library, Princeton NJ; *M. H. Cronberg*, Virginian Hotel, Medicine Bow, WY; *Philip N. Cronenwett*, The Jones Library, Inc., Amherst MA; *Lillian Cronkhite*, Ferguson Museum, Oklahoma City OK; *Virginia R. Crook*, San Luis Obispo City-County Library, San Luis Obispo CA; *Gail Crotty*, Pleasant Valley NY; *Pamela Crowell*, Nevada Division of Historic Preservation and Archeology, Carson City NV; *Eleanor M. Crowley*, Public Library, Ipswich MA; *June A. Culton*, Pineville-Bell County Public Library, Pineville KY; *Charles F. Cummings*, Public Library, Newark NJ; *Lucille Cunningham*, Public Library, Hoboken NJ; *Velma Cunningham*, Butterfield Memorial Library, Cold Spring NY; *Eileen M. Curran*, Colby College, Miller Library, Waterville ME; *Elizabeth Curtis*, Public Library, Hampden MA; *Anne T. Cushman*, Mark Twain Library, Redding CT; Cyrenius Booth Library, Newton CT;

Ann Daley, Public Library, South Hadley MA; *Alice C. Dalligan*, Public Library, Detroit MI; *J. A. Daly*, Public Library, Lincoln Center MA; *William T. Dameron*, Chalmers Memorial Library, Kenyon College, Gambier OH; Danbury Scott-Fanton Museum & Historical Society, Danbury CT; *Jonathan Daniels*, Hilton Head Island NC; Dartmouth College Archives, Hanover NH; *Anita T. Daubenspeck*, Ridgefield Library and Historical Association, Ridgefield CT; Davenport Hotel, Spokane WA; Davie County Library, Mocksville NC; *Clayla Davis*, Public Library, St. Helena CA; *Helen James Davis*, Zane Grey House, Lackawaxen PA; *Jean Davis*, Artz Library, Frederick MD; *Jim Davis*, Idaho State Historical Society, Boise ID; *Mary M. Davis*, Public Library, Colorado Springs CO; *Lori Davisson*, Arizona Historical Society, Tucson AZ; *David R. Day*, Minute Man National Historical Park, Concord MA; *Janet Vaill Day*, Wilton Library Association, Wilton CT; Mayor *Douglas Dayton*, East Hampton NY; *Russell DeBerry*, Buckhorn Museum, Lone Star Brewing Co., Inc. San Antonio TX; *Norma Deck*, Public Library, Racine WI; *Ruth Deffenbaugh*, Public Library, Estes Park CO; *Virginia Deffner*, Rock Island Library, Rock Island IL; *John F. Delahant*, Robert Louis Stevenson's Memorial Cottage, Saranac Lake NY; Deming Public Library, Deming NM; *John Dempsey*, Kingston Area Library, Kingston NY; *Phyllis DeMuth*, Alaska Historical Library, Juneau AK; *Delphine Denesen*, Public Library, St. Paul MN; *Rodney G. Dennis*, The Houghton Library, Harvard University, Cambridge MA; *Ellen DeNooyer*, Wave Hill Center for Environmental Studies, Bronx NY; *Betty Dent*, Public Library, Somerville NJ; *Leonie M. DeRoche*, Alexandria Library, Alexandria VA; De Smet City Library, De Smet, SD; *Babette Deutsch*; *Charlotte M. Devers*, Public Library, Armonk NY; *Judith Devine*, Public Library, St. Paul MN; *Eugene Devlin*, Peck Memorial Library, Kensington CT; *Pearl Devlin*, Nesmeth Library, Windham NH; *Rosalie E. Devlin*, Public Library, San Mateo CA; *Peter De Vries*; *Michael and Frances DeWine*, Whitelaw Reid House, Cedarville OH; *Mary-Thadia d'Hondt*, Seattle Historical Society, Seattle WA; *Emily Dickenson*, Wake County Public Library, Raleigh NC; *James Dickey*; *Mrs. Lewis W. Dickson*, Oregon IL; *Paul di Mauro*, Public Library, Evanston IL; *Don Dinimett*, Public Library, Waterloo IA; *Mary Alice Dirnberger*, Public Library, Hebron OH; District of Columbia Library, Washington DC; *Esther Dittlinger*, Public Library, Anderson IN; Dixie Regional Library, Pontotoc MS; *Timothy Doane*, Essex CT; *Mary Dodd*, Public Library, Weston MA; *Joanne L. Dodds*, Pueblo Regional Library District, Pueblo CO; *Janie A. Dollinger*, Public Library, Lanark IL; *Georgia Donati*, Public Library, Mt. Vernon NY; *Kathryn Dooling*, San Benito County Free Library, Hollister CA; *Honora Dougherty*, Public Library, Larchmont NY; *Ellen Douglas*; *Clifford Dowdey*; *Kenneth E. Dowlin*, Public Library, Colorado Springs CO; *Alice Driscoll*, CEL Regional Library, Savannah GA; *Gertrude Drown*, Goodrich Memorial Library, Newport VT; *Allen Drury*; *Dorothy Dudley*, County Branch Library, West Salem WI; *Robert Dugan*, Public Library, Beverly MA; *Anne T. Dugger*, Public Library, Charlotte NC; *Barbara Dunbar*, McLean County Historical Society, Bloomington IL; *Thomas Duncan*; *Audrey Dunham*, Florida Department of Natural Resources, Tallahassee FL; Dunham Public Library, Whitesboro NY; *Eva Jane Dunn*, Public Library, Galesburg IL; duPont Library, University of the South, Sewanee TN; *Alice DuPuis*, Miami-Dade Public Library System, Miami FL; *Diane Durette*, Public Library, Aurora IL; Durham County Library, Durham NC; *Michael J. Durkan*, Swarthmore College Library, Swarthmore PA; *Margaret Durkin*, Public Library, Somerville MA; *Carol Durrence*, Central Florida Regional Library, Ocala FL; *Muriel Dustin*, State Library, Pierre SD; *Arvilla L. Dyer*, Plainfield Historical Society, Inc., Plainfield MA; *Jane Dyer*, Public Library, Chapel Hill NC; *Richard N. Dyer*, Colby College, Waterville ME; *Alice Evans Dyson*, Mocksville NC;

Yvette Eastman, Gay Head MA; *Betsy Eaton*, Public Library, Littleton NH; *Leona G. Eaton*, Public Library, Lyons NY; *Richard Eberhart*; *Patricia W. Eckels*, Howe Library, Hanover NH; *Martha Eder*, Hamburg Historical Society, Hamburg NY; *Walter Dumaux Edmonds*; *Kathryn F. Edwards*, Burnham Library, Bridgewater CT; *Mary Roy Edwards*, Howard County Library, Columbia MD; *Elizabeth B. Egbert*, Longfellow National Historic Site, Cambridge MA; *Mrs. Amos Eidem*, Burbank SD; *Marc Eisen*, Public Library, East Orange NJ; *Dottie Elder*, Ernie Pyle Memorial, Dana IN; *Kathleen Elder*, Public Library, New Alexandria PA; *Jane A. Eldridge*, Public Library, Eastham MA; *Jean Elkington*, Oregon Historical Society, Portland OR; Elk Point Library, Elk Point SD; El Paso Library, El Paso TX; *David Emerson*, Ralph Waldo Emerson Memorial Association, Boston MA; *B. Emery*, Free Library, Kennebunk ME; *Doreen Emery*, Public Library, Sausalito CA; *Dana Emmons*, Woodstock VT; *Nancy Englander*, MacDowell Colony, Peterborough NH; *Davis Erhardt*, Queens Borough Public Library, Jamaica NY; Erie City and County Library, Erie PA; *Marion Errickson*, Ludington Public Library, Bryn Mawr PA; *Eleanor Estes*; Eugene Field House, St. Louis MO; Eugene Library, Eugene OR; *Ann L. Eustace*, Public Library, Nutley NJ; *Judy Evans*, Altadena Library District, Altadena CA; *Wayne M. Everard*, Public Library, New Orleans LA; *Betty Ewing*, Bartholomew County Library, Columbus IN; Exeter Library, Exeter NH; *John S. Ezell*, University of Oklahoma Library, Norman OK;

Nicholas Falco, Queensborough Public Library, Jamaica NY; *Mrs. John Falkner*, Oxford MS; *Dorothy Fanoele*, Columbus Library, Columbus KS; *Emily C. Farnsworth*, Public Library, Brookline MA; *Howard Fast*; *Stanley P. Fast*, Mark Twain Birthplace, Stoutsville MO; *Frances Faucett*, Jennings County Public Library, North Vernon IN; *Betty Fawcett*, Historic Homes Foundation, Inc., Louisville KY; *Evelyn Fenton*, Hollywood Regional Branch Library, Hollywood CA; *Joan Fenton*, New Brunswick Library, New Brunswick NJ; *Pamelyn Ferguson*, Westfield Memorial Library, Westfield NJ; *David G. Fergusson*, Davie County Public Library, Mocksville NC; *Francis Fergusson*; *Lawrence Ferlinghetti*; *Leslie Fiedler*; *Elizabeth Finch*, Brewster Ladies' Library, Brewster MA; *Vivian Fink*, Public Library, Garnett KS; *Mary L. Finken*, Public Library, Harlan IA; *Doris Finley*, Public Library, St. Joseph MO; *Ann Fisk*, Pigeon Cove MA; *Eleanor B. Fitchen*, Landmarks Preservation Society of Southeast, Brewster NY; *Dorothea Fitzgerald*, Public Library, Boone IA; *Linda R. Fitzgerald*, Public Library, Plymouth MA; *Lee Fletcher*, New Albany–Floyd County Public Library, New Albany IN; *Sadie F. Flint*, Eastham Historical Society, Eastham MA; *John H. Flister*, National Park Service, Andersonville GA; *Julie P. Fogg*, Chelmsford Historical Society, Chelmsford MA; *Shelby Foote*; *Lydia Forbes*, Blackburn College Library, Carlinville IL; *Esther Ford*, Kinsman Library, Kinsman OH; *Janet Ford*, Public Library, Clyde OH; Ford Memorial Library, Ovid NY; Fort Scott Public Library, Fort Scott KS; *Elinor Foster*, Public Library, Riverside CA; *Omer H. Foust*, James Whitcomb Riley Memorial Association, Indianapolis IN; *Betty Fowler*, Village Library, Farmington CT; *Leandra Fox*, Community College of Allegheny County, Pittsburgh PA; *Barbara Fraley*, Johnson County Library, Buffalo WY; *Irma Franklin*, Putnam County Historical Society, Cold Spring NY; *Charlotte Franz*, Public Library, McGregor IA; *Mary H. Frautschi*, Union County Library, Liberty IN; *B. E. Frayne*, Edgar Allan Poe House, Philadelphia PA; *Karen Frederickson*, Public Library, Menlo Park CA; *Donald C. Freeman*, John Greenleaf Whittier Homestead, Haverhill MA; *Mrs. Donald S. French*, Oswego County Historical Society, Oswego NY; Free Library, New Hope PA; *Anne Frese*, Community Library, Niles MI; *Amy Louise Frey*, Public Library, West Hartford CT; *Ramona Friedlander*, New Albany–Floyd County Public Library, New Albany IN; *Hazel R. Friedley*, Palos Park Historical Society, Palos Park IL; *Bertha S. Frothingham*, Windsor VT; *Kent W. Fuhrman*, Sinclair Lewis Foundation, Inc., Sauk Centre MN; *Walline Fuller*, Community Library, Stevenson WA; *John R. Funk*, George Ade Memorial Association, Brook IN;

Aileen Gaddy, Hood River County Library, Hood River OR; *Abner J. Gaines*, University of Rhode Island Library, Kingston RI; *Kevin J. Gallagher*, Adriance Memorial Library, Poughkeepsie NY; *Janice Gallinger*, Lamson Library, Plymouth State College, Plymouth NH; *Suzanne H. Gallup*, Bancroft Library, Berkeley CA; *Robert S. Gamble*, National Parks Service, Washington DC; *Ann Garcia*, Free Library, Keeseville NY; *Paula I. Garner*, Public Library, North Manchester IN; Garnett Library, Garnett KS; *Catharine Garoian*, Free Library, Newton MA; *Margaret E. Gates*, Manhattan Library, Manhattan KS; *Nancy E. Gaudette*, Public Library, Worcester MA; *Eleanor M. Gehres*, Public Library, Denver CO; *Ann Geisel*, Peterborough Library, Peterborough NH; *Kathleen R. Georg*, Gettysburg National Military Park, Gettysburg PA; *Shirley George*, Public Library, Maywood IL; *Virginia Gerhardstein*, Public Library, Mansfield OH; *Isabel W. Gerty*, Ralph Waldo Emerson House, Concord MA; *C. Catherine Gibson*, Indianapolis-Marion County Public Library, Indianapolis IN; *Frank Gibson*, Public Library, Omaha NE; *Elsie Gilbert*, City Library, Frankfort MI; *Marcia Giles*, Public Library, Southington CT; *Harold B. Gill*, Colonial Williamsburg Foundation, Williamsburg VA; Gilroy Historical Museum, Gilroy CA; *Christine Gilson*, Public Library, Lincoln IL; *Allen Ginsberg*; *Mildred T. Giusti*, Public Library, Providence RI; *Ruth E. Godown*, Holland Township Free Public Library, Milford NJ; *Oswald H. Goering*, Lorado Taft Campus, Oregon IL; *Herbert Gold*; *Harry Golden*; *Ruth Golden*, Public Library, Ponca NE; *Magdaline Goodrich*, Town of Southold Historian, Southold NY; *Elizabeth H. Goodwin*, South Berwick ME; *Jayne Gordon*, Orchard House, Concord MA; *William Goyen*; *Linda Graf*, Flagstaff City–Conconino County Public Library, Flagstaff AZ; *Elizabeth E. Graham*, Vachel Lindsay Association, Springfield IL; *Maria Graham*, New Canaan Library, New Canaan CT; *William Granger*, Toledo–Lucas County Public Library, Toledo OH; *Shirley Ann Grau*; *Beth Gray*, Free Library, East Hampton NY; *Dwight E. Gray*, Cosmos Club, Washington DC; *Walter Gray*, Polk County Public Library, Columbus NC; *Diana B. Green*, Jack London Village, Oakland CA; *James Green*, Regenstein Library, University of Chicago, Chicago IL; *Kathleen B. Greenn*, Berkshire County Historical Society, Pittsfield MA; *Paul Green*; *Thomas A. Green*, Public Library, Austin MN; *Stanley Greenberg*, Forbes Library, Northampton MA; *Margaret Greene*, Pettee Memorial Library, Wilmington VT; *Monica Greening*, Public Library, Monrovia CA; Greensboro Free Library, Greensboro VT; Greensboro Library, Greensboro NC; *Thomas B. Greenslade*, Chalmers Memorial Library, Kenyon College, Gambier OH; *Joanne Greenspun*, Public Library, Vineland NJ; Greenville Green Library, Greenville TN; Greenwood–Leflore Library, Greenwood MS; *Hilary Greimann*, Hancock County Historical Society, Britt IA; *James M. Greiner*, Herkimer NY; *Loren Grey*, Northridge CA; Griffith Memorial Library, Danby VT; *Mrs. Ronald Grimes*, Como MS; *Susan Grotyohann*, Public Library, New Brunswick NJ; *Barbara Grosby*, Lithgow Public Library, Augusta ME; *Jean Gross*, Public Library, Vineyard Haven MA; *John Gross*, Salinas Public Library–John Steinbeck Library, Salinas CA; *Raymond E. Gross*, The Courier-Gazette, Rockland ME; *Ruth Gross*, Public Library, Covina CA; Grundy Center Public Library, Grundy IA; *J. Owen Grundy*, Jersey City NJ; *Edgar A. Guest, Jr.*, Troy MI; *C. L. Guild*, Naval Support Facility, Thurmont MD; James Gullickson, Sleepy Hollow Restorations, Tarrytown NY; *June Gustafson*, Springfield–Greene County Library, Springfield MO; Gwinnett Historical Society Inc., Lawrenceville GA;

Bridgie Hackstaff, Public Library, Crawfordsville IN; *Daniel W. Hagelin*, Public Library, Lakewood OH; *Dorothy Hagerman*, Public Library, Galesburg IL; *Richard L. Hagy*, Filson Club, Louisville KY; *David Hahn*, Putnam Publishing Group, New York NY; *Donald Hall*; *Edward Hall*, Public Library of Annapolis & Ann Arundel County, Annapolis MD; *Louise D. Hall*, Public Library, Portland ME; *Rose Hall*, Margaret Mitchell Library, Fayetteville GA; *William H. Hall*, West Hartford Chamber of Commerce, West Hartford CT; *Albert Halper*; *Margaret Halsey*; *Virginia C. Halstead*, Public Library, Stamford CT; *Collin B. Hamer, Jr.*, Public Library, New Orleans LA; *Dwight L. Hamilton*, Rocky Mountain National Park, Estes Park CO; *Marie Hamilton*, Public Library, McLeansboro IL; *Norman Hamilton*, Ellsworth NE; *Rita Hamm*, Parks & Recreation Department, Crawfordsville IN; *Rosemary Hammond*, Boyle County Library, Danville KY; *Elizabeth Hampsten*, University of North Dakota, Grand Forks ND; Hampton Library, Bridgehampton NY; *Rev. Robert W. Haney*, First and Second Church in Boston, Boston MA; *Stanley Haney*, Public Library, Westborough MA; *Patricia Hanna*, Howland Circulating Library, Beacon NY; *William Hanna*, Mississippi Department of Archives and History, Jackson MS; *Olive W. Hannaford*, Berry Memorial Library, Bar Mills ME; Hannibal Library, Hannibal MO; *Charlene Hansen*, Linn County Heritage Society, Cedar Rapids IA; *Constance H. Hansen*, Cross and Sword, St. Augustine FL; *Mrs. Ted Hansen*, Public Library, Anita IA; *Ann Harder*, Public Library, Santa Ana CA; *David J. Hardgrove*, Library of the Chathams, Chatham NJ; *Jonathan P. Harding*, Library of the Boston Athenaeum; *Walter Harding*, The Thoreau Society, Inc., Geneseo NY; *Yvonne H. Hare*, Free Library, New City NY; *Tyrus G. Harmsen*, Occidental College Library, Los Angeles CA; Harnett County Library, Lillington NC; *Steve Harrington*, Thomas Wolfe Memorial, Asheville NC; *C. H. Harris*, Public Library, Jacksonville FL; *Lois Harris*, Vigo County Public Library, Terre Haute IN; *Amalia Harrison*, Alabama Public Library Service, Montgomery AL; Harrison Public Library, Harrison NJ; *Nancy C. Harrison*, Handley Library, Winchester VA; *Fritz Harsdorff*, Times-Picayune Publishing Corp., New Orleans LA; *Nellie A. Hart*, Public Library, Camden ME; *Ruth Hartman*, Ventura County Library, Ventura CA; *Dorothy Harvey*, Sacramento City-County Library, Sacramento CA; *Patricia Haslam*, Stowe VT; *William S. Hass*, Public Library, Westport CT; *John P. Hastings*, Public Library, Paris TX; *Anna R. Hathaway*, Shaw Memorial Library, Plainfield MA; *Rebecca Sue Hatton*, Public Library, Newton KS; *Kathryn Hauck*, Kingsport TN; *Connie M. Haun*, Davy Crockett Tavern Museum, Morristown TN; *Ruth M. Hauser*, Honnold Library, Claremont CA; *Lois Hausman*, Public Library, Grand Rapids MI; *Allan Hauth*, Jackson Metro Library, Jackson MS; *Walter Havighurst*; *Mrs. Frank Hawkins*, Mark Twain Library, Redding CT; *Donna J. Hawthorne*, Taylor Memorial Library, Hampton VA; Hawthorne Community Association, South Casco ME; *Julie Haxton*, University of Oklahoma Library, Norman OK; *Dennis Hayes*, Public Library, Jersey City NJ; *Joseph Hayes*; *Richard Hayes*, Pratt Memorial Library, Cohasset MA; *William F. Hayes*, Public Library & Info Center, Boise ID; *James Hazel*, Forbes Library, Northampton MA; *Howell J. Heaney*, Free Library of Philadelphia, Philadelphia PA; *Mary Glenn Hearne*, Public Library of Nashville and Davidson County, Nashville TN; *Jeanne Heet*, Public Library, Hamilton OH; *Harry Heglund*, Brainerd Community College Library, Brainerd MN; *Davis Heighton*, Grass Valley Branch Library, Grass Valley CA; *Jacqueline L. Heintz*, City of Angels, Angels Camp CA; *Madelyn Helling*, Nevada County Library, Nevada City CA; *Robert Hemmingson*, Public Library, Fergus Falls MN; *Howard A. Henderson*, Matawan Historical Society, Matawan NJ; Hendersonville County Library, Hendersonville NC; *Alice Hendricks*, Auburn-Placer Library, Auburn CA; *Roberta L. Hendrix*, Cazenovia NY; *Ellen W. Henry*, Thomas Balch Library, Leesburg VA; *Cynthia Herman*, Gene Stratton-Porter Cabin, Geneva IN; *Penny Herren*, Public Library, Platteville CO; *Jack W. Herring*, Armstrong Browning Library, Baylor University, Waco TX; *Ruth Herschberger*; *John Hersey*; *Frank Herzog*, Arden DE; *Carolyn Hesketh*, Curtis Memorial Library, Brunswick ME; *Charlotte G. Hewson*, Hawthorne Community Association, South Casco ME; *Gerald R. Hickey*, Public Library, Belleville NJ; *James M. Hicks*, Central Arkansas Library System, Little Rock AR; *Shirley Higginbotham*, Minnesota Valley Regional Library, Mankato MN; *Hannah Hilb*, Burnham Library, Bridgewater CT; *Margaret Hileman*, Public Library, Bozeman MT; *Margaret F. Hinckley*, Public Library, Winthrop MA; *James Hiner*, Minnesota Historical Society, Little Falls MN; Historic Augusta, Inc., Augusta GA; Historic New Orleans Collection, New Orleans LA; Historical Research Cooperative, Leadville CO; Historical Resource Center, Pierre SD; *Edward Hoagland*; *Marcia Hodges*, Brunswick-Glynn County Regional Library, Sea Island GA; *Elliott W. Hoffman*, Noah Webster Foundation, West Hartford CT; *Lydia M. Hoffmann*, Public Library, Hamburg NY; *Nancy Hoffman*, Rollins College, Winter Park FL; *Elizabeth C. Hoke*, Montgomery County Department of Public Libraries, Rockville MD; *W. Kenneth Holditch*, Heritage Tours, New Orleans LA; *John Hollander*; *Pat Hollander*, Public Library, Catskill NY; *Mary P. Holley*, Jefferson Parish Recreation Department, Library, Metaire LA; *Gabrielle L. Holmes*, City Library, Sacramento CA; *John Clellon Holmes*; *Gretchen Holzhauer*, Public Library, Tenafly NJ; *Phoebe Homans*, Free Library, Wayland MA; *Justine L. Hommel*, Free Library, Haines Falls NY; Hood River County Library, Hood River OR; *Helen Hopkins*, Holly Springs MS; *June T. Hopper*, Family Memorial Library, Wrens GA; *Paul Horgan*; *Ann Hornak*, Public Library, Houston TX; *Helen Horstman*, Jennings County Public Library, North Vernon IN; *Henry Beetle Hough*; *Janet R. Houghton*, Woodstock Historical Society, Woodstock VT; *Milo B. Howard, Jr.*, Alabama Department of Archives and History Library, Montgomery AL; *Harriet S. Howe*, Public Library, Sanibel FL; *Barbara Howes*; *John N. M. Howells*, Kittery Point ME; *Jeanne C. Howes*, Melville Society, Georgetown CT; *Lindi Hubbard*, Public Library, Greenville ME; *A. C. Hucksoll*, Greenleaf Memorial Library, Franconia NH; *Corrine P. Hudgins*, Historic Landmarks Commission, Richmond VA; *Michelle Hudson*, Mississippi Department of Archives, Jackson MS; *Elisabeth Huey*, Parish Library, Natchitoches LA; *Carol Huge*, Perry County District Library, New Lexington OH; *Edwin J. Hughes*, Public Library, Oxnard CA; *Mrs. William A. Hughes*, Fort Scott KS; *Carolyn A. Humphries*, Historic Augusta, Inc., Augusta GA; *Charles Hunsberger*, Clark County Library District, Las Vegas NV; *Julie Hunter*, Public Library, Atlanta GA; *Thaddeus B. Hurd*, Clyde Heritage League, Inc., Clyde OH; *Marie V. Hurley*, Public Library, Stamford CT; Huron Library, Huron SD; *Katherine Hurrey*, St. Mary's County Memorial Library, Leonardtown MD; *Louise Hutchison*, Greene County District Library, Xenia OH; *Charlotte M. Hutton*, Lane Memorial Library, Hampton NH; *Aileen A. Hwang*, Bound Brook Memorial Library, Bound Brook NJ;

Idaho Historical Society, Boise ID; Illinois State Historical Library, Springfield IL; Ilsley Library, Middlebury VT; Indiana Library, Indiana PA; *Jeremy Ingalls*; *Cevera Ingraham*, Placer County Museum, Auburn CA; *Charles E. Irvin*, Bartram Trail Regional Library, Washington GA; *Jill Hudson Irvine*, Public Library, Bloomington IL; *Julieanne B. Irving*, Maine Historical Society, Portland ME; *Charles N. Irwin*, Eastern California Museum, Independence CA; *Torrey Isaac*, Chautauqua Library, Chautauqua NY; *Lane Ittelson*, New Mexico Historic Preservation Bureau, Santa Fe NM;

Jackson-George Regional Library, Pascagoula MS; Jacksonville Library, Jacksonville IL; Jacksonville Public Library, Jacksonville FL; *Mary S. Jacobs*, Public Library, Wellsville NY; *Sonia L. Jacobs*, University of Colorado Libraries, Boulder CO; *Charlotte Jacobson*, Norwegian-American Historical Association, Northfield MN; *David Jacobson*, Sinclair Lewis Foundation, Sauk Centre MN; *Ruth B. Jacobson*, Dunn County Historical Society, Menomonie WI; *Brenda James*, Cobb County Public Library System, Marietta GA; *Steve Jansen*, Douglas County Historical Society Museum, Lawrence KS; *Sandi Jarmark*, Public Library, Gardiner ME; *Susanne Javorski*, Wesleyan University, Middletown CT; *Annette R. Jenks*, Public Library, Williamstown MA; *Diane Richmond Jennings*, Franklin County Library, Rocky Mount VA; Jermain Memorial Library, Sag Harbor NY; *Faith Jernigan*, Public Library, Hartford City IN; *William Jerousek*, Public Library, Oak Park IL; *Doro-*

thy Jerse, Vigo County Historical Society, Terre Haute IN; *Ruth A. John*, Public Library, Fairmount IN; John Jay Homestead, Katonah NY; *Mary Johns*, Public Library, Kearney NE; *Carolyn Johnson*, Public Library, Ipswich MA; *Dorothy Johnson*, Mt. Pleasant Public Library, Pleasantville NY; *Gregory A. Johnson*, Alderman Library, University of Virginia, Charlottesville VA; *Joseph J. Johnson*, Public Library, Monterey CA; *Joy A. Johnson*, Community Public Library, St. Mary's OH; *Martha Riley Johnson*, Williams College Library, Williamstown MA; *Thomas B. Johnson, Jr.*, Auburn ME; *Rev. Wayne L. Johnson*, Church of the Holy Comforter, Kenilworth IL; *Helen Johnston*, Birmingham AL; *Betsy Jones*, Charleston Library Society, Charleston SC; *Daisy E. Jones*, Eagle Rock Branch Library, Los Angeles CA; *David L. Jones*, State University of New York, Alfred NY; *Deborah Theresa Jones*, Frederick Douglass Home, Washington DC; *Mrs. James Jones*, Sagaponack NY; *Madison Jones*; *Pearl R. Jones*, Baker County Library, Baker, OR; *Napoleon Jones-Henderson*, Research Institute of African and African Diaspora Arts, Roxbury MA; *Elaine Joselovitz*, The National Theatre, Washington DC; *Alice Joseph*, Public Library, Provincetown MA; *Eleanor Joseph*, East Baton Rouge Parish Library, Baton Rouge LA; *Elizabeth Joslin*, Patterson Library, Patterson NY; *Mrs. Lillis P. Joy*, Northeast Harbor Library, Mount Desert ME; *Patricia L. Joy*, Bronson Library, Waterbury CT; *Ruth D. Jungemann*, Public Library, Olivia MN;

Mrs. Averil J. Kadis, Enoch Pratt Library, Baltimore MD; Kalamazoo County Register of Deeds Office, Kalamazoo MI; *Lois Kamm*, Public Library, Mishawaka IN; *David Kantor*, Volusia County Public Libraries, Daytona Beach FL; *Dorothy Karmiller*, Public Library, Union City NJ; *Leon Karpel*, Mid-Hudson Library System, Poughkeepsie NY; *Frances Katz*, Public Library, Perth Amboy NJ; *Alfred Kazin*; *Margaret Keefe*, Public Library, Oak Park IL; *Joan Keefer*, Public Library, Huntington IN; *Gladys Kehm*, Public Library, Mason City IA; *Helen A. Kehoe*, Public Library, Paterson NJ; *D. Kelleher*, Arlington County Library, Arlington VA; *Margaret Kelleher*, Public Library, Newburyport MA; *Ethel Kelley*, Middletown RI; *Elizabeth Burroughs Kelly*, John Burroughs Memorial Association, West Park NY; *Ruth Kelly*, Public Library, Monterey CA; *Veronica D. Kelly*, Public Library, Seymour CT; *Brent L. Kendrick*, Washington DC; *Mrs. Spencer P. Kennard*, Library Association, Lenox MA; *Johanna Kennell*, Free Library, Riverhead NY; *Margaret T. Kenney*, Public Library, Hopewell NJ; Kentucky Department of Tourism, Frankfort KY; *Susan Kern*, Free Public Library, New Brunswick NJ; *Susan C. Kershner*, Pearl Buck Birthplace, Hillsboro WV; *Jo Anna Kessler*, Joint University Libraries, Nashville TN; *David B. Kesterson*, Nathaniel Hawthorne Society, Denton TX; *David B. Kier*, San Benito County Historical Society, Hollister CA; *Thomas F. Kilfoil*, Public Library, West Hartford CT; *Ruth N. Kimmel*, Free Library, West Nyack NY; Kinderhook Memorial Library, Kinderhook NY; *Errol Kindschy*, West Salem Historical Society, West Salem WI; *Dorothy T. King*, East Hampton Library, East Hampton NY; *Mary Louise King*, New Canaan Historical Society, New Canaan CT; *Patricia L. King*, Public Library, Phoenix AZ; *Russ Kingman*, Glen Ellen CA; *Sidney Kingsley*; *Zerma D. Kinkel*, Montrose County Regional District Library, Montrose CO; *Marjorie Kinney*, Public Library, Kansas City MO; *Jean Kiral*, Teton County Library, Jackson WY; Kirkland Town Library, Clinton NY; *Mary Bess Kirksey*, Public Library, Birmingham AL; *Florence Kirwin*, Pitkin County Library, Aspen CO; *Opal L. Kissinger*, Public Library, Anaheim CA; *Marvin F. Kivett*, Nebraska State Historical Society, Lincoln NE; *Paul Klammer*, Brown County Historical Society, New Ulm MN; *Jeanne Klein*, Public Library, Streator IL; *Stephen C. Klein*, Sutter County Free Library, Yuba City CA; *Jeanette Knapp*, Southern Adirondack Library System, Saratoga Springs NY; *G. R. Wilson Knight*; *James M. Knox*, Green Library, Stanford CA; *Lois R. Koehler*, Rapides Parish Library, Alexandria LA; *Mrs. Manuel Komroff*; *Kathleen L. Kosuda*, Free Library, Gloversville NY; *Geraldine Kozik*, Public Library, Chicopee MA; *Jeanine Krause*, Public Library, Beverly Hills CA; *Sylvia Kropp*, Green Free Library, Wellsboro PA; *Virginia L. Krueger*, Findlay-Hancock County Public Library, Findlay OH; *Peter C. Ku*, Howard Community College Library, Columbia MD; *Katherine Neilson Kurtz*, Landmarks Association of St. Louis, St. Louis MO; *Willa Kurtz*, Encino Tarzana Branch Library, Tarzana CA; *Judith Kuzel*, Public Library, Aurora IL;

La Crosse Library, La Crosse WI; *W. E. R. LaFarge*, Saunderstown RI; Elena Lagorio, Carmel CA; Lake Mills Library, Lake Mills IA; *Claire Lambert*, Jesup Memorial Library, Bar Harbor ME; *Pattie Lambert*, Braswell Memorial Library, Rocky Mount NC; *Corliss Lamont*; *Ruth R. Lampel*, Jamestown Philomenian Library, Jamestown RI; *Carl Landrum*, Quincy IL; *Irene Arnold Lang*, New Hampshire Department of Resources, Concord NH; *Lee C. Lansing, Jr.*, Dumfries VA; *Gail P. Landy*, Dayton and Montgomery County Public Library, Dayton OH; *Stephanie Langenkamp*, Public Library, San Marcos TX; *Nancy Langston*, Public Library, Roswell NM; Lanier Library Association, Tryon NC; *Jane M. Lape*, Fort Ticonderoga Association Library, Ticonderoga NY; *Hans L. Larsen*, Ohlone College, Fremont CA; *Margaret Lasch*, Albany Institute of History and Art, Albany NY; *Shirley Brooks Laseter*, Public Library System, Prattville AL; *Norma Lasley*, Public Library, Muncie IN; *Mr. and Mrs. James Laughlin*, Norfolk CT; Laura Ingalls Wilder Memorial Society, Inc., De Smet SD; *Laure-Abel McNamee*, Historical Preservation Commission, Providence RI; Laurens County Library, Laurens SC; Lauringer Library, Georgetown University, Washington DC; *Betty Lawrence*, Pack Memorial Public Library, Asheville NC; *Jerome Lawrence*; *Patricia A. Lawrence*, Public Library, Sausalito CA; *Tom Lea*; *Margaret Leaf*, Garrett Park MD; *Marcia Learned*, Monroe County Library System, Monroe MI; *James O. Leas*, Crawfordsville IN; Lebanon Advertiser, Lebanon IL; Lebanon City Library, Lebanon OR; *Lauren K. Lee*, Cobb County Public Library System, Marietta GA; *Leola Mae Lee*, San Diego County Library, San Diego CA; *Mary Leen*, The Bostonian Society, Boston MA; *Millie Lehman*, Chamber of Commerce, Oregon City OR; *Catherine F. Leigh*, Amenia NY; *Sara Leishman*, Cambria County Historical Society, Ebensburg PA; *Susan K. Lemke*, United States Military Academy Library, West Point NY; *Margaret Leonard*, Public Library, Osage IA; *Margaret Leonhardt*, Schenectady County Public Library, Schenectady NY; *Ruth Leporsky*, Walpole Historical Society, Walpole NH; *Max Lerner*; *Camille J. Leslie*, Public Library, Bethlehem PA; *S. Lesnick*, Public Library, Jericho NY; *Ann-Ellen Lesser*, Steepletop, Austerlitz NY; *Denise Levertov*; *I. Harry Levin*, Vineland Historical and Antiquarian Society, Vineland NJ; *Stan Levine*, Capital Newspapers Group, Albany NY; *Mary Jo Levy*, City Library, Palo Alto CA; *Nancy Levy*, Public Library, Irvington NY; *Karen R. Lewis*, Harvard University Library, Cambridge MA; *Margot E. Lewis*, Public Library, Kalamazoo MI; *Rabbi Theodore Lewis*, Touro Synagogue, Newport RI; *Irene V. Lichty*, Laura Ingalls Wilder Home, Mansfield MO; *Jerry Dick Lien*, Collegedale TN; *Marina Limoges*, Friske Free Library, Claremont NH; *John Lindahl*, John G. Neihardt Center, Bancroft NE; *Robert A. Linder*, University of Mississippi Library, University MS; Litchfield Historical Society, Litchfield CT; Literature Department, Public Library, San Francisco CA; *Kay Littlefield*, Public Library, Bangor ME; Littleton Area Historical Society, Littleton NH; *Frederick Llewellyn*, Forest Lawn Museum Library, Forest Lawn CA; *Warren F. Lloyd*, Darlington County Libraries, Darlington SC; *Karen Locher*, Public Library, Victoria TX; *Josephine D. Locke*, Charlotte County Free Library, Charlotte Court House VA; *Sharon L. Loe*, Northern Illinois Library System, Rockford IL; *Beatrice M. Loennecke*, Plymouth Church of the Pilgrims, Brooklyn NY; *Donald E. Loker*, Public Library, Niagara Falls NY; Long Beach Independent/Press Telegram, Long Beach CA; *Maryalice Long*, Los Gatos Memorial Library, Los Gatos CA; *Louise Longan*, Public Library, Tulare CA; *Stephen Longstreet*; *Anita Loos*; *William H. Loos*, Buffalo & Erie County Library, Buffalo NY; *Linda Lopez*, Public Library, Corpus Christi TX; *Frank K. Lorenz*, Hamilton College Library, Clinton NY; Louisiana State Library, Baton Rouge, LA; *Ursula Burroughs Love*, John Burroughs Memorial Association, West Park NY; Lowndes County Library System, Columbus MS; *P. Lubin*, Public Library, Groton CT; *Catherine E. Lucas*, Riverside City and County Public Library, Riverside CA; *Clare Boothe Luce*; *Jane Lufkin*, Public Library, St. Paul MN; *Peggy Lugthart*, Public Library, Saginaw MI; *Lucinda Lunceford*, Ray County Library, Richmond MO; *Janel E. Lundgren*, The Vachel Lindsay Association, Springfield IL; *John Lustig*, Public Library, Monrovia CA; *Dr. John E. Lutz*, Woodchuck Lodge, Inc., Roxbury NY; *Robert W. Lyle*, Peabody Library Association, Washington DC; *Ancelin V. Lynch*, Historical Preservation Commission, Providence RI; *Mary Lynch*, Public Library, Colebrook NH; *Sr. Mary Dennis Lynch*, Rosemont College Library, Rosemont PA;

Hubert McAlexander, Jr., University of Georgia, Athens GA; *Elizabeth McAndrew*, Crandall Library, City Park, Glens Falls NY; *Edith H. McCauley*, Public Library, Portland ME; *Robert McCloskey*; *Susan H. McClure*, Public Library, Chapel Hill NC; *Carol McCormick*, Alaska Division of Tourism, Juneau AK; McCormick County Library, McCormick SC; McCoy Library, McLeansboro IL; *Charles H. McCurdy*, Grand Teton National Park, Moose WY; *Evelyn McDaniel*, Blount County Library, Maryville TN; *Christine McDonald*, Free Library, Hyde Park NY; *Edgar MacDonald*, Ellen Glasgow Society, Ashland VA; *Murray McDonald*, Public Library, Dedham MA; *Jeanne McDonnell*, Palo Alto CA; McDowell County Library, Marion NC; *Roberta McGaughran*, Williamsburg Regional Library, Williamsburg VA; *Dorothy McGinniss*, Public Library, Chilmark MA; *Beatrice McGough*, Public Library, Prattville AL; *Shirley H. McGrath*, Memorial Hall Library, Andover MA; *Marion L. McGuire*, Public Library, Rochester NY; *John J. McKay, Jr.*, Middle Georgia Historical Society, Macon GA; *Jeanette McKee*, Public Library, Dedham MA; *Rev. Nelson W. MacKie*, Diocese of Rhode Island, Greenville RI; *Doug McLaughlin*, Public Library, Oxnard CA; *M. R. McPartland*, Free Library, East Greenwich RI; *Bruce D. MacPhail*, Sleepy Hollow Restorations, Tarrytown NY; *Christine McPherson*, Woonsocket Harris Public Library, Woonsocket RI; *Patricia McQueen*, Free Public Library, Summit NJ; *Angela McWhite*, National Park Service, Philadelphia PA; *Patricia McWhorter*, Historic New Orleans Collection, New Orleans LA; *Evangeline Machlin*, Professor Emerita, Boston University, The Frost Place, Franconia NH; *Elsie M. Maddaus*, Public Library, Ballston Spa NY; *Roxie Mahan*, Green County Public Library, Greensburg KY; *Donna Mailloux*, City Library, Lowell MA; *Rose Mailman*, Kellogg-Hubbard Library, Montpelier VT; Maine Historical Society, Portland ME; *John Makonie*, Florence SC; *Bernard Malamud*; Malheur County Library, Ontario OR; *Judy Malott*, Mt. Pleasant Public Library, Pleasantville NY; *Phyllis Manago*, Peabody Library, Columbia City IN; *Chuck Manley*, Washoe County Library, Reno NV; *Gordon Manning*, Oregon Historical Society, Portland OR; *Larry L. Manuel*, Wilmington Library, Wilmington DE; *Lois R. Markey*, Public Library, Concord NH; *Frieda Marion*, John Greenleaf Whittier Home, Amesbury MA; *Bonita C. Marley*, Public Library, Mooresville IN; *R. Marquardt*, Sauk City WI; *Lattie Marr*, Green County Public Library, Greensburg KY; *Margaret Marschner*, Public Library, Conway NH; *Elizabeth C. Marshall*, Henderson County Library, Hendersonville NC; *Susan Marshall*, Monroe County Public Library, Bloomington IN; *Callista Jane Martin*, King's Daughters Library, Del Norte CO; Martin Memorial Library, York PA; *Shirley A. Martin*, Mantor Library, University of Maine, Farmington ME; *Karl Martinez*, Public Library, Salem OR; *Larry B. Massie*, Western Michigan University, Kalamazoo MI; Mathews Memorial Library, Mathews TX; Mattituck Free Library, Mattituck NY; *Laurene L. Mattor*, Patten Free Library, Bath ME; *Leila Mattson*, Great Neck Library, Great Neck NY; *Charles D. Maurer*, Free Library, Rutland VT; *Patrick Max*, University of Notre Dame Libraries, Notre Dame IN; *Rod Maxwell*, Public Library, Sioux City IA; *Henrietta Mayer*, Stonington Free Library Association, Stonington CT; *Luverne C. Mead*, Madison Carnegie Public Library, Madison MN; *Joan Meador*, Tulsa City–County Library System, Tulsa OK; *Helen Medeiros*, Public Library, Taunton MA; *Roderic Meng*, Arkham House Publishers, Inc., Sauk City WI; *Josephine Mentley*, Free Library, Canton NY; *Edward* and *Genevieve Merrill*, Redtop, Belmont MA; *Nancy C. Merrill*, Public Library, Exeter NH; *Judith Michaelson*, Delaware County District Library, Delaware OH; *Michel Michelsen*, Public Library, Danbury CT; *Jean Michie*, Richards Library, Newport NH; *Robert Middleton*, Port of Oakland, Oakland CA; Middletown Free Library, Middletown RI; *Hazel Mikel*, Kendall Young Library, Webster City IA; *Norma Millay*, Steepletop, Austerlitz NY; *Ann B. Miller*, San Simeon State Historical Monument, San Simeon CA; *Doris Miller*, Sullivan County Public Library, Sullivan IN; *Edna Miller*, Public Library, Rockford AL; *Mrs. James Miller*, Westhampton Memorial Library, Westhampton MA; *Lucile Miller*, Public Library, Long Beach CA; *Madelaine Hemingway Miller*, Petoskey MI; *Nancy Miller*, Maryland Historical Trust, Annapolis MD; *Roger Miller*, Free Public Library, Roselle NJ; *Virginia Miller*, Wells Memorial Library, Lafayette IN; *Edna M. Milliken*, Kentucky Department of Library and Archives, Frankfort KY; *Marie C. Mims*, Charleston County Library, Charleston SC; Mitchell Library, Mitchell SD; *Sophy M. Mitchell*, Sebring Historical Society, Sebring FL; *Gerard Mittelstaedt*, McAllen Memorial Library, McAllen TX; *Joan Moak*, Public Library, Newburyport MA; *Louise Moe*, Public Library, Rochester MN; *Jody Moeller*, Champaign County Historical Archives–Urbana Free Library, Urbana IL; *Joan Moench*, Scranton Memorial Library, Madison CT; *Robert Molina*, City Hall, Los Angeles CA; *Gladys Mollart*, Watertown Historical Society, Watertown WI; *Genevieve Moloney*, Public Library, Marblehead MA; *Mary Molyneux*, Public Library, Bedford VA; Monmouth County Historical Association, Freehold NJ; Monroe County Library, Key West FL; *Maria Monters*, Warner Library, Tarrytown NY; Montgomery City Library, Montgomery City MO; *Gail Moon*, Middle Georgia Regional Library, Macon GA; *Brian Moore*; *Ellen Moore*, Public Library, Austin TX; *Jean B. Moore*, Public Library, Paramus NJ; *Mary Moore*, Public Library, Hawthorne FL; *Mildred W. Morgan*, Lanier Library Association, Tryon NC; *Janice Moriarty*, Burlington County Library, Bordentown NJ; *Wright Morris*; *Kenneth D. Morrison*, Mountain Lake Sanctuary, Lake Wales FL; *Wayne E. Morrison, Sr.*, Ovid NY; *Sally Morrison*, Marjorie Kinnan Rawlings Home, Hawthorne FL; *J. E. Moss*, Concord Free Library, Concord MA; *Julia Nail Moss*, Sul Ross State University Library, Alpine TX; Mt. Morris Library, Mt. Morris NY; *Archie Motley*, Chicago Historical Society, Chicago IL; *Raymond A. Mulhern*, Green County Library, Xenia OH; *Nancy A. Mullen*, Ventress Memorial Library, Marshfield MA; *Marguerite Mullennaux*, Albany Library, Albany NY; *Lewis Mumford*; *Louis Mumford*, South Bend Tribune, Niles MI; *Curtis Muncey*, Public Library, Cedar Falls IA; *Winifred F. Munger*, Napa City–County Library, Napa CA; *Ellen Mundell*, Public Library, Brookfield MA; *Ezevel Murphy*, Warren County Library, Monmouth IL; *Meta Murphy*, Hollywood Regional Branch Library, Hollywood CA; *Yolanda Murphy*, Prebie County District Library, Camden OH; Muscatine Library, Muscatine IA; Museum of the City of New York, New York NY; *Emily M. Myers*, University of Oklahoma Library, Norman OK;

Sharon Nantz, Department of Archives, Louisville KY; *N. M. Nee*, Public Library, San Francisco CA; *Ruth B. Neff*, Eleanor I. Jones Archives, Camden OH; *Hope B. Neilson*, Public Library, Kittery ME; *Cordelia M. Neitz*, Cumberland County Historical Society, Carlisle PA; *Dean E. Nelson*, Delaware Hall of Records, Dover DE; *Helen M. Nelson*, Public Library, Oceanside CA; *Marion Nelson*, City Library, Lebanon OR; *Ruth Nelson*, Birmingham Public and Jefferson County Free Library, Birmingham AL; *Howard Nemerov*; *Katherine Nesteby*, Public Library, Bloomington IL; Nevada State Historical Society, Reno NV; New Canaan Historical Society, New Canaan CT; New Castle Library, New Castle DE; New Jersey Reference Division, Public Library, Newark NJ; New Jersey State Park Service, Trenton NJ; *Helen Jean Newman*, Tourism New Brunswick, Fredericton, Canada; Newport Historical Society, Newport RI; *Janet Newton*, Livermore Heritage Guild, Livermore CA; *Sandra B. Neyman*, Dawes Memorial Library, Marietta College, Marietta OH; *Marion Nicholson*, Greenwich Library, Greenwich CT; *Mildred Nilon*, Norlin Library, Boulder CO; *Barbara Nims*, Public Library, Plymouth NH; *David S. Nivison*, Stanford University, Stanford CA; *Douglas R. Noble*, Museum of Arts and Sciences, Macon GA; *Carl R. Nold*, Gadsby's Tavern Museum, Alexandria VA; *Carl R. Nold*, New York State Historical Association, Cooperstown NY; *Donald W. Nolte*, Public Library, Livermore CA; *Dorothy W. Normandy*, Washington GA; *Barbara Norris*, Menlo Park Historical Association, Menlo Park CA; *Jane Northshield*, Croton-on-Hudson NY; *Linda N. Norton*, Public Library, Edgartown MA; *Glenna Nowell*, Public Library, Gardiner ME; *Janice Noyes*, Public Library, Chatham MA; *Ferner Nuhn*, Claremont CA; *Reidun D. Nuquist*, Vermont Historical Society, Montpelier VT;

Oberlin College Archives, Mudd Learning Center, Oberlin OH; *Barbara L. Ochmanek*, Public Library, Morristown NJ; *Elizabeth C. O'Connor, R.S.C.J.*, Manhattanville College, Purchase NY; *Bette O'Dell*, Public Library, Ann Arbor MI; *Thomas F. O'Donnell*, State University College at Brockport, Brockport NY; *Paul H. Oehser*, Cosmos Club, Washington DC; *Dorothy Offensend*, Wells VT; *Hazel Ohl*, Reuben McMillan Free Library Assoc., Youngstown OH; *Steven Olderr*, Public Library, Riverside IL; *Mary Simms Oliphant*, Greenville SC; *Arthur L. Olivas*, Museum of New Mexico, Santa Fe NM; *Helen Oliver*, Free Library, Calais ME; *Evelyn N. Olson*, Public Library, Roselle NJ; Oneida Historical Society, Utica NY; *Gladys O'Neil*, Jesup Memorial Library, Bar Harbor ME; Orford Social Library, Orford NH; *Mary Orsborn*, Public Library, San Antonio TX; *Joseph A. O'Rourke*, Rutgers University, New Brunswick NJ; Osage Chamber of Commerce, Osage IA; *Karen E. Osvald*, Georgia Historical Society, Savannah GA; Oswego County Historical Society Library, Oswego NY; *Eleanor Otten*, Public Library, Evansville IL; *Esau Owens Jr.*, Public Library, Chicago IL; *Margaret Jean Owens*, Central Library, Glendale CA; *Mary Owens*, Washington County Library System, Greenville MS; *William A. Owens*;

Phyllis Paccadolmi, Ridgefield Library and Historical Association, Ridgefield CT; *William E. Padgett*, Office of Historic Preservation, Sacramento CA; Paducah-McCracken County Tourist Commission, Paducah KY; *Thelma Paine*, Public Library, New Bedford MA; *Oloanne Dykeman Palen*, Tulare County Library System, Visalia CA; *Marilyn Paley*, Public Library, Manhasset NY; *Helen Palmer*, Pitkin County Library, Aspen CO; *Leonard J. Panaggio*, Department of Economic Development, Providence RI; *Monique Panaggio*, Preservation Society of Newport County, Newport RI; *Bonnavere Parker*, Public Library, Kingsville OH; *Dorris D. Parker*, Public Library, Blue Hill ME; Parker House, Boston MA; *Elisabeth Scollard Parlon*, Clinton NY; *Elisabeth Patton*, Public Library, New Haven CT; *Paula F. Paul*, Orangeburg County Library, Orangeburg SC; *Marjorie Paulson*, Public Library, Bridgewater MA; *Sally Thomas Pavetti*, Eugene O'Neill Theater Center, Waterford CT; *Barbara Stagg Paylor*, Rugby Restoration Association, Rugby TN; *Ann Pearson*, Public Library, Evansville IN; *Donald Pearson*, Cripple Creek Resort, Ester AK; *Mary L. Pearson*, Public Library, Manistee MI; *Beth Pelle*, Public Library, Rome NY; *Mimi Penchansky*, Queens College, Flushing NY; *Dennis W. Pendleton*, Roanoke Library, Roanoke VA; *Susan C. Penney*, Troy Library, Troy NY; *Mrs. C. J. Penniman*, Essex Literary Association, Essex CT; *Blaine Pennington*, Kingman City–Mohave County Library, Kingman AZ; *Lee Pennington*; *Helen Peppler*, Public Library, Chappaqua NY; *Thelma W. Percy*, Public Library, Mill Valley CA; *Walker Percy*; *Noel Perrin*, Dartmouth College, Hanover NH; Peregrine Press, Old Saybrook CT; *Mary Perkins*, Stewart Free Library, Corinna ME; Petersburg Chamber of Commerce, Petersburg IL; *Keith Petersen*, Latah County Historical Society, Moscow ID; *Max Petersen*, Middlebury College, Middlebury VT; *Loretta Peterson*, Public Library, Carbondale IL; *Marion Peterson*, Public Library, Palm Springs CA; *Inez Pettijohn*, Kern County Library System, Bakersfield CA; *Nancy H. Pettus*, Lawrence County Public Library, Lawrenceburg TN; *Paula F. Peyraud*, Chappaqua Library, Chappaqua NY; *Robert L. Phelps*, Saint Olaf College, Northfield MN; *Lance Phillips*, The Long Islander, Huntington NY; *Mary L. Phillips*, Charlotte and Mecklenburg County Library, Charlotte NC; *Terry Pickens*, Mesa County Public Library, Grand Junction CO; *James H. Pickering*, Public Library, Springfield OH; *Pauline Pierce*, Stockbridge Library Association, Stockbridge MA; *Billie Piercy*, Public Library, Buckeye AZ; *Robert M. Pierson*, McKeldin Library, College Park MD; *Caroline Sandoz Pifer*, Gordon NE; Pigeon Cove Chamber of Commerce, S. Rockport MA; *Thelma Pikett*, Town of Aurora Deputy Historian, Aurora NY; *John Pillsbury*, Robert Frost Foundation, Plymouth NH; *Joyce Y. Pinney*, Public Library, Pasadena CA; *William B. Pinney*, Agency of Development and Community Affairs, Montpelier VT; *Frank Piontek*, Public Library, Beverly Hills CA; Plainfield Public Library, Plainfield NJ; *Dianne Platner*, Searls Historical Library, Nevada City CA; *Steven Plattner*, Cincinnati Historical Society, Cincinnati OH; Pleasant Valley Free Library, Pleasant Valley NY; *Linda J. Plummer*, Salinas Public Library–John Steinbeck Library, Salinas CA; *Gregory J. Plunges*, Monmouth County Historical Society, Freehold NJ; *Michael Plunkett*, Alderman Library, University of Virginia, Charlottesville VA; *Catherine Pollari*, Carrie Rich Memorial Library, Buies Creek NC; *David J. Ponciera*, Adams Library, Chelmsford MA; *Phyllis Pope*, Monroe County Library, Key West FL; *Annis T. Popoff*, Public Library, Glen Ridge NJ; *Paul Porter*; *Lucile Portlock*, Public Library, Norfolk VA; *Dorothy Potter*, Public Library, Jellico TN; *Wanda Potts*, Public Library, Mooresville IN; *Jerry Powell*, Atlanta GA; *Murella Hebert Powell*, Public Library, Biloxi MS; *Robert D. Powell*, Theodore Roosevelt National Park, Medora ND; *Bea Powers*, San Diego County Library, San Diego CA; *James Farl Powers*; *Luella Preminger*, Yellowstone Valley Prints, Billings MT; *Peggy Prescott*, Historical Society, Princeton NJ; *Dickson J. Preston*, Easton MD; *Douglas M. Preston*, Oneida Historical Society, Utica NY; *Mary A. Prien*, Historical Museum, Gilroy CA; *Margaret Priest*, Public Library, Jaffrey NH; *Mary Lou Prieto*, Public Library, Oradell NJ; *Esther Primus*, Public Library, Steamboat Rock IA; *Dolores Pritchard*, Public Library, Corvallis OR; *Mary Proper*, Public Library, Nyack NY; Public Library of Columbus and Franklin County,

Columbus OH; *Craig Pugsley*, Custer State Park, Hermosa SD; *Cheryl A. Pula*, Mid-York Library System, Utica NY; *Verna Pungitore*, Public Library, Plattsburgh NY; *Virginia H. Putnam*, Bridge Memorial Library, Walpole NH;

Liz Quinn, Public Library, Aberdeen SD;

Marita Rack, Free Library, Pawling NY; *Faye L. Raisley*, Public Library, Mars PA; *Louise Ralston*, Public Library, Chagrin Falls OH; *Frances B. Randall*, Peacham Library, Peacham VT; *Samson Raphaelson*; *Ellen Raring*, Public Library, Pottsville PA; *Jane Ratner*, Willard Library, Battle Creek MI; *Elizabeth G. Rawlings*, Committee for Alna History, Alna ME; *Linda Rea*, Public Library, Hastings NE; *George A. Reaves*, Shiloh National Military Park, Shiloh TN; Red Hook Library, Red Hook NY; *Elizabeth A. Reeb*, National Road–Zane Grey Museum, Norwich OH; *Alice Reed*, Nevins Memorial Library, Methuen MA; *Gertrude Reed*, Bryn Mawr College Library, Bryn Mawr PA; *Gary Fuller Reese*, Public Library, Tacoma WA; *Donald G. Reeves*, Public Library, Barrington RI; *Ellie Reichlin*, Society for Preservation of New England Antiquities, Boston MA; *Ruth S. Reid*, Historical Society of Western Pennsylvania, Pittsburgh PA; *Louis Reitz*, St. Mary's Seminary Library, Cantonsville MD; *Larry Remele*, State Historical Society, Bismarck ND; *Virginia J. Renner*, Huntington Library, San Marino CA; *Nelson Lance Reppert*, Public Library, Lawrence KS; *Ada Lois Resseau*, Washington Memorial Library, Macon GA; *M. L. Reynolds*, Frederick County Public Library, Frederick MD; *Kenneth Rexroth*; *Dolores Rhodes*, Public Library, Uniontown PA; *Flora B. Rhodes*, Sanisfield MA; *Rosemarie Rice*, Tompkins County Public Library, Ithaca NY; *Buddy Rich*, Savannah GA; *Jane Richards*, Cary Memorial Library Association, Wayne ME; *Donald S. Richardson*, Public Library, Seminole OK; Richland County Library, Columbia SC; *Marcia Richmond*, Kentucky Heritage Commission, Frankfort KY; Richmond Public Schools Humanities Center, Richmond VA; *Harvena Richter*; Ridgway Library, Ridgway PA; *Eileen Riebe*, Public Library, Girard KS; *Lucy Riegel*, Public Library, Noblesville IN; *Donald E. Riggs*, State University Library, Tempe AZ; *Sally Ripatti*, Public Library, Knoxville TN; *Ellen Robbins*, Public Library, Littleton NH; *Alice Roberts*, Wayne County Public Library, Wooster OH; *David L. Roberts*, The Medicine Bow Post, Medicine Bow WY; *Ellin K. Roberts*, The Woodstock Library, Woodstock NY; *Rubie Roberts*, Field Memorial Library, Conway MA; *William M. Roberts*, Bancroft Library, Berkeley CA; *Evelyn Robertson*, Public Library, Glendale CA; *Jack Robertson*, Hutchinson Library, Hutchinson KS; *Linda Robertson*, Public Library, Wabash IN; *Alfreda Robinson*, McKinley Memorial Library, Niles OH; *Charles Robinson*, Public Library, Towson MD; *Mildred Robinson*, Public Library, Piggott AR; *Sheila Robinson*, McLean County Historical Society, Coleharbor ND; Robinson Memorial Association, Ferrisburgh VT; *Dorothy Rock*, Public Library, Sterling IL; *Arlene R. Rockwood*, Public Library, Butler MO; *Janis Rodman*, Darien Library, Darien CT; *Frank and Rita E. Rogers*, Pendleton IN; *Margaret N. Rogers*, Pugh Library, Senatobia MS; *Samuel B. Rogers*, Convention and Visitors Bureau, Philadelphia PA; *Elizabeth Roland*, Sawyer Free Library, Gloucester MA; *Orlando Romero*, New Mexico State Library, Santa Fe NM; *Amy Roney*, Public Library, Greenfield IN; *Sigmund Roos*, Cambridge Historical Commission, Cambridge MA; *Jean Rosborough*, Public Library, Vestal NY; *A. G. Rose*, Edgar Allan Poe Society, Baltimore MD; *Elizabeth Rose*, Bapst Library, Chestnut Hill MA; *Margaret Rose*, Public Library, Corpus Christi TX; Rosenberg Library, Galveston TX; *Lois Rosenberger*, Public Library, Vevay IN; *Linda Rosenbluth*, Mt. Pleasant Public Library, Pleasantville NY; *Alan Rosenus*, Orion Press, Eugene OR; *Delanie M. Ross*, Memphis-Shelby County Library, Memphis TN; *William B. Rotch*, Milford Cabinet, Milford NH; *Harold L. Roth*, Bryant Library, Roslyn NY; *Philip Roth*; *Carol Ann Rott*, Public Library, Tucson AZ; *Sister Marie Rousek*, College of Saint Elizabeth, Mahoney Library, Convent Station NJ; *Dennis Rowley*, Lee Library, Brigham Young University, Provo UT; *Nancy M. Rubery*, Free Library, Palmyra NY; *Laura Ruble*, District Library, Lancaster OH; *Gerald A. Rudolph*, University of Nebraska, Lincoln NE; *Joseph A. Ruef*, Public Library, Windsor CT; *Virginia Rule*, Ormsby Public Library, Carson City NV; *Norman Runnion*, Brattleboro Daily Reformer, Brattleboro VT; *Marjorie Runyan*, Public Library, Aurora IN; *Patricia A. Rupp*, Public Library, Hudson MI; *Virginia Rust*, Huntington Library, San Marino CA; *Dan Rylance*, University of North Dakota Library, Grand Forks ND;

David Sabsay, Sonoma County Library, Santa Rosa CA; Sacramento Library, Sacramento CA; *Graham H. Sadler*, Denver Public Library, Denver CO; St. Albans Library, St. Albans VT; *Barton St. Armand*, Brown University, Providence RI; St. John's Seminary Library, Brighton MA; *Anne A. Salter*, Historical Society, Atlanta GA; *Sylvia Saltzman*, Peekskill Library, Peekskill NY; *Howard K. Samuelson*, Public Library, Santa Ana CA; *Dorothy C. Sanborn*, Auburn-Placer County Library, Auburn CA; Carl Sandburg Birthplace, Galesburg IL; San Diego Historical Society, San Diego CA; *Edith G. Sanders*, Public Library, Atlanta GA; Sandisfield Free Public Library, Sandisfield MA; *William W. Sannwald*, Ventura County Library Service Agency, Ventura CA; Santa Fe Library, Santa Fe NM; *William Saroyan*; *May Sarton*; *Charlyne Saunders*, Stowell Free Library, Cornish NH; *Robert Saunter*, Public Library, Springfield OH; *Paige Adams Savery*, The Stowe-Day Foundation, Hartford CT; *Joanne M. Sawyer*, Teachout-Price Memorial Library, Hiram College, Hiram OH; *Howard Scammon*, Williamsburg VA; *Jeffrey B. Scanlan*, Greenwich Library, Greenwich CT; *Kathryn W. Scarich*, Public Library, Bethel CT; *Susan Schade*, Noyes Library, Old Lyme CT; *Anne K. Schaller*, Natick Historical Society, South Natick MA; *M. Patricia Schapp*, County of Livingston Historian, Geneseo NY; *Dore Schary*; *Fran Schell*, Tennessee State Library and Archives, Nashville TN; *Leslie Scherer*, Public Library, Wallingford CT; *Bernard Schermetzler*, University of Wisconsin Library, Madison WI; *Vic Schliebs*, Nevada Historical Society, Reno NV; *Catherine Schmiesing*, Sinclair Lewis Foundation, Sauk Centre MN; *John Schmuck*, Memorial Library, Los Gatos CA; *Gail Schneider*, Staten Island Institute of Arts and Sciences, Staten Island NY; *Ralph W. Schneider*, Chicago Public Library, Chicago IL; *Kathleen S. Schoene*, Missouri Historical Society, St. Louis MO; *Merrill Schrader*, Public Library, Colfax IA; *Ruth Schubert*, Public Library, Hingham MA; *Frances Schuh*, Public Library, Kentland IN; *Charles R. Schultz*, Texas A&M University Library, College Station TX; *Leila Schultz*, Public Library, Elyria OH; Schuylkill Haven Library, Schuylkill PA; *Sherry Schwabacher*, Hancock County Library, Bay St. Louis MS; *Carolyn Schwartz*, Public Library, North Caldwell NJ; *Muriel Scoles*, Public Library, Long Branch NJ; *Ann M. Scott*, San Mateo County Library, Belmont CA; *David H. Scott*, Blue Hill ME; *Joan S. Scott*, Public Library, Walden NY; *Pattie J. Scott*, Richmond Public Library, Richmond VA; *Robert Kirk Scott*, Bayou Folk Museum, Cloutierville LA; *Thomas L. Scott*, Central Florida Regional Library, Ocala FL; Scottsdale Library, Scottsdale AZ; *Nancy H. Seamans*, Roanoke County Public Library, Roanoke VA; *George Seldes*; *Patricia Selig*, County Public Library, Madison IN; Selinsgrove Community Center Library, Selinsgrove PA; *Ruth Serber*, Falmouth Memorial Library, Portland ME; *Steven M. Seven*, Abraham Lincoln Birthplace, Hodgenville KY; Sewickley Library, Sewickley PA; *W. N. Seymour*, Lime Rock Press, Salisbury CT; *J. A. Shade*, Pearl S. Buck Foundation, Perkasie PA; *Ellen Shaffer*, Silverado Museum, St. Helena CA; *Mona Shaffer*, County Public Library, Yreka CA; *Susan Shaner*, Hawaii State Archives, Honolulu HI; *Jeanne Sharp*, Free Library, Newburgh NY; *Irwin Shaw*; Shawnee Carnegie Library, Shawnee OK; *Kay Shearier*, Appleton Library, Appleton WI; *Mrs. Raymond Sheldon*, McLean County Historical Society, Washburn ND; *Robert Sheldon*, University of Nebraska, Lincoln NE; *Charles Sheller*, Eldora IA; *Barbara Shelton*, Public Library, Wiscasset ME; *Milo Shepard*, Irving Shepard Trust, Glen Ellen CA; Shepard-Pruden Memorial Library, Edenton NC; *Frances Sherman*, Public Library, Oceanside CA; *Earle G. Shettleworth, Jr.*, Maine Historic Preservation Commission, Augusta ME; *Anna M. Shewbridge*, Old Charles Town Library, Charles Town WV; *Shirley Shisler*, Public Library, Des Moines IA; *Rayma Shrader*, Lincoln City Libraries, Lincoln NE; *M. L. Shuban*, Public Library, Chillicothe OH; *Saundra Shuler*, Metropolitan Library System, Oklahoma City OK; *Helen Silver*, Public Library, Long Beach CA; Silverado Museum, St. Helena CA; Silver Lake NH Chamber of Commerce; *Rebecca H. Siman*, County Historical Society, Burlington NJ; *Georges Simenon*; *Robert Simmons*, Fletcher Library, Westford MA; *Mrs. Clark M. Simms*, Gunn Memorial Library, Washington CT; *Doris Simon*, Public Library, Paducah KY; *Samuel Simons*, Memorial Hall Library, Andover MA; *Wilma Simonsen*, Public Library, Eugene OR; *Ethel C. Simpson*, University of Arkansas Libraries, Fayetteville AR; *William S. Simpson, Jr.*, Public Library, Richmond VA; *Isaac Bashevis Singer*; *Florence Sinsheimer*, Public Library, Scarsdale NY; *L. Sizoo*, Eugene O'Neill Foundation, Danville CA; *Warren Skidmore*, Akron-Summit County Public Library, Akron OH; *Catherine Slaughter*, Regional Library, Newberry SC; *Frank G. Slaughter*, M.D.; *Mrs. Lark Slone*, County Library, Hindman KY; *Bernice Slote*, University of Nebraska, Lincoln NE; *Anita Smith*, Montgomery County Department of History and Archives, Fonda NY; *Jeanette Smith*, Public Library, Duluth MN; *Kathryn Smith*, Public Library, Fullerton CA; *Luetta Smith*, Public Library, Palos Park IL; *Michael D. Smith*, Abraham Lincoln Birthplace, Hodgenville KY; *Nelle Smith*, Lake Park FL; *Rebecca A. Smith*, Historical Association of Southern Florida, Miami FL; *Thomas A. Smith*, Rutherford B. Hayes Library, Fremont OH; *Sherry Smith-Gonzales*, Museum of New Mexico, Santa Fe NM; *Lottie Smolenski*, Public Library, Hillside NJ; *Mrs. Burrill D. Snell*, Hubbard Free Library, Hallowell ME; *W. D. Snodgrass*; *Laurence L. Snook, Jr.*, Burns Library, Jacksonville FL; *C. L. Sonnichsen*, Arizona Historical Society, Tucson AZ; *Scott Sorensen*, Sioux City Public Museum, Sioux City IA; *Gretchen Sorin*, Gadsby Tavern Museum, Alexandria VA; *Mrs. E. Soschin*, Charles Scribner's Sons, New York NY; *Thomas Spencer*, Notre Dame IN; *Milda Sperauskas*, Free Public Library, Mahaw NJ; *Jacqueline Spillane*, Memorial Library, Los Gatos CA; *Robert E. Spiller*; *Dian Spitler*, Public Library, Atlantic City NJ; *Marion Spratt*, Historical Society, LeRoy IL; *Martha Kelsey Squires*, Chemung County Historical Society, Elmira NY; *Eileen Stafiej*, Public Library, Fall River MA; *Mrs. Poyntell C. Staley*, Whitehall Committee, Middletown RI; *Robert W. Stallman*; *Edward W. Stanley*, Clinton NY; *Elsie Stanley*, Public Library, Norridgewock ME; *Pamela J. Stanovich*, Chamber of Commerce, Biloxi MS; *Patricia Bodak Stark*, Yale University Library, New Haven CT; *Mrs. Bill Starnes*, Madisonville TN; *Mae Stanton*, Public Library, Avon-by-the-Sea NJ; *Patricia Stasikelis*, Public Library, Zanesville OH; *Dorothy S. Stavrides*, Public Library, Phoenixville PA; *Mary Stearns*, Public Library, Bethlehem PA; *Louise Stedronsky*, Norfolk Library, Norfolk CT; *Marian R. Steele*, Historical Association of South Jefferson, Adams NY; *Marlene A. Steele*, Willard Library, Battle Creek MI; *Wallace Stegner*; *Michael Steinfeld*, Public Library, Mount Kisco NY; *Kay Stenten*, Lincoln City Libraries, Lincoln NE; *Jerome Stephens*, Public Library, Warren OH; *Lotta R. Stern*, Public Library, Glencoe IL; *Philip Van Doren Stern*; *Christine Sterner*, Public Library, Steubenville OH; *Elaine G. Stetson*, Noah Webster Foundation, West Hartford CT; *Sally W. Stevens*, Town of Ellisburg Historian, Ellisburg NY; *Vaun Stevens*, City-County Library of Missoula, Missoula MT; *Edward M. Stevenson*, House of The Seven Gables, Salem MA; *Elisabeth Steves*, Public Library, Sandwich MA; *Hazel K. Stiebeling*, Luther Place Memorial Church, Washington DC; *William A. Stigler*, Beauvoir, Biloxi MS; *James Still*; *Maye Stills*, County Library, Ontario OR; *Bette Stillwell*, Waioli Tea Room, Honolulu HI; *Jan Stilson*, Northern Illinois University, Oregon IL; *Henry P. Stimpson*, Public Library, Concord NH; *Harriet C. Stoddard*, Blackburn College, Carlinville IL; *Mrs. Walter Stokes*, St. Davids PA; *Mary Stolz*; *Grace Zaring Stone*; *Irving Stone*; *Marjorie Stone*, Public Library, Edgartown MA; *John W. P. Storck*, Public Library, Martins Ferry OH; *Virginia Storey*, Chattahoochee Valley Regional Library, Columbus GA; *Shirley Stowe*, Haverford College Library, Haverford PA; *Donna Strachan*, Merrill Library, Machias ME; *June R. Stratton*, Public Library, South Bend IN; *Bertie M. Strawn*, Kyle TX; *George M. Street*, University of Mississippi, University MS; *Caroline Stride*, Museum of the Concord Antiquarian Society, Concord MA; *Rodger E. Stroup*, Historic Columbia Foundation, Columbia SC; *Oneita Strubinger*, Public Library, Barry IL; *Jewell Stuckert*, Village Library, Katonah NY; *William W. Sturm*, Oakland Library, Oakland CA; *William Styron*; *Edwin Suderow*, Chicago Public Library, Chicago IL; *Audrey Sugarman*, City Library, Santa Clara CA; *Wayne L. Suggs*, Public Library, Albany OR; *Sam A. Suhler*, County Free Library, Fresno CA; *Toby Sulenski*, Las Vegas Library, Las Vegas NV; *Catherine Sullivan*, Montclair NJ; *Sean Sullivan*, Public Library, Waterford NY; *Elizabeth A. Swaim*, Olin Library, Middletown CT; *Marx Swanholm*, Minnesota Historical Society, St. Paul MN; *Clara W. Swann*, County Library, Chattanooga TN; *Cynthia Swanson*, Contra Costa County Library, Berkeley CA; *Beatrice Sweeney*, Office of the City Historian, Saratoga Springs NY; *Margie Sweeney*, Gene Stratton-Porter Cabin, Rome City IN; *Henry Sweets*, Mark Twain Home & Museum, Hannibal MO; *Oliver Swift*, Public Library, White Plains NY; *Sara Carter Swinney*, Sutter County Free Library, Yuba City CA; Syracuse Library, Syracuse NY;

Mae Talbott, Washington County Free Library, Hagerstown MD; *Olive A. Tamborelle*, Public Library, Teaneck NJ; *H. S. Tanner*, Shelburne Falls MA; *Mrs. Homer Tanner*, Tyler Memorial Library, Charlemont MA; *Sheldon Tarakan*, Public Library, Port Washington NY; *Marion Taub*, Saratoga Springs Library, Saratoga Springs NY; *B. Taylor*, Princeton University Library, Princeton NJ; *Bobbie Taylor*, Monroe County Public Library, Bloomington IN; *Maureen Taylor*, Rhode Island Historical Society, Providence RI; *Robert Lewis Taylor*; *Warren E. Taylor*, Public Library, Topeka KS; *William Tema*, Altadena Library District, Altadena CA; *Phyllis Terra*, Public Library, San Jose CA; *Mary Anne Terstegge*, Tulare County Free Library System, Visalia CA; Texas Historical Commission, Austin TX; *Shirley Thayer*, Maine State Library, Augusta ME; The Dalles City–Wasco County Library, The Dalles OR; *Dorothy D. Thews*, Minneapolis Public Library, Minneapolis MN; *Bernice Thibault*, Public Library, Highgate Center VT; *Brother Harold Thibodeau*, Abbey of Gethsemani, Trappist KY; *Paul Thigpen*, CEL Regional Library, Savannah GA; *Dorothy Thomas*, Community Library, Ketchum ID; *Kenneth H. Thomas, Jr.*, Department of Natural Resources, Atlanta GA; *Pauline G. Thompson*, Public Library, Goshen IN; *Rita M. Thormeyer*, Edward Bellamy Memorial Association, Chicopee MA; *Jeanie Thornblom*, Horgan Library, Roswell NM; *Ron Thomson*, Tuskegee Institute National Historic Site, Tuskegee AL; *Elizabeth Thornton*, Cochise Fine Arts, Bisbee AZ; *Nancy Thorson*, Public Library, Red Wing MN; Thurmont Public Library, Thurmont MD; *Viola Tideman*, Public Library, Wichita KS; *Marilyn Tidlow*, Spalding Memorial Library, Athens PA; *Mrs. Freddie Tiemeyer*, Memorial Library, Clifton KS; *Donald Tillett*, Public Library, Keokuk IA; *Natalie Tinkham*, Public Library, Madison WI; *Nancy E. Titcomb*, Thornton W. Burgess Society, Sandwich MA; Tompkins County Library, Ithaca NY; *William H. Toner*, Maine Historical Society, Portland ME; Topeka Library, Topeka KS; *Jean Ann Towle*, Public Library, Tamaqua PA; *Martha Townsend*, Public Library, Santa Monica CA; *Herb Traub*, Pirates' House, Savannah GA; *Robert Trautwein*, Lincoln City Libraries, Lincoln NE; *Margaret Trevathan*, Public Library, Murray KY; *Tom Trice*, Napa City-County Library, Napa CA; *Kathleen Trimble*, Toledo Blade, Toledo OH; *Barbara M. Trott*, Witherle Memorial Library, Castine ME; *Barbara W. Tuchman*; *Louise H. Tullis*, Public Library, Rangely ME; Tulsa City-County Library, Tulsa OK; *Marie L. Tunney*, Public Library, Coronado CA; *Agnes Sligh Turnbull*; *Charles T. Turner*, Pueblo Regional Library District, Pueblo CO; *Essie Turner*, County Clerk's Office, Seminole OK; *Jean-Rae Turner*, Newark NJ; *Joan B. Turner*, Public Library, Westport CT; *Jack M. Tyler*, Public Library, Berkeley CA; *Ober Tyus*, Rollins College, Winter Park FL;

Uncle Remus Museum, Eatonton GA; Union County Library, New Albany MS; University of Wisconsin Library, Madison WI; *John Updike*; Upper Darby Township Library, Upper Darby PA; *Ruth S. Upson*, Public Library, Middlebury CT; Urbana Library, Urbana IL;

Ruth Vanderbloemen, Mann Library, Two Rivers WI; *Viola Van Loo*, Public Library, Brookville IN; *Nancy Smith Van Noppen*, Morganton NC; *James D. Van Trump*, Pittsburgh History & Landmarks Foundation, Pittsburgh PA; *Mrs. Walter Varley*, Bordentown Historical Society, Bordontown NJ; *Colleen Richards Verge*, Public Library, Petoskey MI; *Gore Vidal*; *Ray Vignovich*, Public Library, Appleton WI; Village Library, Cooperstown NY; Vinalhaven Library, Vinalhaven ME; *Raita Vilnins*, Elmhurst Public Library, Wilder Park IL; *Mary Louise Vincent*, Hiram College, Hiram OH; Vineyard Gazette, Edgartown MA; Virginia State Library, Richmond VA; *Rose D. Vitzthum*, New York State Historical Association, Cooperstown NY; *Emily Clark Vogellus*, Sandisfield MA; *Cassandra M. Volpe*, University of Colorado Libraries, Boulder CO; *Susan von Briesen*, Public Library, Portsmouth NH; *Bets Vondrasek*, Walt Whitman Birthplace, Huntington Station NY; *Elizabeth von Oettingen*, Loutit Library, Grand Haven MI;

Waco-McLennan County Library, Waco TX; *Edward Wagenknecht*; *Richard Wagner*, County Library, The Dalles OR; *Dorothy B. Walker*, Public Library, Northport NY; *Mary Faith Walker*, Belfast Free Library, Belfast ME; *Robert B. Wall*, Sacramento City Library, Sacramento CA; *Ellen Wallach*, Perrot Memorial Library, Old Greenwich CT; *W. Walmsley*, Chicago Park District, Chicago IL; *Ronald C. Walrod*, Baker County Public Library, Baker OR; *Bill Walton*, Arlington VA; *Mr. Ward*, Clinton-Essex-Franklin Library Systems, Clinton NY; *Lynn Wardwell*, Public Library, Groton CT; *Cindy Wargo*, Public Library, Evanston IL; *Robert Penn Warren*; Washington County Library, Plymouth NC; *Lucille J. Wasick*, Public Library, Wolcott CT; *Betty Watson*, Town Library, Amherst NH; *Judith Watson*, Phelps Community Memorial Library, Phelps NY; *Warren Watson*, Crane Public Library, Quincy MA; Wayne Township Department of Parks, Wayne NJ; *Marjan Wazeka*, Public Library, Eugene City OR; *Elizabeth Webber*, Beverly Historical Society, Beverly MA; *Warren Weber*, Carl Sandburg National Historic Site, Flat Rock NC; *Solan W. Weeks*, Detroit Historical Museum, Detroit MI; *Robert Wegman*, Public Library, Normal IL; *Jerome Weidman*; *Joanna Weinstock*, Public Library, Highland Falls NY; *Linda Weirather*, Parmly Billings Library, Billings MT; *Maureen Weisblatt*, Public Library, Freehold NJ; *Henry A. Weiss*, Public Library, Palm Springs CA; *Lelia M. Welch*, Community Library, Kennebunkport ME; *Karen Welsh*, University of California at San Diego Library, La Jolla CA; *Jessamyn West*; *Paula West*, Wyoming State Archives, Cheyenne WY; *Shirley L. West*, Public Library, Camden NJ; *Dominique C. Western*, Delaware Department of State, Dover DE; *I. Lois Wetherby*, Public Library, Sebring FL; *Robert G. Wheeler*, Greenfield Village & Henry Ford Museum, Dearborn MI; *Doris L. Whetstone*, Pine Grove PA; *Elizabeth L. White*, Brooklyn Public Library, Brooklyn NY; *Fran White*, Tulsa City-County Library System, Tulsa OK; *Katherine K. White*, Emporia KS; *Livonia E. White*, North Hampton NH; *Paul A. White*, Kinderhook Regional Library, Lebanon MO; *Theodore H. White*; *Wini W. White*, Public Library, Cornwall NY; Public Library, White Plains NY; *Elizabeth P. Whitten*, Public Library, New London CT; *Harvey D. Wickware*, Theodore Roosevelt National Park, Medora ND; *John D. Widdemer*, Gloversville NY; *Wilbur E. Wieprecht*, Nevada Division of Historic Preservation and Archeology, Carson City NV; *Marcella Wignall*, Public Library, Traer IA; *Ronald Wil*, Stark County District Library, Canton OH; *Terry Wilbert*, Public Library, Youngstown OH; *Lowell R. Wilbur*, Iowa Historical Library, Des Moines IA; *Richard Wilbur*; *Bill Wilcox*, Booker T. Washington National Monument, Rocky Mount VA; *Lucy Wilde*, Kansas State University Library, Manhattan KS; *Diane Wilhelm*, Illinois State Historical Library, Springfield IL; *Danby Williams*, Public Library, Winchester KY; *Daniel T. Williams*, Frissell Library, Tuskegee AL; *Elizabeth Williams*, Peabody Institute Library, Danvers MA; *Ethelynn R. Williams*, Public Library, New Rochelle NY; *Joan N. Williams*, Public Library, Elmwood NE; *Philip L. Williams*, Jefferson-Madison Regional Library, Charlottesville VA; *Telva Williams*, Public Library, Arcola IL; *Tennessee Williams*; *William E. Williams*, M.D., Rutherford NJ; *Anne Williamson*, Springfield City Library, Springfield MA; *Anne K. Williamson*, Jones Library, Amherst MA; Williamstown Public Library, Williamstown MA; *Barbara P. Willis*, Wallace Memorial Library, Fredericksburg VA; *Dr. Patricia C. Willis*, Rosenbach Foundation, Philadelphia PA; Will Rogers State Historic Park, Pacific Palisades CA; Wilmington Institute Library, Wilmington DE; *Bonnie Wilson*, Minnesota Historical Society, St. Paul MN; *Sloan Wilson*; *C. Daniel Wilson*, Wilton Library Association, Wilton CT; *Michael E. Wilson*, Rosenberg Library, Galveston TX; Wilton Public and Gregg Free Library, Wilton NH; *Abraham N. Winokur*, Jewish Congregation of Pacific Palisades CA; *Christina Winston*, Public Library, Rushville NE; *Diana Witt*, Public Library, Portsmouth NH; *Martha Leigh Wolf*, Brandywine Conservancy, Chadds Ford PA; *Claire Wood*, West County Historical Association, Girard PA; *Katharine M. Wood*, University of Delaware Library, Newark DE; *Verna L. Wood*, Public Library, Albuquerque NM; *Carol Woodger*, Public Library, Port Chester NY; *Irene Woodhouse*, Weber County Library, Ogden UT; *Thelma Wootton*, Wells Memorial Library, Lafayette IN; *Dorothy Worcester*, Public Library, Dublin NH; *Garnet Workman*, Public Library, Lewiston IL; *Emily Woudenberg*, Harrison Memorial Library, Carmel CA; *Herman Wouk*; *Bonnie Wright*, State Historical Society of Missouri, Columbia MO; *David A. Wright*, Bryan College Library, Dayton TN; *Mildred C. Wright*, Atlanta GA; *Walter W. Wright*, Dartmouth College Library, Hanover NH; *Rhoda Wynn*, Chapel Hill NC;

Daniel A. Yanchisin, Memphis-Shelby County Library, Memphis TN; Yankee Clipper Inn, Rockport MA; *Donald Yannella*, The Melville Society, Glassboro NJ; *Elizabeth Yates*; *Mildred G. Yelverton*, Bowling Library, Marion AL; *Martin I. Yoelson*, Independence National Historical Park, Philadelphia PA;

Marya Zaturenska; *Phyllis Zack*, Public Library, Pittsfield MA; *Louis B. Ziegler*, Municipal Museum, San Jacinto CA; *Robert J. Zietz*, Public Library, Mobile AL; *Maria Zini*, Carnegie Library, Pittsburgh PA; *Kathleen L. Zwicker*, Chamber of Commerce, Brattleboro VT.

Picture Credits

Page 1, Sturges Photo, courtesy of the Committee for Alna History. *Page 2*, Dr. Thomas B. Johnson. *Page 4*, Mrs. Benjamin Butler. *Page 5* (top), Colby College; (bottom), US Department of the Interior. *Page 6*, NY Public Library. *Page 7*, Kennebunkport Historical Society. *Page 8* (top), by permission of the Houghton Library, Harvard University; (bottom), Maine Historical Society. *Page 9*, Rockland *Courier-Gazette*. *Page 10*, W. Fallon, courtesy of The Hawthorne Community Association. *Page 11* (top), E. E. Hubbard, courtesy of The Milford Cabinet; (bottom), William B. Rotch, courtesy of The Milford Cabinet. *Page 12*, NY Public Library. *Page 13*, The Frost Place. *Page 14*, The John Greenleaf Whittier Homestead. *Page 15* (top), The John Greenleaf Whittier Homestead; (bottom), The John Greenleaf Whittier Homestead. *Page 16*, Lotte Jacobi, courtesy of May Sarton. *Page 17* (top), NY Public Library; (bottom) Douglas Armsden, courtesy of Strawberry Banke. *Page 19*, Louis R. Brown, Inc. *Page 20*, University of Vermont. *Page 21* (top), Helen H. Cahill; (bottom left and right), Rowland E. Robinson Memorial Association. *Page 22*, courtesy of Vermont Historical Society. *Page 23*, Middlebury College. *Page 24*, Bertha S. Frothingham. *Page 25*, Stan & Maryjane Bean, courtesy of the John Greenleaf Whittier Home. *Page 26*, NY Public Library. *Page 27*, Edward W. and Geneviève de B. Merrill. *Page 28*, Saint-Gaudens National Historic Site. *Page 29* (top), Robert W. Haney; (bottom), NY Public Library. *Page 30* (left and right), NY Public Library. Page 31 (top), The Thoreau Society. *Page 31* (bottom), NY Public Library. *Page 32* (top), NY Public Library; (bottom), Library of the Boston Athenaeum. *Page 33*, Library of the Boston Athenaeum. *Page 35* (left to right), NY Public Library; M. Shulman & Associates; Stan & Maryjane Bean, courtesy of the John Greenleaf Whittier Home. *Page 36*, The Bostonian Society. *Page 37* (top), Chase News Photo, courtesy Mrs. Conrad Aiken; (bottom), Marvin Beinema. *Page 38*, National Park Service. *Page 39*, National Park Service. *Page 40*, Cambridge Historical Commission. *Page 41*, Cambridge Historical Commission. *Page 42*, Edward Bellamy Memorial Association. *Page 43* (left and right), Marvin Beinema. *Page 44* (top), Concord Free Public Library; (bottom), Ralph Waldo Emerson House, Concord MA. *Page 45*, Marvin Beinema. *Page 47*, Edith Blake, courtesy of Henry Beetle Hough. *Page 48*, John Greenleaf Whittier Homestead. *Page 49*, John Greenleaf Whittier Homestead. *Page 50*, Marvin Beinema. *Page 51* (left and right), Marvin Beinema. *Page 53*, John Greenleaf Whittier Homestead. *Page 55*, Berkshire County Historical Society. *Page 56*, Marvin Beinema. *Page 57* (left and right), Marvin Beinema. *Page 58*, House of Seven Gables. *Page 59* (left), House of Seven Gables; (right), Edward M. Stevenson, House of Seven Gables. *Page 60* (left), The Thornton W. Burgess Society; (right), Marvin Beinema. *Page 61*, Marvin Beinema. *Page 62*, Marvin Beinema. *Page 64*, Marvin Beinema. *Page 65* (left and right), Marvin Beinema. *Page 66*, RI Tourist Promotion Division. *Page 67*, RI Tourist Promotion Division. *Page 68*, RI Department of Economic Development. *Page 69* (top), NY Public Library; (bottom), RI Tourist Promotion Division. *Page 70*, Rhode Island Historical Society. *Page 71*, RI Tourist Promotion Division. *Page 73*, Gérard Cochepin. *Page 74*, Gérard Cochepin. *Page 75*, Gérard Cochepin. *Page 76*, Ken Riehl. *Page 78* (top and bottom), Stowe-Day Foundation, Hartford, Conn. *Page 79* (left), Mark Twain Memorial, Hartford, Conn.; (right), Stowe-Day Foundation, Hartford, Conn. *Page 80* (top and bottom), Mark Twain Memorial, Hartford, Conn. *Page 82*, Photo Morgoli, courtesy of Georges Simenon. *Page 83*, courtesy of The New Canaan Historical Society. *Page 85*, Gérard Cochepin. *Page 87*, Eugene O'Neill Theater Center. *Page 89*, NY Public Library. *Page 90*, Ken Riehl. *Page 91*, Gérard Cochepin. *Page 92*, John J. Trask, from *Stonington Houses*. *Page 93*, Noah Webster Foundation and Historical Society. *Page 94* (left), Arni, courtesy of Eleanor Estes; (right), Thomas Victor, courtesy of Peter DeVries.

Page 97, The Historical Association of South Jefferson. *Page 98* (left and right), Gérard Cochepin. *Page 99*, Ken Riehl. *Page 100*, Marvin Beinema. *Page 101*, Ken Riehl. *Page 102* (top), Joseph Lizzio; (bottom), Gérard Cochepin. *Page 103*, Marvin Beinema. *Page 104* (left), NY Public Library; (right), Ken Riehl. *Page 105* (top and bottom), New York State Historical Association, Cooperstown. *Page 107*, Ken Riehl. *Page 108* (top left and right), Chemung County Historical Society, Elmira, NY; (bottom left), Gérard Cochepin; (bottom right), Chemung County Historical Society, Elmira, NY. *Page 109*, Gérard Cochepin. *Page 110* (top), Valerie Meyer, courtesy of Hortense Calisher; (bottom), Ken Riehl. *Page 111*, Ken Riehl. *Page 113* (top left and right), *The Long-Islander*; (bottom left and right), Ken Riehl. *Page 114*, Joseph Lizzio. *Page 115*, The Rare Book Division, The NY Public Library. *Page 116*, Ken Riehl. *Page 117* (left and right), Ken Riehl. *Page 119* (left and right), NY Public Library. *Page 121*, Marvin Beinema. *Page 122* (top), Culver Pictures; (bottom), UPI. *Page 123* (left), Culver Pictures; (right), Marion Bodine. *Page 124* (top), Culver Pictures; (bottom), Marvin Beinema. *Page 125* (top), Culver Pictures; (bottom), NY Public Library. *Page 126*, Marvin Beinema. *Page 127*, Culver Pictures. *Page 128*, Rodney Smith, courtesy of Robert Anderson. *Page 129*, Culver Pictures. *Page 130*, Culver Pictures. *Page 131*, NY Public Library. *Page 132*, Culver Pictures. *Page 133*, Culver Pictures. *Page 134* (top), courtesy of Charles Scribner's Sons; (bottom), Marion Bodine. *Page 135* (top), Jerry Bauer, courtesy of Hortense Calisher; (bottom), Ezra Stoller, Esto. *Page 136* (left), Lone Star Brewing Co., San Antonio TX; (right), Culver Pictures. *Page 137*, Dith Pran, The New York *Times*. *Page 138*, Culver Pictures. *Page 140*, Marvin Beinema. *Page 141* (left and right), The Richard Gimbel Collection, Free Library of Philadelphia. *Page 142*, Marion Bodine. *Page 143*, NY Public Library. *Page 144* (left), Culver Pictures; (right), UPI. *Page 145*, NY Public Library. *Page 146*, Culver Pictures. *Page 147*, Marion Bodine. *Page 148*, Marion Bodine. *Page 149*, UPI. *Page 150*, BSW Literary Agency. *Page 151*, UPI. *Page 152*, © Irving Penn, Condé Nast Publications. *Page 153* (top), UPI; (bottom left), Marion Bodine; (bottom right), Marvin Beinema. *Page 154*, Philip H. and A. S. W. Rosenbach Foundation, Philadelphia. *Page 155* (top), Hart Crane Papers, Rare Book and Manuscript Library, Columbia University; (bottom left), Marion Bodine; (bottom right), courtesy of Plymouth Church of the Pilgrims. *Page 156*, UPI. *Page 157*, Ken Riehl. *Page 158*, Ken Riehl. *Page 159*, Wave Hill Center for Environmental Studies. *Page 160* (left), Ken Riehl; (right), The Bronx County Historical Society. *Page 161*, Ken Riehl. *Page 162*, Ken Riehl. *Page 163* (top), NY Public Library; (bottom), J. J. Heatley, Inc. *Page 165*, Ken Riehl. *Page 166* (left and right), Ken Riehl. *Page 167*, David B. Carruth. *Page 168* (top and bottom), The Silverado Museum, St. Helena, CA. *Page 169* (top), Raymond Hand, Jr.; (bottom), George S. Bolster. *Page 171* (top and bottom), NY Public Library. *Page 172*, Sleepy Hollow Restorations, Tarrytown, NY. *Page 173* (top left and bottom), Sleepy Hollow Restorations, Tarrytown, NY; (top right), Ken Riehl. *Page 175* (left and right), Ken Riehl. *Page 176* (top and bottom), US Military Academy, West Point, NY. *Page 177*, Ray Godwin, courtesy of Margaret Halsey. *Page 178*, NY Public Library. *Page 179* (left), Bordentown Historical Society; (right), courtesy of the Burlington County Historical Society. *Page 180* (top left), *The Long-Islander*; (top right) NJ Historic Sites, Department of Environmental Protection; (bottom), Richard T. Koles, courtesy of Jean-Rae Turner. *Page 182*, NY Public Library. *Page 183*, The Lawrenceville School. *Page 184*, NY Public Library. *Page 185*, courtesy of Dore Schary. *Page 186*, Columbia University. *Page 188*, St. Peter's Episcopal Church, Perth Amboy, NJ. *Page 189*, Ulli Steltzer, Princeton University Library. *Page 190* (top), NY Public Library; (bottom), Ulli Steltzer, Princeton University Library. *Page 191*, Ulli Steltzer, Princeton University Library. *Page 192*, courtesy of Charles Scribner's Sons. *Page 193*, courtesy of William E. Williams. *Page 194*, Department of Parks and Recreation, Wayne Township. *Page 195*, NY Public Library. *Page 196* (left and right), Marvin Beinema. *Page 197* (left and right), Marvin Beinema. *Page 198*, National Park Service. *Page 199*, Marvin Beinema. *Page 200* (top), NY Public Library; (bottom), Zane Grey Museum, Lackawaxen, PA. *Page 201*, courtesy of Agnes Sligh Turnbull. *Page 202* (top and bottom), Marvin Beinema. *Page 203* (top), Mountain Lake Sanctuary, Lake Wales, FL; (bottom), The John Bartram House Association. *Page 204*, The Historical Society of Pennsylvania. *Page 205*, NY Public Library. *Page 207* (top), Culver Pictures; (bottom), Philip H. and A. S. W. Rosenback Foundation, Philadelphia. *Page 208*, National Park Service. *Page 209* (top), NY Public Library; (bottom), Jonathan Arms, National Park Service. *Page 210*, Culver Pictures. *Page 211* (left and right), Marvin Beinema. *Page 212* (left), Gérard Cochepin; (right), Marvin Beinema. *Page 213* (top and bottom), Marvin Beinema. *Page 214* (top), Gérard Cochepin; (bottom left), Marvin Beinema; (bottom right), Gérard Cochepin. *Page 215*, Mr. and Mrs. Frank Herzog. *Page 216* (top and bottom), Raymond Hand, Jr. *Page 217*, Wilmington Library. *Page 219*, Middle Georgia Historical Society. *Page 220* (left), Manuscripts Division, University of Virginia Library; (right), courtesy of the Edgar Allan Poe Society of Baltimore. *Page 221* (top), NY Public Library; (bottom), courtesy of Katherine Anne Porter. *Page 222*, C. Burr Artz Library, Frederick, MD. *Page 223* (top), Raymond Hand, Jr.; (bottom, left), John W. McGrain; (bottom right), Jennifer Rogers. *Page 224*, Culver Pictures. *Page 225*, National Park Service. *Page 226* (top left and right), Culver Pictures; (bottom) courtesy of Urion Press, Eugene OR. *Page 229*, Manuscripts Division, University of Virginia Library. *Page 230* (top), NY Public Library; (bottom), Virginia Historic Landmarks Commission, courtesy of the Virginia State Library. *Page 231*, Virginia Historic Landmarks Commission, courtesy of the Virginia State Library. *Page 232*, Manuscripts Division, University of Virginia Library. *Page 233*, Manuscripts Division, University of Virginia Library. *Page 234* (left and right), Manuscripts Division, University of Virginia Library. *Page 235*, NY Public Library. *Page 236*, Hilbert H. Campbell, Virginia Polytechnic Institute and State University. *Page 238* (top), NY Public Library; (bottom), West Virginia Department of Culture and History. *Page 239* (left and right), West Virginia Department of Culture and History.

Page 241, William Barnhill Collection, Pack Memorial Library. *Page 242*, Thomas Wolfe Collection, Pack Memorial Library. *Page 243*, Mrs. Byrd Green Cornwell, courtesy of Paul Green. *Page 244* (top), Tom Walters, courtesy of Harry Golden; (bottom), National Park Service. *Page 245*, National Park Service. *Page 246*, Thomas Wolfe Collection, Pack Memorial Library. *Page 247* (top), North Carolina Division of Archives and History, courtesy of the Davis County Public Library; (bottom), courtesy of Jonathan Daniels. *Page 248*, O & O Photography, Columbus NC. *Page 249* (top), courtesy of Mary C. Simms Oliphant; (bottom), Charles N. Bayless. *Page 250*, Charles N. Bayless. *Page 251* (left), Charles N. Bayless; (right), Historic Columbia Foundation, Columbia SC. *Page 252*, Mrs. A. D. Oliphant. *Page 254*, from Private Archives. *Page 255* (top), Historic Augusta, Inc.; (bottom), Margaret Mitchell Collection, Atlanta Public Library. *Page 256*, UPI. *Page 257*, Middle Georgia Historical Society. *Page 258* (all), Middle Georgia Historical Society, *Page 259*, The Pirates House. *Page 260* (left), Savannah *News-Press*; (right), Georgia Historical Society. *Page 261*, Florida Department of Natural Resources. *Page 262*, Monroe County Public Library. *Page 263*, Monroe County Public Library. *Page 265*, Rollins College, Winter Park FL. *Page 266*, Erik Overbey, Mobile Public Library Collection, University of South Alabama Photographic Archives. *Page 267* (left), Mike Thomason; (right), Local History Division, Mobile Public Library. *Page 268*, Ivy Green. *Page 269*, Tuskegee Institute National Historic Site. *Page 270*, Shanks Photo Service, courtesy of Lowndes County Library System. *Page 271*, Mrs. Frank Hopkins. *Page 272*, Mississippi Department of Archives and History. *Page 274* (top), UPI; (bottom), University of Mississippi. *Page 275*, University of Mississippi. *Page 276*, Tommy Covington, Ripley, MS. *Page 277*, B. A. Cohen and Caspaar Blue. *Page 278* (top and bottom), NY Public Library. *Page 279* (left), The Vieux Carre Survey, The Historic New Orleans Collection; (right), *The Times-Picayune States-Item*. *Page 280* (left), NY Public Library; (right), The Vieux Carre Survey, The Historic New Orleans Collection. *Page 281*, The Vieux Carre Survey, The Historic New Orleans Collection. *Page 282* (left), Louisiana State Library; (right), Louisiana State Library. *Page 283* (top), Arkansas Historic Preservation Program; (bottom), University of Arkansas Library. *Page 284*, Arkansas Historic Preservation Program. *Page 285*, Culver Pictures. *Page 286*, Culver Pictures. *Page 287*, Metro Board of Parks and Recreation, Nashville, TN. *Page 288* (top), NY Public Library; (bottom), Rugby Restoration Association. *Page 290*, Brother Patrick Hart. *Page 291*, National Park Service. *Page 292*, Paducah–McCracken County Tourist Commission. *Page 293*, Lee Pennington.

Page 295, NY Public Library. *Page 296*, Greene County Historical Society. *Page 297* (top), NY Public Library; (bottom), from the Collection of the Public Library of Cincinnati and Hamilton County. *Page 298* (top), Culver Pictures; (bottom), from the Collection of the Public Library of Cincinnati and Hamilton County. *Page 299*, Thaddeus B. Hurd. *Page 301* (both), Dayton and Montgomery County Public Library. *Page 302*, Ben Schnall, courtesy of Kenyon College. *Page 303* (top), Culver Pictures; (bottom), Hiram College Library Archives. *Page 304*, Malabar Farm State Park. *Page 305*, Zane Grey Museum, Lackawaxen PA. *Page 306*, Miami University, Oxford OH. *Page 307*, *The Blade*, Toledo OH. *Page 308*, Greene County District Library. *Page 309* (left), George Ade Memorial Association, Inc.; (right), NY Public Library. *Page 310*, Crawfordsville Parks and Recreation Department. *Page 311* (both), Gérard Cochepin. *Page 312* (both), Gérard Cochepin. *Page 313* (all), Gérard Cochepin. *Page 314*, Union County Public Library, Liberty IN. *Page 315*, Vigo County Historical Society. *Page 316*, Braun Photo Service, courtesy of Mr. and Mrs. Max Rosenberger. *Page 317*, Alton *Telegraph*. *Page 319*, Culver Pictures. *Page 320*, Michael C. O'Connor. *Page 321*, Chicago Park District. *Page 322*, Michael C. O'Connor. *Page 323* (top), Culver Pictures; (bottom), Michael C. O'Connor. *Page 324* (left), Culver Pictures; (right), NY Public Library. *Page 325*, Illinois State Historical Library. *Page 326*, Kate Dunham. *Page 327*, Lebanon *Advertiser*. *Page 328*, Illinois State Historical Library. *Page 329*, courtesy of The Vachel Lindsay Association. *Page 330*, Josephine Burtschi. *Page 332* (top), South Bend *Tribune*; (bottom), NY Public Library. *Page 333*, Niles Community Library. *Page 334* (both), Madelaine Hemingway Miller. *Page 335*, Appleton Public Library. *Page 336*, The John M. Russells. *Page 337*, Harold Hume, from the collection of the University of Wisconsin–Madison Archives. *Page 339* (both), The John M. Russells. *Page 340* (top), courtesy Arkham House Publishers; (left and right), West Salem Historical Society. *Page 341*, Ted Miller, Minnesota Historical Society. *Page 342* (top), St. Olaf College; (bottom), Minnesota Historical Society. *Page 343* (top), Coward, McCann & Geoghehan, Inc.; (bottom), Minnesota Historical Society. *Page 344*, Minnesota Historical Society. *Page 345* (top, and bottom right), Sinclair Lewis Foundation; (bottom left), UPI. *Page 346*, courtesy of Thomas Duncan. *Page 348*, Ruth Suckow Memorial Association. *Page 349* (top), Harvard University Press; (bottom), NY Public Library. *Page 351* (left), Musser Public Library, Muscatine IA; (right), courtesy of Paul Corey. *Page 352*, Max D. Maxon, courtesy of Kendall Young Library. *Page 354*, Art Grossmann. *Page 355* (top left and right), Mark Twain Home Board; (bottom), Walker, Missouri Tourism. *Page 356*, John Winkler, courtesy of the Becky Thatcher Book Shop. *Page 358*, Eugene Field House and Toy Museum. *Page 359*, Missouri Historical Society. *Page 360* (left), National Park Service; (right), McLean County Historical Society. *Page 361* (both), Laura Ingalls Wilder Memorial Society, Inc., De Smet SD. *Page 362* (both), Laura Ingalls Wilder Memorial Society, Inc., De Smet SD. *Page 363*, Ron Nicodemus, courtesy of the Nebraska State Historical Society. *Page 364* (top), Caroline Sandoz Piper; (bottom), Nebraska State Historical Society. *Page 365*, Nebraska State Historical Society. *Page 366* Gabriel North Seymour. *Page 367*, Mrs. William L. White. *Page 368*, Manhattan *Mercury*, courtesy of the Manhattan Public Library. *Page 369* (both), Will Rogers Memorial. *Page 370*, © Bern Schwartz. *Page 372*, The Austin-Travis County Collection of the Austin (Texas) Public Library. *Page 373* (top), courtesy William A. Owens; (bottom), El Paso Public Library. *Page 374*, Bertie M. Strawn. *Page 375*, The Lone Star Brewing Co., Inc.

Page 377, Stanford University. *Page 378*, Yellowstone Valley Prints. *Page 379*, Idaho Historical Society. *Page 380* (left), UPI; (right), Lloyd R. Arnold. *Page 381*, Wyoming State Museum. *Page 382*, Medicine Bow *Post*. *Page 383* (both), The Pioneers' Museum. *Page 384* (both), Gérard Cochepin. *Page 385*, Gérard Cochepin. *Page 386*, Rocky Mountain National Park, National Park Service. *Page 388* (left), Tyler Dingee, courtesy of Museum of New Mexico; (right), New Mexico Historic Preservation Bureau. *Page 389* (both), William Clark. *Page 390*, Klickitat County Historical Society. *Page 391* (both), © The Seattle *Times*. *Page 393*, The Davenport Hotel. *Page 394* (both), courtesy of Urion Press. *Page 396*, John Bragg, courtesy of Farrar, Straus and Giroux. *Page 397* (top), Oregon Historical Society; (bottom), Oregon City Historical Society. *Page 400*, Jacqueline L. Heintz. *Page 401*, The Bancroft Library. *Page 402*, The Silverado Museum, St. Helena CA. *Page 403* (left), Harrison Memorial Library; (right), California Department of Parks and Recreation. *Page 404*, California Department of Parks and Recreation. *Page 405*, California Department of Parks and Recreation. *Page 406*, courtesy of William Saroyan. *Page 407*, California Department of Parks and Recreation. *Page 408*, Culver Pictures. *Page 409*, Margaret Reavey. *Page 410*, Long Beach *Independent-Press Telegram*. *Page 411*, UPI. *Page 412*, Culver Pictures. *Page 413* (top), Culver Pictures; (bottom), California Department of Parks and Recreation. *Page 414*, The Viking Press. *Page 415* (top), NY Public Library; (bottom), California Department of Parks and Recreation. *Page 417* (both), Culver Pictures. *Page 418* (both), Public Relations, Port of Oakland. *Page 419*, courtesy of Urion Press. *Page 421*, Sacramento Public Library. *Page 422*, San Diego Historical Society-Title Insurance and Trust Collection. *Page 423*, Larry R. Nicodemo. *Page 424*, Larry R. Nicodemo. *Page 425* (left), Larry R. Nicodemo; (right), City Lights. *Page 426*, The Bancroft Library. *Page 427*, Larry R. Nicodemo. *Page 428* (left), The Bancroft Library; (right), California Department of Parks and Recreation. *Page 429*, International Creative Management. *Page 431*, California Department of Parks and Recreation. *Page 432*, Nevada Historical Society. *Page 434*, Nevada Historical Society. *Page 435* (left), NY Public Library; (right), Brigham Young University Library. *Page 436*, Gérard Cochepin. *Page 437*, Gérard Cochepin. *Page 438*, courtesy of Erskine Caldwell. *Page 439* (top), Donald Pearson; (bottom), Culver Pictures. *Page 441*, R. J. Baker, Bishop Museum. *Page 442*, Milo Shepard. *Page 443*, Waioli Tea Room.

INDEX TO AUTHORS

A

B

C

D

E

F

G

H

I

J

K

L

M

Q

R

T

U

V

W

X-Y

Z

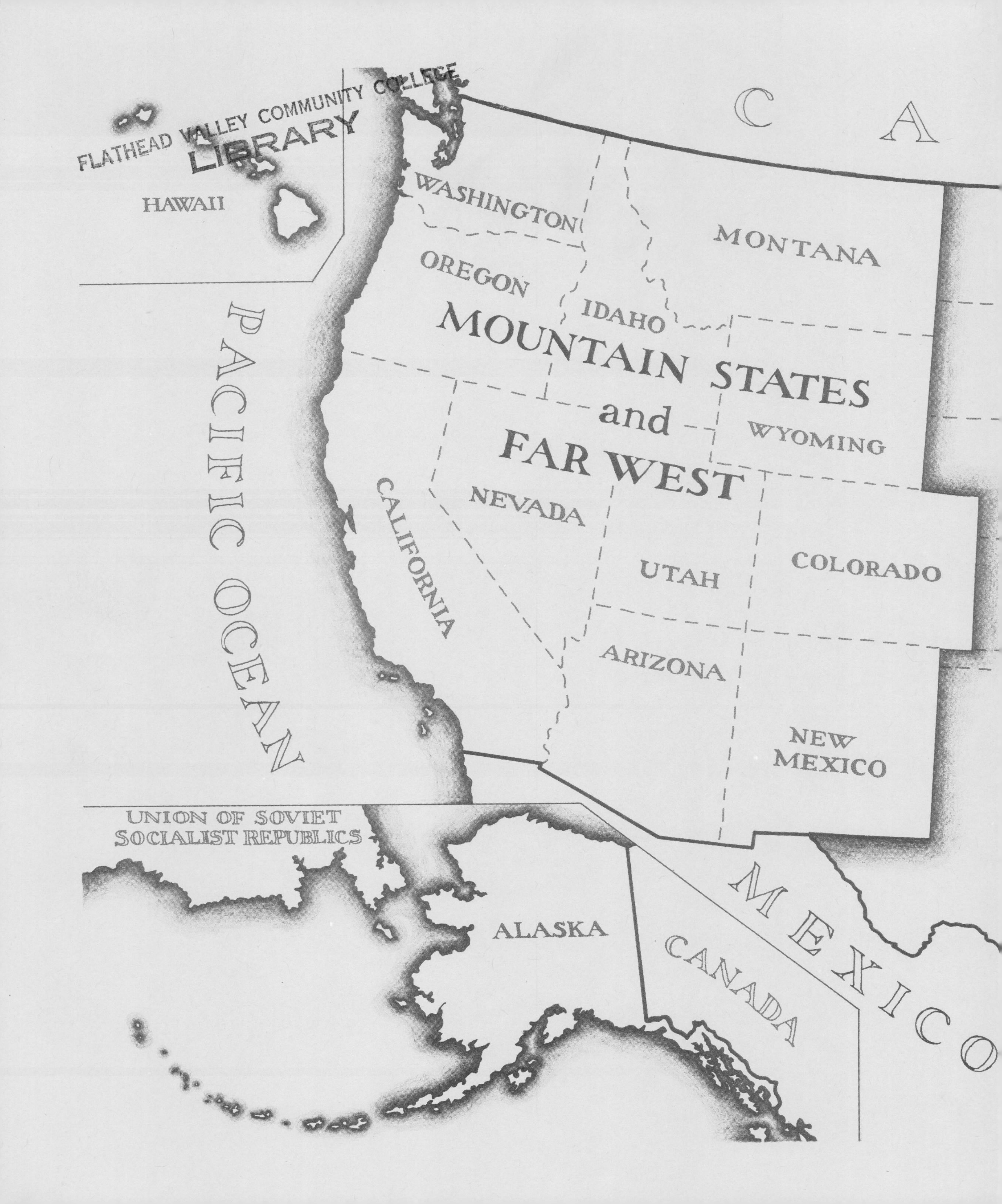

FLATHEAD VALLEY COMMUNITY COLLEGE
LIBRARY
HAWAII
WASHINGTON
MONTANA
OREGON
IDAHO
MOUNTAIN STATES
and
FAR WEST
WYOMING
PACIFIC OCEAN
CALIFORNIA
NEVADA
UTAH
COLORADO
ARIZONA
NEW
MEXICO
UNION OF SOVIET
SOCIALIST REPUBLICS
ALASKA
CANADA
MEXICO
C A